Applications of Differential Equations in Engineering and Mechanics

Applications of Differential Equations in Engineering and Mechanics

K.T. Chau

CRC Press
Taylor & Francis Group
Boca Raton London New York

CRC Press is an imprint of the
Taylor & Francis Group, an informa business

CRC Press
Taylor & Francis Group
6000 Broken Sound Parkway NW, Suite 300
Boca Raton, FL 33487-2742

Printed on acid-free paper

International Standard Book Number-13: 978-0-367-02643-1 (Paperback)
International Standard Book Number-13: 978-1-4987-6697-5 (Hardback)

Library of Congress Cataloging-in-Publication Data

Names: Chau, K. T., author.
Title: Applications of differential equations in engineering and mechanics /
Kam Tim Chau.
Description: Boca Raton : Taylor & Francis, a CRC title, part of the Taylor &
Francis imprint, a member of the Taylor & Francis Group, the academic
division of T&F Informa, plc, [2019] | Includes bibliographical references
and indexes. |
Identifiers: LCCN 2018036501 (print) | LCCN 2018038769 (ebook) | ISBN
9780429894350 (Adobe PDF) | ISBN 9780429894343 (ePub) | ISBN 9780429894336
(Mobipocket) | ISBN 9780367026431 (pbk.) | ISBN 9781498766975 (hardback)
| ISBN 9780429470646 (ebook)
Subjects: LCSH: Engineering mathematics. | Differential equations.
Classification: LCC TA347.D45 (ebook) | LCC TA347.D45 C49 2019 (print) | DDC
620.001/535--dc23
LC record available at https://lccn.loc.gov/2018036501

Visit the Taylor & Francis Web site at
http://www.taylorandfrancis.com

and the CRC Press Web site at
http://www.crcpress.com

To

My wife Lim, son Magnum, and daughter Jaquelee

my late father Chow Yat Wing

and

my late teacher Dr. Yan Sze Kwan

CONTENTS

PREFACE

THE AUTHOR

APPLICATIONS OF DIFFERENTIAL EQUATIONS IN ENGINEERING AND MECHANICS

CHAPTER 1: THEORY OF BEAMS AND COLUMNS 1

1.1 Introduction 1
1.2 Beam Bending 2
1.2.1 Euler-Bernoulli Beam 4
1.2.2 Simply-Supported Beam 7
1.2.3 Cantilever Beam 8
1.2.4 Cable Load 10
1.2.5 Green's Function for Simply-Supported Beams 13
1.3 Beam Vibrations 13
1.3.1 Simply-Supported Beams 14
1.3.2 Orthogonality of the Eigenfunctions 17
1.3.3 Cantilever Beam with Suddenly Removed Point Force 21
1.3.4 Cantilever Beam with a Tip Lump Mass 25
1.3.5 Simply-Supported Beam Subject to an Impulse 27
1.3.6 Seismograph as Vibrations of Rigid Beam 29
1.4 Rocket/Missile Launch Pad as Beam 31
1.5 Beam on Elastic Foundation 33
1.5.1 Formulation 33
1.5.2 Boundary Conditions 34
1.5.3 Infinite Beam under Concentrated Load 35
1.6 Euler's Column Buckling 37
1.7 Vibrations of Beams under Axial Compression 44
1.7.1 Free Vibrations of Cantilever Beams under Axial Compression 45
1.7.2 Orthogonal Approximation 50
1.7.3 Rigorous Approach 51
1.8 Timoshenko Beam Theory 52
1.8.1 Variational Formulation 52
1.8.2 Static Solution for Timoshenko Beam 56
1.8.3 Free Vibrations of Timoshenko Beams 58
1.8.4 Free Vibrations of Simply-Supported Timoshenko Beams 61
1.8.5 Free Vibrations of Cantilever Timoshenko Beams 63
1.8.6 Free Vibrations of Fixed End Timoshenko Beams 65

1.9 Summary and Further Reading 67
1.10 Problems 68

CHAPTER 2: THEORY OF PLATES 79

2.1 Introduction 79
2.2 Kirchhoff Plate Theory 80
2.2.1 Equilibrium Equations 81
2.2.2 Forces and Moments 83
2.2.3 Governing Equations 85
2.2.4 Edge Conditions 86
2.3 Simply-Supported Plates 88
2.3.1 Navier's Solution 88
2.3.2 Levy's Solution 89
2.4 Clamped Rectangular Plates 95
2.4.1 Galerkin Method 95
2.4.2 Approximation for Clamped Plates 96
2.5 Deflection of Circular Plates 98
2.5.1 Clamped Plate with Uniform Load 99
2.5.2 Clamped Plate with Patch Load 100
2.5.3 Plates under Central Point Force 102
2.6 Buckling of Plates 103
2.7 Bending of Anisotropic Plates 107
2.8 Plate on Elastic Foundation 108
2.9 Plate Vibrations 109
2.9.1 Free Vibrations 109
2.9.2 Forced Vibrations 110
2.9.3 Approximation by Rayleigh Quotient 112
2.9.4 Strain Energy of Plates 113
2.9.5 Rayleigh-Ritz Method 113
2.10 Vibrations of Circular Plates 116
2.11 Hertz Problem of Circular Plate under Point Load 122
2.11.1 Series Solution 123
2.11.2 Variational Principle 126
2.11.3 Rayleigh-Ritz Method 127
2.11.4 General Solution in Kelvin Functions 128
2.11.5 Matching of Boundary Condition 130
2.11.6 Wyman's Solution 133
2.12 Summary and Further Reading 134
2.13 Problems 134

CHAPTER 3: THEORY OF SHELLS 137

3.1 Introduction 137
3.2 Stresses, Forces, and Moments in Shells 139
3.3 Membrane Theory for Axisymmetric Shells 141
3.3.1 Dome under Concentrated Apex Load 146
3.3.2 Truncated Dome under Ring Load 147
3.3.3 Compatibility at Ring Foundation 148
3.4 Shell of Revolution under Uniform Load 150
3.4.1 Spherical Shell with Opening 151
3.4.2 Spherical Fluid Container 153
3.4.3 Conical Shells 156
3.5 Membrane Theory for Cylindrical Shells 160
3.5.1 Governing Equations 160
3.5.2 General Solutions for Axisymmetric Case 162
3.5.3 Simply-Supported Tube 163
3.5.4 Circular Tube under Dead Load 164
3.5.5 Membrane Theory versus Beam Theory 166
3.5.6 Pipe Subject to Edge Load 168
3.5.7 Simply-Supported Cylindrical Shell Roof 170
3.6 Bending Theory of Cylindrical Shells 172
3.6.1 Governing Equation for Axisymmetric Cylindrical Shells 172
3.6.2 Deformation Kinematics 174
3.6.3 Shell Bending Theory versus Beam on Elastic Foundation 176
3.6.4 General Solutions 177
3.7 Circular Pipe 178
3.7.1 Semi-Infinite Pipe Subject to End Force 178
3.7.2 Decay of Edge Disturbance 180
3.7.3 Infinite Pipes under Ring Load 181
3.7.4 Effective Length 183
3.8 Buckling of Cylindrical Shell under Axial Load 185
3.9 Bending Theory for Shell of Revolution 187
3.9.1 Force and Moment Equilibrium 189
3.9.2 Hooke's Law 189
3.9.3 Change of Curvature 191
3.9.4 Reissner Formulation 192
3.10 Spherical Shell of Constant Thickness 195
3.10.1 Solution in Terms of Hypergeometric Functions 196
3.10.2 Superposition for Various Boundary Conditions 200
3.11 Thin Spherical Shell 201
3.11.1 Geckeler-Staerman Approximation 201
3.11.2 Hetenyi Approximation 207
3.12 Symmetrical Bending of Thin Shallow Spherical Shell 210
3.12.1 Reissner Formulation 210
3.12.2 Governing Equations for Negligible Self-Weight 213

3.12.3 Solution in Kelvin Functions 214
3.13 Bending of Cylindrical Shell 218
3.13.1 Governing Equations 218
3.13.2 Vlasov's Stress Function 221
3.13.3 Cylindrical Roof Shells 222
3.13.4 Particular Solution 224
3.13.5 Homogeneous Solution 225
3.13.6 General Solution 227
3.13.7 Vertical Load on Shell Surface 228
3.14 Summary and Further Reading 231
3.15 Problems 232

CHAPTER 4: STRUCTURAL DYNAMICS 239

4.1 Introduction 239
4.2 Static Deflection versus Natural Vibration 240
4.3 Single-Story Building 242
4.4 Damped and Undamped Responses 243
4.4.1 Undamped Responses 243
4.4.2 Damped-Free Responses 246
4.4.3 Damping Ratio by Hammer Test 249
4.4.4 Damped Forced Responses 250
4.5 Duhamel Integral for General Ground Motions 253
4.5.1 Formulation of Equation of Motion 254
4.5.2 Duhamel Integral 255
4.6 Response Spectrum 257
4.6.1 Pseudo-Response Spectrum 259
4.6.2 Nonlinear Response Spectrum 261
4.7 Multiple-Story Buildings 263
4.8 Modal Analysis 265
4.8.1 Free Vibrations 266
4.8.2 Decoupling of the Undamped Dynamic System 267
4.8.3 Decoupling of the Damped Dynamic System 268
4.8.4 Rayleigh Damping 268
4.8.5 Caughey and Liu-Gorman Proportional Damping 271
4.8.6 Rayleigh Quotient Technique 274
4.8.7 Response Spectrum Method for MDOF System 275
4.9 Summary and Further Reading 283
4.10 Problems 283

CHAPTER 5: CATENARY AND CABLE-SUPPORTED BRIDGES......... 291

5.1 Introduction 291
5.2 Vibrations of Hanging Chains 293
5.3 Catenary 298
5.4 Inverted Catenary and Arch 301
5.5 Stone Arches 302
5.5.1 Formulation of Stone Arches 302
5.5.2 Inglis Solution 303
5.6 Cable Suspension Bridge 305
5.7 Cable-Stay Bridge 308
5.8 Vibrations of Cable Suspension Bridge 311
5.8.1 Governing Equations for Flexible Deck 311
5.8.2 Symmetric Modes 316
5.8.3 Anti-Symmetric Modes 319
5.8.4 Suspension Bridge with Stiffened Truss 321
5.9 Summary and Further Reading 327
5.10 Problems 328

CHAPTER 6: NONLINEAR BUCKLING 331

6.1 Introduction 331
6.1.1 Column Buckling 332
6.1.2 Plate Buckling 334
6.1.3 Shell Buckling 336
6.2 Lagrangian or Green's Strain 344
6.3 Euler-Bernoulli Beam 346
6.3.1 Strain Energy Function 348
6.3.2 Hamiltonian Principle 350
6.3.3 Calculus of Variations 350
6.3.4 Applied Force versus Applied Displacement 353
6.4 Static Buckling Theory of Beam 355
6.5 Linear Dynamic Stability of Static States 357
6.5.1 Perturbation Method 357
6.5.2 Stability of Straight State 358
6.5.3 Stability of Buckled States 360
6.6 Nonlinear Dynamic Stability 362
6.6.1 Undamped Motions 363
6.6.2 Damped Motions 365
6.7 Multi-Time Perturbation and Stability 368
6.8 Governing Equations of Crooked Beams 375
6.8.1 Lagrangian Strain for Crooked Beams 375
6.8.2 Variational Principle for Crooked Beams 376
6.9 Snap-Through Buckling of Elastic Arches 378
6.9.1 Static Solution under Pressure 378

6.9.2 Linear Dynamic Stability ... 380
6.9.3 Transitions of Snap-Through Buckling ... 388
6.9.4 Linear Dynamic Stability for Unsymmetric State ... 389
6.10 Summary and Further Reading ... 391
6.11 Problems ... 392

CHAPTER 7: TURBULENT DIFFUSIONS IN FLUIDS ... 397

7.1 Introduction ... 397
7.2 Error Function ... 398
7.2.1 Definition ... 398
7.2.2 Relation to Normal Distribution ... 398
7.2.3 Complementary Error Function ... 401
7.2.4 Some Results of Error Function ... 401
7.3 Diffusion of Pollutants in River ... 403
7.4 Ogata and Banks Solution ... 405
7.5 Solution for Decaying Pollutants ... 410
7.6 Dispersion of Decaying Substances ... 412
7.7 Taylor's Point Source Solution ... 414
7.7.1 Taylor's Approach ... 415
7.7.2 Taylor's Solution by Dimensional Analysis ... 416
7.8 Decaying Pollutant in Flowing Fluid ... 418
7.8.1 Point Source Solution ... 418
7.8.2 Continuous Source Solution ... 419
7.9 Diffusion in Higher Dimension ... 423
7.9.1 Two-Dimensional Point Source Solution ... 423
7.9.2 Three-Dimensional Point Source Solution ... 425
7.9.3 Two-Dimensional Line Source ... 427
7.10 Summary and Further Reading ... 428
7.11 Problems ... 429

CHAPTER 8: GEOPHYSICAL FLUID FLOWS ... 433

8.1 Introduction ... 433
8.2 Coriolis Force Due to Rotation ... 434
8.2.1 Coriolis Force for High Altitude ... 436
8.2.2 Coriolis Force for All Altitudes ... 437
8.3 Hydrodynamic Equations for Geophysical Flows ... 440
8.3.1 Continuity Condition ... 440
8.3.2 Momentum Equations ... 441
8.3.3 Mass Conservation ... 442
8.3.4 Constitutive Law ... 443
8.3.5 Energy Equation ... 445
8.3.6 Equation of State ... 445
8.4 System of Equations for Geophysical Flows ... 446

8.4.1	Consideration of Scales	446
8.4.2	Governing Equations	449
8.4.3	Rossby, Ekman and Reynolds Numbers	449
8.5	Storm Surges	451
8.5.1	Storm Surges by Inverse Barometer Effect	451
8.5.2	Storm Surges with Moving Disturbance	453
8.5.3	Wind-Induced Storm Surges	455
8.5.4	Current Profile	457
8.6	Ekman Transport	458
8.6.1	Ekman Transport with No Internal Currents	458
8.6.2	Ekman Transport with Internal Currents	459
8.7	Geostrophic Flows	464
8.7.1	Taylor-Proudman Theorem	464
8.7.2	Homogeneous Geostrophic Flows	464
8.8	2-D Shallow Water Equations	465
8.9	Vorticity and Tornado Dynamics	467
8.9.1	Helmholtz Vorticity Equation	467
8.9.2	Conservation of Angular Momentum	470
8.9.3	Vorticity in Tornadoes	471
8.9.4	Potential Vortex Model	472
8.9.5	Rankine Vortex Model	473
8.9.6	Burgers-Rott Vortex Model	475
8.9.7	Oseen-Lamb Vortex Model	478
8.9.8	Sullivan Vortex Model	482
8.10	Summary and Further Reading	489
8.17	Problems	490

CHAPTER 9: NONLINEAR WAVE AND SOLITONS 495

9.1	Introduction	495
9.2	Nonlinear Transport and Shocks	497
9.3	Dispersive Waves	498
9.4	Shock Waves in Traffic Flow	499
9.5	KdV Equation	502
9.5.1	Formulation of KdV	503
9.5.2	Scale Invariance	505
9.5.3	Physical Interpretation of KdV	506
9.5.4	Dispersion versus Nonlinearity	508
9.5.5	Soliton Solution	508
9.6	Hirota's Direct Method	512
9.6.1	Bilinear Form of KdV Equation	512
9.6.2	One-Soliton Solution	513
9.6.3	Two-Soliton Solution	514
9.6.4	*N*-Soliton Solution	516
9.6.5	Hirota's D-Operator	517

9.7 KdV Equation and Other Nonlinear Equations 518
9.7.1 KdV Equation and mKdV Equation 518
9.7.2 KdV Equation and Boussinesq Equation 519
9.7.3 KdV Equation and Nonlinear Schrödinger Equation 520
9.7.4 KdV Equation and First Painlevé Equation 523
9.7.5 mKdV Equation and Second Painlevé Equation 525
9.8 Conservation Laws of KdV 526
9.9 Nonlinear Schrödinger Equation 529
9.9.1 mKdV Equation and NLSE 529
9.9.2 Bright Soliton 531
9.9.3 Dark Soliton 533
9.9.4 Rogue Waves in Oceans 535
9.10 Other Nonlinear Wave Equations 541
9.11 Summary and Further Reading 543
9.12 Problems 545

CHAPTER 10: MATHEMATICAL THEORY FOR MAXWELL EQUATIONS 551

10.1 Introduction 551
10.2 Microscopic Maxwell Equations 552
10.2.1 Gauss Law for Electric Field 553
10.2.2 Gauss Law for Magnetism 553
10.2.3 Maxwell-Faraday Law 554
10.2.4 Ampere Circuital Law (with Maxwell Correction) 554
10.2.5 Dual Symmetry of Electromagnetic Waves in Vacuum Space 555
10.3 Integral versus Differential Forms 556
10.4 Macroscopic Maxwell Equations 560
10.5 Constitutive Relation and Ohm's Law 562
10.6 Electromagnetic Waves in Vacuum 564
10.7 Maxwell Equations in Gauss Unit 565
10.8 Boundary Conditions 565
10.9 Maxwell's Vector and Scalar Potentials 567
10.10 Gauge Freedom 568
10.10.1 Coulomb Gauge 569
10.10.2 Lorenz Gauge 569
10.10.3 Aharonov-Bohm Effect (Physical Meaning of Wave Potentials) 571
10.11 Solutions of Maxwell Equations: Jefimenko's Equations 571
10.11.1 Gradient Identity of Jefimenko 572
10.11.2 Curl Identity of Jefimenko 574
10.12 Electromagnetic Waves in Materials 576
10.13 Mathematical Theory for Lorenz Gauge 578
10.13.1 Hertz Vector for Electric Field 578
10.13.2 Gauge Invariance of Hertz Vector 580
10.13.3 Hertz Vector for Magnetic Polarization 581

10.13.4 Debye Potential Function for Transverse Magnetic Waves 582
10.13.5 Debye Potential Function for Transverse Electric Waves 587
10.14 Duality and Symmetry 589
10.15 Mathematical Theory for Coulomb Gauge 591
10.15.1 Scalar and Vector Potentials 591
10.15.2 Transverse Waves or Radiation Gauge 591
10.15.3 General Solution for Poisson Equation 593
10.15.4 Single- and Double-Layer Potentials 595
10.16 Kirchhoff Integral Formula for Waves 595
10.17 Summary and Further Reading 600
10.18 Problems 601

CHAPTER 11: QUANTUM MECHANICS AND SCHRÖDINGER EQUATION 603

11.1 Introduction 603
11.2 Black Body Radiation and Quantized Energy 604
11.3 Schrödinger Equation 608
11.3.1 One-Dimensional Schrödinger Equation 608
11.3.2 Three-Dimensional Schrödinger Equation 611
11.3.3 Wave Functions of Particles 612
11.3.4 Expectation Values 613
11.3.5 Stationary State of Energy *E* 615
11.4 Operators and Expectation Values 618
11.5 Classical Mechanics versus Quantum Mechanics 620
11.6 Hydrogen-Like Atom Model 620
11.6.1 Schrödinger Equation in Polar Form 621
11.6.2 Separation of Variables 622
11.6.3 Constraints Imposed by Wavefunctions 625
11.6.4 Laguerre and Associated Laguerre Polynomials 629
11.6.5 Orthogonality of Associated Laguerre Polynomials 635
11.6.6 Admissible Form of the Wavefunctions 640
11.7 Electron Spins 647
11.8 Schrödinger Equation for General Atoms 648
11.9 Radiative Transitions from Atoms 650
11.10 Summary and Further Reading 652
11.11 Problems 653

CHAPTER 12: CELESTIAL MECHANICS AND ASTRODYNAMICS ... 655

12.1 Introduction ... 655
12.2 Equation of Motion for a Rigid Mass ... 657
12.3 Mass under Gravitational Pull ... 658
12.4 Orbital Equations for an Artificial Satellite ... 662
12.5 Orbital Equations in Polar Form ... 664
12.6 Kepler's 1st Law ... 664
12.7 First Escape Velocity (Orbital Speed) ... 666
12.8 Second Escape Velocity (from Earth) ... 668
12.9 Third Escape Velocity (from Our Solar System) ... 671
12.10 Travel to the Moon ... 673
12.11 Kepler's Second Law ... 675
12.12 Kepler's Third Law (Newton's Law) ... 676
12.13 Energy in an Elliptic Orbit ... 679
12.14 Interplanetary Travel ... 683
12.14.1 Hohmann Transfer Orbit ... 684
12.14.2 Launching Time Window ... 688
12.15 Striking Speed of Meteors on Earth ... 691
12.16 Precession of the Perihelion of Mercury ... 692
12.16.1 Schwarzschild Metric for Curved Space-Time ... 694
12.16.2 Energy Term Due to Relativity ... 698
12.16.3 Contribution to Perihelion Precession ... 699
12.17 Motion near the Earth's Surface ... 702
12.18 Rocket and Missile Problem ... 703
12.19 Dynamic of Atmospheric Re-Entry ... 708
12.19.1 Formulation ... 709
12.19.2 Yaroshevsky Solution ... 711
12.20 Restricted Problem of Three Bodies ... 716
12.20.1 Formulation of the Three-Body Problem ... 716
12.20.2 Triangular Lagrangian Points ... 718
12.20.3 Three Collinear Lagrangian Points ... 719
12.20.4 Approximate Solution to Lagrange's Quintic Equation ... 720
12.21 Summary and Further Reading ... 725
12.22 Problems ... 725

CHAPTER 13: FRACTURE MECHANICS AND DYNAMICS ... 731

13.1 Introduction ... 731
13.2 Papkovitch-Neuber Potentials for Axisymmetric Elasticity ... 732
13.3 Mixed Boundary Value Problems as Potential Problems ... 736
13.4 Formulation of Dual Integral Equations ... 736
13.5 Penny-Shaped Crack Problem ... 738
13.5.1 Reduction of Dual Integral Equations to Abel Integral ... 738
13.5.2 Displacement Field Due to Uniform Pressure ... 742
13.5.3 Energy Change Due to Crack Presence ... 743

13.6 Papkovitch-Neuber Potentials for Plane Elasticity 744
13.7 Formulation of Dual Integral Equations 746
13.8 Griffith Crack Problem 747
13.8.1 Reduction of Dual Integral Equations to Abel Integral 747
13.8.2 Solutions 748
13.9 Fracture Dynamics in Wave Equations 749
13.10 Reduction of Wave to Harmonic Problem by Galilean Transform . 751
13.11 Mode I Asymptotic Field at Moving Crack Tip 753
13.11.1 Eigenvalue Problem 755
13.11.2 Asymptotic Fields 756
13.12 Mode II Asymptotic Field at Moving Crack Tip 758
13.12.1 Eigenvalue Problem 760
13.12.2 Asymptotic Fields 761
13.13 Mode III Asymptotic Field at Moving Crack Tip 762
13.13.1 Eigenvalue Problem 763
13.13.2 Asymptotic Fields 764
13.14 Asymptotic Field of Transient Crack Growth 765
13.15 Crack Growth with Intersonic Speed 771
13.15.1 Formulation 771
13.15.2 Mode I 773
13.15.3 Mode II 775
13.15.4 Asymptotic Field for Mode II Crack 777
13.16 Summary and Further Reading 780
13.17 Problems 782

References 785
Author Index 795
Subject Index 799

PREFACE

Studying engineering mathematics may appear as a long arduous journey passing through an "apparently barren and dead" desert. However, there is always an oasis hidden in the middle of a large desert. A journey crossing such arid and barren desert is not easy but it could be very rewarding. Once you discover the oasis, you will be fascinated by the diverse fauna and flora found in it. For example, Havasu Falls within the Havasu Indian Reservation is a breathtaking waterfall in an oasis hidden in the middle of Arizona high desert. To get there, you need to drive 193 km for more than three and one-half hours from the Grand Canyon Visitor Center, followed by 18 km of strenuous hiking in high desert canyons. I went there during my post-doctoral year. Likewise, once you acquire the theories and techniques for solving differential equations, you will be rewarded by your accessibility to "seemingly difficult" theories in engineering and mechanics.

Another analogy is like learning to swim. Once you have learned swimming, you can try platform and springboard diving, water polo, scuba diving, snorkeling, water skiing, surfing, wind surfing, kite surfing, canoeing or kayaking, kayak water polo, dragon boat racing, etc. The opportunities are endless. Once you master the basic mathematical skills in solving differential equations, you can understand and appreciate the mathematical theories behind celestial mechanics, geophysical flows, quantum mechanics, electrodynamics, cable-supported structures, etc. The knowledge accessible by you is endless.

This book consists of thirteen chapters. Our focuses are on mathematical techniques and the associated physical meaning of these thirteen topics, and whenever possible, applications to practical problems will be presented. Chapter 1 considers beams and columns which form the basics of structural mechanics, Chapter 2 discusses the theory and use of plates, Chapter 3 goes into shell theory and the associated method of solutions, Chapter 4 considers structural dynamics, Chapter 5 reviews catenary and cable supported bridges, Chapter 6 introduces nonlinear buckling, and Chapter 7 reviews turbulent diffusion in fluids. These 7 chapters cover some important applications of differential equations and their models in engineering. The next five chapters deviate from traditional topics in engineering. However, they are important topics in mechanics and its applications. Chapter 8 deals with a huge topic in fluid mechanics—geophysical fluid flows. This topic is important in view of the fact that climate change and the rise in sea level has become a hot topic in our society, and recognizing the role of fluid mechanics in it becomes more important. Chapter 9 goes into nonlinear waves and solitons. Although this is a topic originally motivated for civil engineering application (initiated by the solitary wave observations by a civil engineer named John Scott Russell), it has been traditionally studied in the domain of applied mathematics and physics. This is probably the most important and successful mathematical model of nonlinear differential equations that leads to exact solutions and insights into many physical problems (e.g., the nerve impulse in squid is a soliton that led to the Nobel Prize in physiology or medicine in 1963). Chapter 10 deals with the most important mathematical theory of all time—mathematical theory for Maxwell equations. All electronic advances nowadays rely on this successful theory. Traditionally, it is covered in physics and electrical engineering. With breakthroughs in nanotechnologies in recent years, engineers in all disciplines found that it is essential to learn and to become knowledgeable in Maxwell equations and electrodynamics. This chapter mainly looks at the mathematical techniques for solving Maxwell equations from the viewpoint of a mechanician. Chapter 11 covers quantum mechanics and Schrödinger's equation. This is again another huge and important topic in physics and electronics. With the advances in nano-mechanics and their applications in engineering, engineers and mechanicians find that it is essential to have some background in quantum mechanics. This chapter serves as an introduction from a mathematical point of view. Chapter 12 touches upon the oldest mechanics problem and the most classical applications of differential equations—celestial mechanics, and then touches upon astrodynamics or the dynamic problems of spacecrafts. It is also a classical problem with major applications in our daily lives. New technologies such as GPS (global positioning system), remote sensing, and InSAR (Interferometric synthetic aperture radar) rely on artificial satellites and are linked to celestial mechanics and astrodynamics. Finally, Chapter 13 covers fracture mechanics and dynamics, which finds applications in all aspects of engineering and earthquake dynamics.

These topics reflect my academic background and personal interests in engineering and mechanics. As an engineer, I obtained my undergraduate training in civil engineering, my master's degree in structural engineering, and PhD in theoretical and applied mechanics. I learned beams and columns mainly from Drs. S.K. Yan and K.K. Koo at the Hong Kong Baptist College (HKBC) and Profs. D.M. Brotton and W. Kanok-Nukulchai at the Asian Institute of Technology (AIT), plate bending and shell theories from Prof. P. Karasudhi and M. Wieland at AIT, structural dynamics from Dr. K.K. Koo at HKBC and Prof. M. Wieland at AIT, nonlinear buckling from Prof. E.L. Reiss at Northwestern University (NU), and fracture mechanics from Profs. J.W. Rudnicki, J.D. Achenbach, B. Moran, L.M. Keer, T. Mura, and J. Dundurs at NU. My interest in fracture dynamics came from sitting in on lectures on earthquake dynamics delivered by Prof. J.R. Rice at Harvard University. My interest in solitons came from my sitting in on lectures given by W. Kath at NU. My interest in turbulent diffusion in fluids came from my contact with Prof. S. Vongvisessomjai at AIT, where I sat in on his lectures on wave hydrodynamics. My interest in geophysical flows came from my studies on tsunami and storm surges, and my accidental acquisition of the wonderful book by Cushman-Roisin (1994). My interest in tornadoes was aroused by a personal story told by Prof. E.L. Reiss on his encounter of a tornado on a Florida highway (he shared it in his class on bifurcation theory). My knowledge of cable-supported bridges came from my personal studies of the books by Irvine (1981) and Rocard (1957) during the construction of Tsing Ma Bridge in Hong Kong. My self-learning on Maxwell equations and quantum mechanics was aroused by discussions with my son Magnum, who majored in physics at the Hong Kong University of Science and Technology. My interest in celestial mechanics and astrodynamics came from the late Nobel Prize laureate Richard Feynman. In 1988, when I was studying at the Science library at Northwestern University, I came across a newspaper article on Richard Feynman's death and his "investigation" of the space shuttle Challenger disaster. I read both of his books *Surely You're Joking, Mr Feynman!* and *What Do You Care What Other People Think*? They caused me to recall my memory of watching the explosion of the Challenger during takeoff when I was at the Asian Institute of Technology. I also recalled watching the disaster during the atmospheric re-entry of the space shuttle Columbia in 2003.

In this book, I have included a number of interesting examples in engineering and mechanics: the stress estimation of the top spherical shell of the Pantheon in Rome, the prediction of the vibration frequency of Tacoma Narrows Bridge before its failure using linear theory (it agrees surprisingly well with observations), the maximum wind speed in tornadoes, translunar and interplanetary travels and their launching time window, the maximum deceleration of spacecraft during atmospheric re-entry, the firing range of ballistic missiles, the precession of the perihelion of Mercury due to correction from relativity, the quantized energy state of an electron predicted by Schrödinger's equation, comparison of the Peregrine breather solution with the 1995 Draupner platform's freak wave record, estimation of storm surge height in Hong Kong during the 2017 Hato Typhoon, snap-through buckling of a two-bar system, the physical meaning of Rayleigh damping in structural dynamics, the design of a seismograph, and a complete proof of Jefimenko's solution of Maxwell's equations.

When I was writing this book, my beloved father passed away on February 20, 2017 at the age of 103, and two days later Dr. S.K. Yan, who inspired my interest in engineering mathematics and "created" an extra post to hire me as a full-time tutor at the Hong Kong Baptist College, also passed away. This book is dedicated to them.

My father, Chow Yat Wing, came to Hong Kong alone from Canton, China as a teenager with only a few dollars in his pocket. He never had a chance to attend school, either in Hong Kong or in mainland China. When I was in primary school, I would spend all of my time in helping the business at our tiny grocery shop (0.6 m by 3 m) at the corner of Maple Street and Cheung Sha Wan Road after school (I attended A.M. school only). In the late afternoon every day before dusk, my mother would go home to prepare dinner. There were not many customers at those times, and my father would tell me lots of stories. One of these stories still sticks in my mind today, and it was about how my father survived during the Japanese occupation of Hong Kong in the midst of World War II. When the Japanese first invaded Kowloon and the New Territories in December 1941, and the British army was still defending and holding on the Hong Kong Island side, there were ocean liners that got caught in the middle of the war and captains had to abandon their vessels inside the Victoria Harbor. There was fierce fighting across the harbor in the daytime. At night, spotlights were projected from both the Japanese side and the British side onto the harbor. They would fire at anything trying to cross the harbour. A few days before the Japanese successfully crossed Victoria Harbor, my father, who did not know how to swim, and a friend took a sampan to approach these abandoned ocean liners in the harbor. They climbed up the anchor chain of the vessels, carried bags of goods from the vessels back to their sampan, and secretly paddled back to the Japanese side. They would sell them to the local people to earn a living. It was an extremely risky business. If they were spotted by either the Japanese side or the British side, they would be shot. If my father fell into the sea, he would drown. I learned to appreciate things that I had, and to seize the moment when opportunities arose.

I returned to Hong Kong after a 45-day Xinjiang trip (a Silk Road trip) in July 1984. It was my graduation trip and I took off the day after my last final examination. When I returned to Hong Kong, I asked Dr. S.K. Yan, the Head of the Civil Engineering Department at HKBC, about the possibility of getting the full time tutor job at the department. This one-year tutor job would be crucial to me, as it would allow me to save money for my postgraduate studies (my initial plan was to enroll in a one-year master's degree program in the UK). Dr. Yan told me two tutor positions had been filled by my classmates. I applied too late. I felt lost and disappointed. I just didn't know what to do at that time. Every day, I returned to HKBC to play basketball, volleyball and watch others playing in the playground. Evidently, Dr. Yan saw me doing nothing every day except showing up at the playground. After about a week, Dr. Yan approached me in the playground and asked me whether I got a job yet. I told him that I was not looking for any job at all. Then, he said that he had "created" an extra tutor job for me, if I was still interested in it. Oh my god, it was a miracle to me. It turned out that Dr. Yan was able to juggle the budget and came up with half of the budget for another full-time tutor, and then he approached the Dean of the Science Faculty, Dr. Burnett, and persuaded him to support the remaining half of the budget to help the Civil Engineering Department. Without his help, I am sure I could not have the chance today to write this book.

While working on this ambitious book project, I was able to keep my regular regiments of swim training with the PolyU swimming team (serving as the honorable manager, I have the privilege to train with the swimmers), and to continue joining swim competitions with swimmers from the Sea Green Lifesaving and Swimming Club. In the last few years, I am particularly indebted to the training from Coach Mrs. Ngan, Coach Adrian Liu, Coach Dean Chan, Coach Yuen Fong, Coach Pasu Ka Po Chung, and Coach Kasu Andy Li of the PolyU swimming team, and from Coach King Man Lo (Lo Sir), and Coach Peter Chan of the Sea Green Lifesaving and Swimming Club. Inspired by Coach Peter Chan and Coach Eagle Wong, I joined the innovative marathon training by Coach Eagle Wong in October 2017. It has been twenty-something years since I finished the China Coast Marathon in 1994 and the Chicago Oldstyle Marathon in 1991. I eventually finished another full marathon in February of 2018, with only two training sessions per week. This allowed me to have enough time and energy to finish this book project. I appreciated the tolerance and encouragement from Coach Eagle Wong throughout the 4-month training.

My wife Lim continued to cover my back during this book project. Being a student who majored in physics, my son Magnum inspired my interests in electrodynamics and quantum mechanics. Being a student who majored in fine arts, my daughter Jaquelee reminded me to appreciate life.

Special thanks go to Professor James R. Rice of Harvard University, Professor John W. Rudnicki of Northwestern University, and Professor Ken P. Chong of George Washington University for agreeing to write the back cover notes for my books. I am indebted for their generosity and time. This book project was encouraged by Mr. Tony Moore, a senior editor of civil engineering at CRC Press (imprint of Taylor & Francis). The expert assistance from Production Editor Michele Dimont and Editorial Assistant Gabrielle Williams is highly appreciated.

K.T. Chau
The Hong Kong Polytechnic University

THE AUTHOR

Professor K.T. Chau, Ph.D., is the Chair Professor of Geotechnical Engineering of the Department of Civil and Environmental Engineering at the Hong Kong Polytechnic University. He obtained his honors diploma with distinction from Hong Kong Baptist College (Hong Kong), his master's of engineering in structural engineering from the Asian Institute of Technology (Thailand) where he was also awarded the Tim Kendall Memorial Prize (an academic prize for the best graduating student) with straight As, his Ph.D. in Theoretical and Applied Mechanics from Northwestern University (U.S.A.), and an Executive Certificate from the Graduate School of Business of Stanford University.

Dr. Chau worked as a full-time tutor/demonstrator/technician at Hong Kong Baptist College (1984–1985), as a research associate at the Asian Institute of Technology (summer of 1987), a research assistant at Northwestern University (1987–1991), and as a post-doctoral fellow at Northwestern University (1991–1992). At Hong Kong Polytechnic University (PolyU), he has served as a lecturer, an assistant professor, an associate professor, a full professor and a chair professor since 1992. At PolyU, he served as the Associate Dean (Research and Development) of the Faculty of Construction and Environment, the Associated Head of the Department of Civil and Structural Engineering, the Chairman of the Appeals and Grievance Committee, the Alternate Chairman of the Academic Appeals Committee, and the Alternate Chairman of the University Staffing Committee.

Dr. Chau is a fellow of the Hong Kong Institution of Engineers (HKIE), the past Chairman of the Geomechanics Committee (2005–2010) of the Applied Mechanics Division (AMD) of ASME, the Chairman of the Elasticity Committee (2010–2013) of the Engineering Mechanics Institute (EMI) of ASCE, and Chairman of the TC103 of the ISSMGE. He is a recipient of the Distinguished Young Scholar Award of the National Natural Science Foundation, China (2003), the France-Hong Kong Joint Research Scheme (2003–2004) of RGC of Hong Kong, and the Young Professor Overseas Placement Scheme of PolyU. He was a recipient of the Excellent Teaching Award of the Civil and Engineering Department (2014). He is a past president of the Hong Kong Society of Theoretical and Applied Mechanics (2004–2006) after serving as member-at-large and vice president. He also served as a Scientific Advisor of the Hong Kong Observatory of HKSAR Government, an RGC Engineering Panel member of the HKSAR Government for 7 consecutive years, and served as the Vice President of the Hong Kong Institute of Science. He has delivered more than 12 keynote lectures at international/national conferences, served on the advisory committee of 20 international conferences, and on organizing committees of 18 international conferences. He also held visiting positions at Harvard University (USA), Kyoto University (Japan), Polytech-Lille (France), Shandong University (China), Taiyuan

University of Technology (China), the Rock Mechanics Research Center of CSIRO (Australia), and the University of Calgary (Canada).

Dr. Chau's research interests have included geomechanics and geohazards, including bifurcation and stability theories in geomaterials, rock mechanics, fracture and damage mechanics in brittle rocks, three-dimensional elasticity, earthquake engineering and mechanics, landslides and debris flows, tsunami and storm surges, and rockfalls and dynamic impacts, seismic pounding, vulnerability of tall buildings with transfer systems, and shaking-table tests. He is the author of more than 100 journal papers and 200 conference publications. His book *Analytic Methods in Geomechanics* published by CRC Press in 2013 was a major book covering many important topics in geomechanics.

In his leisure time, he enjoys swimming and takes part in master swimming competitions. He is the Honorable Manager of the Hong Kong Polytechnic University Swimming Team. Since 2001, he has competed in Hong Kong Masters Games, the Hong Kong Territory-wise Age-Group Swimming Competition, the Hong Kong Amateur Swimming Association (HKASA) Masters Swimming Championships, and District Swimming Meets of the Leisure and Cultural Services Department (LCSD). He has also participated in international masters swimming competitions, including the Macau Masters Swimming Championship, the Singapore National Masters Swimming, Standard Chartered Asia Pacific Masters Swim Meet, Wisdom-Act International Swimming Championship (Taiwan), Standard Chartered Singapore Masters Swim 2007, Japan Masters Long Distance Swim Meet 2008 (Aichi Meet and Machida Meet), Japan Short Course Masters Swimming Championship 2009 (Kyoto), Marblehead Sprint Classics (USA), the Masters Games Hamilton (New Zealand), Hawaii Senior Olympics, National China Masters Swimming Championships, the Third Annual Hawaii International Masters Swim Meet, and the Fifth Penang Invitational Masters Swimming Championship. By 2007, he had competed in all long-course FINA events (i.e., 50 m, 100 m, 200 m, 400 m, 800 m and 1500 m freestyle; 50 m, 100 m and 200 m butterfly; 50 m, 100 m and 200 m breaststroke; 50 m, 100 m and 200 m backstroke; and 200 m and 400 m individual medley).

He also enjoys jogging and has completed five full marathons, including the Hong Kong International Marathon, the Chicago Oldstyle Marathon, the China Coast Marathon, and the Hong Kong Standard Chartered Marathon, with a personal best of 3 hours, 34 minutes and 15 seconds.

CHAPTER ONE

Theory of Beams and Columns

1.1 INTRODUCTION

Beams and columns have been used for thousands of years in human-made structures. Columns arranged in rectangular patterns were found in many ruins of the ancient Greek temples, showing the popularity of using stone columns in the ancient time. Beams were used in supporting roof structures in these ancient temples, but unfortunately many of these beam-supported roofs did not survive the shaking induced by historical earthquakes.

Galileo Galilei is often credited with the first published theory of the strength of beams in bending. In the "Codex Madrid" by Leonardo da Vinci (published in 1493), the use of beam in hoisting system of weapons is can be found in his hand drawings (see for example Figure 1.1). In the 18th century, it was Jacob Bernoulli first discovered that the curvature of a beam at any point is proportional to the bending moment at that point. The equation of motion for a vibrating beam was first derived by Euler in 1744 and Jacob's nephew, Daniel Bernoulli, in 1751. It was Euler who fully developed the related result for this beam theory, including the celebrated Euler's buckling formula. Therefore, it is normally referred as Euler-Bernoulli beam theory, Bernoulli-Euler beam theory, simple beam theory, or the classical beam theory. A major portion of the present chapter is devoted to this theory. The main shortcoming of this theory is that shear deformation has been neglected in the formulation. In other words, the shear modulus has been assumed infinite. Consequently, the response of Euler-Bernoulli beam is stiffer than the real beam, and for vibrating beams the natural frequency has been overestimated.

In this chapter, we will consider some applications of beam models in mechanics and engineering. Both static and dynamic problems of Euler-Bernoulli beam theory will be considered. The major assumption for the Euler-Bernoulli beam theory is that a vertical plane perpendicular to the neutral axis of the beam before bending remains a plane and normal to the deformed neutral surface. The neutral axis after bending locates on the centroid of the cross-section of the beam. This assumption is found acceptable only for thin beams with aspect ratio larger than ten. Euler buckling formula for columns is also considered. The case of initial imperfection of column buckling is considered, yielding a transition from straight to buckled state. However, more complex case of nonlinear buckling of columns and beams is deferred to Chapter 6. The effect of axial force in beam vibrations is also investigated. Some new results are summarized.

In 1877, Lord Rayleigh introduced the rotatory inertia term in beam vibrations. In 1921, Timoshenko proposed a major improvement to the Euler-Bernoulli beam theory by incorporating shear deformation in terms of an additional rotation at the neutral of the beam cross-section (i.e., the plane after deformation is no longer perpendicular to the neutral surface). In this equivalent process, a factor called shear coefficient is introduced to take care of the cross-section dependence of the shear stress distribution. In Timoshenko beam theory, the assumption that a straight vertical plane before bending remains a plane after bending is preserved. However, the plane is no longer perpendicular to the neutral surface of the deflected beam because of the additional rotation. Timoshenko beam theory is applicable to so-called deep beams with aspect ratio of less than 10. In particular, variational principle is used in deriving Timoshenko beam theory, and both static and dynamic problems are considered. Various applications are considered, including the designs of a rocket/missile launch pad, and seismograph.

Figure 1.1 A sketch of a cantilever beam drawn by da Vinci

1.2 BEAM BENDING

The theory of the flexural strength and stiffness of beams is now attributed to Bernoulli and Euler who developed it almost 400 years ago. Its application to large scale engineering development is more recent, probably becoming more popular when the Effiel Tower and Ferris Wheel was designed and constructed using iron and steel during the industrial revolution. The distribution of bending stress in beams has long fascinated scientists. In addition to da Vinci and Galileo, contributions have been made by Mariotte, Parent, Daniel Bernoulli and Euler. Some early development of beam theory is summarized in Timoshenko (1956) and Heyman (1999). Some experiments were also conducted by Coulomb in 1773 to verify the strength of beam.

We start with the derivation of the Euler-Bernoulli beam theory. Figure 1.2 shows a free body diagram of a beam with a trapezoidal section subject to distributed load $w(x)$, end shear force $V(x)$, and end concentrated bending moment $M(x)$. The deflection and slope of deflection are denoted v and ϕ. Various kinds of supports at the end of the beam will be discussed in the next section. As illustrated in Figure 1.2, the sign convention for the shear force is that all shear forces leading to clockwise moment of the free body are considered as positive, whereas the moments leading to a sagging shape of the beam are taken as positive. For this reason, they are also called sagging moments by engineers. The moments at the end that induce an upward bending of the beam segment is treated as negative and they are called hogging moments.

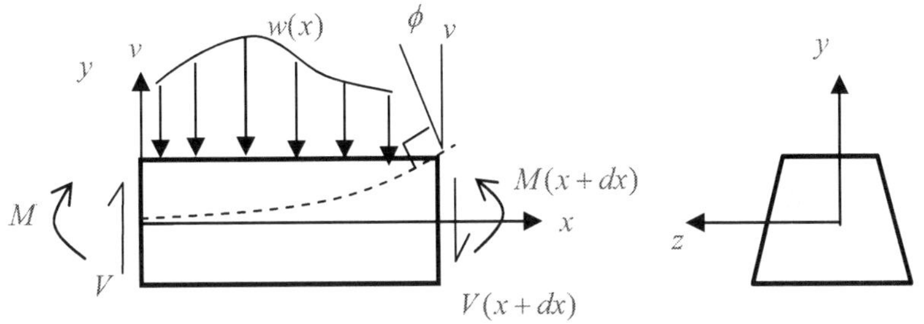

Figure 1.2 A beam segment with applied load of *w(x)* and end moment *M* and shear *V*

The vertical force equilibrium of the free body can be formulated as

$$V-(V+dV)-w(x)dx=0 \tag{1.1}$$

Using the first term in the Taylor's series expansion, we have

$$V+dV=V+\frac{dV}{dx}dx \tag{1.2}$$

Substitution of (1.2) into (1.1) gives the following relation between the disturbed load w and the change of shear force

$$\frac{dV}{dx}=-w(x) \tag{1.3}$$

The moment equilibrium of the beam segment gives

$$M+dM-M-Vdx+(wdx)\cdot\frac{dx}{2}=0 \tag{1.4}$$

For the case of $dx\to 0$, the second order term can be neglected and this assumption leads to the expression linking shear force to the change of the moment along the beam

$$\frac{dM}{dx}=V \tag{1.5}$$

Note that the distributed load $w(x)$ does not appear in (1.5) but only through the calculation of V indirectly given in (1.3). Substitution of (1.5) into (1.3) gives a relation between bending moment and distributed load:

$$\frac{d^2M}{dx^2} = -w(x) \tag{1.6}$$

The so-called Euler-Bernoulli equation will be considered next, which links moment to deflection.

1.2.1 Euler-Bernoulli Beam

In this section, we will derive the simplest type of beam theories which is known to mathematicians as Euler-Bernoulli beam and to engineers as simple beam theory. Approximation has to be made in linking the bending moment with the deflection of the beam. Figure 1.3 shows the situation that the deflected beam can be approximated by an arc of a circle with a radius of ρ. In addition, the plane normal to the neutral axis before bending remains normal to the neutral axis after bending.

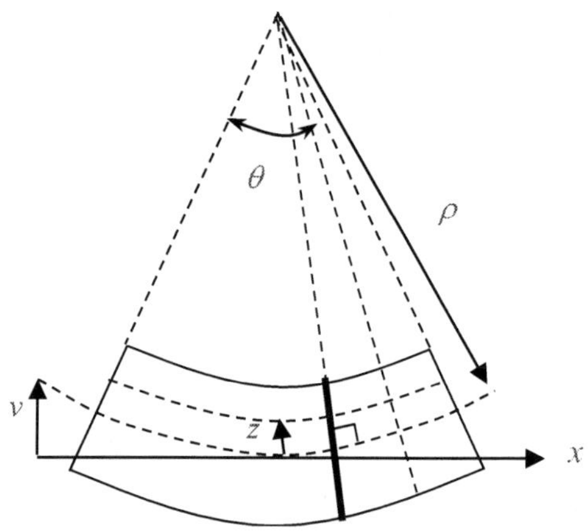

Figure 1.3 The plane-remain-plane assumption in Euler-Bernoulli beam theory

The slope of the deflection of the beam is given by

$$\phi = \frac{dv}{dx} \tag{1.7}$$

The radius of curvature is defined as the change of the slope

$$k = \frac{d\phi}{dx} \tag{1.8}$$

If the beam is relatively long with an aspect ratio of larger than 10, the beam is considered as a simple beam and its deflection curve can be approximated by a circular arc. The radius of curvature can be assumed proportional to the bending moment only

$$k = \frac{1}{\rho} = \frac{M}{EI} = \frac{d^2v}{dx^2} \tag{1.9}$$

The first expression in (1.9) reflects the fact that the radius of curvature is the inverse of the radius ρ shown in Figure 1.3. The assumption that the radius of curvature is proportional to the bending moment implies that a vertical plane on the beam cross-section remains a plane (although inclined) after bending. In addition, this plane remains normal to the neutral axis of the beam after deflection (if this is not true we will have shear deformation in the beam which will be discussed later in the Timoshenko beam theory). This assumption is normally referred as plane-remain-plane assumption. A major consequence of this assumption leads to the linear distribution of the bending stress along the cross-section of the beam. It is also a consequence of neglecting shear deformation in the beam. This is a good approximation only for the case that the depth of the beam d is small compared to the span of the beam L. If the ratio L/d is larger or equal to about 10, the plane-as-plane assumption provides an acceptable result. For deep beams, $L/d << 10$, shear deformation becomes important and there will be shear deformation in the plane after bending. The bending stress is no longer linear in the beam section. Strictly speaking all transverse loading w will induce shear stress and, in turn, shear deformation on the cross-section. Thus, Euler-Bernoulli beam theory is not exact for the beam with transverse load, but it is exact for the case that only constant moment is applied to the beam because there is no shear deformation in this case.

Substitution of (1.9) into (1.5) gives the Euler-Bernoulli beam theory:

$$\frac{d^2M}{dx^2} = \frac{d^2}{dx^2}(EI\frac{d^2v}{dx^2}) = -w(x) \tag{1.10}$$

We will illustrate how to solve this in the next section. Here we first consider the special case that the beam is uniform along x and there is no distributed load:

$$EI\frac{d^4v}{dx^4} = 0 \tag{1.11}$$

Integrating (1.11) four times, we obtaining the solution as

$$v(x) = a_1x^3 + a_2x^2 + a_3x + a_4 \tag{1.12}$$

We now formulate the solution of a fundamental beam element subject to end displacements and end rotations as shown in Figure 1.4.

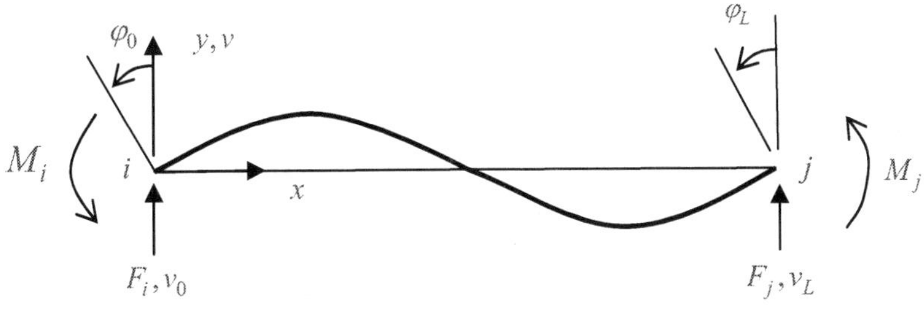

Figure 1.4 A beam subject to end displacements and rotations

For the given end deflection and end rotation at $x = 0$, we require

$$v(0) = v_0 = a_4 \tag{1.13}$$

$$\frac{dv(0)}{dx} = \varphi_0 = a_3 \tag{1.14}$$

For the given end deflection and end rotation at $x = L$, we obtain

$$v(L) = v_L = a_1 L^3 + a_2 L^2 + a_3 L + a_4 \tag{1.15}$$

$$\frac{dv(L)}{dx} = \varphi_L = 3a_1 L^2 + 2a_2 L + a_3 \tag{1.16}$$

This provides a system of four equations for four unknowns, and the solution of this system gives

$$\begin{aligned} v(x) = &\left[\frac{2}{L^3}(v_0 - v_L) + \frac{1}{L^2}(\varphi_0 + \varphi_L)\right]x^3 \\ &+\left[-\frac{3}{L^2}(v_0 - v_L) - \frac{1}{L}(2\varphi_0 + \varphi_L)\right]x^2 + \varphi_0 x + v_0 \end{aligned} \tag{1.17}$$

This solution provides a fundamental result that can be used to formulate the stiffness matrix of beam element used in finite element analysis (see Problem 1.16).

For more general beam problems, four types of end conditions are commonly considered. They are:

(i) <u>Boundary condition I</u>: Fixed end boundary is defined by

$$v(0) = 0, \quad \frac{dv(0)}{dx} = 0 \tag{1.18}$$

Physically, these enforce zero deflection and zero rotation at the support.

(ii) <u>Boundary condition II</u>: Free end boundary is defined by

$$\frac{d}{dx}[EI\frac{d^2 v(0)}{dx^2}] = 0, \quad \frac{d^2 v(0)}{dx^2} = 0 \tag{1.19}$$

Physically, these enforce zero shear and zero moment at the support.

(iii) <u>Boundary condition III</u>: Simply-supported end boundary is defined by

$$v(0) = 0, \quad \frac{d^2 v(0)}{dx^2} = 0 \tag{1.20}$$

Physically, these enforce zero deflection and zero moment at the support.

(iv) <u>Boundary condition IV</u>: Guided end boundary is defined by

$$\frac{d}{dx}[EI\frac{d^2 v(0)}{dx^2}] = 0, \quad \frac{dv(0)}{dx} = 0 \tag{1.21}$$

This enforces zero shear force and zero rotation at the support. These four types of supports are illustrated in Figure 1.5. Problem 1.12 at the end of the chapter considers the more general case of spring supports (both linear translational and rotational springs) for beam formulation. Such spring supports apparently are more important in considering interactions between a beam and the structures that they rest on. For example, the vehicle suspension system is normally modeled by springs. For soil-structure interaction problems, beam on elastic foundations has been formulated using the so-called Winkler spring model (Hetenyi, 1946), which will be considered in a later section. If these springs are nonlinear, such problems become much more complicated mathematically. Bifurcation solution and chaotic response of the beams may be expected. In the context of nonlinear buckling, Example 6.2 will consider the case of a two-beam system with a nonlinear spring at the joint.

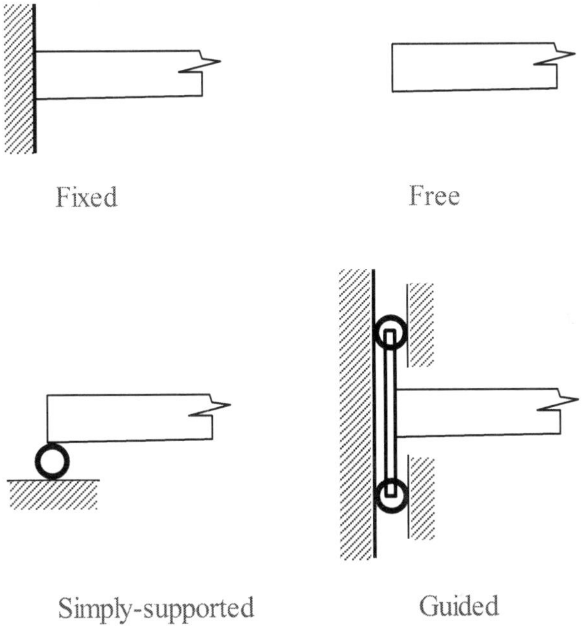

Figure 1.5 Four commonly used beam supports

1.2.2 Simply-Supported Beam

In this section, we return to the case of nonzero distributed loads using a simply-supported beam subject to uniformly distributed load, as shown in Figure 1.6.

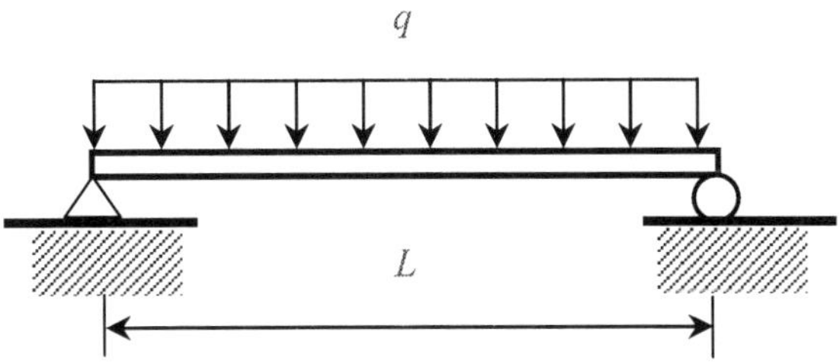

Figure 1.6 A simply-supported beam of length L subject to distributed load q

The governing equation for this case is

$$EI\frac{d^4v}{dx^4} = -q(x) \tag{1.22}$$

Integrating four times, we get

$$v(x) = -\frac{qx^4}{24EI} + a_1x^3 + a_2x^2 + a_3x + a_4 \tag{1.23}$$

The boundary conditions for a simply-supported end at $x = 0$ require

$$a_2 = a_4 = 0 \tag{1.24}$$

The boundary conditions at $x = L$ give two equations for two unknowns as:

$$-\frac{qL^2}{2EI} + 6a_1L = 0, \quad -\frac{qL^4}{24EI} + a_1L^3 + a_3L = 0 \tag{1.25}$$

The solutions of (1.25) are

$$a_1 = \frac{qL}{12EI}, \quad a_3 = -\frac{qL^3}{24EI} \tag{1.26}$$

Finally, substitution of (1.26) into (1.23) yields the deflection as

$$v = -\frac{qx(L^3 + x^3 - 2Lx^2)}{24EI} \tag{1.27}$$

The corresponding slope of deflection, shear force, and bending moment can be found by differentiation.

1.2.3 Cantilever Beam

In this section, we consider cantilever beam problems by considering a point load applied at the tip of the free end, as shown in Figure 1.7.

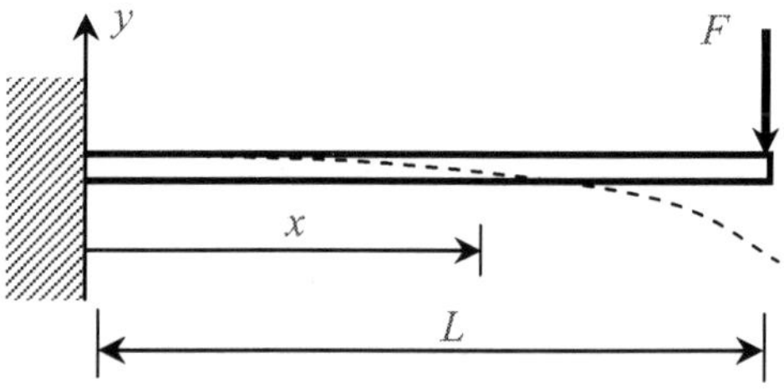

Figure 1.7 A cantilever beam of length *L* subject to an endpoint load *F*

Since there is no distributed load in this case, we have the Euler-Bernoulli beam equation being

$$EI\frac{d^4v}{dx^4} = 0 \tag{1.28}$$

Recall the solution that

$$v(x) = a_1x^3 + a_2x^2 + a_3x + a_4 \tag{1.29}$$

The boundary conditions for the fixed end support are

$$v(0) = 0, \quad \frac{dv}{dx}(0) = 0 \tag{1.30}$$

On the other hand, the free end with a concentrated point force can be formulated as:

$$\frac{d^2v}{dx^2}(L) = 0, \quad EI\frac{d^3v}{dx^3}(L) = -F \tag{1.31}$$

Taking the differentiation of v given in (1.29), we get

$$\frac{dv}{dx} = 3a_1x^2 + 2a_2x + a_3 \tag{1.32}$$

$$\frac{d^2v}{dx^2} = 6a_1x + 2a_2 \tag{1.33}$$

$$\frac{d^3v}{dx^3} = 6a_1 \tag{1.34}$$

The boundary conditions at the fixed end result in

$$a_3 = a_4 = 0 \tag{1.35}$$

The boundary conditions at the free ends give two equations for the remaining unknowns as:

$$6a_1L + 2a_2 = 0, \quad a_1 = -\frac{F}{6EI} \tag{1.36}$$

The solution of (1.36) gives

$$a_1 = -\frac{F}{6EI}, \quad a_2 = \frac{FL}{2EI} \tag{1.37}$$

Substitution of these results into (1.29) gives

$$v(x) = \frac{F}{6EI}(3Lx^2 - x^3) \tag{1.38}$$

The corresponding expressions of slope, bending moment and shear force along the beam can be obtained by differentiation, and they are

$$\phi = \frac{dy}{dx} = \frac{xF}{2EI}(2L - x) \tag{1.39}$$

$$M = EI\frac{d^2y}{dx^2} = F(L - x) \tag{1.40}$$

$$V = EI\frac{d^3y}{dx^3} = -F \tag{1.41}$$

This completes all force and moment analyses for the cantilever beam problem. The distribution of bending stress can be found by using

$$\sigma_{xx} = \frac{My}{I_x} \tag{1.42}$$

where y is the vertical coordinate along the deflection of the beam, and I_x is the moment of inertia of the section for bending about the x-axis.

1.2.4 Cable Load

For a cantilever type of canopy or beam, reinforcing cable force can be used to stabilize the beam. In this section, we consider a beam subject to a cable force at the tip of the free end in a way that, the angle of the cable force makes an angle of α with the vertical axis as shown in Figure 1.8, regardless of the deflection of the beam. This also closely resembles cable-carrying electric transmission towers. It is more convenient to choose the coordinate from the tip of the cantilever beam. Instead of starting from the differential equation of the beam deflection, it is easier to solve this problem by starting from the bending moment equation:

$$EI\frac{d^2y}{dx^2} = M(x) = Py\sin\alpha - Px\cos\alpha \tag{1.43}$$

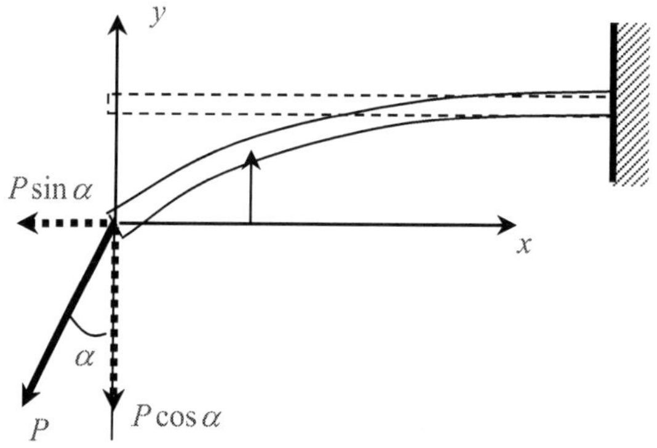

Figure 1.8 A cantilever beam subject to a cable force at the free end

The right-hand side of (1.43) shows the sagging and hopping moments induced by the horizontal and vertical components of the cable force respectively. Rearranging this equation, we get

$$\frac{d^2y}{dx^2} - \frac{P}{EI}\sin\alpha\, y = -\frac{P}{EI}x\cos\alpha \tag{1.44}$$

Since the coefficient of the ODE is constant, the solution is of exponential form (e.g., Chau, 2018). The homogeneous form of the ODE is

$$\frac{d^2y_h}{dx^2} - \frac{P}{EI}\sin\alpha\, y_h = 0 \tag{1.45}$$

Assuming an exponential solution, we find

$$(\lambda^2 - \frac{P}{EI}\sin\alpha)e^{\lambda x} = 0 \tag{1.46}$$

Therefore, we have

$$\lambda = \pm\sqrt{\frac{P}{EI}\sin\alpha} \tag{1.47}$$

Consequently, the homogeneous solution is

$$y_h = A\exp(x\sqrt{\frac{P}{EI}\sin\alpha}) + B\exp(-x\sqrt{\frac{P}{EI}\sin\alpha}) \tag{1.48}$$

For the original ODE with nonhomogeneous term, we can assume

$$y_p = Cx \tag{1.49}$$

Substitution of (1.49) into (1.44) gives

$$C = \frac{\cos\alpha}{\sin\alpha} = \cot\alpha \tag{1.50}$$

Finally, we have the solution as

$$y = y_h + y_p = Ae^{x\sqrt{\frac{P\sin\alpha}{EI}}} + Be^{-x\sqrt{\frac{P\sin\alpha}{EI}}} + x\cot\alpha \tag{1.51}$$

With the chosen coordinate, we can impose the boundary condition as:

$$y(0) = 0, \quad \frac{dy}{dx}(L) = 0 \tag{1.52}$$

Using these boundary conditions, we obtain

$$A = -B \tag{1.53}$$

$$\sqrt{\frac{P\sin\alpha}{EI}}\{A\exp[L\sqrt{\frac{P\sin\alpha}{EI}}] - Be[-L\sqrt{\frac{P\sin\alpha}{EI}}]\} + \cot\alpha = 0 \tag{1.54}$$

Solving for the unknown constants, we find

$$B = -A = \frac{\cot\alpha}{2\sqrt{\frac{P\sin\alpha}{EI}}\cosh(\sqrt{\frac{P\sin\alpha}{EI}}L)} \tag{1.55}$$

Substituting (1.55) into (1.51) gives

$$y = \cot\alpha\{x - \frac{\sinh(\sqrt{\frac{P\sin\alpha}{EI}}x)}{\sqrt{\frac{P\sin\alpha}{EI}}\cosh(\sqrt{\frac{P\sin\alpha}{EI}}L)}\} \tag{1.56}$$

This solution is only valid if the cable is applied on the left of the vertical axis. Figure 1.9 shows the case that the cable force makes an angle β on the right of the vertical axis. Moment equilibrium becomes

$$\frac{d^2y}{dx^2} + \frac{P}{EI}\sin\beta y = \frac{P}{EI}x\cos\beta \tag{1.57}$$

In this case, both cable force components lead to hogging moments. The differential equation changes type. The characteristic equation for the exponential type of solution (since it is again of constant coefficients) becomes

$$(\lambda^2 + \frac{P}{EI}\sin\beta)e^{\lambda x} = 0 \tag{1.58}$$

Therefore, we have

$$\lambda = \pm i\sqrt{\frac{P\sin\beta}{EI}} \tag{1.59}$$

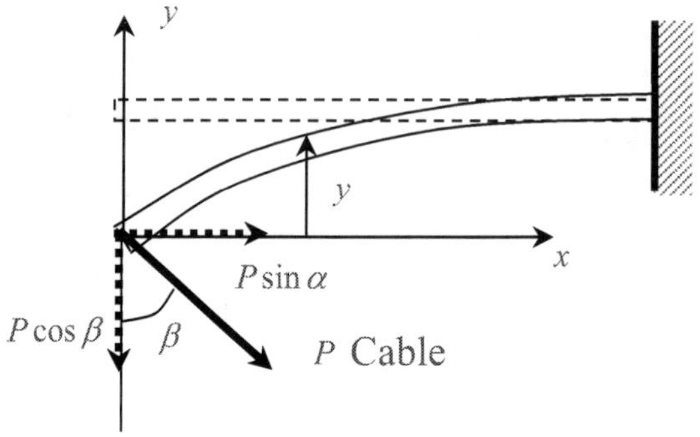

Figure 1.9 A cantilever beam subject to a cable force at the free end forming a hopping moment

Consequently, the homogeneous solution is

$$y_h = A\sin(\sqrt{\frac{P\sin\beta}{EI}}x) + B\cos(\sqrt{\frac{P\sin\beta}{EI}}x) \tag{1.60}$$

For the original ODE with the nonhomogeneous term, we can assume

$$y_p = Cx \tag{1.61}$$

Substitution of (1.61) into (1.57) gives

$$C = -\frac{\cos\beta}{\sin\beta} = -\cot\beta \tag{1.62}$$

Finally, the general solution becomes

$$y = A\sin(\sqrt{\frac{P\sin\beta}{EI}}x) + B\cos(\sqrt{\frac{P\sin\beta}{EI}}x) - x\cot\beta \tag{1.63}$$

Using the boundary conditions given in (1.52), we obtain

$$B = 0 \tag{1.64}$$

$$\sqrt{\frac{P\sin\beta}{EI}}A\cos(\sqrt{\frac{P\sin\beta}{EI}}L) - \cot\beta = 0 \tag{1.65}$$

The solution of (1.65) gives

$$A = \frac{\cot\beta}{\sqrt{\frac{P\sin\beta}{EI}}\cos(\sqrt{\frac{P\sin\beta}{EI}}L)} \tag{1.66}$$

Substituting (1.64) and (1.66) into (1.63), we obtain the final solution as

$$y = \cot\beta\{\sin(\sqrt{\frac{P\sin\beta}{EI}}x)/[\sqrt{\frac{P\sin\beta}{EI}}\cos(\sqrt{\frac{P\sin\beta}{EI}}L)] - x\} \tag{1.67}$$

1.2.5 Green's Function for Simply-Supported Beams

As discussed in Chau (2018), Green's function method is a powerful technique in solving ODEs, although the final solution is normally expressed in the form of an integral, to which numerical integration is inevitable. Without going into the detail, we only summarize the result of the Green's function for beams. In particular, the Green's function G satisfies the following ODE obtained in Example 8.4 of Chau (2018):

$$EI\frac{d^4G}{dx^4} = \delta(x-\xi) \tag{1.68}$$

For the case of the simply-supported beam, boundary conditions are

$$G(0) = G(L) = 0, \quad EI\left.\frac{d^2G}{dx^2}\right|_{x=0} = 0, \quad EI\left.\frac{d^2G}{dx^2}\right|_{x=L} = 0 \tag{1.69}$$

The Green's function was obtained as (Chau, 2018):

$$G_u(x,\xi) = \frac{1}{6EI}[(x-\xi)^3 H(x-\xi) + \frac{L-\xi}{L}(-x^2 + 2\xi L - \xi^2)x] \tag{1.70}$$

The solution of a beam subject to an arbitrary load q can be calculated as:

$$u(x) = \int_0^L G_u(x,\xi)q(\xi)d\xi \tag{1.71}$$

For the case of moment, the Green's function is given as ((8.274) of Chau, 2018)

$$G_M(x,\xi) = M(x,\xi) = (x-\xi)H(x-\xi) - \frac{(L-\xi)}{L}x \tag{1.72}$$

Accordingly, the moment can be expressed in Green's function as:

$$M(x) = \int_0^L G_M(x,\xi)q(\xi)d\xi \tag{1.73}$$

1.3 BEAM VIBRATIONS

When the inertia term is included in the formulation of the equation of motion, we have the following PDE

$$\frac{\partial^4 u}{\partial x^4} + \frac{1}{a^4}\frac{\partial^2 u}{\partial t^2} = 0 \tag{1.74}$$

where

$$a^2 = \sqrt{\frac{EI}{\rho A}} \tag{1.75}$$

The density of the beam is denoted by ρ and the cross-section area of the beam is denoted by A. We are interested in the vibrations set off by the following initial conditions:

$$\left.\frac{\partial u}{\partial t}\right|_{t=0} = g(x), \quad u|_{t=0} = f(x) \tag{1.76}$$

The first of (1.76) is the initial velocity, whereas the second one is the initial deflection. The vibration solutions depend on the type of boundary conditions, which will be discussed in the following sections.

1.3.1 Simply-Supported Beams

Figure 1.10 shows the case that a simply-supported beam is subject to initial conditions given in (1.76). The complete set of initial and boundary conditions are given by

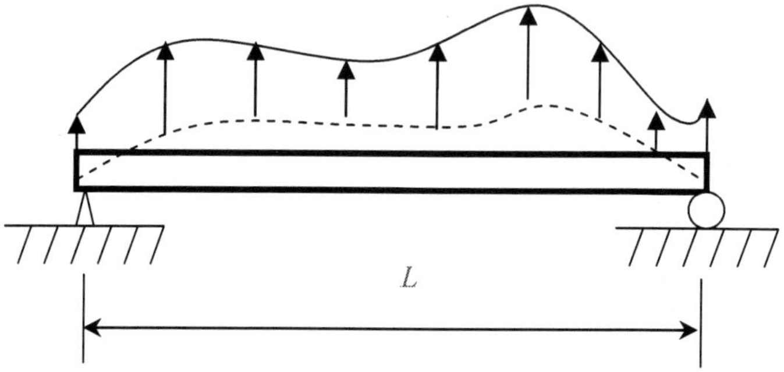

Figure 1.10 A beam subject to initial deflection and velocity

$$u(0,t)=0, \quad \frac{\partial^2 u}{\partial x^2}(0,t)=0, \quad u(L,t)=0, \quad \frac{\partial^2 u}{\partial x^2}(L,t)=0 \tag{1.77}$$

$$u(x,0)=f(x), \quad \left.\frac{\partial u(x,t)}{\partial t}\right|_{t=0}=g(x) \tag{1.78}$$

As discussed by Chau (2018), separation of variables can be applied

$$u(x,t)=X(x)T(t) \tag{1.79}$$

Differentiation of (1.79) gives

$$\frac{\partial^4 u}{\partial x^4}(x,t)=\frac{d^4X(x)}{dx^4}T(t)=X^{(IV)}T \tag{1.80}$$

$$\frac{\partial^2 u}{\partial t^2}(x,t)=X\frac{d^2T(t)}{dt^2}=XT''(t) \tag{1.81}$$

Substitution of (1.80) and (1.81) into (1.74) gives

$$X^{(IV)}T+\frac{1}{a^4}XT''=0 \tag{1.82}$$

Rearranging (1.82), we get

$$\frac{X^{(IV)}}{X}=-\frac{1}{a^4}\frac{T''}{T}=\lambda \tag{1.83}$$

where λ is a constant. The governing equation for X is

$$X^{(IV)} - \lambda X = 0 \tag{1.84}$$

This is a constant coefficient ODE and the standard solution form is (Chau, 2018):

$$X = e^{\alpha x} \tag{1.85}$$

Substitution of (1.85) into (1.84) gives a characteristic equation for α

$$\alpha^4 - \lambda = 0 \tag{1.86}$$

Taking the square root, we find

$$\alpha^2 = \pm\sqrt{\lambda} \tag{1.87}$$

Taking the square root for the second time, we get

$$\alpha = \pm i\sqrt[4]{\lambda}, \pm\sqrt[4]{\lambda} \tag{1.88}$$

The corresponding solution for X is

$$X = A\cos(\sqrt[4]{\lambda}x) + B\sin(\sqrt[4]{\lambda}x) + C\cosh(\sqrt[4]{\lambda}x) + D\sinh(\sqrt[4]{\lambda}x) \tag{1.89}$$

From (1.83), the governing equation for T is

$$T'' + \lambda a^4 T = 0 \tag{1.90}$$

The solution of (1.90) is obviously

$$T = M\cos(\sqrt{\lambda}a^2 t) + N\sin(\sqrt{\lambda}a^2 t) \tag{1.91}$$

The first simply-supported condition given in (1.77) for u implies

$$u(0,t) = 0, \quad \Rightarrow \quad u(0,t) = X(0)T(t) = 0, \quad \Rightarrow \quad X(0) = 0 \tag{1.92}$$

The second simply-supported condition given in (1.77) for u implies

$$\frac{\partial^2 u}{\partial x^2}(0,t) = 0 \quad \Rightarrow \quad X''(0)T(t) = 0, \quad \Rightarrow \quad X''(0) = 0 \tag{1.93}$$

The simply-supported conditions given in (1.77) at $x = L$ give

$$u(L,t) = 0 \quad \Rightarrow \quad u(L,t) = X(L)T(t) = 0, \quad \Rightarrow \quad X(L) = 0 \tag{1.94}$$

$$\left.\frac{\partial^2 u}{\partial x^2}(x,t)\right|_{x=L} = 0 \quad \Rightarrow \quad X''(L)T(t) = 0, \quad \Rightarrow \quad X''(L) = 0 \tag{1.95}$$

The support conditions at $x = 0$ given in (1.92) and (1.93) require

$$X(0) = A + C = 0, \tag{1.96}$$

$$X''(0) = -\sqrt{\lambda}A + \sqrt{\lambda}C = \sqrt{\lambda}(-A + C) = 0 \tag{1.97}$$

This gives

$$A = C = 0 \tag{1.98}$$

The support conditions at $x = L$ given in (1.94) and (1.95) give

$$X(L) = B\sin(\sqrt[4]{\lambda}L) + D\sinh(\sqrt[4]{\lambda}L) = 0 \tag{1.99}$$

$$X''(L) = -\sqrt{\lambda}B\sin(\sqrt[4]{\lambda}L) + \sqrt{\lambda}D\sinh(\sqrt[4]{\lambda}L) = 0 \tag{1.100}$$

Since the sinh function is not zero (except for the trivial solution of $\lambda = 0$) we must have its coefficient D be zero, and this implies that B cannot be zero. Thus, we must have

$$D = 0, \quad \sin(\sqrt[4]{\lambda}L) = 0 \tag{1.101}$$

The eigenvalue is obtained by setting the argument in the sine function to $n\pi$ as:

$$\sqrt[4]{\lambda}L = n\pi, \quad \text{or} \quad \lambda_n = \left(\frac{n\pi}{L}\right)^4 \tag{1.102}$$

The eigenmodes of the spatial variation are

$$X_n = \sin(n\pi x / L), \quad n = 1,2,3,... \tag{1.103}$$

Combining this eigenmode with the time-dependent function, we get

$$u(x,t) = \sum_{n=1}^{\infty}[M_n \cos(\frac{n^2\pi^2 a^2 t}{L^2}) + N_n \sin(\frac{n^2\pi^2 a^2 t}{L^2})]\sin(n\pi x / L) \tag{1.104}$$

For the sake of completeness, note from Chau (2018) that the eigenfunction expansion must contain an infinite number of terms. Differentiation of (1.104) with respect to t gives

$$\frac{\partial u(x,t)}{\partial t} = \sum_{n=1}^{\infty}\frac{n^2\pi^2 a^2}{L^2}[-M_n \sin(\frac{n^2\pi^2 a^2 t}{L^2}) + N_n \cos(\frac{n^2\pi^2 a^2 t}{L^2})]\sin(n\pi x / L) \tag{1.105}$$

Substitution of (1.104) into the first part of (1.78) gives

$$u(x,0) = \sum_{n=1}^{\infty} M_n \sin(n\pi x / L) = f(x) \tag{1.106}$$

Multiplying both sides by the m-th eigenmode and integrating from 0 to L, we obtain

$$\sum_{n=1}^{\infty} M_n \int_0^L \sin(n\pi x / L)\sin(m\pi x / L)dx = \int_0^L f(x)\sin(m\pi x / L)dx \tag{1.107}$$

To find the unknown constant, we use the following orthogonal properties of the sine functions:

$$\int_0^L \sin(n\pi x / L)\sin(m\pi x / L)dx = \begin{cases} 0 & n \neq m \\ L/2 & n = m \end{cases} \tag{1.108}$$

This orthogonal identity can be proved readily by using the sum rules of cosine functions as:

$$\begin{aligned}\int_0^L \sin\frac{m\pi x}{L}\sin\frac{n\pi x}{L}dx &= \frac{1}{2}\int_0^L \{\cos[\frac{(m-n)\pi x}{L}] - \cos[\frac{(m+n)\pi x}{L}]\}dx \\ &= \frac{1}{2}\frac{L}{\pi}\{\frac{\sin[(m-n)\pi x / L]}{(m-n)} - \frac{\sin[(m+n)\pi x / L]}{(m-n)}\}\Bigg|_0^L \\ &= 0\end{aligned} \tag{1.109}$$

$$\begin{aligned}\int_0^L (\sin\frac{m\pi x}{L})^2 dx &= \frac{1}{2}\int_0^L \{1 - \cos(2m\pi x / L)\}dx = \frac{1}{2}\{x - \frac{\sin(2m\pi x / L)}{2m\pi x / L}\}\Bigg|_0^L \\ &= L/2\end{aligned} \tag{1.110}$$

In view of this, we immediately obtain

$$M_n = \frac{2}{L}\int_0^L f(x)\sin(n\pi x / L)dx \tag{1.111}$$

The initial velocity given by the second part of (1.78) gives

$$\left.\frac{\partial u(x,t)}{\partial t}\right|_{t=0} = \sum_{n=1}^{\infty} \frac{n^2\pi^2 a^2}{L^2} N_n \sin(n\pi x / L) = g(x) \tag{1.112}$$

By following the same procedure in obtaining (1.111), we finally obtain

$$N_n = \frac{L^2}{n^2\pi^2 a^2}\left(\frac{2}{L}\right)\int_0^L g(x)\sin(n\pi x / L)dx \tag{1.113}$$

The final solution of the beam vibrations subject to arbitrary initial deflection and initial velocity is therefore

$$\begin{aligned} u(x,t) = \frac{2}{L}\sum_{n=1}^{\infty} [\cos\left(\frac{n^2\pi^2 a^2 t}{L^2}\right)\int_0^L f(\xi)\sin(n\pi\xi / L)d\xi \\ + \frac{L^2}{n^2\pi^2 a^2}\sin\left(\frac{n^2\pi^2 a^2 t}{L^2}\right)\int_0^L g(\xi)\sin(n\pi\xi / L)d\xi]\sin(n\pi x / L) \end{aligned} \tag{1.114}$$

1.3.2 Orthogonality of the Eigenfunctions

In this section, we consider the orthogonal property between two eigenfunctions of two distinct eigenvalues λ_m and λ_n satisfying the following ODEs:

$$X_m^{(IV)} - \lambda_m X_m = 0 \tag{1.115}$$

$$X_n^{(IV)} - \lambda_n X_n = 0 \tag{1.116}$$

Multiplying (1.115) by X_n, multiplying (1.116) by X_m, and subtracting these results, we get

$$X_m X_n^{(IV)} - X_m^{(IV)} X_n - (\lambda_n - \lambda_m) X_n X_m = 0 \tag{1.117}$$

Integrating (1.117) from 0 to L, we have

$$\int_0^L [X_n^{(IV)} X_m - X_n X_m^{(IV)}]dx - (\lambda_n - \lambda_m)\int_0^L X_m X_n dx = 0 \tag{1.118}$$

Carrying out integration by parts on the first term in the first integral of (1.118), we obtain

$$\int_0^L X_n^{(IV)} X_m dx = (X_m X_n''' - X_m' X_n'' + X_m'' X_n' - X_n X_m''')_0^L + \int_0^L X_m^{(IV)} X_n dx \tag{1.119}$$

In view of (1.115), (1.119) can be simplified to

$$\int_0^L X_n^{(IV)} X_m dx = (X_m X_n''' - X_m' X_n'' + X_m'' X_n' - X_n X_m''')_0^L + \lambda_m \int_0^L X_m X_n dx \tag{1.120}$$

Similarly, we also have

$$\int_0^L X_m^{(IV)} X_n dx = (X_n X_m''' - X_n' X_m'' + X_n'' X_m' - X_m X_n''')_0^L + \lambda_n \int_0^L X_n X_m dx \tag{1.121}$$

Substitution of (1.120) and (1.121) into (1.118) gives

$$(-X_n X_m''' + X_n' X_m'' - X_n'' X_m' + X_m X_n''')_0^L - (\lambda_n - \lambda_m)\int_0^L X_m X_n dx = 0 \tag{1.122}$$

For simply-supported beams, we have

$$X_m(0) = X_m''(0) = X_m(L) = X_m''(L) = 0 \tag{1.123}$$

The same is also true for the n-th eigenfunction. We can see that we must have

$$(\lambda_m - \lambda_n)\int_0^L X_m X_n dx = 0 \tag{1.124}$$

Since we start with two distinct eigenvalues, we must have

$$\int_0^L X_m X_n dx = 0 \tag{1.125}$$

This is the orthogonal property and thus the eigenfunctions X_n are also called orthogonal functions.

For fixed-end beams, we have

$$X_m(0) = X_m'(0) = X_m(L) = X_m'(L) = 0 \tag{1.126}$$

Again, all boundary terms are identically zeros; and we arrive at the same orthogonal property obtained in (1.125).

For cantilever beams (one fixed end and one free end), we have

$$X_m(0) = X_m'(0) = X_m''(L) = X_m'''(L) = 0 \tag{1.127}$$

Again, all boundary terms (1.122) vanish; and we get the same orthogonal property derived in (1.125).

Finally, for beams with both ends guided (see Figure 1.5), we have

$$X_m'''(0) = X_m'(0) = X_m'(L) = X_m'''(L) = 0 \tag{1.128}$$

This again leads to (1.125). Note, however, that the beam is unstable for this case as the beam can freely move vertically. The results obtained in this section are not restrict to the eigenfunctions in terms of sine function obtained in the last section but apply to any solution of the differential equation (1.84).

For the case of $m = n$, the integral in (1.125) is not zero but equals

$$\int_0^L X_n^2 dx = \frac{L}{4}\{X_n^2(L) + [X_n''(L)]^2 - 2X_n'(L)X_n'''(L)\} \tag{1.129}$$

where

$$X_n'(\xi) = \frac{dX_n}{d\xi}, \quad \xi = kx = \sqrt[4]{\lambda_n}\, x \tag{1.130}$$

This proof of the result given in (1.129) is not straightforward. To prove (1.129), we first recall from the last section that

$$X_m = X_m(\sqrt[4]{\lambda_m}\, x) = X_m(kx) \tag{1.131}$$

Thus, we have

$$\lambda_m = k^4 \tag{1.132}$$

Assuming that $n > m$, we have

$$\lambda_n = (k + \delta k)^4 = k^4 + 4k^3\delta k + \ldots \tag{1.133}$$

Thus, we have the difference of the eigenvalues as

$$\lambda_n - \lambda_m = 4k^3\delta k + \ldots \tag{1.134}$$

We now expand the eigenfunction in terms of the change of k. Thus, we have

$$X_n = X_m + \frac{dX_m}{dk}\delta k \tag{1.135}$$

Differentiation of this result gives

$$X_n' = X_m' + \frac{dX_m'}{dk}\delta k \tag{1.136}$$

$$X_n'' = X_m'' + \frac{dX_m''}{dk}\delta k \tag{1.137}$$

$$X_n''' = X_m''' + \frac{dX_m'''}{dk}\delta k \tag{1.138}$$

With these results, we find

$$\begin{aligned} X_m X_n''' - X_m''' X_n &= X_m X_m''' + X_m \frac{dX_m'''}{dk}\delta k - X_m''' X_m - X_m''' \frac{dX_m}{dk}\delta k \\ &= X_m \frac{dX_m'''}{dk}\delta k - X_m''' \frac{dX_m}{dk}\delta k \end{aligned} \tag{1.139}$$

Similarly, we have

$$\begin{aligned} -X_n' X_m'' + X_n'' X_m' &= -X_m' X_m'' - X_m' \frac{dX_m''}{dk}\delta k + X_m'' X_m' + X_m'' \frac{dX_m'}{dk}\delta k \\ &= X_m'' \frac{dX_m'}{dk}\delta k - X_m' \frac{dX_m''}{dk}\delta k \end{aligned} \tag{1.140}$$

Substitution of (1.139), (1.140), (1.134) and (1.135) into (1.122) gives

$$\begin{aligned} &\delta k(X_m \frac{dX_m'''}{dk} - X_m''' \frac{dX_m}{dk} + X_m'' \frac{dX_m'}{dk} - X_m' \frac{dX_m''}{dk})_0^L \\ &= 4k^3 \delta k \int_0^L X_m^2 dx + 4k^3 (\delta k)^2 \int_0^L X_m \frac{dX_m}{dk} dx \end{aligned} \tag{1.141}$$

Dropping higher order terms, we get

$$4k^3 \int_0^L X_m^2 dx = (X_m \frac{dX_m'''}{dk} - X_m''' \frac{dX_m}{dk} + X_m'' \frac{dX_m'}{dk} - X_m' \frac{dX_m''}{dk})_0^L \tag{1.142}$$

We now introduce the differentiation with respect to a new variable as:

$$\frac{dX_m}{dx} = \frac{dX_m}{d(kx)} \frac{d(kx)}{dx} = kX_m' \tag{1.143}$$

The prime is now re-defined as differentiation respect to *kx*, instead of with respect to *x*. Similarly, we have

$$\frac{dX_m}{dk} = \frac{dX_m}{d(kx)} \frac{d(kx)}{dk} = xX_m' \tag{1.144}$$

Equation (1.115) can be rewritten as

$$\frac{d^4 X_m}{dx^4} = \lambda_m X_m = k^4 \frac{d^4 X_m}{d\xi^4} = k^4 X_m'''' \tag{1.145}$$

where $\xi = kx$ is defined in (1.130). Using (1.143), we finally have

$$X_m = X_m'''' \tag{1.146}$$

with the prime being defined in (1.143). We will now consider each term on the right-hand side of (1.142):

$$X_m \frac{d}{dk}\left(\frac{d^3 X_m}{dx^3}\right) = X_m \frac{d(k^3 X_m''')}{dk} = X_m 3k^2 X_m''' + X_m k^3 x X_m'''' \\ = 3k^2 X_m X_m''' + xk^3 X_m^2 \tag{1.147}$$

$$\frac{d^3 X_m}{dx^3}\frac{dX_m}{dk} = xk^3 X_m' X_m''' \tag{1.148}$$

$$\frac{d^2 X_m}{dx^2}\frac{d}{dk}\left(\frac{dX_m}{dx}\right) = \frac{d}{dk}(kX_m')k^2 X_m'' = k^2 X_m' X_m'' + xk^3 (X_m'')^2 \tag{1.149}$$

$$\frac{dX_m}{dx}\frac{d}{dk}\left(\frac{d^2 X_m}{dx^2}\right) = kX_m' \frac{d}{dk}(k^2 X_m'') = kX_m'(2kX_m'' + xk^2 X_m''') \tag{1.150}$$

One have to be careful about the definition of prime in (1.142) and in (1.147) to (1.150). Substitution of these results into (1.142) gives

$$4k\int_0^L X_m^2 dx = [3X_m X_m''' + xkX_m^2 - 2xkX_m' X_m''' - X_m' X_m'' + xk(X_m'')^2]_0^L \tag{1.151}$$

In terms of the new variable in the differentiation, it is straightforward to show that the three standard boundary conditions for simply-supported, built-in and free ends are:

Simply-supported end: $\quad X_m = 0, \quad X_m'' = 0 \quad$ (1.152)

Built-in end: $\quad X_m = 0, \quad X_m' = 0 \quad$ (1.153)

Free end: $\quad X_m'' = 0, \quad X_m''' = 0 \quad$ (1.154)

For all these three boundary conditions, we can show that the following boundary terms in (1.151) are identically:

$$X_m X_m''' \equiv 0, \quad X_m' X_m'' \equiv 0 \tag{1.155}$$

With this information, (1.151) is reduced to:

$$4k\int_0^L X_m^2 dx = [xkX_m^2 - 2xkX_m' X_m''' + xk(X_m'')^2]_0^L \tag{1.156}$$

Substituting the boundary values into the right-hand side, we get

$$4\int_0^L X_m^2 dx = \{x[X_m^2 - 2X_m' X_m''' + (X_m'')^2]\}_0^L \\ = L[X_m^2 - 2X_m' X_m''' + (X_m'')^2]_{x=L} \tag{1.157}$$

Finally, the validity of (1.129) is proved

$$\int_0^L X_m^2 dx = \frac{L}{4}[X_m^2 - 2X_m' X_m''' + (X_m'')^2]_{x=L} \tag{1.158}$$

If at the end $x = L$ is free, we have the following simplification:

$$\int_0^L X_m^2 dx = \frac{L}{4}[X_m^2]_{x=L} \tag{1.159}$$

If at the end $x = L$ is fixed, we have the following simplification:

$$\int_0^L X_m^2 dx = \frac{L}{4}[(X_m'')^2]_{x=L} \tag{1.160}$$

If at the end $x = L$ is simply-supported, we have the following simplification:

$$\int_0^L X_m^2 dx = -\frac{L}{2}[X_m' X_m''']_{x=L} \tag{1.161}$$

This completes the discussion of orthogonal properties of the eigenfunctions of beam vibrations.

1.3.3 Cantilever Beam with Suddenly Removed Point Force

In this section, we consider the vibrations of a beam, which was deflected initially by a static point force, and the point force was suddenly removed to set off the vibrations. In practice, it can be used to model the situation of a lump mass suddenly dropped from the tip of the beam. Figure 1.11 shows the problem of a suddenly applied force at time zero. The problem is formulated mathematically as:

$$\frac{\partial^4 u}{\partial x^4} + \frac{1}{a^4}\frac{\partial^2 u}{\partial t^2} = 0 \tag{1.162}$$

$$u(0) = 0, \quad \frac{\partial u}{\partial x}(0) = 0, \quad \frac{\partial^2 u}{\partial x^2}(L) = 0, \quad \frac{\partial^3 u}{\partial x^3}(L) = 0 \tag{1.163}$$

$$u(0) = f(x), \quad \frac{\partial u}{\partial t}(0) = 0 \tag{1.164}$$

where the static deflection of a point force applied at the tip is

$$f(x) = \frac{F}{6EI}(3Lx^2 - x^3) \tag{1.165}$$

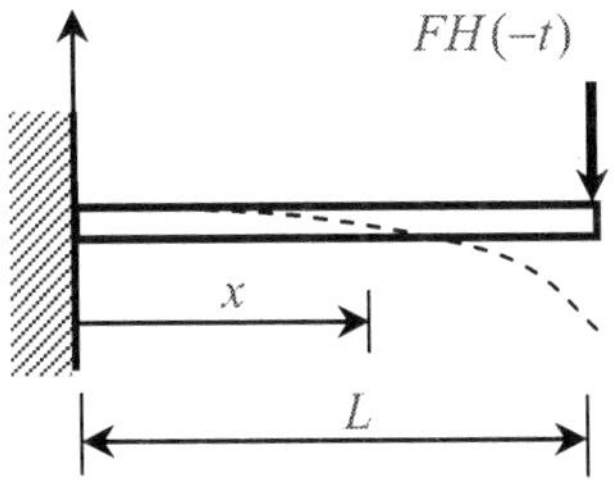

Figure 1.11 A cantilever beam subject to a sudden removal of point force

This solution given in (1.165) has been derived in Section 1.2.3. Recall the result of separation of variables from the last section that

$$u(x) = X(x)T(t) \tag{1.166}$$

$$X^{(IV)} - \lambda X = 0 \tag{1.167}$$

$$T'' + \lambda a^4 T = 0 \tag{1.168}$$

The solutions of this uncoupled system is

$$X = A\cos(\sqrt[4]{\lambda}x) + B\sin(\sqrt[4]{\lambda}x) + C\cosh(\sqrt[4]{\lambda}x) + D\sinh(\sqrt[4]{\lambda}x) \qquad (1.169)$$

$$T = M\cos(\sqrt{\lambda}a^2t) + N\sin(\sqrt{\lambda}a^2t) \qquad (1.170)$$

The first boundary condition of (1.163) gives

$$X(0) = A + C = 0 \qquad (1.171)$$

Differentiation of function X gives

$$X' = \sqrt[4]{\lambda}\{-A\sin(\sqrt[4]{\lambda}x) + B\cos(\sqrt[4]{\lambda}x) + C\sinh(\sqrt[4]{\lambda}x) + D\cosh(\sqrt[4]{\lambda}x)\} \qquad (1.172)$$

The second boundary condition of (1.163) gives

$$X'(0) = \sqrt[4]{\lambda}(B + D) = 0 \qquad (1.173)$$

Thus, the number of unknowns is reduced to two and the solution is first rewritten as

$$X = A[\cos(\sqrt[4]{\lambda}x) - \cosh(\sqrt[4]{\lambda}x)] + B[\sin(\sqrt[4]{\lambda}x) - \sinh(\sqrt[4]{\lambda}x)] \qquad (1.174)$$

Differentiation of X given in (1.174) gives

$$X'' = \sqrt{\lambda}A[-\cos(\sqrt[4]{\lambda}x) - \cosh(\sqrt[4]{\lambda}x)] + \sqrt{\lambda}B[-\sin(\sqrt[4]{\lambda}x) - \sinh(\sqrt[4]{\lambda}x)] \qquad (1.175)$$

$$X''' = \sqrt{\lambda}\sqrt[4]{\lambda}A[\sin(\sqrt[4]{\lambda}x) - \sinh(\sqrt[4]{\lambda}x)] + \sqrt{\lambda}\sqrt[4]{\lambda}B[-\cos(\sqrt[4]{\lambda}x) - \cosh(\sqrt[4]{\lambda}x)] \qquad (1.176)$$

In view of these results, the third and fourth boundary conditions given in (1.163) yield

$$A[\cos(\sqrt[4]{\lambda}L) + \cosh(\sqrt[4]{\lambda}L)] + B[\sin(\sqrt[4]{\lambda}L) + \sinh(\sqrt[4]{\lambda}L)] = 0 \qquad (1.177)$$

$$A[\sin(\sqrt[4]{\lambda}L) - \sinh(\sqrt[4]{\lambda}L)] + B[-\cos(\sqrt[4]{\lambda}L) - \cosh(\sqrt[4]{\lambda}L)] = 0 \qquad (1.178)$$

For nonzero A and B (i.e., nontrivial solutions), we require

$$\begin{aligned}&[\cos(\sqrt[4]{\lambda}L) + \cosh(\sqrt[4]{\lambda}L)]^2 \\ &+[\sin(\sqrt[4]{\lambda}L) + \sinh(\sqrt[4]{\lambda}L)][\sin(\sqrt[4]{\lambda}L) - \sinh(\sqrt[4]{\lambda}L)] = 0\end{aligned} \qquad (1.179)$$

This can be simplified to give the characteristic equation as

$$1 + \cos(\sqrt[4]{\lambda}L)\cosh(\sqrt[4]{\lambda}L) = 0 \qquad (1.180)$$

This can be rewritten as

$$\cos(\gamma_n) = -\frac{1}{\cosh(\gamma_n)}, \quad \gamma_n = \sqrt[4]{\lambda_n}L \qquad (1.181)$$

Figure 1.12 shows that there are infinite roots for the characteristics. The first three roots are also indicated in the figure. The first ten roots of (1.181) were obtained using the root searching subroutine ZBRENT given in Press et al. (1992) and are compiled in Table 1.1.

We see that except for the first root, all other roots can be approximated by:

$$\gamma_n \approx \frac{(2n-1)\pi}{2} \qquad (1.182)$$

Table 1.1 compiles the first ten roots of the eigenvalue equation given in (1.181). The first root differs from the approximation given by (1.182) by 16%, the second root differs from the approximation by less than 0.4%. All subsequent roots are basically indistinguishable from the approximations. Note that the normalized eigenvalue γ_n given in Table 1.1 is independent of the bending stiffness EI of the beam. The eigenvalues can be calculated and estimated as

$$\omega_n = a^2\sqrt{\lambda_n} = \sqrt{\frac{EI}{\rho A}}(\frac{\gamma_n}{L})^2 \approx \sqrt{\frac{EI}{\rho A}}\frac{(2n-1)^2\pi^2}{4L^2}, \quad n = 1,2,3,... \tag{1.183}$$

Returning to eigenfunctions, the constants in (1.174) can be related, by using (1.177) or (1.178), as

$$B = -A\frac{(\cos\gamma_n + \cosh\gamma_n)}{(\sin\gamma_n + \sinh\gamma_n)} \tag{1.184}$$

Using (1.184), we can write the eigenfunctions as

$$\begin{aligned} X_n &= (\sin\gamma_n + \sinh\gamma_n)[\cos(\sqrt[4]{\lambda_n}x) - \cosh(\sqrt[4]{\lambda_n}x)] \\ &\quad -(\cos\gamma_n + \cosh\gamma_n)[\sin(\sqrt[4]{\lambda_n}x) - \sinh(\sqrt[4]{\lambda_n}x)] \end{aligned} \tag{1.185}$$

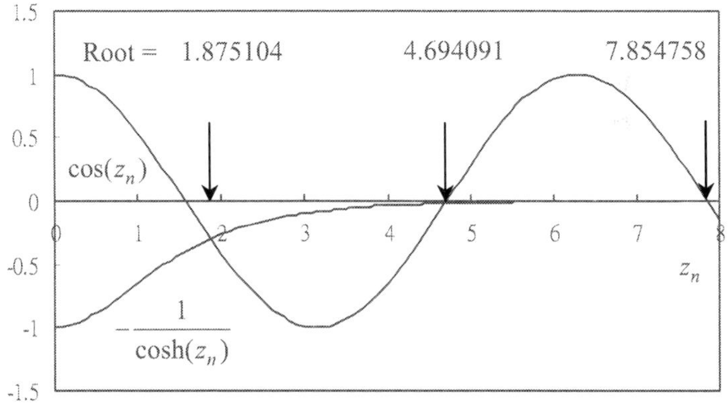

Figure 1.12 Searching characteristic roots for (1.181)

Table 1.1 First ten eigenvalues of natural frequencies for cantilever beams

Number, n	γ_n	$(2n-1)\pi/2$	% error
1	1.875104	1.570796	−16.228843
2	4.694091	4.712389	0.389809
3	7.854758	7.853982	−0.009877
4	10.995541	10.995575	0.000308
5	14.137168	14.137167	−0.000008
6	17.27876	17.27876	0.000003
7	20.420352	20.420353	0.000003
8	23.561945	23.561946	0.000003
9	26.703538	26.703538	0.000003
10	29.84513	29.845131	0.000003

The final solution can be expressed in terms of the eigenfunctions as:

$$u(x,t) = \sum_{n=1}^{\infty} [M_n \cos(\sqrt{\lambda_n} a^2 t) + N_n \sin(\sqrt{\lambda_n} a^2 t)]\{(\sin\gamma_n + \sinh\gamma_n) \\ \times[\cos(\sqrt[4]{\lambda_n} x) - \cosh(\sqrt[4]{\lambda_n} x)] - (\cos\gamma_n + \cosh\gamma_n)[\sin(\sqrt[4]{\lambda_n} x) - \sinh(\sqrt[4]{\lambda_n} x)]\} \tag{1.186}$$

We are now ready to consider the initial conditions. Differentiation of (1.186) with respect to time gives

$$\frac{\partial u}{\partial t}(x,t) = \sum_{n=1}^{\infty} \sqrt{\lambda_n} a^2 [-M_n \sin(\sqrt{\lambda_n} a^2 t) + N_n \cos(\sqrt{\lambda_n} a^2 t)] X_n(x) \tag{1.187}$$

The second of the initial conditions given in (1.164) can be considered as

$$\frac{\partial u}{\partial t}(x,0) = \sum_{n=1}^{\infty} \sqrt{\lambda_n} a^2 N_n X_n(x) = 0 \tag{1.188}$$

Equation (1.188) yields

$$N_n = 0 \tag{1.189}$$

The first initial condition given in (1.164) yields

$$u(x,0) = \sum_{n=1}^{\infty} M_n X_n(x) = f(x) \tag{1.190}$$

Multiplying both sides by the eigenfunction X_m and integrating from 0 to L, we find

$$\sum_{n=1}^{\infty} M_n \int_0^L X_n(x) X_m(x) dx = \int_0^L f(x) X_m(x) dx \tag{1.191}$$

In view of the orthogonal property, we have

$$M_n = \frac{\int_0^L f(x) X_m(x) dx}{\int_0^L X_n^2(x) dx} \tag{1.192}$$

Substitution of (1.189) and (1.192) into (1.186) results in

$$u(x,t) = \sum_{n=1}^{\infty} \frac{\int_0^L f(\xi) X_n(\xi) d\xi}{\int_0^L X_n^2(\xi) d\xi} \cos(\frac{\gamma_n^2 a^2}{L^2} t) X_n(x) \tag{1.193}$$

The integral in the denominator of (1.193) can be further simplified. Recalling from the result of the last section, for cantilever beams we have

$$\int_0^L X_m^2 dx = \frac{L}{4} [X_m^2]_{x=L} \tag{1.194}$$

We first simplify (1.185) as

$$\begin{aligned} X_n(L) &= (\sin\gamma_n + \sinh\gamma_n)(\cos\gamma_n - \cosh\gamma_n) \\ &\quad -(\cos\gamma_n + \cosh\gamma_n)(\sin\gamma_n - \sinh\gamma_n) \\ &= 2(\cos\gamma_n \sinh\gamma_n - \sin\gamma_n \cosh\gamma_n) \end{aligned} \tag{1.195}$$

Substitution of (1.195) into (1.194) gives

$$\int_0^L X_m^2 dx = L(\cos\gamma_n \sinh\gamma_n - \sin\gamma_n \cosh\gamma_n)^2 \tag{1.196}$$

Similarly, the integral in the numerator of (1.193) can be evaluated as

$$\begin{aligned}\int_0^L f(x)X_n(x)dx &= \frac{1}{\lambda_n}\int_0^L f(x)X_n''''(x)dx \\ &= \frac{1}{\lambda_n}\{(X_n''' f - X_n'' f' + X_n' f'' - X_n f''')\Big|_0^L + \int_0^L X_n f^{(IV)} d\xi\} \quad (1.197) \\ &= \frac{1}{\lambda_n}\{\frac{X_n(L)F}{EI}\}\end{aligned}$$

Substitution of (1.196) and (1.197) yields

$$u(x,t) = \frac{4FL^3}{EI}\sum_{n=1}^{\infty}\frac{X_n(x)}{\gamma_n^4 X_n(L)}\cos(\frac{\gamma_n^2 a^2}{L^2}t) \tag{1.198}$$

Finally, the solution can be written as:

$$u(x,t) = \frac{2FL^3}{EI}\sum_{n=1}^{\infty}\frac{X_n(x)\cos(\frac{\gamma_n^2 a^2}{L^2}t)}{\gamma_n^4(\cos\gamma_n \sinh\gamma_n - \sin\gamma_n \cosh\gamma_n)} \tag{1.199}$$

This result agrees with that given on p. 68 of Lebedev et al. (1965).

1.3.4 Cantilever Beam with a Tip Lump Mass

In this section, we will consider the natural vibration frequency of a cantilever beam of mass M having a lump mass M_0 at the free end, as shown in Figure 1.13. The problem is formulated mathematically as:

$$\frac{\partial^4 u}{\partial x^4} + \frac{1}{a^4}\frac{\partial^2 u}{\partial t^2} = 0 \tag{1.200}$$

$$u(0) = 0, \quad \frac{\partial u}{\partial x}(0) = 0, \quad \frac{\partial^2 u}{\partial x^2}(L) = 0, \quad EI\frac{\partial^3 u}{\partial x^3}(L) = M_0\frac{\partial^2 u}{\partial t^2} \tag{1.201}$$

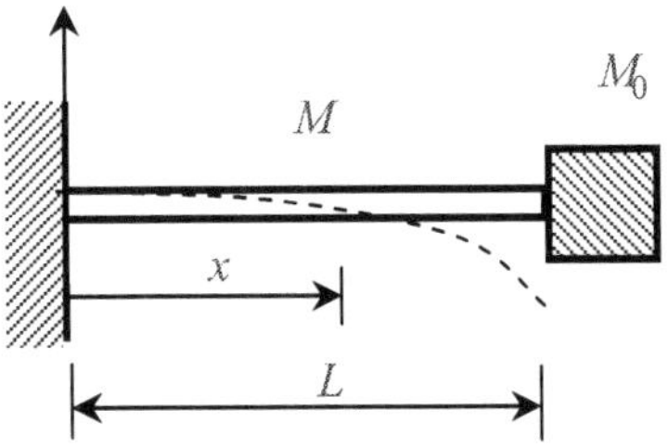

Figure 1.13 A cantilever beam of mass M with a mass M_0 attached to the free end

Since we are looking at the natural vibrations, we can assume the time-dependent function as

$$u(x,t) = v(x)\sin(\omega t + \varphi) \tag{1.202}$$

Substitution of (1.202) into (1.200) gives

$$v^{(IV)} - \frac{\omega^2}{a^4} v = 0 \tag{1.203}$$

The corresponding boundary conditions for v become:

$$v(0) = 0, \quad v'(0) = 0, \quad v''(L) = 0, \quad v'''(L) = -\frac{M_0\omega^2}{EI} v\bigg|_{x=L} \tag{1.204}$$

Similar to our discussion in the last section, we have the solution of v as

$$v(x) = A_1 \cos(\frac{\sqrt{\omega}}{a}x) + B_1 \sin(\frac{\sqrt{\omega}}{a}x) + A_2 \cosh(\frac{\sqrt{\omega}}{a}x) + B_2 \sinh(\frac{\sqrt{\omega}}{a}x) \tag{1.205}$$

The first and second equations of (1.204) give

$$v(0) = A_1 + A_2 = 0 \tag{1.206}$$

$$v'(0) = \frac{\sqrt{\omega}}{a}(B_1 + B_2) = 0 \tag{1.207}$$

With these results, we can first eliminate A_2 and B_2 from the solution as

$$v(x) = A_1[\cos(\frac{\sqrt{\omega}}{a}x) - \cosh(\frac{\sqrt{\omega}}{a}x)] + B_1[\sin(\frac{\sqrt{\omega}}{a}x) - \sinh(\frac{\sqrt{\omega}}{a}x)] \tag{1.208}$$

The third and fourth equations of the boundary conditions in (1.204) give

$$A_1[\cos\gamma + \cosh\gamma] + B_1[\sin\gamma + \sinh\gamma] = 0 \tag{1.209}$$

$$\begin{aligned}&(\frac{\sqrt{\omega}}{a})^3 A_1(-\sin\gamma + \sinh\gamma) + (\frac{\sqrt{\omega}}{a})^3 B_1(\cos\gamma + \cosh\gamma) = \\ &\frac{M_0\omega^2}{EI}\{A_1(\cos\gamma - \cosh\gamma) + B_1(\sin\gamma - \sinh\gamma)\}\end{aligned} \tag{1.210}$$

Equation (1.210) can further be simplified as:

$$\begin{aligned}&A_1(\sin\gamma - \sinh\gamma) + B_1(-\cos\gamma - \cosh\gamma) = \\ &-\frac{M_0\sqrt{\omega}a^3}{EI}\{A_1(\cos\gamma - \cosh\gamma) + B_1(\sin\gamma - \sinh\gamma)\}\end{aligned} \tag{1.211}$$

The coefficient on the right of (1.211) can be simplified as

$$\frac{M_0\sqrt{\omega}a^3}{EI} = M_0(\frac{\gamma_n a}{L})\frac{a^3}{EI} = M_0\frac{\gamma_n}{L}(\frac{1}{EI})\frac{EI}{\rho A} = M_0\frac{\gamma_n}{M} \tag{1.212}$$

where $M = \rho AL$ is the mass of the beam. With this result, (1.211) is reduced to a more compact form:

$$\begin{aligned}&A_1\{\frac{M_0\gamma_n}{M}(\cos\gamma - \cosh\gamma) + (\sin\gamma - \sinh\gamma)\} \\ &+ B_1\{\frac{M_0\gamma_n}{M}(\sin\gamma - \sinh\gamma) - (\cos\gamma + \cosh\gamma)\} = 0\end{aligned} \tag{1.213}$$

Thus, (1.209) and (1.213) provide a system of two equations of two unknowns. For nontrivial solution or for eigenfunctions to exist, we require

$$\begin{aligned}&[\frac{M_0\gamma_n}{M}(\cos\gamma_n-\cosh\gamma_n)+(\sin\gamma_n-\sinh\gamma_n)](\sin\gamma_n+\sinh\gamma_n)\\&\qquad-[\frac{M_0\gamma_n}{M}(\sin\gamma_n-\sinh\gamma_n)-(\cos\gamma_n+\cosh\gamma_n)](\cos\gamma+\cosh\gamma)=0\end{aligned} \tag{1.214}$$

This can be simplified as

$$1+\cos\gamma_n\cosh\gamma_n=\frac{M_0\gamma_n}{M}(\sin\gamma_n\cosh\gamma_n-\cos\gamma_n\sinh\gamma_n) \tag{1.215}$$

If we set $M_0 = 0$, the eigenvalue equation of (1.180) is recovered, as expected. The natural frequency is given by:

$$\omega_n=\frac{a^2\gamma_n^2}{L^2} \tag{1.216}$$

where γ_n are the roots from (1.215). The first ten roots for (1.215) are obtained by using the Fortran subroutines ZBRAC and ZBRENT (Press et al., 1992) for various values of M_0. These results are compiled in Table 1.2 for various values of M_0/M. The first eigenvalue (i.e., $n = 1$) decreases with the increase of M_0/M. For higher eigenvalues (i.e., $n > 1$), the eigenvalues are relatively insensitive to the value of M_0/M.

Table 1.2 The first ten natural frequencies for cantilever beams with tip mass

n	Eigenvalues γ_n				
	$M_0/M = 0$	$M_0/M = 0.2$	$M_0/M = 0.5$	$M_0/M = 1$	$M_0/M = 2$
1	1.875104	1.6164	1.419964	1.247917	1.076196
2	4.694091	4.267062	4.111134	4.03114	3.982574
3	7.854758	7.318373	7.190335	7.134132	7.10265
4	10.995541	10.401563	10.298445	10.256621	10.234015
5	14.137168	13.506702	13.421001	13.387757	13.370122
6	17.27876	16.623354	16.550279	16.522725	16.508278
7	20.420352	19.746859	19.683265	19.659751	19.647518
8	23.561945	22.874754	22.818506	22.798005	22.787398
9	26.703538	26.005618	25.955221	25.93705	25.927689
10	29.84513	29.138587	29.092951	29.076636	29.068259

1.3.5 Simply-Supported Beam Subject to an Impulse

In this section, we consider an impulse applied suddenly at time zero, as shown in Figure 1.14. The problem can be formulated as

$$\frac{\partial^4 u}{\partial x^4}+\frac{1}{a^4}\frac{\partial^2 u}{\partial t^2}=0 \tag{1.217}$$

$$u(x,0) = 0 \tag{1.218}$$

$$\frac{\partial u}{\partial t}(x,0) = g(x) = V\delta(x-c) \tag{1.219}$$

By the analysis of Section 1.3.1, we have the solution for simply-supported beams as

$$u(x,t) = \sum_{n=1}^{\infty}[M_n \cos(\frac{n^2\pi^2 a^2 t}{L^2}) + N_n \sin(\frac{n^2\pi^2 a^2 t}{L^2})]\sin(n\pi x / L) \tag{1.220}$$

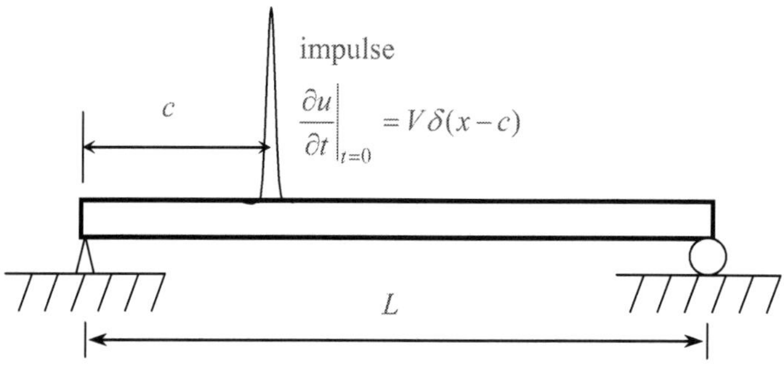

Figure 1.14 A cantilever beam subject to an impulse at $x = c$

The zero initial deflection given in (1.218) leads to

$$M_n = 0 \tag{1.221}$$

The differentiation of (1.220) with respect to time gives

$$\frac{\partial u(x,t)}{\partial t} = \sum_{n=1}^{\infty}\frac{n^2\pi^2 a^2}{L^2}N_n \cos(\frac{n^2\pi^2 a^2 t}{L^2})\sin(n\pi x / L) \tag{1.222}$$

The initial velocity is imposed on (1.222) as

$$\left.\frac{\partial u(x,t)}{\partial t}\right|_{t=0} = \sum_{n=1}^{\infty}\frac{n^2\pi^2 a^2}{L^2}N_n \sin(n\pi x / L) = g(x) \tag{1.223}$$

Using the orthogonality of the eigenfunctions, we get

$$N_n = \frac{L^2}{n^2\pi^2 a^2}(\frac{2}{L})\int_0^L g(x)\sin(n\pi x / L)dx \tag{1.224}$$

Substitution of $g(x)$ from (1.219) into (1.224) gives

$$\int_0^L g(\xi)\sin(n\pi\xi / L)d\xi = \int_0^L V\delta(\xi - c)\sin(n\pi\xi / L)d\xi = V\sin(n\pi c / L) \tag{1.225}$$

The final solution of the beam vibrations subject to an impulse is obtained as

$$u(x,t) = \frac{2VL}{\pi^2}\sqrt{\frac{\rho A}{EI}}\sum_{n=1}^{\infty}\frac{\sin(n\pi c / L)}{n^2}\sin(\frac{n^2\pi^2 a^2 t}{L^2})\sin(\frac{n\pi x}{L}) \tag{1.226}$$

1.3.6 Seismograph as Vibrations of Rigid Beam

One of the beam applications is to use it in a seismograph (see p. 580 of Timoshenko and Young, 1965), as illustrated in Figure 1.15. In this case, the beam is considered as rigid. We are looking for a fine balance among the spring stiffness k, its position a, the length of the beam L, and the weight W at the tip of the beam.

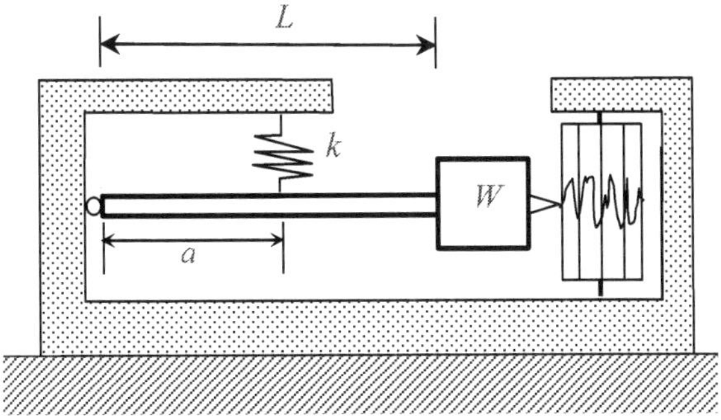

Figure 1.15 A seismograph made of a rigid beam with lump mass at the tip

In particular, we first consider the oscillations of a vertically hanging mass by a spring as shown in Figure 1.15. The equation of motion of the weight W can be formulated as

$$\frac{W}{g}\ddot{x} = W - (k\delta_{st} + kx) = W - (W + kx) = -kx \qquad (1.227)$$

where δ_{st} is the static deformation of the spring due to weight W (i.e., the case of no vibration). Thus, we have an ODE for x as

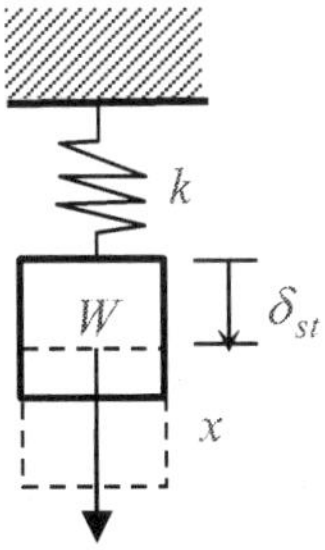

Figure 1.16 A weight W supported by a spring

$$\ddot{x} + \frac{kg}{W}x = \ddot{x} + \omega^2 x = 0 \qquad (1.228)$$

The solution of free vibrations is obviously

$$x = A\cos\omega t + B\sin\omega t \qquad (1.229)$$

where ω is the circular frequency of vibrations. The period can be expressed as:

$$T = \frac{2\pi}{\omega} = 2\pi\sqrt{\frac{W}{kg}} = 2\pi\sqrt{\frac{\delta_{st}}{g}} \tag{1.230}$$

The last of (1.230) shows that once the static deflection of the oscillator can be found, so can the period of natural vibrations. Alternatively, we can find the frequency as

$$f = \frac{1}{2\pi}\sqrt{\frac{g}{\delta_{st}}} \tag{1.231}$$

We now return to the seismograph problem. Consider the static equilibrium of the beam structure as shown in Figure 1.17.

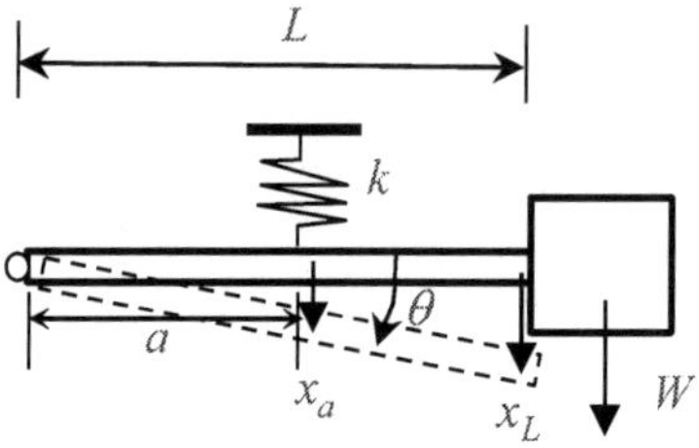

Figure 1.17 A hinged rigid beam with tip mass and spring support

Taking the moment at the left hinge, we have

$$WL = akx_a \tag{1.232}$$

where the deflection at the spring is x_a. Thus, we have

$$W = kx_a\frac{a}{L} \tag{1.233}$$

We define the equivalent stiffness of the beam as

$$W = k_{eq}x_L = k(\frac{x_a a}{L}) \tag{1.234}$$

where the deflection at the beam at the tip is x_L. For rigid beam rotation, we must have

$$x_L = \theta L, \quad x_a = \theta a \tag{1.235}$$

Substitution of (1.235) into (1.234) gives

$$W = k_{eq}\theta L = k(\frac{\theta a^2}{L}) \tag{1.236}$$

Thus, the equivalent stiffness of the system observed by the tip mass is

$$k_{eq} = k(\frac{a}{L})^2 \tag{1.237}$$

At static equilibrium, we have

$$W = k_{eq}\delta_{st} = k(\frac{a}{L})^2\delta_{st} \tag{1.238}$$

In view of (1.238), the static deflection can be calculated as

$$\delta_{st} = \frac{W}{k}(\frac{L}{a})^2 \tag{1.239}$$

Finally, the frequency of natural vibrations of the seismograph is

$$f = \frac{1}{2\pi}\sqrt{\frac{g}{\delta_{st}}} = \frac{1}{2\pi}\frac{a}{L}\sqrt{\frac{kg}{W}} \tag{1.240}$$

We see that the seismograph has its own natural frequency. In order to truly reflect the frequency content of the ground motion during earthquakes, we have to tune a and k such that f given in (1.240) would not overlap with the dominant frequency of the ground motion, which is typically in the order of a few Hertz.

1.4 ROCKET/MISSILE LAUNCH PAD AS BEAM

In this section, we consider a simple model for a rocket or missile launch pad using the Euler-Bernoulli beam theory. As shown in Figure 1.18, during the launch of the rocket, the launch pad is allowed to be released from the vertical position. The vertical truss with a service platform at the top is modeled as a beam with a tip mass M_0 and with a flexible base support, instead of fixed end. The rotational flexibility is modeled by a spring of stiffness k. The equivalent bending stiffness of the beam is EI, and the equivalent mass per length is ρA. It is of interest to estimate the free vibration frequency of the huge truss structure. The vibration analysis is modified from that given in Section 1.3.4.

The problem can be formulated as

$$\frac{\partial^4 u}{\partial x^4} + \frac{1}{a^4}\frac{\partial^2 u}{\partial t^2} = 0 \tag{1.241}$$

$$u(0) = 0, \quad EI\frac{\partial^2 u}{\partial x^2}(0) = k\frac{\partial u}{\partial x}(0), \quad \frac{\partial^2 u}{\partial x^2}(L) = 0, \quad EI\frac{\partial^3 u}{\partial x^3}(L) = M_0\left.\frac{\partial^2 u}{\partial t^2}\right|_{x=L} \tag{1.242}$$

Recalling the solution for free vibrations, we can express the solution as

$$u(x,t) = v(x)\sin(\omega t + \varphi) \tag{1.243}$$

$$v(x) = A_1\cos(\beta x) + A_2\cosh(\beta x) + B_1\sin(\beta x) + B_2\sinh(\beta x) \tag{1.244}$$

where

$$\beta = \frac{\sqrt{\omega}}{a} \tag{1.245}$$

The first equation of (1.242) gives

$$v(0) = A_1 + A_2 = 0 \tag{1.246}$$

The result will be used directly in other boundary conditions (i.e., A_2 will be eliminated from the boundary conditions). The second equation of (1.242) gives

$$(B_1 + B_2) = -(\frac{EI}{k})\frac{\sqrt{\omega}}{a}A_1 \tag{1.247}$$

The third and fourth of the boundary conditions in (1.242) require

$$-A_1[\cos\gamma + \cosh\gamma] - B_1\sin\gamma + B_2\sinh\gamma = 0 \tag{1.248}$$

$$A_1[(\sin\gamma - \sinh\gamma) + \chi(\cos\gamma - \cosh\gamma)] + B_1[-\cos\gamma + \chi\sin\gamma] + B_2[\cosh\gamma + \chi\sinh\gamma] = 0 \tag{1.249}$$

where

$$\gamma = \beta L = \frac{\sqrt{\omega}L}{a} \tag{1.250}$$

$$\chi = \frac{M_0\sqrt{\omega}a^3}{EI} = \frac{M_0 a^3}{EI}\left(\frac{\gamma a}{L}\right) = \left(\frac{M_0 a^4}{EIL}\right)\gamma = \left(\frac{M_0}{\rho AL}\right)\gamma = \frac{M_0}{M}\gamma \tag{1.251}$$

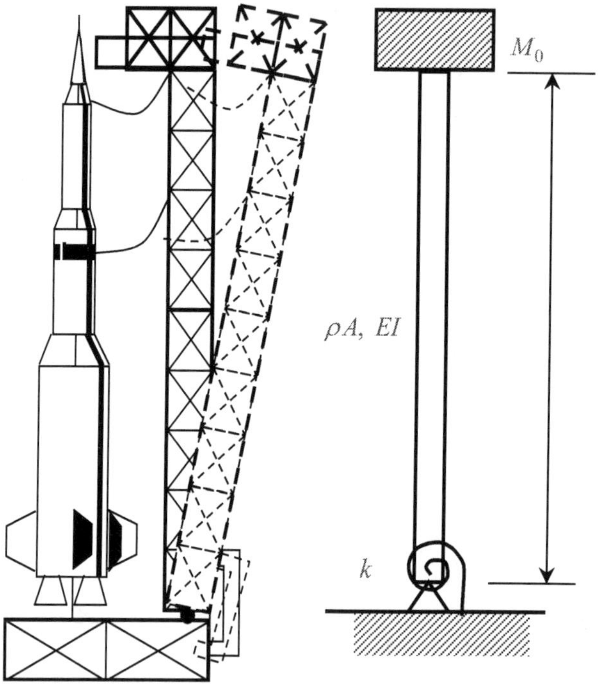

Figure 1.18 A rocket launching pad modeled as a cantilever beam with elastic rotational support

The system of three equations (1.247), (1.248), and (1.249) for three unknowns can be written as

$$\begin{bmatrix} k_R\gamma & 1 & 1 \\ \cos\gamma + \cosh\gamma & \sin\gamma & -\sinh\gamma \\ \sin\gamma - \sinh\gamma + \chi(\cos\gamma - \cosh\gamma) & -\cos\gamma + \chi\cosh\gamma & \cosh\gamma + \chi\sinh\gamma \end{bmatrix} \begin{Bmatrix} A_1 \\ B_1 \\ B_2 \end{Bmatrix} = \begin{Bmatrix} 0 \\ 0 \\ 0 \end{Bmatrix} \tag{1.252}$$

where

$$k_R = \frac{EI}{kL} \tag{1.253}$$

Setting the determinant to zero and simplifying the expression, we find

$$1+\cos\gamma\cosh\gamma=(\frac{M_0}{M}+\frac{1}{2}k_R)\gamma(\sin\gamma\cosh\gamma-\cos\gamma\sinh\gamma) \tag{1.254}$$

For the special case of $k\to\infty$ or $k_R\to 0$, we have

$$1+\cos\gamma\cosh\gamma=\frac{M_0}{M}\gamma(\sin\gamma\cosh\gamma-\cos\gamma\sinh\gamma) \tag{1.255}$$

This of course agrees with the result given in Section 1.3.4.

1.5 BEAM ON ELASTIC FOUNDATION

In the area of railway engineering, rail track is commonly modeled as beams on an elastic foundation. Lintel beams connecting pile caps can also be regarded as beams on an elastic foundation. For soil-structure interactions, the elastic foundation is modeled as a Winkler spring with stiffness or sub-grade reaction k.

1.5.1 Formulation

Figure 1.19 illustrates the problem of a beam resting on an elastic foundation. The only modification needed in the formulation is

$$\frac{dV}{dx}=-w(x)+kv(x) \tag{1.256}$$

where k is the stiffness of the Winkler spring. For the case of the foundation, we have tacitly assumed that the spring can take compression as well as tension. In reality, tension cannot be transmitted through soil, and separation will occur under tension. Equation (1.256) can be rewritten in terms of moment as:

$$\frac{d^2M}{dx^2}=-w(x)+kv(x) \tag{1.257}$$

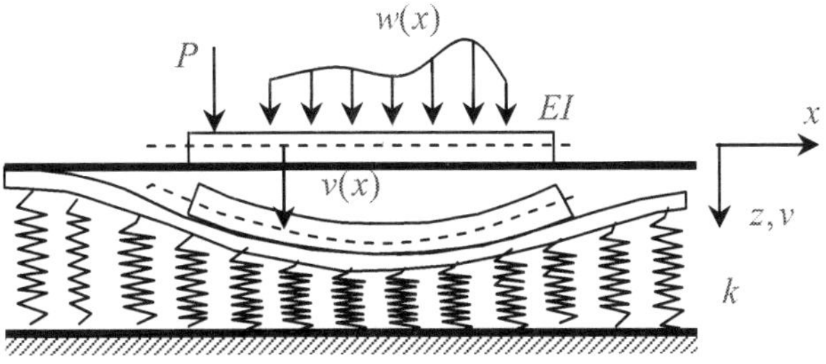

Figure 1.19 A beam on an elastic foundation with Winkler constant *k*

According to the sign convention used in Figure 1.19, (1.256) and (1.257) can be combined to give

$$\frac{d^2}{dx^2}(EI\frac{d^2v}{dx^2})+kv=w(x) \tag{1.258}$$

For prismatic beams, we have

$$\frac{d^4 v}{dx^4} + 4\beta^4 v = \frac{w(x)}{EI} \tag{1.259}$$

where

$$\beta = (\frac{k}{4EI})^{1/4} \tag{1.260}$$

Assuming an exponential solution for v, we have

$$v = e^{\alpha x} \tag{1.261}$$

Substitution of (1.261) into the homogeneous form of (1.259) gives

$$(\alpha^4 + 4\beta^4)e^{\alpha x} = 0 \tag{1.262}$$

Rearranging (1.262), we get

$$\alpha^4 = -4\beta^4 \tag{1.263}$$

Taking the square root on both sides of (1.263), we find

$$\alpha^2 = \pm i2\beta^2 \tag{1.264}$$

Using Euler's formula for the polar form of a complex number (Chau, 2018), we have two scenarios:

$$\alpha^2 = 2e^{i\pi/2}\beta^2, \quad 2e^{-i\pi/2}\beta^2 \tag{1.265}$$

Taking the square root again, we get

$$\alpha = \pm\sqrt{2}e^{i\pi/4}\beta, \quad \pm\sqrt{2}e^{-i\pi/4}\beta \tag{1.266}$$

More explicitly, the four complex roots for α are

$$\alpha = \pm(\beta + i\beta), \quad \pm(\beta - i\beta) \tag{1.267}$$

These can clearly be grouped in pairs as:

$$\alpha = (\beta \pm i\beta), \quad -(\beta \pm i\beta) \tag{1.268}$$

Consequently, the homogeneous solutions are

$$v_h = e^{\beta x}(C_1 \sin\beta x + C_2 \cos\beta x) + e^{-\beta x}(C_3 \sin\beta x + C_4 \cos\beta x) \tag{1.269}$$

For the nonhomogeneous ODE, we can, in general, express the final solution as

$$v = e^{\beta x}(C_1 \sin\beta x + C_2 \cos\beta x) + e^{-\beta x}(C_3 \sin\beta x + C_4 \cos\beta x) + q(w) \tag{1.270}$$

where q is a particular solution depending on the given function w.

1.5.2 Boundary Conditions

In general, two kinds of boundary conditions can be formulated at one end, as shown in Figure 1.20 for the case of a semi-infinite beam on an elastic foundation with no distributed loads. Since the solution must be bounded as $x \to \infty$, we arrive at:

$$C_1 = C_2 = 0 \tag{1.271}$$

For the case of applied moment and point force at $x = 0$, we have

$$EI\left.\frac{d^2 v}{dx^2}\right|_{x=0} = -M_0, \quad EI\left.\frac{d^3 v}{dx^3}\right|_{x=0} = P_0 \tag{1.272}$$

For the case of prescribed deflection and rotation at $x = 0$, we have

$$v\big|_{x=0} = v_0, \quad \frac{dv}{dx}\bigg|_{x=0} = \theta_0 \tag{1.273}$$

If applied moment and point force are prescribed, we have

$$C_3 = \frac{2\beta^2 M_0}{k}, \quad C_4 = \frac{2\beta P_0}{k} - \frac{2\beta^2 M_0}{k} \tag{1.274}$$

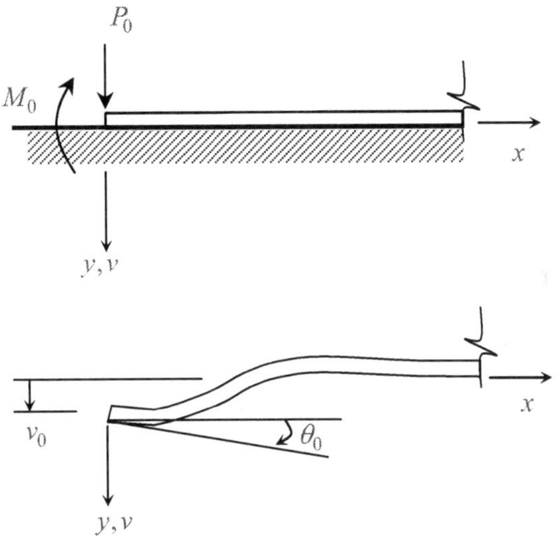

Figure 1.20 A semi-infinite beam on elastic foundation

Substitution of these constants obtained in (1.274) into (1.270) gives

$$v = \frac{2\beta^2 M_0}{k} e^{-\beta x} (\sin\beta x - \cos\beta x) + \frac{2\beta P_0}{k} e^{-\beta x} \cos\beta x \tag{1.275}$$

The slope of the beam can be determined as

$$\theta = \frac{dv}{dx} = \frac{4\beta^3 M_0}{k} e^{-\beta x} \cos\beta x - \frac{2\beta^2 P_0}{k} e^{-\beta x} (\sin\beta x + \cos\beta x) \tag{1.276}$$

1.5.3 Infinite Beam under Concentrated Load

We now consider the case of an infinite beam subject to a point force applied at $x = 0$, as shown in Figure 1.21. This problem can be solved easily based on the solution for a semi-infinite beam given in the last section.

We can chop the infinite beam at $x = 0$. Then, the point force can be split in half and applied to each end of the two chopped semi-infinite beams. That is, by symmetry, we can set

$$P_0 \leftarrow P_0 / 2 \tag{1.277}$$

In addition, the angle of rotation must be zero at $x = 0$:

$$\theta = \left.\frac{dv}{dx}\right|_{x=0} = \frac{4\beta^3 M_0}{k} - \frac{2\beta^2 (P_0/2)}{k} = 0 \tag{1.278}$$

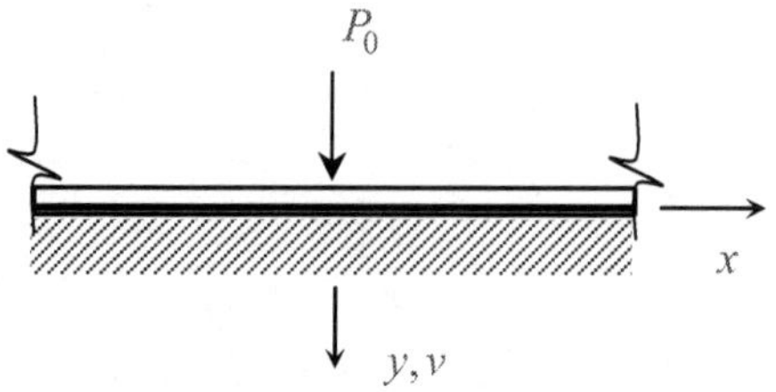

Figure 1.21 An infinite beam resting on an elastic foundation under a point force

This gives

$$M_0 = \frac{P_0}{4\beta} \tag{1.279}$$

Substitution of these constants into (1.269) gives

$$v = \frac{\beta P_0}{2k} e^{-\beta x} (\sin \beta x + \cos \beta x) \tag{1.280}$$

$$\theta = \frac{dv}{dx} = -\frac{\beta^2 P_0}{k} e^{-\beta x} \sin \beta x \tag{1.281}$$

The solution for an infinite beam under a concentrated point force was reconsidered by Weitsman (1970) for the case of a compressive spring that does not take tension. It was concluded that if the beam is weightless, the solution given in this section is only valid for $L \leq \pi/\beta$. If the weight of the beam is taken into account, continuous contact will remain (i.e., no separation occurs) if the following condition is satisfied:

$$-2 \leq \frac{P\beta}{q} \leq 2e^{\pi} \tag{1.282}$$

where q is the weight of the beam per unit meter. Separation will occur if the applied force is given by

$$P > 2e^{\pi} \frac{q}{\beta} \tag{1.283}$$

However, the distance from the origin where separation occurs can only be calculated numerically by solving a system of simultaneous equations. The form of deflection will change type from the contact zone to the separation zone. Our discussion of the nonlinear spring (taking compression only) will end here. Such analysis finds application in train-induced vibrations (Krylov and Ferguson, 1994).

1.6 EULER'S COLUMN BUCKLING

When a column is subject to compression, the column will fail in buckling before the compressive failure. Euler is the first to derive the buckling load. Figure 1.22 shows a free body diagram of a section of a column (oriented horizontally). The main difference between the formulation of the Euler-Bernoulli beam and Euler's column buckling is that we have to formulate the problem in the deformed shape. Such a formulation is known as large deformation formulation. That is, we cannot formulate the problem by assuming the deformed shape is the same as the original shape. Vertical force equilibrium in Figure 1.22 gives

$$qdx + V - (V + dV) = 0 \tag{1.284}$$

This gives the relation between the change of shear force V and the applied transverse loading q

$$\frac{dV}{dx} = q \tag{1.285}$$

Moment equilibrium gives

$$M - Pdv + Vdx + qdx\,dx/2 - (M + dM) = 0 \tag{1.286}$$

Simplification of (1.286) gives

$$V = \frac{dM}{dx} + P\frac{dv}{dx} \tag{1.287}$$

The second term on the right-hand side of (1.287) provides the main contribution of the axial compression. If $P = 0$, the expression for the Euler-Bernoulli beam is recovered.

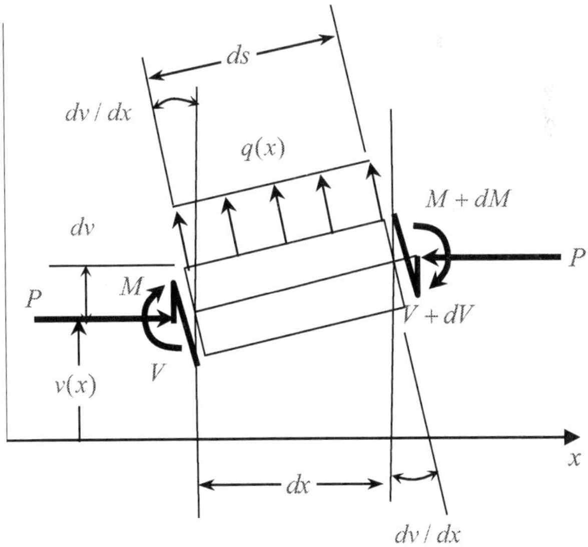

Figure 1.22 A segment of a column subject to end moment, shear force and axial force

Using the result from Section 1.2.1 that bending moment is proportional to the second derivative of deflection v, we have

$$\frac{d^2 v}{dx^2} = \frac{M}{EI} \tag{1.288}$$

Combining (1.285), (1.287) and (1.288), we get

$$\frac{d^2 M}{dx^2} + \lambda^2 M = q \tag{1.289}$$

where

$$\lambda^2 = \frac{P}{EI} \tag{1.290}$$

Substitution of (1.288) into (1.289) gives

$$\frac{d^2}{dx^2}(EI\frac{d^2 v}{dx^2}) + \lambda^2 EI \frac{d^2 v}{dx^2} = q \tag{1.291}$$

For columns with uniform cross-section, (1.291) is simplified to

$$\frac{d^4 v}{dx^4} + \lambda^2 \frac{d^2 v}{dx^2} = \frac{q}{EI} \tag{1.292}$$

Substitution of (1.288) and (1.287) gives

$$V = \frac{d}{dx}(EI\frac{d^2 v}{dx^2}) + P\frac{dv}{dx} \tag{1.293}$$

For the case of zero lateral loading q, (1.292) is reduced to

$$\frac{d^4 v}{dx^4} + \lambda^2 \frac{d^2 v}{dx^2} = 0 \tag{1.294}$$

This is a fourth order ODE with constant coefficient. According to Chau (2018), we can convert (1.294) to a second order ODE by assuming the second derivative of v as the new variable. Using this procedure, it is straightforward to find the solution of v as

$$v = C_1 \sin \lambda x + C_2 \cos \lambda x + C_3 x + C_4 \tag{1.295}$$

Differentiation of v gives

$$v' = \lambda C_1 \cos \lambda x - \lambda C_2 \sin \lambda x + C_3 \tag{1.296}$$

$$v'' = -\lambda^2 C_1 \sin \lambda x - \lambda^2 C_2 \cos \lambda x \tag{1.297}$$

$$v''' = -\lambda^3 C_1 \cos \lambda x + \lambda^3 C_2 \sin \lambda x \tag{1.298}$$

We now illustrate the determination of the buckling load for various boundary conditions.

Example 1.1 Find Euler's buckling load for the simply-supported column shown in Figure 1.23.

Solution: Recalling from (1.294), we have

$$\frac{d^4 v}{dx^4} + \lambda^2 \frac{d^2 v}{dx^2} = 0 \tag{1.299}$$

The deflections at the ends are:

$$v(0) = 0, \quad v(L) = 0 \tag{1.300}$$

The end bending moments are

$$M(0) = EIv''(0) = 0, \quad M(L) = EIv''(L) = 0 \tag{1.301}$$

Substituting (1.295) to (1.297) into (1.300) and (1.301), we obtain four equations for the four unknown constants as

$$\begin{aligned} &v(0) = C_2 + C_4 = 0, \\ &v(L) = C_1 \sin \lambda L + C_2 \cos \lambda L + C_3 L + C_4 = 0, \\ &M(0) = -C_2 EI\lambda^2 = 0, \\ &M(L) = -C_1 EI\lambda^2 \sin \lambda L - C_2 EI\lambda^2 \cos \lambda L = 0 \end{aligned} \tag{1.302}$$

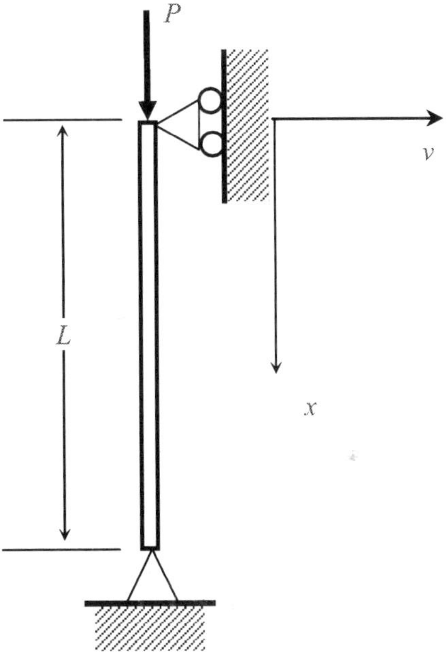

Figure 1.23 A column subject to an axial force *P* with a simply support

This homogeneous system provides an eigenvalue problem. That is, only for certain values of λ can a nonzero solution for the unknown constants satisfy the prescribed boundary conditions. Setting the determinant of the system of equations to zero, we have

$$\begin{vmatrix} 0 & 1 & 0 & 1 \\ \sin \lambda L & \cos \lambda L & L & 1 \\ 0 & -P & 0 & 0 \\ -P \sin \lambda L & -P \cos \lambda L & 0 & 0 \end{vmatrix} = 0 \tag{1.303}$$

The solution for this system must satisfy

$$\sin \lambda L = 0 \tag{1.304}$$

This requires

$$\lambda L = n\pi \tag{1.305}$$

Substitution of the definition of λ in (1.290) gives Euler's buckling formula

$$P_{cr} = \frac{\pi^2 n^2 EI}{L^2} \tag{1.306}$$

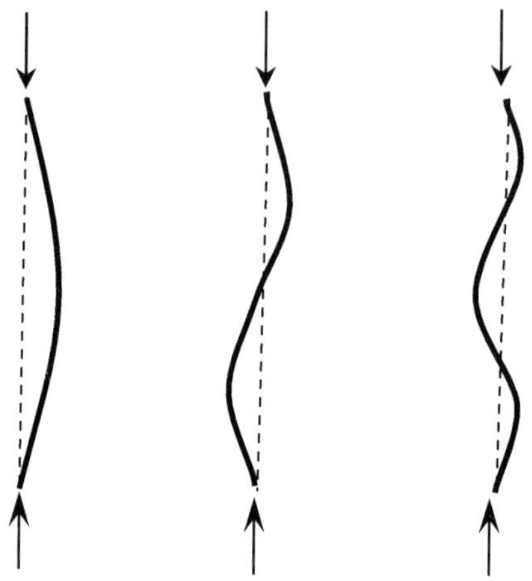

Figure 1.24 The first three buckling modes of columns

This is the Euler buckling formula. The first three modes of buckling are sketched in Figure 1.24 with the corresponding buckling loads of

$$P_{cr} = \frac{\pi^2 EI}{L^2}, \frac{4\pi^2 EI}{L^2}, \frac{9\pi^2 EI}{L^2} \tag{1.307}$$

Note that we cannot find the values of the unknowns in (1.295). That is, the magnitude of deflection is indeterminate. Thus, the column deflection is identically zero before Euler's buckling load is attained. We can only calculate buckling load but not the magnitude of deflection. In reality, we rarely observe buckling that occurs from the state of zero deflection. In a later example (i.e., Example 1.3), we will see that the deflection is nonzero before buckling, if there is an imperfection (e.g., the axial load does not align with the centroid of the cross-section).

Example 1.2 Find the buckling load for the following simply-supported bar subject to concentrated moment and axial force at the ends, as shown in Figure 1.25.

Solution: The deflections at the ends are:

$$v(0) = 0, \quad v(L) = 0 \tag{1.308}$$

The end bending moments are

$$M(0) = -M_0, \quad M(L) = -M_0 \tag{1.309}$$

This results in four equations for the four unknown constants:

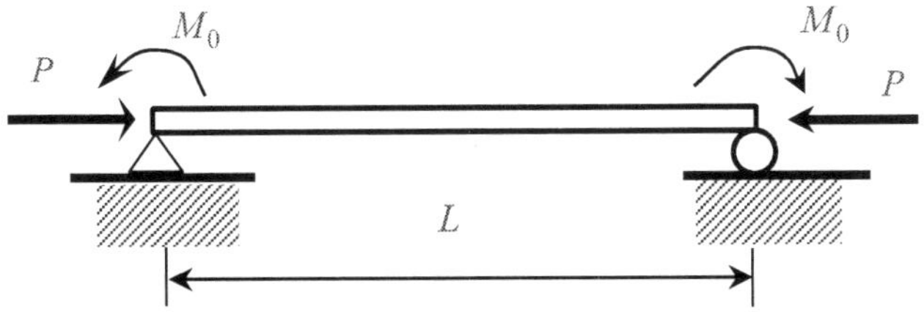

Figure 1.25 A bar subject to concentrated moment and axial force

$$\begin{aligned} &v(0) = C_2 + C_4 = 0, \\ &v(L) = C_1 \sin \lambda L + C_2 \cos \lambda L + C_3 L + C_4 = 0 \\ &M(0) = -C_2 EI\lambda^2 = -M_0 \\ &M(L) = -C_1 EI\lambda^2 \sin \lambda L - C_2 EI\lambda^2 \cos \lambda L = -M_0 \end{aligned} \tag{1.310}$$

The solutions for this system are

$$C_1 = \frac{M_0}{P}\frac{1-\cos\lambda L}{\sin\lambda L}, \quad C_2 = -C_4 = \frac{M_0}{P}, \quad C_3 = 0 \tag{1.311}$$

Substitution of (1.311) into (1.295) gives

$$v = \frac{M_0}{P}\left(\frac{1-\cos\lambda L}{\sin\lambda L}\sin\lambda x + \cos\lambda x - 1\right) \tag{1.312}$$

By symmetry, the maximum deflection must occur at the center of the bar and equals

$$v_{\max} = \frac{M_0}{P}\left[\frac{\sin^2(\lambda L/2)}{\cos(\lambda L/2)} + \cos(\lambda L/2) - 1\right] = \frac{M_0}{P}\left(\sec\frac{\lambda L}{2} - 1\right) \tag{1.313}$$

Similarly, the maximum moment at the center of the bar is

$$M_{\max} = \left|-M_0 - Pv_{\max}\right| = M_0 \sec\left(\frac{\lambda L}{2}\right) = \frac{M_0}{\cos\left(\frac{\lambda L}{2}\right)} \tag{1.314}$$

We can see that both deflection and moment approach infinity (this is what we expect for buckling) if we set

$$\frac{\lambda L}{2} = \frac{n\pi}{2} \Rightarrow v_{\max} \to \infty, \quad M_{\max} \to \infty \tag{1.315}$$

Therefore, the eigenvalues of the buckling problem are

$$\lambda = \frac{n\pi}{L}, \quad n = 1, 2, 3, \ldots \tag{1.316}$$

Substitution of (1.316) into (1.290) yields the buckling load

$$P_n = \frac{n^2\pi^2 EI}{L^2} \tag{1.317}$$

which is the same as (1.306). For the most critical buckling load, we have $n = 1$ or

$$P_{cr} = \frac{\pi^2 EI}{L^2} \quad (1.318)$$

This agrees with the Euler buckling formula. In conclusion, the buckling load does not change even if end moments are applied.

Example 1.3 Consider the case where the axial force does not apply through the centroid of the section, as shown in Figure 1.26. This can be considered as a kind of loading imperfection. The axial compression P is applied at a distance of k from the centroid as shown in Figure 1.26 and the boundary conditions are assumed to be simply-supported. Find the maximum deflection of the bar and the buckling load of the problem.

Solution: Instead of using the ODE for deflection, we use the governing equation for the bending moment since the bending moment can be found easily

$$EI\frac{d^2y}{dx^2} = M(x) = -P(k+y) \quad (1.319)$$

Rearranging (1.319), we have a standard ODE as:

$$\frac{d^2y}{dx^2} + \frac{P}{EI}y = -\frac{Pk}{EI} \quad (1.320)$$

Without going through the details, the general solution of (1.320) can be integrated as:

$$y = A\cos\sqrt{\frac{P}{EI}}x + B\sin\sqrt{\frac{P}{EI}}x - k \quad (1.321)$$

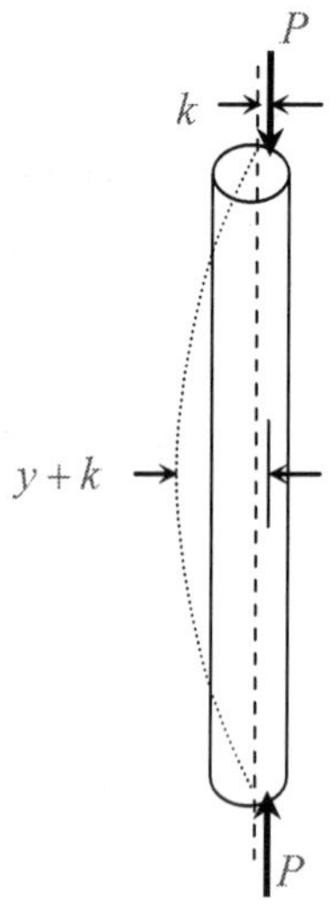

Figure 1.26 A column subject to an eccentric axial force *P*

Note that (1.320) is a second order ODE and we only need two boundary conditions to fix the unknowns. The boundary condition at $x = 0$ is

$$y(0) = 0 \tag{1.322}$$

This boundary condition gives $A = k$ and results in

$$y = k\cos\sqrt{\frac{P}{EI}}x + B\sin\sqrt{\frac{P}{EI}}x - k \tag{1.323}$$

By symmetry, we can set the secondary boundary condition at the center of the bar as

$$\left.\frac{dy}{dx}\right|_{x=L/2} = 0 \tag{1.324}$$

Substitution of (1.323) into (1.324) gives

$$-k\sqrt{\frac{P}{EI}}\sin(\sqrt{\frac{P}{EI}}\frac{L}{2}) + B\sqrt{\frac{P}{EI}}\cos(\sqrt{\frac{P}{EI}}\frac{L}{2}) = 0 \tag{1.325}$$

The unknown constant B can be found as

$$B = k\tan(\sqrt{\frac{P}{EI}}\frac{L}{2}) \tag{1.326}$$

Finally, substitution of (1.326) into (1.323) gives the deflection of the bar

$$y = k\{\sec(\sqrt{\frac{P}{EI}}\frac{L}{2})\cos[\sqrt{\frac{P}{EI}}(x-\frac{L}{2})] - 1\} \tag{1.327}$$

By symmetry, the maximum deflection is clearly at the center and equals

$$y_{\max} = k[\sec(\sqrt{\frac{P}{EI}}\frac{L}{2}) - 1] = k[\frac{1}{\cos(\sqrt{\frac{P}{EI}}\frac{L}{2})} - 1] \tag{1.328}$$

For buckling, we see that the deflection tends to infinity if the cosine function in the denominator tends to zero. This gives

$$\sqrt{\frac{P}{EI}}\frac{L}{2} = 2n\pi + \frac{\pi}{2} \tag{1.329}$$

Rearranging (1.328), we find the buckling load as

$$P_{cr} = \frac{4EI}{L^2}(2n\pi + \frac{\pi}{2})^2 \tag{1.330}$$

The smallest buckling load corresponds to $n = 0$ and equals

$$P_{cr} = \frac{EI\pi^2}{L^2} \tag{1.331}$$

This is precisely Euler's buckling load again. However, the buckling load for the higher modes does not coincide with Euler's result. In fact, Chau (1993) showed by three-dimensional continuum mechanics that Euler's buckling formula remains valid even for the case of anisotropic, compressible pressure-sensitive material if the Young's modulus E is interpreted by tangent Young's modulus.

Timoshenko and Gere (1961) called (1.328) the secant formula. More importantly, we see that we can actually calculate the deflection before buckling occurs. Rearranging (1.327), we find that the load deflection relation before buckling as:

$$\frac{P}{P_{cr}} = \frac{4}{\pi^2}\{\cos^{-1}(\frac{1}{\frac{y_{max}}{k}+1})\}^2 \tag{1.332}$$

Figure 1.27 plots (1.332), and we see that for columns with imperfection, buckling does not occur abruptly but instead the buckling load is approached asymptotically. The solution of Example 1.2 can also be rearranged to yield

$$\frac{P}{P_{cr}} = \frac{4}{\pi^2}\{\cos^{-1}(\frac{1}{\frac{Py_{max}}{M_0}+1})\}^2 \tag{1.333}$$

It is clear that these two formulas are identical if we identify δ as M_0/P. Thus, mathematically the case of eccentric loading can be replaced by equivalent end moments.

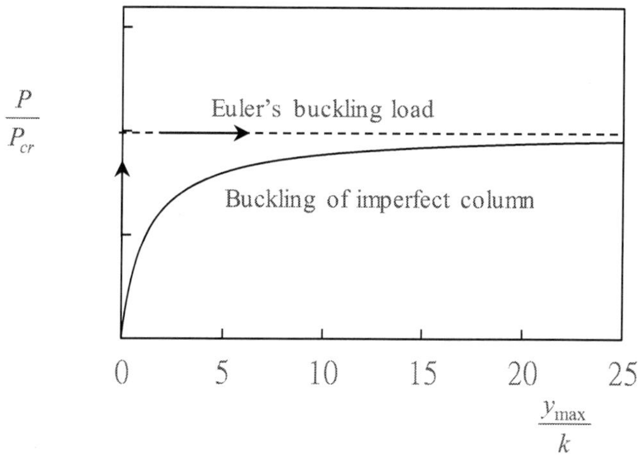

Figure 1.27 Force-displacement curves for columns with and without imperfection

1.7 VIBRATIONS OF BEAMS UNDER AXIAL COMPRESSION

In the previous sections, we have considered vibrations and buckling of beam separately. In this section, we will consider the combined effect of axial force and vibrations. In particular, we consider the case of cantilever beam (i.e., one end built-in and one end free) as shown in Figure 1.28. The boundary conditions of a cantilever can be summarized as:

$$v(0,t) = 0, \quad \frac{\partial v}{\partial x}(0,t) = 0 \tag{1.334}$$

$$\frac{\partial^2 v}{\partial x^2}(L,t)=0,\quad \frac{\partial^3 v}{\partial x^3}(L,t)=0 \tag{1.335}$$

The initial conditions can be expressed as:

$$v(x,0)=f(x),\quad \frac{\partial v}{\partial x}(x,0)=g(x) \tag{1.336}$$

In the next section, we will first find the natural frequency of free vibrations and determine the corresponding eigenfunctions before we consider the solution for vibration solutions induced by initial conditions given in (1.335).

1.7.1 Free Vibrations of Cantilever Beams under Axial Compression

In Figure 1.28, we have included the more general situation of distributed load $q(x,t)$. Equation of motion along the vertical or y-axis is formulated as

$$V(x+dx,t)-V(x,t)+\rho A\frac{\partial^2 v}{\partial t^2}dx=q(x,t)dx \tag{1.337}$$

Using the first term of Taylor's series expansion for the shear force V, we get

$$\frac{\partial V}{\partial x}+\rho A\frac{\partial^2 v}{\partial t^2}=q(x,t) \tag{1.338}$$

Next, we consider the moment equilibrium in the free body of length dx and obtain

$$M(x+dx,t)-M(x,t)+N[v(x+dx,t)-v(x,t)]-V(x+dx,t)dx=0 \tag{1.339}$$

Again, using the usual procedure of retaining the first term of Taylor's series expansion, we arrive at the following relation between moment M and shear force V

$$\frac{\partial M}{\partial x}+N\frac{\partial v}{\partial x}=V \tag{1.340}$$

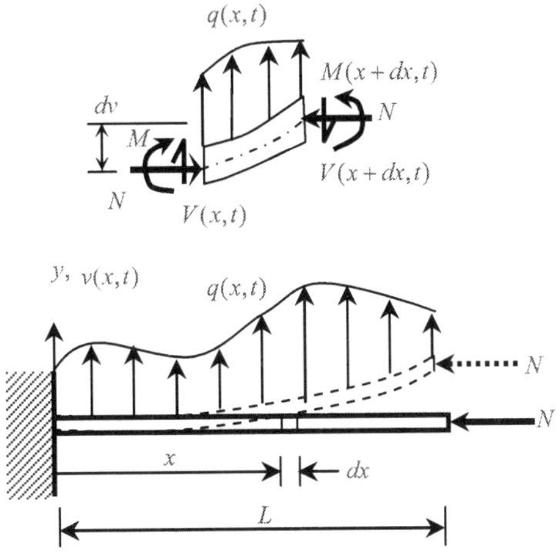

Figure 1.28 Cantilever beam subject to axial force

Substitution of (1.340) into (1.338) gives

$$\frac{\partial^2 M}{\partial x^2}+N\frac{\partial^2 v}{\partial x^2}+\rho A\frac{\partial^2 v}{\partial t^2}=q(x,t) \quad (1.341)$$

Finally, using the relation between moment and deflection, we get a PDE for v

$$\frac{\partial^2}{\partial x^2}(EI\frac{\partial^2 v}{\partial x^2})+N\frac{\partial^2 v}{\partial x^2}+\rho A\frac{\partial^2 v}{\partial t^2}=q(x,t) \quad (1.342)$$

Considering a beam with uniform cross-section (EI is constant) undergoing free vibrations (i.e., setting $q = 0$), we get

$$EI\frac{\partial^4 v}{\partial x^4}+N\frac{\partial^2 v}{\partial x^2}+\rho A\frac{\partial^2 v}{\partial t^2}=0 \quad (1.343)$$

Applying the separation of variables, we assume

$$v=X(x)T(t) \quad (1.344)$$

Substitution of (1.344) into (1.343) gives

$$EI\frac{X^{(iv)}}{X}+N\frac{X''}{X}=-\rho A\frac{T''}{T}=\omega^2\rho A \quad (1.345)$$

where ω is a constant to be determined. Dividing through the whole expression by ρA, we get

$$a^2\frac{X^{(iv)}}{X}+b^2\frac{X''}{X}=-\frac{T''}{T}=\omega^2 \quad (1.346)$$

where

$$a^2=\frac{EI}{\rho A}, \quad b^2=\frac{N}{\rho A} \quad (1.347)$$

The second part of (1.346) gives an ODE for time dependent function T as

$$T''+\omega^2 T=0 \quad (1.348)$$

This is the classic ODE for harmonic oscillators, and the solution is (Chau, 2018):

$$T=A\sin\omega t+B\cos\omega t \quad (1.349)$$

The first part of (1.345) gives another ODE for spatial dependent function X as

$$a^2X^{(iv)}+b^2X''-\omega^2X=0 \quad (1.350)$$

For beams with uniform cross-section (or so-called prismatic beams), the coefficients a and b are constants. Consequently, (1.350) is an ODE with constant coefficients. We expect exponential type of solution (Chau, 2018):

$$X=e^{\alpha x} \quad (1.351)$$

Substitution of (1.351) into (1.350) yields a characteristic equation for α

$$a^2\alpha^4+b^2\alpha^2-\omega^2=0 \quad (1.352)$$

The solution of this quadratic equation is

$$\alpha^2=\frac{-b^2\pm\sqrt{b^4+4a^2\omega^2}}{2a^2}=-\gamma\pm\beta \quad (1.353)$$

Thus, four solutions for α are obtained as:

$$\alpha=\pm\sqrt{\beta-\gamma}, \quad \pm i\sqrt{\beta+\gamma} \quad (1.354)$$

The solution form is the same as those for Euler-Bernoulli beams with no axial force:

$$\begin{aligned} X &= A_1\cos(\sqrt{\beta+\gamma}x) + A_2\sin(\sqrt{\beta+\gamma}x) \\ &+ A_3\sinh(\sqrt{\beta-\gamma}x) + A_4\cosh(\sqrt{\beta-\gamma}x) \end{aligned} \tag{1.355}$$

If we set $N = 0$ (or equivalently $\gamma = 0$), we recover the solution for Euler-Bernoulli beams. Differentiation of (1.355) leads to

$$\begin{aligned} X' &= \sqrt{\beta+\gamma}[-A_1\sin(\sqrt{\beta+\gamma}x) + A_2\cos(\sqrt{\beta+\gamma}x)] \\ &+ \sqrt{\beta-\gamma}[A_3\cosh(\sqrt{\beta-\gamma}x) + A_4\sinh(\sqrt{\beta-\gamma}x)] \end{aligned} \tag{1.356}$$

$$\begin{aligned} X'' &= -(\beta+\gamma)[A_1\cos(\sqrt{\beta+\gamma}x) + A_2\sin(\sqrt{\beta+\gamma}x)] \\ &+ (\beta-\gamma)[A_3\sinh(\sqrt{\beta-\gamma}x) + A_4\cosh(\sqrt{\beta-\gamma}x)] \end{aligned} \tag{1.357}$$

$$\begin{aligned} X''' &= (\beta+\gamma)\sqrt{\beta+\gamma}[A_1\sin(\sqrt{\beta+\gamma}x) - A_2\cos(\sqrt{\beta+\gamma}x)] \\ &+ (\beta-\gamma)\sqrt{\beta-\gamma}[A_3\cosh(\sqrt{\beta-\gamma}x) + A_4\sinh(\sqrt{\beta-\gamma}x)] \end{aligned} \tag{1.358}$$

The built-in boundary conditions given in (1.334) at $x = 0$ lead to

$$A_4 = -A_1, \quad A_3 = -A_2 \tag{1.359}$$

The free boundary conditions given in (1.335) at $x = L$ require

$$\begin{aligned} X''(L) &= -(\beta+\gamma)[A_1\cos(\sqrt{\beta+\gamma}L) + A_2\sin(\sqrt{\beta+\gamma}L)] \\ &+ (\beta-\gamma)[-A_2\sinh(\sqrt{\beta-\gamma}L) - A_1\cosh(\sqrt{\beta-\gamma}L)] = 0 \end{aligned} \tag{1.360}$$

$$\begin{aligned} X'''(L) &= (\beta+\gamma)\sqrt{\beta+\gamma}[A_1\sin(\sqrt{\beta+\gamma}L) - A_2\cos(\sqrt{\beta+\gamma}L)] \\ &+ (\beta-\gamma)\sqrt{\beta-\gamma}[-A_2\cosh(\sqrt{\beta-\gamma}L) - A_1\sinh(\sqrt{\beta-\gamma}L)] = 0 \end{aligned} \tag{1.361}$$

Note that in obtaining (1.360) and (1.361), we have already used the results in (1.359). This gives two equations for two unknowns and this homogeneous system of equations can be recast in matrix form as:

$$\begin{bmatrix} K_{11} & K_{12} \\ K_{21} & K_{22} \end{bmatrix} \begin{Bmatrix} A_1 \\ A_2 \end{Bmatrix} = \begin{Bmatrix} 0 \\ 0 \end{Bmatrix} \tag{1.362}$$

where components of the matrix $\boldsymbol{K}$ can be identified readily from (1.360) and (1.361). For nontrivial solution, we require the determinant of matrix $\boldsymbol{K}$ being zero and this leads to the following eigenvalue equation:

$$K_{11}K_{22} - K_{21}K_{12} = 0 \tag{1.363}$$

where

$$\begin{aligned} K_{11}K_{22} &= (\beta+\gamma)^{5/2}\cos^2(\sqrt{\beta+\gamma}L) + (\beta-\gamma)^{5/2}\cosh^2(\sqrt{\beta-\gamma}L) \\ &+ (\beta-\gamma)(\beta+\gamma)[\sqrt{\beta-\gamma} + \sqrt{\beta+\gamma}]\cos(\sqrt{\beta+\gamma}L)\cosh(\sqrt{\beta-\gamma}L) \end{aligned} \tag{1.364}$$

$$\begin{aligned} -K_{21}K_{12} &= (\beta+\gamma)^{5/2}\sin^2(\sqrt{\beta+\gamma}L) - (\beta-\gamma)^{5/2}\sinh^2(\sqrt{\beta-\gamma}L) \\ &+ (\beta-\gamma)(\beta+\gamma)[-\sqrt{\beta-\gamma} + \sqrt{\beta+\gamma}]\sin(\sqrt{\beta+\gamma}L)\sinh(\sqrt{\beta-\gamma}L) \end{aligned} \tag{1.365}$$

Equation (1.363) can be simplified to

$$
\begin{aligned}
F(\omega) &= (\beta^2-\gamma^2)\{[\sqrt{\beta-\gamma}-\sqrt{\beta+\gamma}]\sin(\sqrt{\beta+\gamma}L)\sinh(\sqrt{\beta-\gamma}L) \\
&-[\sqrt{\beta-\gamma}+\sqrt{\beta+\gamma}]\cos(\sqrt{\beta+\gamma}L)\cosh(\sqrt{\beta-\gamma}L)\} \\
&-(\beta+\gamma)^{5/2}+(\beta-\gamma)^{5/2}=0
\end{aligned}
\tag{1.366}
$$

For the special case of zero axial compression, we have $N = 0$. It leads to $b^2 = 0$, $\gamma = 0$, and $\beta = \omega/a$. This leads to the following eigenvalue equation

$$
\cos(z_n) = -\frac{1}{\cosh(z_n)}, \quad z_n = \sqrt{\frac{\omega}{a}}L \tag{1.367}
$$

This equation has been obtained in Section 1.3.3, at which it was shown that there are infinite roots for (1.367). Thus, we also expect there are also infinite roots in (1.366). For the case of nonzero axial compression, we can solve (1.366) for nonzero N (or equivalently $b \neq 0$). Table 1.3 is obtained for $a = 337$ m^2/s and $b = 69.4$ m/s. These values roughly correspond to a 10 m long steel cantilever beam of rectangular cross-section of 0.23 m by 0.115 m subject to a constant axial compression of 100 kN. The adopted density and Young's modulus are 7850 kg/m^3 and 200 GPa. The first natural circular frequency is found very sensitive to the value of the axial compression. It drops from 11.85 rad/s to 4.25 rad/s. However, for the second and higher natural circular frequencies, the dependence on the axial compression N is negligible. In practice, the first mode of vibrations is most likely to occur compared to higher modes and, thus, the effect of axial force cannot be neglected.

Table 1.3 First ten eigenvalues of natural vibration frequencies for cantilever beam for $a = 337$ (m^2/s) $b = 69.4$ (m/s)

Number, n	ω_n		Approximation	% error
	$N = 0$	$N = 100$ kN	$a[(2n-1)\pi/2L]^2$	
1	11.84897059	4.251728	8.315142	95.57087
2	74.25623237	74.494319	74.83628	0.459042
3	207.9196423	207.755684	207.878554	0.059142
4	407.4394767	407.379274	407.441966	0.015389
5	673.5265792	673.488569	673.526516	0.005634
6	1006.132194	1006.10677	1006.132203	0.002528
7	1405.258914	1405.240798	1405.259027	0.001297
8	1870.906900	1870.893286	1870.906988	0.000732
9	2403.076034	2403.065329	2403.076087	0.000448
10	3001.766114	3001.757651	3001.766324	0.000289

To further investigate the effect of axial force, Figure 1.29 plots the function $F(\omega)$ versus ω for various values of b (or equivalently N). For the case of $b = 77$, the function shift upward such that there is no zero intercept of the function $F(\omega)$. Thus, fundamental mode is highly sensitive to the axial compression or the prestress in the beam. From Table 1.3, we see that a good approximation of the natural frequency can be found as

$$\omega_n = \sqrt{\frac{EI}{\rho A}(\frac{z_n}{L})^2} \approx \sqrt{\frac{EI}{\rho A}\frac{(2n-1)^2\pi^2}{4L^2}}, \quad n = 1,2,3,... \tag{1.368}$$

To investigate the accuracy of (1.367), Table 1.4 compiles the first natural frequency versus the prediction by (1.367) for various values of b and a fixed a. We see that for axial force larger than 50 kN, the accuracy of (1.367) deteriorates to being unacceptable.

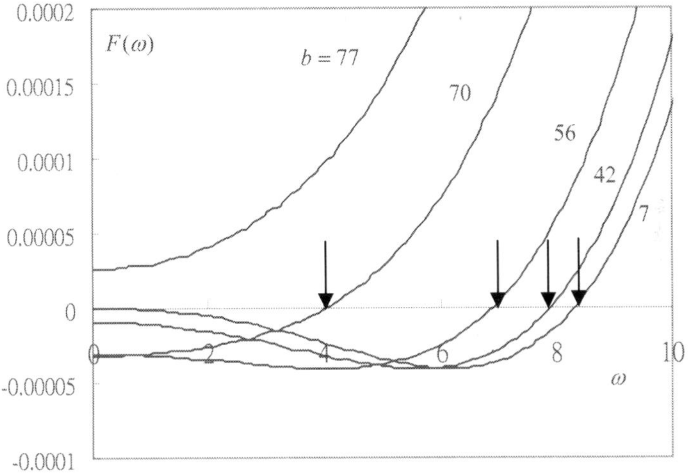

Figure 1.29 The roots of frequency equation $F(\omega)$

Table 1.4 The dependence of first mode as a function of the axial force parameter b (with a = 337 (m^2/s) and b = 7~70 (m/s))

N (kN)	ω_1	$a[\pi/2L]^2$	% error	a	b
10	8.314824	8.315142	0.003828	337	7
20	8.310055	8.315142	0.061215	337	14
30	8.289359	8.315142	0.311042	337	21
40	8.233379	8.315142	0.993065	337	28
50	8.114077	8.315142	2.477978	337	35
60	7.892515	8.315142	5.354788	337	42
70	7.513421	8.315142	10.670517	337	49
80	6.891506	8.315142	20.657836	337	56
90	5.870648	8.315142	41.63925	337	63
100	4.03587	8.315142	106.030956	337	70
110	74.317465	8.315142	-88.811322	337	77

The eigenmodes of vibrations can then be expressed as:

$$X_n = A_n[\cos(\sqrt{\beta_n + \gamma}x) - \cosh(\sqrt{\beta_n - \gamma}x)] + B_n[\sin(\sqrt{\beta_n + \gamma}x) - \sinh(\sqrt{\beta_n - \gamma}x)] \quad (1.369)$$

where

$$\beta_n = \frac{\sqrt{b^4 + 4a^2\omega_n^2}}{2a^2} \quad (1.370)$$

Finally, the eigenmode for v is obtained as

$$\begin{aligned} v_n = &[C_n \sin(\omega t) + D_n \cos(\omega t)]\{[\cos(\sqrt{\beta_n + \gamma}x) - \cosh(\sqrt{\beta_n - \gamma}x)] \\ &+ \frac{B_n}{A_n}[\sin(\sqrt{\beta_n + \gamma}x) - \sinh(\sqrt{\beta_n - \gamma}x)]\} \end{aligned} \quad (1.371)$$

However, it is possible to find the ratio of A_n/B_n using (1.335) as

$$\frac{B_n}{A_n} = -\frac{(\beta_n + \gamma)\cos(\sqrt{\beta_n + \gamma}L) + (\beta_n - \gamma)\cosh(\sqrt{\beta_n - \gamma}L)}{(\beta_n + \gamma)\sin(\sqrt{\beta_n + \gamma}L) + (\beta_n - \gamma)\sinh(\sqrt{\beta_n - \gamma}L)} = Y_n(\beta_n) \quad (1.372)$$

Finally, we can recast the solution in a more compact form

$$\begin{aligned} v(x,t) = &\sum_{n=1}^{\infty}[C_n \sin(\omega t) + D_n \cos(\omega t)]\{[\cos(\sqrt{\beta_n + \gamma}x) - \cosh(\sqrt{\beta_n - \gamma}x)] \\ &+ Y_n[\sin(\sqrt{\beta_n + \gamma}x) - \sinh(\sqrt{\beta_n - \gamma}x)]\} \end{aligned} \quad (1.373)$$

1.7.2 Orthogonal Approximation

In this section, we consider the problem of a beam subject to initial conditions given in (1.336). As discussed in Chau (2018), we need both the completeness and orthogonality of the eigenfunctions X_n in order to solve the problem using eigenfunction expansion. As a first approximation, we assume the following orthogonality of X_n without proof:

$$\int_0^L X_n(x)X_m(x)dx = 0, \quad m \neq n \\ \neq 0, \quad m = n \quad (1.374)$$

In fact, it will be shown in Problem 1.1 that this is not strictly correct for the case of nonzero axial compression (i.e., $b \neq 0$); and this will be further discussed in the next section. Therefore, the following result only provides a first approximation for the case of small b.

Applying the initial conditions, we have

$$v(x,0) = \sum_{n=1}^{\infty} D_n X_n(x) = f(x) \quad (1.375)$$

Multiplying both sides by X_m and integrating from 0 to L, we have

$$\sum_{n=1}^{\infty} D_n \int_0^L X_n(x)X_m(x)dx = \int_0^L f(x)X_m(x)dx \tag{1.376}$$

Assuming the validity of the orthogonal properties given in (1.374), we have

$$D_n = \frac{\int_0^L f(x)X_n(x)dx}{\int_0^L X_n^2(x)dx} \tag{1.377}$$

Similarly, applying the initial condition, we have

$$\frac{\partial v}{\partial t}(x,0) = \sum_{n=1}^{\infty} \omega_n C_n X_n(x) = g(x) \tag{1.378}$$

Multiplying both sides by the m-th eigenmodes and integrating both sides from 0 to L, we have

$$\sum_{n=1}^{\infty} \omega_n C_n \int_0^L X_n(x)X_m(x)dx = \int_0^L g(x)X_m(x)dx \tag{1.379}$$

Using the assumed orthogonal property of the eigenmodes, we have

$$C_n = \frac{\int_0^L g(x)X_n(x)dx}{\omega_n \int_0^L X_n^2(x)dx} \tag{1.380}$$

Finally, substitution of (1.336) and (1.339) gives an approximate solution of the problem.

1.7.3 Rigorous Approach

As shown in Problem 1.6, the correct orthogonal relation is:

$$(\lambda_m - \lambda_n)\int_0^L X_m X_n dx + b^2[X_m(L)X_n'(L) - X_m'(L)X_n(L)] = 0 \tag{1.381}$$

Apparently, this orthogonal relation has not been published in the literature. The only related work was made by Takahashi (1980) for the case of a cantilever with a tip mass supported by a spring subject to axial compression. Other than Takahashi the orthogonal property of beams subject to axial compression has not been reported elsewhere. In conclusion, (1.381) shows that the nice property of orthogonality is destroyed. In layman terms, the unknown constants cannot be evaluated independently for each eigenfunction of n-th order as given in (1.377) and (1.380). Substitution of (1.381) into (1.376) gives

$$\sum_{n=1}^{\infty} \frac{b^2[X_m(L)X_n'(L) - X_m'(L)X_n(L)]}{(\lambda_m - \lambda_n)} D_n = \int_0^L f(x)X_m(x)dx \tag{1.382}$$

This is a system of infinite equations for infinite unknowns. In actual numerical implementation, we cannot solve for a system of infinite equations. Let us consider a truncation of the first N eigenfunctions such that

$$\sum_{n=1}^{N} \psi_{mn} D_n = F_m, \quad m = 1,2,...,N \tag{1.383}$$

where

$$\psi_{mn} = \frac{b^2[X_m(L)X_n'(L) - X_m'(L)X_n(L)]}{(\lambda_m - \lambda_n)}, \quad m \neq n$$
$$= \int_0^L X_m^2 dx, \quad m = n \tag{1.384}$$

$$F_m = \int_0^L f(x) X_m(x) dx, \quad m = 1,2,...,N \tag{1.385}$$

This gives a system of N equations for n unknowns. The evaluation of ψ_{mm} in the second of (1.383) only involves the integration of functions as product of any two of the following functions: sine, cosine, hyperbolic sine, and hyperbolic cosine. Although the calculation may be tedious, it is straightforward.

Similarly, for initial velocity we have

$$\sum_{n=1}^{N} \psi_{mn} \omega_n C_n = G_m, \quad m = 1,2,...,N \tag{1.386}$$

$$G_m = \int_0^L g(x) X_m(x) dx \tag{1.387}$$

This completes the solution. However, we will do attempt to solve any particular initial value problem here.

1.8 TIMOSHENKO BEAM THEORY

In 1921, Timoshenko modified the Euler-Bernoulli theory by allowing for shear deformation. This model is more accurate than the Euler-Bernoulli beam since it is not restrict to beams with aspect ratio of larger than 10 (i.e., long shallow beam).

1.8.1 Variational Formulation

Timoshenko assumed that a plane section remains plane but the plane is no longer perpendicular to the neutral axis after bending by adding an additional shear rotation β, as shown in Figure 1.30. In other words, all the shear deformation effect has been replaced by an equivalent shear rotation angle β, which can be expressed as:

$$\beta = \alpha - \frac{\partial v}{\partial x} \tag{1.388}$$

Clearly, this assumption of an additional rotation does not exactly account for the shear deformation, but this assumption makes the problem mathematically tractable. The strain energy due to bending is:

$$V_b = \frac{1}{2} \int_0^L EI(\alpha')^2 dx \tag{1.389}$$

This is the main term in beam bending problems. The strain energy due to shear deformation is (Timoshenko, 1921):

$$V_s = \frac{1}{2}\int_0^L (\iint_A \tau\gamma dA)dx = \frac{1}{2}\int_0^L \kappa G A \beta^2 dx \tag{1.390}$$

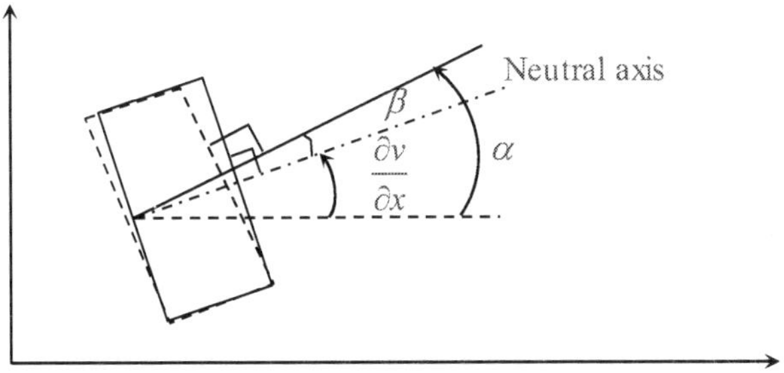

Figure 1.30 Shear deformation represented by the rotation β

where κ is called shear coefficient and has been considered by many, including Timoshenko, Mindlin, Cowper, Stephen, Hutchinson, Kaneko and many others. This shear coefficient is resulted from the fact that the shear deformation effect is replaced by a single parameter β in Timoshenko beam theory. In fact, shear stress must increase from zero at the top surface of the beam and return to zero at the bottom surface of the beam. Clearly, this area average factor κ must be a function of the cross-section (different cross-section will have different shear stress distribution). Depending on the cross-section, Kaneko (1975) suggested the use of the following formulas for κ

$$\begin{aligned} \kappa &= \frac{5(1+\nu)}{6+5\nu}, \quad \text{rectangular} \\ &= \frac{6+12\nu+6\nu^2}{7+12\nu+4\nu^2}, \quad \text{circular} \end{aligned} \tag{1.391}$$

where ν is the Poisson's ratio. They were obtained by Timoshenko by comparing the Timoshenko beam theory with 2-D analysis of rectangular beam and 3-D analysis of circular beam. A comprehensive review on κ was given by Kaneko (1975) and by Hutchinson (2001). The strain energy term due to shear deformation is normally neglected (as in Euler-Bernoulli beam theory). However, for beam with aspect ratio of larger than 0.1 (or deep beam case) shear deformation is no longer negligible. The kinetic energy of the beam is

$$T = \frac{1}{2}\int_0^L \rho A(\dot{v})^2 dx + \frac{1}{2}\int_0^L \rho I(\dot{\alpha})^2 dx \tag{1.392}$$

The second term is due to the rotatory inertia and it has been neglected in Euler-Bernoulli beam theory, and this rotatory inertia was first considered by Lord Rayleigh. The generalized Hamilton principle is

$$\int_{t_1}^{t_2} \delta(T-V)dt + \int_{t_1}^{t_2} \delta W_{nc} dt = 0 \tag{1.393}$$

where δ means taking the variation, and the work done by external non-conservative loading is given by

$$W_{nc} = \int_0^L p_{y'}(x,t)\delta v(x,t)dx \tag{1.394}$$

where δv is the kinetically admissible displacement variations. Note that (1.393) is a generalized form of the Hamilton principle given in (14.10) in Chapter 14 of Chau (2018), including the work done by non-conservative loading. Substitution of (1.389), (1.390), (1.392) and (1.394) into (1.393) gives

$$\begin{aligned}&\frac{1}{2}\int_{t_1}^{t_2}\int_0^L \delta[\rho I(\dot{\alpha})^2 + \rho A(\dot{v})^2 - EI(\alpha')^2 - \kappa GA(\alpha - v')^2]dxdt \\ &+\int_{t_1}^{t_2}\int_0^L p_y(x,t)\delta v(x,t)dxdt = 0\end{aligned} \tag{1.395}$$

The variations v and α satisfy the following conditions

$$\delta v(x,t_1) = \delta v(x,t_2) = 0, \quad \delta\alpha(x,t_1) = \delta\alpha(x,t_2) = 0 \tag{1.396}$$

The following identities of the variations can be established:

$$d\delta v = \frac{d\delta v}{dt}dt = \delta\dot{v}dt = \frac{d\delta v}{dx}dx = \delta v'dx \tag{1.397}$$

$$d\delta\alpha = \frac{d\delta\alpha}{dt}dt = \delta\dot{\alpha}dt = \frac{d\delta\alpha}{dx}dx = \delta\alpha'dx \tag{1.398}$$

$$\delta[\rho A(\dot{v})^2] = 2\rho A\dot{v}\delta\dot{v} \tag{1.399}$$

$$\delta[\rho I(\dot{\alpha})^2] = 2\rho I\dot{\alpha}\delta\dot{\alpha} \tag{1.400}$$

$$\delta[EI(\alpha')^2] = 2EI\alpha'\delta\alpha' \tag{1.401}$$

$$\delta[\kappa GA(\alpha - v')^2] = 2\kappa GA(\alpha - v')(\delta\alpha - \delta v') \tag{1.402}$$

The variation of first term on the left of (1.395) becomes

$$\begin{aligned}\int_{t_1}^{t_2}\delta[\rho I(\dot{\alpha})^2]dt &= 2\rho I\int_{t_1}^{t_2}\dot{\alpha}\delta\dot{\alpha}dt = 2\rho I\int_{t_1}^{t_2}\dot{\alpha}\, d(\delta\alpha) \\ &= 2\rho I\{\dot{\alpha}\,\delta\alpha\Big|_{t_1}^{t_2} - \int_{t_1}^{t_2}\ddot{\alpha}\,\delta\alpha dt\} \\ &= -2\rho I\int_{t_1}^{t_2}\ddot{\alpha}\,\delta\alpha dt\end{aligned} \tag{1.403}$$

We have employed (1.396) in getting the last line of (1.403). The variation of the second term on the left of (1.395) is obtained as

$$\begin{aligned}\int_{t_1}^{t_2}\delta[\rho A(\dot{v})^2]dt &= 2\rho A\int_{t_1}^{t_2}\dot{v}d(\delta v) = 2\rho A\{\dot{v}\delta\, v\Big|_{t_1}^{t_2} - \int_{t_1}^{t_2}\ddot{v}\delta vdt\} \\ &= -2\rho A\int_{t_1}^{t_2}\ddot{v}\delta vdt\end{aligned} \tag{1.404}$$

Again, the boundary terms are zeros in view of (1.396). The variation of the third term on the left of (1.395) gives

$$\int_0^L \delta[EI(\alpha')^2]dx = 2\int_0^L EI\alpha'd\delta\alpha = 2\{EI\alpha'\delta\alpha\Big|_0^L - \int_0^L (EI\alpha')'\delta\alpha dx\} \tag{1.405}$$

Finally, the variation of the last term on the left of (1.395) gives

$$\int_0^L \delta[\kappa GA(\alpha - v')^2]dx = 2\int_0^L \kappa GA(\alpha - v')(\delta\alpha - \delta v')dx$$
$$= 2\int_0^L \kappa GA(\alpha - v')\delta\alpha dx - 2\int_0^L \kappa GA(\alpha - v')\delta v' dx \quad (1.406)$$

The last integral on the last line of (1.406) can be evaluated as

$$\int_0^L \kappa GA(\alpha - v')\delta v' dx = \int_0^L \kappa GA(\alpha - v')d(\delta v)$$
$$= \kappa GA(\alpha - v')\,\delta v\Big|_0^L - \int_0^L [\kappa GA(\alpha - v')]'\delta v dx \quad (1.407)$$

Substitution of all these results into (1.395) gives

$$\int_{t_1}^{t_2}\int_0^L [-\rho A\ddot{v} - [\kappa GA(\alpha - v')]' + p_y]\delta v dx dt$$
$$+\int_{t_1}^{t_2}\int_0^L [(EI\alpha')' - \rho I\ddot{\alpha} - \kappa GA(\alpha - v')]\delta\alpha dx dt \quad (1.408)$$
$$+\int_{t_1}^{t_2} \left[\kappa GA(\alpha - v')\delta v\right]_0^L dt - \int_0^L [EI\alpha'\delta\alpha]_0^L dt = 0$$

Since the variations $\delta\alpha \neq 0$ and $\delta v \neq 0$, we have the following Euler-Lagrange equations for Timoshenko beam theory:

$$\rho A\frac{\partial^2 v}{\partial t^2} + \frac{\partial}{\partial x}[\kappa GA(\alpha - \frac{\partial v}{\partial x})] = p_y \quad (1.409)$$

$$\kappa GA(\alpha - \frac{\partial v}{\partial x}) - \frac{\partial}{\partial x}(EI\frac{\partial \alpha}{\partial x}) + \rho I\frac{\partial^2 \alpha}{\partial t^2} = 0 \quad (1.410)$$

The boundary conditions are

$$\left[\kappa GA(\alpha - v')\delta v\right]_0^L = 0, \quad [EI\alpha'\delta\alpha]_0^L = 0 \quad (1.411)$$

The Timoshenko beam theory is described in (1.409) and (1.410). For deformation of non-uniform Timoshenko beams, we have to resort to numerical methods in solving the system given in (1.409) and (1.410). When finite element technique is applied, we have to model the deflection v and rotation α independently (i.e., treating them as two independent variables). For thin beams, a phenomenon called shear locking, leading to unrealistically stiff behavior, has been observed. Beam element with linear interpolation or shape functions appears to be more susceptible to such numerical problems.

If the beam is uniform, we can eliminate α to get a single uncoupled PDE for v:

$$EI\frac{\partial^4 v}{\partial x^4} + \rho A\frac{\partial^2 v}{\partial t^2} - p_y - \rho I\frac{\partial^4 v}{\partial t^2 \partial x^2} + \frac{EI}{\kappa GA}\frac{\partial^2}{\partial t^2}(p_y - \rho A\frac{\partial^2 v}{\partial t^2})$$
$$-\frac{\rho I}{\kappa GA}\frac{\partial^2}{\partial t^2}(p_y - \rho A\frac{\partial^2 v}{\partial t^2}) = 0 \quad (1.412)$$

The first two terms forms the Euler-Bernoulli beam theory, the fourth term is the rotatory inertia, the fifth term is caused by shear deformation, and the sixth term is due to the combined effect of rotatory inertia and shear deformation. For the case of free vibrations, (1.412) can be simplified as

$$EI\frac{\partial^4 v}{\partial x^4}+\rho A\frac{\partial^2 v}{\partial t^2}-\rho A r_G^2(1+\frac{E}{\kappa G})\frac{\partial^4 v}{\partial t^2 \partial x^2}+\frac{\rho^2 A r_G^2}{\kappa G}\frac{\partial^4 v}{\partial t^4}=0 \tag{1.413}$$

where we have used the radius of gyration defined as

$$I = A r_G^2 \tag{1.414}$$

After v is obtained from solving (1.413), the angle of rotation can be evaluated as

$$\frac{\partial \alpha}{\partial x}=\frac{\partial^2 v}{\partial x^2}-\frac{\rho}{\kappa G}\frac{\partial^2 v}{\partial t^2} \tag{1.415}$$

Alternatively, we can also eliminate v to get a differential equation for α (see Problem 1.7). For the case of free vibrations, we note that the differential equation for v given in (1.413) also applies to α (see Problem 1.8).

If the shear deformation is negligible or $G \to \infty$ (i.e., $\beta \to 0$), we recover the Rayleigh beam theory:

$$EI\frac{\partial^4 v}{\partial x^4}+\rho A\frac{\partial^2 v}{\partial t^2}-\rho I\frac{\partial^4 v}{\partial t^2 \partial x^2}=p_y \tag{1.416}$$

This theory was considered by Lord Rayleigh in 1877 in Chapter 8 of his classic book "*The Theory of Sound*".

1.8.2 Static Solution for Timoshenko Beam

In this section, we consider the static problem of a simply-supported Timoshenko beam subject to a uniform load p. First, we specify the time independent form of the governing equations given in (1.409) and (1.410) as

$$\frac{d}{dx}[\kappa GA(\alpha-\frac{dv}{dx})]=p \tag{1.417}$$

$$\kappa GA(\alpha-\frac{dv}{dx})-\frac{d}{dx}(EI\frac{d\alpha}{dx})=0 \tag{1.418}$$

Differentiating of (1.418) with respect to x and using (1.417), we find

$$\frac{d^2}{dx^2}(EI\frac{d\alpha}{dx})-p=0 \tag{1.419}$$

Assuming the beam is uniform in cross-section and integrating (1.419), we find

$$EI\frac{d\alpha}{dx}=\frac{1}{2}px^2+C_1x+C_2 \tag{1.420}$$

Further integrating (1.420), we get

$$\alpha=\frac{px^3}{6EI}+\frac{x^2}{2EI}C_1+\frac{x}{EI}C_2+C_3 \tag{1.421}$$

Because of the nonzero shear deformation, the boundary conditions to be imposed at the supports of the beam need to be modified. In particular, the moment of the beam is evaluated as the x derivative of the slope and, therefore, is defined as:

$$M = EI\frac{d\alpha}{dx} \tag{1.422}$$

The shear force can be found by employing (1.422) into (1.418)

$$V = \frac{dM}{dx} = \kappa GA(\alpha - \frac{dv}{dx}) \tag{1.423}$$

The boundary conditions for simply-supported end at $x = 0$ are

$$v(0) = 0, \quad EI\frac{d\alpha}{dx}(0) = 0 \tag{1.424}$$

The boundary conditions for fixed end at $x = 0$ are

$$v(0) = 0, \quad \alpha(0) = 0 \tag{1.425}$$

The boundary conditions for free end at $x = 0$ are

$$V(0) = \kappa G(\alpha - \frac{dv}{dx})\bigg|_{x=0} = 0, \quad M(0) = EI\frac{d\alpha}{dx}\bigg|_{x=0} = 0 \tag{1.426}$$

The boundary conditions for guided end at $x = 0$ are

$$V(0) = \kappa G(\alpha - \frac{dv}{dx})\bigg|_{x=0} = 0, \quad \alpha(0) = 0 \tag{1.427}$$

In summary, for simply-supported beams, we have

$$v(0) = 0, \quad EI\frac{d\alpha}{dx}(0) = 0, \quad v(L) = 0, \quad EI\frac{d\alpha}{dx}(L) = 0 \tag{1.428}$$

Applying (1.421) to the second and fourth conditions of (1.428), we have

$$C_2 = 0, \quad C_1 = -\frac{pL}{2} \tag{1.429}$$

Combining (1.420) and (1.421), we get

$$\frac{d}{dx}(EI\frac{d\alpha}{dx}) = px + C_1 = \frac{p}{2}(2x - L) \tag{1.430}$$

Substitution of (1.420) and (1.430) into (1.418) gives

$$\frac{p}{2}(2x - L) - \kappa GA(\frac{px^3}{6EI} - \frac{pLx^2}{4EI} + C_3 - \frac{dv}{dx}) = 0 \tag{1.431}$$

Solving for the deflection *dv/dx*, we find

$$\frac{dv}{dx} = -\frac{p}{2\kappa GA}(2x - L) + (\frac{px^3}{6EI} - \frac{pLx^2}{4EI} + C_3) \tag{1.432}$$

Integration of (1.432) gives

$$v = -\frac{p}{2\kappa GA}(x^2 - xL) + \frac{px^4}{24EI} - \frac{pLx^3}{12EI} + C_3 x + C_4 \tag{1.433}$$

The first of (1.428) gives

$$C_4 = 0 \tag{1.434}$$

Finally, substitution of (1.433) into the third condition of (1.428) leads to

$$C_3 = \frac{pL^3}{24EI} \tag{1.435}$$

In summary, the solutions for v and α are

$$v = \frac{p(xL - x^2)}{2\kappa GA} + \frac{px^4}{24EI} - \frac{pLx^3}{12EI} + \frac{pL^3x}{24EI} \tag{1.436}$$

$$\alpha = \frac{px^3}{6EI} - \frac{pLx^2}{4EI} + \frac{pL^3}{24EI} \tag{1.437}$$

Substitution of (1.436) and (1.437) into (1.388) gives the shear-induced rotation β as

$$\beta = -\frac{p(L-2x)}{2\kappa GA} \tag{1.438}$$

Substitution of these solutions into (1.422) and (1.423) gives the moment and shear force as

$$M = \frac{px}{2}(x - L) \tag{1.439}$$

$$V = \frac{p}{2}(2x - L) \tag{1.440}$$

Assuming a rectangular section and isotropic material, we have the moment of inertia related to the cross-section area A and shear modulus to Young's modulus as

$$I = \frac{bh^3}{12} = \frac{Ah^2}{12}, \quad G = \frac{E}{2(1+\nu)} \tag{1.441}$$

Substituting (1.441) into (1.436), we can rewrite the deflection in a more informative form:

$$v = \frac{pL^4}{24EI}\left\{ \left(\frac{x^4}{L^4} - \frac{2x^3}{L^3} + \frac{x}{L}\right) + \frac{2}{\kappa}(1+\nu)\left(\frac{h}{L}\right)^2\left(\frac{x}{L} - \frac{x^2}{L^2}\right) \right\} \tag{1.442}$$

The aspect ratio h/L appears explicitly in the formula. Therefore, Euler-Bernoulli beam theory can be interpreted as a first order term of 3-D elastic solution, whereas Timoshenko beam theory provides a second order refinement in terms of $(h/L)^2$. As expected, for long beam (say $h/L < 0.1$) the second term in the left of (1.442) becomes negligible. Thus, shear deformation is important mainly for deep beams, as remarked earlier. For circular cross-section, the deflection becomes

$$v = \frac{pL^4}{24EI}\left\{ \left(\frac{x^4}{L^4} - \frac{2x^3}{L^3} + \frac{x}{L}\right) + \frac{3}{2\kappa}(1+\nu)\left(\frac{d}{L}\right)^2\left(\frac{x}{L} - \frac{x^2}{L^2}\right) \right\} \tag{1.443}$$

where d is the diameter of the circular beam. The nonzero shear deformation also leads to a major change in bending stress. Figure 1.31 illustrates qualitatively how the bending stress change from linear variations of compressive and tensile stresses in Euler-Bernoulli to a highly nonlinear bending stress distribution (predicted by FEM using plane stress assumption). The tensile bending stress is more severe in deep beams than in thin beams.

1.8.3 Free Vibrations of Timoshenko Beams

In the section, we consider the natural frequency of free vibrations of Timoshenko beams. Let us consider the usual separation of variables:

$$v(x,t) = X(x)T(t) \tag{1.444}$$

Substitution of (1.444) into (1.413) gives

$$EI\frac{X^{(IV)}}{X} + \rho A\frac{\ddot{T}}{T} - \rho A r_G^2 (1+\frac{E}{\kappa G})\frac{X''}{X}\frac{\ddot{T}}{T} + \frac{\rho^2 A r_G^2}{\kappa G}\frac{\ddddot{T}}{T} = 0 \tag{1.445}$$

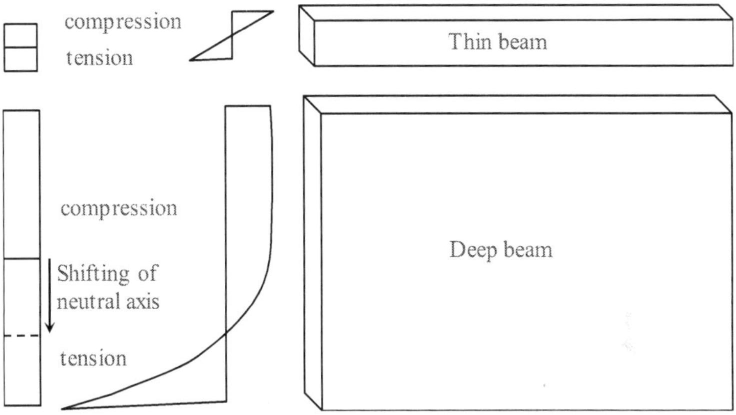

Figure 1.31 Illustration of bending stress for thin and deep beams

We find that we cannot put all functions of x to one side and all functions of t to the other. That is, as illustrated in Problem 7.13 of Chau (2018) for biharmonic equation in Cartesian coordinates, not all PDEs are separable. The systematic technique discussed in Chapter 7 of Chau (2018) is not applicable here.

However, by intuition we know that the time dependent solutions for natural vibrations should be expressible in sine and cosine functions (i.e., oscillating functions of time with finite magnitudes). In particular, we assume

$$v(x,t) = V(x)[C\sin\omega t + D\cos\omega t] \tag{1.446}$$

Substitution of (1.446) into (1.413) gives

$$EIV^{(IV)} + \rho A r_G^2 (1+\frac{E}{\kappa G})\omega^2 V'' + \rho A\omega^2 (\frac{\rho r_G^2}{\kappa G}\omega^2 - 1)V = 0 \tag{1.447}$$

We introduce the following variables:

$$a^4 = \frac{EI}{\rho A}, \quad \lambda^2 = \frac{\omega}{a^2} \tag{1.448}$$

In addition, we note that

$$\frac{\rho r_G^2}{\kappa G}\omega^2 = \frac{\omega^2}{a^4}(\frac{EI}{\rho A})\frac{\rho r_G^2}{\kappa G} = \frac{\omega^2}{a^4}\frac{E r_G^4}{\kappa G} = \lambda^4 \frac{E r_G^4}{\kappa G} \tag{1.449}$$

In view of (1.449), (1.447) can be expressed as:

$$V^{(IV)} + \lambda^4 r_G^2 (1+\frac{E}{\kappa G})V'' + \lambda^4 (\lambda^4 \frac{E r_G^4}{\kappa G} - 1)V = 0 \tag{1.450}$$

This is a linear fourth order ODE with constant coefficients, and as discussed in Chapter 3 of Chau (2018) we can assume an exponential function in x. In particular, we look for

$$V(x) = e^{rx} \tag{1.451}$$

Substitution of (1.451) into (1.450) leads to the following characteristic equation for r:

$$r^4 + \lambda^4 r_G^2 (1 + \frac{E}{\kappa G}) r^2 + \lambda^4 (\lambda^4 \frac{E r_G^4}{\kappa G} - 1) = 0 \tag{1.452}$$

This is a quadratic equation for r^2 and the roots are:

$$r^2 = \frac{1}{2}\left\{ -\lambda^4 r_G^2 (1 + \frac{E}{\kappa G}) \pm \sqrt{\lambda^8 r_G^4 (1 + \frac{E}{\kappa G})^2 - 4\lambda^4 (\lambda^4 \frac{E r_G^4}{\kappa G} - 1)} \right\} \tag{1.453}$$

$$= -\gamma \pm \beta$$

Before we write out the solution for V, we consider the following scenarios. If

$$\lambda^4 \frac{E r_G^4}{\kappa G} > 1 \tag{1.454}$$

The square root term in (1.453) can be potentially imaginary. However, this is very unlikely, since we need to have $G \to 0$ in (1.453) or require very soft shear stiffness. In the Rayleigh beam limit, we require $G \to \infty$ or very stiff response against shear (i.e. shear strain can be neglected). The roots for Rayleigh beam are

$$r^2 = \frac{1}{2}\left\{ -\lambda^4 r_G^2 \pm \sqrt{\lambda^8 r_G^4 + 4\lambda^4} \right\} = -\gamma \pm \beta \tag{1.455}$$

Clearly, for this case the roots for r become

$$r = \pm i\sqrt{\gamma + \beta}, \quad \pm\sqrt{\beta - \gamma} \tag{1.456}$$

The solution for Rayleigh beam is

$$\begin{aligned} V(x) &= A_1 \sin(\sqrt{\gamma + \beta} x) + A_2 \sinh(\sqrt{\beta - \gamma} x) \\ &+ B_1 \cos(\sqrt{\gamma + \beta} x) + B_2 \cosh(\sqrt{\beta - \gamma} x) \end{aligned} \tag{1.457}$$

where β and γ are given by (1.455).

For the subsequent discussions, we restrict to the case of finite G (i.e., G not too small) such that (1.453) remains valid for Timoshenko beam provided that:

$$\lambda^4 \frac{E r_G^4}{\kappa G} < 1 \tag{1.458}$$

Since λ is a function of frequency, the validity of (1.458) can be assumed only for the case of lower vibration modes. For higher modes, λ is large and (1.458) is likely to be violated even for finite values of G. In summary, we assume the solution for lower vibration modes of Timoshenko beam being

$$\begin{aligned} V(x) &= A_1 \sin(\sqrt{\gamma + \beta} x) + A_2 \sinh(\sqrt{\beta - \gamma} x) \\ &+ B_1 \cos(\sqrt{\gamma + \beta} x) + B_2 \cosh(\sqrt{\beta - \gamma} x) \end{aligned} \tag{1.459}$$

where β and γ are given by (1.453). For small G and higher modes, the sinh and cosh functions in (1.459) may evolve to sine and cosine functions. Using the definitions of λ and a, we find that the mode shifting occurs if the natural frequency satisfies

$$\omega^2 > \omega_{cr}^2 = \frac{\kappa G}{\rho r_G^2} \tag{1.460}$$

For a steel rectangular beam of depth 0.231 m and width 0.115 m, this critical value circular frequency equals 44873 rad/s (or 7142 Hz), which is unusually high for most engineering applications. Thus, possibility of mode shifting will not be considered here.

1.8.4 Free Vibrations of Simply-supported Timoshenko Beams

For the case of simply-supported beams shown in Figure 1.32, we have

$$V(0) = 0, \quad EI\frac{\partial \alpha}{\partial x}(0,t) = 0, \quad V(L) = 0, \quad EI\frac{\partial \alpha}{\partial x}(L,t) = 0 \tag{1.461}$$

By virtue of (1.415) for the case of free vibrations, we have

$$\frac{\partial \alpha}{\partial x} = \frac{\partial^2 v}{\partial x^2} - \frac{\rho}{\kappa G}\frac{\partial^2 v}{\partial t^2} \tag{1.462}$$

Substitution of (1.462) into the second condition in (1.461) gives

$$\frac{\partial \alpha}{\partial x}(0,t) = \{\frac{d^2V(0)}{dx^2} + \frac{\rho\omega^2}{\kappa G}V(0)\}[C\sin\omega t + D\cos\omega t] = 0 \tag{1.463}$$

Therefore, the zero moment conditions lead to

$$\frac{d^2V(0)}{dx^2} = 0, \quad V(0) = 0 \tag{1.464}$$

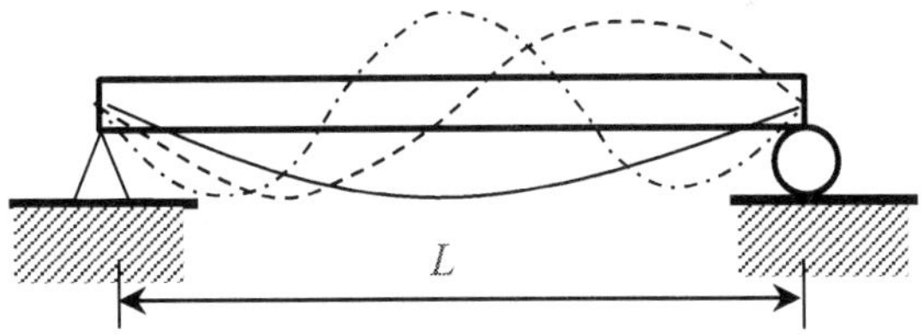

Figure 1.32 Free vibrations of a Timoshenko beam

The second of (1.464) overlaps with the first of (1.461). Therefore, the boundary conditions for simply-supported Timoshenko beams can be rewritten as:

$$V(0) = 0, \quad \frac{d^2V(0)}{dx^2} = 0, \quad V(L) = 0, \quad \frac{d^2V(L)}{dx^2} = 0 \tag{1.465}$$

Differentiation and integration of the solution form given in (1.459) gives

$$\begin{aligned} V'(x) &= \sqrt{\gamma+\beta}\{A_1\cos(\sqrt{\gamma+\beta}x) - B_1\sin(\sqrt{\gamma+\beta}x)\} \\ &+ \sqrt{\beta-\gamma}\{A_2\cosh(\sqrt{\beta-\gamma}x) + B_2\sinh(\sqrt{\beta-\gamma}x)\} \end{aligned} \tag{1.466}$$

$$V''(x) = (\gamma+\beta)\{-A_1 \sin(\sqrt{\gamma+\beta}x) - B_1 \cos(\sqrt{\gamma+\beta}x)\} \\ +(\beta-\gamma)\{A_2 \sinh(\sqrt{\beta-\gamma}x) + B_2 \cosh(\sqrt{\beta-\gamma}x)\} \tag{1.467}$$

$$\int V(x)dx = \frac{1}{\sqrt{\gamma+\beta}}\{-A_1 \cos(\sqrt{\gamma+\beta}x) + B_1 \sin(\sqrt{\gamma+\beta}x)\} \\ +\frac{1}{\sqrt{\beta-\gamma}}\{A_2 \cosh(\sqrt{\beta-\gamma}x) + B_2 \sinh(\sqrt{\beta-\gamma}x)\} \tag{1.468}$$

Substitution of (1.459) into the first condition of (1.465) gives

$$B_2 = -B_1 \tag{1.469}$$

Substitution of (1.467) into the second condition of (1.465) gives

$$B_2 = 0 \tag{1.470}$$

Using the third and fourth conditions of (1.465) gives two equations for A_1 and A_2 as

$$A_1 \sin(\sqrt{\gamma+\beta}L) + A_2 \sinh(\sqrt{\beta-\gamma}L) = 0 \tag{1.471}$$

$$-(\gamma+\beta)A_1 \sin(\sqrt{\gamma+\beta}L) + (\beta-\gamma)A_2 \sinh(\sqrt{\beta-\gamma}L) = 0 \tag{1.472}$$

For nonzero constants, we have to set the determinant of the system to zero and this leads to

$$2\beta \sin(\sqrt{\gamma+\beta}L)\sinh(\sqrt{\beta-\gamma}L) = 0 \tag{1.473}$$

Therefore, we require

$$\sin(\sqrt{\gamma+\beta}L) = 0, \quad \text{or} \quad \sqrt{\gamma+\beta}L = n\pi, \quad n = 1,2,3,... \tag{1.474}$$

Using the definitions of γ and β given in (1.453), we have

$$(\frac{n\pi}{L})^2 = \frac{1}{2}\left\{-\lambda^4 r_G^2(1+\frac{E}{\kappa G}) \pm \sqrt{\lambda^8 r_G^4(1+\frac{E}{\kappa G})^2 - 4\lambda^4(\lambda^4\frac{Er_G^4}{\kappa G}-1)}\right\} \tag{1.475}$$

Clearly, this eigenvalue equation can be rewritten as

$$(\frac{n\pi}{L})^4 + \lambda^4 r_G^2(1+\frac{E}{\kappa G})(\frac{n\pi}{L})^2 + \lambda^4(\lambda^4\frac{Er_G^4}{\kappa G}-1) = 0 \tag{1.476}$$

Solving for λ, we get

$$\lambda^4 = \frac{\kappa G}{2r_G^2 E}\{-r_G^2(1+\frac{E}{\kappa G})(\frac{n\pi}{L})^2 + 1 \\ \pm\sqrt{\{r_G^2(1+\frac{E}{\kappa G})(\frac{n\pi}{L})^2 - 1\}^2 - 4\frac{Er_G^4}{\kappa G}(\frac{n\pi}{L})^4}\} \tag{1.477}$$

Back substitution the definition of λ in (1.448) gives

$$\omega^2 = \frac{I\kappa G}{2\rho A r_G^4}\{-r_G^2(1+\frac{E}{\kappa G})(\frac{n\pi}{L})^2 + 1 \\ \pm\sqrt{\{r_G^2(1+\frac{E}{\kappa G})(\frac{n\pi}{L})^2 - 1\}^2 - 4\frac{Er_G^4}{\kappa G}(\frac{n\pi}{L})^4}\} \tag{1.478}$$

Note that for higher modes (i.e., $n \to \infty$), the natural frequency may be complex. Therefore, in a sense, the shear deformation suppresses high modes of vibrations.

1.8.5 Free Vibrations of Cantilever Timoshenko Beams

The boundary conditions for cantilever Timoshenko beams shown in Figure 1.33 can be rewritten as:

$$V(0)=0, \quad \alpha(0)=0, \quad \alpha'(L)=0, \quad \alpha(L)-\frac{\partial v}{\partial x}(L)=0 \tag{1.479}$$

Again by virtue of (1.446) for the case of free vibrations, (1.463) can be written as

$$\frac{\partial \alpha}{\partial x}(x,t)=[\frac{d^2V}{dx^2}+\frac{\omega^2\rho}{\kappa G}V][C\sin\omega t+D\cos\omega t] \tag{1.480}$$

Integrating once, we get

$$\begin{aligned}\alpha(x,t)&=\Psi(x)[C\sin\omega t+D\cos\omega t]\\ &=[\frac{dV}{dx}+\frac{\omega^2\rho}{\kappa G}\int Vdx][C\sin\omega t+D\cos\omega t]\end{aligned} \tag{1.481}$$

The fourth boundary condition given in (1.479) becomes

$$\Psi(L)-\frac{dV}{dx}(L)=\frac{\omega^2\rho}{\kappa G}\int Vdx\bigg|_{x=L} \tag{1.482}$$

In view of these, the boundary conditions become

$$\begin{aligned}&V(0)=0, \quad \left[\frac{dV}{dx}+\frac{\omega^2\rho}{\kappa G}\int Vdx\right]_{x=0}=0,\\ &\left[\frac{d^2V}{dx^2}+\frac{\omega^2\rho}{\kappa G}V\right]_{x=L}=0, \quad \frac{\omega^2\rho}{\kappa G}\int Vdx\bigg|_{x=L}=0\end{aligned} \tag{1.483}$$

The first condition of (1.483) gives

$$B_2=-B_1 \tag{1.484}$$

The second condition of (1.483) gives

$$\sqrt{\gamma+\beta}A_1-\frac{\omega^2\rho}{\kappa G}\frac{1}{\sqrt{\gamma+\beta}}A_1+\sqrt{\beta-\gamma}A_2+\frac{\omega^2\rho}{\kappa G}\frac{1}{\sqrt{\beta-\gamma}}A_2=0 \tag{1.485}$$

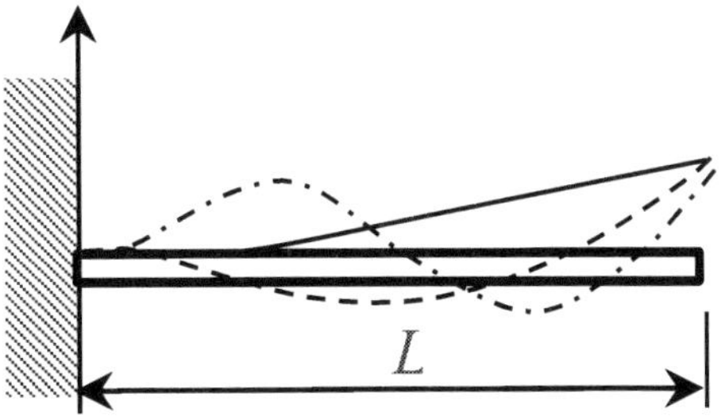

Figure 1.33 Free vibrations of a cantilever Timoshenko beam

This can be simplified to

$$A_2 = \sqrt{\frac{\beta-\gamma}{\gamma+\beta}}\left[\frac{\omega^2\rho - \kappa G(\beta+\gamma)}{\omega^2\rho + \kappa G(\beta-\gamma)}\right]A_1 = \chi A_1 \tag{1.486}$$

We can see that both A_2 and B_2 can be expressed in terms of A_1 and B_1. The third and fourth conditions in (1.483) lead to two equations for A_1 and B_1 (in view of (1.486) and (1.484)):

$$K_{11}A_1 + K_{12}A_2 = 0 \tag{1.487}$$

$$K_{21}A_1 + K_{22}A_2 = 0 \tag{1.488}$$

where

$$K_{11} = \frac{\chi}{\sqrt{\beta-\gamma}}\cosh(\sqrt{\beta-\gamma}L) - \frac{1}{\sqrt{\gamma+\beta}}\cos(\sqrt{\gamma+\beta}L) \tag{1.489}$$

$$K_{12} = \frac{1}{\sqrt{\gamma+\beta}}\sin(\sqrt{\gamma+\beta}L) - \frac{1}{\sqrt{\beta-\gamma}}\sinh(\sqrt{\beta-\gamma}L) \tag{1.490}$$

$$K_{21} = \left[\frac{\omega^2\rho}{\kappa G} - (\gamma+\beta)\right]\sin(\sqrt{\gamma+\beta}L) + \chi\left[\frac{\omega^2\rho}{\kappa G} + (\beta-\gamma)\right]\sinh(\sqrt{\beta-\gamma}L) \tag{1.491}$$

$$K_{22} = \left[\frac{\omega^2\rho}{\kappa G} - (\gamma+\beta)\right]\cos(\sqrt{\gamma+\beta}L) - \left[\frac{\omega^2\rho}{\kappa G} + (\beta-\gamma)\right]\cosh(\sqrt{\beta-\gamma}L) \tag{1.492}$$

Setting the determinant of the system of equations to zero, we get

$$K_{11}K_{22} - K_{12}K_{21} = 0 \tag{1.493}$$

Substitution (1.489) to (1.492) into (1.493) gives the following eigenvalue equation

$$\begin{aligned}
&\left[\frac{\omega^2\rho}{\kappa G} - (\gamma+\beta)\right]\left\{-\frac{1}{\sqrt{\gamma+\beta}} + \frac{\chi}{\sqrt{\beta-\gamma}}\cosh(\sqrt{\beta-\gamma}L)\cos(\sqrt{\gamma+\beta}L)\right. \\
&\left. + \frac{1}{\sqrt{\beta-\gamma}}\sinh(\sqrt{\beta-\gamma}L)\sin(\sqrt{\gamma+\beta}L)\right\} \\
&+\left[\frac{\omega^2\rho}{\kappa G} + (\beta-\gamma)\right]\left\{-\frac{\chi}{\sqrt{\beta-\gamma}} + \frac{1}{\sqrt{\gamma+\beta}}\cosh(\sqrt{\beta-\gamma}L)\cos(\sqrt{\gamma+\beta}L)\right. \\
&\left. - \frac{\chi}{\sqrt{\gamma+\beta}}\sinh(\sqrt{\beta-\gamma}L)\sin(\sqrt{\gamma+\beta}L)\right\} = 0
\end{aligned} \tag{1.494}$$

where χ was defined in (1.486) and γ and β are defined in (1.453). For the case of Rayleigh beam, we can set $G \to \infty$

$$
\begin{aligned}
&-(\gamma+\beta)\left\{-\frac{1}{\sqrt{\gamma+\beta}}+\frac{\chi}{\sqrt{\beta-\gamma}}\cosh(\sqrt{\beta-\gamma}L)\cos(\sqrt{\gamma+\beta}L)\right. \\
&\left.+\frac{1}{\sqrt{\beta-\gamma}}\sinh(\sqrt{\beta-\gamma}L)\sin(\sqrt{\gamma+\beta}L)\right\} \\
&+(\beta-\gamma)\left\{-\frac{\chi}{\sqrt{\beta-\gamma}}+\frac{1}{\sqrt{\gamma+\beta}}\cosh(\sqrt{\beta-\gamma}L)\cos(\sqrt{\gamma+\beta}L)\right. \\
&\left.-\frac{\chi}{\sqrt{\gamma+\beta}}\sinh(\sqrt{\beta-\gamma}L)\sin(\sqrt{\gamma+\beta}L)\right\}=0
\end{aligned}
\tag{1.495}
$$

where

$$
\chi=-\sqrt{\frac{\beta+\gamma}{\beta-\gamma}}, \quad \gamma=-\frac{1}{2}\lambda^4 r_G^2, \quad \beta=\frac{1}{2}\sqrt{\lambda^8 r_G^4+4\lambda^4} \tag{1.496}
$$

For Euler-Bernoulli beams, we can further set $r_G = 0$ to get

$$
1+\cosh(\lambda L)\cos(\lambda L)=0 \tag{1.497}
$$

This result, of course, agrees with our solution obtained in earlier section.

1.8.6 Free Vibrations of Fixed End Timoshenko Beams

Note that to separate anti-symmetric modes from symmetric modes, we shift the origin of the coordinate to the center of the beam of length $2L$, as shown in Figure 1.34. The boundary conditions for both end fixed are

$$
\begin{aligned}
&V(-L)=0, \quad \left[\frac{dV}{dx}+\frac{\omega^2\rho}{\kappa G}\int V dx\right]_{x=-L}=0, \\
&V(L)=0, \quad \left[\frac{dV}{dx}+\frac{\omega^2\rho}{\kappa G}\int V dx\right]_{x=L}=0
\end{aligned}
\tag{1.498}
$$

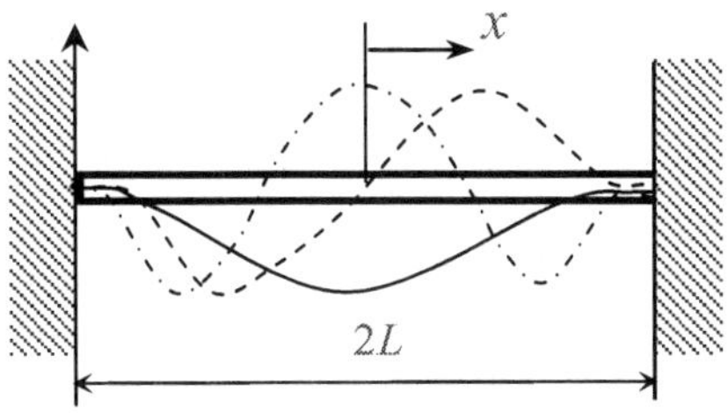

Figure 1.34 Vibrations of a fixed end Timoshenko beam with origin at center

The first and third conditions of (1.498) lead to

$$
-A_1\sin(\sqrt{\gamma+\beta}L)-A_2\sinh(\sqrt{\beta-\gamma}L)+B_1\cos(\sqrt{\gamma+\beta}L)+B_2\cosh(\sqrt{\beta-\gamma}L)=0
$$

$$(1.499)$$

$$A_1 \sin(\sqrt{\gamma+\beta}L) + A_2 \sinh(\sqrt{\beta-\gamma}L) + B_1 \cos(\sqrt{\gamma+\beta}L) + B_2 \cosh(\sqrt{\beta-\gamma}L) = 0 \tag{1.500}$$

Adding (1.498) and (1.499), we find

$$B_1 \cos(\sqrt{\gamma+\beta}L) + B_2 \cosh(\sqrt{\beta-\gamma}L) = 0 \tag{1.501}$$

Subtracting (1.498) from (1.499) gives

$$A_1 \sin(\sqrt{\gamma+\beta}L) + A_2 \sinh(\sqrt{\beta-\gamma}L) = 0 \tag{1.502}$$

The first and third conditions of (1.498) result in another set of two equations

$$\begin{aligned} &\sqrt{\gamma+\beta}\{A_1 \cos(\sqrt{\gamma+\beta}L) + B_1 \sin(\sqrt{\gamma+\beta}L)\} \\ &+\sqrt{\beta-\gamma}\{A_2 \cosh(\sqrt{\beta-\gamma}L) - B_2 \sinh(\sqrt{\beta-\gamma}L)\} \\ &+\frac{\omega^2\rho}{\kappa G}\{\frac{1}{\sqrt{\gamma+\beta}}[-A_1 \cos(\sqrt{\gamma+\beta}x) - B_1 \sin(\sqrt{\gamma+\beta}x)] \\ &+\frac{1}{\sqrt{\beta-\gamma}}[A_2 \cosh(\sqrt{\beta-\gamma}x) - B_2 \sinh(\sqrt{\beta-\gamma}x)]\} = 0 \end{aligned} \tag{1.503}$$

$$\begin{aligned} &\sqrt{\gamma+\beta}\{A_1 \cos(\sqrt{\gamma+\beta}L) - B_1 \sin(\sqrt{\gamma+\beta}L)\} \\ &+\sqrt{\beta-\gamma}\{A_2 \cosh(\sqrt{\beta-\gamma}L) + B_2 \sinh(\sqrt{\beta-\gamma}L)\} \\ &+\frac{\omega^2\rho}{\kappa G}\{\frac{1}{\sqrt{\gamma+\beta}}[-A_1 \cos(\sqrt{\gamma+\beta}L) + B_1 \sin(\sqrt{\gamma+\beta}L)] \\ &+\frac{1}{\sqrt{\beta-\gamma}}[A_2 \cosh(\sqrt{\beta-\gamma}L) + B_2 \sinh(\sqrt{\beta-\gamma}L)]\} = 0 \end{aligned} \tag{1.504}$$

Adding and subtracting, we get two equations

$$\begin{aligned} &-\frac{1}{\sqrt{\gamma+\beta}}[\frac{\omega^2\rho}{\kappa G} - (\gamma+\beta)]\cos(\sqrt{\gamma+\beta}L)A_1 \\ &+\frac{1}{\sqrt{\beta-\gamma}}[\frac{\omega^2\rho}{\kappa G} + (\beta-\gamma)]\cosh(\sqrt{\beta-\gamma}L)A_2 = 0 \end{aligned} \tag{1.505}$$

$$\begin{aligned} &\frac{1}{\sqrt{\gamma+\beta}}[\frac{\omega^2\rho}{\kappa G} - (\gamma+\beta)]\sin(\sqrt{\gamma+\beta}L)B_1 \\ &+\frac{1}{\sqrt{\beta-\gamma}}[\frac{\omega^2\rho}{\kappa G} + (\beta-\gamma)]\sinh(\sqrt{\beta-\gamma}L)B_2 = 0 \end{aligned} \tag{1.506}$$

For anti-symmetric modes, we have

$$B_1 = B_2 = 0 \tag{1.507}$$

Equations (1.502) and (1.505) provides two equations for two unknowns, and for nontrivial solutions, we obtain the following eigenvalue equation

$$[\frac{\omega^2\rho}{\kappa G\sqrt{\beta-\gamma}} + \sqrt{\beta-\gamma}]\tan(\sqrt{\gamma+\beta}L) + [\frac{\omega^2\rho}{\kappa G\sqrt{\gamma+\beta}} - \sqrt{\gamma+\beta}]\tanh(\sqrt{\beta-\gamma}L) = 0$$

(1.508)

For symmetric modes, we have

$$A_1 = A_2 = 0 \quad (1.509)$$

Equations (1.501) and (1.506) provides two equations for two unknowns, and for nontrivial solutions, we obtain the following eigenvalue equation

$$[\frac{\omega^2 \rho}{\kappa G\sqrt{\beta-\gamma}}+\sqrt{\beta-\gamma}]\tanh(\sqrt{\beta-\gamma}L)-[\frac{\omega^2 \rho}{\kappa G\sqrt{\gamma+\beta}}-\sqrt{\gamma+\beta}]\tan(\sqrt{\beta+\gamma}L)=0$$

(1.510)

For the case of Rayleigh beam, we can set $G \to \infty$ to get

$$\sqrt{\beta-\gamma}\tan(\sqrt{\gamma+\beta}L)-\sqrt{\gamma+\beta}\tanh(\sqrt{\beta-\gamma}L)=0, \quad \text{anti-symmetric} \quad (1.511)$$

$$\sqrt{\beta-\gamma}\tanh(\sqrt{\beta-\gamma}L)+\sqrt{\gamma+\beta}]\tan(\sqrt{\beta+\gamma}L)=0, \quad \text{symmetric} \quad (1.512)$$

For Euler-Bernoulli beams, we can set $G \to \infty$ and $r_G \to 0$ to recover

$$\tan(\lambda L)-\tanh(\lambda L)=0, \quad \text{anti-symmetric} \quad (1.513)$$

$$\tanh(\lambda L)+\tan(\lambda L)=0, \quad \text{symmetric} \quad (1.514)$$

These eigenvalue equations for Euler-Bernoulli beams have been obtained in Problems 1.1 and 1.2.

1.9 SUMMARY AND FURTHER READING

The development of the theory for beams and columns is a major triumph in mechanics and engineering when differential equation was first developed by Bernoulli and Euler. It relates to our daily lives in a subtle way because nearly all present day man-made structures involves beams and columns. In this chapter, we introduce the mathematical theories of Euler-Bernoulli and of Timoshenko beams. We cover both static and dynamic problems related to beams subject to various kinds of support. Although Euler-Bernoulli beam theory was developed more than 200 years ago whereas Timoshenko beam theory was developed for about 100 years, it remains an area of active research. More recently, beam theory finds important applications to functional graded materials, nano-scale tubes, sensors, actuators, transistors, probes, and resonators. For beams with semi-rigid joints, we can model the joints by adding rotational and translational springs (see Problems 1.9 and 1.12). For engineering structures, components of bridges, tall buildings, spacecraft, ships, nuclear power plants, and hanging beams can be modeled by bean theory (e.g., Yokoyama, 1990). A topic of recent interests is the effect of axial force to the vibration behavior of beams. We have introduced the Winkler model for beam on elastic foundations. However, for some classes of material, we need to use an additional spring that interacts with the original spring. Such foundation is normally referred as Pasternak-type foundation, which was proposed by Pasternak in 1954. Similar idea has actually been proposed by Filonenko-Borodich in 1945 and Hetenyi in 1946. This is, however, is out of the scope of the present chapter. For beams on foundation, we refer the readers to the books by Hetenyi (1946) and by Vlasov and Leontev (1966).

1.10 PROBLEMS

Problem 1.1 Find the natural vibration frequency of a beam of length $2L$ with both ends being fixed, as shown in Figure 1.35. Assume that the free vibration is symmetric.

Solution:

$$\omega_n = \sqrt{\frac{EI}{\rho A}}(\frac{\gamma_n}{L})^2, \quad n = 1,2,3,... \tag{1.515}$$

where γ_n satisfies the following characteristic equation:

$$\tan\gamma + \tanh\gamma = 0 \tag{1.516}$$

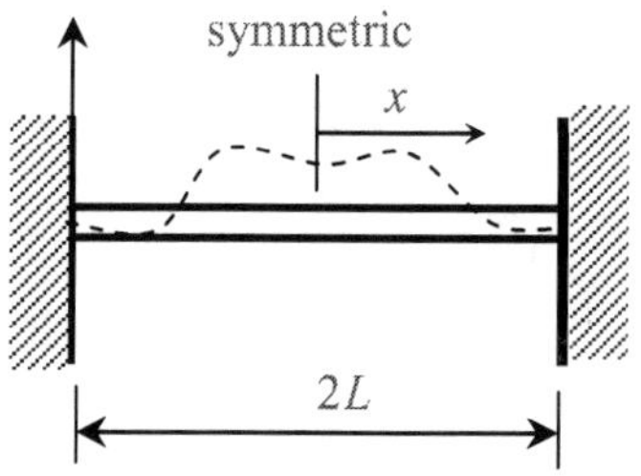

Figure 1.35 Symmetric free vibrations of a fixed end beam

Problem 1.2 Find the natural vibration frequency of a beam $2L$ with both ends being fixed, as shown in Figure 1.36. Assume that the free vibration is anti-symmetric.

Solution:

$$\omega_n = \sqrt{\frac{EI}{\rho A}}(\frac{\gamma_n}{L})^2, \quad n = 1,2,3,... \tag{1.517}$$

where γ_n satisfies the following characteristic equation:

$$\tan\gamma - \tanh\gamma = 0 \tag{1.518}$$

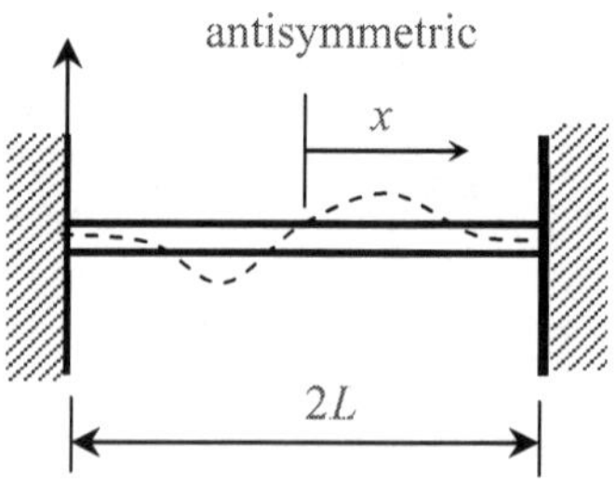

Figure 1.36 Anti-symmetric free vibrations of a fixed end beam

Problem 1.3 Show that the eigenfunctions for symmetric and anti-symmetric modes of Problems 1.1 and 1.2 are given by:

$$X_n^{(s)} = \cosh\gamma_n \cos(\sqrt[4]{\lambda_n}x) - \cos\gamma_n \cosh(\sqrt[4]{\lambda_n}x), \quad n = 1,2,3,... \tag{1.519}$$

$$X_n^{(a)} = \cosh\gamma_n \sin(\sqrt[4]{\lambda_n}x) - \cos\gamma_n \sinh(\sqrt[4]{\lambda_n}x), \quad n = 1,2,3,... \tag{1.520}$$

Problem 1.4 Find the solution of the following problem with given boundary and initial conditions, shown in Figure 1.37:

$$\frac{\partial^4 u}{\partial x^4} + \frac{1}{a^4}\frac{\partial^2 u}{\partial t^2} = 0 \tag{1.521}$$

$$u(-L) = 0, \quad \frac{\partial u}{\partial x}(-L) = 0, \quad u(L) = 0, \quad \frac{\partial u}{\partial x}(L) = 0 \tag{1.522}$$

$$u(x,0) = f(x), \quad \frac{\partial u}{\partial t}(x,0) = 0 \tag{1.523}$$

where $f(x)$ is symmetric in x.

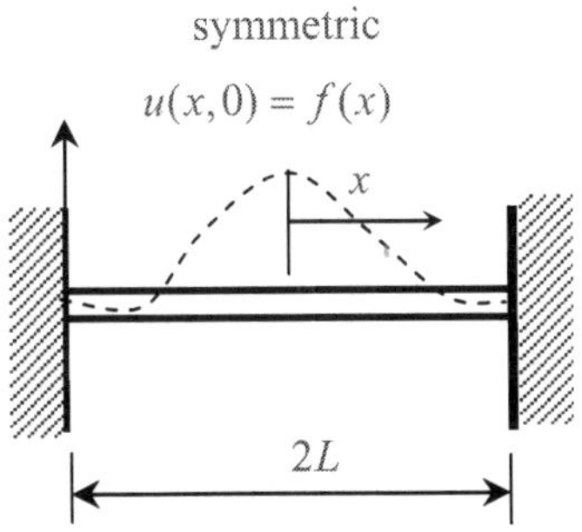

Figure 1.37 Symmetric vibrations set off by an initial deflection

Solution:

$$u(x,t) = \frac{1}{L}\sum_{n=1}^{\infty} \frac{\cos(\frac{\gamma_n^2 a^2 t}{L^2})}{\cosh^2\gamma_n \cos^2\gamma_n} \int_0^L f(\xi)X_n(\xi)d\xi X_n(x) \tag{1.524}$$

$$X_n(x) = \cosh\gamma_n \cos(\frac{\gamma_n}{L}x) - \cos\gamma_n \sinh(\frac{\gamma_n}{L}x) \tag{1.525}$$

where γ_n is the n-th root of

$$\tan\gamma + \tanh\gamma = 0 \tag{1.526}$$

Problem 1.5 Find the solution of the following problem with given boundary and initial conditions shown in Figure 1.38:

$$\frac{\partial^4 u}{\partial x^4} + \frac{1}{a^4}\frac{\partial^2 u}{\partial t^2} = 0 \tag{1.527}$$

$$u(-L) = 0, \quad \frac{\partial u}{\partial x}(-L) = 0, \quad u(L) = 0, \quad \frac{\partial u}{\partial x}(L) = 0 \tag{1.528}$$

$$u(x,0)=0,\quad \frac{\partial u}{\partial t}(x,0)=g(x) \tag{1.529}$$

$g(x)$ is anti-symmetric.

Solution:

$$u(x,t)=4L^5\sqrt{\frac{\rho A}{EI}}\sum_{m=1}^{\infty}\frac{\sin(\frac{\gamma_m^2 a^2 t}{L^2})\int_0^L g(\xi)X_m(\xi)d\xi}{\gamma_n^6[\sin\gamma_m\cosh\gamma_m-\cos\gamma_m\sinh\gamma_m]^2}X_m(x) \tag{1.530}$$

$$X_m(x)=\cosh\gamma_m\sin(\sqrt[4]{\lambda_m}x)-\cos\gamma_m\sinh(\sqrt[4]{\lambda_m}x) \tag{1.531}$$

where γ_m is the m-th root of

$$\tan\gamma-\tanh\gamma=0 \tag{1.532}$$

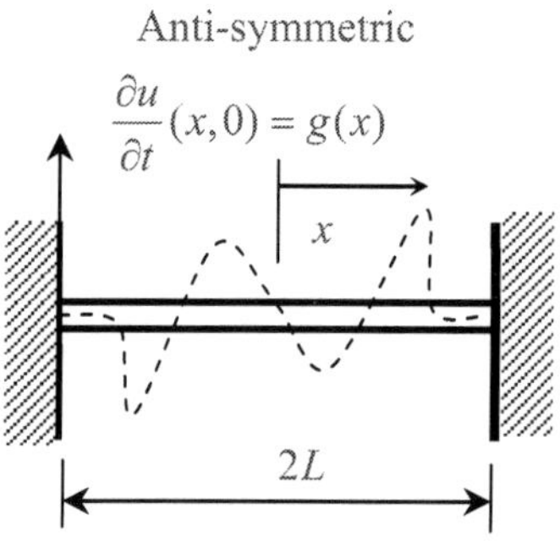

Figure 1.38 Anti-symmetric vibrations set off by an initial velocity

Problem 1.6 We will demonstrate in this problem that orthogonal properties of eigenfunctions given in (1.374) for the case of beam vibrations with nonzero axial compression is not strictly correct. In particular, consider two eigenfunctions as:

$$a^2X_m^{(IV)}+b^2X_m''-\lambda_m X_m=0 \tag{1.533}$$

$$a^2X_n^{(IV)}+b^2X_n''-\lambda_n X_n=0 \tag{1.534}$$

(i) Multiply the first by X_n and the second by X_m, subtract these results, and integrate along the beam to show

$$a^2\int_0^L[X_n^{(IV)}X_m-X_nX_m^{(IV)}]dx+b^2\int_0^L[X_n''X_m-X_nX_m'']dx \\ +(\lambda_m-\lambda_n)\int_0^L X_mX_n dx=0 \tag{1.535}$$

(ii) Use integration by parts to show

$$\int_0^L X_n^{(IV)}X_m dx=(X_mX_n'''-X_m'X_n''+X_m''X_n'-X_nX_m''')_0^L+\int_0^L X_m^{(IV)}X_n dx \tag{1.536}$$

$$\int_0^L X_n''X_m dx=(X_mX_n'-X_m'X_n)_0^L+\int_0^L X_m''X_n dx \tag{1.537}$$

(iii) Use the result in Parts (i) and (ii) to prove that

$$a^2(X_m X_n''' - X_m' X_n'' + X_m'' X_n' - X_n X_m''')_0^L + b^2(X_m X_n' - X_m' X_n)_0^L + (\lambda_m - \lambda_n)\int_0^L X_m X_n dx = 0 \tag{1.538}$$

(iv) Use the boundary conditions for the cantilever beam to show

$$(\lambda_m - \lambda_n)\int_0^L X_m X_n dx + b^2[X_m(L)X_n'(L) - X_m'(L)X_n(L)] = 0 \tag{1.539}$$

(v) Show that the following orthogonal property is obtained only for the case of zero axial compression (i.e., $N = 0$):

$$\int_0^L X_m X_n dx = 0, \quad m \neq n \tag{1.540}$$

Problem 1.7 By eliminating v from (1.409) and (1.410), show that the governing differential equation for α is

$$EI\frac{\partial^4 \alpha}{\partial x^4} + \rho A\frac{\partial^2 \alpha}{\partial t^2} - \rho I(1+\frac{E}{\kappa G})\frac{\partial^4 \alpha}{\partial t^2 \partial x^2} + \frac{\rho^2 I}{\kappa G}\frac{\partial^4 \alpha}{\partial t^4} = \frac{\partial p_y}{\partial x} \tag{1.541}$$

Problem 1.8 Continued from Problem 1.7, show that the differential equation for free vibrations for the angle rotation of the centroid line α is

$$EI\frac{\partial^4 \alpha}{\partial x^4} + \rho A\frac{\partial^2 \alpha}{\partial t^2} - \rho A r_G^2(1+\frac{E}{\kappa G})\frac{\partial^4 \alpha}{\partial t^2 \partial x^2} + \frac{\rho^2 A r_G^2}{\kappa G}\frac{\partial^4 \alpha}{\partial t^4} = 0 \tag{1.542}$$

which is the same as the differential equation for v given in (1.413).

Problem 1.9 Find the eigenvalue equation of the Euler-Bernoulli beam shown in Figure 1.39. One end is simply-supported and the other was support by an elastic spring (i.e., the shear force at the tip is proportional to the stiffness k). More specifically, the boundary condition is

$$u(0) = 0, \quad u''(0) = 0, \quad u''(L) = 0, \quad EIu'''(L) = ku(L) \tag{1.543}$$

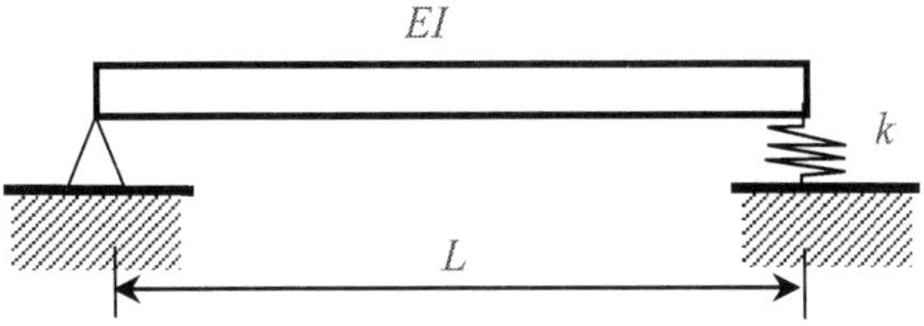

Figure 1.39 Vibration of an Euler-Bernoulli beam with an elastic support on the right

Ans:

$$\coth \beta L - \cot \beta L = \frac{2k}{EI\beta^3}, \quad \beta = \sqrt{w}(\frac{\rho A}{EI})^{1/4} \tag{1.544}$$

Problem 1.10 For the case of free vibrations of Timoshenko beam subject to axial compression N, it was shown by Abramovich and Elishakoff in 1990 and by Sato in 1991 that the governing equations become

$$\rho A\frac{\partial^2 v}{\partial t^2}+\frac{\partial}{\partial x}[\kappa GA(\alpha-\frac{\partial v}{\partial x})]+N\frac{\partial^2 v}{\partial x^2}=0 \tag{1.545}$$

$$\kappa GA(\alpha-\frac{\partial v}{\partial x})-\frac{\partial}{\partial x}(EI\frac{\partial \alpha}{\partial x})+\rho I\frac{\partial^2 \alpha}{\partial t^2}=0 \tag{1.546}$$

(i) Show that the PDE for v becomes

$$EI(1-\frac{N}{\kappa GA})\frac{\partial^4 v}{\partial x^4}+N\frac{\partial^2 v}{\partial x^2}+\rho A\frac{\partial^2 v}{\partial t^2}-\rho A r_G^2[(1-\frac{N}{\kappa GA})+\frac{E}{\kappa G}]\frac{\partial^4 v}{\partial t^2\partial x^2}$$
$$+\frac{\rho^2 A r_G^2}{\kappa G}\frac{\partial^4 v}{\partial t^4}=0 \tag{1.547}$$

(ii) For the case of simply-supported beams, show that the natural frequency is given by

$$\omega^2=\frac{I\kappa G}{2\rho A r_G^4}\{r_G^2[(1-\frac{N}{\kappa AG})+\frac{E}{\kappa G}](\frac{n\pi}{L})^2+1$$
$$\pm\sqrt{\{r_G^2[(1-\frac{N}{\kappa GA})+\frac{E}{\kappa G}](\frac{n\pi}{L})^2+1\}^2-4\frac{E r_G^4}{\kappa G}(\frac{n\pi}{L})^2[(\frac{n\pi}{L})^2-\frac{N}{EI}]}\} \tag{1.548}$$

(iii) Show that the result in (1.478) is recovered as a special when $N = 0$.

Remarks: Derivations of (1.545) and (1.546) are given in Problem 1.16.

Problem 1.11 Find the buckling load of the column with one free end and one fixed end subject to eccentric axial force, as shown in Figure 1.40.

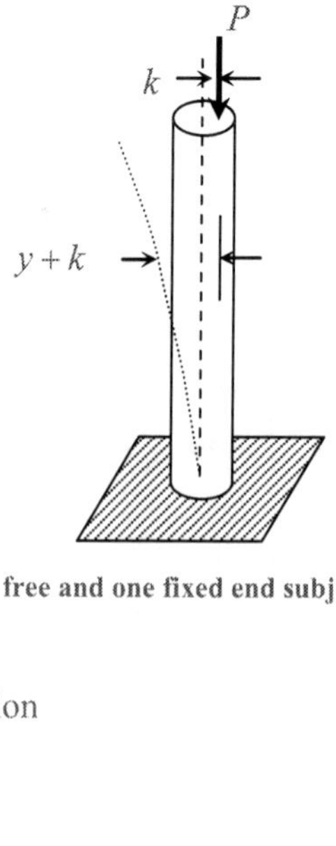

Figure 1.40 A column with one free and one fixed end subject to eccentric axial force

Hint: Look for symmetric solution

Ans:

$$y_{\max} = k[\sec(\sqrt{\frac{P}{EI}}L) - 1] = k[\frac{1}{\cos(\sqrt{\frac{P}{EI}}L)} - 1] \tag{1.549}$$

Problem 1.12 Express mathematically the following boundary conditions of a beam shown in Figure 1.41.

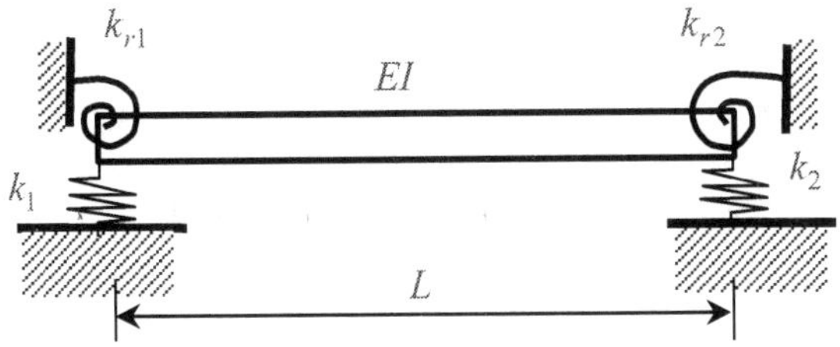

Figure 1.41 A beam with elastic translational and rotational supports

Ans:

$$\begin{aligned} &EI\frac{\partial^3 v}{\partial v^3}\bigg|_{x=0} = k_1 v\big|_{x=0}, \quad EI\frac{\partial^2 v}{\partial v^2}\bigg|_{x=0} = k_{r1}\frac{\partial v}{\partial v}\bigg|_{x=0}, \\ &EI\frac{\partial^3 v}{\partial v^3}\bigg|_{x=L} = k_2 v\big|_{x=L}, \quad EI\frac{\partial^2 v}{\partial v^2}\bigg|_{x=L} = k_{r2}\frac{\partial v}{\partial v}\bigg|_{x=L} \end{aligned} \tag{1.550}$$

Problem 1.13 Find the eigenvalue equation for the natural frequency of free vibrations of the following Euler-Bernoulli beam with free-free boundary condition, as shown in Figure 1.42.

Ans:

$$\cosh\gamma\cos\gamma = 1 \tag{1.551}$$

where

$$\gamma = \frac{\sqrt{\omega}L}{a} = (\frac{EI}{\rho A})\sqrt{\omega}L \tag{1.552}$$

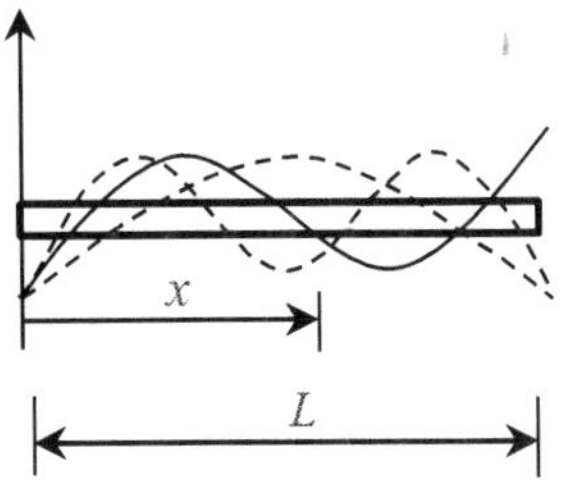

Figure 1.42 Free vibrations of Euler-Bernoulli beam with free-free ends

Problem 1.14 Derive the Euler-Lagrange equation of a Timoshenko beam vibrating under its own weight, as shown in Figure 1.43. In particular, the self-weight of the beam induced a non-uniform axial force $N(x)$ in the beam and its corresponding strain energy V as

$$V=\frac{1}{2}\int_0^L EI(\alpha')^2 dx+\frac{1}{2}\int_0^L \kappa GA\beta^2 dx+\frac{1}{2}\int_0^L N(x)(v')^2 dx \tag{1.553}$$

$$N(x)=\rho Ag(L-x) \tag{1.554}$$

Note that the kinetic energy T is same as that given in Section 1.8.1.

(i) Following the procedure of Hamilton's principle used in Section 1.8.1, show that the governing equations are

$$\rho A\frac{\partial^2 v}{\partial t^2}+\frac{\partial}{\partial x}[\kappa GA(\alpha-\frac{\partial v}{\partial x})]-\frac{\partial}{\partial x}[N(x)\frac{\partial v}{\partial x}]=0 \tag{1.555}$$

$$\kappa GA(\alpha-\frac{\partial v}{\partial x})-\frac{\partial}{\partial x}(EI\frac{\partial \alpha}{\partial x})+\rho I\frac{\partial^2 \alpha}{\partial t^2}=0 \tag{1.556}$$

Note: These results are the same as (1.545) and (1.546) by noting the set for Problem 1.10 is for axial compression whereas the present case is for tension.

(ii) Show that the results derived in (i) can be normalized as:

$$r\frac{\partial^2 \eta}{\partial \tau^2}+s\frac{\partial}{\partial \xi}(\alpha-\frac{\partial \eta}{\partial \xi})-g\frac{\partial}{\partial \xi}[(1-\xi)\frac{\partial \eta}{\partial \xi}]=0 \tag{1.557}$$

$$s(\alpha-\frac{\partial \eta}{\partial \xi})-\frac{\partial^2 \alpha}{\partial \xi^2}+\frac{\partial^2 \alpha}{\partial \tau^2}=0 \tag{1.558}$$

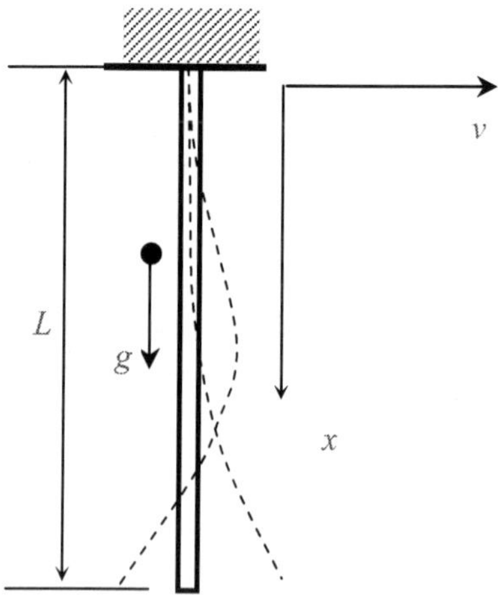

Figure 1.43 A hanging beam under self weight

where

$$s=\frac{\kappa GAL^2}{EI},\quad r=(\frac{L}{r_G})^2,\quad g=\frac{\rho AgL^2}{EI},\quad \xi=\frac{x}{L},\quad \tau=\sqrt{\frac{E}{\rho}}\frac{t}{L},\quad \eta=\frac{v}{L} \tag{1.559}$$

(iii) Show that α can be eliminated from (1.558) and (1.559) to give an ODE for η

$$\begin{aligned}&\frac{\partial^4\eta}{\partial\xi^4}-(1+\frac{r}{s})\frac{\partial^4\eta}{\partial\tau^2\partial\xi^2}+\frac{r}{s}\frac{\partial^4\eta}{\partial\tau^4}+r\frac{\partial^2\eta}{\partial\tau^2}\\&+\frac{g}{s}(\frac{\partial^2}{\partial\xi^2}-\frac{\partial^2}{\partial\tau^2}-s)\{\frac{\partial}{\partial\xi}[(1-\xi)\frac{\partial\eta}{\partial\xi}]\}=0\end{aligned} \tag{1.560}$$

(iv) Show that the governing equation for the free vibrations of a hanging Euler-Bernoulli beam under gravity is

$$\frac{\partial^4\eta}{\partial\xi^4}+r\frac{\partial^2\eta}{\partial\tau^2}-g\frac{\partial}{\partial\xi}[(1-\xi)\frac{\partial\eta}{\partial\xi}]=0 \tag{1.561}$$

Remarks: This equation finds it application in testing light weight flexible space structures (such as spacecraft and manned space station). When these flexible space structures are tested on earth under 1g condition, its dynamic behavior can be quite different from that in orbit under 0g condition. The hanging beam model under gravity provides a simple model for calibrating earth 1g testing for beam-like space structures (Yokoyama, 1990). A set of series solutions for an equation similar to (1.561) has been solved by Naguleswara (2004) using Frobenius series.

Problem 1.15 In the formulation given in Problem 1.16, we have tacitly assumed that the axial compressive force always aligns with the axis of a Timoshenko beam. Equivalently, the additional shear force due to a constant axial force N is given

Case 1:
$$V=\frac{\partial M}{\partial x}=\kappa GA(\alpha-\frac{\partial v}{\partial x})+N\frac{\partial v}{\partial x} \tag{1.562}$$

If the axial force is always normal to the shear force instead of parallel to the axis of bending, we have (1.562) being revised as:

Case 2:
$$V=\frac{dM}{dx}=\kappa GA(\alpha-\frac{dv}{dx})+N\alpha \tag{1.563}$$

Actually, these two different assumptions regarding the effect of axial force have been discussed briefly on p. 135 of Timoshenko and Gere (1961). For Case 2, the following equations for free vibrations are obtained (Esmailzadeh and Ohadi, 2000; Arboleda-Monsalve et al., 2007):

$$\rho A\frac{\partial^2 v}{\partial t^2}+\frac{\partial}{\partial x}[\kappa GA(\alpha-\frac{\partial v}{\partial x})]+N\frac{\partial\alpha}{\partial x}=0 \tag{1.564}$$

$$(\kappa GA+N)(\alpha-\frac{\partial v}{\partial x})-\frac{\partial}{\partial x}(EI\frac{\partial\alpha}{\partial x})+\rho I\frac{\partial^2\alpha}{\partial t^2}=0 \tag{1.565}$$

Note: These are different from (1.555) and (1.556) obtained in Problem 1.14.

(i) Show that these two equations can be written in normalized forms:

$$r\frac{\partial^2\eta}{\partial\tau^2}+s\frac{\partial}{\partial\xi}(\alpha-\frac{\partial\eta}{\partial\xi})+a\frac{\partial\alpha}{\partial\xi}=0 \tag{1.566}$$

$$(s+a)(\alpha-\frac{\partial\eta}{\partial\xi})-\frac{\partial^2\alpha}{\partial\xi^2}+\frac{\partial^2\alpha}{\partial\tau^2}=0 \tag{1.567}$$

where all normalized parameters have been defined in (1.559) and

$$a=\frac{NL^2}{EI} \tag{1.568}$$

(ii) Prove that η satisfies the following 4-th order PDE:

$$\frac{\partial^4\eta}{\partial\xi^4}-(1+\frac{r}{s})\frac{\partial^4\eta}{\partial\tau^2\partial\xi^2}+\frac{r}{s}\frac{\partial^4\eta}{\partial\tau^4}+\frac{r}{s}(s+a)\frac{\partial^2\eta}{\partial\tau^2}+\frac{a(s+a)}{s}\frac{\partial^2\eta}{\partial\xi^2}=0 \tag{1.569}$$

Problem 1.16 This problem formulates the stiffness matrix for the Euler-Bernoulli beam element with nodal deflections and rotations.

(i) Show that the result given in (1.17) can written in matrix form as:

$$v(x)=[N]\{d\} \tag{1.570}$$

where

$$[N]=[N_1,N_2,N_3,N_4],\quad \{d\}=\begin{Bmatrix} v_0\\ \varphi_0\\ v_L\\ \varphi_L\end{Bmatrix}, \tag{1.571}$$

$$\begin{aligned} N_1&=\frac{2x^3}{L^3}-\frac{3x^2}{L^2}+1,\quad N_2=\frac{x^3}{L^2}-\frac{2x^2}{L}+x,\\ N_3&=-\frac{2x^3}{L^3}+\frac{3x^2}{L^2},\quad N_4=\frac{x^3}{L^2}-\frac{x^2}{L}\end{aligned} \tag{1.572}$$

(ii) Derive the following formulas for the shear forces and moments at the end of the beam:

$$f_{0y}=V=EI\frac{d^3v}{dx^3}\bigg|_{x=0}=\frac{EI}{L^3}(12v_0+6L\varphi_0-12v_L+6L\varphi_L) \tag{1.573}$$

$$f_{Ly}=-V=-EI\frac{d^3v}{dx^3}\bigg|_{x=L}=\frac{EI}{L^3}(-12v_0-6L\varphi_0+12v_L-6L\varphi_L) \tag{1.574}$$

$$m_0=-M=-EI\frac{d^2v}{dx^2}\bigg|_{x=0}=\frac{EI}{L^3}(6Lv_0+4L^2\varphi_0-6Lv_L+2L^2\varphi_L) \tag{1.575}$$

$$m_L=M=EI\frac{d^2v}{dx^2}\bigg|_{x=L}=\frac{EI}{L^3}(6Lv_0+2L^2\varphi_0-6Lv_L+4L^2\varphi_L) \tag{1.576}$$

(iii) Find the stiffness matrix K defined as:

$$\{f\} = [K]\{d\} \tag{1.577}$$

where

$$\{f\} = \begin{Bmatrix} f_{0y} \\ m_0 \\ f_{Ly} \\ m_L \end{Bmatrix}, \quad \{d\} = \begin{Bmatrix} v_0 \\ \varphi_0 \\ v_L \\ \varphi_L \end{Bmatrix}, \tag{1.578}$$

Ans: The following stiffness forms the basics of finite element analysis for Euler-Bernoulli beams

$$[K] = \frac{EI}{L^3}\begin{bmatrix} 12 & 6L & -12 & 6L \\ 6L & 4L^2 & -6L & 2L^2 \\ -12 & -6L & 12 & -6L \\ 6L & 2L^2 & -6L & 4L^2 \end{bmatrix} \tag{1.579}$$

Problem 1.17 Find the stiffness matrix for a Timoshenko beam element in terms of the nodal deflection and rotation at the ends.

Ans: The following stiffness forms the basics of finite element analysis for Timoshenko beams

$$[K] = \frac{EI}{L^3(1+\phi)}\begin{bmatrix} 12 & 6L & -12 & 6L \\ 6L & (4+\phi)L^2 & -6L & (2-\phi)L^2 \\ -12 & -6L & 12 & -6L \\ 6L & (2-\phi)L^2 & -6L & (4+\phi)L^2 \end{bmatrix}, \quad \phi = \frac{12EI}{\kappa AGL^2} \tag{1.580}$$

CHAPTER TWO

Theory of Plates

2.1 INTRODUCTION

The theory of plates finds many applications in civil, mechanical, aeronautical, and marine engineering, such as flat slabs in structures, waffle slabs, bridge decks, raft foundations, shear walls, pavements, folded plates, and sheet pile retaining walls. The development of plate bending theory occupied a central place in the history of engineering mechanics. Euler in 1776 considered plate vibration using membrane theory, and Jacques Bernoulli in 1786 derived the governing equation of plate bending, but missed the cross derivative terms in the biharmonic operator. Chladni in 1809 presented his experiments of vibrating plates at the French Academy of Science; Napoleon was very impressed and suggested setting up a prize to reward the mathematical formulation of plate theory. In 1811, the first call for competition for the prize resulted in only one candidate, Sophie Germain, who made a mistake in her calculation and was not awarded the prize. Being one of the judges, Lagrange in 1813 corrected a term in Sophie Germain's formulation. Sophie Germain entered the competition again at the second call for the prize in 1813 and the judges requested justification of her assumption in formulating the strain energy. She was again unsuccessful. In 1816, Sophie Germain was finally awarded the prize when the competition was called for the third time, despite the fact that the judges were not fully satisfied with her work. Poisson in 1814 further developed the plate bending formulation and successfully formulated the strain energy but obtained an incorrect rigidity constant being proportional to thickness squared, instead of to thickness cubed. In 1823, Navier derived the plate theory with the correct rigidity constant for the case of Poisson's ratio being 0.25, and obtained the solution of a simply-supported rectangular plate in Fourier series. In 1829, Poisson asserted that there are three boundary conditions on the edge of the plates. A major breakthrough in plate theory came from Kirchhoff's thesis in 1850, which made two major hypotheses: (i) a vertical plane before deflection remains plane and perpendicular to the neutral axis, and (ii) no stretching on the mid-plane of the plates. Kirchhoff derived the strain energy and derived the correct rigidity for an arbitrary Poisson's ratio and correctly found that only two boundary conditions need to be satisfied on the edges of the plates. Kirchhoff's plate theory is also referred to as the classical plate theory for thin plates. For thick plates with shear deformation (say in-plane dimension is at least 20 times the thickness), Kirchhoff's plate theory underestimates deflection but overestimates vibration frequency and buckling loads. Employing the Saint-Venant principle, Kelvin and Tate in 1883 showed that the twisting moments on the edges can be represented by a pair of shear forces, which completely resolved the dispute between Poisson's three boundary conditions and Kirchhoff's two boundary conditions. Plate bending with large deflections was considered by Kirchhoff in 1877, Clebsh in 1862, and

Foppl in 1907. In 1899, Levy proposed solutions for plates with two parallel simple supports. Thick plate theories were proposed by Reissner in 1945 and Mindlin in 1951. As remarked by Chau (2018), the Galerkin method was originally proposed in Russia to solve plate bending problems. Other major Russian contributors include Krylov and his student Bubnov. Reinforced concrete plates and wooden plates are inherently anisotropic, and anisotropic plate theory was considered by Gehring in 1877, Boussinesq in 1879, Huber in 1929, and Lekhnitski in 1968. Plate buckling and its post-buckling strength are important and were studied by Navier, Dinnik, Nadai, Meissner and many others. Vibration of plates was considered by Kirchhoff, Poisson, Voight, Ritz, and many others.

In this chapter, we will only cover a discussion of the classical plate theory of Kirchhoff, with no elaboration on thick plate theory. However, both buckling and vibration of plates are included. Both Navier and Levy's series solution for simply-supported rectangular plates are presented, whereas the Galerkin method is used to obtain an approximation for the bending problem of a clamped rectangular plate. Clamped circular plates under uniform load and patch loads are discussed, and simply-supported circular plates under central point load are examined. The classical solution for rectangular plate buckling is also considered. Bending of anisotropic plates, a plate on an elastic foundation, and forced and free vibrations of plates are other topics covered in this chapter. Both the Rayleigh quotient and Rayleigh-Ritz method are used in solving plate vibration problems. The boundary and initial value problem of circular plate vibrations are formulated and solved. Finally, Hertz's problem of a circular plate under central point load is formulated and solved using the Rayleigh-Ritz method and Kelvin functions.

2.2 KIRCHHOFF PLATE THEORY

The following nonhomogeneous biharmonic equation was first derived by Lagrange in 1811

$$\nabla^2\nabla^2 w = \nabla^4 w = \frac{p(x,y)}{D} \tag{2.1}$$

where the plate bending rigidity D is defined as

$$D = \frac{Eh^3}{12(1-\nu^2)} \tag{2.2}$$

The two-dimensional Laplacian is defined as

$$\nabla^2(..) = \frac{\partial^2(..)}{\partial x^2} + \frac{\partial^2(..)}{\partial y^2} \tag{2.3}$$

The plate deflection is denoted by w. In addition, E is the Young's modulus, ν is the Poisson's ratio, and $p(x,y)$ is the vertical loads applied normally to the plate surface (x-y plane). The biharmonic equation is considered in Section 7.9 of Chau (2018), Green's function for the biharmonic equation is discussed in Section 8.13.4 of Chau (2018), the boundary integral equation for the biharmonic equation is covered in Section 8.9 of Chau (2018), and variational methods for the biharmonic equation is given in Section 14.12 of Chau (2018). The derivation of (2.1) will be considered in the next two sections.

2.2.1 Equilibrium Equations

To derive (2.1), let us consider the correspondence between bending moments, twisting moments, axial forces, shear forces and the normal and shear stresses as shown in Figure 2.1. The membrane forces are denoted by N_x and N_y, bending moments by M_x and M_y, twisting moments by M_{xy} and M_{yx}, and shear forces by Q_x, Q_y, N_{xy} and N_{yx}. The corresponding normal stresses are σ_x and σ_y, and shear stresses τ_{xy}, τ_{yx}, τ_{zx}, τ_{xz}, τ_{yz} and τ_{zy}. If the plate is only subject to in-plane loading, only membrane stresses exist. That is, both σ_x and σ_y are constants and shear stresses τ_{xy} and τ_{yx} are also constants and $\tau_{zx} = \tau_{xz} = \tau_{yz} = \tau_{zy} = \sigma_z = 0$. Actually, these are the plane stress conditions discussed in Chau (2013). However, for the case of loading applied perpendicularly to the middle plane of the plate, we have flexural stresses only but with $\sigma_z = 0$.

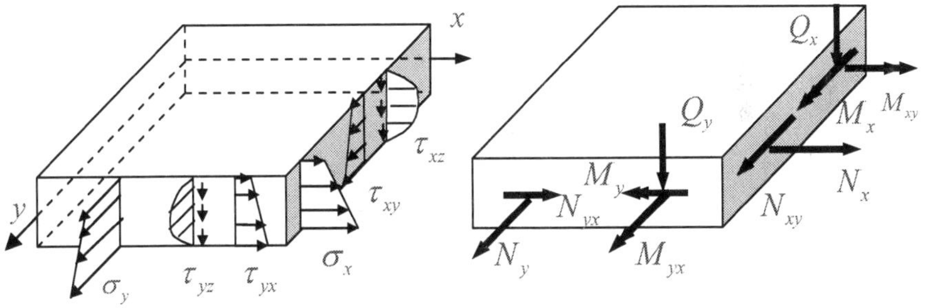

Figure 2.1 Two-dimensional stress states corresponding to bending moments, twisting moments and normal forces

We start with the force equilibriums of a three-dimensional elastic body expressed in terms of stresses (Chau, 2013):

$$\frac{\partial \sigma_x}{\partial x} + \frac{\partial \tau_{xy}}{\partial y} + \frac{\partial \tau_{xz}}{\partial z} + X = 0 \tag{2.4}$$

$$\frac{\partial \tau_{yx}}{\partial x} + \frac{\partial \sigma_y}{\partial y} + \frac{\partial \tau_{yz}}{\partial z} + Y = 0 \tag{2.5}$$

$$\frac{\partial \tau_{zx}}{\partial x} + \frac{\partial \tau_{zy}}{\partial y} + \frac{\partial \sigma_z}{\partial z} + Z = 0 \tag{2.6}$$

where X, Y, and Z are body forces along the x-, , y-, and z-axes respectively.

By assuming zero body forces, multiplying (2.4) by z and integrating from the bottom surface of the plate ($-h/2$) to the top surface of the plate ($h/2$) gives

$$\int_{-h/2}^{h/2} \frac{\partial \sigma_x}{\partial x} z dz + \int_{-h/2}^{h/2} \frac{\partial \tau_{xy}}{\partial y} z dz + \int_{-h/2}^{h/2} \frac{\partial \tau_{xz}}{\partial z} z dz = 0 \tag{2.7}$$

Application of integration by parts to the last term of (2.7) gives

$$\frac{\partial M_x}{\partial x} + \frac{\partial M_{xy}}{\partial y} + \tau_{xz} z \Big|_{-h/2}^{h/2} - \int_{-h/2}^{h/2} \tau_{xz} dz = 0 \tag{2.8}$$

where

$$M_x = \int_{-h/2}^{h/2} \sigma_x z dz, \quad M_{xy} = \int_{-h/2}^{h/2} \tau_{xy} z dz \tag{2.9}$$

Physically, the first part of (2.9) is the resultant moment on the edge with constant x, whereas the second term of (2.9) is the resulting twisting moment shown in Figure 2.1. Since both the top and bottom surfaces are traction-free and the shear force is defined as

$$Q_x = \int_{-h/2}^{h/2} \tau_{xz} dz \tag{2.10}$$

Finally, (2.8) is reduced to

$$\frac{\partial M_x}{\partial x} + \frac{\partial M_{xy}}{\partial y} - Q_x = 0 \tag{2.11}$$

Similarly, we can multiply (2.5) by z and integrate it from the bottom surface of the plate ($-h/2$) to the top surface of the plate ($h/2$) to give

$$\frac{\partial M_{xy}}{\partial x} + \frac{\partial M_x}{\partial y} - Q_y = 0 \tag{2.12}$$

Integration of (2.6) from $-h/2$ to $h/2$ gives

$$\int_{-h/2}^{h/2} \frac{\partial \tau_{zx}}{\partial x} dz + \int_{-h/2}^{h/2} \frac{\partial \tau_{zy}}{\partial y} dz + \int_{-h/2}^{h/2} \frac{\partial \sigma_z}{\partial z} dz = 0 \tag{2.13}$$

Noting that

$$Q_x = \int_{-h/2}^{h/2} \tau_{zx} dz, \quad Q_y = \int_{-h/2}^{h/2} \tau_{zy} dz \tag{2.14}$$

We have

$$\frac{\partial Q_x}{\partial x} + \frac{\partial Q_y}{\partial y} + \left[\sigma_z\right]_{-h/2}^{h/2} = 0 \tag{2.15}$$

By referring to Figure 2.2, the difference between normal tractions on the bottom and top surfaces of the plate is

$$\left[\sigma_z\right]_{-h/2}^{h/2} = \sigma_z(h/2) - \sigma_z(-h/2) = p(x, y) \tag{2.16}$$

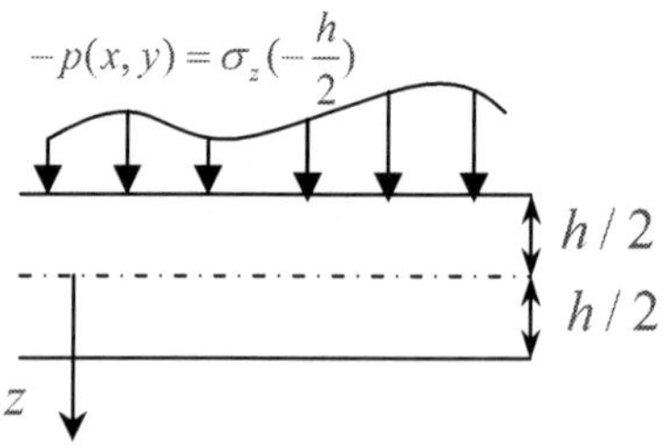

Figure 2.2 Loading applied to the top of the plate

Substitution of (2.16) into (2.15) gives the final equation:

$$\frac{\partial Q_x}{\partial x} + \frac{\partial Q_y}{\partial y} + p = 0 \tag{2.17}$$

Substitution of (2.11) and (2.12) into (2.17) gives

$$\frac{\partial^2 M_x}{\partial x^2} + 2\frac{\partial^2 M_{xy}}{\partial x \partial y} + \frac{\partial^2 M_y}{\partial y^2} + p = 0 \tag{2.18}$$

Clearly, one differential equation is not sufficient to solve for three unknown moments. Additional conditions are needed to solve the problem.

2.2.2 Forces and Moments

We now consider the displacement-strain relation for plate deflection w. Figure 2.3 shows the deflection w of a typical section of a plate. The point M on the mid-surface of the plate moves with nonzero deflection w but with zero deflections u and v, whereas the point P, which measures z from the mid-surface, will undergo a displacement of nonzero u_z, v_z and w_z. Suppose that a vertical plane containing points M and P as shown in Figure 2.3 remains a plane after bending (comparing the plane-remains-plane condition of Euler-Bernoulli beam theory discussed in Chapter 1). Using this condition and referring to Figure 2.3, we have

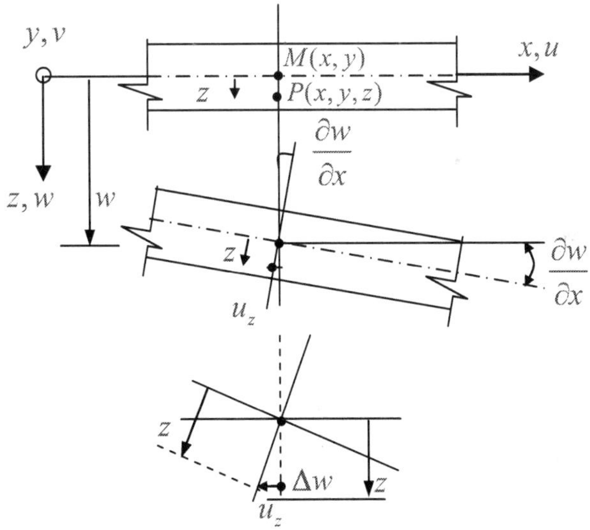

Figure 2.3 Kinematics of the deflection of a plate

$$u_z = -z\sin(\frac{\partial w}{\partial x}) \approx -z\frac{\partial w}{\partial x} \tag{2.19}$$

$$w_z = w - \Delta w = w - z[1-\cos(\frac{\partial w}{\partial x})] \approx w - \frac{z}{2}(\frac{\partial w}{\partial x})^2 \approx w \tag{2.20}$$

Note that Figure 2.3 is restricted to deflection on the x-z plane. Similarly, we can consider deflection on the y-z plane to yield

$$v_z = -z\frac{\partial w}{\partial y} \tag{2.21}$$

Therefore, correcting to the first order of magnitude of the deflection derivative, we have the deflection at point P as

$$u_z = -z\frac{\partial w}{\partial x}, \quad v_z = -z\frac{\partial w}{\partial y}, \quad w_z = w \tag{2.22}$$

The corresponding strain can be evaluated in terms of w as:

$$\varepsilon_x = \frac{\partial u_z}{\partial x} = -\frac{\partial^2 w}{\partial x^2} z = -\kappa_x z \tag{2.23}$$

$$\varepsilon_y = \frac{\partial v_z}{\partial y} = -\frac{\partial^2 w}{\partial y^2} z = -\kappa_y z \tag{2.24}$$

$$\gamma_{xy} = \frac{\partial v_z}{\partial x} + \frac{\partial u_z}{\partial y} = -2\frac{\partial^2 w}{\partial y \partial x} z = -2\kappa_{xy} z \tag{2.25}$$

Figure 2.4 illustrates the physical meaning of the curvatures κ_x, κ_y, and κ_{xy}. To model the arbitrary curved surface of shells, we need to use principal curvatures, Gaussian curvature, mean curvature, Lame's parameters, Codazzi's condition, and Gauss's condition in the context of differential geometry, which are however out of the scope of the present chapter. Hooke's law can now be used to relate stress and strain as

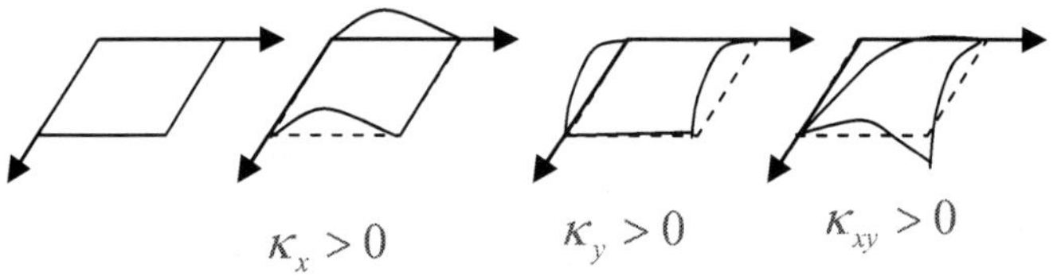

Figure 2.4 The physical meaning of nonzero curvatures

$$\varepsilon_x = \frac{1}{E}(\sigma_x - \nu\sigma_y), \quad \varepsilon_y = \frac{1}{E}(\sigma_y - \nu\sigma_x), \quad \gamma_{xy} = \frac{1}{G}\tau_{xy} \tag{2.26}$$

In view of (2.23) to (2.25), the stress components can thus be evaluated in terms of the curvature of deflection as

$$\sigma_x = \frac{E}{1-\nu^2}[\varepsilon_x + \nu\varepsilon_y] = -\frac{Ez}{1-\nu^2}[\kappa_x + \nu\kappa_y] \tag{2.27}$$

$$\sigma_y = \frac{E}{1-\nu^2}[\varepsilon_y + \nu\varepsilon_x] = -\frac{Ez}{1-\nu^2}[\kappa_y + \nu\kappa_x] \tag{2.28}$$

$$\tau_{xy} = G\gamma_{xy} = -\frac{Ez}{1-\nu^2}(1-\nu)\kappa_{xy} \tag{2.29}$$

Equations (2.27) to (2.29) show that for constant curvatures, we have the stresses being proportional to z. Substitution of (2.27) into the moment defined in (2.9) leads to

$$M_x = \int_{-h/2}^{h/2} \sigma_x z dz = -\frac{E}{1-\nu^2}(\kappa_x + \nu\kappa_y)\int_{-h/2}^{h/2} z^2 dz = -D(\kappa_x + \nu\kappa_y) \quad (2.30)$$

where

$$D = \frac{Eh^3}{12(1-\nu^2)} \quad (2.31)$$

This provides the proof of plate rigidity D. Similarly, we can find the moment on constant y as

$$M_y = -D(\kappa_y + \nu\kappa_x) \quad (2.32)$$

Substitution of (2.29) into the twisting moment defined in (2.9) leads to

$$M_{xy} = \int_{-h/2}^{h/2} \tau_{xy} z dz = -\frac{E}{1-\nu^2}(1-\nu)\kappa_{xy}\int_{-h/2}^{h/2} z^2 dz = -D(1-\nu)\kappa_{xy} \quad (2.33)$$

By virtue of (2.23) to (2.25), the bending moments and twisting moment can be expressed in terms of deflection w as:

$$M_x = -D(\kappa_x + \nu\kappa_y) = -D(\frac{\partial^2 w}{\partial x^2} + \nu\frac{\partial^2 w}{\partial y^2}) \quad (2.34)$$

$$M_y = -D(\kappa_y + \nu\kappa_x) = -D(\frac{\partial^2 w}{\partial y^2} + \nu\frac{\partial^2 w}{\partial x^2}) \quad (2.35)$$

$$M_{xy} = -D(1-\nu)\kappa_{xy} = -D(1-\nu)\frac{\partial^2 w}{\partial x \partial y} \quad (2.36)$$

Substitution of (2.34) and (2.36) into (2.11) gives

$$Q_x = -D\frac{\partial}{\partial x}\nabla^2 w \quad (2.37)$$

Substitution of (2.35) and (2.36) into (2.12) gives

$$Q_y = -D\frac{\partial}{\partial y}\nabla^2 w \quad (2.38)$$

Therefore, all bending moments and shear forces can be related to the plate deflection w.

2.2.3 Governing Equations

Finally, substitution of (2.34) to (2.36) into (2.18) gives

$$\nabla^2\nabla^2 w = \nabla^4 w = \frac{p}{D} \quad (2.39)$$

This completes the proof of (2.1). This formula was first derived by Lagrange in 1811. Proper boundary conditions for plate bending are considered next.

Along a boundary with a constant value of $x = x_r$, the boundary conditions of a plate can be expressed in terms of M_x, M_y, and M_{xy} (static boundary conditions) or w, and $\partial w/\partial x$ (kinematic boundary conditions). On each boundary, the prescribed boundary condition can be of static type or of kinematic type. Equation (2.39) is a fourth partial differential equation and we need only two boundary conditions on each boundary of the 2-D domain. For free edges, we have three moments being prescribed instead of two. This inconsistency was considered and solved by Kirchhoff. Thus, the theory of plate bending without considering shear deformation is normally referred as Kirchhoff plate theory.

2.2.4 Edge Conditions

For a *clamped boundary*, the kinematic conditions for an edge parallel to the x-axis are

$$w(x_r) = 0, \quad \frac{\partial w}{\partial x}(x_r) = 0 \tag{2.40}$$

where x_r is the coordinate on the edge. More generally, if the edge is not parallel to either x-axis or y-axis, the second kinematic condition given in (2.40) becomes:

$$\frac{\partial w}{\partial n}(x_r) = 0 \tag{2.41}$$

For a *simply-supported edge*, zero deflection and zero moment conditions give

$$w(x_r) = 0, \quad M_x(x_r) = -D\left(\frac{\partial^2 w}{\partial x^2} + \nu \frac{\partial^2 w}{\partial y^2}\right) = 0 \tag{2.42}$$

Clearly, for a simply-supported edge on x = constant, we must have the $w = 0$ or $\partial^2 w/\partial y^2 = 0$. These can be rewritten as

$$w(x_r) = 0, \quad \frac{\partial^2 w}{\partial x^2}(x_r) = 0 \tag{2.43}$$

More generally, if the edge is not parallel to either the x-axis or y-axis, the second kinematic condition becomes:

$$\nabla^2 w(x_r) = 0 \tag{2.44}$$

Note that $\nabla^2 w$ is an invariant with respect to any straight boundary. For a *free edge*, Poisson proposed setting the bending moment, twisting moment and shear force to zero:

$$M_{xy}(x_r) = 0, \quad M_x(x_r) = 0, \quad Q_x(x_r) = 0 \tag{2.45}$$

As remarked earlier, these conditions can be reduced to two and this was done by Kelvin and Tate in 1883. In particular, M_{xy} and Q_x can be combined to give a supplementary shear force V_x.

To examine the effect of twisting moments, we consider in Figure 2.5 the additional shear force T_x induced by distributed twisting moments. In particular, by referring to Figure 2.5 we can define T_x as

$$\begin{aligned} T_x dy &= S_x(y+dy)dy - S_x(y)dy \\ &= M_{xy} + \frac{\partial M_{xy}}{\partial y}dy - M_{xy} \end{aligned} \tag{2.46}$$

Simplification of (2.46) gives

$$T_x = \frac{\partial M_{xy}}{\partial y} \tag{2.47}$$

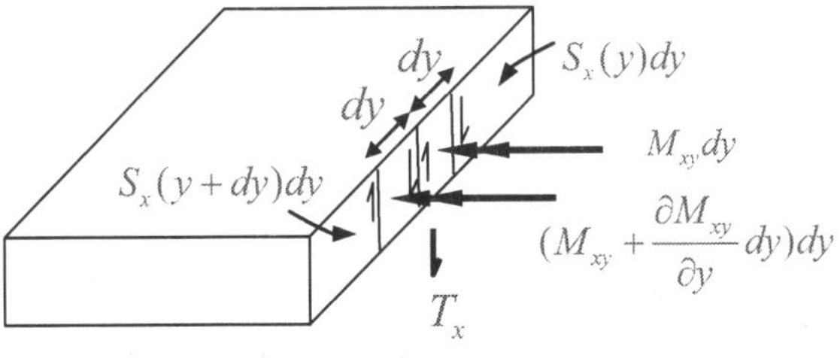

Figure 2.5 Shear force induced by twisting moments

In view of this addition of shear forces, supplementary shear forces can be defined as

$$V_x = Q_x + T_x = -D\frac{\partial}{\partial x}\nabla^2 w + \frac{\partial M_{xy}}{\partial y} = -D\frac{\partial}{\partial x}\nabla^2 w - D(1-\nu)\frac{\partial^3 w}{\partial x \partial y^2} \tag{2.48}$$

$$V_y = Q_y + T_y = -D\frac{\partial}{\partial y}\nabla^2 w + \frac{\partial M_{xy}}{\partial x} = -D\frac{\partial}{\partial y}\nabla^2 w - D(1-\nu)\frac{\partial^3 w}{\partial x^2 \partial y} \tag{2.49}$$

In addition, by symmetry we have the following condition at the corner

$$M_{xy} = M_{yx} \tag{2.50}$$

At the corner, an additional force results from the twisting moments. In particular, Figure 2.6 shows the two shear forces induced at the corner and the resultant force R can be summed as

$$R = S_x dy + S_y dx = M_{xy} + M_{yx} = 2M_{xy}, \quad x = a, \quad y = b \tag{2.51}$$

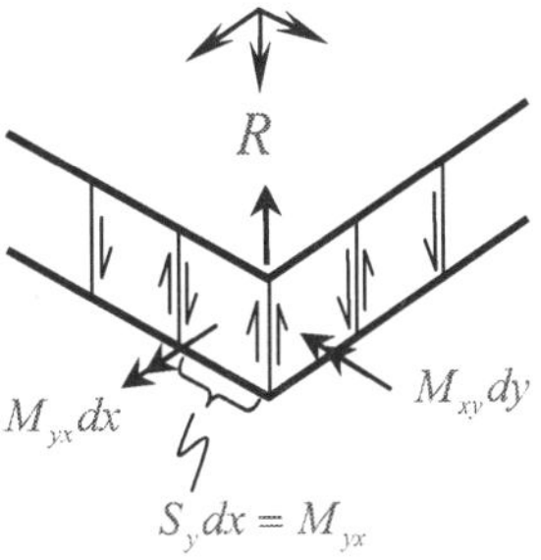

Figure 2.6 Resultant corner force induced by the twisting moment

This reactive force R at the corner is induced by the twisting moment. The tendency of uplift at the corners of flat slabs is likely to induce damage of the plate near the corners.

2.3 SIMPLY-SUPPORTED PLATES

Two analytical solutions for simply-supported plates will be considered in this section, and they are Navier's solution and Levy's solution.

2.3.1 Navier's Solution

For simply-supported boundary conditions, it is straightforward to see that the following solution form proposed by Navier in 1820 satisfies the boundary condition (2.42) identically:

$$w(x,y)=\sum_{i=1}^{\infty}\sum_{k=1}^{\infty}c_{ik}\sin k\pi\frac{x}{a}\sin i\pi\frac{y}{b} \tag{2.52}$$

where c_{ik} are unknown constants to be determined. Similarly, it is assumed that the transverse loading applied to the plate can also be expanded in double Fourier series expansion as

$$p(x,y)=\sum_{i=1}^{\infty}\sum_{k=1}^{\infty}b_{ik}\sin k\pi\frac{x}{a}\sin i\pi\frac{y}{b} \tag{2.53}$$

where

$$b_{ik}=\frac{4}{ab}\int_0^a\int_0^b p(x,y)\sin k\pi\frac{x}{a}\sin i\pi\frac{y}{b}dxdy \tag{2.54}$$

Substitution of (2.52) and (2.53) into (2.1) gives

$$\sum_{i=1}^{\infty}\sum_{k=1}^{\infty}c_{ik}\pi^4\left(\frac{k^2}{a^2}+\frac{i^2}{b^2}\right)^2\sin k\pi\frac{x}{a}\sin i\pi\frac{y}{b}=\frac{1}{D}\sum_{i=1}^{\infty}\sum_{k=1}^{\infty}b_{ik}\sin k\pi\frac{x}{a}\sin i\pi\frac{y}{b} \tag{2.55}$$

Balancing term by term on both sides of (2.55), we get

$$c_{ik}=\frac{b_{ik}}{D\pi^4\left(\frac{k^2}{a^2}+\frac{i^2}{b^2}\right)^2} \tag{2.56}$$

Substitution of (2.56) into (2.52) gives the deflection as

$$w(x,y)=\sum_{i=1}^{\infty}\sum_{k=1}^{\infty}\frac{b_{ik}}{D\pi^4\left(\frac{k^2}{a^2}+\frac{i^2}{b^2}\right)^2}\sin k\pi\frac{x}{a}\sin i\pi\frac{y}{b} \tag{2.57}$$

This is Navier's solution, and the key problem of this solution is the convergence of the double infinite series obtained in (2.57). For concentrated loads, we have the following convergence of the summation

$$w\sim\sum\frac{1}{k^4},\quad M\sim\sum\frac{1}{k^2},\quad Q\sim\sum\frac{(-1)^k}{k} \tag{2.58}$$

For the nonhomogeneous biharmonic equation with the unknown being deflection, the Navier solution is more accurate for w and less accurate for moment M, whilst the shear force has the slowest convergence and is thus most inaccurate.

2.3.2 Levy's Solution

For the problems with two parallel simply-supported edges on $x = 0$ and $x = a$, Levy proposed in 1899 the following solution form:

$$w(x,y) = w_0(x,y) + w_1(x,y) \tag{2.59}$$

This, in essence, is expressing the general solution in terms of the homogeneous and the particular solution (see Section 3.3.2 of Chau, 2018) such that

$$\nabla^4 w_0(x,y) = \frac{p}{D}, \quad \nabla^4 w_1(x,y) = 0 \tag{2.60}$$

with both w_0 and w_1 satisfying the edge conditions on $x = 0$ and $x = a$. We first introduce the following separation of variables

$$w_1(x,y) = X(x)Y(y) \tag{2.61}$$

To satisfy the simply-supported conditions on the two edges, we can look for the following solution form:

$$w_1(x,y) = \sum_{k=1}^{\infty} Y_k(y) \sin k\pi \frac{x}{a} \tag{2.62}$$

Substitution of (2.62) into the second part of (2.60) gives

$$\nabla^4 w_1(x,y) = \sum_{k=1}^{\infty} [\frac{k^4\pi^4}{a^4} Y_k - \frac{2k^2\pi^2}{a^2}\frac{d^2Y_k}{dy^2} + \frac{d^4Y_k}{dy^4}] \sin k\pi \frac{x}{a} = 0 \tag{2.63}$$

For arbitrary x, (2.63) requires

$$\frac{k^4\pi^4}{a^4} Y_k - \frac{2k^2\pi^2}{a^2}\frac{d^2Y_k}{dy^2} + \frac{d^4Y_k}{dy^4} = 0 \tag{2.64}$$

This is a fourth order ODE with constant coefficients. As usual, we can assume an exponential solution

$$Y_k = e^{\lambda y} \tag{2.65}$$

Substitution of (2.65) into (2.64) leads to

$$F(\lambda) = \lambda^4 - \frac{2k^2\pi^2}{a^2}\lambda^2 + \frac{k^4\pi^4}{a^4} = 0 \tag{2.66}$$

Although (2.66) is a fourth order algebraic equation, λ only appears as a quadratic function of λ^2. Thus, (2.66) can be factorized easily as

$$[\lambda^2 - (\frac{k\pi}{a})^2]^2 = 0 \tag{2.67}$$

Consequently, the four roots are obtained as

$$\lambda_{1,2} = \frac{k\pi}{a}, \quad \lambda_{3,4} = -\frac{k\pi}{a} \tag{2.68}$$

Figure 2.7 plots F against λ and it is apparent that the roots for λ are indeed given by (2.68). For this case of repeated roots, the solution form can be expressed as (Chau, 2018):

$$Y_k(y) = (A_k + B_k \frac{k\pi y}{a}) \sinh k\pi \frac{y}{a} + (C_k + D_k \frac{k\pi y}{a}) \cosh k\pi \frac{y}{a} \tag{2.69}$$

For finite rectangular plates, the general solution form can be expressed as:

$$w_1(x,y)=\sum_{k=1}^{\infty}[(A_k+B_k\frac{k\pi y}{a})\sinh k\pi\frac{y}{a}+(C_k+D_k\frac{k\pi y}{a})\cosh k\pi\frac{y}{a}]\sin k\pi\frac{x}{a} \tag{2.70}$$

where the unknown constants have to be determined.

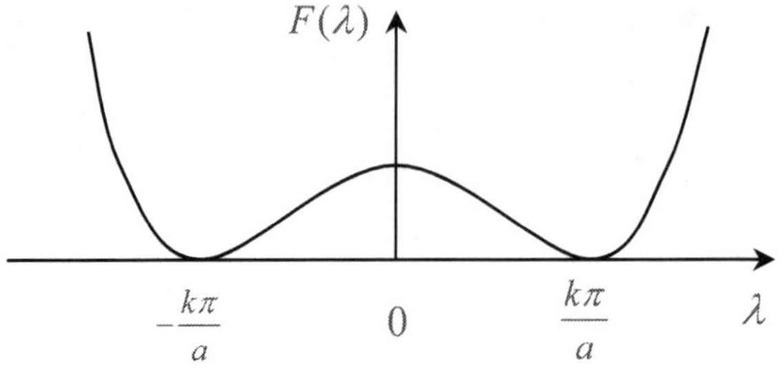

Figure 2.7 The roots of $F(\lambda)$ defined in (2.66)

In particular, the remaining unknown constants need to be determined from the boundary conditions of the other edges. For infinite plates in the shape of a strip, it is more convenient to express the solution in exponential form in view of the decay condition

$$w_1(x,y)=\sum_{k=1}^{\infty}[(A_k+B_k\frac{k\pi y}{a})e^{k\pi\frac{y}{a}}+(C_k+D_k\frac{k\pi y}{a})e^{-k\pi\frac{y}{a}}]\sin k\pi\frac{x}{a} \tag{2.71}$$

As $y\to\infty$, we must have

$$A_k=B_k=0 \tag{2.72}$$

For plates with four simply-supported edges subject to uniform load shown in Figure 2.8, we have the particular solution for (2.60) being

$$w_0(x,y)=\frac{p}{24D}(x^4-2ax^3+a^3x) \tag{2.73}$$

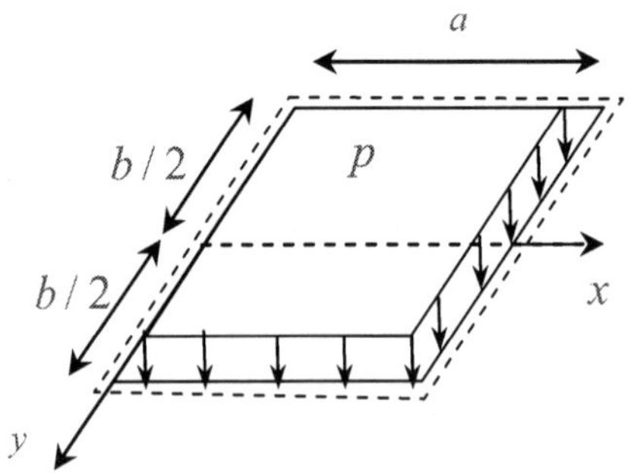

Figure 2.8 Simply-supported rectangular plates under uniform load

It is straightforward to check that

$$\nabla^4 w_0(x, y) = \frac{p}{D} \tag{2.74}$$

In particular, differentiation of (2.73) with respect to x twice gives

$$\frac{\partial^2 w_0}{\partial x^2} = \frac{p}{24D}(12x^2 - 12ax) \tag{2.75}$$

Therefore, we can see that (2.73) and (2.75) satisfy the following simply-supported boundary conditions

$$w = 0, \quad \frac{\partial^2 w}{\partial y^2} = 0, \quad y = \pm b/2 \tag{2.76}$$

We now expand the right-hand side of (2.73) into eigenfunction expansion (i.e., sine function), and in particular, we have

$$q(x) = \sum_{m=1}^{\infty} b_m \sin\frac{m\pi x}{a} = \frac{p}{24D}(x^4 - 2ax^3 + a^3 x) \tag{2.77}$$

$$b_m = \frac{2}{a}\frac{p}{24D}\int_0^a (x^4 - 2ax^3 + a^3 x)\sin\frac{m\pi x}{a}dx \tag{2.78}$$

To evaluate the integral, by applying integration by parts, we have

$$\int_0^a x^4 \sin\frac{m\pi x}{a}dx = \frac{1}{\beta_m}[a^4 + 4\int_0^a x^3 \cos\frac{m\pi x}{a}dx] \tag{2.79}$$

The formula from Equation 14.372 of Spiegel (1964) is useful for the present integration

$$\int x^3 \cos ax dx = (\frac{3x^2}{a^2} - \frac{6}{a^4})\cos ax + (\frac{x^3}{a} - \frac{6x}{a^3})\sin ax \tag{2.80}$$

Applying (2.79) and (2.80), we obtain

$$\int_0^a x^3 \cos\frac{m\pi x}{a}dx = -\frac{3a^4}{m^2\pi^2}(1 - \frac{4}{m^2\pi^2}), \quad m = 1, 2, 3, \ldots \tag{2.81}$$

$$\int_0^a x^4 \sin\frac{m\pi x}{a}dx = \frac{a^5}{m\pi}[1 - \frac{12}{m^2\pi^2}(1 - \frac{4}{m^2\pi^2})] \tag{2.82}$$

Similarly, we find that

$$\int_0^a x^3 \sin\frac{m\pi x}{a}xdx = \frac{a^4}{m\pi}(1 - \frac{6}{m^2\pi^2}) \tag{2.83}$$

$$\int_0^a x\sin\frac{m\pi x}{a}xdx = \frac{a^2}{m\pi} \tag{2.84}$$

Using all these results found in (2.79) to (2.84), we have from (2.78)

$$b_m = \frac{4pa^4}{m^5\pi^5 D} \tag{2.85}$$

Thus, we have the following eigenfunction expansion for the particular solution

$$\frac{p}{24D}(x^4 - 2ax^3 + a^3 x) = \frac{4pa^4}{\pi^5 D}\sum_{m=1,3,5\ldots}^{\infty}\frac{1}{m^5}\sin\frac{m\pi x}{a} \tag{2.86}$$

On the other hand, in view of symmetry, the other function w_1 given in (2.71) is reduced to the following form (retaining only the even functions of y):

$$w_1(x,y)=\sum_{k=1}^{\infty}[B_k\frac{k\pi y}{a}\sinh k\pi\frac{y}{a}+C_k\cosh k\pi\frac{y}{a}]\sin k\pi\frac{x}{a} \tag{2.87}$$

Finally, the general solution of this rectangular plate can be expressed as

$$\begin{aligned} w(x,y)&=\frac{p}{24D}(x^4-2ax^3+a^3x)\\ &+\frac{pa^4}{D}\sum_{m=1}^{\infty}[A_m\cosh\frac{m\pi y}{a}+B_m\frac{m\pi y}{a}\sinh\frac{m\pi y}{a}]\sin\frac{m\pi x}{a} \end{aligned} \tag{2.88}$$

Note that the unknowns have been renamed and scaled. Applying (2.86), we obtain

$$w(x,y)=\frac{pa^4}{D}\sum_{m=1,3,5,...}^{\infty}[\frac{4}{\pi^5m^5}+A_m\cosh\frac{m\pi y}{a}+B_m\frac{m\pi y}{a}\sinh\frac{m\pi y}{a}]\sin\frac{m\pi x}{a} \tag{2.89}$$

Differentiating (2.89), we have

$$\frac{\partial^2 w}{\partial y^2}=\frac{pa^4}{D}\sum_{m=1,3,5,...}^{\infty}(\frac{m\pi}{a})^2[(A_m+2B_m)\cosh\frac{m\pi y}{a}+B_m\frac{m\pi y}{a}\sinh\frac{m\pi y}{a}]\sin\frac{m\pi x}{a} \tag{2.90}$$

Substitution of (2.90) into the second boundary condition given in (2.76) leads to

$$(A_m+2B_m)\cosh\alpha_m+B_m\alpha_m\sinh\alpha_m=0 \tag{2.91}$$

where

$$\alpha_m=\frac{m\pi b}{2a} \tag{2.92}$$

Substitution of (2.89) into the first boundary condition given in (2.76) leads to another equation. In summary, we have two equations from the two boundary conditions given in (2.76):

$$\frac{4}{\pi^5m^5}+A_m\cosh\alpha_m+B_m\alpha_m\sinh\alpha_m=0 \tag{2.93}$$

$$(A_m+2B_m)\cosh\alpha_m+B_m\alpha_m\sinh\alpha_m=0 \tag{2.94}$$

The solutions of (2.93) and (2.94) are

$$A_m=-\frac{2(\alpha_m\tanh\alpha_m+2)}{\pi^5m^5\cosh\alpha_m},\quad B_m=\frac{2}{\pi^5m^5\cosh\alpha_m} \tag{2.95}$$

Substitution of (2.95) into (2.89) yields the final result

$$\begin{aligned} w(x,y)&=\frac{4pa^4}{\pi^5D}\sum_{m=1,3,5,...}^{\infty}\frac{1}{m^5}[1-\frac{\alpha_m\tanh\alpha_m+2}{2\cosh\alpha_m}\cosh\frac{2\alpha_m y}{b}\\ &+\frac{\alpha_m}{2\cosh\alpha_m}(\frac{2y}{b})\sinh\frac{2\alpha_m y}{b}]\sin\frac{m\pi x}{a} \end{aligned} \tag{2.96}$$

The maximum deflection occurs at $y=0$ and $x=a/2$ is

$$w_{\max} = \frac{4pa^4}{\pi^5 D} \sum_{m=1,3,5,...}^{\infty} \frac{(-1)^{(m-1)/2}}{m^5} (1 - \frac{\alpha_m \tanh \alpha_m + 2}{2\cosh \alpha_m}) \tag{2.97}$$

By dropping the second term in the series, we have the deflection in the middle of a uniformly loaded strip being ($b \to 0$)

$$w_{\max} = \frac{4pa^4}{\pi^5 D} \sum_{m=1,3,5,...}^{\infty} \frac{(-1)^{(m-1)/2}}{m^5} = \frac{5pa^4}{384D} \tag{2.98}$$

For the special case of a square plate with $a = b$, we have

$$w_{\max} = \frac{5pa^4}{384D} - \frac{4pa^4}{\pi^5 D} \sum_{m=1,3,5,...}^{\infty} \frac{(-1)^{(m-1)/2}}{m^5} \frac{\alpha_m \tanh \alpha_m + 2}{2\cosh \alpha_m} \tag{2.99}$$

This solution converges much faster than the Navier solution, and we see that the first two terms gives

$$\frac{\frac{\pi}{2}\tanh(\frac{\pi}{2}) + 2}{2\cosh(\frac{\pi}{2})} = 0.685614743 \tag{2.100}$$

$$-\frac{1}{3^5} \frac{\frac{3\pi}{2}\tanh(\frac{3\pi}{2}) + 2}{2\cosh(\frac{3\pi}{2})} = -0.0602876 \times \frac{1}{3^5} = -0.00024805973109... \tag{2.101}$$

There will be only 0.036% error if we retained only the first term

$$w_{\max} = \frac{5pa^4}{384D} - \frac{4pa^4}{\pi^5 D}(0.68562 - 0.00025 + ...) = 0.00406 \frac{pa^4}{D} \tag{2.102}$$

The bending moment due to these deflections can be determined as

$$M_x = M_x' + M_x'', \quad M_y = M_y' + M_y'' \tag{2.103}$$

where

$$M_x' = \frac{px(a-x)}{2}, \quad M_y' = \nu \frac{px(a-x)}{2} \tag{2.104}$$

$$M_x'' = (1-\nu) pa^2 \pi^2 \sum_{m=1,3,...}^{\infty} m^2 [A_m \cosh \frac{m\pi y}{a} + B_m (\frac{m\pi y}{a} \sinh \frac{m\pi y}{a} - \frac{2\nu}{1-\nu} \cosh \frac{m\pi y}{a})] \sin \frac{m\pi x}{a} \tag{2.105}$$

$$M_y'' = -(1-\nu) pa^2 \pi^2 \sum_{m=1,3,...}^{\infty} m^2 [A_m \cosh \frac{m\pi y}{a} + B_m (\frac{m\pi y}{a} \sinh \frac{m\pi y}{a} + \frac{2}{1-\nu} \cosh \frac{m\pi y}{a})] \sin \frac{m\pi x}{a} \tag{2.106}$$

The moment along the x-axis at $y = 0$ will be

$$(M_x)_{y=0} = \frac{px(a-x)}{2} - pa^2\pi^2 \sum_{m=1,3,...}^{\infty} m^2[2\nu B_m - (1-\nu)A_m]\sin\frac{m\pi x}{a} \quad (2.107)$$

$$(M_y)_{y=0} = \frac{\nu px(a-x)}{2} - pa^2\pi^2 \sum_{m=1,3,...}^{\infty} m^2[2B_m + (1-\nu)A_m]\sin\frac{m\pi x}{a} \quad (2.108)$$

These infinite series converge very fast because of the cosh term in the denominator of (2.96). To evaluate the shear force, we first calculate the Laplacian of w as:

$$\nabla^2 w = \frac{\partial^2 w}{\partial x^2} + \frac{\partial^2 w}{\partial y^2} = -\frac{px(a-x)}{2D} + \frac{2\pi^2 pa^2}{D} \sum_{m=1,3,...}^{\infty} m^2 B_m \cosh\frac{m\pi y}{a}\sin\frac{m\pi x}{a} \quad (2.109)$$

Then, we have

$$Q_x = \frac{p(a-2x)}{2} - 2\pi^3 pa \sum_{m=1,3,...}^{\infty} m^3 B_m \cosh\frac{m\pi y}{a}\cos\frac{m\pi x}{a} \quad (2.110)$$

$$Q_y = -2\pi^3 pa \sum_{m=1,3,...}^{\infty} m^3 B_m \sinh\frac{m\pi y}{a}\sin\frac{m\pi x}{a} \quad (2.111)$$

Along the edge $x = 0$, we have

$$\begin{aligned}(Q_x)_{x=0} &= \frac{pa}{2} - 2\pi^3 pa \sum_{m=1,3,...}^{\infty} m^3 B_m \cosh\frac{m\pi y}{a} \\ &= \frac{pa}{2} - \frac{4pa}{\pi^2} \sum_{m=1,3,...}^{\infty} \frac{\cosh\frac{m\pi y}{a}}{m^2\cosh\alpha_m}\end{aligned} \quad (2.112)$$

Along the edge $y = -b/2$, we have

$$\begin{aligned}(Q_y)_{y=-b/2} &= 2\pi^3 pa \sum_{m=1,3,...}^{\infty} m^3 B_m \sinh\alpha_m \sin\frac{m\pi x}{a} \\ &= \frac{4pa}{\pi^2} \sum_{m=1,3,...}^{\infty} \frac{\tanh\alpha_m}{m^2}\sin\frac{m\pi x}{a}\end{aligned} \quad (2.113)$$

The reactive force along the edge $x = 0$ can be determined as

$$\begin{aligned}V_x &= (Q_x - \frac{\partial M_{xy}}{\partial y})_{x=0} = \frac{pa}{2} - \frac{4pa}{\pi^2} \sum_{m=1,3,...}^{\infty} \frac{\cosh\frac{m\pi y}{a}}{m^2\cosh\alpha_m} \\ &+ \frac{2(1-\nu)pa}{\pi^2} \sum_{m=1,3,...}^{\infty} \frac{1}{m^2\cosh^2\alpha_m}(\alpha_m\sinh\alpha_m\cosh\frac{m\pi y}{a} \\ &- \frac{m\pi y}{a}\cosh\alpha_m\sinh\frac{m\pi y}{a})\end{aligned} \quad (2.114)$$

The reactive force at the center of the edge $x = 0$ is (see Figure 2.8)

$$(V_x)_{x=0,y=0} = pa[\frac{1}{2} - \frac{4}{\pi^2} \sum_{m=1,3,...}^{\infty} \frac{1}{m^2 \cosh \alpha_m} + \frac{2(1-\nu)}{\pi^2} \sum_{m=1,3,...}^{\infty} \frac{\alpha_m \sinh \alpha_m}{m^2 \cosh^2 \alpha_m}] \quad (2.115)$$

The reactive force at the corner $x = a$, $y = b/2$ is

$$R = 2(M_{xy})_{x=a,y=b/2} = 2D(1-\nu)(\frac{\partial^2 w}{\partial x \partial y})_{x=a,y=b/2} \quad (2.116)$$

$$R = \frac{4(1-\nu)pa^2}{\pi^3} \sum_{m=1,3,...}^{\infty} \frac{1}{m^3 \cosh \alpha_m} [(1+\alpha_m \tanh \alpha_m)\sinh \alpha_m - \alpha_m \cosh \alpha_m] \quad (2.117)$$

This provides the complete Levy solution for the simply-supported plates.

2.4 CLAMPED RECTANGULAR PLATES

In this section, solutions for clamped rectangular plates are considered by using the Galerkin method.

2.4.1 Galerkin Method

According to Section 14.12 of Chau (2018), the Galerkin method for plate bending can be expressed as:

$$(\delta V)_1 = \iint p \delta w dxdy \quad (2.118)$$

$$(\delta V)_2 = D \iint \nabla^2 \nabla^2 w \delta w \, dxdy \quad (2.119)$$

The biharmonic equation can be satisfied globally as

$$(\delta V)_1 = (\delta V)_2 \quad (2.120)$$

The deflection w can be approximated by

$$w = a_1 \varphi_1(x,y) + a_2 \varphi_2(x,y) + a_3 \varphi_2(x,y) + ... + a_n \varphi_n(x,y) \quad (2.121)$$

Note that these approximate functions φ_i must satisfy both the essential and natural boundary conditions (see Section 14.10 of Chau, 2018). The variations of the deflection can be expressed as

$$\delta w_1 = \delta a_1 \varphi_1(x,y) \quad (2.122)$$

$$\delta w_2 = \delta a_2 \varphi_2(x,y) \quad (2.123)$$

$$\delta w_3 = \delta a_3 \varphi_2(x,y) \quad (2.124)$$

$$\delta w_n = \delta a_n \varphi_n(x,y) \quad (2.125)$$

The Galerkin method gives the following n equations for determining n unknown constants for the approximate functions assumed in (2.121) as

$$\iint (\nabla^2 \nabla^2 w - \frac{p}{D}) \varphi_i dxdy = 0, \quad i = 1, 2, ..., n \quad (2.126)$$

Special approximated functions for (2.121) will be considered in the next section for the case of clamped plates.

2.4.2 Approximation for Clamped Plates

We now consider the case of clamped plate subject to uniform load p, as shown in Figure 2.9. Let us consider a one-term expansion in (2.121) as

$$w = \sum_{i=1}^{n} c_i \varphi_i \approx c_1 \varphi_1 = c_1 (x^2 - a^2)(y^2 - b^2)^2 \tag{2.127}$$

It is clear that (2.127) satisfies the following boundary conditions:

$$w = \frac{\partial w}{\partial x} = 0, \quad x = \pm a \tag{2.128}$$

$$w = \frac{\partial w}{\partial y} = 0, \quad y = \pm b \tag{2.129}$$

For this case, (2.126) is reduced to one equation with one unknown:

$$c_1 \iint (\nabla^2 \nabla^2 \varphi_1) \varphi_1 dxdy = \frac{1}{D} \iint p \varphi_1 dxdy \tag{2.130}$$

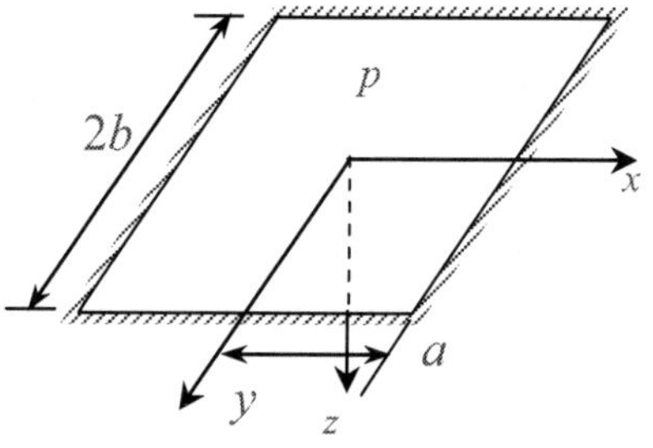

Figure 2.9 Rectangular plate with all fixed edges subject to uniform load *p*

The unknown constant c_1 is found equal to

$$c_1 = \frac{b_1}{a_{11}} \tag{2.131}$$

where

$$a_{11} = D \int_0^a \int_0^b \left(\frac{\partial^4 \varphi_1}{\partial x^4} \varphi_1 + 2 \frac{\partial^4 \varphi_1}{\partial x^2 \partial y^2} \varphi_1 + \frac{\partial^4 \varphi_1}{\partial y^4} \varphi_1 \right) dxdy \tag{2.132}$$

$$b_1 = p \int_0^a \int_0^b \varphi_1 dxdy = p \int_0^a \int_0^b (x^2 - a^2)(y^2 - b^2)^2 dxdy = \frac{64p}{225} a^5 b^5 \tag{2.133}$$

The integral on the right-hand side of (2.132) can be evaluated one-by-one as

$$\int_0^a \int_0^b \frac{\partial^4 \varphi_1}{\partial x^4} \varphi_1 dxdy = \int_0^a \int_0^b 24(x^2 - a^2)^2 (y^2 - b^2)^4 dxdy = \frac{8192}{1575} a^5 b^9 \tag{2.134}$$

$$\int_0^a \int_0^b \frac{\partial^4 \varphi_1}{\partial y^4} \varphi_1 dxdy = \int_0^a \int_0^b 24(x^2 - a^2)^4 (y^2 - b^2)^2 dxdy = \frac{8192}{1575} a^9 b^5 \tag{2.135}$$

$$\int_0^a \int_0^b 2\frac{\partial^4 \varphi_1}{\partial x^2 \partial y^2}\varphi_1 dxdy$$

$$= \int_0^a \int_0^b 32(3x^2 - a^2)(3y^2 - b^2)(x^2 - a^2)^2(y^2 - b^2)^2 dxdy \tag{2.136}$$

$$= \frac{32768}{11025}a^7 b^7$$

Substitution of (2.134) to (2.136) into (2.132) gives

$$a_{11} = 4D(\frac{8192}{1575}a^5 b^9 + \frac{32768}{11025}a^7 b^7 + \frac{8192}{1575}a^9 b^5) \tag{2.137}$$

Finally, substitution of (2.133) and (2.137) into (2.131) leads to

$$c_1 = \frac{7}{128}\frac{1}{(b^4 + \frac{4}{7}a^2 b^2 + a^4)}(\frac{p}{D}) \tag{2.138}$$

The first approximation given by (2.127) becomes

$$w \approx c_1\varphi_1 = \frac{7}{128}\frac{1}{(b^4 + \frac{4}{7}a^2 b^2 + a^4)}(\frac{p}{D})(x^2 - a^2)^2(y^2 - b^2)^2 \tag{2.139}$$

The maximum deflection at the center of the plate is

$$w_{\max} = \frac{7pa^4 b^4}{128D(b^4 + \frac{4}{7}a^2 b^2 + a^4)} \tag{2.140}$$

For the special case of a square plate (i.e., $a = b$), the maximum deflection occurs

$$w_{\max} = \frac{49}{2304}(\frac{pa^4}{D}) = 0.02127\frac{pa^4}{D} \tag{2.141}$$

Table 35 of Timoshenko and Woinowsky-Krieger (1959) gives a more accurate result by solving a system of equations:

$$w_{\max} = 0.02016\frac{pa^4}{D} \tag{2.142}$$

We see that the error by the Galerkin method with a single term is about 5.5%. Equation (2.139) is considered very accurate in terms of the amount of computation cost. The moment M_x can be evaluated as

$$M_x = -D\frac{\partial^2 w}{\partial x^2} = -4Dc_1(3x^2 - a^2)(y^2 - b^2)^2 \tag{2.143}$$

The maximum moment that occurs at the edge is

$$(M_x)_{\max} = (M_x)_{y=0, x=b/2} = -8Dc_1 a^6 = -\frac{49}{288}pa^2 = -0.17014pa^2 \tag{2.144}$$

Table 32 of Timoshenko and Woinowsky-Krieger (1959) gives a more accurate result as:

$$(M_x)_{\max} = -0.2052pa^2 \tag{2.145}$$

The error of the moment is 16.3%, which is much larger than that of the deflection. More terms need to be included if a more accurate moment is needed.

Nevertheless, we have demonstrated that an approximate solution for clamped plates can be obtained easily by using the Galerkin method.

2.5 DEFLECTION OF CIRCULAR PLATES

In this section, bending of circular plates is considered. In polar form, the nonhomogeneous biharmonic equation for plate bending becomes

$$\nabla^2\nabla^2 w = \nabla^4 w = \frac{p}{D} \tag{2.146}$$

where

$$\left(\frac{\partial^2}{\partial r^2} + \frac{1}{r}\frac{\partial}{\partial r} + \frac{1}{r}\frac{\partial^2}{\partial \theta^2}\right)\left(\frac{\partial^2}{\partial r^2} + \frac{1}{r}\frac{\partial}{\partial r} + \frac{1}{r}\frac{\partial^2}{\partial \theta^2}\right) w = \frac{p}{D} \tag{2.147}$$

The corresponding formulas for bending moments and shear forces in polar form are

$$M_r = -D\left[(1-\nu)\frac{\partial^2 w}{\partial r^2} + \nu\nabla^2 w\right] \tag{2.148}$$

$$M_\theta = -D\left[\nabla^2 w - (1-\nu)\frac{\partial^2 w}{\partial r^2}\right] \tag{2.149}$$

$$M_{r\theta} = -D(1-\nu)\frac{\partial}{\partial r}\left(\frac{1}{r}\frac{\partial w}{\partial \theta}\right) \tag{2.150}$$

$$Q_r = -D\frac{\partial}{\partial r}\nabla^2 w, \quad Q_\theta = -D\frac{\partial}{r\partial \theta}\nabla^2 w \tag{2.151}$$

$$V_r = Q_r + \frac{\partial M_{r\theta}}{r\partial \theta} \tag{2.152}$$

$$V_\theta = Q_\theta + \frac{\partial M_{r\theta}}{\partial r} \tag{2.153}$$

For axisymmetric loading and bending of circular plates, the biharmonic equation is simplified to

$$\frac{1}{r}\frac{d}{dr}\left\{r\frac{d}{dr}\left[\frac{1}{r}\frac{d}{dr}\left(r\frac{dw}{dr}\right)\right]\right\} = \frac{p}{D} \tag{2.154}$$

Integrating (2.154) once, we have

$$r\frac{d}{dr}\left[\frac{1}{r}\frac{d}{dr}\left(r\frac{dw}{dr}\right)\right] = \frac{pr^2}{2D} + 4B \tag{2.155}$$

Dividing both sides of (2.155) by r and integrating one more time, we get

$$\frac{1}{r}\frac{d}{dr}\left(r\frac{dw}{dr}\right) = \frac{pr^2}{4D} + 4B\ln r + 4(B+C) \tag{2.156}$$

Rearranging and integrating (2.156), we obtain

$$\frac{dw}{dr} = \frac{pr^3}{16D} + 2Br\left(\ln r - \frac{1}{2}\right) + 2r(B+C) + \frac{A}{r} \tag{2.157}$$

Finally, integrating (2.157) once more, we get

$$w = \frac{pr^4}{64D} + Cr^2 + Br^2 \ln r + A \ln r + F \tag{2.158}$$

The first term on the right of (2.158) is the particular solution, whereas the last four terms of (2.158) are the homogeneous solutions. There are four unknown constants to be determined. The case of clamped circular plates will be considered next.

2.5.1 Clamped Plate with Uniform Load

For clamped plates subject to uniform loads shown in Figure 2.10, the boundary conditions are

$$w(a) = 0, \quad \frac{dw}{dr}(a) = 0 \tag{2.159}$$

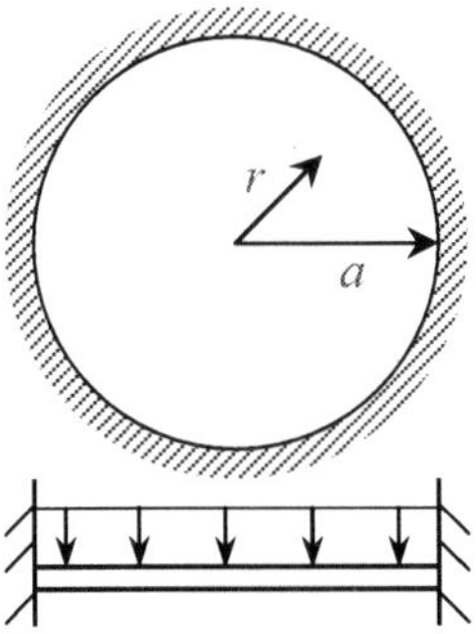

Figure 2.10 Clamped circular plate subject to uniform load

It is clear that the deflection shown in Figure 2.10 is axisymmetric and the solution given in (2.158) can be employed. The boundedness condition of the plate as $r \to 0$ requires

$$w(0), \quad M(0), \quad \frac{dw}{dr}(0), \quad Q_r(0), \quad \text{are finite} \tag{2.160}$$

The boundedness condition for w gives

$$A = 0 \tag{2.161}$$

The shear force boundedness condition requires that the following term is finite

$$Q_r = -D\frac{d}{dr}[\frac{1}{r}\frac{d}{dr}(r\frac{dw}{dr})] = -\frac{pr^2}{2} - \frac{4BD}{r} \tag{2.162}$$

Therefore, we must have

$$B = 0 \tag{2.163}$$

In view of (2.161) and (2.163), the displacement given in (2.158) is reduced to

$$w = \frac{pr^4}{64D} + Cr^2 + F \tag{2.164}$$

Finally, using the clamped edge conditions given in (2.159), we arrive at the following two equations

$$\frac{pa^4}{64D} + Ca^2 + F = 0 \tag{2.165}$$

$$\frac{pa^3}{16D} + 2Ca = 0 \tag{2.166}$$

The solutions of (2.165) and (2.166) are

$$C = -\frac{pa^3}{32D}, \quad F = \frac{pa^4}{64D} \tag{2.167}$$

Finally, the deflection becomes

$$w = \frac{pa^4}{64D}\left(\frac{r^2}{a^2} - 1\right)^2 \tag{2.168}$$

The maximum deflection at the center is

$$w_{\max} = \frac{pa^4}{64D} \tag{2.169}$$

Substitution of (2.1698) into (2.148), (2.149) and (2.151) gives the corresponding non-zero moments and shear force

$$M_r = -D\left[\frac{d^2 w}{dr^2} + \frac{v}{r}\frac{dw}{dr}\right] = -\frac{pa^2}{16}\left[(3+v)\frac{r^2}{a^2} - (1+v)\right] \tag{2.170}$$

$$M_\theta = -D\left[\frac{1}{r}\frac{dw}{dr} + v\frac{d^2 w}{dr^2}\right] = -\frac{pa^2}{16}\left[(1+3v)\frac{r^2}{a^2} - (1+v)\right] \tag{2.171}$$

$$Q_r = -D\frac{\partial}{\partial r}\nabla^2 w = -\frac{pr}{2} \tag{2.172}$$

This completes the analysis for axisymmetric circular plates with clamped edges.

2.5.2 Clamped Plate with Patch Load

If the loading is applied in the form of a circular patch as shown in Figure 2.11, the plate has to be divided into sub-domains. The general solution in regions I and II are

$$w_I = w_0 + w_{hI}, \quad b > r > 0 \tag{2.173}$$

$$w_{II} = w_{hII}, \quad a > r > b \tag{2.174}$$

Each of these solutions has four unknown constants (see (2.158)), or a total of eight unknown constants. There are eight conditions to be satisfied. There are two boundary conditions at the edge for solution II:

$$w_{II}(a) = \frac{dw_{II}}{dr}(a) = 0 \tag{2.175}$$

There are four continuity conditions at the interface between regions I and II shown in Figure 2.11:

$$w_I(b) = w_{II}(b) \tag{2.176}$$

$$\frac{dw_I}{dr}(b) = \frac{dw_{II}}{dr}(b) \tag{2.177}$$

$$M_{rI}(b) = M_{rII}(b) \tag{2.178}$$

$$Q_{rI}(b) = Q_{rII}(b) \tag{2.179}$$

Finally, we also have the boundedness condition for $r \to 0$:

$$w_I(0), M_{rII}(0) \quad \text{are finite} \tag{2.180}$$

This results in eight equations for eight unknown constants. It is very complicated mathematically and we not try to solve the system here. Instead, we will consider the vertical force equilibrium shown in the free body diagram in Figure 2.11

$$2\pi r Q_r + \int_0^r p(r) 2\pi r dr = 0 \tag{2.181}$$

Rearranging (2.181), we find that the shear force is

$$Q_r = -\frac{1}{r}\int_0^r p(r) r dr \tag{2.182}$$

If the loading is constant, we have

$$\begin{aligned} p(r) &= p \quad r \le b \\ &= 0 \quad r > b \end{aligned} \tag{2.183}$$

The shear force becomes

$$\begin{aligned} Q_r &= -\frac{pr}{2}, \quad r \le b \\ &= -\frac{P}{2\pi r}, \quad r > b \end{aligned} \tag{2.184}$$

where

$$P = \int_0^b p 2\pi r dr = p\pi b^2 \tag{2.185}$$

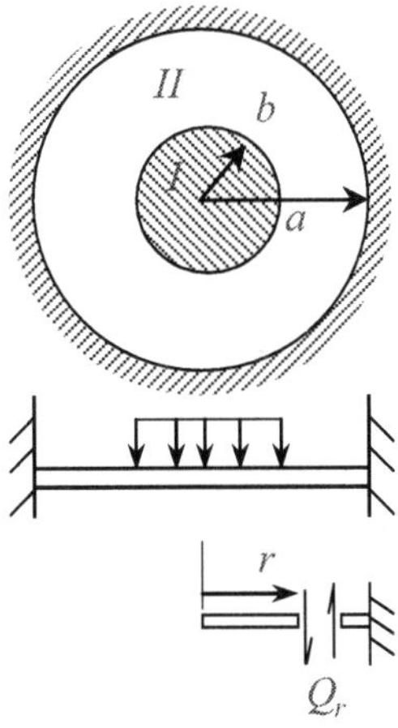

Figure 2.11 Clamped circular plate with central patch load

For $b > r$ (or outside the patch load), we have

$$Q_r = -D\frac{\partial}{\partial r}\nabla^2 w = -\frac{pr}{2} - \frac{4BD}{r} = -\frac{4BD}{r} \tag{2.186}$$

Note that we have $p = 0$ outside the loading patch. Therefore, we have

$$B = \frac{P}{8\pi D} \tag{2.187}$$

This result will be used in the next section for the case of applied central point force.

2.5.3 Plates under Central Point Force

Let us consider the limiting case of a concentrated load from (2.184) as

$$\lim_{b\to 0} Q_r(r) = -\frac{P}{2\pi r}, \quad \lim_{\substack{b\to 0\\ p\to\infty}} \pi b^2 p = P \tag{2.188}$$

The solution given in (2.158) becomes

$$w(r) = \frac{P}{8\pi D} r^2 \ln r + Cr^2 + F \tag{2.189}$$

(i) For simply-supported edges, we have

$$w(a) = M_r(a) = 0 \tag{2.190}$$

Substituting (2.189) into (2.190), we find

$$w(r) = \frac{P}{8\pi D}[r^2 \ln\frac{r}{a} + \frac{3+\nu}{1+\nu}(\frac{a^2 - r^2}{2})] \tag{2.191}$$

The corresponding moments and shear force can be obtained as

$$M_r = \frac{-P}{4\pi}[(1+\nu)\ln\frac{r}{a}] \tag{2.192}$$

$$M_\theta = \frac{-P}{4\pi}[(1+\nu)\ln\frac{r}{a} - (1+\nu)] \tag{2.193}$$

$$Q_r = -\frac{P}{2\pi r} \tag{2.194}$$

(ii) For clamped edges, we have

$$w(a) = \frac{dw}{dr}(a) = 0 \tag{2.195}$$

Substituting (2.189) into (2.195), we find

$$w(r) = \frac{P}{8\pi D}[r^2 \ln\frac{r}{a} + \frac{a^2 - r^2}{2}] \tag{2.196}$$

The corresponding moments and shear force can be obtained as

$$M_r = \frac{-P}{4\pi}[(1+\nu)\ln\frac{r}{a} + 1] \tag{2.197}$$

$$M_\theta = \frac{-P}{4\pi}[(1+\nu)\ln\frac{r}{a} + \nu] \tag{2.198}$$

$$Q_r = -\frac{P}{2\pi r} \tag{2.199}$$

Note the moment and shear forces tend to infinity as $r \to 0$, as expected.

2.6 BUCKLING OF PLATES

In this section, buckling of plates will be considered. Figure 2.12 shows the vertical components induced from the shear forces along the side of a rectangular plate because of the angle of rotation from the non-uniform deflection. Figure 2.13 shows the vertical components of the axial force N_x because of the plate deflection. In particular, the vertical force due to the change of axial force N_x is

$$\begin{aligned} F_{z,x} &= (N_x + \frac{\partial N_x}{\partial x}dx)dy(\frac{\partial w}{\partial x} + \frac{\partial^2 w}{\partial x^2}dx) - N_x dy\frac{\partial w}{\partial x} \\ &= (N_x \frac{\partial^2 w}{\partial x^2} + \frac{\partial N_x}{\partial x}\frac{\partial w}{\partial x})dxdy \end{aligned} \tag{2.200}$$

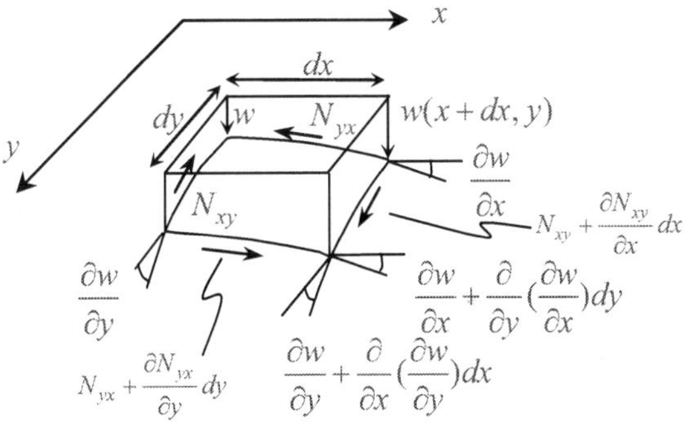

Figure 2.12 Vertical force induced by rotation from non-uniform deflection

Similarly, the vertical force induced by the change of N_y is

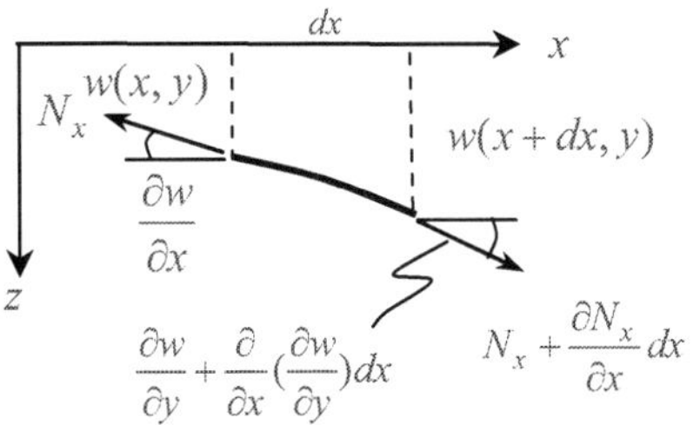

Figure 2.13 Vertical force induced by axial force and plate deflection

$$F_{z,y} = (N_y \frac{\partial^2 w}{\partial y^2} + \frac{\partial N_y}{\partial y}\frac{\partial w}{\partial y})dxdy \tag{2.201}$$

By referring to Figure 2.12, the vertical force due to changes in shear force N_{yx} is

$$\begin{aligned} F_{z,yx} &= (N_{yx} + \frac{\partial N_{yx}}{\partial y}dy)dx(\frac{\partial w}{\partial x} + \frac{\partial^2 w}{\partial x \partial y}dy) - N_{yx}dx\frac{\partial w}{\partial x} \\ &= (N_{yx}\frac{\partial^2 w}{\partial x \partial y} + \frac{\partial N_{yx}}{\partial y}\frac{\partial w}{\partial x})dxdy \end{aligned} \tag{2.202}$$

Similarly, the vertical force induced by N_{xy} is

$$F_{z,xy} = (N_{xy}\frac{\partial^2 w}{\partial x \partial y} + \frac{\partial N_{xy}}{\partial x}\frac{\partial w}{\partial y})dxdy \tag{2.203}$$

Therefore, the total vertical force induced by the axial and shear membrane forces of the plate is

$$\begin{aligned} F_z &= F_{z,x} + F_{z,y} + F_{z,xy} + F_{z,yx} \\ &= (N_x \frac{\partial^2 w}{\partial x^2} + \frac{\partial N_x}{\partial x}\frac{\partial w}{\partial x} + N_y\frac{\partial^2 w}{\partial y^2} + \frac{\partial N_y}{\partial y}\frac{\partial w}{\partial y} + N_{yx}\frac{\partial^2 w}{\partial x \partial y} \\ &\quad + \frac{\partial N_{yx}}{\partial y}\frac{\partial w}{\partial x} + N_{xy}\frac{\partial^2 w}{\partial x \partial y} + \frac{\partial N_{xy}}{\partial x}\frac{\partial w}{\partial y})dxdy \end{aligned} \tag{2.204}$$

Grouping terms, we obtain

$$\begin{aligned} F_z &= [N_x\frac{\partial^2 w}{\partial x^2} + 2N_{xy}\frac{\partial^2 w}{\partial x \partial y} + N_y\frac{\partial^2 w}{\partial y^2} + (\frac{\partial N_x}{\partial x} + \frac{\partial N_{yx}}{\partial y})\frac{\partial w}{\partial x} + (\frac{\partial N_y}{\partial y} + \frac{\partial N_{xy}}{\partial x})\frac{\partial w}{\partial y}]dxdy \\ &= [N_x\frac{\partial^2 w}{\partial x^2} + 2N_{xy}\frac{\partial^2 w}{\partial x \partial y} + N_y\frac{\partial^2 w}{\partial y^2}]dxdy \end{aligned} \tag{2.205}$$

The last two terms on the first line of (2.205) are zeros due to in-plane equilibrium. Adding this additional vertical force to the normal force equilibrium, we get

$$D\nabla^4 w - N_x\frac{\partial^2 w}{\partial x^2} - 2N_{xy}\frac{\partial^2 w}{\partial x \partial y} - N_y\frac{\partial^2 w}{\partial y^2} = p(x,y) \tag{2.206}$$

This is the well-known von Karman equation for plate buckling. The cause of buckling comes from the membrane forces, which can be separated from bending moments. Within the framework of plane stress, they can be expressed in terms of the Airy stress function as:

$$N_x = \frac{\partial^2 \Phi}{\partial y^2}, \quad N_y = \frac{\partial^2 \Phi}{\partial x^2}, \quad N_{xy} = -\frac{\partial^2 \Phi}{\partial x \partial y} \tag{2.207}$$

It can be shown that the Airy stress function satisfies the biharmonic equation (Chau, 2013)

$$\nabla^4 \Phi = 0 \tag{2.208}$$

Consider the special case of axial compression $N_x = -p$ as shown in Figure 2.14 without normal applied load. Thus, we have

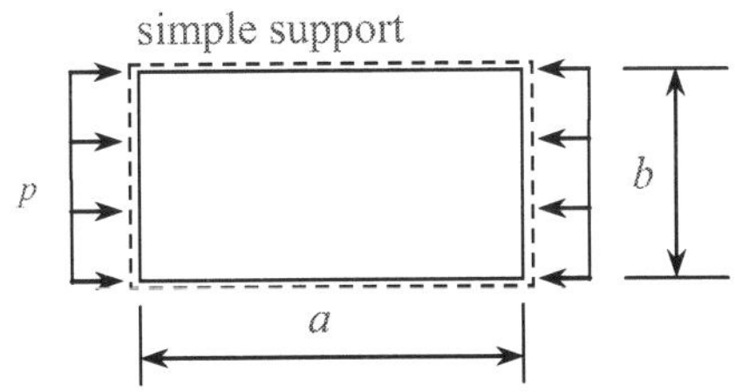

Figure 2.14 Buckling of a simply-supported rectangular plate subject to compression *p*

$$D\nabla^4 w + p\frac{\partial^2 w}{\partial x^2} = 0 \tag{2.209}$$

The boundary conditions for the simply-supported edges are

$$w = 0, \quad M_n = 0 \tag{2.210}$$

The trivial or unbuckled state is obviously given by $w = 0$. To satisfy the boundary conditions, we can seek a nontrivial solution of the following form:

$$w_{ik}(x, y) = C_{ik} \sin\frac{k\pi x}{a} \sin\frac{i\pi y}{b} \tag{2.211}$$

Substitution of (2.211) into (2.209) yields

$$C_{ik} \sin\frac{k\pi x}{a} \sin\frac{i\pi x}{a}[D\pi^4(\frac{k^2}{a^2} + \frac{i^2}{b^2})^2 - p\pi^2\frac{k^2}{a^2}] = 0 \tag{2.212}$$

Thus, the buckling load can be written as

$$p_{crit} = \frac{D\pi^2 a^2}{k^2}(\frac{k^2}{a^2} + \frac{i^2}{b^2})^2 = \frac{D\pi^2}{b^2}(\frac{kb}{a} + \frac{i^2 a}{kb})^2 = \frac{\pi^2 D}{b^2}\kappa \tag{2.213}$$

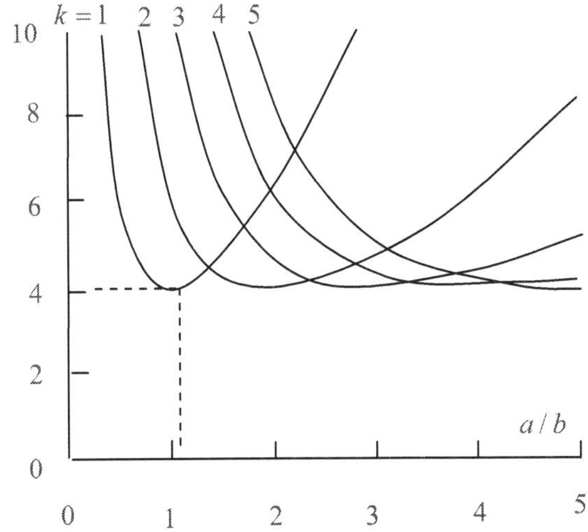

Figure 2.15 Buckling load factor *κ* versus *a/b* for rectangular plates of length *a* and width *b*

Figure 2.15 plots the variations of the factor κ versus a/b for $i = 1$ and $k = 1,2,3,4,5$. It is clear that the minimum value of κ occurs at $a = b$ and also some higher values of a/b. Therefore, for $a/b = 1$, the minimum buckling load is

$$(p_{crit})_{\min} = \frac{4\pi^2 D}{b^2} = \frac{Eh^3\pi^2}{3(1-\nu^2)b^2} \tag{2.214}$$

For the case of $i = 1$ and $k = 2$, the deflection pattern of the buckling mode is shown in Figure 2.16.

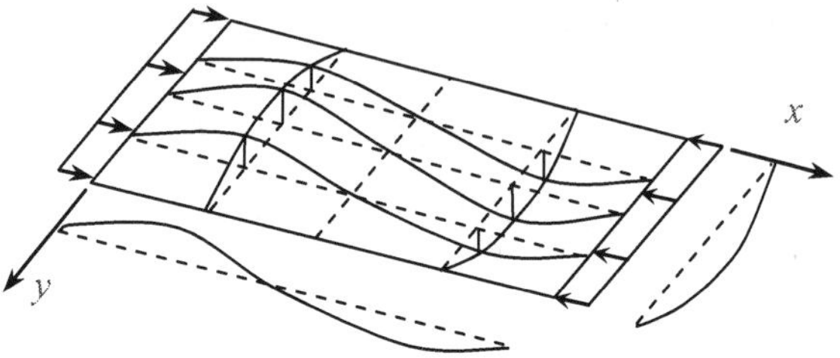

Figure 2.16 Buckling mode for $i = 1$ and $k = 2$

This result for buckling load can also be used to design stiffener in preventing lower modes of buckling to occur. This idea is illustrated in Figure 2.17. Consider the case of

$$N_x = N, \quad N_y = N_{xy} = 0 \tag{2.215}$$

Substitution of (2.215) into (2.206) gives

$$D\nabla^4 w = p(x,y) + N\frac{\partial^2 w}{\partial x^2} \tag{2.216}$$

We first expand the loading $p(x,y)$ in double Fourier series expansion as

$$p(x,y) = \sum_{i=1}^{\infty}\sum_{k=1}^{\infty} a_{ik} \sin\frac{k\pi x}{a}\sin\frac{i\pi y}{b} \tag{2.217}$$

where

$$a_{ik} = \frac{4}{ab}\int_0^a\int_0^b p(x,y)\sin\frac{k\pi x}{a}\sin\frac{i\pi y}{b}\,dxdy \tag{2.218}$$

The plate deflection w is also expanded in double Fourier series

$$w(x,y) = \sum_{i=1}^{\infty}\sum_{k=1}^{\infty} C_{ik} \sin\frac{k\pi x}{a}\sin\frac{i\pi y}{b} \tag{2.219}$$

Substitution of (2.217) and (2.219) into (2.216) gives

$$C_{ik} = \frac{a_{ik}}{D\pi^4[(\frac{k^2}{a^2}+\frac{i^2}{b^2})^2 + \frac{Nk^2}{\pi^2 Da^2}]} \tag{2.220}$$

Therefore, we can see that the magnitude of plate deflection decreases with a tension pre-tensioning N (as demonstrated in the right diagram in Figure 2.17). If

$N < 0$ or is compressive, we have a softening effect as C_{ik} in (2.220) increases and, consequently, the deflection w in (2.219) also increases.

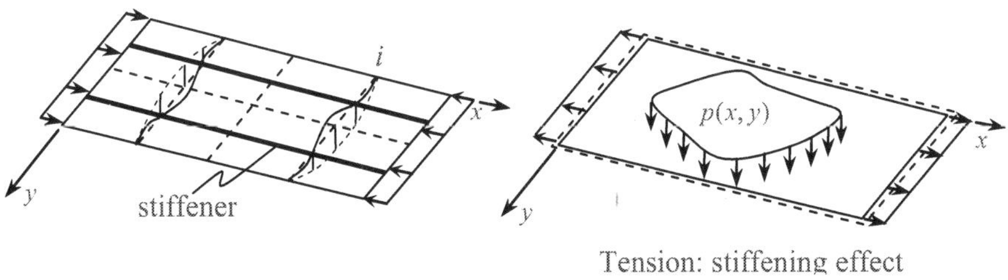

Figure 2.17 Stiffener in the plate for increasing the mode number and controlling deflection

2.7 BENDING OF ANISOTROPIC PLATES

For real plate applications, plates are often made of RC slabs, fiberglass panels, plywood slabs, and composite sandwich slabs. These construction materials are anisotropic. For orthotropic materials, the plate bending theory has to be revised as

$$D_x \frac{\partial^4 w}{\partial x^4} + 2H \frac{\partial^4 w}{\partial x^2 \partial y^2} + D_y \frac{\partial^4 w}{\partial y^4} = p(x, y) \tag{2.221}$$

where

$$D_x = \frac{E_x h^3}{12}, \quad D_y = \frac{E_y h^3}{12}, \quad D_1 = \frac{\bar{E} h^3}{12}, \quad H = D_1 + 2D_{xy}, \quad D_{xy} = \frac{Gh^3}{12} \tag{2.222}$$

There are four independent elastic constants. The proof of (2.221) is sketched in Problem 2.2. For the isotropic case, we have

$$D_x = D_x = H = D = \frac{Eh^3}{12(1-\nu^2)} \tag{2.223}$$

This is equivalent to having the following special cases

$$E_x = E_x = \frac{E}{1-\nu^2}, \quad \bar{E} = \frac{\nu E}{1-\nu^2}, \quad G = \frac{E}{2(1+\nu)} \tag{2.224}$$

For reinforced concrete slabs with two-way reinforcement, the following form of rigidities are assumed by Huber in 1914 (see Article 86 of Timoshenko and Woinowsky-Krieger, 1959):

$$\begin{gathered} D_x = \frac{E_c}{1-\nu_c^2}[I_{cx} + (n-1)I_{sx}], \quad D_y = \frac{E_c}{1-\nu_c^2}[I_{cy} + (n-1)I_{sy}], \\ D_1 = \nu_c \sqrt{D_x D_y}, \quad D_{xy} = \frac{1-\nu_c}{2}\sqrt{D_x D_y} \end{gathered} \tag{2.225}$$

where ν_c is the Poisson ratio of concrete (≈ 0.16), $n = E_s/E_c$ or the ratio of Young's modulus of steel to concrete, I_{cx} and I_{sx} are the moment of inertia of concrete and steel sections with respect to the x-axis. With these assumptions, the rigidity H can be simplified to

$$H = D_1 + 2D_{xy} = \sqrt{D_x D_y} \tag{2.226}$$

With (2.226), we have (2.221) simplified to

$$D_x \frac{\partial^4 w}{\partial x^4} + 2\sqrt{D_x D_y} \frac{\partial^4 w}{\partial x^2 \partial y^2} + D_y \frac{\partial^4 w}{\partial y^4} = p(x, y) \tag{2.227}$$

To further simplify (2.227), we introduce the following change of variables

$$\xi = y(D_x / D_y)^{1/4} \tag{2.228}$$

Thus, the PDE becomes

$$D_x \overline{\nabla}^4 w = p \tag{2.229}$$

where

$$\overline{\nabla}^4 = \frac{\partial^4}{\partial x^4} + 2\frac{\partial^4}{\partial x^2 \partial \xi^2} + \frac{\partial^4}{\partial \xi^4} \tag{2.230}$$

Mathematically, this is the biharmonic equation in the transformed space. All mathematical techniques discussed so far apply to these anisotropic plates by identifying

$$D \leftarrow D_x, \quad y \leftarrow \xi \tag{2.231}$$

The solution is then $w(x, \xi)$. However, there is no free lunch. For example, a circular orthotropic plate is mathematically equivalent to solving an elliptic isotropic plate, which is mathematically more difficult to solve than solving circular plates.

For the case of simply-supported anisotropic rectangular plates, Navier's solution can be used to solve the problem by expanding the lateral loads in series expansion:

$$p(x, \xi) = \sum_{i=1}^{\infty}\sum_{k=1}^{\infty} a_{ik} \sin\frac{k\pi x}{a} \sin\frac{i\pi\xi}{\overline{b}} \tag{2.232}$$

where

$$a_{ik} = \frac{4}{a\overline{b}} \int_0^a \int_0^{\overline{b}} p(x, \xi) \sin\frac{k\pi x}{a} \sin\frac{i\pi\xi}{\overline{b}} \, dx d\xi \tag{2.233}$$

$$\overline{b} = b(D_x / D_y)^{/4} \tag{2.234}$$

The deflection w is also expressed as a Fourier series

$$w(x, \xi) = \sum_{i=1}^{\infty}\sum_{k=1}^{\infty} C_{ik} \sin\frac{k\pi x}{a} \sin\frac{i\pi\xi}{\overline{b}} \tag{2.235}$$

Substitution of (2.232) and (2.234) into (2.229) gives

$$C_{ik} = \frac{a_{ik}}{\pi^4 [\frac{k^2}{a^2} D_x + 2\frac{k^2 i^2}{a^2 \overline{b}^2} \sqrt{D_x D_y} + \frac{i^4}{\overline{b}^4} D_y]} \tag{2.236}$$

2.8 PLATE ON ELASTIC FOUNDATION

A plate on an elastic foundation is normally modeled by the so-called Winkler foundation, which was proposed in 1867, in studying soil-structure interaction, as shown in Figure 2.18. The total loading can be expressed as the difference between the vertical load and the foundation reaction, or normally referred to as a sub-grade reaction in the context of soil mechanics

$$\nabla^4 w = \frac{p_t}{D} = \frac{p - kw}{D} \tag{2.237}$$

where k is the foundation modulus. This can be rearranged as

$$D\nabla^4 w + kw = p(x, y) \tag{2.238}$$

More discussions of plates on an elastic foundation can be found in Vlasov and Leontev (1966) and Ventsel and Krauthammer (2001). This topic is out of the scope of the present chapter. However, a problem mathematically related to plates on an elastic foundation is called the Hertz problem, and will be discussed in detail in Section 2.11.

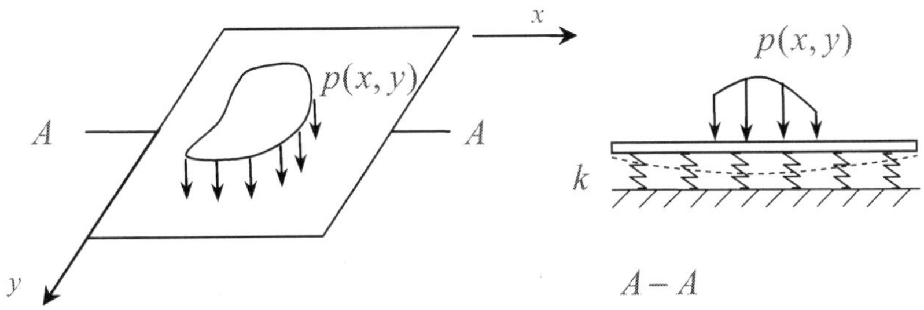

Figure 2.18 Plate on an elastic foundation

2.9 PLATE VIBRATIONS

For the case of plate vibrations, the effect of inertia force can be applied like the case of plates on an elastic foundation:

$$\nabla^4 w = \frac{p_t}{D} = \frac{p - m\ddot{w}}{D} \tag{2.239}$$

Thus, the governing equation of a vibrating plate can be expressed as

$$D\nabla^4 w + m\frac{\partial^2 w}{\partial t^2} = p(x, y, t) \tag{2.240}$$

where m is the mass of the plate.

2.9.1 Free Vibrations

For free vibrations, we have $p = 0$ and (2.240) becomes

$$D\nabla^4 w + m\frac{\partial^2 w}{\partial t^2} = 0 \tag{2.241}$$

We seek the following vibration modes

$$w(x,y,t)=W(x,y)\cos\kappa t \tag{2.242}$$

Substitution of (2.242) into (2.241) gives

$$D\nabla^4 W - m\kappa^2 W = 0 \tag{2.243}$$

For simply-supported edges, we can assume a solution form for W as

$$W_{ik}(x,y)=C_{ik}\sin\frac{k\pi x}{a}\sin\frac{i\pi y}{b} \tag{2.244}$$

Substitution of (2.244) into (2.243) yields

$$D\pi^4\left(\frac{k^2}{a^2}+\frac{i^2}{b^2}\right)^2 - m\kappa^2 = 0 \tag{2.245}$$

The eigenfrequency is then

$$\kappa_{ik}=\pi^2\left(\frac{k^2}{a^2}+\frac{i^2}{b^2}\right)\sqrt{\frac{D}{m}} \tag{2.246}$$

The fundamental frequency for $k = 1$ and $i = 1$ is

$$\kappa_{\min}=\pi^2\left(\frac{1}{a^2}+\frac{1}{b^2}\right)\sqrt{\frac{D}{m}} \tag{2.247}$$

2.9.2 Forced Vibrations

We now consider a forced vibration with zero initial conditions:

$$w(x,y,0)=w_0(x,y)=0, \quad \frac{\partial w}{\partial t}(x,y,0)=v_0(x,y)=0 \tag{2.248}$$

The forcing term p is assumed to be expressible as

$$p(x,y,t)=\sum_{m=1}^{\infty}\sum_{n=1}^{\infty} f_{mn}(t)W_{mn}(x,y) \tag{2.249}$$

where W_{mn} is the mode shape, which is a function of the boundary conditions. The unknown deflection w is also expanded as

$$w=\sum_{m=1}^{\infty}\sum_{n=1}^{\infty} F_{mn}(t)W_{mn}(x,y) \tag{2.250}$$

Substitution of (2.250) and (2.249) into (2.240) and in view of (2.243) results in

$$\ddot{F}_{mn}+\kappa_{mn}^2 F_{mn}=\frac{1}{\rho h}f_{mn} \tag{2.251}$$

where m in (2.243) has been set to ρh in order not to confuse with the mode number. The general solution is the sum of the homogeneous and particular solutions

$$F_{mn}(t)=A_{mn}\cos\kappa_{mn}t+B_{mn}\sin\kappa_{mn}t+F_{mn}^{p}(t) \tag{2.252}$$

For the case of simply-supported rectangular plates under harmonic loading, the loading can be expanded as

$$p(x,y,t) = p_0(x,y)\cos\Omega t = \cos\Omega t \sum_{m=1}^{\infty}\sum_{n=1}^{\infty} C_{mn} \sin\alpha_m x \sin\beta_n y \tag{2.253}$$

where

$$C_{mn} = \frac{4}{ab}\int_0^a\int_0^b p_0(x,y)\sin\alpha_m x \sin\beta_n y\, dx dy \tag{2.254}$$

$$\alpha_m = \frac{m\pi}{a}, \quad \beta_n = \frac{n\pi}{b} \tag{2.255}$$

Let the particular solution be

$$F_{mn}^p(t) = D_{mn}\cos\Omega t \tag{2.256}$$

Substitution of (2.256) into (2.151) yields

$$(\kappa_{mn}^2 - \Omega^2) D_{mn}\cos\Omega t = \frac{1}{\rho h}\cos\Omega t \tag{2.257}$$

The solution for D_{mn} can be obtained from (2.257) as

$$D_{mn} = \frac{1}{\rho h(\kappa_{mn}^2 - \Omega^2)} \tag{2.258}$$

The particular solution assumed in (2.256) becomes

$$F_{mn}^p(t) = \frac{\cos\Omega t}{\rho h(\kappa_{mn}^2 - \Omega^2)} \tag{2.259}$$

Applying the initial conditions given in (2.248) to (2.252) gives

$$A_{mn} = -\frac{1}{\rho h(\kappa_{mn}^2 - \Omega^2)}, \quad B_{mn} = 0 \tag{2.260}$$

Finally, the deflection is obtained as

$$w = \frac{1}{\rho h}\sum_{m=1}^{\infty}\sum_{n=1}^{\infty}\frac{C_{mn}}{(\kappa_{mn}^2 - \Omega^2)}(\cos\Omega t - \cos\kappa_{mn} t)\sin\alpha_m x \sin\beta_n y \tag{2.261}$$

For the special case of the applied frequency approaching the natural frequency or

$$\Omega \to \kappa_{mn}, \tag{2.262}$$

we can see that (2.261) is of the indeterminate form of 0/0. Applying L'Hôpital's rule, we have

$$w = \frac{1}{\rho h}\sum_{m=1}^{\infty}\sum_{n=1}^{\infty}\frac{C_{mn}}{2\kappa_{mn}} t \sin\kappa_{mn} t \sin\alpha_m x \sin\beta_n y \tag{2.263}$$

It is obvious that the solution increases to infinity as time approaches infinity. Since there is no damping incorporated into the analysis, the forcing term applied with the natural frequency will induce the so-called resonance.

For the special case of constant applied load of p_0 being a constant, we have from (2.254)

$$C_{mn} = \frac{16 p_0}{mn\pi^2}, \quad m,n = 1,3,5,... \tag{2.264}$$

The vibrations of the rectangular plates are

$$w = \frac{16 p_0}{\rho h \pi^2} \sum_{m=1,3,\ldots}^{\infty} \sum_{n=1,3,\ldots}^{\infty} \frac{\cos \Omega t - \cos \kappa_{mn} t}{mn(\kappa_{mn}^2 - \Omega^2)} \sin \alpha_m x \sin \beta_n y \tag{2.265}$$

The first three mode shapes of the vibrations are illustrated in Figure 2.19.

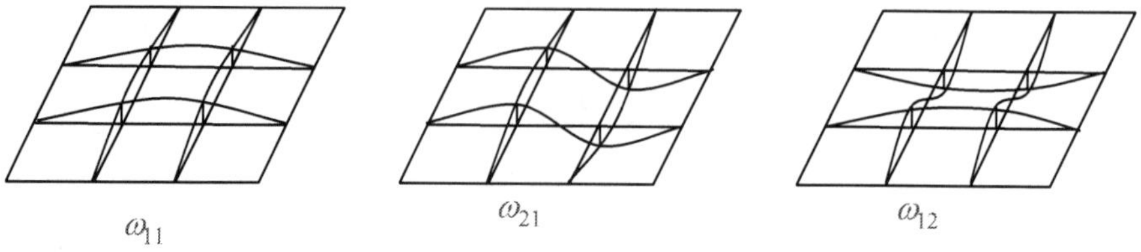

Figure 2.19 First three mode shapes of plate vibrations

2.9.3 Approximation by Rayleigh Quotient

Rayleigh's principle can be expressed as a form of conservation of energy

$$U_{\max} = K_{\max}, \tag{2.266}$$

that is, the maximum potential energy equals the maximum kinetic energy. The kinetic energy can be determined as

$$K = \frac{1}{2} \iint_A \rho h [\frac{\partial w(x, y, t)}{\partial t}]^2 dxdy \tag{2.267}$$

For harmonic motions discussed in the last section, we have

$$w(x, y, t) = W(x, y) \sin \omega t \tag{2.268}$$

Thus, we have

$$K = \frac{\omega^2}{2} \cos^2 \omega t \iint_A \rho h W^2(x, y) dxdy \tag{2.269}$$

Therefore, the maximum kinetic energy is

$$K_{\max} = \frac{\omega^2}{2} \iint_A \rho h W^2(x, y) dxdy \tag{2.270}$$

Substitution of (2.270) into (2.266) yields

$$\omega^2 = \frac{2K_{\max}}{\iint_A \rho h W^2(x, y) dxdy} = \frac{2U_{\max}}{\iint_A \rho h W^2(x, y) dxdy} \tag{2.271}$$

where (see (14.178) of Chau, 2018)

$$U_{\max} = \frac{1}{2} \iint_A D\{(\nabla^2 W)^2 + 2(1-\nu)[(\frac{\partial^2 W}{\partial x \partial y})^2 - \frac{\partial^2 W}{\partial x^2} \frac{\partial^2 W}{\partial y^2}]\} dxdy \tag{2.272}$$

The proof of (2.272) will be given in the next section. The right-hand side of (2.271) is known as Rayleigh's quotient, which was proposed by Rayleigh in

1877. To show the accuracy of (2.271), we consider its application in Example 2.1 below.

2.9.4 Strain Energy of Plates

The strain energy can be evaluated as (Chau, 2013)

$$\begin{aligned} U &= \frac{1}{2}\iint_A \sigma : \varepsilon dV \\ &= \frac{1}{2}\iint_A [\frac{1}{2E}(\sigma_x^2 + \sigma_x^2 - 2\nu\sigma_x\sigma_y) + \frac{1+\nu}{E}\tau_{xy}^2]dV \end{aligned} \tag{2.273}$$

Substitution of (2.23) to (2.29) into (2.273) gives

$$U = \frac{1}{2}\iint_A D[(\frac{\partial^2 w}{\partial x^2})^2 + (\frac{\partial^2 w}{\partial y^2})^2 + 2\nu\frac{\partial^2 w}{\partial x^2}\frac{\partial^2 w}{\partial y^2} + 2(1-\nu)(\frac{\partial^2 w}{\partial x\partial y})^2]dA \tag{2.274}$$

Rearranging (2.274), we have

$$U = \frac{1}{2}\iint_A D\{(\frac{\partial^2 w}{\partial x^2} + \frac{\partial^2 w}{\partial y^2})^2 - 2(1-\nu)[\frac{\partial^2 w}{\partial x^2}\frac{\partial^2 w}{\partial y^2} - (\frac{\partial^2 w}{\partial x\partial y})^2]\}dA \tag{2.275}$$

In view of (2.268), we arrive at the maximum strain energy as (2.272).

2.9.5 Rayleigh-Ritz Method

In the Rayleigh-Ritz method, we assume an admissible function as (see Section 14.8 of Chau, 2018)

$$W(x,y) = \sum_{i=1}^{n} c_i W_i(x,y) \tag{2.276}$$

Note that W_i needs to satisfy the essential boundary condition or the deflection boundary condition on the edges for the present case of plate bending. The unknown constants need to be determined from the minimum total energy principle or

$$\frac{\partial(U_{\max} - K_{\max})}{\partial c_i} = 0, \quad i = 1, 2, \ldots, n \tag{2.277}$$

This provides a system of n equations for n unknowns.

Example 2.1 Find the first fundamental vibration frequency of a clamped plate shown in Figure 2.20 using the following trial function by using Rayleigh's quotient:

$$W(x,y) = c(x^2 - a^2)^2(y^2 - b^2)^2 \tag{2.278}$$

where c is an unknown constant.

Solution: It is straightforward to see that

$$W = \frac{\partial W}{\partial x} = 0, \quad \text{on } x = \pm a \tag{2.279}$$

$$W = \frac{\partial W}{\partial y} = 0, \quad \text{on } y = \pm b \tag{2.280}$$

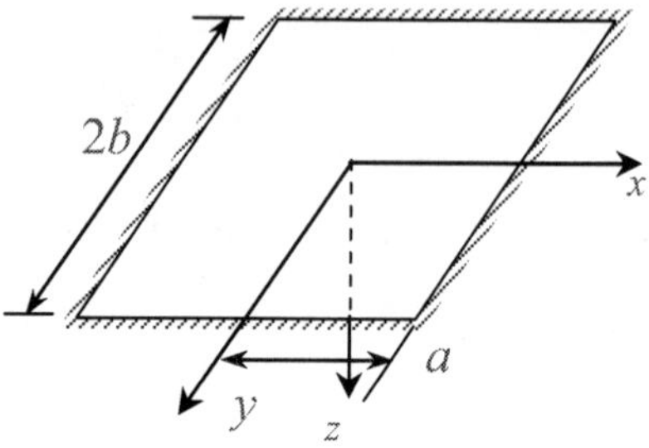

Figure 2.20 Free vibrations of a clamped rectangular plate

Differentiating (2.278), we find

$$\frac{\partial W}{\partial x} = 4cx(x^2 - a^2)(y^2 - b^2)^2 \tag{2.281}$$

$$\frac{\partial W}{\partial y} = 4cy(x^2 - a^2)^2(y^2 - b^2) \tag{2.282}$$

$$\frac{\partial^2 W}{\partial y \partial x} = 16cxy(x^2 - a^2)(y^2 - b^2) \tag{2.283}$$

$$\frac{\partial^2 W}{\partial x^2} = 4c(3x^2 - a^2)(y^2 - b^2)^2 \tag{2.284}$$

$$\frac{\partial^2 W}{\partial y^2} = 4c(3y^2 - b^2)(x^2 - a^2)^2 \tag{2.285}$$

Adding (2.284) and (2.285) and squaring the result, we get

$$\begin{aligned}(\nabla^2 W)^2 = 16c^2[&(x^2 - a^2)^2(y^2 - b^2)^2 + 2x^2(y^2 - b^2)^2 \\ &+ 2y^2(x^2 - a^2)^2 + (y^2 - b^2)(x^2 - a^2)^2]^2\end{aligned} \tag{2.286}$$

Squaring (2.278), we find

$$W^2 = c^2(x^2 - a^2)^4(y^2 - b^2)^4 \tag{2.287}$$

Multiplying (2.284) and (2.285), we obtain

$$\frac{\partial^2 W}{\partial x^2}\frac{\partial^2 W}{\partial y^2} = 16c^2(3x^2 - a^2)(3y^2 - b^2)(y^2 - b^2)^2(x^2 - a^2)^2 \tag{2.288}$$

Integration of these functions shows:

$$\int_0^a \int_0^b \frac{\partial^2 W}{\partial x^2}\frac{\partial^2 W}{\partial y^2} dxdy = \frac{16384 a^7 b^7 c^2}{11025} \tag{2.289}$$

$$\int_0^a \int_0^b (\nabla^2 W)^2 dxdy = \frac{16(3584a^5b^9 + 2048a^7b^7 + 3584a^9b^5)c^2}{11025} \quad (2.290)$$

$$\int_0^a \int_0^b (\frac{\partial^2 W}{\partial y \partial x})^2 dxdy = \frac{16384a^7b^7c^2}{11025} \quad (2.291)$$

$$\int_0^a \int_0^b W^2 dxdy = \frac{1}{9}(\frac{16384a^9b^9c^2}{11025}) \quad (2.292)$$

The maximum potential energy becomes

$$U_{\max} = \frac{8D}{11025}(3584a^5b^9 + 2048a^7b^7 + 3584a^9b^5) \quad (2.293)$$

Substitution of (2.289) to (2.293) into (2.272) and the result into (2.271) gives

$$\begin{aligned}\omega_{11} &= \{\frac{72}{8192}(\frac{3584}{a^4} + \frac{2048}{a^2b^2} + \frac{3584}{b^4})\}^{1/2} \sqrt{\frac{D}{\rho h}} \\ &= 3\{\frac{7}{2}(\frac{1}{a^4} + \frac{4}{7}\frac{1}{a^2b^2} + \frac{1}{b^4})\}^{1/2} \sqrt{\frac{D}{\rho h}}\end{aligned} \quad (2.294)$$

In the literature, the width of the plate is denoted as a instead of $2a$ as shown in Figure 2.20, and for such terminology we can modify (2.294) as

$$\omega_{11} = 12\{\frac{7}{2}(\frac{1}{a^4} + \frac{4}{7}\frac{1}{a^2b^2} + \frac{1}{b^4})\}^{1/2} \sqrt{\frac{D}{\rho h}} \quad (2.295)$$

This formula was first obtained by Galin in 1947. For the special case of a square plate with size $a \times a$, we have the fundamental vibration being:

$$\omega_{11} = \frac{36}{a^2} \sqrt{\frac{D}{\rho h}} \quad (2.296)$$

This is within 1% error of the value compiled in Table 4.22 of Leissa (1969) of 35.9866. This result will be further compared to various formulas:

Warburton 1954

$$\omega_{11} = \frac{36.1327}{a^2} \sqrt{\frac{D}{\rho h}} \quad (2.297)$$

Jenich in 1962

$$\omega_{11} = \frac{37.22}{a^2} \sqrt{\frac{D}{\rho h}} \quad (2.298)$$

Tomotika 1936

$$35.984 < \omega_{11} a^2 \sqrt{\frac{\rho h}{D}} < 36.09 \quad (2.299)$$

Young 1950

$$\omega_{11} = \frac{35.99}{a^2} \sqrt{\frac{D}{\rho h}} \quad (2.300)$$

Bolotin 1961

$$\omega_{11} = \frac{35.09193}{a^2}\sqrt{\frac{D}{\rho h}} \tag{2.301}$$

This demonstrates that Rayleigh's quotient provides an accurate result compared to the results from other methods.

2.10 VIBRATIONS OF CIRCULAR PLATES

We now consider axisymmetric vibrations of a clamped circular plate subject to arbitrary initial conditions, shown in Figure 2.21. This problem was considered by Reid (1962). The governing equation in cylindrical coordinates can be expressed as

$$\frac{1}{r}\frac{\partial}{\partial r}\{r\frac{\partial}{\partial r}[\frac{1}{r}\frac{\partial}{\partial r}(r\frac{\partial u}{\partial r})]\} + \frac{1}{b^4}\frac{\partial^2 u}{\partial t^2} = 0 \tag{2.302}$$

where

$$D = \frac{Eh^3}{12(1-\nu^2)}, \quad b^2 = \sqrt{\frac{D}{\rho h}} \tag{2.303}$$

The initial conditions are given as

$$\left.\frac{\partial u}{\partial t}\right|_{t=0} = g(r), \quad u\big|_{t=0} = f(r) \tag{2.304}$$

The edge boundary conditions for a clamped plate at $r = a$ are

$$\left.\frac{\partial u}{\partial r}\right|_{r=a} = 0, \quad u\big|_{r=a} = 0 \tag{2.305}$$

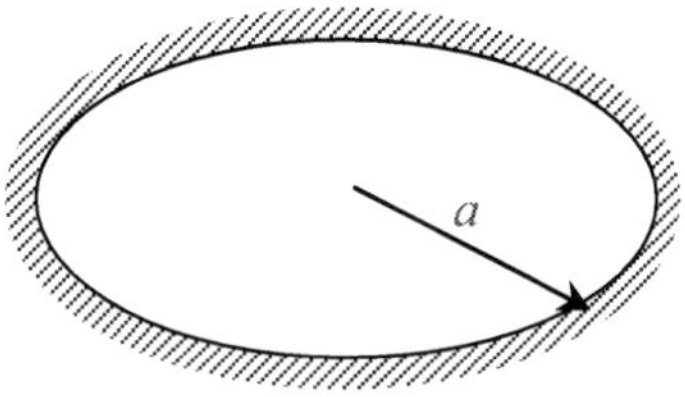

Figure 2.21 Vibrations of circular plates

Consider the standard separation of variables

$$u(r,t) = R(r)T(t) \tag{2.306}$$

Substitution of (2.306) into (2.302) gives

$$\frac{1}{r}\frac{d}{dr}\{r\frac{d}{dr}[\frac{1}{r}\frac{d}{dr}(r\frac{dR}{dr})]\}T + \frac{1}{b^4}\frac{d^2T}{dt^2}R = 0 \tag{2.307}$$

Rearrangement of (2.307) yields

$$\frac{1}{r}\frac{d}{dr}\{r\frac{d}{dr}[\frac{1}{r}\frac{d}{dr}(r\frac{dR}{dr})]\}\frac{1}{R}=-\frac{1}{b^4}\frac{T''}{T}=\lambda \tag{2.308}$$

where λ is a constant to be determined. For axisymmetric vibrations, we have the Laplacian being

$$\frac{1}{r}\frac{d}{dr}(r\frac{dR}{dr})=\frac{1}{r}\frac{dR}{dr}+\frac{d^2R}{dr^2}=\nabla^2 R \tag{2.309}$$

Using (2.309) and (2.308), we obtain the following governing equation for R

$$\nabla^2\nabla^2 R-\lambda R=0 \tag{2.310}$$

This can be factorized as (see 3.3.5 of Chau, 2018)

$$(\nabla^2-\sqrt{\lambda})(\nabla^2+\sqrt{\lambda})R=(\nabla^2+\sqrt{\lambda})(\nabla^2-\sqrt{\lambda})R=0 \tag{2.311}$$

Thus, the solution of (2.310) is equivalent to the solution of the following system

$$(\nabla^2+\sqrt{\lambda})R=0, \quad (\nabla^2-\sqrt{\lambda})R=0 \tag{2.312}$$

From the last part of (2.308), the governing equation for T is

$$T''+\lambda b^4 T=0 \tag{2.313}$$

The solution of T is, of course, sine and cosine

$$T=A\sin(\sqrt{\lambda}b^2 t)+B\cos(\sqrt{\lambda}b^2 t) \tag{2.314}$$

Let us consider the first of (2.312)

$$(\nabla^2+\sqrt{\lambda})R=0 \tag{2.315}$$

In explicit form, (2.315) can be written as

$$\frac{1}{r}\frac{dR}{dr}+\frac{d^2R}{dr^2}+\sqrt{\lambda}R=0 \tag{2.316}$$

Referring to Chapter 4 of Chau (2018), this is a Bessel equation of zero order and the solution is

$$R=AJ_0(\sqrt[4]{\lambda}r)+BY_0(\sqrt[4]{\lambda}r) \tag{2.317}$$

In view of Figure 4.6 of Chau (2018), the Bessel function of the second kind is unbounded at $r=0$. Thus, the boundedness condition requires

$$r\to 0, \quad u=finite, \quad B=0 \tag{2.318}$$

The second of (2.312) is

$$(\nabla^2-\sqrt{\lambda})R=0 \tag{2.319}$$

In explicit form, (2.319) can be expressed as

$$\frac{1}{r}\frac{dR}{dr}+\frac{d^2R}{dr^2}-\sqrt{\lambda}R=0 \tag{2.320}$$

Recalling the differential equation for modified Bessel function (see (4.331) of Chau, 2018), we have

$$x^2y''+xy'-(x^2+n^2)y=0 \tag{2.321}$$

The solution of (2.321) is

$$y=AI_n(x)+BK_n(x) \tag{2.322}$$

Evidently, (2.320) and (2.321) are mathematically the same, and thus the solution for R is

$$R=AI_0(\sqrt[4]{\lambda}r)+BK_0(\sqrt[4]{\lambda}r) \tag{2.323}$$

The modified Bessel function of the first kind of zero order is

$$I_0(x) = J_0(ix) = 1 + \frac{x^2}{2^2} + \frac{x^4}{2^2 \cdot 4^2} + \frac{x^6}{2^2 \cdot 4^2 \cdot 6^2} + \dots \tag{2.324}$$

The modified Bessel function of the second kind of zero order is

$$\begin{aligned} K_0(x) &= \frac{\pi}{2}\{iJ_0(ix) - Y_0(ix)\} = -[\ln(\frac{x}{2}) + \gamma]I_0(x) \\ &+ \frac{2}{\pi}[\frac{x^2}{2^2} + \frac{x^4}{2^2 \cdot 4^2}(1+\frac{1}{2}) + \frac{x^6}{2^2 \cdot 4^2 \cdot 6^2}(1+\frac{1}{2}+\frac{1}{3}) + \dots] \end{aligned} \tag{2.325}$$

Again, the boundedness condition of u at $r \to 0$ gives $B = 0$ in (2.323). Thus, the solution of R can be expressed as

$$R = AJ_0(\sqrt[4]{\lambda} r) + BI_0(\sqrt[4]{\lambda} r) \tag{2.326}$$

The second boundary condition given in (2.305) yields

$$AJ_0(\sqrt[4]{\lambda} a) + BI_0(\sqrt[4]{\lambda} a) = 0 \tag{2.327}$$

The first boundary condition given in (2.305) results in

$$A\frac{dJ_0(\sqrt[4]{\lambda} r)}{dr}\bigg|_{r=a} + B\frac{dI_0(\sqrt[4]{\lambda} r)}{dr}\bigg|_{r=a} = 0 \tag{2.328}$$

The determinant of the system of equations formed by (2.327) and (2.328) must be zero, and this gives

$$\begin{bmatrix} J_0 & I_0 \\ \dfrac{dJ_0}{dr} & \dfrac{dI_0}{dr} \end{bmatrix} \begin{Bmatrix} A \\ B \end{Bmatrix} = \begin{Bmatrix} 0 \\ 0 \end{Bmatrix} \tag{2.329}$$

We now define the following function

$$R'(r) = I_0(\gamma_n)\frac{dJ_0}{dr}(\frac{\gamma_n r}{a}) - J_0(\gamma_n)\frac{dI_0}{dr}(\frac{\gamma_n r}{a}) \tag{2.330}$$

where "prime" denotes differentiation with respect to r. Evaluating this function at $r = a$, we have from (2.329)

$$R'(a) = I_0(\gamma_n)\frac{dJ_0}{dr}(\gamma_n) - J_0(\gamma_n)\frac{dI_0}{dr}(\gamma_n) = 0 \tag{2.331}$$

where

$$\gamma_n = \sqrt[4]{\lambda_n} a, \quad \lambda_n = (\frac{\gamma_n}{a})^4 \tag{2.332}$$

Equation (2.331) is exactly the same as setting the determinant of (2.329) to zero, and, thus, it is the characteristic equation for the eigenvalue λ_n. In addition, B can relate to A by (2.327) as:

$$B = -A\frac{J_0(\gamma_n)}{I_0(\gamma_n)} \tag{2.333}$$

We can define R_n as

$$R_n(r) = I_0(\gamma_n)J_0(\frac{\gamma_n}{a}r) - J_0(\gamma_n)I_0(\frac{\gamma_n}{a}r) \tag{2.334}$$

In view of (2.333) and (2.326), we can see that (2.334) is actually the radial eigenfunctions of R. Thus, the general solution of R is expressible as

$$R(r)=\sum_{n=1}^{\infty}C_nR_n(r)=\sum_{n=1}^{\infty}C_n\{I_0(\gamma_n)J_0(\frac{\gamma_n}{a}r)-J_0(\gamma_n)I_0(\frac{\gamma_n}{a}r)\} \quad (2.335)$$

Before we attempt to consider the initial conditions given in (2.304), we will first consider the orthogonal property of the Bessel functions. In particular, for the m-th eigenmode and n-th eigenmode, we have

$$\frac{1}{r}\frac{d}{dr}\{r\frac{d}{dr}[\frac{1}{r}\frac{d}{dr}(r\frac{dR_m}{dr})]\}-\lambda_mR_m=0 \quad (2.336)$$

$$\frac{1}{r}\frac{d}{dr}\{r\frac{d}{dr}[\frac{1}{r}\frac{d}{dr}(r\frac{dR_n}{dr})]\}-\lambda_nR_n=0 \quad (2.337)$$

We subtract the results of multiplying (2.336) by $rR_n(r)$ and of multiplying (2.337) by $rR_m(r)$, and integrate this difference from 0 to a to get

$$\begin{aligned}&(\lambda_n-\lambda_m)\int_o^a rR_nR_mdr+\int_o^a R_n\frac{d}{dr}\{r\frac{d}{dr}[\frac{1}{r}\frac{d}{dr}(r\frac{dR_m}{dr})]\}dr\\&-\int_o^a R_m\frac{d}{dr}\{r\frac{d}{dr}[\frac{1}{r}\frac{d}{dr}(r\frac{dR_n}{dr})]\}dr=0\end{aligned} \quad (2.338)$$

The second integral on the left-hand side of (2.338) can be evaluated by applying integration by parts four times to get

$$\begin{aligned}I_2&=\int_o^a R_n\frac{d}{dr}\{r\frac{d}{dr}[\frac{1}{r}\frac{d}{dr}(r\frac{dR_m}{dr})]\}dr\\&=\{R_nr\frac{d}{dr}[\frac{1}{r}\frac{d}{dr}(r\frac{dR_m}{dr})]-\frac{d}{dr}(r\frac{dR_m}{dr})R_n'\\&+R_m'\frac{d}{dr}(r\frac{dR_n}{dr})-R_mr\frac{d}{dr}[\frac{1}{r}\frac{d}{dr}(r\frac{dR_n}{dr})]\}_0^a+\lambda_n\int_o^a rR_nR_mdr\end{aligned} \quad (2.339)$$

We note that from (2.334) we have

$$R_n(a)=R_m(a)=0 \quad (2.340)$$

We also note the following identity:

$$R_m'\frac{d}{dr}(r\frac{dR_n}{dr})-\frac{d}{dr}(r\frac{dR_m}{dr})R_n'=r(R_m'R_n''-R_m''R_n') \quad (2.341)$$

Clearly, the boundary values of (2.341) are identically zero by noting (2.331). In view of (2.340) and (2.341), we have all the boundary terms in (2.339) vanish and we have

$$I_2=\lambda_n\int_o^a rR_nR_mdr \quad (2.342)$$

Similar analysis can also be applied to the third integral on the left-hand side of (2.338) and the third integral reduces to

$$I_3=\lambda_m\int_o^a rR_nR_mdr \quad (2.343)$$

Substitution of (2.342) and (2.343) into (2.338) gives

$$2(\lambda_n - \lambda_m)\int_o^a rR_nR_m dr = 0 \tag{2.344}$$

Since the eigenvalues are distinct, we must have the following orthogonal condition

$$\int_o^a rR_nR_m dr = 0 \tag{2.345}$$

We are now ready to consider the initial conditions. In particular, the deflection and its time derivative are

$$u(r,t) = \sum_{n=1}^{\infty} R_n(r)[A_n \sin(\frac{\gamma_n^2 b^2}{a^2} t) + B_n \cos(\frac{\gamma_n^2 b^2}{a^2} t)] \tag{2.346}$$

$$\frac{\partial u}{\partial t}(r,t) = \sum_{n=1}^{\infty} \frac{\gamma_n^2 b^2}{a^2} R_n(r)[A_n \cos(\frac{\gamma_n^2 b^2}{a^2} t) - B_n \sin(\frac{\gamma_n^2 b^2}{a^2} t)] \tag{2.347}$$

Applying the initial condition given in the second part of (2.304), we have

$$u(r,0) = f(r) = \sum_{n=1}^{\infty} R_n B_n \tag{2.348}$$

Multiplying by rR_m on both sides of (2.348), integrating the product from 0 to a, and using the orthogonal property of R found in (2.345), we get

$$\int_0^a \rho f(\rho) R_m(\rho) d\rho = \sum_{n=1}^{\infty} B_n \int_0^a \rho R_n(\rho) R_m(\rho) d\rho \tag{2.349}$$

Setting $n = m$, we have

$$B_n = \frac{\int_0^a \rho f(\rho) R_n(\rho) d\rho}{\int_0^a \rho R_n^2(\rho) d\rho} \tag{2.350}$$

Similarly, we can apply the first condition given in (2.304) to get

$$u_t(r,0) = g(r) = \sum_{n=1}^{\infty} \frac{\gamma_n^2 b^2}{a^2} R_n(r) A_n \tag{2.351}$$

Applying an orthogonal condition similar to the one that we used in obtaining (2.350), we obtain A_n as

$$A_n = \frac{a^2}{\gamma_n^2 b^2} \frac{\int_0^a \rho g(\rho) R_n(\rho) d\rho}{\int_0^a \rho R_n^2(\rho) d\rho} \tag{2.352}$$

Finally substituting (2.352) and (2.350) into (2.346), we obtain

$$u(r,t)=\sum_{n=1}^{\infty}\frac{R_n(r)}{\int_0^a \rho R_n^2(\rho)d\rho}\left[\frac{a^2}{\gamma_n^2 b^2}\sin\left(\frac{\gamma_n^2 b^2}{a^2}t\right)\int_0^a \rho g(\rho)R_n(\rho)d\rho\right.$$
$$\left.+\cos\left(\frac{\gamma_n^2 b^2}{a^2}t\right)\int_0^a \rho f(\rho)R_n(\rho)d\rho\right] \tag{2.353}$$

This gives the complete solution except that we still need to evaluate the integral in the denominator of the solution given in (2.353). The determination of that integral is not straightforward. In particular, we first introduce two functions Q and R such that

$$\sqrt{\lambda_n}Q=\frac{1}{r}\frac{d}{dr}\left(r\frac{dR}{dr}\right) \tag{2.354}$$

$$\sqrt{\lambda_n}R=\frac{1}{r}\frac{d}{dr}\left(r\frac{dQ}{dr}\right) \tag{2.355}$$

If we substitute Q from (2.354) into (2.355), we get (2.337) or (2.338) for R. Similarly, we can reverse the roles for R and Q to give (2.337) or (2.338) for Q. Thus, (2.354) and (2.355) are somehow equivalent to our differential equation for R. The introduction of (2.354) and (2.355) is the most critical and tricky step. Multiplying (2.354) by rQ and multiplying (2.355) by rR, and integrating these products from 0 to a, subtracting these two integrals, and applying integration by parts, we obtain

$$\sqrt{\lambda_n}\int_0^a rQ^2 dr=\sqrt{\lambda_n}\int_0^a rR^2 dr+\left[rQ\frac{dR}{dr}-rR\frac{dQ}{dr}\right]_0^a \tag{2.356}$$

Similarly, we can add the product of multiplying (2.354) by $2r^2dQ/dr$ to the product of multiplying (2.355) by $2r^2dR/dr$, and we obtain

$$\sqrt{\lambda_n}r^2\frac{d}{dr}(Q^2+R^2)=2\frac{d}{dr}\left(r^2\frac{dR}{dr}\frac{dQ}{dr}\right) \tag{2.357}$$

We then integrate both sides of (2.357) from 0 to a and use (2.356) to give

$$\int_0^a rR^2 dr=\psi(a) \tag{2.358}$$

where

$$\psi(r)=\frac{1}{4\sqrt{\lambda_n}}\left\{\sqrt{\lambda_n}r^2(Q^2+R^2)-2r^2\frac{dQ}{dr}\frac{dR}{dr}-2rQ\frac{dR}{dr}+2rR\frac{dQ}{dr}\right\} \tag{2.359}$$

More steps in deriving (2.358) are given in Problem 2.4. For the special case of fixed edges (i.e., $R(a)=R'(a)=0$, and $Q(a)=R''(a)$), we can simplify (2.358) as

$$\int_0^a rR^2 dr=\frac{a^2}{4}[R''(a)]^2 \tag{2.360}$$

where

$$R''=J_0''I_0-I_0''J_0 \tag{2.361}$$

Equation (2.360) can further be simplified by using a Bessel equation and a modified Bessel equation:

$$\xi^2 J_0''+\xi J_0'+\xi^2 J_0=0 \tag{2.362}$$

$$\xi^2 I_0'' + \xi I_0' - \xi^2 I_0 = 0 \tag{2.363}$$

Rearranging these equations, we find

$$J_0'' = -\frac{1}{\xi} J_0' - J_0, \quad I_0'' = -\frac{1}{\xi} I_0' + I_0 \tag{2.364}$$

Substituting (2.361) into (2.360) and taking the square of it, in view of (2.364) and (2.331) we obtain

$$(R'')^2 = (J_0'' I_0 - I_0'' J_0)^2 = 4 I_0^2 J_0^2 \tag{2.365}$$

Finally, substituting (2.365) into (2.360), we obtain the final result

$$\int_0^a r R^2 dr = a^2 I_0^2(\gamma_n) J_0^2(\gamma_n) \tag{2.366}$$

Using (2.366), the solution of u given in (2.353) can be summarized as

$$\begin{aligned} u(r,t) = \frac{1}{a^2} \sum_{n=1}^{\infty} \frac{R_n(r)}{I_0^2(\gamma_n) J_0^2(\gamma_n)} [&\frac{a^2}{\gamma_n^2 b^2} \sin(\frac{\gamma_n^2 b^2}{a^2} t) \int_0^a \rho g(\rho) R_n(\rho) d\rho \\ &+ \cos(\frac{\gamma_n^2 b^2}{a^2} t) \int_0^a \rho f(\rho) R_n(\rho) d\rho] \end{aligned} \tag{2.367}$$

where

$$R_n(r) = I_0(\gamma_n) J_0(\frac{\gamma_n}{a} r) - J_0(\gamma_n) I_0(\frac{\gamma_n}{a} r) \tag{2.368}$$

$$R'(a) = I_0(\gamma_n) \frac{dJ_0}{dr}(\gamma_n) - J_0(\gamma_n) \frac{dI_0}{dr}(\gamma_n) = 0 \tag{2.369}$$

This gives the final solution of the initial boundary value problem defined in (2.302) to (2.305).

2.11 HERTZ PROBLEM OF CIRCULAR PLATE UNDER POINT LOAD

In 1884, Hertz considered the problem of a large circular plate floating on a fluid and subjected to a point force at the origin of the cylindrical coordinates. The problem relates to aircraft landing on ice sheets and to oil drilling on ice sheets at the North Pole. The problem can be mathematically modeled by

$$\nabla^2 \nabla^2 w + \frac{kw}{D} = (\frac{d^2}{dr^2} + \frac{1}{r}\frac{d}{dr})(\frac{d^2 w}{dr^2} + \frac{1}{r}\frac{dw}{dr}) + \frac{kw}{D} = q \tag{2.370}$$

$$q(r) = \delta(r) P \tag{2.371}$$

where k is the unit weight of water. The first term is the normal biharmonic term in the plate bending equation. The second term results from the fact that the water pressure from the water beneath the plate depends on the deflection of the plate, and it is this term that changes the type of the governing equation of this plate bending problem. Note that (2.370) is mathematically equivalent to (2.237) given for plates on an elastic foundation.

Except at the origin, we can normalize and rewrite (2.370) as

$$l^4\left(\frac{d^2}{dr^2}+\frac{1}{r}\frac{d}{dr}\right)\left(\frac{d^2w}{dr^2}+\frac{1}{r}\frac{dw}{dr}\right)+w=0 \tag{2.372}$$

where

$$l^4=\frac{D}{k} \tag{2.373}$$

We note that the unit of l is length. Thus, l can be interpreted as a characteristic length of the problem. Both the radial coordinate and deflection of the plate can be normalized as:

$$z=\frac{w}{l},\quad x=\frac{r}{l} \tag{2.374}$$

The resulting equation can be rewritten as

$$\left(\frac{d^2}{dx^2}+\frac{1}{x}\frac{d}{dx}\right)\left(\frac{d^2z}{dx^2}+\frac{1}{x}\frac{dz}{dx}\right)+z=0 \tag{2.375}$$

Or equivalently, it can be expressed as:

$$\nabla^2\nabla^2 z+z=0 \tag{2.376}$$

where

$$\nabla^2(.)=\frac{d^2(.)}{dx^2}+\frac{1}{x}\frac{d(.)}{dx} \tag{2.377}$$

Since it is a fourth order differential equation, there must be four independent solutions. Thus, we have

$$z=A_1X_1(x)+A_2X_2(x)+A_3X_3(x)+A_4X_4(x) \tag{2.378}$$

2.11.1 Series Solution

We will now seek a series solution for $X(x)$:

$$X_i(x)=a_nx^n \tag{2.379}$$

Taking the Laplacian of (2.379), we have

$$\nabla^2a_nx^n=n(n-1)a_nx^{n-2}+na_nx^{n-2}=n^2a_nx^{n-2} \tag{2.380}$$

Taking the Laplacian one more time, we have

$$\nabla^2\nabla^2a_nx^n=n^2(n-2)^2a_nx^{n-4} \tag{2.381}$$

Substituting (2.381) and (2.379) into (2.376), we have

$$n^2(n-2)^2a_nx^{n-4}+a_{n-4}x^{n-4}=0 \tag{2.382}$$

Thus, we obtain a recursive relation between the coefficients:

$$a_n=-\frac{a_{n-4}}{n^2(n-2)^2} \tag{2.383}$$

The final term in the recursive formula given in (2.383) is

$$a_4=-\frac{a_0}{4^2 2^2} \tag{2.384}$$

Note also that

$$\nabla^2\nabla^2(a_0)=0, \quad \nabla^2\nabla^2(a_2x^2)=0 \tag{2.385}$$

Thus, two series can be formulated as

$$X_1(x)=1-\frac{x^4}{2^2\cdot 4^2}+\frac{x^8}{2^2\cdot 4^2\cdot 6^2\cdot 8^2}-\frac{x^{12}}{2^2\cdot 4^2\cdot 6^2\cdot 8^2\cdot 10^2\cdot 12^2}+... \tag{2.386}$$

$$X_2(x)=x^2-\frac{x^6}{4^2\cdot 6^2}+\frac{x^{10}}{4^2\cdot 6^2\cdot 8^2\cdot 10^2}-\frac{x^{14}}{4^2\cdot 6^2\cdot 8^2\cdot 10^2\cdot 12^2\cdot 14^2}+... \tag{2.387}$$

However, both of these series are finite at the origin. Since a concentrated load is applied at the origin, the bending moment there would be infinite. Thus, we seek a singular solution of the form:

$$X_3(x)=X_1\ln x+F_3(x) \tag{2.388}$$

Applying the biharmonic operator on X_3, we find

$$\nabla^2\nabla^2X_3=\frac{4}{x}\frac{d^3X_1}{dx^3}+\ln x\nabla^2\nabla^2X_1+\nabla^2\nabla^2F_3(x) \tag{2.389}$$

Recall that X_3 must satisfy

$$\nabla^2\nabla^2X_3+X_3=0 \tag{2.390}$$

Combining (2.388) and (2.390), we obtain

$$\nabla^2\nabla^2X_3=-X_1\ln x-F_3(x) \tag{2.391}$$

Substitution of (2.391) into (2.389) gives

$$\frac{4}{x}\frac{d^3X_1}{dx^3}+\ln x(\nabla^2\nabla^2X_1+X_1)+\nabla^2\nabla^2F_3(x)+F_3(x)=0 \tag{2.392}$$

Note that the second term is zero since X_1 satisfies (2.376). Thus, substituting (2.386) into (2.392), we have

$$\begin{aligned}\nabla^2\nabla^2F_3(x)+F_3(x)&=-\frac{4}{x}\frac{d^3X_1}{dx^3}\\ &=-4\left(-\frac{4\cdot 3\cdot 2}{2^2\cdot 4^2}+\frac{6\cdot 7\cdot 8}{2^2\cdot 4^2\cdot 6^2\cdot 8^2}x^4-\frac{10\cdot 11\cdot 12}{2^2\cdot 4^2\cdot 6^2\cdot 8^2\cdot 10^2\cdot 12^2}x^8+...\right)\end{aligned} \tag{2.393}$$

In view of the power series on the right-hand side of (2.393), we assume F_3 as:

$$F_3(x)=b_4x^4+b_8x^8+b_{12}x^{12}+... \tag{2.394}$$

Taking the biharmonic operator on the first term on the right-hand side of (2.394), we have

$$\nabla^2\nabla^2(b_4x^4)=4^2\cdot 2^2b_4 \tag{2.395}$$

Thus, balancing the coefficients on both sides of (2.393), we have

$$b_4=\frac{2\cdot 3\cdot 4^2}{2^4\cdot 4^4}=\frac{3}{128} \tag{2.396}$$

Similarly, the second coefficient can be found as

$$b_8=-\frac{25}{1{,}769{,}472} \tag{2.397}$$

More generally, the coefficient b_n can be evaluated as

$$b_n = (-1)^{n/4-1} \frac{1}{n^2(n-2)^2}[b_{n-4} + \frac{n(n-1)(n-2)}{2^2 \cdot 4^2 \cdot 6^2 \cdots n^2}] \tag{2.398}$$

Thus, the third solution becomes

$$X_3 = X_1 \ln x + \frac{3}{128}x^4 - \frac{25}{1,769,472}x^8 + ... \tag{2.399}$$

Similarly, the fourth solution can be obtained in a similar manner as

$$\begin{aligned} X_4 &= X_2 \ln x + F_4(x) \\ &= X_2 \ln x + 4\frac{4 \cdot 5 \cdot 6}{4^4 \cdot 6^4}x^6 - \frac{1}{10^2 \cdot 8^2}(4\frac{4 \cdot 5 \cdot 6}{4^4 \cdot 6^4} + \frac{10 \cdot 9 \cdot 8}{4^2 \cdot 6^2 \cdots 10^2})x^{10} + ... \end{aligned} \tag{2.400}$$

Finally, the series solution becomes

$$\begin{aligned} z &= A_1(1 - \frac{x^4}{2^2 \cdot 4^2} + \frac{x^8}{2^2 \cdot 4^2 \cdot 6^2 \cdot 8^2} - ...) \\ &+ A_2(x^2 - \frac{x^6}{4^2 \cdot 6^2} + \frac{x^{10}}{4^2 \cdot 6^2 \cdot 8^2 \cdot 10^2} - ...) \\ &+ A_3[(1 - \frac{x^4}{2^2 \cdot 4^2} + \frac{x^8}{2^2 \cdot 4^2 \cdot 6^2 \cdot 8^2} - ...) \ln x + \frac{3}{128}x^4 - \frac{25}{1,769,472}x^8 + ...] \\ &+ A_4[(x^2 - \frac{x^6}{4^2 \cdot 6^2} + \frac{x^{10}}{4^2 \cdot 6^2 \cdot 8^2 \cdot 10^2} - ...) \ln x + \frac{5}{3,456}x^6 - \frac{1,054 \cdot 10^{-4}}{442,368}x^{10} + ...] \end{aligned} \tag{2.401}$$

A shortcut to obtain the solution of (2.375) is to recognize that it was the Kelvin equation discussed in (4.443) of Chau (2018). More details of this approach will be discussed in a later section.

For the case of the free edge of the circular plate, both the moment and shear force are zeros:

$$M_n = 0, \quad Q_r = 0 \tag{2.402}$$

These conditions can be expressed in terms of w as

$$(\frac{d^2w}{dr^2} + \nu\frac{1}{r}\frac{dw}{dr})_{r=a} = 0, \quad \frac{d}{dr}(\frac{d^2w}{dr^2} + \frac{1}{r}\frac{dw}{dr})_{r=a} = 0 \tag{2.403}$$

The deflection must be finite at the origin ($r = 0$), we have

$$A_3 = 0 \tag{2.404}$$

The vertical force equilibrium requires

$$\left(\int_0^{2\pi} Q_r r d\theta\right)_{r=\varepsilon} + P = 0 \tag{2.405}$$

where ε is the radius of an infinitesimal small cylinder embracing the origin. Substituting the expression of Q_r in terms of w into (2.405), we have

$$-kl^4 \frac{d}{dr}\left(\frac{d^2w}{dr^2} + \frac{1}{r}\frac{dw}{dr}\right)_{r=\varepsilon} 2\pi\varepsilon + P = 0 \tag{2.406}$$

In view of (2.401), (2.406) becomes

$$-kl^4 \frac{4A_4}{l\varepsilon} 2\pi\varepsilon + P = 0 \tag{2.407}$$

Finally, we can obtain the unknown constant as

$$A_4 = \frac{P}{8\pi k l^3} \tag{2.408}$$

There are two equations given by boundary conditions in (2.402) for the remaining constants A_1 and A_2. However, we will not go into the details of this analysis.

2.11.2 Variational Principle

When shear deformation of the plate is neglected, the strain energy of a plate can be evaluated by integrating the strain energy induced by bending moments and twisting moments for an infinitesimal plate element over the whole plate as

$$I = \iint \left(\frac{D}{2} \left\{ \left(\frac{\partial^2 w}{\partial x^2} + \frac{\partial^2 w}{\partial y^2}\right)^2 - 2(1-\nu)\left[\frac{\partial^2 w}{\partial x^2}\frac{\partial^2 w}{\partial y^2} - \left(\frac{\partial^2 w}{\partial y \partial x}\right)^2\right] \right\} - wq \right) dxdy \tag{2.409}$$

This equation was first obtained by Kirchhoff in 1850. The polar form of this strain energy for circular plate is

$$\begin{aligned} I = \iint [\frac{D}{2}\{ & \left(\frac{\partial^2 w}{\partial r^2} + \frac{1}{r}\frac{\partial w}{\partial r} + \frac{1}{r^2}\frac{\partial^2 w}{\partial \theta^2}\right)^2 - 2(1-\nu)\frac{\partial^2 w}{\partial r^2}\left(\frac{1}{r}\frac{\partial w}{\partial r} + \frac{1}{r^2}\frac{\partial^2 w}{\partial \theta^2}\right) \\ & + 2(1-\nu)\left(\frac{1}{r}\frac{\partial^2 w}{\partial r \partial \theta} - \frac{1}{r^2}\frac{\partial w}{\partial \theta}\right)^2\} - wq] r dr d\theta \end{aligned} \tag{2.410}$$

For the case of axisymmetric bending, it can be simplified as

$$I = \pi \int \{D[\left(\frac{d^2 w}{dr^2} + \frac{1}{r}\frac{dw}{dr}\right)^2 - 2(1-\nu)\frac{d^2 w}{dr^2}\left(\frac{1}{r}\frac{dw}{dr}\right)] - wq\} r dr \tag{2.411}$$

The variational principle is based on the introduction of certain types of approximations that satisfy boundary conditions and at the same time minimize the strain energy. The most popular choice of variational methods is called the Rayleigh-Ritz method. Before we consider this method, we would further consider some special forms of the strain energy. Let us consider the following integration:

$$\iint \left[\frac{\partial^2 w}{\partial x^2}\frac{\partial^2 w}{\partial y^2} - \left(\frac{\partial^2 w}{\partial y \partial x}\right)^2\right] dxdy = \iint \left[\frac{\partial}{\partial x}\left(\frac{\partial w}{\partial x}\frac{\partial^2 w}{\partial y^2}\right) - \frac{\partial}{\partial y}\left(\frac{\partial w}{\partial x}\frac{\partial^2 w}{\partial y \partial x}\right)\right] dxdy \tag{2.412}$$

Note from Green's identity that

$$\iint \left[\frac{\partial}{\partial x} P(x,y) - \frac{\partial}{\partial y} Q(x,y)\right] dxdy = \int [Q(x,y)dx + P(x,y)dy] \tag{2.413}$$

Application of (2.412) to (2.411) gives

$$\iint \left[\frac{\partial}{\partial x}\left(\frac{\partial w}{\partial x}\frac{\partial^2 w}{\partial y^2}\right) - \frac{\partial}{\partial y}\left(\frac{\partial w}{\partial x}\frac{\partial^2 w}{\partial y \partial x}\right)\right] dxdy = \int \left[\frac{\partial w}{\partial x}\frac{\partial^2 w}{\partial y^2} dy + \frac{\partial w}{\partial x}\frac{\partial^2 w}{\partial y \partial x} dx\right] \tag{2.414}$$

For a clamped boundary, we have $\partial w/\partial n = 0$ for any normal vector $\boldsymbol{n}$ pointing out of the boundary and (2.409) can be simplified to

$$I = \iint [\frac{D}{2}(\nabla^2 w)^2 - wq]dxdy \tag{2.415}$$

In particular, if the boundary is clamped, we have

$$dx = 0, \quad \frac{\partial w}{\partial x} = 0 \tag{2.416}$$

and

$$dy = 0, \quad \frac{\partial^2 w}{\partial x \partial y} = 0 \tag{2.417}$$

Similarly, for circular plates with a clamped boundary (or $dw/dr = 0$) we have

$$I = \pi D \int [r(\frac{d^2 w}{dr^2})^2 + \frac{1}{r}(\frac{dw}{dr})^2]dr \tag{2.418}$$

2.11.3 Rayleigh-Ritz Method

In the Rayleigh-Ritz method, we seek an approximation that

$$w = a_1\varphi_1(x,y) + a_2\varphi_2(x,y) + ... + a_n\varphi_n(x,y) \tag{2.419}$$

$$\frac{\partial I}{\partial a_1} = 0, \quad \frac{\partial I}{\partial a_2} = 0, \quad ..., \quad \frac{\partial I}{\partial a_n} = 0 \tag{2.420}$$

where I is the functional given in Section 2.11.2. This provides a system of n equations for n unknowns.

Let us consider a particular solution based on the Rayleigh-Ritz method. First, we assume that

$$w = A + Br^2 \tag{2.421}$$

Thus, we have

$$\frac{dw}{dr} = 2Br, \quad \frac{1}{r}\frac{dw}{dr} = 2B, \quad \frac{d^2 w}{dr^2} = 2B \tag{2.422}$$

The strain energy of the plate is

$$\begin{aligned} V_1 &= \pi \int_0^a \{D[(\frac{d^2 w}{dr^2} + \frac{1}{r}\frac{dw}{dr})^2 - 2(1-\nu)\frac{d^2 w}{dr^2}(\frac{1}{r}\frac{dw}{dr})]rdr \\ &= \pi D \int_0^a \{16B^2 - 2(1-\nu)4B^2]rdr \\ &= 4B^2 D\pi(1+\nu)a^2 \end{aligned} \tag{2.423}$$

where a is the radius of the circular plate. Assuming that the water resistance can be modeled by a spring stiffness k, the strain energy stored in the water is

$$V_2 = 2\pi \int_0^a \{\frac{kw^2}{2} rdr = \pi k(\frac{1}{2}A^2 a^2 + \frac{1}{2}ABa^4 + \frac{1}{6}B^2 a^6) \tag{2.424}$$

The work done by the point force at the center is $-PA$ (note $w = A$ at the origin from (2.421)). Thus, from (2.411) we have the functional as

$$I = 4B^2 D\pi(1+\nu)a^2 + \pi k(\frac{1}{2}A^2a^2 + \frac{1}{2}ABa^4 + \frac{1}{6}B^2a^6) - PA \tag{2.425}$$

To find A and B, by following (2.420) we differentiate (2.425) with respect to A and B respectively to give two equations

$$\frac{\partial I}{\partial A} = 0, \quad \frac{\partial I}{\partial B} = 0 \tag{2.426}$$

More explicitly, they are

$$A + Ba^2[\frac{2}{3} + \frac{16D(1+\nu)}{ka^4}] = 0, \quad A + \frac{1}{2}Ba^2 = \frac{P}{\pi ka^2} \tag{2.427}$$

The solutions of A and B are

$$A = \frac{4P[ka^4 + 24D(1+\nu)]}{\pi ka^2[ka^4 + 96D(1+\nu)]}, \quad B = -\frac{6P}{\pi[ka^4 + 96D(1+\nu)]} \tag{2.428}$$

Substitution of (2.428) into (2.421) gives the approximation as

$$w = \frac{4P[ka^4 + 24D(1+\nu)]}{\pi ka^2[ka^4 + 96D(1+\nu)]} - \frac{6r^2P}{\pi[ka^4 + 96D(1+\nu)]} \tag{2.429}$$

If the stress singularity is needed under the point load, we need to add a singular term as

$$w = A + Br^2 + C\frac{P}{8\pi D}r^2 \ln r \tag{2.430}$$

2.11.4 General Solution in Kelvin Functions

Another way to solve this problem is to rewrite the governing equation into a form with recognizable solutions. Let us consider the following change of variables

$$\xi = x\sqrt{i}, \quad \xi^2 = ix^2 \tag{2.431}$$

With this new variable, differentiation becomes

$$\frac{d^2(.)}{dx^2} = i\frac{d^2(.)}{d\xi^2}, \quad \frac{1}{x}\frac{d(.)}{dx} = i\frac{1}{\xi}\frac{d(.)}{d\xi} \tag{2.432}$$

Using this new variable, (2.376) becomes

$$\nabla^2\nabla^2 z + z = i^2\overline{\nabla}^2\overline{\nabla}^2 z + z = -\overline{\nabla}^2\overline{\nabla}^2 z + z = 0 \tag{2.433}$$

where

$$\overline{\nabla}^2(.) = \frac{d^2(.)}{d\xi^2} + \frac{1}{\xi}\frac{d(.)}{d\xi} \tag{2.434}$$

To factorize the last of (2.433), we observe that

$$\overline{\nabla}^2\overline{\nabla}^2 z - z = \overline{\nabla}^2\overline{\nabla}^2 z + \overline{\nabla}^2 z - \overline{\nabla}^2 z - z = \overline{\nabla}^2(\overline{\nabla}^2 z + z) - (\overline{\nabla}^2 z + z) = 0 \tag{2.435}$$

$$\overline{\nabla}^2\overline{\nabla}^2 z - z = \overline{\nabla}^2\overline{\nabla}^2 z - \overline{\nabla}^2 z + \overline{\nabla}^2 z - z = \overline{\nabla}^2(\overline{\nabla}^2 z - z) + (\overline{\nabla}^2 z - z) = 0 \tag{2.436}$$

In view of (2.435) and (2.436), (2.433) can be factorized as

$$(\overline{\nabla}^2 - 1)(\overline{\nabla}^2 z + z) = 0 \tag{2.437}$$

$$(\bar{\nabla}^2+1)(\bar{\nabla}^2 z - z) = 0 \tag{2.438}$$

Therefore, solving (2.433) is equivalent to solving the following PDEs

$$(\bar{\nabla}^2 z + z) = 0, \quad (\bar{\nabla}^2 z - z) = 0 \tag{2.439}$$

The solutions of these equations are Bessel functions and modified Bessel functions with complex arguments:

$$z = B_1 I_0(x\sqrt{i}) + B_3 K_0(x\sqrt{i}) \tag{2.440}$$

$$z = \bar{B}_2 J_0(x\sqrt{i}) + \bar{B}_4 Y_0(x\sqrt{i}) \tag{2.441}$$

Note the following identities of the Bessel function and the modified Bessel function (Abramowitz and Stegun, 1964)

$$I_\nu(\zeta) = e^{3\nu\pi i/2} J_\nu(\zeta e^{-3\pi i/2}) \tag{2.442}$$

$$K_\nu(\zeta) = -\frac{1}{2}\pi i e^{-\nu\pi i/2} H_\nu^{(2)}(\zeta e^{-\pi i/2}) \tag{2.443}$$

$$H_\nu^{(2)}(\zeta) = J_\nu(\zeta) - iY_\nu(\zeta) \tag{2.444}$$

for $\pi/2 < \arg z \leq \pi$. For the special case that $\nu = 0$, we note the following identities

$$\zeta = xi\sqrt{i} = xe^{i\pi/2}e^{i\pi/4} = xe^{i3\pi/4} \tag{2.445}$$

Thus, (2.442) and (2.443) can be written as

$$I_0(\zeta) = J_0(\zeta e^{-3\pi i/2}) \tag{2.446}$$

$$K_0(\zeta) = -\frac{1}{2}\pi i H_0^{(2)}(\zeta e^{-\pi i/2}) \tag{2.447}$$

Consider that ζ is purely imaginary

$$I_0(xi\sqrt{i}) = J_0[xi\sqrt{i}(-i)] = J_0(x\sqrt{i}) \tag{2.448}$$

$$Y_0(x\sqrt{i}) = \frac{2}{\pi}K_0(xi\sqrt{i}) + iI_0(xi\sqrt{i}) \tag{2.449}$$

Employing these results, we have the second solution given in (2.441) being

$$z = \bar{B}_2 J_0(x\sqrt{i}) + \bar{B}_4 Y_0(x\sqrt{i}) = B_2 I_0(xi\sqrt{i}) + B_4 K_0(xi\sqrt{i}) \tag{2.450}$$

Finally, the solution for (2.433) can be written as

$$z = B_1 I_0(x\sqrt{i}) + B_2 I_0(xi\sqrt{i}) + B_3 K_0(x\sqrt{i}) + B_4 K_0(xi\sqrt{i}) \tag{2.451}$$

However, all these functions are complex. In 1890, Kelvin proposed to single out the real part of these complex functions when he encountered similar functions in solving electrical problems. These functions are called Kelvin functions (Abramowitz and Stegun, 1964). More specially, Kelvin functions of the first kind are defined as

$$I_0(x\sqrt{\pm i}) = J_0(xi\sqrt{\pm i}) = \mathrm{ber}(x) \pm i\,\mathrm{bei}(x) \tag{2.452}$$

The terms "ber" and "bei" were originally proposed by Kelvin in the hope that they would have significance similar to "sin" and "cos." In terms of infinite series, Kelvin functions of the first kind can be expressed as

$$\mathrm{ber}(x) = 1 - \frac{(\frac{1}{2}x)^4}{(2!)^2} + \frac{(\frac{1}{2}x)^8}{(4!)^2} - \dots \tag{2.453}$$

$$\operatorname{bei}(x)=\frac{(\frac{1}{2}x)^2}{(1!)^2}-\frac{(\frac{1}{2}x)^6}{(3!)^2}+\frac{(\frac{1}{2}x)^{10}}{(5!)^2}-.. \tag{2.454}$$

For complex Bessel functions of higher order, these Kelvin functions had been extended to higher order by Russell in 1909 and by Whitehead in 1911:

$$\operatorname{ber}_\nu(z)\pm i\operatorname{bei}_\nu(z)=J_\nu(ze^{\pm 3\pi i/4}) \tag{2.455}$$

For modified Bessel functions of the second kind, the following Kelvin function of the second kind had been defined by Russell

$$K_0(x\sqrt{\pm i})=\operatorname{ker}(x)\pm i\operatorname{kei}(x) \tag{2.456}$$

These functions had been extended to higher order ν by Whitehead as:

$$\operatorname{ker}_\nu(x)+i\operatorname{kei}_\nu(x)=e^{-\pi\nu i/2}K_\nu(xe^{\pi i/4}) \tag{2.457}$$

Kelvin functions of the third kind can be defined as

$$\operatorname{her}(x)\pm i\operatorname{hei}(x)=H_\nu^{(1)}(xe^{\pm 3\pi i/4}) \tag{2.458}$$

The general solutions of

$$x^2w''+xw'-(ix^2+\nu^2)w=0 \tag{2.459}$$

are Kelvin functions

$$w(x)=A[\operatorname{ber}_\nu(x)+i\operatorname{bei}_\nu(x)]+B[\operatorname{ker}_\nu(x)+i\operatorname{kei}_\nu(x)] \tag{2.460}$$

where A and B are complex constants. Alternatively, Kelvin functions are also solutions of the following real differential equation

$$x^4w^{(IV)}+2x^3w'''-(1+2\nu^2)(x^2w''-xw')-(\nu^4-4\nu^2+x^4)w=0 \tag{2.461}$$

$$w(x)=A\operatorname{ber}_\nu(x)+B\operatorname{bei}_\nu(x)+C\operatorname{ker}_\nu(x)+D\operatorname{kei}_\nu(x) \tag{2.462}$$

Thus, we can rewrite the complex Bessel functions in terms of Kelvin functions as

$$K_0(x\sqrt{\pm i})=\operatorname{ker}(x)\pm i\operatorname{kei}(x) \tag{2.463}$$

$$I_0(x\sqrt{\pm i})=\operatorname{ber}(x)\pm i\operatorname{bei}(x) \tag{2.464}$$

The solution of z given in (2.233) becomes

$$\begin{aligned}z&=B_1[\operatorname{ber}(x)+i\operatorname{bei}(x)]+B_2[\operatorname{ber}(x)-i\operatorname{bei}(x)]+B_3[\operatorname{ker}(x)+i\operatorname{kei}(x)]\\&+B_4[\operatorname{ker}(x)-i\operatorname{kei}(x)]\end{aligned} \tag{2.465}$$

This solution can be recast in real form as

$$w=C_1\operatorname{ber}(x)+C_2\operatorname{bei}(x)+C_3\operatorname{ker}(x)+C_4\operatorname{kei}(x) \tag{2.466}$$

where the unknown constants are all real.

2.11.5 Matching of Boundary Condition

We now try to solve the Hertz problem in terms of Kelvin's function (e.g., Wyman, 1950). The governing equation in polar form is

$$D\frac{1}{r}\frac{d}{dr}\{r\frac{d}{dr}[\frac{1}{r}\frac{d}{dr}(r\frac{dw}{dr})]\}+\rho g w=-P\delta(0) \tag{2.467}$$

where ρ is the density of water and P is the downward applying point force. If we exclude the origin, this ODE can be factorized as

$$(\frac{d^2}{d\bar{r}^2}+\frac{1}{\bar{r}}\frac{d}{d\bar{r}}+i)(\frac{d^2}{d\bar{r}^2}+\frac{1}{\bar{r}}\frac{d}{d\bar{r}}-i)w=0 \tag{2.468}$$

where

$$\bar{r}=\frac{r}{l}=r(\frac{\rho g}{D})^{1/4} \tag{2.469}$$

In terms of Kelvin's functions, the solution of (2.467) is

$$w=C_1\text{ber}(\frac{r}{l})+C_2\text{bei}(\frac{r}{l})+C_3\text{kei}(\frac{r}{l})+C_4\text{ker}(\frac{r}{l}) \tag{2.470}$$

For decay conditions, we observe the following behaviors of Kelvin functions for a large argument:

$$\text{ber}(\frac{r}{l}),\quad \text{bei}(\frac{r}{l}),\quad \text{ker}(\frac{r}{l})\to\infty,\quad \text{as}\, r\to\infty \tag{2.471}$$

Thus, by boundedness condition, we have

$$w=C_3\text{kei}(\frac{r}{l}) \tag{2.472}$$

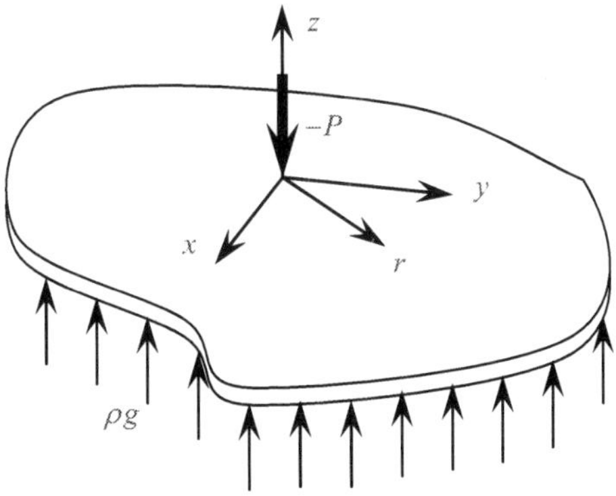

Figure 2.22 Hertz problem: An infinite plate resting on water under a point force

Considering vertical force equilibrium, we have

$$\int_0^\infty \rho g w 2\pi r dr=-P \tag{2.473}$$

Substituting (2.472) into (2.473), we obtain

$$\int_0^\infty \rho g 2\pi r C_3\text{kei}(\frac{r}{l})dr=-P \tag{2.474}$$

We now introduce a new variable as

$$\xi=\frac{r}{l} \tag{2.475}$$

In this new variable, (2.474) can be written as

$$\rho g 2\pi C_3 l^2 \int_0^\infty \xi \text{kei}(\xi) d\xi = -P \tag{2.476}$$

Applying formula (4.472) of Chau (2018) for the integration of the Kelvin function, we find

$$\int_0^\infty \xi \text{kei}(\xi) d\xi = -\xi \,\text{ker}'(\xi)\Big|_0^\infty = -\frac{\xi}{\sqrt{2}}(\text{ker}_1 \xi + \text{kei}_1 \xi)\Bigg|_0^\infty \tag{2.477}$$

The last part of (2.477) is a result of applying (4.475) of Chau (2018). To evaluate the upper limit in (2.477), we note the following asymptotic form of Kelvin's functions as $\xi \to \infty$ (Abramowitz and Stegun, 1964):

$$\xi \,\text{ker}_1 \xi = \sqrt{\frac{\pi\xi}{2}} e^{-\xi/\sqrt{2}} \{\cos[\frac{\xi}{\sqrt{2}} + \frac{5\pi}{8}]\} \tag{2.478}$$

$$\xi \text{kei}_1 \xi = \sqrt{\frac{\pi\xi}{2}} e^{-\xi/\sqrt{2}} \{-\sin[\frac{\xi}{\sqrt{2}} + \frac{5\pi}{8}]\} \tag{2.479}$$

The sine and cosine functions are bounded between ± 1 as $\xi \to \infty$, and we only need to consider the following limit:

$$\lim_{\xi\to\infty} \frac{\sqrt{\xi}}{e^{\xi/\sqrt{2}}} = \lim_{\xi\to\infty} \frac{1}{\sqrt{2}} \frac{1}{\sqrt{\xi} e^{\xi/\sqrt{2}}} = 0 \tag{2.480}$$

We have applied L'Hôpital's rule to the first of (2.480) since it is of the form ∞/∞. In view of (2.478) to (2.480), the upper limit of (2.477) approaches zero:

$$\frac{\xi}{\sqrt{2}}(\text{ker}_1 \xi + \text{kei}_1 \xi) \to 0, \quad \xi \to \infty \tag{2.481}$$

To evaluate the lower limit in (2.477), we note the following asymptotic form of Kelvin's functions as $\xi \to 0$ (Abramowitz and Stegun, 1964):

$$\begin{aligned} \xi \,\text{ker}_1 \xi &= \cos(\frac{3\pi}{4}) - \xi \ln(\frac{\xi}{2}) \text{ber}_1 \xi + \frac{1}{4} \xi \pi \text{bei}_1 \xi \\ &+ \frac{\xi}{2}(\frac{\xi}{2}) \sum_{k=0}^\infty \cos\{(\frac{3}{4} + \frac{k}{2})\pi\} \frac{\psi(k+1)+\psi(k+2)}{k!(k+1)!} (\frac{\xi^2}{4})^k \end{aligned} \tag{2.482}$$

$$\begin{aligned} \xi \text{kei}_1 \xi &= -\sin(\frac{3\pi}{4}) - \xi \ln(\frac{\xi}{2}) \text{bei}_1 \xi - \frac{1}{4} \xi \pi \text{ber}_1 \xi \\ &+ \frac{\xi}{2}(\frac{\xi}{2}) \sum_{k=0}^\infty \sin\{(\frac{3}{4} + \frac{k}{2})\pi\} \frac{\psi(k+1)+\psi(k+2)}{k!(k+1)!} (\frac{\xi^2}{4})^k \end{aligned} \tag{2.483}$$

Application of (2.481) to (2.483) yields the following limit in (2.477):

$$\int_0^\infty \xi \text{kei}(\xi) d\xi = \frac{\xi}{\sqrt{2}}(\text{ker}_1 \xi + \text{kei}_1 \xi)\Big|_0^\infty = 0 - \frac{1}{\sqrt{2}}(-\frac{1}{\sqrt{2}} - \frac{1}{\sqrt{2}}) = -1 \tag{2.484}$$

Substitution of (2.484) into (2.476) results in

$$w = \frac{P}{2\pi\rho g l^2} \text{kei}(\frac{r}{l}) \tag{2.485}$$

This is Wyman's solution obtained in 1950. However, the steps in obtaining this solution are not provided by Wyman. The maximum deflection of the floating plate is thus

$$w_{\max} = \frac{P}{2\pi\rho g l^2}\mathrm{kei}(0) = -\frac{P}{2\pi\rho g l^2}\frac{\pi}{4} = -\frac{P}{8\rho g l^2} \tag{2.486}$$

2.11.6 Wyman's Solution

For a more general loading as shown in Figure 2.23, Wyman (1950) also gave the following solution in the form of Green's function method:

$$w(r) = \frac{1}{2\pi\rho g l^2}\iint_A F(r',\theta)\mathrm{kei}(R/l)r'dr'd\theta \tag{2.487}$$

where

$$R^2 = r'^2 + r^2 - 2rr'\cos\theta \tag{2.488}$$

If the load is applied to a circular area of area a, the solution can be obtained as (Wyman, 1950):

$$w(r) = \begin{cases} -\dfrac{P}{\pi\rho g}\{\dfrac{1}{a^2} + \dfrac{1}{al}[\mathrm{ker}'(a/l)\mathrm{ber}(r/l) - \mathrm{kei}'(a/l)\mathrm{bei}(r/l)]\}, & 0 \le r \le a \\ -\dfrac{P}{\pi\rho g a l}[\mathrm{ber}'(a/l)\mathrm{ker}(r/l) - \mathrm{bei}'(a/l)\mathrm{kei}(r/l)]\}, & r > a \end{cases} \tag{2.489}$$

where

$$P = \int_0^a \int_0^{2\pi} F(r,\theta)d\theta dr \tag{2.490}$$

The maximum plate deflection at $r = 0$ can be evaluated as

$$w_{\max} = -\frac{P}{\pi\rho g l^2}[(\frac{l}{a})^2 + (\frac{l}{a})\mathrm{ker}'(\frac{a}{l})] \tag{2.491}$$

This result is obvious if we recall from (4.479) of Chau (2018) that

$$\mathrm{ber}(0) = 1, \quad \mathrm{bei}(0) = 0 \tag{2.492}$$

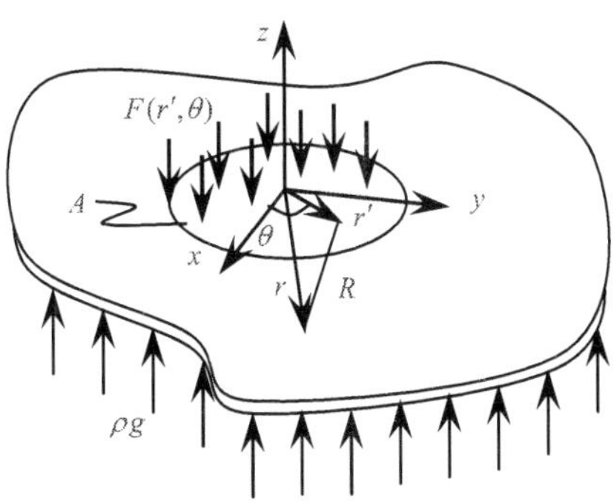

Figure 2.23 An infinite plate on water subject to circular loads

2.12 SUMMARY AND FURTHER READING

In this chapter, we have demonstrated how to apply various theories for ODEs and PDEs to plate bending problems. First of all, we derived the deflection of plate bending obeying a nonhomogeneous fourth order PDE, or to be precise the nonhomogeneous biharmonic equation, which has been introduced and discussed in Section 7.9 of Chau (2018). Plate bending was mentioned briefly in Section 7.9.3 of Chau (2018). Various problems of plate bending are discussed by solving the biharmonic equation. In the present chapter, we discuss the deflection of rectangular plates both with simple supports (Navier's solution and Levy's solution) or with fixed support (by Galerkin's method). Clamped circular plates under uniform, patch, and point loads are considered. Buckling load of rectangular plates was derived for the simplest case of uniaxial compression. It was demonstrated that bending of anisotropic plates can be solved by methods developed for isotropic plates if a proper change of variables is applied. Both free and forced vibrations of plates are considered, using the Rayleigh quotient and Rayleigh-Ritz method. The initial boundary value problem of plate vibrations induced by initial deflection and initial velocity is solved in full detail. The Hertz problem is a circular plate floating on fluid subject to a point force. The problem finds application in aircraft landing on ice sheets at the North Pole. The problem is solved by series solution, by Rayleigh-Ritz method, and by Kelvin function.

For further reading, the general theory of plate bending is referred to in Timoshenko and Woinkowsky-Krieger (1959), vibrations of elastic plates are referred to in Mindlin (2001), and buckling and post-buckling of thin plates is referred to in Bloom and Coffin (2001). Leissa (1969) compiled and compared extensively various solutions for plate vibrations, and it is a very resourceful book containing a lot of formulas, numerical tables, and graphs. The more recent textbook of Ventsel and Krauthammer (2001) includes the bending of elliptical, sector-shaped, semi-circular, triangular, and skew plates. For plates on an elastic foundation (somehow related to the Hertz problem considered in Section 2.11), we referred to Vlasov and Leontev (1966).

2.13 PROBLEMS

Problem 2.1 Consider a rectangular plate subject to compression along both x- and y- directions, as shown in Figure 2.24. Show that the buckling forces N_x and N_y satisfy the following equation:

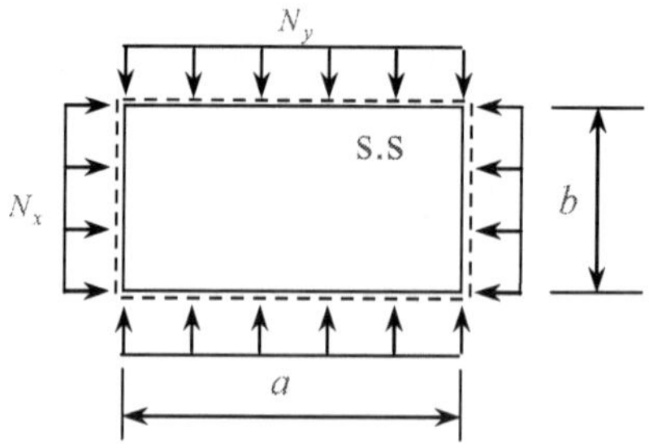

Figure 2.24 Buckling of rectangular plates subject to biaxial compression

$$\frac{m^2 N_x}{\pi^2 a^2 D} + \frac{n^2 N_y}{\pi^2 b^2 D} = (\frac{m^2}{a^2} + \frac{n^2}{b^2})^2 \tag{2.493}$$

Hints: Assume the following solution modes

$$w_{mn}(x, y) = C_{mn} \sin\frac{m\pi x}{a} \sin\frac{n\pi x}{b} \tag{2.494}$$

Problem 2.2 It was given (2.221) that the governing equation of anisotropic plate bending is governed by

$$D_x \frac{\partial^4 w}{\partial x^4} + 2H \frac{\partial^4 w}{\partial x^2 \partial y^2} + D_y \frac{\partial^4 w}{\partial y^4} = p(x, y) \tag{2.495}$$

To prove this, we have to answer the following questions:

(i) For orthotropic materials, we have three orthogonal symmetric planes and for the plane stress problem we have $\sigma_z = \tau_{zx} = \tau_{zy} = 0$. The Hooke's law is given as

$$\sigma_x = E_x \varepsilon_x + \bar{E} \varepsilon_y \tag{2.496}$$

$$\sigma_y = E_y \varepsilon_y + \bar{E} \varepsilon_x \tag{2.497}$$

$$\tau_{xy} = G \gamma_{xy} \tag{2.498}$$

Using (2.24) to (2.38), show that the bending moment and shear forces are

$$M_x = -(D_x \kappa_x + D_y \kappa_y) \tag{2.499}$$

$$M_y = -(D_1 \kappa_x + D_y \kappa_y) \tag{2.500}$$

$$M_{xy} = -2 D_{xy} \kappa_{xy} \tag{2.501}$$

$$Q_x = -\frac{\partial}{\partial x}[D_x \kappa_x + H \kappa_y] \tag{2.502}$$

$$Q_y = -\frac{\partial}{\partial y}[D_y \kappa_y + H \kappa_x] \tag{2.503}$$

(ii) Substitute these results into (2.18) to show that

$$D_x \frac{\partial^2 w}{\partial x^2} + 2H \frac{\partial^2 w}{\partial x \partial y} + D_y \frac{\partial^2 w}{\partial y^2} = p(x, y) \tag{2.504}$$

Problem 2.3 This problem uses Rayleigh's quotient to find the fundamental frequency of a circular plate with clamped support. Figure 2.25 shows such a circular plate with radius R. The following deflection function is found to satisfy both clamped boundary conditions:

$$W = c(R^2 - r^2)^2 \tag{2.505}$$

(i) For the axisymmetric case, the maximum potential energy can be evaluated as:

$$U_{\max} = \frac{D}{2} \int_0^{2\pi} \int_0^R (\frac{\partial^2 W}{\partial r^2} + \frac{1}{r}\frac{\partial W}{\partial r} + \frac{1}{r^2}\frac{\partial^2 W}{\partial \varphi^2})^2 r dr d\varphi \tag{2.506}$$

Show that for the deflection given in (2.505) it is given by

$$U_{\max} = \frac{32\pi DR^6}{3} \tag{2.507}$$

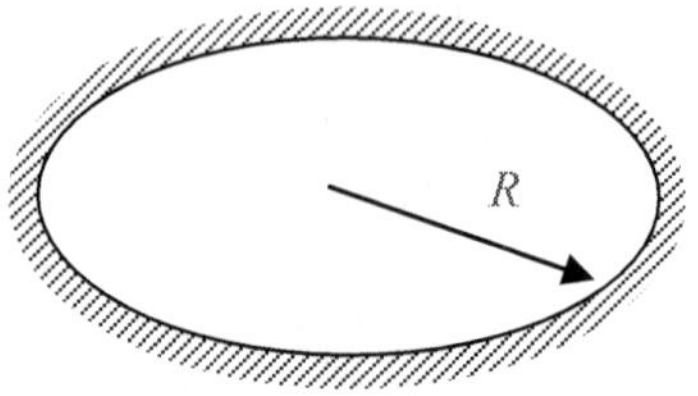

Figure 2.25 Vibrations of clamped circular plates

(ii) For the axisymmetric case, the maximum kinetic energy can be evaluated as:

$$K_{\max} = \frac{\omega^2}{2}\int_0^{2\pi}\int_0^R \rho h W^2(r,\varphi) r dr d\varphi \tag{2.508}$$

Show that for the deflection given in (2.505) $K_{\max}$ is given by

$$K_{\max} = \frac{\omega^2 \pi\rho h R^{10}}{10} \tag{2.509}$$

(iii) Find the fundamental frequency ω_{11} using Rayleigh's quotient, and estimate the error of the present approximation by comparing to the following formula given in Table 2.1 of Leissa (1969).

$$\omega_{11} = 10.2158\frac{1}{R^2}\sqrt{\frac{D}{\rho h}} \tag{2.510}$$

Ans: (iii) The fundamental frequency is

$$\omega_{11} = 8\sqrt{\frac{5}{3}}\left(\frac{1}{R^2}\sqrt{\frac{D}{\rho h}}\right) \tag{2.511}$$

and the estimated error is 1.098%.

Problem 2.4 (i) Show that

$$\sqrt{\lambda_n} r^2 \frac{d}{dr}(Q^2 + R^2) = 2\frac{d}{dr}\left(r^2 \frac{dR}{dr}\frac{dQ}{dr}\right) \tag{2.512}$$

(ii) Integrating (2.512) from 0 to a and applying integration by parts to show that

$$\sqrt{\lambda_n}[r^2(Q^2 + R^2)] - 2\sqrt{\lambda_n}\int_0^a rR^2 dr - 2\sqrt{\lambda_n}\int_0^a rQ^2 dr = 2\left[r^2 \frac{dR}{dr}\frac{dQ}{dr}\right]_0^a \tag{2.513}$$

(iii) Substituting (2.356) into the third term on the left of (2.513), show that

$$\int_0^a rR^2 dr = \frac{1}{4\sqrt{\lambda_n}}\left[\sqrt{\lambda_n} r^2(Q^2 + R^2) - 2r^2\frac{dQ}{dr}\frac{dR}{dr} - 2rQ\frac{dR}{dr} + 2rR\frac{dQ}{dr}\right]_0^a \tag{2.514}$$

CHAPTER THREE

Theory of Shells

3.1 INTRODUCTION

The word "shell" was originally used for describing the hard rigid calcareous covering of an animal, like a snail, mussel, abalone, oyster, clam, turtle, etc. It is also the standard term for the covering of eggs. The cover of nuts and human skulls are also shell structures. In nature, it provides the strongest three-dimensional cover in terms of the most optimum use of materials. Shell structures have been adopted and used in many man-made structures, such as aircraft, automobiles, pressure pipes, silos, water tanks, ships, submarines, rockets, domes, stadiums, tents, chapels, planetariums, opera theatres, cooling towers of power plants, and many others. By definition, the thickness of a shell is thin and is in the order of 400 or 500 times smaller than the span or equivalent radius of the shell structures. For reinforced concrete structures, the shell thickness is in the order of 0.1 m and thus it gives a typical span of about 50 m. Therefore, it is ideal for building sports and storage facilities. The main advantage of shell structures is that there is no internal support, creating an open, unobstructed interior space inside the structure. To increase the stability, some of the shell structures are supported by tension steel cables. Architects admire the elegant simplicity and aesthetic look of shell structures, and examples include the Sydney Opera House in Australia, the Rolex Library at EPFL in Lausanne Switzerland, and the Hong Kong Exhibition and Convention Center. The disadvantages in building shell structures are its cost of construction and the issue of quality control. The centering of shells (or the building of temporary scaffolding in casting or positioning the shell structures) can be very expensive because of the unique design of each shell structure. Leakage of rain water through a shell-type roof is another problem that needs to be addressed properly in the design. Condensation of moisture indoors due to the temperature gradient inside and outside the shells is another pragmatic issue.

In terms of structural mechanics, shells can be classified as lattice-type shell structures (like the geodesic dome or hyperboloid structures) and membrane-type shell structures (like the barrel vault roof and cooling tower). Shell structures can be singly curved or doubly curved. Sometimes, conoid-type shells are also used, in which the shell surface is formed by a system of rotated straight lines such as those used in the hyperbolic paraboloid that we see in the Tokyo TV tower and the Guangzhou TV tower.

Shell structures have been used for thousands of years. Some of the most famous examples include the Pantheon in Rome which was completed in AD 125 and was rebuilt a few times, and the Basilica of the Hagia Sophia in Istanbul built in AD 537. The Pantheon is a spherical shell structure with a diameter of 43.3 m and with a hole opening at the top of the dome of diameter of 7.8 m, allowing sunlight to enter. The interior rises 37 m from the ground to the roof. It remains the world's largest unreinforced dome. To reduce the self-weight of the unreinforced dome, the thickness of the dome was reduced as the top was approached. The lower part of the dome was built using bricks and pumice, and the upper part was made by pumice only. The modern era of shell construction started with the thin concrete shell of the Zeiss planetarium, which was built in 1925 in Germany. In the 1950s to 1970s, a lot of shell structures were built. The largest shell structure in the world is CNIT at Paris built in 1958 with a main span of 219 m, which used a double-skin system connected by cellular shear transferring diaphragms. However, somehow shell construction declined at the end of the 1970s. This may relate to a number of shell structure damages and collapses. For example, on November 1, 1965, a 115-m-tall RC cooling tower of a coal-fired- power station in Ferry Bridge UK collapsed. On September 14, 1970, the Tucker High School gymnasium in the United States, which is in a form of hypar shell, collapsed suddenly and, luckily, there was no fatality but only some injuries. On May 21, 1980, the West Berlin Congress Hall in Germany collapsed due to a corrosion problem and killed one and injured five others. In 1980, cracking was observed in the Kresge Auditorium at the Massachusetts Institute of Technology (MIT) built in 1955 because of a settlement problem of the edge beam.

Theoretically, the first shell theory was derived by Love, the author of the classic book *A Treatise of the Mathematical Theory of Elasticity*, in 1888. Love adopted the same assumption that leads to the Euler-Bernoulli beam theory and the Kirchhoff plate theory; that is, a plane normal to the neutral axis before bending remains a plane and normal to the deformed axis. In other words, no shear deformation is allowed for. Love's theory is referred as the thin shell theory or Love's first approximation theory. Second order theories of thick shells were considered by Lure, Flugge, Byrne, Novozhilov, Vlasov, Reissner and Naghdi. The geometric nonlinear theory of shells has been considered by Naghdi, Nordgren, Sanders, Koiter and Vlasov. For thin shells, Love identified that the membrane solution applied to most parts of the thin shell and the bending solution becomes important in regions near the supports, where membrane forces alone could not satisfy equilibrium and the boundary conditions. In 1912, H. Reissner discovered that bending moments can be calculated separately and superimposed onto the membrane solution near the boundaries. Zoelly in 1915 in his PhD study discovered that unlike plates and columns, there is a great discrepancy between theory and experiments for shells. It turns out that buckling of shell structures is particularly sensitive to imperfections. Cylindrical shell theory was developed by Lloyd Donnell in 1934 and Leslie Morley in 1959. Other contributors to shell theory include Wilhelm Flugge, Ralph Byrne, Eric Reissner, Paul Naghdi, Stephen Timoshenko, V. Vlasov, Lyell Sanders and Warner Koiter. The Sanders-Koiter 21 equations for thin shells were derived based on tensor theory and appear to be very general and included all spherical and cylindrical shells as special cases, but they are not popular among engineers due to the use of tensor.

This chapter serves as an introduction to membrane shell theory and bending shell theory and their applications to engineering structures. Shell theory is difficult to solve analytically, and thus various simplifications are adopted to make the problem mathematically tractable. Reissner's formulation of bending theory for shells of revolution will be solved using hypergeometric functions; and for thin shells, the solution is obtained using the Geckeler-Staerman approximation and Hetenyi approximation. The eighth order PDE of the Vlasov stress function for the bending theory of cylindrical shells will be solved using factorization.

3.2 STRESSES, FORCES, AND MOMENTS IN SHELLS

Figure 3.1 shows the typical geometry, shear and normal forces, and bending and twisting moments of a shell element. The dimension of the element is dx by dy and the thickness of the shell element is h. The surface of the shell can be characterized by two curvatures denoted by $1/r_x$ and $1/r_y$. Another related term is called Gaussian curvature, which is defined as $\kappa = 1/(r_x r_y)$. Using the Kirchhoff-Love hypothesis of plane-remain-plane and the zero axial strain assumption of the mid-surface, we can evaluate the resultant forces per unit length of the normal section by integrating across the thickness as:

$$N_x = \int_{-h/2}^{h/2} \sigma_x (1 - \frac{z}{r_y}) dz, \quad N_y = \int_{-h/2}^{h/2} \sigma_y (1 - \frac{z}{r_x}) dz \tag{3.1}$$

$$N_{xy} = \int_{-h/2}^{h/2} \tau_{xy} (1 - \frac{z}{r_y}) dz, \quad N_{yx} = \int_{-h/2}^{h/2} \tau_{yx} (1 - \frac{z}{r_x}) dz \tag{3.2}$$

$$Q_x = \int_{-h/2}^{h/2} \tau_{xz} (1 - \frac{z}{r_y}) dz, \quad Q_y = \int_{-h/2}^{h/2} \tau_{yz} (1 - \frac{z}{r_x}) dz \tag{3.3}$$

where z is the vertical distance from the mid-surface of the shell, which is positive below the mid-surface and negative above the mid-surface. Note that the side surfaces shown in Figure 3.1 are of trapezoidal shape because of the curvature. Consequently, a correction term is included in the brackets of each of these equations in (3.1) to (3.3) to take into account the effect of curvature of the shell.

Although the shear parallel to the two edges are equal due to complementary shear, the result shear forces are not:

$$\tau_{xy} = \tau_{yx}, \quad N_{xy} \neq N_{yx} \tag{3.4}$$

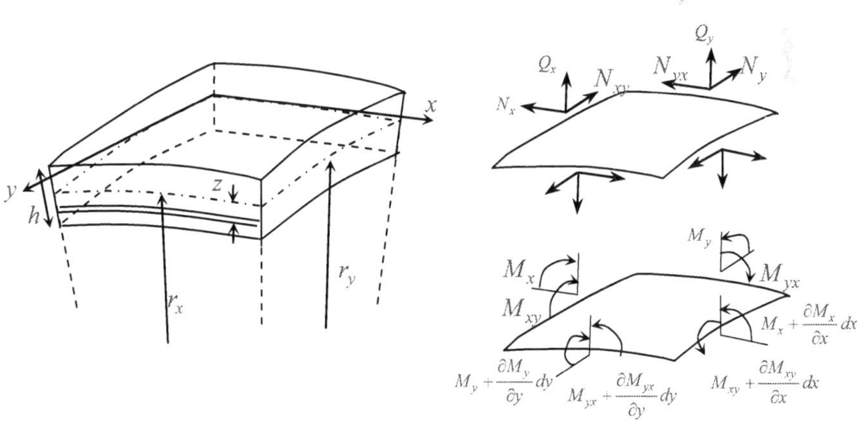

Figure 3.1 Definitions of curvatures, forces and moments on a shell element

Similarly, the bending moments and twisting moments can also be determined by integrating across the thickness:

$$M_x = \int_{-h/2}^{h/2} \sigma_x z(1-\frac{z}{r_y})dz, \quad M_y = \int_{-h/2}^{h/2} \sigma_y z(1-\frac{z}{r_x})dz \tag{3.5}$$

$$M_{xy} = -\int_{-h/2}^{h/2} \tau_{xz} z(1-\frac{z}{r_y})dz, \quad M_{yx} = \int_{-h/2}^{h/2} \tau_{yz} z(1-\frac{z}{r_x})dz \tag{3.6}$$

Figure 3.2 shows the elongation of a fiber of original length l_1 if the neutral surface is subject to an axial strain ε_1. The original length and deformed length of the fiber measured z from the neutral axis are

$$l_1 = (1-\frac{z}{r_x})ds, \quad l_2 = (1+\varepsilon_1)(1-\frac{z}{r'_x})ds \tag{3.7}$$

where r'_x is the radius of curvature after deformation. Using (3.7), the axial strain of the fiber can be evaluated as

$$\varepsilon_x = \frac{l_2 - l_1}{l_1} = \frac{\varepsilon_1}{1-z/r_x} - \frac{z}{1-z/r_x}[\frac{1}{r'_x} - \frac{1}{r_x}] \tag{3.8}$$

Typically, thick shells are having $h/a > 1/20$, whereas thin shells are having $1/1000 < h/a < 1/20$.

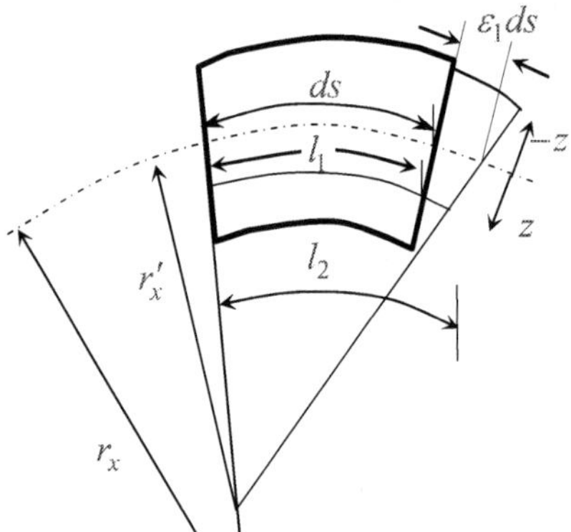

Figure 3.2 Axial strains of the neutral axis and of a length element in bending

If the shell is thin or z is small compared to the radius of curvature, we can set

$$z/r_x, z/r_y \to 0 \tag{3.9}$$

Substitution of (3.9) into (3.8) gives

$$\varepsilon_x = \varepsilon_1 - z(\frac{1}{r'_x} - \frac{1}{r_x}) = \varepsilon_1 - z\chi_x \tag{3.10}$$

where χ represents the change of curvature. A similar expression can also be derived for the axial strain along the y-direction:

$$\varepsilon_y = \varepsilon_2 - z(\frac{1}{r'_y} - \frac{1}{r_y}) = \varepsilon_2 - z\chi_y \tag{3.11}$$

Substitution of these strains into Hooke's law yields

$$\sigma_x = \frac{E}{1-\nu^2}[\varepsilon_1 + \nu\varepsilon_2 - z(\chi_x + \nu\chi_y)] \tag{3.12}$$

$$\sigma_y = \frac{E}{1-\nu^2}[\varepsilon_2 + \nu\varepsilon_1 - z(\chi_y + \nu\chi_x)] \tag{3.13}$$

In view of (3.9), substitution of (3.12) and (3.13) into (3.1) and (3.5) results in the following expressions

$$N_x = \frac{Eh}{1-\nu^2}[\varepsilon_1 + \nu\varepsilon_2], \quad N_y = \frac{Eh}{1-\nu^2}[\varepsilon_2 + \nu\varepsilon_1] \tag{3.14}$$

$$M_x = -D(\chi_x + \nu\chi_y), \quad M_y = -D(\chi_y + \nu\chi_x) \tag{3.15}$$

where

$$D = \frac{Eh^3}{12(1-\nu^2)} \tag{3.16}$$

This bending stiffness D has been derived for plate bending in Chapter 2. More generally, if there is shear stress acting on the lateral sides of the shell element as shown in Figure 3.1, we have

$$\tau_{xy} = (\gamma - 2z\chi_{xy})G \tag{3.17}$$

where γ is the shear strain on the mid-surface of the shell and the second term in (3.17) can be derived as we did for the plate bending in (2.25) of Chapter 2. Substitution of (3.17) into (3.2) and (3.6) gives

$$N_{xy} = N_{yx} = \frac{\gamma Eh}{2(1+\nu)} \tag{3.18}$$

$$M_{xy} = -M_{yx} = D(1-\nu)\chi_{xy} \tag{3.19}$$

For thin shells, the bending effects are not important and membrane forces dominate the deformation of the shell. Thus, membrane theory for shells will be considered in the next section, and bending theory will be postponed to Section 3.6.

3.3 MEMBRANE THEORY FOR AXISYMMETRIC SHELL

Before we consider the membrane theory, we first make the assumption that the curvature terms z/r_x and z/r_y are negligible in the calculation of membrane forces. Consequently, we can simplify the membrane forces as

$$N_x = \int_{-h/2}^{h/2} \sigma_x dz, \quad N_y = \int_{-h/2}^{h/2} \sigma_y dz \tag{3.20}$$

$$N_{xy} = N_{yx} = \int_{-h/2}^{h/2} \tau_{yx} dz \tag{3.21}$$

Being consistent with (3.20) and (3.21), the moment and vertical shear forces are also simplified as

$$M_x = \int_{-h/2}^{h/2} \sigma_x z dz, \quad M_y = \int_{-h/2}^{h/2} \sigma_y z dz \tag{3.22}$$

$$M_{xy} = M_{yx} = \int_{-h/2}^{h/2} \tau_{yz} z dz \tag{3.23}$$

$$Q_x = \int_{-h/2}^{h/2} \tau_{xz} dz, \quad Q_y = \int_{-h/2}^{h/2} \tau_{yz} dz \tag{3.24}$$

Figure 3.3 shows the shear and normal stress on the side surface with normal vector parallels to the x-axis. We see that there are two components of the normal stress σ_x, one from M_x and one from N_x. Similarly, this is also true for τ_{xy}, which depends on both N_{xy} and M_{xy}. On the other hand, τ_{xz} depends only on Q_x. Mathematically, these stress distributions can be expressed as:

$$\sigma_x = \frac{N_x}{h} + \frac{M_x}{h^3/12} z \tag{3.25}$$

$$\tau_{xy} = \frac{N_{xy}}{h} + \frac{M_{xy}}{h^3/12} z \tag{3.26}$$

$$\tau_{xz} = \frac{3}{2}\frac{Q_x}{h}(1 - \frac{4z^2}{h^2}) \tag{3.27}$$

As shown in Figure 3.3, (3.25) and (3.26) are linear functions of depth and (3.27) is a parabolic distribution with depth. That is, once the forces and moments are obtained, we can find the corresponding stress in the shell according to (3.25) to (3.27).

In the membrane shell theory, only the membrane forces given in (3.20) and (3.21) are nonzero, whereas all resultant moment and shear force given in (3.22) to (3.24) are neglected.

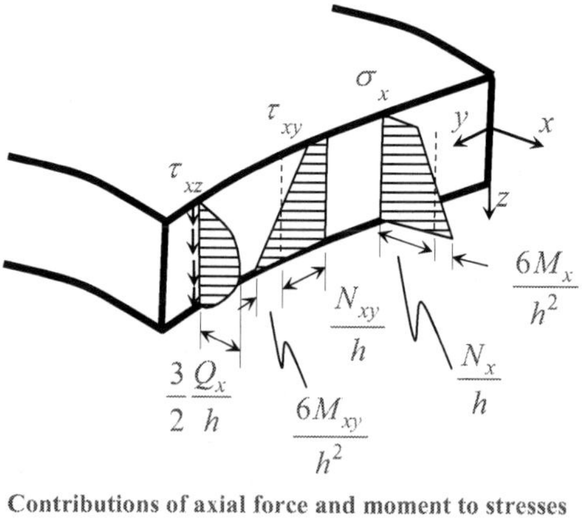

Figure 3.3 Contributions of axial force and moment to stresses

Next, we consider the case that the shells are in the form of a surface of revolution and the loading is axisymmetric with respect to the axis, as shown in Figure 3.4. Shells formed by surfaces of revolution find extensive applications, such as tanks, domes, and other containers. As illustrated in Figure 3.4, the membrane forces on a shell element can be expressed in polar components and the shell element is measured r horizontally from the axis of symmetry. The radius of curvature is r_2, which measures an angle φ from the axis of symmetry. The curve swept by the increasing φ is called a meridian. Thus, the shell element is cut out by two meridians and two parallel circles.

Referring to Figure 3.4, the radius r can be found as

$$r = r_2 \sin\varphi \tag{3.28}$$

The length ds of the shell element is

$$ds = r_1 d\varphi \tag{3.29}$$

The horizontal projection of ds is dr and is given by

$$dr = ds\cos\varphi = r_1 \cos\varphi d\varphi \tag{3.30}$$

Combining (3.29) and (3.30), we get

$$\frac{dr}{d\varphi} = \frac{ds}{d\varphi}\cos\varphi = r_1 \cos\varphi \tag{3.31}$$

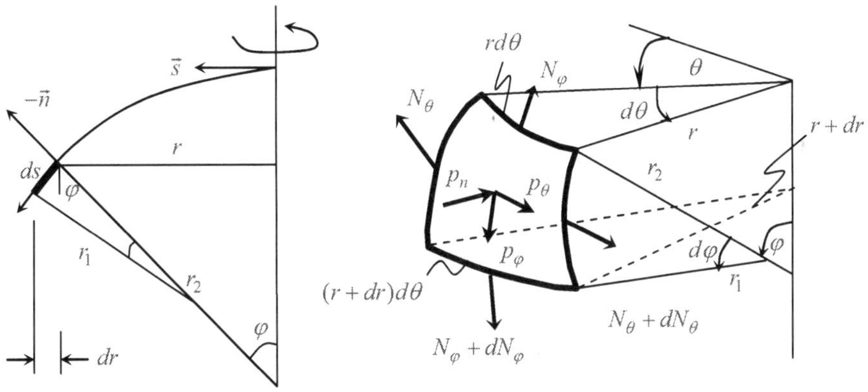

Figure 3.4 Forces in membrane theory for shell of revolution

For the symmetry requirement, we must have

$$dN_\theta = \frac{\partial N_\theta}{\partial \theta} d\theta = 0 \tag{3.32}$$

The loadings on the shell element are also axisymmetric and must be expressible as:

$$p_\theta = 0, \quad p_\varphi = p_\varphi(\varphi), \quad p_n = p_n(\varphi) \tag{3.33}$$

Consequently, the axial forces must be expressible as

$$N_{\varphi\theta} = N_{\theta\varphi} = 0, \quad N_\varphi = N_\varphi(\varphi), \quad N_\theta = N_\theta(\varphi) \tag{3.34}$$

As shown in Figure 3.5, the tangential forces must be balanced by a horizontal resultant force R_r acting along the r-direction and its value is

$$R_r = N_\theta d\theta ds = N_\theta r_1 d\theta d\varphi \tag{3.35}$$

This horizontal resultant force can be decomposed into two components, one normal to the shell element and the other parallel to the meridian direction. By referring to Figure 3.6, we have

$$R_n = R_r \sin\varphi = N_\theta r_1 \sin\varphi d\theta d\varphi \tag{3.36}$$

$$R_\varphi = R_r \cos\varphi = N_\theta r_1 \cos\varphi d\theta d\varphi \tag{3.37}$$

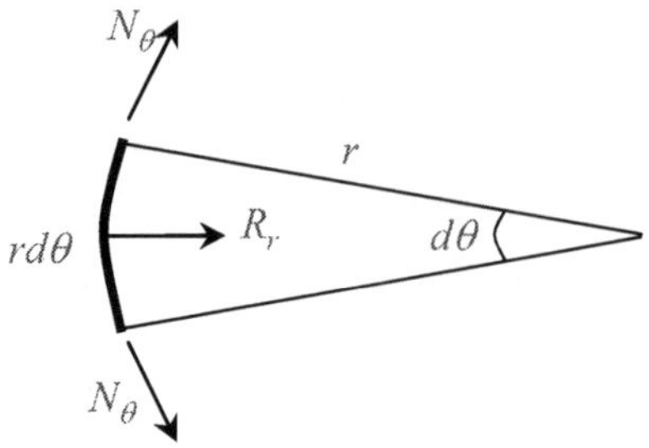

Figure 3.5 Force components of R_r

The force equilibrium along the φ-direction gives

$$-R_\varphi - N_\varphi r d\theta + (N_\varphi + \frac{dN_\varphi}{d\varphi} d\varphi)(r + dr)d\theta + p_\varphi r r_1 d\theta d\varphi = 0 \tag{3.38}$$

Substituting (3.37) into (3.38) and in view of (3.29), we obtain

$$N_\varphi r_1 \cos\varphi + \frac{dN_\varphi}{d\varphi} r - N_\theta r_1 \cos\varphi + p_\varphi r r_1 = 0 \tag{3.39}$$

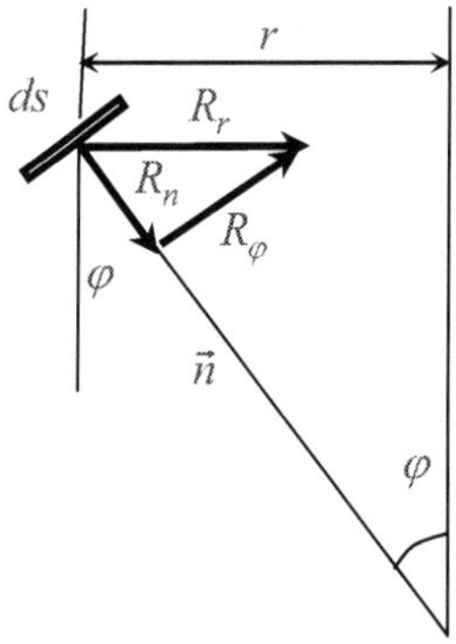

Figure 3.6 Projection of the radial force R_r

On the other hand, we note that

$$\frac{d(N_\varphi r)}{d\varphi} = r\frac{dN_\varphi}{d\varphi} + N_\varphi \frac{dr}{d\varphi} = r\frac{dN_\varphi}{d\varphi} + N_\varphi r_1 \cos\varphi \tag{3.40}$$

The last of (3.40) results from (3.31). Applying (3.40) into (3.39), we arrive at

$$\frac{d}{d\varphi}(N_\varphi r) - N_\theta r_1 \cos\varphi + p_\varphi r r_1 = 0 \tag{3.41}$$

The equilibrium along the n-direction is

$$R_n + N_\varphi r d\theta d\varphi + p_n r r_1 d\theta d\varphi = 0 \tag{3.42}$$

Substitution of (3.36) into (3.42) yields

$$N_\theta r_1 \sin\varphi + N_\varphi r + p_n r r_1 = 0 \tag{3.43}$$

In view of (3.28), (3.43) can be simplified as

$$\frac{N_\theta}{r_2} + \frac{N_\varphi}{r_1} = -p_n \tag{3.44}$$

In summary, for axisymmetric shells, the two governing equations for two unknowns N_φ and N_θ are given in (3.41) and (3.44).

We now proceed to solve this system of equations. Adding the result of (3.41), multiplying by $\sin\varphi$ to the result of (3.43), and multiplying by $\cos\varphi$, we get

$$\frac{d}{d\varphi}(N_\varphi r \sin\varphi) = -r_1 r_2 (p_\varphi \sin^2\varphi + p_n \sin\varphi\cos\varphi) \tag{3.45}$$

Equation (3.45) can be readily integrated to get

$$N_\varphi = -\frac{1}{r_2\sin^2\varphi}\{\int r_1 r_2 (p_\varphi \sin^2\varphi + p_n \sin\varphi\cos\varphi)d\varphi + C\} = \frac{-R_V(\varphi)}{2\pi r\sin\varphi} \tag{3.46}$$

where

$$R_V(\varphi) = 2\pi\{\int r_1 r_2 (p_\varphi \sin^2\varphi + p_n \sin\varphi\cos\varphi)d\varphi + C\} \tag{3.47}$$

The integration on the right of (3.47) results from the distributed loads and the integration constant relates to concentrated line load. Once (3.46) is obtained, we can use (3.44) to determine N_θ as:

$$N_\theta = -(p_n + \frac{N_\varphi}{r_1})r_2 \tag{3.48}$$

Physically, the term R_V equals the total vertical force. To see this, Figure 3.7 considers the forces applied on a truncated dome. By referring to the free body diagram in Figure 3.7, we have the vertical force component from the membrane force N_φ as

$$R_\varphi = 2\pi r N_\varphi^V = 2\pi r_2 \sin\varphi N_\varphi^V = N_\varphi 2\pi r_2 \sin^2\varphi \tag{3.49}$$

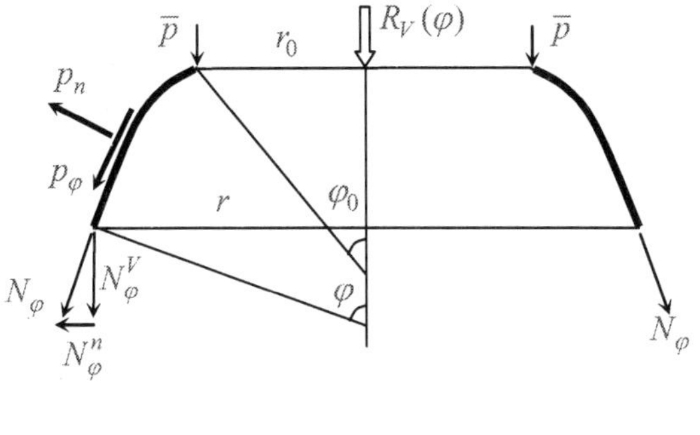

Figure 3.7 Physical meaning of $R_V(\varphi)$

Vertical force equilibrium requires:

$$R_V(\varphi) + R_\varphi = 0 \tag{3.50}$$

Substitution of (3.49) into (3.50) yields

$$N_{\varphi}=\frac{-R_V(\varphi)}{2\pi r_2\sin^2\varphi}=\frac{-R_V(\varphi)}{2\pi r\sin\varphi} \tag{3.51}$$

If the vertical force from above the shell is compressive, from (3.51) the membrane force N_{φ} is also compressive.

If the vertical force from above is applied as a ring line load at the upper surface of the truncated dome shown in Figure 3.7, we have

$$R_V(\varphi_0)=2\pi C=2\pi r_0\bar{p} \tag{3.52}$$

This gives the integration constant C in (3.47) and also gives the value of R_V. Equations (3.51) and (3.52) are general expressions that are applicable to conical shells, spherical shells, shallow shells, and open shells (or truncated shells). A number of special cases will be considered in the following sections.

3.3.1 Dome under Concentrated Apex Load

In this section, we consider a dome shell subject to a vertical point force at the tip of the dome, as shown in Figure 3.8. If the self-weight of the shell is negligible, we have

$$p_{\theta}=p_{\varphi}=p_n=0 \tag{3.53}$$

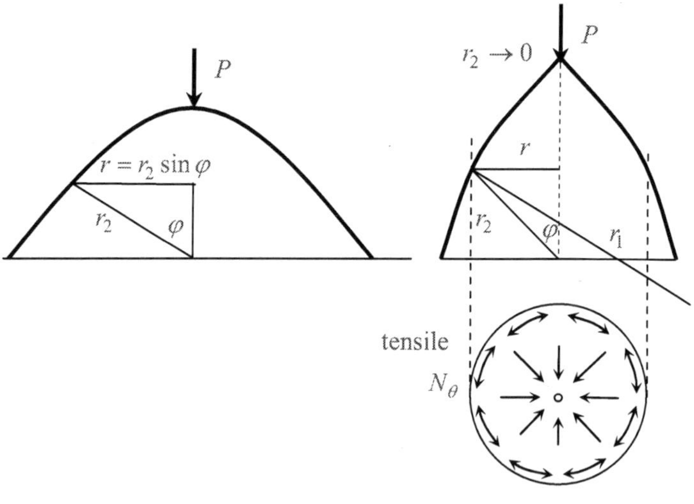

Figure 3.8 Dome under concentrated apex load

Obviously, the vertical force from above the shell is thus

$$R_V(\varphi)=P \tag{3.54}$$

Substitution of (3.54) into (3.51) gives

$$N_\varphi = -\frac{P}{2\pi r_2 \sin^2 \varphi} \tag{3.55}$$

Substitution of (3.55) into (3.48) gives the other tangential membrane force

$$N_\theta = -\frac{P}{2\pi r_1 \sin^2 \varphi} \tag{3.56}$$

The lower diagram in Figure 3.8 illustrates the mechanism of why tensile N_θ is generated. If the apex of the dome is approached, we have the following singularity behavior of the membrane force

$$N_\theta \approx N_\varphi \propto \frac{1}{\varphi^2}, \quad \varphi \to 0 \tag{3.57}$$

It is of singularity of second order near the apex. Infinite membrane forces also lead to infinite bending moments at the apex. For the case of a pointed dome shown in the right diagram of Figure 3.8, φ is not zero and we have the following singularity:

$$N_\varphi \propto \frac{1}{r_2}, \quad r_2 \to 0 \tag{3.58}$$

In this case, N_φ is singular but the meridian membrane force N_θ is finite.

3.3.2 Truncated Dome under Ring Load

We see from the last section that the membrane force becomes singular, and this is clearly undesirable from a practical point of view. Therefore, it is more often that a truncated dome is used and the vertical load from structural components above the dome is applied through a ring load, as shown in Figure 3.9. The axial stiffness $E_r A_r$ of the support ring is given whilst the support ring is assumed to have no torsional stiffness.

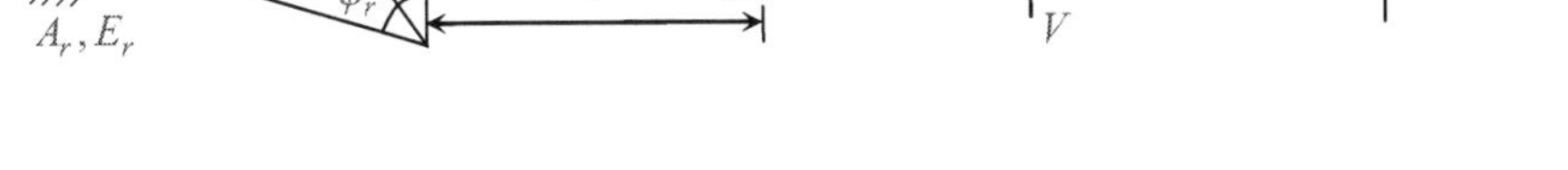

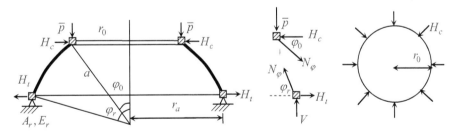

Figure 3.9 Spherical dome under ring loads

The free body diagram at the top compression ring gives the following vertical force equilibrium:

$$\bar{p} + N_\varphi \sin \varphi_0 = 0 \tag{3.59}$$

From Figure 3.9, the horizontal force equilibrium of the top ring gives the radial force per length as

$$H_c = -N_\varphi \cos\varphi_0 = \cos\varphi_0 \frac{\overline{p}}{\sin\varphi_0} = \overline{p}\cot\varphi_0 \tag{3.60}$$

From Problem 3.1, we can prove that the compressive force in the ring can be evaluated as

$$S_c = \overline{p} r_0 \cot\varphi_0 \tag{3.61}$$

For the lower compression ring at the support, horizontal force equilibrium gives

$$H_t = -N_\varphi \cos\varphi_r \tag{3.62}$$

Vertical force equilibrium at the support ring gives

$$V + N_\varphi \sin\varphi_r = 0 \tag{3.63}$$

Substitution of (3.63) into (3.62) yields

$$H_t = \frac{V}{\sin\varphi_r}\cos\varphi_r = V\cot\varphi_r \tag{3.64}$$

From the vertical force equilibrium of the whole truncated dome, the vertical reaction force at the support relates to the ring load as

$$V = \frac{R_V(\varphi)}{2\pi r_a} = \frac{2\pi r_0 \overline{p}}{2\pi r_a} = \frac{r_0 \overline{p}}{r_a} \tag{3.65}$$

Applying the result of Problem 3.1, we obtain the tensile force in the lower support ring as

$$S_t = \overline{p} r_0 \cot\varphi_r \tag{3.66}$$

In conclusion, we need to install a compressive ring at the top of the truncated dome whereas a tensile ring is needed at the support. The design of the support ring is considered next.

3.3.3 Compatibility at Ring Foundation

If the Young's modulus and cross-sectional area of the support ring are known, the tensile strain of the ring can be determined as

$$\varepsilon_r = \frac{S_t}{E_r A_r} \tag{3.67}$$

On the other hand, the strain in the shell can be determined from Hooke's law as:

$$\varepsilon_\theta = \frac{1}{E}(\sigma_\theta - \nu\sigma_\varphi) = \frac{1}{Eh}(N_\theta - \nu N_\varphi) \tag{3.68}$$

In general, the tangential strain of the ring does not match the tangential strain of the shell

$$\varepsilon_\theta \neq \varepsilon_r \tag{3.69}$$

Thus, the strain is not compatible. To see this, substituting (3.66) into (3.67) we get

$$\varepsilon_r = \frac{\overline{p} r_0 \cot\varphi_r}{E_r A_r} \tag{3.70}$$

Consider the special case of a spherical dome, which by the way is a popular choice for dome structures. Thus, we have for a spherical shell $r_1 = r_2 = a$, and (3.55) gives

$$N_\varphi(\varphi_r) = -\frac{P}{2\pi r_2 \sin^2 \varphi_r} = -\frac{2\pi a \bar{p} \sin \varphi_0}{2\pi a \sin^2 \varphi_r} = -\frac{\bar{p} \sin \varphi_0}{\sin^2 \varphi_r} \tag{3.71}$$

For spherical shells, (3.44) is simplified to

$$N_\theta = -N_\varphi - a p_n = \frac{\bar{p} \sin \varphi_0}{\sin^2 \varphi_r} \tag{3.72}$$

Substitution of (3.71) and (3.72) into (3.68) gives the tangential strain of the shell

$$\varepsilon_\theta = \frac{1}{Eh}(N_\theta - \nu N_\varphi) = \frac{\bar{p}(1+\nu)\sin \varphi_0}{Eh \sin^2 \varphi_r} = \varepsilon_{shell} \tag{3.73}$$

The membrane forces of the shell are

$$N_\theta = -N_\varphi = \frac{\bar{p} \sin \varphi_0}{\sin^2 \varphi}, \quad \varphi_0 < \varphi < \varphi_r \tag{3.74}$$

This incompatibility between the ring and shell will lead to a boundary disturbance such that a moment will be induced near the support. In particular, the torsional stiffness of the ring cannot be neglected and it leads to both horizontal reaction and moment at the support. In short, membrane theory of a shell alone cannot model the behavior of shells near the support. However, more detailed analyses have shown that if the shell is thin, the moments are only nonzero near the support, whereas most of the shell is still under membrane forces only.

For non-spherical shells, we have

$$R_V(\varphi_0) = 2\pi r_0 \bar{p} \tag{3.75}$$

The corresponding membrane forces are

$$N_\varphi(\varphi_r) = -\frac{R_V(\varphi)}{2\pi r_2 \sin^2 \varphi} = -\frac{r_0 \bar{p}}{r_2 \sin^2 \varphi_r} \tag{3.76}$$

$$N_\theta = -r_2 p_n - \frac{r_2}{r_1} N_\varphi = \frac{r_0 \bar{p}}{r_1 \sin^2 \varphi_r} \tag{3.77}$$

In general, as discussed in Article 131 of Timoshenko and Woinowsky-Krieger (1959), the edge ring can be designed according to the following compatibility conditions:

$$\delta_{ring} = \delta_{shell}, \quad \left(\frac{dw}{d\varphi}\right)_{ring} = \left(\frac{dw}{d\varphi}\right)_{shell} \tag{3.78}$$

Such analysis will not be discussed here. This problem was solved by Girkmann in 1956 for a particular shallow shell. Figure 276 of Timoshenko and Woinowsky-Krieger (1959) presented Girkmann's numerical result of M_φ, which shows that the bending moment appears only near the support ring.

We should discuss a little more about the "Girkmann problem" here. In 2008, Pitkaranta, Babuska and Szabo proposed a challenge in the Bulletin of the IACM (the International Association for Computational Mechanics), *Expressions*, to develop a FEM model with less than 5% error from the Girkmann problem. They then reported in 2009 *Expressions* that among the FEM numerical solutions submitted from 15 responses, only the four p-version FEM models were accurate, whereas most of the eleven h-version FEM models were not accurate. The bending moment at the support ring was wildly off from the Girkmann solution. Thus, they concluded that the Girkmann problem should serve as a good benchmark problem for calibrating the accuracy of commercial FEM programs. Indeed, analytic solutions are important in guiding the use or avoiding the misuse of FEM-type numerical models.

3.4 SHELL OF REVOLUTION UNDER UNIFORM LOAD

In the last section, we have assumed that the shell is thin and the self-weight is negligible. In this section, a shell of revolution under uniform self-weight is considered, as shown in Figure 3.10. The radius of the spherical shell is a and the thickness of the shell is h. For shells under self-weight, we have the uniform loads being

$$p_n = g\cos\varphi, \quad p_\varphi = g\sin\varphi \tag{3.79}$$

where g is the unit dead load per length (not to be confused with the gravitational constant).

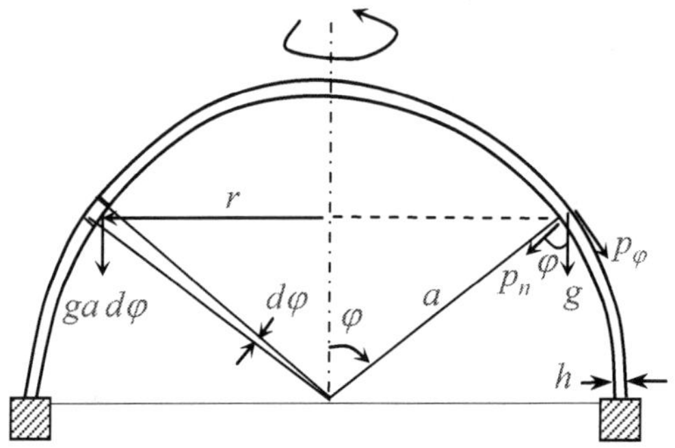

Figure 3.10 Spherical dome with support ring under self-weight

In addition, we have

$$r_1 = r_2 = a, \quad r = a\sin\varphi \tag{3.80}$$

Substitution of (3.79) into (3.47) gives

$$R_V(\varphi) = \int_0^\varphi 2\pi rgad\varphi = 2\pi a^2 g\int_0^\varphi \sin\varphi d\varphi = 2\pi a^2 g(1-\cos\varphi) \tag{3.81}$$

Consequently, the membrane forces can be determined from (3.46) and (3.48)

$$N_\varphi(\varphi_r) = -\frac{R_V(\varphi)}{2\pi a\sin^2\varphi} = -\frac{ag(1-\cos\varphi)}{\sin^2\varphi} = -\frac{ag}{1+\cos\varphi} \tag{3.82}$$

$$N_\theta = -ap_n - N_\varphi = \frac{ag}{1+\cos\varphi} - ag\cos\varphi = ag[\frac{1}{1+\cos\varphi} - \cos\varphi] \tag{3.83}$$

The tangential membrane force N_θ is zero at a particular value of φ. To see that, we set

$$N_\theta = 0 \tag{3.84}$$

Substitution of (3.84) into (3.83) leads to

$$\cos\varphi = \frac{1}{1+\cos\varphi} \tag{3.85}$$

Rearranging (3.85), we get

$$\cos^2\varphi + \cos\varphi - 1 = 0 \tag{3.86}$$

The solution is readily obtained as

$$\cos\varphi = \frac{-1 \pm \sqrt{5}}{2} \tag{3.87}$$

The critical angle is therefore determined as

$$\varphi_1 = \cos^{-1}\left(\frac{-1+\sqrt{5}}{2}\right) = 51.8273^\circ \tag{3.88}$$

This result had been known to scientists and mathematicians for a long time, at least since the time of Mascheroni (1750-1800). Figure 3.11 plots both the membrane forces N_θ and N_φ versus φ. As expected from (3.88), N_φ is tensile for $\varphi < \varphi_1$.

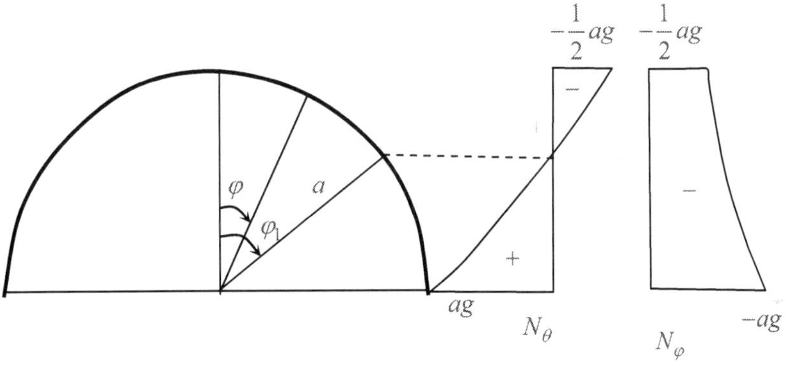

Figure 3.11 Membrane forces for a spherical shell under self-weight

For RC shells, we should avoid the concrete shell being subject to tensile force because of the weak tensile strength of concrete. Tensile cracking of concrete will cause eventual rusting of reinforcement and leakage of water into the structure. Therefore, it is a normal practice to chop off the upper part of the spherical shell to form a shell with an opening to avoid the tensile N_θ. Therefore, a spherical shell with an opening is considered next.

3.4.1 Spherical Shell with Opening

For the case of a spherical shell with an opening, Figure 3.12 shows a spherical dome subject to self-weight and ring load. First, we consider the case of zero ring load and we have for this case

$$r = a\sin\varphi \tag{3.89}$$

$$p_n = g\cos\varphi \tag{3.90}$$

Equation (3.81) is revised as:

$$R_V(\varphi) = g\int_{\varphi_0}^{\varphi} 2\pi r a d\varphi = 2\pi a^2 g\int_{\varphi_0}^{\varphi} \sin\varphi d\varphi = 2\pi a^2 g(\cos\varphi_0 - \cos\varphi) \tag{3.91}$$

Subsequently, (3.82) and (3.83) become

$$N_\varphi(\varphi) = -\frac{R_V(\varphi)}{2\pi a\sin^2\varphi} = -\frac{ag}{\sin^2\varphi}(\cos\varphi_0 - \cos\varphi) \tag{3.92}$$

$$N_\theta = -ap_n - N_\varphi = ag[\frac{\cos\varphi_0 - \cos\varphi}{\sin^2\varphi} - \cos\varphi] \tag{3.93}$$

Similar to the case of no opening, the membrane force can also be negative. That is, we can set

$$N_\theta = 0 \tag{3.94}$$

Substitution of (3.94) into (3.93) gives

$$\cos\varphi_0 - 2\cos\bar{\varphi} + \cos^3\bar{\varphi} = 0 \tag{3.95}$$

The solution of (3.95) gives the critical value of φ that yields zero tangential membrane force

$$\bar{\varphi} = \varphi_{crit} \tag{3.96}$$

For the case of $\varphi_0 = \pi/4$, we have $\varphi_{crit} = 0.3849\pi$ (or 69.28°). Note that the critical angle obtained in (3.96) differs from that for a spherical shell without an opening. The actual value of this critical value is a function of φ_0.

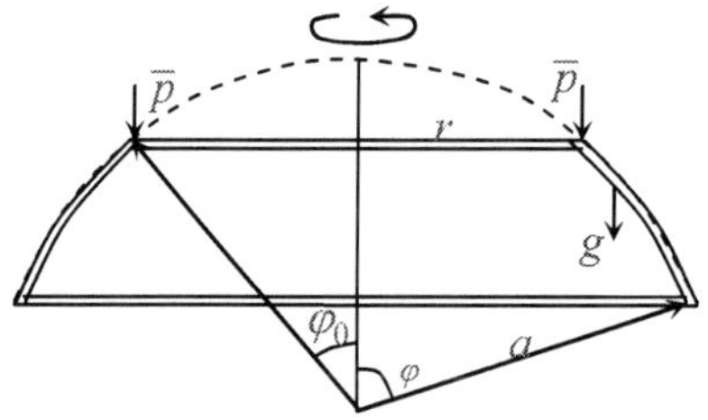

Figure 3.12 Spherical shells with an opening under self-weight and ring load

For nonzero ring load, (3.91) is replaced by

$$R_V(\varphi) = 2\pi a^2 g\int_{\varphi_0}^{\varphi} \sin\varphi d\varphi + 2\pi r\bar{p}$$
$$= 2\pi a^2 g(\cos\varphi_0 - \cos\varphi) + 2\pi a\sin\varphi_0\bar{p} \tag{3.97}$$

Subsequently, the membrane forces are updated as

$$N_\varphi(\varphi) = -\frac{ag}{\sin^2\varphi}(\cos\varphi_0 - \cos\varphi) - \frac{\bar{p}\sin\varphi_0}{\sin^2\varphi} \tag{3.98}$$

$$N_\theta = ag[\frac{\cos\varphi_0 - \cos\varphi}{\sin^2\varphi} - \cos\varphi] + \frac{\bar{p}\sin\varphi_0}{\sin^2\varphi} \quad (3.99)$$

3.4.2 Spherical Fluid Container

This section considers the case of a spherical fluid tank resting on a support ring as illustrated in Figure 3.13. Spherical tanks are commonly used in wastewater treatment plants and liquid gas containers. If the container stores gas, the shell is subjected to a uniform pressure p_0. If the container stores liquid, it is subject to a pressure that increases linearly with depth.

(i) Membrane forces under constant pressure

For the case of uniform pressure, Figure 3.14 shows the force equilibrium of the membrane force and the uniform pressure. More specifically, we have

$$\pi a^2 p_0 = 2\pi a N_\varphi \quad (3.100)$$

Solving for the membrane force from (3.100), we obtain

$$N_\varphi = \frac{a p_0}{2} \quad (3.101)$$

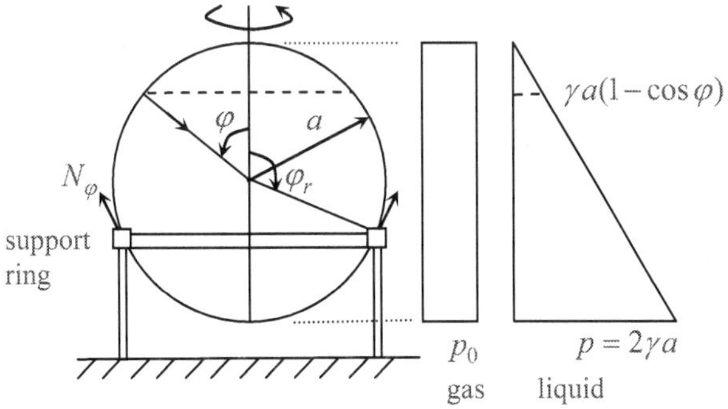

Figure 3.13 A spherical shell under gas and liquid pressures

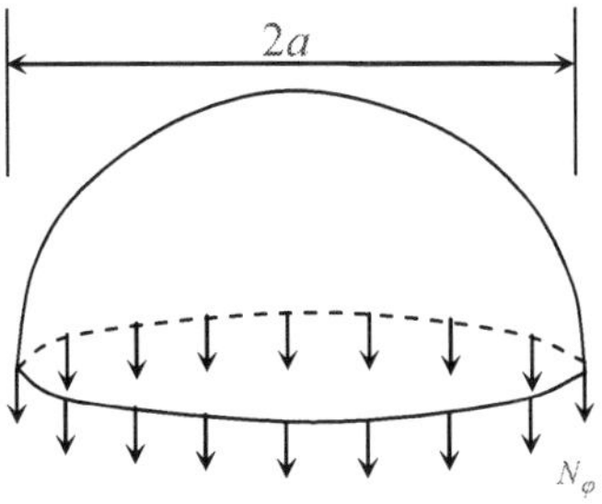

Figure 3.14 For equilibrium for spherical shell under gas pressure

Due to the symmetry of the problem, we have

$$N_\theta = N_\varphi = \frac{ap_0}{2} \tag{3.102}$$

(ii) Linearly varying pressure

For spherical tanks, we have

$$r_1 = r_2 = a \tag{3.103}$$

For the case of linearly increasing pressure, the loading is given by

$$p_\varphi = 0, \quad p_n = -\gamma_w h = -\gamma_w a(1-\cos\varphi) \tag{3.104}$$

where the normal pressure is a linearly increasing function of depth h, which in turn can be expressed in terms of φ. Substitution of (3.104) into (3.46) gives

$$N_\varphi = \frac{\gamma_w a^2}{\sin^2\varphi}[\int (1-\cos\varphi)\sin\varphi\cos\varphi d\varphi + C] \tag{3.105}$$

Vertical force equilibrium at any angle φ is given by

$$2\pi r_0 N_\varphi \sin\varphi + R_V = 0 \tag{3.106}$$

Substitution of (3.105) into (3.106) gives

$$\begin{aligned} R_V(\varphi) &= -2\pi a^3\gamma_w \int_0^\varphi (1-\cos\varphi)\sin\varphi\cos\varphi d\varphi \\ &= -2\pi a^3\gamma_w[\frac{1}{6} - \frac{1}{2}\cos^2\varphi(1-\frac{2}{3}\cos\varphi)], \quad 0<\varphi<\varphi_r \end{aligned} \tag{3.107}$$

In obtaining the last expression in (3.107), we have used the following integration formulas:

$$\int \sin\varphi\cos^2\varphi d\varphi = -\frac{1}{3}\cos^3\varphi \tag{3.108}$$

$$\int \sin\varphi\cos\varphi d\varphi = -\frac{1}{2}\cos^2\varphi \tag{3.109}$$

The total weight of the water in the tank can be determined as

$$W = \frac{4\pi a^3\gamma_w}{3} \tag{3.110}$$

For $\varphi > \varphi_r$, the reaction from the support ring needs to be taken into consideration. Thus, the vertical reaction is revised as

$$\begin{aligned} R_V(\varphi) &= -\frac{4\pi a^3\gamma_w}{3} - 2\pi a^3\gamma_w[\frac{1}{6} - \frac{1}{2}\cos^2\varphi(1-\frac{2}{3}\cos\varphi)] \\ &= -2\pi a^3\gamma_w[\frac{5}{6} - \frac{1}{2}\cos^2\varphi(1-\frac{2}{3}\cos\varphi)], \quad \varphi_r<\varphi<\pi \end{aligned} \tag{3.111}$$

Recall from (3.82) that

$$N_\varphi(\varphi) = -\frac{R_V(\varphi)}{2\pi a\sin^2\varphi} \tag{3.112}$$

Substitution of (3.107) and (3.111) into (3.112) gives

$$
\begin{aligned}
N_\varphi(\varphi) &= \frac{a^2\gamma_w}{\sin^2\varphi}\left[\frac{1}{6}-\frac{1}{2}\cos^2\varphi(1-\frac{2}{3}\cos\varphi)\right], \quad 0<\varphi<\varphi_r \\
&= \frac{a^2\gamma_w}{\sin^2\varphi}\left[\frac{5}{6}-\frac{1}{2}\cos^2\varphi(1-\frac{2}{3}\cos\varphi)\right], \quad \varphi_r<\varphi<\pi
\end{aligned}
\tag{3.113}
$$

The first part of (3.113) can be simplified as

$$
\begin{aligned}
N_\varphi(\varphi) &= \frac{\gamma_w a^2}{6\sin^2\varphi}[1-\cos^2\varphi(3-2\cos\varphi)] \\
&= \frac{\gamma_w a^2}{6(1-\cos^2\varphi)}[1-\cos^2\varphi+2\cos^2\varphi(\cos\varphi-1)] = \frac{\gamma_w a^2}{6}\left[1-\frac{2\cos^2\varphi}{1+\cos\varphi}\right]
\end{aligned}
\tag{3.114}
$$

The second part of (3.113) can be reduced to

$$
\begin{aligned}
N_\varphi(\varphi) &= \frac{a^2\gamma_w}{6\sin^2\varphi}[5-\cos^2\varphi(3-2\cos\varphi)] \\
&= \frac{\gamma_w a^2}{6(1-\cos^2\varphi)}[5(1-\cos^2\varphi)+2\cos^2\varphi(1+\cos\varphi)] \\
&= \frac{\gamma_w a^2}{6}\left[5+\frac{2\cos^2\varphi}{1-\cos\varphi}\right]
\end{aligned}
\tag{3.115}
$$

Consequently, (3.113) is simplified to

$$
\begin{aligned}
N_\varphi(\varphi) &= \frac{\gamma_w a^2}{6}\left[1-\frac{2\cos^2\varphi}{1+\cos\varphi}\right], \quad 0<\varphi<\varphi_r \\
&= \frac{\gamma_w a^2}{6}\left[5+\frac{2\cos^2\varphi}{1-\cos\varphi}\right], \quad \varphi_r<\varphi<\pi
\end{aligned}
\tag{3.116}
$$

The other membrane force N_θ can be found from (3.83) as

$$
\begin{aligned}
N_\theta(\varphi) &= \frac{\gamma_w a^2}{6}\left[5-6\cos\varphi+\frac{2\cos^2\varphi}{1+\cos\varphi}\right], \quad 0<\varphi<\varphi_r \\
&= \frac{\gamma_w a^2}{6}\left[1-6\cos\varphi-\frac{2\cos^2\varphi}{1-\cos\varphi}\right], \quad \varphi_r<\varphi<\pi
\end{aligned}
\tag{3.117}
$$

Therefore, there is a jump in N_θ and N_φ at $\varphi = \varphi_r$. These membrane forces are illustrated in Figure 3.15. The support ring clearly induces an unwanted disturbance of the membrane forces that will further induce bending moment. A simple way to minimize this support disturbance is to use a support truss with a reaction parallel to the tangential membrane locally at the support, as shown in Figure 3.16. If the tank is not fully filled, there is another force discontinuity at the water surface. Figure 3.16 also illustrates the unwanted effect of non-zero bending moment at the location of the liquid level.

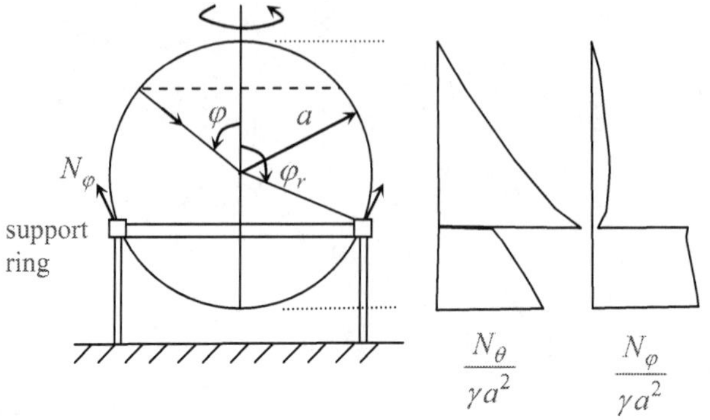

Figure 3.15 Membrane forces for spherical shell under liquid pressure

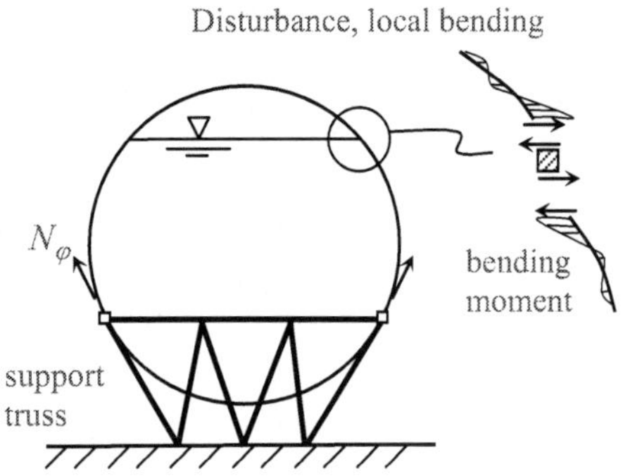

Figure 3.16 Local bending for partially filled spherical tank

3.4.3 Conical Shells

In this section, the special case of a conical shell is considered. The two examples discussed in this section are a conical shell subject to an apex point load and a liquid tank with a conical bottom.

(i) Concentrated apex load

For conical shells, as shown in Figure 3.17 we have

$$r_1 \to \infty, \quad \varphi = \alpha = const. \tag{3.118}$$

The tension in the support ring at the base of the conical shell can be evaluated as (see Problem 3.1):

$$S_t = |H| r_{\max} = r_{\max} |N\varphi(r_{\max})| \cos\alpha \tag{3.119}$$

The horizontal force H at the support can be found by horizontal force equilibrium, as shown in Figure 3.17. Vertical force equilibrium gives

$$\left|N\varphi(r_{\max})\right| = \frac{P}{2\pi r_{\max}\sin\varphi} \tag{3.120}$$

Substitution of (3.120) into (3.119) results in

$$S_t = \frac{P\sin\alpha}{2\pi\sin\varphi} = \frac{P}{2\pi}\cot\alpha \tag{3.121}$$

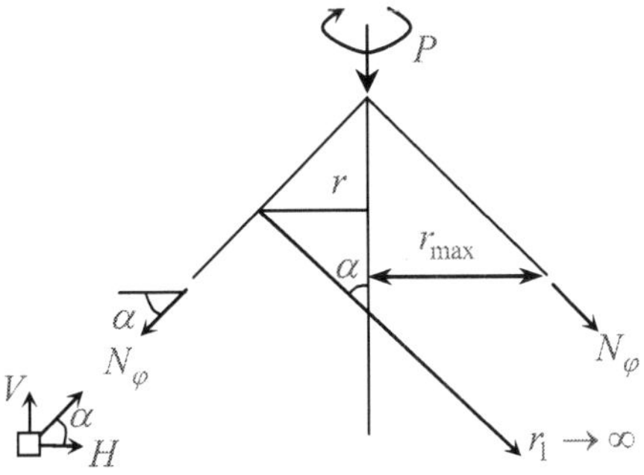

Figure 3.17 Force equilibrium at the support ring of a conical shell subject to a concentrated apex force *P*

The membrane forces can be readily found as

$$N_\varphi = -\frac{R_V(\varphi)}{2\pi r\sin\varphi} = -\frac{P}{2\pi r\sin\alpha} \tag{3.122}$$

$$N_\theta = 0 \tag{3.123}$$

The tangential membrane is zero since $r_1 \to \infty$ and $p_n = 0$. At the apex, $r \to 0$, the membrane tends to infinity (i.e., $N_\varphi \to -\infty$). Thus, local yielding is expected to occur. In practice, as long as the "point" load is actually distributed, the membrane force will be finite.

(ii) Liquid storage tank

The second example of a conical shell is shown in Figure 3.18. The base of a liquid storage container is an inverted conical shell. The reactions of the support ring are denoted as H and V as shown.

Note that the angle φ relates to α by

$$\varphi = \frac{\pi}{2} + \alpha \tag{3.124}$$

Consequently, using the sum rule of sine function, we have

$$\sin\varphi = \cos\alpha \tag{3.125}$$

In view of (3.125) and Figure 3.18, we have

$$p_n = -\gamma z - p_0, \quad r_2 = \frac{r}{\cos\alpha}, \quad r_1 \to \infty \tag{3.126}$$

where the pressure p_0 is the water pressure resulting from liquid above the inverted conical shell. The angle α can be expressed in terms of r, h, and z as

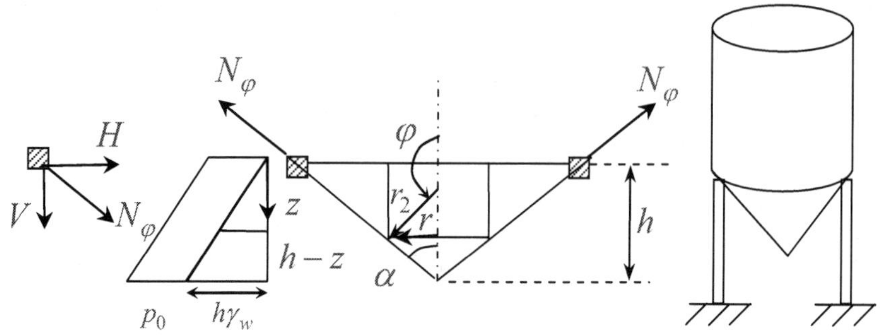

Figure 3.18 An inverted conical shell at the base of a liquid storage tank

$$\tan\alpha = \frac{r}{h-z} \tag{3.127}$$

The water force acting below z can be calculated as

$$G = \gamma\pi r^2[z + \frac{1}{3}(h-z)] + p_0\pi r^2 = R_V(z) \tag{3.128}$$

where the first term in the bracket is induced by the cylindrical liquid of radius r and length z, whereas the second term in the bracket is induced by the conical liquid below z. Substitution of (3.128) into (3.122) gives

$$N_\varphi = -\frac{R_V(\varphi)}{2\pi r\sin\varphi} = -\frac{\gamma r(2z+h)}{6\cos\alpha} + \frac{p_0 r}{2\cos\alpha} \tag{3.129}$$

In view of (3.127), (3.129) can be rewritten as

$$N_\varphi = -\frac{\gamma(2z+h)(h-z)\sin\alpha}{6\cos^2\alpha} + \frac{p_0(h-z)\sin\alpha}{2\cos^2\alpha} \tag{3.130}$$

The other membrane force is (in view of (3.126)):

$$N_\theta = -r_2 p_n - \frac{r_2}{r_1}N_\varphi = -\frac{r}{\cos\alpha}(-\gamma z - p_0) \tag{3.131}$$

Substitution of (3.127) into (3.131) gives

$$N_\theta = \frac{\sin\alpha}{\cos^2\alpha}(\gamma z + p_0)(h-z) \tag{3.132}$$

For the special case of $p_0 = 0$, we have the maximum membrane forces occurring at different depths:

$$\frac{dN_\theta}{dz} = 0 \quad \Rightarrow z = h/4 \tag{3.133}$$

$$\frac{dN_\varphi}{dz} = 0 \quad \Rightarrow z = h/2 \tag{3.134}$$

Note that the location of the maximum membrane forces is independent of the value of α. This is illustrated in Figure 3.19. The corresponding values of the membrane forces are

$$N_{\varphi,\max} = \frac{3\gamma h^2}{16}\frac{\sin\alpha}{\cos^2\alpha} \tag{3.135}$$

$$N_{\theta,\max} = \frac{\gamma h^2}{4}\frac{\sin\alpha}{\cos^2\alpha} \tag{3.136}$$

Thus, we always have $N_{\theta,\max} > N_{\varphi,\max}$. Next, we will consider the compression in the support ring. Global vertical force equilibrium requires

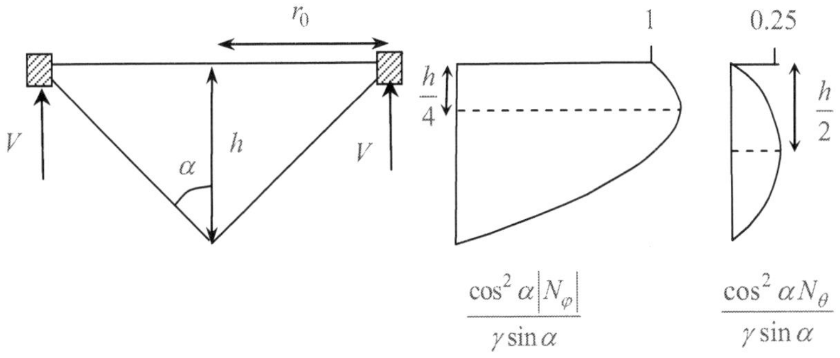

Figure 3.19 Location of maximum forces in conical tank

$$\frac{1}{3}\gamma\pi r_0^2 h = 2\pi r_0 V \tag{3.137}$$

This can be used to solve for the vertical reaction force per length V as

$$V = \frac{1}{6}r_0 h\gamma = \frac{1}{6}\gamma h^2\tan\alpha \tag{3.138}$$

On the other hand, the force equilibrium at the support ring gives

$$V = N_\varphi\cos\alpha \tag{3.139}$$

$$H = N_\varphi\sin\alpha \tag{3.140}$$

Substitution of (3.138) into (3.139) gives

$$N_\varphi = \frac{V}{\cos\alpha} = \frac{\gamma h^2}{6}\frac{\sin\alpha}{\cos^2\alpha} \tag{3.141}$$

Then, from (3.141) we find the horizontal force in (3.140) as

$$H = \frac{\gamma h^2}{6}\tan^2\alpha \tag{3.142}$$

Again, the ring load can be found by using the result derived in Problem 3.1 as

$$S_c = Hr_0 = \frac{\gamma h^2}{6} r_0 \tan^2 \alpha = \frac{1}{6} \gamma h^3 \tan^3 \alpha \tag{3.143}$$

The last of (3.143) is obtained in view of the following identity

$$r_0 = h \tan \alpha \tag{3.144}$$

Similar to the earlier discussions, the lateral strains of the ring and the shell are not compatible at the support. Therefore, boundary disturbance at the ring will manifest in the form of bending moment. In the next section, we will consider the membrane theory for cylindrical shells.

3.5 MEMBRANE THEORY FOR CYLINDRICAL SHELLS

In this section, membrane theory for cylindrical shells will be considered. For a special case that one of the radii of curvature becomes infinite, we have a cylindrical shell, as shown in Figure 3.20.

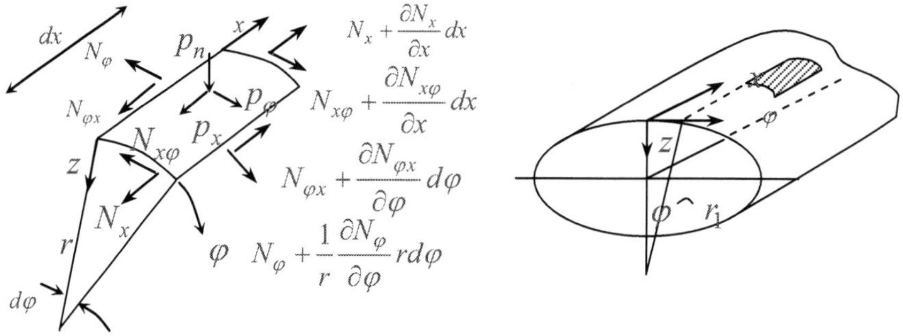

Figure 3.20 Membrane theory for cylindrical shells

3.5.1 Governing Equations

For thin shells, we can ignore the effect of curvature on the stresses and, consequently, we have

$$N_{\varphi x} = N_{x\varphi} \tag{3.145}$$

Referring to Figure 3.20, the force equilibrium along the x-axis requires

$$(N_x + \frac{\partial N_x}{\partial x} dx) r d\varphi + \frac{\partial N_{\varphi x}}{\partial \varphi} d\varphi dx - N_x r d\varphi + p_x r d\varphi dx = 0 \tag{3.146}$$

Similarly, the force equilibrium along the φ-direction requires

$$\frac{\partial N_\varphi}{\partial \varphi} d\varphi dx + \frac{\partial N_{x\varphi}}{\partial x} r d\varphi dx + p_\varphi r d\varphi dx = 0 \tag{3.147}$$

Finally, to consider the force equilibrium along the z-axis, we refer to Figure 3.21. By neglecting the higher order terms, we obtain

$$N_\varphi d\varphi dx + p_n r d\varphi dx = 0 \tag{3.148}$$

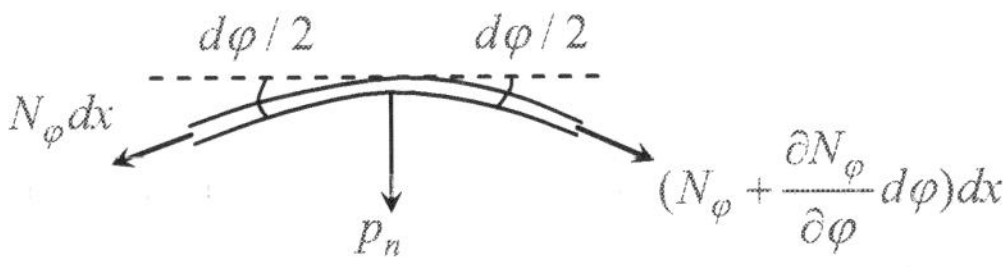

Figure 3.21 Force equilibrium in the z-direction

Equations (3.146) to (3.148) can be simplified as

$$\frac{\partial N_x}{\partial x} + \frac{1}{r}\frac{\partial N_{\varphi x}}{\partial \varphi} = -p_x \tag{3.149}$$

$$\frac{\partial N_{x\varphi}}{\partial x} + \frac{1}{r}\frac{\partial N_\varphi}{\partial \varphi} = -p_\varphi \tag{3.150}$$

$$N_\varphi = -p_n r \tag{3.151}$$

Equation (3.151) provides a very important limitation of the membrane theory, and Figure 3.22 illustrates the limitation of membrane theory because of this prediction.

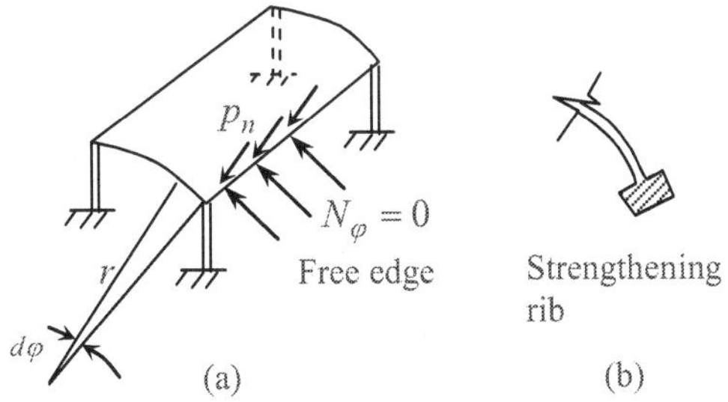

Figure 3.22 Free edge condition of cylindrical shells

In particular, as shown in Figure 3.22(a), on the lateral free edge there is no membrane force N_φ. For a shell with a non-zero normal load p_n, (3.151) cannot be satisfied on the lateral edge. Mathematically, that is

$$N_\varphi = 0 \quad \Rightarrow p_n = 0 \tag{3.152}$$

In other words, membrane forces cannot satisfy the boundary conditions. Actually none of the open shell problems can be solved by membrane theory exactly (i.e., boundary condition on φ = constant cannot be prescribed). In reality, similar to previous discussions, bending stress will develop due to boundary disturbance. The edge disturbance will weaken the shell. One simple way to solve the problem is to install a strengthening rib at the edge as shown in Figure 3.22(b). However, for the closed shell shown on the right diagram of Figure 3.20, we do not have such a problem.

3.5.2 General Solutions for Axisymmetric Case

For a shell with loading and geometric shape independent of x, we have

$$p_x = p_x(\varphi), \quad p_\varphi = p_\varphi(\varphi), \quad p_n = p_n(\varphi), \quad r = r(\varphi) \tag{3.153}$$

From (3.151) and (3.153), we see that the membrane force is only a function of φ:

$$N_\varphi = -p_n(\varphi)r(\varphi) = f(\varphi) \tag{3.154}$$

In view of (3.154), we can integrate (3.150) to give

$$N_{x\varphi} = -\int [p_\varphi + \frac{1}{r}\frac{\partial N_\varphi}{\partial \varphi}]dx + f_1(\varphi) \tag{3.155}$$

Therefore, the shear force can, in general, be rewritten as

$$N_{x\varphi} = -xF(\varphi) + f_1(\varphi) \tag{3.156}$$

Differentiating (3.156) with respect to φ, we obtain

$$\frac{\partial N_{\varphi x}}{\partial \varphi} = -x\frac{dF(\varphi)}{d\varphi} + \frac{df_1(\varphi)}{d\varphi} \tag{3.157}$$

On the other hand, (3.149) can be expressed as

$$\frac{\partial N_x}{\partial x} = -\frac{1}{r}\frac{\partial N_{\varphi x}}{\partial \varphi} - p_x \tag{3.158}$$

Substitution of (3.157) into (3.158) gives

$$N_x = -\int [\frac{1}{r}(-x\frac{dF(\varphi)}{d\varphi} + \frac{df_1(\varphi)}{d\varphi}) + p_x]dx + f_2(\varphi) \tag{3.159}$$

More specifically, we can integrate (3.159) to give

$$N_x = \frac{x^2}{2r}\frac{dF(\varphi)}{d\varphi} - \frac{x}{r}\frac{df_1(\varphi)}{d\varphi} - x\,p_x(\varphi) + f_2(\varphi) \tag{3.160}$$

In summary, the general solution can be expressed in the following form

$$N_\varphi = -p_n r \tag{3.161}$$

$$N_{x\varphi} = N_{\varphi x} = -xF(\varphi) + f_1(\varphi) \tag{3.162}$$

$$N_x = \frac{x^2}{2r}\frac{dF(\varphi)}{d\varphi} - \frac{x}{r}\frac{df_1(\varphi)}{d\varphi} - x\,p_x(\varphi) + f_2(\varphi) \tag{3.163}$$

where

$$F(\varphi) = p_\varphi + \frac{1}{r}\frac{\partial N_\varphi}{\partial \varphi} = p_\varphi - \frac{dp_n}{d\varphi} - \frac{p_n}{r}\frac{dr}{d\varphi} \tag{3.164}$$

In these equations, f_1 and f_2 are unknown integration functions depending on the boundary conditions.

Figure 3.23 illustrates that the membrane force $N_{x\varphi}$ at the end of the cylindrical shells has to be carried by a crosswall or stiffening ring, otherwise flexural edge disturbance will soften the shell.

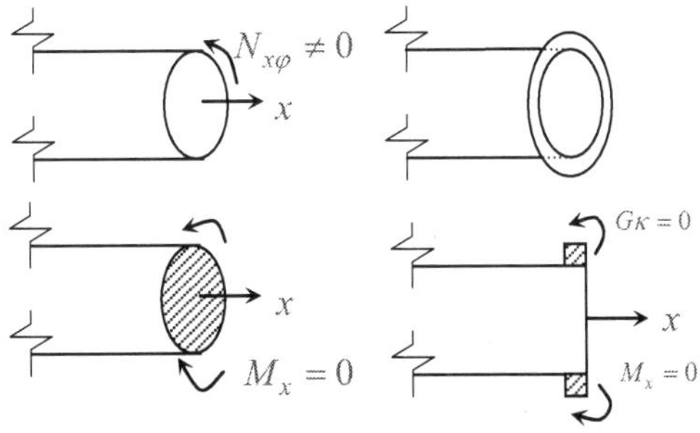

Figure 3.23 Cylindrical shells with end crosswall or stiffening ring

3.5.3 Simply-Supported Tube

One of the simplest problems of cylindrical shells is a simply-supported tube of thin wall. As shown in Figure 3.24, the axial membrane force of the tube at the supports is zero:

$$N_x(L/2) = \frac{L^2}{8r}\frac{dF}{d\varphi} - \frac{L}{2r}\frac{df_1}{d\varphi} - \frac{L}{2}p_x + f_2(\varphi) = 0 \tag{3.165}$$

$$N_x(-L/2) = \frac{L^2}{8r}\frac{dF}{d\varphi} + \frac{L}{2r}\frac{df_1}{d\varphi} + \frac{L}{2}p_x + f_2(\varphi) = 0 \tag{3.166}$$

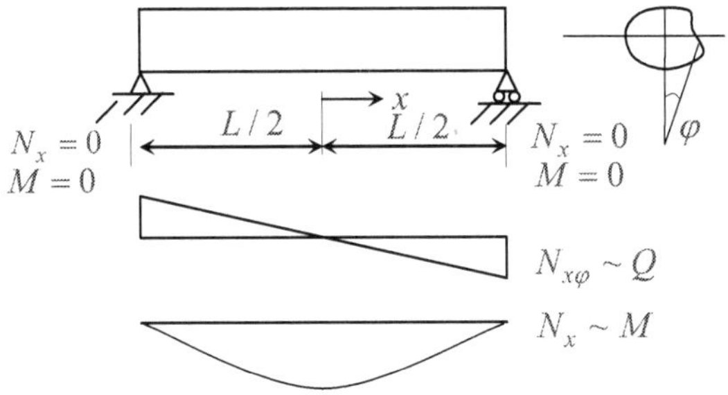

Figure 3.24 Variation of membrane forces and its similarity to simply-supported beams

Adding (3.165) to (3.166), we obtain a relation between F and f_2:

$$f_2 = -\frac{L^2}{8r}\frac{dF}{d\varphi} \tag{3.167}$$

Subtracting (3.165) from (3.166), we obtain

$$f_1 = -\int r p_x d\varphi + C \tag{3.168}$$

Consider the special case that no loading is applied along the edge direction or

$$p_x = 0 \tag{3.169}$$

Then, (3.168) is simplified to

$$f_1(\varphi) = C \tag{3.170}$$

If no end torsional moment is applied at the support, we have

$$f_1(\varphi) = 0 \tag{3.171}$$

In summary, the membrane forces of a simply-supported tube are summarized as

$$N_x = -\frac{1}{8r}(L^2 - 4x^2)\frac{dF}{d\varphi} \tag{3.172}$$

$$N_{x\varphi} = -xF(\varphi) \tag{3.173}$$

$$N_\varphi = -rp_n(\varphi) \tag{3.174}$$

where

$$F(\varphi) = p_\varphi - \frac{p_n}{r}\frac{dr}{d\varphi} - \frac{dp_n}{d\varphi} \tag{3.175}$$

Figure 3.24 plots the variation of $N_{x\varphi}$ and N_x along the span of the simply-supported tube. We see that the variations of $N_{x\varphi}$ closely resemble that of shear force in a Bernoulli-Euler beam, and variations of N_x closely resemble that of bending the moment of a simply-supported beam.

3.5.4 Circular Tube under Dead Load

We now specify the results of cylindrical shells to the case of a circular tube under dead load. The loads are

$$p_\varphi = g\sin\varphi, \quad p_n = g\cos\varphi \tag{3.176}$$

where g is the unit dead load per length. For circular tubes, we have

$$N_\varphi = -rp_n = -ag\cos\varphi \tag{3.177}$$

Substitution of (3.177) into (3.175) gives

$$F(\varphi) = p_\varphi + \frac{1}{r}\frac{\partial N_\varphi}{\partial \varphi} = g\sin\varphi + \frac{1}{a}ag\sin\varphi = 2g\sin\varphi \tag{3.178}$$

Differentiation of (3.178) with respect to φ gives

$$\frac{dF}{d\varphi} = 2g\cos\varphi \tag{3.179}$$

Employing (3.178) and (3.179) in (3.172) to (3.174), we obtain

$$N_x = -\frac{1}{8a}(L^2 - 4x^2)2g\cos\varphi \tag{3.180}$$

$$N_{x\varphi} = -2gx\sin\varphi \tag{3.181}$$

$$N_\varphi = -ag\cos\varphi \tag{3.182}$$

Referring to Figure 3.25, z can relate to φ as

$$z = a - a\cos\varphi = a(1-\cos\varphi) \tag{3.183}$$

In view of (3.183), we can rewrite (3.180) to (3.182) as

$$N_x = -\frac{1}{8a}(L^2 - 4x^2)2g(1 - \frac{z}{a}) \tag{3.184}$$

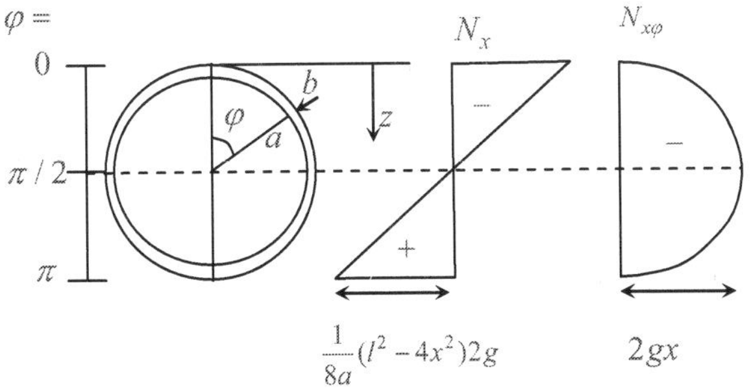

Figure 3.25 Bending force and shear force in a circular tube

$$N_{x\varphi} = -2gx\sqrt{1 - (1 - \frac{z}{a})^2} \tag{3.185}$$

$$N_\varphi = -ag(1 - \frac{z}{a}) \tag{3.186}$$

The variations of (3.184) and (3.185) with depth z are showed in Figure 3.25. These membrane force distributions again closely resemble bending stress and shear stress for beam bending.

Example 3.1 Find the axial and shear forces for the cantilever tube beam shown in Figure 3.26.

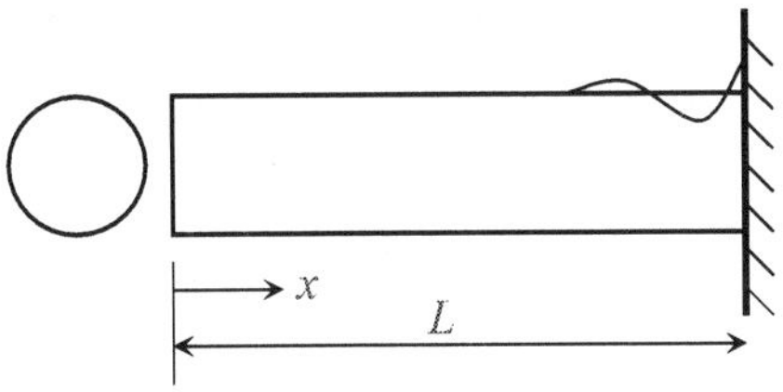

Figure 3.26 Bending of a cantilever circular tube beam

Solution: The boundary conditions at the free ends are

$$N_x = N_{x\varphi} = 0, \quad x = 0 \tag{3.187}$$

Thus, we have from (3.162) and (3.163) that

$$N_{x\varphi}(0) = f_1(\varphi) = 0 \tag{3.188}$$

$$N_x(0) = f_2(\varphi) = 0 \tag{3.189}$$

The membrane forces given in (3.160) and (3.156) become

$$N_x = \frac{x^2}{2r}\frac{dF}{d\varphi} \tag{3.190}$$

$$N_{x\varphi} = N_{\varphi x} = -xF(\varphi) \tag{3.191}$$

These axial forces and shear forces closely reflect the variation of bending moment and shear force of simple beam theory (see Figure 3.27).

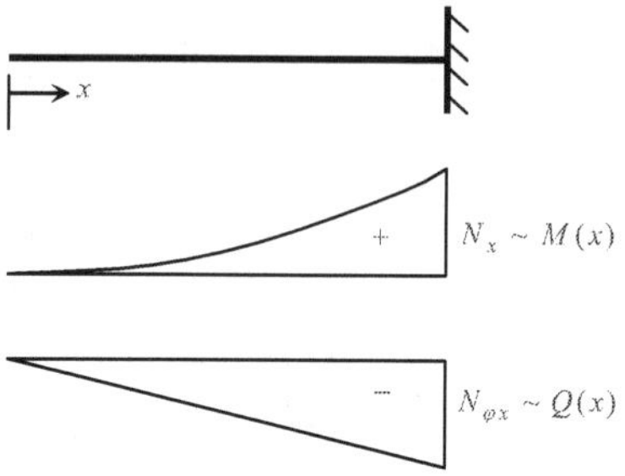

Figure 3.27 Axial and shear forces of a cantilever circular tube beam

These calculations of N_x and $N_{x\varphi}$ are exact. However, the membrane theory of shells will lead to nonzero values of $u(L)$, $v(L)$, and $w(L)$ at the fixed end. That is, the fixed end boundary conditions are not satisfied exactly. This observation can be generalized to conclude that membrane theory of shells can only satisfy the static boundary condition (or force boundary), but not the geometric boundary conditions.

3.5.5 Membrane Theory versus Beam Theory

We mentioned repeatedly in the last two sections that a close similarity between simply-supported circular tubes and simply-supported beams exists. In this section, we will compare the predictions of a simply-supported beam to that of a circular pipe. We first recall from Chapter 1 that according to beam theory, the bending moment and shear force of a simply-supported beam subject to uniformly distributed load are those given in Figure 3.28.

We first recall from the results of the membrane theory of shells derived in the last section:

$$N_x = -\frac{1}{8a}(L^2 - 4x^2)2g\cos\varphi \tag{3.192}$$

$$N_{x\varphi} = -2gx\sin\varphi \tag{3.193}$$

$$N_{\varphi} = -ag\cos\varphi \tag{3.194}$$

The corresponding results of the beam theory can be found as

$$M(x) = \frac{q}{8}(L^2 - 4x^2) = \frac{2\pi ag}{8}(L^2 - 4x^2) \tag{3.195}$$

$$Q(x) = qx = 2\pi agx \tag{3.196}$$

The bending stress and shear stress in the pipe can be calculated as (see p. 278 of Gere and Timoshenko, 1990)

$$\sigma = \frac{My}{I}, \quad \tau = \frac{Qs}{I\bar{b}} \tag{3.197}$$

where

$$I = \frac{\pi}{4}[(a+\frac{b}{2})^4 - (a-\frac{b}{2})^4] = \pi a^3 b + O(b^2) \tag{3.198}$$

$$\bar{b} = 2b \tag{3.199}$$

$$s(y) = 2\int y dA = 2\int_0^{\varphi} ba^2 \cos\varphi d\varphi = 2ba^2 \sin\varphi \tag{3.200}$$

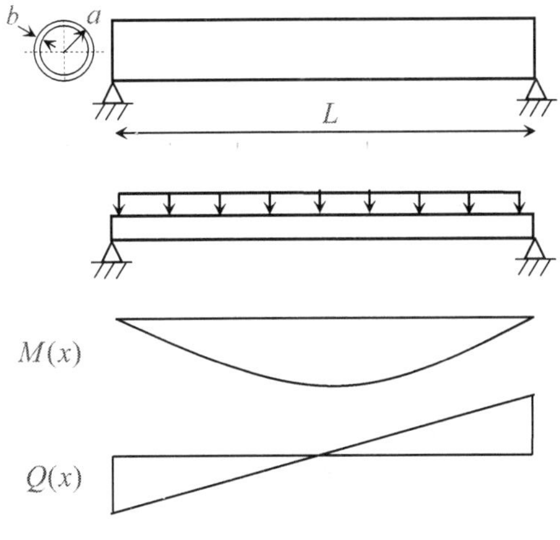

Figure 3.28 Axial and shear forces of a cantilever circular tube beam

The integration of (3.200) is conducted by referring to Figure 3.29. The shear and bending stress can be found as

$$\tau(x) = \frac{(2\pi agx)(2ba^2 \sin\varphi)}{\pi a^3 b(2b)} = \frac{2gx\sin\varphi}{b} = \frac{|N_{x\varphi}|}{b} \tag{3.201}$$

$$\sigma(x) = \frac{2\pi ag\frac{1}{8}(L^2 - 4x^2)a\cos\varphi}{\pi a^3 b} = \frac{(L^2 - 4x^2)2g\cos\varphi}{8ab} = \frac{|N_x|}{b} \tag{3.202}$$

Clearly, these bending and shear stresses relate to the magnitudes of shear and axial membrane forces. These results are the same as those from membrane theory of shells that we obtained in the last section. Therefore, the relation between membrane theory and beam theory is formally established.

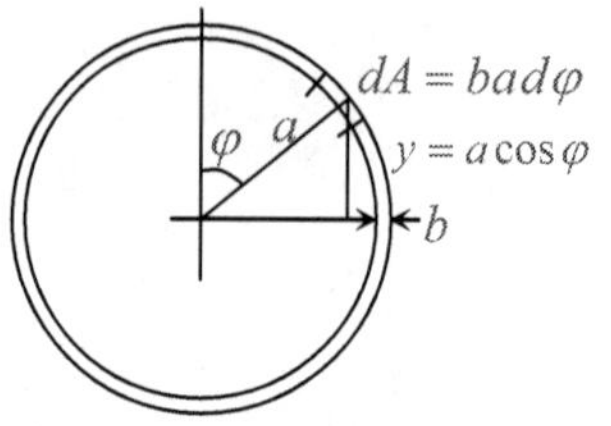

Figure 3.29 The integration of the first moment defined in (3.200)

3.5.6 Pipe Subject to Edge Load

In this section, we consider the case that a pipe is subject to nonzero edge loading of N_x and $N_{x\varphi}$. For a pipe or a cylindrical shell subject to edge loads only, we have

$$p_x = p_n = p_\varphi = 0 \tag{3.203}$$

From (3.155), we have the meridian membrane force being identically zero:

$$N_\varphi = -p_n r = 0 \tag{3.204}$$

In view of (3.164), (3.157) and (3.164) can be simplified to

$$\begin{aligned} N_{x\varphi} &= -xF(\varphi) + f_1(\varphi) \\ &= -x\{p_\varphi + \frac{1}{r}\frac{\partial N_\varphi}{\partial \varphi}\} + f_1(\varphi) = f_1(\varphi) \end{aligned} \tag{3.205}$$

$$N_x = -\frac{x}{r}\frac{df_1}{d\varphi} + f_2(\varphi) \tag{3.206}$$

In summary, we have membrane forces as

$$N_x = -\frac{x}{r}\frac{df_1}{d\varphi} + f_2(\varphi) \tag{3.207}$$

$$N_\varphi = -p_n r = 0 \tag{3.208}$$

$$N_{x\varphi} = f_1(\varphi) \tag{3.209}$$

The static boundary conditions for simple supports are:

$$N_x(0) = N_x(L) = 0 \tag{3.210}$$

Application of (3.210) to (3.207) gives

$$f_2(\varphi) = 0 \tag{3.211}$$

$$-\frac{L}{r}\frac{df_1}{d\varphi} + f_2(\varphi) = 0 \tag{3.212}$$

These two equations yield

$$f_1 = C_1 \tag{3.213}$$

$$N_x = N_\varphi = 0 \tag{3.214}$$

$$N_{x\varphi} = f_1 = const. \tag{3.215}$$

Physically, $N_{x\varphi}$ corresponds to the shear flow of the section. This result will be employed in solving the torsion problem in the next example.

Example 3.2 Find the shear stress for a simply-supported tube of arbitrary cross-section under end torsion T as shown in Figure 3.30.

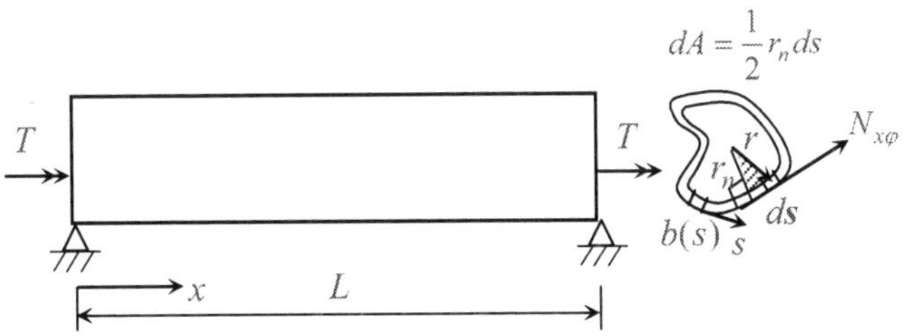

Figure 3.30 Torsion acting on a simply-supported tube beam

Solution: Torsion equilibrium

$$T = \oint N_{x\varphi} r_n ds = N_{x\varphi} \oint r_n ds = 2A_0 N_{x\varphi} \tag{3.216}$$

where A_0 is the total area enclosed by the contour of the section. Strictly speaking, we should use the area enclosed by the middle surface of the thickness, but for thin shells, this approximation is accurate for practical purposes. Rearranging (3.216), we obtain

$$N_{x\varphi} = \frac{T}{2A_0} \tag{3.217}$$

Note that T is a constant and the area A_0 is also a constant once the section is given. Thus, as expected from (3.217), $N_{x\varphi}$ is also a constant. From (3.201), the shear stress is therefore

$$\tau(s) = \frac{N_{x\varphi}}{b(s)} = \frac{T}{2A_0} \tag{3.218}$$

Example 3.3 Find the membrane forces in a circular tube under pre-stressing force distribution $R_x(\varphi)$, which is illustrated in Figure 3.31.

Solution: Pre-stressing force distribution in the shell can be modeled by the following system. If there is no torsion applied, we have $N_{x\varphi} = 0$ and the membrane forces are:

$$N_x = -\frac{x}{r}\frac{df_1}{d\varphi} + f_2(\varphi) = -R_x(\varphi) \tag{3.219}$$

$$N_\varphi = 0 \tag{3.220}$$

From the boundary conditions on $x = 0, L$, we get two equations from $N_x(0)$ and $N_x(L)$. Subtracting them, we find that $df_1/d\varphi = 0$ or f_1 equals constant. Physically, this constant must equal the end torsion. For the present case of zero end torsion, we have

$$N_{x\varphi} = f_1(\varphi) = 0 \tag{3.221}$$

Therefore, only the axial membrane force is nonzero. This situation is relevant to pre-stressed RC shells tied to anchorage at the boundary. Pre-stressing can be intentionally applied to obtain the more favorable condition of initial compressive stress for reinforced concrete shells. Concrete is strong in compression but weak in tension. Pre-compression can prevent tensile cracking in concrete when it is subject to loading.

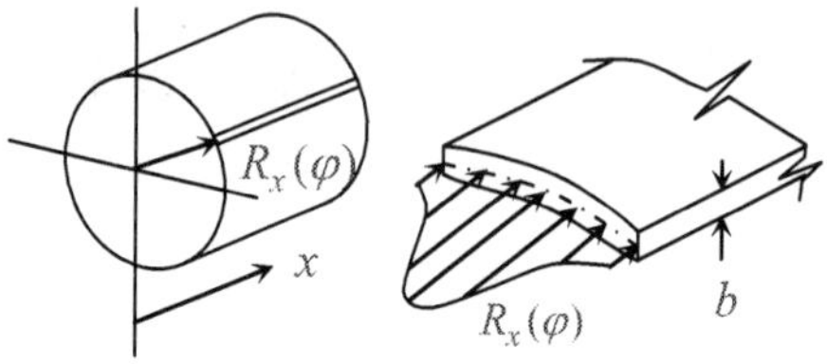

Figure 3.31 Circular tube subject to pre-stressing

3.5.7 Simply-Supported Cylindrical Shell Roof

So far, we have only considered closed cylindrical shells in the form of a tube. In this section, open cylindrical shells are considered as shell roofs, as shown in Figure 3.32. For cylindrical shell roofs subject to vertical loading, we have

$$p_\varphi = g \sin\varphi, \quad p_n = g\cos\varphi, \quad p_x = 0 \tag{3.222}$$

Simply-supported boundary conditions are

$$N_x(\pm L/2) = 0 \tag{3.223}$$

Substitution of (3.222) into (3.172) to (3.175) gives

$$N_\varphi = -ap_n = -ag\cos\varphi \tag{3.224}$$

$$N_x = -\frac{1}{8a}(L^2 - 4x^2)\frac{dF}{d\varphi} \tag{3.225}$$

$$N_{x\varphi} = -xF(\varphi) \tag{3.226}$$

where (see (3.222))

$$F(\varphi) = p_\varphi - \frac{dp_n}{d\varphi} = 2g\sin\varphi \tag{3.227}$$

$$\frac{dF}{d\varphi} = 2g\cos\varphi \tag{3.228}$$

It is straightforward to see that (3.223) is identically satisfied. In summary, the membrane forces are

$$N_\varphi = -ag\cos\varphi \tag{3.229}$$

$$N_x = -\frac{g\cos\varphi}{4a}(L^2 - 4x^2) \tag{3.230}$$

$$N_{x\varphi} = -2gx\sin\varphi \tag{3.231}$$

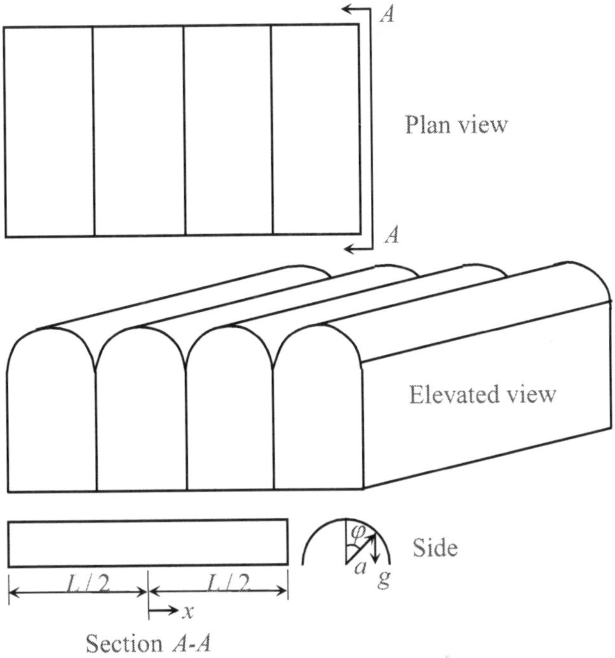

Figure 3.32 Cylindrical shell roof (simply-supported)

If the shell roof spanned an angle of $2\varphi_r$ as shown in Figure 3.33, the total axial force in the x-direction can be evaluated as

$$R(N_x) = \int_{-\varphi_r}^{\varphi_r} N_x(x,\varphi)a d\varphi \tag{3.232}$$

On the other hand, the total reaction force from the shear force on the two parallel edges can be determined as

$$R(N_x) = -[Z(\varphi_r) + Z(-\varphi_r)] \tag{3.233}$$

where

$$Z(\varphi_r) = -\int_x^{L/2} N_{\varphi x}(x,\varphi_r)dx = 2g\sin\varphi_r \int_x^{L/2} x dx = g\sin\varphi_r\left(\frac{L^2}{4} - x^2\right) \tag{3.234}$$

Substitution of (3.234) into (3.233) gives

$$R(N_x) = -\frac{g}{4}(L^2 - 4x^2)\int_{-\varphi_r}^{\varphi_r} \cos\varphi d\varphi = -2g\sin\varphi_r\left(\frac{L^2}{4} - x^2\right) \tag{3.235}$$

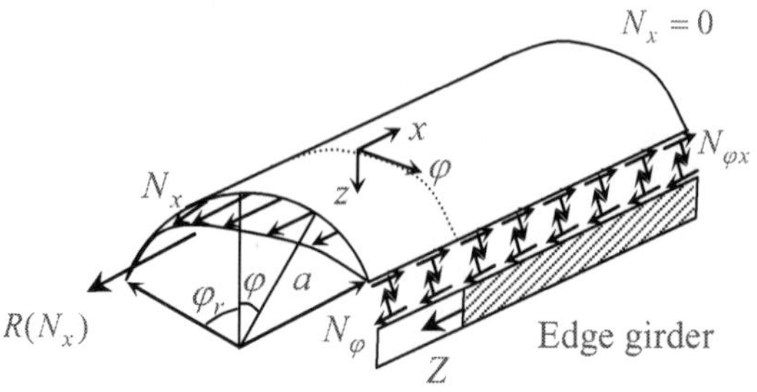

Figure 3.33 Cylindrical shell roof with lateral edge girder (simply-supported)

We can see that (3.235) is always negative or is always under tension. Thus, as shown in Figure 3.33, a girder is needed to take up the shear force and normal force at the two parallel edges. To ensure the edge girder will carry N_φ and $N_{\varphi x}$, we require the following compatibility

$$\varepsilon_{x,shell}(\varphi_r) = \varepsilon_{x,girder}(\varphi_r) \tag{3.236}$$

If there is no girder at the parallel edges, the boundary conditions become

$$N_{\varphi x}(\varphi_r) = 0, \quad N_\varphi(\varphi_r) = 0 \tag{3.237}$$

It is obvious that these boundary conditions cannot be satisfied by the membrane theory. Thus, a boundary disturbance will induce an unwanted bending moment. The bending theory of shells is needed for more accurate analysis near the edge, which is the topic to be covered in the next few sections.

3.6 BENDING THEORY OF CYLINDRICAL SHELLS

We have seen that boundary conditions at free edges cannot be satisfied if bending moments are ignored. In this section, the bending theory of shells will be introduced.

3.6.1 Governing Equation for Axisymmetric Cylindrical Shell

For the case of axisymmetric bending of a cylindrical shell, we have

$$N_{\varphi x} = M_{yx} = Q_r = 0 \tag{3.238}$$

Referring to Figure 3.34, the force equilibrium along the x-direction gives

$$\frac{dN_x}{dx} dxad\varphi + p_x dxad\varphi = 0 \tag{3.239}$$

Force equilibrium along the z-direction gives

$$N_\varphi dxd\varphi + \frac{dQ_x}{dx} dxad\varphi + p_n adxd\varphi = 0 \tag{3.240}$$

Finally, moment equilibrium about the φ-direction requires

$$\frac{dM_x}{dx}dx\,ad\varphi - Q_x dx\,ad\varphi = 0 \tag{3.241}$$

Note that the force equilibrium and moment equilibrium in the x-direction are satisfied automatically. Equations (3.239) to (3.241) can be simplified as

$$\frac{dN_x}{dx} = -p_x \tag{3.242}$$

$$\frac{dM_x}{dx} - Q_x = 0 \tag{3.243}$$

$$\frac{dQ_x}{dx} + \frac{N_\varphi}{a} = -p_a \tag{3.244}$$

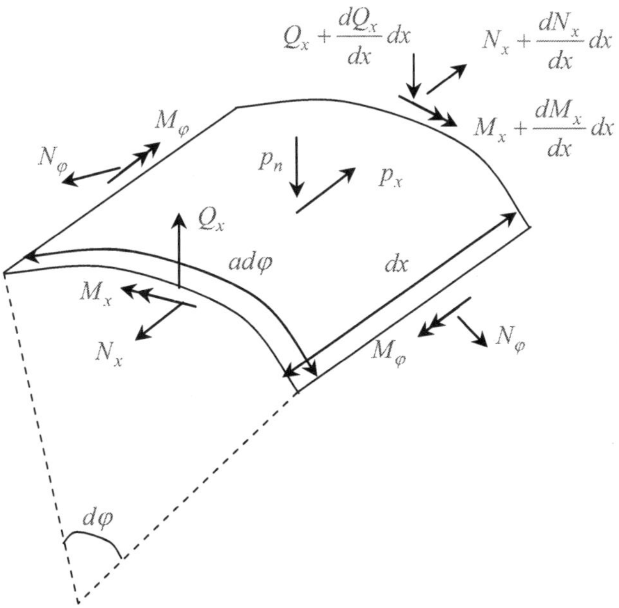

Figure 3.34 Membrane forces and bending moments for a cylindrical shell

In summary, we have two force equilibria and one moment equilibrium that need to be satisfied. Differentiation of (3.243) with respect to x and substitution of the result into (3.244) gives

$$\frac{d^2 M_x}{dx^2} + \frac{N_\varphi}{a} = -p_n \tag{3.245}$$

We will see from the next section that this theory closely resembles the theory for a beam on an elastic foundation. If there is no load being applied along the x-direction, equation (3.242) implies that N_x is a constant. That is,

$$p_x = 0, \quad \Rightarrow N_x = const. \tag{3.246}$$

For the case of nonzero load along the x-direction, we have

$$p_x \neq 0, \quad \Rightarrow N_x = -\int p_x dx + C \tag{3.247}$$

More importantly, we observe that in either case the membrane force N_x does not interact with the bending moment. This observation is, however, true only for axisymmetric cases. Equations (3.242) and (3.245) provide a system of two equations for three unknowns N_x, N_φ, and M_x. Clearly, we need one more equation, and this requires the consideration from deformation kinematics.

3.6.2 Deformation Kinematics

We have seen from the last section that for the bending theory of axisymmetric shells, we have two equations for three unknowns: N_x, N_φ and M_x. One more equation should be considered from deformation kinematics (see Figure 3.35). This idea is similar to that of indeterminate structure. In particular, Figure 3.35 shows the length of a line element measured z from the mid-surface and its length after bending. The local radius of curvature is a and thickness is h. The positive z-axis is pointing downward and due to symmetry, we have displacement along the φ direction being zero (or $v = 0$).

The normal strain along the φ-direction can be determined from

$$\varepsilon_\varphi = \frac{ds' - ds}{ds} = -\frac{w}{a - z} \approx -\frac{w}{a} \tag{3.248}$$

In obtaining the last expression in (3.248), we have assumed that $h/a \approx 0$, typical for a thin shell. A normality condition similar to the Kirchhoff condition for plate bending leads to

$$u_A = u - z\frac{dw}{dx} \tag{3.249}$$

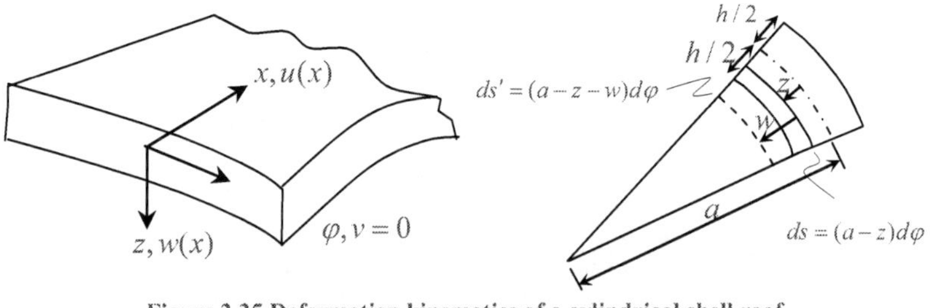

Figure 3.35 Deformation kinematics of a cylindrical shell roof

where Point A is defined in Figure 3.36. The undeformed points are denoted by A and A_0, whilst the deformed points are denoted as A^* and A^*_0. As shown in Figure 3.36, the corresponding deflections of Point A_0 to Point A^*_0 are translated by displacements u and w, whereas those for Point A to Point A^* are u_A and w. The application of this normality condition to shell formulation was first proposed by Love. Differentiation of (3.249) with respect to x gives the bending strain as

$$\varepsilon_x = \frac{\partial u_A}{\partial x} = \frac{\partial u}{\partial x} - z\frac{d^2 w}{dx^2} \tag{3.250}$$

Equation (3.250) shows that the axial strain of any fiber at level z results from both axial strain and the curvature of deflection of the middle surface of the shell.

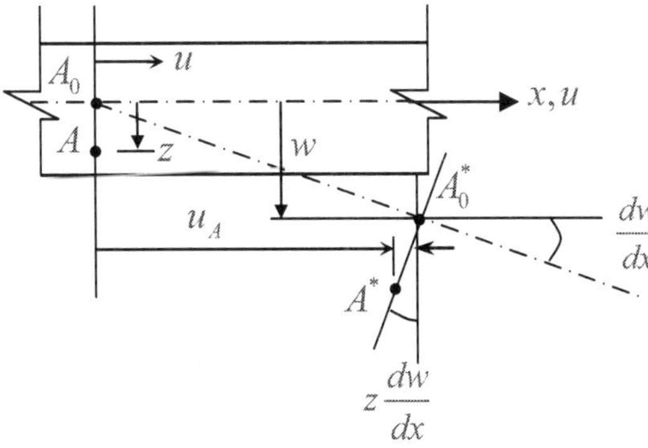

Figure 3.36 Bending deflection of point A measured z from the neutral surface

Applying Hooke's law for isotropic solids, we have

$$\sigma_x = \frac{E}{1-\nu^2}(\varepsilon_x + \nu\varepsilon_\varphi) \tag{3.251}$$

$$\sigma_\varphi = \frac{E}{1-\nu^2}(\varepsilon_\varphi + \nu\varepsilon_x) \tag{3.252}$$

For axisymmetric shell bending, w is not a function of z. Thus, the resultant forces can be readily integrated as

$$N_x = \int_{-h/2}^{h/2} \sigma_x dz = \frac{E}{1-\nu^2}(\frac{du}{dx} - \nu\frac{w}{a}) \tag{3.253}$$

$$N_\varphi = \int_{-h/2}^{h/2} \sigma_\varphi dz = \frac{E}{1-\nu^2}(-\frac{w}{a} + \nu\frac{du}{dx}) \tag{3.254}$$

Similarly, the bending moment can be determined by integrating the bending stresses as

$$M_x = \int_{-h/2}^{h/2} \sigma_x zdz = -\frac{E}{1-\nu^2}\frac{d^2w}{dx^2}\int_{-h/2}^{h/2} z^2 dz = -D\frac{d^2w}{dx^2} \tag{3.255}$$

$$M_\varphi = \int_{-h/2}^{h/2} \sigma_\varphi zdz = -\frac{E\nu}{1-\nu^2}\frac{d^2w}{dx^2}\int_{-h/2}^{h/2} z^2 dz = -\nu D\frac{d^2w}{dx^2} = \nu M_x \tag{3.256}$$

If there is no axial force along the x-axis of the cylindrical shell or $N_x = 0$, we have from (3.253)

$$\frac{du}{dx} = \nu\frac{w}{a} \tag{3.257}$$

Substitution of (3.257) into (3.254) gives

$$N_\varphi = \frac{E}{1-\nu^2}(-\frac{w}{a} + \nu\frac{du}{dx}) = -\frac{Eh}{a}w \tag{3.258}$$

Substitution of (3.255) and (3.258) into (3.245) yields

$$\frac{d^2}{dx^2}(D\frac{d^2w}{dx^2})+\frac{Eh}{a^2}w=p_n \tag{3.259}$$

For the case of a constant section, (3.259) is simplified to

$$D\frac{d^4w}{dx^4}+\frac{Eh}{a^2}w=p_n \tag{3.260}$$

Mathematically, (3.260) is equivalent to the governing equation of a beam on an elastic foundation (see Chapter 1):

$$EI\frac{d^4w}{dx^4}+kw=p \tag{3.261}$$

More detailed comparison is given in the next section.

3.6.3 Shell Bending Theory versus Beam on Elastic Foundation

Table 3.1 compiles the results of the last two sections together with the governing equations of a beam on an elastic foundation. Obviously, there is a close resemblance between the bending theory of an axisymmetric shell and the beam theory on an elastic foundation (see for example Hetenyi, 1946). We see that for the special case of N_φ is zero, the two theories are actually identical. The first three equations of the bending theory of shells given in the first three rows in Table 3.1 were derived in Section 3.6.1, and the last one was derived in Section 3.6.2.

Apparently, the similarity of these two theories had led von Karman and co-authors to model the buckling of cylindrical shells as a beam on an elastic spring (more discussion of this will be given in Chapter 6). Figure 3.37 depicts the physical interpretation of why axisymmetric bending of cylindrical shells bears a close resemblance to a beam on an elastic foundation. The shell surface can be imagined as comprising a series of vertical elastic strips supported by elastic rings. By the solution of the ring subject to uniform pressure considered in Problem 3.1, the ring tensile can be calculated and equated to (3.258) as

Table 3.1 Comparison of cylindrical shell bending theory and beam on foundation theory

Equilibrium in shell	Cylindrical shell bending theory	Beam on elastic foundation
φ-direction moment	$\frac{dM_x}{dx}-Q_x=0$	$\frac{dM_x}{dx}-Q_x=0$
z-direction force	$\frac{dQ_x}{dx}+\frac{N_\varphi}{a}=-p_n$	$\frac{dQ_x}{dx}=-p$
Combining rows 1 and 2	$\frac{d^2M_x}{dx^2}+\frac{N_\varphi}{a}=-p_n$	$\frac{d^2M_x}{dx^2}=-p$
Governing equation of w	$D\frac{d^4w}{dx^4}+\frac{Eh}{a^2}w=p_n$	$EI\frac{d^4w}{dx^4}+kw=p$

$$N_{ring} = -p_n a = N_\varphi = -\frac{Eh}{a} w \tag{3.262}$$

Equation (3.262) gives

$$p_n = \frac{Eh}{a^2} w \tag{3.263}$$

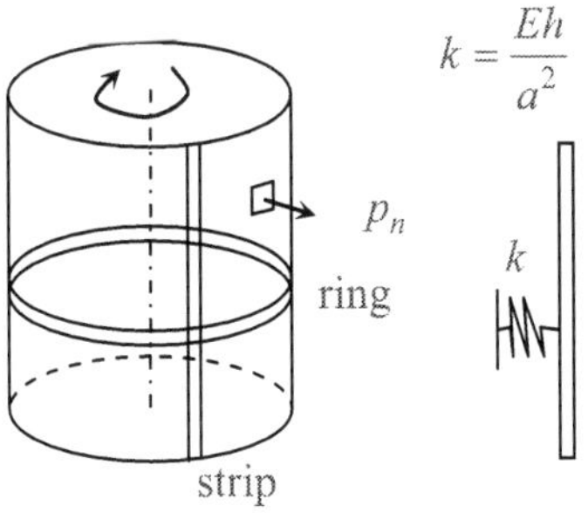

Figure 3.37 Physical interpretation of a shell as a beam on an elastic foundation

The ring stiffness can be defined as

$$k = \frac{p_n}{w} = \frac{Eh}{a^2} \tag{3.264}$$

The last part of (3.264) is obvious in view of (3.263).

3.6.4 General Solutions

We now proceed to consider the solution of the bending theory. First, we rewrite the homogeneous form of (3.260) as

$$\frac{d^4 w}{dx^4} + \frac{4\beta^4}{a^4} w = 0 \tag{3.265}$$

where

$$\beta^4 = \frac{Eha^2}{4D} \tag{3.266}$$

From the standard solution procedure discussed in Chau (2018), we note that the solutions of ODEs with constant coefficients can be expressed as

$$w_h = Ae^{\lambda x/a} \tag{3.267}$$

Substitution of (3.267) into (3.265) yields the following characteristic equation

$$\lambda^4 + 4\beta^4 = 0 \tag{3.268}$$

To solve the roots for λ, we first note from Euler's formula (see (1.122) of Chau, 2018) that

$$\lambda^4 = -4\beta^4 = 4\beta^4 e^{i[\pi + 2(n-1)\pi]} \tag{3.269}$$

The root is obtained as

$$\lambda = \sqrt{2}\beta e^{i[\pi+2(n-1)\pi]/4} \tag{3.270}$$

By setting $n = 0,1,2,3$, we obtain four roots for λ as

$$\lambda_{1,2,3,4} = \pm(1 \pm i)\beta \tag{3.271}$$

The corresponding homogeneous solution is

$$w_h = e^{-\beta x/a}(A_1 \cos\beta\frac{x}{a} + A_2 \sin\beta\frac{x}{a}) + e^{\beta x/a}(A_3 \cos\beta\frac{x}{a} + A_4 \sin\beta\frac{x}{a}) \tag{3.272}$$

This solution can be rewritten differently by introducing the following variable

$$\bar{x} = L - x \tag{3.273}$$

With this new variable, (3.272) can be expressed as

$$w_h = e^{-\beta x/a}(A_1 \cos\beta\frac{x}{a} + A_2 \sin\beta\frac{x}{a}) + e^{-\beta\bar{x}/a}(\bar{A}_3 \cos\beta\frac{\bar{x}}{a} + \bar{A}_4 \sin\beta\frac{\bar{x}}{a}) \tag{3.274}$$

In a sense, the first term is written in terms of a coordinate from the left boundary $x = 0$, whereas the second term is in terms of a coordinate from the left boundary $L = 0$. The particular solution of (3.261) can easily be obtained as

$$w_p = p_n \frac{a^2}{Eh} \tag{3.275}$$

There are two boundary conditions at $x = 0$, and two boundary conditions at $x = L$, leading to a system of four equations for four unknowns. For a very long tube, the end conditions are approximately uncoupled using the form of (3.274). That is, two equations for two unknowns for each end. Some special cases of closed shells in the form of circular tubes are considered next.

3.7 CIRCULAR PIPE

In Section 3.5, we have considered the membrane theory for a simply-supported tube. In this section, we will consider a similar problem using bending theory.

3.7.1 Semi-Infinite Pipe Subject to End Force

First, we consider a semi-infinite pipe, subject to end moment and end shear force, as shown in Figure 3.38. For this case, we have

$$p_n = 0 \tag{3.276}$$

The boundary conditions are

$$M_x(0) = M_0, \quad Q_x(0) = -Q_0 \tag{3.277}$$

At the far ends (i.e., $x \to \infty$), we must have both moment and shear force decaying to zero:

$$M_x(\infty) \to 0, \quad Q_x(\infty) \to 0 \tag{3.278}$$

For this case, (3.274) can be simplified to:

$$w = e^{-\beta x/a}(A_1 \cos\beta\frac{x}{a} + A_2 \sin\beta\frac{x}{a}) \tag{3.279}$$

Substitution of (3.279) into (3.255) gives

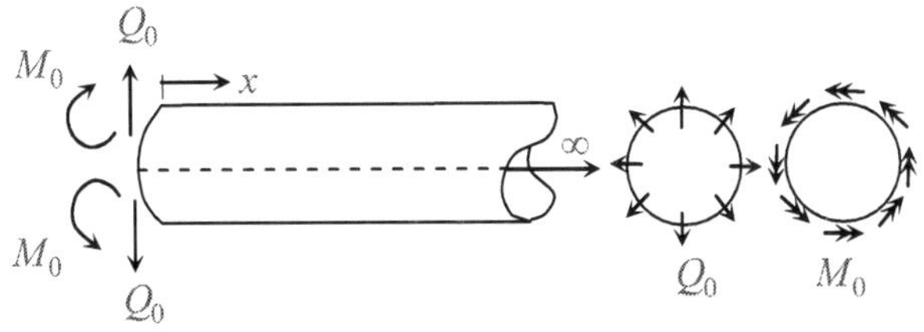

Figure 3.38 Semi-infinite pipe with end force Q_0 and M_0

$$M_x = -D\frac{d^2w}{dx^2} = -\frac{D}{a^2}2\beta^2 e^{-\beta x/a}(A_1 \sin\beta\frac{x}{a} - A_2\cos\beta\frac{x}{a}) \quad (3.280)$$

Differentiating this moment, we get the shear force as

$$Q_x = \frac{dM_x}{dx} = -\frac{D}{a^3}2\beta^3 e^{-\beta x/a}[(A_1 + A_2)\cos\beta\frac{x}{a} - (A_1 - A_2)\sin\beta\frac{x}{a})] \quad (3.281)$$

The first boundary condition given in (3.277) gives

$$M_x(0) = M_0 = \frac{D}{a^2}2\beta^2 A_2 \quad (3.282)$$

The second boundary condition given in (3.277) leads to

$$Q_x(0) = Q_0 = -\frac{D}{a^3}2\beta^3(A_1 + A_2) \quad (3.283)$$

Thus, the unknown constants can be found as

$$A_1 = -\frac{a^2}{2D\beta^2}(M_0 + \frac{a}{\beta}Q_0), \quad A_2 = \frac{a^2}{2D\beta^2}M_0 \quad (3.284)$$

Finally, substitution of these constants found in (3.284) into (3.279) gives

$$w = \frac{a^2}{2D\beta^2}e^{-\beta x/a}[M_0(\sin\beta\frac{x}{a} - \cos\beta\frac{x}{a}) - \frac{a}{\beta}Q_0\cos\beta\frac{x}{a}] \quad (3.285)$$

This solution can be recast in a more compact form as:

$$w = -\frac{a^2}{2D\beta^2}[M_0 f_2(\beta\xi) + \frac{a}{\beta}Q_0 f_3(\beta\xi)] \quad (3.286)$$

where

$$f_1(\beta\xi) = e^{-\beta\xi}(\cos\beta\xi + \sin\beta\xi) \quad (3.287)$$

$$f_2(\beta\xi) = e^{-\beta\xi}(\cos\beta\xi - \sin\beta\xi) \quad (3.288)$$

$$f_3(\beta\xi) = e^{-\beta\xi}\cos\beta\xi \quad (3.289)$$

$$f_4(\beta\xi) = e^{-\beta\xi}\sin\beta\xi \quad (3.290)$$

$$\xi = \frac{x}{a} \quad (3.291)$$

The other two functions defined in (3.287) and (3.290) appear naturally in the following derivatives of w

$$\frac{dw}{dx} = \frac{a}{2D\beta}[2M_0 f_3(\beta\xi) + \frac{a}{\beta} Q_0 f_1(\beta\xi)] \tag{3.292}$$

$$\frac{d^2 w}{dx^2} = -\frac{1}{D}[M_0 f_1(\beta\xi) + \frac{a}{\beta} Q_0 f_4(\beta\xi)] \tag{3.293}$$

$$\frac{d^3 w}{dx^3} = \frac{\beta}{aD}[2M_0 f_4(\beta\xi) - \frac{a}{\beta} Q_0 f_2(\beta\xi)] \tag{3.294}$$

Using these results, the resultant forces and moments can be found as

$$N_\varphi = -\frac{Eh}{a} w = \frac{6(1-\nu^2)a}{\beta^2 h^2}[M_0 f_2 + \frac{a}{\beta} Q_0 f_3] \tag{3.295}$$

$$M_x = M_0 f_1 + \frac{a}{\beta} Q_0 f_4 \tag{3.296}$$

$$M_\varphi = \nu[M_0 f_1 + \frac{a}{\beta} Q_0 f_4] \tag{3.297}$$

$$Q_x = -D\frac{d^3 w}{dx^3} = -\frac{\beta}{a}[2M_0 f_4(\beta\xi) - \frac{a}{\beta} Q_0 f_2(\beta\xi)] \tag{3.298}$$

Note that these expressions are valid only for constant thickness h. In the next section, we will consider the decay distance of these bending moments.

3.7.2 Decay of Edge Disturbance

The solution that we have just obtained clearly concentrates near the end $x = 0$. To consider the decay of the edge disturbance, we set a distance defined as

$$\beta\xi = \beta\frac{x}{a} = \pi \tag{3.299}$$

$$e^{-\pi} = 0.043 \tag{3.300}$$

Therefore, at a normalized distance of π/β, all deformation, moment, and shear force induced at the end reduce to less than 5%. Let us recall the definition of β:

$$\beta^4 = \frac{Eha^2}{4D} = \frac{3(1-\nu^2)a^2}{h^2} \tag{3.301}$$

where β physically can be interpreted as a decay parameter of an edge effect. Substitution of (3.301) into (3.299) leads to:

$$\beta\frac{x}{a} = \pi = \frac{x}{a}\sqrt[4]{3(1-\nu^2)\frac{a^2}{h^2}} \tag{3.302}$$

Solving for x, we obtain

$$x = \frac{\pi}{\sqrt[4]{3(1-\nu^2)}}\sqrt{ha} \tag{3.303}$$

This critical distance x for moments and forces attenuating to 5% of their edge values depends on Poisson's ratio, tube radius, and tube thickness. For examples, for $\nu = 0$, 0.16 and 0.5, we have $x = 2.387(ah)^{1/2}$, $2.403(ah)^{1/2}$, and $2.565(ah)^{1/2}$ respectively. The following example shows the zone of influence of the bending moment in the vicinity of the edge boundary of cylindrical water tanks, made of both concrete and steel.

Example 3.4 Taking 5% of the edge disturbance, find the boundary zone of a cylindrical water tank of radius of 5 m (i.e., $a = 5$ m) made of (i) concrete with $\nu = 0.16$ and $h = 0.2$ m and (ii) steel with $\nu = 0.3$ and $h = 0.012$ m.

Solution: For concrete tanks, we have from (3.303) the decay zone being

$$x = \frac{\pi}{\sqrt[4]{3(1-0.16^2)}}\sqrt{0.2\times 5} = 2.403\text{m} \tag{3.304}$$

For a steel tank, the thickness is much smaller (16.6 times thinner than that of concrete) because of its high value of Young's modulus. Thus, although Young's modulus E does not enter the equation given in (3.301), it enters the calculation indirectly through the thickness h of the water tank. More specifically, we have from (3.303) the influence zone for the steel tank:

$$x = \frac{\pi}{\sqrt[4]{3(1-0.3^2)}}\sqrt{0.012\times 5} = 0.5987\text{m} \tag{3.305}$$

These results are illustrated in Figure 3.39. Thus, the disturbance zone is smaller in steel tanks than in concrete tanks.

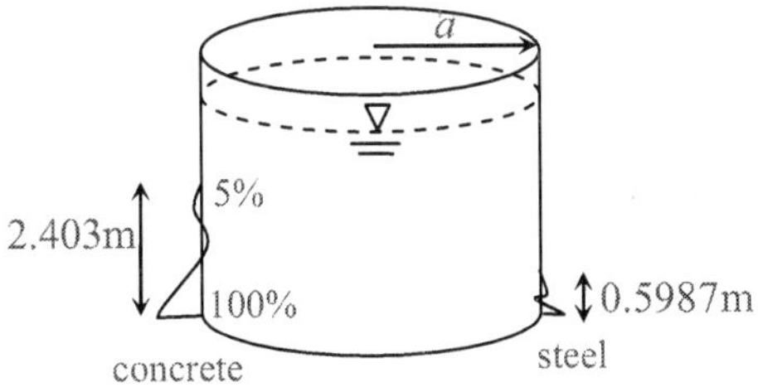

Figure 3.39 The decay zones in RC and steel circular tanks

3.7.3 Infinite Pipes under Ring Load

In this section, we consider the case of ring load on an infinitely long circular pipe, as illustrated in Figure 3.40. Vertical force equilibrium for the free body cutting out the ring load (see the lower diagram in Figure 3.40) gives

$$\bar{p} + Q_0^+ - Q_0^- = 0 \tag{3.306}$$

By referring to Figure 3.40, symmetry also requires

$$Q_0^+ = -Q_0^- \tag{3.307}$$

$$\left.\frac{dw}{dx}\right|_{x=0} = 0 \tag{3.308}$$

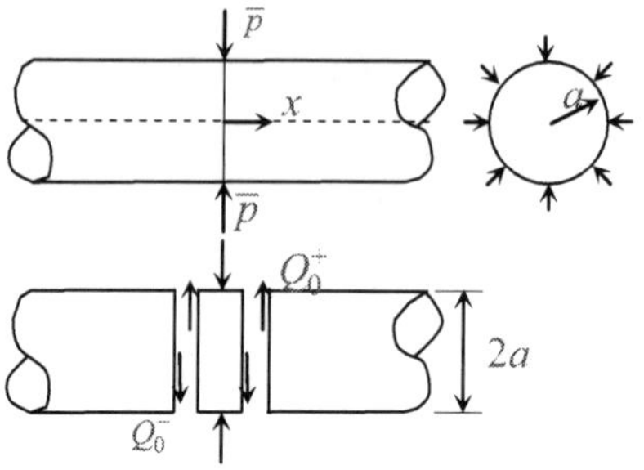

Figure 3.40 An infinitely long pipe subject to ring load

Substitution of (3.307) into (3.306) gives

$$Q_0^+ = -\frac{1}{2}\bar{p} = Q_0 \tag{3.309}$$

At the boundary (i.e., $x = 0^+$), we have from (3.287) and (3.289)

$$f_1 = f_3 = 1, \quad x = 0^+ \tag{3.310}$$

Substitution of (3.308) into (3.292) leads to (in view of (3.310)):

$$M_0 = -\frac{a}{2\beta}Q_0 = \frac{a}{4\beta}\bar{p} \tag{3.311}$$

Thus, the deflection can be determined as

$$\begin{aligned} w &= -\frac{a^2}{2D\beta^2}[M_0 f_2(\beta\xi) + \frac{a}{\beta}Q_0 f_3(\beta\xi)] \\ &= -\frac{a^3\bar{p}}{8D\beta^3}[f_2(\beta\xi) - 2f_3(\beta\xi)] = \frac{a^3\bar{p}}{8D\beta^3}f_1(\beta\xi) \end{aligned} \tag{3.312}$$

Differentiation of this deflection gives

$$\frac{dw}{dx} = -\frac{a^2\bar{p}}{4D\beta^2}f_4(\beta\xi) \tag{3.313}$$

$$\frac{d^2w}{dx^2} = -\frac{a\bar{p}}{4D\beta}f_2(\beta\xi) \tag{3.314}$$

$$\frac{d^3w}{dx^3} = \frac{\bar{p}}{2D}f_3(\beta\xi) \tag{3.315}$$

With these expressions, the corresponding forces and moments are obtained as

$$N_\varphi = -\frac{Eh}{a}w = -\frac{Eha^2\bar{p}}{8D\beta^3}f_1(\beta\xi) \tag{3.316}$$

$$M_x = -D\frac{d^2w}{dx^2} = \frac{a\bar{p}}{4\beta} f_2(\beta\xi) \tag{3.317}$$

$$M_\varphi = \nu M_x = \frac{a\bar{p}\nu}{4\beta} f_2(\beta\xi) \tag{3.318}$$

$$Q_x = -D\frac{d^3w}{dx^3} = -\frac{\bar{p}}{2} f_3(\beta\xi) \tag{3.319}$$

These solutions can be normalized as

$$\tilde{N}_\varphi = \frac{8D\beta^3}{Eha^2\bar{p}} N_\varphi, \quad \tilde{M}_x = \frac{4\beta}{a\bar{p}} M_x, \quad \tilde{M}_\varphi = \frac{4\beta}{a\bar{p}} M_\varphi, \quad \tilde{Q}_x = \frac{2}{\bar{p}} Q_x \tag{3.320}$$

Figure 3.41 plots the variations of these normalized moments and membrane forces versus x/a.

We can see that all solutions are localized within a zone of $x/a < 0.2$. This agrees with our earlier intuitive argument that the bending stresses are localized in the vicinity of the edge, and in this case in the vicinity of the concentrated load. Now, let us interpret these results in the context of designing an airplane fuselage. The radius, thickness, and Poisson's ratio of an airplane fuselage are assumed as 2 m, 0.012 m, and 0.3. With these typical values, we have $\beta = 4.62$ m, and this leads to a decay distance of 2.72 m for having 5% of the bending moment from the ring load. This zone may be too big for designing a fuselage using membrane theory. Therefore, a stiffening ring is normally used, and this is the topic of the next section.

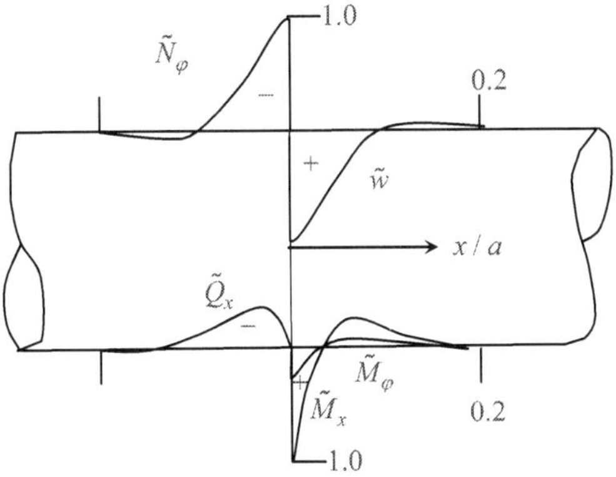

Figure 3.41 Forces and moments due to ring load on pipe

3.7.4 Effective Length

The idea of using effective width has been employed in connection with the design of stiffening ribs in plates by von Karman in 1933 and in shells by Thurlimann in

1950. For the case of an infinite pipe considered in the last section, we can define the effective width b by

$$bN_\varphi(0) = \int_{-\infty}^{\infty} N_\varphi(x)dx \tag{3.321}$$

The physical meaning of b is illustrated in Figure 3.42. Equivalently, we have

$$b = \frac{\int_{-\infty}^{\infty} N_\varphi(x)dx}{N_\varphi(0)} \tag{3.322}$$

Note, however, that N_φ is proportional to w (see (3.316)). Let us recall our solution obtained in (3.312)

$$w = \frac{a^3 p}{8D\beta^3} f_1(\beta\xi) \tag{3.323}$$

Substitution of (3.323) into (3.322) gives

$$b = \frac{2\int_0^{\infty} w(x)dx}{w(0)} = \frac{2\int_0^{\infty} f_1(\beta \frac{x}{a})dx}{f_1(0)} = \frac{2a}{\beta}\int_0^{\infty} f_1(\xi)dx \tag{3.324}$$

More specifically, we have from (3.287) that

$$b = \frac{2a}{\beta}\int_0^{\infty} e^{-\xi}(\cos\xi + \sin\xi)d\xi = \frac{2a}{\beta} e^{-\xi}\cos\xi\Big|_0^{\infty} = \frac{2a}{\beta} \tag{3.325}$$

From (3.266), we finally get the effective width as

$$b = \frac{2a}{\beta} = \frac{2\sqrt{ah}}{\sqrt[4]{3(1-\nu^2)}} \approx 1.520\sqrt{ah} \tag{3.326}$$

The last of (3.326) is for $\nu \to 0$. Stiffening rings will be installed for every distance of the effective width. For the case of the airplane fuselage that we considered in the last section, we have $b = 0.241$ m.

The axial and hoop stresses at the origin (i.e., $x = 0$) can be evaluated as

$$\sigma_x\Big|_{x=0} = \frac{M_x(0)}{h^2/6} = \frac{6ap}{h^2 4\beta} = \frac{3p\sqrt{ah}}{2h^2\sqrt[4]{3(1-\nu^2)}} \tag{3.327}$$

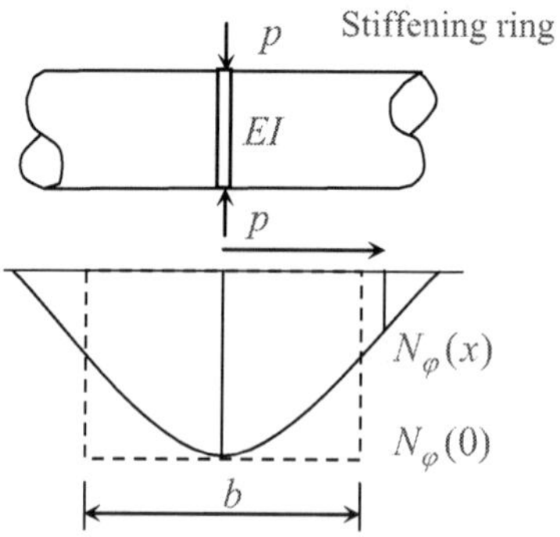

Figure 3.42 Effective width of cylindrical shell

$$\left|\sigma_\varphi\right|_{x=0} = \left|\frac{N_\varphi(0)}{h}\right| + \nu\left|\frac{M_x(0)}{h^2/6}\right| = \frac{Eh^2 p}{8\beta^3 D} + \nu\sigma_x \tag{3.328}$$

Substitution of (3.327) into (3.328) gives

$$\left|\sigma_\varphi\right|_{x=0} = \frac{p\sqrt{ah}}{2h^2\sqrt[4]{3(1-\nu^2)}}[3\nu + \sqrt{3(1-\nu^2)}] \tag{3.329}$$

Finally, we have the axial to hoop stress ratio at the origin being

$$\left|\frac{\sigma_x}{\sigma_\varphi}\right|_{x=0} = \frac{3}{3\nu + \sqrt{3(1-\nu^2)}} \tag{3.330}$$

For the special case of $\nu = 0$, we have the stress ratio of about 1.732:

$$\left|\frac{\sigma_x}{\sigma_\varphi}\right|_{x=0} = \sqrt{3} = 1.732, \quad \nu = 0 \tag{3.331}$$

The design of stiffener in cylindrical shells relates to the designs of liquid tanks, an arched roof, airplane fuselages, pressure vessels, pressure pipelines, ships, automobiles, and submarines. Stiffeners are supposed to be installed to take up bending moments.

3.8 BUCKLING OF CYLINDRICAL SHELL UNDER AXIAL LOAD

Buckling of cylindrical shells is one of the most important topics in shell analysis since pretty much all shells fail in buckling in reality. However, buckling analysis can be quite involved in mathematical techniques and tedious. In this section, we will only discuss one of the simplest buckling problems of a cylindrical shell subject to axial loads. Recall from (3.260) that the deflection of a cylindrical shell with no axial load is governed by

$$D\frac{d^4 w}{dx^4} + \frac{Eh}{a^2}w = p_n \tag{3.332}$$

If an axial force N_x is applied as shown in Figure 3.43, the governing equation can be revised as

$$D\frac{d^4 w}{dx^4} - N_x\frac{d^2 w}{dx^2} + \frac{Eh}{a^2}w = p_n \tag{3.333}$$

For simplicity, we ignore the effect of external load p_n. For such case, the deflection mode can be assumed as

$$w = A\sin n\pi\frac{x}{L} \tag{3.334}$$

Substitution of (3.334) into the homogeneous form of (3.333) gives

$$[\frac{n^4\pi^4}{L^4} + \frac{Eh}{a^2 D} + \frac{N_x}{D}\frac{n^2\pi^2}{L^2}]\sin n\pi\frac{x}{L} = 0 \tag{3.335}$$

Since the sine function is not identically zero, we must have the bracket term be zero, leading to

$$N_x = -D(\frac{n^2\pi^2}{L^2} + \frac{EhL^2}{Da^2n^2\pi^2}) \tag{3.336}$$

where n is the mode number of the buckled deflection. We see that the buckling load is both a function of n and L.

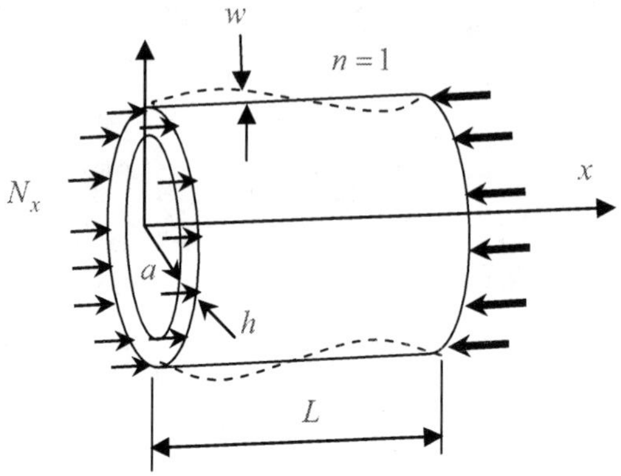

Figure 3.43 Buckling of a cylindrical shell subject to axial force N_x

Considering the lowest mode of $n = 1$, we have the buckling load being

$$N_x = -D(\frac{\pi^2}{L^2} + \frac{EhL^2}{Da^2\pi^2}) \tag{3.337}$$

Clearly, this buckling load is a function of the length of the cylindrical shell L. Naturally, we expect there is a critical value of the buckling length that gives the minimum buckling load. Differentiating (3.337) with respect to L, we obtain

$$\frac{dN_x}{dL} = 0 \tag{3.338}$$

This results in the critical length for the first mode:

$$L_{crit} = \pi \sqrt[4]{\frac{a^2h^2}{12(1-\nu^2)}} \tag{3.339}$$

Substitution of (3.339) into (3.337) gives the critical buckling load as

$$N_{x,crit} = -\frac{Eh^2}{a\sqrt{3(1-\nu^2)}} \tag{3.340}$$

The critical applied stress at buckling can be determined as

$$\sigma_{x,cr} = \frac{N_{x,cr}}{h} = -\frac{Eh}{a\sqrt{3(1-\nu^2)}} \tag{3.341}$$

In practice, slight imperfection will lead to much lower buckling load than this prediction. This is a main observation made by Koiter. Figure 3.44 plots (3.336) versus L for various n. We see that all the buckling loads for different n appear to be the same value as long as the critical value of L is determined accordingly. In fact, it can be shown that the most critical buckling load for all modes is the same, as suggested in Figure 3.44. Problem 3.13 provides the proof of this observation. For $n \to \infty$, the corresponding wavelength of the buckling approaches zero. This physically corresponds to a short wavelength limit and it can also be interpreted as surface instabilities (see Chau, 1994 and 1995).

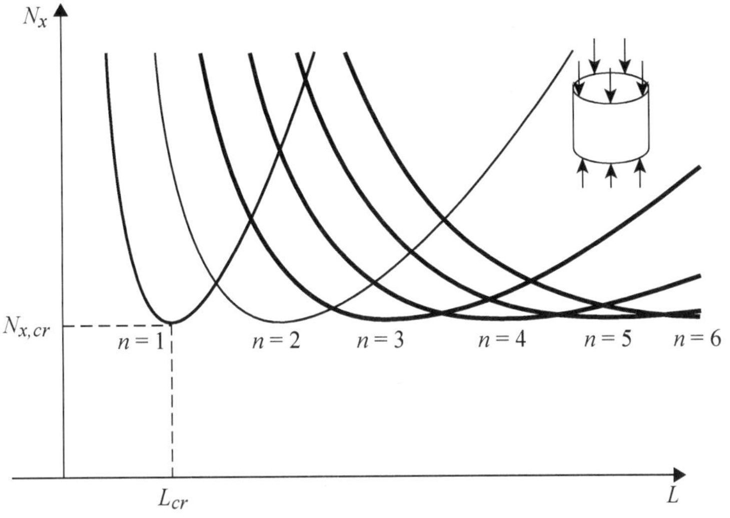

Figure 3.44 Buckling loads of cylindrical shell versus length

3.9 BENDING THEORY FOR SHELL OF REVOLUTION

We now consider the bending theory for shells with surface of revolution subject to symmetric loads. The z-axis points inward to the shell whilst the meridian direction is labeled as the y-axis and the tangential direction is labeled as the x-axis, as shown in Figure 3.45. The definitions of M_θ, M_φ, N_θ, N_φ, and Q_φ are also defined in the figure. The loads per area parallel to these axes are defined as X, Y, and Z (see Figure 3.46). For axisymmetric loading, we must have

$$F_x = X = 0, \quad N_\theta \neq N_\theta(\theta), \quad M_\theta \neq M_\theta(\theta) \tag{3.342}$$

The area of the shell element shown in Figure 3.46 can be written as

$$dA = r_0 r_1 d\theta d\varphi = r_2 r_1 \sin\varphi d\theta d\varphi \tag{3.343}$$

Consequently, the forces tangential to the meridian and normal to the shell surface are obtained as

$$F_\varphi = Y r_2 r_1 \sin\varphi d\theta d\varphi \tag{3.344}$$

$$F_z = Z r_2 r_1 \sin\varphi d\theta d\varphi \tag{3.345}$$

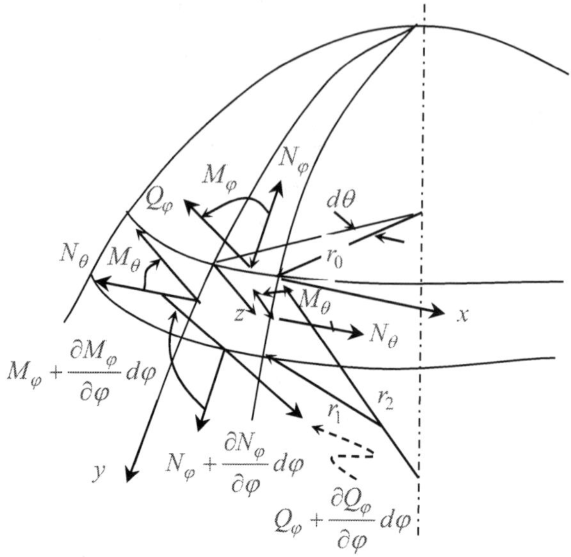

Figure 3.45 Forces and moments for shell of revolution

The axial force, resultant moment, and resultant shear on the side of the shell element normal to the meridian direction can be evaluated as

$$Force = N_\varphi r_2 \sin\varphi d\theta \tag{3.346}$$

$$Moment = M_\varphi r_2 \sin\varphi d\theta \tag{3.347}$$

$$Shear = Q_\varphi r_2 \sin\varphi d\theta \tag{3.348}$$

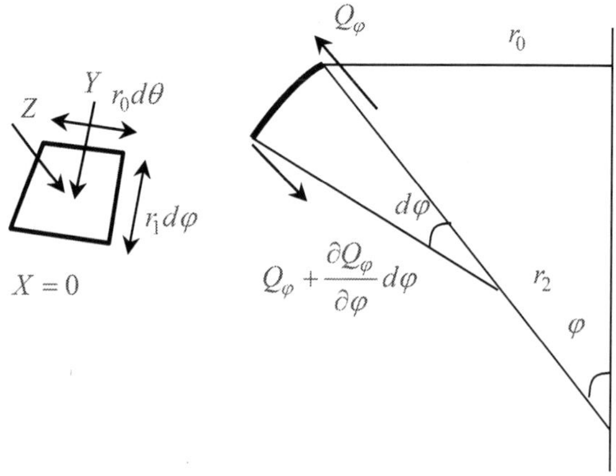

Figure 3.46 Contribution of Q_φ to the φ-direction force equilibrium

3.9.1 Force and Moment Equilibrium

The force equilibrium tangential to the meridian direction has been considered for the membrane theory in (3.41). We only need to add the contribution of the shear force Q_φ on the lower side of the shell element shown in Figure 3.46. By referring to Figure 3.46 and neglecting the higher order terms, the additional contribution of Q_φ leads to an additional force of

$$\begin{aligned} \text{Change of force on } \varphi-\text{direction} &= -Q_\varphi r_0 d\theta(\sin d\varphi) \\ &\approx -Q_\varphi r_0 d\theta d\varphi \end{aligned} \tag{3.349}$$

Consequently, (3.41) can be revised as

$$\frac{d(N_\varphi r_0)}{d\varphi} - N_\theta r_1 \cos\varphi - r_0 Q_\varphi + r_0 r_1 Y = 0 \tag{3.350}$$

Similarly, there is an additional inward force induced by Q_φ. The force equilibrium in the z-direction given in (3.43) can be revised as

$$N_\varphi r_0 + N_\theta r_1 \sin\varphi + \frac{d(r_0 Q_\varphi)}{d\varphi} + r_0 r_1 Z = 0 \tag{3.351}$$

Moment equilibrium leads to

$$\begin{aligned} &(M_\varphi + \frac{dM_\varphi}{d\varphi}d\varphi)(r_0 + \frac{dr_0}{d\varphi}d\varphi) - M_\varphi r_0 d\varphi \\ &-M_\theta r_1 \cos\varphi d\varphi d\theta - Q_\varphi r_2 \sin\varphi r_1 d\varphi d\theta = 0 \end{aligned} \tag{3.352}$$

By dropping the higher order terms, we can simplify (3.352) as

$$\frac{d(r_0 M_\varphi)}{d\varphi} - M_\theta r_1 \cos\varphi - Q_\varphi r_0 r_1 = 0 \tag{3.353}$$

There are three equations (i.e., (3.350), (3.351) and (3.353)) but there are five unknowns (N_φ, N_θ, M_θ, M_φ and Q_φ). For such an indeterminate system, we must consider the deformation of the shell to get two additional equations.

3.9.2 Hooke's Law

Hooke's law in polar form can be written as

$$N_\varphi = \frac{Eh}{1-\nu^2}[\varepsilon_\varphi + \nu\varepsilon_\theta] \tag{3.354}$$

$$N_\theta = \frac{Eh}{1-\nu^2}[\varepsilon_\theta + \nu\varepsilon_\varphi] \tag{3.355}$$

Figure 3.47 shows the deformation kinematics on the y-z plane. The shell segment AB is deformed to $A'B'$. The displacements v and w induce both axial strain and change of curvature. In particular, the axial strain along the meridian direction is

$$\varepsilon_\varphi = \frac{[v + \frac{dv}{d\varphi}d\varphi - v] + [(r_1 - w)d\varphi - r_1 d\varphi]}{r_1 d\varphi} \tag{3.356}$$

The first term in the numerator is caused by displacement v, whereas the second by displacement w. Equation (3.356) can be simplified to

$$\varepsilon_{\varphi} = \frac{1}{r_1}\left(\frac{dv}{d\varphi} - w\right) \tag{3.357}$$

Referring to the lower diagram given in Figure 3.47, we have

$$\varepsilon_{\theta} = \frac{(r_0 + dr_0)d\theta - r_0 d\theta}{r_0 d\theta} \tag{3.358}$$

Note from Figure 3.47 that

$$dr_0 = v\cos\varphi - w\sin\varphi \tag{3.359}$$

Substitution of (3.359) into (3.358) leads to

$$\varepsilon_{\theta} = \frac{v\cos\varphi - w\sin\varphi}{r_0} = \frac{1}{r_2}\frac{(v\cos\varphi - w\sin\varphi)}{\sin\varphi} = \frac{1}{r_2}(v\cot\varphi - w) \tag{3.360}$$

Combining Hooke's law (i.e., (3.354) and (3.355)) and the strain-displacement relation (i.e., (3.357) and (3.360)), we have

$$N_{\varphi} = \frac{Eh}{1-\nu^2}\left[\frac{1}{r_1}\left(\frac{dv}{d\varphi} - w\right) + \nu\frac{1}{r_2}(v\cot\varphi - w)\right] \tag{3.361}$$

$$N_{\theta} = \frac{Eh}{1-\nu^2}\left[\frac{1}{r_2}(v\cot\varphi - w) + \nu\frac{1}{r_1}\left(\frac{dv}{d\varphi} - w\right)\right] \tag{3.362}$$

Next, we have to derive the bending moments in terms of the displacements.

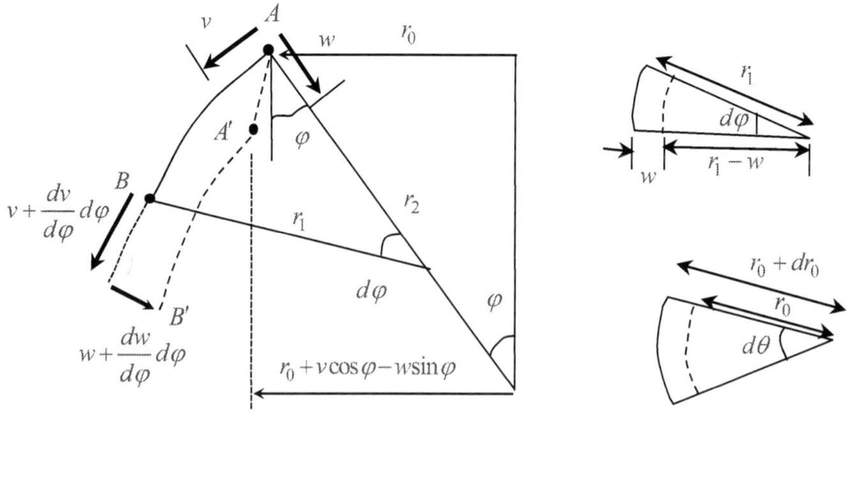

Figure 3.47 Deformation kinematics on the y-z plane

3.9.3 Change of Curvature

To relate the bending moments to displacements v and w, we need to derive the change of curvature. Referring to Figure 3.47, the change of curvature due to the rotation caused by w of the upper side A is

$$\psi_2 = \frac{w+dw-w}{r_1 d\varphi} = \frac{dw}{d\varphi}d\varphi\frac{1}{r_1 d\varphi} = \frac{1}{r_1}\frac{dw}{d\varphi} \tag{3.363}$$

In obtaining (3.363), the mid-surface elongation or shortening has been neglected. Adding the rotation caused by v on the upper side A (i.e., v/r_1), we have the total rotation of the upper side being

$$\psi_u = \psi_1 + \psi_2 = \frac{v}{r_1} + \frac{1}{r_1}\frac{dw}{d\varphi} \tag{3.364}$$

In general, the rotation of the lower side B can be expressed as

$$\psi_l = \frac{v}{r_1} + \frac{1}{r_1}\frac{dw}{d\varphi} + \frac{d}{d\varphi}\left(\frac{v}{r_1} + \frac{1}{r_1}\frac{dw}{d\varphi}\right)d\varphi \tag{3.365}$$

The change of curvature in the meridian direction is

$$\chi_\varphi = \frac{\psi_l - \psi_u}{r_1 d\varphi} = \frac{1}{r_1}\frac{d}{d\varphi}\left(\frac{v}{r_1} + \frac{1}{r_1}\frac{dw}{d\varphi}\right) \tag{3.366}$$

The change of curvature in the θ-direction is shown in Figure 3.48. When $\varphi \to \pi/2$, we see from Figure 3.48 that the y-direction is perpendicular to the tangential θ-direction. When $\varphi \to 0$, the angle between y and the θ-direction is $\pi/2-d\theta$. Therefore, for any arbitrary value of φ, the angle between y and the θ-direction becomes $\pi/2-\cos\varphi d\theta$. As shown in the right diagram of Figure 3.48, the angle rotation of (3.364) in the meridian plane now has a projection in the θ-direction as

$$d\theta(B') - d\theta(A') = \left(\frac{v}{r_1} + \frac{1}{r_1}\frac{dw}{d\varphi}\right)\cos\varphi d\theta \tag{3.367}$$

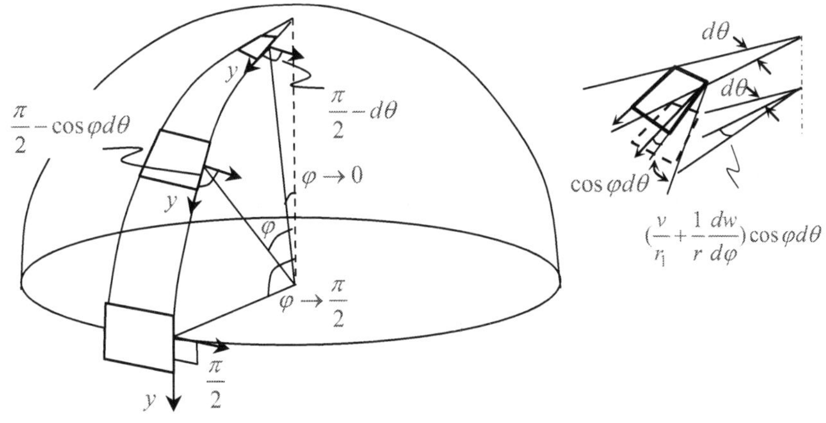

Figure 3.48 The change of curvature in the θ-direction

Thus, the change of curvature in the θ-direction is

$$\chi_\theta = (\frac{v}{r_1} + \frac{1}{r_1}\frac{dw}{d\varphi})\frac{\cos\varphi d\theta}{r_0 d\theta} = (\frac{v}{r_1} + \frac{1}{r_1}\frac{dw}{d\varphi})\frac{\cot\varphi}{r_2} \tag{3.368}$$

Substitution of (3.366) and (3.368) into the polar form of (3.15) gives

$$M_\varphi = -D[\frac{1}{r_1}\frac{d}{d\varphi}(\frac{v}{r_1} + \frac{1}{r_1}\frac{dw}{d\varphi}) + \nu(\frac{v}{r_1} + \frac{1}{r_1}\frac{dw}{d\varphi})\frac{\cot\varphi}{r_2}] \tag{3.369}$$

$$M_\theta = -D[(\frac{v}{r_1} + \frac{1}{r_1}\frac{dw}{d\varphi})\frac{\cot\varphi}{r_2} + \frac{\nu}{r_1}\frac{d}{d\varphi}(\frac{v}{r_1} + \frac{1}{r_1}\frac{dw}{d\varphi})] \tag{3.370}$$

Substitution of N_φ, N_θ, M_θ, and M_φ, found in (3.361), (3.362), (3.369), and (3.370) into three equilibrium equations (3.350), (3.351) and (3.353) gives three governing equations for u, w and Q_φ. However, they are still not easy to solve. Various frameworks have been proposed in obtaining the solution and are considered next.

3.9.4 Reissner Formulation

To simplify the problem mathematically, H. Reissner in 1912 introduced the following change of variables:

$$V = \frac{1}{r_1}(v + \frac{dw}{d\varphi}) \tag{3.371}$$

$$U = r_2 Q_\varphi \tag{3.372}$$

If $Y = Z = 0$, the vertical force equilibrium in Figure 3.49 gives

$$2\pi r_0 N_\varphi \sin\varphi + 2\pi r_0 Q_\varphi \cos\varphi = 0 \tag{3.373}$$

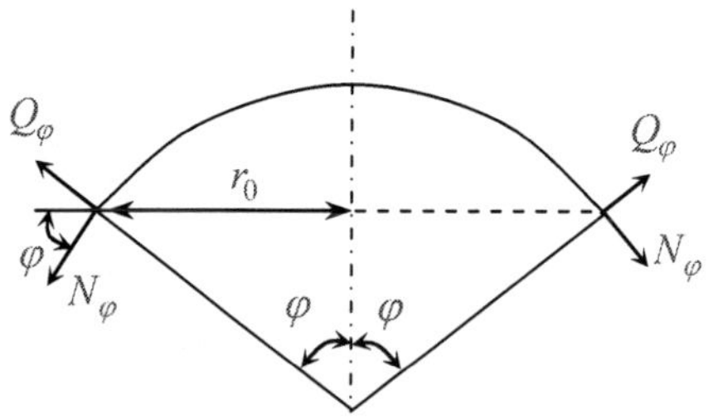

Figure 3.49 Vertical force equilibrium for a spherical shell

Equation (3.373) can be used to solve N_φ

$$N_\varphi = -Q_\varphi \cot\varphi = -\frac{1}{r_2}U\cot\varphi \tag{3.374}$$

From (3.351), we have

$$N_\theta r_1 \sin\varphi = -\frac{d(Q_\varphi r_0)}{d\varphi} - r_0 N_\varphi \tag{3.375}$$

Substitution of (3.374) into (3.375) and in view of (3.374) gives

$$N_\theta = \frac{1}{r_1 \sin\varphi}\{-\frac{d(Q_\varphi r_2 \sin\varphi)}{d\varphi} - r_0 Q_\varphi \cot\varphi\} \tag{3.376}$$

Note that $r_0 = r_2 \sin\varphi$ and the definition given in (3.372), (3.376) can be simplified as

$$N_\theta = -\frac{1}{r_1}\frac{dU}{d\varphi} \tag{3.377}$$

Equations (3.374) and (3.377) show that both membrane forces are functions of U. Combining (3.361) and (3.362), we get

$$N_\varphi - \nu N_\theta = \frac{Eh}{1-\nu^2}[\frac{1}{r_1} - \frac{\nu^2}{r_1}](\frac{dv}{d\varphi} - w) \tag{3.378}$$

$$N_\theta - \nu N_\varphi = \frac{Eh}{1-\nu^2}[\frac{1}{r_2} - \frac{\nu^2}{r_2}](v\cot\varphi - w) \tag{3.379}$$

Rearranging (3.378) and (3.379), we obtain

$$\frac{dv}{d\varphi} - w = \frac{r_1}{Eh}(N_\varphi - \nu N_\theta) \tag{3.380}$$

$$v\cot\varphi - w = \frac{r_2}{Eh}(N_\theta - \nu N_\varphi) \tag{3.381}$$

Subtracting (3.381) from (3.380), we have

$$\frac{dv}{d\varphi} - v\cot\varphi = \frac{1}{Eh}[(r_1 + \nu r_2)N_\varphi - (r_2 + \nu r_1)N_\theta] \tag{3.382}$$

Differentiating (3.381), we have

$$\frac{dv}{d\varphi}\cot\varphi - \frac{v}{\sin^2\varphi} - \frac{dw}{d\varphi} = \frac{d}{d\varphi}[\frac{r_2}{Eh}(N_\theta - \nu N_\varphi)] \tag{3.383}$$

Eliminating $dv/d\varphi$ from (3.382) and (3.383), we obtain

$$-v\cot^2\varphi + \frac{v}{\sin^2\varphi} + \frac{dw}{d\varphi} = \frac{\cot\varphi}{Eh}[(r_1 + \nu r_2)N_\varphi - (r_2 + \nu r_1)N_\theta] - \frac{d}{d\varphi}[\frac{r_2}{Eh}(N_\theta - \nu N_\varphi)] \tag{3.384}$$

The first two terms on the left hand side of (3.384) can be simplified to

$$-v\cot^2\varphi + \frac{v}{\sin^2\varphi} = v(-\frac{\cos^2\varphi}{\sin^2\varphi} + \frac{1}{\sin^2\varphi}) = v \tag{3.385}$$

Finally, in view of the definition of V given in (3.371), (3.384) is reduced to

$$v + \frac{dw}{d\varphi} = r_1 V = \frac{\cot\varphi}{Eh}[(r_1 + \nu r_2)N_\varphi - (r_2 + \nu r_1)N_\theta] - \frac{d}{d\varphi}[\frac{r_2}{Eh}(N_\theta - \nu N_\varphi)] \tag{3.386}$$

Substituting (3.374) and (3.377) into (3.386) and simplifying the result, we eventually find

$$EhV = \frac{r_2}{r_1^2}\frac{d^2U}{d\varphi^2} + \frac{1}{r_1}\left[\frac{d}{d\varphi}\left(\frac{r_2}{r_1}\right) + \frac{r_2}{r_1}\cot\varphi - \frac{r_2}{r_1 h}\frac{dh}{d\varphi}\right]\frac{dU}{d\varphi} - \frac{1}{r_1}\left[\frac{r_1}{r_2}\cot^2\varphi - \nu - \frac{\nu}{h}\frac{dh}{d\varphi}\cot\varphi\right]U \tag{3.387}$$

This is the first governing equation for U and V. On the other hand, using Reissner variable V given in (3.371), (3.369) and (3.370) become

$$M_\varphi = -D\left[\frac{1}{r_1}\frac{dV}{d\varphi} + \nu V\frac{\cot\varphi}{r_2}\right] \tag{3.388}$$

$$M_\theta = -D\left[V\frac{\cot\varphi}{r_2} + \frac{\nu}{r_1}\frac{dV}{d\varphi}\right] \tag{3.389}$$

We first rewrite (3.388) as

$$\begin{aligned} M_\varphi r_0 &= -D\sin\varphi\left[\frac{r_2}{r_1}\frac{dV}{d\varphi} + \nu V\cot\varphi\right] \\ &= -D\left[\frac{r_2}{r_1}\frac{dV}{d\varphi}\sin\varphi + \nu V\cos\varphi\right] \end{aligned} \tag{3.390}$$

Differentiation of (3.390) gives

$$\frac{d(M_\varphi r_0)}{d\varphi} = -D\left[\frac{d}{d\varphi}\left(\frac{r_2}{r_1}\right)\frac{dV}{d\varphi}\sin\varphi + \frac{r_2}{r_1}\frac{d^2V}{d\varphi^2}\sin\varphi + \frac{r_2}{r_1}\frac{dV}{d\varphi}\cos\varphi + \nu\frac{dV}{d\varphi}\cos\varphi - \nu V\sin\varphi\right] - \frac{3D}{h}\frac{dh}{d\varphi}\left[\frac{r_2}{r_1}\frac{dV}{d\varphi}\sin\varphi + \nu V\cos\varphi\right] \tag{3.391}$$

Substitution of (3.391) into (3.353) and in view of (3.389) and (3.372) gives

$$\frac{r_2}{r_1}\frac{d^2V}{d\varphi^2} + \frac{1}{r_1}\left[\frac{d}{d\varphi}\left(\frac{r_2}{r_1}\right) + \frac{r_2}{r_1}\cot\varphi + \frac{3}{h}\frac{dh}{d\varphi}\left(\frac{r_2}{r_1}\right)\right]\frac{dV}{d\varphi} - \frac{1}{r_1}\left[\nu - \frac{3\nu\cot\varphi}{h}\frac{dh}{d\varphi} + \frac{r_1}{r_2}\cot^2\varphi\right]V = -\frac{U}{D} \tag{3.392}$$

If h is constant, we have

$$\frac{dh}{d\varphi} = 0 \tag{3.393}$$

Using this result, we can further simplify (3.391) and (3.392). Finally, we have two coupled ODEs for U and V as

$$\frac{r_2}{r_1^2}\frac{d^2U}{d\varphi^2} + \frac{1}{r_1}\left[\frac{d}{d\varphi}\left(\frac{r_2}{r_1}\right) + \frac{r_2}{r_1}\cot\varphi\right]\frac{dU}{d\varphi} + \frac{1}{r_1}\left[-\frac{r_1}{r_2}\cot^2\varphi + \nu\right]U = EhV \tag{3.394}$$

$$\frac{r_2}{r_1^2}\frac{d^2V}{d\varphi^2} + \frac{1}{r_1}\left[\frac{d}{d\varphi}\left(\frac{r_2}{r_1}\right) + \frac{r_2}{r_1}\cot\varphi\right]\frac{dV}{d\varphi} - \frac{1}{r_1}\left[\frac{r_1}{r_2}\cot^2\varphi + \nu\right]V = -\frac{U}{D} \tag{3.395}$$

These equations can be recast into the classic form

$$L(U) + \frac{\nu}{r_1}U = EhV \tag{3.396}$$

$$L(V)-\frac{\nu}{r_1}V=-\frac{U}{D} \tag{3.397}$$

where

$$L(..)=\frac{r_2}{r_1^2}\frac{d^2(..)}{d\varphi^2}+\frac{1}{r_1}\left[\frac{d}{d\varphi}\left(\frac{r_2}{r_1}\right)+\frac{r_2}{r_1}\cot\varphi\right]\frac{d(..)}{d\varphi}-\frac{r_1\cot^2\varphi}{r_2r_1}(..) \tag{3.398}$$

Finally, we can eliminate U or V to get a single fourth order ODE for U or V as:

$$LL(U)+\nu L\left(\frac{U}{r_1}\right)-\frac{\nu}{r_1}L(U)-\frac{\nu^2}{r_1^2}U=-\frac{Eh}{D}U \tag{3.399}$$

$$LL(V)-\nu L\left(\frac{V}{r_1}\right)+\frac{\nu}{r_1}L(V)-\frac{\nu^2}{r_1^2}V=-\frac{Eh}{D}V \tag{3.400}$$

For a spherical shell ($r_1 = a$) or a conical shell ($r_1 \to \infty$), we can simplify these ODEs as

$$LL(V)+\mu^4V=0 \tag{3.401}$$

$$LL(U)+\mu^4U=0 \tag{3.402}$$

where

$$\mu^4=\frac{Eh}{D}-\frac{\nu^2}{r_1^2} \tag{3.403}$$

These two equations for U and V are identical. These fourth order ODEs for U and V can actually be reduced to two second order ODEs if we allow the use of an ODE with complex coefficient (note that Schrödinger's equation in quantum mechanics is also an ODE with a complex coefficient and it will be discussed in Chapter 11). To see this, we note that

$$L[L(U)+i\mu^2U]-i\mu^2[L(U)+i\mu^2U]=LL(U)+\mu^4U=0 \tag{3.404}$$

$$L[L(U)-i\mu^2U]+i\mu^2[L(U)-i\mu^2U]=LL(U)+\mu^4U=0 \tag{3.405}$$

As discussed in Section 3.3.5 of Chau (2018), these expressions can be factorized as

$$(L-i\mu^2)[L(U)+i\mu^2U]=(L+i\mu^2)[L(U)-i\mu^2U]=LL(U)+\mu^4U=0 \tag{3.406}$$

Since the factorization is commutative, the solution of (3.402) is equivalent to the solution of the following equations

$$L(U)\pm i\mu^2U=0 \tag{3.407}$$

The method of solution for these second order ODEs will be considered for two special cases in the following sections. This is Reissner's formulation.

3.10 SPHERICAL SHELL OF CONSTANT THICKNESS

For spherical shells, we have r_1 and r_2 being a. The operator L is reduced to

$$L(U)=\frac{1}{a}\left[\frac{d^2U}{d\varphi^2}+\cot\varphi\frac{dU}{d\varphi}-\cot^2\varphi U\right] \tag{3.408}$$

We can rewrite (3.372) as

$$U = aQ_\varphi \tag{3.409}$$

Consequently, (3.407) can also be rewritten as

$$\frac{d^2Q_\varphi}{d\varphi^2} + \cot\varphi \frac{dQ_\varphi}{d\varphi} - \cot^2\varphi Q_\varphi + 2i\rho^2 Q_\varphi = 0 \tag{3.410}$$

where

$$\rho^2 = \frac{a\mu^2}{2} = \sqrt{\frac{3a^2(1-\nu^2)}{h^2} - \frac{\nu^2}{4}} \tag{3.411}$$

3.10.1 Solution in Terms of Hypergeometric Functions

To solve (3.410), we use the following change of variables introduced by Meissner in 1913:

$$x = \sin^2\varphi \tag{3.412}$$

This change of variables gives

$$dx = 2\cos\varphi\sin\varphi d\varphi \tag{3.413}$$

Differentiation using the chain rule gives

$$\frac{dQ_\varphi}{d\varphi} = 2\cos\varphi\sin\varphi \frac{dQ_\varphi}{dx} \tag{3.414}$$

Applying the chain rule one more time, we obtain

$$\frac{d^2Q_\varphi}{d\varphi^2} = \frac{d}{d\varphi}\left(\frac{dQ_\varphi}{d\varphi}\right) = \frac{d}{d\varphi}\left(2\cos\varphi\sin\varphi \frac{dQ_\varphi}{dx}\right) \tag{3.415}$$

Differentiating (3.415), we get

$$\frac{d^2Q_\varphi}{d\varphi^2} = 2(1-2\sin^2\varphi)\frac{dQ_\varphi}{dx} + 4\sin^2\varphi\cos^2\varphi \frac{d^2Q_\varphi}{dx^2} \tag{3.416}$$

Substitution of these derivatives into (3.410) gives

$$4\sin^2\varphi\cos^2\varphi \frac{d^2Q_\varphi}{dx^2} + 2(1-2\sin^2\varphi)\frac{dQ_\varphi}{dx} + 2\cos^2\varphi \frac{dQ_\varphi}{dx} - \cot^2\varphi Q_\varphi + 2i\rho^2 Q_\varphi = 0 \tag{3.417}$$

Note the following identities between x and φ

$$\cos^2\varphi = 1 - \sin^2\varphi = 1 - x \tag{3.418}$$

$$\cot^2\varphi = \frac{\cos^2\varphi}{\sin^2\varphi} = \frac{1-x}{x} \tag{3.419}$$

Using these identities, we can rewrite (3.417) as

$$4x(1-x)\frac{d^2Q_\varphi}{dx^2} + 2(2-3x)\frac{dQ_\varphi}{dx} + \left(2i\rho^2 - \frac{1-x}{x}\right)Q_\varphi = 0 \tag{3.420}$$

We introduce another round of change of variable to (3.420) as

$$z = \frac{Q_\varphi}{\sqrt{x}} \tag{3.421}$$

Applying the chain rule, we have

$$\frac{dQ_\varphi}{dx} = \frac{dz}{dx}\sqrt{x} + \frac{1}{2}\frac{z}{\sqrt{x}} \tag{3.422}$$

$$\frac{d^2Q_\varphi}{dx^2} = \frac{d^2z}{dx^2}\sqrt{x} + \frac{1}{\sqrt{x}}\frac{dz}{dx} - \frac{z}{4x\sqrt{x}} \tag{3.423}$$

Finally, (3.420) is converted to the following second order ODE with complex constant:

$$x(x-1)\frac{d^2z}{dx^2} + (\frac{5}{2}x - 2)\frac{dz}{dx} + \frac{1-2i\rho^2}{4}z = 0 \tag{3.424}$$

Recalling from Section 4.12 of Chau (2018), we obtain the following differential equation for the hypergeometric function

$$x(x-1)\frac{d^2y}{dx^2} + [\gamma - (\alpha+\beta+1)x]\frac{dy}{dx} - \alpha\beta y = 0 \tag{3.425}$$

Comparison of (3.424) and (3.425) gives

$$\gamma = 2, \quad \alpha = \frac{3 \pm \sqrt{5+8i\rho^2}}{4}, \quad \beta = \frac{3 \mp \sqrt{5+8i\rho^2}}{4} \tag{3.426}$$

It is straightforward to see that

$$\alpha + \beta + 1 = \frac{5}{2}, \quad \alpha\beta = \frac{1}{4}[1 - 2i\rho^2] \tag{3.427}$$

The solutions of (3.425) are hypergeometric functions

$$y_1 = F(\alpha,\beta,\gamma,x) = 1 + \frac{\alpha\beta}{1\cdot\gamma}x + \frac{\alpha(\alpha+1)\beta(\beta+1)}{1\cdot 2\cdot\gamma(\gamma+1)}x^2 + \dots \tag{3.428}$$

$$y_2 = x^{1-\gamma}F(\alpha-\gamma+1, \beta-\gamma+1, 2-\gamma, x) \tag{3.429}$$

for

$$-1 < x < 1, \quad \gamma - (\alpha+\beta) > -1 \tag{3.430}$$

Using (3.426), we have

$$\gamma - (\alpha+\beta) = \frac{1}{2} > -1 \tag{3.431}$$

Therefore, the hypergeometric functions are the valid solutions for (3.424)

$$z = Az_1 + Bz_2 = AF(\alpha,\beta,2,x) + \frac{1}{x}BF(\alpha-1,\beta-1,0,x) \tag{3.432}$$

We see that the second solution tends to infinity if we include the point $x = 0$. This point corresponds to the zenith of the spherical shell. Thus, we should not involve the second solution in (3.432) for the case of no hole at the top of the spherical shell. Note that the largest of the unreinforced spherical dome structure in the world, the Pantheon in Rome, actually has a hole at the top of the spherical shell.

To simplify later analyses, we introduce another parameter δ such that

$$\delta^2 = 5 + 8i\rho^2 = 5 + 4i\sqrt{\frac{12a^2(1-\nu^2)}{h^2} - \nu^2} = 5 + i\kappa \tag{3.433}$$

$$1 - 2i\rho^2 = \frac{9-\delta^2}{4} \tag{3.434}$$

$$\alpha\beta = \frac{1}{4}(1 - 2i\rho^2) = \frac{9-\delta^2}{16} \tag{3.435}$$

$$(\alpha+1)(\beta+1) = \frac{7^2-\delta^2}{16} \tag{3.436}$$

$$(\alpha+2)(\beta+2) = \frac{11^2-\delta^2}{16} \tag{3.437}$$

Recalling from (3.421), we have

$$z_1 = S_1 + iS_2 = \frac{Q_\varphi}{\sqrt{x}} = \frac{Q_\varphi}{\sin\varphi} \tag{3.438}$$

Using the hypergeometric function given in (3.428), we obtain

$$z_1 = 1 + \frac{3^2 - 5 - i\kappa}{16\cdot 1\cdot 2}x + \frac{(3^2-5-i\kappa)(7^2-5-i\kappa)}{16^2\cdot 1\cdot 2\cdot 2\cdot 3}x^2 + ... \tag{3.439}$$

Collecting the real and imaginary parts of (3.439) and comparing with (3.438), we find

$$S_1 = 1 + \frac{3^2-5}{16\cdot 1\cdot 2}x + \frac{(3^2-5)(7^2-5)-\kappa^2}{16^2\cdot 1\cdot 2\cdot 2\cdot 3}x^2 + .. \tag{3.440}$$

$$S_2 = -\frac{\kappa}{16\cdot 1\cdot 2}x - \frac{(3^2-5)\kappa + (7^2-5)\kappa}{16^2\cdot 1\cdot 2\cdot 2\cdot 3}x^2 + .. \tag{3.441}$$

Recalling from (3.409) and (3.421), we have

$$U_1 = a\sin\varphi z_1 = I_1 + iI_2 \tag{3.442}$$

Substituting (3.439) into (3.442), we find

$$I_1 = a\sin\varphi[1 + \frac{3^2-5}{16\cdot 1\cdot 2}\sin\varphi + \frac{(3^2-5)(7^2-5)-\kappa^2}{16^2\cdot 1\cdot 2\cdot 2\cdot 3}\sin^2\varphi + ...] \tag{3.443}$$

$$I_2 = -a\sin^2\varphi[\frac{\kappa}{16\cdot 1\cdot 2} + \frac{(3^2-5)\kappa + (7^2-5)\kappa}{16^2\cdot 1\cdot 2\cdot 2\cdot 3}\sin\varphi + ...] \tag{3.444}$$

Note that the original ODE given in (3.402) for U is real and fourth order. The solution for U can be taken as

$$U = aQ_\varphi = AI_1 + BI_2 \tag{3.445}$$

Taking the operator L on U, we have

$$L(U) = AL(I_1) + BL(I_2) \tag{3.446}$$

Substitution of (3.442) into (3.407) gives

$$L(I_1 + iI_2) = -i\mu^2(I_1 + iI_2) \tag{3.447}$$

Since L is a linear operator, we have

$$L(I_1)+iL(I_2)=-i\mu^2 I_1+\mu^2 I_2 \tag{3.448}$$

Comparing the real and imaginary parts, we obtain

$$L(I_1)=\mu^2 I_2, \quad L(I_2)=-\mu^2 I_1 \tag{3.449}$$

The general solution of V can now be expressed in terms of I_1 and I_2 using (3.396)

$$\begin{aligned} EhaV &= aL(U)+\nu U = aAL(I_1)+aBL(I_2)+\nu AI_1+\nu BI_2 \\ &= (A\nu - Ba\mu^2)I_1+(Aa\mu^2+B\nu)I_2 \end{aligned} \tag{3.450}$$

Once U and V are obtained in terms of I_1 and I_2, we can find the forces and moments in the shell as

$$N_\varphi = -\frac{1}{a}U\cot\varphi \tag{3.451}$$

$$N_\theta = -\frac{1}{a}\frac{dU}{d\varphi} \tag{3.452}$$

$$M_\varphi = -\frac{D}{a}[\frac{dV}{d\varphi}+\nu V\cot\varphi] \tag{3.453}$$

$$M_\theta = -\frac{D}{a}[V\cot\varphi+\nu\frac{dV}{d\varphi}] \tag{3.454}$$

The strain can be found by inverting Hooke's law given in (3.354) and (3.355) as

$$\varepsilon_\varphi = \frac{1}{Eh}(N_\varphi - \nu N_\theta) \tag{3.455}$$

$$\varepsilon_\theta = \frac{1}{Eh}(N_\theta - \nu N_\varphi) \tag{3.456}$$

Subsequently, the strain-displacement relation can be used to obtain

$$\begin{aligned} a(\varepsilon_\varphi-\varepsilon_\theta) &= \frac{dv}{d\varphi} - v\cot\varphi = \frac{a}{Eh}[N_\varphi(1+\nu)-N_\theta(1+\nu)] \\ &= \frac{(1+\nu)}{Eh}(\frac{dU}{d\varphi}-U\cot\varphi) = f(\varphi) \end{aligned} \tag{3.457}$$

This is a nonhomogeneous first order ODE for v, and can be solved as

$$v=\sin\varphi[\int\frac{f(\varphi)}{\sin\varphi}d\varphi + C] \tag{3.458}$$

If the membrane forces N_φ and N_θ are known in (3.457), v can be solved from (3.457) and ε_θ cannot be obtained from (3.456). Recalling from (3.360), we have

$$\varepsilon_\theta = \frac{1}{a}(v\cot\varphi - w) \tag{3.459}$$

This can be rearranged as

$$w = v\cot\varphi - a\varepsilon_\theta \tag{3.460}$$

Thus, both displacement components w and v can be determined. In practical considerations, the displacement in the horizontal parallel circles can be defined as

$$\delta_r = v\cos\varphi - w\sin\varphi \tag{3.461}$$

Physically, this δ_r also represents the increase of the radius r_0 of the parallel circles. In view of this, we can express it in terms of U easily

$$\begin{aligned}\delta_r &= a\sin\varphi\varepsilon_\theta = \frac{a\sin\varphi}{Eh}(N_\theta - \nu N_\varphi) \\ &= -\frac{\sin\varphi}{Eh}(\frac{dU}{d\varphi} - \nu U\cot\varphi)\end{aligned} \tag{3.462}$$

Spherical shells with forces and couples uniformly distributed along the edge can be represented in terms of I_1 and I_2. The convergence of these series depends primarily on the value of ρ defined in (3.411):

$$\rho = \sqrt[4]{\frac{3a^2(1-\nu^2)}{h^2} - \frac{\nu^2}{4}} \approx \sqrt[4]{3}\sqrt{\frac{a}{h}} \tag{3.463}$$

The last of (3.463) is an approximation for small ν^2. Numerical calculations by Bolle in 1915 showed that the convergence of I_1 and I_2 is quite rapid if $\rho < 10$ (see Article 128 Of Timoshenko and Woinowsky-Krieger, 1959). In general, the convergence of I_1 and I_2 decreases as a/h becomes larger, which is typical for thin shells with small h and large a. A special consideration for the case of "shallow" shells will be considered in Section 3.11.

3.10.2 Superposition for Various Boundary Conditions

In this section, we demonstrate how to satisfy various boundary conditions. Figure 3.50 shows two types of boundary conditions.

Simply-supported Case

The left diagram in Figure 3.50 is a shell with simple supports. The membrane stresses for a simply-supported spherical shell under uniform compression has been solved and we recall from Section 3.4.2 that

$$\sigma_\varphi = \sigma_\theta = -\frac{pa}{2h} \tag{3.464}$$

$$(N_\varphi)_{\varphi=\alpha} = -\frac{pa}{2} \tag{3.465}$$

$$H = \frac{pa}{2}\cos\alpha \tag{3.466}$$

The boundary conditions at the simple support give

$$(N_\varphi)_{\varphi=\alpha} = H\cos\alpha, \quad (M_\varphi)_{\varphi=\alpha} = 0 \tag{3.467}$$

Since the membrane force and moment are expressible in terms of U and V, and in turn in constants A and B (defined in (3.445)), this results in two equations for two unknowns. The results are derived and given in Problem 3.3.

Fixed-edge Case

Superposition can be used to solve the problem of fixed end support. In particular, the solution of the simply-supported case just obtained can be added to the solution of the following problem:

$$(\varepsilon_\theta)_{\varphi=\alpha} = 0, \quad (V)_{\varphi=\alpha} = 0 \tag{3.468}$$

These two conditions correspond to fixed edges with zero lateral strain and no angle of rotation at the support. Again, these two conditions provide two equations for unknown constants A and B. The calculations are rather tedious and will not be given here. Such analyses can also be extended to the case of thermal difference between the inside and outside of the spherical shells with fixed support similar to the discussion given in Article 14 of Timoshenko and Woinowsky-Krieger (1959).

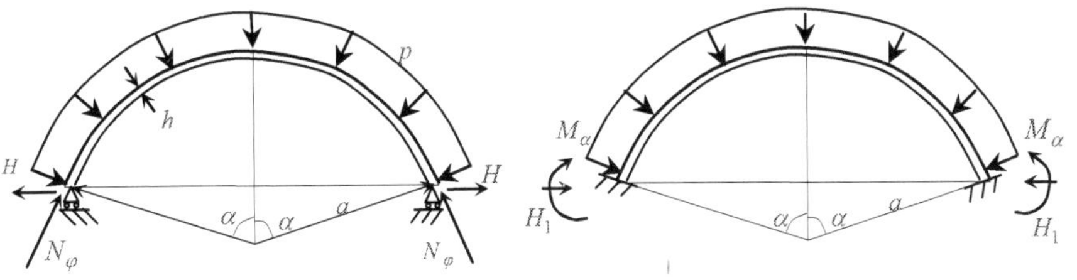

Figure 3.50 Spherical shell under uniform compression with simple support and fixed support

3.11 THIN SPHERICAL SHELL

When the thickness of the shell is small, a/h becomes very large. For such case, the solution given in terms of the hypergeometric series discussed in the last section converges very slowly, as we mentioned earlier. Two approximations have been proposed and they are introduced in this section.

3.11.1 Geckeler-Staerman Approximation

The first approximation was proposed by Staerman in 1924 and by Geckeler in 1926. For spherical shells, we first recall from (3.394), (3.395) and (3.409) that

$$U = aQ_\varphi \tag{3.469}$$

$$\frac{d^2 Q_\varphi}{d\varphi^2} + \cot\varphi \frac{dQ_\varphi}{d\varphi} - (\cot^2\varphi - \nu)Q_\varphi = EhV \tag{3.470}$$

$$\frac{d^2 V}{d\varphi^2} + \cot\varphi \frac{dV}{d\varphi} - (\cot^2\varphi + \nu)V = -\frac{a^2 Q_\varphi}{D} \tag{3.471}$$

The main approximation in this approach is based on the fact that Q_φ and V damp out rapidly from the edge such that we assume

$$\frac{d^2 Q_\varphi}{d\varphi^2} >> \frac{dQ_\varphi}{d\varphi}, \quad \frac{d^2 V}{d\varphi^2} >> \frac{dV}{d\varphi} \tag{3.472}$$

This approximation is quite natural in the sense that the highest derivative term always controls the solution form of the ODE. In effect, we can drop the second and third terms on the left of (3.470) and (3.471) to get

$$\frac{d^2 Q_\varphi}{d^2 \varphi} = EhV \tag{3.473}$$

$$\frac{d^2 V}{d\varphi^2} = -\frac{a^2 Q_\varphi}{D} \tag{3.474}$$

The unknown V (rotation of a tangent with respect to a meridian) can be eliminated from (3.473) and (3.474) as

$$\frac{d^4 Q_\varphi}{d\varphi^4} + 4\lambda^4 Q_\varphi = 0 \tag{3.475}$$

where

$$\lambda^4 = 3(1-\nu^2)(\frac{a}{h})^2 \tag{3.476}$$

Equation (3.475) is a fourth order ODE with constant coefficient, and its general solution can be written as

$$Q_\varphi = e^{\lambda\varphi}(C_1 \cos\lambda\varphi + C_2 \sin\lambda\varphi) + e^{-\lambda\varphi}(C_3 \cos\lambda\varphi + C_4 \sin\lambda\varphi) \tag{3.477}$$

We now introduce a new variable (see Figure 3.51(b)):

$$\psi = \alpha - \varphi \tag{3.478}$$

Thus, (3.477) can be rewritten as

$$\begin{aligned} Q_\varphi &= e^{\alpha\varphi}(C_1 \cos\lambda\alpha + C_2 \sin\lambda\alpha)\cos\lambda\psi e^{-\lambda\psi} \\ &+ e^{\alpha\varphi}(C_1 \sin\lambda\alpha - C_2 \cos\lambda\alpha)\sin\lambda\psi e^{-\lambda\psi} \end{aligned} \tag{3.479}$$

To rewrite the shear force in a more concise manner, we assume

$$C\sin\gamma = C_1 \cos\lambda\alpha + C_2 \sin\lambda\alpha \tag{3.480}$$

$$C\cos\gamma = C_1 \sin\lambda\alpha - C_2 \cos\lambda\alpha \tag{3.481}$$

It is straightforward to show the following identities

$$(C\cos\lambda\alpha)^2 + (C\sin\lambda\alpha)^2 = C_1^2 + C_2^2 = C^2 \tag{3.482}$$

$$\tan\gamma = \frac{C\sin\gamma}{C\cos\gamma} = \frac{C_1\cos\lambda\alpha + C_2\sin\lambda\alpha}{C_1\sin\lambda\alpha - C_2\cos\lambda\alpha} = \frac{C_1 + C_2\tan\lambda\alpha}{C_1\tan\lambda\alpha - C_2} \tag{3.483}$$

It is obvious from (3.483) that the new constant γ can be expressed in terms of the other constants as

$$\gamma = \tan^{-1}\{\frac{C_1 + C_2\tan\lambda\alpha}{C_1\tan\lambda\alpha - C_2}\} \tag{3.484}$$

Consequently, the shear force in (3.479) can now be simplified greatly to

$$Q_\varphi = Ce^{-\lambda\psi}\sin(\lambda\psi + \gamma) \tag{3.485}$$

The membrane forces can be found by combining (3.451), (3.452), and (3.485) as

$$N_\varphi = -Q_\varphi \cot\varphi = -C\cot(\alpha - \psi)e^{-\lambda\psi}\sin(\lambda\psi + \gamma) \tag{3.486}$$

$$N_\theta = -\frac{dQ_\varphi}{d\varphi} = \frac{dQ_\varphi}{d\psi} = -\lambda Ce^{-\lambda\psi}\sin(\lambda\psi + \gamma) + \lambda Ce^{-\lambda\psi}\cos(\lambda\psi + \gamma) \tag{3.487}$$

The equation for the membrane force N_θ can be further simplified by observing the following identity

$$\sin(\lambda\psi+\gamma-\frac{\pi}{4}) = \sin(\lambda\psi+\gamma)\cos\frac{\pi}{4} - \cos(\lambda\psi+\gamma)\sin\frac{\pi}{4}$$
$$= \frac{1}{\sqrt{2}}[\sin(\lambda\psi+\gamma) - \cos(\lambda\psi+\gamma)] \quad (3.488)$$

Using (3.488), we can simplify (3.487) as

$$N_\theta = -\lambda\sqrt{2}Ce^{-\lambda\psi}\sin(\lambda\psi+\gamma-\frac{\pi}{4}) \quad (3.489)$$

The corresponding bending moments are

$$M_\varphi = -\frac{D}{a}(\frac{dV}{d\varphi}+\nu\cot\varphi V) \approx -\frac{D}{a}\frac{dV}{d\varphi}$$
$$= \frac{aC}{\sqrt{2}\lambda}e^{-\lambda\psi}\sin(\lambda\psi+\gamma+\frac{\pi}{4}) \quad (3.490)$$

$$M_\theta = \nu M_\varphi = \frac{a\nu}{\lambda\sqrt{2}}Ce^{-\lambda\psi}\sin(\lambda\psi+\gamma+\frac{\pi}{4}) \quad (3.491)$$

Note that we have dropped the term proportional to V in (3.490). Finally, the angle of rotation of the deflection of the shell can be found as

$$V = \frac{1}{Eh}\frac{d^2Q_\varphi}{d\varphi^2} = -\frac{2\lambda^2C}{Eh}e^{-\lambda\psi}\cos(\lambda\psi+\gamma) \quad (3.492)$$

$$\delta_r = -\frac{\sin\varphi}{Eh}(\frac{dU}{d\varphi}-\nu U\cot\varphi) \approx -a\frac{\sin\varphi}{Eh}\frac{dU}{d\varphi}$$
$$= -\frac{\lambda a}{Eh}\sin(\alpha-\psi)\sqrt{2}Ce^{-\lambda\psi}\sin(\lambda\psi+\gamma-\frac{\pi}{4}) \quad (3.493)$$

Again, the term proportional to U in (3.493) has been dropped. This method will be employed in the next three examples to solve the problem of a shallow spherical shell subject to uniform pressure with fixed edge support.

Example 3.5 Figure 3.51 shows the two problems to be solved in this and the next example. In particular, a spherical shell is subject to uniform edge moment M_α shown in Figure 3.51(a) and is subject to the horizontal edge force H shown in Figure 3.51(b). The solution of these solutions will be used in Example 3.7 later to solve the problem of a spherical shell with a fixed edge under uniform distributed pressure. In particular, solve the problem defined in Figure 3.51(a).

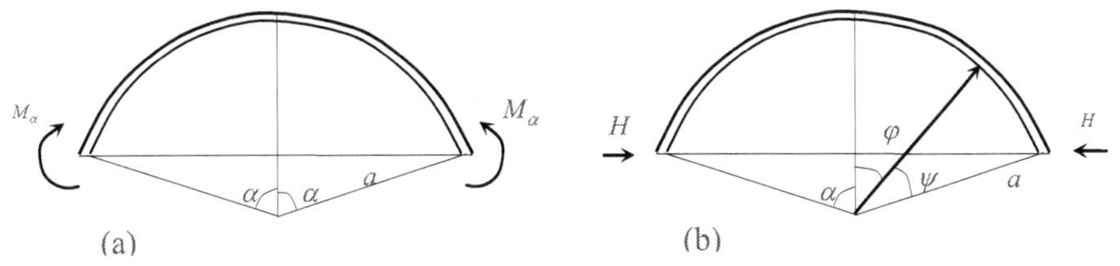

Figure 3.51 Spherical shell subject to (a) edge moment and (b) horizontal edge force

Solution: The boundary conditions for Figure 3.51(a) are

$$(N_\varphi)_{\varphi=\alpha} = 0, \quad (M_\varphi)_{\varphi=\alpha} = M_\alpha \tag{3.494}$$

At the boundary $\varphi = \alpha$, we have $\psi = 0$. We see from (3.486) that

$$(N_\varphi)_{\varphi=\alpha} = -C \cot\alpha \sin(\gamma) = 0 \tag{3.495}$$

This gives

$$\gamma = 0 \tag{3.496}$$

The second boundary condition in (3.494) leads to

$$(M_\varphi)_{\varphi=\alpha} = M_\alpha = \frac{aC}{2\lambda} \tag{3.497}$$

Rearranging (3.497), we find C as

$$C = \frac{2\lambda M_\alpha}{a} \tag{3.498}$$

Therefore, the resulting forces and moments are obtained as

$$N_\varphi = -\frac{2\lambda M_\alpha}{a} \cot(\alpha - \psi) e^{-\lambda\psi} \sin(\lambda\psi) \tag{3.499}$$

$$N_\theta = -\frac{2^{3/2}\lambda^2 M_\alpha}{a} e^{-\lambda\psi} \sin(\lambda\psi - \frac{\pi}{4}) \tag{3.500}$$

$$M_\varphi = \sqrt{2} M_\alpha e^{-\lambda\psi} \sin(\lambda\psi + \frac{\pi}{4}) \tag{3.501}$$

$$M_\theta = \nu M_\varphi = \nu\sqrt{2} M_\alpha e^{-\lambda\psi} \sin(\lambda\psi + \frac{\pi}{4}) \tag{3.502}$$

The angle of rotation and deflection are

$$V = -\frac{4\lambda^3}{Eha} M_\alpha e^{-\lambda\psi} \cos\lambda\psi \tag{3.503}$$

$$\delta_r = -\frac{2^{3/2}\lambda^2 M_\alpha}{Eh} \sin(\alpha - \psi) e^{-\lambda\psi} \sin(\lambda\psi - \frac{\pi}{4}) \tag{3.504}$$

Finally, we can specify the rotation and deflection at the edge (i.e., $\psi = 0$) as

$$(V)_{\psi=0} = -\frac{4\lambda^3}{Eha} M_\alpha, \quad (\delta_r)_{\psi=0} = \frac{2\lambda^2 \sin\alpha}{Eh} M_\alpha \tag{3.505}$$

These results will be used in solving the problem considered in Example 3.7.

Example 3.6 Solve the problem of a shallow shell subject to horizontal edge force as shown in Figure 3.51(b).

Solution: The boundary conditions for the problem shown in Figure 3.51(b) are

$$(N_\varphi)_{\varphi=\alpha} = -H\cos\alpha, \quad (M_\varphi)_{\varphi=\alpha} = 0 \tag{3.506}$$

At the boundary $\varphi = \alpha$, we have $\psi = 0$. We see from (3.490) and the second of (3.506) that

$$(M_\varphi)_{\psi=0} = \frac{aC}{\sqrt{2}\lambda}\sin(\gamma+\frac{\pi}{4}) = 0 \tag{3.507}$$

This results in the following value for γ

$$\gamma = -\frac{\pi}{4} \tag{3.508}$$

The first boundary condition given in (3.506) leads to

$$(N_\varphi)_{\psi=0} = -H\cos\alpha = C\cot\alpha\frac{1}{\sqrt{2}} \tag{3.509}$$

Thus, finally we have

$$C = -\sqrt{2}H\sin\alpha \tag{3.510}$$

Therefore, the resulting forces, moments, deflection and angle of rotation are

$$N_\varphi = \sqrt{2}H\sin\alpha\cot(\alpha-\psi)e^{-\lambda\psi}\sin(\lambda\psi-\frac{\pi}{4}) \tag{3.511}$$

$$N_\theta = \lambda 2H\sin\alpha e^{-\lambda\psi}\sin(\lambda\psi-\frac{\pi}{2}) \tag{3.512}$$

$$M_\theta = -\frac{a\nu}{\lambda}H\sin\alpha e^{-\lambda\psi}\sin\lambda\psi \tag{3.513}$$

$$M_\varphi = -\frac{a}{\lambda}H\sin\alpha e^{-\lambda\psi}\sin\lambda\psi \tag{3.514}$$

$$V = \frac{2^{3/2}\lambda^2}{Eh}H\sin\alpha e^{-\lambda\psi}\cos(\lambda\psi-\frac{\pi}{4}) \tag{3.515}$$

$$\delta_r = \frac{2\lambda aH\sin\alpha}{Eh}\sin(\alpha-\psi)e^{-\lambda\psi}\sin(\lambda\psi-\frac{\pi}{2}) \tag{3.516}$$

The rotation and deflection at the edge (i.e., $\psi=0$) are

$$(V)_{\psi=0} = \frac{2\lambda^2\sin\alpha}{Eh}H, \quad (\delta_r)_{\psi=0} = -\frac{2\lambda a\sin^2\alpha}{Eh}H \tag{3.517}$$

This result and the result from Example 3.5 will be employed in the next example to consider the problem of a spherical shell subject to uniform pressure and with a fixed base.

Note that the coefficient of H of the second equation of (3.517) is the same as the coefficient of M_α of the second equation of (3.505). This follows from the Maxwell-Rayleigh reciprocity theorem.

Example 3.7 Solve the problem of a shallow shell subject to uniform load with a fixed edge as shown in Figure 3.52(a).

Solution: As depicted in Figure 3.52, the original problem considered in (a) is the sum of the solutions for Problems (b) and (c). The solution for the problem shown in (b) can be solved from membrane theory as (see Example 3.5):

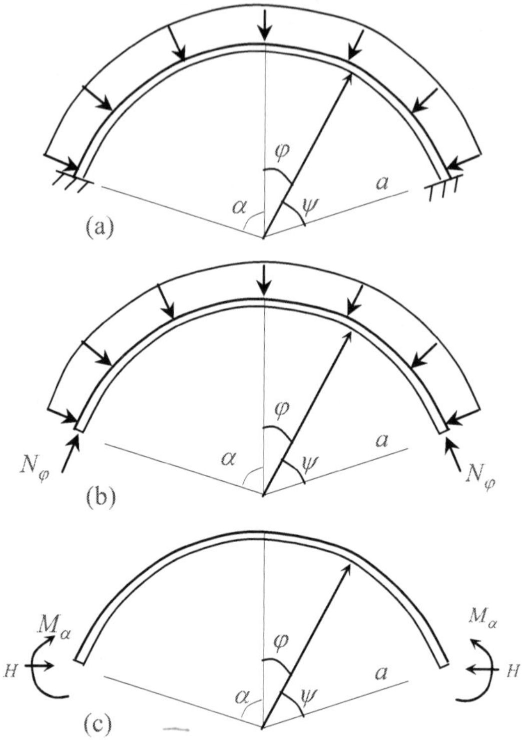

Figure 3.52 Spherical shell subject to normal pressure with a fixed edge. Problem (a) is decomposed into Problems (b) and (c)

$$N_\theta = N_\varphi = -\frac{pa}{2} \tag{3.518}$$

The corresponding deflection at the edge or support is

$$\delta_r = \frac{a\sin\alpha}{Eh}(N_\theta - \nu N_\varphi) = -\frac{pa^2(1-\nu)}{Eh}\sin\alpha \tag{3.519}$$

The last result obtained in (3.519) results from the substitution of (3.518). There is no rotation for this Problem (b). The solution of Problem (c) can be derived from the following boundary conditions for Problem (a):

$$(V)_{\psi=0} = 0, \quad (\delta_r)_{\psi=0} = 0 \tag{3.520}$$

Physically, we need to adjust H and M_α such that (3.520) will be satisfied. In particular, the edge deflection caused by H and M_α will balance those from Problem (b) given in (3.519) to satisfy the boundary conditions given in (3.520) for Problem (a). By superposition, we require that the angle rotations and deflection due to H given in (3.517) and M_α given in (3.505), given in Examples 3.5 and 3.6, are balanced to give

$$-\frac{4\lambda^3}{Eha}M_\alpha+\frac{2\lambda^2\sin\alpha}{Eh}H=0 \tag{3.521}$$

$$\frac{2\lambda^2\sin\alpha}{Eh}M_\alpha-\frac{2\lambda a\sin^2\alpha}{Eh}H=\frac{pa^2(1-\nu)}{2Eh}\sin\alpha \tag{3.522}$$

The solutions of H and M_α of (3.521) and (3.522) are

$$M_\alpha=-\frac{pa^2(1-\nu)}{4\lambda^2}=-\frac{pah}{4}\sqrt{\frac{1-\nu}{3(1+\nu)}} \tag{3.523}$$

$$H=\frac{2\lambda}{a\sin\alpha}M_\alpha=-\frac{pa(1-\nu)}{2\lambda\sin\alpha} \tag{3.524}$$

Note that the directions of H and M_α are opposite to those shown in Figure 3.52(c). This gives the Geckeler-Staerman approximation for the problem of the spherical shell subject to uniform pressure with a fixed support. The corresponding forces and moments in the shells can be determined by superimposing the results of Problems (b) and (c) on H and M_α obtained in (3.523) and (3.524) (see Problem 3.16).

3.11.2 Hetenyi Approximation

A more accurate approximation was proposed by Hetenyi in 1938. First, we note that in view of (3.409), (3.470) and (3.471) can be rewritten as

$$\frac{d^2Q_\varphi}{d\varphi^2}+\cot\varphi\frac{dQ_\varphi}{d\varphi}-\cot^2\varphi Q_\varphi+\nu Q_\varphi=EhV \tag{3.525}$$

$$\frac{d^2V}{d\varphi^2}+\cot\varphi\frac{dV}{d\varphi}-\cot^2\varphi V-\nu V=-\frac{a^2Q_\varphi}{D} \tag{3.526}$$

In particular, the following change of variables is proposed by Blumenthal in 1914:

$$Q_1=Q_\varphi\sqrt{\sin\varphi},\quad V_1=V\sqrt{\sin\varphi} \tag{3.527}$$

We find that the first order derivative terms in (3.525) and (3.526) can be eliminated by these changes of variables (instead of neglecting them as we did in the last section for the Geckeler-Staerman Approximation). In particular, differentiation of (3.527) gives

$$\frac{dQ_\varphi}{d\varphi}=\frac{1}{\sqrt{\sin\varphi}}\frac{dQ_1}{d\varphi}-\frac{Q_1\cos\varphi}{2\sin^{3/2}\varphi} \tag{3.528}$$

$$\frac{d^2Q_\varphi}{d\varphi^2}=\frac{1}{\sqrt{\sin\varphi}}\frac{d^2Q_1}{d\varphi^2}-\frac{\cos\varphi}{\sin^{3/2}\varphi}\frac{dQ_1}{d\varphi}+(\frac{\sin^2\varphi}{2}+\frac{3\cos^2\varphi}{4})\frac{Q_1}{\sin^{5/2}\varphi} \tag{3.529}$$

Substitution of (3.528) and (3.529) into (3.525) gives

$$\frac{d^2Q_1}{d\varphi^2}+(\frac{5}{4}+\nu-\frac{1}{\sin^2\varphi})Q_1=EhV_1 \tag{3.530}$$

The mathematical structures of both equations in (3.525) and (3.526) are the same. Consequently, the corresponding equation for V_1 can be obtained from (3.528) and (3.529) by replacing Q_1 by V_1. Similarly, we obtain the following governing equation for V_1

$$\frac{d^2V_1}{d\varphi^2}+(\frac{5}{4}-\nu-\frac{1}{\sin^2\varphi})V_1=-\frac{a^2Q_1}{D} \tag{3.531}$$

After this round of change of variables, we now drop the linear order terms in (3.530) and (3.531) by assuming

$$\frac{d^2Q_1}{d\varphi^2}>>Q_1,\quad \frac{d^2V_1}{d\varphi^2}>>V_1 \tag{3.532}$$

This is a reasonable first approximation since the highest order derivative term in an ODE is always the dominant term. Using (3.532), we can simplify (3.530) and (3.531) to give

$$\frac{d^2Q_1}{d\varphi^2}=EhV_1 \tag{3.533}$$

$$\frac{d^2V_1}{d\varphi^2}=-\frac{a^2Q_1}{D} \tag{3.534}$$

We see that the mathematical structure is now exactly the same as that found in the Geckeler-Staerman Approximation. Without repeating the procedure, we can quote the solutions as (Timoshenko and Woinowsky-Krieger, 1959):

$$Q_\varphi=C\frac{e^{-\lambda\psi}}{\sqrt{\sin(\alpha-\psi)}}\sin(\lambda\psi+\gamma) \tag{3.535}$$

$$N_\varphi=-C\cot(\alpha-\psi)\frac{e^{-\lambda\psi}}{\sqrt{\sin(\alpha-\psi)}}\sin(\lambda\psi+\gamma) \tag{3.536}$$

$$N_\theta=C\frac{\lambda e^{-\lambda\psi}}{2\sqrt{\sin(\alpha-\psi)}}[2\cos(\lambda\psi+\gamma)-(k_1+k_2)\sin(\lambda\psi+\gamma)] \tag{3.537}$$

The corresponding shear and moment are

$$M_\varphi=\frac{aC}{2\lambda}\frac{e^{-\lambda\psi}}{\sqrt{\sin(\alpha-\psi)}}[k_1\cos(\lambda\psi+\gamma)+\sin(\lambda\psi+\gamma)] \tag{3.538}$$

$$M_\theta=\frac{a}{4\nu\lambda}C\frac{e^{-\lambda\psi}}{\sqrt{\sin(\alpha-\psi)}}\{[(1+\nu^2)(k_1+k_2)-2k_2]\cos(\lambda\psi+\gamma)+2\nu^2\sin(\lambda\psi+\gamma)\} \tag{3.539}$$

Finally, the angle of rotation of deflection of the shell can be found as

$$V=-\frac{2\lambda^2}{Eh}C\frac{e^{-\lambda\psi}}{\sqrt{\sin(\alpha-\psi)}}\cos(\lambda\psi+\gamma) \tag{3.540}$$

$$\delta_r=\frac{a\sin(\alpha-\psi)}{Eh}C\frac{\lambda e^{-\lambda\psi}}{\sqrt{\sin(\alpha-\psi)}}[\cos(\lambda\psi+\gamma)-k_2\sin(\lambda\psi+\gamma)] \tag{3.541}$$

where

$$k_1 = 1 - \frac{1-2\nu}{2\lambda}\cot(\alpha - \psi) \tag{3.542}$$

$$k_2 = 1 - \frac{1+2\nu}{2\lambda}\cot(\alpha - \psi) \tag{3.543}$$

(i) Spherical shell subject to edge moment

Consider the case of Figure 3.51(a) for applying edge moment:

$$(N_\varphi)_{\varphi=\alpha} = 0, \quad (M_\varphi)_{\varphi=\alpha} = M_\alpha \tag{3.544}$$

Substitution of (3.536) and (3.538) into (3.544) yields

$$(N_\varphi)_{\varphi=\alpha} = -\cot\alpha C \frac{\sin\gamma}{\sqrt{\sin\alpha}} = 0 \tag{3.545}$$

$$(M_\varphi)_{\varphi=\alpha} = \frac{aC}{2\lambda}\frac{1}{\sqrt{\sin\alpha}}(k_1\cos\gamma + \sin\gamma) = M_\alpha \tag{3.546}$$

Solving for C and γ, we obtain

$$C = \frac{2\lambda\sqrt{\sin\alpha}}{ak_1} M_\alpha, \quad \gamma = 0 \tag{3.547}$$

(ii) Spherical shell subject to edge force H

Consider the case of Figure 3.51(b) for applying edge horizontal force H:

$$(N_\varphi)_{\varphi=\alpha} = -H\cos\alpha, \quad (M_\varphi)_{\varphi=\alpha} = 0 \tag{3.548}$$

Substitution of (3.536) and (3.538) into (3.548) yields

$$(N_\varphi)_{\varphi=\alpha} = -\cot\alpha C \frac{\sin\gamma}{\sqrt{\sin\alpha}} = -H\cos\alpha \tag{3.549}$$

$$(M_\varphi)_{\varphi=\alpha} = \frac{aC}{2\lambda}\frac{1}{\sqrt{\sin\alpha}}(k_1\cos\gamma + \sin\gamma) = 0 \tag{3.550}$$

Solving for C and γ, we obtain

$$C = -\sin^{3/2}\alpha \frac{\sqrt{1+k_1^2}}{k_1} H, \quad \tan\gamma = -k_1 \tag{3.551}$$

(iii) Spherical shell subject to uniform compression with fixed support

From (3.540), (3.541), and (3.547), the edge rotation and deflection for shallow spherical shells subject to edge moment are

$$(V)_{\psi=0} = -\frac{4\lambda^3}{Eahk_1} M_\alpha, \quad (\delta_r)_{\psi=0} = \frac{2\lambda^2 \sin\alpha}{Ehk_1} M_\alpha \tag{3.552}$$

From (3.540), (3.541), and (3.551), the edge rotation and deflection for shallow spherical shells subject to horizontal edge force are

$$(V)_{\psi=0} = \frac{2\lambda^2 \sin\alpha}{Ehk_1} H, \quad (\delta_r)_{\psi=0} = -\frac{\lambda a \sin^2\alpha}{Eh}(k_2 + \frac{1}{k_1})H \tag{3.553}$$

Subsequently, these results can be used for solving the problem of fixed edge spherical shells under uniform pressure. In particular, superposition can be used to solve Problem (c) defined in Figure 3.52(c) as:

$$(\delta_r)^H_{\psi=0} + (\delta_r)^M_{\psi=0} = -\frac{pa^2(1-\nu)}{Eh}\sin\alpha \tag{3.554}$$

$$(V)^H_{\psi=0} + (V)^M_{\psi=0} = 0 \tag{3.555}$$

where the superscripts H and M denote those contributions from horizontal force H and from bending moment M_α. Subsequently, these can be used for solving the problem of fixed edge spherical shells under uniform pressure. In particular, superposition can be used to solve Problem (c) defined in Figure 3.52(c) as:

$$-\frac{4\lambda^3}{Ehak_1} M_\alpha + \frac{2\lambda^2 \sin\alpha}{Ehk_1} H = 0 \tag{3.556}$$

$$\frac{2\lambda^2 \sin\alpha}{Ehk_1} M_\alpha - \frac{\lambda a \sin^2\alpha}{Eh}(k_2 + \frac{1}{k_1})H = \frac{pa^2(1-\nu)}{2Eh}\sin\alpha \tag{3.557}$$

The solutions of H and M_α from (3.556) and (3.557) are

$$M_\alpha = -\frac{pah}{4k_2}\sqrt{\frac{1-\nu}{3(1+\nu)}} \tag{3.558}$$

$$H = -\frac{p\sqrt{ah}}{2\sin\alpha k_2}\sqrt[4]{\frac{(1-\nu)^3}{3(1+\nu)}} \tag{3.559}$$

Note that if $k_2 = 1$ in (3.558) and (3.559), we recover the results by Geckeler-Staerman Approximation. Timoshenko and Woinowsky-Krieger (1959) showed that the approximation of Hetenyi is much closer to the exact series solution in terms of the hypergeometric function than the Geckeler-Staerman approximation.

3.12 SYMMETRICAL BENDING OF THIN SHALLOW SPHERICAL SHELL

We now consider the special case of a spherical shell that the rise of the shell z_0 is much smaller than the horizontal radius l of the parallel circle. In general, for $z_0/l <$ 5, the shell can be considered as shallow, and $r/a \to 0$. The loading is assumed to be symmetric and thus is only a function of r. This theory was proposed by E. Reissner in 1946 and 1947.

3.12.1 Reissner Formulation

From Figure 3.53, we have the following geometric compatibility

$$z = \sqrt{a^2 - r^2} - (a - z_0) \tag{3.560}$$

The slope of deflection can be approximated as

$$\frac{dz}{dr} = -\frac{r}{\sqrt{a^2 - r^2}} \approx -\frac{r}{a} \tag{3.561}$$

For spherical shells, we have

$$r_1 = r_2 = a, \quad r = a \sin\varphi \tag{3.562}$$

For symmetrical bending of spherical shells, we have the following approximations as $\varphi \to 0$:

$$Q_r = Q_\varphi \cos\varphi \approx Q_\varphi, \quad N_r = N_\varphi \cos\varphi \approx N_\varphi, \quad M_r = M_\varphi \cos\varphi \approx M_\varphi \tag{3.563}$$

These approximations greatly simplify the subsequent analysis. In view of $r_0 = r = r_1 \sin\varphi = a \sin\varphi$, the force and moment equilibrium equations given in (3.350), (3.351) and (3.353) can be rewritten as

$$\frac{d}{dr}(rN_r) - N_\theta - \frac{r}{a}Q_r + rp_r = 0 \tag{3.564}$$

$$\frac{r}{a}(N_r + N_\theta) + \frac{d(rQ_r)}{dr} + rp = 0 \tag{3.565}$$

$$\frac{d}{dr}(rM_r) - M_\theta - Q_r r = 0 \tag{3.566}$$

The strain-displacement relation can be obtained through Hooke's law as:

$$\varepsilon_r = \frac{1}{Eh}(N_r - \nu N_\theta) = \frac{dv}{dr} - \frac{w}{a} \tag{3.567}$$

$$\varepsilon_\theta = \frac{1}{Eh}(N_\theta - \nu N_r) = \frac{v}{r} - \frac{w}{a} \tag{3.568}$$

The moment-displacement relation can be obtained through the curvature-moment relation as:

$$M_r = -D(\chi_r + \nu\chi_\theta) = -D(\frac{d^2 w}{dr^2} + \frac{\nu}{r}\frac{dw}{dr}) \tag{3.569}$$

$$M_\theta = -D(\chi_\theta + \nu\chi_r) = -D(\frac{1}{r}\frac{dw}{dr} + \nu\frac{d^2 w}{dr^2}) \tag{3.570}$$

Assuming that the radial load is derivable from a potential function Ω, we define

$$p_r = -\frac{d\Omega}{dr} \tag{3.571}$$

For thin shallow shells, the shear force Q_r can be neglected, and the equilibrium equation given in (3.564) becomes

$$\frac{d}{dr}(rN_r) - N_\theta - r\frac{d\Omega}{dr} = 0 \tag{3.572}$$

We see that (3.572) is identically satisfied if the following Reissner's stress function is introduced:

$$N_r = \frac{1}{r}\frac{dF}{dr} + \Omega, \quad N_\theta = \frac{d^2 F}{dr^2} + \Omega \tag{3.573}$$

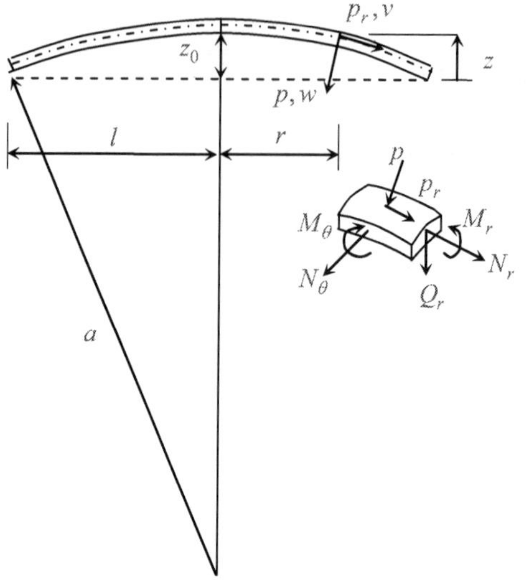

Figure 3.53 Symmetrical bending of thin shallow spherical shells

The validity of this claim is left as an exercise for the reader (see Problem 3.5). The compatibility of the strains can be expressed as (see Problem 3.6):

$$\frac{1}{r^2}\frac{d}{dr}(r^2\frac{d\varepsilon_\theta}{dr})-\frac{1}{r}\frac{d\varepsilon_r}{dr}+\frac{1}{a}\nabla^2 w=0 \tag{3.574}$$

Substitution of (3.573) into (3.567) and (3.568) and then the results into (3.574) results in

$$\nabla^2\nabla^2 F+\frac{Eh}{a}\nabla^2 w=-(1-\nu)\nabla^2\Omega \tag{3.575}$$

For the details of the proof of (3.75), the reader can refer to Problem 3.7. Equation (3.566) can be rearranged as:

$$Q_r r=\frac{d}{dr}(rM_r)-M_\theta \tag{3.576}$$

Differentiating (3.576) with respect to r and using (3.565), we obtain

$$\frac{d}{dr}[\frac{d}{dr}(rM_r)-M_\theta]+\frac{r}{a}(N_r+N_\theta)+rp=0 \tag{3.577}$$

Using (3.573), we have

$$N_r+N_\theta=\frac{d^2F}{dr^2}+\frac{1}{r}\frac{dF}{dr}+2\Omega=\nabla^2 F+2\Omega \tag{3.578}$$

Substitution of (3.569) and (3.570) into the first term in (3.577) results in

$$\frac{1}{r}\frac{d}{dr}[\frac{d}{dr}(rM_r)-M_\theta]=-D\{\frac{d^4w}{dr^4}+\frac{2}{r}\frac{d^3w}{dr^3}-\frac{1}{r^2}\frac{d^2w}{dr^2}+\frac{1}{r^3}\frac{dw}{dr}\} \tag{3.579}$$

$$=-D\nabla^2\nabla^2 w$$

Substitution of (3.578) and (3.579) into (3.577) gives

$$\nabla^2\nabla^2 w-\frac{1}{Da}\nabla^2 F=\frac{p}{D}+\frac{2\Omega}{Da} \tag{3.580}$$

Substitution of (3.575) and (3.580) provides a system of two equations for two unknowns w and F for any given p and Ω. For formulating the edge conditions, the vertical boundary shear and horizontal boundary displacement are useful and they are evaluated as

$$Q_v=Q_r+\frac{r}{a}N_r,\quad \delta_h=v-\frac{r}{a}w \tag{3.581}$$

where

$$Q_r=-D\frac{d}{dr}(\nabla^2 w) \tag{3.582}$$

This shear-displacement equation (3.582) is similar to that for plate bending.

3.12.2 Governing Equations for Negligible Self-Weight

For the case of $p=\Omega=0$, we can introduce a parameter λ and combine (3.575) and (3.580) to get a simple PDE as:

$$\nabla^2\nabla^2(w-\lambda F)-\frac{\lambda Eh}{a}\nabla^2(w+\frac{1}{\lambda hDE}F)=0 \tag{3.583}$$

To further simply this PDE, we can set

$$\lambda=-\frac{1}{\lambda hDE} \tag{3.584}$$

Using (3.584), we obtain

$$\nabla^2\nabla^2(w-\lambda F)-\frac{\lambda Eh}{a^2}\nabla^2(w-\lambda F)=0 \tag{3.585}$$

Equation (3.584) implies

$$\lambda^2=-(\frac{1}{hE})\frac{12(1-v^2)}{Eh^3} \tag{3.586}$$

Thus, we have

$$\lambda=\frac{i}{Eh^2}\sqrt{12(1-v^2)} \tag{3.587}$$

Let us introduce a new characteristic length parameter L as

$$\frac{\lambda Eh}{a}=\frac{i}{L^2} \tag{3.588}$$

where

$$L = \frac{\sqrt{ah}}{\sqrt[4]{12(1-\nu^2)}} \tag{3.589}$$

Substitution of (3.588) into (3.585) gives

$$\nabla^2\nabla^2(w-\lambda F) - \frac{i}{L^2}\nabla^2(w-\lambda F) = 0 \tag{3.590}$$

We further assume that

$$w-\lambda F = \Phi + \Psi \tag{3.591}$$

Substitution of (3.591) into (3.590) leads to

$$\nabla^2(\nabla^2\Phi - \frac{i}{L^2}\Phi) + \nabla^2(\nabla^2\Psi - \frac{i}{L^2}\Psi) = 0 \tag{3.592}$$

If we set Φ as a harmonic function, we have

$$\nabla^2\Phi = 0 \tag{3.593}$$

Thus, (3.592) is reduced to

$$\nabla^2\Psi - \left(\frac{\sqrt{i}}{L}\right)^2 \Psi = 0 \tag{3.594}$$

3.12.3 Solution in Kelvin Functions

In explicit form (3.594) can be written as

$$L^2\frac{d^2\Psi}{dr^2} + \frac{L^2}{r}\frac{d\Psi}{dr} - i\Psi = 0 \tag{3.595}$$

Absorbing L into the variable in (3.595), we have

$$\frac{d^2\Psi}{dx^2} + \frac{1}{x}\frac{d\Psi}{dx} - i\Psi = 0 \tag{3.596}$$

where

$$x = \frac{r}{L}, \quad L = \frac{\sqrt{ah}}{\sqrt[4]{12(1-\nu^2)}} \tag{3.597}$$

Now the physical meaning of characteristic length for L is clear. This solution of (3.596) is the Kelvin function (see Section 4.9 of Chau, 2018). We can try the following solutions for Φ and Ψ:

$$\Phi = A_1 + A_2 \ln x \tag{3.598}$$

$$\Psi = A_3[Ber(x) + iBei(x)] + A_4[Ker(x) + iKei(x)] \tag{3.599}$$

where the unknown constants are complex. We see from Section 8.3 of Chau (2018) that (3.598) is in fact the Green's function of the axisymmetric case of the Laplace equation. Substituting these solution forms into (3.591), we have

$$
\begin{aligned}
& w-\lambda F \\
& =\Phi+\Psi \\
& =C_1 Ber(x)+C_2 Bei(x)+C_3 Ker(x)+C_4 Kei(x)+C_5+C_7 \ln x \\
& +i\{-C_2 Ber(x)+C_1 Ber(x)-C_4 Ker(x)+C_3 Kei(x)+C_6 \ln x+C_8\}
\end{aligned} \tag{3.600}
$$

where all the constants are now real and are defined as

$$C_1=\mathrm{Re}[A_3],\quad C_2=-\mathrm{Im}[A_3],\quad C_3=\mathrm{Re}[A_4],\quad C_4=-\mathrm{Im}[A_4] \tag{3.601}$$

$$C_6=\mathrm{Im}[A_2],\quad C_7=\mathrm{Re}[A_2],\quad C_5=\mathrm{Re}[A_1],\quad C_8=\mathrm{Im}[A_1] \tag{3.602}$$

Thus, we have

$$w=C_1 Ber(x)+C_2 Bei(x)+C_3 Ker(x)+C_4 Kei(x)+C_5 \tag{3.603}$$

$$F=-\frac{Eh^2}{\sqrt{12(1-\nu^2)}}\{-C_2 Ber(x)+C_1 Ber(x)-C_4 Ker(x)+C_3 Kei(x)+C_6 \ln x\} \tag{3.604}$$

Note that the boundedness of w for $x \to 0$ requires $C_7 = 0$, whereas we can set the $C_8 = 0$ without changing the membrane forces (see the definition of F given in (3.573)).

Example 3.8 Consider the problem of a shallow spherical shell subject to a point force P at $r = 0$, as shown in Figure 3.54.

Solution: Vertical force equilibrium gives

$$P+2\pi r Q_V=0 \tag{3.605}$$

Using the definitions of x and L given in (3.597), we can be rewrite (3.603) as

$$Q_v=-\frac{P}{2\pi xL} \tag{3.606}$$

We expect that w, dw/dr, N_θ, and N_r must be finite at $r = 0$. We also expect bending to occur near the point load only, and bending decays to zero as $r \to \infty$. Substitution of (3.606) into the first condition of (3.581) gives

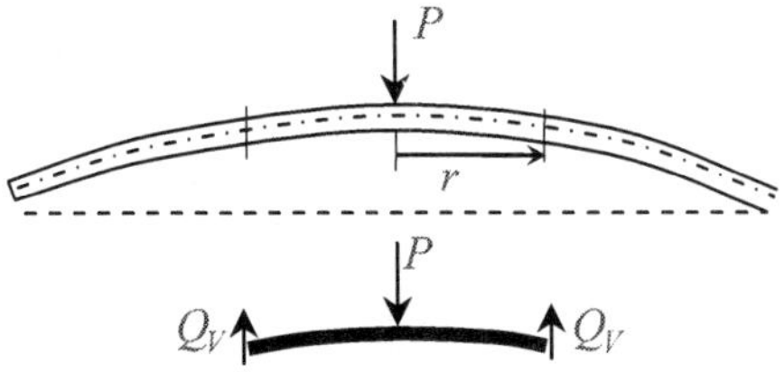

Figure 3.54 Shallow spherical shell subject to a point force

$$Q_v=Q_r+\frac{r}{a}N_r=-\frac{P}{2\pi xL} \tag{3.607}$$

The membrane force N_r in (3.573) can be evaluated as

$$N_r = \frac{1}{r}\frac{dF}{dr} = \frac{1}{xL^2}\frac{dF}{dx}$$
$$= -\frac{Eh^2}{xL^2\sqrt{12(1-\nu^2)}}\{-C_2 Bei'(x) + C_1 Bei'(x) - C_4 Ker'(x) + C_3 Kei'(x) + \frac{C_6}{x}\} \tag{3.608}$$

To find Q_r in (3.582), we first evaluate

$$\frac{1}{L^2}\bar{\nabla}^2 w = C_1\bar{\nabla}^2 Ber(x) + C_2\bar{\nabla}^2 Bei(x) + C_3\bar{\nabla}^2 Ker(x) + C_4\bar{\nabla}^2 Kei(x) \tag{3.609}$$

where

$$\bar{\nabla}^2(.) = \frac{d^2(.)}{dx^2} + \frac{1}{x}\frac{d(.)}{dx} \tag{3.610}$$

Note that the differential equation for Ψ is

$$\frac{d^2\Psi}{dx^2} + \frac{1}{x}\frac{d\Psi}{dx} - i\Psi = \bar{\nabla}^2\Psi - i\Psi = 0 \tag{3.611}$$

Since one of the solutions of Ψ is the Kelvin function, we have

$$\Psi = A\{Ber(x) + iBei(x)\} \tag{3.612}$$

Substitution of (3.600) into (3.599) gives

$$\bar{\nabla}^2\Psi - i\Psi = A\bar{\nabla}^2 Ber(x) + iA\bar{\nabla}^2 Bei(x) - iA\{Ber(x) + iBei(x)\} = 0 \tag{3.613}$$

Both the real and imaginary parts of (3.613) must be zero, and this yields

$$\bar{\nabla}^2 Ber(x) + Bei(x) = 0 \tag{3.614}$$

$$\bar{\nabla}^2 Bei(x) - Ber(x) = 0 \tag{3.615}$$

Similarly, the second independent solution of (3.607) must also satisfy (3.611). This leads to

$$\bar{\nabla}^2 Ker(x) + Kei(x) = 0 \tag{3.616}$$

$$\bar{\nabla}^2 Kei(x) - Ker(x) = 0 \tag{3.617}$$

Substitution of (3.614) to (3.617) into (3.609) gives

$$\frac{1}{L^2}\bar{\nabla}^2 w = \nabla^2 w = -C_1 Bei(x) + C_2 Ber(x) - C_3 Kei(x) + C_4 Ker(x) \tag{3.618}$$

Subsequently, the shear force can be evaluated as

$$Q_r = -D\frac{d}{dr}\nabla^2 w = -\frac{D}{L^3}[-C_1 Bei'(x) + C_2 Ber'(x) - C_3 Kei'(x) + C_4 Ker'(x)] \tag{3.619}$$

Substitution of (3.608) and (3.619) into (3.607) yields

$$\frac{Eh^2}{12L^3(1-\nu^2)}[-C_1 Bei'(x) + C_2 Ber'(x) - C_3 Kei'(x) + C_4 Ker'(x)]$$
$$+\frac{Eh^2}{aL\sqrt{12(1-\nu^2)}}\{-C_2 Ber'(x) + C_1 Ber'(x) - C_4 Ker'(x) + C_3 Kei'(x) + \frac{C_6}{x}\} \tag{3.620}$$
$$= \frac{P}{2\pi xL}$$

This equilibrium must be satisfied even for $r \to 0$ or $x \to 0$. From the table of Nevel (1959) for the Kelvin function, we find that

$$Bei' \to 0, \quad Ber' \to 0, \quad Kei' \to 0, \quad Ker' \to -\infty, \quad x \to 0 \tag{3.621}$$

Using this information, (3.620) can be written as

$$\frac{Eh^2}{12L^3(1-\nu^2)}C_4 - \frac{Eh^2}{aL\sqrt{12(1-\nu^2)}}[C_4 + C_6] = -\frac{P}{2\pi L} \tag{3.622}$$

This can be solved by choosing

$$C_4 + C_6 = 0, \quad C_4 = -\frac{Pa}{2\pi}\frac{\sqrt{12(1-\nu^2)}}{Eh^2} \tag{3.623}$$

The validity of these choices can be proved by noting from (3.597) that

$$L = \frac{\sqrt{ah}}{\sqrt[4]{12(1-\nu^2)}} \tag{3.624}$$

In summary, we have the unknowns being

$$C_4 = -C_6 = -\frac{Pa}{2\pi}\frac{\sqrt{12(1-\nu^2)}}{Eh^2} \tag{3.625}$$

$$C_1 = C_2 = C_3 = C_5 = 0 \tag{3.626}$$

Using these constants, the solutions can be summarized as

$$w = -\frac{Pa}{2\pi}\frac{\sqrt{12(1-\nu^2)}}{Eh^2}\operatorname{kei}(x) \tag{3.627}$$

$$F = -\frac{Pa}{2\pi}[\operatorname{ker}(x) + \ln x] \tag{3.628}$$

$$N_r = \frac{1}{r}\frac{dF}{dr} = \frac{Pa}{2\pi L^2 x}[\operatorname{ker}'(x) + \frac{1}{x}] \tag{3.629}$$

$$N_\theta = \frac{d^2F}{dr^2} = -\frac{Pa}{2\pi L^2}[\operatorname{ker}''(x) - \frac{1}{x^2}] \tag{3.630}$$

$$M_r = -D(\frac{d^2w}{dr^2} + \frac{\nu}{r}\frac{dw}{dr}) = \frac{Pah}{2\pi L^2\sqrt{12(1-\nu^2)}}[\operatorname{kei}''(x) + \frac{\nu}{x}\operatorname{kei}'(x)] \tag{3.631}$$

$$M_\theta = -D(\frac{1}{r}\frac{dw}{dr} + \nu\frac{d^2w}{dr^2}) = \frac{Pah}{2\pi L^2\sqrt{12(1-\nu^2)}}[\frac{1}{x}\operatorname{kei}'(x) + \nu\operatorname{kei}''(x)] \tag{3.632}$$

This completes the solution. The solutions for thin shallow spherical shells under axisymmetric loading are expressible in Kelvin functions. We have seen in Chapter 2 that the solutions for the Hertz problem of a circular plate floating on water and subject to central point load are also expressible in Kelvin functions. In fact, the solution for a cylindrical water tank with non-uniform thickness can also be solved in terms of Kelvin functions, but such analysis is out of the scope of the present chapter due to space limitation.

3.13 BENDING OF CYLINDRICAL SHELL

3.13.1 Governing Equations

Recall from an earlier section that membrane theory for a cylindrical shell cannot satisfy all boundary conditions and only provides a first approximation. The bending theory for cylindrical shells presented in Section 3.6.1 is restricted to the axisymmetric case. In this section, we will consider the effect of bending in a cylindrical shell under more general loadings, as shown in Figure 3.55. If the change of curvature of the shell can be neglected, all products of derivatives of displacement with axial forces and moments can be ignored in the equilibrium formulations. For this case, Timoshenko and Woinowsky-Krieger (1959) show that the force equilibrium equations are (see Article 101 of Timoshenko and Woinowsky-Krieger, 1959):

$$a\frac{\partial N_x}{\partial x}+\frac{\partial N_{\varphi x}}{\partial \varphi}=0 \tag{3.633}$$

$$\frac{\partial N_\varphi}{\partial \varphi}+a\frac{\partial N_{x\varphi}}{\partial x}-Q_\varphi=0 \tag{3.634}$$

$$a\frac{\partial Q_x}{\partial x}+\frac{\partial Q_\varphi}{\partial \varphi}+N_\varphi+qa=0 \tag{3.635}$$

where q is the normal pressure on the shell. Equations of moment equilibrium are

$$a\frac{\partial M_{x\varphi}}{\partial x}-\frac{\partial M_\varphi}{\partial \varphi}+aQ_\varphi=0 \tag{3.636}$$

$$\frac{\partial M_{\varphi x}}{\partial \varphi}+a\frac{\partial M_x}{\partial x}-aQ_x=0 \tag{3.637}$$

We can use (3.634) and (3.635) to eliminate Q_x and Q_φ from (3.636) and (3.637). These expressions along with (3.633) give three equations

$$a\frac{\partial N_x}{\partial x}+\frac{\partial N_{\varphi x}}{\partial \varphi}=0 \tag{3.638}$$

$$\frac{\partial N_\varphi}{\partial \varphi}+a\frac{\partial N_{x\varphi}}{\partial x}+\frac{\partial M_{x\varphi}}{\partial x}-\frac{1}{a}\frac{\partial M_\varphi}{\partial \varphi}=0 \tag{3.639}$$

$$N_\varphi+\frac{\partial^2 M_{\varphi x}}{\partial x\partial \varphi}+a\frac{\partial^2 M_x}{\partial x^2}-\frac{\partial^2 M_{x\varphi}}{\partial x\partial \varphi}+\frac{1}{a}\frac{\partial^2 M_\varphi}{\partial \varphi^2}+qa=0 \tag{3.640}$$

Now we recall the following Hooke's law:

$$N_x=\frac{Eh}{1-\nu^2}(\varepsilon_x+\nu\varepsilon_\varphi),\quad N_\varphi=\frac{Eh}{1-\nu^2}(\varepsilon_\varphi+\nu\varepsilon_x) \tag{3.641}$$

$$M_x=-D(\chi_x+\nu\chi_\varphi),\quad M_\varphi=-\nu(\chi_\varphi+\nu\chi_x) \tag{3.642}$$

$$N_{x\varphi}=N_{\varphi x}=\frac{Eh}{2(1+\nu)}\gamma_{x\varphi},\quad M_{x\varphi}=-M_{\varphi x}=D(1-\nu)\chi_{x\varphi} \tag{3.643}$$

These expressions can be obtained from (3.14) and (3.15) by replacing y by φ. The displacement strain relations are

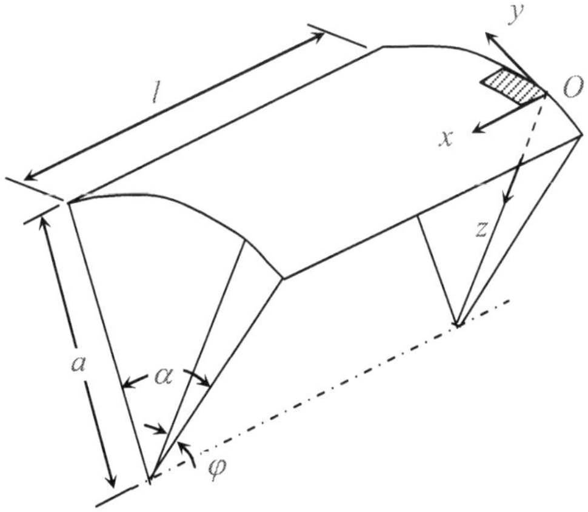

Figure 3.55 Bending of a cylindrical shell

$$\varepsilon_x = \frac{\partial u}{\partial x}, \quad \varepsilon_\varphi = \frac{1}{a}(\frac{\partial v}{\partial \varphi} - w), \quad \gamma_{x\varphi} = \frac{1}{a}\frac{\partial u}{\partial \varphi} + \frac{\partial v}{\partial x} \tag{3.644}$$

These relations can be obtained from the strain-displacement relation (see equation (1.7) of Chau, 2013). The curvature changes become

$$\chi_x = \frac{\partial^2 w}{\partial x^2}, \quad \chi_\varphi = \frac{1}{a^2}(\frac{\partial v}{\partial \varphi} + \frac{\partial^2 w}{\partial \varphi^2}), \quad \chi_{x\varphi} = \frac{1}{a}(\frac{\partial v}{\partial x} + \frac{\partial^2 w}{\partial x \partial \varphi}) \tag{3.645}$$

The first equation of (3.645) can be visualized immediately by referring to the proof of (1.19) in Chapter 1 for beam bending, the second one has been proved in (3.366) and the last one is proved with the help of Figure 3.56. In particular, Figure 3.56(a) illustrates twisting above the y-axis and Figure 3.56(b) illustrates twisting above the z-axis.

According to these diagrams, we have the change of curvature as

$$-\chi_{x\varphi} a d\varphi = -\frac{\partial^2 w}{\partial x \partial \varphi} d\varphi - (\frac{\partial v}{\partial x} + \frac{\partial^2 v}{\partial x \partial \varphi} d\varphi) d\varphi \tag{3.646}$$

By neglecting the higher order term in (3.646), we obtain

$$\chi_{x\varphi} = \frac{1}{a}(\frac{\partial v}{\partial x} + \frac{\partial^2 w}{\partial x \partial \varphi}) \tag{3.647}$$

Note that (3.644) and (3.645) are only valid if the effect of strain in the middle surface on the curvature is negligible. Substitution of (3.641) to (3.645) into (3.638) to (3.640) results in the following equations

$$\frac{\partial^2 u}{\partial \xi^2} + \frac{1-\nu}{2}\frac{\partial^2 u}{\partial \varphi^2} + \frac{1+\nu}{2}\frac{\partial^2 v}{\partial \xi \partial \varphi} - \nu \frac{\partial w}{\partial \xi} = 0 \tag{3.648}$$

$$\frac{1+\nu}{2}\frac{\partial^2 u}{\partial x\partial\varphi}+\frac{1}{a}\frac{\partial^2 v}{\partial\varphi^2}+\frac{a(1-\nu)}{2}\frac{\partial^2 v}{\partial x^2}-\frac{1}{a}\frac{\partial w}{\partial\varphi}+\frac{h^2}{12a}(\frac{\partial^3 u}{\partial x^2\partial\varphi}+\frac{\partial^3 v}{a^2\partial\varphi^3})$$
$$+\frac{h^2}{12a}[(1-\nu)\frac{\partial^2 v}{\partial x^2}+\frac{\partial^2 v}{a^2\partial\varphi^2}]=0 \tag{3.649}$$

$$\nu\frac{\partial u}{\partial x}+\frac{1}{a}\frac{\partial v}{\partial\varphi}-\frac{w}{a}-\frac{h^2}{12}(a\frac{\partial^4 w}{\partial x^4}+\frac{2}{a}\frac{\partial^4 w}{\partial x^2\partial\varphi^2}+\frac{1}{a^3}\frac{\partial^4 w}{\partial x^4})$$
$$-\frac{h^2}{12}(\frac{2-\nu}{a}\frac{\partial^3 v}{\partial x^2\partial\varphi}+\frac{\partial^3 v}{a^3\partial\varphi^3})=-\frac{aq(1-\nu^2)}{Eh} \tag{3.650}$$

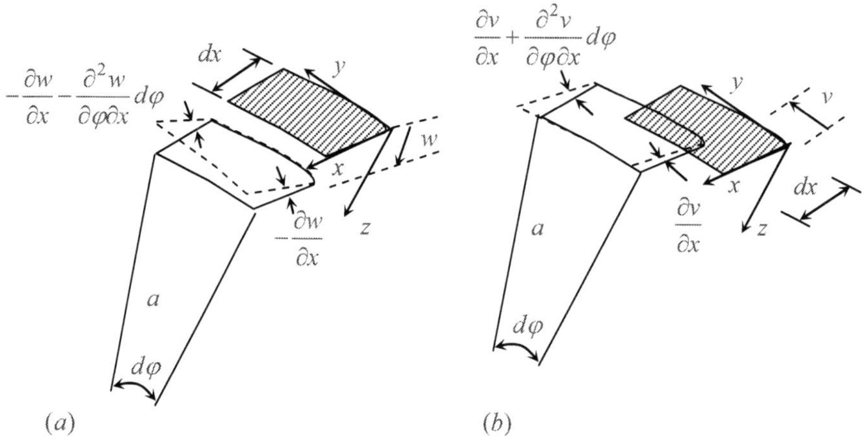

Figure 3.56 Twisting of a cylindrical shell

Vlasov in 1949 showed that the last two terms in (3.649) are small quantities that are comparable to those terms that have been neglected in deriving results obtained in Section 3.2, and the same also applies to the last term on the left of (3.650). To be consistent, with our assumption on linear stress distribution through the thickness made in Section 3.2, (3.648) to (3.650) should be simplified as

$$\frac{\partial^2 u}{\partial\xi^2}+\frac{1-\nu}{2}\frac{\partial^2 u}{\partial\varphi^2}+\frac{1+\nu}{2}\frac{\partial^2 v}{\partial\xi\partial\varphi}-\nu\frac{\partial w}{\partial\xi}=0 \tag{3.651}$$

$$\frac{1+\nu}{2}\frac{\partial^2 u}{\partial x\partial\varphi}+\frac{1}{a}\frac{\partial^2 v}{\partial\varphi^2}+\frac{a(1-\nu)}{2}\frac{\partial^2 v}{\partial x^2}-\frac{1}{a}\frac{\partial w}{\partial\varphi}=0 \tag{3.652}$$

$$\nu\frac{\partial u}{\partial x}+\frac{1}{a}\frac{\partial v}{\partial\varphi}-\frac{w}{a}-\frac{h^2}{12}(a\frac{\partial^4 w}{\partial x^4}+\frac{2}{a}\frac{\partial^4 w}{\partial x^2\partial\varphi^2}+\frac{1}{a^3}\frac{\partial^4 w}{\partial x^4})=-\frac{aq(1-\nu^2)}{Eh} \tag{3.653}$$

Now consider the following change of variables

$$c^2=\frac{h^2}{12a^2},\quad \xi=\frac{x}{a},\quad \nabla^2=\frac{\partial^2}{\partial\xi^2}+\frac{\partial^2}{\partial\varphi^2} \tag{3.654}$$

In addition, if all external loadings are included, (3.651) to (3.653) can be generalized to

$$\frac{\partial^2 u}{\partial \xi^2}+\frac{1-\nu}{2}\frac{\partial^2 u}{\partial \varphi^2}+\frac{1+\nu}{2}\frac{\partial^2 v}{\partial \xi \partial \varphi}-\nu\frac{\partial w}{\partial \xi}=-\frac{(1-\nu^2)a^2}{Eh}X \tag{3.655}$$

$$\frac{1+\nu}{2}\frac{\partial^2 u}{\partial \xi \partial \varphi}+\frac{\partial^2 v}{\partial \varphi^2}+\frac{1-\nu}{2}\frac{\partial^2 v}{\partial \xi^2}-\frac{\partial w}{\partial \varphi}=-\frac{(1-\nu^2)a^2}{Eh}Y \tag{3.656}$$

$$\nu\frac{\partial u}{\partial \xi}+\frac{\partial v}{\partial \varphi}-w-c^2\nabla^2\nabla^2 w=-\frac{(1-\nu^2)a^2}{Eh}Z \tag{3.657}$$

To solve this system of fourth order PDEs, Vlasov in 1949 ingeniously proposed a single stress function in uncoupling the system, which will be presented in the next section.

3.13.2 Vlasov's Stress Function

In particular, Vlasov proposed a stress function F defined as

$$u=\frac{\partial^3 F}{\partial \xi \partial \varphi^2}-\nu\frac{\partial^3 F}{\partial \xi^3}+u_0 \tag{3.658}$$

$$v=-\frac{\partial^3 F}{\partial \varphi^3}-(2+\nu)\frac{\partial^3 F}{\partial \xi^2 \partial \varphi}+v_0 \tag{3.659}$$

$$w=-\nabla^2\nabla^2 F+w_0 \tag{3.660}$$

For the special case that $X = Y = Z = 0$, we have $u_0 = v_0 = w_0 = 0$. In view of (3.655) to (3.657), the function F satisfies the following governing equation

$$\nabla^2\nabla^2\nabla^2\nabla^2 F+\frac{(1-\nu^2)}{c^2}\frac{\partial^4 F}{\partial \xi^4}=0 \tag{3.661}$$

Equation (3.661) is a multi-harmonic equation, which has been briefly discussed in Section 8.13.5 of Chau (2018). The proof of (3.661) is left as an exercise for the readers. This is the highest order PDE that we have encountered so far in this book. Equation (3.661) is first rewritten as

$$(\nabla^2)^4 F+4\gamma^4\frac{\partial^4 F}{\partial \xi^4}=0 \tag{3.662}$$

where

$$\gamma=\sqrt[4]{\frac{3(1-\nu^2)a^2}{h^2}} \tag{3.663}$$

Note from the earlier section that Reissner's variables V and U can be factorized if complex constants are allowed. The same technique can be applied here to factorize (3.662). In particular, we see that

$$[\nabla^2 - \gamma(1+i)\frac{\partial}{\partial \xi}][\nabla^2 F + \gamma(1+i)\frac{\partial F}{\partial \xi}]$$

$$= (\nabla^2)^2 F + \gamma(1+i)\frac{\partial}{\partial \xi}\nabla^2 F - \gamma(1+i)\frac{\partial}{\partial \xi}\nabla^2 F - \gamma^2(1+i)^2 \frac{\partial^2 F}{\partial \xi^2} \tag{3.664}$$

$$= (\nabla^2)^2 F - \gamma^2(1+i)^2 \frac{\partial^2 F}{\partial \xi^2}$$

Similarly, we have

$$[\nabla^2 - \gamma(1-i)\frac{\partial}{\partial \xi}][\nabla^2 F + \gamma(1-i)\frac{\partial F}{\partial \xi}] = (\nabla^2)^2 F - \gamma^2(1-i)^2 \frac{\partial^2 F}{\partial \xi^2} \tag{3.665}$$

It is straightforward to show that

$$[(\nabla^2)^2 - \gamma^2(1-i)^2 \frac{\partial^2}{\partial \xi^2}][(\nabla^2)^2 F - \gamma^2(1+i)^2 \frac{\partial^2 F}{\partial \xi^2}]$$

$$= (\nabla^2)^4 F + \gamma^4(1+i)^2(1-i)^2 \frac{\partial^4 F}{\partial \xi^4} = (\nabla^2)^4 F + 4\gamma^4 \frac{\partial^4 F}{\partial \xi^4} \tag{3.666}$$

Therefore, according to Section 3.3.5 of Chau (2018) the solution to (3.662) is equivalent to the solution of the following four governing equations:

$$\nabla^2 F_n \pm \gamma(1 \pm i)\frac{\partial F_n}{\partial \xi} = 0, \quad n = 1,2,3,4 \tag{3.667}$$

where $i = (-1)^{1/2}$. These can further be simplified by letting

$$F_1 = e^{-\gamma(1+i)\xi/2}\Phi_1 \tag{3.668}$$

$$F_2 = e^{\gamma(1+i)\xi/2}\Phi_2 \tag{3.669}$$

$$F_3 = e^{-\gamma(1-i)\xi/2}\Phi_3 \tag{3.670}$$

$$F_4 = e^{\gamma(1-i)\xi/2}\Phi_4 \tag{3.671}$$

where $i = (-1)^{1/2}$. Consequently, (3.662) is converted to

$$\nabla^2 \Phi_n + i\mu_n \Phi_n = 0 \tag{3.672}$$

where

$$\mu_1 = \mu_2 = -\frac{a}{2h}\sqrt{3(1-\nu^2)}, \quad \mu_3 = \mu_4 = \frac{a}{2h}\sqrt{3(1-\nu^2)} \tag{3.673}$$

All membrane forces and bending moments can be evaluated in terms of Vlasov's stress function F (see Problem 3.9). We will apply this bending theory to consider the case of cylindrical roof shells in the next section.

3.13.3 Cylindrical Roof Shells

Consider the multi-bay cylindrical roof shown in Figure 3.57. For most cylindrical shells, the width of the bay is typically half of the span l. The rise of the roof is typically one-fifth of the width. As discussed before, the membrane forces cannot satisfy the edge condition. The shell edge along the span direction needs to be reinforced by either a tensile rod or by a stiffening beam.

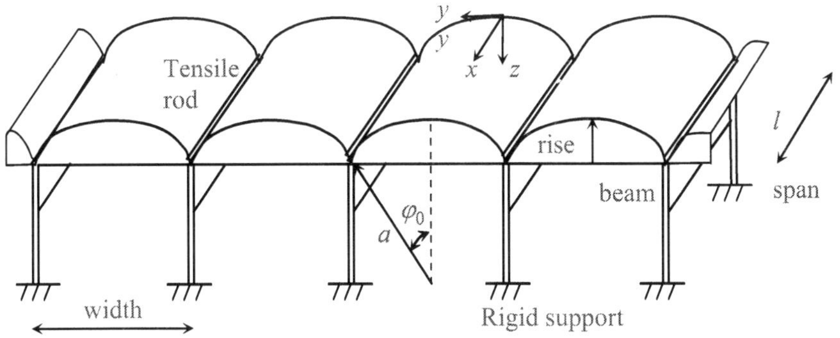

Figure 3.57 Multi-bay cylindrical roof shells

Figure 3.58 shows various types of stiffening beams that are commonly used in supporting cylindrical shells. They are the downstand edge beam, upstand edge beam, dropped edge beam, and valley beam. Note that the middle two cases in Figure 3.58 are popular for tropical regions as they can act as rain channels. Consider a general loading case X, Y, and Z as

$$X = \sum_{m=1}^{\infty} X_m(\varphi)\cos\frac{\lambda_m x}{a} \tag{3.674}$$

$$Y = \sum_{m=1}^{\infty} Y_m(\varphi)\sin\frac{\lambda_m x}{a} \tag{3.675}$$

$$Z = \sum_{m=1}^{\infty} Z_m(\varphi)\sin\frac{\lambda_m x}{a} \tag{3.676}$$

where

$$\lambda_m = \frac{m\pi a}{l} \tag{3.677}$$

The boundary conditions at the simply-supported edge on $x = 0, l$ are

$$v = 0, \quad w = 0, \quad M_x = 0, \quad N_x = 0 \tag{3.678}$$

For the free edge, the boundary conditions are

$$M_x = 0, \quad N_x = 0, \quad S_x = N_{x\varphi} + \frac{M_{x\varphi}}{a} = 0, \quad T_x = Q_x + \frac{M_{x\varphi}}{a} = 0 \tag{3.679}$$

3.13.4 Particular Solution

The particular solution can be expressed as

$$u_0 = \sum_{m=1}^{\infty} U_{0m}(\varphi)\cos\frac{\lambda_m x}{a} \tag{3.680}$$

$$v_0 = \sum_{m=1}^{\infty} V_{0m}(\varphi)\sin\frac{\lambda_m x}{a} \tag{3.681}$$

$$w_0 = \sum_{m=1}^{\infty} W_{0m}(\varphi)\sin\frac{\lambda_m x}{a} \tag{3.682}$$

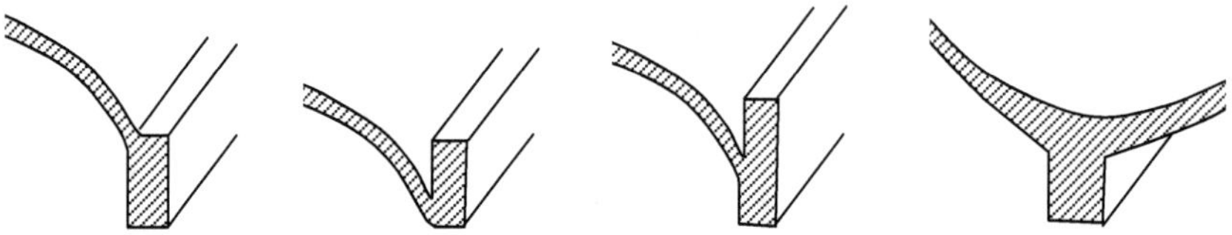

Figure 3.58 Four types of edge beams

At $\varphi = \varphi_0$, we have the following boundary values on $x = 0$

$$u_0 = \sum U_{0m}(\varphi_0) \tag{3.683}$$

$$v_0 = 0 \tag{3.684}$$

$$w_0 = 0 \tag{3.685}$$

It is straightforward to show that the boundary condition on $x = l$ also leads to these conditions. From the expressions obtained in Problem 3.9, we have

$$N_x = \frac{K}{a}[\frac{\partial u_0}{\partial \xi} + \nu(\frac{\partial v_0}{\partial \varphi} - w_0)] \tag{3.686}$$

$$M_x = -\frac{D}{a^2}(\frac{\partial^2 w_0}{\partial \xi^2} + \nu\frac{\partial^2 w_0}{\partial \varphi^2}) \tag{3.687}$$

Substitution of (3.680) to (3.682) into (3.686) and (3.687) gives

$$N_x = \frac{K}{a}\sum_{m=1}^{\infty}[-U_{0m}\lambda_m + \nu(V_{0m}' - W_{0m})]\sin\lambda_m\xi \tag{3.688}$$

$$M_x = -\frac{D}{a^2}\sum_{m=1}^{\infty}(-\lambda_m^2 W_{0m} + \nu W_{0m}'')\sin\lambda_m\xi \tag{3.689}$$

where $\xi = x/a$. Thus, the boundary conditions on $x = 0$, l given in (3.678) are satisfied identically by choosing (3.680) to (3.682).

3.13.5 Homogeneous Solution

For the homogeneous case, we have

$$X = Y = Z = 0 \tag{3.690}$$

We seek a homogeneous solution for the Vlasov stress function F of the following form:

$$F_m = e^{\alpha\varphi} \sin \frac{\lambda_m x}{a} \tag{3.691}$$

Taking the Laplacian of (3.691) repeatedly, we have

$$\nabla^2 F_m = (-\lambda_m^2 + \alpha^2) e^{\alpha\varphi} \sin \frac{\lambda_m x}{a} \tag{3.692}$$

$$\nabla^2 \nabla^2 F_m = (\alpha^2 - \lambda_m^2)^2 e^{\alpha\varphi} \sin \frac{\lambda_m x}{a} \tag{3.693}$$

$$\nabla^2 \nabla^2 \nabla^2 \nabla^2 F_m = (\alpha^2 - \lambda_m^2)^4 e^{\alpha\varphi} \sin \frac{\lambda_m x}{a} \tag{3.694}$$

$$\frac{\partial^4 F}{\partial \xi^4} = \lambda_m^4 e^{\alpha\varphi} \sin \frac{\lambda_m x}{a} \tag{3.695}$$

Using these results, (3.662) gives

$$(\alpha^2 - \lambda_m^2)^4 + \frac{(1-\nu^2)}{c^2} \lambda_m^4 = 0 \tag{3.696}$$

Taking square root of (3.696) repeatedly, we have

$$(\alpha^2 - \lambda_m^2)^2 = \pm i \sqrt{\frac{(1-\nu^2)}{c^2}} \lambda_m^2 \tag{3.697}$$

$$(\alpha^2 - \lambda_m^2) = \pm \frac{1}{\sqrt{2}} (1 \pm i) \sqrt[4]{\frac{(1-\nu^2)}{c^2}} \lambda_m \tag{3.698}$$

Thus, α can be solved as

$$\alpha = \pm \sqrt{\lambda_m^2 \pm \frac{1}{\sqrt{2}} (1 \pm i) \sqrt[4]{\frac{(1-\nu^2)}{c^2}} \lambda_m} \tag{3.699}$$

The following polar form is proposed for the term inside the square root sign on the right hand side of (3.699):

$$re^{i\theta} = \lambda_m^2 \pm \frac{1}{\sqrt{2}} (1 \pm i) \sqrt[4]{\frac{(1-\nu^2)}{c^2}} \lambda_m = r(\cos\theta + i \sin\theta) \tag{3.700}$$

The square of the magnitude of r is

$$r^2 = \left[\lambda_m^2 \pm \frac{1}{\sqrt{2}} \sqrt[4]{\frac{(1-\nu^2)}{c^2}} \lambda_m \right]^2 + \frac{1}{2} \sqrt{\frac{(1-\nu^2)}{c^2}} \lambda_m^2 \tag{3.701}$$

The angle θ is given by

$$\theta = \tan^{-1} \{ \frac{\frac{1}{\sqrt{2}} \sqrt[4]{\frac{(1-\nu^2)}{c^2}} \lambda_m}{\lambda_m^2 \pm \frac{1}{\sqrt{2}} \sqrt[4]{\frac{(1-\nu^2)}{c^2}} \lambda_m} \} \tag{3.702}$$

This can be rewritten as

$$r^2 = \rho^4 (\sigma^2 \pm \sqrt{2}\sigma + 1) = \frac{\rho^4}{2} [(1 \pm \sigma\sqrt{2})^2 + 1] \tag{3.703}$$

where

$$\rho^2 = \lambda_m \sqrt[4]{\frac{1-\nu^2}{c^2}}, \quad \sigma = \frac{\lambda_m^2}{\rho^2} = \frac{\lambda_m}{\sqrt[4]{\frac{1-\nu^2}{c^2}}} \tag{3.704}$$

With this notation of ρ, we have

$$\begin{aligned} \alpha &= r^{1/2} e^{i\theta/2} = r^{1/2} (\cos\frac{\theta}{2} + i \sin\frac{\theta}{2}) \\ &= r^{1/2} (\sqrt{\frac{\cos\theta + 1}{2}} + i \sqrt{\frac{1-\cos\theta}{2}}) \end{aligned} \tag{3.705}$$

where

$$\cos\theta = \frac{\sqrt{2}\sigma \pm 1}{\sqrt{1 + (1 \pm \sqrt{2}\sigma)^2}} \tag{3.706}$$

The sine and cosine of the half angle can be evaluated as

$$\sin\frac{\theta}{2} = \sqrt{\frac{1-\cos\theta}{2}} = \frac{1}{\sqrt{2}} \frac{\{\sqrt{1+(1\pm\sqrt{2}\sigma)^2} - (\sqrt{2}\sigma \pm 1)\}^{1/2}}{\sqrt[4]{1+(1\pm\sqrt{2}\sigma)^2}} \tag{3.707}$$

$$\cos\frac{\theta}{2} = \sqrt{\frac{1+\cos\theta}{2}} = \frac{1}{\sqrt{2}} \frac{\{\sqrt{1+(1\pm\sqrt{2}\sigma)^2} + (\sqrt{2}\sigma \pm 1)\}^{1/2}}{\sqrt[4]{1+(1\pm\sqrt{2}\sigma)^2}} \tag{3.708}$$

Substituting (3.703), (3.707) and (3.708) into (3.705), we find

$$\alpha = \gamma + i\beta \tag{3.709}$$

where

$$\gamma = \frac{\rho}{\sqrt[4]{2}\sqrt{2}} \sqrt{\sqrt{1+(1\pm\sqrt{2}\sigma)^2} + (\sqrt{2}\sigma \pm 1)} \tag{3.710}$$

$$\beta = \frac{\rho}{\sqrt[4]{2}\sqrt{2}} \sqrt{\sqrt{1+(1\pm\sqrt{2}\sigma)^2} - (\sqrt{2}\sigma \pm 1)} \tag{3.711}$$

We can show that β can be related to ρ and γ as (see Problem 3.10)

$$\beta = \frac{\rho^2}{\sqrt{8}\gamma} \tag{3.712}$$

The eight roots for α can now be expressed as

$$\alpha_1 = \gamma_1 + i\beta_1, \quad \alpha_5 = -\alpha_1 \tag{3.713}$$

$$\alpha_2 = \gamma_1 - i\beta_1, \quad \alpha_6 = -\alpha_2 \tag{3.714}$$

$$\alpha_3 = \gamma_2 + i\beta_2, \quad \alpha_7 = -\alpha_3 \tag{3.715}$$

$$\alpha_4 = \gamma_2 - i\beta_2, \quad \alpha_8 = -\alpha_4 \tag{3.716}$$

where

$$\gamma_1 = \frac{\rho}{\sqrt[4]{8}}\sqrt{\sqrt{1+(1+\sqrt{2}\sigma)^2} + \sqrt{2}\sigma + 1} \tag{3.717}$$

$$\gamma_2 = \frac{\rho}{\sqrt[4]{8}}\sqrt{\sqrt{1+(1-\sqrt{2}\sigma)^2} - (1-\sqrt{2}\sigma)} \tag{3.718}$$

$$\beta_1 = \frac{1}{\gamma_1}\frac{\rho^2}{\sqrt{8}} \tag{3.719}$$

$$\beta_2 = \frac{1}{\gamma_2}\frac{\rho^2}{\sqrt{8}} \tag{3.720}$$

Note that there is a typo in equation (h) on p. 526 of Timoshenko and Woinowsky-Krieger (1959), which is corrected in (3.717) and (3.718). Finally, the homogeneous solution becomes

$$F = \sum_{m=1}^{\infty} f_m(\varphi)\sin\frac{\lambda_m x}{a} \tag{3.721}$$

where

$$f_m(\varphi) = C_{1m}e^{\alpha_1\varphi} + C_{2m}e^{\alpha_2\varphi} + \ldots + C_{8m}e^{\alpha_8\varphi} \tag{3.722}$$

3.13.6 General Solution

Combining (3.721) to (3.658), (3.659), and (3.660), we get the displacements as

$$u = \frac{\partial^3 F}{\partial\xi\partial\varphi^2} - \nu\frac{\partial^3 F}{\partial\xi^3} + u_0 = \sum_{m=1}^{\infty}(\lambda_m f_m'' + \nu\lambda_m^3 f_m + U_{0m})\cos\frac{\lambda_m x}{a} \tag{3.723}$$

$$v = -\frac{\partial^3 F}{\partial\varphi^3} - (2+\nu)\frac{\partial^3 F}{\partial\xi^2\partial\varphi} + v_0 = \sum_{m=1}^{\infty}[(2+\nu)\lambda_m^2 f_m' - f_m''' + V_{0m})\sin\frac{\lambda_m x}{a} \tag{3.724}$$

$$w = -\nabla^2\nabla^2 F + w_0 = \sum_{m=1}^{\infty}(2\lambda_m^2 f_m'' - f_m^{(IV)} - \lambda_m^4 f_m + W_{0m})\sin\frac{\lambda_m x}{a} \tag{3.725}$$

Substitution of (3.723) to (3.725) into (3.644) and (3.645) gives the expressions for strains and changes of curvature. Further substitution of these strains and changes of curvature into (3.641) and (3.642) gives the expressions for axial forces and moments. Since there are eight unknowns, we need four boundary conditions at $\varphi = \varphi_0$ and another four at $\varphi = -\varphi_0$.

3.13.7 Vertical Load on Shell Surface

For the case of vertical uniform loadings on the cylindrical roof, we have

$$X = 0, \quad Y = p\sin\varphi, \quad Z = p\cos\varphi \tag{3.726}$$

The coefficients of expansions for these external loadings are

$$X_m = \frac{2}{l}\int_0^l X\cos\frac{\lambda_m x}{a}dx = 0 \tag{3.727}$$

$$Y_m = \frac{2}{l}\int_0^l Y\sin\frac{\lambda_m x}{a}dx = \frac{4p}{m\pi}\sin\varphi \tag{3.728}$$

$$Z_m = \frac{2}{l}\int_0^l Z\sin\frac{\lambda_m x}{a}dx = \frac{4p}{m\pi}\cos\varphi \tag{3.729}$$

where λ_m is defined in (3.677). An appropriate form of the particular solution is

$$U_{0m}(\varphi) = A_{0m}\cos\varphi \tag{3.730}$$

$$V_{0m}(\varphi) = B_{0m}\sin\varphi \tag{3.731}$$

$$W_{0m} = C_{0m}\cos\varphi \tag{3.732}$$

Substitution of these particular solutions into (3.655) to (3.657) results in three equations for constants A_{0m}, B_{0m}, and C_{0m}. The solutions can be obtained readily and some details are given in Problems 3.11 and 3.12.

Since the problem is symmetric with respect to $\varphi = 0$, we can rewrite f_m in (3.722) as

$$\begin{aligned} f_m(\varphi) = {} & A_{1m}\cos\beta_1\varphi\cosh\gamma_1\varphi + A_{2m}\sin\beta_1\varphi\sinh\gamma_1\varphi \\ & + A_{3m}\cos\beta_2\varphi\cosh\gamma_2\varphi + A_{4m}\sin\beta_2\varphi\sinh\gamma_2\varphi \end{aligned} \tag{3.733}$$

The vertical and horizontal components of the edge displacement at the support are (see Figure 3.59(a)):

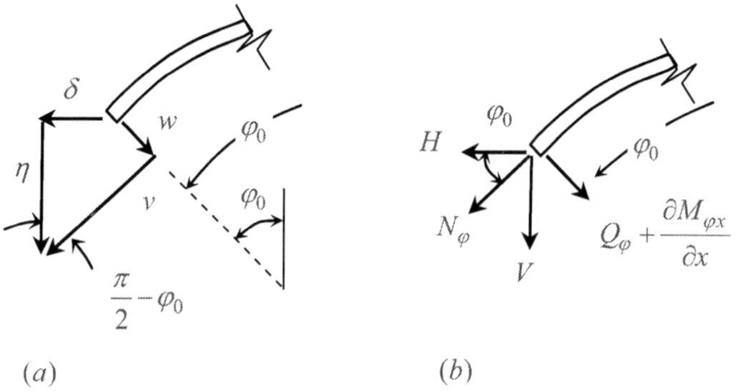

Figure 3.59 Forces and deformations at the support

$$\eta = v\sin\varphi_0 + w\cos\varphi_0 \tag{3.734}$$

$$\delta = v\cos\varphi_0 - w\sin\varphi_0 \tag{3.735}$$

The vertical and horizontal reactions at the support are (see Figure 3.59(b)):

$$V = N_\varphi \sin\varphi_0 + (Q_\varphi + \frac{\partial M_{\varphi x}}{\partial x})\cos\varphi_0 \tag{3.736}$$

$$H = N_\varphi \cos\varphi_0 - (Q_\varphi + \frac{\partial M_{\varphi x}}{\partial x})\sin\varphi_0 \tag{3.737}$$

The rotation of the shell with respect to the edge line along the x-axis is

$$\chi = \frac{v}{a} + \frac{1}{a}\frac{\partial w}{\partial \varphi} \tag{3.738}$$

Three types of cylindrical shell roofs are considered next to illustrate the application of the theory.

(i) Roof with perfectly flexible tensile rods

We first consider the case where the roof edge is supported by a flexible tensile rod without the edge beam, as shown in Figure 3.57 (in the figure, both tensile rod and beam are shown simultaneously). Due to symmetry, the horizontal deflection of the edge and the twisting angle must be zero at $\varphi = \varphi_0$, leading to:

$$\chi = 0, \quad \delta = 0 \tag{3.739}$$

In view of (3.735) and (3.738), these conditions become

$$v\cos\varphi_0 - w\sin\varphi_0 = 0 \tag{3.740}$$

$$v + \frac{\partial w}{\partial \varphi} = 0 \tag{3.741}$$

In addition to (3.736), we must also have

$$2V = Q_0 \tag{3.742}$$

where Q_0 is the weight of the tensile rod per unit length. Using Fourier expansion, we can expand the uniform weight of the tensile rod as (Timoshenko and Woinowsky-Krieger, 1959)

$$Q_0 = \frac{4Q_0}{\pi} \sum_{m=1,3,5,...}^{\infty} \frac{1}{m} \sin\frac{\lambda_m x}{a} \tag{3.743}$$

Deformation compatibility requires that the axial strain of the shell on the edge and at $\varphi = \varphi_0$ must equal the strain of the tensile rod, and this results in:

$$\frac{1}{E_0 A_0}\int_0^x 2N_{\varphi x}\, dx = \frac{\partial u}{\partial x} \tag{3.744}$$

where A_0 and E_0 are the cross-section area and Young's modulus of the tensile rod. We can calculate the four unknown constants A_{1m}, A_{2m} , ... A_{4m} by using conditions given in (3.740), (3.741), (3.742) and (3.744) for each m =1,3,5,... Then, the Vlasov stress function F is calculated from (3.733) and (3.721). Subsequently, the displacements can be determined from (3.723) to (3.725). All other strain, force and moment can then be found. In short, we need four conditions to solve the problem.

(ii) Roof with multi-bay, stiffened by beam

This situation is shown in Figure 3.57 (without the tensile rod). The support conditions given in (3.740) and (3.741) remain the same. The third condition comes from the deflection of the stiffened beam, and its governing equation is

$$E_0 I_0 \frac{d^4\eta}{dx^4} = Q_0 - 2V + 2\frac{h_0}{2}\frac{\partial N_{\varphi x}}{\partial x} \tag{3.745}$$

Deformation compatibility becomes

$$\frac{2}{E_0 A_0}\int_0^x N_{\varphi x} dx + \frac{h_0}{2}\frac{d^2\eta}{dx^2} = \frac{\partial u}{\partial x} \tag{3.746}$$

The second term on the left results from the curvature of the beam. Thus, again we have four conditions for four unknown constants.

(iii) One-bay roof, stiffened by beam

The vertical deflection of the shell on the edge is governed by the following equilibrium expressed in terms of the deflection η of the stiffening beam

$$E_0 I_0 \frac{d^4\eta}{dx^4} = Q_0 - V + \frac{h_0}{2}\frac{\partial N_{\varphi x}}{\partial x} \tag{3.747}$$

Since this is a single-bay cylindrical roof, the vertical force that acts on the beam is only half of that in (3.745). The horizontal deflection of the shell at the edge is reduced by the flexural rigidity of the beam in the horizontal plane as

$$E_0 \bar{I}_0 \frac{d^4}{dx^4}(\delta - \chi\frac{h_0}{2}) = -H \tag{3.748}$$

The equilibrium of the twisting moment of the beam above its axis is

$$\frac{dM_t}{dx} - \frac{Hh_0}{2} + M_\varphi = 0 \tag{3.749}$$

The twisting moment can be related to the twisting angle θ, which can be expressed in terms of the change of curvature as:

$$M_t = C_0 \frac{d\chi}{dx} \tag{3.750}$$

where C_0 is the torsional rigidity of the beam. Combining (3.749) and (3.750) leads to

$$C_0 \frac{d^2\chi}{dx^2} - \frac{Hh_0}{2} + M_\varphi = 0 \tag{3.751}$$

Finally, the elongation compatibility between the stiffening beam and the shell at the side edge leads to

$$\frac{1}{E_0 A_0}\int_0^x N_{\varphi x} dx + \frac{h_0}{2}\frac{d^2\eta}{dx^2} = \frac{\partial u}{\partial x} \tag{3.752}$$

The main difference between the case of multi-bay roof and a single-bay roof results in a factor of 2 in the integral terms. In summary, the four conditions for determining the unknown constants A_{1m}, A_{2m} , ... A_{4m} become (3.747) and (3.748), (3.751) and (3.752).

3.14 SUMMARY AND FURTHER READING

In this chapter, we introduce the mathematical problems of shell bending. The membrane theory of shells is formulated and applied to axisymmetric shells, including a dome under concentrated apex load, a truncated dome under ring load, and a compatibility condition at the support ring. Shells of revolution, including spherical shells with an opening and without an opening, and conical shells, are considered. Membrane theory for cylindrical shells is formulated and applied to a circular tube subjected to various loads. Mathematical similarity between the membrane theory of cylindrical shells and beam theory is summarized. Bending theory of axisymmetric cylindrical shells, and its mathematical similarity with a beam on an elastic foundation is noted. Circular pipe problems are reconsidered using bending theory of shells. The idea of effective length is discussed in the context of designing a stiffening rib. The buckling load of cylindrical shells under uniform axial compression is considered. Reissner's bending theory of shells of revolution is formulated, and its solution is shown to be expressible in terms of hypergeometric series. Approximate solutions by Geckeler-Staerman and by Hetenyi are discussed for thin spherical shells subject to radial pressure and with fixed support. A solution of symmetric bending of thin shallow shells is shown to be expressible in terms of Kelvin functions. Finally, a bending theory of a cylindrical shell is formulated, Vlasov's stress function is introduced, and the resulting eighth order PDE is factorized by allowing for complex constants. Cylindrical roofs subject to self-weight are considered as examples.

In this chapter, we do not consider the general formulation of shell theory following the Sanders and Koiter approach, in which differential geometry is used to model the curvature of shell bending in terms of Lame parameters, mean curvature, and Gauss curvature. The equation of Gauss and the Codazzi condition are important equations governing the curvatures. In the Sanders-Koiter formulation, a total of 21 equations are formulated and all formulations considered in the present chapter can be considered as certain special cases. The Sanders-Koiter equations also include plate bending as a special case. Another important topic that we do not discuss in detail is the buckling strength of shells (only buckling of a cylindrical shell under uniform axial force is considered in Section 13.8). In the 1930s, extensive experiments were conducted on the buckling of steel cylinders in relation to the design of better airplanes. It turned out that buckling strength is extremely sensitive to residual stress, temperature stress, inhomogeneities, eccentricity of loading, and geometric imperfection (like dimples). Contributors to shell buckling include Lorentz, Southwell, von Mises, Timoshenko, Flugge, Donnel, von Karman, Tsien, Koiter and Calladine. In particular, during the Second World War, Koiter in his PhD thesis was able to quantify a small geometric imperfection using perturbation analysis in 1945. However, English translation of Koiter's thesis was only available in 1967. Koiter's half-power law was found accurate even for finite amplitude of imperfection. Hutchinson extended the perturbation analysis to 2-mode initial imperfection in 1965. More reading can be found in Timoshenko and Woinowsky-

Krieger (1959), Flugge (1960, 1962), Timoshenko and Gere (1961), Ventsel and Krauthammar (2001), Calladine (1983), Shilkrut (2002), Vlasov (1951), Blaauwendraad and Hoeffaker (2014), and Vlasov and Leontev (1966).

3.15 PROBLEMS

Problem 3.1 This problem provides the proof of (3.61) in the text. Refer to Figure 3.60, and show that the tension in the ring subject to radial force can be evaluated as

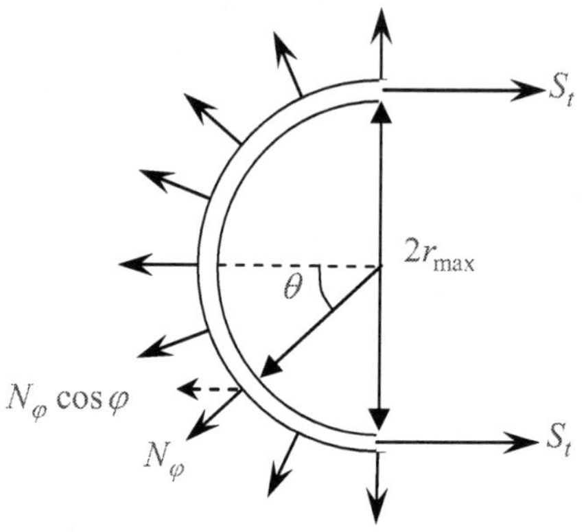

Figure 3.60 Ring subject to uniform pressure

(i) The tensile force S_t can be equated to the radial pressure as

$$2S_t = \int_{-\pi/2}^{\pi/2} N\varphi(r_{max}) \cos\alpha \cos\theta r_{max} d\theta \tag{3.753}$$

(ii) Integrate (3.753) to show that

$$S_t = |H| r_{max} = r_{max} |N\varphi(r_{max})| \cos\alpha \tag{3.754}$$

Problem 3.2 Using the variables introduced by H. Reissner in 1912, U and V, we eventually show that they are expressible in terms of hypergeometric series:

$$U = AI_1 + BI_2 \tag{3.755}$$

$$V = \frac{1}{Eha}\{(Av - Ba\mu^2)I_1 + (Aa\mu^2 + Bv)I_2\} \tag{3.756}$$

where A and B are unknown constants.

(i) Show that the two boundary conditions given in (3.467) lead to two equations:

$$AI_1 + BI_2 = -aH \tag{3.757}$$

$$A(vK_1 + a\mu^2 K_2) + B(vK_2 - a\mu^2 K_1) = 0 \tag{3.758}$$

where

$$K_1 = \frac{dI_1}{d\varphi}(\alpha) + \nu\cos\alpha I_1(\alpha), \quad K_2 = \frac{dI_2}{d\varphi}(\alpha) + \nu\cos\alpha I_2(\alpha) \tag{3.759}$$

The definitions of I_1 and I_2 are given in (3.443) and (3.444).

(ii) Prove that the constants A and B in a spherical shell subject to uniform compression considered in Section 3.10.2 are

$$A = \frac{(\nu K_2 - a\mu^2 K_1)aH}{I_2(\nu K_1 + a\mu^2 K_2) - I_1(\nu K_2 - a\mu^2 K_1)} \tag{3.760}$$

$$B = -\frac{(\nu K_1 + a\mu^2 K_2)aH}{I_2(\nu K_1 + a\mu^2 K_2) - I_1(\nu K_2 - a\mu^2 K_1)} \tag{3.761}$$

Problem 3.3 Derive the following expressions for H and M_α of the Hetenyi approximation derived in 1914 for a thin spherical shell subject to uniform pressure:

$$M_\alpha = -\frac{pah}{4k_2}\sqrt{\frac{1-\nu}{3(1+\nu)}} \tag{3.762}$$

$$H = -\frac{p\sqrt{ah}}{2\sin\alpha k_2}\sqrt[4]{\frac{(1-\nu)^3}{3(1+\nu)}} \tag{3.763}$$

Similar to the observation for the Geckeler-Staerman approximation, note that the directions of H and M_α are opposite to those shown in Figure 3.52(c).

Problem 3.4 Show that the following equation

$$\frac{d}{dr}(rN_r) - N_\theta - r\frac{d\Omega}{dr} = 0 \tag{3.764}$$

is identically satisfied by the following choices of N_θ and N_r:

$$N_r = \frac{1}{r}\frac{dF}{dr} + \Omega, \quad N_\theta = \frac{d^2F}{dr^2} + \Omega \tag{3.765}$$

Problem 3.5 Recall the definitions of the tangential and radial strains and Laplacian for symmetrical bending of shallow spherical shells:

$$\varepsilon_r = \frac{dv}{dr} - \frac{w}{a} \tag{3.766}$$

$$\varepsilon_\theta = \frac{v}{r} - \frac{w}{a} \tag{3.767}$$

$$\nabla^2 w = \frac{d^2w}{dr^2} + \frac{1}{r}\frac{dw}{dr} \tag{3.768}$$

Prove the following compatibility equation

$$\frac{1}{r^2}\frac{d}{dr}(r^2\frac{d\varepsilon_\theta}{dr}) - \frac{1}{r}\frac{d\varepsilon_r}{dr} + \frac{1}{a}\nabla^2 w = 0 \tag{3.769}$$

Problem 3.6 For shallow spherical shells considered in Section 3.12, prove the following identities:

(i)
$$\nabla^2\nabla^2 F = (\frac{d^2}{dr^2} + \frac{1}{r}\frac{d}{dr})(\frac{d^2 F}{dr^2} + \frac{1}{r}\frac{dF}{dr}) = \frac{d^4 F}{dr^4} + \frac{2}{r}\frac{d^3 F}{dr^3} - \frac{1}{r^2}\frac{d^2 F}{dr^2} + \frac{1}{r^3}\frac{dF}{dr} \tag{3.770}$$

(ii)
$$\frac{1}{r^2}\frac{d}{dr}(r^2\frac{d\varepsilon_\theta}{dr}) = \frac{1}{Eh}\{\frac{d^4 F}{dr^4} + \frac{2-\nu}{r}\frac{d^3 F}{dr^3} + \frac{2(1-\nu)}{r}\frac{d\Omega}{dr} + (1-\nu)\frac{d^2\Omega}{dr^2}\} \tag{3.771}$$

(iii)
$$-\frac{1}{r}\frac{d\varepsilon_r}{dr} = \frac{1}{Eh}\{\frac{\nu}{r}\frac{d^3 F}{dr^3} - \frac{1}{r^2}\frac{d^2 F}{dr^2} + \frac{1}{r^3}\frac{dF}{dr} - \frac{1-\nu}{r}\frac{d\Omega}{dr}\} \tag{3.772}$$

(iv) Substitute these identities into (3.561) and prove

$$\nabla^2\nabla^2 F + \frac{Eh}{a}\nabla^2 w = -(1-\nu)\nabla^2\Omega \tag{3.773}$$

Problem 3.7 Prove the validity of the following governing equation for Vlasov stress function F:

$$\nabla^2\nabla^2\nabla^2\nabla^2 F + \frac{(1-\nu^2)}{c^2}\frac{\partial^4 F}{\partial\xi^4} = 0 \tag{3.774}$$

Problem 3.8 By neglecting the effect of u and v on the bending and twisting moments, show the following axial forces and moments in terms of Vlasov's stress functions:

$$N_x = \frac{K}{a}[\frac{\partial u}{\partial\xi} + \nu(\frac{\partial v}{\partial\varphi} - w)] = \frac{Eh}{a}\frac{\partial^4 F}{\partial\varphi^2\partial\xi^2} \tag{3.775}$$

$$N_\varphi = \frac{K}{a}[\frac{\partial v}{\partial\varphi} - w + \nu\frac{\partial u}{\partial\xi}] = \frac{Eh}{a}\frac{\partial^4 F}{\partial\xi^4} \tag{3.776}$$

$$N_{x\varphi} = \frac{K(1-\nu)}{2a}(\frac{\partial u}{\partial\varphi} + \frac{\partial v}{\partial\xi}) = -\frac{Eh}{a}\frac{\partial^4 F}{\partial\xi^3\partial\varphi} \tag{3.777}$$

$$M_x = -\frac{D}{a^2}(\frac{\partial^2 w}{\partial\xi^2} + \nu\frac{\partial^2 w}{\partial\varphi^2}) = \frac{D}{a^2}(\frac{\partial^2}{\partial\xi^2} + \nu\frac{\partial^2}{\partial\varphi^2})\nabla^2\nabla^2 F \tag{3.778}$$

$$M_\varphi = -\frac{D}{a^2}(\frac{\partial^2 w}{\partial\varphi^2} + \nu\frac{\partial^2 w}{\partial\xi^2}) = \frac{D}{a^2}(\frac{\partial^2}{\partial\varphi^2} + \nu\frac{\partial^2}{\partial\xi^2})\nabla^2\nabla^2 F \tag{3.779}$$

$$M_{x\varphi} = -M_{\varphi x} = \frac{D(1-\nu)}{a^2}\frac{\partial^2 w}{\partial\xi\partial\varphi} = -\frac{D}{a^2}(1-\nu)\frac{\partial^2}{\partial\xi\partial\varphi}\nabla^2\nabla^2 F \tag{3.780}$$

$$Q_x = -\frac{D}{a^3}\frac{\partial}{\partial\xi}\nabla^2 w = \frac{D}{a^3}\frac{\partial}{\partial\xi}\nabla^2\nabla^2\nabla^2 F \tag{3.781}$$

$$Q_{\varphi} = -\frac{D}{a^3}\frac{\partial}{\partial \varphi}\nabla^2 w = \frac{D}{a^3}\frac{\partial}{\partial \varphi}\nabla^2\nabla^2\nabla^2 F \tag{3.782}$$

where

$$K = \frac{Eh}{1-\nu^2}, \quad D = \frac{Eh^3}{12(1-\nu^2)} \tag{3.783}$$

Problem 3.9 In Section 3.13.4, the real and imaginary parts of the root α are found as

$$\gamma = \frac{\rho}{\sqrt[4]{2}\sqrt{2}}\sqrt{\sqrt{1+(1\pm\sqrt{2}\sigma)^2}+(\sqrt{2}\sigma\pm 1)} \tag{3.784}$$

$$\beta = \frac{\rho}{\sqrt[4]{2}\sqrt{2}}\sqrt{\sqrt{1+(1\pm\sqrt{2}\sigma)^2}-(\sqrt{2}\sigma\pm 1)} \tag{3.785}$$

Prove the following relation between β and γ

$$\beta = \frac{\rho^2}{\sqrt{8}\gamma} \tag{3.786}$$

Problem 3.10 This problem considers cylindrical shells subject to vertical load as discussed in Section 3.13.7.

(i) Substituting (3.730) to (3.732) into (3.680) to (3.682) to show that the particular solution can be expressed as:

$$u_0 = \sum_{m=1}^{\infty} A_{0m}\cos\varphi\cos\lambda_m\xi \tag{3.787}$$

$$v_0 = \sum_{m=1}^{\infty} B_{0m}\sin\varphi\sin\lambda_m\xi \tag{3.788}$$

$$w_0 = \sum_{m=1}^{\infty} C_{0m}\cos\varphi\sin\lambda_m\xi \tag{3.789}$$

where $\xi = x/a$.

(ii) Substituting of the results in Part (i) and (3.727) to (3.729) into (3.655) to (3.657) to show that the unknown constants A_{0m}, B_{0m}, and C_{0m} satisfy the following equations:

$$-\lambda_m^2 A_{0m} + [(\frac{1+\nu}{2})\lambda_m - (\frac{1-\nu}{2})]B_{0m} - \nu\lambda_m C_{0m} = 0 \tag{3.790}$$

$$(\frac{1+\nu}{2})\lambda_m A_{0m} + [1+\lambda_m^2(\frac{1-\nu}{2})]B_{0m} + C_{0m} = -\frac{4a^2 p(1-\nu^2)}{Ehm\pi} \tag{3.791}$$

$$-\nu\lambda_m A_{0m} + B_{0m} - [1+c^2(1+\lambda_m^2)^2]C_{0m} = -\frac{4a^2 p(1-\nu^2)}{Ehm\pi} \tag{3.792}$$

Problem 3.11 Show that the solutions for the unknown constants A_{0m}, B_{0m}, and C_{0m} given in Problem 3.10 are

$$\frac{A_{0m}}{\overline{p}} = \{2\lambda_m^3 \nu^2 - 2\nu[(1+\lambda_m)(c^2\lambda_m^4 + c^2 + 2) + \lambda_m^3(1+2c^2) + 2c^2\lambda_m^2]$$
$$+2(1-\lambda_m)[c^2(\lambda_m^2+1)^2 + 2]\}/d \tag{3.793}$$

$$\frac{B_{0m}}{\overline{p}} = 2\lambda_m^2[\nu(3\nu+1) - 2c^2\lambda_m^2(\lambda_m^2-2) - 2(c^2+2)]/d \tag{3.794}$$

$$\frac{C_{0m}}{\overline{p}} = \lambda_m\{3\nu^2(\lambda_m+1) - 2\nu(\lambda_m^3 - 2\lambda_m + 1) + (2\lambda_m^3 + \lambda_m - 1)\}/d \tag{3.795}$$

where

$$d = 2\lambda_m^4\nu^3 + \lambda_m\nu^2[c^2\lambda_m^4(1+\lambda_m) + 2\lambda_m^3(c^2-1) + 2c^2\lambda_m^2 + (c^2-7)\lambda_m$$
$$+c^2 - 1] + 2\lambda_m\nu\{\lambda_m(1+\lambda_m^2)(c^2 - c^2\lambda_m^4 - 1) + 1\} + \lambda_m^5 c^2(2\lambda_m^3 + 9\lambda_m - 1) \tag{3.796}$$
$$+2\lambda_m^3[\lambda_m(6c^2+1) - c^2] + \lambda_m^2(5c^2+9) - \lambda_m(c^2+1)$$

$$\overline{p} = \frac{4a^2 p(1-\nu^2)}{Ehm\pi} \tag{3.797}$$

Problem 3.12 Prove that the effective width b for a semi-infinite pipe subject to a ring load of p at $x = 0$:

$$b_{semi} = \frac{1}{4} b_\infty = 0.38\sqrt{ah} \tag{3.798}$$

where b_∞ is the effective width of the infinite tube discussed in Section 3.7.3.

Hints:

(i) The boundary condition is not the same as the infinite tube. The proper boundary conditions are

$$Q_0 = -p, \quad M_0 = 0, \quad \text{on } x = 0 \tag{3.799}$$

(ii) Show that the deflection for this case is

$$w = \frac{a^3 p}{2D\beta^3} f_3(\beta\xi) \tag{3.800}$$

(iii) For the semi-infinite case, the effective width should be defined as

$$b = \frac{\int_0^\infty w(x)dx}{w(0)} \tag{3.801}$$

Problem 3.13 Consider the general buckling modes (i.e., n = 1,2,3, ...) for cylindrical shells under axial force of N_x.

(i) Show that

$$L_{crit} = n\pi \sqrt[4]{\frac{a^2 h^2}{12(1-\nu^2)}} \tag{3.802}$$

(ii) Show that for any value of n, the buckling load remains the same and equals

$$N_{x,crit} = -\frac{Eh^2}{a\sqrt{3(1-\nu^2)}} \tag{3.803}$$

Problem 3.14 Show that for the shallow spherical shell shown in Figure 3.53, the radius of curvature of the spherical shell can be expressed in terms of the rise (or sagitta) z_0 and the horizontal radius l as:

$$a = \frac{l^2}{8z_0} + \frac{z_0}{2} \tag{3.804}$$

Problem 3.15 Half of the cross-section of the Pantheon in Rome is shown in Figure 3.61. Assume that the dome with an opening at the top of the Pantheon can be modeled by a spherical shell of uniform thickness of 1.2 m and diameter of 21.7 m. Also assume that the unit weight of the pumice and of the brick mixture is 8.829 kN/m^3. The compressive and tensile strengths of the pumice mixture are 4000 kPa and 400 kPa.

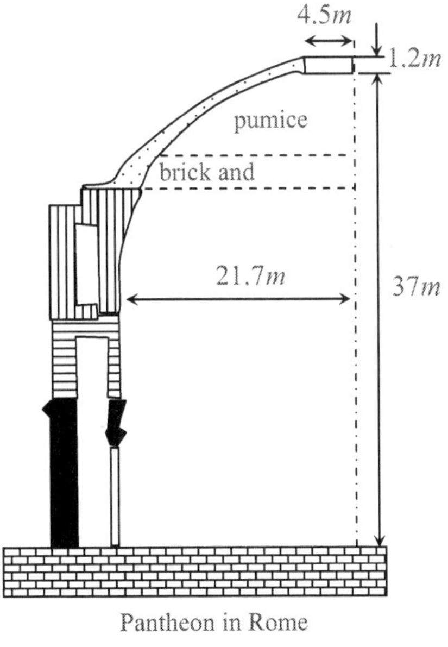

Figure 3.61 Half cross-section of the Pantheon in Rome

(i) Find the total weight of the spherical shell roof.

(ii) Find the maximum and minimum membrane forces N_θ and N_φ.

(iii) Find the factors of safety against compressive failure and tensile failure of the unreinforced spherical shell roof.

Ans:
(i) weight = 30.675 MN
(ii) $N_{\theta\max} = 224.98$kN/m, $N_{\theta\min} = -224.98$kN/m; $N_{\varphi\max} = 0$, $N_{\varphi\min} = -224.98$kN/m
(iii) 17.78 and 1.778

Problem 3.16 Show that with the Geckeler-Staerman approximation, solutions for membrane forces and bending moments for a spherical shell under uniform compression shown in Figure 3.52(a) are:

$$N_\varphi = -\frac{pa}{2} - \frac{2\lambda M_\alpha}{a}\cot(\alpha-\psi)e^{-\lambda\psi}\sin(\lambda\psi) + \sqrt{2}H\sin\alpha\cot(\alpha-\psi)e^{-\lambda\psi}\sin(\lambda\psi-\frac{\pi}{4}) \tag{3.805}$$

$$N_\theta = -\frac{pa}{2} - \frac{2^{3/2}\lambda^2 M_\alpha}{a}e^{-\lambda\psi}\sin(\lambda\psi-\frac{\pi}{4}) + 2\lambda H\sin\alpha e^{-\lambda\psi}\sin(\lambda\psi-\frac{\pi}{2}) \tag{3.806}$$

$$M_\varphi = \sqrt{2}M_\alpha e^{-\lambda\psi}\sin(\lambda\psi+\frac{\pi}{4}) - \frac{a}{\lambda}H\sin\alpha e^{-\lambda\psi}\sin\lambda\psi \tag{3.807}$$

$$M_\theta = \nu M_\varphi = \nu\sqrt{2}M_\alpha e^{-\lambda\psi}\sin(\lambda\psi+\frac{\pi}{4}) - \frac{a\nu}{\lambda}H\sin\alpha e^{-\lambda\psi}\sin\lambda\psi \tag{3.808}$$

where

$$M_\alpha = -\frac{pah}{4}\sqrt{\frac{1-\nu}{3(1+\nu)}} \tag{3.809}$$

$$H = -\frac{p\sqrt{ah}}{2\sin\alpha}\sqrt[4]{\frac{(1-\nu)^3}{3(1+\nu)}} \tag{3.810}$$

These are the solutions for the Geckeler-Staerman approximation.

CHAPTER FOUR

Structural Dynamics

4.1 INTRODUCTION

Structural dynamics is considered a difficult topic for most college engineering students, but it is in action in every aspect of our daily lives, from riding on bicycles or automobiles, flying on an airplane, hiding in a house from a typhoon or hurricane, or starting a refrigerator or air conditioner. During the lifespan of real structures, including building structures, bicycles, or automobiles, they are constantly subject to time-dependent excitations. If the change of this excitation is "fast," the structures will move in an unsteady dynamic manner and inertia effects cannot be ignored. The study of structural dynamics is particularly important in the design of buildings against winds and earthquakes, machines or automobiles against oscillations, airplanes against self-excited fluttering vibrations from the wings, cable suspension bridges against wind-induced vibrations, magnetic levitated trains against dynamic instability, and speedboats against vibrations from fluid-structure interactions.

Dynamics is one of the classical topics tackled by great scientists and mathematicians during the development of differential equations. They include Aristole, Galieo Galilei, Newton, Euler, Daniel Bernoulli, Lagrange, D'Alembert, Laplace, Poincare, Rayleigh, and Sophus Lie. In terms of n-degree-of-freedom dynamical systems, it was mainly developed by Lagrange, Navier, Poisson, and Plana. More recent development of dynamic instability includes the work of Liapunov, Routh, Menabrea, and Rayleigh. A major piece of work on dynamics of structures is *The Theory of Sound* by Lord Rayleigh. Kneser's theorem deals with the condition that an oscillating solution exists for a second order ODE, and another similar theorem is called the Sturm-Picone comparison theorem. The vibrations of beams have been considered extensively by Stokes, Philips, Saint-Venant, Boussinesq, Krall, Kirchhoff, Bernerd, Rayleigh, Kussner, and Pochhammer. Axial impact on beams or bars was considered by Young, Navier, Cauchy, Poisson, Clebsch, Babinet, Philips, and Boussinesq. Flexural impacts on beams were considered Hodgkinson. Kirchhoff considered the vibrations of beams with varying cross-section, Kussner considered the effect of axial loads on vibrations, and Pochhammer combined torsional and longitudinal vibrations. Traveling loads on beams were considered by Philips, Stokes, Resal, Melan, Zimmermann, Kriloff, Bleich, Prager, Timoshenko and Krall. The main contributors of the vibrations of plates include Germain, Kirchhoff, Bernerd, Rayleigh, Lamb, Southwell, Klotter, Reissner, Schmidt, and Grammel.

In terms of civil engineering applications, the structural dynamics of trusses and frames was considered by H. Reissner, of bridges by F. Bleich, and of buildings by G. Alfani. The dynamics actions on soils were considered by A.

Sommerfeld, A. Hertwig, and H. Lorentz. In mechanical engineering, vibrations of machinery, engines and cars have been considered by I. Radinger, A. Stodola, I. Heun, H. Lorentz, W. Hort, and R. von Mises. Dynamics of shock absorbers was considered A. Foppl and K. Klotter. Dynamics of ships was considered by W. Froude and A. Krylov.

Structural dynamics can be, at least, classified into a steady harmonic vibration from a conservative force (small vibrations of continuous and stable oscillations) and a non-steady vibration from non-conservative forces (large chaotic vibrations that may cause instability). Dynamics analysis for harmonic motions had been considered since the time of Euler, Lagrange, and D'Alembert, whereas dynamic stability of structures for non-conservative systems (external forces are not derivatives of a potential) was studied in detail by Routh and Hurwitz. For example, when the wind speed attains a critical value, suspension bridges may undergo unstable oscillations because wind-induced forces are non-conservative.

In this chapter, we focus our discussion on single-degree-of-freedom oscillators and multi-degree-of-freedom oscillators, with particular relevance to single- and multi-story buildings with rigid slabs. Ground-induced motions are addressed through discussions of response spectrum and modal superposition.

4.2 STATIC DEFLECTION VERSUS NATURAL VIBRATION

In this section, we consider a vibrating system of a single-degree-of-freedom oscillator and its application to consider the suspension system for the wheel. In particular, as shown in Figure 4.1, the vertical force equilibrium for the single-degree-of-freedom oscillator is

$$m\frac{d^2z}{dt^2} = -k(z - z_0) + mg \tag{4.1}$$

For the special case of static deflection, we can set the left hand side of (4.1) to zero resulting in

$$\Delta z = (z_1 - z_0) = \frac{mg}{k} \tag{4.2}$$

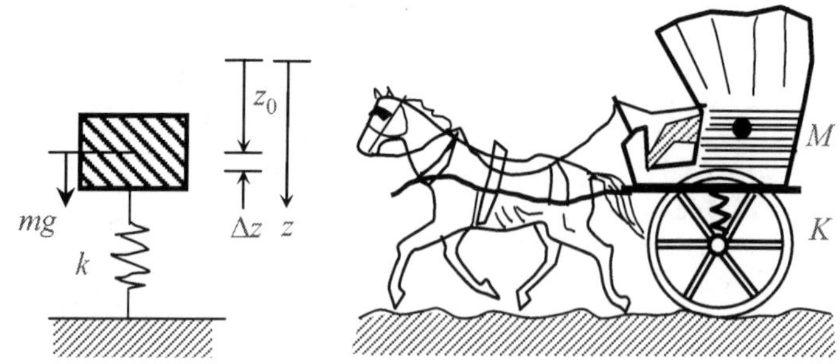

Figure 4.1 Mechanical model for a carriage

For free oscillations, by subtracting the dead loads, we can consider the system as

$$\frac{d^2 z_2}{dt^2} + \frac{k}{m} z_2 = 0 \tag{4.3}$$

For this harmonic oscillator, the solutions are sine and cosine functions as

$$z_2 = A\cos(\omega_0 t) + B\sin(\omega_0 t) \tag{4.4}$$

where the fundamental natural circular frequency of the oscillator is defined as

$$\omega_0 = \sqrt{\frac{k}{m}} \tag{4.5}$$

Once the oscillator is excited, the mass m will vibrate according to this circular frequency. The period and frequency of this oscillation can be expressed in terms of ω_0 as:

$$f = \frac{\omega_0}{2\pi} = \frac{1}{2\pi}\sqrt{\frac{k}{m}} \tag{4.6}$$

$$T = \frac{1}{f} = \frac{2\pi}{\omega_0} = 2\pi\sqrt{\frac{m}{k}} \tag{4.7}$$

This solution can be recast as

$$z_2 = R\sin(\omega_0 t + \varphi) \tag{4.8}$$

Applying the sum rule of the sine function, we can rewrite the right-hand side of (4.8) as

$$R\sin(\omega_0 t + \varphi) = R\sin\omega_0 t\cos\varphi + R\sin\varphi\cos\omega_0 t \tag{4.9}$$

Comparison of (4.9) with (4.4) gives

$$A = R\cos\varphi, \quad B = R\sin\varphi \tag{4.10}$$

Solving for R and φ, we have

$$R = \sqrt{A^2 + B^2}, \quad \varphi = \tan^{-1}(\frac{B}{A}) \tag{4.11}$$

Physically, R is the amplitude of the maximum displacement and φ is the phase or phase angle (dimensionless). Therefore, (4.4) and (4.8) are equivalent. In fact, we can also rewrite (4.4) in a single cosine function instead of the sine function (see Problem 4.1).

For human beings, the walking frequency is about 1.3 steps per second. Naturally, this frequency can also be considered as the human comfort frequency. Therefore, we can adjust the stiffness on the carriage on the right diagram of Figure 4.1, such that the static deflection will satisfy the following constraint:

$$\frac{K}{M} = \frac{k}{m} = \omega_0^2 = \frac{g}{\Delta z} = \frac{4\pi^2}{T^2} \tag{4.12}$$

Equivalently, we have

$$\Delta z = \frac{g}{\omega_0^2} \tag{4.13}$$

In particular, we have the circular frequency being

$$\omega_0 = 2\pi f = (1.3)(2\pi) = 8.168 \text{ s}^{-1} \tag{4.14}$$

The static deflection of the carriage such that it will oscillate with the comfort circular frequency is

$$\Delta z = \frac{g}{\omega_0^2} = \frac{9.81}{(8.168)^2} = 0.147 \text{ m} = 14.7 \text{ cm} \tag{4.15}$$

If the total mass M to be carried by the carriage is fixed, we can easily select a stiffness K such that the desirable static deflection can be achieved. If the carriage is for tourists, the mass of the passenger may vary considerably. For such case, a spring suspension system with tunable stiffness is needed to achieve the human comfort frequency. If the natural frequency is too low, the magnitude of displacement will be large and young children will suffer from motion sickness since their intestines are still movable. If the natural frequency is too high, older people with ossified joints may suffer backache. We need to strike a delicate balance on this choice.

4.3 SINGLE-STORY BUILDING

Figure 4.2 shows a model of a single-story building. The mass of the story is m and the mass of the columns is ignored (this is the so-called lump mass assumption versus the case of distributed mass). There is only one unknown in this problem, the horizontal displacement $u(t)$. This is the reason that it is normally referred to as a single-degree-of-freedom (SDOF) oscillator. Of course, in real single-story structures, the building can sway in two independent horizontal directions. If the columns are not rigid axially, vertical displacement can potentially occur, but it is normally much smaller than the translational displacements. In addition, torsional twisting may also appear if twisting moment is applied. If an asymmetric structure is excited under earthquake motions, torsional response will set in as well. Nevertheless, we restrict our discussion to the SDOF system in this section. The dashpot is used to model the material damping of the structure, which is used to reflect the fact that vibrations from initial excitation will decay as a function of time. The bending stiffness of the column provides the restoring force to return the story to the original position under external loading.

Force equilibrium of the free body diagram given in Figure 4.3 is

$$f_I + f_D + f_S = F(t) \tag{4.16}$$

where f_I is the inertia force, f_D is the damping force by the dashpot, f_S is the restoring force due to the columns, and $F(t)$ is the external applied force at the story level. They are defined as

$$f_I = m\ddot{u}_t, \quad f_D = c\dot{u}, \quad f_s = ku \tag{4.17}$$

In (4.17), the damping force is normally assumed to be proportional to velocity and it is called viscous damping. Substitution of (4.17) into (4.16) gives

$$m\ddot{u} + c\dot{u} + ku = F(t) \tag{4.18}$$

where the stiffness k can be estimated as

$$k = \frac{24EI}{L^3} \tag{4.19}$$

This is twice of the bending of each column (see for example p. 545 of Gere and Timoshenko, 1990). The solution will be considered in the next section.

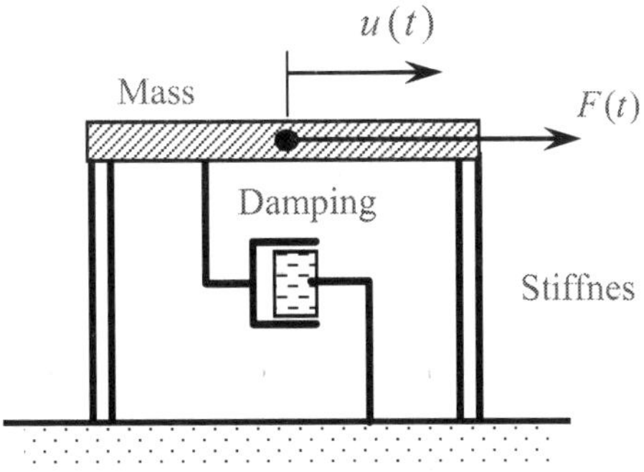

Figure 4.2 Model for a single-story building

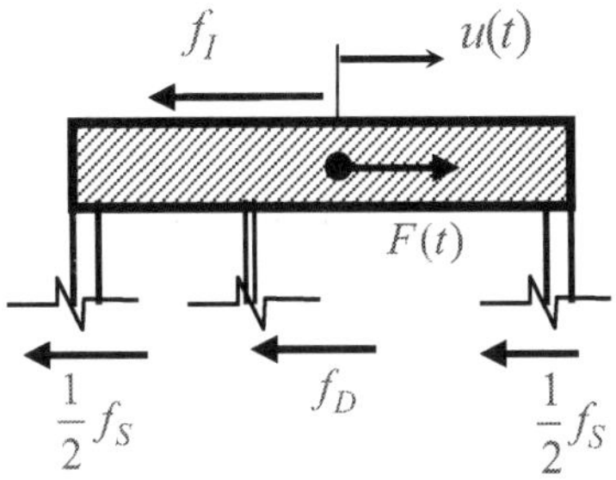

Figure 4.3 Free body diagram for a single-story building

4.4 DAMPED AND UNDAMPED RESPONSES

4.4.1 Undamped Responses

Let us consider the undamped situation first (i.e., c = 0). Consider the external force as a harmonic function:

$$m\ddot{u}(t) + ku(t) = F_0 \cos \omega t \tag{4.20}$$

The homogeneous solution is

$$u_h(t) = c_1 \cos \omega_0 t + c_2 \sin \omega_0 t \tag{4.21}$$

where

$$\omega_0 = \sqrt{\frac{k}{m}} \tag{4.22}$$

If $\omega \neq \omega_0$, we can assume the particular solution as

$$u_p(t) = A_1 \cos \omega t + A_2 \sin \omega t \tag{4.23}$$

Substitution of (4.23) into (4.20) leads to

$$A_1(k-m\omega^2)\cos\omega t + A_2(k-m\omega^2)\sin\omega t = F_0\cos\omega t \tag{4.24}$$

Equating the coefficients of sine and cosine functions on both sides, we obtain two equations

$$A_1(k-m\omega^2) = F_0, \quad A_2(k-m\omega^2) = 0 \tag{4.25}$$

Since $\omega \neq \omega_0$, we must have

$$A_1 = \frac{F_0}{m(\omega_0^2-\omega^2)}, \quad A_2 = 0 \tag{4.26}$$

Finally, the general solution is

$$u(t) = c_1\cos\omega_0 t + c_2\sin\omega_0 t + \frac{F_0}{m(\omega_0^2-\omega^2)}\cos\omega t \tag{4.27}$$

Example 4.1 Consider the case that an undamped structure is modeled by (4.20) and subject to the following forcing term and initial conditions:

$$F(t) = F_0\cos\omega t \tag{4.28}$$

$$u(0) = u_0, \quad \dot{u}(0) = v_0 \tag{4.29}$$

Solution: Differentiating (4.27) with respect to time, we find

$$\dot{u}(t) = -\omega_0 c_1\sin\omega_0 t + \omega_0 c_2\cos\omega_0 t - \frac{\omega F_0}{m(\omega_0^2-\omega^2)}\sin\omega t \tag{4.30}$$

Using the initial conditions of (4.29), we have

$$u(0) = c_1 + \frac{F_0}{m(\omega_0^2-\omega^2)} = u_0 \tag{4.31}$$

$$\dot{u}(0) = \omega_0 c_2 = v_0 \tag{4.32}$$

Thus, the constants are

$$c_1 = u_0 - \frac{F_0}{m(\omega_0^2-\omega^2)} \tag{4.33}$$

$$c_2 = \frac{v_0}{\omega_0} \tag{4.34}$$

Substitution of (4.33) and (4.34) into (4.27) gives

$$u(t) = [u_0 - \frac{F_0}{m(\omega_0^2-\omega^2)}]\cos\omega_0 t + \frac{v_0}{\omega_0}\sin\omega_0 t + \frac{F_0}{m(\omega_0^2-\omega^2)}\cos\omega t \tag{4.35}$$

Now consider the special case of the forced vibrations with no initial excitations (i.e., $u_0 = v_0 = 0$). The solution becomes

$$u(t) = \frac{F_0}{m(\omega_0^2-\omega^2)}(\cos\omega t - \cos\omega_0 t) \tag{4.36}$$

This equation can be written in a form from which the physical meaning is more apparent. First, we note the following identities:

$$\cos\omega_0 t = \cos(\frac{\omega_0+\omega}{2})\cos(\frac{\omega_0-\omega}{2}) - \sin(\frac{\omega_0+\omega}{2})\sin(\frac{\omega_0-\omega}{2}) \tag{4.37}$$

$$\cos\omega t = \cos(\frac{\omega_0+\omega}{2})\cos(\frac{\omega_0-\omega}{2}) + \sin(\frac{\omega_0+\omega}{2})\sin(\frac{\omega_0-\omega}{2}) \tag{4.38}$$

Substitution of (4.38) and (4.39) into (4.37) results in

$$u(t) = \left[\frac{2F_0}{m(\omega_0^2-\omega^2)}\sin\frac{(\omega_0-\omega)t}{2}\right]\sin\frac{(\omega_0+\omega)t}{2} \tag{4.39}$$

We observe that there are two frequencies in (4.39). The frequency of oscillation inside the square bracket is smaller than the oscillation frequency of the last term. Therefore, a higher frequency of oscillations has a sinusoidal amplitude as an envelope. Figure 4.4 illustrates a typical case of undamped forced vibrations with a sinusoidal amplitude.

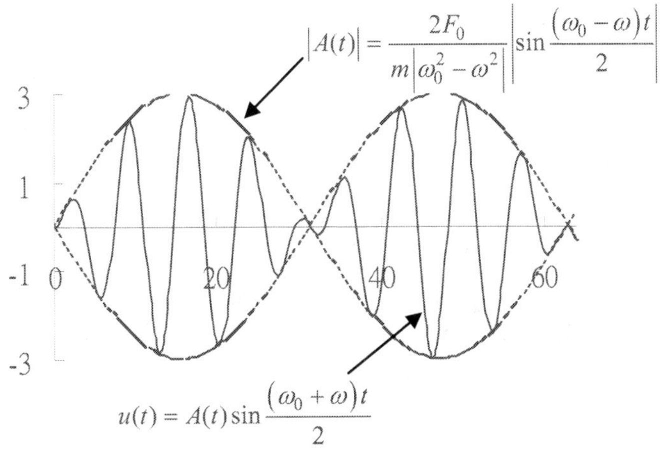

Figure 4.4 Forced oscillations of an SDOF showing beats

If the frequency of the forcing harmonic term is the same as the natural frequency (i.e., $\omega = \omega_0$), (4.20) becomes

$$m\ddot{u}(t) + ku(t) = F_0\cos\omega_0 t \tag{4.40}$$

Using the method of undetermined coefficients, we have to assume the particular solution as (see Section 3.3.3 of Chau, 2018)

$$u_p(t) = (A_1\cos\omega_0 t + A_2\sin\omega_0 t)t \tag{4.41}$$

Note that the forced frequency ω_0 coincides with the natural frequency of the system found in (4.22). Substitution of (4.41) into (4.40) leads to

$$2m\omega_0(-A_1\sin\omega_0 t + A_2\cos\omega_0 t) + [k - m\omega_0^2](A_1\cos\omega_0 t + A_2\sin\omega_0 t)t = F_0\cos\omega_0 t \tag{4.42}$$

In view of (4.22), we have from (4.42) the unknown constant being

$$A_1 = 0, \quad A_2 = \frac{F_0}{2m\omega_0} \tag{4.43}$$

Thus, the solution becomes

$$u(t) = c_1 \cos \omega_0 t + c_2 \sin \omega_0 t + \frac{F_0}{2m\omega_0} t \sin \omega_0 t \tag{4.44}$$

Example 4.2 Consider the case that an undamped structure is modeled by (4.40) and subject to the following forcing term and initial conditions:

$$F(t) = F_0 \cos \omega_0 t \tag{4.45}$$

$$u(0) = u_0, \quad \dot{u}(0) = v_0 \tag{4.46}$$

Solution: Differentiating (4.44) with respect to time, we find

$$\dot{u}(t) = \omega_0(-c_1 \sin \omega_0 t + c_2 \cos \omega_0 t) + \frac{F_0}{2m} t \cos \omega_0 t + \frac{F_0}{2m\omega_0} \sin \omega_0 t \tag{4.47}$$

Using the initial conditions of (4.46), we have

$$u(0) = c_1 = u_0 \tag{4.48}$$

$$\dot{u}(0) = \omega_0 c_2 = v_0 \tag{4.49}$$

Substitution of (4.48) and (4.49) into (4.44) gives

$$u(t) = u_0 \cos \omega_0 t + \frac{v_0}{\omega_0} \sin \omega_0 t + \frac{F_0}{2m\omega_0} t \sin \omega_0 t \tag{4.50}$$

We see from (4.44) and (4.50) that when the excitation is applied with $\omega = \omega_0$, the solution is proportional to time. Thus, the solution blows up and resonance occurs because the forced excitation frequency equals the natural frequency of the system. This is the reason why soldiers are not supposed to march across a bridge. If the natural frequency coincides with the marching frequency, resonance may occur. Soldier-marching-induced damages were observed at the Broughton Bridge in England in 1831 and Ostrawitza River bridge in Austria in 1886. Figure 4.5 illustrates a typical solution with an ever-increasing amplitude with time when the applied frequency of the harmonic forcing equals the natural frequency of an oscillator.

4.4.2 Damped-Free Responses

In this section, we will include damping effects of structures. It is clear from (4.18) that it is an ODE with constant coefficients, and thus, the solution must be in exponential form (Section 3.3.1 of Chau, 2018):

$$u(t) = e^{rt} \tag{4.51}$$

Substitution of (4.51) into the homogeneous form of (4.18) gives a characteristic equation for r

$$mr^2 + cr + k = 0 \tag{4.52}$$

The roots for r are

$$r_1, r_2 = \frac{-c \pm \sqrt{c^2 - 4mk}}{2m} = \frac{c}{2m}\left[-1 \pm \sqrt{1 - \frac{4mk}{c^2}}\right] \quad (4.53)$$

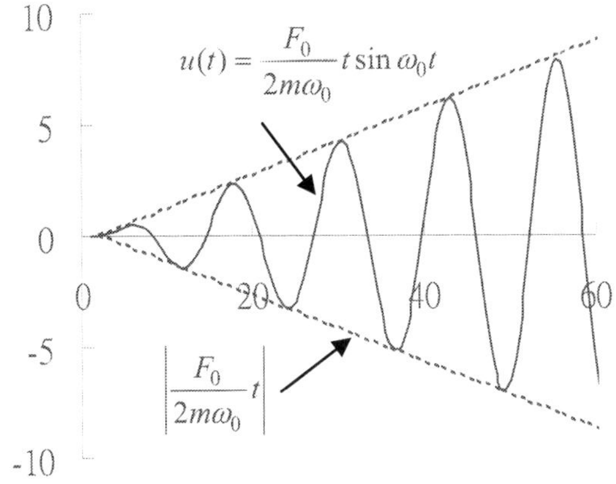

Figure 4.5 Forced oscillations imposed at the natural frequency of an SDOF oscillator

There are three different scenarios:

(i) $c^2 - 4mk > 0$:

The roots given in (4.53) are real negative and distinct. The solution is given by (see (3.363) of Chau, 2018):

$$u(t) = Ae^{r_1 t} + Be^{r_2 t}, \text{ where } r_1 < 0,\ r_2 < 0; \quad (4.54)$$

(ii) $c^2 - 4mk = 0$:

The roots given in (4.53) are repeated real negative roots. The solution is given by (see (3.370) of Chau, 2018):

$$u(t) = (A + Bt)e^{-ct/2m}, \text{ where } c/2m > 0; \quad (4.55)$$

(iii) $c^2 - 4mk < 0$:

For this case, the roots given in (4.53) are a complex conjugate pair. The solution is given by (see (3.375) of Chau, 2018):

$$u(t) = e^{-\omega_0 \zeta t}(A \cos \omega_D t + B \sin \omega_D t) \quad (4.56)$$

where the so-called quasi-frequency or damped frequency is defined as

$$\omega_D = \frac{\sqrt{4mk - c^2}}{2m} = \sqrt{\frac{k}{m}}\sqrt{1 - \frac{c^2}{4mk}} = \omega_0\sqrt{1 - \frac{c^2}{4m^2\omega_0^2}} = \omega_0\sqrt{1-\zeta^2} \tag{4.57}$$

where the damping ratio is defined as

$$\zeta = \frac{c}{2m\omega_0} \tag{4.58}$$

However, for real structures the value of damping c is normally smaller than k and m such that case (iii) defined above prevails. It is straightforward to see that the magnitude of (4.56) is given by

$$|u(t)| \le \sqrt{A^2 + B^2}\, e^{-\omega_0\zeta t} \tag{4.59}$$

Thus, for a damped oscillator, the magnitude tends to zero as $t \to \infty$. Let us examine the quasi-frequency in more detail. Using Taylor's series expansion, we can approximate the damped frequency and damped period as

$$\frac{\omega_D}{\omega_0} = \sqrt{1-\zeta^2} = 1 - \frac{\zeta^2}{2} + \dots \tag{4.60}$$

$$\frac{T_d}{T} = \frac{2\pi/\omega_D}{2\pi/\omega_0} = \frac{\omega_0}{\omega_D} = \frac{1}{\sqrt{1-\zeta^2}} = 1 + \frac{\zeta^2}{2} + \dots \tag{4.61}$$

From (4.60) and (4.61), we see that damping reduces the frequency of oscillations and elongates the period of oscillations. For the extreme case, when $\zeta \to 1$, we have

$$\lim_{\zeta\to 1}\frac{\omega_D}{\omega_0} = 0, \quad \lim_{\zeta\to 1}\frac{T_d}{T} = \infty \tag{4.62}$$

Therefore, a critical damping can be defined as

$$c = 2m\omega_0 \tag{4.63}$$

For $c > 2m\omega_0$ (or $\zeta > 1$), the motion of the structure damps out extremely fast, as shown in (4.54). Therefore, it is normally called overdamped.

Example 4.3 Consider the case that a damped structure is modeled by (4.18) and subject to the following forcing term and initial conditions:

$$F(t) = 0 \tag{4.64}$$

$$u(0) = u_0, \quad \dot{u}(0) = v_0 \tag{4.65}$$

Solution: Differentiating (4.56) with respect to time, we find

$$\begin{aligned} \dot{u}(t) &= -\omega_0\zeta e^{-\omega_0\zeta t}(A\cos\omega_D t + B\sin\omega_D t) \\ &+ e^{-\omega_0\zeta t}(-A\omega_D \sin\omega_D t + B\omega_D\cos\omega_D t) \end{aligned} \tag{4.66}$$

Using the initial conditions of (4.65), we have

$$u(0) = A = u_0 \tag{4.67}$$

$$\dot{u}(0) = -\omega_0\zeta A + \omega_D B = v_0 \tag{4.68}$$

Substitution of the results of A and B obtained from (4.67) and (4.68) into (4.56) gives

$$u(t) = e^{-\zeta\omega_0 t}\left[u_0\cos(\omega_D t) + \frac{v_0 + \zeta\omega_0 u_0}{\omega_D}\sin(\omega_D t)\right] \tag{4.69}$$

4.4.3 Damping Ratio by Hammer Test

The solution given in (4.69) provides a simple way to find the damping ratio in structures or in structure models in the laboratory. This technique is called the hammer test. If the structure is set in motion after an initial excitation, like a hammer blow, the structure will oscillate. If a displacement transducer is used to capture the motions of the attenuating vibrations, we can record the magnitude of two successive peak values in the damped-free vibrations (as illustrated in Figure 4.6).

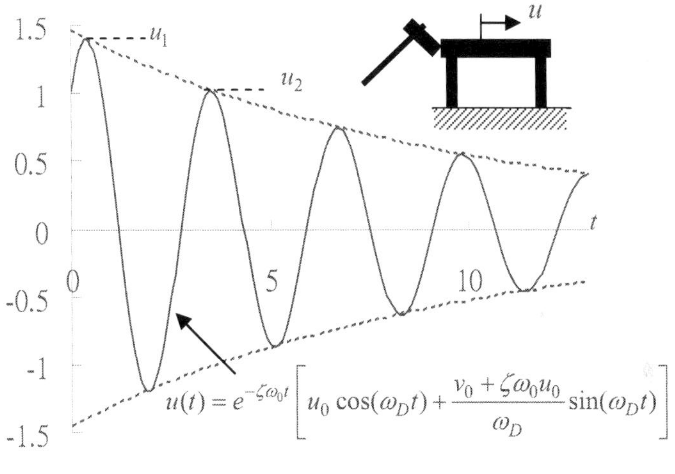

Figure 4.6 Illustration of a hammer test on damped-free vibrations

The first peak is assumed to be measured at time t_1 and the magnitude can be written as

$$u_1(t_1) = e^{-\zeta\omega_0 t_1}\sqrt{u_0^2 + \left(\frac{v_0 + \zeta\omega_0 u_0}{\omega_D}\right)^2} \tag{4.70}$$

Then, at the next peak we can express the measurement as

$$u_2(t_1 + T_D) = e^{-\zeta\omega_0 (t_1 + T_D)}\sqrt{u_0^2 + \left(\frac{v_0 + \zeta\omega_0 u_0}{\omega_D}\right)^2} \tag{4.71}$$

Taking the ratio of these two values, we have

$$\frac{u_1}{u_2} = \frac{e^{-\zeta\omega_0 t_1}}{e^{-(\zeta\omega_0 t_1 + T_d)}} = e^{\zeta\omega_0 T_D} \tag{4.72}$$

Solving for the damping ratio, we get

$$\zeta = \frac{\delta}{\sqrt{(2\pi)^2 + \delta^2}}. \tag{4.73}$$

where

$$\delta = \ln(\frac{u_2}{u_1}) \tag{4.74}$$

This is a very useful and powerful technique in estimating the damping ratio of a structure. Table 4.1 compiles some typically assumed values of the damping ratio for various types of structures, including steel, reinforced concrete, pre-stressed concrete, and timber. In engineering applications, the most commonly assumed value of ζ is 5% or 0.05.

Table 4.1 Typical values of damping ratios

Structures	Stress level	Damping ratio %
Welded steel	0.5 yield stress	2-3
	Yield stress	5-7
Pre-stressed concrete	0.5 yield stress	2-3
	Yield stress	5-7
Reinforced Concrete	0.5 yield stress	2-3
Reinforced Concrete with considerable cracking	0.5 yield stress	3-5
	Yield stress	7-10
Bolted or riveted steel	0.5 yield stress	5-7
	Yield stress	10-15
Timber	0.5 yield stress	5-7
Bolted timber	Yield stress	10-15
Nailed timber	Yield stress	15-20

Another method in determining ζ is called the half-power bandwidth method and is given in Problem 4.6.

4.4.4 Damped Forced Responses

Consider the case that a harmonic forcing function is applied

$$m\ddot{u} + c\dot{u} + ku = F_0 \cos(\omega t) \tag{4.75}$$

Noting the introduction of a damping ratio in (4.58), we can first rewrite (4.75) as

$$\ddot{u} + 2\omega_0\zeta\dot{u} + \omega_0^2 u = \frac{F_0}{m}\cos(\omega t) \tag{4.76}$$

As discussed in the last section, for damped structures, the damping ratio is typically less than 10% and is not expected to be larger than 20% even for the most extreme cases (see Table 4.1). Thus, we can restrict our consideration to case (iii) of Section 4.4.2 only. That is, the homogeneous solution of (4.75) is

$$u_h(t) = e^{-\omega_0 \zeta t}(c_1 \cos \omega_D t + c_2 \sin \omega_D t) \tag{4.77}$$

For the case that $\omega \neq \omega_D$ in the forcing term, the particular solution can be assumed as:

$$u_p(t) = A \cos \omega t + B \sin \omega t \tag{4.78}$$

Substitution of (4.78) into (4.76) leads to

$$(\omega_0^2 - \omega^2)(A \cos \omega t + B \sin \omega t) + 2\omega_0 \omega \zeta(-A \sin \omega t + B \cos \omega t) = \frac{F_0}{m} \cos \omega t \tag{4.79}$$

Collecting coefficients for the sine and cosine, we obtain two equations for the two unknown constants:

$$(\omega_0^2 - \omega^2)A + 2\omega_0 \omega \zeta B = \frac{F_0}{m} \tag{4.80}$$

$$(\omega_0^2 - \omega^2)B - 2\omega_0 \omega \zeta A = 0 \tag{4.81}$$

Solving for A and B, we have

$$A = \frac{F_0(1-\beta^2)}{k[(1-\beta^2)^2 + (2\beta\zeta)^2]} \tag{4.82}$$

$$B = \frac{2\beta\zeta F_0}{k[(1-\beta^2)^2 + (2\beta\zeta)^2]} \tag{4.83}$$

where

$$\beta = \frac{\omega}{\omega_0}, \quad \zeta = \frac{c}{2m\omega_0} \tag{4.84}$$

Substituting (4.82) and (4.83) into (4.78) and adding the results to the homogeneous solution given in (4.77), we obtain the general solution as

$$\begin{aligned} u(t) &= e^{-\omega_0 \zeta t}(c_1 \cos \omega_D t + c_2 \sin \omega_D t) \\ &+ \frac{F_0}{k[(1-\beta^2)^2 + (2\beta\zeta)^2]}\{(1-\beta^2)\cos \omega t + 2\beta\zeta \sin \omega t\} \end{aligned} \tag{4.85}$$

Example 4.4 Consider a damped structure modeled by the following ODE and initial conditions:

$$\ddot{u} + 2\omega_0 \zeta \dot{u} + \omega_0^2 u = \frac{F_0}{m} \cos(\omega t) \tag{4.86}$$

$$u(0) = u_0, \quad \dot{u}(0) = v_0 \tag{4.87}$$

Solution: Differentiating (4.85) with respect to time, we find

$$\dot{u}(t) = -\omega_0 \zeta e^{-\omega_0 \zeta t}(c_1 \cos \omega_D t + c_2 \sin \omega_D t) + \omega_D e^{-\omega_0 \zeta t}(-c_1 \sin \omega_D t + c_2 \cos \omega_D t)$$
$$+ \frac{\omega F_0}{k[(1-\beta^2)^2 + (2\beta\zeta)^2]}\{-(1-\beta^2)\sin \omega t + 2\beta\zeta \cos \omega t\} \tag{4.88}$$

Using the initial conditions of (4.87), we have

$$u(0) = c_1 + \frac{F_0(1-\beta^2)}{k[(1-\beta^2)^2 + (2\beta\zeta)^2]} = u_0 \tag{4.89}$$

$$\dot{u}(0) = -\omega_0 \zeta c_1 + \omega_D c_2 + \frac{2\beta\zeta\omega F_0}{k[(1-\beta^2)^2 + (2\beta\zeta)^2]} = v_0 \tag{4.90}$$

The solutions for (4.89) and (4.90) are

$$c_1 = u_0 - \frac{F_0(1-\beta^2)}{k[(1-\beta^2)^2 + (2\beta\zeta)^2]} \tag{4.91}$$

$$c_2 = \frac{1}{\sqrt{1-\zeta^2}}\{\frac{v_0}{\omega_0} + u_0\zeta + (\frac{\zeta F_0}{k})\frac{2\beta^2 - (1-\beta^2)}{(1-\beta^2)^2 + (2\beta\zeta)^2}\} \tag{4.92}$$

Substitution of (4.91) and (4.92) into (4.85) gives the final solution as

$$u(t) = e^{-\zeta\omega_0 t}\{[u_0 - \frac{F_0(1-\beta^2)}{k[(1-\beta^2)^2 + (2\beta\zeta)^2]}]\cos(\omega_D t)$$
$$+ \frac{1}{\sqrt{1-\zeta^2}}[\frac{v_0}{\omega_0} + u_0\zeta + (\frac{\zeta F_0}{k})(\frac{2\beta^2 - (1-\beta^2)}{(1-\beta^2)^2 + (2\beta\zeta)^2})]\sin(\omega_D t)\} \tag{4.93}$$
$$+ \frac{F_0}{k}\{\frac{(1-\beta^2)\cos \omega t + 2\beta\zeta \sin \omega t}{(1-\beta^2)^2 + (2\beta\zeta)^2}\}$$

If the forcing term is a sine function, instead of a cosine function, we can follow a similar procedure to get the particular solution (see Problems 4.2 and 4.3).

We now take the special case of steady case (i.e., $t \to \infty$), and we have

$$u(t) = \frac{F_0}{k}\{\frac{(1-\beta^2)\cos \omega t + 2\beta\zeta \sin \omega t}{(1-\beta^2)^2 + (2\beta\zeta)^2}\} \tag{4.94}$$

We first rewrite this solution as

$$u(t) = R\cos(\omega t - \delta) \tag{4.95}$$

Expanding this, we have

$$u(t) = R\cos \omega t \cos \delta + R \sin \omega t \sin \delta \tag{4.96}$$

Comparison of (4.94) and (4.96) gives

$$R\cos \delta = \frac{F_0}{k}\{\frac{(1-\beta^2)}{(1-\beta^2)^2 + (2\beta\zeta)^2}\} \tag{4.97}$$

$$R\sin\delta = \frac{F_0}{k}\{\frac{2\beta\zeta}{(1-\beta^2)^2+(2\beta\zeta)^2}\} \tag{4.98}$$

Squaring (4.97) and (4.98) and adding the results, we get

$$R = \frac{F_0}{k}[\frac{1}{\sqrt{(1-\beta^2)^2+(2\beta\zeta)^2}}] \tag{4.99}$$

This is the amplitude of the long-term damped forced vibrations. Differentiating (4.99) with respect to β and set the result to zero, we have

$$\beta_{\max} = \sqrt{1-2\zeta^2} \tag{4.100}$$

Note that maximum response is not at β = 1, unless the damping ratio is zero (i.e., ζ = 0). With this critical normalized frequency, the corresponding maximum amplitude becomes

$$R_{\max} = \frac{F_0}{k}[\frac{1}{4\zeta^2(1-\zeta^2)}] \tag{4.101}$$

Therefore, the maximum amplitude appears at the critical damping. Figure 4.7 plots the amplitude of the steady-state response R versus the normalized frequency β for various values of damping ratio ζ.

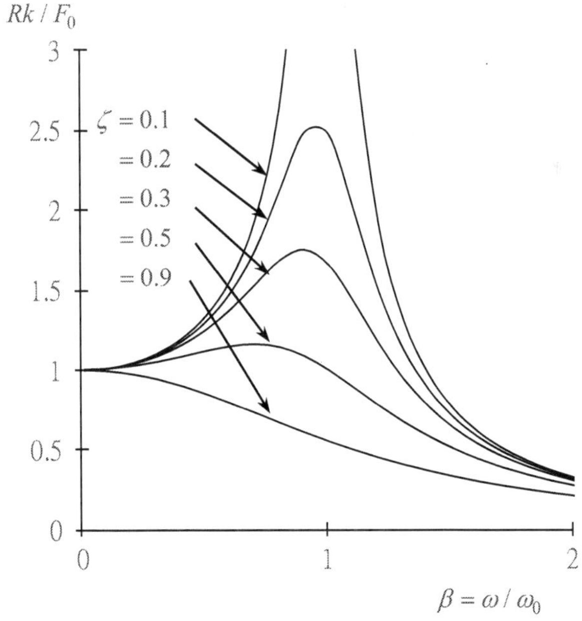

Figure 4.7 The amplitude function versus frequency

4.5 DUHAMEL INTEGRAL FOR GENERAL GROUND MOTIONS

In this section, we consider a structure subject to ground shaking induced by earthquakes. The estimation of the amplitude and frequency of ground shaking due to earthquakes is not our concern here, and this topic is highly technical in the area of seismology and huge uncertainty is involved. We refer the readers to the book by Aki and Richards (1980). In this section, we restrict our discussion to structural responses during earthquakes.

4.5.1 Formulation of Equation of Motion

Figure 4.8 shows the effect of ground shaking on an SDOF structure. The total displacement at the story level comprises two components, namely the ground displacement u_g and the relative displacement u between the ground and the structure.

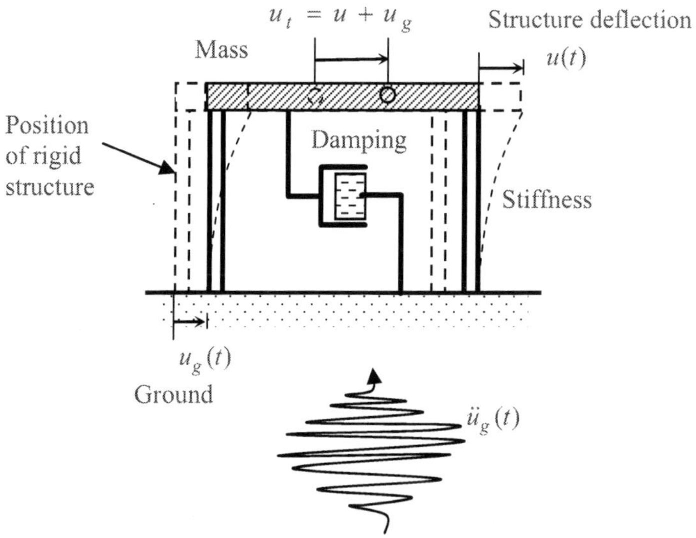

Figure 4.8 A single-story building subject to ground shaking

Note that the inertia force is proportional to the total displacement u_t, whereas both the column restoring force and damping force are proportional to the relative displacement u. The equation of motions of the structure shown in Figure 4.8 is

$$m\ddot{u}_t + c\dot{u} + ku = 0 \tag{4.102}$$

Referring to Figure 4.8, the total displacement is defined as

$$u_t = u + u_g \tag{4.103}$$

Substitution of (4.103) into (4.102) gives

$$m\ddot{u} + c\dot{u} + ku = -m\ddot{u}_g(t) \tag{4.104}$$

Similar to the previous section, we can rewrite (4.102) as

$$\ddot{u} + 2\zeta\omega_0\dot{u} + \omega_0^2 u = -\ddot{u}_g(t) \tag{4.105}$$

where the natural circular frequency and damping ratio are

$$\omega_0 = \sqrt{\frac{k}{m}}, \quad \zeta = \frac{c}{2m\omega_0} \tag{4.106}$$

When the earthquake first strikes, the structure is motionless. The initial conditions for earthquake problems are

$$u(0) = u_0 = 0, \quad \dot{u}(0) = v_0 = 0 \tag{4.107}$$

For an arbitrary ground shaking $u_g(t)$, the solution can be expressed in terms of the Duhamel integral.

4.5.2 Duhamel Integral

Figure 4.9 illustrates the idea of using superposition via the Duhamel integral. A continuous input function of the ground is divided into a number of Heaviside step functions, and the solution due to each step function can be found by multiplying the fundamental response function. The final solution is calculated as a sum of each of these response functions. The summation can, of course, be replaced by an integral when the increment in time is small. In particular, the solution can be expressed as

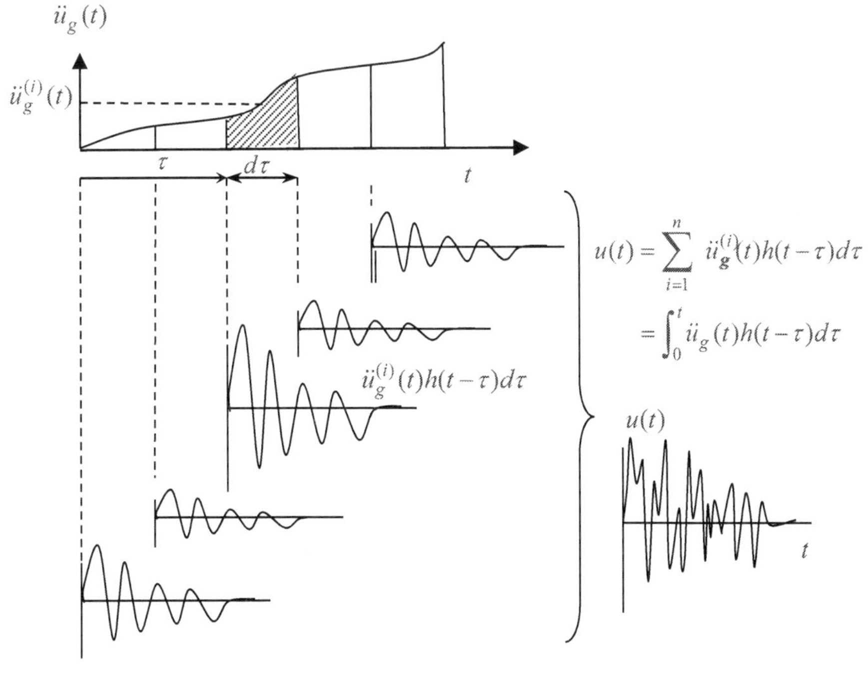

Figure 4.9 Superposition in terms of the Duhamel integral

$$u(t) = -\int_0^t \ddot{u}_g(\tau)h(t-\tau)d\tau \tag{4.108}$$

where

$$h(t) = \frac{1}{\omega_D} e^{-\zeta\omega t}\sin(\omega_D t), \quad \omega_D = \omega_0\sqrt{1-\zeta^2} \tag{4.109}$$

To prove the validity of the Duhamel integral given in (4.108), we differentiate (4.108) as

$$\dot{u}(t) = -h(0)\ddot{u}_g(t) - \int_0^t \ddot{u}_g(\tau)h'(t-\tau)d\tau \tag{4.110}$$

In obtaining (4.110), we have applied the following Leibniz's rule of differentiation under integral sign (see Section 1.3.4 of Chau, 2018 for its proof):

$$\frac{\partial}{\partial t}\int_{f(t)}^{g(t)} h(t,\zeta)d\zeta = g'(t)h[t,g(t)] - f'(t)h[t,f(t)] + \int_{f(t)}^{g(t)} \frac{\partial h(t,\zeta)}{\partial t}d\zeta \tag{4.111}$$

Applying the Leibniz's rule again, we find

$$\ddot{u}(t) = -h(0)\dddot{u}_g(t) - h'(0)\ddot{u}_g(t) - \int_0^t \ddot{u}_g(\tau)h''(t-\tau)d\tau \tag{4.112}$$

Substitution of (4.108), (4.110) and (4.112) into (4.105) gives

$$\int_0^t \ddot{u}_g(\tau)(h''+2\zeta\omega_0 h'+h\omega_0^2)d\tau + \dddot{u}_g(t)h_0 + \ddot{u}_g(t)[h_0' + 2\zeta\omega_0 h_0 - 1] = 0 \tag{4.113}$$

where

$$h' = \frac{dh(\xi)}{d\xi} \tag{4.114}$$

$$h_0 = h(0), \quad h'(0) = h_0' \tag{4.115}$$

Since the ground acceleration and its derivative are not identically zero, (4.113) gives an ODE for h and its initial conditions

$$\frac{d^2h}{d\xi^2} + 2\zeta\omega_0\frac{dh}{d\xi} + \omega_0^2 h = 0, \quad h(0) = 0, \quad \frac{dh}{d\xi}(0) = 1 \tag{4.116}$$

Following the solution technique employed in the last section, we have the solution of (4.116) as

$$h(\xi) = e^{-\omega_0\zeta\xi}(c_1\cos\omega_D\xi + c_2\sin\omega_D\xi) \tag{4.117}$$

Differentiating (4.117), we obtain

$$\begin{aligned}\frac{dh}{d\xi} &= -\omega_0\zeta e^{-\omega_0\zeta\xi}(c_1\cos\omega_D\xi + c_2\sin\omega_D\xi) \\ &+\omega_D e^{-\omega_0\zeta\xi}(-c_1\sin\omega_D\xi + c_2\cos\omega_D\xi)\end{aligned} \tag{4.118}$$

By applying the initial conditions given in (4.116), we find

$$h(0) = c_1 = 0 \tag{4.119}$$

$$\frac{dh}{d\xi}(0) = \omega_D c_2 = 1, \text{ or } c_2 = \frac{1}{\omega_D} \tag{4.120}$$

Using these results, the solution for h is

$$h(\xi) = \frac{1}{\omega_D} e^{-\omega_0 \zeta \xi} \sin \omega_D \xi \tag{4.121}$$

This is the solution given for h in (4.109). This completes the proof of (4.108) and (4.109). Once the ground acceleration is given, we can integrate (4.109) numerically to give the solution. Similarly, we can also use (4.110) and (4.112) to evaluate the velocity or acceleration of the SDOF structure subject to any ground motion. However, real ground acceleration induced by earthquakes does not appear to have any clear pattern, and the resulting structural response is consequently not easy to interpret. In the next section, we will discuss a technique called the response spectrum to interpret characteristics of apparently chaotic ground motions.

4.6 RESPONSE SPECTRUM

In this section, we will discuss the effect of ground motions on buildings in terms of the response spectrum. The analysis of structures against earthquake-induced ground motions is also called seismic analysis. There are at least three different approaches in seismic analysis. They are quasi-static analysis, response spectrum analysis, and time history analysis. Quasi-static analysis replaces ground-induced motion by equivalent static, lateral loads. This method is cheap and quick, but only gives a rough approximation. Response spectrum analysis considers structural response subject to a particular ground shaking by considering the response of a spectrum of single-degree-of-freedom oscillators. Time history analysis employs a real or synthetic time history of ground motion as input to carry out step-by-step time integration of the PDE of a realistic structural model (such as those from finite element methods). As discussed in Section 13.3.1 and 13.3.2 of Chau (2018), the Wilson θ method and the Newmark β method are two commonly employed schemes for numerical integration of structural dynamics problems. It can take nonlinear behavior of a structure into account, but it is very time consuming and expensive. Normally, time history analysis is only required for important structures, such as fire stations, and nuclear power plants. Quasi-static analysis is cheap but inaccurate, and time history analysis is accurate but expensive. Thus, response spectrum analysis provides a better approximation at a reasonable cost. It was proposed by M.A. Biot in 1932. It was a part of Biot's PhD thesis at Caltech under the supervision of von Karman. The technique was subsequently further developed and popularized by Housner and Newmark. However, its application in seismic design is not realized before the popular use of personal computers in the 1970s and the availability of enough strong ground motions after the 1971 San Fernando earthquake. The numerical integration of the digitally measured data of input ground acceleration in (4.108) is a tedious process without a computer. Biot was a Belgium-born American scientist and mathematician. Biot was trained as an electrical and mining engineer. Biot had published two seminal papers on poro-elastic solids and both of them had been cited over 8,000 times by early 2018. The most amazing fact is that Biot's prolific research in diversified areas was all done by himself without the help of any graduate students. Biot received ASME's Timoshenko medal in 1962 and ASCE has awarded Biot's medal since 2003 in honor of him.

The idea is illustrated in Figures 4.10 and 4.11. In particular, SDOF oscillators of different damped natural fundamental frequencies are subjected to the input ground motion, with the damping ratio ζ as a parameter. The response of each oscillator is calculated by integrating (4.108). Figure 4.10 illustrates the displacement time histories of three oscillators with different circular natural frequencies, indicating the values of the corresponding maximum displacement. These maximum values are plotted in Figure 4.11 against the circular natural frequencies of each oscillator at a fixed damping ratio. Mathematically, the response spectrum S_r can be defined symbolically as

$$S_r(\omega,\zeta) = \max_t \left| r(t;\omega,\zeta) \right| \tag{4.122}$$

where the function r can be any type of response such as displacement, velocity or acceleration. That is, we have

$$r = u(t), \dot{u}(t), \ddot{u}_t(t) \tag{4.123}$$

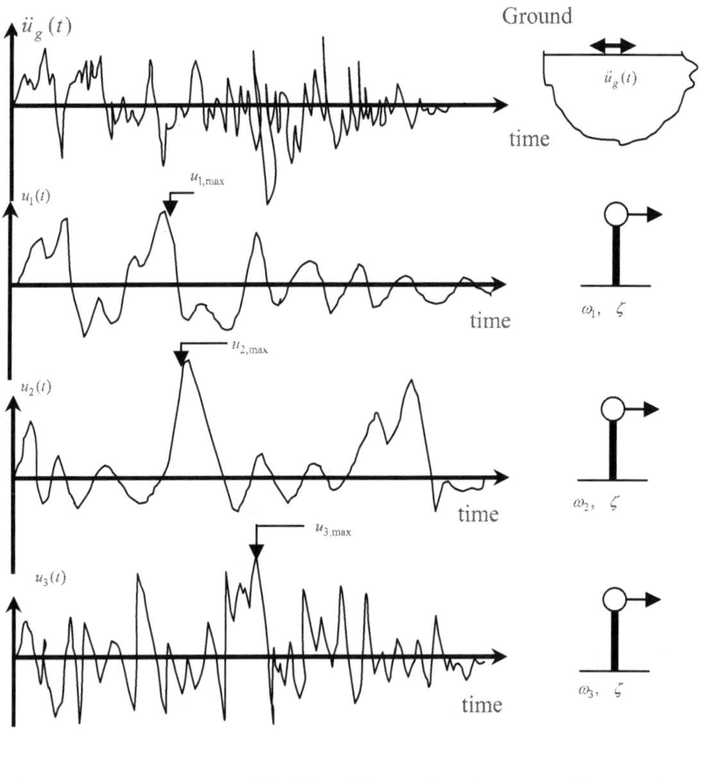

Figure 4.10 Displacement responses of SDOF oscillators of various natural frequencies subject to the same ground excitation

Since Figure 4.10 plots the displacement response, the response spectrum given in Figure 4.11 is for the displacement response spectrum. Thus, the response spectrum represents the maximum response of a SDOF oscillator subject to a given ground motion. In fact, Biot proposed in 1932 in his PhD thesis the idea of taking the average of the response spectra of some actual ground accelerations recorded during earthquakes for a seismic zone to get a "design" response spectrum for seismic design. The idea was apparently inspired by Professor Kyoji Suyehiro's lecture delivered in the United States in 1931. Such analyses became possible in 1971 after a lot of acceleration records were obtained during the San Fernando earthquake when 241 accelerographs were recorded. The shapes of the response spectrum were studied extensively by Newmark, who also proposed a scaled "design" peak acceleration.

More specifically, we can define various response spectra as:

Relative displacement spectrum

$$S_d(\omega,\zeta) = \max_t |u(t;\omega,\zeta)| \tag{4.124}$$

Relative velocity spectrum

$$S_v(\omega,\zeta) = \max_t |\dot{u}(t;\omega,\zeta)| \tag{4.125}$$

Absolute acceleration spectrum

$$S_a(\omega,\zeta) = \max_t |\ddot{u}_t(t;\omega,\zeta)| \tag{4.126}$$

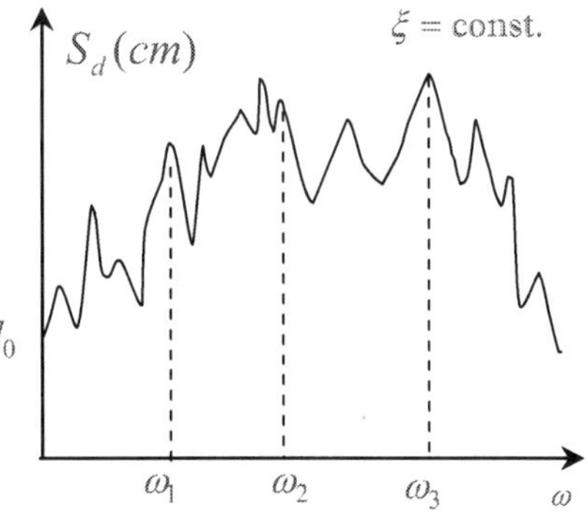

Figure 4.11 Extraction of maximum displacement values from Figure 4.10 to form the response spectrum

4.6.1 Pseudo-Response Spectrum

The pseudo-response spectrum is an approximate technique relating displacement, velocity and acceleration response spectra. To define the pseudo-response spectrum, we first we recall the equation of motion for an undamped oscillator

$$m\ddot{u} + ku = 0 \tag{4.127}$$

Clearly, the solution can be expressed as

$$u = \phi(x)\sin(\omega t + \delta) \tag{4.128}$$

Then, the velocity and acceleration are

$$\dot{u} = \omega\phi(x)\cos(\omega t + \delta) \tag{4.129}$$

$$\ddot{u} = -\omega^2\phi(x)\sin(\omega t + \delta) = -\omega^2 u(t) \tag{4.130}$$

From the harmonic-free vibrations, we have

$$\dot{u}_{\max} = \omega|u_{\max}| \tag{4.131}$$

$$\ddot{u}_{\max} = \omega^2|u_{\max}| \tag{4.132}$$

Clearly, actual earthquake-induced ground motion is not of the harmonic type. Nevertheless, pseudo-response spectra are defined based on the observations from (4.131) and (4.132). In particular, the pseudo-velocity spectrum is defined as

$$S_{pv}(\omega,\zeta) = \omega S_d(\omega,\zeta) \tag{4.133}$$

The pseudo-absolute acceleration spectrum is defined as

$$S_{pa}(\omega,\zeta) = \omega^2 S_d(\omega,\zeta) \tag{4.134}$$

The prefix "p" stands for pseudo and it clearly indicates that it is not the true spectrum. However, for long-duration ground motions with small damping, we have

$$S_{pv}(\omega,\zeta) \approx S_v(\omega,\zeta), \quad \omega > \omega_0 \tag{4.135}$$

where ω_0 is related to the earthquake duration T as

$$\omega_0 = \frac{2\pi}{T} \tag{4.136}$$

The pseudo-response spectrum is an approximate but convenient way to use the response spectrum. The advantage of using the approximate pseudo-spectra rather than the exact spectra is that all the maximum responses of S_d, S_{pv}, and S_{pa} of a given earthquake can be represented by just one spectrum. That is, if S_d is known, all the other spectra can be estimated by (4.133) and (4.134):

$$S_{pv}(\omega,\zeta) = \frac{1}{\omega}S_{pa}(\omega,\zeta) = \omega S_d(\omega,\zeta) \tag{4.137}$$

Taking the logarithm of the first pair in (4.137), we obtain

$$\log_{10} S_{pv}(\omega,\zeta) = -\log_{10}\omega + \log_{10} S_{pa}(\omega,\zeta) \tag{4.138}$$

Similarly, the logarithm of the second pair of (4.137) gives

$$\log_{10} S_{pv}(\omega,\zeta) = \log_{10}\omega + \log_{10} S_d(\omega,\zeta) \tag{4.139}$$

These two equations can be rewritten as

$$y = -x + z \tag{4.140}$$

$$y = x + z' \tag{4.141}$$

where

$$y = \log_{10} S_{pv}, \quad x = \log_{10}\omega, \quad z = \log_{10} S_{pa}, \quad z' = \log_{10} S_d \tag{4.142}$$

Equations (4.140) and (4.141) are illustrated in Figure 4.12 for various constant values of z and z'. By combining the two plots of Figure 4.12, we have the four-way logarithmic diagram shown in Figure 4.13. The envelope of thick lines is the response spectrum of ground motion defined by parameters a_0, d_0 and v_0. (i.e., the maximum ground acceleration, displacement and velocity). The units of the pseudo-acceleration, pseudo-velocity, and displacement response spectra are g, m/s and m, respectively. The scale is a logarithmic scale along four axes. When the damping is small, the spectra calculated from the recorded ground motions exhibit a number of peaks. Typically, the maximum spectra occur at different peaks, as shown in Figure 4.13. The frequencies at the maximum displacement, maximum pseudo-velocity and maximum pseudo-acceleration are also indicated in Figure 4.13, and they do not occur at the same frequency. If we have a very stiff structure or $\omega \rightarrow \infty$ (or the period goes to zero), we have that the pseudo-acceleration spectrum approaches the maximum ground acceleration:

$$\lim_{\omega \to \infty} S_{pa}(\omega) = a_0 \tag{4.143}$$

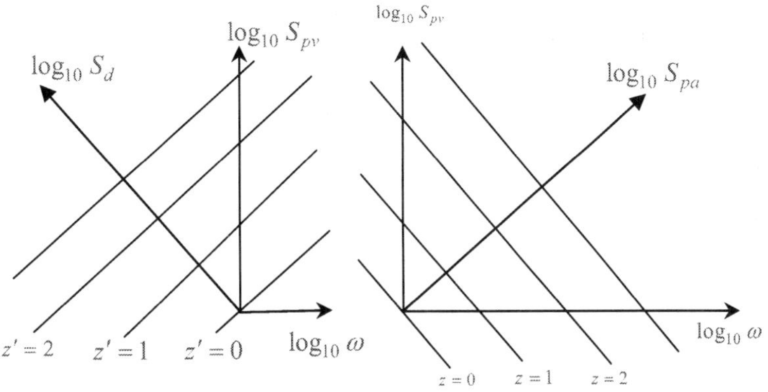

Figure 4.12 Relation between pseudo-response spectra and the relative displacement spectrum

It is shown in the right lower corner of Figure 4.13 that the response spectrum approaches the maximum ground acceleration a_0. For a very soft structure, we have $\omega \rightarrow 0$ (or the period goes to infinity), and the displacement response spectrum approaches the maximum ground displacement:

$$\lim_{\omega \to 0} S_d(\omega) = d_0 \tag{4.144}$$

It can be seen in the left lower corner of Figure 4.13 that the response spectrum approaches the maximum ground acceleration d_0. The use of this response spectrum is easy. All we need to do is to find the parameters of an equivalent SDOF oscillator with a specific damping ratio. Once the natural frequency ω of the equivalent SDOF oscillator is found, we can use Figure 4.13 to read out the S_d, S_{pv} and S_{pa} from Figure 4.13. It is also possible to find the envelope of the various scaled response spectra of ground records to generate a "design response spectrum," and one of these is the Newmark-Hall design response spectrum.

4.6.2 Nonlinear Response Spectrum

The idea of the response spectrum can also be extended to consider a nonlinear SDOF oscillator, for which the equation of motion becomes

$$m\ddot{u}_t + f(u, \dot{u}) = 0 \tag{4.145}$$

where f is a nonlinear function of displacement and velocity. Expanding f in Taylor's series expansion, we obtain

$$f(u,\dot{u}) = f(0,0) + \left.\frac{\partial f}{\partial u}\right|_{(0,0)} u + \left.\frac{\partial f}{\partial \dot{u}}\right|_{(0,0)} \dot{u} + ... \tag{4.146}$$

where the tangential stiffness and tangential damping constant can be interpreted as

$$k_t = \left.\frac{\partial f}{\partial u}\right|_{(0,0)}, \quad c_t = \left.\frac{\partial f}{\partial \dot{u}}\right|_{(0,0)} \tag{4.147}$$

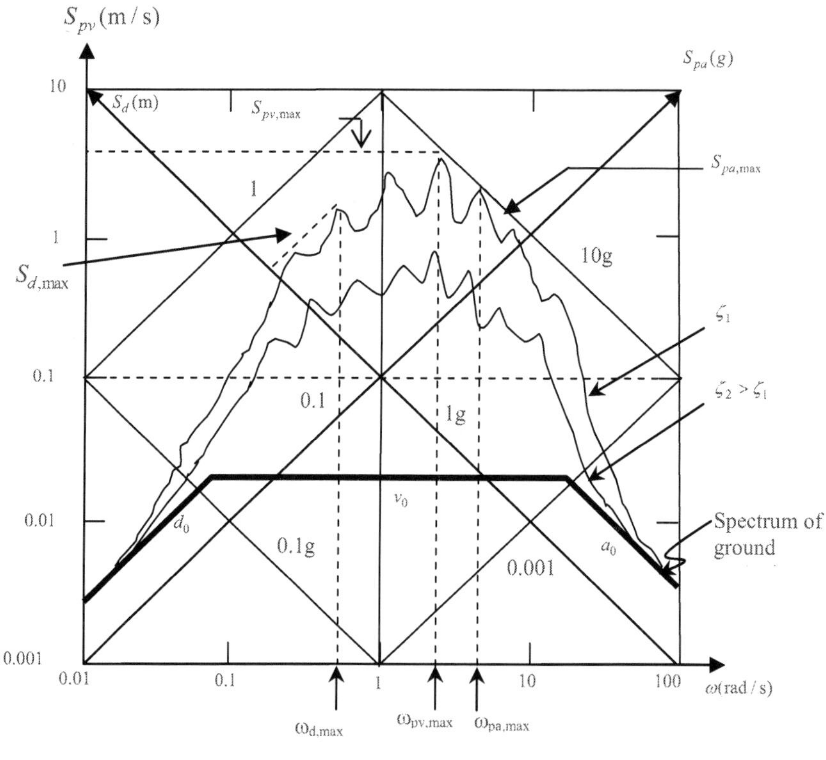

Figure 4.13 Four-way pseudo-response spectra

If the amplitude of oscillations is small, we can find the frequency and damping ratio as:

$$\omega = \sqrt{\frac{k_t}{m}} = \sqrt{\frac{1}{m}(\partial f / \partial u)|_{(0,0)}} \tag{4.148}$$

$$\zeta = \frac{c}{2m\omega} = \frac{1}{2}\frac{(\partial f / \partial \dot{u})|_{(0,0)}}{\sqrt{m(\partial f / \partial u)|_{(0,0)}}} \tag{4.149}$$

For nonlinear structures, we can evaluate the ductility factor as

$$\mu = \frac{u_{\max}}{u_0} \tag{4.150}$$

However, for nonlinear structure, we cannot use the pseudo-response spectra that we discussed in the last section.

4.7 MULTI-STORY BUILDINGS

Multi-story buildings are normally modeled as a multi-degree-of-freedom oscillator. Assuming only one horizontal displacement u per story (i.e., rigid slab assumption), we can model a building as shown in Figure 4.14. The masses of columns are neglected such that story mass can be considered as a lump mass model. The stiffness of the column can be estimated by considering the stiffness of a clamp-clamp beam due to support movement.

For the case of earthquake-induced ground motions, we can formulate the model as a system of ODEs in matrix form as:

$$\boldsymbol{M}\ddot{\boldsymbol{u}}_t(t) + \boldsymbol{C}\dot{\boldsymbol{u}} + \boldsymbol{K}\boldsymbol{u} = \boldsymbol{0} \tag{4.151}$$

where

$$\boldsymbol{M} = \begin{bmatrix} m_1 & 0 & \vdots & 0 \\ 0 & m_2 & \vdots & 0 \\ \cdots & \cdots & \ddots & \cdots \\ 0 & 0 & \vdots & m_n \end{bmatrix}, \quad \boldsymbol{C} \begin{bmatrix} c_{11} & c_{12} & \vdots & c_{1n} \\ c_{21} & c_{22} & \vdots & c_{12} \\ \cdots & \cdots & \ddots & \cdots \\ c_{n1} & c_{n2} & \vdots & c_{nn} \end{bmatrix}, \tag{4.152}$$

$$\boldsymbol{K} = \begin{bmatrix} k_1 + k_2 & -k_1 & \vdots & 0 \\ -k_1 & k_2 + k_3 & \vdots & 0 \\ \cdots & \cdots & \ddots & \cdots \\ 0 & 0 & \vdots & k_n \end{bmatrix}, \tag{4.153}$$

where these matrices are of dimension $n \times n$. The displacement and force vectors are defined as:

$$\ddot{\boldsymbol{u}}_t(t) = \begin{Bmatrix} 1 \\ 1 \\ \vdots \\ 1 \end{Bmatrix} \ddot{u}_g(t) + \begin{Bmatrix} \ddot{u}_1 \\ \ddot{u}_2 \\ \vdots \\ \ddot{u}_n \end{Bmatrix} = \boldsymbol{e}\ddot{u}_g(t) + \ddot{\boldsymbol{u}}(t), \quad \boldsymbol{F}_t(t) = \begin{Bmatrix} F_1 \\ F_2 \\ \vdots \\ F_n \end{Bmatrix} \tag{4.154}$$

Using (4.154), we can rewrite the system given in (4.151) as

$$\boldsymbol{M}\ddot{\boldsymbol{u}}(t) + \boldsymbol{C}\dot{\boldsymbol{u}} + \boldsymbol{K}\boldsymbol{u} = -\boldsymbol{M}\boldsymbol{e}\ddot{u}_g(t) \tag{4.155}$$

For more general external applied loads, we can formulate the model as a system of ODEs in matrix form as:

$$\boldsymbol{M}\ddot{\boldsymbol{u}}(t) + \boldsymbol{C}\dot{\boldsymbol{u}}(t) + \boldsymbol{K}\boldsymbol{u}(t) = \boldsymbol{F}(t) \tag{4.156}$$

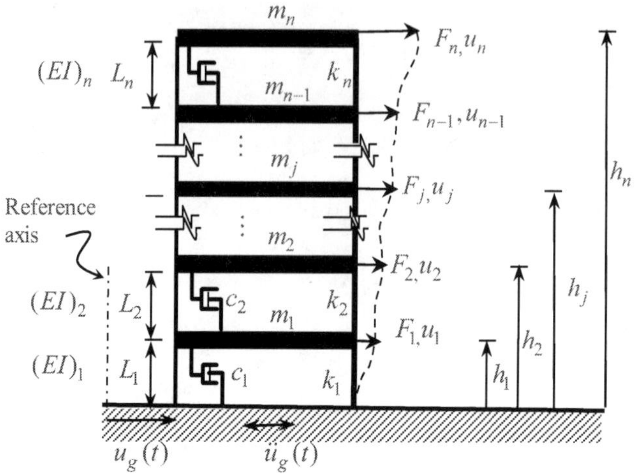

Figure 4.14 Multi-story buildings with rigid floors

where

$$F^T(t) = \{F, F_2, ..., F_n\} \tag{4.157}$$

Note that, in general, we have assumed a full damping matrix. The formulation of the stiffness matrix is illustrated in the following example. The determination of the damping matrix is less straightforward and will be discussed later.

Example 4.5 (i) Derive the general stiffness matrix of a three-story building as a special case of Figure 4.15. (ii) Give the special form of the stiffness matrix for the following values of column stiffness

$$E_1I_1 = 3EI, \quad E_2I_2 = 2EI, \quad E_3I_3 = EI \tag{4.158}$$

$$L_1 = \frac{3}{2}L, \quad L_2 = L, \quad L_3 = L \tag{4.159}$$

Solution: (i) Figure 4.15 shows the case of a three-story building, together with the definition of each component of the stiffness matrix. They are the reactions at the slab levels for a unit displacement for each degree of freedom.

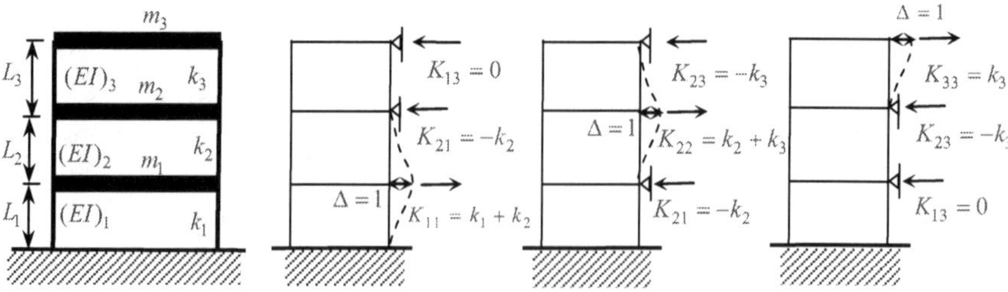

Figure 4.15 Three-story buildings subject to unit displacement at each level

For the stiffness components obtained in Figure 4.15, the stiffness matrix is formulated as:

$$\boldsymbol{K}=\begin{bmatrix} K_{11} & K_{12} & K_{13} \\ K_{21} & K_{22} & K_{23} \\ K_{31} & K_{32} & K_{33} \end{bmatrix}=\begin{bmatrix} k_1+k_2 & -k_2 & 0 \\ -k_2 & k_2+k_3 & -k_3 \\ 0 & -k_3 & k_3 \end{bmatrix} \quad (4.160)$$

where the column stiffnesses are (see p. 545 of Gere and Timoshenko, 1990)

$$k_1=2\times(\frac{12E_1I_1}{L_1^3})=\frac{24E_1I_1}{L_1^3} \quad (4.161)$$

$$k_2=2\times(\frac{12E_2I_2}{L_2^3})=\frac{24E_2I_2}{L_2^3} \quad (4.162)$$

$$k_3=2\times(\frac{12E_3I_3}{L_3^3})=\frac{24E_3I_3}{L_3^3} \quad (4.163)$$

The factor two before the bracket takes into account the two columns (since the 2-D case here has only 2 two columns).

(ii) Using (4.159) and (4.160), we have

$$k_1=\frac{24E_1I_1}{L_1^3}=3(\frac{2}{3})^3\frac{24EI}{L^3}=\frac{64}{3}(\frac{EI}{L^3}) \quad (4.164)$$

$$k_2=\frac{24E_2I_2}{L^3}=2\frac{24EI}{L^3}=48(\frac{EI}{L^3}) \quad (4.165)$$

$$k_3=\frac{24E_3I_3}{L_3^3}=24(\frac{EI}{L^3}) \quad (4.166)$$

Using these values, we obtain the stiffness matrix as

$$\boldsymbol{K}=\begin{bmatrix} k_1+k_2 & -k_2 & 0 \\ -k_2 & k_2+k_3 & -k_3 \\ 0 & -k_3 & k_3 \end{bmatrix}=\frac{8EI}{3L^3}\begin{bmatrix} 26 & -18 & 0 \\ -18 & 27 & -9 \\ 0 & -9 & 9 \end{bmatrix} \quad (4.167)$$

4.8 MODAL ANALYSIS

The matrix equation given in (4.155) and (4.156) can be integrated numerically using the Runge-Kutta method as discussed in Chapter 13 of Chau (2018). However, for a proportional type of damping, we can uncouple the matrix equation to become a scalar equation for an SDOF oscillator using the modes of free undamped vibrations. This technique is called modal analysis.

4.8.1 Free Vibrations

Modal analysis is based on the natural modes of vibrations of a multi-degree-of-freedom oscillator. It can only apply to the case of a linear oscillator, and damping is so-called proportional (such as the Rayleigh damping and Caughey damping to be discussed in later sections). We first recall the governing equations of free vibration of a multi-degree-of-freedom oscillator as:

$$\boldsymbol{M}\ddot{\boldsymbol{u}}(t)+\boldsymbol{K}\boldsymbol{u}(t)=\boldsymbol{0} \tag{4.168}$$

For structural dynamics problems, we seek for a harmonic function in time:

$$\boldsymbol{u}(t)=\boldsymbol{\phi}e^{i\omega t} \tag{4.169}$$

Substitution of (4.169) into (4.168) results in

$$-\omega^2\boldsymbol{M}\boldsymbol{\phi}+\boldsymbol{K}\boldsymbol{\phi}=\boldsymbol{0} \tag{4.170}$$

Note that we can also assume a sine or cosine for time, the resulting equation is the same as (4.170). Rearranging (4.170), we get

$$(\boldsymbol{K}-\omega^2\boldsymbol{M})\boldsymbol{\phi}=\boldsymbol{0} \tag{4.171}$$

This is an eigenvalue problem and the vector $\boldsymbol{\phi}$ is the eigenvector. It is well known that (4.171) cannot be satisfied by an arbitrary value of circular frequency ω. For a multi-story system of n stories, the number of eigenvalues can be shown to be n as well. Since (4.168) is a homogeneous system, for a nonzero eigenvector, we must require the determinant of the square matrix to be zero:

$$\left|\boldsymbol{K}-\omega^2\boldsymbol{M}\right|=0 \tag{4.172}$$

This gives an n-th order algebraic equation, and by Gauss's theorem we must have n roots. It is straightforward to see that if $\boldsymbol{M}^{-1}\boldsymbol{K}$ is symmetric, all the eigenvalues are real (Hermitian matrix as discussed in Section 5.5.1 of Chau (2018). From (4.152) and (4.153), we see that $\boldsymbol{K}$ is symmetric and $\boldsymbol{M}$ is diagonal, thus, we have that $\boldsymbol{M}^{-1}\boldsymbol{K}$ is indeed symmetric. Thus, this formulation is in fact similar to the theory for a system of first order ODEs discussed in Chapter 5 of Chau (2018). For the case of two degrees of freedom, the explicit forms for the natural frequencies are given in Problem 4.14. The eigenvalues and their corresponding eigenvectors (or vibration mode shapes) are

$$\omega_j,\boldsymbol{\phi}_j,\quad j=1,2,...,n \tag{4.173}$$

which satisfy the following equations

$$\omega_j^2\boldsymbol{M}\boldsymbol{\phi}_j=\boldsymbol{K}\boldsymbol{\phi}_j \tag{4.174}$$

$$\omega_k^2\boldsymbol{M}\boldsymbol{\phi}_k=\boldsymbol{K}\boldsymbol{\phi}_k \tag{4.175}$$

As discussed in Chapter 5 of Chau (2018), real symmetric matrices are also known as Hermitian matrices (or self-adjoint) and their associated eigenvalues must be real. Taking the transpose of (4.174) gives

$$(\omega_j^2\boldsymbol{M}\boldsymbol{\phi}_j)^T=\omega_j^2\boldsymbol{\phi}_j^T\boldsymbol{M}^T=(\boldsymbol{K}\boldsymbol{\phi}_j)^T=\boldsymbol{\phi}_j^T\boldsymbol{K}^T \tag{4.176}$$

Thus, (4.176) becomes

$$\omega_j^2\boldsymbol{\phi}_j^T\boldsymbol{M}=\boldsymbol{\phi}_j^T\boldsymbol{K} \tag{4.177}$$

We can now turn (4.174) into a scalar equation by multiplying $\boldsymbol{\phi}_k$ from the right to give

$$\omega_j^2 \boldsymbol{\phi}_j^T \boldsymbol{M} \boldsymbol{\phi}_k = \boldsymbol{\phi}_j^T \boldsymbol{K} \boldsymbol{\phi}_k \tag{4.178}$$

On the other hand, we multiply the transpose of $\boldsymbol{\phi}_j$ to (4.178) to give

$$\omega_k^2 \boldsymbol{\phi}_j^T \boldsymbol{M} \boldsymbol{\phi}_k = \boldsymbol{\phi}_j^T \boldsymbol{K} \boldsymbol{\phi}_k \tag{4.179}$$

Comparison of (4.178) and (4.179) gives

$$(\omega_j^2 - \omega_k^2) \boldsymbol{\phi}_j^T \boldsymbol{M} \boldsymbol{\phi}_k = 0 \tag{4.180}$$

Since $\omega_j \neq \omega_k$, we must have

$$\boldsymbol{\phi}_j^T \boldsymbol{M} \boldsymbol{\phi}_k = 0 \tag{4.181}$$

This is the orthogonality of the eigenvectors on the mass matrix. For either (4.178) or (4.179), the left hand side is zero because of (4.181), and consequently, we must have the right-hand side of them be zero as well. This leads to the second orthogonality condition

$$\boldsymbol{\phi}_j^T \boldsymbol{K} \boldsymbol{\phi}_k = 0 \tag{4.182}$$

With these two orthogonal conditions, we are ready to uncouple the system.

4.8.2 Decoupling of the Undamped Dynamic System

We now consider the most important step of the mode superposition technique. Following the idea of eigenfunction expansion as discussed in Chapter 10 of Chau (2018), we assume that the solution $\boldsymbol{u}$ can be expanded in series expansion of the eigenvectors as

$$\boldsymbol{u} = \sum_{i=1}^{n} y_i(t) \boldsymbol{\phi}_i = \boldsymbol{\Phi} \boldsymbol{y} \tag{4.183}$$

where the fundamental matrix is defined as

$$\boldsymbol{\Phi} = [\boldsymbol{\phi}_1, \boldsymbol{\phi}_2, ..., \boldsymbol{\phi}_n], \quad \boldsymbol{y}^T = [y_1, y_2, ..., y_n] \tag{4.184}$$

Note that $\boldsymbol{\Phi}$ is an $n \times n$ matrix. For the case of an undamped system, we can simplify (4.155) as

$$\boldsymbol{M}\ddot{\boldsymbol{u}}(t) + \boldsymbol{K}\boldsymbol{u} = -\boldsymbol{M}\boldsymbol{e}\ddot{u}_g(t) \tag{4.185}$$

Substituting (4.183) into (4.185), we get the following system

$$\sum_{i=1}^{n} [\boldsymbol{M}\boldsymbol{\phi}_i \ddot{y}_i + \boldsymbol{K}\boldsymbol{\phi}_i y_i] = -\boldsymbol{M}\boldsymbol{e}\ddot{u}_g(t) \tag{4.186}$$

Multiplying the transpose of $\boldsymbol{\phi}_k$ by (4.186), we obtain

$$\sum_{i=1}^{n} [\boldsymbol{\phi}_k^T \boldsymbol{M}\boldsymbol{\phi}_i \ddot{y}_i + \boldsymbol{\phi}_k^T \boldsymbol{K}\boldsymbol{\phi}_i y_i] = -\boldsymbol{\phi}_k^T \boldsymbol{M}\boldsymbol{e}\ddot{u}_g(t) \tag{4.187}$$

Using the orthogonal conditions derived in (4.181) and (4.182), we have

$$\boldsymbol{\phi}_k^T \boldsymbol{M}\boldsymbol{\phi}_k \ddot{y}_k + \boldsymbol{\phi}_k^T \boldsymbol{K}\boldsymbol{\phi}_k y_k = -\boldsymbol{\phi}_k^T \boldsymbol{M}\boldsymbol{e}\ddot{u}_g(t) \tag{4.188}$$

Equation (4.188) is actually a scalar equation and can be rewritten as

$$M_k \ddot{y}_k + K_k y_k = -L_k \ddot{u}_g(t) \tag{4.189}$$

where

$$M_k = \boldsymbol{\phi}_k^T \boldsymbol{M} \boldsymbol{\phi}_k \tag{4.190}$$

$$K_k = \boldsymbol{\phi}_k^T \boldsymbol{K} \boldsymbol{\phi}_k \tag{4.191}$$

$$L_k = \boldsymbol{\phi}_k^T \boldsymbol{M}\boldsymbol{e} \tag{4.192}$$

Sometimes, it is more convenient to normalize the eigenvector with respect to the mass matrix:

$$\tilde{\boldsymbol{\phi}}_k^T \boldsymbol{M} \tilde{\boldsymbol{\phi}}_k = 1, \quad \tilde{\boldsymbol{\phi}}_k = \frac{1}{\sqrt{M_k}} \boldsymbol{\phi}_k \tag{4.193}$$

Therefore, we have uncoupled the system, and we can solve each degree of freedom independently using the technique discussed in Sections 4.4 and 4.5. For the general case of dynamic loads, similarly we obtain

$$M_k \ddot{y}_k + K_k y_k = L_k(t) \tag{4.194}$$

where

$$L_k = \boldsymbol{\phi}_k^T \boldsymbol{F}(t) \tag{4.195}$$

For this case, the mathematical technique given in Section 4.4 for an SDOF oscillator can be applied to solve the uncoupled equation.

4.8.3 Decoupling of the Damped Dynamic System

In reality, damping exists in real structures or machines. Retaining the damping term, we have

$$\boldsymbol{M}\ddot{\boldsymbol{u}}(t) + \boldsymbol{C}\dot{\boldsymbol{u}} + \boldsymbol{K}\boldsymbol{u} = -\boldsymbol{M}\boldsymbol{e}\ddot{u}_g(t) \tag{4.196}$$

Substituting (4.183) into (4.196) and multiplying the transpose of $\boldsymbol{\phi}_k$ to its result, we obtain

$$\sum_{i=1}^{n} [\boldsymbol{\phi}_k^T \boldsymbol{M} \boldsymbol{\phi}_i \ddot{y}_i + \boldsymbol{\phi}_k^T \boldsymbol{C} \boldsymbol{\phi}_i \dot{y}_i + \boldsymbol{\phi}_k^T \boldsymbol{K} \boldsymbol{\phi}_i y_i] = -\boldsymbol{\phi}_k^T \boldsymbol{M}\boldsymbol{e}\ddot{u}_g(t) \tag{4.197}$$

We have seen from the last section that both the first and third terms are orthogonal and can be uncoupled. The key issue here is whether the second term in (4.197) is diagonal. This issue will be discussed in the next section.

4.8.4 Rayleigh Damping

Rayleigh (1877) proposed that the damping matrix can be assumed to be proportional to both the mass and stiffness matrices as:

$$\boldsymbol{C} = \beta_0 \boldsymbol{K} + \beta_1 \boldsymbol{M} \tag{4.198}$$

This is called Rayleigh damping. Physically, Rayleigh damping can be interpreted that each story is interconnected to the dashpot with a damping coefficient proportional to the column stiffness, and each story is also connected to the ground by a separate dashpot with a damping coefficient proportional to the mass. More details of this physical meaning are given in Problem 4.11. Thus, the second term in (4.197) becomes

$$\boldsymbol{\phi}_k^T \boldsymbol{C}\boldsymbol{\phi}_i \dot{y}_i = (\beta_0 \boldsymbol{\phi}_k^T \boldsymbol{K}\boldsymbol{\phi}_i + \beta_1 \boldsymbol{\phi}_k^T \boldsymbol{M}\boldsymbol{\phi}_i)\dot{y}_i \tag{4.199}$$

Substitution of (4.199) into (4.197) gives

$$\sum_{i=1}^{n} [\boldsymbol{\phi}_k^T \boldsymbol{M}\boldsymbol{\phi}_i \ddot{y}_i + (\beta_0 \boldsymbol{\phi}_k^T \boldsymbol{K}\boldsymbol{\phi}_i + \beta_1 \boldsymbol{\phi}_k^T \boldsymbol{M}\boldsymbol{\phi}_i)\dot{y}_i + \boldsymbol{\phi}_k^T \boldsymbol{K}\boldsymbol{\phi}_i y_i] = -\boldsymbol{\phi}_k^T \boldsymbol{M}\boldsymbol{e}\ddot{u}_g(t) \tag{4.200}$$

Thus, we have

$$M_k \ddot{y}_k + C_k \dot{y}_k + K_k y_k = -L_k \ddot{u}_g(t) \tag{4.201}$$

where

$$C_k = \beta_0 K_k + \beta_1 M_k \tag{4.202}$$

This can be solved by using the technique for an SDOF oscillator. In particular, we can rewrite (4.201) as

$$\ddot{y}_k + 2\xi_k \omega_k \dot{y}_k + \omega_k^2 y_k = -\alpha_k \ddot{u}_g(t) \tag{4.203}$$

where

$$\omega_k = \sqrt{\frac{K_k}{M_k}} \tag{4.204}$$

$$\xi_k = \frac{C_k}{2\omega_k M_k} \tag{4.205}$$

$$\alpha_k = \frac{L_k}{M_k} = \frac{\boldsymbol{\phi}_k^T \boldsymbol{M}\boldsymbol{e}}{\boldsymbol{\phi}_k^T \boldsymbol{M}\boldsymbol{\phi}_k} \tag{4.206}$$

The factor α_k is known as the mode participation factor of the k-th mode. For most engineering applications, the summation assumed in (4.183) needs only to sum the first four to six terms for sufficient accuracy. The displacement of the structure can be expressed as:

$$y_k(t) = -\alpha_k \int_0^t \ddot{u}_g(\tau) h_k(t-\tau) d\tau = \frac{\alpha_k}{\omega_{Dk}} V_k(t) \tag{4.207}$$

where

$$h_k(t) = \frac{1}{\omega_{Dk}} e^{-\xi_k \omega_k t} \sin(\omega_{Dk} t), \quad \omega_{Dk} = \omega_k \sqrt{1-\xi_k^2} \tag{4.208}$$

Note that ξ_k has been defined in (4.205) and physically V_k is the velocity. Therefore, the solution is

$$\boldsymbol{u} = \sum_{k=1}^{n} y_k(t) \boldsymbol{\phi}_k = \sum_{k=1}^{n} \boldsymbol{\phi}_k \frac{\alpha_k}{\omega_{Dk}} V_k(t) \tag{4.209}$$

The elastic force can be calculated as

$$\boldsymbol{f}_s(t) = \boldsymbol{K}\boldsymbol{u} = \sum_{k=1}^{n} \boldsymbol{K}\boldsymbol{\phi}_k \frac{\alpha_k}{\omega_{Dk}} V_k(t) \tag{4.210}$$

From (4.174), this elastic restoring force can be written as

$$\boldsymbol{f}_s(t) = \sum_{k=1}^{n} \boldsymbol{M}\boldsymbol{\phi}_k \alpha_k \frac{\omega_k^2}{\omega_{Dk}} V_k(t) \tag{4.211}$$

The base shear at ground level can be calculated as

$$H_{tot} = \sum_{k=1}^{n} f_{si}(t) = \boldsymbol{e}^T \boldsymbol{f}_s = \sum_{k=1}^{n} \boldsymbol{e}^T \boldsymbol{M}\boldsymbol{\phi}_k \alpha_k \frac{\omega_k^2}{\omega_{Dk}} V_k(t) \tag{4.212}$$

Recalling from (4.206) that

$$\alpha_k = \frac{\boldsymbol{\phi}_k^T \boldsymbol{Me}}{M_k} \tag{4.213}$$

Since this is a scalar equation, we can take the transpose without changing its value:

$$\alpha_k = \frac{(\boldsymbol{\phi}_k^T \boldsymbol{Me})^T}{M_k} = \frac{\boldsymbol{e}^T \boldsymbol{M}\boldsymbol{\phi}_k}{M_k} \tag{4.214}$$

The last part of (4.214) can be substituted into (4.212) to get

$$H_{tot} = \sum_{k=1}^{n} M_k \alpha_k^2 \frac{\omega_k^2}{\omega_{Dk}} V_k(t) \tag{4.215}$$

The overturning moment of structure at the ground level is

$$M_0(t) = (h_1, h_2, ..., h_n)\boldsymbol{f}_s(t) = \boldsymbol{h}^T \boldsymbol{f}_s(t) = \sum_{k=1}^{n} \boldsymbol{h}^T \boldsymbol{M}\boldsymbol{\phi}_k \alpha_k \frac{\omega_k^2}{\omega_{Dk}} V_k(t) \tag{4.216}$$

where h_i is defined in Figure 4.14. For most engineering applications, the main contribution comes from the first four to six modes. For example, in considering a ten-story building, we only need to solve for the first four to six modes. In actual numerical calculations, we can stop the summation whenever the contribution from the next higher mode is smaller than a prescribed percentage.

To find the constants in Rayleigh damping, we can first rewrite the damping matrix in diagram form by multiplying the fundamental matrix defined in (4.184) as

$$\boldsymbol{C}_g = \boldsymbol{\Phi}^T \boldsymbol{C}\boldsymbol{\Phi} = \beta_1 \begin{bmatrix} M_1 & 0 & \vdots & 0 \\ 0 & M_2 & \vdots & 0 \\ \cdots & \cdots & \ddots & 0 \\ 0 & 0 & 0 & M_n \end{bmatrix} + \beta_0 \begin{bmatrix} \omega_1^2 M_1 & 0 & \vdots & 0 \\ 0 & \omega_2^2 M_2 & \vdots & 0 \\ \cdots & \cdots & \ddots & 0 \\ 0 & 0 & 0 & \omega_n^2 M_n \end{bmatrix} \tag{4.217}$$

On the other hand, from (4.205), we can identify the diagonal matrix as

$$\boldsymbol{C}_g = \begin{bmatrix} 2\xi_1\omega_1 M_1 & 0 & \vdots & 0 \\ 0 & 2\xi_2\omega_2 M_2 & \vdots & 0 \\ \cdots & \cdots & \ddots & 0 \\ 0 & 0 & 0 & 2\xi_n\omega_n M_n \end{bmatrix} \tag{4.218}$$

Taking first two modes, we have

$$2\xi_1\omega_1 M_1 = M_1(\beta_0 + \omega_1^2 \beta_1) \tag{4.219}$$

$$2\xi_2\omega_2 M_2 = M_2(\beta_0 + \omega_2^2 \beta_2) \tag{4.220}$$

These equations can be rewritten as

$$\xi_1 = \frac{1}{2}(\frac{\beta_0}{\omega_1} + \omega_1\beta_1) \tag{4.221}$$

$$\xi_2 = \frac{1}{2}(\frac{\beta_0}{\omega_2} + \omega_2\beta_2) \tag{4.222}$$

The solutions of (4.194) and (4.195) are

$$\beta_0 = \frac{2\omega_1\omega_2(\xi_1\omega_2 - \xi_2\omega_1)}{\omega_2^2 - \omega_1^2} \tag{4.223}$$

$$\beta_1 = \frac{2(\omega_2\xi_2 - \omega_1\xi_1)}{\omega_2^2 - \omega_1^2} \tag{4.224}$$

There is a major problem in this evaluation of β_0 and β_1. If we pick two other arbitrary natural frequencies, we will obtain a different set of β_0 and β_1. There are many different choices of calculating the damping coefficients. A recent analysis by Song and Su (2017) showed, for the case of a hydropower house, that different ways of evaluating β_0 and β_1 lead to quite different Rayleigh damping coefficients. A more systematic approach is proposed to use a weighted least square approach (similar to the Gauss least square method for linear regression):

$$E = \sum_{i=1}^{n} W(m_i, \omega_i)(\frac{\beta_0}{\omega_i} + \omega_i\beta_1 - \xi_i)^2 \tag{4.225}$$

$$\frac{dE}{d\beta_0} = 0, \quad \frac{dE}{d\beta_1} = 0 \tag{4.226}$$

If the weighting function W is one, we have the classical linear regression approach. Another way to mitigate this problem will be discussed in the next section. The results derived in this section are valid as long as Rayleigh damping is a good approximation of the actual system. Rayleigh damping is also called proportional damping, or classical damping. Rayleigh proposed (4.198) in his famous book *The Theory of Sound* in 1877. This assumption makes the mathematics particularly attractive, but Rayleigh did emphasize that such assumption needs to be verified by experiments. A number of researchers, including Lord Kelvin, Rayleigh, Kimball, Lovell, Becker, Foppl, Kussner and Kassner, did conduct experiments and found that it is not satisfied by all dynamics systems. Hasselsman (1976) found that Rayleigh damping is a good approximation when there is an adequate frequency separation between the natural modes. Example 4.6 given in a later section illustrates how to find β_0 and β_1.

In the next section, we will consider a more general proportional damping.

4.8.5 Caughey and Liu-Gorman Proportional Damping

Rayleigh damping is not the only type of proportional damping. Caughey (1960) showed that a damping matrix can be assumed as proportional damping if $\boldsymbol{M}^{-1}\boldsymbol{C}$ can be expanded in a series of $\boldsymbol{M}^{-1}\boldsymbol{K}$. More specifically, Caughey and O'Kelly (1965) proposed the following proportional damping:

$$C = M\sum_{k=0}^{n-1}\beta_k[M^{-1}K]^k \tag{4.227}$$

where the power of the square bracket term is defined as

$$[M^{-1}K]^k = \underset{1}{[M^{-1}K]}\underset{2}{[M^{-1}K]}\underset{\cdots}{\cdots}\underset{k}{[M^{-1}K]} \tag{4.228}$$

We can see that if we take only the first two terms, we recover Rayleigh damping as a special case:

$$C = \beta_0 M + M\beta_1 M^{-1}K + ... = \beta_0 M + \beta_1 K + ... \tag{4.229}$$

To see the validity of (4.227), we can first consider the following damped-free vibrations:

$$M\ddot{u}(t) + C\dot{u} + Ku = 0 \tag{4.230}$$

Multiplying (4.230) by the inverse of M, we find

$$\ddot{u}(t) + M^{-1}C\dot{u} + M^{-1}Ku = 0 \tag{4.231}$$

The problem can be uncoupled if we can express the following

$$M^{-1}C = f(M^{-1}K) \tag{4.232}$$

However, in matrix theory, the Cayley-Hamilton theorem states that (e.g., Lin, 1966)

$$f(M^{-1}K) = \sum_{k=0}^{n-1}\beta_k[M^{-1}K]^k \tag{4.233}$$

Substituting (4.233) into (4.232) and multiplying M by the result gives (4.227). Alternatively, multiplying (4.230) by the inverse of K, we have

$$K^{-1}M\ddot{u}(t) + K^{-1}C\dot{u} + u = 0 \tag{4.234}$$

Similarly, we can look for the following possibility:

$$K^{-1}C = f(K^{-1}M) \tag{4.235}$$

Again, we can apply the Cayley-Hamilton theorem to obtain the following expansion

$$f(K^{-1}M) = \sum_{k=0}^{n-1}\beta_k[K^{-1}M]^k \tag{4.236}$$

Substitution of (4.236) into (4.235) and multiplying K by the result gives an alternative series to that of Caughey:

$$C = K\sum_{k=0}^{n-1}\beta_k[K^{-1}M]^k \tag{4.237}$$

Apparently, this particular series form was first given by Liu and Gorman (1995). Further result of using this series is given in Problems 4.7 to 4.9.

More generally, Adhikari (2006) showed that the damping matrix can be expressed as:

$$C = Mf_1(M^{-1}K, K^{-1}M) + Kf_2(M^{-1}K, K^{-1}M) \tag{4.238}$$

The expansion of the function f_1 and f_2 given in (4.238) will lead to the Caughey series or the Liu-Gorman series as special cases. Two special cases of (4.238) are

$$C = \beta_0 M + \alpha_1 K + \beta_2 K M^{-1} K \quad (4.239)$$

$$C = \beta_{-1} M K^{-1} M + \beta_0 M + \beta_1 K \quad (4.240)$$

These are the first three terms of the Caughey and Liu-Gorman series, respectively. If we have an n-story building, we can take n terms in Caughey's power series defined in (4.227). In contrast to Rayleigh damping, we have n equations for n coefficients compared to n equations for two unknowns. Similarly, we can also use n terms in Liu-Gorman's power series to give n equations for n unknowns.

Now, let us derive the equations for the damping coefficients for the case of Caughey's series. Recalling the eigenvalue problem, we have:

$$\omega_n^2 M \phi_n = K \phi_n \quad (4.241)$$

We can multiply (4.241) by a carefully selected vector such that (4.241) becomes a scalar equation as:

$$\phi_n^T K M^{-1} (\omega_n^2 M \phi_n) = \phi_n^T K M^{-1} (K \phi_n) \quad (4.242)$$

Multiplying the matrix through, we get

$$\omega_n^2 \phi_n^T K \phi_n = \phi_n^T K M^{-1} K \phi_n \quad (4.243)$$

However, it is straightforward to see from (4.241) and (4.190) that

$$\phi_n^T K \phi_n = \omega_n^2 \phi_n^T M \phi_n = \omega_n^2 M_n \quad (4.244)$$

Substitution of (4.244) into (4.243) gives

$$\phi_n^T K M^{-1} K \phi_n = \omega_n^4 M_n \quad (4.245)$$

We can continue the line of analysis used in obtaining (4.245) to get

$$\phi_n^T K M^{-1} K M^{-1} (\omega_n^2 M \phi_n) = \phi_n^T K M^{-1} K M^{-1} (K \phi_n) \quad (4.246)$$

Simplification of (4.246) gives

$$\phi_n^T K M^{-1} K M^{-1} K \phi_n = \phi_n^T K M^{-1} K \phi_n \omega_n^2 = \omega_n^6 M_n \quad (4.247)$$

The last part of (4.247) results from the substitution of (4.245) in the second-to-last part of (4.247) (i.e., the result from the previous step). Clearly this procedure can continue forever. More importantly, the matrix within the two eigenvectors in (4.245) and (4.247) are those that appear in the following equation, which was obtained by multiplying the series (4.227) by the transpose of the eigenvector from the front and by the eigenvector from the back, as

$$\phi_n^T C \phi_n = \beta_0 \phi_n^T M \phi_n + \beta_1 \phi_n^T K \phi_n + \beta_2 \phi_n^T K M^{-1} K \phi_n + \beta_2 \phi_n^T K M^{-1} K M^{-1} K \phi_n + \ldots \quad (4.248)$$

From (4.248), we obtain the following explicit form for the damping ratio

$$\begin{aligned} C_n = 2\xi_n \omega_n M_n &= \beta_0 M_n + \beta_1 K_n + \beta_2 \omega_n^4 M_n + \beta_3 \omega_n^6 M_n + \ldots \\ &= \sum_{k=0}^{n-1} \beta_k \omega_n^{2k} M_n \end{aligned} \quad (4.249)$$

Thus, we have n equations for n unknowns. Let us consider the explicit form for the case of $n = 3$:

$$\xi_1 \omega_1 M_1 = \beta_0 M_1 + \beta_1 \omega_1^2 M_1 + \beta_2 \omega_1^4 M_1 \quad (4.250)$$

$$\xi_2 \omega_2 M_2 = \beta_0 M_2 + \beta_1 \omega_2^2 M_2 + \beta_2 \omega_2^4 M_2 \quad (4.251)$$

$$\xi_3\omega_3 M_3 = \beta_0 M_3 + \beta_1\omega_3^2 M_3 + \beta_2\omega_3^4 M_3 \tag{4.252}$$

Then, the coefficients for Caughey's proportional damping can be found uniquely, without using the least square method discussed in (4.225).

Similar expressions can also be developed for the Liu-Gorman series. The readers can refer to Problems 4.7 to 4.9 for details. For n equal or larger than four, we can also pick some terms from Caughey's series and some from the Liu-Gorman series. We will not discuss such possibility here.

4.8.6 Rayleigh Quotient Technique

To uncouple the system of ODEs and to make the modal analysis applicable, we have seen in the last section that proportional damping has to be assumed. In fact, if we can diagonalize the damping matrix, decoupling can also be achieved. To do that we can use a technique called the Rayleigh quotient. First, we multiply the equation of motions given in (4.155) by the inverse of the fundamental matrix from the front and substitute (4.183) for $\boldsymbol{u}$ to get:

$$\boldsymbol{\Phi}^T \boldsymbol{M\Phi}\ddot{\boldsymbol{y}}(t) + \boldsymbol{\Phi}^T \boldsymbol{C\Phi}\dot{\boldsymbol{y}} + \boldsymbol{\Phi}^T \boldsymbol{K\Phi y} = -\boldsymbol{\Phi}^T \boldsymbol{Me}\ddot{u}_g(t) \tag{4.253}$$

$$\boldsymbol{M}_g\ddot{\boldsymbol{y}}(t) + \boldsymbol{C}_g\dot{\boldsymbol{y}} + \boldsymbol{K}_g\boldsymbol{y} = -\boldsymbol{\Phi}^T \boldsymbol{Me}\ddot{u}_g(t) \tag{4.254}$$

where

$$\boldsymbol{M}_g = \boldsymbol{\Phi}^T \boldsymbol{M\Phi} = \begin{bmatrix} M_1 & 0 & \vdots & 0 \\ 0 & M_2 & \vdots & 0 \\ \cdots & \cdots & \ddots & \vdots \\ 0 & 0 & \cdots & M_n \end{bmatrix} \tag{4.255}$$

$$\boldsymbol{K}_g = \boldsymbol{\Phi}^T \boldsymbol{K\Phi} = \begin{bmatrix} M_1\omega_1^2 & 0 & \vdots & 0 \\ 0 & M_2\omega_2^2 & \vdots & 0 \\ \cdots & \cdots & \ddots & \vdots \\ 0 & 0 & \cdots & M_n\omega_n^2 \end{bmatrix} \tag{4.256}$$

$$\boldsymbol{C}_g = \boldsymbol{\Phi}^T \boldsymbol{C\Phi} = \begin{bmatrix} C_{11} & C_{12} & \vdots & C_{1n} \\ C_{21} & C_{21} & \vdots & C_{2n} \\ \cdots & \cdots & \ddots & \vdots \\ C_{n1} & C_{n2} & \cdots & C_{nn} \end{bmatrix} \tag{4.257}$$

In general, the system of ODEs is not uncoupled because of $\boldsymbol{C}_g$ given in (4.257), unless the damping matrix is diagonalized. We replace (4.257) by the following diagonal matrix

$$\boldsymbol{C}_g^{RQ} = \begin{bmatrix} C_1^{RQ} & 0 & \vdots & 0 \\ 0 & C_2^{RQ} & \vdots & 0 \\ \cdots & \cdots & \ddots & \vdots \\ 0 & 0 & \cdots & C_n^{RQ} \end{bmatrix} \tag{4.258}$$

where the diagonal term is approximated by

$$C_n^{RQ} = \frac{\boldsymbol{\phi}_n^T \boldsymbol{C} \boldsymbol{\phi}_n}{\boldsymbol{\phi}_n^T \boldsymbol{\phi}_n} \tag{4.259}$$

$$M_k \ddot{y}_k(t) + C_k^{RQ} \dot{y}_k + K_k y_k = -\boldsymbol{\phi}^T \boldsymbol{M} \boldsymbol{e} \ddot{u}_g(t) \tag{4.260}$$

This gives n SDOF equations and again can be solved independently.

4.8.7 Response Spectrum Method for MDOF System

For the case of proportional damping, modal analysis allows decoupling of the MDOF system. We can use the response spectrum method to find the maximum response for each mode. In particular, we can find the peak values of response of the k-th mode as:

$$y_{k,\max} = \frac{\alpha_k}{\omega_{Dk}} \max_t |V_k(t)| = \frac{\alpha_k}{\omega_{Dk}} S_{pv}(\omega_k, \zeta_k) = \alpha_k S_d(\omega_k, \zeta_k) \tag{4.261}$$

The maximum displacement of the k-th mode is

$$\boldsymbol{u}_{k,\max} = \boldsymbol{\phi}_k y_{k,\max} = \boldsymbol{\phi}_k \frac{\alpha_k}{\omega_{Dk}} S_{pv}(\omega_k, \zeta_k) = \boldsymbol{\phi}_k \alpha_k S_d(\omega_k, \zeta_k) \tag{4.262}$$

Similarly, the maximum elastic force, the maximum base shear, and the maximum overturning moment are

$$(\boldsymbol{f}_{s,k})_{\max} = \boldsymbol{M} \boldsymbol{\phi}_k \alpha_k \frac{\omega_k^2}{\omega_{Dk}} S_{pv}(\omega_k, \zeta_k) \tag{4.263}$$

$$(H_{tot.k})_{\max} = M_k \alpha_k^2 \frac{\omega_k^2}{\omega_{Dk}} S_{pv}(\omega_k, \zeta_k) \tag{4.264}$$

$$(M_{0,k})_{\max} = \boldsymbol{h}^T \boldsymbol{M} \boldsymbol{\phi}_k \alpha_k \frac{\omega_k^2}{\omega_{Dk}} S_{pv}(\omega_k, \zeta_k) \tag{4.265}$$

Now, the main issue is how to add the contribution from each mode, because the modal maxima do not appear at the same time. Simple summation will not work. That is, we have

$$\boldsymbol{u}_{\max} \neq \sum_{k=1}^{n} \boldsymbol{u}_{k,\max} \tag{4.266}$$

In addition, we cannot derive the maximum elastic force from the maximum displacement. There is no unique or exact way of adding the maximum. We will only discuss three approaches here. For the most conservative approach, we can use the absolute sum (AS) rule:

$$r_a = \sum_{k=1}^{N} |r_k| \tag{4.267}$$

where r_a is the total response and r_k is the maximum response of the k-th mode. The AS was proposed by M.A. Biot when he introduced the concept of the response spectrum method. This approach, of course, overestimates the maximum response. The most widely accepted rule is the square root sum of squares (SRSS):

$$r_a = [\sum_{k=1}^{N} r_k^2]^{1/2} \tag{4.268}$$

This method gives good results provided that the natural frequencies are well separated. The third approach is the so-called complete quadratic combination (CQC) rule:

$$r_a = [\sum_{j=1}^{N} \sum_{k=1}^{N} S_{jk} r_j r_k]^{1/2} \tag{4.269}$$

where

$$S_{jk} = \frac{8\sqrt{\zeta_j \zeta_k}(\zeta_j + \gamma\zeta_k)\gamma^{3/2}}{(1-\gamma^2)^2 + 4\zeta_j\zeta_k\gamma(1+\gamma^2) + 4(\zeta_j^2 + \zeta_k^2)\gamma^2} \tag{4.270}$$

$$\gamma = \frac{\omega_k}{\omega_j} \tag{4.271}$$

This combination rule includes all cross-modal terms, taking into consideration the interactions between modes. For $\omega_{n+1}/\omega_n < 1.5$, CQC should be used because mode interaction cannot be neglected. It was also found that when the mode is not adequately separated, the use of Rayleigh damping is not accurate. The next example illustrates the application of modal analysis to the case of a two-story building.

Example 4.6 Seismic design is applied to a two-story building as shown in Figure 4.14, by using modal analysis and the response spectrum. The masses are m_1 =60,000 kg and m_2 = 50,000 kg. The equivalent column stiffnesses are $k_1 = 5\times10^4$ kN/m and $k_2 = 3\times10^4$ kN/m. It is assumed that the structure obeys Rayleigh damping. It was given that the damping ratios for both modes 1 and 2 are 0.05 or 5%. For 5% damping, assume that the pseudo-response spectrum of the design ground motion is prescribed by the upper spectrum curve in Figure 4.13.

(i) Find the natural frequencies of the structure.

(ii) Plot the corresponding mode shapes.

(iii) Find the maximum displacement of modes 1 and 2 using the response spectrum method.

(iv) Find the maximum restoring forces in the columns for both modes 1 and 2.

(v) Find the total displacement using the AS rule, SRSS rule and CQC rule.

(vi) Find the damping coefficients (β_0 and β_1) in Rayleigh damping.

(vii) Find the corresponding damping matrix.

Solution: (i) The stiffness matrix of the two-story structure shown in Figure 4.14 is

$$\boldsymbol{K} = \begin{bmatrix} k_1 + k_2 & -k_2 \\ -k_2 & k_2 \end{bmatrix} = \begin{bmatrix} 8 & -3 \\ -3 & 3 \end{bmatrix} \times 10^4 \text{kN/m} \tag{4.272}$$

The mass matrix of the two-story structure is

$$\boldsymbol{M} = \begin{bmatrix} m_1 & 0 \\ 0 & m_2 \end{bmatrix} = \begin{bmatrix} 60 & 0 \\ 0 & 50 \end{bmatrix} \times 10^3 \text{kg} \tag{4.273}$$

The eigenvalue or the natural frequency can be determined from

$$\left|\boldsymbol{K} - \omega^2 \boldsymbol{M}\right| = \begin{vmatrix} K_{11} - m_1\omega^2 & K_{12} \\ K_{21} & K_{22} - m_2\omega^2 \end{vmatrix} = 0 \tag{4.274}$$

Substitution of (4.272) and (4.273) into (4.274) gives

$$\begin{vmatrix} 8\times10^4 - 60\omega^2 & -3\times10^4 \\ -3\times10^4 & 3\times10^4 - 50\omega^2 \end{vmatrix} = 0 \tag{4.275}$$

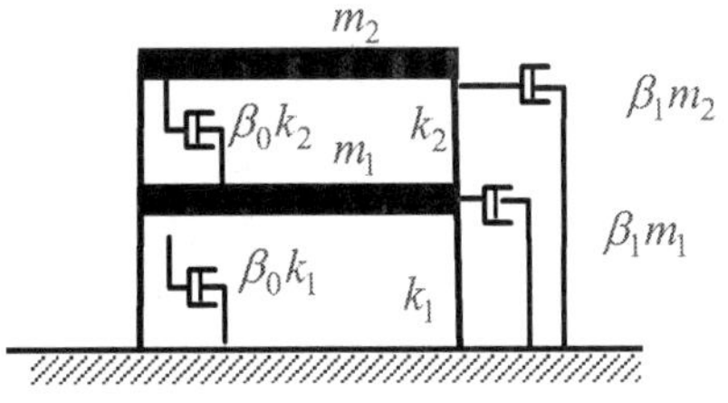

Figure 4.16 Two-story building with Rayleigh damping

Expanding (4.275) gives a quadratic equation for ω^2:

$$(8\times10^4 - 60\omega^2)(3\times10^4 - 50\omega^2) - 9\times10^8 = 0 \tag{4.276}$$

Rearranging terms in (4.276), we have

$$0.00003\omega^4 - 0.058\omega^2 + 15 = 0 \tag{4.277}$$

The solutions for (4.277) are

$$\omega_1^2 = 307.5, \quad \omega_2^2 = 1625.8 \tag{4.278}$$

Finally, we have the fundamental frequencies being

$$\omega_1 = 17.54 \text{ rad/s}, \quad \omega_2 = 40.32 \text{ rad/s} \tag{4.279}$$

(ii) The mode shape of the first natural frequency is (see Problem 4.15 for the following formula)

$$\phi_1 = \begin{Bmatrix} \dfrac{K_{12}}{m_1\omega_1^2 - K_{11}} \\ 1 \end{Bmatrix} = \begin{Bmatrix} \dfrac{-3\times 10^4}{60\times 307.5 - 8\times 10^4} \\ 1 \end{Bmatrix} = \begin{Bmatrix} 0.4874 \\ 1 \end{Bmatrix} \tag{4.280}$$

For the second mode, we have

$$\phi_2 = \begin{Bmatrix} \dfrac{K_{12}}{m_1\omega_2^2 - K_{11}} \\ 1 \end{Bmatrix} = \begin{Bmatrix} \dfrac{-3\times 10^4}{60\times 1625.8 - 8\times 10^4} \\ 1 \end{Bmatrix} = \begin{Bmatrix} -1.7096 \\ 1 \end{Bmatrix} \tag{4.281}$$

These mode shapes are plotted in Figure 4.17.

(iii) Recalling from (4.190), we have

$$M_k = \phi_k^T \boldsymbol{M} \phi_k \tag{4.282}$$

Substitution of (4.273), (4.280), and (4.281) into (4.282) gives

$$M_1 = \phi_1^T \boldsymbol{M} \phi_1 = \{0.4874 \quad 1\} \begin{bmatrix} 60 & 0 \\ 0 & 50 \end{bmatrix} \begin{Bmatrix} 0.4874 \\ 1 \end{Bmatrix} = 64.25 \tag{4.283}$$

$$M_2 = \phi_2^T \boldsymbol{M} \phi_2 = \{-1.7096 \quad 1\} \begin{bmatrix} 60 & 0 \\ 0 & 50 \end{bmatrix} \begin{Bmatrix} -1.7096 \\ 1 \end{Bmatrix} = 225.36 \tag{4.284}$$

From (4.191), we have

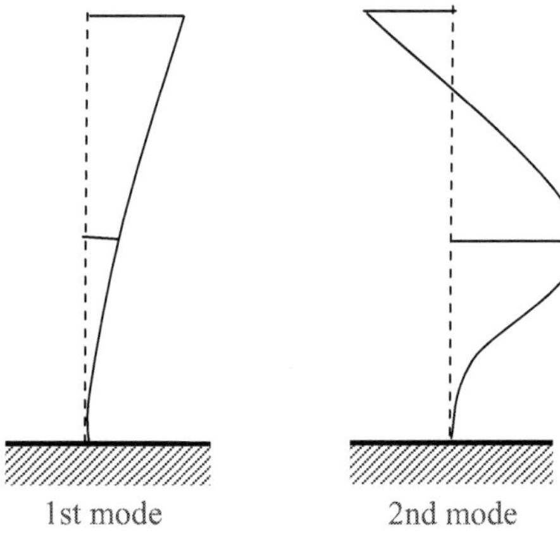

Figure 4.17 Mode shapes for the two-story building given in Figure 4.16

$$K_k = \phi_k^T \boldsymbol{K} \phi_k \tag{4.285}$$

Substitution of (4.273), (4.280), and (4.281) into (4.285) gives

$$K_1 = \phi_1^T \boldsymbol{K} \phi_k = \{0.4874 \quad 1\} \begin{bmatrix} 8\times 10^4 & -3\times 10^4 \\ -3\times 10^4 & 3\times 10^4 \end{bmatrix} \begin{Bmatrix} 0.4874 \\ 1 \end{Bmatrix} = 19760 \tag{4.286}$$

$$K_2 = \phi_2^T K \phi_2 = \{-1.7096 \quad 1\}\begin{bmatrix} 8\times10^4 & -3\times10^4 \\ -3\times10^4 & 3\times10^4 \end{bmatrix}\begin{Bmatrix} -1.7096 \\ 1 \end{Bmatrix} = 366395 \quad (4.287)$$

From (4.192), we have

$$L_k = \phi_k^T Me \quad (4.288)$$

Substitution of (4.273), (4.280), and (4.281) into (4.288) gives

$$L_1 = \phi_1^T Me = \{0.4874 \quad 1\}\begin{bmatrix} 60 & 0 \\ 0 & 50 \end{bmatrix}\begin{Bmatrix} 1 \\ 1 \end{Bmatrix} = 79.244 \quad (4.289)$$

$$L_2 = \phi_2^T Me = \{-1.7096 \quad 1\}\begin{bmatrix} 60 & 0 \\ 0 & 50 \end{bmatrix}\begin{Bmatrix} 1 \\ 1 \end{Bmatrix} = -52.576 \quad (4.290)$$

Finally, we have the mode participation factors

$$\alpha_1 = \frac{L_1}{M_1} = \frac{79.244}{64.25} = 1.2334 \quad (4.291)$$

$$\alpha_2 = \frac{L_2}{M_2} = -\frac{52.576}{225.36} = -0.2333 \quad (4.292)$$

By using the response spectrum method, the maximum displacement can be evaluated from (4.262)

$$u_{k,\max} = \phi_k \frac{\alpha_k}{\omega_{Dk}} S_{pv}(\omega_k, \zeta_k) \quad (4.293)$$

More specifically, we have

$$u_{1,\max} = \phi_1 \frac{\alpha_1}{\omega_{D1}} S_{pv}(\omega_1, \zeta_1) \quad (4.294)$$

$$u_{2,\max} = \phi_2 \frac{\alpha_2}{\omega_{D2}} S_{pv}(\omega_2, \zeta_2) \quad (4.295)$$

Since the damping ratio for both modes is given as 5%, we have

$$\omega_{D1} = \omega_1\sqrt{1-\zeta_1^2} = 17.54\sqrt{1-0.05^2} = 17.52 \text{ rad/s} \quad (4.296)$$

$$\omega_{D2} = \omega_2\sqrt{1-\zeta_2^2} = 40.32\sqrt{1-0.05^2} = 40.27 \text{ rad/s} \quad (4.297)$$

Finally, we can look up the spectrum value from Figure 4.13, as shown in Figure 4.18

$$S_{pv}(\omega_1, \zeta_1) = 0.27 \text{ m/s}, \quad S_{pv}(\omega_2, \zeta_2) = 0.05 \text{ m/s} \quad (4.298)$$

Employing these values of the pseudo-velocity spectrum, we find

$$\begin{aligned} u_{1,\max} &= \phi_1 \frac{\alpha_1}{\omega_{D1}} S_{pv}(\omega_1, \zeta_1) \\ &= \begin{pmatrix} 0.4874 \\ 1 \end{pmatrix}\frac{1.2334}{17.52}\times 0.27 \\ &= \begin{pmatrix} 0.00926 \\ 0.019 \end{pmatrix} m \end{aligned} \quad (4.299)$$

$$\begin{aligned}\boldsymbol{u}_{2,\max} &= \boldsymbol{\phi}_2 \frac{\alpha_2}{\omega_{D2}} S_{pv}(\omega_2, \zeta_2) \\ &= \begin{pmatrix} -1.7096 \\ 1 \end{pmatrix} \frac{-0.2333}{40.27} \times 0.05 \\ &= \begin{pmatrix} 0.000495 \\ -0.000289 \end{pmatrix} m\end{aligned} \tag{4.300}$$

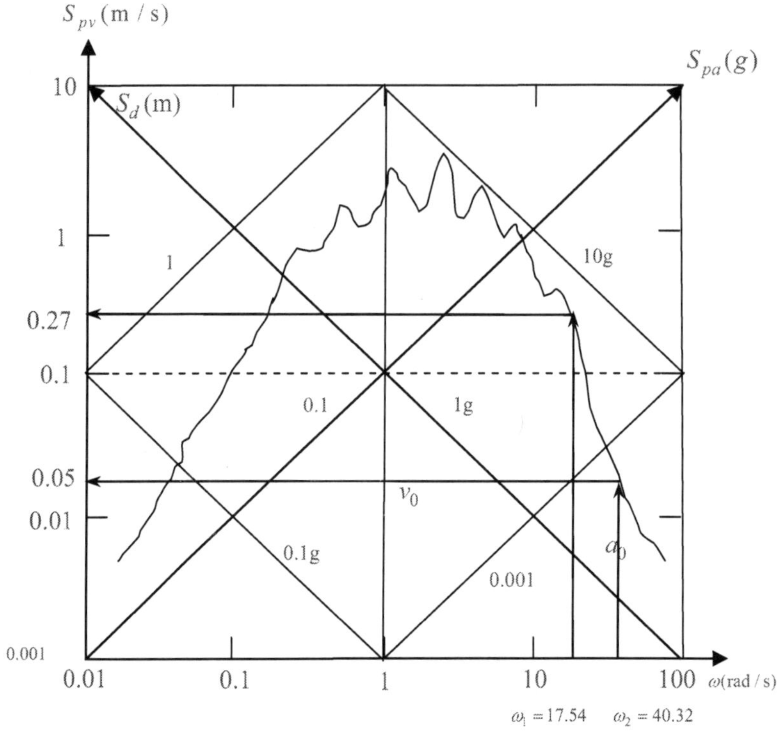

Figure 4.18 The values of the pseudo-velocity response spectrum for modes 1 and 2 given in Figure 4.17

(iv) The maximum restoring forces in the columns for both modes 1 and 2 are

$$(\boldsymbol{f}_{s,n})_{\max} = \boldsymbol{M}\boldsymbol{\phi}_n \alpha_n \frac{\omega_n^2}{\omega_{Dn}} S_{pv}(\omega_n, \zeta_n) \tag{4.301}$$

For the present two-story structure, we have

$$(\boldsymbol{f}_{s,1})_{\max} = \boldsymbol{M}\boldsymbol{\phi}_1\alpha_1 \frac{\omega_1^2}{\omega_{D1}} S_{pv}(\omega_1,\zeta_1)$$
$$= \begin{bmatrix} 60 & 0 \\ 0 & 50 \end{bmatrix} \begin{Bmatrix} 0.4874 \\ 1 \end{Bmatrix} \times 1.2333 \times \frac{307.5}{17.52} \times 0.27$$
$$= \begin{pmatrix} 29.244 \\ 50 \end{pmatrix} \times 1.2333 \times \frac{307.5}{17.52} \times 0.27 \tag{4.302}$$
$$= \begin{pmatrix} 170.93 \\ 292.25 \end{pmatrix} kN$$

$$(\boldsymbol{f}_{s,2})_{\max} = \boldsymbol{M}\boldsymbol{\phi}_2\alpha_2 \frac{\omega_2^2}{\omega_{D2}} S_{pv}(\omega_2,\zeta_2)$$
$$= \begin{bmatrix} 60 & 0 \\ 0 & 50 \end{bmatrix} \begin{Bmatrix} -1.7096 \\ 1 \end{Bmatrix} \times (-0.2333) \times \frac{1625.8}{40.27} \times 0.05$$
$$= \begin{pmatrix} -102.58 \\ 50 \end{pmatrix} \times (-0.2333) \times \frac{1625.8}{40.27} \times 0.05 \tag{4.303}$$
$$= \begin{pmatrix} 48.31 \\ -23.55 \end{pmatrix} kN$$

(v) The total displacement can be determined by using the AS rule, SRSS rule and CQC rule. By the AS rule, we have

$$\boldsymbol{u}_{\max} = \sum_{k=1}^{n} \left| \boldsymbol{u}_{k,\max} \right|$$
$$= \begin{bmatrix} 9.26 \\ 19 \end{bmatrix} + \begin{bmatrix} 0.495 \\ 0.289 \end{bmatrix} = \begin{bmatrix} 9.755 \\ 19.289 \end{bmatrix} mm \tag{4.304}$$

By the SRSS rule, we have

$$\boldsymbol{u}_{\max} = \begin{bmatrix} (9.26^2 + 0.495^2)^{1/2} \\ (19^2 + 0.289^2)^{1/2} \end{bmatrix}$$
$$= \begin{bmatrix} 9.273 \\ 19.002 \end{bmatrix} mm \tag{4.305}$$

By the CQC rule, we have to first evaluate the coefficient of the quadratic terms given in (4.270)

$$S_{11} = S_{22} = \frac{8(0.05)(2)(0.05)}{4(0.05)^2(2) + 4(2)(0.05)^2} = 1 \tag{4.306}$$

$$S_{12} = \frac{8(0.05)^2(1+2.299)(2.299)^{3/2}}{[1-(2.299)^2]^2 + 4(0.05)^2(2.299)(1+2.299^2) + 4(2)(0.05)^2(2.299)^2} \tag{4.307}$$
$$= 0.01235$$

$$S_{21}=\frac{8(0.05)^2(1+0.435)(0.435)^{3/2}}{[1-(0.435)^2]^2+4(0.05)^2(0.435)(1+0.435^2)+4(2)(0.05)^2(0.435)^2} \tag{4.308}$$
$$=0.29058$$

Using these results, we have

$$u_{1,\max}=[(9.26)^2+(0.29058+0.01235)(-9.26)(0.494)+(0.494)^2]^{1/2} \tag{4.309}$$
$$=9.198mm$$

$$u_{2,\max}=[(19)^2+(0.29058+0.01235)(19)(0.289)+(0.289)^2]^{1/2} \tag{4.310}$$
$$=19.050mm$$

Finally, we get the total displacement predicted by the CQC rule

$$\boldsymbol{u}_{\max}=\begin{bmatrix}9.198\\19.050\end{bmatrix}mm \tag{4.311}$$

As expected, the AS rule is most conservative, the SRSS rule is a popular choice for buildings, whereas the CQC rule allows mode interactions. Since for the present case the $\omega_2/\omega_1 > 1.5$, the mode interactions are not very severe for the present case.

(vi) Since the damping ratios for both modes 1 and 2 are given as 5%, (4.223) and (4.224) can be used to find the Rayleigh damping coefficients

$$\beta_0=\frac{2\omega_1\omega_2(\xi_1\omega_2-\xi_2\omega_1)}{\omega_2^2-\omega_1^2}$$
$$=\frac{2(17.54)(40.32)(40.32-17.54)(0.05)}{1625.8-307.5} \tag{4.312}$$
$$=1.2158\,s$$

$$\beta_1=\frac{2(\omega_2\xi_2-\omega_1\xi_1)}{\omega_2^2-\omega_1^2}$$
$$=\frac{2(40.32-17.54)(0.05)}{1625.8-307.5} \tag{4.313}$$
$$=0.001728\,s^{-1}$$

(vii) Employing the results obtained in (4.312) and (4.313), we can find the damping matrix as follows:

$$\boldsymbol{C}=\beta_0\boldsymbol{K}+\beta_1\boldsymbol{M}$$
$$=1.222\times\begin{bmatrix}8\times10^4 & -3\times10^4\\-3\times10^4 & 3\times10^4\end{bmatrix}+0.001728\times\begin{bmatrix}60 & 0\\0 & 50\end{bmatrix} \tag{4.314}$$
$$=\begin{bmatrix}9.776\times10^4 & -3.666\times10^4\\-3.666\times10^4 & 3.666\times10^4\end{bmatrix}\text{kN s m}^{-1}$$

4.9 SUMMARY AND FURTHER READING

In conclusion, we have given a brief introduction to structural dynamics in the context of structures under earthquake actions. First, a single-degree-of-freedom oscillator is considered, including free undamped oscillations, free damped oscillations, forced undamped oscillations, and forced damped oscillations. The idea of the response spectrum is then introduced, including the use of the pseudo-response spectrum. For multi-degree-of-freedom (MDOF) systems, we introduce the concept of modal analysis for cases of proportional damping. Rayleigh, Caughey and Liu-Gorman damping are discussed. Modal analysis was originally introduced for an undamped system, but can be extended to the case of proportional damping by Rayleigh. If a dynamic oscillator involves contact force and an actively controlled system during oscillations, the assumption of proportional damping is not appropriate. The idea of analyzing an inelastic structure using the response spectrum method is introduced. The response spectrum method using modal analysis is discussed. The summation rules for maximum response from each mode are discussed, including the absolute sum (AS) rule, square root sum of squares (SRSS) rule, and complete quadratic combination (CQC) rule.

In the current chapter, our focus has been on linear structures with proportional damping. The idea of proportional damping can also be extended to the case of a nonlinear problem by using the updated tangential stiffness matrix of the system:

$$\boldsymbol{C}(t) = \boldsymbol{M}\sum_{k=0}^{n-1} \beta_k [\boldsymbol{M}^{-1}\boldsymbol{K}_{\tan}(t)]^k \tag{4.315}$$

This can be considered as an extension of Caughey's damping. Similarly, we can also express the time-dependent $\boldsymbol{C}$ as

$$\boldsymbol{C}(t) = \boldsymbol{K}_{\tan}\sum_{k=0}^{n-1} \beta_k [\boldsymbol{K}_{\tan}^{-1}(t)\boldsymbol{M}^{-1}]^k \tag{4.316}$$

This is an extension of Liu-Gorman series damping. This idea of using tangential stiffness for nonlinear structures was proposed by Charney (2008) for the case of Rayleigh damping. In addition, damping arising from frictional effect, active dampers, and wind-induced damping might not be proportional. However, it is still an active area of research and is out of the scope of the present chapter.

Clough and Penzien (2003) provided comprehensive coverage of structural dynamics, and this book has been widely used in the world as one of the standard textbooks.

4.10 PROBLEMS

Problem 4.1 The solution given in (4.4) can be rewritten as a cosine function

$$z_2 = R\cos(\omega_0 t - \delta) \tag{4.317}$$

Find R and δ in terms of A and B defined in (4.4).

Ans:

$$R = \sqrt{A^2 + B^2}, \quad \delta = \tan^{-1}(\frac{B}{A}) \tag{4.318}$$

Problem 4.2 Consider a damped structure modeled by the following nonhomogeneous ODE:

$$\ddot{u} + 2\omega_0 \zeta \dot{u} + \omega_0^2 u = \frac{F_0}{m}\sin(\omega t) \tag{4.319}$$

Find the particular solution of this ODE.

Ans:

$$u_p(t) = \frac{F_0}{k[(1-\beta^2)^2 + (2\beta\zeta)^2]}\{-2\beta\zeta\cos\omega t + (1-\beta^2)\sin\omega t\} \tag{4.320}$$

Problem 4.3 Find the general solution for the following damped structure subject to the following harmonic excitation and initial conditions:

$$\ddot{u} + 2\omega_0 \zeta \dot{u} + \omega_0^2 u = \frac{F_0}{m}\sin(\omega t) \tag{4.321}$$

$$u(0) = u_0, \quad \dot{u}(0) = v_0 \tag{4.322}$$

Ans:

$$u(t) = e^{-\zeta\omega_0 t}\{[u_0 + \frac{2\beta\zeta F_0}{k[(1-\beta^2)^2 + (2\beta\zeta)^2]}]\cos(\omega_D t)$$
$$+ \frac{1}{\sqrt{1-\zeta^2}}[\frac{v_0}{\omega_0} + u_0\zeta + (\frac{\beta F_0}{k})(\frac{2\zeta^2 - (1-\beta^2)}{(1-\beta^2)^2 + (2\beta\zeta)^2})]\sin(\omega_D t)\} \tag{4.323}$$
$$+ \frac{F_0}{k}[\frac{-2\beta\zeta\cos\omega t + (1-\beta^2)\sin\omega t}{(1-\beta^2)^2 + (2\beta\zeta)^2}]$$

Problem 4.4 When a man of mass m stands still on a beam of length L and of bending stiffness EI, the static deflection at the center of a beam is δ_s (see Figure 4.19). Derive a formula for the natural frequency f (in Hertz) of free vibrations for a flexible simply-supported beam in terms of δ_s when the man jumps off the beam.

Hints: You can assume the beam with the man is a single-degree-of-freedom oscillator of vibrating mass m without damping and the weight of the beam is much smaller than the weight of the man.

Ans:

$$f = \frac{1}{2\pi}\sqrt{\frac{g}{\delta_s}} \tag{4.324}$$

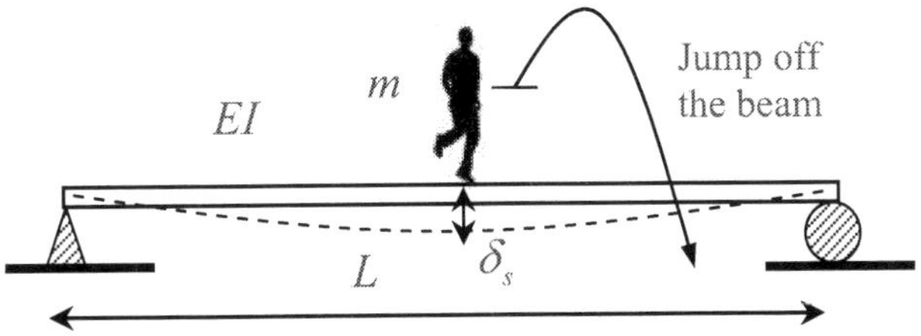

Figure 4.19 Simple vibration model of a man jumping off a beam

Problem 4.5 This problem continues the analysis of Problem 4.4.

(i) Show that the bending stiffness EI of the beam can be estimated in terms of the mass m of the man and the static deflection at the center of a beam δ_s as:

$$EI = \frac{mgL^3}{48\delta_s} \tag{4.325}$$

(ii) Using the result from part (i), show that the natural frequency of vibration can be estimated as:

$$f = \frac{2}{\pi}\sqrt{\frac{3EI}{mL^3}} \tag{4.326}$$

(iii) It is given that the exact solution of the natural frequency of the vibration of a beam alone is (see Section 12.8 of Timoshenko and Young, 1965):

$$f_{exact} = \frac{\pi}{2}\sqrt{\frac{EI}{ML^3}} \tag{4.327}$$

where M is the total mass of the beam. Show that the vibrations of a beam of mass M (the total mass of a beam with distributed mass) can be replaced by a beam vibrating under a concentrated mass of m at the center, where

$$m = \frac{48M}{\pi^4} = 0.4927M \tag{4.328}$$

Problem 4.6 In Section 4.4.3, we see from the hammer test that two successive peaks in damped-free vibrations for a structure is sufficient to estimate the damping ratio. However, if more comprehensive forced vibrations were conducted using various input frequencies such that the amplitude response curve versus force harmonic frequency is available, we can use the half-power bandwidth method to find the damping ratio. In particular, Figure 4.20 shows an experimentally obtained steady-state response function defined in (4.95) as R versus β.

(i) Show that

$$R_1 = \frac{1}{\sqrt{2}}\frac{F_0/k}{2\zeta\sqrt{1-\zeta^2}} \tag{4.329}$$

(ii) Show that β_1 and β_2 satisfy the following equation

$$\beta^4 + 2(\zeta^2 - 1)\beta^2 + 1 - 8\zeta^2(1 - \zeta^2) = 0 \tag{4.330}$$

(iii) Show that if $\zeta^2 << 1$

$$\zeta \approx \frac{1}{2}(\beta_2 - \beta_1) \tag{4.331}$$

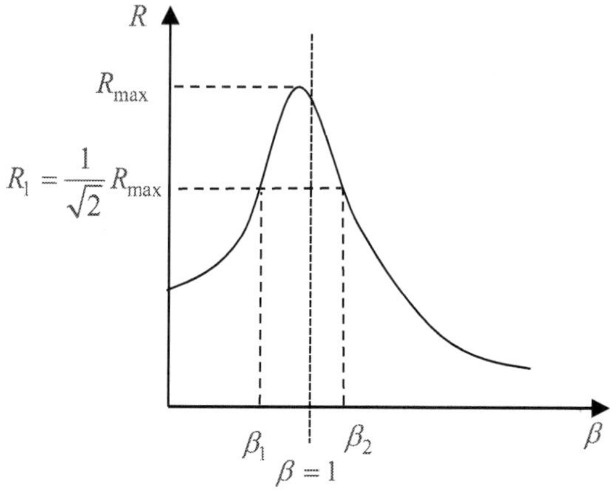

Figure 4.20 Half-power bandwidth method in determining damping ratio

Problem 4.7 Instead of using Caughey's series of proportional damping, we use the Liu-Gorman series as:

$$\boldsymbol{C} = \boldsymbol{K}\sum_{k=0}^{n-1} \beta_k [\boldsymbol{K}^{-1}\boldsymbol{M}]^k \tag{4.332}$$

(i) Following the technique employed in the text, prove the following formula

$$\boldsymbol{\phi}_n^T \boldsymbol{M}\boldsymbol{K}^{-1}\boldsymbol{M}\boldsymbol{\phi}_n = \frac{K_n}{\omega_n^4} \tag{4.333}$$

$$\boldsymbol{\phi}_n^T \boldsymbol{M}\boldsymbol{K}^{-1}\boldsymbol{M}\boldsymbol{K}^{-1}\boldsymbol{M}\boldsymbol{\phi}_n = \frac{K_n}{\omega_n^6} \tag{4.334}$$

(ii) Use the results in Part (i) to prove the following formula

$$C_n = \sum_{k=0}^{n-1} \beta_k \frac{K_n}{\omega_n^{2k}} \tag{4.335}$$

Problem 4.8 For the Liu-Gorman series, show that the damping constant can be calculated as

$$\zeta_n = \frac{1}{2} M_n \sum_{k=0}^{n-1} \beta_k \frac{K_n}{\omega_n^{2k+1}} \tag{4.336}$$

Problem 4.9 For the case of $n = 3$ in the Liu-Gorman series, show that three coefficients can be evaluated from the following equations

$$\zeta_1 = \frac{1}{2}\{\beta_0\omega_1 + \frac{\beta_1}{\omega_1} + \frac{\beta_2}{\omega_1^3}\} \tag{4.337}$$

$$\zeta_2 = \frac{1}{2}\{\beta_0\omega_2 + \frac{\beta_1}{\omega_2} + \frac{\beta_2}{\omega_2^3}\} \tag{4.338}$$

$$\zeta_3 = \frac{1}{2}\{\beta_0\omega_3 + \frac{\beta_1}{\omega_3} + \frac{\beta_2}{\omega_3^3}\} \tag{4.339}$$

Problem 4.10 Example 4.5 formulates coefficients of the stiffness matrix by imposing a unit displacement at each level. This formulation is called the stiffness method. In fact, we can also formulate the equation of motion using the flexibility method by imposing a unit force at each degree of freedom (see Figure 4.21). In particular, for the case of undamped free vibrations of a three-story structure, the flexibility method can be formulated as:

$$\boldsymbol{\delta}\boldsymbol{M}\ddot{\boldsymbol{u}}(t) + \boldsymbol{u}(t) = \boldsymbol{0} \tag{4.340}$$

where

$$\boldsymbol{\delta} = \begin{bmatrix} \delta_{11} & \delta_{12} & \delta_{13} \\ \delta_{21} & \delta_{22} & \delta_{23} \\ \delta_{31} & \delta_{32} & \delta_{33} \end{bmatrix}, \quad \boldsymbol{M} = \begin{bmatrix} m_1 & 0 & 0 \\ 0 & m_2 & 0 \\ 0 & 0 & m_3 \end{bmatrix} \tag{4.341}$$

The determination of the flexibility matrix can be found by applying a unit force at each story level, as shown in Figure 4.13. Show that if the columns are uniform

$$E_1I_1 = E_2I_2 = E_3I_3 = EI \tag{4.342}$$

$$L_1 = L_2 = L_3 = L \tag{4.343}$$

the flexibility matrix is

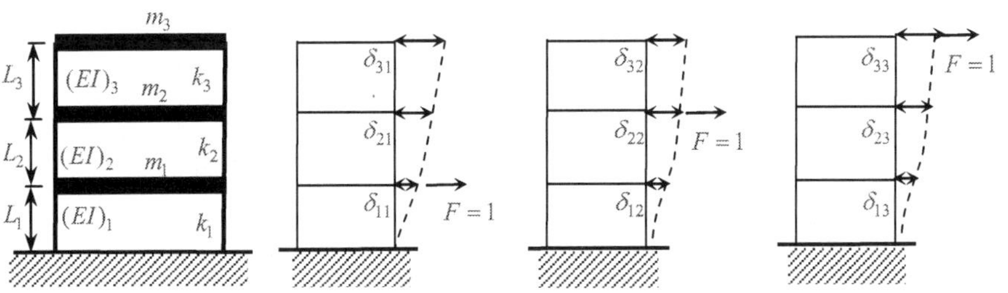

Figure 4.21 Flexibility method for a three-story structure

$$\boldsymbol{\delta} = \frac{L^3}{24EI}\begin{bmatrix} 3 & 2 & 1 \\ 2 & 2 & 1 \\ 1 & 1 & 1 \end{bmatrix} \tag{4.344}$$

Hints: First show that for unit columns the stiffness matrix is

$$K = \frac{EI}{L^3}\begin{bmatrix} 24 & -24 & 0 \\ -24 & 48 & -24 \\ 0 & -24 & 48 \end{bmatrix} \tag{4.345}$$

Then, compare (4.340) with the special case of (4.168) as

$$\boldsymbol{M}\ddot{\boldsymbol{u}}(t) + \boldsymbol{K}\boldsymbol{u}(t) = \boldsymbol{0} \tag{4.346}$$

Problem 4.11 Rayleigh damping has been very popular in the literature because of its simplicity. Figure 4.22 shows the physical meaning of Rayleigh damping.

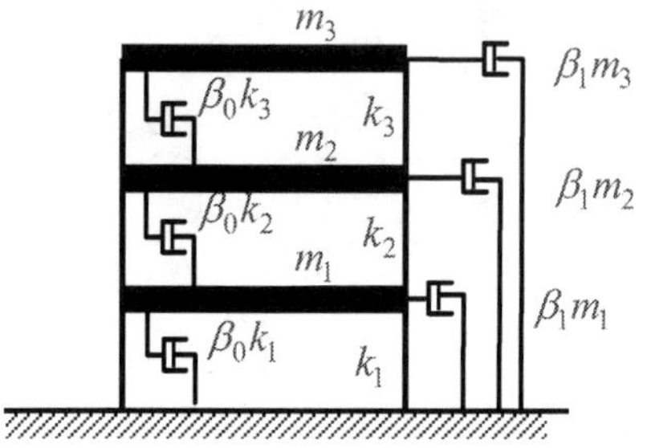

Figure 4.22 Rayleigh damping for a three-story building

(i) Show that the equation of motion for each mass is

$$m_1\ddot{u}_1 + (\beta_0 k_1 + \beta_0 k_2 + \beta_1 m_1)\dot{u}_1 - \beta_0 k_2 \dot{u}_2 + (k_1 + k_2)u_1 - k_2 u_2 = 0 \tag{4.347}$$

$$\begin{aligned} &m_2\ddot{u}_2 + (\beta_0 k_2 + \beta_0 k_3 + \beta_1 m_2)\dot{u}_2 - \beta_0 k_2 \dot{u}_1 - \beta_0 k_3 \dot{u}_3 \\ &+(k_3 + k_2)u_2 - k_2 u_1 - k_3 u_3 = 0 \end{aligned} \tag{4.348}$$

$$m_3\ddot{u}_3 + (\beta_0 k_3 + \beta_1 m_3)\dot{u}_3 - \beta_0 k_3 \dot{u}_2 - k_3 u_2 + k_3 u_3 = 0 \tag{4.349}$$

(ii) Show that the equation of motion can be put in a matrix form as

$$\boldsymbol{M}\ddot{\boldsymbol{u}}(t) + \boldsymbol{C}\dot{\boldsymbol{u}}(t) + \boldsymbol{K}\boldsymbol{u}(t) = \boldsymbol{0} \tag{4.350}$$

where

$$\boldsymbol{C} = \beta_0 \boldsymbol{K} + \beta_1 \boldsymbol{M} \tag{4.351}$$

Remarks: This problem provides a physical meaning for Rayleigh's damping.

Problem 4.12 Show that the uncoupled equation for the case of Caughey damping is

$$\ddot{y}_k + 2\zeta_k \omega_k \dot{y}_k + \omega_k^2 y_k = -\alpha_k \ddot{u}_g(t) \tag{4.352}$$

where

$$\omega_k = \sqrt{\frac{\phi_k^T \boldsymbol{K} \phi_k}{\phi_k^T \boldsymbol{M} \phi_k}}, \quad \zeta_k = \frac{1}{2}\sum_{k=0}^{k-1} \beta_k \omega_k^{2j-1}, \quad \alpha_k = \frac{\phi_k^T \boldsymbol{M}\boldsymbol{e}}{\phi_k^T \boldsymbol{M} \phi_k} \tag{4.353}$$

Problem 4.13 Show that the uncoupled equation for the case of Liu-Gorman damping is

$$\ddot{y}_k + 2\zeta_k \omega_k \dot{y}_k + \omega_k^2 y_k = -\alpha_k \ddot{u}_g(t) \tag{4.354}$$

where

$$\omega_k = \sqrt{\frac{\boldsymbol{\phi}_k^T \boldsymbol{K} \boldsymbol{\phi}_k}{\boldsymbol{\phi}_k^T \boldsymbol{M} \boldsymbol{\phi}_k}}, \quad \zeta_k = \frac{1}{2}\sum_{j=0}^{k-1} \beta_j \frac{1}{\omega_k^{2j-1}}, \quad \alpha_k = \frac{\boldsymbol{\phi}_k^T \boldsymbol{M}\boldsymbol{e}}{\boldsymbol{\phi}_k^T \boldsymbol{M} \boldsymbol{\phi}_k} \tag{4.355}$$

Problem 4.14 Find the explicit form for the square of the natural frequency for a two-degree-of-freedom oscillator from the following equation

$$\left| \boldsymbol{K} - \omega^2 \boldsymbol{M} \right| = 0 \tag{4.356}$$

Ans:

$$\omega^2 = \frac{1}{2}\left(\frac{K_{11}}{m_1} + \frac{K_{22}}{m_2}\right) \pm \sqrt{\left[\frac{1}{2}\left(\frac{K_{11}}{m_1} + \frac{K_{22}}{m_2}\right)\right]^2 - \frac{K_{11}K_{22} - K_{12}K_{21}}{m_1 m_2}} \tag{4.357}$$

Problem 4.15 Show that the mode shapes for the two-degree-of-freedom oscillator can be written as

$$\boldsymbol{\phi}_1 = \begin{Bmatrix} \phi_{11} \\ \phi_{12} \end{Bmatrix} = \begin{Bmatrix} K_{12} \\ m_1\omega_1^2 - K_{11} \end{Bmatrix}, \quad \boldsymbol{\phi}_2 = \begin{Bmatrix} \phi_{21} \\ \phi_{22} \end{Bmatrix} = \begin{Bmatrix} K_{12} \\ m_1\omega_2^2 - K_{11} \end{Bmatrix} \tag{4.358}$$

Problem 4.16 Prove the validity of (4.217) for the overturning moment.

CHAPTER FIVE

Catenary and Cable-Supported Bridges

5.1 INTRODUCTION

The catenary occupies a central place in the development of differential equations. It was derived by Johann Bernoulli in 1691 during the first stage of the development of differential equations. It is the shape of a hanging chain across two supports of the same height and is also the shape of a hanging spider web. In a catenary, uniform mass is distributed along the length of the hanging chain. Catenary shape can be expressed mathematically as a hyperbolic cosine function. The shape of the catenary also inspires the shape of arches and vaults. They are called inverted catenaries. Instead of tension, arches are subjected to compression. Inglis showed in 1863 that the optimum shape of a stone arch bridge is in the form of an inverted catenary (see details discussed in a later section). The Gateway Arch at St. Louis in Missouri is a typical example of an inverted catenary, as shown in Figure 5.1. The height and span of the Gateway Arch is 192 m. The eye-catching monument was designed by Eero Saarinen and is the tallest man-made monument in the United States.

Figure 5.1 Gateway Arch in St. Louis, Missouri (copyright 2008 by Daniel Schwen)

A related problem is the cable suspension bridge for which a uniform mass is distributed along the horizontal distance of the span. Ancient cable suspension bridges were found in China and South America. However, the first modern suspension bridge was built in Pennsylvania in 1796 by James Finley. Theoretical analyses of cable suspension bridges have been contributed by famous mechanicians and engineers, including M. Navier, W.J.M. Rankine, M. Levy, A.A. Jakkula, W. Ritter, J. Melan, R.V. Southwell, L.S. Moisseiff, S.P. Timoshenko, H.H. Bleich, K. Kloppel, S.O. Asplund, A. Selberg, O..H. Amman, T. von Karman, H. Reissner, F. Bleich, D.B. Steinman, A.G. Pugsley, H.M. Irvine, R.H. Scalan, and D.B. Botton. It is normally assumed that the mass of hanging cable is negligible compared to that of the cable, such that the cable for suspension bridges is in the shape of a parabola. The major advance in suspension bridge analysis unfortunately resulted from the collapse of the Tacoma Narrows Bridge on November 7, 1940. The wind is not particularly strong at about 67 km per hour, but noticeable torsion vibrations of extremely large amplitude were observed before its collapse. The whole process of collapse was captured in motion pictures. As a result, a 430-page mathematical analysis of vibration in suspension bridge was published in 1950 by the Bureau of Public Roads and was jointly financed by the Oregon State Highway Commission, the American Institute Steel Construction, and the Bureau of Public Roads.

In fact, the Tacoma Narrows Bridge failure is by no means the first of its kind in the 200-year development of suspension bridges. As compiled by Bleich et al. (1950), failures and damages of suspension bridges were reported repeatedly before the collapse of the Tacoma Narrows Bridge. For example, wind-induced damage or collapse occurred at the Drybourgh Abbey footbridge in Scotland in 1817, Union Bridge in England in 1820, Brighton Chain Pier in England in 1833, Nassau Bridge in Germany in 1833, Montrose Bridge in Scotland in 1837, Roche-Bernard Bridge in France in 1852, Wheeling Bridge in the United States in 1854, and International Bridge at the Niagara River on the US-Canada border in 1863. Collapse caused by vibrations induced by marching of soldiers was reported for little Broughton Bridge in England in 1831 and Ostrawitza River Bridge in Austria in 1886. The Menai Bridge in Wales built by Telford was damaged three times by wind from 1826 to 1939.

In the last few decades, with the advances of construction technology, the cable suspension bridge has been commonly used for long span bridges. Notable examples include the Akashi Kaiyo Bridge in Japan (main span 1991 m), Great Belt East Bridge in Denmark (main span 1624 m), Humber Bridge in the UK (main span 1410 m), Jing Yin Yangtze River Bridge in mainland China (main span 1385 m), Tsing Ma Bridge in Hong Kong (main span 1377 m), and Golden Gate Bridge in the United States (main span 1280 m). The Tsing Ma Bridge in Hong Kong is shown in Figure 5.2, and there are also two side spans of suspension bridge from each tower to the anchor blocks of the main cables. As shown in Figure 5.2, vertical hangers were used to transfer the weight of the bridge deck to the two main cables. One of the major problems for suspension bridges is wind-induced vibrations. The interaction of the wind and bridge (or fluid-structure interaction) is believed to be caused by "von Karman vortex street." However, we will not detour into such problems. Instead, the vibration frequencies of cable suspension bridges will be considered in this chapter. Cable-stay bridges will also be covered briefly.

Figure 5.2 Tsing Ma Bridge in Hong Kong with a main span of 1377 m (copyright 2017 by K.T. Chau)

5.2 VIBRATIONS OF HANGING CHAINS

Before we consider the problems of the catenary and cable suspension bridges, it is informative to first consider the case of vibrating cables or chains. Figure 5.3 shows the first four modes of vibrations of hanging chains or cables after the cable is displaced initially.

The boundary and initial conditions of the hanging cable problem can be formulated as

$$u(x,0) = u_0(x), \quad \frac{\partial u}{\partial t}(x,0) = 0 \tag{5.1}$$

$$u(L,t) = 0, \quad u(0,t) = \textit{finite} \tag{5.2}$$

The origin of the coordinate is set at the lower tip of the cable. The first equation of (5.1) imposed an initial displacement of the cable, while there is no initial velocity being imposed. Equation (5.2) indicates that the cable is fixed at the top and the vibrations of the cable must be finite at all times.

As shown in Figure 5.4, the horizontal component of the tension force in the chain can be approximated as:

$$F(x) = W \tan\alpha \approx W \frac{\partial u}{\partial x} \tag{5.3}$$

The deflection of the cable is relatively small compared to the length of the cable or

$$\left|\sqrt{x^2 + u^2} - x\right| \approx 0 \tag{5.4}$$

Using this approximation, we have

$$\begin{aligned} F(x+\Delta x) &= F(x) + \frac{\partial F(x)}{\partial x}\Delta x \\ &= F(x) + \frac{\partial}{\partial x}(W\frac{\partial u}{\partial x})\Delta x \end{aligned} \tag{5.5}$$

The weight of the chain can be estimated as

$$W = \rho g x \tag{5.6}$$

where ρ is the mass per length of the chain and g is the gravitational constant. The equation of motion of the vibrating chain is

$$m\frac{\partial^2 u}{\partial t^2} = \rho \Delta x \frac{\partial^2 u}{\partial t^2} = F(x+\Delta x) - F(x) \tag{5.7}$$

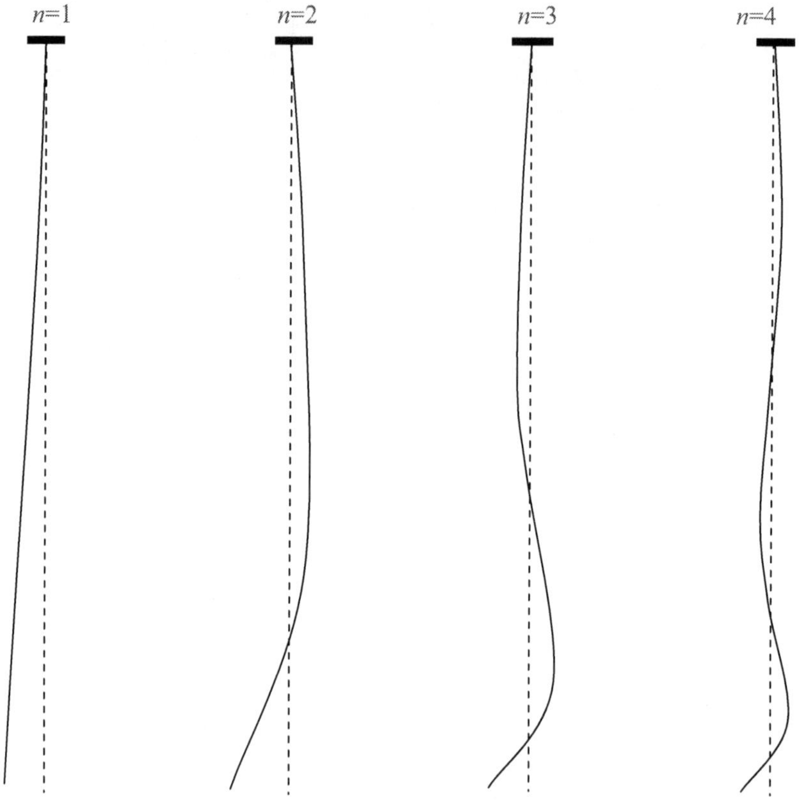

Figure 5.3 Vibration of hanging chains

Combining (5.5) and (5.7), we find the following equation of motion of the vibrating chain

$$\frac{\partial}{\partial x}(\rho g x \frac{\partial u}{\partial x}) = \rho \frac{\partial^2 u}{\partial t^2} \tag{5.8}$$

This can be expanded as

$$\frac{\partial^2 u}{\partial t^2} = g(\frac{\partial u}{\partial x} + x \frac{\partial^2 u}{\partial x^2}) \tag{5.9}$$

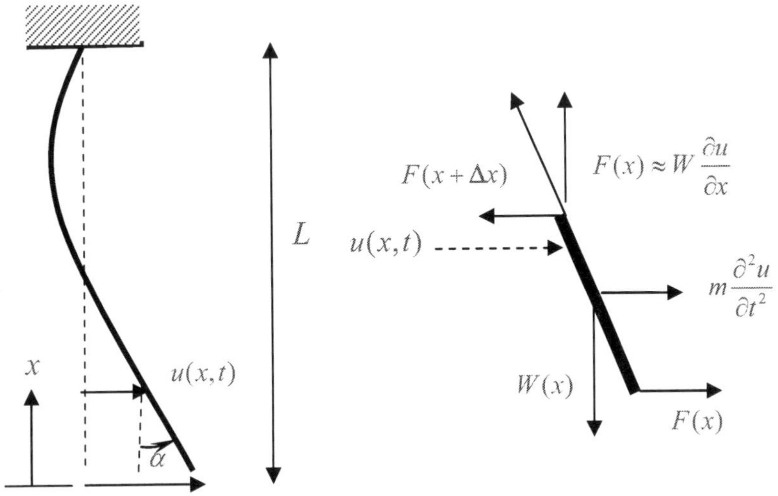

Figure 5.4 Free body diagram of a vibrating hanging chain

This is a second order linear PDE, and the standard approach of separation of variables can be applied

$$u(x,t) = F(x)G(t) \tag{5.10}$$

Substitution of (5.10) into (5.9) gives

$$F\ddot{G} = gF'G + gxF''G \tag{5.11}$$

Dividing through by u, we get

$$\frac{\ddot{G}}{G} = \frac{g(F' + xF'')}{F} = -\omega^2 \tag{5.12}$$

Since a function of t cannot equal a function of x, the first and second terms in (5.12) must be constant, which is represented by $-\omega^2$. We will shortly see why we must use a negative constant on the right-hand side of (5.12) and why we use the symbol ω for the constant. Equation (5.12) leads to two ODEs. The ODE for G is

$$\ddot{G} + \omega^2 G = 0 \tag{5.13}$$

This is an equation of a harmonic oscillator, and the general solution is (see Section 2.5 of Chau, 2018)

$$G = A\cos\omega t + B\sin\omega t \tag{5.14}$$

It is now obvious that ω is the circular frequency of the vibration and this is the reason that we have chosen the symbol ω. In addition, if we use a positive constant on the right-hand side of (5.12), we will end up with an ODE of the type (2.13) of Chau (2018) and the corresponding solution becomes hyperbolic sine and cosine functions. This is clearly unphysical as the solution becomes infinity at infinite time. This is the reason why we must use a negative constant in (5.12). In view of (5.12), the ODE for F is

$$xF'' + F' + \frac{\omega^2}{g} F = 0 \tag{5.15}$$

The corresponding initial condition for G and boundary condition for F are

$$\frac{\partial u}{\partial t}(x,0) = F(x)\dot{G}(0) = 0, \Rightarrow \quad \dot{G}(0) = 0 \tag{5.16}$$

$$u(L,t) = F(L)G(t) = 0, \quad \Rightarrow \quad F(L) = 0 \tag{5.17}$$

To obtain the general solution for F, we introduce the following change of variables

$$z^2 = \frac{4x}{g} \tag{5.18}$$

Differentiations of F in terms of the new variables are

$$\frac{dF}{dx} = \frac{dF}{dz}\frac{dz}{dx} = \frac{2}{zg}\frac{dF}{dz} \tag{5.19}$$

$$\frac{d^2F}{dx^2} = \frac{dF}{dz}\left(\frac{2}{zg}\frac{dF}{dz}\right)\frac{dz}{dx} = \frac{2}{g}\left(-\frac{1}{z^2}\frac{dF}{dz} + \frac{1}{z}\frac{d^2F}{dz^2}\right)\frac{2}{zg} \tag{5.20}$$

Substitution of (5.19) and (5.20) into (5.15) gives the following ODE

$$z^2\frac{d^2F}{dz^2} + z\frac{dF}{dz} + z^2\omega^2 F = 0 \tag{5.21}$$

From Section 4.6 of Chapter 4 of Chau (2018), we recognize that this is the Bessel equation of zero order and the general solution is

$$F(z) = AJ_0(\omega z) + BY_0(\omega z) \tag{5.22}$$

The second condition of (5.2) requires the solution be bounded at all locations and this leads to $B = 0$. Thus, the solution becomes

$$F(z) = AJ_0(\omega z) \tag{5.23}$$

The first condition of (5.2) requires that

$$F(L) = AJ_0\left(2\omega_n\sqrt{\frac{L}{g}}\right) = 0, \quad n = 1,2,3,... \tag{5.24}$$

Note that we have replaced ω by ω_n because there will be more than one root for (5.24) (see Figure 4.5 of Chau, 2018). Since the constant A cannot be zero (otherwise the solution becomes trivial), we must require

$$J_0\left(2\omega_n\sqrt{\frac{L}{g}}\right) = 0 \tag{5.25}$$

The first two roots of (5.25) are

$$\omega_1 = \frac{\bar{\omega}_1}{2}\sqrt{\frac{g}{L}} = 1.202415\sqrt{\frac{g}{L}}, \quad \omega_2 = \frac{\bar{\omega}_2}{2}\sqrt{\frac{g}{L}} = 2.276004\sqrt{\frac{g}{L}},... \tag{5.26}$$

There are infinite natural frequencies of vibrations. Because of the constraint imposed by the first condition of (5.2), only discrete values of ω can satisfy the boundary condition. In terms of the eigenfunction of u_n, the general solution is (see Chapter 10 of Chau, 2018)

$$u(x,t)=\sum_{n=1}^{\infty}u_n(x,t)=\sum_{n=1}^{\infty}J_0(2\omega_n\sqrt{\frac{x}{g}})[A_n\cos\omega_n t+B_n\sin\omega_n t] \tag{5.27}$$

The first four eigenmodes are illustrated in Figure 5.3. Finally, we impose the initial conditions given in (5.1) in resulting

$$\frac{\partial u}{\partial t}(x,0)=0, \quad \Rightarrow B_n=0 \tag{5.28}$$

$$u(x,0)=u_0(x)=\sum_{n=1}^{\infty}J_0(2\omega_n\sqrt{\frac{x}{g}})A_n \tag{5.29}$$

Multiplying both sides of (5.29) by the Bessel function for ω_m and integrating from 0 to L, we obtain

$$\int_0^L u_0(x)J_0(2\omega_m\sqrt{\frac{x}{g}})dx=\sum_{n=1}^{\infty}A_n\int_0^L J_0(2\omega_n\sqrt{\frac{x}{g}})J_0(2\omega_m\sqrt{\frac{x}{g}})dx \tag{5.30}$$

Introduce the following change of variables

$$\xi=2\sqrt{\frac{x}{g}}, \quad dx=\frac{g}{2}\xi d\xi \tag{5.31}$$

Subsequently, (5.30) becomes

$$\int_0^{2\sqrt{\frac{L}{g}}}\xi u_0(\frac{g\xi^2}{4})J_0(\omega_m\xi)d\xi=\sum_{n=1}^{\infty}A_n\int_0^{2\sqrt{\frac{L}{g}}}\xi J_0(\omega_n\xi)J_0(\omega_m\xi)d\xi \tag{5.32}$$

It can be further simplified to

$$\int_0^{2\sqrt{\frac{L}{g}}}\xi u_0(\frac{g\xi^2}{4})J_0(\bar{\omega}_m\frac{\xi}{2}\sqrt{\frac{g}{L}})d\xi=\sum_{n=1}^{\infty}A_n\int_0^{2\sqrt{\frac{L}{g}}}\xi J_0(\bar{\omega}_n\frac{\xi}{2}\sqrt{\frac{g}{L}})J_0(\bar{\omega}_m\frac{\xi}{2}\sqrt{\frac{g}{L}})d\xi \tag{5.33}$$

In view of the orthogonal properties of the Bessel function derived in (10.281) of Chau (2018), we have the right-hand side of (5.33) equal to zero if $n \neq m$. Thus, find

$$A_n=\frac{\int_0^{2\sqrt{\frac{L}{g}}}\xi u_0(\frac{g\xi^2}{4})J_0(\bar{\omega}_n\frac{\xi}{2}\sqrt{\frac{g}{L}})d\xi}{\int_0^{2\sqrt{\frac{L}{g}}}\xi[J_0(\bar{\omega}_n\frac{\xi}{2}\sqrt{\frac{g}{L}})]^2d\xi} \tag{5.34}$$

The general solution of the problem becomes

$$u(x,t)=\sum_{n=1}^{\infty}\frac{\int_0^{2\sqrt{\frac{L}{g}}}\xi u_0(\frac{g\xi^2}{4})J_0(\bar{\omega}_n\frac{\xi}{2}\sqrt{\frac{g}{L}})d\xi}{\int_0^{2\sqrt{\frac{L}{g}}}\xi[J_0(\bar{\omega}_n\frac{\xi}{2}\sqrt{\frac{g}{L}})]^2d\xi}J_0(2\omega_n\sqrt{\frac{x}{g}})\cos\omega_n t \tag{5.35}$$

where ω_n satisfies (5.25). If an initial velocity is imposed, we can also solve the problem using the same approach (see Problem 5.1).

5.3 CATENARY

As remarked in the introduction, the catenary appears in nature as the shape of a hanging chain between two supports. The hanging chain was coined as "catenaria" by Leibniz and Huygens in 1691, which was derived from the Latin word "catena," which means chain. The English term "catenary" apparently was first used by Thomas Jefferson, the third US president, when he received a copy of the treatise on the arch by Mascheroni. It is one of the most fundamental problems solved using differential equation. In particular, Figure 5.5 shows a free body diagram of a suspending cable of weight per unit length as μ. Considering the force equilibrium in the vertical direction, we get

$$(T+dT)\sin(\theta+d\theta)=T\sin\theta+\mu ds \tag{5.36}$$

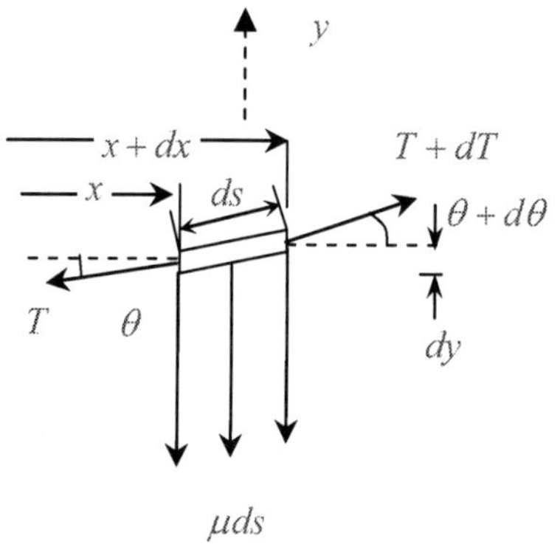

Figure 5.5 Force equilibrium for suspending cable

By sum rule for sine functions, we have the following approximation

$$\begin{aligned}\sin(\theta+d\theta)&=\sin\theta\cos(d\theta)+\sin(d\theta)\cos\theta\\&\approx\sin\theta+d\theta\cos\theta\end{aligned} \tag{5.37}$$

For small deflection, the slope or θ is also small such that we have

$$\cos(d\theta)\approx 1,\quad \sin(d\theta)\approx d\theta,\quad d\theta\to 0 \tag{5.38}$$

Substitution of (5.38) into the left-hand side of (5.36) gives

$$\begin{aligned}(T+dT)\sin(\theta+d\theta)&=T\sin\theta+\sin\theta dT+T\cos\theta d\theta+\cos\theta d\theta dT\\&\approx T\sin\theta+\sin\theta dT+T\cos\theta d\theta\end{aligned} \tag{5.39}$$

Substitution of (5.39) into (5.36) gives

$$T\cos\theta d\theta+dT\sin\theta=\mu ds \tag{5.40}$$

This can be rewritten as

$$d(T\sin\theta)=\mu ds \tag{5.41}$$

This result will be used later. Next, we can consider the force equilibrium in the horizontal direction:

$$(T+dT)\cos(\theta+d\theta)=T\cos\theta \tag{5.42}$$

Using the sum rule for cosine functions, we have the following approximation

$$\begin{aligned}\cos(\theta + d\theta) &= \cos\theta\cos d\theta - \sin\theta\sin d\theta \\ &\approx \cos\theta - d\theta\sin\theta\end{aligned} \tag{5.43}$$

Substitution of (5.43) into the left-hand side of (5.42) yields

$$\begin{aligned}(T + dT)(\cos\theta - \sin\theta d\theta) &= T\cos\theta + dT\cos\theta - T\sin\theta d\theta - \sin\theta dT d\theta \\ &\approx T\cos\theta + dT\cos\theta - T\sin\theta d\theta\end{aligned} \tag{5.44}$$

Substitution of (5.44) into (5.42) yields

$$-T\sin\theta d\theta + dT\cos\theta = 0 \tag{5.45}$$

This evidently can be written as a total differential as

$$d(T\cos\theta) = 0 \tag{5.46}$$

Integration of (5.46) gives

$$T\cos\theta = C = T_0 \tag{5.47}$$

where T_0 is an unknown constant. Using (5.47), we can rewrite (5.41) as

$$\frac{d}{ds}(T_0\frac{\sin\theta}{\cos\theta}) = \frac{d}{ds}(T_0\tan\theta) = \mu \tag{5.48}$$

Note that the tangent equals the slope of the catenary, and we have

$$\frac{d}{ds}(\frac{dy}{dx}) = \frac{\mu}{T_0} \tag{5.49}$$

Integrating with respect to *s*, we get

$$\frac{dy}{dx} = \frac{\mu}{T_0}s + C \tag{5.50}$$

Alternatively, we can also apply the chain rule to (5.49) as

$$\frac{d}{ds}(\frac{dy}{dx}) = \frac{d}{dx}[(\frac{dy}{dx})]\frac{dx}{ds} = \frac{\mu}{T_0} \tag{5.51}$$

The last equation of (5.51) can be rearranged as

$$\frac{d^2y}{dx^2} = \frac{\mu}{T_0}\frac{ds}{dx} \tag{5.52}$$

Using the Pythagorean theorem in Figure 5.5, we have

$$(ds)^2 = (dx)^2 + (dy)^2 \tag{5.53}$$

This can be rewritten as

$$(\frac{ds}{dx})^2 = 1 + (\frac{dy}{dx})^2 \tag{5.54}$$

Solving for *ds*/*dx* and substituting the result into (5.52), we get

$$\frac{d^2y}{dx^2} = \frac{\mu}{T_0}\sqrt{1 + (\frac{dy}{dx})^2} \tag{5.55}$$

This is a nonlinear second order ODE, but it is of autonomous type (i.e., variable *x* does not appear explicitly in the ODE). Using the change of variables suggested in Section 3.3.6 of Chau (2018), we assume

$$p = \frac{dy}{dx} \tag{5.56}$$

Using this new variable, we can rewrite (5.55) as

$$\frac{dp}{\sqrt{1+p^2}} = \frac{\mu}{T_0} dx \tag{5.57}$$

This is separable and can be integrated as (see Formula of 14.182 of Spiegel, 1968)

$$\ln(p + \sqrt{1+p^2}) = \frac{\mu}{T_0} x + C \tag{5.58}$$

Taking an exponential function on both sides, we have

$$p + \sqrt{1+p^2} = e^{\mu x/T_0 + C} \tag{5.59}$$

Rearranging and squaring, we get

$$1 + p^2 = (e^{\mu x/T_0 + C} - p)^2 = e^{2(\mu x/T_0 + C)} - 2pe^{\mu x/T_0 + C} + p^2 \tag{5.60}$$

Solving for p and substituting the result into its definition given in (5.56), we finally obtain:

$$\begin{aligned} \frac{dy}{dx} &= p = (e^{\mu x/T_0 + C} - e^{-\mu x/T_0 - C})/2 \\ &= \sinh(\mu x / T_0 + C) \end{aligned} \tag{5.61}$$

Now, we can set the origin of the coordinate at the lowest point of the catenary as shown in Figure 5.6. The slope at the origin must be zero or

$$y'(0) = 0 \tag{5.62}$$

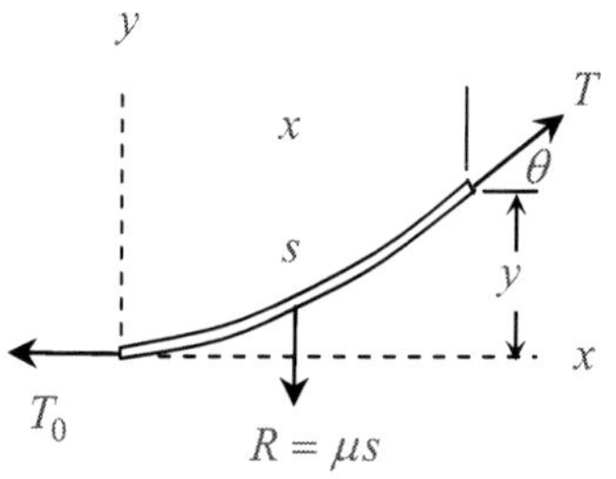

Figure 5.6 Force equilibrium for a suspending cable

Substitution of (5.62) into (5.61) gives

$$C = 0 \tag{5.63}$$

Integrating (5.61), we get

$$y = \frac{T_0}{\mu} \cosh(\frac{\mu x}{T_0}) + K \tag{5.64}$$

where K is an unknown constant. At the origin shown in Figure 5.6, the deflection is zero or

$$y(0) = 0 \tag{5.65}$$

Applying this boundary condition to (5.64), we find the unknown constant K as

$$K = -\frac{T_0}{\mu} \tag{5.66}$$

Finally, we find the final catenary

$$y = \frac{T_0}{\mu}[\cosh(\frac{\mu x}{T_0}) - 1] \tag{5.67}$$

As expected, the catenary is in the form of a hyperbolic cosine. To find the length of the deflected catenary s, we first recall (5.50)

$$\frac{dy}{dx} = \frac{\mu}{T_0} s + C_1 \tag{5.68}$$

Substituting (5.61) in the left-hand side gives

$$s = \frac{T_0}{\mu}[\sinh(\frac{\mu x}{T_0}) - C_1] \tag{5.69}$$

The constant must be zero due to the following boundary condition for s:

$$s(0) = 0 \tag{5.70}$$

Consequently, the length of the chain or cable is obtained as

$$s = \frac{T_0}{\mu} \sinh(\frac{\mu x}{T_0}) \tag{5.71}$$

The tension at any location of the catenary can be found using the force equilibrium in Figure 5.6. The three forces shown in Figure 5.6 must form a right-angle triangle and, thus, we have

$$T^2 = \mu^2 s^2 + T_0^2 \tag{5.72}$$

Substitution of (5.71) into (5.72) results in

$$T^2 = T_0^2 (1 + \sinh^2 \frac{\mu x}{T_0}) = T_0^2 \cosh^2 (\frac{\mu x}{T_0}) \tag{5.73}$$

Taking square root, the tension is a hyperbolic cosine function of x:

$$T = T_0 \cosh(\frac{\mu x}{T_0}) \tag{5.74}$$

However, (5.67) can be rearranged as

$$\mu y + T_0 = T_0 \cosh(\frac{\mu x}{T_0}) \tag{5.75}$$

Equating (5.74) and (5.75), we find another expression for the tension in terms of y instead of x:

$$T = T_0 + \mu y \tag{5.76}$$

This completes the discussion of the catenary.

5.4 INVERTED CATENARY AND ARCH

The Gateway Arch in St. Louis is now considered as an example. The arch is 192 m high, and the shape is an inverted catenary:

$$y = A(\cosh \frac{Cx}{L} - 1) \tag{5.77}$$

where A = 20.96 m and C = 3.0022 (see Figure 5.1). The Taq-i Kisra in Cresiphon, Iraq, the world's tallest barrel vault of 37 m, is in the shape of an inverted catenary. The hallway of the Casa Mila in Barcelona, Spain, was designed by Spanish artist Antoni Gaudi and is also in the shape of an inverted catenary. The use of the inverted catenary as a stone arch has long been recognized in the scientific community. For example, Robert Hooke in 1675 wrote "As hangs the flexible chain so, inverted, stand the touching pieces of an arch."

5.5 STONE ARCHES

5.5.1 Formulation of Stone Arches

The optimum shape of a stone arch bridge was the subject of research for centuries and Hooke in 1675 stated that the shape of an arch is the inverted shape of a hanging flexible cable. It was known that the line of thrust must pass through the mid-depth of the ring of stones of the arch. However, the complete mathematical solution of an arch bridge with a level ground was found by Inglis in 1863. The governing equation for the shape of the arch and its solution will be considered next. Review of stone arches can be found in Heyman (1999).

Let us consider a shallow stone arch (or called voussoirs, meaning "ring of stones") with the rise of the arch being d, the overburden at the mid-span being c, and the span of the arch being l (see Figure 5.7). The horizontal force of the thrust in the arch is denoted as H, which is a constant along the span l. Vertical force equilibrium requires

$$(T+\frac{dT}{dx}dx)\sin(\theta+\frac{d\theta}{dx}dx)-T\sin\theta=qdx \tag{5.78}$$

The intensity of the loading per unit length can be expressed as

$$q=-\gamma(c+d-z) \tag{5.79}$$

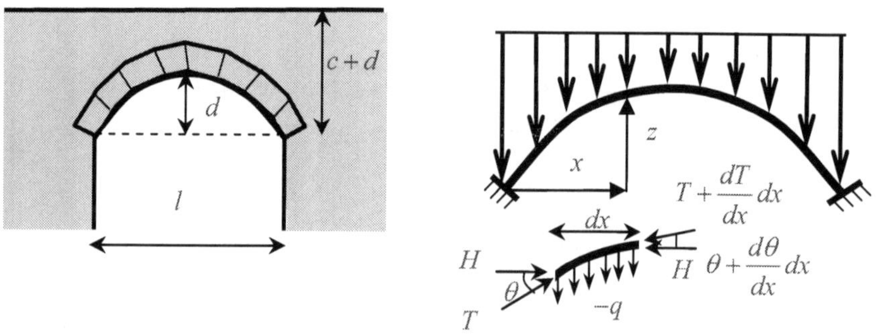

Figure 5.7 Stone arch bridge and the force equilibrium of an arch element

Since we assume that the arch is shallow, and thus the inclination θ is small such that we can approximate

$$\sin\theta \approx \theta, \quad \cos\theta \approx 1 \tag{5.80}$$

The vertical force equilibrium gives

$$\frac{dT}{dx}dx\,\theta + T\frac{d\theta}{dx}dx = qdx \tag{5.81}$$

Dividing through by *dx*, we find

$$\frac{dT}{dx}\theta + T\frac{d\theta}{dx} = \frac{d(T\theta)}{dx} = q \tag{5.82}$$

Replacing θ by the slope, we obtain

$$\frac{d}{dx}(T\frac{dz}{dx}) = q \tag{5.83}$$

The horizontal force equilibrium gives

$$T\cos\theta = H \tag{5.84}$$

For shallow arches, we can assume

$$T \approx H, \quad \theta \to 0 \tag{5.85}$$

On the other hand, there is no applied horizontal force within the arch element, and so we must have

$$\frac{dH}{dx} = 0 \tag{5.86}$$

In other words, the horizontal component of the arch thrust T must be constant. Substitution of (5.79) and (5.85) into (5.83) gives the governing equation for z

$$H\frac{d^2z}{dx^2} = -\gamma(c+d-z) \tag{5.87}$$

This is a linear second order nonhomogeneous ODE. We will consider in the next section the solution of this differential equation with appropriate boundary conditions.

5.5.2 Inglis Solution

We first rearrange (5.87) as

$$H\frac{d^2z}{dx^2} - \gamma z = -\gamma(c+d) \tag{5.88}$$

Dividing through by the constant thrust *H*, we have the standard form of (2.13) of Chau (2018) with a nonhomogeneous term:

$$\frac{d^2z}{dx^2} - kz = -k(c+d), \quad k = \frac{\gamma}{H} \tag{5.89}$$

Referring to Figure 5.7, we have the boundary conditions at the fixed support and zero slope at the top of the arch as

$$z(0) = \frac{dz}{dx}(\frac{l}{2}) = 0 \tag{5.90}$$

The homogeneous solution of (5.89) is given in (2.15) of Chau (2018) as

$$z_h = A\sinh\sqrt{k}x + B\cosh\sqrt{k}x \tag{5.91}$$

It is straightforward to see that particular solution is

$$z_p = c + d \tag{5.92}$$

Adding (5.91) and (5.92), we obtain the general solution as:

$$z = z_h + z_p = A\sinh\sqrt{k}x + B\cosh\sqrt{k}x + c + d \tag{5.93}$$

The first boundary condition in (5.90) gives

$$z(0) = B + c + d = 0 \tag{5.94}$$

Thus, we find B as

$$B = -(c+d) \tag{5.95}$$

Differentiating (5.93), we get

$$\frac{dz}{dx} = A\sqrt{k}\cosh(\sqrt{k}x) + B\sqrt{k}\sinh(\sqrt{k}x) \tag{5.96}$$

The second boundary condition in (5.90) leads to

$$A\cosh(\sqrt{k}\frac{l}{2}) + B\sinh(\sqrt{k}\frac{l}{2}) = 0 \tag{5.97}$$

Using the value of B obtained in (5.95) and solving for A, we obtain

$$A = (c+d)\tanh(\sqrt{k}\frac{l}{2}) \tag{5.98}$$

Back-substituting these constants into (5.93) yields

$$z = (c+d)\left\{\tanh(\sqrt{k}\frac{l}{2})\sinh(\sqrt{k}x) - \cosh(\sqrt{k}x) + 1\right\} \tag{5.99}$$

Although γ is a constant defined in the second equation of (5.89), the value of k still depends on the arch thrust H, which is still an unknown. To find H, we use the following condition

$$z(\frac{l}{2}) = d \tag{5.100}$$

Substituting (5.100) into (5.99), we find

$$(c+d)\left\{\tanh(\sqrt{k}\frac{l}{2})\sinh(\sqrt{k}\frac{l}{2}) - \cosh(\sqrt{k}\frac{l}{2}) + 1\right\} = d \tag{5.101}$$

This is the solution of the shape of the stone arch bridge. However, recall that H is still an unknown. To find H, we apply another condition at mid-span

$$(c+d)\sinh^2(\sqrt{k}\frac{l}{2}) = (c+d)\cosh^2(\sqrt{k}\frac{l}{2}) - c\cosh(\sqrt{k}\frac{l}{2}) \tag{5.102}$$

Recalling (1.226) of Chau (2018), we have

$$\cosh(\sqrt{k}\frac{l}{2}) = \frac{c+d}{c} \tag{5.103}$$

Taking arc cosh, we get

$$\sqrt{k}\frac{l}{2} = \sqrt{\frac{\gamma}{H}}\frac{l}{2} = \cosh^{-1}(\frac{c+d}{c}) \tag{5.104}$$

Squaring and rearranging, we obtain the arch thrust as a function of the span, unit weight, the rise of the arch, and the thickness of the overburden:

$$H = \frac{\gamma l^2}{4} \frac{1}{[\cosh^{-1}(\frac{c+d}{c})]^2} \tag{5.105}$$

We first rearrange (5.99) as

$$\begin{aligned} z &= (c+d) - (c+d)\left\{\cosh(\sqrt{k}x) - \tanh(\sqrt{k}\frac{l}{2})\sinh(\sqrt{k}x)\right\} \\ &= c+d) - \frac{(c+d)}{\cosh(\sqrt{k}\frac{l}{2})}\left\{\cosh(\sqrt{k}\frac{l}{2})\cosh(\sqrt{k}x) - \sinh(\sqrt{k}\frac{l}{2})\sinh(\sqrt{k}x)\right\} \end{aligned} \tag{5.106}$$

The sum rule of hyperbolic cosine given in (1.228) of Chau (2018) leads to

$$\cosh[\sqrt{k}\frac{l}{2} - \sqrt{k}x] = \cosh(\sqrt{k}\frac{l}{2})\cosh(\sqrt{k}x) - \sinh(\sqrt{k}\frac{l}{2})\sinh(\sqrt{k}x) \tag{5.107}$$

With this formula, (5.106) reduces to

$$\begin{aligned} z &= c + d - c\cosh(\sqrt{k}\frac{l}{2} - \sqrt{k}x) = d + c\{1 - \cosh(\sqrt{k}\frac{l}{2} - \sqrt{k}x)\} \\ &= d + c\{1 - \cosh[\sqrt{k}\frac{l}{2}(1 - \frac{2x}{l})]\} \end{aligned} \tag{5.108}$$

Substituting (5.104) into (5.108), we finally obtain the Inglis solution as

$$z = d + c\{1 - \cosh[(1 - \frac{2x}{l})\cosh^{-1}(\frac{c+d}{c})]\} \tag{5.109}$$

The solution is no longer a function of H, and this is the so-called Inglis solution derived by Inglis in 1863. This solution was given in Irvin (1981) without proof. This is clear that the stone arch bridge is in the shape of an inverted catenary.

5.6 CABLE SUSPENSION BRIDGE

In this section, we will derive the cable force and length of the cable in a cable suspension bridge as a function of the weight per length of the deck, the span and the sag. Figure 5.8 shows a cable suspension bridge with towers of unequal heights, together with a free body diagram for a cable element. Note that the main difference between the cable suspension bridge and the catenary considered in Figures 5.5 and 5.6 lies on the vertical load calculation (μ per length in catenary versus w per horizontal distance in suspension bridge). In fact, in the formulation of a suspension bridge, the weight of cable is negligible when comparing the weight of the bridge deck and its live loads. Thus, the vertical load on the cable element is proportional to the horizontal distance instead of the length of the cable. Thus, (5.36) can be modified as

$$(T + dT)\sin(\theta + d\theta) = T\sin\theta + wdx \tag{5.110}$$

Neglecting the second order term and assuming a small value of $d\theta$, we have

$$T\cos\theta d\theta + dT\sin\theta = wdx \tag{5.111}$$

The left-hand side can be written as a total differential as

$$d(T\sin\theta) = wdx \tag{5.112}$$

The horizontal force equilibrium is exactly the same as that for the catenary, and, thus, we have

$$T\cos\theta = C = T_0 \tag{5.113}$$

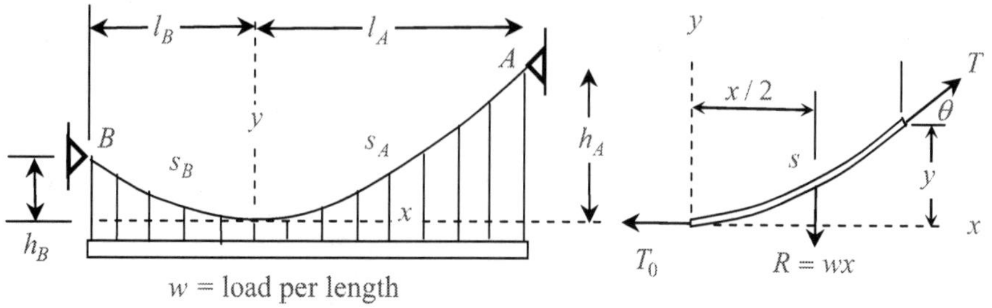

Figure 5.8 Force equilibrium for cable suspension bridge

Substitution of (5.113) into (5.112) gives

$$\frac{d}{dx}(T_0\frac{\sin\theta}{\cos\theta}) = T_0\frac{d}{dx}(\tan\theta) = T_0\frac{d^2y}{dx^2} = w \tag{5.114}$$

The last of (5.114) can be rewritten as

$$\frac{d^2y}{dx^2} = \frac{w}{T_0} \tag{5.115}$$

Integrating twice, we get

$$y = \frac{wx^2}{2T_0} + Cx + D \tag{5.116}$$

The deflection and slope at the origin are zeros:

$$y(0) = 0,\ y'(0) = 0 \tag{5.117}$$

Application of (5.117) to (5.116) gives both integration as zeros, and the deflection then becomes

$$y = \frac{wx^2}{2T_0} \tag{5.118}$$

This is an equation of a parabola. To find the tension at the lowest point of the cable, we substitute the following coordinate at the right support

$$x = l_A,\quad y = h_A \tag{5.119}$$

Consequently, we find

$$T_0 = \frac{wl_A^2}{2h_A} \tag{5.120}$$

From the geometry of the cable and loading, we can find the tension in the cable regardless of the mechanical properties of the cable. Substitution of (5.120) into (5.118) gives

$$y = h_A(\frac{x}{l_A})^2 \tag{5.121}$$

Referring to the right diagram in Figure 5.8, we find the tension T as

$$T = \sqrt{T_0^2 + w^2x^2} \tag{5.122}$$

Evidently, the maximum tension is at the support and is given by (5.122) and (5.120) as

$$T_{\max} = wl_A\sqrt{1+(\frac{l_A}{2h_A})^2} \tag{5.123}$$

The length of the cable can be found by integrating (5.54) as

$$\int_0^{s_A} ds = \int_0^{l_A} \sqrt{1+(dy/dx)^2}\,dx = \int_0^{l_A} \sqrt{1+(wx/T_0)^2}\,dx \tag{5.124}$$

Recalling the binomial theorem from (1.4) of Chau (2018), we note

$$(1+x)^n = 1 + nx + \frac{n(n-1)}{2!}x^2 + \frac{n(n-1)(n-2)}{3!}x^3 + \ldots \tag{5.125}$$

Expanding the square root term in (5.124) by employing (5.125), we have the following approximation

$$s_A = \int_0^{l_A} (1 + \frac{w^2x^2}{2T_0^2} - \frac{w^4x^4}{8T_0^4} + \ldots)dx = l_A\left[1 + \frac{2}{3}(\frac{h_A}{l_A})^2 - \frac{2}{5}(\frac{h_A}{l_A})^4 + \ldots\right] \tag{5.126}$$

Alternatively, we can integrate (5.124) exactly as

$$s_A = \frac{L_a\sqrt{(wL_a/T_0)^2+1}}{2} + \frac{1}{2}(\frac{T_0}{w})\ln[L_a + \sqrt{L_a^2 + (T_0/w)^2}\,] - \frac{1}{2}(\frac{T_0}{w})\ln(\frac{T_0}{w}) \tag{5.127}$$

For the cable on the left, we can find the tension, profile and cable length by making the following substitution

$$x = l_B, \quad y = h_B \tag{5.128}$$

In reality, the two towers are always equal in height, as shown in Figure 5.9. For this case, we can define the span of the bridge and length of the cable as

$$L = 2l_A, \quad S = 2s_A \tag{5.129}$$

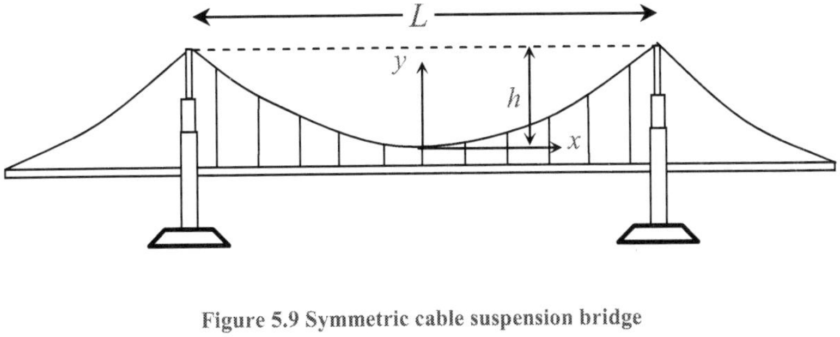

Figure 5.9 Symmetric cable suspension bridge

Using (5.129), we have the tension and cable length for a symmetric suspension bridge being

$$T_0 = \frac{wL^2}{8h} \tag{5.130}$$

$$T_{max} = \frac{wL}{2}\sqrt{1+(\frac{L}{4h})^2} \tag{5.131}$$

$$S = L\left[1+\frac{8}{3}(\frac{h}{L})^2 - \frac{32}{5}(\frac{h}{L})^4 + ...\right] \tag{5.132}$$

These formulas are illustrated for the Tsing Ma Bridge in the following example.

Example 5.1 Tsing Ma Bridge is a cable suspension bridge with the longest main span in the world that carries both motorway and railway (see Figure 5.2). The main span is 1377 m, the sag is 112 m, and the dead weight is 272.52 kN/m. Find the horizontal component of the cable force, maximum cable force, and length of the cable.

Hints: There are two main cables carrying the deck.

Solution: For the horizontal component of the tension, (5.120) can be used:

$$T_0 = (\frac{1}{2})\frac{wl_A^2}{2h_A} = \frac{272.52\times 688.5^2}{2\times 2\times 112} = 288.35MN \tag{5.133}$$

Since the loading is taken by two main cables, we have divided the prediction by two in (5.133). The maximum tension in the cable is half of that of (5.131)

$$\begin{aligned} T_{max} &= (\frac{1}{2})\frac{wL}{2}\sqrt{1+(\frac{L}{4h})^2} = \frac{(272.52)(1377)}{(2)(2)}\sqrt{1+(\frac{1377}{4\times 112})^2} \\ &= 303.23MN \end{aligned} \tag{5.134}$$

Finally, (5.132) can be used to find the total length of the main cable between the two towers

$$\begin{aligned} S &= L\left[1+\frac{8}{3}(\frac{h}{L})^2 - \frac{32}{5}(\frac{h}{L})^4 + ...\right] \\ &= (1377)[1+\frac{8}{3}(\frac{112}{1377})^2 - \frac{32}{5}(\frac{112}{1377})^4 + ...] \\ &= (1377)[1+0.0176415-0.00028+...] \\ &= 1400.9m \end{aligned} \tag{5.135}$$

5.7 CABLE-STAY BRIDGE

The earliest construction of cable-stay bridges can be traced back to the primitive bridges built more than 2000 years ago in Borneo and Laos, and they were supported by inclined vines attached to trees from both sides of the crossing. The first ever cable-stay system similar to the modern design (including tower, deck and stay) is probably the wooden bridges designed and built by German carpenter C.J. Loscher in 1784. In 1784-1824, a number of cable-stay bridges were built in Europe. Two notable collapses of cable-stay bridges were the 79-m pedestrian Tweed River Bridge near Dryburg-Abbey England in 1818 and the 78-m Saale River Bridge in Nienburg Germany in 1824 (overloaded due to excessive crowd joining a river festival). After examining failures of both types of bridges, C.L. Navier in 1823 concluded in his famous report that a cable suspension bridge is preferable. After World War II, it was the German engineer, F. Dischinger, in 1949 who proposed building the first modern cable-stay bridge over River Rhine. However, the first modern cable-stay bridge was the Stromsund Bridge in Sweden completed in 1955. A series of cable-stay bridges were then completed in Germany, including the North Bridge in 1958, the Severin Bridge in 1960, and the Leverkusen Bridge in 1964.

Two different types of cable-stayed bridges are shown in Figures 5.10 and 5.11. In the first type, the weight of any segment of the bridge deck is carried by two cable systems, one connected to the left tower and the other connected to the right tower. In the second type, only one cable system is used for carrying each segment of bridge deck. The segments on the left of the mid-span are carried by cables connected to the left tower and those on the right of the mid-span are supported by the right tower. Let the span of the bridge be l, the height of towers above the bridge deck is d, and the uniform dead weight of the deck per unit length is mg, where m is the mass of deck per length.

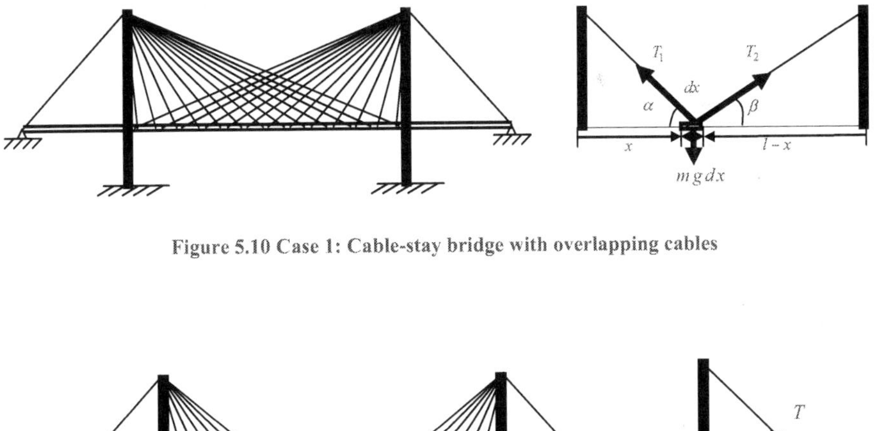

Figure 5.10 Case 1: Cable-stay bridge with overlapping cables

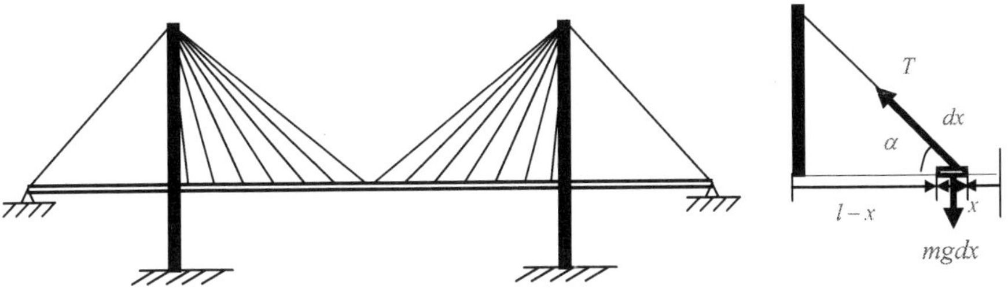

Figure 5.11 Case 2: Cable-stay bridge with non-overlapping cables

For the first case of overlapping cables, the vertical force equilibrium shown in Figure 5.10 gives

$$T_1 \sin\alpha + T_2 \sin\beta = mgdx \tag{5.136}$$

where angles α and β are shown in Figure 5.10. Horizontal force equilibrium requires

$$T_1 \cos\alpha = T_2 \cos\beta \tag{5.137}$$

This can be rewritten as

$$T_1 = T_2 \frac{\cos\beta}{\cos\alpha} \tag{5.138}$$

Substitution of (5.138) into (5.136) gives

$$T_2 = \frac{mgdx}{(\tan\alpha\cos\beta + \sin\beta)} \tag{5.139}$$

The total horizontal force that needs to be taken up by the right tower backstay can be found by integration as

$$H_1 = \int_0^l T_2 \cos\beta dx \tag{5.140}$$

Substitution of (5.139) into (5.140) gives

$$\begin{aligned} H_1 &= \int_0^l T_2 \cos\beta dx = mg\int_0^l \frac{dx}{(\tan\alpha + \tan\beta)} = mg\int_0^l \frac{dx}{[d/x + d/(l-x)]} \\ &= \frac{mg}{ld}\int_0^l x(l-x)dx = \frac{mgl^2}{6d} \end{aligned} \tag{5.141}$$

By symmetry, the total horizontal force of the backstay on the left of the left tower must also be given by the same result. The total cable force of the backstay depends on the inclination of the cable. A suitable size of anchor block must be provided to balance the cable force from the mid-span. In addition, since the horizontal force of the left and right cables of each deck element is in equilibrium, there is no initial compressive force on the deck section.

For the case of the non-overlapping cable-stay bridge shown in Figure 5.11, the weight of each deck segment is carried by one set of cables. The vertical force equilibrium on the free-body diagram requires

$$T \sin\alpha = mgdx \tag{5.142}$$

Or, the cable force is

$$T = \frac{mg}{\sin\alpha} dx \tag{5.143}$$

The total horizontal force of the backstay can be integrated as

$$\begin{aligned} H_2 &= \int_0^{l/2} T\cos\beta dx = \int_0^{l/2} \frac{mg}{\sin\beta}\cos\beta dx = \int_0^{l/2} mg\cot\beta dx \\ &= mg\int_0^{l/2} \left(\frac{l}{2d} - \frac{x}{d}\right)dx = \frac{mgl^2}{8d} \end{aligned} \tag{5.144}$$

The horizontal force component will induce compressive force on the deck of the bridge. Therefore, there is an initial compressive force for the case 2 configuration of the cable-stay bridge shown in Figure 5.11. This initial compression attains a minimum at the mid-span and approaches a maximum towards the tower. Special consideration has to be made to ensure that unwanted buckling of the bridge deck section will not occur.

Let the total cross-section area of the wire strands in the backstay for case 1 shown in Figure 5.10 be A_1 and that for case 2 shown in Figure 5.10 be A_2. Then, for the same working stress for the backstay cable, the size of the strands for the cable-stay of layout 1 and layout 2 are

$$H_1 = A_1\sigma = \frac{mgl^2}{6d}, \quad H_2 = A_2\sigma = \frac{mgl^2}{8d} \tag{5.145}$$

where σ is the working tensile stress in the cable. Thus, the ratio of the size of the cable in the backstay for layout 1 to layout 2 is

$$\frac{A_1}{A_2} = \frac{8}{6} = \frac{4}{3} \tag{5.146}$$

Thus, it is clear that layout 2 is more efficient, but care needs to be taken in preventing buckling due to the initial compressive stress. Apparently, M.C. Tang was the first to consider the buckling of a cable-stay bridge.

5.8 VIBRATIONS OF CABLE SUSPENSION BRIDGE

So far, we have only considered the static force equilibrium in calculating the tension in the cables. As summarized in the Introduction, wind-induced vibrations are the main reason for suspension bridge collapse, such as the failure of the Tacoma Narrows Bridge. In this section, we will consider the vibration frequency of cable suspension bridges.

5.8.1 Governing Equations for Flexible Deck

Consider the case that the deck of the cable suspension bridge is quite flexible, such as the case of the Tacoma Narrows Bridge. That is, the bending stiffness can be neglected. Let us shift the origin of the coordinate up to the top of the tower and positive y pointing downward (see Figure 5.12). The main span is L, the side spans are L_1, the sag is h, and the weight per unit length of the deck is w. For this coordinate system, we can rewrite the results for static equilibrium as

$$y = h[1 - (\frac{2x}{L})^2] \tag{5.147}$$

$$\frac{dy}{dx} = -\frac{wx}{T_0} = -\frac{8hx}{L^2} \tag{5.148}$$

$$\frac{d^2y}{dx^2} = -\frac{w}{T_0} = -\frac{8h}{L^2} \tag{5.149}$$

The horizontal component of cable force can be obtained from (5.149):

$$T_0 = \frac{wL^2}{8h} \tag{5.150}$$

Now, suppose that some time-dependent excitations are applied to the bridge, such as wind-induced excitations. The bridge will then undergo vibrations. The horizontal component of cable tension will vary and may be expressed as $T_0+\Delta T_0$, where the additional horizontal tension force ΔT_0 is a function of time and depends on the nature of excitations. As long as the horizontal inertia can be neglected, this additional horizontal tension cannot be a function of x. The additional deflection due to the vibrations is expressed as η with respect to the cable profile, and, thus, the instantaneous position of the cable is $y+\eta$ (see Figure 5.13). The equilibrium equation given in (5.149) can be revised as

$$(T_0 + \Delta T_0)(\frac{d^2y}{dx^2} + \frac{\partial^2\eta}{\partial x^2}) = -w + \frac{w}{g}\frac{\partial^2\eta}{\partial t^2} \tag{5.151}$$

The last term is the inertia force due to the time-dependent deflection η. Applying the static equilibrium given in (5.149) to (5.151), we have the equation of motion as

$$\Delta T_0 \frac{d^2y}{dx^2} + (T_0 + \Delta T_0)\frac{\partial^2\eta}{\partial x^2} = m\frac{\partial^2\eta}{\partial t^2} \tag{5.152}$$

where m is the mass of the deck per unit length. This is a nonlinear PDE since we expect that ΔT_0 is a function of η as the additional tension in cable must result from the vibrations.

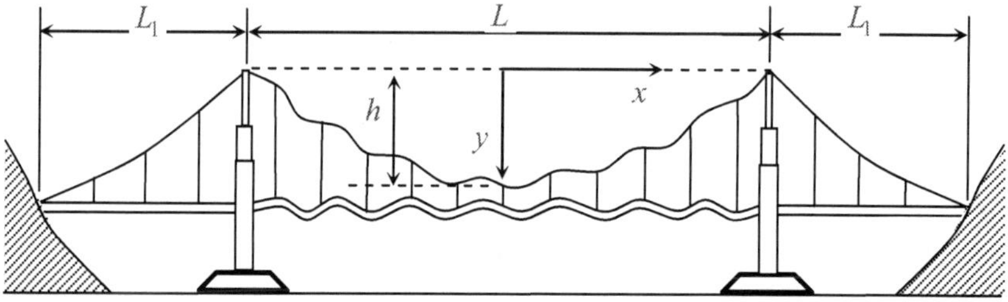

Figure 5.12 Anti-symmetric mode vibrations in a cable suspension bridge

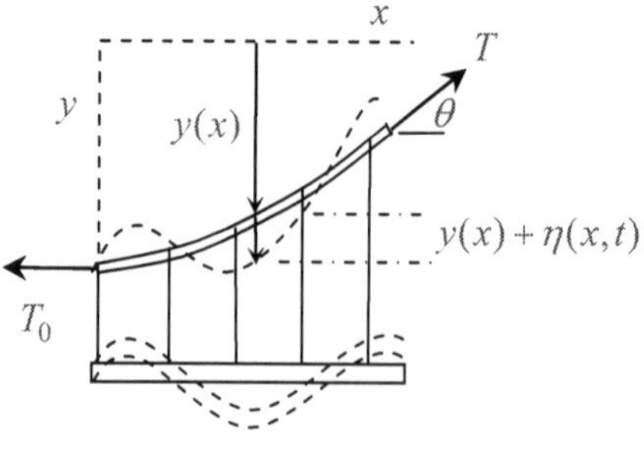

Figure 5.13 The definition of vibrations η measured from the cable profile y

In reality, the vibration amplitude η is expected to be much smaller than y and its derivative is also much smaller than the derivative of y. Thus, we assume

$$\eta \ll y, \quad \left|\frac{\partial^2 \eta}{\partial x^2}\right| \ll \left|\frac{d^2 y}{dx^2}\right| = \frac{8h}{L^2} \tag{5.153}$$

This assumption can linearize the PDE given in (5.152) and simplify the mathematical analysis. For vibrations of higher mode, the second condition given in (5.153) may not be valid. To see this, we can assume a deflection mode as

$$\eta = \eta_0 \sin(2n\pi \frac{x}{L}) f(t) \tag{5.154}$$

Differentiating twice with respect to x, we have

$$\frac{\partial^2 \eta}{\partial x^2} = -\frac{4n^2\pi^2}{L^2}\eta \tag{5.155}$$

Substitution of (5.155) into the second condition given in (5.153) results in

$$\eta \ll \frac{2h}{\pi^2 n^2} \tag{5.156}$$

It is clear that for large n, (5.156) may be violated. For the case of Tsing Ma Bridge and with nine nodes between the two towers or ten half sine waves on the span (i.e., $n = 5$), we have

$$\eta \ll \frac{2h}{\pi^2 n^2} = \frac{2(112.48)}{\pi^2 (5)^2} = 0.9117m \tag{5.157}$$

Therefore, the vibration amplitude in the bridge has to be much less than 0.9 m. Otherwise, the assumption given in (5.153) is invalidated, and all subsequent results derived in this section are not valid. For the case of the Tacoma Narrows Bridge with nine nodes (indeed the bridge underwent an eight-node oscillation on the main span before it turned into a single wave and finally collapsed on November 7, 1940):

$$\eta \ll \frac{2h}{\pi^2 n^2} = \frac{2(70.7)}{\pi^2 (5)^2} = 0.573m \tag{5.158}$$

From the 16-mm film taken before the collapse of the Tacoma Narrows Bridge, the amplitude of vibrations clearly exceeds 0.573 m before failure. Thus, a fully nonlinear mathematical model at the onset of failure is still required. This is, however, out of the scope of the present chapter. We will see later that linear theory yields surprisingly accurate result on the vibration frequency.

Nevertheless, assuming the linear theory is valid, we expect, for a small amplitude of oscillations, the additional horizontal cable tension is proportional to η as a first approximation:

$$\Delta T_0 \sim \eta \tag{5.159}$$

Therefore the following term appearing in (5.152) is much smaller than the other terms

$$\Delta T_0 \frac{\partial^2 \eta}{\partial x^2} << \frac{\partial^2 \eta}{\partial x^2} \sim O(\eta) \tag{5.160}$$

Applying (5.160), we can linearize (5.152) as

$$\Delta T_0 \frac{d^2 y}{dx^2} + T_0 \frac{\partial^2 \eta}{\partial x^2} = m \frac{\partial^2 \eta}{\partial t^2} \tag{5.161}$$

Substitution of (5.149) and (5.150) from the static solution, we have the following equation of motion:

$$m \frac{\partial^2 \eta}{\partial t^2} - \frac{mgL^2}{8h} \frac{\partial^2 \eta}{\partial x^2} + \frac{8h}{L^2} \Delta T_0 = 0 \tag{5.162}$$

Assuming a sinusoidal oscillation in time, we have

$$\eta(x,t) = \bar{\eta}(x) \sin \omega t \tag{5.163}$$

$$\Delta T_0 = H \sin \omega t \tag{5.164}$$

where H is an unknown constant to be determined. We have tacitly assumed that the additional cable tension is in phase with the vibration amplitude. Substitution of (5.163) and (5.164) into (5.162) yields

$$-m\omega^2 \bar{\eta} - \frac{mgL^2}{8h} \frac{d^2 \bar{\eta}}{dx^2} + \frac{8hH}{L^2} = 0 \tag{5.165}$$

This can be written as

$$\frac{d^2 \bar{\eta}}{dx^2} + (\frac{2\mu}{L})^2 \bar{\eta} = \frac{8Hh}{T_0 L^2} = \frac{64Hh^2}{wL^4} \tag{5.166}$$

where

$$\mu = \omega \sqrt{\frac{2h}{g}} \tag{5.167}$$

To determine the natural frequency of the oscillations, we need to consider the global condition of the entire bridge which is constrained by the anchoring condition at the side spans. In particular, referring to Figure 5.12 the cable is anchored at the ends of the side spans as

$$\eta = 0, \quad x = \pm(\frac{L}{2} + L_1) \tag{5.168}$$

Let us assume the towers are rigid and the cable is allowed to slide on the top of the towers, in a way that the horizontal cable tension is the same on both sides of the tower. The total change in length of the cable due to the vibrations must be zero if the cable is inextensible, or must be compatible with the elastic elongation of the main cable.

Case 1: Inextensible cable

Let us consider an element of length ds that is defined as

$$ds^2 = dx^2 + dy^2 \tag{5.169}$$

Assuming that the horizontal and vertical deflections at any point of the cable can be expressed as ξ and η, we can take the variation of (5.169) as

$$ds\delta ds = dxd\xi + dyd\eta \tag{5.170}$$

The condition of cable inextensibility (setting $\delta ds = 0$) leads to

$$d\xi = -\frac{dy}{dx} d\eta \tag{5.171}$$

Since the cable is anchored with a fixed total length, we require

$$\int d\xi = 0 \tag{5.172}$$

Substitution of (5.171) into (5.172) and integration by parts gives

$$\int d\xi = -\int \frac{dy}{dx}\frac{d\eta}{dx}dx = -\left[\eta \frac{dy}{dx}\right]_{-L/2-L_2}^{L/2+L_1} + \int \eta \frac{d^2y}{dx^2}dx = 0 \tag{5.173}$$

In view of (5.168) and (5.149), we must have the following additional constraint on η

$$\int_{-L/2-L_2}^{L/2+L_1} \eta dx = 0 \tag{5.174}$$

This constraint will be used to determine the frequency of vibrations.

Case 2: Extensible cable

For the case of extensible cable (set $\delta ds \neq 0$), we can apply Hooke's law as

$$\Delta T = EA\frac{\delta ds}{ds} \tag{5.175}$$

where E is the Young's modulus of the cable, A is the cross-section area of the cable, and ΔT is the tension induced by cable extension. The horizontal component of this additional tension must, of course, equal H assumed in (5.164), resulting in

$$\Delta T = \frac{H}{\cos\theta} = H\frac{ds}{dx} \tag{5.176}$$

Substitution of (5.176) into (5.175) gives

$$\delta ds = \frac{H}{EA}(\frac{ds}{dx})^2 dx \tag{5.177}$$

The horizontal deflection can be determined from (5.170) as

$$d\xi = \frac{ds}{dx}\delta ds - \frac{dy}{dx}d\eta = \frac{H}{EA}(\frac{ds}{dx})^3 dx - \frac{dy}{dx}\frac{d\eta}{dx}dx \tag{5.178}$$

For a fixed horizontal distance between the anchorages, we have

$$\int_{-L/2-L_2}^{L/2+L_1} d\xi = \frac{H}{EA}\int_{-L/2-L_2}^{L/2+L_1}(\frac{ds}{dx})^3 dx - \int_{-L/2-L_2}^{L/2+L_1}\frac{dy}{dx}\frac{d\eta}{dx}dx = 0 \tag{5.179}$$

Thus, we obtain the following compatibility condition

$$\frac{H}{EA}\int_{-L/2-L_2}^{L/2+L_1}(\frac{ds}{dx})^3 dx = \frac{HL_s}{EA} = \frac{8h}{L^2}\int_{-L/2-L_2}^{L/2+L_1}\eta dx \tag{5.180}$$

where L_s has been used to denote the result of the first integral in (5.180). The last term in (5.180) results from integration by parts similar to that in deriving (5.173).

5.8.2 Symmetric Modes

When the vibration mode is symmetric with respect to $x = 0$, we have

$$\eta(x) = \eta(-x) \tag{5.181}$$

The homogeneous solution for (5.166) is

$$\bar{\eta}_h = A\cos(\frac{2\mu x}{L}) + B\sin(\frac{2\mu x}{L}) \tag{5.182}$$

The particular solution for (5.166) is

$$\bar{\eta}_p = \frac{2Hh}{T_0\mu^2} \tag{5.183}$$

The general solution is the sum of the homogeneous solution and the particular solution (see Section 3.3.2 of Chau, 2018)

$$\bar{\eta} = \bar{\eta}_h + \bar{\eta}_p = A\cos(\frac{2\mu x}{L}) + B\sin(\frac{2\mu x}{L}) + \frac{2Hh}{T_0\mu^2} \tag{5.184}$$

Due to symmetry, applying (5.184) to (5.181), we have

$$B = 0 \tag{5.185}$$

For the main span, there is no deflection at the top of the tower:

$$\bar{\eta} = 0, \quad x = \pm\frac{L}{2} \tag{5.186}$$

Applying boundary condition (5.186) to (5.184), we get

$$A = -\frac{1}{\cos\mu}\frac{2Hh}{T_0\mu^2} \tag{5.187}$$

For the right side span, we have the following boundary conditions:

$$\bar{\eta} = 0, \quad x = \frac{L}{2}, \frac{L}{2} + L_1 \tag{5.188}$$

In view of the boundary condition, we have to modify the cosine function as

$$\bar{\eta} = A\cos[\mu(1+\alpha-\frac{2x}{L})] + \frac{2Hh}{T_0\mu^2} \tag{5.189}$$

where

$$\alpha = \frac{L_1}{L} \tag{5.190}$$

Substitution of (5.189) into the first part of (5.188) gives

$$A = -\frac{1}{\cos\mu\alpha}\frac{2Hh}{T_0\mu^2} \tag{5.191}$$

Thus, we have the deflection of the right side span being

$$\bar{\eta} = \frac{2Hh}{T_0\mu^2}[1 - \frac{\cos\mu(1+\alpha-2x/L)}{\cos\alpha\mu}] \tag{5.192}$$

Following a similar procedure, we can see that the solution for the left side span is

$$\bar{\eta} = \frac{2Hh}{T_0\mu^2}[1 - \frac{\cos\mu(1+\alpha+2x/L)}{\cos\alpha\mu}] \tag{5.193}$$

In summary, the solutions for symmetric modes are divided into the following three segments:

$$\bar{\eta} = \frac{2Hh}{T_0\mu^2}[1 - \frac{\cos(2\mu x/L)}{\cos\mu}], \quad -\frac{L}{2} < x < \frac{L}{2} \tag{5.194}$$

$$\bar{\eta} = \frac{2Hh}{T_0\mu^2}[1 - \frac{\cos\mu(1+\alpha-2x/L)}{\cos\alpha\mu}], \quad \frac{L}{2} < x < \frac{L}{2} + L_1 \tag{5.195}$$

$$\bar{\eta} = \frac{2Hh}{T_0\mu^2}[1 - \frac{\cos\mu(1+\alpha+2x/L)}{\cos\alpha\mu}], \quad -\frac{L}{2} - L_1 < x < -\frac{L}{2} \tag{5.196}$$

For inextensible cable, we can substitute this solution into (5.174) to yield

$$\begin{aligned}\int_{-L/2-L_1}^{L/2+L_1} \bar{\eta}\, dx &= \int_{-L/2-L_1}^{-L/2} \bar{\eta}\, dx + \int_{-L/2}^{L/2} \bar{\eta}\, dx + \int_{L/2}^{L/2+L_1} \bar{\eta}\, dx = I_1 + I_2 + I_3 = 0 \\ &= \frac{2Hh}{T_0\mu^2}\int_{-L/2-L_1}^{-L/2} [1 - \frac{\cos\mu(1+\alpha-2x/L)}{\cos\alpha_2\mu}]dx \\ &+ \frac{2Hh}{T_0\mu^2}\int_{-L/2}^{L/2} [1 - \frac{\cos(2\mu x/L)}{\cos\mu}]\eta dx \\ &+ \frac{2Hh}{T_0\mu^2}\int_{L/2}^{L/2+L_1} [1 - \frac{\cos\mu(1+\alpha+2x/L)}{\cos\alpha_2\mu}]dx\end{aligned} \tag{5.197}$$

It is straightforward to integrate these integrals in (5.197):

$$\begin{aligned}I_1 &= \frac{2Hh}{T_0\mu^2}\int_{-L/2-L_1}^{-L/2} [1 - \frac{\cos\mu(1+\alpha+2x/L)}{\cos\alpha\mu}]dx \\ &= \frac{2Hh}{T_0\mu^2}\{L_1 - \frac{L}{2\mu\cos\alpha\mu}[\sin\mu(1+\alpha+\frac{2x}{L})]_{-L/2-L_1}^{-L/2}\} = \frac{2Hh}{T_0\mu^2}[L_1 - \frac{L\tan\alpha\mu}{\mu}]\end{aligned} \tag{5.198}$$

$$\begin{aligned}I_2 &= \frac{2Hh}{T_0\mu^2}\int_{-L/2}^{L/2} [1 - \frac{\cos(2\mu x/L)}{\cos\mu}]\eta dx = \frac{2Hh}{T_0\mu^2}\{L - \frac{L}{2\mu}\frac{1}{\cos\mu}[\sin(2\mu\frac{x}{L})]_{-L/2}^{L/2}\} \\ &= \frac{2Hh}{T_0\mu^2}[L - \frac{L}{\mu}\tan\mu]\end{aligned} \tag{5.199}$$

$$\begin{aligned}I_3 &= \frac{2Hh}{T_0\mu^2}\int_{L/2}^{L/2+L_1} [1 - \frac{\cos\mu(1+\alpha-2x/L)}{\cos\alpha\mu}]dx \\ &= \frac{2Hh}{T_0\mu^2}\{L_1 + \frac{L}{2\mu\cos\alpha\mu}[\sin\mu(1+\alpha-\frac{2x}{L})]_{L/2}^{L/2+L_1}\} = \frac{2Hh}{T_0\mu^2}[L_1 - \frac{L\tan\alpha\mu}{\mu}]\end{aligned} \tag{5.200}$$

Substitution of (5.198) to (5.200) to (5.197) gives

$$2\tan\alpha\mu + \tan\mu = (1+2\alpha)\mu \tag{5.201}$$

Recall from (5.167) that μ is a function of vibration frequency ω. This eigenvalue equation needs to be solved by numerical methods. Numerical results show that the roots are very close to

$$\alpha\mu \approx \frac{\pi}{2}(2n+1), \quad \mu \approx \frac{\pi}{2}(2n+1), \quad n = 0,1,2,... \tag{5.202}$$

Suppose that the side span is about 1/3 of the main span or α = 1/3 in (5.190) (this is a typical value for cable suspension bridge). We have the fundamental frequency of the cable

$$\omega \approx \frac{\pi}{2}\sqrt{\frac{g}{2h}} \tag{5.203}$$

Equation (5.203) shows that the natural frequency of the cable is independent of the weight of the bridge deck. The circular frequency for a pendulum of length h (the sag of the main cable in the suspension bridge) is (see Section 2.7 of Chau, 2018):

$$\omega_P = \frac{2\pi}{T} = \sqrt{\frac{g}{h}} \tag{5.204}$$

Therefore, the fundamental frequency of a cable suspension bridge is about 1.11 times the circular frequency of a pendulum of length h (the sag of the cable):

$$\omega \approx \frac{\pi}{2\sqrt{2}}\sqrt{\frac{g}{h}} = \frac{\pi}{2\sqrt{2}}\omega_p = 1.1107\omega_p \tag{5.205}$$

For the special case of α = 1/3, Table 5.1 compiles the first six fundamental frequencies for the bending modes of the vibrations of the Tacoma Narrows Bridge (h = 70.7 m). The lowest fundamental frequency is 0.4137 rad/s and this corresponds to half a wave in the main span (or there is no node observed in the main span). Here is an excerpt from Rocard (1957):

> "From 8 to 10 a.m., the (Tacoma Narrows) bridge was seen to vibrate *vertically in bending* with a moderate amplitude and a frequency of 36 oscillations per minute, 8 nodes being formed on the main span..." (p. 98 of Rocard, 1957)

Table 5.1 Fundamental frequencies of Tacoma Narrows Bridge in bending modes

$m =$	$n = (2m+1)/2$ wave number	$\mu_m \approx \pi(2m+1)/2$	$\omega_m = \mu_m \sqrt{(g/h)}$ (rad/s)
0	1/2	$\pi/2$	0.4137
1	3/2	$3\pi/2$	1.2412
2	5/2	$5\pi/2$	2.0687
3	7/2	$7\pi/2$	2.8962
4	9/2	$9\pi/2$	3.7237
5	11/2	$11\pi/2$	4.5512

If there are eight nodes in the span, there will be 4.5 complete waves. In fact, only four and one-half waves indeed reflects symmetric modes of vibration. For the main span, the deflection can be expressed as

$$\bar{\eta} = \frac{2Hh}{T_0\mu^2}[1 - \frac{\cos(2\mu\frac{x}{L})}{\cos\mu}] = \frac{2Hh}{T_0\mu^2}[1 - \frac{\cos(2n\pi\frac{x}{L})}{\cos\mu}] \tag{5.206}$$

According to Table 5.1, what we observed before the failure of the Tacoma Narrows Bridge corresponds to m = 4 or 4.5 complete waves, and the natural circular frequency is 3.7237 rad/s. This value agrees amazing well with the observed frequency:

$$\omega_{obs} = 2\pi f = \frac{2\times\pi\times 36}{60} = 3.7699 \text{ rad/s} \tag{5.207}$$

This corresponds to only 1.22% error, despite that nonlinearity should have been incorporated in the view of the magnitude of oscillations, which is in the order of 1.07 m (Farquharson, 1950). Nevertheless, this observed error gives credence to the present linear analysis. In the next section, we will consider the case of anti-symmetric vibration modes.

5.8.3 Anti-Symmetric Modes

By definition, the anti-symmetric modes are defined by

$$\eta(x) = -\eta(-x) \tag{5.208}$$

As shown in Figure 5.14, the suspension bridge is originally symmetric with respect to O, but cable is displaced laterally to O' due to the anti-symmetric vibrations. For anti-symmetric vibrations, the cable will slide longitudinally with respect to the deck. Two possibilities can be identified:

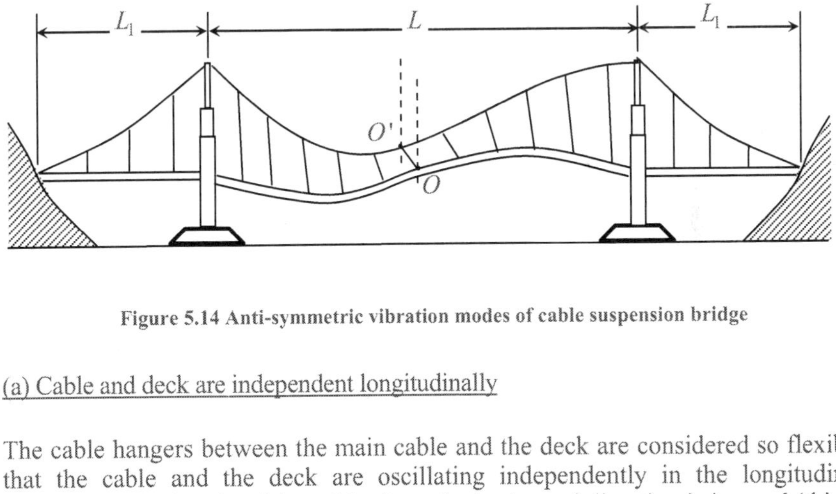

Figure 5.14 Anti-symmetric vibration modes of cable suspension bridge

<u>(a) Cable and deck are independent longitudinally</u>

The cable hangers between the main cable and the deck are considered so flexible that the cable and the deck are oscillating independently in the longitudinal direction. If the inertia of the cable along the horizontal direction is ignored (this is just since the cable is not very heavy and the horizontal acceleration is relatively small), there will be no additional horizontal component of cable tension being induced by the anti-symmetric vibrations. The change of cable tension on the left of O' cancels the change of cable tension on the right of O' due to vibrations. Thus, we have

$$\Delta T_0 = 0 \tag{5.209}$$

This, in turn, gives $H = 0$ (see (5.164)) and (5.166) becomes

$$\frac{d^2\bar{\eta}}{dx^2}+(\frac{2\mu}{L})^2\bar{\eta}=0 \tag{5.210}$$

The general solution of (5.210) is, of course, consisting of sine and cosine functions. However, the anti-symmetric condition imposed by (5.208) leads to a sine function only. The following solution form can be assumed:

$$\bar{\eta}=\eta_0\sin(2\mu\frac{x}{L}),\quad -\frac{L}{2}<x<\frac{L}{2} \tag{5.211}$$

$$\bar{\eta}=\eta_0\sin\frac{2\mu}{L}(x-\frac{L}{2}),\quad \frac{L}{2}<x<\frac{L}{2}+L_1 \tag{5.212}$$

$$\bar{\eta}=\eta_0\sin\frac{2\mu}{L}(x+\frac{L}{2}),\quad -\frac{L}{2}-L_1<x<-\frac{L}{2} \tag{5.213}$$

Substitution of this solution form into the boundary condition at the top of the towers ($x=\pm L/2$) results in

$$\sin\mu=0 \tag{5.214}$$

Thus, we must have $\mu=n\pi$, $n=1,2,3,...$ The anchor condition of the side span at $x=\pm(L/2+L_1)$, given by (5.168), requires

$$\sin(2\mu\alpha)=0 \tag{5.215}$$

This leads to the natural frequency of vibrations as $\mu=n\pi/(2\alpha)$ for $n=1,2,3,...$ for the side span vibrations given in (5.212) and (5.213). Substitution of (5.211) to (5.213) into the left-hand side of (5.174) gives

$$\int_{-L/2-L_1}^{L/2+L_1}\bar{\eta}dx=I_1+I_2+I_3 \tag{5.216}$$

where

$$I_1=\eta_0\int_{-L/2-L_1}^{-L/2}\sin\{\frac{2\mu}{L}(x+\frac{L}{2})\}dx=-\frac{\eta_0 L}{2\mu}\{1-\cos 2\mu\alpha\} \tag{5.217}$$

$$I_2=\eta_0\int_{-L/2}^{L/2}\sin(\frac{2\mu}{L}x)dx=-\frac{\eta_0 L}{2\mu}\{\cos\mu-\cos(\mu)\}=0 \tag{5.218}$$

$$I_3=\eta_0\int_{L/2}^{L/2+L_1}\sin\{\frac{2\mu}{L}(x-\frac{L}{2})\}dx=-\frac{\eta_0 L}{2\mu}\{\cos 2\mu\alpha-1\} \tag{5.219}$$

Substitution of (5.217) to (5.219) into (5.216) results in zero, and (5.174) is identically satisfied. Thus, the cable inextensibility constraint is fulfilled by the choice of (5.211) to (5.213).

(b) Cable is braced to the center of the deck

For a given value of sag, the level of the bridge deck can be set at any level before the main cable. A common practice is to reduce the cost by setting the deck level as close to the main cable as possible. Sometimes, the main cable is braced to the center of the deck to avoid vibrations. The installation of bracing will bar the longitudinal sliding relative to the deck so anti-symmetric vibrations are not possible.

5.8.4 Suspension Bridge with Stiffened Truss

In the previous section, we have assumed that the deck is flexible, such as the case of the Tacoma Narrows Bridge. It turns out that a suspension bridge with a flexible deck is highly conducive to wind-induced vibrations (such as the Tacoma Narrows Bridge disaster), and is highly undesirable. Subsequently, the decks of newer suspension bridges are intentionally designed to be stiffer. For modern suspension bridges, we need to account for the bending stiffness of the deck and we have the equation of motion of the bridge deck as

$$m\frac{\partial^2 \eta}{\partial t^2}+EI\frac{\partial^4 \eta}{\partial x^4}-\frac{mgL^2}{8h}\frac{\partial^2 \eta}{\partial x^2}+\frac{8h}{L^2}\Delta T_0=0 \tag{5.220}$$

where EI is the bending stiffness of the bridge deck (typically a truss). The last two terms are considered as the external forces added on the beam by the cables. In setting up (5.220), we have tacitly assumed that there is no loosening of the hangers that connect the deck to the cable. For very large amplitude of vibrations, this assumption may be violated, making the present linear analysis invalid. Similar to the analysis used in the previous section, we assume

$$\eta(x,t)=\bar{\eta}(x)\sin\omega t \tag{5.221}$$

$$\Delta T_0 = H\sin\omega t \tag{5.222}$$

Substitution of (5.221) and (5.222) into (5.220) gives

$$-m\omega^2\bar{\eta}+EI\frac{d^4\bar{\eta}}{dx^4}-\frac{mgL^2}{8h}\frac{d^2\bar{\eta}}{dx^2}+\frac{8Hh}{L^2}=0 \tag{5.223}$$

The ODE can be rewritten as

$$\frac{d^4\bar{\eta}}{dx^4}-(\frac{2K}{L})^2\frac{d^2\bar{\eta}}{dx^2}-(\frac{4\mu K}{L^2})^2\bar{\eta}=-\frac{8Hh}{T_0L^2}(\frac{2K}{L})^2 \tag{5.224}$$

where

$$\mu=\omega\frac{L}{2}\sqrt{\frac{m}{T_0}}=\omega\sqrt{\frac{2h}{g}} \tag{5.225}$$

$$\frac{T_0}{EI}=(\frac{2K}{L})^2 \tag{5.226}$$

The general solution of (5.244) is of exponential form

$$\bar{\eta}=\exp(\lambda x) \tag{5.227}$$

Substitution of (5.227) into the homogeneous form of (5.224) gives the following characteristic equation

$$\lambda^4-(\frac{2K}{L})^2\lambda^2-(\frac{4\mu K}{L^2})^2=0 \tag{5.228}$$

$$\lambda^2=\frac{1}{2}\{(\frac{2K}{L})^2\pm\sqrt{(\frac{2K}{L})^4+4(\frac{4\mu K}{L^2})^2}\} \tag{5.229}$$

This can be further simplified as

$$\lambda^2=\frac{4K^2}{L^2}\{\frac{1}{2}\pm\sqrt{(\frac{\mu}{K})^2+\frac{1}{4}}\} \tag{5.230}$$

Taking the square root again, we have

$$\lambda_{1,2} = \pm\frac{2}{L}K_2, \quad \lambda_{3,4} = \pm i\frac{2}{L}\mu_2 \tag{5.231}$$

where

$$K_2^2 = K^2\left[\sqrt{(\frac{\mu}{K})^2+\frac{1}{4}}+\frac{1}{2}\right] \tag{5.232}$$

$$\mu_2^2 = K^2\left[\sqrt{(\frac{\mu}{K})^2+\frac{1}{4}}-\frac{1}{2}\right] \tag{5.233}$$

It can be shown that these coefficients are related by (see Problem 5.5)

$$\mu_2 K_2 = \mu K, \quad K_2^2 = \mu_2^2 + K^2, \quad \mu^2 = \mu_2^2\left[(\frac{\mu_2}{K})^2+1\right] \tag{5.234}$$

The homogeneous solution of (5.224) is expressible in the following form

$$\bar{\eta}_h = C_1 e^{2K_2 x/L} + C_2 e^{-2K_2 x/L} + C_3 e^{2i\mu_2 x/L} + \bar{C}_3 e^{-2i\mu_2 x/L} \tag{5.235}$$

As shown in Section 1.11 of Chau (2018), C_3 must be a complex constant and the superimposed bar means its complex conjugate. Since the cable vibration η is a real function, it is more preferable to express the homogeneous solution as

$$\begin{aligned}\bar{\eta}_h = &A\cosh(2K_2 x/L) + C\sinh(2K_2 x/L) + B\cos(2\mu_2 x/L)\\ &+D\sin(2\mu_2 x/L)\end{aligned} \tag{5.236}$$

It is straightforward to show that the particular solution of (5.224) is

$$\bar{\eta}_p = \frac{2Hh}{T_0\mu^2} \tag{5.237}$$

The general solution to (5.224) becomes

$$\begin{aligned}\bar{\eta} = &A\cosh(2K_2 x/L) + C\sinh(2K_2 x/L) + B\cos(2\mu_2 x/L)\\ &+D\sin(2\mu_2 x/L) + \frac{2Hh}{T_0\mu^2}\end{aligned} \tag{5.238}$$

We now consider a very general situation that the side spans are of unequal span. The load per unit length of the bridge deck is m (see Figure 5.15). Note that the origins of the coordinate systems for the side spans now locate at their mid-span, as shown in Figure 5.15.

Symmetric modes

We now consider the symmetric modes, which require

$$\bar{\eta}(x) = \bar{\eta}(-x) \tag{5.239}$$

In view of this requirement, we have the symmetric modes of (5.238) as

$$\bar{\eta} = A\cosh\left(2K_2\frac{x}{L}\right) + B\cos\left(2\mu_2\frac{x}{L}\right) + \frac{2Hh}{T_0\mu^2} \tag{5.240}$$

If the deck of the bridge is hinged on both towers, both deflection and bending moment are zeros leading to

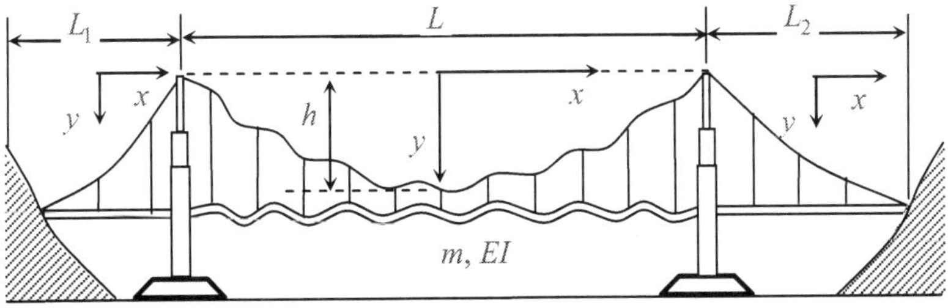

Figure 5.15 Symmetric vibration modes of a cable suspension bridge with unequal side spans and suspending different loads

$$\bar{\eta} = 0, \quad x = \pm\frac{L}{2} \tag{5.241}$$

$$\frac{d^2\bar{\eta}}{dx^2} = 0, \quad x = \pm\frac{L}{2} \tag{5.242}$$

Applying (5.241) to (5.240), we have

$$A\cosh(K_2) + B\cos(\mu_2) + \frac{2Hh}{T_0\mu^2} = 0 \tag{5.243}$$

Applying (5.242) to (5.240), we get

$$(\frac{2K_2}{L})^2 A\cosh(K_2) - (\frac{2\mu_2}{L})^2 B\cos(\mu_2) = 0 \tag{5.244}$$

Equations (5.243) and (5.244) provide two equations for the two unknown constants. The solutions are

$$A = -\frac{2Hh}{T_0\mu^2}\frac{\mu_2^2}{\mu_2^2 + K_2^2}\frac{1}{\cosh K_2} \tag{5.245}$$

$$B = -\frac{2Hh}{T_0\mu^2}\frac{K_2^2}{\mu_2^2 + K_2^2}\frac{1}{\cos \mu_2} \tag{5.246}$$

Substitution of (5.245) and (5.246) into (5.240) gives the deflection for the main span

$$\bar{\eta} = \frac{2Hh}{T_0\mu^2}\{1 - \frac{1}{\mu_2^2 + K_2^2}[K_2^2\frac{\cos(2\mu_2 x/L)}{\cos\mu_2} + \mu_2^2\frac{\cosh(2K_2 x/L)}{\cosh K_2}]\}, \quad -\frac{L}{2} < x < \frac{L}{2} \tag{5.247}$$

(a) Suspended side spans

The solution can be modified from (5.247) to that for side spans as

$$\bar{\eta} = \frac{2Hh}{T_0\mu^2}\{1 - \frac{1}{\mu_2^2 + K_2^2}[K_2^2\frac{\cos(2\mu_2 x/L)}{\cos\alpha_1\mu_2} + \mu_2^2\frac{\cosh(2K_2 x/L)}{\cosh\alpha_1 K_2}]\}, \quad -\frac{L_1}{2} < x < \frac{L_1}{2} \tag{5.248}$$

$$\bar{\eta} = \frac{2Hh}{T_0 \mu^2}\{1 - \frac{1}{\mu_2^2 + K_2^2}[K_2^2 \frac{\cos(2\mu_2 x / L)}{\cos \alpha_2 \mu_2} + \mu_2^2 \frac{\cosh(2K_2 x / L)}{\cosh \alpha_2 K_2}]\}, \quad -\frac{L_2}{2} < x < \frac{L_2}{2} \tag{5.249}$$

where

$$\alpha_1 = \frac{L_1}{L}, \quad \alpha_2 = \frac{L_2}{L} \tag{5.250}$$

Substitution of these solutions into (5.180) gives

$$\frac{HL_s}{EA}(\frac{L^2}{8h}) = \int_{-L/2-L_2}^{L/2+L_1} \bar{\eta}\, dx = I_0 + I_1 + I_2 \tag{5.251}$$

where the sub-integrals I_0, I_1, and I_2 for the main and side spans are defined in the first line of (5.197). Carrying out the integral, we have

$$I_0 = \frac{2HhL}{T_0 \mu^2}\{1 - \frac{\mu_2^2}{\mu_2^2 + K_2^2}(\frac{\tanh K_2}{K_2}) - \frac{K_2^2}{\mu_2^2 + K_2^2}(\frac{\tan \mu_2}{\mu_2})\} \tag{5.252}$$

$$I_1 = \frac{2HhL}{T_0 \mu^2}\{\alpha_1 - \frac{\mu_2^2}{\mu_2^2 + K_2^2}[\frac{\tanh(\alpha_1 K_2)}{K_2}] - \frac{K_2^2}{\mu_2^2 + K_2^2}[\frac{\tan(\alpha_1 \mu_2)}{\mu_2}]\} \tag{5.253}$$

$$I_2 = \frac{2HhL}{T_0 \mu^2}\{\alpha_2 - \frac{\mu_2^2}{\mu_2^2 + K_2^2}[\frac{\tanh(\alpha_2 K_2)}{K_2}] - \frac{K_2^2}{\mu_2^2 + K_2^2}[\frac{\tan(\alpha_2 \mu_2)}{\mu_2}]\} \tag{5.254}$$

Substitution of (5.252) to (5.254) into (5.251) gives

$$\begin{aligned} \tan \mu_2 + 2\tan(\mu_2 \alpha_1) &= (1 + \alpha_1 + \alpha_2)\mu_2[(\frac{\mu_2}{K_2})^2 + 1] \\ &-(\frac{\mu_2}{K_2})^3[\tanh K_2 + \tanh(\alpha_2 K_2) + \tanh(\alpha_1 K_2)] - C\mu_2^3(\frac{\mu_2^2 + K_2^2}{K^2}) \end{aligned} \tag{5.255}$$

where

$$C = \frac{T_0 L}{16h^2}(\frac{L_s}{EA}) \tag{5.256}$$

When we take the special case of equal side spans that $\alpha_1 = \alpha_2 = \alpha$, we find that (5.255) converts to the box equation on p. 180 of Rocard (1957).

(b) Unsuspended side spans

For the case of unsuspended side spans, there is no interaction between the deck and the cable and, thus, the bending stiffness does not go into the equation. However, the tension from the main span does transfer to the cable tension through T_0 and ΔT_0. The weights acting on the unit length of the left- and right-side cables are assumed to be $m_1 g$ and $m_2 g$, respectively. The governing equations for the cable deflection on the side spans are

$$\frac{d^2\bar{\eta}}{dx^2} + \frac{m_1 \omega^2}{T_0}\bar{\eta} = \frac{H}{T_0^2} m_1 g, \quad -\frac{L_1}{2} < x < \frac{L_1}{2} \tag{5.257}$$

$$\frac{d^2\bar{\eta}}{dx^2}+\frac{m_2\omega^2}{T_0}\bar{\eta}=\frac{H}{T_0^2}m_2 g,\quad -\frac{L_2}{2}<x<\frac{L_2}{2} \tag{5.258}$$

where H is the magnitude is ΔT_0 defined in (5.222). These equations can be rewritten as

$$\frac{d^2\bar{\eta}}{dx^2}+\beta_1\frac{m\omega^2}{T_0}\bar{\eta}=\frac{H}{T_0^2}\beta_1 mg \tag{5.259}$$

$$\frac{d^2\bar{\eta}}{dx^2}+\beta_2\frac{m\omega^2}{T_0}\bar{\eta}=\frac{H}{T_0^2}\beta_2 mg \tag{5.260}$$

where

$$\beta_1=\frac{m_1}{m},\quad \beta_2=\frac{m_2}{m} \tag{5.261}$$

The deflections of the side span are then

$$\bar{\eta}=\frac{2hH}{T_0\mu^2}\left[1-\frac{\cos(2\sqrt{\beta_1}\,\mu x/L)}{\cos(\sqrt{\beta_1}\,\alpha_1\mu)}\right],\quad -\frac{L_1}{2}<x<\frac{L_1}{2}, \tag{5.262}$$

$$\bar{\eta}=\frac{2hH}{T_0\mu^2}\left[1-\frac{\cos(2\sqrt{\beta_2}\,\mu x/L)}{\cos(\sqrt{\beta_2}\,\alpha_2\mu)}\right],\quad -\frac{L_2}{2}<x<\frac{L_2}{2}, \tag{5.263}$$

The solution for the main span has been given in (5.247). Substitution of these into the cable length constraint gives

$$\begin{aligned}\int_{-L/2-L_1}^{L/2+L_2}\bar{\eta}\,dx&=\int_{-L_1/2}^{L_1/2}\bar{\eta}\,dx+\int_{-L/2}^{L/2}\bar{\eta}\,dx+\int_{-L_2/2}^{L_2/2}\bar{\eta}\,dx=\frac{HL_s}{EA}\left(\frac{L^2}{8h}\right),\\&=I_1+I_0+I_2\end{aligned} \tag{5.264}$$

More specifically, we have

$$I_1=\int_{-L_1/2}^{L_1/2}\eta\,dx=\frac{2HhL}{T_0\mu^2}\left[\alpha_1-\left(\frac{1}{\sqrt{\beta_1}\,\mu}\right)\tan\left(\sqrt{\beta_1}\,\alpha_1\mu\right)\right] \tag{5.265}$$

$$I_2=\int_{-L_1/2}^{L_1/2}\eta\,dx=\frac{2HhL}{T_0\mu^2}\left[\alpha_2-\left(\frac{1}{\sqrt{\beta_2}\,\mu}\right)\tan\left(\sqrt{\beta_2}\,\alpha_2\mu\right)\right] \tag{5.266}$$

$$I_0=\frac{2HhL}{T_0\mu^2}\left[1-\left(\frac{K_2^2}{\mu_2^2+K_2^2}\right)\frac{\tan\mu_2}{\mu_2}-\left(\frac{\mu_2^2}{\mu_2^2+K_2^2}\right)\frac{\tanh K_2}{K_2}\right] \tag{5.267}$$

Note that I_0 given in (5.267) is the same as I_0 given in (5.252). Substitution of (5.265) to (5.267) into (5.264) gives the eigenvalue equation as

$$1-\left(\frac{K_2^2}{\mu_2^2+K_2^2}\right)\frac{1}{\mu_2}\tan\mu_2-\left(\frac{\mu_2^2}{\mu_2^2+K_2^2}\right)\frac{1}{K_2}\tanh K_2$$
$$+[\alpha_1-(\frac{1}{\sqrt{\beta_1}\mu})\tan\left(\sqrt{\beta_1}\alpha_1\mu\right)]+[\alpha_2-(\frac{1}{\sqrt{\beta_2}\mu})\tan\left(\sqrt{\beta_2}\alpha_2\mu\right)] \quad (5.268)$$
$$=\frac{L_s}{EA}\left(\frac{T_0L}{16h^2}\right)\mu^2$$

Rearrangement of (5.268) gives the frequency equation as

$$\tan\mu_2+\frac{K(\mu_2^2+K_2^2)}{K_2^3}\left[\frac{1}{\sqrt{\beta_1}}\tan(\sqrt{\beta_1}\alpha_1\frac{\mu_2K_2}{K})+\frac{1}{\sqrt{\beta_2}}\tan(\sqrt{\beta_2}\alpha_2\frac{\mu_2K_2}{K})\right]$$
$$=(1+\alpha_1+\alpha_2)(\mu_2+\frac{\mu_2^3}{K_2^2})-(\frac{\mu_2}{K_2})^3\tanh K_2-C\frac{\mu_2^3}{K^2}(\mu_2^2+K_2^2) \quad (5.269)$$

where C has been defined in (5.256). For the special case $\alpha_1 = \alpha_2 = \alpha$, and $\beta_1 = \beta_2 = \beta$, (5.269) should reduce to the first box equation on p. 180 of Rocard (1957). However, we find that there is a typo in the formula shown on p. 180 of Rocard (1957). See Problem 5.7 for more details.

Anti-symmetric modes

We now consider the anti-symmetric modes, which require

$$\bar{\eta}(x)=-\bar{\eta}(-x) \quad (5.270)$$

Similar to the earlier discussions in Section 5.8.3, for the case of anti-symmetric modes, the horizontal motion of cable and deck are independent. Thus, there is no additional tension induced (i.e., $H = 0$). For the mid-span, (5.238) becomes

$$\bar{\eta}=A\sinh\left(2K_2\frac{x}{L}\right)+B\sin\left(2\mu_2\frac{x}{L}\right) \quad (5.271)$$

The boundary conditions are

$$\bar{\eta}=0,\quad x=\pm\frac{L}{2} \quad (5.272)$$

$$\frac{d^2\bar{\eta}}{dx^2}=0,\quad x=\pm\frac{L}{2} \quad (5.273)$$

In view of (5.272) and (5.273), we have $A = 0$ and

$$B\sin(\mu_2)=0 \quad (5.274)$$

Since for nontrivial solution, we require $B \neq 0$, we must have

$$\mu_2=n\pi,\quad n=1,2,3,... \quad (5.275)$$

Therefore, the eigenmode is

$$\bar{\eta}=B\sin(2n\pi\frac{x}{L}) \quad (5.276)$$

To get the frequency equation, we can substitute (5.271) into the homogeneous form of (5.224) or the following

$$\frac{d^4\bar{\eta}}{dx^4}-(\frac{2K}{L})^2\frac{d^2\bar{\eta}}{dx^2}-(\frac{4\mu K}{L^2})^2\bar{\eta}=0 \tag{5.277}$$

This results in

$$(\frac{2n\pi}{L})^4+(\frac{2K}{L})^2(\frac{2n\pi}{L})^2-(\frac{4\mu K}{L^2})^2=0 \tag{5.278}$$

Using the definition of K and μ defined in (5.225) and (5.226), we can simplify (5.278) as

$$\omega^2=\frac{\pi^2 g(2n)^2}{8h}+\frac{\pi^4 EI(2n)^4}{mgL^4} \tag{5.279}$$

In obtaining (5.279), we have used (5.510) and $w = mg$. This equation agrees with the first box equation on p. 181 of Rocard (1957), except for a minor typo in the second term.

5.9 SUMMARY AND FURTHER READING

In this chapter, we have summarized the theories of the catenary, inverted catenary, Inglis solution of the stone arch, the cable-stay bridge, and the cable suspension bridge. Both static and vibration analyses have been considered for cable suspension bridges. For the case of a flexible bridge deck, we derive the natural frequency of both symmetric and anti-symmetric bending modes. Observed oscillations of the Tacoma Narrows Bridge before its collapse reveals that the mode shape consists of four and one-half complete waves and a natural frequency of 36 oscillations per minute or a circular frequency of 3.7699 rad/s while the theoretical prediction of frequency for the same bending mode is 3.7237 rad/s, with an error of only 1.22%. This lends credence to the use of linear theory. For modern cable suspension bridges, symmetric and anti-symmetric modes are considered taking into consideration the bridge deck stiffness, with equal and unequal side spans, and with suspended and unsuspended side spans. Our formulas are more general than those derived by Rocard (1957). For the special cases considered by Rocard (1957), we find some typos in the formulas of Rocard (1957).

For general discussion of cable structures, the excellent textbook by Irvine (1981) is recommended. Comprehensive theoretical considerations of cable suspension bridges are given by the classic works of Bleich et al. (1950), Rocard (1957), and Pugsley (1968). For example, Bleich et al. (1950) discussed the use of the energy method in analyzing the suspension bridge. The effects of flexural rigidity and torsional rigidity of towers, wind-induced vibrations, damping, towers with cable stays, torsional oscillations, and the main cable braced to the bridge deck were considered by both Bleich et al. (1950) and Rochard (1957). General review of modern suspension bridges can be found in Chapter 18 of the *Bridge Engineering Handbooks* edited W.F. Chen and L. Duan (Okukawa et al., 2000), Fujino (2002), and Xu (2013).

Further readings on the cable-stay bridge can be found in Troitsky (1988), and Chapter 19 of the *Bridge Engineering Handbooks* by Tang (2000) and Kanok-Nukulchai et al. (1992). Design methods of cable-stay bridges can be referred to

Walther et al. (1999). Gimsing and Georgakis (2011) covered the design aspects of both cable-stay and cable-suspension bridges.

5.10 PROBLEMS

Problem 5.1 Solve the vibration problem of a hanging chain subject to an initial velocity field as

$$u(x,0)=0, \quad \frac{\partial u}{\partial t}(x,0)=v_0(x) \tag{5.280}$$

$$u(L,t)=0, \quad u(0,t)=\textit{finite} \tag{5.281}$$

Ans:

$$u(x,t)=\sum_{n=1}^{\infty}\frac{\int_0^{2\sqrt{\frac{L}{g}}}\xi v_0(\frac{g\xi^2}{4})J_0(\bar{\omega}_n\frac{\xi}{2}\sqrt{\frac{g}{L}})d\xi}{\omega_n\int_0^{2\sqrt{\frac{L}{g}}}\xi[J_0(\bar{\omega}_n\frac{\xi}{2}\sqrt{\frac{g}{L}})]^2d\xi}J_0(2\omega_n\sqrt{\frac{x}{g}})\sin\omega_n t \tag{5.282}$$

Problem 5.2 Extend the analysis in Section 5.8.2 to the case of unequal side spans, as shown in Figure 5.16. Show that the symmetric modes can be formulated as:

$$\eta=\frac{2Hh}{T_0\mu^2}[1-\frac{\cos(2\mu x/L)}{\cos\mu}], \quad -\frac{L}{2}<x<\frac{L}{2} \tag{5.283}$$

$$\eta=\frac{2Hh}{T_0\mu^2}[1-\frac{\cos\mu(1+\alpha_1+2x/L)}{\cos\alpha_1\mu}], \quad -L_1-\frac{L}{2}<x<-\frac{L}{2} \tag{5.284}$$

$$\eta=\frac{2Hh}{T_0\mu^2}[1-\frac{\cos\mu(1+\alpha_2-2x/L)}{\cos\alpha_2\mu}], \quad \frac{L}{2}<x<\frac{L}{2}+L_2 \tag{5.285}$$

where

$$\alpha_1=\frac{L_1}{L}, \quad \alpha_2=\frac{L_2}{L} \tag{5.286}$$

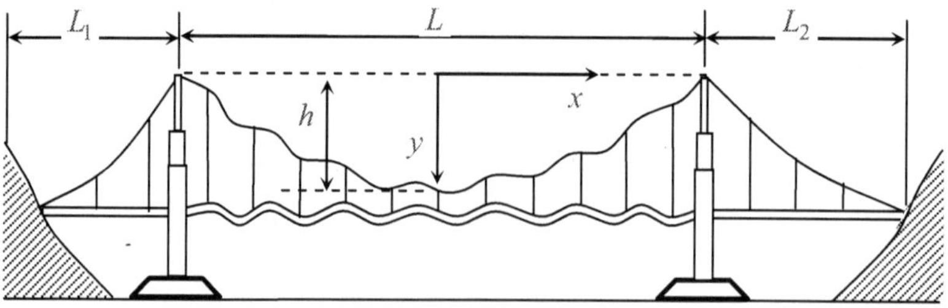

Figure 5.16 Symmetric vibration modes of a cable suspension bridge with unequal side spans

Problem 5.3
(i) Show that the vibration frequency for the cable suspension bridge with unequal side spans shown in Figure 5.16 satisfies the following eigenvalue equation:

$$\tan(\alpha_1\mu)+\tan(\alpha_2\mu)+\tan\mu=(1+\alpha_1+\alpha_2)\mu \tag{5.287}$$

(ii) Show that (5.201) can be recovered as a special case of (5.287) if we set $\alpha_1 = \alpha_2 = \alpha$.

Problem 5.4 Consider the case that the deck was not suspended by the side cable, as shown in Figure 5.17. The mass per unit length of the side span is m_1, instead of m in the main span.

(i) Show that the governing equation for the left side span can be written as (compare with (5.165) for the main span):

$$-m_1\omega^2\bar{\eta}-\frac{m_1 g L_1^2}{8h}\frac{d^2\bar{\eta}}{dx^2}+\frac{8hH}{L_1^2}=0 \tag{5.288}$$

(ii) Show that this ODE can be rewritten as

$$\frac{d^2\bar{\eta}}{dx^2}+\frac{m\omega^2}{\alpha_1^2 T_0}\bar{\eta}=\frac{H}{\beta\alpha_1^4 T_0^2}mg \tag{5.289}$$

where

$$\beta=\frac{m_1}{m},\quad \alpha_1=\frac{L_1}{L} \tag{5.290}$$

(iii) Similarly, show that the ODE for the right side span can be rewritten as

$$\frac{d^2\bar{\eta}}{dx^2}+\frac{m\omega^2}{\alpha_2^2 T_0}\bar{\eta}=\frac{H}{\beta\alpha_2^4 T_0^2}mg \tag{5.291}$$

where

$$\beta=\frac{m_1}{m},\quad \alpha_2=\frac{L_2}{L} \tag{5.292}$$

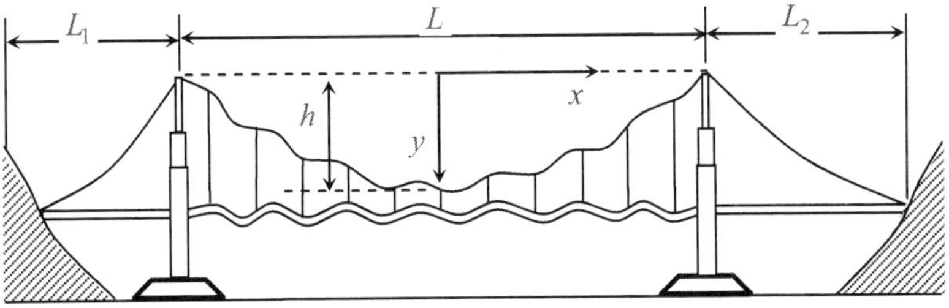

Figure 5.17 Symmetric vibration modes of a cable suspension bridge with unsuspended side spans

Problem 5.5 Prove the following identities

$$\mu_2 K_2=\mu K,\quad K_2^2=\mu_2^2+K^2,\quad \mu^2=\mu_2^2\left[\left(\frac{\mu_2}{K}\right)^2+1\right] \tag{5.293}$$

where μ, μ_2, K, and K_2 are defined in (5.232) and (5.233).

Problem 5.6

(i) Show that the side span vibrations of the case of a suspension bridge with stiffened truss discussed in Section 5.8.5 can also be formulated as

Left span

$$\eta(x) = \eta_0 \sin[\frac{n\pi}{\alpha_1 L}(x - \frac{L_1}{2})] \tag{5.294}$$

Right span

$$\eta(x) = \eta_0 \sin[\frac{n\pi}{\alpha_2 L}(x - \frac{L_2}{2})] \tag{5.295}$$

(ii) Show that the following boundary conditions of the left and right spans

$$\eta(\pm\frac{L_1}{2}) = 0, \quad \eta(\pm\frac{L_2}{2}) = 0 \tag{5.296}$$

are satisfied by (5.294) and (5.295).

Problem 5.7

(i) Show that for $\alpha_1 = \alpha_2 = \alpha$, and $\beta_1 = \beta_2 = \beta$, (5.269) becomes

$$\tan \mu_2 + \frac{2K(\mu_2^2 + K_2^2)}{K_2^3} \frac{1}{\sqrt{\beta}} \tan(\sqrt{\beta}\alpha \frac{\mu_2 K_2}{K}) = (1+2\alpha)(\mu_2 + \frac{\mu_2^3}{K_2^2})$$
$$-(\frac{\mu_2}{K_2})^3 \tanh K_2 - C\frac{\mu_2^3}{K^2}(\mu_2^2 + K_2^2) \tag{5.297}$$

(ii) Show that there is a typo in the frequency equation given on p. 180 of Rocard (1957).

CHAPTER SIX

Nonlinear Buckling

6.1 INTRODUCTION

The classical theory of elasticity assumes small strain and small deformation. Small strain will preclude plastic or inelastic deformation (material nonlinearity), and small deformation will preclude buckling (geometric nonlinearity). Thus, we need to incorporate either large strain or large deformation to investigate the nonlinear response of structures. Material nonlinearity occurs for the case of finite strain (finite in the sense that the strain is no longer infinitesimal) such that linear stress-strain behavior is not valid, whereas the deformation can remain small. Geometric nonlinearity, on the other hand, implies a linear stress-strain response but the deflection of the solid can be large. This normally occurs for an elastic solid with a "thin" dimension, such as for the cases of bars, plates or shells. We cannot develop a general nonlinear theory for the buckling of thin structures, because all theories developed for bars, plates, and shells are special theories that, are derived based upon on various degree of idealizations. As we have seen in Chapter 1, popular beam theories include the Bernoulli-Euler beam, Rayleigh beam, and Timoshenko beam. Bernoulli-Euler beams assume a plane normal to the neutral axis (along which there is no bending strain) of the beam remaining a plane and normal to the deformed neutral axis after bending, the Rayleigh beam allows for torsional inertia of the beam, and the Timoshenko beam allows the plane not to be normal to the deformed neutral axis after bending to account for shear deformation. They are all special theories. In terms of buckling, the elastic theory of Euler was developed as the elastic theory for column buckling, the von-Karman nonlinear plate theory was developed for buckling of elastic plates, and the Donnell-von-Karman-Marguerre theory was developed for buckling of elastic shells. The present chapter mainly considers the nonlinear buckling of beams using bifurcation theory via perturbation analysis. The present chapter mainly considers the buckling of one-dimensional solids, including columns, beams, and arches. Material nonlinearity will not be included in the present chapter. Only geometric nonlinearity is considered.

This chapter evolves from a course on nonlinear buckling offered by E.L. Reiss at Northwestern University in 1988 and from Reiss (1980b), Reiss (1984), and Reiss and Matkowsky (1971). This chapter consists of three main parts: (i) elastic buckling of straight columns; (ii) dynamic buckling of straight columns; and (iii) dynamic buckling of elastic arches. The main focus of the present chapter is on the case of columns or arches subject to end shortening, instead of subject to end compressive force. End shortening results in a nonlinear PDE which was first derived by Hoff in 1951. Both linear and nonlinear stability analyses are considered. For nonlinear dynamic stability, multi-scale perturbation technique is employed. The mathematical techniques of asymptotic expansion and perturbation analysis, calculus of variations, and variational methods covered in Chapters 12 to 14 of Chau (2018) are essential before the reader continues studying the present chapter. Section 2.5 of Chau (2013) on elastic materials undergoing large deformations is also useful. Before we go into the mathematical formulation, the next three subsections summarize buckling of columns, plates, and shells.

6.1.1 Column Buckling

The buckling of a column subject to compression had been considered by Euler. Figure 6.1 depicts the buckling of a column as a bifurcation process. At the critical load P_{crit}, the straight state of the column (i.e., $\delta = 0$) may yield to a buckled state through bifurcation (i.e., $\delta \neq 0$). As shown in Figure 6.1, beyond the bifurcation point, both the straight state and buckled state are possible. Whether buckling will occur, we need to examine the dynamic stability of each of these equilibrium solutions. Since there are two solutions for the same P, the problem cannot be modeled by linear PDE, which can only yield one solution for any given P. In real columns with imperfections, the deformation will not appear as an abrupt process of bifurcation, but instead appears as a continuous smooth transition to the buckled state (see the dotted line shown in Figure 6.1).

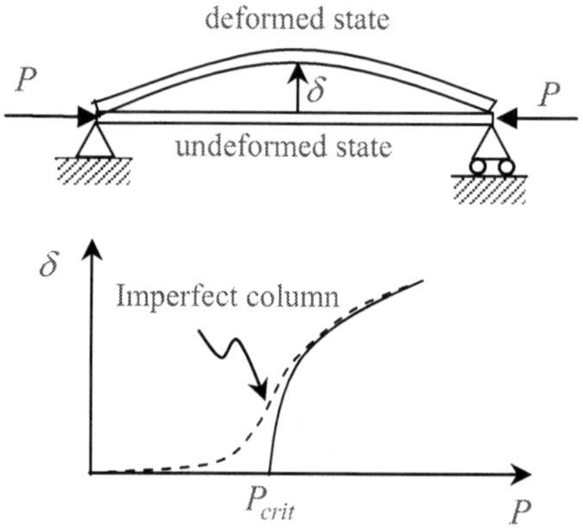

Figure 6.1 Buckling of column

If the imperfection of the column appears as a crooked column, the force-displacement curve will appear as a smooth continuous curve as illustrated in Figure 6.2. There is no bifurcation in this case as the unbuckled state deformed smoothly to the buckled state. The growth of δ increases rapidly as the theoretical buckling load P_{crit} is approached. If the applied load is a function of time (i.e., $P = P(t)$), the buckling response may be completely different from static buckling, and it is called dynamic buckling. This kind of dynamic buckling has not been studied as extensively as static buckling. Dynamic buckling will be considered in later sections of this chapter. In particular, stability for damped and undamped cases is considered separately. Various forms of initial disturbances will be considered using multi-time perturbation analysis for the case of snap-through buckling of arches.

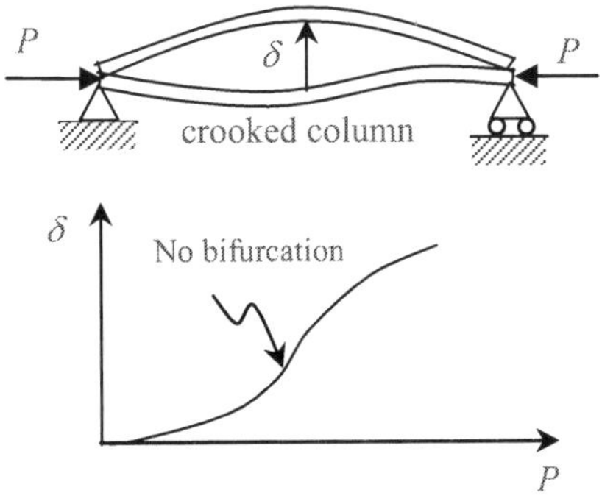

Figure 6.2 Deformation of a crooked column

If a beam is built curved upwards as shown in Figure 6.3, it is known as an arch. The theory of the stone arch has been considered in Chapter 4 under the context of a cable structure. The buckling of an arch appears in a very different manner called snap-through buckling. Figure 6.3 shows that it is a dynamic jump phenomenon (e.g., Reiss, 1980b). The transition segment between the lower curve and the upper curve of the zip-zag load deflection is unstable. Experiments show that the dynamic transition from the lower curve to the upper symmetric buckled state is, in fact, unsymmetric as shown in Figure 6.3. More detailed discussions on arch buckling will be given in later sections. The dynamic transition from $P = P_u$ is a snap-through buckling. Similarly, for the case of unloading, the symmetric state will snap back to the lower curve through buckling at $P = P_l$. These values P_l and P_u are called limit points. This kind of snap-through buckling also appears in the buckling of shallow spherical and cylindrical shells. In principle, the whole force-displacement shown in Figure 6.3 can be traced numerically from the arc-length control technique discussed in Chapter 15 of Chau (2018) (see Section 15.7.3).

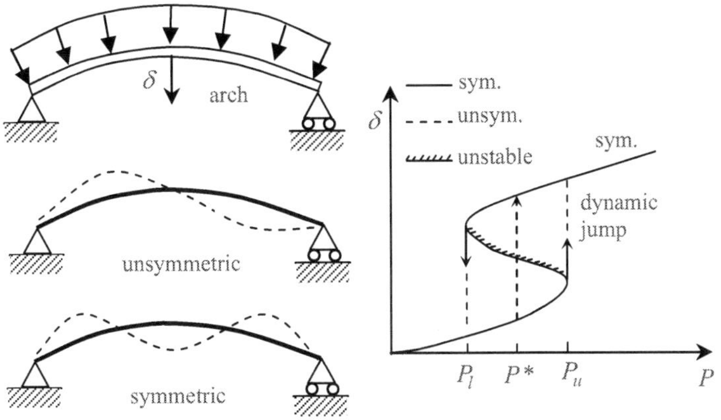

Figure 6.3 Deformation of a crooked column

6.1.2 Plate Buckling

If a plate is sufficiently thin, under axial compression large deformation can develop in the plates without much material nonlinearity. However, if the plate is moderately thick, plastic hinge or damage that develops in plates cannot be neglected. For thin plates undergoing large deformation, strain is small but can be nonlinear due to rotation from large deformation. From plane strain equilibrium, Hooke's law and strain compatibility, Kirchhoff in 1877 developed the large-deflection theory, Foppl in 1907 employed the use of the Airy stress function, and von Karman in 1910 derived the following system of equations for plate buckling for the out-plane deflection w (p. 347 of Timoshenko and Gere, 1961; p. 417 of Timoshenko and Woinowsky-Krieger, 1959):

$$\nabla^4 F = Eh[(\frac{\partial^2 w}{\partial x \partial y})^2 - \frac{\partial^2 w}{\partial x^2}\frac{\partial^2 w}{\partial y^2}] \tag{6.1}$$

$$D\nabla^4 w = p + (\frac{\partial^2 F}{\partial y^2}\frac{\partial^2 w}{\partial x^2} + \frac{\partial^2 F}{\partial x^2}\frac{\partial^2 w}{\partial y^2} - 2\frac{\partial^2 F}{\partial y \partial x}\frac{\partial^2 w}{\partial y \partial x}) \tag{6.2}$$

where F is the Airy stress function, h is the thickness of the plate, E is the Young's modulus of the plate, p is the applied normal load, and D is the flexural rigidity of the plate defined as

$$D = \frac{Eh^2}{12(1-\nu^2)} \tag{6.3}$$

This is known as the von Karman-Foppl theory (Bazant and Cedolin, 1991) or the Foppl-von Karman theory (Coman and Bassom, 2016). The exact solution for this system is not easy to obtain. Timoshenko and Woinowsky-Krieger (1959) gave an approximate solution for the case of circular and rectangular plates subject to uniform compression using the variational method for extremely thin plates with negligible bending stiffness (i.e., $D \approx 0$ in the membrane theory).

The qualitative illustration of primary and secondary buckling of a circular plate subject to uniform compression T is shown in Figure 6.4. This problem leads to the development of the so-called boundary layer theory in plate buckling. In particular, Friedrichs and Stoker (1941) in their landmark paper proposed the idea of boundary layer analysis discovering that tensile stress in the central part and compressive stress near the edge can co-exist in the plate under radial compression because of buckling. This technique eventually evolves as the boundary layer analysis. As discussed in Section 12.4 of Chau 2018, the method is also called matched asymptotic expansion (MMAE) or "singular perturbations." In fact, the term "singular perturbations" was coined by Friedrichs and Wasow in 1946 in a paper related to nonlinear vibrations. The primary buckling manifests as a pop-up of the central part, whereas secondary buckling appears as plate wrinkles along the edge of the plates. After primary buckling, the pop-up buckling will lead to larger compression near the edge as T further increases. With increasing T, the width of this compressive zone near the edge will decrease. This will eventually lead to wrinkle buckling as a secondary bifurcation. These wrinkles are formed as unsymmetrical buckling in the form of a ring of waves around the edge while the central part remains roughly axisymmetric. This quantitative result can only be obtained numerically (Cheo and Reiss, 1973, 1974), which is out of the scope of the present chapter. The dotted line on the force-deflection plate shows the smooth transition to the buckled state for imperfect plates.

Figure 6.5 illustrates the buckling of a rectangular plate subject to uniform compression T. The force-deflection curve closely resembles that shown in Figure 6.4 for circular plates. A primary buckling occurs at T_c, followed by a secondary buckling at T_s. The analysis of these secondary bifurcations for plate buckling is more complicated than those for columns and arches, and will not be considered in this chapter.

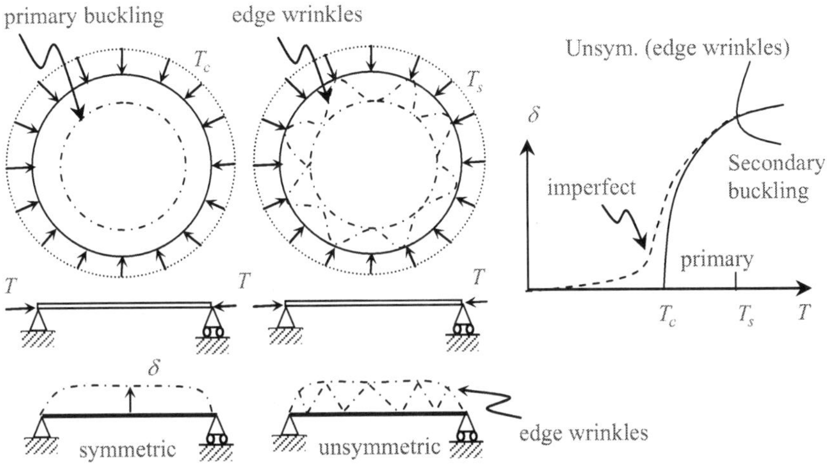

Figure 6.4 Deformation of a crooked column

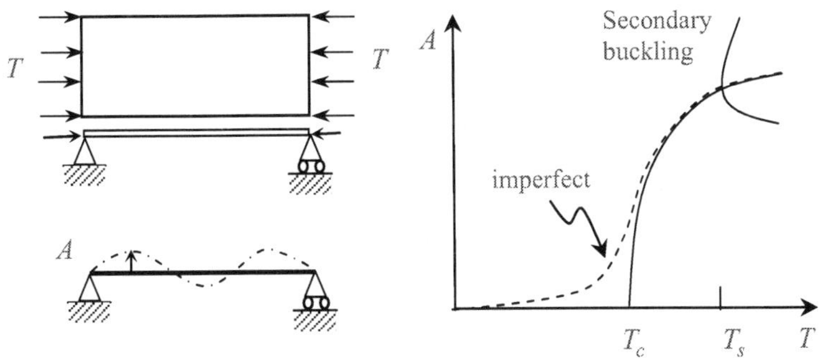

Figure 6.5 Buckling of a rectangular plate

6.1.3 Shell Buckling

Figure 6.6 illustrates the buckling of a cylindrical shell under axial compression and of a shallow spherical cap under radial compression. The force and displacement response is similar for these two cases. See, for example, Figure 1 of Bauer et al. (1967) for the case of spherical cap buckling. For the case of a cylindrical shell, the linearized Donnell equation is developed. In particular, the deflection w of the shell is governed by the following equation:

$$D\nabla^4\nabla^4 w - \nabla^4\left(N_{xx}^0 \frac{\partial^2 w}{\partial x^2} + 2N_{xy}^0 \frac{\partial^2 w}{\partial y \partial x} + N_{yy}^0 \frac{\partial^2 w}{\partial y^2}\right) + \frac{Eh}{a^2}\frac{\partial^4 w}{\partial x^4} = 0 \quad (6.4)$$

where h is the thickness of the shell, a is the radius of the shallow shell, and E is Young's modulus. The initial forces within the shells are denoted with a superscript 0. If the shell wall is thin and the number of buckles is large, the critical stress at buckling is found equal to (p. 472 of Timoshenko and Gere, 1961):

$$\sigma_{cr} = \frac{N_{xx}^{cr}}{h} = \frac{Eh}{a\sqrt{3(1-\nu^2)}} \quad (6.5)$$

This result has been derived in Section 3.8. This is the buckling stress for the short wavelength limit, and it can be interpreted as surface instability of surface wrinkling. This solution is valid for

$$Z = \frac{L^2}{ah}\sqrt{1-\nu^2} \geq 2.85 \quad (6.6)$$

where Z is called the Batdorf parameter.

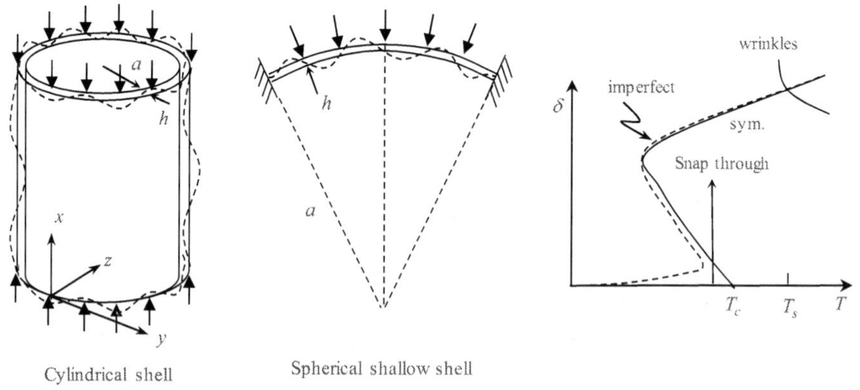

Figure 6.6 Snap-through buckling of cylindrical and spherical shells

Snap-through buckling and wrinkling of shallow spherical shells is illustrated in Figure 6.7. For the case of symmetrical buckling of spherical shallow shells, the governing equation is (see Section 3.12.1)

$$\nabla^2\nabla^2 F + \frac{Eh}{a}\nabla^2 w = 0 \tag{6.7}$$

$$\nabla^2\nabla^2 w - \frac{1}{Da}\nabla^2 F = 0 \tag{6.8}$$

If the wavelength of the buckles λ in the spherical shell is much smaller than the radius a of the spherical shell, the buckling stress is also obtained as (6.5) (see p. 517 of Timoshenko and Gere, 1961). The discussion on shell buckling will not be covered in the present chapter.

Before we formulate the problems of nonlinear buckling, it is instructive to consider two simple examples. Example 6.1 illustrates the existence of snap-through buckling in a two-bar system, and Example 6.2 illustrates the buckling of a two-bar system jointed by a deflection spring and a rotational spring. The mathematical structures of these two simple problems closely resemble the results of the more involved analyses in subsequent sections.

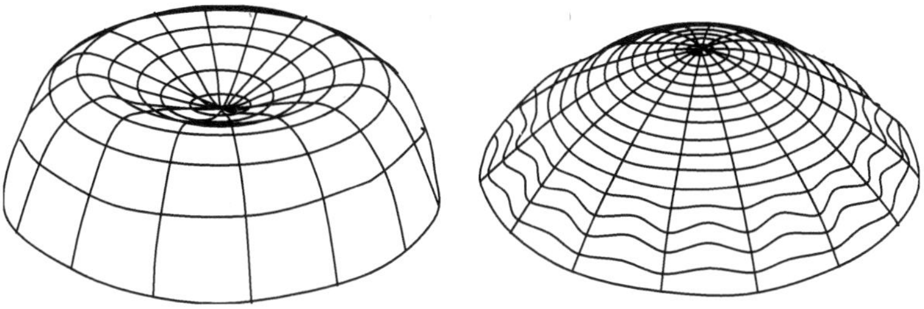

Figure 6.7 Snap-through buckling and edge wrinkling of a spherical shell

---

Example 6.1 We have seen from Figures 6.3 and 6.6 that snap-through buckling is possible for shallow arches as well as shallow spherical caps. In this example, an analytic solution of snap-through buckling for a two-bar system is considered. As shown in Figure 6.8, two bars inclined at an angle of 18° from horizontal are hinged in the middle. This example is modified from Bathe (1982). A vertical downward force R is applied at the hinge. The length and stiffness of the bar are denoted by L and k. Find the load-deflection curve of this problem.

Solution: By compatibility of the deformation in the free body diagram given in Figure 6.8, we have

$$(L-\delta)\cos\beta = L\cos(18^\circ) \tag{6.9}$$

$$(L-\delta)\sin\beta = L\sin(18^\circ) - \Delta \tag{6.10}$$

Taking the square of both (6.9) and (6.10), and adding these results, we find

$$\delta = L - \sqrt{L^2 - 2L\Delta\sin(18^\circ) + \Delta^2} \tag{6.11}$$

By considering force equilibrium of the free body diagram shown in Figure 6.8, we obtain

$$2F\sin\beta = R \tag{6.12}$$

The force F leads to a bar shortening of δ and they are related by

$$F = k\delta \tag{6.13}$$

Substitution of (6.13) into (6.12) gives

$$2k\delta \sin\beta = R \tag{6.14}$$

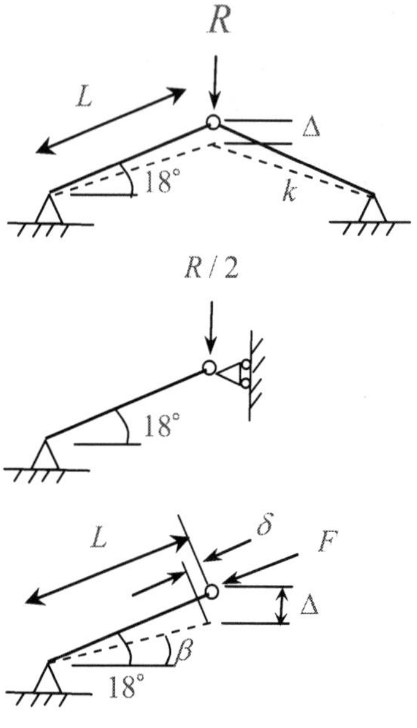

Figure 6.8 The free body diagram of half of the two-bar system with a vertical roller

Using (6.10) to obtain sinβ, we can rewrite (6.14) as

$$2k\delta[\frac{L\sin(18^\circ)-\Delta}{L-\delta}] = R \tag{6.15}$$

Rearranging (6.15), we obtain

$$\frac{R}{2Lk} = [\frac{\sin(18^\circ)-\Delta/L}{1/(\frac{\delta}{L})-1}] \tag{6.16}$$

Equation (6.11) can be rearranged as

$$\delta/L = 1-\sqrt{1-2(\Delta/L)\sin(18^\circ)+(\Delta/L)^2} = 1-\sqrt{\alpha} \tag{6.17}$$

In view of (6.17), the denominator of (6.16) can be expressed as

$$\frac{1}{1/(\frac{\delta}{L})-1} = \frac{\delta/L}{1-\delta/L} = \frac{1-\sqrt{\alpha}}{\sqrt{\alpha}} = (-1+\frac{1}{\sqrt{\alpha}}) \tag{6.18}$$

Substitution of (6.18) into (6.16) gives

$$\frac{R}{2Lk}=\left\{\frac{1}{\sqrt{1-2(\Delta/L)\sin(18^{\circ})+(\Delta/L)^{2}}}-1\right\}[\sin(18^{\circ})-\Delta/L] \tag{6.19}$$

This gives the force-displacement curve for the problem, which is plotted in Figure 6.9. This simple yet illustrative example shows the physical reality of snap-through buckling. A more complicated case of snap-through buckling of a shallow arch of sinusoidal shape subject to sinusoidal pressure will be investigated by using perturbation theory in Section 6.4.

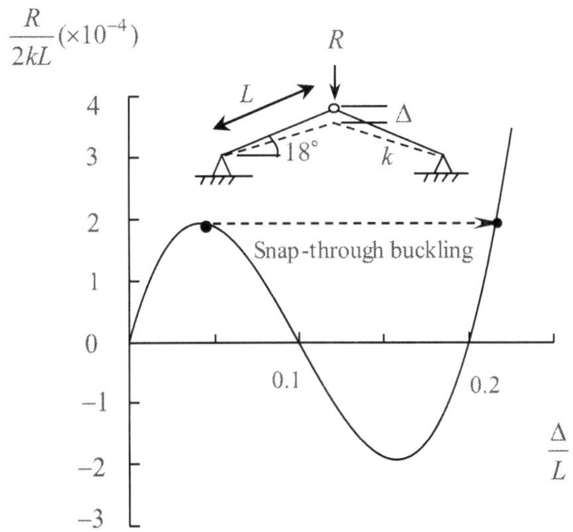

Figure 6.9 Snap-through buckling of a two-bar system

Example 6.2 In this second example, we will illustrate the buckling of two rigid rods connected by a lateral nonlinear spring and a linear rotation spring, subjected to axial compression (see Figure 6.10). This example was provided by Stoker (1950) and gave considerable insight into the nonlinear dynamic stability analysis to be considered in later sections. (i) Find the buckled load of the system and the corresponding solutions (straight state and buckled states); and (ii) examine the dynamic stabilities of these solutions.

Solution: (i) The nonlinear spring force and linear spring moment are modeled by:

$$f(x)=\alpha x+\beta x^{3},\quad M=k\gamma=2k\theta \tag{6.20}$$

where the angle of rotation γ at the joint can be interpreted as 2θ from Figure 6.10 (i.e., exterior angle of a triangle). Vertical force equilibrium of the free body leads to

$$P = V \tag{6.21}$$

Horizontal force equilibrium of the free body results in

$$H = \frac{F}{2} \tag{6.22}$$

Moment equilibrium about the upper support gives

$$-\frac{M}{2} + VL\sin\theta - \frac{FL}{2}\cos\theta = 0 \tag{6.23}$$

In view of (6.21) and (6.22), (6.23) can be rewritten as

$$F = -\frac{2k\theta}{L\cos\theta} + 2P\frac{\sin\theta}{\cos\theta} \tag{6.24}$$

By the compatibility of the deformation, we have

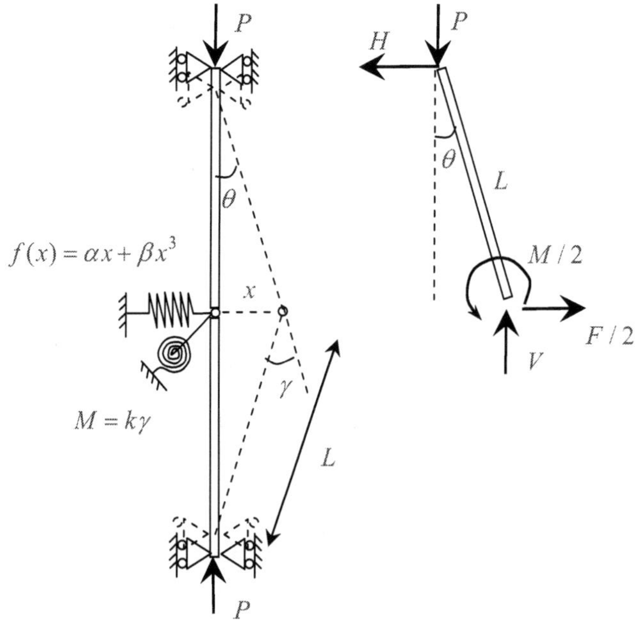

Figure 6.10 Nonlinear buckling of a rigid two-bar system

$$\cos\theta = \frac{\sqrt{L^2 - x^2}}{L}, \quad \sin\theta = \frac{x}{L}, \quad \theta = \tan^{-1}\left(\frac{x}{\sqrt{L^2 - x^2}}\right) \tag{6.25}$$

Substitution of (6.25) into (6.24) leads to

$$F = -\frac{2k}{\sqrt{L^2 - x^2}} \tan^{-1}(\frac{x}{\sqrt{L^2 - x^2}}) + 2P\frac{x}{\sqrt{L^2 - x^2}} \tag{6.26}$$

Assuming x is small compared to L, we can expand the arc tangent function by the following Taylor's series expansion:

$$\tan^{-1}(\frac{x}{\sqrt{L^2 - x^2}}) = \frac{x}{\sqrt{L^2 - x^2}} - \frac{x^3}{3(L^2 - x^2)^{3/2}} + ... \tag{6.27}$$

In view of (6.27), (6.26) can be approximated as

$$F = -\frac{2k}{\sqrt{L^2 - x^2}} \{\frac{x}{\sqrt{L^2 - x^2}} - \frac{x^3}{3(L^2 - x^2)^{3/2}}\} + 2P\frac{x}{\sqrt{L^2 - x^2}} \tag{6.28}$$

We can further expand the denominator using the following formulas (Spiegel, 1968):

$$\frac{1}{1-x} = 1 + x + x^2 + ... \tag{6.29}$$

$$\frac{1}{(1-x)^{1/2}} = 1 + \frac{1}{2}x + \frac{3}{8}x^2 + ... \tag{6.30}$$

Applying (6.29) and (6.30) to (6.28), we obtain

$$F = \frac{2Px}{L} + \frac{Px^3}{L^3} - \frac{2kx}{L^2} - \frac{4k}{3}\frac{x^3}{L^4} + ... \tag{6.31}$$

The equation of motion in the x-direction gives

$$m\ddot{x} + f(x) - F = 0 \tag{6.32}$$

where m is the mass of the bar system. Using (6.20) and (6.31) retaining up to the third order terms only, we can rewrite (6.32) as

$$m\ddot{x} + (\alpha - \frac{2P}{L} + \frac{2k}{L^2})x + (\beta - \frac{P}{L^3} + \frac{4k}{3L^4})x^3 = 0 \tag{6.33}$$

Mathematically, we can rewrite (6.33) as

$$m\ddot{x} + a_1 x + a_3 x^3 = 0 \tag{6.34}$$

where

$$a_1 = \alpha - \frac{2P}{L} + \frac{2k}{L^2} \tag{6.35}$$

$$a_3 = \beta - \frac{P}{L^3} + \frac{4k}{3L^4} \tag{6.36}$$

Note that in this formulation, we have only retained the first nonlinear term in the formulation. In addition, this third order nonlinear term has nothing to do with whether we use the nonlinear or linear spring in Figure 6.10. To see this, we can set $\beta = 0$ in (6.36), and the third nonlinear term remains.

For static buckling, we can set the acceleration term to zero:

$$\ddot{x} = 0 \tag{6.37}$$

Thus, (6.34) becomes

$$a_1 x + a_3 x^3 = x(a_1 + a_3 x^2) = 0 \tag{6.38}$$

The three roots for (6.38) are obviously

$$x_1 = 0, \quad x_{2,3} = \pm\sqrt{-\frac{a_1}{a_3}} \tag{6.39}$$

The first root in (6.39) corresponds to the straight unbuckled state, whereas the other two are the buckled states. The linearized equation of (6.38) becomes

$$xa_1 = 0 \tag{6.40}$$

For buckling to occur, x cannot be zero. Thus, we must have $a_1 = 0$, leading to

$$P_{crit} = \frac{\alpha L}{2} + \frac{k}{L} \tag{6.41}$$

Since x can be of any value, and, thus, this is the buckling load of the system. For $a_1 < 0$, we have

$$\alpha - \frac{2P}{L} + \frac{2k}{L^2} < 0 \tag{6.42}$$

Rearranging (6.42), we have

$$P > \frac{\alpha L}{2} + \frac{k}{L} = P_{crit} \tag{6.43}$$

and, for $a_1 < 0$, the second and third roots given in (6.39) are real. Similarly, for $a_1 > 0$, we have

$$P < \frac{\alpha L}{2} + \frac{k}{L} = P_{crit} \tag{6.44}$$

The results can be summarized as:

$$\begin{aligned} &a_1 > 0: \quad P < P_{crit}, \quad x = 0 \\ &a_1 < 0: \quad P > P_{crit}, \quad x = \pm\sqrt{-\frac{a_1}{a_3}} \end{aligned} \tag{6.45}$$

The static buckling problem is solved.

(ii) The more in-depth question is whether these solutions obtained in Part (i) are stable. To address the stability of the solution, we have to consider the dynamic stability analysis. For the unbuckled or straight state $x = 0$, we consider the neighborhood of $x \to 0$. In particular, for small x, we have the system approximated by

$$m\ddot{x} + a_1 x = 0 \tag{6.46}$$

Initial disturbance is imposed as:

$$\dot{x}(0) = d_0, \quad x(0) = v_0 \tag{6.47}$$

The solution of (6.46) is

$$x = c_1 \cos(\sqrt{\frac{a_1}{m}}t) + c_2 \sin(\sqrt{\frac{a_1}{m}}t) \tag{6.48}$$

Employing the initial conditions given in (6.47), we find

$$x = d_0 \cos(\sqrt{\frac{a_1}{m}}t) + \sqrt{\frac{m}{a_1}}v_0 \sin(\sqrt{\frac{a_1}{m}}t) \tag{6.49}$$

For a given initial disturbance, (6.49) gives the locus of an ellipse, and the size of the ellipse depends on the initial disturbance or the energy level after excitations.

Since no damping is taken into the formulation, the dynamic motion of the column is a stable center in the terminology of Poincaré (see Section 5.9 of Chau, 2018).

For more general cases, we can solve (6.34) by multiplying it with the velocity as:

$$m\dot{x}\ddot{x} + a_1\dot{x}x + a_3\dot{x}x^3 = 0 \tag{6.50}$$

This can be grouped as

$$\frac{d}{dt}\{\frac{1}{2}m\dot{x}^2 + \frac{1}{2}a_1x^2 + \frac{1}{4}a_3x^4\} = 0 \tag{6.51}$$

Thus, we can find the first integral as (i.e., the equation after integrating once)

$$\frac{1}{2}m\dot{x}^2 + \frac{1}{2}a_1x^2 + \frac{1}{4}a_3x^4 = E \tag{6.52}$$

where E is a constant and physically the energy of the system. Physically, the first term on the left-hand side is the kinetic energy, and the second and third terms are the potential energy stored in the springs. Equation (6.52) can be rewritten as

$$\frac{1}{2}m\dot{x}^2 = E - \varphi(x) \tag{6.53}$$

where

$$\varphi(x) = \frac{1}{2}a_1x^2 + \frac{1}{4}a_3x^4 \tag{6.54}$$

We next examine the maximum and the minimum of this potential energy function φ. Differentiating (6.54) with respect to x and setting the result to zero, we get

$$\varphi'(x) = a_1x + a_3x^3 = x(a_1 + a_3x^2) = 0 \tag{6.55}$$

Therefore, the maximum and minimum potential energy occurs at

$$x = 0, \quad x = \pm\sqrt{-\frac{a_1}{a_3}} \tag{6.56}$$

which are exactly the same as the solution obtained in (6.39). There is a local maximum at $x = 0$ and two minima at $x = \pm(-a_1/a_3)^{1/2}$. The minimum value of the potential φ is

$$\varphi(\sqrt{-\frac{a_1}{a_3}}) = -\frac{a_1}{2}(\frac{a_1}{a_3}) + \frac{a_3}{4}(\frac{a_1}{a_3})^2 = -\frac{a_1^2}{4a_3} \tag{6.57}$$

The potential φ and its relation with velocity and displacement can be illustrated in Figure 6.11. From the energy point of view, the unbuckled state is unstable. Imagine that a ball initially located at $x = 0$ will fall into one of the two energy wells when it is subject to a small disturbance.

The framework outlined here can be employed for later analysis for column buckling subject to end compression.

Remarks: A simplified version of the two-bar system without rotational spring shown in Figure 6.10 has been found useful in preliminary buckling analysis of cylindrical shell by Karman, Dun and Tsien in 1940 (see Hutchinson and Budiansky, 1966).

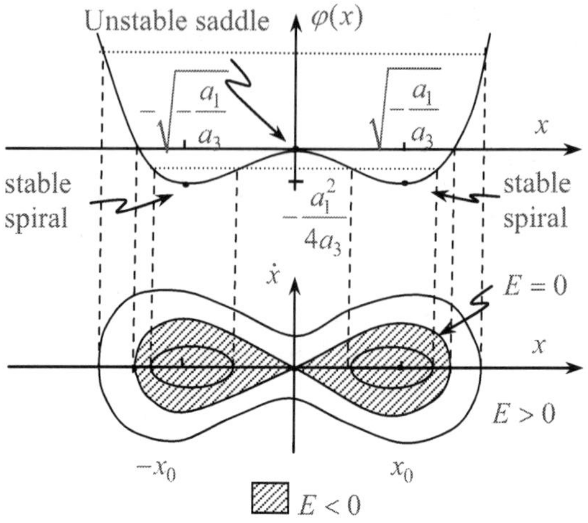

Figure 6.11 Sectional and contour plots of the potential energy of the two-bar system

6.2 LAGRANGIAN OR GREEN'S STRAIN

In this section, we will formulate nonlinear strain due to large deformation but small strain, in terms of Lagrangian or Green's strain. Figure 6.12 shows an initial body *I* and a deformed body *D* together with the definition of displacement vector *U*. The position vectors of the initial undeformed and current deformed bodies are denoted by

$$\boldsymbol{X} = (X_1, X_2, X_3) \tag{6.58}$$

$$\boldsymbol{x} = (x_1, x_2, x_3) \tag{6.59}$$

Note that we must distinguish the initial and current bodies to allow for large deformation, and without such formulation, buckling is not possible. The new deformed position equals

$$\boldsymbol{x} = \boldsymbol{X} + \boldsymbol{U}(\boldsymbol{X}) \tag{6.60}$$

In component form, we have

$$x_i = X_i + U_i \tag{6.61}$$

Similarly, an initial element of length *dS* will deform to *ds*. The component of the element can be expressed in terms of the incremental displacement as

$$\begin{aligned} dx_i &= dX_i + dU_i(\boldsymbol{X}) \\ &= dX_i + U_{i,j} dX_j \\ &= (\delta_{ij} + U_{i,j}) dX_j \end{aligned} \tag{6.62}$$

where δ_{ij} is the Dirac delta function. We have used tensor notation in (6.62), and recall from Chau (2018) that the repeated indices imply summation. An infinitesimal line element in the original body can be mapped to the deformed body by using the deformation gradient $\boldsymbol{F}$ as

$$dx_i = \frac{\partial x_i}{\partial X_j} dX_j = F_{ij} dX_j \tag{6.63}$$

In tensor form, (6.63) can be written as

$$d\boldsymbol{x} = \boldsymbol{F} \cdot d\boldsymbol{X} \tag{6.64}$$

The length of the line element in the original body and the deformed body can be determined as a dot product of the line element vectors:

$$(dS)^2 = d\boldsymbol{X} \cdot d\boldsymbol{X} = \delta_{ij} dX_i dX_j \tag{6.65}$$

$$(ds)^2 = d\boldsymbol{x} \cdot d\boldsymbol{x} = dx_i dx_i = x_{k,j} dX_j x_{k,l} dX_l = C_{jl} dX_j dX_l \tag{6.66}$$

where the right Cauchy-Green's deformation tensor $\boldsymbol{C}$ can be defined as

$$\boldsymbol{C} = \boldsymbol{F}^T \cdot \boldsymbol{F} \tag{6.67}$$

The difference between the square of the line element before and after deformation is

$$(ds)^2 - (dS)^2 = (C_{jl} - \delta_{jl}) dX_i dX_j \tag{6.68}$$

We can define the Lagrangian or Green's strain as

$$E_{ij}(\boldsymbol{X}) = \frac{1}{2}(C_{ij} - \delta_{ij}) \tag{6.69}$$

Employing (6.62), (6.63), and (6.66), we obtain

$$\begin{aligned} C_{ij} &= x_{k,i} x_{k,j} = (\delta_{ki} + U_{k,i})(\delta_{kj} + U_{k,j}) \\ &= \delta_{ij} + U_{j,i} + U_{i,j} + U_{k,i} U_{k,j} \end{aligned} \tag{6.70}$$

Substitution of (6.70) into (6.69) leads to

$$E_{ij} = \frac{1}{2}(U_{j,i} + U_{i,j} + U_{k,i} U_{k,j}) \tag{6.71}$$

This result agrees with Section 2.5 of Chau (2018). In Cartesian coordinates, we can set

$$U_1 = U, \quad U_2 = V, \quad U_3 = W \tag{6.72}$$

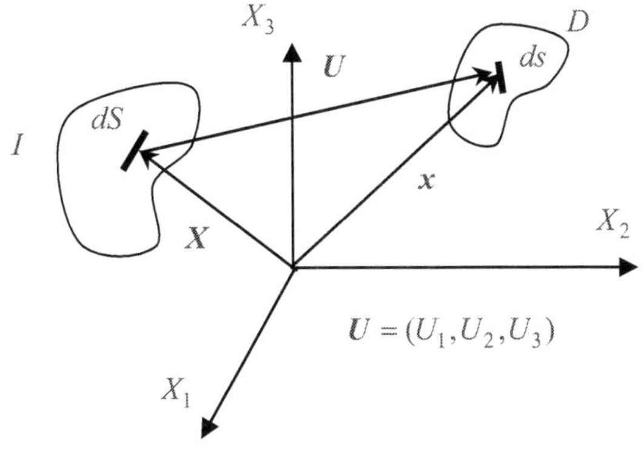

Figure 6.12 Initial and deformed bodies with definitions of displacement vector U

Taking the normal strain in plane with normal along the X_1-direction, we have the axial strain being

$$E_{11} = e_X = U_X + \frac{1}{2}(U_X^2 + V_X^2 + W_X^2) \tag{6.73}$$

This formula will be employed in the next section. This result can be found in Novozhilov (1953).

6.3 EULER-BERNOULLI BEAM

In this section, we will consider the Euler-Bernoulli theory. To make this theory valid, the depth of the beam D must be much smaller than the length of the beam L (say $L/D > 10$), and, thus, the beam can be considered thin. As shown in Figure 6.13, only two displacement components are nonzero, namely U and W. The coordinate is chosen at the centroid of the section such that

$$\iint Z dY dZ = 0 \tag{6.74}$$

All stresses and strains are negligible except for those along the X direction, and using the 1-D Hooke's law, we get:

$$\sigma_X = E e_X \tag{6.75}$$

From the result derived in (6.73), we have the following nonlinear axial strain

$$e_X = U_X + \frac{1}{2}\left[U_X^2 + V_X^2 + W_X^2\right] \tag{6.76}$$

For the case of a 1-D bar, we can assume

$$|U_X|, |V_X| << 1 \tag{6.77}$$

Using the approximation given in (6.77), we simplify the axial strain to

$$e_X = U_X + \frac{1}{2} W_X^2 \tag{6.78}$$

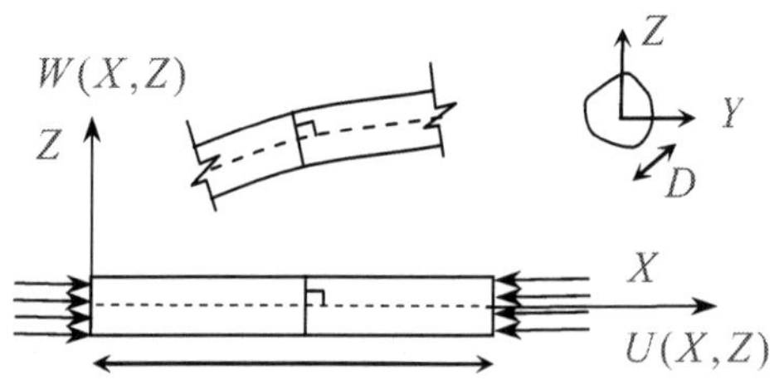

Figure 6.13 Initial and deformed bodies with definitions of displacement vector U

Another important assumption of the Euler-Bernoulli theory is the so-called plane-remain-plane assumption. That is, a plane originally normal to the neutral axis of the beam will remain a plane and normal to the deflected neutral axis after bending (see Figure 6.13). This idea of plane-remain-plane had also been applied to plate formulation by Kirchhoff in 1850 and to shell formulation by Love in 1888. Physically, this assumption also implies zero shear strain:

$$e_{XZ}(X,0)=0 \tag{6.79}$$

A more refined theory taking shear strain effect approximately had been proposed by Timoshenko (see Chapter 1). For thin beams, we further assume that the unknown displacements can be expanded as

$$U(X,Z)=u(X)+Z\hat{U}(X) \tag{6.80}$$

$$W(X,Z)=w(X) \tag{6.81}$$

The second term on the right-hand side of (6.80) relates to bending, which is positive above the centroid and negative below the centroid. For large deformation, (6.71) can be used to express the shear strain as

$$2e_{XZ}=U_Z+W_X+U_XU_Z+V_XV_Z+W_XW_Z \tag{6.82}$$

The axial deformation along the X-direction is negligible, and the variations of U and W along the X-direction are small. Thus, we have

$$V\equiv 0, \quad |U_X|,|W_X|<<1 \tag{6.83}$$

In view of (6.83), (6.82) can be simplified as

$$2e_{XZ}\approx U_Z+W_X \tag{6.84}$$

Substituting (6.80), (6.81) and (6.84) into (6.79), we obtain

$$2e_{XZ}=\widehat{U}(X)+w_X(X)=0 \tag{6.85}$$

Thus, we have

$$\widehat{U}(X)=-w_X(X) \tag{6.86}$$

Substituting (6.86) into (6.80), we have

$$U(X,Z)=u(X)-Zw_X(X) \tag{6.87}$$

The negative sign on the right-hand side of (6.87) indicates that sagging type bending will lead to tensile displacement below the centroid. Employing (6.87) and (6.81), we can evaluate (6.78) as

$$e_X=U_X+\frac{1}{2}W_X{}^2=u_X-Zw_{XX}+\frac{1}{2}w_X{}^2=e_X^0-Zw_{XX} \tag{6.88}$$

where

$$e_X^0=u_X+\frac{1}{2}w_X{}^2 \tag{6.89}$$

The curvature of the deformed center line can be estimated as (see Section 18.5 of Britton et al., 1966)

$$\kappa=\frac{w_{XX}}{\left[1+w_X^2\right]^{3/2}}\approx w_{XX} \tag{6.90}$$

The axial stress in the column can be obtained by substituting (6.88) into (6.75) as

$$\sigma_X=Ee_X=E\left[e_X^0-Zw_{XX}\right] \tag{6.91}$$

Once the stress is obtained, the strain energy function can be determined.

6.3.1 Strain Energy Function

Since only axial stress is retained, we can evaluate the strain energy function as

$$\mathcal{E} = \iiint\limits_{beam} \frac{1}{2}\sigma_X e_X dXdYdZ \tag{6.92}$$

The factor 1/2 takes into account of the fact that e_X increases linearly with σ_X from zero (i.e., for linear elastic solid). For example, see Section 2.14 of Chau (2013). The cross-section area and the moment of inertia about the X-axis are defined as

$$A \equiv \iint dYdZ \tag{6.93}$$

$$I \equiv \iint\limits_{A} Z^2 dYdZ \tag{6.94}$$

Using (6.88) and (6.91), we find

$$\begin{aligned}
\mathcal{E} &= \frac{E}{2}\iiint e_X^2 dXdYdZ \\
&= \frac{E}{2}\iiint \left(e_X^0 - Zw_{XX}\right)^2 dXdYdZ \\
&= \frac{E}{2}\iint\limits_{A}\int_0^L \left[\left(e_X^0\right)^2 - 2Ze_X^0 w_{XX} + Z^2 w_{XX}^2\right] dXdYdZ
\end{aligned} \tag{6.95}$$

Note that the integration of the second term on the right-hand side is identically zero (since we have chosen the centroid as the coordinate system as shown in (6.74)). In view of (6.93) and (6.94), we find

$$\mathcal{E} = \frac{E}{2}\int_0^L A\left(e_X^0\right)^2 dX + \frac{E}{2}\int_0^L Iw_{XX}^2 dX \tag{6.96}$$

The first term on the right-hand side is the stretching energy and the second term is the bending energy. The simple support boundary condition can be expressed as

$$w = M_X = 0, \quad X = 0, L \tag{6.97}$$

The bending moment about the X-axis can be determined as

$$M_X = \iint\limits_{A} Z\sigma_X dYdZ = E\iint\limits_{A} Z\left[e_X^0 - Zw_{XX}\right] dYdZ \tag{6.98}$$

In view of (6.74) and (6.94), we have

$$M_X = -EIw_{XX} \tag{6.99}$$

This result agrees with the classical result that we got for beam bending in (1.92) of Chapter 1. The simple support condition becomes

$$w = w_{XX} = 0, \quad X = 0, L \tag{6.100}$$

For axial compression, loading can applied through two different ways. It is well known in experimental mechanics that axial compression can be applied using force control or displacement control. In force control experiments, we need a servo-control system to get the feedback of the bar. Once a prescribed force is applied, the bar will deform and the initially prescribed force will drop. The signal from the servo-control sensor is then used to adjust the actuator pressure to achieve the prescribed axial force. Intuitively, it is not difficult to imagine that a relatively rigid bar only deforms very little before a prescribed force is maintained, whereas a soft bar will undergo substantial deformation before a required force can be maintained. If the bar is too soft, it may even yield or buckle before a required load can be sustained. In displacement control experiments, we simply add a force continuously until a prescribed displacement is achieved. Thus, in displacement control, we need a load cell to record the load applied through the actuator. As a general principle in mechanics, if we apply force, we need to measure the deformation of the bar (which cannot be controlled because it depends on the stiffness of the bar), whereas if we impose displacement, we have to measure the force (which cannot be controlled because again it depends on the stiffness of the bar). Motivated by experimental set-up, the boundary condition 2 below has been formulated by Hoff (1951). The nonlinearity induced by this end condition is also the main focus of the present chapter.

<u>End boundary condition 1</u>: Force control

If an axial stress of magnitude H is applied, the boundary condition is

$$\sigma_X(0,Z) = \sigma_X(L,Z) = -H \tag{6.101}$$

The work done by this applied stress is

$$\begin{aligned} \mathcal{W}_1 &= \iint_A HU(0,Z)dYdZ - \iint_A HU(L,Z)dYdZ \\ &= H\iint_A [u(0) - Zw_X(0)]dYdZ - H\iint_A [u(L) - Zw_X(L)]dYdZ \end{aligned} \tag{6.102}$$

In view of (6.74), we obtain

$$\mathcal{W}_1 = HA[u(o) - u(L)] \tag{6.103}$$

<u>End boundary condition 2</u>: Displacement control

If an axial compression of magnitude C is applied at the ends, we have

$$U\left(0,Z\right) = u(0) - Zw_x(0) = \mathrm{C}, \tag{6.104}$$

$$U\left(L,Z\right) = u(L) - Zw_x(L) = -\mathrm{C} \tag{6.105}$$

Note that most of the compression machines in the laboratory actually use displacement control (the present case), and only a compression machine equipped with a servo-control sensor can apply a load control condition (case 1 above). There is no work done by the external force for this case:

$$\mathcal{W}_2 = \iint_A U(0,Z)dYdZ = 0 \tag{6.106}$$

If we integrate (6.104) and (6.105) over the cross-section, we have

$$u\left(0\right) = \mathrm{C}, \quad u\left(L\right) = -\mathrm{C} \tag{6.107}$$

The kinetic energy of the column is

$$\mathcal{K} = \iiint \frac{1}{2}\rho\left[U_T{}^2 + W_T{}^2\right]dXdYdZ \tag{6.108}$$

where ρ is the density of the beam. Substitution of (6.87) into (6.108) yields

$$\mathcal{K} = \frac{\rho}{2}\iiint\left[\left(u_T - Zw_{XT}\right)^2 + w_T{}^2\right]dXdYdZ \tag{6.109}$$

Assuming $Zw_{XT} << u_T$, we have

$$\mathcal{K} = \frac{\rho}{2}\iiint\left[u_T^2 + w_T^2\right]dXdYdZ = \frac{\rho A}{2}\int_0^L (u_T{}^2 + w_T^2)dX \tag{6.110}$$

The horizontal kinetic energy is normally negligible compared to the lateral kinetic energy, and thus we have

$$\mathcal{K} = \frac{\rho}{2}\iiint w_T^2 dXdYdZ \tag{6.111}$$

With all the strain energy, kinetic energy and external work done as known factors, we can now apply the Hamiltonian principle or variational principle to find the governing equation.

6.3.2 Hamiltonian Principle

According to Section 14.5 of Chau (2018), we can formulate the Hamiltonian principle for a deformable body as

$$V[u,w] \equiv \int_{T_0}^{T_1}[\mathcal{E} - \mathcal{W} - \mathcal{K}]dT \tag{6.112}$$

Substitution of (6.96), (6.103), and (6.110) into (6.112) gives

$$V = \int_{T_0}^{T_1}\{\int_0^L\left[\frac{AE}{2}\left(e_X^0\right)^2 + \frac{EI}{2}w_{XX}^2 - \frac{A\rho}{2}\left(u_T^2 + w_T^2\right)\right]dX \\ -HA\left[u(0,T) - u(L,T)\right]\}dT \tag{6.113}$$

Expressing the axial strain in terms of displacements, we get

$$V[u,w] = \int_{T_0}^{T_1}\{\int_0^L\left[\frac{AE}{2}(u_X + \frac{1}{2}w_X^2)^2 + \frac{EI}{2}w_{XX}^2 - \frac{\rho A}{2}\left(u_T^2 + w_T^2\right)\right]dX \\ -HA\left[u(0,T) - u(L,T)\right]\}dT \tag{6.114}$$

According to the Hamilton principle, the actual motion that occurs makes V stationary.

6.3.3 Calculus of Variations

Following Chapter 13 of Chau (2018), we can take the calculus of variations of V to ensure stationary conditions. In particular, we impose on the admissible function an admissible variation such that

$$V\left[u + \varepsilon\bar{u}, w + \varepsilon\bar{w}\right] = V\left[u, \bar{u}, w, \bar{w}\right] \tag{6.115}$$

where ε is a small parameter. Then, we require for all admissible variations satisfying

$$V_1\left[u, w; \bar{u}, \bar{w}\right] = \left.\frac{dV}{d\varepsilon}\right|_{\varepsilon=0} = 0 \tag{6.116}$$

More explicitly, (6.115) can be expressed as

$$V\left[u+\varepsilon\bar{u},w+\varepsilon\bar{w}\right]=\int_{T_0}^{T_1}\{\int_0^L[\frac{AE}{2}\left(u_X+\varepsilon\bar{u}_X+\frac{1}{2}(w_X+\varepsilon\bar{w}_X)^2\right)^2$$
$$+\frac{EI}{2}\left(w_{XX}+\varepsilon\bar{w}_{XX}\right)^2 \tag{6.117}$$
$$-\frac{\rho A}{2}\left((u_T+\varepsilon\bar{u}_T)^2+\left(w_T+\varepsilon\bar{w}_T\right)^2\right)]dX$$
$$-HA\left[u\left(0,T\right)+\varepsilon\bar{u}(0,T)-u\left(L,T\right)-\varepsilon\bar{u}(L,T)\right]\}dT$$

Taking the differentiation of (6.117) with respect to ε, we have

$$\frac{dV}{d\varepsilon}=\int_{T_0}^{T_1}\{\int_0^L[AE\left(u_X+\varepsilon\bar{u}_X+\frac{1}{2}\left(w_X+\varepsilon\bar{w}_X\right)^2\right)[\bar{u}_X+\left(w_X+\varepsilon\bar{w}_X\right)\bar{w}_X]$$
$$+EI\left(w_{XX}+\varepsilon\bar{w}_{XX}\right)\bar{w}_{XX} \tag{6.118}$$
$$-\rho A\left(\left(u_T+\varepsilon\bar{u}_T\right)\bar{u}_T+\left(w_T+\varepsilon\bar{w}_T\right)\bar{w}_T\right)]dX$$
$$-HA\left[\bar{u}(0,T)-\bar{u}(L,T)\right]\}dT$$

Setting $\varepsilon = 0$, we find

$$\left.\frac{dV}{d\varepsilon}\right|_{\varepsilon=0}=\int_{T_0}^{T_1}\{\int_0^L[AEe_X^0\left(\bar{u}_X+w_X\bar{w}_X\right)+EIw_{XX}\bar{w}_{XX}$$
$$-\rho A\left(u_T\bar{u}_T+w_T\bar{w}_T\right)]dX \tag{6.119}$$
$$-HA\left[\bar{u}(0,T)-\bar{u}(L,T)\right]\}dT$$

Using integration by parts, we have

$$\int_0^L e_X^0\bar{u}_X dX=e_X^0\bar{u}\Big|_0^L-\int_0^L e_{X,X}\bar{u}dX \tag{6.120}$$

$$\int_0^L\left(e_Xw_X\right)\bar{w}_XdX=\left(e_Xw_X\bar{w}_X\right)\Big|_0^L-\int_0^L(e_Xw_X)_{,X}\bar{w}_XdX \tag{6.121}$$

$$\int_0^L w_{XX}\bar{w}_{XX}dX=w_{XX}\bar{w}_X\Big|_0^L-\int_0^L w_{XXX}\bar{w}_XdX$$
$$=-w_{XXX}\bar{w}\Big|_0^L-\int_0^L w_{XXXX}\bar{w}dX \tag{6.122}$$

$$\int_{T_0}^T u_T\bar{u}_TdT=u_T\bar{u}\Big|_{T_0}^T-\int_{T_0}^T u_{TT}\bar{u}dT \tag{6.123}$$

Substitution of (6.120) to (6.123) into (6.119) results in

$$V_1\left[u,w;\bar{u},\bar{w}\right]=\int_{T_0}^{T_1}\{[\int_0^L\left(-AEe_{X,X}^0+\rho Au_{TT}\right)\bar{u}dX$$
$$+\int_0^L\left(-AE(e_X^0w_X)_{,X}+EIw_{XXXX}+\rho Aw_{TT}\right)\bar{w}\,dX] \tag{6.124}$$
$$+AEe_X^0\,\bar{u}\Big|_0^L+AH\,\bar{u}\Big|_0^L\}dT$$

Considering the axial stress at the centroid, we can write (6.91) as

$$\sigma_X^0 = Ee_X^0 = \sigma(X,T) = E\left[u_X + \frac{1}{2}w_X^2\right] \tag{6.125}$$

Recalling (6.116), we obtain the following Euler-Lagrange equations

$$\frac{\partial \sigma}{\partial X} = \rho u_{TT} \tag{6.126}$$

$$EIw_{XXXX} - A\left(\sigma w_X\right)_{,X} + \rho A w_{TT} = 0 \tag{6.127}$$

$$\bar{u}(0,T) = \bar{u}(L,T) = 0 \tag{6.128}$$

$$\sigma = -H, \quad X = 0, L \tag{6.129}$$

Note that the second term of (6.127) is nonlinear since σ is a function of u and w as shown in (6.125). The linearized form is

$$\sigma_{,X} = \rho u_{TT} \tag{6.130}$$

$$\sigma = Eu_X \tag{6.131}$$

$$EIw_{XXXX} + \rho A w_{TT} = 0 \tag{6.132}$$

Substitution of (6.131) into (6.130) yields the classical Euler-Bernoulli vibrating beam

$$u_{XX} - \frac{\rho}{E}u_{TT} = 0 \tag{6.133}$$

$$EIw_{XXXX} + \rho A w_{TT} = 0 \tag{6.134}$$

This set of PDEs admits the following traveling waves in the beam

$$u = e^{ik(X-\gamma_1 T)}, \quad w = e^{ik(X-\gamma_2 T)} \tag{6.135}$$

Substituting (6.135) into (6.133) and (6.134), we obtain

$$-k^2 + \frac{\gamma_1^2 k^2}{E}\rho = 0 \tag{6.136}$$

$$k^4 - \gamma_2^2 k^2 \frac{\rho A}{EI} = 0 \tag{6.137}$$

Substituting (6.135) into (6.133) and (6.134), we obtain

$$\gamma_1^2 = \frac{E}{\rho}, \quad \gamma_2^2 = \frac{EI}{\rho A}k^2 = \gamma_1^2 \alpha\left(hk\right)^2 \tag{6.138}$$

where γ_1 is the axial wave speed, and γ_2 is the flexural wave speed and h is the thickness of the beam. For rectangular columns, we have α being

$$\alpha = \frac{1}{12} \tag{6.139}$$

Thus, α is a constant order term even for $h \to 0$. Since physically k is the wave number, and thus is inversely proportional to the wavelength:

$$hk = \frac{2\pi h}{\lambda} \tag{6.140}$$

If hk is much smaller than γ_1, the bending wave speed γ_2 is smaller than the axial wave speed (see (6.138)). For the thin beam theory to be valid, we require that hk is a small number. When the column starts to buckle laterally, many axial waves will have occurred and the axial vibrations would have died down because of damping.

Anyhow, when axial inertia is neglected, the equation of motion along the X-direction given in (6.126) becomes

$$\sigma_{,X} = 0 \tag{6.141}$$

This assumption is essential for later formulation. Thus, the axial stress is constant and equals

$$\sigma = E\left[u_X + \frac{1}{2}\omega_X^{\ 2}\right] \tag{6.142}$$

The flexural motions are governed by

$$\frac{EI}{A} w_{XXXX} - (\sigma w_X)_X + \rho w_{TT} = 0 \tag{6.143}$$

$$w = w_{XX} = 0, \quad X = 0, L \tag{6.144}$$

The end boundary conditions for Case 1 are

$$\sigma(0,T) = \sigma(L,T) = -H \tag{6.145}$$

The end boundary conditions for Case 2 are

$$u(0,T) = C, \quad u(L,T) = -C \tag{6.146}$$

6.3.4 Applied Force versus Applied Displacement

Case 1: Applied stress

For end boundary condition 1, since (6.141) requires a constant axial stress, it must equal $-H$ induced by the end boundary. Consequently, (6.143) is reduced to

$$\frac{EI}{A} w_{XXXX} + H w_{XX} + \rho w_{TT} = 0 \tag{6.147}$$

For steady-state deflection, we have

$$\frac{EI}{A} w_{XXXX} + H w_{XX} = 0 \tag{6.148}$$

As discussed in Section 3.5.8 of Chau (2018), we can assume

$$\phi(X) = w_{XX} \tag{6.149}$$

Applying (6.149), we can simplify (6.148) as

$$\phi'' + \mu^2 \phi = 0, \quad \phi(0) = \phi(L) = 0 \tag{6.150}$$

where

$$\mu^2 = \frac{HA}{EI} \tag{6.151}$$

The solution of (6.150) is (see Section 1.11 of Chau, 2018):

$$\phi = C_1 \cos \mu X + C_2 \sin \mu X \tag{6.152}$$

The first boundary condition given in (6.150) leads to

$$\phi(0) = 0, \quad \Rightarrow C_1 = 0 \tag{6.153}$$

The second boundary condition given in (6.150) results in

$$\phi(L) = 0, \quad \Rightarrow \sin \mu L = 0 \tag{6.154}$$

Thus, we must have

$$\mu L = n\pi = \mu_n, \quad n = 1, 2, \ldots \tag{6.155}$$

This gives the eigenvalues of the homogeneous problem defined in (6.150) (compare Section 10.4 of Chau, 2018). The corresponding eigenfunction is

$$\phi(X) = C_2 \sin \frac{\mu_n}{L} X \tag{6.156}$$

Integrating (6.156) once, we have

$$w_X = \int \phi(X) dX + C = -\frac{C_2 L}{\mu_n} \cos \frac{\mu_n}{L} X + C \tag{6.157}$$

Integrating (6.157) one more time, we have

$$w = -\frac{C_2 L^2}{\mu_n^2} \sin \frac{\mu_n}{L} X + CX + D \tag{6.158}$$

The boundary conditions at the support lead to

$$w(0) = D = 0 \tag{6.159}$$

$$w(L) = CL = 0 \tag{6.160}$$

Thus, we have $C = D = 0$. Substituting $\sigma = -H$ into (6.142), we get

$$u_X = -\frac{H}{E} - \frac{1}{2} w_X^2 = -\frac{H}{E} - \frac{1}{2} \frac{C_2^2 L^2}{\mu_n^2} \cos^2 (\mu_n \frac{X}{L}) \tag{6.161}$$

Integrating (6.161), we find the axial deformation as

$$\begin{aligned} u &= -\frac{H}{E} X - \frac{1}{2} \frac{C_2^2 L^3}{\mu_n^3} \{\frac{1}{4} \sin(2\mu_n \frac{X}{L}) + \frac{1}{2} \mu_n \frac{X}{L}\} \\ &= -(\frac{H}{E} + \frac{1}{4} \frac{C_2^2 L^2}{\mu_n^2}) X - \frac{1}{8} \frac{C_2^2 L^3}{\mu_n^3} \sin(2\mu_n \frac{X}{L}) \end{aligned} \tag{6.162}$$

The buckling load can be found (6.151) and (6.155) as

$$H_n = \frac{EI}{A} (n\pi)^2, \quad n = 1, 2, \ldots \tag{6.163}$$

The problem is completely solved.

Case 2: Applied end displacement

For the case of applied end compression C with damping, the PDE for lateral motion w is

$$w_{XXXX} - \frac{A}{EI} \sigma w_{XX} + \frac{\rho A}{EI} w_{TT} + \frac{\Gamma}{EI} w_T = 0 \tag{6.164}$$

where Γ is the damping coefficient. As we have seen from (6.141), the stress is a constant along the beam. This fact allows us to integrate the axial strain given in (6.142) from 0 to L and set the result to σL:

$$\sigma L = \int_0^L E\left[u_X + \frac{1}{2}w_X^2\right]dX$$
$$= E[u(L) - u(0) + \frac{1}{2}\int_0^L w_X^2 dX] \tag{6.165}$$

$$\sigma L = E[-2C + \frac{1}{2}\int_0^L w_X^2 dX] \tag{6.166}$$

We now apply the following change of variables

$$x = \frac{X}{L}, \quad t = T\sqrt{\frac{E}{\rho L^2 k}}, \quad \hat{w} = \frac{1}{L}w(X,T), \quad k = \frac{L^2 A}{I} \tag{6.167}$$

Substitution of (6.167) to (6.164) leads to (after dropping the "∧")

$$w_{xxxx} + \lambda w_{xx} + w_{tt} + 2\gamma w_t = 0 \tag{6.168}$$

where

$$\lambda(t) = -\frac{k}{E}\sigma(t), \quad \gamma = \frac{\Gamma L^2 k}{2EA}\sqrt{\frac{E}{\rho L^2 k}} \tag{6.169}$$

$$\lambda(t) = k[2c - \frac{1}{2}\int_0^1 w_x^2 dx] \tag{6.170}$$

$$c = \frac{C}{L} \tag{6.171}$$

The normalized term λ can be interpreted as the normalized axial stress, which is a function of time; and c is the normalized end compression. Equation (6.168) is nonlinear because (6.170) shows that λ is a function of the unknown w. It is this nonlinearity leading to the possibility of bifurcation, which is the main concern for the rest of this chapter. Apparently, this nonlinear PDE was first obtained by Hoff in 1951.

6.4 STATIC BUCKLING THEORY OF BEAM

For the static case, we can ignore all time-dependent terms. Equation (6.168) is reduced to

$$w_{xxxx} + \lambda w_{xx} = 0 \tag{6.172}$$

where λ is a function of w and is equal to

$$\lambda(t) = k[2c - \frac{1}{2}\int_0^1 w_x^2 dx] \tag{6.173}$$

The boundary conditions for simply-supported ends are

$$w = w_{xx} = 0, \quad x = 0,1 \tag{6.174}$$

The trivial solution is

$$w = 0, \quad \lambda = 2kc \tag{6.175}$$

For the nontrivial solution, we can reduce the order of (6.172) by introducing

$$f(x) = w_{xx} \tag{6.176}$$

Substitution of (6.176) into (6.174) gives

$$f'' + \lambda f = 0 \tag{6.177}$$

The homogeneous boundary conditions for f are

$$f(0) = f(1) = 0 \tag{6.178}$$

This eigenvalue problem has been solved in the last section and the general solution of f can be expanded in its eigenfunction expansion:

$$f = A_n \phi_n(x) \tag{6.179}$$

$$\phi_n(x) = \sqrt{2}\sin(n\pi x), \quad \lambda_n = n\pi \tag{6.180}$$

Integrating (6.176) twice, we obtain

$$w(x) = -\frac{A_n \phi_n}{(n\pi)^2} + a_n x + b_n \tag{6.181}$$

The boundary conditions given in (6.174) for w result in

$$a_n = b_n = 0 \tag{6.182}$$

Consequently, (6.181) becomes

$$w(x) = -\frac{A_n \phi_n}{(n\pi)^2} \tag{6.183}$$

Substitution of (6.183) into (6.173) gives

$$\begin{aligned} \lambda_n &= k\{2c - \frac{1}{2}\int_0^1 (-\frac{A_n}{n\pi}\sqrt{2})^2[\cos^2(n\pi x)]dx\} \\ &= k\{2c - \frac{A_n^2}{2(n\pi)^2}\} \end{aligned} \tag{6.184}$$

Solving for c, we obtain the post-buckling solution as

$$c = \frac{\lambda_n}{2k} + \frac{A_n^2}{4\lambda_n^2} = c_n + \frac{A_n^2}{4\lambda_n^2} \tag{6.185}$$

where the eigenvalue is defined as

$$c_n = \frac{\lambda_n}{2k} = \frac{n\pi I}{2L^2 A} \tag{6.186}$$

Figure 6.14 plots (6.185), showing a parabolic post-buckling curve of c versus A_n.

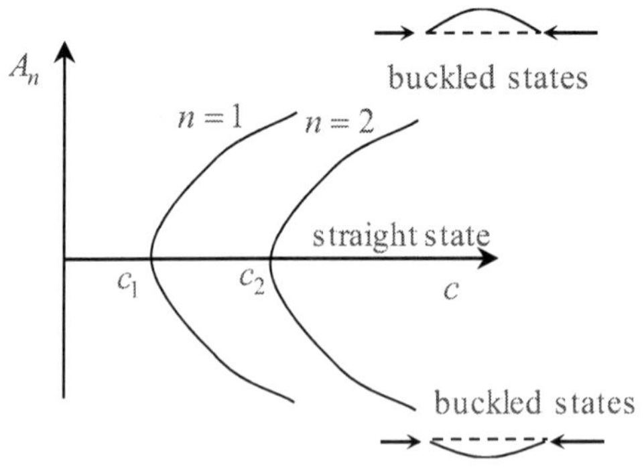

Figure 6.14 Buckling of column under end displacements

6.5 LINEAR DYNAMIC STABILITY OF STATIC STATES

To check the linear dynamic stability of the infinite solution (one trivial and infinite nontrivial solutions for λ_n, $n = 1,2,3...$), we have to linearize the nonlinear PDE about the static solution obtained in the last section and examine its stability by adding a perturbation. First, the nonlinear dynamic theory is summarized as

$$w_{xxxx} + \lambda w_{xx} + w_{tt} + 2\gamma w_t = 0 \tag{6.187}$$

$$\lambda(t) = k[2c - \frac{1}{2}\int_0^1 w_x^2 dx] \tag{6.188}$$

$$w = w_{xx} = 0, \quad x = 0,1 \tag{6.189}$$

Let us assume the static solution obtained in Section 6.4 is w^0. To check the dynamic stability, we need to apply disturbances to the system about the static solution:

$$w(x,0) = \delta f(x), \quad w_t(x,0) = \delta g(x) \tag{6.190}$$

where δ is a small parameter.

6.5.1 Perturbation Method

To linearize the system, we assume the following perturbation

$$w(x,t,\delta) = w^0(x,t) + \delta w^1(x,t) + \frac{\delta^2}{2} w^2(x,t) + ... \tag{6.191}$$

where we require

$$w^1(x,t) = \left.\frac{\partial w(x,t,\delta)}{\partial \delta}\right|_{\delta=0} \tag{6.192}$$

Differentiation with respect to δ is represented as

$$\dot{w} = \frac{\partial w(x,t,\delta)}{\partial \delta} \tag{6.193}$$

In view of (6.191), we differentiate the nonlinear PDE given in (6.187) with respect to δ to get

$$\dot{w}_{xxxx} + \lambda \dot{w}_{xx} + \dot{\lambda} w_{xx} + \dot{w}_{tt} + 2\gamma \dot{w}_t = 0 \tag{6.194}$$

$$\dot{\lambda}(t) = -\frac{k}{2}\int_0^1 2\dot{w}_x w_x dx \tag{6.195}$$

The boundary condition becomes

$$\dot{w} = \dot{w}_{xx} = 0, \quad x = 0,1 \tag{6.196}$$

$$\dot{w}(x,0) = f(x), \quad \dot{w}_t(x,0) = g(x) \tag{6.197}$$

Note that these initial disturbances (in terms of both displacement and velocity) are imposed to check whether the system will return to the equilibrium solution when it is disturbed slightly. If yes, the corresponding solution is stable; if not, it is unstable. Let us make the following identification:

$$\lambda_0 = \lambda\big|_{\delta=0}, \quad \dot{\lambda}_0 = \left.\frac{\partial \lambda}{\partial \delta}\right|_{\delta=0} \tag{6.198}$$

Substituting (6.191) into the system and setting δ = 0 in (6.194) to (6.197), we obtain

$$w^1_{xxxx} + \lambda_0 w^1_{xx} + \dot{\lambda}_0 w^0_{xx} + w^1_{tt} + 2\gamma w^1_t = 0 \tag{6.199}$$

$$\dot{\lambda}_0 = -k\int_0^1 w^0_x w^1_x dx \tag{6.200}$$

$$\lambda_0 = k[2c - \frac{1}{2}\int_0^1 (w^0_x)^2 dx] \tag{6.201}$$

$$w^1 = w^1_{xx} = 0, \quad x = 0,1 \tag{6.202}$$

$$w^1(x,0) = f(x), \quad w^1_t(x,0) = g(x) \tag{6.203}$$

We look for the solution of w^1 governed by the system of equations given in (6.199) to (6.203). The linear dynamic stability of the equilibrium state w^0 depends on whether w^1 grows or decays with time when subject to initial excitation given in (6.203).

6.5.2 Stability of Straight State

To test the dynamic stability of the straight state or the trivial solution, we have the equilibrium solution given by

$$w^0 = 0, \quad \lambda_0 = 2kc \tag{6.204}$$

Substitution of this trivial solution into (6.199) gives

$$w^1_{xxxx} + 2kcw^1_{xx} + w^1_{tt} + 2\gamma w^1_t = 0 \tag{6.205}$$

$$w^1 = w^1_{xx} = 0, \quad x = 0,1 \tag{6.206}$$

$$w^1(x,0) = f(x), \quad w^1_t(x,0) = g(x) \tag{6.207}$$

This is a linear fourth order PDE with the unknown being w^1. To solve for this linearized system, we assume the following eigenfunction expansion for the prescribed function $f(x)$ and $g(x)$ as (see Chapter 10 of Chau, 2018):

$$f(x) = \sum_{n=1}^{\infty} f_n \phi_n(x), \quad g(x) = \sum_{n=1}^{\infty} g_n \phi_n(x) \tag{6.208}$$

where $\phi_n(x)$ is defined in (6.180). We have assumed in (6.208) the whole spectrum of Fourier series expansion in the initial disturbances; thus, it can model any arbitrary initial disturbances. Once the functions $f(x)$ and $g(x)$ are given, the Fourier coefficients can be found:

$$f_n = \frac{2}{L}\int_0^L f(x)\phi_n(x)dx, \quad g_n = \frac{2}{L}\int_0^L g(x)\phi_n(x)dx \tag{6.209}$$

Similarly, we also expand the unknown function w^1 in Fourier series expansion (or eigenfunction expansion) as:

$$w^1(x,t) = \sum_{n=1}^{\infty} S_n(t)\phi_n(x) \tag{6.210}$$

Substitution of (6.210) into (6.205) gives

$$\sum_{n=1}^{\infty} [S_n'' + 2\gamma S_n' - 2kc(n\pi)^2 S_n + (n\pi)^4 S_n]\phi_n = 0 \tag{6.211}$$

This requires for each n

$$S_n'' + 2\gamma S_n' + (n\pi)^2[(n\pi)^2 - 2kc]S_n = 0 \tag{6.212}$$

From the last section, we have the eigenvalue for c as

$$c_n = \frac{(n\pi)^2}{2k} \tag{6.213}$$

Substitution of (6.208) and (6.210) into (6.207) gives

$$S_n(0) = f_n, \quad S_n'(0) = g_n \tag{6.214}$$

Equation (6.212) is a linear ODE for $S_n(t)$ with constant coefficients. As shown by Section 3.3.1 of Chau (2018), the solution can be expressed as the exponential function

$$S_n(t) = e^{\sigma_n t} \tag{6.215}$$

Substitution of (6.215) into (6.212) gives

$$\sigma_n^2 + 2\gamma\sigma_n + (n\pi)^2 2k(c_n - c) = 0 \tag{6.216}$$

This is a quadratic equation and the roots are

$$\sigma_n^{\pm} = -\gamma \pm \sqrt{\gamma^2 - (n\pi)^2 2k(c_n - c)} \tag{6.217}$$

Thus, the solution $S_n(t)$ becomes

$$S_n(t) = H_n^- e^{\sigma_n^- t} + H_n^+ e^{\sigma_n^+ t} \tag{6.218}$$

Applying the initial conditions given in (6.214), we have

$$S_n(0) = H_n^- + H_n^+ = f_n \tag{6.219}$$

$$S_n'(0) = H_n^- \sigma_n^- + H_n^+ \sigma_n^+ = g_n \tag{6.220}$$

Solving for the unknown constants, we find

$$H_n^{\pm} = \pm \frac{\sigma_n^{\mp} f_n - g_n}{\sigma_n^+ - \sigma_n^-} \tag{6.221}$$

Consider the special case of $c = c_n$, the solution of $\sigma_n(t)$ becomes

$$\sigma_n^+ = 0, \quad \sigma_n^- = -2\gamma \tag{6.222}$$

Subsequently, the unknown constants become

$$H_n^+ = -\frac{g_n}{2\gamma}, \quad H_n^- = f_n + \frac{g_n}{2\gamma} \tag{6.223}$$

The time-dependent function of w^1 is obtained as

$$S_n(t) = f_n + \frac{g_n}{2\gamma}[1 - e^{-2\gamma t}] \tag{6.224}$$

For the case of no damping (i.e., $\gamma = 0$), the solution becomes

$$S_n(t) = f_n + g_n t \tag{6.225}$$

This solution is unstable because $S_n(t) \to \infty$ as $t \to \infty$. For damped motions, we have (6.217) that

$$c < c_1 \quad \Rightarrow \quad \mathrm{Re}[\sigma_n^{\pm}] < 0, \quad n = 1, 2, \ldots \tag{6.226}$$

$$c > c_1 \quad \Rightarrow \quad \mathrm{Re}[\sigma_1^+] > 0 \tag{6.227}$$

Thus, we have $S_n(t) \to 0$ as $t \to \infty$ for $c < c_1$. Thus, the straight state is stable before the buckling induced by the end compression occurs. However, we have $S_n(t) \to \infty$ as $t \to \infty$ for $c > c_1$, or the straight state is unstable after the buckling is possible. This means that the column must buckle after the bifurcation point at $c = c_1$. At $c = c_1$, $S_n(t)$ does not go to zero as $t \to \infty$ and it is also unstable. In conclusion, the straight state is stable for $c < c_1$ but becomes unstable for $c \geq c_1$.

6.5.3 Stability of Buckled States

The equilibrium solution of the buckled states obtained in Section 6.4 is

$$w^0 = -\frac{A_n}{(n\pi)^2}\phi_n, \quad \lambda_0 = \lambda_n = (n\pi)^2 \tag{6.228}$$

where again $\phi_n(x)$ is defined in (6.180). Substitution of the equilibrium solution given in (6.228) into (6.199) gives

$$w_{xxxx}^1 + \lambda_n w_{xx}^1 + w_{tt}^1 + 2\gamma w_t^1 = -\dot{\lambda}_0 w_{xx}^0 \tag{6.229}$$

$$\dot{\lambda}_0 = -k\int_0^1 w_x^0 w_x^1 dx \tag{6.230}$$

$$w^1 = w_{xx}^1 = 0, \quad x = 0, 1 \tag{6.231}$$

Following the same procedure used in the last section, we can expand the initial conditions and the unknown function w^1 in terms of the eigenfunction as

$$w^1(x,0) = f(x) = \sum_{n=1}^{\infty} f_n\phi_n, \quad w_t^1(x,0) = g(x) = \sum_{n=1}^{\infty} g_n\phi_n \tag{6.232}$$

Following the same procedure used in the last section, we can expand the initial conditions and the unknown function w^1 in terms of the eigenfunction as

$$w^1(x,t) = \sum_{p=1}^{\infty} S_p(t)\phi_p(x) \tag{6.233}$$

Substitution of (6.228) and (6.233) into (6.229) gives

$$\begin{aligned}
&\sum_{p=1}^{\infty} [S_p'' + 2\gamma S_p' + (p\pi)^2[(p\pi)^2 - (n\pi)^2]S_p]\phi_p \\
&= -k\int_0^1 \{\frac{A_n}{n\pi}\sqrt{2}\cos n\pi x \sum_{p=1}^{\infty} \sqrt{2}S_p(p\pi)\cos p\pi x\} dx A_n\phi_n \\
&= -2\frac{A_n^2 k}{n\pi}\{\sum_{p=1}^{\infty} S_p p\pi\int_0^1 \cos n\pi x \cos p\pi x dx\}\phi_n
\end{aligned} \tag{6.234}$$

Next, we note the following orthogonal identity for the cosine function (see Section 10.5 of Chau, 2018):

$$\int_0^1 \cos n\pi x \cos p\pi x dx = 0 \qquad p \neq n$$
$$= \frac{1}{2} \qquad p = n \tag{6.235}$$

Application of (6.235) to (6.234) results in

$$\sum_{p=1}^{\infty} [S_p'' + 2\gamma S_p' + (p\pi)^2[(p\pi)^2 - (n\pi)^2]S_p]\phi_p = -kA_n^2 S_n \phi_n \tag{6.236}$$

Therefore, we have two ODEs, one for $p \neq n$ and one for $p = n$:

$$S_p'' + 2\gamma S_p' + (p\pi)^2[(p\pi)^2 - (n\pi)^2]S_p = 0, \quad p \neq n \tag{6.237}$$

$$S_n'' + 2\gamma S_n' + kA_n^2 S_n = 0, \qquad p = n \tag{6.238}$$

Since these ODEs have constant coefficients, the solutions are again in exponential form:

$$S_p(t) = A_p^+ e^{a_p^+ t} + A_p^- e^{a_p^- t}, \quad p = 1, 2, \ldots \tag{6.239}$$

where the exponential coefficients are

$$a_p^{\pm} = -\gamma \pm \sqrt{\gamma^2 - (p\pi)^2 \pi^2 (p^2 - n^2)}, \quad p \neq n \tag{6.240}$$

$$a_n^{\pm} = -\gamma \pm \sqrt{\gamma^2 - kA_n^2}, \quad p = n \tag{6.241}$$

It is straightforward to see that for $n = 1$

$$\text{Re}[a_p^{\pm}] < 0, \quad p = 2, 3, \ldots \tag{6.242}$$

From (6.239), we find that $S_p \to 0$ as $t \to \infty$, and therefore the buckled state is stable. For higher modes with $n > 1$ with $p = 1$, we see that

$$\text{Re}[a_1^{\pm}] > 0 \tag{6.243}$$

Therefore, we find that all buckled states with $n > 1$ are unstable. Recall from (6.232) and (6.233) that p is the mode of the buckled equilibrium state and n indicates the mode shape of the initial disturbance. That is, initial excitations of higher mode shapes will not lead to the buckling of lower modes. This is illustrated in Figure 6.15.

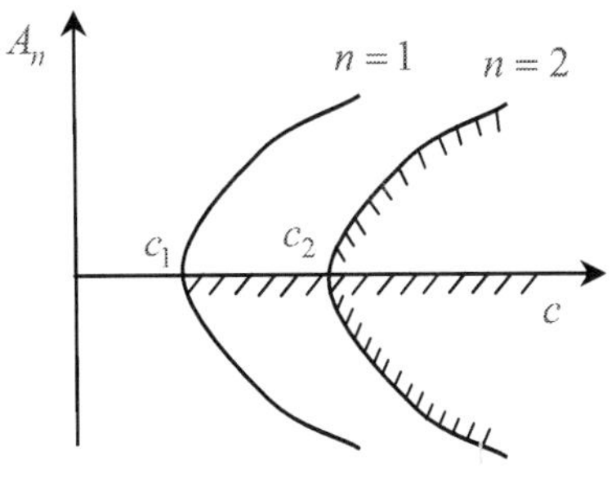

Figure 6.15 Unstable and stable solution for different ranges of *c*

6.6 NONLINEAR DYNAMIC STABILITY

To study the transition from the unstable straight state to a buckled state as c increases above the bifurcation point c_n, we must use nonlinear dynamic stability. The presentation in this section follows closely that of Reiss and Matkowsky (1971). For $n > 1$ with $p = 1$, we have

$$w_{xxxx} + \lambda w_{xx} + w_{tt} + 2\gamma w_t = 0 \tag{6.244}$$

$$\lambda(t) = k[2c - \frac{1}{2}\int_0^1 w_x^2 dx] \tag{6.245}$$

$$w = w_{xx} = 0, \quad x = 0,1 \tag{6.246}$$

We consider the following "special" initial condition as the disturbance:

$$w(x,0) = F_n\phi_n(x), \quad w_t(x,0) = G_n\phi_n(x) \tag{6.247}$$

Note that this initial disturbance differs essentially from the arbitrary initial disturbances considered in the last two sections (see and compare (6.208) and (6.232)). In (6.247), only a particular mode of disturbance is imposed.

In view of the initial conditions imposed in (6.247), we seek the following solution form

$$w(x,t) = B(t)\phi_n(x) \tag{6.248}$$

where ϕ_n is the eigenfunction of the problem given in (6.180) and is defined as:

$$\phi_n(x) = \sqrt{2}\sin(n\pi x) \tag{6.249}$$

Applying the initial conditions to (6.247), we obtain the following initial condition for function B:

$$B(0) = F_n, \quad B'(0) = G_n \tag{6.250}$$

Substitution of (6.248) into (6.244) yields

$$B'' + 2\gamma B' + [(n\pi)^4 - \lambda(n\pi)^2]B = 0 \tag{6.251}$$

where

$$\begin{aligned} \lambda(t) &= k[2c - \frac{B^2}{2}\int_0^1 2(n\pi)^2\cos^2(n\pi x)dx] \\ &= k[2c - \frac{(n\pi)^2}{2}B^2] \end{aligned} \tag{6.252}$$

More explicitly, (6.251) can be written as

$$B'' + 2\gamma B' + \{(n\pi)^4 - k(n\pi)^2[2c - \frac{(n\pi)^2}{2}B^2]\}B = 0 \tag{6.253}$$

This is a nonlinear second order ODE with unknown B, and can be recast as

$$B'' + 2\gamma B' + K[\beta B + B^3] = 0 \tag{6.254}$$

where

$$K = \frac{k\lambda_n^2}{2}, \quad \beta = \frac{4}{\lambda_n}(c_n - c) \tag{6.255}$$

where λ_n is defined in (6.228). Equation (6.254) is known as the homogeneous Duffing equation (e.g., Minorsky, 1962), and as discussed in Section 12.5 of Chau (2018), it is this equation that leads to the development of the multi-scale perturbation method. This is a nonlinear ODE known to exhibit chaotic behavior.

The mathematical structure of this nonlinear ODE will be examined qualitatively in the next section.

6.6.1 Undamped Motions

For the case of undamped columns, we have $\gamma = 0$ and the ODE given in (6.254) becomes

$$B'' + K[\beta B + B^3] = 0 \tag{6.256}$$

This is another form of the Duffing equation (see p. 217 of Stoker, 1950). To solve this, we can multiply (6.256) by B' to get

$$B'B'' + K[\beta BB' + B^3 B'] = 0 \tag{6.257}$$

We recognize that this can be rewritten as

$$\frac{d}{dt}\{\frac{(B')^2}{2} + K[\beta\frac{B^2}{2} + \frac{B^4}{4}]\} = 0 \tag{6.258}$$

Integrating once, we get

$$(B')^2 + K[\beta B^2 + \frac{B^4}{2}] = E \tag{6.259}$$

where E is an integration constant, and physically it actually represents the energy of the column. On the boundary, the energy becomes

$$E = G_n^2 + K[\beta F_n^2 + \frac{F_n^4}{2}] \tag{6.260}$$

However, for the case of no damping (i.e., $\gamma = 0$), we have the conservation of energy. That is, the left-hand side of (6.259) must equal the right-hand side of (6.260). This observation agrees with Eq. (5.4) of Reiss and Matkowsky (1971).

The equilibrium state can be obtained by setting $B'' = 0$. Thus, we have

$$\beta B + B^3 = 0 \tag{6.261}$$

Factorizing (6.261), we have

$$B(\beta + B^2) = 0 \tag{6.262}$$

There are three roots for B. The straight state is represented by

$$B = 0 \tag{6.263}$$

Recalling (6.25), we have the buckled state

$$B^2 = -\beta = \frac{4}{\lambda_n}(c - c_n) \tag{6.264}$$

The stability of the solution depends on the value of the end compression c. Two possibilities are considered.

Case 1: $c < c_n$

When $c < c_n$ in (6.264), we must have $\beta > 0$. Equation (6.264) shows that there is no real root for $B = (-\beta)^{1/2}$. Thus, only the straight state (i.e., $B = 0$) is an equilibrium point for $\beta > 0$.

For the case of the straight state subject to small initial disturbances (or excitations), undamped motion of the column is, in essence, a so-called stable center (in the sense of Poincaré's stability classification) (e.g., see Section 5.9 of Chau, 2018). In particular, if the disturbances are small and, thus, $B \to 0$, we can simplify (6.256) to

$$B'' + K\beta B = 0 \tag{6.265}$$

This is the classical harmonic equation and its solution is

$$B = a_1 \sin\sqrt{K\beta}t + a_2 \cos\sqrt{K\beta}t \tag{6.266}$$

Thus, as long as the initial excitations are small such that the magnitude of B is small and the nonlinear term is negligible, the vibrations induced by the excitations are stable centers. Figure 6.16 plots the phase diagram of this solution. Therefore, the motions induced by the initial excitations are nearly an ellipse in the phase diagram.

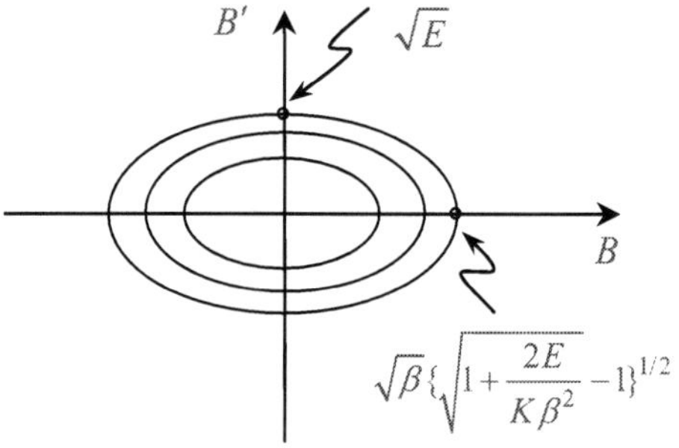

Figure 6.16 Initially excited vibration of an undamped bar

<u>Case 2: $c > c_n$</u>

When $c > c_n$ in (6.264), we must have $\beta < 0$. The equilibrium solution of the buckled states are:

$$B = \pm\sqrt{-\beta} = \pm B_0 \tag{6.267}$$

Since we have $\beta < 0$, B_0 is real. For small values of B (i.e., small initial excitation), the governing equation is reduced to

$$B'' + K\beta B = 0 \tag{6.268}$$

Since $\beta < 0$, the solution is in exponential form:

$$B = c_1 e^{\sqrt{-K\beta}t} + c_2 e^{-\sqrt{-K\beta}t} \tag{6.269}$$

According to Poincaré's stability classification, this solution is a stable saddle (see Section 5.9 of Chau, 2018). To see this, we can rewrite (6.259) as

$$(B')^2 = E - K[\beta B^2 + \frac{B^4}{2}] = E - \varphi(B) \tag{6.270}$$

where

$$\varphi(B) = K[\beta B^2 + \frac{B^4}{2}] \tag{6.271}$$

It is straightforward to see that

$$\varphi(0) = 0 \tag{6.272}$$

Differentiating (6.271) and setting the result to zero, we obtain the maximum and minimum points of the function φ as:

$$\varphi'(B) = K[2\beta B + 2B^3] = 2KB(\beta + B^2) = 0 \tag{6.273}$$

Thus, the maximum and minimum points are

$$B = 0, \quad B = \pm\sqrt{-\beta} = \pm B_0 \tag{6.274}$$

These points are precisely the equilibrium solutions. Figure 6.17 plots the function φ and the corresponding phase diagram of B'-B space. It is clear from the figure that the system is unstable for $E \geq 0$ and is stable when $E < 0$. For a damped system (i.e., $\gamma \neq 0$), we have the same equilibrium states as for the undamped case (i.e., $\gamma = 0$). However, there is no first integral (in terms of energy conservation) or we could not find the conservative law for the damped system.

6.6.2 Damped Motions

In this section, we sketch the stability analysis for the case of damped motions. For this case, an exact analytical solution is not possible. The first integral in terms of energy conservation does not exist. For damped motion, we recall from (6.254) as

$$B'' + 2\gamma B' + K[\beta B + B^3] = 0 \tag{6.275}$$

This ODE can be simplified by using the following change of variables

$$B' = v, \quad B'' = v' \tag{6.276}$$

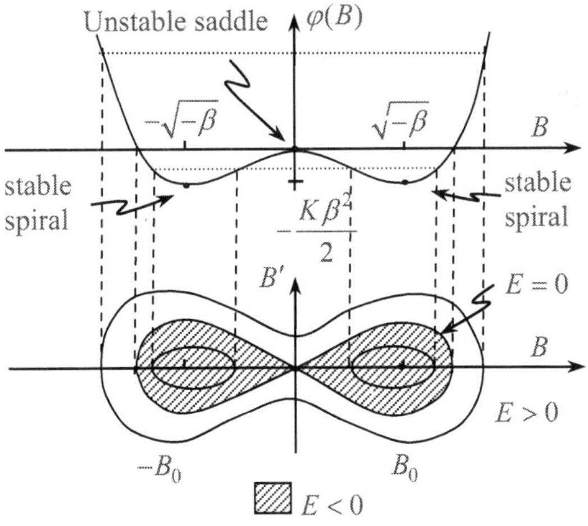

Figure 6.17 Unstable saddle and stable spiral of a damped column subject to an initial condition consisting of a single eigenmode

This is the standard approach discussed in Section 3.3.6 of Chau (2018) for an autonomous system. With this change of variables, we have

$$\frac{dv}{dt}=\frac{dv}{dB}(\frac{dB}{dt})=v\frac{dv}{dB} \tag{6.277}$$

Substitution of (6.277) and (6.276) into (6.275) gives

$$\frac{dv}{dB}=-\frac{2\gamma v+K[\beta B+B^3]}{v} \tag{6.278}$$

This first order ODE is not separable and needs to be solved by a numerical method, such as the Runge-Kutta method discussed in Section 15.5 of Chau (2018). A numerical calculation was made by Reiss and Matkowsky (1971) and is reproduced in Figure 6.18. Any initial condition that falls in the gray color zone will end up at the left buckled state (i.e., $-B_0$), whereas any initial condition that falls in the white color zone will end up at the right buckled state (i.e., B_0),

This is an autonomous system (i.e., it is of constant coefficients and variable t does not appear explicitly in the ODE), and thus the solution form is exponential:

$$B=e^{\alpha t} \tag{6.279}$$

Substitution of (6.279) into the linearized form of (6.275) leads to

$$\alpha^2+2\gamma\alpha+K\beta=0 \tag{6.280}$$

where

$$-\beta=\frac{4}{\lambda_n}(c-c_n) \tag{6.281}$$

Case 1: $c<c_n$

The roots of α in (6.280) are

$$\alpha=-\gamma\pm\sqrt{\gamma^2-K\beta} \tag{6.282}$$

For $c < c_n$, from (6.281), we have $\beta > 0$. There are two scenarios for the straight state solution:

(i) Stable node

If the argument within the square root sign is positive, the real part of α is smaller than zero:

$$\gamma^2-K\beta>0, \quad \Rightarrow \quad \mathrm{Re}[\alpha]<0, \quad \mathrm{Im}[\alpha]=0 \tag{6.283}$$

The solution is a stable node.

(ii) Stable spiral point

If the argument within the square root sign is negative, the real part of α is smaller than zero with a nonzero imaginary part:

$$\gamma^2-K\beta<0, \quad \Rightarrow \quad \alpha=-\gamma\pm i\sqrt{|\gamma^2-K\beta|}, \quad \mathrm{Re}[\alpha]<0 \tag{6.284}$$

Thus, the straight state solution before buckling is stable no matter what the value of the damping is.

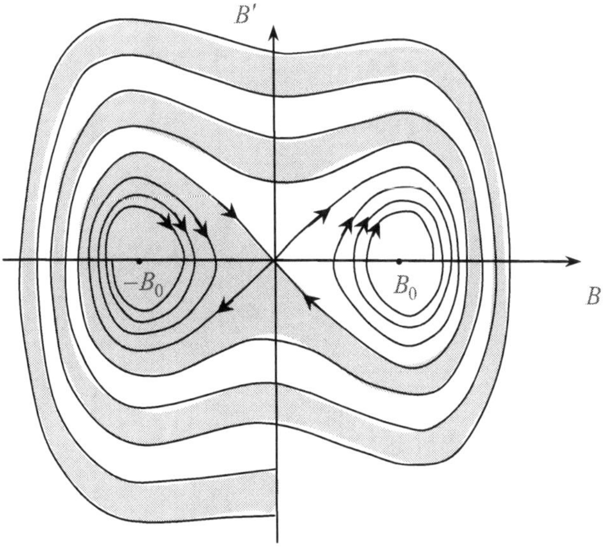

Figure 6.18 Phase diagram of a damped column showing the separatrix, dividing the zone of two "attractors" (after Reiss and Matkowsky, 1971)

Case 2: $c > c_n$

For $c > c_n$, we have three possible equilibrium states

$$B = 0, \quad B = \pm B_0 = \pm\sqrt{|\beta|} \tag{6.285}$$

For this case, note that from the definitions given in (6.255)

$$K = \frac{k\lambda_n^2}{2} > 0, \quad \beta = \frac{4}{\lambda_n}(c_n - c) < 0 \tag{6.286}$$

Near the straight state, we can set $B \to 0$, and the linearized form of (6.275) is

$$B'' + 2\gamma B' + K\beta B = 0 \tag{6.287}$$

For $c > c_n$, it is obvious from (6.288) that $K\beta < 0$, and the root α becomes

$$\alpha = -\gamma \pm \sqrt{\gamma^2 + |K\beta|} \tag{6.288}$$

The real part of one of the roots of α is less than zero while that of the other is larger than zero:

$$\text{Re}[\alpha_1] = \text{Re}[-\gamma + \sqrt{\gamma^2 + |K\beta|}] > 0 \tag{6.289}$$

$$\text{Re}[\alpha_2] = \text{Re}[-\gamma - \sqrt{\gamma^2 + |K\beta|}] < 0 \tag{6.290}$$

Therefore, the straight state $B = 0$ is an unstable saddle point. Near the buckled solution $B = \pm B_0$, and we can write B as

$$B = B_0 + b \tag{6.291}$$

Substitution of (6.291) into (6.275) leads to

$$\begin{aligned} & B''+2\gamma B'+K[\beta B+B^3] \\ & = b''+2\gamma b'+K[\beta(B_0+b)+(B_0+b)^3] \\ & = b''+2\gamma b'+Kb^3+3KB_0b^2+(K\beta+3KB_0^2)b+K(B_0^3+\beta B_0) \\ & = b''+2\gamma b'+Kb^3+3KB_0b^2+(K\beta+3KB_0^2)b=0 \end{aligned} \tag{6.292}$$

Note that in obtaining the last part of (6.292), we have used the following condition for B_0

$$B_0^2=-\beta \tag{6.293}$$

The linearized form of (6.292) is

$$b''+2\gamma b'+(K\beta+3KB_0^2)b=0 \tag{6.294}$$

For linear stability analysis, we seek an exponent solution for b

$$b=e^{\alpha t} \tag{6.295}$$

Substitution of (6.295) into (6.294) gives the following characteristic equation

$$\alpha^2+2\gamma\alpha+(K\beta+3KB_0^2)=0 \tag{6.296}$$

Thus, the roots are

$$\begin{aligned} \alpha &= \frac{-2\gamma\pm\sqrt{4\gamma^2-4(K\beta+3KB_0^2)}}{2} \\ &= -\gamma\pm\sqrt{\gamma^2+2K\beta} \\ &= -\gamma\pm\sqrt{\gamma^2-2|K\beta|} \end{aligned} \tag{6.297}$$

Again, we have used (6.286) and (6.293) in obtaining the result in (6.297). There are two scenarios for the buckled state solution:

(i) Stable node

$$\gamma^2-2|K\beta|\geq 0, \quad \Rightarrow \quad \mathrm{Re}[\alpha]<0, \quad \mathrm{Im}[\alpha]=0 \tag{6.298}$$

(ii) Stable spiral point

$$\gamma^2-2|K\beta|<0, \quad \Rightarrow \quad \alpha=-\gamma\pm i\sqrt{\left|\gamma^2-2|K\beta|\right|}, \quad \mathrm{Re}[\alpha]<0 \tag{6.299}$$

Therefore, the buckled state is stable regardless of the value of γ. In conclusion, for $c > c_n$ (i.e., $B = 0$) the straight state $B = 0$ is an unstable saddle point and the buckled states $B = \pm B_0$ are either a stable node or a stable spiral point (depending on the value of γ).

6.7 MULTI-TIME PERTURBATION AND STABILITY

In this section, we will consider the case that the initial conditions are expressed in a finite number of eigenstates only (i.e., arbitrary initial conditions). Before we proceed to consider the multi-time perturbation method, it is instructive to consider the asymptotic expansion of the solution around c_1 (the compressive displacement leading to the first buckled state). Recall from (6.269)

$$B=c_1e^{\sqrt{-K\beta}t}+c_2e^{-\sqrt{-K\beta}t} \tag{6.300}$$

Now substituting the definition of β, we have the exponential index in (6.300) as

$$\sqrt{-K\beta} = \sqrt{\frac{4K}{\lambda_n}(c-c_1)} \tag{6.301}$$

The result derived in (6.185) and depicted in Figure 6.14 suggests that we should expect a quadratic form in the neighborhood of c_1 (see Figure 6.19). Thus, we assume

$$c = c_1 + \varepsilon^2 \tag{6.302}$$

where ε is a small parameter.

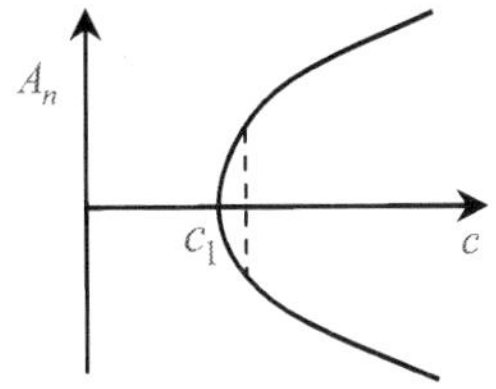

Figure 6.19 The quadratic behavior of the buckled state near bifurcation point c_1

Substitution of (6.302) into (6.301) gives

$$\sqrt{-K\beta} = \sqrt{\frac{4K}{\lambda_n}(c-c_1)} = \sqrt{\frac{4K}{\lambda_n}}\varepsilon \tag{6.303}$$

This suggests a second slower time scale of

$$\tau = \varepsilon t \tag{6.304}$$

where ε is a small parameter. With this insight from the asymptotic expansion around the bifurcation point, we now return to our problem of dynamic stability analysis of the straight state for $c > c_1$ with small damping, which is assumed to be proportional to ε (see (6.316)). In particular, we rewrite the problem defined in (6.187) to (6.189) as

$$w_{xxxx} + \lambda w_{xx} + w_{tt} + 2\tilde{\gamma}\varepsilon w_t = 0 \tag{6.305}$$

$$\lambda(t) = k[2c - \frac{1}{2}\int_0^1 w_x^2 dx] \tag{6.306}$$

$$w = w_{xx} = 0, \quad x = 0,1 \tag{6.307}$$

We consider the following small initial conditions as disturbances:

$$w(x,0) = \varepsilon f(x) = \varepsilon \sum_{n=1}^{N} f_n \phi_n, \quad w_t(x,0) = \varepsilon g(x) = \varepsilon \sum_{n=1}^{N} g_n \phi_n \tag{6.308}$$

The key difference between these initial conditions and those considered in earlier sections lies in the fact that there are only a finite number of terms N in the initial conditions in (6.308) and their proportionality to a small parameter ε. This set of equations of (6.305) to (6.308) is labeled as Problem 0, or the zero-th order problem. In short, we have assumed all small damping and small initial disturbances are proportional to ε, which was introduced in (6.302) and (6.304) for the slow time τ. Following the analysis of Reiss and Matkowsky (1971), we assume the following expansion of w in the series of ε

$$w(x,t,\varepsilon)=w^1(x,t,\tau)\varepsilon+w^2(x,t,\tau)\frac{\varepsilon^2}{2}+w^3(x,t,\tau)\frac{\varepsilon^3}{3!}+\dots \tag{6.309}$$

We assume that w^1 is nonzero and all w^i are bounded functions. Differentiating w with respect to time, we have

$$\frac{\partial w}{\partial t}(x,t,\varepsilon)=\frac{\partial w^1}{\partial t}(x,t,\tau)\varepsilon+\frac{\partial w^2}{\partial t}(x,t,\tau)\frac{\varepsilon^2}{2}+\frac{\partial w^3}{\partial t}(x,t,\tau)\frac{\varepsilon^3}{3!}+\dots \tag{6.310}$$

However, w^1 is now expressed as a function of t and τ, and in turn τ is a function of t. Thus, we have to apply the chain rule as

$$\frac{\partial w^1}{\partial t}=w_t^1+w_\tau^1\frac{d\tau}{dt}=w_t^1+w_\tau^1\varepsilon \tag{6.311}$$

In view of (6.311), we have (6.310) being

$$w_t=[w_t^1+w_\tau^1\varepsilon]\varepsilon+[w_t^2+w_\tau^2\varepsilon]\frac{\varepsilon^2}{2}+[w_t^3+w_\tau^3\varepsilon]\frac{\varepsilon^3}{3!}+\dots \tag{6.312}$$

Collecting terms in (6.312), we obtain

$$w_t=w_t^1\varepsilon+(w_t^2+2w_\tau^1)\frac{\varepsilon^2}{2}+(w_t^3+3w_\tau^3)\frac{\varepsilon^3}{3!}+\dots \tag{6.313}$$

Similarly, the second derivative of w with respect to t is

$$\begin{aligned}w_{tt}&=[w_{tt}^1+w_{t\tau}^1\varepsilon]\varepsilon+[w_{tt}^2+w_{t\tau}^2\varepsilon+2w_{t\tau}^1+2w_{\tau\tau}^1\varepsilon]\frac{\varepsilon^2}{2}\\&+[w_{tt}^3+w_{t\tau}^3\varepsilon+3w_{t\tau}^2+3w_{\tau\tau}^2\varepsilon]\frac{\varepsilon^3}{3!}+\dots\end{aligned} \tag{6.314}$$

Collecting terms in (6.314), we obtain

$$w_{tt}=w_{tt}^1\varepsilon+(w_{tt}^2+4w_{t\tau}^1)\frac{\varepsilon^2}{2}+(w_{tt}^3+6w_{t\tau}^2+6w_{t\tau}^1)\frac{\varepsilon^3}{3!}+\dots \tag{6.315}$$

We are going to differentiate Problem 0 with respect to ε, and in doing so, the following formulas are useful:

$$\gamma=\varepsilon\tilde{\gamma},\quad \dot{\gamma}=\frac{d\gamma}{d\varepsilon}=\tilde{\gamma},\quad \ddot{\gamma}=0,\quad \gamma(0)=0 \tag{6.316}$$

$$c(\varepsilon)=c_1+\varepsilon^2,\quad \dot{c}=\frac{dc}{d\varepsilon}=2\varepsilon,\quad \ddot{c}=2,\quad c(0)=c_1,\quad \dot{c}(0)=0 \tag{6.317}$$

$$\lambda(\varepsilon)=\lambda^0+\lambda^1\varepsilon+\lambda^2\frac{\varepsilon^2}{2}+\dots,\quad \dot{\lambda}(\varepsilon)=\lambda^1+\lambda^2\varepsilon+\dots \tag{6.318}$$

$$\lambda(0)=\lambda^0,\quad \dot{\lambda}(0)=\lambda^1 \tag{6.319}$$

For the differentiation of w, the following formulas are useful

$$\dot{w}(x,t,\varepsilon)=w^1+w^2\varepsilon+\dots,\quad \dot{w}(x,t,0)=w^1,\quad \dot{w}=\frac{\partial w}{\partial\varepsilon} \tag{6.320}$$

$$w_{xxxx}(x,t,\varepsilon)=w^1_{xxxx}\varepsilon+w^2_{xxxx}\frac{\varepsilon^2}{2}+\ldots \tag{6.321}$$

$$\dot{w}_{xxxx}(x,t,\varepsilon)=w^1_{xxxx}+w^2_{xxxx}\varepsilon+\ldots,\quad \dot{w}_{xxxx}(x,t,0)=w^1_{xxxx} \tag{6.322}$$

$$\dot{w}(x,t,\varepsilon)=w^1+w^2\varepsilon+\ldots,\quad \dot{w}(x,t,0)=w^1 \tag{6.323}$$

$$\dot{w}_t(x,t,\varepsilon)=w^1_t+w^2_\tau\varepsilon+2w^1_\tau\varepsilon+\ldots,\quad \dot{w}_t(x,t,0)=w^1_t \tag{6.324}$$

$$w_t(x,t,0)=0,\quad w_{tt}(x,t,0)=0, \tag{6.325}$$

$$w_x(x,t,\varepsilon)=w^1_x(x,t,\tau)\varepsilon+w^2_x(x,t,\tau)\frac{\varepsilon^2}{2}+\ldots,\quad w_x(x,t,0)=0 \tag{6.326}$$

$$\dot{w}_x(x,t,\varepsilon)=w^1_x(x,t,\tau)+w^2_x(x,t,\tau)\varepsilon+\ldots,\quad \dot{w}_x(x,t,0)=w^1_x(x,t,\tau) \tag{6.327}$$

$$\dot{w}_{tt}=w^1_{tt}+(w^2_{tt}+4w^1_{t\tau})\varepsilon+(w^3_{tt}+6w^2_{t\tau}+6w^1_{t\tau}]\frac{\varepsilon^2}{2!}+\ldots,\dot{w}_{tt}(x,t,0)=w^1_{tt} \tag{6.328}$$

We differentiate Problem 0 given from (6.305) to (6.308), and we obtain

$$\dot{w}_{xxxx}+\lambda\dot{w}_{xx}+\dot{\lambda}w_{xx}+\dot{w}_{tt}+2\gamma\dot{w}_t+2\dot{\gamma}w_t=0 \tag{6.329}$$

$$\dot{\lambda}(t)=k[2\dot{c}-\frac{1}{2}\int_0^1 w_x\dot{w}_x dx] \tag{6.330}$$

$$\dot{w}=\dot{w}_{xx}=0,\quad x=0,1 \tag{6.331}$$

$$\dot{w}(x,0)=f(x),\quad \dot{w}_t(x,0)=g(x) \tag{6.332}$$

Setting ε = 0 and using the formulas derived above from (6.316) to (6.328), we obtain Problem I or the first order problem:

$$L[w^1]=w^1_{xxxx}+\lambda^0 w^1_{xx}+w^1_{tt}=0 \tag{6.333}$$

$$\dot{\lambda}(0)=\lambda^1=0 \tag{6.334}$$

$$w^1=w^1_{xx}=0,\quad x=0,1 \tag{6.335}$$

$$w^1(x,0)=\sum_{n=1}^{N}f_n\phi_n,\quad w^1_t(x,0)=\sum_{n=1}^{N}g_n\phi_n \tag{6.336}$$

To show the validity of (6.334), we see that

$$\begin{aligned}\dot{\lambda}(0)&=k[2\dot{c}(0)-\frac{1}{2}\int_0^1 w_x(x,t,0)\dot{w}_x dx]\\&=k[2\times 0-\frac{1}{2}\int_0^1 0\times\dot{w}_x dx]=0\end{aligned} \tag{6.337}$$

To find the solution of (6.333), we seek the following solution form

$$w^1(x,t,\tau)=\sum_{n=1}^{N}A^1_n(t,\tau)\phi_n(x) \tag{6.338}$$

Substitution of (6.338) into (6.333) gives

$$\sum_{n=1}^{N}\{\lambda_n^2A^1_n-\lambda_n\lambda_1A^1_n+A^1_{n,tt})\}\phi_n(x)=0 \tag{6.339}$$

Therefore, the amplitude of the unknown satisfies

$$A_{n,tt}^1 + \omega_n^2 A_n^1 = 0 \tag{6.340}$$

where

$$\omega_n = \sqrt{\lambda_n(\lambda_n - \lambda_1)} \tag{6.341}$$

$$A_n^1(0,0) = f_n, \quad A_{n,t}^1(0,0) = g_n \tag{6.342}$$

For $n = 1$, we have

$$\omega_1 = \sqrt{\lambda_1(\lambda_1 - \lambda_1)} = 0 \tag{6.343}$$

Therefore, the amplitude of the unknown satisfies

$$A_{1,tt}^1 = 0 \tag{6.344}$$

Integrating twice, we get

$$A_1^1 = a_1^1(\tau)t + b_1^1(\tau) \tag{6.345}$$

From the boundedness requirement as $t \to \infty$, we require

$$a_1^1(\tau) = 0 \tag{6.346}$$

For $n = 2,3,4,...$, we have the general solutions being

$$A_n^1 = a_n^1(\tau)\sin\omega_n t + b_n^1(\tau)\cos\omega_n t \tag{6.347}$$

$$A_1^1 = b(\tau), \quad b(0) = f_1 \tag{6.348}$$

From (6.345), we get

$$A_{1,t}^1(0,0) = 0, \quad \Rightarrow \quad g_1 = 0 \tag{6.349}$$

Thus, the current multi-time scale cannot be used to solve the initial value problem that contains data of g_1 (imposing velocity of mode with $n = 1$). A revised slow time scale proportional to $\varepsilon^{1/2}$ is found necessary (Reiss, 1980a). However, we will not consider such a complication in this chapter. Using the initial conditions, we have for $n > 1$

$$A_n^1(0,0) = b_n^1(0) = f_n \tag{6.350}$$

$$A_{n,t}^1(0,0) = \omega_n a_n^1(0) = g_n \tag{6.351}$$

Finally, the solution in (6.338) can be expressed as

$$w^1(x,t,\tau) = b(\tau)\phi_1(x) + \sum_{n=2}^{N}[a_n^1(\tau)\sin\omega_n t + b_n^1(\tau)\cos\omega_n t]\phi_n(x) \tag{6.352}$$

To find the amplitude functions, we have to go to a higher order problem (this procedure is a standard approach in perturbation analysis). The process is quite tedious and we will not provide the full details here. More details can be found in Reiss and Matkowsky (1971).

In particular, if we differentiate Problem I defined in (6.329) to (6.332) one more time with respect to ε and then set ε to zero, we obtain

$$L[w^2] = w_{xxxx}^2 + \lambda^0 w_{xx}^2 + w_{tt}^2 = -4[w_{t\tau}^1 + \tilde{\gamma} w_t^1] \tag{6.353}$$

$$w^2 = w_{xx}^2 = 0, \quad x = 0,1 \tag{6.354}$$

$$w^2(x,0,0) = 0, \quad w_t^2(x,0,0) + 2w_\tau^1(x,0,0) = 0 \tag{6.355}$$

This is Problem II, and some steps in obtaining Problem II will be given in Problems 6.2 and 6.4 at the back of the chapter. By solving this problem, we can obtain the following solution (see Problems 6.5 and 6.6):

$$w^1(x,t,\tau) = b(\tau)\phi_1(x) + e^{-\tilde{\gamma}\tau}\sum_{n=2}^{N}[(\frac{g_n}{\omega_n})\sin\omega_n t + f_n\cos\omega_n t]\phi_n(x) \tag{6.356}$$

where $b(\tau)$ satisfies the following problem (Reiss and Matkowsky, 1971):

$$b_{\tau\tau} + 2\tilde{\gamma}b_\tau + K(\mu b + b^3) = 0 \tag{6.357}$$

$$K = \frac{k\lambda_1^2}{2}, \quad \mu = \frac{4}{\lambda_1}\{-1 + \frac{1}{8}e^{-2\tilde{\gamma}\tau}\sum_{n=2}^{N}\lambda_n[f_n^2 + (\frac{g_n}{\omega_n})^2]\} \tag{6.358}$$

$$b(0) = f_1, \quad b_\tau(0) = 0 \tag{6.359}$$

Note that we need to go to a higher level problem to obtain (6.357) to (6.359) for $b(\tau)$ (see Problems 6.7-6.9 for some details of the proof). The first term in the summation on the right of (6.358) depends on the primary data f_1 through the first equation of (6.359) and is independent of t (with $g_1 = 0$). Note that τ is much slower than t. Thus, this first term is a slow-time standing wave motion and constitutes the primary motion. The square term in the summation on the right of (6.358) depends on the fast time t and, thus, is a high-frequency "noise" and depends on the secondary data of the excitation (i.e., f_n and g_n). Reiss and Matkowsky (1971) defined the following:

$$\begin{aligned} &\mu \geq 0, \quad \text{noisy initial data} \\ &\mu < 0, \quad \text{quiet initial data} \end{aligned} \tag{6.360}$$

More importantly, the mathematical form of (6.357) for b is the same as that for B given in (6.275). Thus, the discussions given in Section 6.6.1 and 6.6.2 for undamped and damped motions apply equally here.

More specially, for undamped motions,

$$b_{\tau\tau} + K(\mu b + b^3) = 0 \tag{6.361}$$

where

$$\mu = \frac{4}{\lambda_1}\{-1 + \frac{1}{8}\sum_{n=2}^{N}\lambda_n[f_n^2 + (\frac{g_n}{\omega_n})^2]\} \tag{6.362}$$

As discussed in Section 6.6.1, the first integral of (6.361) equals the energy conservation and is

$$(b_\tau)^2 + K[\mu b^2 + \frac{b^4}{2}] = E \tag{6.363}$$

Figure 6.20 plots the potential function and the phase plot of b.

For noisy data (i.e., $\mu \geq 0$), $b = 0$ is the only real solution as $b_0 = \pm i(\mu)$ is not real. Thus, the unbuckled state is nonlinearly stable. For quiet data (i.e., $\mu < 0$), we also have $E < 0$, referring to Figure 6.20, if

$$|f_1| < f_1^* = \sqrt{-2\mu} \tag{6.364}$$

Then, the primary motion is a slow time periodic wave and polarized around $b = b_0$ if $f_1 > 0$ and polarized around $b = -b_0$ if $f_1 < 0$. For $f_1 > f_1^*$, we have $E > 0$, the solution sways around but the unbuckled state is considered nonlinearly stable by Reiss and Matkowsky (1971) according to their definition of nonlinear stability (see their definition 5.1). If $|f_1| = f_1^*$, the motion approaches the unbuckled state for $\tau \to \infty$. In short, if there is no damping, the motions will not settle to the buckled states and thus the unbuckled state is considered as nonlinearly stable. This conclusion differs from that of the linear stability analysis.

For damped motions (i.e., $\gamma \neq 0$), as $t \to \infty$, we have (as the higher modes in the secondary motions decay to zero as predicted by (6.358))

$$w(x,t,\tau) = [b(\tau)\phi_1(x)]\varepsilon + O(\varepsilon^2) \tag{6.365}$$

$$b_{\tau\tau} + 2\tilde{\gamma} b_\tau + K(\mu b + b^3) = 0 \tag{6.366}$$

$$\mu = -\frac{4}{\lambda_1} \tag{6.367}$$

$$b(0) = f_1, \quad b_\tau(0) = 0 \tag{6.368}$$

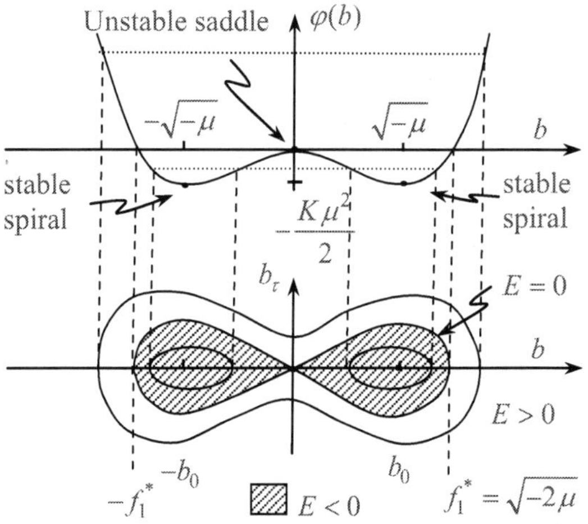

Figure 6.20 Nonlinear dynamic stability of a damped column subject to initial conditions consisting of a finite number of eigenmodes

The solution of this system is somewhat similar to that shown in Figure 6.18. If the initial data fall into the shaded zone in Figure 6.21, the motion of the column will settle to the buckled state at $b = -b_0$. Likewise, the initial motion in the white zone will settle to $b = -b_0$. If $|f_1| < p$, the motion of the column is polarized and damped to the buckled state closer to f_1. If the initial data $|f_1| > p$, the column will sway back and forth a few times between the two buckled states, and eventually settle into one of these states.

As we demonstrated here, the dynamic nonlinear stability analysis subject to arbitrary initial disturbance is not an easy subject.

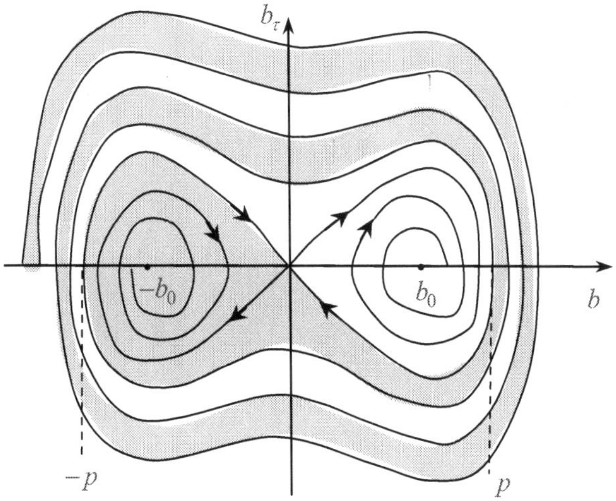

Figure 6.21 Nonlinear stability analysis of a damped column subject to initial conditions

6.8 GOVERNING EQUATIONS OF CROOKED BEAMS

In this section, we will modify the analysis given in Section 6.2 for a column with initial imperfections (or a crooked beam or an arch). As shown in Figure 6.22, the initial profile is given as w^0. Although we show in Figure 6.22 the case of applied compression P, the formulation for end compression will also be considered. Note that an arch can be considered as a special case of initial imperfections.

6.8.1 Lagrangian Strain for Crooked Beams

Recall the result derived in Section 6.2 that the axial strain in the X-axis can be expressed as

$$e_X = U_X + \frac{1}{2} w_X^2 = u_X - Z w_{XX} + \frac{1}{2} w_X^2 = e_X^0 - Z w_{XX} \tag{6.369}$$

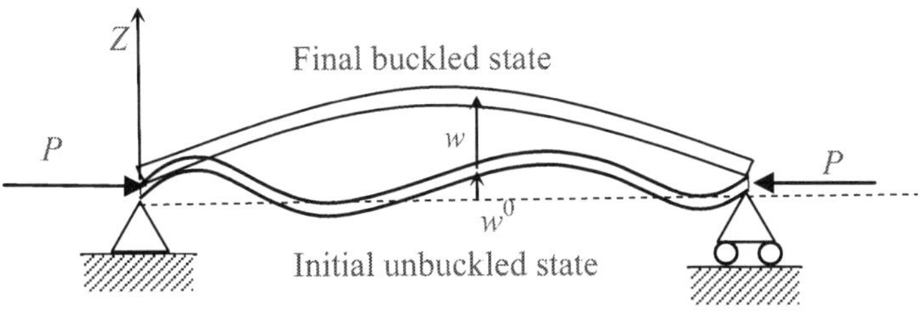

Figure 6.22 Buckling of a crooked beam

The current position vector can be expressed in terms of displacement $\boldsymbol{u}$ as

$$\boldsymbol{x} = \boldsymbol{X} + \boldsymbol{u}(\boldsymbol{X}, T) \tag{6.370}$$

Taking the total differential of (6.370), the length of a line element dX after deformation can be evaluated as

$$dx = dX + u_X(\boldsymbol{X}, T)dX \tag{6.371}$$

As shown in Figure 6.22, the vertical position of the beam is

$$z = w^0(\boldsymbol{X}, T) + w(\boldsymbol{X}, T) \tag{6.372}$$

Taking total differential of (6.372), we have

$$dz = (w_X^0 + w_X)dX \tag{6.373}$$

The arc length of the initial state can be determined as

$$(dS)^2 = [1 + (w_X^0)^2](dX)^2 \tag{6.374}$$

Note that if $w^0 = 0$, we will recover the expected result for a straight beam. An initial element of length dS will deform to ds. Employing (6.373) and (6.371), we find the following result for ds

$$\begin{aligned}(ds)^2 &= (dx)^2 + (dz)^2 \\ &= [(1 + u_X)^2 + (w_X^0 + w_X)^2](dX)^2\end{aligned} \tag{6.375}$$

Subtracting (6.375) from (6.376), we have

$$\begin{aligned}(ds)^2 - (dS)^2 &= \{1 + 2u_X + u_X^2 + (w_X^0)^2 + 2w_X^0 w_X + (w_X)^2 - 1 - (w_X^0)^2\}(dX)^2 \\ &= 2[u_X + \frac{1}{2}u_X^2 + w_X^0 w_X + \frac{1}{2}(w_X)^2](dX)^2\end{aligned} \tag{6.376}$$

By dropping the second term on the right-hand side of (6.376), we obtain

$$e_X^0 = u_X + w_X^0 w_X + \frac{1}{2}(w_X)^2 \tag{6.377}$$

If the column is straight, the second term will be zero and (6.89) is recovered as a special case.

6.8.2 Variational Principle for Crooked Beams

Recall the Hamiltonian principle

$$V_0[u, w] \equiv \int_{T_0}^{T_1} [\mathcal{E} - \mathcal{W} - \mathcal{K}] dT \tag{6.378}$$

where

$$\mathcal{E} = \frac{E}{2}\int_0^L A\left(e_X^0\right)^2 dX + \frac{E}{2}\int_0^L I w_{XX}^2 dX \tag{6.379}$$

$$\mathcal{K} = \frac{\rho A}{2}\int_0^L w_T^2 dX \tag{6.380}$$

In this section, we consider only the boundary condition 2 (i.e., displacement boundary condition), and we have $\mathcal{W} = 0$. The difference between a crooked beam and a straight beam mainly arises from the axial strain term (or the first term in (6.379)). We will only examine this term (as all other terms are similar to the analyses used in Section 6.2):

$$V_0[u,w]=\int_{T_0}^{T_1}\{\int_0^L \frac{AE}{2}[u_X+w_X^0 w_X+\frac{1}{2}(w_X)^2]^2 dXdT \tag{6.381}$$

Substituting u and w with their admissible solutions plus admissible variations, we have

$$\begin{aligned} &V_0[u+\varepsilon\bar{u},w+\varepsilon\bar{w}] \\ &=\int_{T_0}^{T_1}\{\int_0^L \frac{AE}{2}[u_X+\varepsilon\bar{u}_X+w_X^0(w_X+\varepsilon\bar{w}_X)+\frac{1}{2}(w_X+\varepsilon\bar{w}_X)^2]^2 dXdT \end{aligned} \tag{6.382}$$

Differentiating (6.382) with respect to ε, we have

$$\begin{aligned} \frac{dV_0}{d\varepsilon}=&\int_{T_0}^{T_1}\{\int_0^L AE[u_X+\varepsilon\bar{u}_X+w_X^0(w_X+\varepsilon\bar{w}_X)+\frac{1}{2}(w_X+\varepsilon\bar{w}_X)^2] \\ &\times[\bar{u}_X+w_X^0\bar{w}_X+(w_X+\varepsilon\bar{w}_X)\bar{w}_X]dX\}dT \end{aligned} \tag{6.383}$$

Setting $\varepsilon = 0$, we finally obtain

$$\left.\frac{dV_0}{d\varepsilon}\right|_{\varepsilon=0}=\int_{T_0}^{T_1}\{\int_0^L AEe_X^0[\bar{u}_X+w_X^0\bar{w}_X+w_X\bar{w}_X]dXdT \tag{6.384}$$

The second term in (6.384) is the additional term induced by imperfection, and its integral can be evaluated by integration by parts as

$$\int_0^L e_X^0 w_X^0 \bar{w}_X dX=e_X^0 w_X^0\,\bar{w}\Big|_0^L-\int_0^L (e_X^0 w_X^0)_{,X}\bar{w}dX \tag{6.385}$$

We note that the boundary terms in (6.385) are zero because

$$\bar{w}(0)=\bar{w}(L)=0 \tag{6.386}$$

Applying (6.385) and (6.386) to (6.384), we obtain

$$\left.\frac{dV_0}{d\varepsilon}\right|_{\varepsilon=0}=\int_{T_0}^{T_1}\{\int_0^L -AEe_{X,X}^0\bar{u}dX+\int_0^L[-AE(e_X^0 w_X)_{,X}-AE(e_X^0 w_X^0)_{,X}]\bar{w}dX\}dT \tag{6.387}$$

In obtaining this result, we have applied the following boundary conditions for the admissible variations

$$\bar{u}(0,T)=\bar{u}(L,T)=0 \tag{6.388}$$

Without going into the details, we obtain the following Euler equation for w

$$EIw_{XXXX}-A\left(\sigma w_X+\sigma w_X^0\right)_{,X}+\rho Aw_{TT}=0 \tag{6.389}$$

Integrating the axial stress along the beam, we have

$$\begin{aligned} \sigma L&=\int_0^L E[u_X+w_X^0 w_X+\frac{1}{2}(w_X)^2]dX \\ &=E[u(L)-u(0)+\int_0^L w_X^0 w_X dX+\frac{1}{2}\int_0^L (w_X)^2 dX] \end{aligned} \tag{6.390}$$

Applying end compression given in (6.146) to (6.390), we get

$$\sigma L=E[-2C+\int_0^L w_X^0 w_X dX+\frac{1}{2}\int_0^L (w_X)^2 dX] \tag{6.391}$$

The normalized stress λ becomes

$$\lambda=-\frac{k\sigma}{E}=k[2c-\int_0^1 w_x^0 w_x dx-\frac{1}{2}\int_0^1 w_x^2 dx] \tag{6.392}$$

where the normalized parameters x, λ, and w are defined in (6.167), (6.169), (6.170), and (6.171). As expected, the previous result for λ is recovered if w^0 is zero. This revised formula will be used to model the arch problem in the next section.

6.9 SNAP-THROUGH BUCKLING OF ELASTIC ARCHES

In this section, nonlinear buckling of a shallow elastic arch subject to distributed load will be considered, as shown in Figure 6.23. The span of the arch is L and the initial amplitude of the arch is a. Snap-through buckling of shallow or flat arches were studied by Timoshenko in 1935 under distributed load and by Biezeno in 1938 under point load. Other contributors on snap-through buckling of shallow arches include Fung, Kaplan, Marguerre, Hoff, Bruce, Masur, Lo, Humphreys, Reiss and Lock. Snap-through buckling also appears in flat cylindrical shells, shallow spherical shells, and tied trusses. For example, the standard wall switch of electric lights and Bazant's safety ski binding are also designed based on the idea of snap-through buckling (e.g., Bazant and Cedolin, 1991). The presentation in this section mainly follows that of Reiss (1980b) and Reiss (1984).

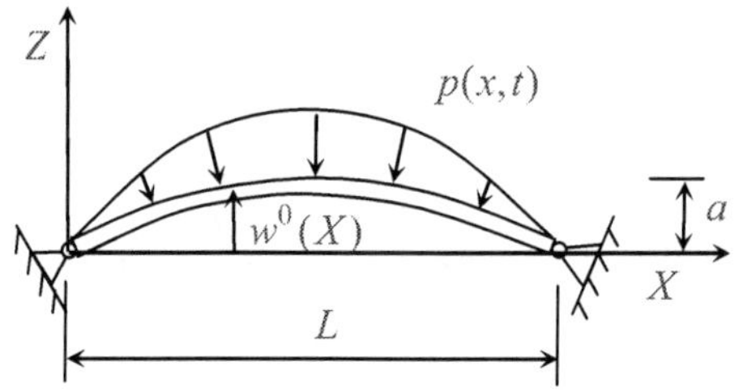

Figure 6.23 Arch with an initial shape w^0 subject to compressive pressure $p(x,t)$

In normalized form, (6.389), and its associated conditions can be rewritten as

$$w_{xxxx} + \lambda w_{xx} + w_{tt} + 2\Gamma w_t = -\lambda w^0_{xx} + p(x,t) \tag{6.393}$$

$$w = w_{xx} = 0, \quad x = 0,1 \tag{6.394}$$

$$\lambda = -k[\frac{1}{2}\int_0^1 (w_x)^2 dx + \int_0^1 w^0_x w_x dx] \tag{6.395}$$

$$k = \frac{L^2 A}{I} \tag{6.396}$$

where w^0 is the initial shape of the arch. As expected, the previous formulation is recovered if w^0 is zero.

6.9.1 Static Solution under Pressure

Consider an arch in the shape of a sine function subject to an applied pressure of a sine function also. In particular, we have

$$p = p(x) = -q\pi^4 \sin \pi x \tag{6.397}$$

$$w^0(x) = a \sin \pi x \tag{6.398}$$

Naturally, we are looking for a solution as a sine function with an unknown amplitude of A.

$$w(x) = -aA \sin \pi x \tag{6.399}$$

Using (6.398) and (6.399), the integrals defined in (6.395) can be evaluated as

$$\int_0^1 w_x^0 w_x dx = -a^2\pi^2 A \int_0^1 \cos^2 \pi x dx = -\frac{a^2\pi^2 A}{2} \tag{6.400}$$

$$\int_0^1 (w_x)^2 dx = a^2 A^2 \pi^2 \int_0^1 \cos^2 \pi x dx = \frac{a^2 A^2 \pi^2}{2} \tag{6.401}$$

Substituting (6.400) and (6.401) into (6.395), we obtain

$$\lambda = -\frac{ka^2 A\pi^2}{2}(-1 + \frac{A}{2}) \tag{6.402}$$

For a static solution, we can drop the time derivative terms in (6.393) and substitute (6.397), (6.398), and (6.399) into it, and eventually, we have

$$q\pi^4 = aA\pi^4 - \lambda a\pi^2 (A-1) \tag{6.403}$$

In view of (6.402), (6.403) can be rewritten as

$$q = aA + \frac{ka^3 A}{4}(A^2 - 3A + 2) \tag{6.404}$$

Equation (6.404) can be recast as

$$Q = A[A^2 - 3A + 2 + K] \tag{6.405}$$

where Q and K are defined as

$$Q = \frac{4q}{ka^3}, \quad K = \frac{4}{ka^2} \tag{6.406}$$

Physically, Q is the loading parameter and K is the geometric parameter. To examine the maximum and minimum of (6.405), we differentiate it with respect to A and set the result to zero:

$$\frac{dQ}{dA} = 3A^2 - 6A + 2 + K = 0 \tag{6.407}$$

The maximum and minimum of Q occurs at

$$A = 1 \pm \sqrt{1-B} \tag{6.408}$$

where

$$B = \frac{2+K}{3} \tag{6.409}$$

The second derivative of Q is

$$\frac{d^2Q}{dA^2} = 6(A-1) \tag{6.410}$$

We see that the Q-A curve changes its curvature at $A = 1$ and from (6.405) $Q = K$. If $B > 1$, the Q-A curve has no inflexion point and it is a monotonically increasing curve. If $B = 1$, the Q-A curve has a vertical slope at $A = 1$ and $Q = K$. If $B < 1$, the Q-A curve has two vertical slopes: one at A_u and one at A_L. These three scenarios are demonstrated in Figure 6.24. The more interesting scenario is for $B < 1$, which indicates the possibility of jump in the solution.

In Figure 6.24, the two limit points are defined as:

$$A_u = 1 - \sqrt{1-B} \tag{6.411}$$

$$A_L = 1 + \sqrt{1-B} \tag{6.412}$$

The corresponding loading parameters can be derived as

$$\begin{aligned} Q_u &= A_u[A_u^2 - 3A_u + 2 + K] \\ &= (1-\sqrt{1-B})K + (1-\sqrt{1-B})[A_u^2 - 3A_u + 2] \\ &= (1-\sqrt{1-B})K + B\sqrt{1-B} \end{aligned} \tag{6.413}$$

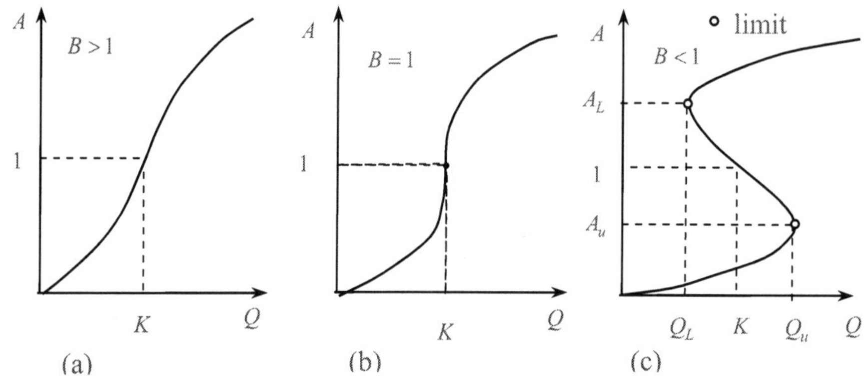

Figure 6.24 The load-deflection curves for arch buckling $B > 1$, $B = 0$ and $B < 1$ (static solution): (a) $K = 2$; (b) $K = 1$; (c) and $K = 0.4$

$$\begin{aligned} Q_L &= A_L[A_L^2 - 3A_L + 2 + K] \\ &= (1+\sqrt{1-B})[K + \sqrt{1-B}(\sqrt{1-B} - 1)] \end{aligned} \tag{6.414}$$

In deriving (6.413), we have used the identity derived in Problem 6.10. The linear dynamic stability of the static solution is considered next.

6.9.2 Linear Dynamic Stability

We first denote the equilibrium state as

$$w(x,t), \quad \lambda(w) = \lambda_0 \tag{6.415}$$

To check the dynamic stability, we consider a solution which is the sum of the equilibrium solution and a perturbation:

$$\overline{w} = w + v, \quad \overline{\lambda} = \lambda_0 + \Lambda \tag{6.416}$$

Recall the following differential equation for arch deflection

$$\bar{w}_{xxxx}+\bar{\lambda}\bar{w}_{xx}+\bar{w}_{tt}+2\Gamma\bar{w}_t=-\bar{\lambda}w^0_{xx}+p(x,t) \tag{6.417}$$

Substitution of (6.416) into (6.417) gives

$$w_{xxxx}+v_{xxxx}+\bar{\lambda}w_{xx}+\bar{\lambda}v_{xx}+w_{tt}+v_{tt}+2\Gamma w_t+2\Gamma v_t=-\bar{\lambda}w^0_{xx}+p(x,t)-\Lambda w^0_{xx} \tag{6.418}$$

Recall the fact that w is an equilibrium solution, and it satisfies the following PDE

$$w_{xxxx}+\lambda_0 w_{xx}+w_{tt}+2\Gamma w_t=-\lambda_0 w^0_{xx}+p(x,t) \tag{6.419}$$

Applying (6.419) to (6.418), we obtain

$$v_{xxxx}+\Lambda w_{xx}+\lambda_0 v_{xx}+\Lambda v_{xx}+v_{tt}+2\Gamma v_t=-\Lambda w^0_{xx} \tag{6.420}$$

To find λ, we substitute the second equation of (6.416) into (6.395) to get

$$\begin{aligned}\bar{\lambda}&=-k[\frac{1}{2}\int_0^1(w_x+v_x)^2dx+\int_0^1 w^0_x(w_x+v_x)dx]\\&=-k[\frac{1}{2}\int_0^1(w_x)^2dx+\int_0^1 w^0_x w_x dx]\\&-k\{\int_0^1[w_x v_x+\frac{1}{2}(v_x)^2]dx+\int_0^1 w^0_x v_x dx\}\end{aligned} \tag{6.421}$$

Clearly, the first term on the right-hand side of (6.421) is λ_0, and the extra term is a function of v. Thus, (6.421) can be rewritten as

$$\bar{\lambda}=\lambda(w)+\Lambda(v)=\lambda_0+\Lambda \tag{6.422}$$

where

$$\Lambda=-k\{\int_0^1[w_x v_x+\frac{1}{2}(v_x)^2]dx+\int_0^1 w^0_x v_x dx\} \tag{6.423}$$

Recall the equilibrium equation for v

$$v_{xxxx}+\Lambda w_{xx}+\lambda_0 v_{xx}+\Lambda v_{xx}+v_{tt}+2\Gamma v_t=-\Lambda w^0_{xx} \tag{6.424}$$

In view of (6.423), we see that the fourth term on the left-hand side of (6.424) is a nonlinear function of the unknown v. Linearization of (6.424) gives

$$v_{xxxx}+\Lambda w_{xx}+\lambda_0 v_{xx}+v_{tt}+2\Gamma v_t=-\Lambda w^0_{xx} \tag{6.425}$$

$$v=v_{xx}=0,\quad x=0,1 \tag{6.426}$$

$$\Lambda=-k\{\int_0^1 w_x v_x dx+\int_0^1 w^0_x v_x dx\} \tag{6.427}$$

To check linear dynamic stability, we impose the following initial perturbations as

$$v(x,0)=\sum_{n=1}^{\infty}f_n\sin n\pi x,\quad v_t(x,0)=\sum_{n=1}^{\infty}g_n\sin n\pi x \tag{6.428}$$

Naturally, we seek the following solution form:

$$v(x,t)=\sum_{n=1}^{\infty}v_n(t)\sin n\pi x \tag{6.429}$$

Recall the equilibrium solution from (6.399)

$$w(x)=-aA\sin\pi x \tag{6.430}$$

With (6.429) and (6.430), we can now evaluate the new stress parameter Λ defined in (6.427). In particular, the following integrals reduce to orthogonality of cosine functions (see Section 10.5 of Chau, 2018):

$$\int_0^1 w_x^0 v_x dx = nav_n\pi^2 \sum_{n=1}^{\infty} \int_0^1 \cos n\pi x \cos \pi x dx = \frac{nav_1\pi^2}{2} \tag{6.431}$$

$$\int_0^1 w_x v_x dx = -Ana\pi^2 \sum_{n=1}^{\infty} \int_0^1 \cos n\pi x \cos \pi x dx = -\frac{Anav_1\pi^2}{2} \tag{6.432}$$

Substitution of (6.431) and (6.432) into (6.427) gives the final expression for Λ

$$\Lambda = -\frac{k}{2} na\pi^2 (1-A) v_1 \tag{6.433}$$

Substitution of (6.398), (4.429) and (6.430) into (6.424) results in

$$\begin{aligned}
&\sum_{n=1}^{\infty} [(n\pi)^4 v_n - (n\pi)^2 \lambda_0 v_n + v_n'' + 2\Gamma v_n'] \sin n\pi x \\
&= (-aA\Lambda\pi^2 + \Lambda\pi^2 a) \sin \pi x \\
&= -\frac{ka^2\pi^4}{2} (1-A) v_1 \sin \pi x
\end{aligned} \tag{6.434}$$

Balancing the sine terms on both sides of (6.434), we obtain two equations

$$v_n'' + 2\Gamma v_n' + n^2\pi^4 [n^2 + \frac{ka^2 A}{4}(A-2)] v_n = 0, \quad n \ge 2 \tag{6.435}$$

$$v_1'' + 2\Gamma v_1' + \pi^4 [1 + \frac{ka^2 A}{4}(A-2) + \frac{ka^2}{2}(1-A)] v_1 = 0 \tag{6.436}$$

Recalling the definition of K from (6.406), (6.436) can be further simplified to

$$v_1'' + 2\Gamma v_1' + \frac{\pi^4}{K} [K + A^2 - 4A + 2] v_1 = 0 \tag{6.437}$$

In summary, we have two second order ODEs

$$v_1'' + 2\Gamma v_1' + D_1 v_1 = 0 \tag{6.438}$$

$$v_n'' + 2\Gamma v_n' + D_n v_n = 0, \quad n \ge 2 \tag{6.439}$$

where

$$D_1 = \frac{\pi^4}{K} [A^2 - 4A + 2 + K] \tag{6.440}$$

$$D_n = n^2\pi^4 [n^2 + \frac{ka^2 A}{4}(A-2)] = \frac{n^2\pi^4}{K} [Kn^2 + A(A-2)] \tag{6.441}$$

The right-hand side of (6.441) can be factorized by observing the roots of the following quadratic equation

$$A^2 - 2A + Kn^2 = 0 \tag{6.442}$$

$$A_n^{\pm} = 1 \pm \sqrt{1 - Kn^2} \tag{6.443}$$

Employing (6.442) and (6.443), we can factorize (6.441) as

$$D_n = \frac{n^2\pi^4}{K}(A - A_n^+)(A - A_n^-) \tag{6.444}$$

Figure 6.25 illustrates that D_n is negative between the two roots of A given in (6.443), and this is of profound importance for later discussion.

Following the same procedure, we can also factorize D_1 as

$$D_1 = \frac{\pi^4}{K}[A^2 - 4A + 2 + K] = \frac{\pi^4}{K}[(A - A_1^-)(A - A_1^+)] \tag{6.445}$$

where

$$A_1^{\pm} = 2 \pm \sqrt{2 - K} \tag{6.446}$$

The general solution for (6.438) and (6.439) is clearly in exponential form

$$v_n = V_n e^{\sigma t} \tag{6.447}$$

Substitution of (6.447) into (6.438) or (6.439) leads to

$$\sigma^2 + 2\Gamma\sigma + D_n = 0 \tag{6.448}$$

The roots of this quadratic equation are

$$\sigma^{\pm} = -\Gamma \pm \sqrt{\Gamma^2 - D_n} \tag{6.449}$$

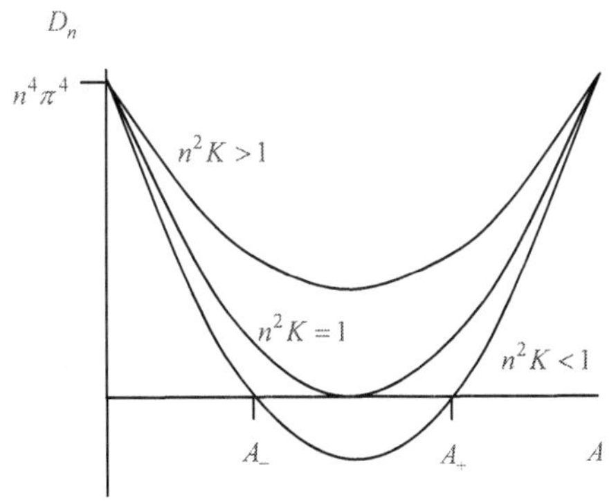

Figure 6.25 The possibilities of having two real roots for D_n

Depending on the sign of D_n, three scenarios of dynamic stability can be identified. If $D_n < 0$, (6.449) shows that $\sigma^+ > 0$. Consequently, the corresponding equilibrium state is unstable. If all $D_n > 0$, the corresponding equilibrium state is asymptotically stable. If $D_n = 0$ for some n, the corresponding equilibrium state is not asymptotically stable, but instead it suggests the possibilities of a singular point (either a bifurcation or a limit point). These rules are used for later discussions. Note that $D_n = 0$ if A satisfies (6.442).

<u>Case 1</u>: $B > 1$ and $K > 1$,

For $B > 1$, (6.409) shows that $K > 1$ as well. As demonstrated in Figure 6.25, all D_n larger than zero for $n^2K > 1$ (since n is an integer). For D_1 and D_n larger than zero, (6.449) shows that all the roots for σ are smaller than zero. That is, the solution is asymptotically stable, and the solution returns to an equilibrium state when it is subject to dynamic disturbance. In summary, we have

$$D_1 > 0, \quad D_n > 0, \quad \sigma^{\pm} < 0, \quad n = 2,3,4,... \tag{6.450}$$

This is the case with no bifurcation point or no buckling as shown in Figure 6.24(a). Recall from (6.406) that

$$K = \frac{4}{a^2 k} = \frac{4I}{a^2 L^2 A} = \frac{4r_g^2}{\alpha^2 L^4} \tag{6.451}$$

where r_g is the radius of gyration and α is the shallowness parameter:

$$r_g^2 = \frac{I}{A}, \quad \alpha = \frac{a}{L} \tag{6.452}$$

Therefore, the load-deflection curve is stable and no snap-through buckling is expected for a short-span shallow arch with a strong section against bending (i.e., $K > 1$).

Case 2: $B = 1$ and $K = 1$,

Case 2 is the transition from stable to unstable behavior (or from no buckling to snap-through buckling case). For $B = 1$, we have

$$D_1 = 0, \quad D_n > 0, \quad \sigma^{+} = 0, \quad \sigma^{-} = -2\Gamma < 0, \quad n = 2,3,4,... \tag{6.453}$$

The equilibrium state is not asymptotically stable. This suggests a bifurcation point or limit point, as shown in Figure 6.24(b).

Case 3: $B < 1$ and $K < 1$

Case 3 is the most interesting scenario of snap-through buckling, as illustrated in Figure 6.24(c). The actual shape of the bifurcation mode depends on the actual value K (see Cases 3A, 3B and 3C below). We first examine the possibility of mode 1 buckling (or $n = 1$). To find the range of negative D_1 on the Q-A plot by noting (6.409), we first rewrite (6.440) as

$$D_1 = \frac{\pi^4}{K}[A^2 - 4A + 3B] \tag{6.454}$$

We want to find the value of A on the Q-A curve such that $D_1 = 0$. Setting the right-hand side of (6.454) to zero, we have

$$3B = 4A_* - A_*^2 \tag{6.455}$$

Substitution of (6.455) into (6.405) together with (6.409) yields

$$\begin{aligned} Q &= A_*^3 - 3A_*^2 + 3BA_* \\ &= A_*^2 \end{aligned} \tag{6.456}$$

Therefore, we get

$$A_* = \sqrt{Q} \tag{6.457}$$

This curve provides the boundary between $D_1 > 0$ and $D_1 < 0$ (i.e., between the stable and unstable equilibrium states). Three scenarios are considered below:

Case 3A: $1 > K > 1/4$

Figure 6.26 shows the unstable zone for the buckling modes of $n = 1$. The dashed line is the plot for (6.457) or $D_1 = 0$. Note that Figure 6.26 plots Q against A, instead of A versus Q as in Figure 6.24. The solution is unstable shortly after Q_u is passed. The diagram on the right of Figure 6.26 shows that the jump to the next stable point on the Q-A curve requires a much higher loading Q at a much larger deflection A. This is quite difficult to achieve in real experiments, and, thus, buckling with $n = 1$ is unlikely.

For $n > 1$ and $K > 1/4$, we have $n^2K > 1$, and from Figure 6.25 we have $D_n > 0$, leading to (see (6.449)):

$$D_n > 0, \quad \sigma^{\pm} < 0, \quad n = 2,3,4,... \tag{6.458}$$

Therefore, the higher mode of buckling is asymptotically stable.

<u>Case 3B</u>: $K = 1/4$

For this case $K = 1/4$, we have seen in Case 3A that $n = 1$ is unstable. The Q-A plot for this case is shown in Figure 6.27. For $n = 2$, we have from (6.441)

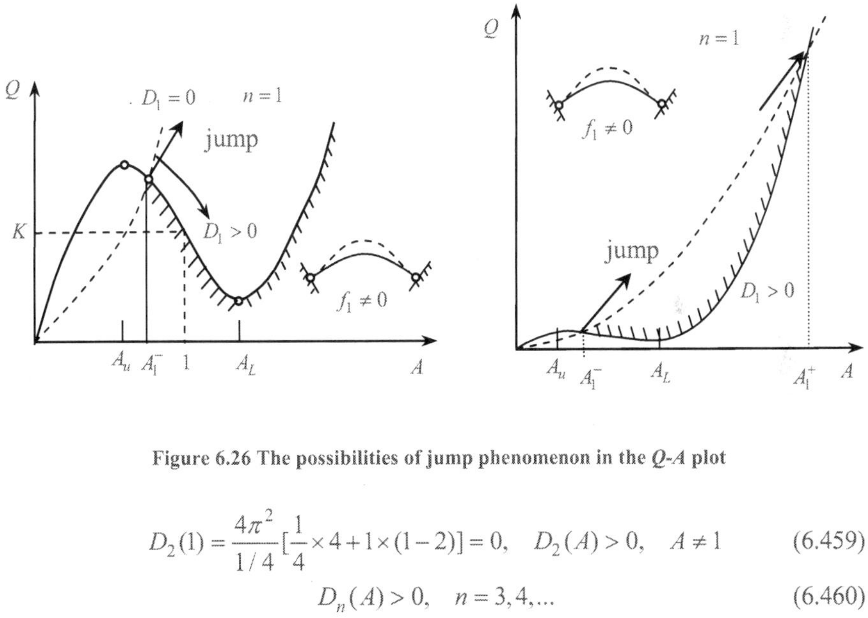

Figure 6.26 The possibilities of jump phenomenon in the Q-A plot

$$D_2(1) = \frac{4\pi^2}{1/4}[\frac{1}{4}\times 4 + 1\times(1-2)] = 0, \quad D_2(A) > 0, \quad A \neq 1 \tag{6.459}$$

$$D_n(A) > 0, \quad n = 3,4,... \tag{6.460}$$

For the deformation mode of $n = 2$, the equilibrium state is stable for A, except for $A = 1$ and $K = 1/4$. At this point, we have one of the roots of (6.449) being zero, and, thus, it is a possible bifurcation point for snap-through buckling, as illustrated in Figure 6.24. Note from Figure 6.27 that for this case $Q = 0$ for $A = A_L$.

<u>Case 3C</u>: $1/9 < K < 1/4$

For this range of K, the roots for A equal

$$A_2^{\pm} = 1 \pm \sqrt{1-4K} \tag{6.461}$$

which is clearly real. Therefore, real roots for A exist and $D_2 < 0$ for

$$A_2^- < A < A_2^+ \tag{6.462}$$

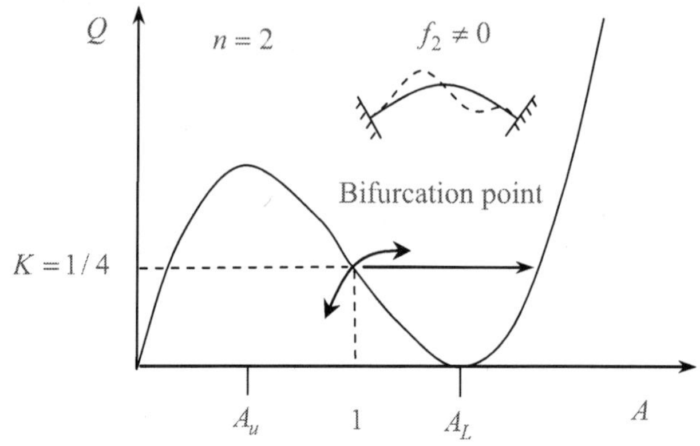

Figure 6.27 The possibilities of bifurcation at $K = 1/4$ and $A = 1$ for $n = 2$

In summary, we have

$$D_2(A_2^{\pm}) = 0, \quad D_2(A) < 0, \quad A_2^- < A < A_2^+ \tag{6.463}$$

$$D_n(A) > 0, \quad n = 3, 4, ... \tag{6.464}$$

According to the rule, there is a possible pair of bifurcation points at $A = A^+$ and A^-. The situation is summarized in Figure 6.28. The equilibrium states marked by hatched lines on the Q-A curve are unstable. The bifurcation points are shown by curved arrows. For $K = 1/4$, there is only one possible bifurcation point. As K decreases, the possible bifurcation point is split into two bifurcation points. At $K = 2/11$, we have the bifurcation points coincide with the limit points (A_u and A_L) (see Problem 6.11 for proof). At $K = 1/9$, the unstable zone spreads further beyond the limit points, as shown. Thus, snap-through buckling would not occur for mode 2 (i.e., $n = 2$), whilst all higher modes would be stable.

This procedure can be continued for considering snap-through buckling for higher modes with $K = 1/n^2$. Therefore, snap-through buckling of higher modes is more likely to occur if the arch is deeper and of weaker moment of inertia, subject to corresponding initial perturbations. At each higher mode, the possible bifurcation starts at $A = 1$ and $Q = K$.

Physically, Bazant and Cedolin (1991) provides an insightful explanation of the snap-through buckling phenomena for the case of a bifurcation point being lower than the first peak Q_u shown in Figure 6.28 for $K = 1/9$. In particular, the lowering of bifurcation force can be interpreted as a consequence that the kinetic energy of the initial disturbance of the corresponding mode equals the energy hump required for the equilibrium state to move to a nearby lower energy state. This concept is illustrated in Figure 6.29. In terms of the potential energy of the equilibrium states, we can at least consider two situations. The left figure illustrates the case of a dynamic jump with no energy loss and the right figure illustrates the case of a loss of potential energy as an energy hump is overcome by the input kinetic energy of initial disturbances. The situation depicted on the right is referred to as meta-stable, in the sense that a finite disturbance leads to an unstable state whereas the state is stable if the disturbance is infinitesimally small or $f_2 \to 0$. As a first approximation, it can be shown that change of the force is proportional to 2/3 power of the required kinetic energy:

$$\Delta Q \propto (\Delta T)^{2/3} \tag{6.465}$$

Equivalently, the change of the force is roughly proportional to 4/3 power of the disturbance velocity:

$$\Delta Q \propto v^{4/3} \tag{6.466}$$

The proofs of these (6.465) and (6.466) are left for readers as problems (Problems 6.12 and 6.13). In the next two sections, we are going to show that the transition of the bifurcation on the left figure of Figure 6.29 can be modeled by an unsymmetric state, which itself is unstable.

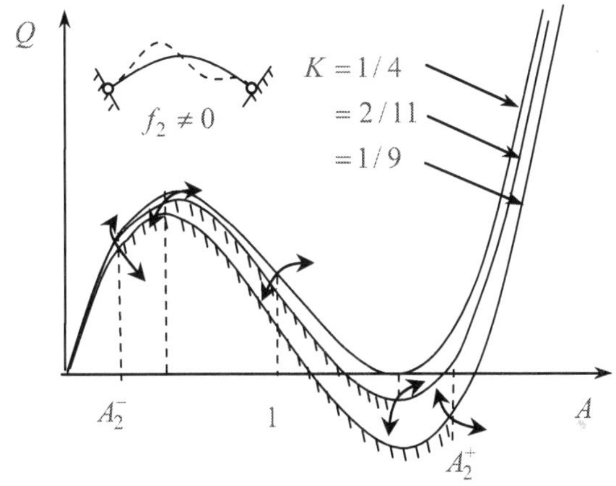

Figure 6.28 A pair of bifurcation points for 1/9< *K* < 1/4

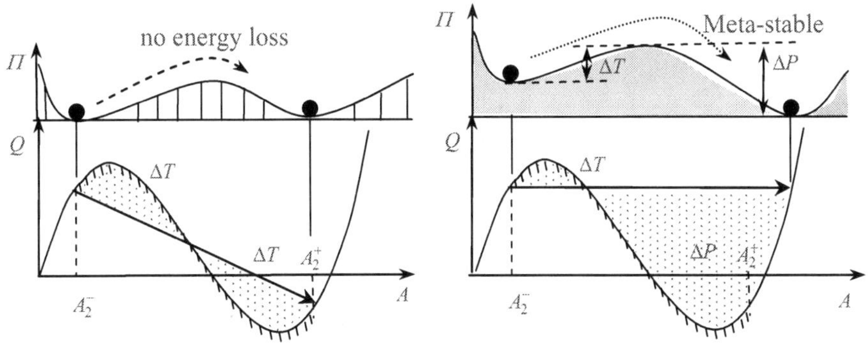

Figure 6.29 Energy hump required for an equilibrium state to move to a nearby stable state

6.9.3 Transitions of Snap-Through Buckling

To find the possible solution connecting the equilibrium states shown in the left diagram of Figure 6.29, we consider the following two-mode solution of the nonlinear static theory:

$$w(x) = -a[A\sin \pi x + B\sin 2\pi x] \tag{6.467}$$

$$p(x,t) = p(x) = -q\pi^4 \sin \pi x \tag{6.468}$$

Note that the $\sin 2\pi x$ is an unsymmetric deformation mode with respect to the center line at the crest of the arch. The steady-state arch problem with zero damping can be expressed as

$$w_{xxxx} + \lambda w_{xx} = -\lambda w_{xx}^0 + p(x,t) \tag{6.469}$$

$$w = w_{xx} = 0, \quad x = 0,1 \tag{6.470}$$

$$\lambda = -k[\frac{1}{2}\int_0^1 (w_x)^2 dx + \int_0^1 w_x^0 w_x dx] \tag{6.471}$$

Substitution of (6.398) and (6.467) into (6.471) gives

$$\begin{aligned} \lambda &= -k[\frac{1}{2}\int_0^1 (w_x)^2 dx + \int_0^1 w_x^0 w_x dx] \\ &= -\frac{ka^2\pi^2}{2}(\frac{A^2}{2} - A + 2B^2) \end{aligned} \tag{6.472}$$

Substitution of (6.398), (6.467), (6.468) and (6.472) into (6.469) leads to

$$(-aA\pi^4 + a\lambda A\pi^2)\sin \pi x + (-16aB\pi^4 + 4aB\lambda\pi^2)\sin 2\pi x = (a\lambda\pi^2 - q\pi^4)\sin \pi x \tag{6.473}$$

Setting the coefficients of each sine function to zero independently, we obtain a set of two coupled equations for A and B:

$$A + \frac{ka^2}{2}(\frac{A^2}{2} - A + 2B^2)(A-1) = \frac{q}{a} \tag{6.474}$$

$$B[4 + \frac{ka^2}{2}(\frac{A^2}{2} - A + 2B^2)] = 0 \tag{6.475}$$

Considering the special case of $B = 0$, we recover the symmetric solution given in (6.399). For $B \neq 0$, we have

$$4 + \frac{ka^2}{2}(\frac{A^2}{2} - A + 2B^2) = 0 \tag{6.476}$$

Substituting (6.476) into (6.474) yields

$$A - 4(A-1) = \frac{q}{a} \tag{6.477}$$

Using the definitions of Q and K given in (6.406), we can rewrite (6.477) as

$$Q = K(4 - 3A) \tag{6.478}$$

On the other hand, (6.476) can be rearranged as

$$(A-1)^2 + 4B^2 = 1 - 4K \tag{6.479}$$

This is the equation of an ellipse in the B-A plot. This is demonstrated in Figure 6.30. Thus, B is related to A through K. At $B = 0$, we have

$$A = A_2^{\mp} = 1 \pm \sqrt{1-4K} \tag{6.480}$$

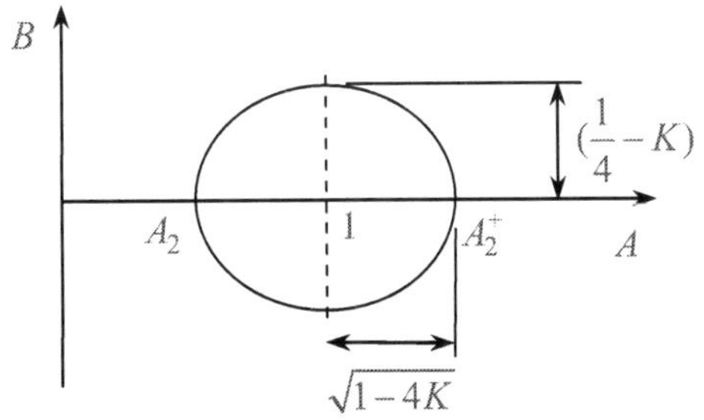

Figure 6.30 The elliptic locus in the plot of amplitudes of first and second modes defined in (6.467)

Equation (6.467) is an unsymmetric solution connecting the two bifurcation points of the symmetric solution that we obtained in the last section.

6.9.4 Linear Dynamic Stability for Unsymmetric State

To check the dynamic stability of the solution found in the last section, we consider a perturbation analysis around the unsymmetric state obtained in the last section:

$$\overline{w} = w + v \tag{6.481}$$

where

$$w(x) = -a[A\sin\pi x + B\sin 2\pi x] \tag{6.482}$$

Following the procedure that we used in Section 6.9.2, we obtain the linearized problem as

$$v_{xxxx} + \Lambda w_{xx} + \lambda_0 v_{xx} + v_{tt} + 2\Gamma v_t = -\Lambda w_{xx}^0 \tag{6.483}$$

where

$$\lambda_0 = -\frac{ka^2\pi^2}{2}\left(\frac{A^2}{2} - A + 2B^2\right) \tag{6.484}$$

We now seek for the following solution form:

$$v(x,t) = \sum_{n=1}^{\infty} v_n(t)\sin n\pi x \tag{6.485}$$

Using (6.398), (6.482) and (6.485), we can determine Λ as

$$\begin{aligned}\Lambda &= -k\left\{\int_0^1 w_x v_x dx + \int_0^1 w_x^0 v_x dx\right\} \\ &= -a\pi^2 k\left[\frac{1}{2}(1-A)v_1 - 2Bv_2\right]\end{aligned} \tag{6.486}$$

The derivation of (6.486) is left as a problem for the reader (see Problem 6.13). Substituting (6.398), (6.482), (6.485), and (6.486) into (6.483), we find

$$\sum_{n=1}^{\infty}\{[n^4\pi^4-\frac{ka^2\pi^2}{2}(\frac{A^2}{2}-A+2B^2)n^2\pi^2]v_n+v_n''+2\Gamma v_n'\}\sin n\pi x$$
$$=ka\pi^2[\frac{a\pi^2}{2}v_1(1-A)-2a\pi^2Bv_2][(A+1)\sin\pi x+4B\sin 2\pi x] \tag{6.487}$$

Collecting terms for each sine function, we obtain three ODEs for different values of n

$$v_1''+2\Gamma v_1'+\pi^4[-3+\frac{ka^2}{2}(1-A)^2]v_1+2k\pi^4a^2B(1-A)v_2=0 \tag{6.488}$$

$$v_2''+2\Gamma v_2'-2k\pi^4a^2B(1-A)v_1+8ka^2\pi^4Bv_2=0 \tag{6.489}$$

$$v_n''+2\Gamma v_n'+[n^4\pi^4-\frac{ka^2\pi^2}{2}(\frac{A^2}{2}-A+2B^2)n^2\pi^2]v_n=0,\quad n\geq 3 \tag{6.490}$$

Our focus here is on the first two coupled ODEs for v_1 and v_2. Equations (6.488) and (6.489) are ODEs with constant coefficients, and thus we seek a solution of the form

$$v_1=C_1e^{\sigma t},\quad v_2=C_2e^{\sigma t} \tag{6.491}$$

Substitution of (6.491) into (6.488) and (6.489) gives

$$\{\sigma^2+2\Gamma\sigma+\pi^4[-3+\frac{ka^2}{2}(1-A)^2]\}C_1+2k\pi^4a^2B(1-A)C_2=0 \tag{6.492}$$

$$2k\pi^4a^2B(A-1)C_1+\{\sigma^2+2\Gamma\sigma+8ka^2\pi^4B^2\}C_2=0 \tag{6.493}$$

For a non-trivial solution, we set the determinant of this system of equations to zero, and this leads to

$$P(\sigma)=\sigma^4+a_1\sigma^3+a_2\sigma^2+a_3\sigma+a_4=0 \tag{6.494}$$

where

$$a_1=4\Gamma>0 \tag{6.495}$$

$$a_2=4\Gamma^2+\pi^4\{-3+\frac{ka^2}{2}[16B^2+(1-A)^2]\} \tag{6.496}$$

$$a_3=2\Gamma\pi^4\{-3+\frac{ka^2}{2}[16B^2+(1-A)^2]\} \tag{6.497}$$

$$a_4=-24\pi^8ka^2B^2\leq 0 \tag{6.498}$$

To investigate the possibility that any roots of σ have a positive real part, we need to study the roots of the fourth order algebraic equation found in (6.494). Since the roots must appear as complex conjugate pairs, we can assume

$$\sigma_1=\alpha_1+i\beta_1,\quad \sigma_2=\alpha_1-i\beta_1,\quad \sigma_3=\alpha_3+i\beta_3,\quad \sigma_4=\alpha_3-i\beta_3 \tag{6.499}$$

The polynomials can be factorized using these roots as

$$\begin{aligned}P(\sigma)&=(\sigma-\sigma_1)(\sigma-\sigma_2)(\sigma-\sigma_3)(\sigma-\sigma_4)\\&=\sigma^4-2(\alpha_1+\alpha_3)\sigma^3+(\alpha_1^2+\beta_1^2+\alpha_3^2+\beta_3^2+4\alpha_1\alpha_3)\sigma^2\\&\quad-2[\alpha_1(\alpha_3^2+\beta_3^2)+\alpha_3(\alpha_1^2+\beta_1^2)]\sigma\\&\quad+(\alpha_3^2+\beta_3^2)(\alpha_1^2+\beta_1^2)\end{aligned} \tag{6.500}$$

Comparison of (6.500) with (6.494) gives

$$a_1 = -2(\alpha_1 + \alpha_3) \tag{6.501}$$

$$a_2 = (\alpha_1^2 + \beta_1^2 + \alpha_3^2 + \beta_3^2 + 4\alpha_1\alpha_3) \tag{6.502}$$

$$a_3 = -2[\alpha_1(\alpha_3^2 + \beta_3^2) + \alpha_3(\alpha_1^2 + \beta_1^2)] \tag{6.503}$$

$$a_4 = (\alpha_3^2 + \beta_3^2)(\alpha_1^2 + \beta_1^2) \tag{6.504}$$

For a stable solution, we require that all real parts of the roots given in (6.499) be negative or

$$\alpha_1 < 0, \quad \alpha_3 < 0 \tag{6.505}$$

Substitution of (6.505) into (6.501) to (6.504) gives

$$a_1 > 0, \quad a_2 > 0, \quad a_3 > 0, \quad a_4 > 0 \tag{6.506}$$

Conversely, if (6.506) is satisfied, all roots of σ have negative real parts or the solution is stable. However, it is clear from (6.498) that $a_4 < 0$, and, consequently, at least one of the roots of σ has a positive real part. Therefore, the unsymmetric solution from the unbuckled to the buckled states is unstable. Physically, this suggests that the transition in terms of unsymmetric deformation appears in a dynamic manner.

6.10 SUMMARY AND FURTHER READING

In this chapter, we have demonstrated the use of perturbation analysis to study buckling problems of straight columns and shallow arches. Two types of axial compression of columns or arches are possible, and they are the load and displacement controls at the end of the columns or arches. It turns out that the case of applied end displacement leads to a much more meaningful bifurcation scenario. The nonlinearity of the PDE arises from the fact that axial stress due to end compressions depends on the unknown deflection of the columns or arches. A static equilibrium solution is first obtained, and its stability is examined by using both linear and nonlinear dynamic stability analyses. It demonstrates that the mathematical structure of the straight column, subject to end compression, is the same as that for a simple example of two rigid bars connected by a deflection spring and a rotational spring. Stability can be interpreted by visualizing the existence of energy basins around the buckled states. Stability for damped and undamped cases are considered separately. Three particular forms of initial disturbances are considered: (i) an infinite series of the eigenfunctions; (ii) a particular n-th mode of the eigenfunction; and (iii) a finite number of series of eigenfunctions. The third case requires the use of multi-time perturbation analysis and is the most complicated scenario. Transitions from the straight to the buckled states are also presented. The role of dynamic disturbance is examined in detail. For the case of snap-through buckling, a simple example of two elastic bars connected by a hinge is considered. For arch buckling, we consider the elastic buckling of an arch under sinusoidal pressure. Mathematically, we find that the problem is convertible to the Duffing equation, which consists of a third order nonlinear term. The stability of the snap-through buckling is considered by perturbations. A transition solution connecting two ends of the snap-through buckling is found in terms of an unsymmetric solution, which is proved to be unstable.

Our buckling analyses on a straight column under end displacement are based on and expanded from the work of Reiss (1969), and Reiss and Matkowsky (1971), whereas our buckling analysis of shallow arches expanded on the work of Reiss (1980b) and Reiss (1984). For further reading on nonlinear buckling, we refer readers to the classical textbooks on elastic stability of structures including Timoshenko and Gere (1961), and Bazant and Cedolin (1991). Buckling of bars, plates and shells was considered by Brush and Almroth (1975). Wang et al. (2005) compiled exact solutions of buckling of structures, including columns, plates and shells. For the general theory of elastic stability, we refer to Bolotin (1963, 1964).

6.11 PROBLEMS

Problem 6.1. Show that the sizes of the elliptic orbit given in Figure 6.16 are correct. That is,

$$B = \sqrt{\beta}\{\sqrt{1+\frac{2E}{K\beta^2}}-1\}^{1/2}, \quad for\ B'=0 \tag{6.507}$$

$$B' = \sqrt{E}, \quad for\ B=0 \tag{6.508}$$

Problem 6.2. Show that the differentiation of Problem I defined in (6.329) to (6.332) one more time with respect to ε leads to

$$\ddot{w}_{xxxx} + 2\dot{\lambda}\dot{w}_{xx} + \lambda\ddot{w}_{xx} + \ddot{\lambda}w_{xx} + \ddot{w}_{tt} + 4\dot{\gamma}\dot{w}_t + 2\gamma\ddot{w}_t + 2\ddot{\gamma}w_t = 0 \tag{6.509}$$

$$\ddot{\lambda} = k\{2\ddot{c} - \int_0^1 [(\dot{w}_x)^2 + w_x\ddot{w}_x]dx\} \tag{6.510}$$

$$\ddot{w} = \ddot{w}_{xx} = 0, \quad x = 0,1 \tag{6.511}$$

$$\ddot{w}(x,t,0) = 0, \quad \ddot{w}_t(x,t,0) = 0 \tag{6.512}$$

Problem 6.3. Show the following identities

$$\ddot{w}_{xxxx}(x,t,0) = w^2_{xxxx}, \quad \ddot{w}_{xx}(x,t,0) = w^2_{xx} \tag{6.513}$$

$$\ddot{w}_t(x,t,0) = 2w^1_\tau + w^2_t, \quad \ddot{w}_{tt}(x,t,0) = 4w^1_{t\tau} + w^2_{tt} \tag{6.514}$$

Problem 6.4. Use the results derived in Problems 6.2 and 6.3 to derive Problem II given (6.353) to (6.355).

Problem 6.5. Substitute (6.352) into (6.353) to (6.355) and derive the following system:

$$L[w^2] = -4\sum_2^N \omega_n[(a^1_{n,\tau} + \tilde{\gamma}a^1_n)\cos\omega_n t - (b^1_{n,\tau} + \tilde{\gamma}b^1_n)\sin\omega_n t]\phi_n(x) \tag{6.515}$$

$$w^2 = w^2_{xx} = 0, \quad x = 0,1 \tag{6.516}$$

$$w^2(x,0,0) = 0, \tag{6.517}$$

$$w_t^2(x,0,0) = -2[b_\tau(0)\phi_1 + \sum_{n=2}^{N} b_{n,\tau}^1(0)\phi_n] \tag{6.518}$$

Problem 6.6. To solve the system derived in Problem 6.5, we assume the following solution form:

$$w^2(x,t,\tau) = \sum_{n=1}^{N} A_n^2(t,\tau)\phi_n(x) \tag{6.519}$$

(i) Prove that

$$A_{n,tt}^2 + \omega_n^2 A_n^2 = -4\omega_n[(a_{n,\tau}^1 + \tilde{\gamma} a_n^1)\cos\omega_n t - (b_{n,\tau}^1 + \tilde{\gamma} b_n^1)\sin\omega_n t] \tag{6.520}$$

$$A_n^2(0,0) = 0, \quad A_{n,t}^2(0,0) = -2b_{n,\tau}^1(0) \tag{6.521}$$

where

$$b_1^1(\tau) = b(\tau), \quad \omega_n = \sqrt{\lambda_n(\lambda_n - \lambda_1)} \tag{6.522}$$

(ii) Note that the homogeneous solution of (6.520) coincides with the nonhomogeneous term on the right-hand side of (6.398), and the solution must be bounded. Show that

$$a_{n,\tau}^1 = \frac{g_n}{\omega_n} e^{-\tilde{\gamma}\tau}, \quad b_n^1 = f_n e^{-\tilde{\gamma}\tau} \tag{6.523}$$

Hints: (6.351) and (6.352) are useful.

(a) Consider the case of $n = 1$, we have

$$A_{1,tt}^2 = 0 \tag{6.524}$$

(b) Show that the solution of (a) is

$$A_1^2 = a_1^2(\tau)t + b_1^2(\tau) \tag{6.525}$$

(c) By the boundedness condition and (6.392), show that

$$A_{1,\tau}^2(0,0) = -2b_\tau(0) = 0 \tag{6.526}$$

(d) Finally, show the validity of (6.357).

Problem 6.7. Show that differentiation of Problem II, defined in (6.509) to (6.512), one more time with respect to ε leads to the following Problem III

$$\dddot{w}_{xxxx} + 3\ddot{\lambda}\dot{w}_{xx} + \lambda\dddot{w}_{xx} + +\dddot{w}_{tt} + 6\dot{\gamma}\ddot{w}_t = 0 \tag{6.527}$$

$$\dddot{\lambda} = k[2\dddot{c} - \int_0^1 (3\dot{w}_x\ddot{w}_x + w_x\dddot{w}_x)dx] \tag{6.528}$$

$$\dddot{w} = \dddot{w}_{xx} = 0, \quad x = 0,1 \tag{6.529}$$

$$\dddot{w}(x,t,0) = 0, \quad \dddot{w}_t(x,t,0) = 0 \tag{6.530}$$

Problem 6.8. Substituting $\varepsilon = 0$ in Problem 6.7, we get Problem III as:

$$L[w^3] = w_{xxxx}^3 + \lambda^0 w_{xx}^3 + w_{tt}^3 = -3\lambda^2 w_{xx}^1 - 6w_{\tau t}^2 - 6w_{\tau\tau}^1 - 6\tilde{\gamma}(w_\tau^2 + 2w_\tau^1) \tag{6.531}$$

$$w^3 = w_{xx}^3 = 0, \quad x = 0,1 \tag{6.532}$$

$$w^3(x,0,0)=0, \quad w_t^3(x,0,0)=-3w_\tau^2(x,0,0) \tag{6.533}$$

$$\lambda^2 = 2k[2-\frac{1}{2}\int_0^1 (w_x^1)^2 dx] \tag{6.534}$$

Problem 6.9. To solve Problem III, derived in Problem 6.8, we seek the following solution:

$$w^3(x,t,\tau)=\sum_{n=1}^{N} A_n^3(t,\tau)\phi_n(x) \tag{6.535}$$

(i) Show that the amplitude of (6.535) satisfies

$$\begin{aligned}
&\sum_{n=1}^{N}(A_{n,tt}^3+\omega_n^2 A_n^3)= \\
&3\lambda^2\{\lambda_1 b\phi_1 + e^{-\tilde{\gamma}\tau}\sum_{n=2}^{N}\lambda_n[\frac{g_n}{\omega_n}\sin\omega_n t + f_n\cos\omega_n t]\phi_n\} \\
&-6\sum_{n=2}^{N}\omega_n[a_{n,\tau}^2\cos\omega_n t - b_{n,\tau}^2\sin\omega_n t]\phi_n \\
&-6\{b_{\tau\tau}\phi_1+\tilde{\gamma}^2 e^{-\tilde{\gamma}\tau}\sum_{n=2}^{N}[\frac{g_n}{\omega_n}\sin\omega_n t + f_n\cos\omega_n t]\phi_n\} \\
&-6\tilde{\gamma}\sum_{n=2}^{N}\omega_n[a_n^2\cos\omega_n t - b_n^2\sin\omega_n t]\phi_n \\
&-12\tilde{\gamma}\{b_\tau\phi_1 - \tilde{\gamma}e^{-\tilde{\gamma}\tau}\sum_{n=2}^{N}(\frac{g_n}{\omega_n}\sin\omega_n t + f_n\cos\omega_n t)\phi_n\}
\end{aligned} \tag{6.536}$$

(ii) For $n=1$, show that

$$A_{1,tt}^3 = -6(b_{\tau\tau}+2\tilde{\gamma}b_\tau - \frac{1}{2}\lambda^2\lambda_1 b) \tag{6.537}$$

where

$$\lambda^2 = 2k\{2-\frac{1}{2}b^2\lambda_1 - \frac{1}{2}e^{-\tilde{\gamma}\tau}\sum_{n=2}^{N}\lambda_n[(\frac{g_n}{\omega_n})^2+f_n^2]\sin^2(\omega_n t+\gamma_p)\} \tag{6.538}$$

(iii) Show that the homogeneous solution of (6.415) is

$$A_1^3 = a_1^3(\tau)t + b_1^3(\tau) \tag{6.539}$$

(iv) By boundedness as $t\to\infty$, show that

$$A_{1,tt}^3 = 0 \tag{6.540}$$

(v) By considering the fact that the long-term motion (i.e., $t\to\infty$) should not have a constant term, show from (6.537) and (6.538) that

$$b_{\tau\tau}+2\tilde{\gamma}b_\tau + K(\mu b + b^3)=0 \tag{6.541}$$

where

$$K = \frac{k\lambda_1^2}{2}, \quad \mu = \frac{4}{\lambda_1}\{-1 + \frac{1}{8}e^{-\bar{\gamma}\tau}\sum_{n=2}^{N}\lambda_n[f_n^2 + (\frac{g_n}{\omega_n})^2]\} \tag{6.542}$$

Remark: This result was given by Reiss and Matkowsky (1971) without proof.

Problem 6.10. Prove the following identity

$$(1-\sqrt{1-B})[A_u^2 - 3A_u + 2] = B\sqrt{1-B} \tag{6.543}$$

Problem 6.11. Prove for dynamic stability analysis of a shallow arch and $n = 2$, that the bifurcation points coincide with the limit points when $K = 2/11$ as reported in Section 6.9.2.

Problem 6.12. Prove the validity of (6.465) and (6.466) by assuming that the Q-A curve near the peak at Q_u can be approximated by a parabolic curve.

Problem 6.13. Show the validity of (6.494).

CHAPTER SEVEN

Turbulent Diffusions in Fluids

7.1 INTRODUCTION

Recent rapid urban and industrial developments in coastal cities around the peripheries of estuaries and gulfs in both developed and developing countries have created major concern regarding the capability of their waters to act as a buffer zone for receiving waste effluents and subsequent dispersion of waste in offshore waters. Unwanted effluents may come from exploration of coastal waters for oil and mineral deposits. The dynamics of the dispersion and diffusion mechanism in the estuaries and gulf are complex but are crucial in affecting the water quality in local waters. Flow patterns near shore are complex, depending on meteorological conditions, bathymetry, and boundary conditions. The nonlinear interactions of waves, current, and the discharge of polluted fluids from factories or domestic sewage can be quite complex, but this topic is out of the scope of the present chapter. The coastal water quality depends on the pollutant concentration, discharge rate, waste characteristics, diffusion, and dispersion processes. Only analytic solutions will be discussed in the present chapter. Although numerical methods seem inevitably to simulate more realistic prediction of pollutant dispersion and diffusion under various initial conditions, analytic solutions to be discussed are found very useful in providing insights to the understanding of the mechanisms of pollutant transports in fluids.

In this chapter, error function and complementary error function are first reviewed. Most of the solutions that appear later in the chapter are expressible in error functions. Fick's first and second laws are derived for modeling turbulent eddy-type diffusion. A steady-state solution for dispersion in one-dimensional situations is first discussed. The Ogata and Banks (1961) solution is discussed and this solution is then extended to the case of a fluid with nonzero flow speed and with decaying behavior of the pollutants. The classical solution of Taylor (1954) for a point source for a one-dimensional problem is discussed and the solution is extended to the case of a fluid with nonzero flow speed and with decaying behavior of the pollutants. Such a one-dimensional solution is relevant to the dumping of wastewater into a river by factories. Point source solutions in two-dimensional and three-dimensional cases are then derived and the superposition principle is used to derive the solution for a continuous line source of pollutants. This line source solution in an infinite two-dimensional domain is relevant to the situation of some cities releasing their treated wastewater into deeper waters off the coastline through submarine pipelines.

7.2 ERROR FUNCTION

7.2.1 Definition

A very important function that appears naturally in diffusion problems is called the error function. The error function is defined as (Abramowitz and Stegun, 1964)

$$erf(x) = \frac{2}{\sqrt{\pi}} \int_0^x e^{-\eta^2} d\eta \tag{7.1}$$

The function inside the integral actually represents a normal distribution in probability theory or the Gaussian normal distribution. This integral relates to the error estimation in probability theory, the theory of heat conduction (Carslaw and Jaeger, 1959), fluid pressure in a porous medium (Chau, 2013), and the Rayleigh impulsive flow problem (Segel, 1987). Although this function is normally considered as a kind of special function, there is normally not much discussion on this function in most books in special functions. The purpose of this section is to summarize some essential results of the error function before its appearance in later sections on diffusion of pollutants.

7.2.2 Relation to Normal Distribution

The probability of finding a certain quantity of a sample in a population is essential to engineering applications, related to mass production, and product inspection for defective items. For example, it is prohibitively expensive and time-consuming to inspect every product before its delivery to the market. Therefore, we normally select some samples for inspection and then interpret the possibility of having defects using the data obtained from the samples. These samples are supposed to be selected randomly from among all the products, depending on the size of the samples, the size of population, and the distribution of the defects or errors. Various kinds of distributions have been proposed. For example, if we plot the compressive strength of tested concrete specimens against its occurrence, we normally observe the distribution of the strength around a mean value. The spread of the strength data on the plots can be reflected in its standard deviation. From these plots for different types of data, it is normally assumed that the observed and unobserved variables follow a certain distribution function. Binomial, Poisson, hypergeometric, lognormal, Weibull, and Gaussian normal distributions are among the popular choices. If the observed data is characterized by success or failure, we can use the binomial or Bernoulli distribution to represent the probability of observing success x times (say p) among n samples (it is like a binomial expansion of the success and failure probabilities). A limiting case for the binomial distribution is called the Poisson distribution, and it can be interpreted as the case that the probability of success is small ($p \to 0$) among infinite samples of data ($n \to \infty$) whilst the mean equals the product of np approaching a finite value. For example, it is useful in estimating the probability of observing a certain level of ground shaking in n years. This is useful in analyzing probabilistic seismic hazards for buildings. If we want to find the probability of having a defect in products without replacement, hypergeometric distribution is used instead of the binomial distribution. For the case that the probability of success p is close to 50%

while $n \to \infty$, the binomial distribution can be approximated by the so-called normal distribution (Johnson et al., 1994).

In particular, scientists in the eighteenth century observed an astonishing degree of regularity in errors of measurement. They found that the distribution of errors can be represented by a bell-shaped curve:

$$F(x) = \frac{1}{\sigma\sqrt{2\pi}} \int_{-\infty}^{x} \exp[-\frac{1}{2}(\frac{\eta-\mu}{\sigma})^2] d\eta \tag{7.2}$$

where σ is the standard deviation, and μ is the mean. Let us introduce the following change of variables

$$u = \frac{\eta - \mu}{\sigma} \tag{7.3}$$

Applying (7.3) to (7.2), we have

$$F(x) = \Phi(\frac{x-\mu}{\sigma}) = \Phi(z) = \frac{1}{\sqrt{2\pi}} \int_{-\infty}^{z} \exp(-\frac{u^2}{2}) du \tag{7.4}$$

It has been well known to most engineering and science students that this probability integral cannot be evaluated using an elementary method. We always resort to looking up the values of this integral as a function of z from tables in textbooks on probability. In fact, this integral can be expressed in terms of the error function defined in (7.1). In particular, we first decompose the integral in (7.4) as

$$\begin{aligned} \Phi(z) &= \frac{1}{\sqrt{2\pi}} \int_{-\infty}^{z} \exp(-\frac{u^2}{2}) du \\ &= \frac{1}{\sqrt{2\pi}} \{\int_{-\infty}^{0} \exp(-\frac{u^2}{2}) du + \int_{0}^{z} \exp(-\frac{u^2}{2}) du\} \\ &= \frac{1}{\sqrt{2\pi}} \{-\int_{\infty}^{0} \exp(-\frac{u^2}{2}) du + \int_{0}^{z} \exp(-\frac{u^2}{2}) du\} \\ &= \frac{1}{\sqrt{2\pi}} \{\int_{0}^{\infty} \exp(-\frac{u^2}{2}) du + \int_{0}^{z} \exp(-\frac{u^2}{2}) du\} \end{aligned} \tag{7.5}$$

We further simplify (7.5) by applying the following change of variables

$$\frac{u^2}{2} = \eta^2 \tag{7.6}$$

Applying this change of variables, we get

$$\Phi(z) = \frac{1}{\sqrt{\pi}} \{\int_{0}^{\infty} \exp(-\eta^2) d\eta + \int_{0}^{z/\sqrt{2}} \exp(-\eta^2) d\eta\} \tag{7.7}$$

The first integral on the right-hand side of (7.7) is the Laplace/Gauss integral (see Section 1.4.6 of Chau, 2018), and the second integral on the right can be expressed in terms of the error function defined in (7.1). Thus, we have

$$\Phi(z)=\frac{1}{\sqrt{\pi}}\{\frac{\sqrt{\pi}}{2}+\frac{\sqrt{\pi}}{2}erf(\frac{z}{\sqrt{2}})\}$$
$$=\frac{1}{2}+\frac{1}{2}erf(\frac{z}{\sqrt{2}}) \tag{7.8}$$

Some properties of the error function are considered in the following examples.

Example 8.1 Show that the following antisymmetric property of error function is true:

$$erf(-x)=-erf(x) \tag{7.9}$$

Solution: Using the definition of the error function given in (7.1), we have

$$erf(-x)=\frac{2}{\sqrt{\pi}}\int_0^{-x}e^{-\eta^2}d\eta \tag{7.10}$$

Let us rewrite (7.10) as

$$erf(-x)=-\frac{2}{\sqrt{\pi}}\int_{-x}^{0}e^{-\eta^2}d\eta$$
$$=-\frac{2}{\sqrt{\pi}}\{\int_{-\infty}^{0}e^{-\eta^2}d\eta-\int_{-\infty}^{-x}e^{-\eta^2}d\eta\} \tag{7.11}$$

We can use the following change of variables as

$$\eta=-\zeta \tag{7.12}$$

Consequently, we have

$$erf(-x)=-\frac{2}{\sqrt{\pi}}\{-\int_{\infty}^{0}e^{-\zeta^2}d\zeta+\int_{\infty}^{x}e^{-\zeta^2}d\zeta\}$$
$$=-\frac{2}{\sqrt{\pi}}\{\int_{0}^{\infty}e^{-\zeta^2}d\zeta-\int_{x}^{\infty}e^{-\zeta^2}d\zeta\} \tag{7.13}$$
$$=-\frac{2}{\sqrt{\pi}}\int_{0}^{x}e^{-\zeta^2}d\zeta=-erf(x)$$

This completes the proof.

The series expansion of the error function is considered in the following example.

Example 8.2 Show that the series expansion of the error function is

$$erf(x)=\frac{2}{\sqrt{\pi}}\{x-\frac{x^3}{1!\times 3}+\frac{x^5}{2!\times 5}-\frac{x^7}{3!\times 7}+..\} \tag{7.14}$$

Solution: We can apply Taylor's series expansion for the exponential function inside the integral as

$$erf(x)=\frac{2}{\sqrt{\pi}}\int_0^{x}e^{-\eta^2}d\eta=\frac{2}{\sqrt{\pi}}\int_0^{x}\sum_{n=0}^{\infty}\frac{(-1)^n\eta^{2n}}{n!}d\eta \tag{7.15}$$

We can reverse the order of integration and summation as

$$erf(x) = \frac{2}{\sqrt{\pi}} \int_0^x e^{-\eta^2} d\eta = \frac{2}{\sqrt{\pi}} \sum_{n=0}^{\infty} \frac{(-1)^n}{n!} \int_0^x \eta^{2n} d\eta \tag{7.16}$$

The integration can be carried out term by term as

$$\begin{aligned} erf(x) &= \frac{2}{\sqrt{\pi}} \int_0^x e^{-\eta^2} d\eta = \frac{2}{\sqrt{\pi}} \sum_{n=0}^{\infty} \frac{(-1)^n x^{2n+1}}{(2n+1)n!} \\ &= \frac{2}{\sqrt{\pi}} \{x - \frac{x^3}{1!\times 3} + \frac{x^5}{2!\times 5} - \frac{x^7}{3!\times 7} + ...\} \end{aligned} \tag{7.17}$$

This completes the proof.

7.2.3 Complementary Error Function

Another function closely associated with the error function is called the complementary error function and is defined as

$$erfc(x) = \frac{2}{\sqrt{\pi}} \int_x^{\infty} e^{-\eta^2} d\eta \tag{7.18}$$

This function can be related to the error function as

$$erfc(x) = \frac{2}{\sqrt{\pi}} \{\int_0^{\infty} e^{-\eta^2} d\eta - \int_0^x e^{-\eta^2} d\eta\} \tag{7.19}$$

Using the Laplace/Gauss integral discussed in Section 1.4.6 of Chau (2018) and the definition given in (7.1), we obtain

$$erfc(x) = \frac{2}{\sqrt{\pi}} \frac{\sqrt{\pi}}{2} - erf(x) = 1 - erf(x) \tag{7.20}$$

For the negative argument, we have

$$\begin{aligned} erfc(-x) &= 1 - erf(-x) = 1 + erf(x) \\ &= 2 - erfc(x) \end{aligned} \tag{7.21}$$

7.2.4 Some Results of Error Function

We list some limiting values of the error function here:

$$erf(\infty) = \frac{2}{\sqrt{\pi}} \int_0^{\infty} e^{-\eta^2} d\eta = \frac{2}{\sqrt{\pi}} \frac{\sqrt{\pi}}{2} = 1 \tag{7.22}$$

$$erf(-\infty) = \frac{2}{\sqrt{\pi}} \int_0^{-\infty} e^{-\eta^2} d\eta = -\frac{2}{\sqrt{\pi}} \int_0^{\infty} e^{-\eta^2} d\eta = -\frac{2}{\sqrt{\pi}} \frac{\sqrt{\pi}}{2} = -1 \tag{7.23}$$

$$erfc(\infty) = 1 - erf(\infty) = 1 - 1 = 0 \tag{7.24}$$

$$erfc(-\infty) = 1 - erf(-\infty) = 1 + 1 = 2 \tag{7.25}$$

By direct differentiation, we obtain the following derivatives of the error function

$$\frac{d}{dx} erf(x) = \frac{2}{\sqrt{\pi}} e^{-x^2} \tag{7.26}$$

$$\frac{d^2}{dx^2} erf(x) = -\frac{4}{\sqrt{\pi}} x e^{-x^2} \tag{7.27}$$

$$\frac{d^3}{dx^3} erf(x) = -\frac{4}{\sqrt{\pi}} e^{-x^2} + \frac{8}{\sqrt{\pi}} x^2 e^{-x^2} \tag{7.28}$$

The integral of the error function can be proved equal to

$$\int erf(x)dx = x\, erf(x) + \frac{1}{\sqrt{\pi}} e^{-x^2} + C \tag{7.29}$$

A generalized error function can be defined as

$$E_n(x) = \frac{1}{\Gamma\{(n+1)/n\}} \int_0^x \exp(-\xi^n) d\xi \tag{7.30}$$

The error function is recovered for $n = 2$.

Example 8.3 Find the solution of the following ODE with the given boundary conditions in terms of the error function

$$\frac{d^2F}{d\eta^2} + 2\eta \frac{dF}{d\eta} = 0, \quad 0 < \eta < \infty \tag{7.31}$$

$$F(0) = 0, \quad F(\infty) = 1 \tag{7.32}$$

Solution: We can first reduce the order of the ODE by assuming

$$G = \frac{dF}{d\eta} \tag{7.33}$$

Applying (7.33), we can reduce the ODE given in (7.31) to

$$\frac{dG}{d\eta} + 2\eta G = 0 \tag{7.34}$$

This is a separable first order ODE and can be integrated directly to give

$$G(\eta) = Ce^{-\eta^2} \tag{7.35}$$

Substitution of (7.35) into (7.33) gives

$$\frac{dF}{d\eta} = G = Ce^{-\eta^2} \tag{7.36}$$

Integration of (7.36) gives

$$F(\eta) = C\int_0^{\eta} e^{-s^2} ds \tag{7.37}$$

It is obvious that the first condition of (7.32) is satisfied automatically. The second boundary condition of (7.30) gives

$$F(\infty) = C\int_0^{\infty} e^{-s^2} ds = C\frac{\sqrt{\pi}}{2} = 1 \tag{7.38}$$

Thus, the constant C is

$$C = \frac{2}{\sqrt{\pi}} \tag{7.39}$$

The solution of F is

$$F(\eta) = erf(\eta) \tag{7.40}$$

If the boundary conditions given in (7.32) are interchanged, the problem is called the Rayleigh impulsive flow problem (see Problem 7.4).

7.3 DIFFUSION OF POLLUTANTS IN RIVER

Figure 7.1 shows a flux of substance of concentration c flowing in and out of a control volume along the x-direction. Using conservation of mass, we have

$$\rho cudydz + \frac{\partial}{\partial x}(\rho cudydz)dx - \rho cudydz = \frac{\partial(\rho c)}{\partial t} dxdydz \tag{7.41}$$

where c is the concentration of the pollutant defined by

$$c = \frac{\text{mass of pollutant}}{\text{mass of water}} \tag{7.42}$$

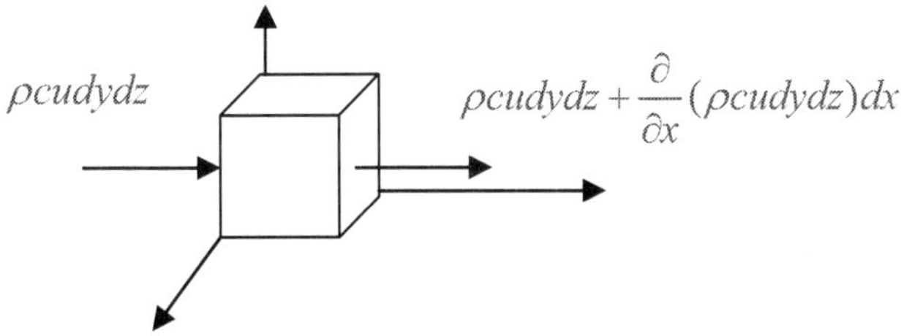

Figure 7.1 Flux of substance c through a fluid element

Simplification of (7.1) leads to the following one-dimensional conservation of mass

$$\frac{\partial(\rho cu)}{\partial x} = \frac{\partial(\rho c)}{\partial t} \tag{7.43}$$

If the flux of the pollutant of concentration c also exists in both the y- and z-directions, the conservation of mass can be extended to the following more general form

$$\frac{\partial(\rho c)}{\partial t} + \frac{\partial(\rho cu)}{\partial x} + \frac{\partial(\rho cv)}{\partial y} + \frac{\partial(\rho cw)}{\partial z} = 0 \tag{7.44}$$

The concept of turbulence is now introduced to interpret the diffusion process in fluid. When the flow velocity is fairly large such that the Reynolds number is not small, the velocity of flow in fluid appears to fluctuate in an extremely irregular manner with time at each point. This fluctuating velocity can be considered to comprise two parts: a mean velocity field plus an irregularly varying velocity field. This fluctuating part of velocity characterizes the turbulence in fluid. No complete quantitative theory of turbulence has yet been discovered. This is one of the long-standing unsolved problems in fluid mechanics. Nevertheless, some conceptual models have generally been accepted. These irregular motions are expressed as turbulent eddies. As flow velocity increases, laminar flows will yield to turbulent flows. Turbulent eddies of both small and large scales will evolve. The scale of eddies is the distance over which the eddy velocity varies appreciably. The largest eddies can be of the order of the size of the region where the flows take place. Naturally, the largest eddies have the largest amplitudes. For the largest eddies, the viscosity effect is negligible, whereas dissipation of energy mainly occurs at the smallest scales. The smallest eddies are also of the highest frequencies (velocity divided by the length scale). The change in the concentration of pollutants in water can come from two mechanisms; the first is mechanical mixing of fluids due to a flowing motion of the fluid (which is essentially non-dissipative), and the second is of diffusion type due to turbulent eddies. Figures 7.2 and 7.3 illustrate the turbulence-induced diffusions in rivers and the smallest eddy turbulence, respectively.

In particular, the total velocity field is the sum of the mean velocity plus the velocity fluctuation:

$$u = \bar{u} + u', \; v = \bar{v} + v', \; w = \bar{w} + w' \tag{7.45}$$

Similarly, the concentration can also be expressed as

$$c = \bar{c} + c' \tag{7.46}$$

Substitution of (7.45) and (7.46) into (7.44) gives

$$\frac{\partial(\bar{c}+c')}{\partial t} + \frac{\partial(\bar{c}+c')(\bar{u}+u')}{\partial x} + \frac{\partial(\bar{c}+c')(\bar{v}+v')}{\partial y} + \frac{\partial(\bar{c}+c')(\bar{w}+w')}{\partial z} = 0 \tag{7.47}$$

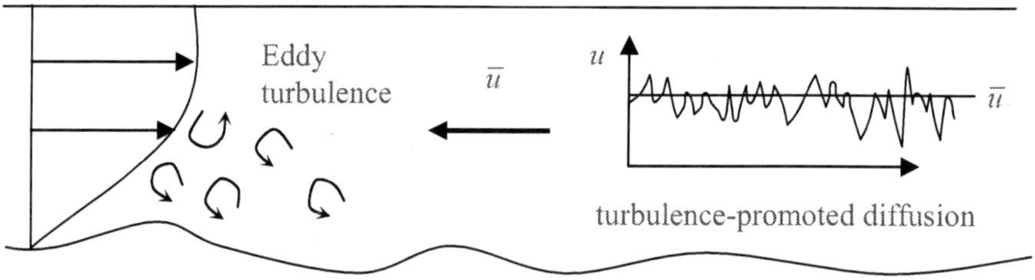

Figure 7.2 Eddy turbulence-induced diffusion

If we take the time average over a certain period T, the time average of the product of an average quantity and velocity variation must be zero:

$$\frac{1}{T}\int_0^T \bar{c}u'dt = \frac{\bar{c}}{T}\int_0^T u'dt = 0 \tag{7.48}$$

Note that the fluctuation of velocity of turbulences over time is zero by definition. Similar expressions can also be written down for the other two components of the fluctuating velocity fields. Considering the time average value of (7.47) and using (7.46), we have

$$\frac{\partial(\bar{c})}{\partial t}+\bar{u}\frac{\partial(\bar{c})}{\partial x}+\bar{v}\frac{\partial(\bar{c})}{\partial y}+\bar{w}\frac{\partial(\bar{c})}{\partial z}+\frac{\partial(\overline{c'u'})}{\partial x}+\frac{\partial(\overline{c'v'})}{\partial y}+\frac{\partial(\overline{c'w'})}{\partial z}=0 \quad (7.49)$$

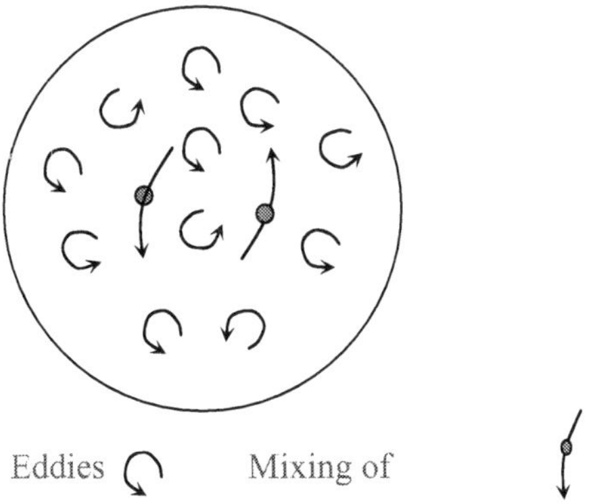

Figure 7.3 Eddy-enhanced mixing of pollutants

The time-average term of the product of pollutant concentration and the velocity field can be approximated by Fick's first diffusion law as:

$$(\overline{c'u'})=-D_x\frac{\partial(\bar{c})}{\partial x},\quad (\overline{c'v'})=-D_y\frac{\partial(\bar{c})}{\partial x},\quad (\overline{c'w'})=-D_z\frac{\partial(\bar{c})}{\partial x} \quad (7.50)$$

where D_x, D_y, and D_z are the effective turbulent eddy diffusion coefficients along different directions. The negative sign indicates that the direction of the flow of matter is toward the region of decreasing pollutant concentration. Note that Fick's first law is applied here to diffusion resulting from turbulent eddies, instead of the conventional interpretation of molecular diffusion. Substitution of (7.50) into (7.49) results in

$$\frac{\partial(\bar{c})}{\partial t}+\bar{u}\frac{\partial(\bar{c})}{\partial x}+\bar{v}\frac{\partial(\bar{c})}{\partial y}+\bar{w}\frac{\partial(\bar{c})}{\partial z}=\frac{\partial}{\partial x}(D_x\frac{\partial\bar{c}}{\partial x})+\frac{\partial}{\partial y}(D_y\frac{\partial\bar{c}}{\partial y})+\frac{\partial}{\partial z}(D_z\frac{\partial\bar{c}}{\partial z}) \quad (7.51)$$

This equation is also known as Fick's second law. These two laws for molecular diffusion were proposed by Adolph Fick, a German physiologist, in 1855. He recognized that the diffusion is equivalent to the Fourier law of heat conduction. The three-dimensional unsteady-state diffusion equation given in (7.51) needs to be solved by numerical methods. In general, we can also add a decay term to model the decay of concentration of the pollutant in the model (see later section).

7.4 OGATA AND BANKS SOLUTION

The special case of one-dimensional diffusion of (7.51) was solved by Ogata and Banks (1961). Banks considered himself an applied mathematician and published two interesting books, focusing on using applied mathematics in solving daily problems (Banks, 1998, 1999). More about Banks will be given in the summary section at the end of the chapter. First of all, it is straightforward to show that for the one-dimensional case shown in Figure 7.4, (7.51) is reduced to

$$\frac{\partial c}{\partial t} = D\frac{\partial^2 c}{\partial x^2} - u\frac{\partial c}{\partial x} \tag{7.52}$$

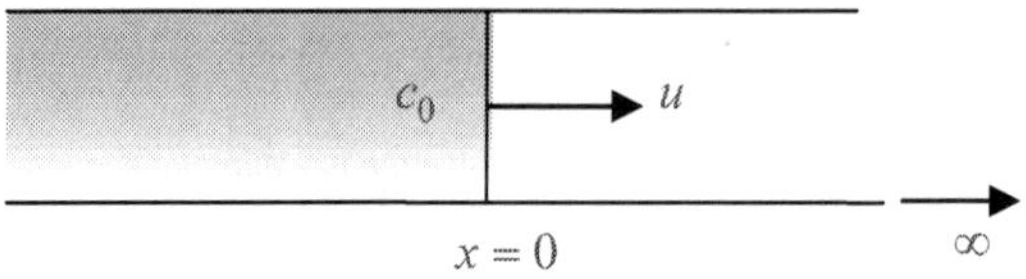

Figure 7.4 Ogata-Banks Problem: One-dimensional diffusion

For the sake of simplicity, we have dropped the superimposed bar for c and dropped the subscript for D. The Ogata-Banks problem is subjected to the following initial and boundary conditions:

$$\begin{aligned} &c(0,t) = c_0, \quad t \ge 0, \\ &c(x,0) = 0, \quad x \ge 0, \\ &c(\infty,t) = 0, \quad t \ge 0, \end{aligned} \tag{7.53}$$

To solve this problem, we first propose the following change of variables as

$$c(x,t) = \Gamma(x,t)\exp(\frac{ux}{2D} - \frac{u^2 t}{4D}) \tag{7.54}$$

As shown in Problem 7.1, the Ogata-Banks problem can be converted to

$$\frac{\partial \Gamma}{\partial t} = D\frac{\partial^2 \Gamma}{\partial x^2} \tag{7.55}$$

with the following conditions for Γ

$$\begin{aligned} &\Gamma(0,t) = \phi(t) = c_0 \exp(\frac{u^2 t}{4D}), \quad t \ge 0, \\ &\Gamma(x,0) = 0, \quad x \ge 0, \\ &\Gamma(\infty,t) = 0, \quad t \ge 0 \end{aligned} \tag{7.56}$$

To solve this problem, we first propose the following problem of applying a unit step function in time:

$$\begin{aligned} &\Gamma(0,t) = H(t), \quad t \ge 0, \\ &\Gamma(x,0) = 0, \quad x \ge 0, \\ &\Gamma(\infty,t) = 0, \quad t \ge 0 \end{aligned} \tag{7.57}$$

Once this problem is solved, we can use the Duhamel integral to solve (7.55) subject to condition (7.56). In particular, we can define the Laplace transform as (see Section 11.2.1 of Chau, 2018):

$$\bar{\Gamma}(x,s)=\int_0^{\infty}e^{-st}\Gamma(x,t)dt \tag{7.58}$$

Applying the Laplace transform to (7.55), we obtain

$$\frac{s}{D}\bar{\Gamma}=\frac{d^2\bar{\Gamma}}{dx^2} \tag{7.59}$$

Application of the Laplace transform to the initial and boundary conditions in (7.57) gives

$$\bar{\Gamma}(0,s)=\frac{1}{s},\quad \bar{\Gamma}(x,\infty)=0,\quad \bar{\Gamma}(\infty,s)=0 \tag{7.60}$$

The general solution of (7.59) is

$$\bar{\Gamma}=Ae^{-qx}+Be^{qx} \tag{7.61}$$

where

$$q=\sqrt{s/D} \tag{7.62}$$

The first and third equations given in (7.60) require

$$x\to\infty,\quad B=0 \tag{7.63}$$

$$x=0,\qquad A=\frac{1}{s} \tag{7.64}$$

The solution for (7.59) becomes

$$\bar{\Gamma}=\frac{1}{s}e^{-qx} \tag{7.65}$$

The final solution in the time domain can be found by using the following Laplace transform inversion

$$L^{-1}\{\frac{e^{-a\sqrt{s}}}{s}\}=erfc(\frac{a}{2\sqrt{t}}) \tag{7.66}$$

where the complementary error function was defined in (7.18). Application of (7.66) to (7.65) gives the fundamental solution for the unit step excitation prescribed in (7.55) and (7.57)

$$\Gamma(x,t)=erfc\{\frac{x}{2\sqrt{Dt}}\}=\frac{2}{\sqrt{\pi}}\int_{\frac{x}{2\sqrt{Dt}}}^{\infty}e^{-\eta^2}d\eta=1-erf\{\frac{x}{2\sqrt{Dt}}\} \tag{7.67}$$

The solution of the original problem defined in (7.55) and (7.56) can be expressed using Duhamel's theorem as

$$\Gamma(x,t)=\int_0^t\phi(\tau)\frac{\partial}{\partial t}\left[\frac{2}{\sqrt{\pi}}\int_{\frac{x}{2\sqrt{D(t-\tau)}}}^{\infty}e^{-\eta^2}d\eta\right]d\tau \tag{7.68}$$

Using Leibniz's rule of differentiation under an integral, we find the following differentiation of (7.68)

$$\frac{\partial}{\partial t}\left[\frac{2}{\sqrt{\pi}}\int_{\frac{x}{2\sqrt{D(t-\tau)}}}^{\infty}e^{-\eta^2}d\eta\right]=\frac{x}{2\sqrt{\pi D}(t-\tau)^{3/2}}e^{-\frac{x^2}{4D(t-\tau)}} \tag{7.69}$$

In view of (7.68), we find the solution as

$$\Gamma(x,t) = \frac{x}{2\sqrt{\pi D}} \int_0^t \phi(\tau) \frac{e^{-x^2/[4D(t-\tau)]}}{(t-\tau)^{3/2}} d\tau \tag{7.70}$$

Introduce the following change of variables

$$\lambda = \frac{x}{2\sqrt{D(t-\tau)}} \tag{7.71}$$

The solution for Γ becomes

$$\Gamma(x,t) = \frac{2}{\sqrt{\pi}} \int_{\frac{x}{2\sqrt{Dt}}}^{\infty} \phi(t - \frac{x^2}{4D\lambda^2}) e^{-\lambda^2} d\lambda \tag{7.72}$$

Substitution of the first condition in (7.56) into (7.72) gives

$$\Gamma(x,t) = c_0 \frac{2}{\sqrt{\pi}} e^{-u^2 t/(4D)} \int_{\frac{x}{2\sqrt{Dt}}}^{\infty} \exp[-\frac{u^2 x^2}{(4D\lambda)^2}] e^{-\lambda^2} d\lambda \tag{7.73}$$

We now introduce two new variables

$$\alpha = \frac{x}{2\sqrt{Dt}}, \quad \varepsilon = \frac{ux}{4D} \tag{7.74}$$

With these new variables, the integration in (7.73) can be rewritten as

$$\Gamma(x,t) = c_0 \frac{2}{\sqrt{\pi}} e^{-u^2 t/(4D)} \left\{ \int_0^{\infty} \exp[-(\lambda^2 + \frac{\varepsilon^2}{\lambda^2})] d\lambda - \int_0^{\alpha} \exp[-(\lambda^2 + \frac{\varepsilon^2}{\lambda^2})] d\lambda \right\} \tag{7.75}$$

By employing the formula given in Formula 3.325 of Gradshteyn and Ryzhik (1980), we obtain

$$\int_0^{\infty} \exp[-(\lambda^2 + \frac{\varepsilon^2}{\lambda^2})] d\lambda = \frac{\sqrt{\pi}}{2} e^{-2\varepsilon} \tag{7.76}$$

It is straightforward to see the validity of the following identity

$$-(\lambda + \frac{\varepsilon}{\lambda})^2 + 2\varepsilon = -\lambda^2 - \frac{\varepsilon^2}{\lambda^2} = -(\lambda - \frac{\varepsilon}{\lambda})^2 - 2\varepsilon \tag{7.77}$$

The main success of the Ogata-Banks solution lies on this identity. Employing the identity in (7.77), we have the second integral in (7.75) being

$$\begin{aligned} I &= \int_0^{\alpha} \exp[-(\lambda^2 + \frac{\varepsilon^2}{\lambda^2})] d\lambda \\ &= \frac{1}{2}\{e^{2\varepsilon} \int_0^{\alpha} \exp[-(\lambda + \frac{\varepsilon}{\lambda})^2] d\lambda + e^{-2\varepsilon} \int_0^{\alpha} \exp[-(\lambda - \frac{\varepsilon}{\lambda})^2] d\lambda\} \\ &= \frac{1}{2}(I_1 + I_2) \end{aligned} \tag{7.78}$$

This technique is a clever way to make the second integration in (7.75) mathematically tractable. For the first integral I_1, we apply the following change of variables

$$z = \frac{\varepsilon}{\lambda} \tag{7.79}$$

Using (7.79), we can simplify the first integral in (7.78) as

$$I_1 = e^{2\varepsilon}\int_{\infty}^{\varepsilon/\alpha} \exp[-(z+\frac{\varepsilon}{z})^2](-\frac{\varepsilon}{z^2})dz$$
$$= -e^{2\varepsilon}\int_{\varepsilon/\alpha}^{\infty} (1-\frac{\varepsilon}{z^2})\exp[-(z+\frac{\varepsilon}{z})^2]dz + e^{2\varepsilon}\int_{\varepsilon/\alpha}^{\infty} \exp[-(z+\frac{\varepsilon}{z})^2]dz \quad (7.80)$$

We employ another change of variable for the first integral in (7.80)

$$\beta = \frac{\varepsilon}{z} + z \quad (7.81)$$

Thus, (7.80) becomes

$$I_1 = -e^{2\varepsilon}\int_{\alpha+\varepsilon/\alpha}^{\infty} e^{-\beta^2} d\beta + e^{2\varepsilon}\int_{\varepsilon/\alpha}^{\infty} \exp[-(z+\frac{\varepsilon}{z})^2]dz \quad (7.82)$$

For the second integral defined in (7.78), we again apply the following change of variables

$$z = \frac{\varepsilon}{\lambda} \quad (7.83)$$

Following a similar procedure, the second integral in (7.78) becomes

$$I_2 = e^{-2\varepsilon}\int_{\varepsilon/\alpha-\alpha}^{\infty} e^{-\beta^2} d\beta - e^{-2\varepsilon}\int_{\varepsilon/\alpha}^{\infty} \exp[-(\frac{\varepsilon}{z}-z)^2]dz \quad (7.84)$$

The proof of (7.84) is left as Problem 7.2 for readers. It is straightforward to prove that (see Problem 7.3):

$$\int_{\varepsilon/\alpha}^{\infty} \exp[-(\frac{\varepsilon}{z}+z)^2+2\varepsilon]dz = \int_{\varepsilon/\alpha}^{\infty} \exp[-(\frac{\varepsilon}{z}-z)^2-2\varepsilon]dz \quad (7.85)$$

Back-substituting (7.82) and (7.84) into (7.78), we obtain

$$I = \frac{1}{2}\{-e^{2\varepsilon}\int_{\alpha+\varepsilon/\alpha}^{\infty} e^{-\beta^2} d\beta + e^{2\varepsilon}\int_{\varepsilon/\alpha}^{\infty} \exp[-(z+\frac{\varepsilon}{z})^2]dz$$
$$+e^{-2\varepsilon}\int_{\varepsilon/\alpha-\alpha}^{\infty} e^{-\beta^2} d\beta - e^{-2\varepsilon}\int_{\varepsilon/\alpha}^{\infty} \exp[-(\frac{\varepsilon}{z}-z)^2]dz\} \quad (7.86)$$
$$= \frac{1}{2}\{e^{-2\varepsilon}\int_{\varepsilon/\alpha-\alpha}^{\infty} e^{-\beta^2} d\beta - e^{2\varepsilon}\int_{\alpha+\varepsilon/\alpha}^{\infty} e^{-\beta^2} d\beta\}$$

The last part of (7.86) is obtained in view of (7.85). Finally, substitution of (7.76) and (7.86) into (7.75) gives

$$\Gamma(x,t) = c_0 \frac{2}{\sqrt{\pi}} e^{u^2 t/(4D)} \left\{ \frac{\sqrt{\pi}}{2} e^{-2\varepsilon} - \frac{1}{2}[e^{-2\varepsilon}\int_{\varepsilon/\alpha-\alpha}^{\infty} e^{-\beta^2} d\beta - e^{2\varepsilon}\int_{\alpha+\varepsilon/\alpha}^{\infty} e^{-\beta^2} d\beta] \right\} \quad (7.87)$$

The first integral on the right-hand side of (7.87) is now expressible in terms of the complementary error function as

$$e^{-2\varepsilon}\int_{\varepsilon/\alpha-\alpha}^{\infty} e^{-\beta^2} d\beta = e^{-2\varepsilon}\frac{\sqrt{\pi}}{2} erfc\{\frac{\varepsilon}{\alpha}-\alpha\} \quad (7.88)$$

Similarly, the second integral on the right-hand side of (7.87) can also be expressed in terms of the complementary error function. Finally, we obtain

$$\Gamma(x,t) = c_0 \frac{2}{\sqrt{\pi}} \exp(\frac{u^2 t}{4D}) \{ \frac{\sqrt{\pi}}{2} e^{-2\varepsilon} - \frac{1}{2} [e^{-2\varepsilon} \frac{\sqrt{\pi}}{2} erfc(\frac{\varepsilon}{\alpha} - \alpha)$$
$$- e^{2\varepsilon} \frac{\sqrt{\pi}}{2} erfc(\frac{\varepsilon}{\alpha} + \alpha)]\} \tag{7.89}$$

Using (7.21), we have

$$\Gamma(x,t) = c_0 \exp(\frac{u^2 t}{4D}) \left\{ e^{-2\varepsilon} - \frac{1}{2} [e^{-2\varepsilon} (1 + erf(\alpha - \frac{\varepsilon}{\alpha})) - e^{2\varepsilon} erfc(\frac{\varepsilon}{\alpha} + \alpha)] \right\} \tag{7.90}$$

This can be further simplified as

$$\Gamma(x,t) = c_0 \exp(\frac{u^2 t}{4D}) \left\{ \frac{1}{2} e^{-2\varepsilon} - \frac{1}{2} e^{-2\varepsilon} erf(\alpha - \frac{\varepsilon}{\alpha}) + \frac{1}{2} e^{2\varepsilon} erfc(\frac{\varepsilon}{\alpha} + \alpha)] \right\} \tag{7.91}$$

Finally, we have the solution in a more compact form

$$\Gamma(x,t) = c_0 \exp(\frac{u^2 t}{4D}) \left\{ \frac{1}{2} e^{-2\varepsilon} erfc(\alpha - \frac{\varepsilon}{\alpha}) + \frac{1}{2} e^{2\varepsilon} erfc(\frac{\varepsilon}{\alpha} + \alpha)] \right\} \tag{7.92}$$

Recalling the following definitions

$$\alpha = \frac{x}{2\sqrt{Dt}}, \quad \varepsilon = \frac{ux}{4D} \tag{7.93}$$

and converting Γ back to c, we get the finally solution of the problem:

$$c(x,t) = \Gamma(x,t) \exp\{ux/(2D) - u^2 t/(4D)\}$$
$$= \frac{c_0}{2} \left\{ erfc(\frac{x - ut}{2\sqrt{Dt}}) + e^{ux/D} erfc(\frac{x + ut}{2\sqrt{Dt}}) \right\} \tag{7.94}$$

This analytical result has been validated using experimental results on glass beads and sands in fluids (see Figure 2 of Ogata and Banks, 1961). For the steady-state equation, for $t \to \infty$, we have

$$c(x,\infty) = \frac{c_0}{2} \left\{ erfc(-\infty) + e^{ux/D} erfc(\infty) \right\} = \frac{c_0}{2} \{2 + 0\} = c_0 \tag{7.95}$$

In obtaining the last of (7.95), we have employed (7.24) and (7.25). As expected, the whole river is diffused with the prescribed concentration c_0 in the long term.

7.5 SOLUTION FOR DECAYING POLLUTANTS

In the formulation in the previous section, we have assumed that the total amount of pollutant remains constant. In reality, many pollutants may decay with time due to their chemical reaction with oxygen or other compounds in water, or due to their consumption by microorganisms in water. If such effects can be modeled by a decay term which is proportional to its concentration, we have

$$\frac{\partial c}{\partial t} = D \frac{\partial^2 c}{\partial x^2} - u \frac{\partial c}{\partial x} - Kc \tag{7.96}$$

where K is the linear decay coefficient of the pollutant. The decay rate is assumed linearly proportional to the concentration. An example for decaying c is the non-conservative biochemical oxygen demand (BOD) in sewage effluent. The initial and boundary conditions are the same as the Ogata-Banks problem

$$c(0,t)=c_0,\quad t\ge 0,$$
$$c(x,0)=0,\quad x\ge 0, \tag{7.97}$$
$$c(\infty,t)=0,\quad t\ge 0,$$

To solve this problem, we need to propose a different change of variables as

$$c(x,t)=\Gamma(x,t)\exp\{\frac{ux}{2D}-\frac{(u^2+4DK)t}{4D}\} \tag{7.98}$$

If we set $K = 0$, we will recover the change of variables proposed in (7.54). Differentiation of Γ gives

$$\frac{\partial c}{\partial t}=\frac{\partial \Gamma}{\partial t}\exp[\frac{ux}{2D}-\frac{(u^2+4DK)t}{4D}]-\frac{(u^2+4DK)}{4D}\Gamma\exp[\frac{ux}{2D}-\frac{(u^2+4DK)t}{4D}] \tag{7.99}$$

$$\frac{\partial c}{\partial x}=\frac{\partial \Gamma}{\partial x}\exp[\frac{ux}{2D}-\frac{(u^2+4DK)t}{4D}]+\frac{u}{2D}\Gamma\exp[\frac{ux}{2D}-\frac{(u^2+4DK)t}{4D}] \tag{7.100}$$

$$\begin{aligned}\frac{\partial^2 c}{\partial x^2}&=\frac{u}{D}\frac{\partial \Gamma}{\partial x}\exp[\frac{ux}{2D}-\frac{(u^2+4DK)t}{4D}]\\&+\frac{u^2}{4D^2}\Gamma\exp[\frac{ux}{2D}-\frac{(u^2+4DK)t}{4D}]+\frac{\partial^2 \Gamma}{\partial x^2}\exp[\frac{ux}{2D}-\frac{(u^2+4DK)t}{4D}]\end{aligned} \tag{7.101}$$

Substitution of (7.99) to (7.101) into (7.96) gives

$$\frac{\partial \Gamma}{\partial t}=D\frac{\partial^2 \Gamma}{\partial x^2} \tag{7.102}$$

Note that by assuming an appropriate form of (7.98) we are able to eliminate the decay term in the ODE and arrive at the same ODE as in the Ogata-Banks problem given in (7.52). This is the key to the successful application of the Ogata-Banks solution procedure in the present problem. The following initial and boundary conditions for Γ can be obtained

$$\Gamma(0,t)=\phi(t)=c_0\exp\{\frac{(u^2+4DK)t}{4D}\},\quad t\ge 0,$$
$$\Gamma(x,0)=0,\quad x\ge 0, \tag{7.103}$$
$$\Gamma(\infty,t)=0,\quad t\ge 0,$$

The method of solution is exactly the same as that used in the last section. Therefore, we will simply outline the main difference in the analysis only. The solution in terms of Duhamel's theorem for the initial condition given in (7.101) becomes

$$\Gamma(x,t)=\frac{2}{\sqrt{\pi}}\int_{\frac{x}{2\sqrt{Dt}}}^{\infty}\phi(t-\frac{x^2}{4D\lambda^2})e^{-\lambda^2}d\lambda \tag{7.104}$$

where ϕ is defined in (7.103). Substitution of the first condition in (7.103) into (7.104) gives

$$\Gamma(x,t) = c_0 \frac{2}{\sqrt{\pi}} \exp\{\frac{(u^2+4DK)t}{4D}\} \int_{\frac{x}{2\sqrt{Dt}}}^{\infty} \exp[-\frac{(u^2+4DK)x^2}{(4D\lambda)^2} - \lambda^2] d\lambda \quad (7.105)$$

We now introduce two new variables

$$\alpha = \frac{x}{2\sqrt{Dt}}, \quad \varepsilon = \frac{\sqrt{u^2+4DK}x}{4D} \quad (7.106)$$

With these new variables, the integration in (7.105) can be rewritten as

$$\Gamma(x,t) = c_0 \frac{2}{\sqrt{\pi}} \exp\{\frac{(u^2+4DK)t}{4D}\} \int_{\alpha}^{\infty} \exp[-(\lambda^2 + \frac{\varepsilon^2}{\lambda^2})] d\lambda \quad (7.107)$$

Following the solution procedure discussed in the last section, we obtain

$$\Gamma(x,t) = c_0 \exp\{\frac{(u^2+4DK)t}{4D}\} \left\{ \frac{1}{2} e^{-2\varepsilon} erfc(\alpha - \frac{\varepsilon}{\alpha}) + \frac{1}{2} e^{2\varepsilon} erfc(\frac{\varepsilon}{\alpha} + \alpha)] \right\} \quad (7.108)$$

Recalling the definition of Γ, we get the finally solution of the problem:

$$c(x,t) = \frac{c_0}{2} \exp(\frac{ux}{2D}) \{ \exp(-\frac{\sqrt{u^2+4DK}x}{2D}) erfc(\frac{x - \sqrt{u^2+4DK}t}{\sqrt{4Dt}}) + \exp(\frac{\sqrt{u^2+4DK}x}{2D}) erfc(\frac{x + \sqrt{u^2+4DK}t}{\sqrt{4Dt}}) \} \quad (7.109)$$

If we set $K = 0$, the solution by Ogata and Banks (1961) is covered as a special case. This special case of $K = 0$ was also given in (2.65) of Fischer et al. (1979) without proof. For a steady-state solution ($t \to \infty$), we have, in view of (7.24) and (7.25),

$$\begin{aligned} c(x,\infty) &= \frac{c_0}{2} \exp(\frac{ux}{2D}) 2 \exp(-\frac{\sqrt{u^2+4DK}x}{2D}) \\ &= c_0 \exp[\frac{x}{2D}(u - \sqrt{u^2+4DK})] \end{aligned} \quad (7.110)$$

For the special case that $K = 0$, we have $c(x, \infty) = c_0$. This result is, of course, the same as that for the Ogata-Banks solution. More generally, the present solution shows a decay of c_0.

7.6 DISPERSION OF DECAYING SUBSTANCES

Taking the steady case of (7.96), we have

$$u \frac{dc}{dx} = D \frac{d^2 c}{dx^2} - Kc \quad (7.111)$$

Consider the case of a constant influx of pollutant over a line source of length X_m. It is assumed that a steady state of the spatial distribution of the pollutant has been attained (see Figure 7.5). The length scale of the mixing zone is assumed to be implicitly shorter than the line source, such that thorough mixing occurs. The concentration c satisfies the conservation of mass as:

$$\int_{X_m}^{\infty} Kc(x)A_0 dx = \dot{M}(X_m) \tag{7.112}$$

$$c(\infty) = 0 \tag{7.113}$$

For u, D, and K being constant, (7.111) is an ODE of constant coefficient. Thus, we seek a solution in exponential form:

$$c = \exp(\lambda x) \tag{7.114}$$

Substitution of (7.114) into (7.111) gives

$$D\lambda^2 - u\lambda - K = 0 \tag{7.115}$$

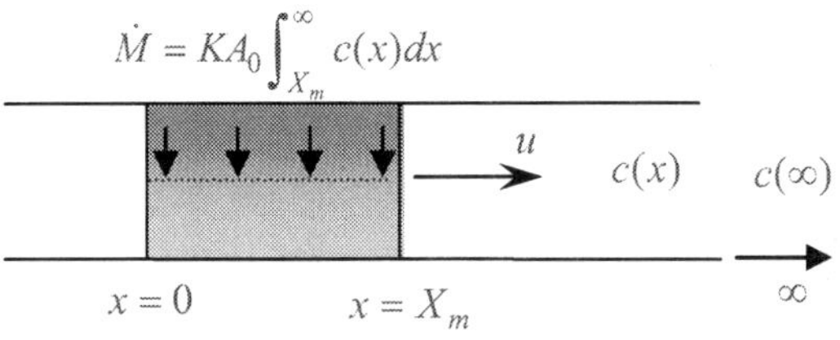

Figure 7.5 Constant influx of pollutants over a line source

The roots for λ are

$$\lambda_{1,2} = \frac{u \pm \sqrt{u^2 + 4KD}}{2D} \tag{7.116}$$

It is obvious that

$$\lambda_1 = \frac{u + \sqrt{u^2 + 4KD}}{2D} > 0, \quad \lambda_2 = \frac{u - \sqrt{u^2 + 4KD}}{2D} < 0 \tag{7.117}$$

The solution for c is

$$c = C_1 \exp(\lambda_1 x) + C_2 \exp(\lambda_2 x) \tag{7.118}$$

The boundedness condition given in (7.113) requires $C_1 = 0$, and thus, the solution becomes

$$c = C_2 \exp(\frac{u - \sqrt{u^2 + 4KD}}{2D} x) \tag{7.119}$$

Substitution of (7.119) into (7.112) gives

$$C_2 KA_0 \int_{X_m}^{\infty} \exp(-\frac{u}{2D}[\sqrt{1+\alpha} - 1]x)dx = \dot{M}(X_m) \tag{7.120}$$

where

$$\alpha = \frac{4KD}{u^2} \tag{7.121}$$

We now introduce the following change of variables

$$\zeta = \frac{u}{2D}[\sqrt{1+\alpha}-1]x \tag{7.122}$$

Applying (7.122), we have (7.120) becoming

$$\frac{2DC_2KA_0}{u(\sqrt{1+\alpha}-1)}\int_{\frac{u}{2D}(\sqrt{1+\alpha}-1)X_m}^{\infty}\exp(-\zeta)d\zeta = \dot{M} \tag{7.123}$$

The unknown constant is obtained as

$$C_2 = \frac{\dot{M}u}{2DKA_0}(\sqrt{1+\alpha}-1)\exp[-\frac{u}{2D}(\sqrt{1+\alpha}-1)X_m] \tag{7.124}$$

Using the definition given in (7.121) and the flow rate Q (= uA), we obtain

$$C_2 = \frac{2\dot{M}}{Q\alpha}(\sqrt{1+\alpha}-1)\exp[-\frac{u}{2D}(\sqrt{1+\alpha}-1)X_m] \tag{7.125}$$

Finally, we obtain the steady-state solution

$$c(x) = \frac{2\dot{M}}{Q\alpha}(\sqrt{1+\alpha}-1)\exp\{-\frac{u}{2D}(\sqrt{1+\alpha}-1)(X_m+x)\} \tag{7.126}$$

This illustrates the effect of dispersion. It is clear that as $x \to \infty$, the concentration of the pollutant drops to zero. Thus, the boundedness condition given in (7.113) is identically satisfied. For a non-decaying pollutant, we have $K \to 0$ or $\alpha \to 0$. We note that

$$\lim_{\alpha\to 0}\frac{\sqrt{1+\alpha}-1}{\alpha} = \lim_{\alpha\to 0}\frac{1}{\alpha}[1+\frac{1}{2}\alpha-1+O(\alpha^2)] = \frac{1}{2} \tag{7.127}$$

Using this result, we finally have

$$c(x) = \frac{\dot{M}}{Q} \tag{7.128}$$

It is obvious to see that the unit of $c(x)$ is kg/m^3. Thus, the steady-state solution is a constant function for the case of no decay. This is because there is a continuous supply of pollutant in the line source.

7.7 TAYLOR'S POINT SOURCE SOLUTION

For one-dimensional diffusion analysis, the fundamental solution was first considered by G.I. Taylor in 1954. Mathematically, it is defined by Fick's second law:

$$\frac{\partial c}{\partial t} = D\frac{\partial^2 c}{\partial x^2}, \quad -\infty < x < \infty \tag{7.129}$$

subjected to the following impulsive initial condition

$$c(x,0) = \frac{M}{A_0}\delta(x) \tag{7.130}$$

where A_0 is the cross-section area of the river and c is in the unit of kg/m^3. It is illustrated in Figure 7.6. By recognizing the property of the Dirac delta function (Chau, 2018), initial condition (7.130) for an impulsive point source can be rewritten as

$$\int_{-\infty}^{\infty} c(x,t)dx = \int_{-\infty}^{\infty} c(x,0)dx = \int_{-\infty}^{\infty} \frac{M}{A_0}\delta(x)dx = \frac{M}{A_0}\int_{-\infty}^{\infty} \delta(x)dx = \frac{M}{A_0} \tag{7.131}$$

The first part of (7.131) is a result of mass conservation. Two different approaches of solving this problem will be discussed next.

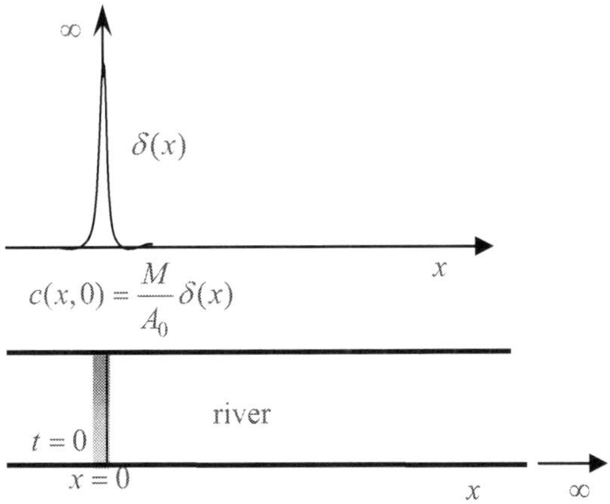

Figure 7.6 One-dimensional diffusion of pollutants from a source point

7.7.1 Taylor's Approach

Taylor (1954) proposed the following solution form for the impulsive point source solution prescribed by (7.129) and (7.130) or (7.131):

$$c(x,t) = \frac{A}{t^{1/2}}\exp(-\frac{x^2}{4Dt}) \tag{7.132}$$

where A is an unknown constant. Differentiation of (7.132) gives

$$\frac{\partial c}{\partial t} = [-\frac{A}{2t^{3/2}} + \frac{Ax^2}{4Dt^{5/2}}]\exp(-\frac{x^2}{4Dt}) \tag{7.133}$$

$$D\frac{\partial^2 c}{\partial x^2} = [-\frac{A}{2t^{3/2}} + \frac{Ax^2}{4Dt^{5/2}}]\exp(-\frac{x^2}{4Dt}) \tag{7.134}$$

Combining (7.133) and (7.134) leads to

$$\frac{\partial c}{\partial t} = D\frac{\partial^2 c}{\partial x^2} = [-\frac{A}{2t^{3/2}} + \frac{Ax^2}{4Dt^{5/2}}]\exp(-\frac{x^2}{4Dt}) \tag{7.135}$$

Therefore, it demonstrates that (7.132) is indeed a solution to (7.129). To find the unknown constant A, we substitute (7.132) in (7.132) to give

$$\frac{A}{t^{1/2}}\int_{-\infty}^{\infty} \exp(-\frac{x^2}{4Dt})dx = \frac{M}{A_0} \tag{7.136}$$

We introduce the following change of variables:

$$\xi^2 = \frac{x^2}{4Dt}, \quad dx = \sqrt{4Dt}\, d\xi \tag{7.137}$$

Substitution of (7.137) into (7.136) gives

$$2\sqrt{D}A\int_{-\infty}^{\infty} \exp(-\xi^2) d\xi = \frac{M}{A_0} \tag{7.138}$$

Recalling the Laplace/Gauss integral discussed in Section 1.4.6 of Chau (2018), we have

$$2\sqrt{D}A\sqrt{\pi} = \frac{M}{A_0} \tag{7.139}$$

Thus, the constant is

$$A = \frac{M}{A_0\sqrt{4\pi D}} \tag{7.140}$$

Finally, we obtain Taylor's fundamental point source solution

$$c(x,t) = \frac{M}{A_0\sqrt{4\pi Dt}} \exp(-\frac{x^2}{4Dt}) \tag{7.141}$$

This is the result obtained by G.I. Taylor (1954).

7.7.2 Taylor's Solution by Dimensional Analysis

Instead of seeing (7.132) by intuition, a proper functional form of the pollutant concentration c (kg/m^3) can be identified by dimensional analysis. From the formulation, we see that the unknown concentration c must be a function of x, t, D, A_0, and M. In particular, we see that the length variable x must be normalized by a proper length scale of the problem. However, there is no physical length for a point source in an infinite 1-D domain. The only length scale is from the square root of the product of diffusivity constant D (m^2/s) and time t (s) such that a dimensionless quantity can be defined as

$$\Pi_1 = \frac{x}{\sqrt{Dt}} \tag{7.142}$$

The unit for c should be in mass per unit volume, and thus the second dimensionless number must be

$$\Pi_2 = \frac{M}{cA_0\sqrt{Dt}} \tag{7.143}$$

Thus, the solution for c must be in the following functional form:

$$c = \frac{M}{A_0\sqrt{Dt}} f(\frac{x}{\sqrt{Dt}}) \tag{7.144}$$

We now introduce a new variable η as

$$\eta = \frac{x}{\sqrt{Dt}} \tag{7.145}$$

Using the chain rule, we have

$$\frac{\partial c(\eta,t)}{\partial t}=\frac{\partial c(\eta,t)}{\partial t}+\frac{\partial c(\eta,t)}{\partial \eta}\frac{\partial \eta}{\partial t}=-\frac{1}{2t}(\frac{M}{A_0\sqrt{Dt}})f+\frac{M}{A_0\sqrt{Dt}}\frac{df}{d\eta}\frac{\partial \eta}{\partial t}$$
$$=-\frac{M}{2A_0t\sqrt{Dt}}(f+\eta\frac{df}{d\eta}) \tag{7.146}$$

$$\frac{\partial^2 c}{\partial x^2}=\frac{\partial}{\partial x}\{\frac{\partial}{\partial x}(\frac{M}{A_0\sqrt{Dt}}f(\eta)]\}=\frac{M}{DtA_0\sqrt{Dt}}\frac{d^2 f}{d\eta^2} \tag{7.147}$$

Substitution of (7.146) and (7.147) into (7.129) gives

$$\frac{d^2 f}{d\eta^2}+\frac{1}{2}(f+\eta\frac{df}{d\eta})=0 \tag{7.148}$$

This equation can be rewritten as

$$\frac{d^2 f}{d\eta^2}+\frac{1}{2}\frac{d(\eta f)}{d\eta}=\frac{d}{d\eta}[\frac{df}{d\eta}+\frac{1}{2}\eta f]=0 \tag{7.149}$$

Integrating once, we get

$$\frac{df}{d\eta}+\frac{1}{2}\eta f=C_0 \tag{7.150}$$

Substituting (7.144) into (7.130), we obtain the following condition for f

$$\frac{M}{A_0}=\int_{-\infty}^{\infty}\frac{M}{A_0\sqrt{Dt}}f(\frac{x}{\sqrt{Dt}})dx=\int_{-\infty}^{\infty}\frac{M}{A_0\sqrt{Dt}}f(\eta)\sqrt{Dt}d\eta \tag{7.151}$$

This leads to

$$\int_{-\infty}^{\infty}f(\eta)d\eta=1 \tag{7.152}$$

In summary, the original problem can be expressed equivalently as

$$\frac{df}{d\eta}+\frac{1}{2}\eta f=C_0,\quad \int_{-\infty}^{\infty}f(\eta)d\eta=1 \tag{7.153}$$

Clearly, the particular solution for the first equation in (7.153) must contain a constant term. If this solution is substituted into the second equation of (7.153), the integral will become infinity instead of unity on the right-hand side. Consequently, we must set $C_0 = 0$. Thus, the problem is reduced to:

$$\frac{df}{d\eta}+\frac{1}{2}\eta f=0,\quad \int_{-\infty}^{\infty}f(\eta)d\eta=1 \tag{7.154}$$

The first of (7.154) is a separable first order ODE, and direct integration gives

$$\ln f=-\frac{1}{4}\eta^2+\bar{C}_1 \tag{7.155}$$

Thus, we have

$$f=C_1\exp(-\frac{1}{4}\eta^2) \tag{7.156}$$

Substitution of (7.156) into the second equation of (7.154) leads to

$$C_1\int_{-\infty}^{\infty}\exp(-\frac{1}{4}\eta^2)d\eta=1 \tag{7.157}$$

Making the following change of variables

$$\xi^2 = \frac{1}{4}\eta^2, \tag{7.158}$$

we arrive at

$$2C_1 \int_{-\infty}^{\infty} \exp(-\xi) d\xi = 1 \tag{7.159}$$

Recalling the Laplace/Gauss integral discussed in Section 1.4.6 of Chau (2018), we have

$$C_1 = \frac{1}{\sqrt{4\pi}} \tag{7.160}$$

Combining all these results, we obtain the final solution

$$c = \frac{M}{A_0\sqrt{4D\pi t}} \exp(-\frac{x^2}{4Dt}) \tag{7.161}$$

This, of course, agrees with (7.141) derived in the last section. This solution suggests that the penetration distance of any pollutant is roughly proportional to the square root of time (see also Problem 7.5).

7.8 DECAYING POLLUTANT IN FLOWING FLUID

7.8.1 Point Source Solution

In this section, Taylor's (1954) solution will be extended to the case of a decaying pollutant in a flowing river. In particular, the mathematical formulation of the problem is

$$\frac{\partial c}{\partial t} = D\frac{\partial^2 c}{\partial x^2} - u\frac{\partial c}{\partial x} - Kc \tag{7.162}$$

$$c(x,0) = \frac{M}{A_0}\delta(x) \tag{7.163}$$

Taylor's solution form is revised as

$$c(x,t) = \frac{A}{t^{1/2}} \exp[-\frac{(x-ut)^2}{4Dt} - Kt] \tag{7.164}$$

Differentiation of (7.164) gives

$$\frac{\partial c}{\partial t} = \exp[-\frac{(x-ut)^2}{4Dt} - Kt]\{-\frac{A}{2t^{3/2}} + \frac{A(x-ut)^2}{4Dt^{5/2}} + \frac{Au(x-ut)}{2Dt^{3/2}} - \frac{KA}{t^{1/2}}\} \tag{7.165}$$

$$\frac{\partial c}{\partial x} = -\frac{A(x-ut)}{2Dt^{3/2}} \exp[-\frac{(x-ut)^2}{4Dt} - Kt] \tag{7.166}$$

$$\frac{\partial^2 c}{\partial x^2} = \{-\frac{A}{2Dt^{3/2}} + \frac{A(x-ut)^2}{4D^2t^{5/2}}\} \exp[-\frac{(x-ut)^2}{4Dt} - Kt] \tag{7.167}$$

Using (7.164), (7.166) and (7.167), we find

$$D\frac{\partial^2 c}{\partial x^2}-u\frac{\partial c}{\partial x}-Kc=\{-\frac{A}{2t^{3/2}}+\frac{A(x-ut)^2}{4Dt^{5/2}}+\frac{Au(x-ut)}{2Dt^{3/2}}-\frac{KA}{t^{1/2}}\}\exp[-\frac{(x-ut)^2}{4Dt}-Kt] \tag{7.168}$$

This equals precisely the right-hand side of (7.165). Thus, we have shown that (7.164) satisfies (7.162). To find the unknown constant, we substitute (7.164) into (7.131) to get

$$\frac{A}{t^{1/2}}\int_{-\infty}^{\infty}\exp[-\frac{(x-ut)^2}{4Dt}-Kt]dx=\frac{M}{A_0} \tag{7.169}$$

We now introduce the following change of variables

$$\xi^2=\frac{(x-ut)^2}{4Dt}+Kt \tag{7.170}$$

The integral in (7.169) becomes

$$2\sqrt{D}A\int_{-\infty}^{\infty}\exp[-\xi^2]d\xi=\frac{M}{A_0} \tag{7.171}$$

The unknown constant becomes

$$A=\frac{M}{A_0\sqrt{4\pi D}} \tag{7.172}$$

$$c(x,t)=\frac{M}{A_0\sqrt{4\pi Dt}}\exp[-\frac{(x-ut)^2}{4Dt}-Kt] \tag{7.173}$$

This is the extension of Taylor's (1954) solution, and Taylor's solution is recovered if we set $K = 0$ and $u = 0$ in (7.173).

7.8.2 Continuous Source Solution

The solution for a point source can readily be used to solve the problem of a non-steady continuous source injected to a flowing river at $x = 0$, as shown in Figure 7.7. The problem can be formulated as

$$\frac{\partial c}{\partial t}=D\frac{\partial^2 c}{\partial x^2}-u\frac{\partial c}{\partial x}-Kc \tag{7.174}$$

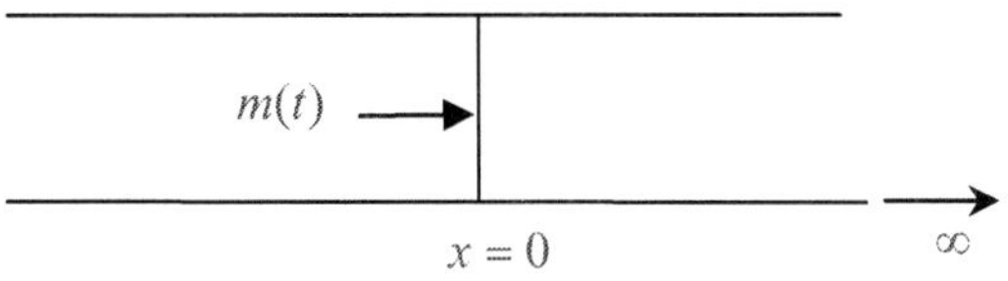

Figure 7.7 One-dimensional diffusion of a pollutant from a continuous source point

A continuous supply of pollutant at a rate of m (kg/s) is injected to the river at $x = 0$. The solution can be expressed as

$$\begin{aligned} c(x,t) &= \int_0^t dc = \int_0^t \frac{dM}{A_0\sqrt{4D\pi(t-\tau)}} \exp[-\frac{[x-u(t-\tau)]^2}{4D(t-\tau)} - K(t-\tau)] \\ &= \int_0^t \frac{m(\tau)}{A_0\sqrt{4D\pi(t-\tau)}} \exp[-\frac{[x-u(t-\tau)]^2}{4D(t-\tau)} - K(t-\tau)]d\tau \end{aligned} \tag{7.175}$$

Note that $dM = m(\tau)d\tau$. In general, for an arbitrary function of $m(\tau)$ this integral needs to be solved by numerical integration. We now consider a special case of constant injection rate that (7.175) becomes

$$c(x,t) = \frac{m}{A_0\sqrt{4D\pi}} \int_0^t \frac{1}{\sqrt{(t-\tau)}} \exp\{-\frac{[x-u(t-\tau)]^2}{4D(t-\tau)} - K(t-\tau)\}d\tau \tag{7.176}$$

For this case, the integration can be carried out exactly. We first introduce the following change of variables

$$s = t - \tau \tag{7.177}$$

Using (7.177), we have

$$c(x,t) = \frac{m}{A_0\sqrt{4D\pi}} \int_0^t \frac{1}{\sqrt{s}} \exp\{-\frac{(x-us)^2}{4Ds} - Ks\}ds \tag{7.178}$$

We note that

$$\begin{aligned} -\frac{(x-us)^2}{4Ds} - Ks &= -(\frac{x^2}{4D})\frac{1}{s} + \frac{2ux}{4D} - (\frac{u^2}{4D} + K)s \\ &= -\frac{1}{4D}[\frac{x^2}{s} - 2ux + (u^2 + 4DK)s] \end{aligned} \tag{7.179}$$

This can further be rewritten as

$$\begin{aligned} -[\frac{(x-us)^2}{4Ds} + Ks] &= -\frac{1}{4D}[\frac{x^2}{s} - 2ux + \lambda^2 s] \\ &= \frac{ux}{2D} - \frac{1}{4D}(\frac{x^2}{s} + \lambda^2 s) \end{aligned} \tag{7.180}$$

where

$$\lambda^2 = u^2 + 4DK \tag{7.181}$$

Using (7.180), we can simplify (7.178) to

$$c(x,t) = \frac{m}{A_0\sqrt{4D\pi}} \exp(\frac{ux}{2D}) \int_0^t \frac{1}{\sqrt{s}} \exp\{-\frac{1}{4D}(\frac{x^2}{s} + \lambda^2 s)\}ds \tag{7.182}$$

Following the procedure used by Ogata and Banks (1961) discussed in Section 7.4, we note the following identity

$$\begin{aligned} -\frac{1}{4D}(\frac{x^2}{s} + \lambda^2 s) &= -\frac{1}{4D}(\frac{x^2}{s} + \lambda^2 s) - \frac{\lambda x}{2D} + \frac{\lambda x}{2D} \\ &= -\frac{1}{4Ds}(x + \lambda s)^2 + \frac{\lambda x}{2D} \\ &= -\frac{1}{4Ds}(x - \lambda s)^2 - \frac{\lambda x}{2D} \end{aligned} \tag{7.183}$$

In view of (7.183), the integral in (7.182) can be expressed as

$$\begin{aligned} I &= \int_0^t \frac{1}{\sqrt{s}} \exp\{-\frac{1}{4D}(\frac{x^2}{s}+\lambda^2 s)\} ds \\ &= \frac{1}{2}\{\exp(\frac{\lambda x}{2D})\int_0^t \frac{1}{\sqrt{s}} \exp[-\frac{1}{4Ds}(x+\lambda s)^2] ds \\ &+\exp(-\frac{\lambda x}{2D})\int_0^t \frac{1}{\sqrt{s}} \exp[-\frac{1}{4Ds}(x-\lambda s)^2] ds\} \\ &= \frac{1}{2}\{I_1+I_2\} \end{aligned} \tag{7.184}$$

The first integral in (7.184) can be further rewritten as

$$I_1 = \exp(\frac{\lambda x}{2D})\int_0^t \frac{1}{\sqrt{s}} \exp[-\frac{\lambda^2}{4D}(\frac{x}{\lambda\sqrt{s}}+\sqrt{s})^2] ds \tag{7.185}$$

We introduce another round of change of variables as

$$z = \frac{x}{\lambda\sqrt{s}} \tag{7.186}$$

Employing (7.186), (7.185) is converted to

$$I_1 = 2\exp(\frac{\lambda x}{2D})\int_{\frac{x}{\lambda\sqrt{t}}}^{\infty} \frac{x}{\lambda z^2} \exp[-\frac{\lambda^2}{4D}(z+\frac{x}{\lambda z})^2] dz \tag{7.187}$$

We then add and subtract an integral to (7.187) to rewrite it as

$$\begin{aligned} I_1 &= -2\exp(\frac{\lambda x}{2D})\int_{\frac{x}{\lambda\sqrt{t}}}^{\infty} (1-\frac{x}{\lambda z^2}) \exp[-\frac{\lambda^2}{4D}(z+\frac{x}{\lambda z})^2] dz \\ &+2\exp(\frac{\lambda x}{2D})\int_{\frac{x}{\lambda\sqrt{t}}}^{\infty} \exp[-\frac{\lambda^2}{4D}(z+\frac{x}{\lambda z})^2] dz \\ &= -2\exp(\frac{\lambda x}{2D})\int_{\frac{x}{\lambda\sqrt{t}}}^{\infty} (1-\frac{x}{\lambda z^2}) \exp[-\frac{\lambda^2}{4D}(z+\frac{x}{\lambda z})^2] dz + I_{10} \end{aligned} \tag{7.188}$$

where I_{10} can be identified from (7.188) readily. The integral on the right-hand side of (7.188) can be simplified by introducing the following variables

$$\beta = z + \frac{x}{\lambda z} \tag{7.189}$$

Note that

$$d\beta = (1-\frac{x}{\lambda z^2}) dz \tag{7.190}$$

We see that this form appears exactly in the integral of (7.188) and, thus, (7.188) becomes

$$I_1 = -2\exp(\frac{\lambda x}{2D})\int_{\frac{x+\lambda t}{\lambda\sqrt{t}}}^{\infty} \exp[-\frac{\lambda^2}{4D}\beta^2] d\beta + I_{10} \tag{7.191}$$

We can further absorb the constant in the exponential function by

$$\eta^2 = \frac{\lambda^2}{4D}\beta^2 \tag{7.192}$$

Applying (7.192), we can rewrite (7.191) as

$$\begin{aligned} I_1 &= -2(\frac{\sqrt{4D}}{\lambda})\exp(\frac{\lambda x}{2D})\int_{\frac{x+\lambda t}{\sqrt{4Dt}}}^{\infty} \exp[-\eta^2]d\eta + I_{10} \\ &= -2(\frac{\sqrt{4D}}{\lambda})\frac{\sqrt{\pi}}{2}\exp(\frac{\lambda x}{2D})erfc(\frac{x+\lambda t}{\sqrt{4Dt}}) + I_{10} \end{aligned} \tag{7.193}$$

Following the same procedure, the second integral in (7.184) can be converted to

$$\begin{aligned} I_2 &= 2(\frac{\sqrt{4D}}{\lambda})\exp(-\frac{\lambda x}{2D})\int_{\frac{x-\lambda t}{\sqrt{4Dt}}}^{\infty} \exp[-\eta^2]d\eta - I_{20} \\ &= 2(\frac{\sqrt{4D}}{\lambda})\frac{\sqrt{\pi}}{2}\exp(-\frac{\lambda x}{2D})erfc(\frac{x-\lambda t}{\sqrt{4Dt}}) - I_{20} \end{aligned} \tag{7.194}$$

where

$$I_{20} = 2\exp(-\frac{\lambda x}{2D})\int_{\frac{x}{\lambda\sqrt{t}}}^{\infty} \exp[-\frac{\lambda^2}{4D}(z-\frac{x}{\lambda z})^2]dz \tag{7.195}$$

Now we note the following identity:

$$-\frac{\lambda^2}{4D}(z-\frac{x}{\lambda z})^2 - \frac{\lambda x}{2D} = -\frac{\lambda^2}{4D}(z^2+\frac{x^2}{\lambda^2 z^2}) = -\frac{\lambda^2}{4D}(z+\frac{x}{\lambda z})^2 + \frac{\lambda x}{2D} \tag{7.196}$$

In view of this identity, it is straightforward to show that

$$I_{10} = I_{20} \tag{7.197}$$

Substituting (7.193) and (7.194) into (7.184) and observing (7.197), we finally obtain the following elegant solution

$$\begin{aligned} c(x,t) = \frac{m}{2A_0\sqrt{u^2+4DK}}\exp(\frac{ux}{2D})\{-\exp(\frac{\sqrt{u^2+4DK}x}{2D})erfc(\frac{x+\sqrt{u^2+4DK}t}{\sqrt{4Dt}}) \\ +\exp(-\frac{\sqrt{u^2+4DK}x}{2D})erfc(\frac{x-\sqrt{u^2+4DK}t}{\sqrt{4Dt}})\} \end{aligned} \tag{7.198}$$

The success of this approach lies on the beautiful identity derived in (7.196) and the cancellation of I_{10} and I_{20}. This result agrees with that given in (2.156) on p. 69 of TRACOR (1971) by Professor Harleman, which was given for modeling a uniform estuary without proof. For the steady-state solution $t \to \infty$, we have

$$c(x,t) = \frac{m}{A_0\sqrt{u^2+4DK}}\exp\{\frac{x}{2D}(u-\sqrt{u^2+4DK})\} \tag{7.199}$$

For a non-decaying pollutant ($K = 0$), the steady-state solution is further reduced to

$$c(x,t) = \frac{m}{A_0 u} \tag{7.200}$$

If the flow speed u is high compared to the injection rate m, the pollutant will be washed away and c diminishes.

7.9 DIFFUSION IN HIGHER DIMENSIONS

7.9.1 Two-Dimensional Point Source Solution

To protect water quality in a harbor, sometimes an underwater pipeline is built to release pollutant or treated sewage at an offshore location from the coastline. For simplicity, the source is assumed to be uniform in the vertical direction. Such problem can be modeled as a point source release in a two-dimensional domain, as shown in Figure 7.8.

For the two-dimensional case, Fick's second law can be expressed

$$\frac{\partial c}{\partial t} = D_x \frac{\partial^2 c}{\partial x^2} + D_y \frac{\partial^2 c}{\partial y^2} - u\frac{\partial c}{\partial x} - v\frac{\partial c}{\partial y} - Kc \tag{7.201}$$

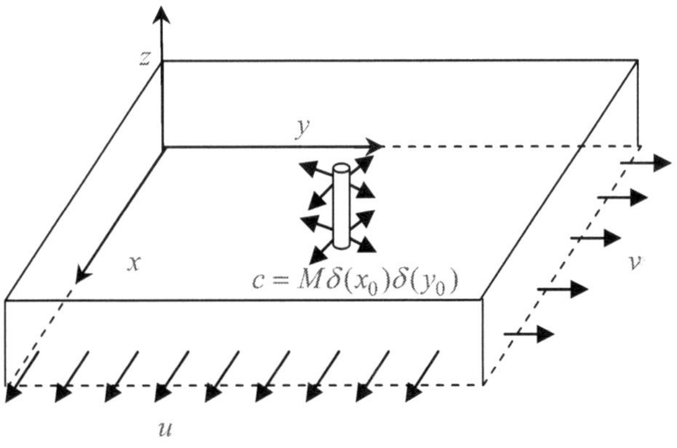

Figure 7.8 Point source in a two-dimensional ocean

The initial condition is given by

$$c(x, y, 0) = \frac{M}{H}\delta(x_0)\delta(y_0) \tag{7.202}$$

where H is the depth of the ocean as shown in Figure 7.8. Similar to the separation of variables discussed in Section 2.5 of Chau (2018), we assume the concentration can be expressed as:

$$c(x, y, t) = c_1(x,t)c_2(y,t) \tag{7.203}$$

Substitution of (7.203) into each term in (1.201) gives

$$c_1\frac{\partial c_2}{\partial t} + c_2\frac{\partial c_1}{\partial t} = c_2 D_x \frac{\partial^2 c_1}{\partial x^2} + c_1 D_y \frac{\partial^2 c_2}{\partial y^2} - uc_2\frac{\partial c_1}{\partial x} - vc_1\frac{\partial c_2}{\partial y} - Kc_1c_2 \tag{7.204}$$

Grouping terms, we obtain

$$c_1\{\frac{\partial c_2}{\partial t} - D_y\frac{\partial^2 c_2}{\partial x^2} + v\frac{\partial c_2}{\partial y}\} + c_2\{\frac{\partial c_1}{\partial t} - D_x\frac{\partial^2 c_1}{\partial x^2} + u\frac{\partial c_1}{\partial x}\} + Kc_1c_2 = 0 \quad (7.205)$$

Dividing through by c, we get

$$\frac{1}{c_2}\{\frac{\partial c_2}{\partial t} - D_y\frac{\partial^2 c_2}{\partial y^2} + v\frac{\partial c_2}{\partial y}\} + \frac{1}{c_1}\{\frac{\partial c_1}{\partial t} - D_x\frac{\partial^2 c_1}{\partial x^2} + u\frac{\partial c_1}{\partial x}\} + K = 0 \quad (7.206)$$

Since K is a constant, the first two terms of (7.206) must be constant as well because a function of x cannot cancel a function of y. In view of this, we can set

$$\frac{1}{c_2}\{\frac{\partial c_2}{\partial t} - D_y\frac{\partial^2 c_2}{\partial y^2} + v\frac{\partial c_2}{\partial y}\} = -K_2 \quad (7.207)$$

$$\frac{1}{c_1}\{\frac{\partial c_1}{\partial t} - D_x\frac{\partial^2 c_1}{\partial x^2} + u\frac{\partial c_1}{\partial x}\} = -K_1 \quad (7.208)$$

It is straightforward to see that

$$K = K_1 + K_2 \quad (7.209)$$

Rearranging (7.207) and (7.208), we obtain the following PDEs for c_1 and c_2:

$$\frac{\partial c_2}{\partial t} - D_y\frac{\partial^2 c_2}{\partial y^2} + v\frac{\partial c_2}{\partial y} + K_2c_2 = 0 \quad (7.210)$$

$$\frac{\partial c_1}{\partial t} - D_x\frac{\partial^2 c_1}{\partial x^2} + u\frac{\partial c_1}{\partial x} + K_1c_1 = 0 \quad (7.211)$$

This is precisely Fick's one-dimensional second law. Thus, the solution that we obtained in the previous section can be used to obtain

$$c_1(x,t) = \frac{A_1}{\sqrt{4\pi D_x t}}\exp[-\frac{(x - x_0 - ut)^2}{4D_x t} - K_1 t] \quad (7.212)$$

$$c_2(y,t) = \frac{A_2}{\sqrt{4\pi D_y t}}\exp[-\frac{(y - y_0 - vt)^2}{4D_y t} - K_2 t] \quad (7.213)$$

where A_1 and A_2 are unknown constants. Putting (7.212) and (7.213) into (7.203), we get

$$c = \frac{A_1A_2}{4\pi t\sqrt{D_xD_y}}\exp[-\frac{(x - x_0 - ut)^2}{4D_x t} - \frac{(y - y_0 - vt)^2}{4D_y t} - (K_1 + K_2)t] \quad (7.214)$$

By conservation of mass, we have from (7.202)

$$\frac{M}{H} = \int_{-\infty}^{\infty}\int_{-\infty}^{\infty} c\,dxdy$$

$$= \frac{A_1A_2}{4\pi t\sqrt{D_xD_y}}\int_{-\infty}^{\infty}\exp[-\frac{(x - x_0 - ut)^2}{4D_x t} - K_1 t]dx\int_{-\infty}^{\infty}\exp[-\frac{(y - y_0 - vt)^2}{4D_y t} - K_2 t]dy \quad (7.215)$$

For the first integral on the right-hand side of (7.215), we introduce the following change of variables

$$\xi^2 = \frac{(x-x_0-ut)^2}{4D_x t} - K_1 t, \quad d\xi = \frac{dx}{\sqrt{4D_x t}} \tag{7.216}$$

Thus, we have

$$\int_{-\infty}^{\infty} \exp[-\frac{(x-x_0-ut)^2}{4D_x t} - K_1 t]dx = \sqrt{4D_x t}\int_{-\infty}^{\infty} \exp(-\xi^2)d\xi = \sqrt{4\pi D_x t} \tag{7.217}$$

Evidently, the second integral on the right-hand side of (7.215) can be evaluated as

$$\int_{-\infty}^{\infty} \exp[-\frac{(y-y_0-vt)^2}{4D_y t} - K_2 t]dy = \sqrt{4D_y t}\int_{-\infty}^{\infty} \exp(-\xi^2)d\xi = \sqrt{4\pi D_y t} \tag{7.218}$$

Using the results in (7.217) and (7.218), we can evaluate (7.215) as

$$\frac{M}{H} = A_1 A_2 \tag{7.219}$$

Finally, we obtain the solution as

$$c = \frac{(M/H)}{4\pi t\sqrt{D_x D_y}} \exp[-\frac{(x-x_0-ut)^2}{4D_x t} - \frac{(y-y_0-vt)^2}{4D_y t} - Kt] \tag{7.220}$$

At any section of constant x, we obtain a Gaussian distribution of the pollutant along the y-direction, and vice versa. If $D_y > D_x$, the cloud of pollutant grows faster along the y-direction, and vice versa. A special case of (7.220) was given in Fischer et al. (1979). However, the definition of M in Fischer et al. (1979) differs from the present one as M in the present chapter always has a unit of kg.

7.9.2 Three-Dimensional Point Source Solution

The analysis given in the last section can be easily extended to the three-dimensional case. For the 3-D point source problem, Fick's second law with initial condition is

$$\frac{\partial c}{\partial t} = D_x \frac{\partial^2 c}{\partial x^2} + D_y \frac{\partial^2 c}{\partial y^2} + D_z \frac{\partial^2 c}{\partial z^2} - u\frac{\partial c}{\partial x} - v\frac{\partial c}{\partial y} - w\frac{\partial c}{\partial z} - Kc \tag{7.221}$$

$$c(x,y,z,0) = M\delta(x_0)\delta(y_0)\delta(z_0) \tag{7.222}$$

Equivalently, the initial condition can be expressed as:

$$M = \int_{-\infty}^{\infty}\int_{-\infty}^{\infty}\int_{-\infty}^{\infty} c\,dxdydz \tag{7.223}$$

Similarly, we can assume a separation of variables as

$$c(x,y,z,t) = c_1(x,t)c_2(y,t)c_3(z,t) \tag{7.224}$$

Substitution of (7.224) into (7.221) gives

$$\begin{aligned} &c_1c_3\{\frac{\partial c_2}{\partial t} - D_y\frac{\partial^2 c_2}{\partial x^2} + v\frac{\partial c_2}{\partial y}\} + c_1c_2\{\frac{\partial c_3}{\partial t} - D_z\frac{\partial^2 c_3}{\partial z^2} + w\frac{\partial c_3}{\partial x}\} \\ &+c_3c_2\{\frac{\partial c_1}{\partial t} - D_x\frac{\partial^2 c_1}{\partial x^2} + u\frac{\partial c_1}{\partial x}\} + Kc_1c_2c_3 = 0 \end{aligned} \tag{7.225}$$

Dividing through by c, we arrive at three PDEs for c_1, c_2, and c_3

$$\frac{1}{c_2}\{\frac{\partial c_2}{\partial t}-D_y\frac{\partial^2 c_2}{\partial y^2}+v\frac{\partial c_2}{\partial y}\}=-K_2 \tag{7.226}$$

$$\frac{1}{c_1}\{\frac{\partial c_1}{\partial t}-D_x\frac{\partial^2 c_1}{\partial x^2}+u\frac{\partial c_1}{\partial x}\}=-K_1 \tag{7.227}$$

$$\frac{1}{c_3}\{\frac{\partial c_3}{\partial t}-D_z\frac{\partial^2 c_3}{\partial z^2}+w\frac{\partial c_3}{\partial z}\}=-K_3 \tag{7.228}$$

It is straightforward to see that

$$K=K_1+K_2+K_3 \tag{7.229}$$

As in Section 7.9.1, the solutions for c_1, c_2, and c_3 for the point source are

$$c_1(x,t)=\frac{A_1}{\sqrt{4\pi D_x t}}\exp[-\frac{(x-x_0-ut)^2}{4D_x t}-K_1 t] \tag{7.230}$$

$$c_2(y,t)=\frac{A_2}{\sqrt{4\pi D_y t}}\exp[-\frac{(y-y_0-vt)^2}{4D_y t}-K_2 t] \tag{7.231}$$

$$c_3(z,t)=\frac{A_3}{\sqrt{4\pi D_z t}}\exp[-\frac{(z-z_0-wt)^2}{4D_z t}-K_3 t] \tag{7.232}$$

Combining these results, we obtain the solution for c as

$$c=\frac{A_1A_2A_3}{(4\pi t)^{3/2}\sqrt{D_xD_yD_z}}\exp[-\frac{(x-x_0-ut)^2}{4D_x t}-\frac{(y-y_0-vt)^2}{4D_y t}-\frac{(z-z_0-wt)^2}{4D_z t}-(K_1+K_2+K_3)t] \tag{7.233}$$

Substituting (7.233) into (7.223), we obtain

$$M=\int_{-\infty}^{\infty}\int_{-\infty}^{\infty}\int_{-\infty}^{\infty}c\,dxdydz=\frac{A_1A_2A_3}{(4\pi t)^{3/2}\sqrt{D_xD_yD_z}}\int_{-\infty}^{\infty}\exp[-\frac{(x-x_0-ut)^2}{4D_x t}-K_1 t]dx$$
$$\times\int_{-\infty}^{\infty}\exp[-\frac{(y-y_0-vt)^2}{4D_y t}-K_2 t]dy$$
$$\times\int_{-\infty}^{\infty}\exp[-\frac{(z-z_0-wt)^2}{4D_z t}-K_2 t]dz \tag{7.234}$$

Applying the technique used in the last section, we finally get

$$M=A_1A_2A_3 \tag{7.235}$$

The final 3-D solution is

$$c=\frac{M}{(4\pi t)^{3/2}\sqrt{D_xD_yD_z}}\exp[-\frac{(x-x_0-ut)^2}{4D_x t}-\frac{(y-y_0-vt)^2}{4D_y t}-\frac{(z-z_0-wt)^2}{4D_z t}-Kt] \tag{7.236}$$

For the special case of $u = v = w = K = 0$ and $x_0 = y_0 = z_0 = 0$, we recover (2.54) of Fischer et al. (1979) as a limiting case. However, no proof was given by Fischer et al. (1979).

7.9.3 Two-Dimensional Line Source

In this section, we will illustrate the method of superposition in generating a solution for another source. Figure 7.9 illustrates a line source parallel to the y-axis. The mass injection along the line source can be modeled as:

$$m = \frac{M}{HL} \tag{7.237}$$

The concentration c can be evaluated using (7.220) and superposition in terms of integration as

$$c = \int_{\eta_1}^{\eta_2} dc = \frac{m}{4\pi t\sqrt{D_x D_y}} \exp[-\frac{(x-x_0-ut)^2}{4D_x t} - Kt]\int_{\eta_1}^{\eta_2} \exp[-\frac{(y-\eta-vt)^2}{4D_y t}]d\eta \tag{7.238}$$

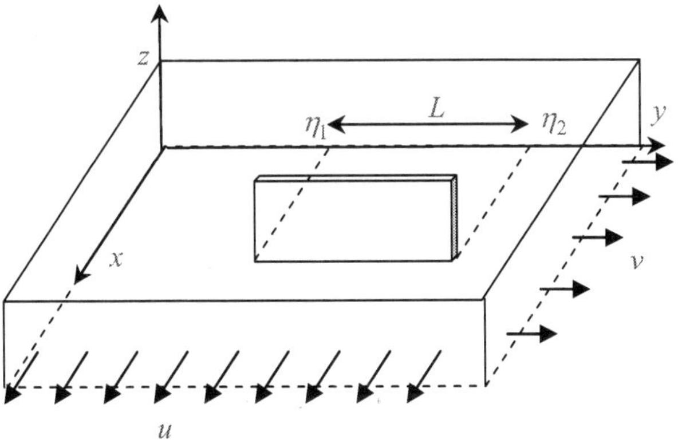

Figure 7.9 Line source in a two-dimensional ocean

We now introduce the following change of variables

$$\xi^2 = \frac{(y-\eta-vt)^2}{4D_y t}, \quad d\eta = -\sqrt{4D_y t}\, d\xi \tag{7.239}$$

The solution given in (7.238) becomes

$$c=\frac{m}{2\pi\sqrt{D_x t}}\exp[-\frac{(x-x_0-ut)^2}{4D_x t}-Kt]\{-\int_{\frac{y-vt}{\sqrt{4D_y t}}}^{\frac{y-\eta_2-vt}{\sqrt{4D_y t}}}\exp(-\xi^2)d\xi$$
$$+\int_{\frac{y-vt}{\sqrt{4D_y t}}}^{\frac{y-\eta_1-vt}{\sqrt{4D_y t}}}\exp(-\xi^2)d\xi\} \tag{7.240}$$

This can be rewritten as

$$c=\frac{m}{2\pi\sqrt{D_x t}}\exp[-\frac{(x-x_0-ut)^2}{4D_x t}-Kt]\{-\int_0^{\frac{y-\eta_2-vt}{\sqrt{4D_y t}}}\exp(-\xi^2)d\xi$$
$$+\int_0^{\frac{y-\eta_1-vt}{\sqrt{4D_y t}}}\exp(-\xi^2)d\xi\} \tag{7.241}$$

Finally, using the definition of the error function, we obtain the final solution as

$$c=\frac{m}{4\sqrt{\pi D_x t}}\exp[-\frac{(x-x_0-ut)^2}{4D_x t}-Kt]\{erf(\frac{y-\eta_1-vt}{\sqrt{4D_y t}})-erf(\frac{y-\eta_2-vt}{\sqrt{4D_y t}})\} \tag{7.242}$$

The maximum value of pollutant concentration locates at

$$x=x_0+ut,\quad y=\frac{1}{2}(\eta_1+\eta_2)+vt, \tag{7.243}$$

It can be shown that the maximum of c is

$$c_{\max}=\frac{M}{LH\sqrt{4\pi D_x t}}\exp(-Kt)erf(\frac{L}{4\sqrt{D_y t}}) \tag{7.244}$$

where L is the length of the line source. The proof is left for readers as an exercise (see Problem 7.8). Problem 7.9 gives the solution for the case where the line source is parallel to the x-axis.

7.10 SUMMARY AND FURTHER READING

For an introduction of the diffusion and dispersion of pollutant, readers can refer to TRACOR (1971) *Estuarine Modeling: An Assessment.* It was a technical review done by TRACOR Incorporation commissioned by the US Environmental Protection Agency. Part II of this review contained nearly 100 pages of hydrodynamic models, which were contributed Professor D.W. Pritchard of John Hopkins University and D.R.F. Harleman of MIT. More generally, classic analytic solutions relevant to diffusion are discussed in Carslaw and Jaeger (1946), Ozisik (1968), Parkus (1962), and Crank (1975). A good book is Banks (1994) on the mathematical framework in solving growth and diffusion problems. Problems with non-constant diffusivity coefficients are also discussed by John Crank in his excellent book of *The Mathematics of Diffusion* (Crank, 1975). Actually, we mentioned the name of John Crank in Section 15.2.3 of Chau (2018) noting that the central difference scheme is also known as the Crank-Nicolson scheme. In fact, the Crank-Nicolson finite difference scheme was derived to solve the heat diffusion equation by John Crank and Phyllis Nicolson. More one-dimensional, two-dimensional and three-dimensional analytical solutions are complied in Wexler (1992) in the context of solute transport in ground water.

The physical problems of pollutant transport in rivers, lakes, and coastal waters are discussed in the comprehensive book by Fischer et al. (1979). In the present chapter, we have ignored the potential of the pollutants having chemical reactions with other constituents in the stream or in the sea. We also have ignored the possibility of having pre-existing pollutants deposited previously at seabed or riverbed and their entrainment into the stream.

To end this chapter, we tell an interesting story about Robert Banks. In addition to the Ogata and Banks solution (1961) that we covered in this chapter, Banks (1994) provided a formal mathematical framework for solving diffusion problems. Robert Banks was also the author of two interesting books on "adventures in applied mathematics" (Banks, 1998, 1999). When he served as the president of the Asian Institute of Technology (AIT) in Thailand, he published some of the interesting problems covered in these two books in the *AIT Bulletin* as challenge problems, offering 100 Baht (Thai currency) for anyone coming up with the correct answer. What an academician at heart! We all do something for a reason. When Banks was a first-year graduate student majoring in environmental engineering, his supervisor asked him to enroll in an undergraduate course on bacteriology to learn more life science as a starter. Since Banks was the only engineering student and the only graduate student in the whole class, the professor of the subject asked him to write an additional term paper. Banks eventually submitted a paper entitled *Applications of mathematics in bacteriology*, including the coverage of nonlinear differential equations for population growth. The professor evidently liked the paper and Banks got an "A." This positive experience clearly led Banks to his later ambition of publishing the two interesting books on adventures in applied mathematics. Another motivation probably came from his wife, Gunta. When Banks was working very hard on his first technical book *Growth and Diffusion Phenomena: Mathematical Frameworks and Applications* (Banks, 1994) which appeared as Vol. 14 of the Texts in Applied Mathematics edited by F. John, J.E. Marsdon, L. Sirovich, M. Golubitsky and W. Jäger, his wife, Gunta, helped to type his manuscripts. She occasionally and tactfully expressed her view that the market would be quite limited without a somewhat better plot and a good deal more passion and excitement in the subject matter. Banks may not have been very successful in his 1994 book, but clearly excelled in his endeavor in publishing the two later books *Towing Icebergs, Falling Dominoes, and Other Adventures in Applied Mathematics* (Banks, 1998) and *Slicing Pizzas, Racing Turtles and Further Adventures in Applied Mathematics* (Banks, 1999).

7.11 PROBLEMS

Problem 7.1 The Ogata-Banks problem is defined as

$$\frac{\partial c}{\partial t} = D\frac{\partial^2 c}{\partial x^2} - u\frac{\partial c}{\partial x} \tag{7.245}$$

with the following initial and boundary conditions:

$$c(0,t)=c_0, \quad t\geq 0,$$
$$c(x,0)=0, \quad x\geq 0, \tag{7.246}$$
$$c(\infty,t)=0, \quad t\geq 0,$$

(i) Introduce the following change of variables

$$c(x,t)=\Gamma(x,t)\exp(\frac{ux}{2D}-\frac{u^2t}{4D}) \tag{7.247}$$

and show the following

$$\frac{\partial c}{\partial t}=\frac{\partial\Gamma}{\partial t}\exp(\frac{ux}{2D}-\frac{u^2t}{4D})-\frac{u^2}{4D}\Gamma\exp(\frac{ux}{2D}-\frac{u^2t}{4D}) \tag{7.248}$$

$$\frac{\partial c}{\partial x}=\frac{\partial\Gamma}{\partial x}\exp(\frac{ux}{2D}-\frac{u^2t}{4D})+\frac{u}{2D}\Gamma\exp(\frac{ux}{2D}-\frac{u^2t}{4D}) \tag{7.249}$$

$$\frac{\partial^2 c}{\partial x^2}=\frac{u}{D}\frac{\partial\Gamma}{\partial x}\exp(\frac{ux}{2D}-\frac{u^2t}{4D})+\frac{u^2}{4D^2}\Gamma\exp(\frac{ux}{2D}-\frac{u^2t}{4D})+\frac{\partial^2\Gamma}{\partial x^2}\exp(\frac{ux}{2D}-\frac{u^2t}{4D}) \tag{7.250}$$

(ii) Substitute these results into (7.245) to show that

$$\frac{\partial\Gamma}{\partial t}=D\frac{\partial^2\Gamma}{\partial x^2} \tag{7.251}$$

(iii) Finally, show that the initial and boundary conditions for Γ are

$$\Gamma(0,t)=\phi(t)=c_0\exp(\frac{u^2t}{4D}), \quad t\geq 0,$$
$$\Gamma(x,0)=0, \quad x\geq 0, \tag{7.252}$$
$$\Gamma(\infty,t)=0, \quad t\geq 0,$$

Problem 7.2 Prove the following identity

$$\int_0^\alpha \exp[-(\lambda-\frac{\varepsilon}{\lambda})^2]d\lambda=\int_{\varepsilon/\alpha-\alpha}^\infty e^{-\beta^2}d\beta-\int_{\varepsilon/\alpha}^\infty \exp[-(\frac{\varepsilon}{z}-z)^2]dz \tag{7.253}$$

Problem 7.3 Prove the following identity given in (7.85):

$$\int_{\varepsilon/\alpha}^\infty \exp[-(\frac{\varepsilon}{z}+z)^2+2\varepsilon]dz=\int_{\varepsilon/\alpha}^\infty \exp[-(\frac{\varepsilon}{z}-z)^2-2\varepsilon]dz \tag{7.254}$$

Problem 7.4 Find the solution of the following ODE with the given boundary conditions in terms of the complementary error function

$$\frac{d^2F}{d\eta^2}+2\eta\frac{dF}{d\eta}=0 \tag{7.255}$$

$$F(0)=1, \quad F(\infty)=0 \tag{7.256}$$

Hints: This is known as the Rayleigh impulsive flow problem (Segel, 1987).

Ans:

$$F(\eta) = erfc(\eta) \tag{7.257}$$

Problem 7.5 Show that for a 1-D point source for an infinite river (Taylor's 1954 problem) the penetration distance of a fixed concentration c_0 can be expressed as:

$$x = \sqrt{4Dt}\{\ln[\frac{M}{A_0 c_0 \sqrt{4D\pi t}}]\}^{1/2} \tag{7.258}$$

Problem 7.6 As discussed in Section 7.5, the PDE for 1-D diffusion of a pollutant in a flowing river is

$$\frac{\partial c}{\partial t} = D\frac{\partial^2 c}{\partial x^2} - u\frac{\partial c}{\partial x} \tag{7.259}$$

Show that this PDE can be converted to the following Fick's second law

$$\frac{\partial c}{\partial \tau} = D\frac{\partial^2 c}{\partial \xi^2} \tag{7.260}$$

where the moving coordinates ξ and τ are defined as

$$\tau = t, \quad \xi = x - ut \tag{7.261}$$

Problem 7.7 Evaluate the following integral in terms of the complementary error function:

$$I = \frac{1}{\sqrt{4\pi}}\int_0^\infty \exp[-\frac{(s-z)^2}{4}]ds \tag{7.262}$$

Ans:

$$I = \frac{1}{2} erfc(-\frac{z}{2}) \tag{7.263}$$

Problem 7.8 Show the validity of (7.244).

Problem 7.9 Repeat the analysis in Section 7.9.3 for the case that the line source is parallel to the x-axis and is imposed between ξ_1 and ξ_2. Show that the solution is

$$c = \frac{m}{4\sqrt{\pi D_y t}}\exp[-\frac{(y-y_0-vt)^2}{4D_y t} - Kt]\{erf(\frac{x-\xi_1-ut}{\sqrt{4D_x t}}) - erf(\frac{x-\xi_2-ut}{\sqrt{4D_x t}})\} \tag{7.264}$$

Problem 7.10 Solve the following problem of finite time injection of pollutants into a river:

$$\frac{\partial c}{\partial t} = D\frac{\partial^2 c}{\partial x^2} - u\frac{\partial c}{\partial x} - Kc \tag{7.265}$$

$$c(0,t) = c_0[H(t) - H(t-\tau)], \quad t \geq 0,$$
$$c(x,0) = 0, \quad x \geq 0, \tag{7.266}$$
$$c(\infty,t) = 0, \quad t \geq 0,$$

where H is the Heaviside step function which is defined in (8.181) of Chau (2018).

Ans:

For $t < \tau$:

$$c(x,t) = \frac{c_0}{2}\exp(\frac{ux}{2D})\{-\exp(\frac{\sqrt{u^2+4DK}\,x}{2D})erfc(\frac{x+\sqrt{u^2+4DK}\,t}{\sqrt{4Dt}})$$
$$+\exp(-\frac{\sqrt{u^2+4DK}\,x}{2D})erfc(\frac{x-\sqrt{u^2+4DK}\,t}{\sqrt{4Dt}})\} \tag{7.267}$$

For $t > \tau$:

$$c(x,t) = \frac{c_0}{2}\exp(\frac{ux}{2D})\{-\exp(\frac{\sqrt{u^2+4DK}\,x}{2D})$$
$$\times[erfc(\frac{x+\sqrt{u^2+4DK}\,t}{\sqrt{4Dt}}) - erfc(\frac{x+\sqrt{u^2+4DK}\,(t-\tau)}{\sqrt{4D(t-\tau)}})]$$
$$+\exp(-\frac{\sqrt{u^2+4DK}\,x}{2D})$$
$$\times[erfc(\frac{x-\sqrt{u^2+4DK}\,t}{\sqrt{4Dt}}) - erfc(\frac{x-\sqrt{u^2+4DK}\,(t-\tau}{\sqrt{4D(t-\tau)}})]\} \tag{7.268}$$

CHAPTER EIGHT

Geophysical Fluid Flows

8.1 INTRODUCTION

There are two main fluids found on our Planet Earth, namely water and air. Their appearances on Earth's surface make our planet habitable for living organisms. Air and water control both climate and weather on Earth's surface, through atmospheric and oceanic flows. The former one relates to meteorology and the latter one relates to oceanography. With growing concerns on climate change, sea-level rise, drought, flooding, and more extreme weather occurrences (such as the increasing intensity of hurricanes, tropical storms, and typhoons), the understanding and prediction of geophysical fluid flows become more important. This chapter focuses on the fundamental fluid dynamics used in modeling geophysical flows. As shown in Chau (2018), fluid flows are governed by Navier-Stokes equations. For geophysical flows, we need to incorporate the rotational effects of the Earth. On the global scale, Rossby waves, or planetary waves, are upper atmospheric cold fronts that influence the mid-latitude weather in the Northern Hemisphere in winter. Heat transfer in the atmosphere from the equatorial areas to the polar areas occurs often through violent weather systems in the form of hurricanes or tropical cyclones. Such extreme weather has been making profound impacts on all kinds of human activities, including explorations, trading, travels, and fisheries. Mathematical treatment of the Navier-Stokes equations under various simplified conditions leads to analytical solutions, providing meaningful insights for weather prediction and forecast. Meaningful analytical models for hurricanes or tropical cyclones are still being developed. In this chapter, after reviewing the governing equations for geophysical flows, we discuss the prediction of storm surges considering the inverse barometer effect, moving center of low pressure, and wind-induced effects. Ekman transport is then discussed in explaining the orthogonal drifting of icebergs with respect to the wind. Various vortex models are discussed in detail and their implications on the wind speed of tornadoes. Although these solutions for tornadoes are on much smaller scales than that of hurricanes, they do provide insight to the structures of hurricanes.

In particular, in Section 8.2, the concept of Coriolis force is introduced, and the special case of high altitude approximation is considered. Section 8.3 considers the governing equations of geophysical flows, including the continuity equation, momentum equations, energy equations, and equations of state. Section 8.4 discusses scaling of force terms and their importance in terms of Rossby, Ekman, and Reynolds numbers. Section 8.5 discusses storm surges induced by atmospheric pressure (inverse barometer effect), moving pressure disturbance, and wind. Section 8.6 derives the Ekman transport with and without internal currents. Section

8.7 presents geostrophic flows, while the 2-D shallow water equation is discussed in Section 8.8. Vorticity is discussed in the context of tornado dynamics in Section 8.9, including the potential vortex, Rankine vortex, Burgers-Rott vortex, Oseen-Lamb vortex, and Sullivan vortex.

8.2 CORIOLIS FORCE DUE TO ROTATION

All fluid motions on the Earth's surface can, in principle, be formulated with respect to a fixed frame of reference, such as a distant star. However, it is well-known that the Earth is revolving around the Sun with a period of one year, and the Earth itself is rotating about its own axis with a period of one day. We are observing all geophysical fluid flows on Earth's surface, which is a rotating frame of reference. Naturally, governing equations for fluid flows should be formulated on a reference rotating with our planet. However, a fictitious force, called the Coriolis force, is observed in a rotating frame of reference. This Coriolis force plays an important role in geophysical fluids, and it was named after French military engineer G. Coriolis (1792-1843) who first recognized its importance. The best way to visualize Coriolis acceleration and force is to consider a man standing at the center of a merry-go-round and throwing a ball outward from the center of the rotating disk. Another man who serves as an observer while standing on the ground records the path, velocity, and acceleration of the flying ball. The situation is depicted in Figure 8.1. The rotating frame is labeled as S' whereas the stationary frame is labeled as S. The motion of the ball appears to be along a straight line O'A' in the rotating frame S'; whereas the ball appears to move from A to B in the stationary frame S. Thus, the ball moves outward as well as in the transverse direction with a velocity of $v_\theta = r\omega$. The change of velocity in the transverse direction can be evaluated as

$$\delta v_\theta = [\omega(r+\delta r)\cos(\frac{\delta\theta}{2}) + v_r \sin(\frac{\delta\theta}{2})] - [\omega r \cos(\frac{\delta\theta}{2}) - v_r \sin(\frac{\delta\theta}{2})] \tag{8.1}$$

Note that the change of the transverse velocity is always perpendicular to the outward motion. The first term inside the first square bracket term on the right-hand side is due to rotation, whereas the second term in the square bracket is due to change of direction of v_r. The first bracket is evaluated at Point B while the second bracket is evaluated at Point A. For small rotation, we have

$$\delta\theta \to 0, \quad \cos(\frac{\delta\theta}{2}) \to 1, \quad \sin(\frac{\delta\theta}{2}) \to \frac{\delta\theta}{2} \tag{8.2}$$

Substitution of (8.2) into (8.1) gives

$$\begin{aligned} \delta v_\theta &\approx [\omega(r+\delta r) + v_r \frac{\delta\theta}{2}] - [\omega r - v_r \frac{\delta\theta}{2}] \\ &= \omega\delta r + v_r \delta\theta \end{aligned} \tag{8.3}$$

Dividing both sides of (8.3) by the change of time δt, we get

$$a_\theta = \frac{\delta v_\theta}{\delta t} = \omega\frac{\delta r}{\delta t} + v_r \frac{\delta\theta}{\delta t} = \omega v_r + v_r \omega = 2\omega v_r \tag{8.4}$$

The transverse force is then

$$F_\theta = m a_\theta = 2m\omega v_r \tag{8.5}$$

This is a real force in the stationary frame S but this force is inferred in the rotating frame S' as Coriolis force as:

$$F'_\theta = -2m\omega v'_r \tag{8.6}$$

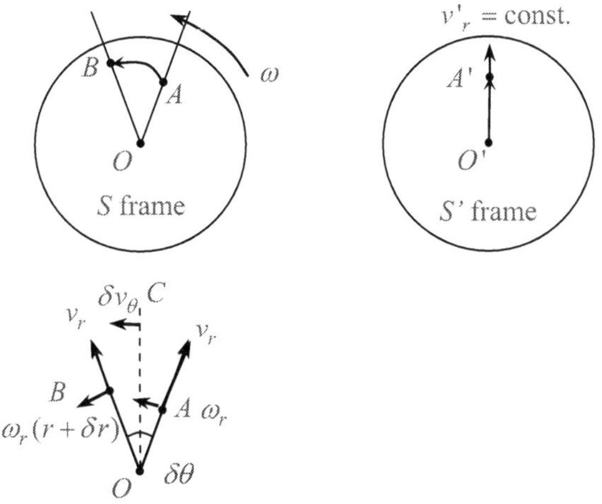

Figure 8.1 A ball is falling outward from the center of a rotating disk observed from a stationary frame *S* and a rotating frame *S'*

Note that the radial velocity is the same for both rotating and stationary frames (i.e., $v_r = v'_r$). Equation (8.6) implies that the Coriolis force is felt in the opposite direction of positive θ. For the case shown in Figure 8.1, the Coriolis force acts to deflect the flying ball to the right with respect to O'A'. In fact, the Coriolis force always acts as a deflecting force perpendicular to the direction of motion as observed in the rotating frame. The Coriolis force is very real from the viewpoint of the rotating frame. If you try to walk outward or inward in a merry-go-round, you would feel this real "fictitious force," trying to push you off balance.

The transverse acceleration experience in the rotating frame given in (8.4) can be generalized as

$$\boldsymbol{a}'_\theta = -2\boldsymbol{\omega} \times \boldsymbol{v}_{rel} \tag{8.7}$$

where the subscript "rel" of $\boldsymbol{v}$ indicates relative velocity. That is, the acceleration due to the Coriolis effect can be represented by the cross-product of angular rotation with the velocity. The cross-product also reflects the fact that Coriolis force always acts perpendicular to the motion in the direction opposite to the prediction of right-hand rule. This is known as Coriolis acceleration and represents the difference of the acceleration of Point A relative to O measured from the non-rotating frame and from the rotating frame. It is named after the French military engineer G. Coriolis who first studied this force.

8.2.1 Coriolis Force for High Altitude

For the case of rotations of the Earth with a rotation rate of ω_s, an eastward-moving air or water mass P will experience a southward acceleration toward the equator, as shown in Figure 8.2. In this formulation, we have assumed that ϕ is not close to zero (or near equator). For the case of zero eastward velocity of u, the outward acceleration due to the Earth's rotation is

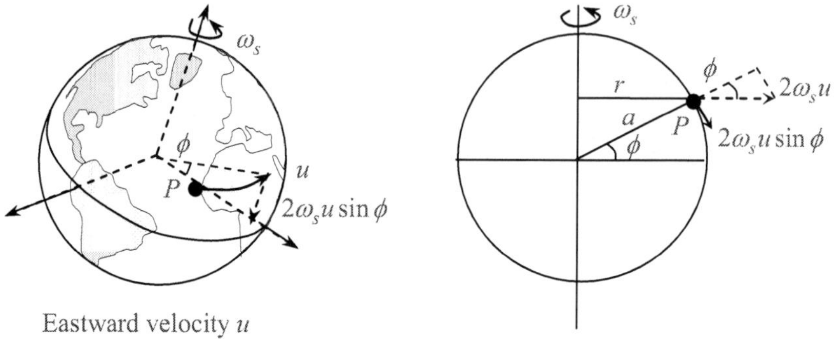

Figure 8.2 Coriolis force on Earth's surface on a point *P* moving eastward

$$\boldsymbol{a}_{r0} = \frac{v_r^2}{r}\boldsymbol{e}_r = \frac{(\omega_s r)^2}{r}\boldsymbol{e}_r = \omega_s^2 r\boldsymbol{e}_r \tag{8.8}$$

For the case of nonzero u, the outward acceleration due to the Earth's rotation and eastward velocity becomes

$$\boldsymbol{a}_{r1} = \frac{(\omega_s r + u)^2}{r}\boldsymbol{e}_r = (\omega_s^2 r + 2\omega_s u + \frac{u^2}{r})\boldsymbol{e}_r \tag{8.9}$$

For air or ocean currents on Earth's surface, the velocity u is much smaller than that induced by Earth's rotation. That is, we have

$$\boldsymbol{a}_{r1} = (\omega_s^2 r + 2\omega_s u)\boldsymbol{e}_r \tag{8.10}$$

The net outward acceleration for the case of eastward motion can be found by subtracting (8.8) from (8.10) as

$$\boldsymbol{a}_r = \boldsymbol{a}_{r1} - \boldsymbol{a}_{r0} = (\omega_s^2 r + 2\omega_s u - \omega_s^2 r)\boldsymbol{e}_r = 2\omega_s u\boldsymbol{e}_r \tag{8.11}$$

Referring to the diagram on the left of Figure 8.2, the Coriolis acceleration acts southward as

$$a_{Coriolis} = 2\omega_s u \sin\phi = fu \tag{8.12}$$

where

$$f = 2\omega_s \sin\phi \tag{8.13}$$

The parameter f is called the Coriolis parameter and is a function of the latitude ϕ. Next, we consider the Coriolis force for a point moving northward, as shown in Figure 8.3. For this case, we can see that r decreases with time t. To conserve angular momentum, we must have u being increased. The angular momentum at point P at rest must balance that at the new position after time change Δt movement:

$$\begin{aligned}\omega_s r^2 &= [\omega_s(r-\Delta r)+\Delta u](r-\Delta r) \\ &= \omega_s r^2 - 2\omega_s(\Delta r)r + r\Delta u + \omega_s(\Delta r)^2 - \Delta u \Delta r\end{aligned} \tag{8.14}$$

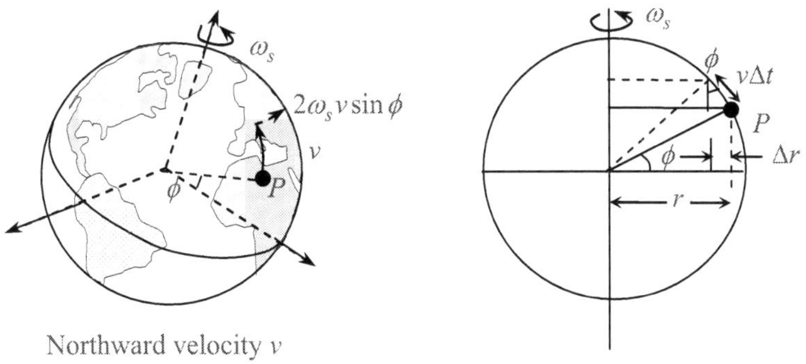

Figure 8.3 Coriolis force on Earth's surface on a point *P* moving northward

Cancelling the term on the left, we can simplify (8.14) as

$$\Delta u = \{\frac{2\omega_s r - \omega_s \Delta r}{r-\Delta r}\}(\Delta r) \tag{8.15}$$

Taking the limit of a small change of radial distance r, we obtain

$$\lim_{\Delta r\to 0} \Delta u = 2\omega_s \Delta r \tag{8.16}$$

From the left diagram in Figure 8.3, we have

$$\sin\phi = \frac{\Delta r}{v\Delta t} \tag{8.17}$$

The apparent eastward acceleration is

$$a_{Coriolis} = \frac{\Delta u}{\Delta t} = \frac{2\omega_s \Delta r}{\Delta r} v\sin\phi = 2\omega_s v\sin\phi \tag{8.18}$$

We can rewrite it as

$$a_{Coriolis} = 2\omega_s v\sin\phi = fv \tag{8.19}$$

where f has been defined in (8.13).

8.2.2 Coriolis Force for All Altitudes

To relate the quantities between a stationary frame and a rotating frame, Figure 8.4 shows a position vector $\boldsymbol{r}$ in terms of both coordinates as:

$$\boldsymbol{r} = x\boldsymbol{i} + y\boldsymbol{j} = X\boldsymbol{I} + Y\boldsymbol{J} \tag{8.20}$$

where $\boldsymbol{I}$ and $\boldsymbol{J}$ are the base vectors of the stationary coordinate (X,Y) and $\boldsymbol{i}$ and $\boldsymbol{j}$ are those for rotating coordinate (x,y). The base vectors between these two coordinates can be established according to Figure 8.4:

$$\boldsymbol{I} = \boldsymbol{i}\cos\Omega t - \boldsymbol{j}\sin\Omega t, \quad \boldsymbol{J} = \boldsymbol{i}\sin\Omega t + \boldsymbol{j}\cos\Omega t \tag{8.21}$$

Substitution of (8.21) into (8.20) gives

$$x = X\cos\Omega t + Y\sin\Omega t, \quad y = -X\sin\Omega t + Y\cos\Omega t \tag{8.22}$$

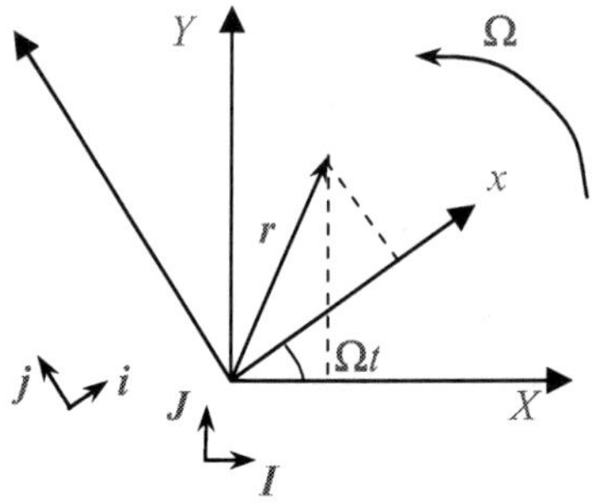

Figure 8.4 Transformation between a stationary frame and a rotating frame

Differentiating (8.22) with respect to time, we obtain

$$u = \frac{dx}{dt} = \frac{dX}{dt}\cos\Omega t + \frac{dY}{dt}\sin\Omega t - \Omega X\sin\Omega t + \Omega Y\cos\Omega t \tag{8.23}$$

$$v = \frac{dy}{dt} = -\frac{dX}{dt}\sin\Omega t + \frac{dY}{dt}\cos\Omega t - \Omega X\cos\Omega t - \Omega Y\sin\Omega t \tag{8.24}$$

The velocity in the rotating frame and the stationary frame are defined as

$$\boldsymbol{u} = \frac{dx}{dt}\boldsymbol{i} + \frac{dy}{dt}\boldsymbol{j} = u\boldsymbol{i} + v\boldsymbol{j} \tag{8.25}$$

$$\boldsymbol{U} = \frac{dX}{dt}\boldsymbol{I} + \frac{dY}{dt}\boldsymbol{J} \tag{8.26}$$

Substitution of (8.21) into (8.26) gives

$$\boldsymbol{U} = (\frac{dX}{dt}\cos\Omega t + \frac{dY}{dt}\sin\Omega t)\boldsymbol{i} + (-\frac{dX}{dt}\sin\Omega t + \frac{dY}{dt}\cos\Omega t)\boldsymbol{j} = U\boldsymbol{i} + V\boldsymbol{j} \tag{8.27}$$

Substitution of (8.23) and (8.24) into (8.27) gives

$$\begin{aligned} U &= \frac{dX}{dt}\cos\Omega t + \frac{dY}{dt}\sin\Omega t = u + \Omega X\sin\Omega t - \Omega Y\cos\Omega t \\ &= u - \Omega y \end{aligned} \tag{8.28}$$

$$\begin{aligned} V &= -\frac{dX}{dt}\sin\Omega t + \frac{dY}{dt}\cos\Omega t = v + \Omega X\cos\Omega t + \Omega Y\sin\Omega t \\ &= v + \Omega x \end{aligned} \tag{8.29}$$

The rotation can be expressed in vector form as

$$\boldsymbol{\Omega} = \Omega\boldsymbol{k} \tag{8.30}$$

where $\boldsymbol{k}$ is the base vector normal to the plane shown in Figure 8.4.

Using vector form, (8.29) and (8.30) can be generalized to

$$\frac{d\boldsymbol{R}}{dt} = \boldsymbol{U} = \boldsymbol{u} + \boldsymbol{\Omega}\times\boldsymbol{r} = \left(\frac{d}{dt} + \boldsymbol{\Omega}\times\right)\boldsymbol{r} \tag{8.31}$$

Therefore, the derivative of a vector $\boldsymbol{R}$ in the reference frame with respect to time is equivalent to the operator in the bracket of (8.31) on $\boldsymbol{r}$ in the rotating frame. Using this information, we can find the acceleration as

$$\begin{aligned} \boldsymbol{A} &= \frac{d^2\boldsymbol{R}}{dt^2} = \left(\frac{d}{dt} + \boldsymbol{\Omega}\times\right)\left(\frac{d\boldsymbol{r}}{dt} + \boldsymbol{\Omega}\times\boldsymbol{r}\right) = \frac{d^2\boldsymbol{r}}{dt^2} + 2\boldsymbol{\Omega}\times\frac{d\boldsymbol{r}}{dt} + \boldsymbol{\Omega}\times(\boldsymbol{\Omega}\times\boldsymbol{r}) \\ &= \frac{d\boldsymbol{u}}{dt} + 2\boldsymbol{\Omega}\times\boldsymbol{u} + \boldsymbol{\Omega}\times(\boldsymbol{\Omega}\times\boldsymbol{r}) \end{aligned} \tag{8.32}$$

Employing the definition of acceleration in the rotating frame, we can rewrite (8.32) as

$$\boldsymbol{A} = \boldsymbol{a} + 2\boldsymbol{\Omega}\times\boldsymbol{u} + \boldsymbol{\Omega}\times(\boldsymbol{\Omega}\times\boldsymbol{r}) \tag{8.33}$$

where the last term is caused by centrifugal acceleration. We now apply (8.33) to the situation of the rotating Earth, as shown in Figure 8.5. That is, the rotation can be expressed as:

$$\boldsymbol{\Omega} = \Omega\cos\phi\boldsymbol{j} + \Omega\sin\phi\boldsymbol{k} \tag{8.34}$$

where the base vectors $\boldsymbol{i}$, $\boldsymbol{j}$, and $\boldsymbol{k}$ align with the east, north, and up direction respectively. At the equator, we have the special case that $\phi = 0$, and (8.34) becomes

$$\boldsymbol{\Omega} = \Omega\boldsymbol{j} \tag{8.35}$$

At the north pole, we have the special case that $\phi = \pi/2$, and (8.34) becomes

$$\boldsymbol{\Omega} = \Omega\boldsymbol{k} \tag{8.36}$$

Substitution of (8.34) into (8.32) gives

$$\begin{aligned} \boldsymbol{\Omega}\times\boldsymbol{u} &= (\Omega\cos\phi\boldsymbol{j} + \Omega\sin\phi\boldsymbol{k})\times(u\boldsymbol{i} + v\boldsymbol{j} + w\boldsymbol{k}) \\ &= -\Omega u\cos\phi\boldsymbol{k} + \Omega w\cos\phi\boldsymbol{i} + \Omega u\sin\phi\boldsymbol{j} - \Omega v\sin\phi\boldsymbol{i} \\ &= (\Omega w\cos\phi - \Omega v\sin\phi)\boldsymbol{i} + \Omega u\sin\phi\boldsymbol{j} - \Omega u\cos\phi\boldsymbol{k} \end{aligned} \tag{8.37}$$

Thus, the absolute acceleration minus the centrifugal component $\boldsymbol{\Omega}\times(\boldsymbol{\Omega}\times\boldsymbol{r})$ are

$$\boldsymbol{A} - \boldsymbol{\Omega}\times(\boldsymbol{\Omega}\times\boldsymbol{r}) = \left(\frac{du}{dt} + 2f_*w - 2fv\right)\boldsymbol{i} + \left(\frac{dv}{dt} + 2fu\right)\boldsymbol{j} + \left(\frac{dw}{dt} - 2f_*u\right)\boldsymbol{k} \tag{8.38}$$

where

$$f_* = \Omega\cos\phi, \quad f = \Omega\sin\phi \tag{8.39}$$

The terms f and f_* defined in (8.39) are called the Coriolis parameter and the reciprocal Coriolis parameter. We see that $f > 0$ in the Northern Hemisphere, $f < 0$ in the Southern Hemisphere and $f = 0$ at equator, whereas $f_* > 0$ everywhere, except $f_* = 0$ at the North Pole and South Pole. The rotation of the Earth can be estimated as:

$$\Omega = \frac{2\pi}{24\times60\times60} + \frac{2\pi}{365.24\times24\times60\times60} = 7.2921\times10^{-5}\,s^{-1} \tag{8.40}$$

This is the rotation rate of the Earth in a *sidereal day*, which equals 23 hours, 56 minutes and 1.7 seconds. This is the period of time spanning the moment when a fixed distance star is seen one day and the moment on the next day when it is seen at the same angle from the same point on Earth. This is shorter than 24 hours because the Earth's orbital motion about the Sun makes it rotate slightly faster than one full turn with respect to a fixed distance star.

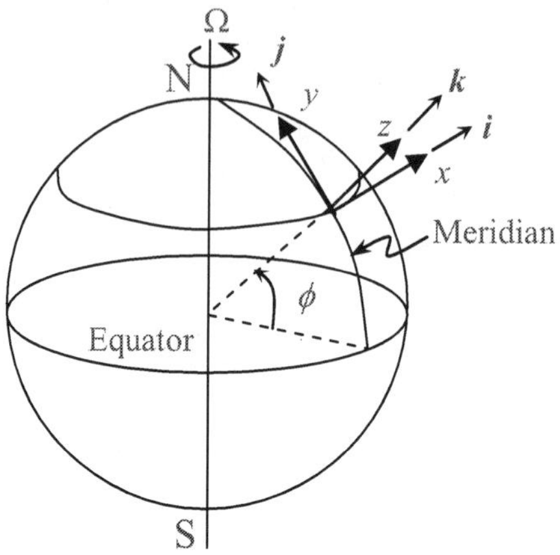

Figure 8.5 Definition of a local Cartesian coordinate for a rotating Earth

8.3 HYDRODYNAMIC EQUATIONS FOR GEOPHYSICAL FLOWS

In this section, we will derive the basic equations governing geophysical flows. Geophysical flows normal to the spherical surface of the Earth will not be considered. First, we will derive the continuity in oceans.

8.3.1 Continuity Condition

Consider a sea level fluctuation ζ being displaced from the mean water depth of h, as shown in Figure 8.2. A column of water with height of $h+\zeta$ is shown with horizontal inflows and outflows along both x and y axes. Subtracting the inflows from the outflows and adding the change of the mass inside the column, we have that the net inflow equals the change of fluid mass inside the column:

$$\rho\{dxdy\frac{\partial}{\partial x}[(h+\zeta)u]\}+\rho\{dxdy\frac{\partial}{\partial y}[(h+\zeta)v]\}+\frac{\partial}{\partial t}[\rho dxdy(h+\zeta)]=0 \quad (8.41)$$

If the density is constant and the change in water level is small compared to the depth of the water, we have

$$\frac{\partial \zeta}{\partial t}+\frac{\partial (hu)}{\partial x}+\frac{\partial (hv)}{\partial y}=0 \quad (8.42)$$

This is the conservation of mass. If the sea bottom is relatively flat, (8.9) can be simplified as

$$\frac{\partial \zeta}{\partial t} + h(\frac{\partial u}{\partial x} + \frac{\partial v}{\partial y}) = 0 \tag{8.43}$$

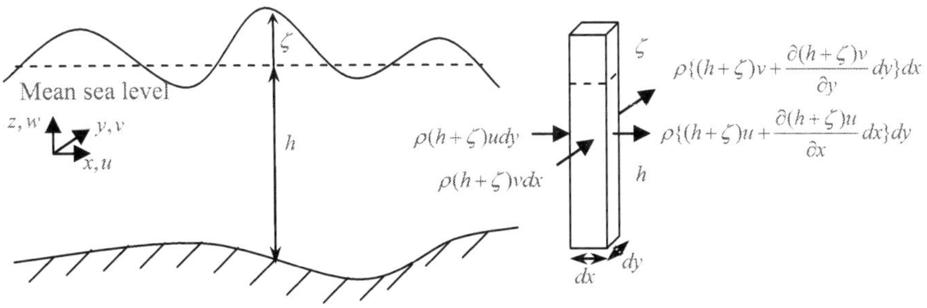

Figure 8.6 The conservation of mass in a column of water subject to two-dimensional flows

8.3.2 Momentum Equations

In fluid mechanics, Eulerian formulation employs the idea of a control volume, as shown in Figure 8.7. The fluid velocity u is a function of three variables. Thus, taking the total differential of u, we get

$$du(x,y,z,t) = \frac{\partial u}{\partial t}dt + \frac{\partial u}{\partial x}dx + \frac{\partial u}{\partial y}dy + \frac{\partial u}{\partial z}dz \tag{8.44}$$

Dividing (8.44) by dt, we have the material time derivative as

$$\frac{du}{dt} = \frac{\partial u}{\partial t} + \frac{\partial u}{\partial x}\frac{dx}{dt} + \frac{\partial u}{\partial y}\frac{dy}{dt} + \frac{\partial u}{\partial z}\frac{dz}{dt} = \frac{\partial u}{\partial t} + u\frac{\partial u}{\partial x} + v\frac{\partial u}{\partial y} + w\frac{\partial u}{\partial z} \tag{8.45}$$

Referring to Figure 8.7, the force equilibrium from the pressure difference and shear stress difference from the top and bottom surfaces leads to

$$\begin{aligned} F_x &= -[p + \frac{\partial p}{\partial x}dx]dydz + pdydz + \tau_{xz}dxdy - (\tau_{xz} - \frac{\partial \tau_{xz}}{\partial z}dz)dxdy \\ &= -\frac{\partial p}{\partial x}dxdydz + \frac{\partial \tau_{xz}}{\partial z}dzdxdy \end{aligned} \tag{8.46}$$

where the last term is due to body force (say tidal gravitational potential) per mass. The force equilibrium per unit mass is

$$\frac{F_x}{m} = -\frac{1}{\rho}(\frac{\partial p}{\partial x} - \frac{\partial \tau_{xz}}{\partial z}) \tag{8.47}$$

Combining (8.45), (8.38) and (8.47), the Coriolis acceleration in Section 8.2.2, we get

$$\frac{\partial u}{\partial t} + u\frac{\partial u}{\partial x} + v\frac{\partial u}{\partial y} + w\frac{\partial u}{\partial z} + f_* w - fv = -\frac{1}{\rho}(\frac{\partial p}{\partial x} - \frac{\partial \tau_{xz}}{\partial z}) \tag{8.48}$$

In Figure 8.7, only the stresses from the top and bottom surface of the control volume have been considered. We can generalize the stresses on the other surface as

$$\frac{\partial u}{\partial t}+u\frac{\partial u}{\partial x}+v\frac{\partial u}{\partial y}+w\frac{\partial u}{\partial z}+f_*w-fv=-\frac{1}{\rho}\left(\frac{\partial p}{\partial x}-\frac{\partial \tau_{xx}}{\partial x}-\frac{\partial \tau_{xy}}{\partial y}-\frac{\partial \tau_{xz}}{\partial z}\right) \quad (8.49)$$

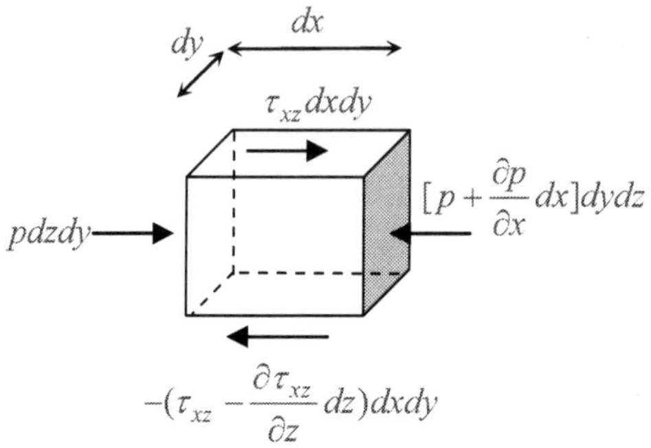

Figure 8.7 Horizontal momentum in a fluid element along the *x*-direction

Similarly, we can also consider the momentum equation along the *y*- and *z*-directions as

$$\frac{\partial v}{\partial t}+u\frac{\partial v}{\partial x}+v\frac{\partial v}{\partial y}+w\frac{\partial v}{\partial z}+fu=-\frac{1}{\rho}\left(\frac{\partial p}{\partial y}-\frac{\partial \tau_{yx}}{\partial x}-\frac{\partial \tau_{yy}}{\partial y}-\frac{\partial \tau_{yz}}{\partial z}\right) \quad (8.50)$$

$$\frac{\partial w}{\partial t}+u\frac{\partial w}{\partial x}+v\frac{\partial w}{\partial y}+w\frac{\partial w}{\partial z}-f_*u=-\frac{1}{\rho}\left(\frac{\partial p}{\partial z}-\frac{\partial \tau_{xz}}{\partial x}-\frac{\partial \tau_{zy}}{\partial y}-\frac{\partial \tau_{zz}}{\partial z}\right) \quad (8.51)$$

The first term on the left of (8.50) is the time derivative of velocity, the next three terms on the left are the advective terms, the fifth term on the left is the Coriolis force, the first term on the right is the pressure gradient term, and the next three terms are due to the stresses. The physical meaning of each term in (8.51) is similar to those for (8.50), except for the reciprocal Coriolis parameter f_*.

Further simplification of these momentum equations is possible by considering the mass conservation as well as the constitutive law.

8.3.3 Mass Conservation

Following the procedure for the proof of (8.42), if we replace the surface level ζ by density ρ and consider the changes from all three directions, we can derive the following continuity equation as:

$$\frac{\partial \rho}{\partial t}+\nabla\cdot(\rho v)=\frac{\partial \rho}{\partial t}+\frac{\partial(\rho u)}{\partial x}+\frac{\partial(\rho v)}{\partial y}+\frac{\partial(\rho w)}{\partial z}=0 \quad (8.52)$$

If spherical geometry of the Earth is considered, there will be additional curvature terms. However, for most geophysical flows, the length scales of flows are much smaller than the global scale and such curvature effects can be neglected. The inverted triangle followed by a dot is called the divergence (see Chau, 2018). In geophysical flows, the density does not change greatly from a mean value. For example, the variation of density in seawater rarely exceeds 2% whether due to density variations within an ocean basin or in estuaries for mixing of fresh and seawater. Within the first 10 km from the surface of the Earth (within which the weather pattern is confined), the variations of the density of air are no more than 5%. Therefore, it is justifiable to consider a small variation of density about the mean value as:

$$\rho = \rho_0 + \bar{\rho}(\boldsymbol{x}) \tag{8.53}$$

Substitution of (8.53) into (8.52) gives the following continuity equation.

$$\frac{\partial(\rho_0 + \bar{\rho})}{\partial t} + \nabla \cdot [(\rho_0 + \bar{\rho})\boldsymbol{v}] = \rho_0 \nabla \cdot \boldsymbol{v} + \bar{\rho}\nabla \cdot \boldsymbol{v} + [\frac{\partial \bar{\rho}}{\partial t} + \boldsymbol{v} \cdot \nabla \bar{\rho}] \tag{8.54}$$

To further simplify (8.54), we can assume that the time and spatial derivatives of the density variation are much smaller than the mean density such that the following approximations are valid

$$\bar{\rho} \ll \rho_0, \quad \boldsymbol{v} \cdot \nabla \bar{\rho} \ll \rho_0 \nabla \cdot \boldsymbol{v}, \quad \frac{\partial \bar{\rho}}{\partial t} \ll \rho_0 \nabla \cdot \boldsymbol{v} \tag{8.55}$$

In view of (8.55), we arrive at the following continuity equation

$$\nabla \cdot \boldsymbol{v} = \frac{\partial u}{\partial x} + \frac{\partial v}{\partial y} + \frac{\partial w}{\partial z} = 0 \tag{8.56}$$

This equation is also known as the reduced continuity equation. Physically, in essence (8.56) shows that the conservation of mass becomes the conservation of volume.

8.3.4 Constitutive Law

In a viscous fluid, we normally assume the stresses are proportional to the velocity gradient (or this is called the Newtonian fluid):

$$\begin{aligned} &\tau_{xx} = 2\mu \frac{\partial u}{\partial x}, \quad \tau_{xy} = \mu(\frac{\partial u}{\partial y} + \frac{\partial v}{\partial x}), \quad \tau_{xz} = \mu(\frac{\partial u}{\partial z} + \frac{\partial w}{\partial x}), \\ &\tau_{yy} = 2\mu \frac{\partial v}{\partial y}, \quad \tau_{zy} = \mu(\frac{\partial w}{\partial y} + \frac{\partial v}{\partial z}), \quad \tau_{zz} = 2\mu \frac{\partial w}{\partial z} \end{aligned} \tag{8.57}$$

where μ is the dynamic viscosity. The stress gradient terms in (8.49) are

$$\frac{\partial \tau_{xx}}{\partial x} + \frac{\partial \tau_{xy}}{\partial y} + \frac{\partial \tau_{xz}}{\partial z} = 2\mu \frac{\partial^2 u}{\partial x^2} + \mu \frac{\partial^2 u}{\partial y^2} + \mu \frac{\partial^2 v}{\partial x \partial y} + \mu \frac{\partial^2 u}{\partial z^2} + \mu \frac{\partial^2 w}{\partial x \partial z} \tag{8.58}$$

By the conservation of volume derived in (8.56), we have

$$\frac{\partial^2 u}{\partial x^2} = -\frac{\partial^2 v}{\partial y \partial x} - \frac{\partial^2 w}{\partial x \partial z} \tag{8.59}$$

Using (8.59), we can convert all velocity gradient terms on the right of (8.58) to depend on u only:

$$\frac{\partial \tau_{xx}}{\partial x}+\frac{\partial \tau_{xy}}{\partial y}+\frac{\partial \tau_{xz}}{\partial z}=2\mu\frac{\partial^2 u}{\partial x^2}+\mu\frac{\partial^2 u}{\partial y^2}+\mu\frac{\partial^2 u}{\partial z^2}-\mu\frac{\partial^2 u}{\partial x^2}$$
$$=\mu(\frac{\partial^2 u}{\partial x^2}+\frac{\partial^2 u}{\partial y^2}+\frac{\partial^2 u}{\partial z^2})=\mu\nabla^2 u \tag{8.60}$$

The normal notation of the Laplacian operator is adopted in the last part of (8.60). Following a similar procedure, it is straightforward to show that the stress gradient terms on the right of (8.50) and (8.51) can also be expressed as

$$\frac{\partial \tau_{yx}}{\partial x}+\frac{\partial \tau_{yy}}{\partial y}+\frac{\partial \tau_{yz}}{\partial z}=\mu\nabla^2 v \tag{8.61}$$

$$\frac{\partial \tau_{xz}}{\partial x}+\frac{\partial \tau_{zy}}{\partial y}+\frac{\partial \tau_{zz}}{\partial z}=\mu\nabla^2 w \tag{8.62}$$

The pressure in the fluid can be expressed as a sum of a mean value plus the variation of the pressure as:

$$p=p_0(z)+\overline{p}(\boldsymbol{x},t) \tag{8.63}$$

The first term on the right of (8.63) is the hydrostatic pressure related to the mean density of the fluid

$$p_0(z)=P_0-\rho_0 gz \tag{8.64}$$

where P_0 is a constant. Substituting (8.64) into (8.63) and differentiating the result with respect to z, we obtain

$$\frac{\partial p}{\partial z}=\frac{\partial p_0}{\partial z}+\frac{\partial \overline{p}}{\partial z}=-\rho_0 g+\frac{\partial \overline{p}}{\partial z} \tag{8.65}$$

Substituting (8.60) into (8.62) and (8.65) into the momentum equations given in (8.49) to (8.51), we finally obtain

$$\frac{du}{dt}+f_* w-fv=-\frac{1}{\rho_0}\frac{\partial \overline{p}}{\partial x}+\nu\nabla^2 u \tag{8.66}$$

$$\frac{dv}{dt}+fu=-\frac{1}{\rho_0}\frac{\partial \overline{p}}{\partial y}+\nu\nabla^2 v \tag{8.67}$$

$$\frac{dw}{dt}-f_* u=-\frac{1}{\rho_0}\frac{\partial \overline{p}}{\partial z}+\nu\nabla^2 w-\frac{\overline{\rho}}{\rho_0}g \tag{8.68}$$

where the kinematic viscosity and the material time derivative are defined as

$$\nu=\frac{\mu}{\rho_0} \tag{8.69}$$

$$\frac{d}{dt}=\frac{\partial}{\partial t}+u\frac{\partial}{\partial x}+v\frac{\partial}{\partial y}+w\frac{\partial}{\partial z} \tag{8.70}$$

In view of the first part of (8.53), it seems that we can drop the last term in (8.68), however, this term accounts for the weight of the fluid and this term contributes to the hydrostatic pressure and cannot be neglected. In the next section, the energy equation will be derived and discussed.

8.3.5 Energy Equation

We now consider the fact that the internal energy of a fluid parcel obeys a balanced budget. According to the first law of thermodynamics, the change of internal energy *de* equals the heat *Q dt* that it receives minus the mechanical work *pdv* that it does:

$$\frac{de}{dt} = Q - p\frac{dv}{dt} \tag{8.71}$$

where v is the specific volume (or $1/\rho$). The internal energy can be expressed as:

$$e = C_v T \tag{8.72}$$

where C_v is the heat capacity at constant volume and T is the absolute temperature measured in Kelvin. Using the Fourier law of heat diffusion, we have

$$\rho Q = k\nabla^2 T \tag{8.73}$$

where k is the thermal conductivity. The derivative terms involved in (8.71) can be rewritten as

$$p\frac{dv}{dt} = p\frac{d}{dt}\left(\frac{1}{\rho}\right) = -\frac{p}{\rho^2}\frac{d\rho}{dt} \tag{8.74}$$

$$\frac{de}{dt} = C_V\frac{dT}{dt} \tag{8.75}$$

Combining (8.71), (8.73), (8.74) and (8.75), we find

$$\rho C_V\frac{dT}{dt} - \frac{p}{\rho}\frac{d\rho}{dt} = k\nabla^2 T \tag{8.76}$$

From (8.52), we have

$$-\frac{p}{\rho}\frac{\partial\rho}{\partial t} = p\left(\frac{\partial u}{\partial x} + \frac{\partial v}{\partial y} + \frac{\partial w}{\partial z}\right) \tag{8.77}$$

Substitution of (8.77) into (8.76) gives

$$\rho C_V\frac{dT}{dt} + p\left(\frac{\partial u}{\partial x} + \frac{\partial v}{\partial y} + \frac{\partial w}{\partial z}\right) = k\nabla^2 T \tag{8.78}$$

This is the energy equation that involves the temperature T.

8.3.6 Equation of State

For any fluid, the density is a function of temperature and pressure and this is known as the equation of state. We now have to consider air and water separately since water is nearly incompressible or independent of pressure. In addition, for seawater the density is a function of salinity while for fresh water it is not.

For dry air, we assume the ideal gas law as:

$$\rho = \frac{p}{RT} \tag{8.79}$$

where R is 287 m^2/s^2K and C_V = 718 m^2/s^2K. For water, we can assume a linear expression of state as:

$$\rho = \rho_0\{1-\alpha(T-T_0)+\beta(S-S_0)\} \tag{8.80}$$

where α is the coefficient of thermal expansion and β is the coefficient of saline contraction. For fresh water, we have β = 0. For typical seawater, we have

$$\begin{aligned} &\rho_0 = 1028kg/m^3, \quad S_0 = 35\%, \quad T_0 = 10^\circ C, \\ &\alpha = 1.7\times10^{-4}K^{-1}, \quad \beta = 7.6\times10^{-4}, \quad C_V = 4000m^2/s^2K \end{aligned} \tag{8.81}$$

For seawater, the local salinity satisfies the following diffusion law:

$$\frac{dS}{dt} = k_s\nabla^2 S \tag{8.82}$$

where k_s is the coefficient of salt diffusion.

In conclusion, for air, we have the unknowns being u, v, w, p, ρ, and T with the governing equations being (8.66), (8.67), (8.68), (8.52), (8.78) and (8.79). For seawater, we have the unknowns being u, v, w, p, ρ, T, and S with the governing equations being (8.66), (8.67), (8.68), (8.52), (8.78), (8.80) and (8.82). For fresh water, we have the same governing equations as seawater except (8.82) and $\beta = 0$ (8.80).

8.4 SYSTEM OF EQUATIONS FOR GEOPHYSICAL FLOWS

8.4.1 Consideration of Scales

In geophysical flows, it is important to recognize the length and time scales of the flows and the identification of the controlling terms in the governing equations. The scales of spatial variables, velocity fields, pressure and density are summarized in Table 8.1.

Table 8.1 Typical scales of atmospheric and oceanic flows

Variables	Scale	Atmospheric	Oceanic
x, y	L	100 km	10 km
z	H	1 km	100 m
t	T	$\geq 1/2\text{day} \approx 4\times10^4\text{s}$	$\geq 1\text{day} \approx 9\times10^4\text{s}$
u, v	U	10 m/s	0.1 m/s
w	W	$W<<U$	$W<<U$
p	P		
ρ	$\Delta\rho$	1% of ρ_0	0.1% of ρ_0

In terms of the time scale, we expect the time is larger than the following:

$$T \geq \frac{1}{\Omega} = 3.8\text{ hours} \tag{8.83}$$

in which we have employed the value of Ω given in (8.40). If a particle with velocity U covers a distance of L in a time longer than or comparable to the Earth's rotational period, we expect the trajectory of the particle to be influenced by the ambient rotation of the Earth. This leads to the following ratio:

$$\varepsilon = \frac{\text{time of one revolution}}{\text{time taken by particle to cover distance L at speed U}} = \frac{2\pi / \Omega}{L / U} = \frac{2\pi U}{\Omega L} \tag{8.84}$$

Thus, the threshold of the velocity can be evaluated as ($\varepsilon \leq 1$)

$$U \leq \frac{\Omega L}{2\pi} \tag{8.85}$$

For the atmospheric scale of geophysical flows given in Table 8.1 (L = 100 km), we have $U \leq$ 1.2 m/s for the Coriolis force to become important; for oceanic scale (L = 10 km), we have $U \leq$ 12 cm/s for the Coriolis force to become important. This provides a simple criterion for whether we can drop the terms due to Coriolis force.

Let us consider the relative scale of the momentum equations. Recall (8.66) and the relative scales of each term are listed below as:

$$\begin{gathered} \frac{\partial u}{\partial t} + u\frac{\partial u}{\partial x} + v\frac{\partial u}{\partial y} + w\frac{\partial u}{\partial z} + f_* w - fv = -\frac{1}{\rho_0}\frac{\partial p}{\partial x} + \nu(\frac{\partial^2 u}{\partial x^2} + \frac{\partial^2 u}{\partial y^2} + \frac{\partial^2 u}{\partial z^2}) \\ \frac{U}{T}, \quad \frac{U^2}{L}, \quad \frac{U^2}{L}, \quad \frac{WU}{H}, \quad \Omega W, \ \Omega U, \quad \frac{P}{\rho_0 L}, \quad \frac{\nu U}{L^2}, \quad \frac{\nu U}{L^2}, \quad \frac{\nu U}{H^2} \end{gathered} \tag{8.86}$$

For geophysical flows, the atmospheric layer that determines the weather is only about 10 km thick, but hurricanes, cyclones, and extra-cyclones (or called anticyclones) are in the scale of hundreds to thousands of kilometers; oceanic currents are confined to the upper hundreds of meters of the sea, but they span over tens of kilometers. Thus, we can assume

$$H << L \tag{8.87}$$

In view of this, the second and third terms on the right-hand side of (8.86) can be neglected. The scales of the continuity equation given in (8.56) are

$$\begin{gathered} \frac{\partial u}{\partial x} + \frac{\partial v}{\partial y} + \frac{\partial w}{\partial z} = 0 \\ \frac{U}{L}, \quad \frac{U}{L}, \quad \frac{W}{H} \end{gathered} \tag{8.88}$$

Therefore, the vertical velocity must be constrained by:

$$W \leq \frac{H}{L}U \tag{8.89}$$

In view of (8.87), (8.89) implies

$$W << U \tag{8.90}$$

It is clear now that we can drop the reciprocal Coriolis term in (8.86) except at the equator because $f \to 0$ and f^* approaches its maximum value. However, for extreme weather systems, like hurricanes, tropical cyclones, and tornadoes, the Coriolis force has to be big enough to induce spiralling wind. Thus, tropical cyclones were only observed beyond 5° north and south of the equator (or farther than 500 km north or south of the equator) (see, Lighthill, 1998). In this chapter, we restrict ourselves to the cases in which the reciprocal Coriolis term can be neglected. The scales for the momentum equation in the vertical direction are

$$\frac{\partial w}{\partial t}+u\frac{\partial w}{\partial x}+v\frac{\partial w}{\partial y}+w\frac{\partial w}{\partial z}-f_*u=-\frac{1}{\rho_0}\frac{\partial \overline{p}}{\partial z}+\nu(\frac{\partial^2 w}{\partial x^2}+\frac{\partial^2 w}{\partial y^2}+\frac{\partial^2 w}{\partial z^2})-\frac{\overline{\rho}}{\rho_0}g$$
$$\frac{W}{T},\quad \frac{UW}{L},\quad \frac{UW}{L},\quad \frac{W^2}{H},\quad \Omega U,\quad \frac{P}{\rho_0 H},\quad \frac{\nu W}{L^2},\quad \frac{\nu W}{L^2},\quad \frac{\nu W}{H^2},\quad \frac{g\Delta\rho}{\rho_0} \tag{8.91}$$

Note from (8.83) and (8.90) that

$$\frac{W}{T} << \frac{U}{T} \leq \Omega U \tag{8.92}$$

Therefore, the first term is negligible compared to the fifth term. From (8.90), we see that the second to the fourth terms are much smaller than the fifth term. By virtue of (8.87), the second and third terms on the right of (8.91) are much smaller than the fourth term on the right. Comparing the last term on the left- and right-hand sides of (8.86), we see that

$$\frac{\nu U}{H^2} \approx \Omega U \tag{8.93}$$

Thus, we have

$$\frac{\nu W}{H^2} \approx \Omega W << \Omega U \tag{8.94}$$

Therefore, the fourth term on the right is again negligible compared to the last term on the left of (8.91). Finally, the ratio of the last terms on the left to the last term on the right of (8.91) is

$$\frac{\rho_0 \Omega U}{g\Delta\rho} \tag{8.95}$$

which, according to the values given in Table 8.1, ranges from 10^{-2} (for atmosphere) to 10^{-3} (for ocean). Therefore, all terms on the left can be neglected. In short, (8.91) is reducible to

$$0=-\frac{1}{\rho_0}\frac{\partial \overline{p}}{\partial z}-\frac{\overline{\rho}}{\rho_0}g \tag{8.96}$$

Therefore, for geophysical flows, the pressure variation is also hydrostatic in terms of the density variation, and, thus, even in the presence of substantial motions, the pressure is fully hydrostatic.

For seawater, we can combine (8.56) and (8.78) to give

$$\frac{dT}{dt}=\frac{k}{\rho C_V}\nabla^2 T = k_T \nabla^2 T \tag{8.97}$$

We are going to show that the heat diffusion given in (8.97) can be combined with the salinity diffusion given (8.82) to convert the equation of state to a diffusion equation for density. In particular, we can rewrite the equation of state given in (8.80)

$$\overline{\rho}=\rho-\rho_0=-\alpha\rho_0(T-T_0)+\beta\rho_0(S-S_0) \tag{8.98}$$

Taking differentiation of (8.98) with respect to time and taking Laplacian of (8.98) gives

$$\frac{d\overline{\rho}}{dt}=-\alpha\rho_0\frac{dT}{dt}+\beta\rho_0\frac{dS}{dt} \tag{8.99}$$

$$\nabla^2 \bar{\rho} = -\alpha \rho_0 \nabla^2 T + \beta \rho_0 \nabla^2 S \tag{8.100}$$

Substitution of (8.97) and (8.82) into (8.99) gives

$$\begin{aligned} \frac{d\bar{\rho}}{dt} &= -\alpha \rho_0 k_T \nabla^2 T + \beta \rho_0 k_s \nabla^2 S \\ &= k(-\alpha \rho_0 \nabla^2 T + \beta \rho_0 \nabla^2 S) \\ &- k \nabla^2 \bar{\rho} \end{aligned} \tag{8.101}$$

where

$$k_T = k_s = k \tag{8.102}$$

In obtaining the last part of (8.101), we have employed (8.100). For small-scale diffusion at the molecular level, we cannot assume the validity of (8.102). For large-scale diffusion, both temperature and salinity diffusion are regulated by turbulence. Thus, k can be interpreted as eddy diffusivity. We will justify this in a later section. The diffusion for density can further be simplified in view of (8.87) and by scaling consideration as

$$\frac{d\bar{\rho}}{dt} = k\left(\frac{\partial^2 \bar{\rho}}{\partial x^2} + \frac{\partial^2 \bar{\rho}}{\partial y^2} + \frac{\partial^2 \bar{\rho}}{\partial z^2}\right) \approx k \frac{\partial^2 \bar{\rho}}{\partial z^2} \tag{8.103}$$

It has been argued that (8.103) is also applicable to the case of air (Cushman-Roisin, 1994).

8.4.2 Governing Equations

In summary, the governing equations after the scaling consideration are simplified to

$$\frac{\partial u}{\partial t} + u\frac{\partial u}{\partial x} + v\frac{\partial u}{\partial y} + w\frac{\partial u}{\partial z} - fv = -\frac{1}{\rho_0}\frac{\partial \bar{p}}{\partial x} + \nu \frac{\partial^2 u}{\partial z^2} \tag{8.104}$$

$$\frac{\partial v}{\partial t} + u\frac{\partial v}{\partial x} + v\frac{\partial v}{\partial y} + w\frac{\partial v}{\partial z} + fu = -\frac{1}{\rho_0}\frac{\partial \bar{p}}{\partial y} + \nu \frac{\partial^2 v}{\partial z^2} \tag{8.105}$$

$$0 = -\frac{1}{\rho_0}\frac{\partial \bar{p}}{\partial z} - \frac{\bar{\rho}}{\rho_0} g \tag{8.106}$$

$$\frac{\partial u}{\partial x} + \frac{\partial v}{\partial y} + \frac{\partial w}{\partial z} = 0 \tag{8.107}$$

$$\frac{\partial \bar{\rho}}{\partial t} + u\frac{\partial \bar{\rho}}{\partial x} + v\frac{\partial \bar{\rho}}{\partial y} + w\frac{\partial \bar{\rho}}{\partial z} = k\frac{\partial^2 \bar{\rho}}{\partial z^2} \tag{8.108}$$

This provides a system of five equations for five unknowns, u, v, w, p, and $\bar{\rho}$.

8.4.3 Rossby, Ekman and Reynolds Numbers

The scaling of both (8.104) and (8.105) leads to the following sequence

$$\frac{U}{T},\quad \frac{U^2}{L},\quad \frac{U^2}{L},\quad \frac{WU}{H},\quad \Omega U,\quad \frac{P}{\rho_0 L},\quad \frac{\nu U}{H^2} \tag{8.109}$$

Normalizing this sequence by the fifth term, we get

$$\frac{1}{\Omega T},\quad \frac{U}{\Omega L},\quad \frac{U}{\Omega L},\quad \frac{W}{\Omega H},\quad 1,\quad \frac{P}{\Omega U \rho_0 L},\quad \frac{\nu}{\Omega H^2} \tag{8.110}$$

These ratios give the relative magnitude of various kinds of force. The first one is called the temporal Rossby number:

$$Ro_T = \frac{1}{\Omega T} \tag{8.111}$$

This is a ratio of the time rate of change of velocity to the Coriolis force. Equation (8.83) suggests that Ro_T is of the order of unity or less. The second ratio is the Rossby number

$$Ro = \frac{U}{\Omega L} \tag{8.112}$$

which is the ratio of the advection to the Coriolis force. According to (8.85), Ro is again of unity or less. The values of the Rossby numbers generally control the characteristics of geophysical flows. The next ratio in (8.110) can be written as

$$\frac{W}{\Omega H} = \frac{U}{\Omega L}\frac{WL}{UH} = Ro\frac{WL}{UH} \tag{8.113}$$

By virtue of (8.89), this ratio is of one order or less than the Rossby number. The last number is the Ekman number defined as

$$Ek = \frac{\nu}{\Omega H^2} \tag{8.114}$$

This number measures the relative importance of viscous force compared to the Coriolis force. For geophysical flows, the Ekman number is extremely small. To examine the effect of the boundary layer, this stress must be retained. The Ekman layer will be considered in Section 8.6.

In nonrotating fluid flows, the ratio of inertial and viscous forces is called the Reynolds number and it can be expressed in terms of the Rossby number and the Ekman number:

$$\mathrm{Re} = \frac{UL}{\nu} = \frac{U}{\Omega L}\frac{\Omega H^2}{\nu}\frac{L^2}{H^2} = \frac{Ro}{Ek}\left(\frac{L}{H}\right)^2 \tag{8.115}$$

Since the Rossby number is in the order of unity whereas the Ekman number and geometric ratio H/L are very small, the Reynolds number is extremely large for geophysical flows. Thus, the flows are turbulent and this is the reason why we can replace the molecular diffusion by eddy diffusion. This justifies the validity of (8.102) in Section 8.4.1.

The final number is the ratio between the pressure force and the Coriolis force. Since the Coriolis force is an important component in geophysical flows, and the force that arose from the pressure variation must somehow be in the same order of the Coriolis force. This provides a scale for the dynamic pressure (an additional part from the hydrostatic pressure) as:

$$P = \rho_0 \Omega L U \tag{8.116}$$

This dynamic pressure is normally very small compared to the hydrostatic pressure in the context of geophysical flows, and will be neglected.

8.5 STORM SURGES

8.5.1 Storm Surges by Inverse Barometer Effect

Tropical cyclones and hurricanes are devastating not only in terms of strong wind but also in terms of the storm surges associated with them. The highest storm surge ever recorded was 14.6 m at Bathurst Bay in Australia during the 1899 Cyclone Mahina. Fish and dolphins were reported found on top of 15-m cliffs. Other recent examples include the 2005 hurricane Katrina and its storm surge (leading to 1836 of deaths in New Orleans), the 2014 tropical cyclone Haiyin and its storm surge (leading to six thousand deaths in the Philippines), and the more recent 2017 Hato and its storm surge (leading to 22 deaths in Macau and mainland China). In fact, storm surge is more deadly than the strong wind. Historical records show that exceptionally high storm surges are localized, probably within tens of kilometers, and drop off quickly away from the point of landfall. For example, during the 2017 Hato storm surge, the highest run-up in Macau is significantly larger than that of Hong Kong, despite of their close proximity. The location of the highest storm surge very often does not coincide with the actual location of the tide gauge, and thus it is very unlikely being recorded. Even in the event that the location of a tide gauge happens to coincide with the highest surge, very often it is damaged by the surge. The surges may further be amplified by the resonance of the seas and basins themselves, coupled with high tides, and interaction with river runout or ocean currents. Small variations in the strength, path, and angle of attack of tropical cyclones also influence the surge.

Suppose that the sea reaches equilibrium subject to a change in the atmospheric pressure such that there is no current. Then, (8.104) and (8.105) can be simplified to

$$\frac{\partial p}{\partial y} = 0, \quad \frac{\partial p}{\partial x} = 0 \tag{8.117}$$

where p is the total pressure because there is no current. Combining (8.63) and (8.64), we have

$$p = p_0 - \rho_0 g z + \overline{p} \tag{8.118}$$

where p_0 is the atmospheric pressure. Assuming that the sea level variation ζ is associated with the pressure variation, we get

$$\overline{p} = \rho_0 g \zeta \tag{8.119}$$

Substituting (8.119) into (8.118) gives

$$p = p_0 - \rho_0 g(z - \zeta) \tag{8.120}$$

where p_a is the change in the atmospheric pressure. Substituting (8.120) into the first and second parts of (8.117) gives

$$\frac{\partial p}{\partial x} = \frac{\partial p_0}{\partial x} + \rho_0 g \frac{\partial \zeta}{\partial x} = 0 \tag{8.121}$$

$$\frac{\partial p}{\partial y}=\frac{\partial p_0}{\partial y}+\rho_0 g\frac{\partial \zeta}{\partial y}=0 \tag{8.122}$$

This suggests that

$$p_0+\rho_0 g\zeta = const. \tag{8.123}$$

For a given pressure variation, the variation in sea level can be evaluated as

$$\Delta\zeta=-\frac{\Delta p_0}{\rho_0 g} \tag{8.124}$$

For seawater, we can set ρ_0 = 1026 kg/m^3 and g = 9.81m/s^2. Using these values, we can rewrite (8.124) as

$$\Delta\zeta=-0.000099\Delta p_0 \tag{8.125}$$

This formula is known as the inverted or inverse barometer effect. That is, the rise in sea level is due to pressure change alone. For a Category 5 hurricane according to the Saffir-Simpson scale, the central pressure may drop to as low as 885 mb recorded during the 2005 Hurricane Wilma from a maximum of 1013 mb (i.e., a potential drop of pressure of 128 mb = 12800 Pa), where mb is milli-bar or 0.1 kPa. Substitution of this extreme value into (8.125) gives a surge due to pressure variation as 1.2672 m, and this value is much smaller than the historical record of 14.6 m that was observed at Bathurst Bay in Australia. For the 2017 Hato, pressure recorded at a remote island along the passage of the tropical cyclone in the south of the Pearl River Delta suggested that the central pressure was 950 mb (a super typhoon or a Category 3 hurricane). For tropical cyclone Hato, the pressure drop was about 6383 Pa or a sea level rise of 0.6320 m, which is again much smaller than the storm surge observed at Macau.

In fact, this inverse barometer effect only gives a static amplification for the surge. It typically accounts for only 10% to 15% of the final surges. This pressure surge is illustrated in Fig. 8.8. The dynamic effect will be examined next.

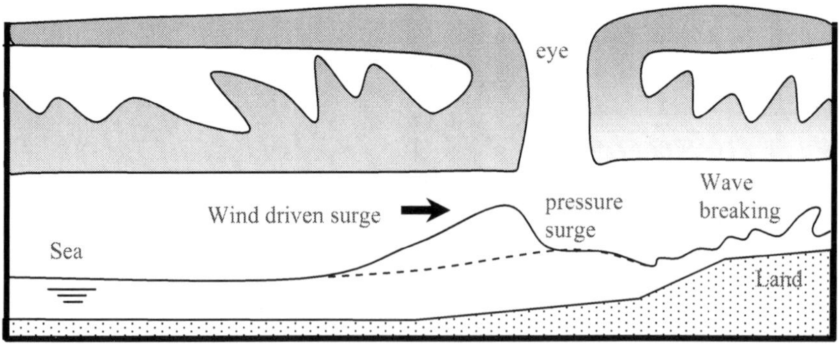

Figure 8.8 Storm surge due to pressure, wind, and wave shoaling

8.5.2 Storm Surges with Moving Disturbance

A simple but elegant analytical solution was derived by Proudman in 1953 for the water level due to a moving disturbance traveling in the x-direction. In particular, by omitting the advective force, Coriolis force, and viscous force, we can reduce the momentum equation given in (8.104) to

$$\frac{\partial u}{\partial t}=-\frac{1}{\rho_0}\frac{\partial p}{\partial x} \tag{8.126}$$

In obtaining (8.126), we have replaced the pressure variation by total pressure because of zero current. Differentiating (8.120) with respect to x, we get

$$\frac{\partial p}{\partial x}=\frac{\partial p_0}{\partial x}+\rho_0 g\frac{\partial \zeta}{\partial x} \tag{8.127}$$

Substitution of (8.127) into (8.126) gives

$$\frac{\partial u}{\partial t}=-\frac{1}{\rho_0}\frac{\partial p_0}{\partial x}-g\frac{\partial \zeta}{\partial x} \tag{8.128}$$

Assuming a given depression of pressure is moving along the x-axis with velocity c_a, we have

$$p_0=A(x-c_a t) \tag{8.129}$$

where A is an arbitrary function and its argument is in the form of characteristics. We are seeking a surge ζ moving with the same velocity but with an unknown amplitude:

$$\zeta=\alpha A(x-c_a t) \tag{8.130}$$

where α is an amplification factor to be determined. For a one-dimensional wave along the x-direction, (8.42) can be simplified as

$$\frac{\partial \zeta}{\partial t}+h(\frac{\partial u}{\partial x})=0 \tag{8.131}$$

Substitution of (8.130) into (8.131) yields

$$\frac{\partial u}{\partial x}=\frac{\alpha c_a}{h}A'(x-c_a t) \tag{8.132}$$

Integrating with respect to x, we obtain

$$u=\frac{\alpha c_a}{h}A(x-c_a t)+C \tag{8.133}$$

Since tropical cyclone-induced surge is a local disturbance, we must have $u\to 0$ as $x\to\infty$. This yields $C=0$. Differentiating (8.130) with respect to x, we get

$$\frac{\partial \zeta}{\partial x}=\alpha A'(x-c_a t) \tag{8.134}$$

Differentiating (8.129) with respect to x, we have

$$\frac{\partial p_0}{\partial x}=A'(x-c_a t) \tag{8.135}$$

Differentiating (8.133) with respect to t, we obtain

$$\frac{\partial u}{\partial t}=-\frac{\alpha c_a^2}{h}A'(x-c_a t) \tag{8.136}$$

Substitution of (8.134) to (8.136) into (8.128) yields

$$-\frac{\alpha c_a^2}{h} A'(x-c_a t) = -\frac{1}{\rho_0} A'(x-c_a t) - g\alpha A'(x-c_a t) \tag{8.137}$$

Solving for α, we finally get the magnitude of the surge due to a moving pressure depression

$$\alpha = -\frac{1}{\rho_0 g[1-c_a^2/(gh)]} \tag{8.138}$$

The corresponding dynamic surge ζ_D can be expressed in terms of the static surge ζ_S as

$$\zeta_D = -\frac{p_a}{\rho g[1-c_a^2/(gh)]} = -\frac{\zeta_s}{[1-c_a^2/(gh)]} = -F\zeta_s \tag{8.139}$$

where the dynamic surge factor is denoted by F. If $C_a = 0$, we recover the static surge obtained in (8.124) (with a slightly different notation). This is the classic result obtained by Proudman. We see that $F \to \infty$ as C_a approaches $(gh)^{1/2}$.

Example 8.1 Consider a tropical cyclone with a pressure depression moving at 20 km/h. (i) Find the dynamic surge amplification at a depth of 50 m and 25 m. (ii) Find the critical depth at which the dynamic surge amplification becomes infinite.

Solution: First of all, we can convert the propagating speed to m/s as

$$c_a = \frac{20\times 10^3}{60\times 60} = 5.56 m/s \tag{8.140}$$

At $h = 50$ m, we have

$$F = \frac{1}{[1-c_a^2/(gh)]} = \frac{1}{[1-5.56^2/(9.81\times 50)]} = 1.067 \tag{8.141}$$

At $h = 25$ m, we have

$$F = \frac{1}{[1-c_a^2/(gh)]} = \frac{1}{[1-5.56^2/(9.81\times 25)]} = 1.144 \tag{8.142}$$

For $F \to \infty$, we have the critical depth as

$$h_{crit} = c_a^2/g = 5.56^2/9.81 = 3.15m \tag{8.143}$$

Thus, for a steadily moving tropical cyclone, the surge increases drastically only at a very shallow water depth. Once F becomes sufficiently large, nonlinear effects set in and the analytical solution by Proudman, of course, breaks down.

Nevertheless, we see that the dynamic amplification cannot explain the big difference between the prediction and observed surge water level. Various kinds of empirical formulas have been proposed in predicting storm surge. These formulas share a common form that:

$$\zeta = \frac{W^2 L}{d_1} f\left(\frac{d_1}{d_2}\right) \tag{8.144}$$

where L is the wind fetch (length of the wind blowing on the sea surface), W is the maximum wind speed in the air, d_1 is the depth at the edge of the shelf, and d_2 is the depth of water near the coast. To see the physical basis of this kind of empirical, wind-induced flow will be considered in the next section. However, for more accurate estimation of the storm surge, numerical analysis of the system given in (8.104) to (8.108) has to be conducted. Typically, the system will be simplified to a two-dimensional system for average depth velocity fields.

8.5.3 Wind-Induced Storm Surges

So far, we have not incorporated the effect of wind-driven surges, which will be considered in the present section. The stress induced on the water surface by the blowing wind in air can be approximated by a power law

$$\tau_s = C_D \rho_a^m W^n \tag{8.145}$$

where C_D is the dimensionless wind drag coefficient, W is the maximum wind speed, and ρ_a is the density of air. The indices m and n can be found by dimensional analysis. The unit of the stress on the left-hand side of (8.145) is

$$[\tau_s] = ML^{-1}T^{-2} \tag{8.146}$$

where the square bracket represents the unit of the physical parameter inside it. The units of mass, length and time are represented symbolically as M, L, and T (for SI units they are kilogram, meter and second). On the other hand, the unit of the right-hand side of (8.145) must be

$$[C_D \rho_a^m W^n] = (ML^{-3})^m (LT^{-1})^n \tag{8.147}$$

By dimensional requirement, the unit on both sides must balance and this leads to

$$m = 1, \quad -1 = -3m + n, \quad -2 = -n \tag{8.148}$$

Thus, by dimensional analysis we must have the wind-induced stress expressed in the following form:

$$\tau_s = C_D \rho_a W^2 \tag{8.149}$$

This formula applies regardless of the nature of the drag force. For example, the same formula applies to estimate drag force applied on a solid by a fluid, such as wind-induced drag on a cable suspension bridge deck. For the case of wind-induced surge on water, experimental observation suggests that the wind drag coefficient C_D is a linear function of the wind speed:

$$C_D = 10^{-3}(0.63 + 0.066 W_{10}), \quad 2.5 m/s < W_{10} < 21 m/s \tag{8.150}$$

where W_{10} is the maximum wind speed measured in m/s at a height of 10 m above the sea level. This formula was proposed by Smith and Banke in 1975.

For steady wind-induced traction, we can neglect the inertia force, advective force, and Coriolis force, and, consequently, (8.48) can be simplified as:

$$\frac{\partial p}{\partial x} = \frac{\partial \tau_{xz}}{\partial z} = \frac{\partial \tau_s}{\partial z} \tag{8.151}$$

Integrating (8.151) over the depth h of the sea, we have

$$\frac{1}{h}\int_{-h}^{0} \frac{\partial \tau_s}{\partial z} dz = \frac{1}{h}(\tau_S - \tau_B) \tag{8.152}$$

where the surface and bottom stresses are denoted by subscript S and B. The right-hand side of (8.152) can be interpreted as the depth average stress. Differentiation of (8.120) and substituting the result into (8.151) gives

$$\frac{\partial p}{\partial x}=\frac{\partial p_0}{\partial x}+\rho_0 g\frac{\partial \zeta}{\partial x}=\frac{\partial \tau_s}{\partial z} \tag{8.153}$$

Assuming p_0 is not a function of x and using the result of (8.149), we can simplify (8.153) as

$$\frac{\partial \zeta}{\partial x}=\frac{1}{\rho_0 g}\frac{\partial \tau_s}{\partial z}=\frac{C_D\rho_a W^2}{\rho_0 g h} \tag{8.154}$$

For the case of a uniform wind of fetch L blowing on a sea with uniform depth h, we have

$$\zeta_W=\frac{C_D\rho_a W^2 L}{\rho_0 g h} \tag{8.155}$$

This formula agrees with the structural form given in the empirical formula of (8.144), and thus provides a theoretical base for (8.144).

A simple way to estimate the sea level induced by storm surge is to sum (8.139) and (8.155) as

$$\zeta=\frac{p_a}{\rho g[1-c_a^2/(gh)]}+\frac{C_D\rho_a W^2 L}{\rho_0 g h} \tag{8.156}$$

In the following example, we will demonstrate the estimation of storm surge in Hong Kong during the 2017 tropical cyclone Hato.

Example 8.2 Consider the storm surge induced by the super typhoon Hato at Hong Kong which made a landfall on August 23, 2017 in Zhuhai City next to Macau. The maximum wind speed measured is up to 67 m/s or 240 km/h (gust). Hato moved at about 20 km/h with a ten-minute wind speed average up to 175 km/h and a central depression of pressure of 950 mb. Assume that the wind fetch is about the width of the continental shelf in the South China Sea or about 172.5 km. The average wind speed in the continental shelf of 10 m can be taken as 110 km/h or 30 m/s. The average depth of the continental shelf is 60 m whereas the average water depth in the Victoria Harbour is about 10 m. Density of air can be taken as 1.2 kg/m^3.

Solution: First of all, we can convert the propagating speed of Hato to m/s as

$$c_a=\frac{20\times 10^3}{60\times 60}=5.56m/s \tag{8.157}$$

In the Hong Kong water of about $h = 10$ m, we have

$$F=\frac{1}{[1-c_a^2/(gh)]}=\frac{1}{[1-5.56^2/(9.81\times 10)]}=1.4601 \tag{8.158}$$

The inverse barometer effect induces

$$\zeta_s=\frac{6383}{(1026)(9.81)}=0.634m \tag{8.159}$$

For wind driven surge, the drag coefficient can be estimated as

$$C_D = 10^{-3}(0.63 + 0.066 \times 30) = 2.61 \times 10^{-3} \quad (8.160)$$

Note that we have assumed the validity of (8.150) for W_{10} = 30 m/s. The wind driven surge is

$$\zeta_W = \frac{C_D \rho_a W^2 L}{\rho_0 g h} = \frac{(2.61\times10^{-3})(1.2)(30)^2(172.5\times10^3)}{(1026)(9.81)(60)} = 0.8049m \quad (8.161)$$

Finally, the storm surge due to inverse barometer effective, moving depression and wind is

$$\zeta = 0.634 + 1.4601 \times 0.634 + 0.8049 = 2.36m \quad (8.162)$$

According to the tide gauge records, the storm surge at Tsim Bei Tsui was 2.48 m (4.56 m including high tide), at Tai Po Kau was 1.65 m (4.09 m including tide) and at Quarry Bay was 1.2 m (3.5 m including tide). Of course, we should not expect the estimation given in (8.162) to be very accurate because we do not include the geometric effect of the coastline and the detailed bathymetry near the tide gauge. Nevertheless, (8.162) at least provides the correct order of magnitude of surge in terms of water level.

8.5.4 Current Profile

We now examine how the wind-induced current decreases with depth. Using dimensional analysis, we assume that the rate of speed change with depth can be expressed as

$$\frac{\partial u}{\partial z} = (\frac{1}{\kappa})\tau_s^n \rho^m z^l \quad (8.163)$$

where κ is called the von Karman constant. Balancing the dimension on both sides of (8.163), we have

$$[LT^{-1}/L] = [ML^{-1}T^{-2}]^n [ML^{-3}]^m [L]^l \quad (8.164)$$

This leads to

$$n = \frac{1}{2}, \quad m = -\frac{1}{2}, \quad l = -1 \quad (8.165)$$

The solution of (8.165) gives

$$\frac{\partial u}{\partial z} = \frac{1}{\kappa z}(\frac{\tau_s}{\rho})^{1/2} \quad (8.166)$$

The rate of speed decrease becomes

$$\frac{\partial u}{\partial z} = \frac{1}{\kappa z}(\frac{\tau_s}{\rho})^{1/2} = \frac{u_*}{\kappa z} \quad (8.167)$$

where u^* can be interpreted as frictional velocity, which is proportional to the wind speed W. In obtaining (8.167), we have employed (8.149). We cannot integrate this shear flow profile from the water surface (z = 0), where (8.167) breaks down. Instead, the concept of a roughness length z_0 has been proposed and this corresponds to a current speed of u_0. Integrating (8.167) and applying the boundary condition at a depth of roughness length, we find

$$u = u_0 - \frac{1}{\kappa}(\frac{\tau_s}{\rho})^{1/2} \ln(\frac{z}{z_0}) \tag{8.168}$$

The following values have been proposed by Pugh (1987): (κ = 0.4, u^* = $0.0012W_{10}$, $u_0 = 0.03W_{10}$ and z_0 = 0.0015 m. For the case of W_{10} = 30 m/s (see Example 8.2), we find that the surface current is in the order of 1 m and the wind-driven current drops to 0.2 m/s at about 4-m depth.

8.6 EKMAN TRANSPORT

8.6.1 Ekman Transport with No Internal Currents

The Norwegian scientist and explorer Nansen discovered in 1893-1896 from his ship *Fram* that polar-ice drifts to the right of the direction of the surface wind, instead of drifting along the wind direction. This phenomenon was explained by Vagn Walfrid Ekman in 1902 in his PhD thesis using hydrodynamics. This is called Ekman drift. Suppose that the depth is constant, sea bottom friction is negligible, density is constant, the wind blowing on the surface is steady, and there is no horizontal pressure gradient. Equations (8.49) and (8.50) can be recast as

$$-fv = -\frac{1}{\rho}\frac{\partial F}{\partial z} \tag{8.169}$$

$$fu = -\frac{1}{\rho}\frac{\partial G}{\partial z} \tag{8.170}$$

where F and G are the applied tractions. In arriving at (8.169) and (8.170), we have assumed that the flows are generally slow such that time derivative can be ignored, and the slow velocity over long distance allows neglecting the nonlinear advection. Integrating (8.169) and (8.170) from the bottom to the surface and assuming the wind blows along the x-direction, we find

$$-f\overline{V} = -\frac{1}{\rho}F_s \tag{8.171}$$

$$f\overline{U} = 0 \tag{8.172}$$

where U and $\overline{V}$ are volume transports per meter along the x- and y-directions, and F_s is the surface wind stress. Noting that f is a function of the latitude, we have

$$\overline{V} = -\frac{F_s}{\rho f} = -\frac{F_s}{2\rho\Omega\sin\phi}, \quad \overline{U} = 0 \tag{8.173}$$

This is called the Ekman volume transport per meter per section (in the unit of m^2/s). We see that when the wind is blowing in the x-direction, and the fluid is flowing in the negative y-direction. That is, the fluid is flowing to the right of the wind direction. This agrees with the observation made from the *Fram*. The wind-induced Ekman layer suggested that the transport is at 90° to the right of the wind direction. At the equator, we have $\phi = 0$, and, thus, Ekman's theory breaks down.

8.6.2 Ekman Transport with Internal Currents

In this section, we consider the case that a steady interior flow field in deep water is subjected to steady surface wind-driven stress (see Figure 8.9). Equations (8.104) and (8.105) can be written as

$$-f(v-\overline{v}) = \nu \frac{\partial^2 u}{\partial z^2} \tag{8.174}$$

$$f(u-\overline{u}) = \nu \frac{\partial^2 v}{\partial z^2} \tag{8.175}$$

The boundary conditions on the surface are

$$\rho_0 \nu \frac{\partial u}{\partial z} = \tau_x, \quad \rho_0 \nu \frac{\partial v}{\partial z} = \tau_y \tag{8.176}$$

The far field boundary condition at the bottom of the deep sea is

$$u = \overline{u}, \quad v = \overline{v}, \quad z \to -\infty \tag{8.177}$$

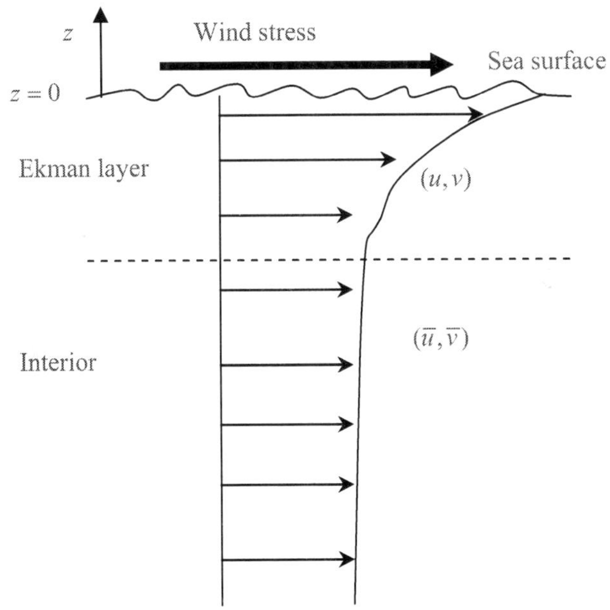

Figure 8.9 Ekman layer induced by wind stress

Let us look for exponential solutions as:

$$u-\overline{u} = A\exp(\lambda z), \quad v-\overline{v} = B\exp(\lambda z) \tag{8.178}$$

Substitution of (8.178) into (8.174) and (8.175) gives

$$\nu\lambda^2 A + fB = 0 \tag{8.179}$$

$$fA - \nu\lambda^2 B = 0 \tag{8.180}$$

These equations can be put in matrix form as

$$\begin{bmatrix} \nu\lambda^2 & f \\ f & -\nu\lambda^2 \end{bmatrix} \begin{Bmatrix} A \\ B \end{Bmatrix} = \begin{Bmatrix} 0 \\ 0 \end{Bmatrix} \tag{8.181}$$

Since this is a homogeneous system, for a nontrivial solution to exist, we require the determinant for the coefficient matrix to zero

$$\nu^2\lambda^4 + f^2 = 0 \tag{8.182}$$

The roots for λ^2 are

$$\lambda^2 = \pm\sqrt{-\frac{f^2}{\nu^2}} = \pm i\frac{f}{\nu} \tag{8.183}$$

Taking the square root again, we have

$$\lambda = \pm\sqrt{\pm i}\sqrt{\frac{f}{\nu}} \tag{8.184}$$

By virtue of Euler's formula, the square root of i can be determined as

$$\sqrt{\pm i} = e^{\pm i\pi/4} = \cos\frac{\pi}{4} \pm i\sin\frac{\pi}{4} = \frac{1}{\sqrt{2}}(1 \pm i) \tag{8.185}$$

Thus, the roots for λ are

$$\lambda = \pm(1 \pm i)\frac{1}{d} \tag{8.186}$$

where d is the Ekman depth

$$d = \sqrt{\frac{2\nu}{f}} \tag{8.187}$$

Therefore, the Ekman depth is a function of latitude. Near the equator, the Ekman depth is thicker than that in the higher attitude. It can be considered as a boundary layer near the sea surface. The solution becomes

$$u = \bar{u} + e^{\frac{z}{d}}\{A_1\cos\frac{z}{d} + B_1\sin\frac{z}{d}\} + e^{-\frac{z}{d}}\{C_1\cos\frac{z}{d} + D_1\sin\frac{z}{d}\} \tag{8.188}$$

$$v = \bar{v} + e^{\frac{z}{d}}\{A_2\cos\frac{z}{d} + B_2\sin\frac{z}{d}\} + e^{-\frac{z}{d}}\{C_2\cos\frac{z}{d} + D_2\sin\frac{z}{d}\} \tag{8.189}$$

The boundary condition given in (8.177) requires the second terms in both (8.188) and (8.189) to vanish, resulting in

$$u = \bar{u} + e^{\frac{z}{d}}\{A_1\cos\frac{z}{d} + B_1\sin\frac{z}{d}\} \tag{8.190}$$

$$v = \bar{v} + e^{\frac{z}{d}}\{A_2\cos\frac{z}{d} + B_2\sin\frac{z}{d}\} \tag{8.191}$$

Differentiation of (8.190) and (8.191) gives

$$\frac{\partial u}{\partial z} = \frac{1}{d} e^{\frac{z}{d}} (A_1 \cos\frac{z}{d} + B_1 \sin\frac{z}{d}) + \frac{1}{d} e^{\frac{z}{d}} (-A_1 \sin\frac{z}{d} + B_1 \cos\frac{z}{d}) \qquad (8.192)$$

$$\frac{\partial^2 u}{\partial z^2} = \frac{1}{d^2} e^{\frac{z}{d}} (2B_1 \cos\frac{z}{d} - 2A_1 \sin\frac{z}{d}) \qquad (8.193)$$

$$\frac{\partial v}{\partial z} = \frac{1}{d} e^{\frac{z}{d}} (A_2 \cos\frac{z}{d} + B_2 \sin\frac{z}{d}) + \frac{1}{d} e^{\frac{z}{d}} (-A_2 \sin\frac{z}{d} + B_2 \cos\frac{z}{d}) \qquad (8.194)$$

$$\frac{\partial^2 v}{\partial z^2} = \frac{1}{d^2} e^{\frac{z}{d}} (2B_2 \cos\frac{z}{d} - 2A_2 \sin\frac{z}{d}) \qquad (8.195)$$

Applying the boundary condition given in (8.176) on the water surface, we get

$$\frac{\rho_0 \nu}{d}\{A_1 + B_1\} = \tau_x \qquad (8.196)$$

$$\frac{\rho_0 \nu}{d}\{A_2 + B_2\} = \tau_y \qquad (8.197)$$

There are four unknown constants in (8.196) and (8.197), and they appear to be unsolvable. However, substitution of (8.194) and (8.191) into (8.174) gives

$$-\frac{fd^2}{\nu}\{A_2 \cos\frac{z}{d} + B_2 \sin\frac{z}{d}\} = 2B_1 \cos\frac{z}{d} - 2A_1 \sin\frac{z}{d} \qquad (8.198)$$

In view of (8.187), (8.198) provides two additional relations for the unknown constants

$$A_2 = -B_1, \quad B_2 = A_1 \qquad (8.199)$$

If we substitute (8.195) and (8.190) into (8.175), we will arrive at the same conditions given in (8.199). Therefore, we have

$$\frac{\rho_0 \nu}{d}\{A_1 + B_1\} = \tau_x \qquad (8.200)$$

$$\frac{\rho_0 \nu}{d}\{-B_1 + A_1\} = \tau_y \qquad (8.201)$$

The solutions of (8.200) and (8.201) for A_1 and B_1 are

$$A_1 = \frac{d}{2\rho_0 \nu}(\tau_x + \tau_y) \qquad (8.202)$$

$$B_1 = \frac{d}{2\rho_0 \nu}(\tau_x - \tau_y) \qquad (8.203)$$

Substitution of (8.199), (8.202) and (8.203) into (8.190) and (8.191) gives the final solutions as

$$u = \bar{u} + e^{\frac{z}{d}} \frac{d}{2\rho_0 \nu}\{(\tau_x + \tau_y)\cos\frac{z}{d} + (\tau_x - \tau_y)\sin\frac{z}{d}\} \qquad (8.204)$$

$$v = \bar{v} + e^{\frac{z}{d}} \frac{d}{2\rho_0 \nu}\{-(\tau_x - \tau_y)\cos\frac{z}{d} + (\tau_x + \tau_y)\sin\frac{z}{d}\} \qquad (8.205)$$

These solutions can be rewritten slightly in view of the following identities:

$$\cos\frac{z}{d} + \sin\frac{z}{d} = \sqrt{2}\cos(\frac{z}{d} - \frac{\pi}{4}) \tag{8.206}$$

$$\cos\frac{z}{d} - \sin\frac{z}{d} = -\sqrt{2}\sin(\frac{z}{d} - \frac{\pi}{4}) \tag{8.207}$$

and (8.204) and (8.205) are reduced to

$$u = \bar{u} + \frac{\sqrt{2}}{\rho_0 fd}\{\tau_x \cos(\frac{z}{d} - \frac{\pi}{4}) - \tau_y \sin(\frac{z}{d} - \frac{\pi}{4})\}e^{\frac{z}{d}} \tag{8.208}$$

$$v = \bar{v} + \frac{\sqrt{2}}{\rho_0 fd}\{\tau_x \sin(\frac{z}{d} - \frac{\pi}{4}) + \tau_y \cos(\frac{z}{d} - \frac{\pi}{4})\}e^{\frac{z}{d}} \tag{8.209}$$

On the surface z = 0, we see from (8.208) and (8.209) that the surface wind-driven current on the top of the Ekman layer is offset by 45° to the right of the wind direction. The wind-driven horizontal transport in the Ekman layer close to the surface can be evaluated as:

$$U = \int_{-\infty}^{0} (u - \bar{u})dz = \frac{\tau_y}{\rho_0 f} \tag{8.210}$$

$$V = \int_{-\infty}^{0} (v - \bar{v})dz = -\frac{\tau_x}{\rho_0 f} \tag{8.211}$$

In obtaining these results we have used the following integrals (see p. 85 of Spiegel, 1968)

$$\int_{-\infty}^{0} e^{z/d}\sin\frac{z}{d}dz = \frac{d}{z}e^{z/d}(\sin\frac{z}{d} - \cos\frac{z}{d})\Big|_{-\infty}^{0} = -\frac{d}{2} \tag{8.212}$$

$$\int_{-\infty}^{0} e^{z/d}\cos\frac{z}{d}dz = \frac{d}{z}e^{z/d}(\cos\frac{z}{d} + \sin\frac{z}{d})\Big|_{-\infty}^{0} = \frac{d}{2} \tag{8.213}$$

For the case of wind blowing along the x-axis only, we have only nonzero V, which is perpendicular to the wind direction. The negative sign for V in (8.211) indicates that the flow is to the right of the wind direction. This is consistent with the result obtained in the previous section without internal current and with the observation of iceberg movements by Nansen in 1893-1896. If the wind is blowing along the y-axis, we have only nonzero τ_y and, from (8.210), results in U only.

Figure 8.10 plots three different scenarios of surface wind and the associated phase diagram plot in the velocity space. In Figure 8.10, we plot the phase diagrams for: (a) nonzero τ_x only, (b) nonzero τ_y only; (c) $\tau_x = \tau_y$; and (d) the three-dimensional current pattern in the Ekman layer. These phase diagram plots are known as the Ekman spirals. In these plots, we have restricted the case of $f > 0$, or they are for the Northern Hemisphere only. For the Southern Hemisphere, we have $f < 0$, and, thus, the current shifts to the left of the wind direction, instead of the right direction in the Northern Hemisphere. As remarked after Equation (8.209), the surface current shifts 45° to the right from the wind direction, whereas the overall horizontal Ekman transport is diverged by 90° (see Fig. 8.10(a)-(c)). This provides an additional insight that was not available from the treatment given in Section 8.6.1.

For currents in the Ekman layer, the divergence of the flows can be evaluated as

$$\int_{-\infty}^{0}\left(\frac{\partial u}{\partial x}+\frac{\partial v}{\partial y}\right)dz=\frac{1}{\rho_0 f}\left(\frac{\partial \tau_y}{\partial x}-\frac{\partial \tau_x}{\partial y}\right) \tag{8.214}$$

The proof of this formula is left as a problem for the reader (see Problem 8.3). We see that the divergence of the Ekman layer depends on the wind-stress curl. If the applied wind creates a nonzero wind-stress curl, this divergence must be accompanied by a vertical flow. Integrating the continuity given in (8.107), we obtain the vertical flow as

$$\overline{w}=\int_{-\infty}^{0}\left(\frac{\partial u}{\partial x}+\frac{\partial v}{\partial y}\right)dz=\frac{1}{\rho_0 f}\left(\frac{\partial \tau_y}{\partial x}-\frac{\partial \tau_x}{\partial y}\right) \tag{8.215}$$

This is called Ekman pumping. In the North Hemisphere (i.e., $f > 0$), a clockwise wind pattern with nonzero curl generates a downwelling whereas a counterclockwise wind pattern generates an upwelling. For the case of a tropical cyclone, the counterclockwise wind will create upwelling in addition to the storm surge considered in the previous sections. This unfavorable effect on storm surge apparently has not been pursued seriously. Further studies are needed.

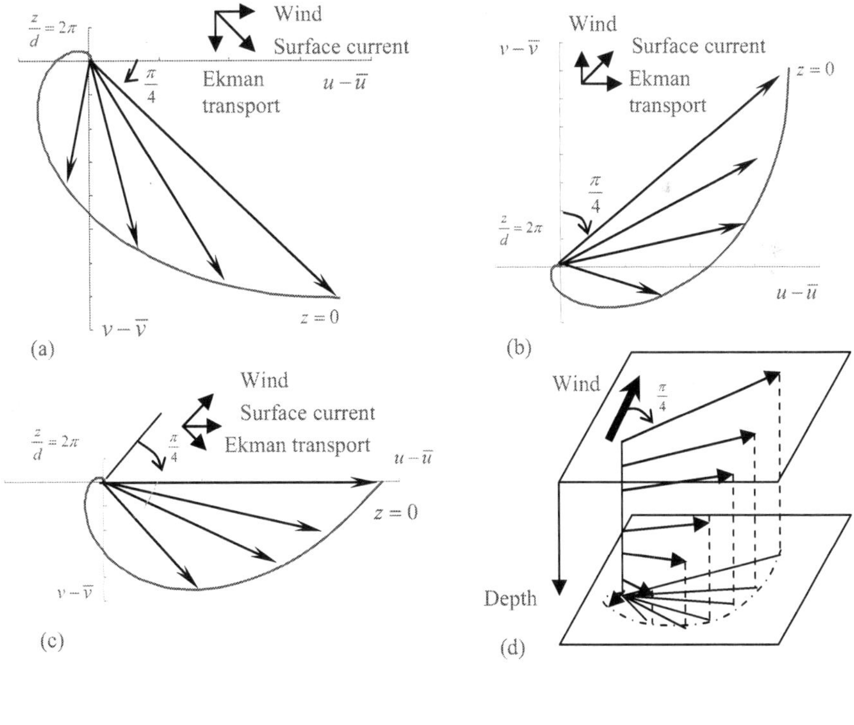

Figure 8.10 The velocity spiral of the Ekman layer for three different directions

8.7 GEOSTROPHIC FLOWS

8.7.1 Taylor-Proudman Theorem

In this section, we consider a very special kind of geophysical flow, and it is the geostrophic flows. When the rotational effect in geophysical flows is the most important factor, we can neglect all terms in the governing equations except the Coriolis force and the pressure gradient. In other words, we have

$$-fv = -\frac{1}{\rho_0}\frac{\partial \overline{p}}{\partial x} \tag{8.216}$$

$$fu = -\frac{1}{\rho_0}\frac{\partial \overline{p}}{\partial y} \tag{8.217}$$

$$0 = -\frac{1}{\rho_0}\frac{\partial \overline{p}}{\partial z} \tag{8.218}$$

$$\frac{\partial u}{\partial x} + \frac{\partial v}{\partial y} + \frac{\partial w}{\partial z} = 0 \tag{8.219}$$

$$\frac{\partial \overline{\rho}}{\partial t} + u\frac{\partial \overline{\rho}}{\partial x} + v\frac{\partial \overline{\rho}}{\partial y} + w\frac{\partial \overline{\rho}}{\partial z} = k\frac{\partial^2 \overline{\rho}}{\partial z^2} \tag{8.220}$$

Let us take differentiation of (8.216) and (8.217) with respect to z to get

$$-f\frac{\partial v}{\partial z} = -\frac{1}{\rho_0}\frac{\partial}{\partial z}(\frac{\partial \overline{p}}{\partial x}) = -\frac{1}{\rho_0}\frac{\partial}{\partial x}(\frac{\partial \overline{p}}{\partial z}) \tag{8.221}$$

$$f\frac{\partial u}{\partial z} = -\frac{1}{\rho_0}\frac{\partial}{\partial z}(\frac{\partial \overline{p}}{\partial y}) = -\frac{1}{\rho_0}\frac{\partial}{\partial y}(\frac{\partial \overline{p}}{\partial z}) \tag{8.222}$$

The reversal of the order of partial differentiation is guaranteed by the Clairaut theorem (see Section 1.3.6 of Chau, 2018). However, substitution of (8.218) into (8.221) and (8.222) gives

$$\frac{\partial u}{\partial z} = \frac{\partial v}{\partial z} = 0 \tag{8.223}$$

This leads to both vertical differentiations of the horizontal flow velocity being zero. This is known as the Taylor-Proudman theorem. Physically, the flow is truly two-dimensional and there is no variation of the flow with the height. This is also known as vertical rigidity.

8.7.2 Homogeneous Geostrophic Flows

If the pressure is given, the velocity field in (8.216) and (8.217) can be evaluated as:

$$v = \frac{1}{\rho_0 f}\frac{\partial \overline{p}}{\partial x}, \quad u = -\frac{1}{\rho_0 f}\frac{\partial \overline{p}}{\partial y} \tag{8.224}$$

Physically, this shows that the flow is not perpendicular to the pressure gradient but instead parallel to it. For non-rotating flows (like incompressible potential flow), the flows are always from high pressure to low pressure, but for rotating flows dominated by Coriolis force, the flow is along the isobar (contour lines of constant pressure). Thus, no work is done by the fluid or on the fluid. Once this kind of flow is initiated, the fluid can persist even without a continuous energy supply. In this case, the pressure is balanced precisely by Coriolis force, and the flow is called geostrophic flow. Meteorological diagrams of weather systems on Earth do indicate parallelism of wind and pressure or indicate geostrophic flow.

Another consequence of the geostrophic flow can be seen from small-scale geophysical flows. If the flow field does not span too wide in the meridional span, we can assume f is constant. Thus, we have

$$\frac{\partial u}{\partial x}+\frac{\partial v}{\partial y}=-\frac{\partial}{\partial x}\left(\frac{1}{\rho_0 f}\frac{\partial \overline{p}}{\partial y}\right)+\frac{\partial}{\partial y}\left(\frac{1}{\rho_0 f}\frac{\partial \overline{p}}{\partial x}\right)=0 \tag{8.225}$$

By the continuity equation, we must have

$$\frac{\partial w}{\partial z}=0 \tag{8.226}$$

That is, the vertical flow velocity is independent of the depth. For weather systems over flat surfaces (such as a level sea or flat ground), we have no vertical air flow from the sea surface and ground, thus, the flow is strictly two-dimensional.

8.8 2-D SHALLOW WATER EQUATIONS

We now extend the geostrophic flows discussed in the last section to a more general case that only the viscous force can be neglected. More specifically, (8.104) to (8.107) can be reduced to

$$\frac{\partial u}{\partial t}+u\frac{\partial u}{\partial x}+v\frac{\partial u}{\partial y}+w\frac{\partial u}{\partial z}-fv=-\frac{1}{\rho_0}\frac{\partial \overline{p}}{\partial x} \tag{8.227}$$

$$\frac{\partial v}{\partial t}+u\frac{\partial v}{\partial x}+v\frac{\partial v}{\partial y}+w\frac{\partial v}{\partial z}+fu=-\frac{1}{\rho_0}\frac{\partial \overline{p}}{\partial y} \tag{8.228}$$

$$0=-\frac{1}{\rho_0}\frac{\partial \overline{p}}{\partial z} \tag{8.229}$$

$$\frac{\partial u}{\partial x}+\frac{\partial v}{\partial y}+\frac{\partial w}{\partial z}=0 \tag{8.230}$$

Let us consider the case that horizontal flows are initially depth independent (see (8.229)). Since the Coriolis force and pressure force (see (8.227)) are also depth independent, the first terms on the left-hand sides of (8.227) and (8.228) must also be depth independent. That is, the time derivatives of the horizontal flows are also depth independent. In other words, the horizontal flows must remain depth independent at all subsequent times. Thus, we have

$$\frac{\partial u}{\partial t}+u\frac{\partial u}{\partial x}+v\frac{\partial u}{\partial y}-fv=-\frac{1}{\rho_0}\frac{\partial \overline{p}}{\partial x} \tag{8.231}$$

$$\frac{\partial v}{\partial t}+u\frac{\partial v}{\partial x}+v\frac{\partial v}{\partial y}+fu=-\frac{1}{\rho_0}\frac{\partial \overline{p}}{\partial y} \tag{8.232}$$

We now integrate the continuity equation (8.230) over the entire column of water from bottom to top as

$$(\frac{\partial u}{\partial x}+\frac{\partial v}{\partial y})\int_{b}^{b+h} dz + [w]_{b}^{b+h} = 0 \tag{8.233}$$

where b is the bottom elevation of the fluid and h is the instantaneous depth of the fluid, as shown in Figure 8.11.

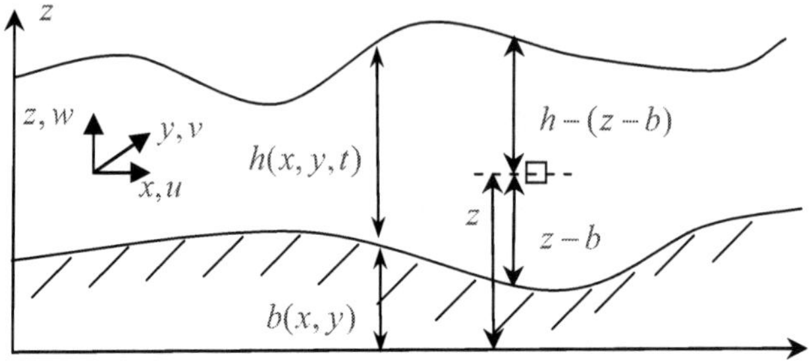

Figure 8.11 Shallow water flow with non-constant depth h

In a more explicit form, we have

$$(\frac{\partial u}{\partial x}+\frac{\partial v}{\partial y})h + w(b+h) - w(b) = 0 \tag{8.234}$$

However, from the kinematic condition, the vertical velocity at the surface is

$$w(b+h) = \left.\frac{d(b+h)}{dt}\right|_{z=b+h} = \frac{\partial(b+h)}{\partial t} + u\frac{\partial(b+h)}{\partial x} + v\frac{\partial(b+h)}{\partial y} \tag{8.235}$$

Similarly, the vertical velocity at the bottom must satisfy

$$w(b) = u\frac{\partial b}{\partial x} + v\frac{\partial b}{\partial y} \tag{8.236}$$

Substitution of (8.235) and (8.236) into (8.234) gives

$$\frac{\partial h}{\partial t} + \frac{\partial(hu)}{\partial x} + \frac{\partial(hv)}{\partial y} = 0 \tag{8.237}$$

This expression eliminates the vertical velocity altogether. In addition, the dynamic pressure can be expressed in terms of h as

$$p_d = p - (P_0 - \rho_0 g z) = [h - (z-b)]\rho_0 g - P_0 + \rho_0 g z = \rho_0 g(h+b) - P_0 \tag{8.238}$$

In summary, we have

$$\frac{\partial u}{\partial t} + u\frac{\partial u}{\partial x} + v\frac{\partial u}{\partial y} - fv = -\frac{1}{\rho_0}\frac{\partial(h+b)}{\partial x} \tag{8.239}$$

$$\frac{\partial v}{\partial t} + u\frac{\partial v}{\partial x} + v\frac{\partial v}{\partial y} + fu = -\frac{1}{\rho_0}\frac{\partial(h+b)}{\partial y} \tag{8.240}$$

$$\frac{\partial h}{\partial t} + \frac{\partial(hu)}{\partial x} + \frac{\partial(hv)}{\partial y} = 0 \tag{8.241}$$

Once the bottom profile b is given, (8.239) to (8.241) provide a system of three equations for three unknowns u, v, and h. This system is known as the shallow-water model. This model is of fundamental importance in the simulation of tsunami and storm surges. However, numerical simulations are out of the scope of the present chapter.

8.9 VORTICITY AND TORNADO DYNAMICS

Various vortex models for tornadoes are considered in this section. For tornado modeling, it is more convenient to work on a polar coordinate system. The governing equations for geophysical flows can first be rewritten in polar form as (Wu et al., 2015):

$$\frac{\partial u}{\partial t}+u\frac{\partial u}{\partial r}+\frac{v}{r}\frac{\partial u}{\partial \theta}+w\frac{\partial u}{\partial z}-\frac{v^2}{r}=-\frac{1}{\rho}\frac{\partial p}{\partial r}+\nu(\nabla^2 u-\frac{u}{r^2}-\frac{2}{r^2}\frac{\partial v}{\partial \theta}) \tag{8.242}$$

$$\frac{\partial v}{\partial t}+u\frac{\partial v}{\partial r}+\frac{v}{r}\frac{\partial v}{\partial \theta}+w\frac{\partial v}{\partial z}+\frac{vu}{r}=-\frac{1}{\rho}\frac{\partial p}{r\partial \theta}+\nu(\nabla^2 v-\frac{v}{r^2}+\frac{2}{r^2}\frac{\partial u}{\partial \theta}) \tag{8.243}$$

$$\frac{\partial w}{\partial t}+u\frac{\partial w}{\partial r}+\frac{v}{r}\frac{\partial w}{\partial \theta}+w\frac{\partial w}{\partial z}=-\frac{1}{\rho}\frac{\partial p}{\partial z}+\nu\nabla^2 w \tag{8.244}$$

$$\frac{\partial (ru)}{\partial r}+\frac{\partial v}{\partial \theta}+\frac{\partial (rw)}{\partial z}=0 \tag{8.245}$$

where

$$\nabla^2=\frac{\partial^2}{\partial r^2}+\frac{1}{r}\frac{\partial}{\partial r}+\frac{1}{r^2}\frac{\partial^2}{\partial \theta^2}+\frac{\partial^2}{\partial z^2} \tag{8.246}$$

Since for tornadoes, the velocity speed is high and the length scale is relative small, Coriolis force has been neglected in (8.242) to (8.244).

8.9.1 Helmholtz Vorticity Equation

In this section, vorticity will be considered and it is defined as

$$\boldsymbol{\omega}=\nabla\times\boldsymbol{u} \tag{8.247}$$

where

$$\nabla=\boldsymbol{e}_r\frac{\partial}{\partial r}+\boldsymbol{e}_\theta\frac{1}{r}\frac{\partial}{\partial \theta}+\boldsymbol{e}_z\frac{\partial}{\partial z},\quad \boldsymbol{u}=\boldsymbol{e}_r u_r+\boldsymbol{e}_\theta u_\theta+\boldsymbol{e}_z u_z \tag{8.248}$$

The physical component of the vorticity can be explicitly written as

$$\omega_r=\frac{1}{r}\frac{\partial w}{\partial \theta}-\frac{\partial v}{\partial z} \tag{8.249}$$

$$\omega_\theta=\frac{\partial u}{\partial z}-\frac{\partial w}{\partial r} \tag{8.250}$$

$$\omega_z=\frac{1}{r}\frac{\partial (rv)}{\partial r}-\frac{1}{r}\frac{\partial u}{\partial \theta} \tag{8.251}$$

The governing equation for vorticity can be obtained by starting with the Navier-Stokes (NS) equation. In particular, we recall from (2.148) of Chau (2018) that

$$\frac{D\boldsymbol{u}}{Dt} = -\frac{1}{\rho}\nabla p + \nu\nabla^2\boldsymbol{u} \tag{8.252}$$

This particular form of NS equation is for incompressible flow (which is consistent with (8.107)). By the result of (2.121) of Chau (2018), we can rewrite (8.252) as

$$\frac{D\boldsymbol{u}}{Dt} = \frac{\partial \boldsymbol{u}}{\partial t} + \nabla(\frac{1}{2}|\boldsymbol{u}|^2) - \boldsymbol{u}\times\nabla\times\boldsymbol{u} = -\frac{1}{\rho}\nabla p + \nu\nabla^2\boldsymbol{u} \tag{8.253}$$

where ν must be interpreted as the effective turbulent eddy viscosity for the case of a tornado. This can be rearranged as

$$\frac{\partial \boldsymbol{u}}{\partial t} + \boldsymbol{\omega}\times\boldsymbol{u} = \nabla(-\frac{p}{\rho} + \frac{1}{2}|\boldsymbol{u}|^2) + \nu\nabla^2\boldsymbol{u} \tag{8.254}$$

In obtaining (8.254), we have used the following identity

$$\boldsymbol{\omega}\times\boldsymbol{u} = -\boldsymbol{u}\times\boldsymbol{\omega} \tag{8.255}$$

Taking the curl of (8.254), we get

$$\frac{\partial(\nabla\times\boldsymbol{u})}{\partial t} + \nabla\times(\boldsymbol{\omega}\times\boldsymbol{u}) = \nabla\times\nabla(-\frac{p}{\rho} + \frac{1}{2}|\boldsymbol{u}|^2) + \nu\nabla\times(\nabla^2\boldsymbol{u}) \tag{8.256}$$

Using the vector identities from (1.346) and (1.360) of Chau (2018), we have

$$\frac{\partial \boldsymbol{\omega}}{\partial t} + \nabla\times(\boldsymbol{\omega}\times\boldsymbol{u}) = \nu\nabla^2\boldsymbol{\omega} \tag{8.257}$$

This is the Helmholtz vorticity equation. Another popular form of the Helmholtz vorticity equation is

$$\frac{\partial \boldsymbol{\omega}}{\partial t} + \boldsymbol{u}\cdot\nabla\boldsymbol{\omega} = \boldsymbol{\omega}\cdot\nabla\boldsymbol{u} + \nu\nabla^2\boldsymbol{\omega} \tag{8.258}$$

This can be proved easily by using the identity derived in Example 8.3 to be followed. The first term on the left-hand side of (8.258) shows the time evolution of the vorticity, the second term on the left-hand side illustrates the convective transport of vorticity, the first term on the right-hand side indicates vortex stretching and turning, and the second term on the right-hand side controls the diffusion of the vorticity. Therefore, (8.258) governs the evolution of vorticity.

Note that the pressure term is not involved in (8.258), suggesting that pressure change has no consequence on vortex dynamics. Similarly, (8.258) also does not involve the gravity term, and, thus, gravity also has no effect on the vortex dynamics. Vorticity cannot be generated inside the fluid, and therefore, it must be generated from the boundary.

We also note for two-dimensional flows that the first term on the right-hand side must be zero:

$$\boldsymbol{\omega}\cdot\nabla\boldsymbol{u} = 0 \tag{8.259}$$

Thus, we have for two-dimensional flows

$$\frac{\partial \boldsymbol{\omega}}{\partial t} + \boldsymbol{u}\cdot\nabla\boldsymbol{\omega} = \nu\nabla^2\boldsymbol{\omega} \tag{8.260}$$

Therefore, vorticity stretching is intrinsically a three-dimensional phenomenon. The diffusion term in (8.260) relates to viscous dissipation of energy. Thus, dissipation is most effective at small scales.

Actually, by the Helmholtz theorem, we can decompose the velocity of fluid into two parts (see Section 4.2.1 of Chau, 2013):

$$\boldsymbol{u} = \nabla\phi + \nabla\times\psi \tag{8.261}$$

Taking the divergence of (8.261), we have

$$\nabla\cdot\boldsymbol{u} = \nabla\cdot\nabla\phi + \nabla\cdot\nabla\times\psi = \nabla^2\phi = 0 \tag{8.262}$$

Recalling that the divergence of the curl of a vector must be zero (see (1.347) of Chau, 2018), we arrive at the last (8.261). Therefore, the irrotational part of the velocity is

$$\boldsymbol{u}_{irr} = \nabla\phi \tag{8.263}$$

The potential ϕ satisfies the Laplace equation (see (8.262)). As shown in Section 9.7 of Chau (2018), incompressible and inviscid flow can be modeled by the potential function and the stream function, and both of them satisfy the Laplace equation. It is also known as potential flow. For irrotational or potential flows, the flow is steady or time-independent and, thus, the momentum equation is not needed in solving the flow field. The flow can be modeled by the Bernoulli equation or Laplace equation. For rotational flow, taking the curl of (8.261), we have

$$\nabla\times\boldsymbol{u} = \nabla\times\nabla\phi + \nabla\times\nabla\times\psi = \nabla\times\nabla\times\psi = \boldsymbol{\omega} \tag{8.264}$$

Thus, the rotational flow can be expressed as

$$\boldsymbol{u}_{rot} = \nabla\times\psi \tag{8.265}$$

Vorticity implies the local rotation of a fluid element. When circular or swirling motion is apparent from the flow, we call it a vortex. The strength of the vortex is normally expressed in terms of the so-called circulation Γ. For inviscid two-dimensional rotating flows, the circulation in the fluid conserves. This is also known as the Helmholtz-Kelvin theorem. When a fluid is squeezed laterally, its vorticity must increase to conserve circulation.

Example 8.3 Prove the following vector identity for vorticity, which has been used in obtaining (8.258):

$$\nabla\times(\boldsymbol{\omega}\times\boldsymbol{u}) = \boldsymbol{u}\cdot\nabla\boldsymbol{\omega} - \boldsymbol{\omega}\cdot\nabla\boldsymbol{u} \tag{8.266}$$

Solution: To prove (8.266), we first write

$$\nabla\times(\boldsymbol{\omega}\times\boldsymbol{u}) = (\frac{\partial}{\partial x_n}\boldsymbol{e}_n)\times(\omega_i u_j e_{mij}\boldsymbol{e}_m) \tag{8.267}$$

The left-hand side of (8.266) can further be written as

$$\nabla\times(\boldsymbol{\omega}\times\boldsymbol{u}) = (\omega_{i,n}u_j + \omega_i u_{j,n})e_{lnm}e_{mij}\boldsymbol{e}_l \tag{8.268}$$

Recalling the e-δ identity from (1.419) of Chau (2018), (8.268) is simplified to

$$\begin{aligned}\nabla\times(\boldsymbol{\omega}\times\boldsymbol{u}) &= (\omega_{i,n}u_j + \omega_i u_{j,n})(\delta_{il}\delta_{jn} - \delta_{in}\delta_{jl})\boldsymbol{e}_l \\ &= (\omega_{l,n}u_n - \omega_{n,n}u_l + \omega_l u_{n,n} - \omega_n u_{l,n})\boldsymbol{e}_l \\ &= \boldsymbol{u}\cdot\nabla\boldsymbol{\omega} - \boldsymbol{u}(\nabla\cdot\boldsymbol{\omega}) + \boldsymbol{\omega}(\nabla\cdot\boldsymbol{u}) - \boldsymbol{\omega}\cdot\nabla\boldsymbol{u}\end{aligned} \tag{8.269}$$

Physically, the divergence of the vorticity must be zero (this is the fundamental definition of the rotation field):

$$\nabla\cdot\boldsymbol{\omega} = \nabla\cdot(\nabla\times\boldsymbol{u}) = 0 \tag{8.270}$$

The last result is from the famous vector identity given in (1.347) of Chau (2018). The continuity equation given in (8.107) can be expressed in tensor form as

$$\nabla \cdot \boldsymbol{u} = 0 \tag{8.271}$$

Substitution of (8.270) and (8.271) into (8.269) gives

$$\nabla \times (\boldsymbol{\omega} \times \boldsymbol{u}) = \boldsymbol{u} \cdot \nabla \boldsymbol{\omega} - \boldsymbol{\omega} \cdot \nabla \boldsymbol{u} \tag{8.272}$$

This completes the proof of (8.266).

For the case of two-dimensional axisymmetric rotating flows, we have

$$\frac{d\boldsymbol{\omega}}{dt} = \frac{\partial \boldsymbol{\omega}}{\partial t} + \boldsymbol{u} \cdot \nabla \boldsymbol{\omega} = \nu \nabla^2 \boldsymbol{\omega} \tag{8.273}$$

For the special case of $u = w = 0$ and $v = v(\theta)$, the Helmholtz vorticity equation given in (8.273) can be simplified as

$$\frac{\partial \omega_z}{\partial t} = \nu \left(\frac{\partial^2 \omega_z}{\partial r^2} + \frac{1}{r} \frac{\partial \omega_z}{\partial r} \right) \tag{8.274}$$

Problem 8.4 gives another proof of (8.274) at the end of the chapter. It can be shown that a similarity solution for the Helmholtz vorticity equation given in (8.274) can be found for ω_z in terms of Laguerre polynomials (see Problem 8.10).

8.9.2 Conservation of Angular Momentum

The conservation of angular momentum of rotating flows can be expressed as (Lewellen, 1976)

$$\frac{\partial (rv)}{\partial t} + u \frac{\partial (rv)}{\partial r} + \frac{v}{r} \frac{\partial (rv)}{\partial \theta} + w \frac{\partial (rv)}{\partial z} = \nu r \left\{ \frac{\partial}{\partial r} \left(\frac{1}{r} \frac{\partial (rv)}{\partial r} \right) + \frac{\partial^2 v}{\partial z^2} \right\} \tag{8.275}$$

To prove this equation given by Lewellen (1976), we carry out the differentiation in (8.275) as

$$r \frac{\partial v}{\partial t} + uv + ur \frac{\partial v}{\partial r} + v \frac{\partial v}{\partial \theta} + rw \frac{\partial v}{\partial z} = \nu r \left\{ \frac{\partial}{\partial r} \left(\frac{1}{r} \frac{\partial (rv)}{\partial r} \right) + \frac{\partial^2 v}{\partial z^2} \right\} \tag{8.276}$$

Dividing through by r, we get

$$\frac{\partial v}{\partial t} + u \frac{\partial v}{\partial r} + \frac{v}{r} \frac{\partial v}{\partial \theta} + w \frac{\partial v}{\partial z} + \frac{uv}{r} = \nu \left\{ \frac{\partial}{\partial r} \left(\frac{1}{r} \frac{\partial (rv)}{\partial r} \right) + \frac{\partial^2 v}{\partial z^2} \right\} \tag{8.277}$$

The right-hand side can be expanded as

$$\frac{\partial}{\partial r} \left(\frac{1}{r} \frac{\partial (rv)}{\partial r} \right) + \frac{\partial^2 v}{\partial z^2} = \frac{\partial^2 v}{\partial r^2} + \frac{1}{r} \frac{\partial v}{\partial r} - \frac{v}{r^2} + \frac{\partial^2 v}{\partial z^2} \tag{8.278}$$

Note from (8.246) that, for axisymmetric flow, we have

$$\nabla^2 v - \frac{v}{r^2} = \frac{\partial^2 v}{\partial r^2} + \frac{1}{r} \frac{\partial v}{\partial r} + \frac{\partial^2 v}{\partial z^2} - \frac{v}{r^2} \tag{8.279}$$

It is straightforward to see that (8.278) and (8.279) are the same. Substitution of (8.279) into (8.277) gives

$$\frac{\partial v}{\partial t}+u\frac{\partial v}{\partial r}+\frac{v}{r}\frac{\partial v}{\partial \theta}+w\frac{\partial v}{\partial z}+\frac{uv}{r}=\nu\{\nabla^2 v-\frac{v}{r^2}\} \tag{8.280}$$

This is exactly the same as (8.243) if we set $p \neq p(\theta)$ and $u \neq u(\theta)$ (these are obviously true for axisymmetric case). Thus, the validity of (8.275) is demonstrated.

8.9.3 Vorticity in Tornadoes

For the axisymmetric case, the velocity field in the horizontal plane can be expressed in terms of the Stokes stream function ψ as

$$u=-\frac{1}{r}\frac{\partial \psi}{\partial z}, \quad w=\frac{1}{r}\frac{\partial \psi}{\partial r} \tag{8.281}$$

In addition, for uniform swirling motion, we can define the circulation of a vortex as

$$\Gamma = rv \tag{8.282}$$

Note that some authors define the circulation in a slightly different way by adding an extra factor 2π on the right-hand side of (8.282). Adopting (8.281) and (8.282), (8.249) to (8.251) can be expressed as

$$\omega_r = -\frac{1}{r}\frac{\partial \Gamma}{\partial z}, \quad \omega_z = \frac{1}{r}\frac{\partial \Gamma}{\partial r} \tag{8.283}$$

$$\omega_\theta = -[\frac{\partial}{\partial r}(\frac{1}{r}\frac{\partial \psi}{\partial r})+\frac{1}{r}\frac{\partial^2 \psi}{\partial z^2}] \tag{8.284}$$

In the following sections, we will consider some popular models for axisymmetric vortices that have been found useful in explaining tornadoes. Strictly speaking, these vortex models apply only to flow regime "I" or the funnel cloud in Figure 8.12. The flow solutions for other regimes, especially the corner regime III that is normally associated with debris clouds near the ground, are more complicated. Their full descriptions need numerical simulations that are out of the scope of the present chapter.

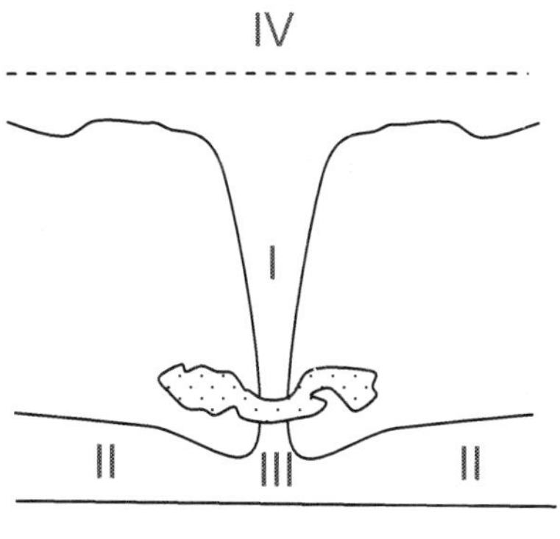

Figure 8.12 Flow regimes in tornadoes

8.9.4 Potential Vortex Model

As the first approximation, the funnel cloud in a tornado can be modeled as a vertical vortex with swirling motion around a vertical axis. The flow field can be approximated by a time-independent (or steady-state) vortex with only an r-dependent tangential component of velocity v (see (8.282)):

$$v = \frac{\Gamma}{r} \tag{8.285}$$

The momentum equation given in (8.242) and (8.243) can be greatly simplified to

$$-\frac{v^2}{r} + \frac{1}{\rho}\frac{\partial p}{\partial r} = 0 \tag{8.286}$$

$$\nabla^2 v - \frac{v}{r^2} = 0 \tag{8.287}$$

In explicit form, (8.287) can be written as

$$\frac{d^2 v}{dr^2} + \frac{1}{r}\frac{dv}{dr} - \frac{v}{r^2} = 0 \tag{8.288}$$

Taking differentiation of (8.282) with respect to r, we get

$$\frac{d\Gamma}{dr} = v + r\frac{dv}{dr} \tag{8.289}$$

$$r\frac{d^2\Gamma}{dr^2} = r^2\frac{d^2 v}{dr^2} + 2r\frac{dv}{dr} \tag{8.290}$$

Subtracting (8.289) from (8.290), we obtain

$$\frac{1}{r}\frac{d^2\Gamma}{dr^2} - \frac{1}{r^2}\frac{d\Gamma}{dr} = \frac{d^2 v}{dr^2} + \frac{1}{r}\frac{dv}{dr} - \frac{v}{r^2} = 0 \tag{8.291}$$

The last of (8.291) results from (8.288). Therefore, the governing equation for Γ is

$$\frac{1}{r}\frac{d^2\Gamma}{dr^2} - \frac{1}{r^2}\frac{d\Gamma}{dr} = 0 \tag{8.292}$$

This can be solved by reducing the order of the ODEs as discussed in Section 3.3.6 in Chau (2018). The solution is

$$\Gamma = \frac{1}{2}Ar^2 + B \tag{8.293}$$

The proof of (8.293) is left for the reader. Substituting (8.293) into (8.285) gives

$$v = \frac{1}{2}Ar + \frac{B}{r} \tag{8.294}$$

In order to have a finite solution for $r \to \infty$, we require $A = 0$ and $B = \Gamma_0/2\pi$. Thus, the tangential velocity becomes

$$v = \frac{\Gamma_0}{2\pi r} \tag{8.295}$$

where $\Gamma_0/2\pi$ is the strength of the circulation at far field. To find the pressure around the vortex, we can substitute (8.295) into (8.286) to give

$$\frac{1}{\rho}\frac{\partial p}{\partial r} = \frac{v^2}{r} = \frac{\Gamma_0^2}{4\pi^2 r^3} \tag{8.296}$$

Integrating (8.296), we have

$$p = -\frac{\rho\Gamma_0^2}{8\pi^2 r^2} + C \tag{8.297}$$

If the far field pressure is p_∞, the pressure profile can be obtained as

$$p = p_\infty - \frac{\rho\Gamma_0^2}{8\pi^2 r^2} \tag{8.298}$$

This model is called the potential vortex. The main problem of this model is that both the tangential velocity and the pressure are singular at $r = 0$. Therefore, it is physically impossible.

8.9.5 Rankine Vortex Model

To remove the undesirable singular behavior at the origin, Rankine in 1882 proposed a two-regime vortex model by assuming the following circulation

$$\begin{aligned} \Gamma &= \Omega_0 r^2, && 0 < r < r_{\max} \\ &= \Gamma_\infty, && r_{\max} < r < \infty \end{aligned} \tag{8.299}$$

In addition, in view of (8.283) and (8.285) and (8.299), we have a jump in vorticity as

$$\begin{aligned} \omega_z &= 2\Omega_0, && 0 < r < r_{\max} \\ &= 0, && r_{\max} < r < \infty \end{aligned} \tag{8.300}$$

Recalling (8.283), the velocity field of the Rankine vortex model is

$$\begin{aligned} v &= \Omega_0 r, && 0 < r < r_{\max} \\ &= \frac{\Gamma_\infty}{r}, && r_{\max} < r < \infty \end{aligned} \tag{8.301}$$

Physically, we can see that the Rankine vortex model uses the first term in (8.294) for the near-field region and use the second term in (8.294) for the far-field region. For a continuous velocity field, equating v at r_{max}, we find

$$r_{\max} = (\frac{\Gamma_\infty}{\Omega_0})^{1/2} \tag{8.302}$$

The tangential velocity increases linearly with r from zero at the center of the vortex to a maximum at $r = r_{max}$, then it decreases inversely with r. The corresponding pressure distribution for the Rankine vortex can be evaluated using (8.286) as

$$\frac{v^2}{r} = \frac{1}{\rho}\frac{\partial p}{\partial r} \tag{8.303}$$

Integrating (8.303), we have

$$p = \rho\Omega_0^2 \frac{r^2}{2} + g(z), \qquad 0 < r < r_{\max}$$
$$= -\frac{\rho\Gamma_\infty^2}{2}\left[\frac{1}{r^2} - \frac{1}{r_{\max}^2}\right] + g(z), \qquad r_{\max} < r < \infty \tag{8.304}$$

By hydrostatic balance of the vertical momentum, we can approximate the pressure gradient as

$$\frac{\partial p}{\partial z} = -\rho g \tag{8.305}$$

Integration of (8.305) gives

$$p = -\rho g z + f(r) \tag{8.306}$$

Finally, comparing (8.304) and (8.306), we obtain the pressure distribution as

$$p = \rho\Omega_0^2 \frac{r^2}{2} - \rho g z, \qquad 0 < r < r_{\max}$$
$$= -\frac{\rho\Gamma_\infty^2}{2}\left[\frac{1}{r^2} - \frac{1}{r_{\max}^2}\right] - \rho g z, \qquad r_{\max} < r < \infty \tag{8.307}$$

One main objection to the Rankine vortex model is that the velocity distribution is not smooth at $r = r_{max}$, and, consequently, vorticity is not continuous. Thus, the Rankine vortex does not obey the Navier-Stokes equation. To apply the model, we must impose Ω_0 and Γ_∞. In real applications to tornado modeling, we can observe the maximum tangential velocity around a tornado and its corresponding radius of occurrence. Then, we have

$$r_{\max} v_{\max} = \Gamma_\infty \tag{8.308}$$

Subsequently, we use (8.302) to find Ω_0. Observations from real tornadoes suggest that $\Gamma_\infty \approx 10^4$ m^2/s and $\nu \approx 10$ m^2/s. If we assume that the maximum velocity occurs at about 80 m, we have from (8.308)

$$v_{\max} = \frac{\Gamma_\infty}{r_{\max}} = \frac{10^4}{80} = 125 \text{ m/s} \tag{8.309}$$

This value is comparable to the acceptable values of 112 to 123 m/s for maximum velocity in tornadoes by meteorologists and engineers (Minor, 1976). By using photogrammetry, Forbes (1976) estimated a maximum tangential velocity of up to 127 m/s in suction vortices, which is a term employed by Professor Fujita, who was a professor at the University of Chicago. By examining the ground marks on a corn field left behind by tornadoes, Fujita interpreted that many strong tornadoes (typically of diameter of 50 m or more), are actually composed of a number of suction vortices with a vorticity feeder, and only small tornadoes are composed of a single suction vortex (typically of diameter of 10 m or less). Fujita also proposed that there are four systems of geophysical vortices, namely hurricanes or tropical cyclones (diameters of 50 km), tornado cyclones (diameters of 5 km), tornadoes (diameters of 50 m), and suction vortices (diameters of 5 m).

In the discipline of tornado studies, the Fujita scale is normally adopted. Table 8.2, which is compiled from Fujita and Pearson (1976), shows six scales of tornadoes, their wind speed, path length, and path width. The distribution of wind-speed intensity for tornadoes, waterspouts, and dust devils are also compiled in

Table 8.2. We can see that it is rare for tornadoes to have wind speed of up to 125 m/s, as estimated in (8.309).

8.9.6 Burgers-Rott Vortex Model

The two-regime type Rankine vortex model discussed in the last section does not satisfy the governing equations (8.242) to (8.245). In this section, we will see that it is possible to find other vortex solutions that can satisfy the Navier-Stokes equation. One of such solutions is the Burgers-Rott vortex model, which was derived by Burgers in 1948 and Rott in 1958. In this model, all three velocity components are nonzero, and we seek a balance between radial advection and radial diffusion. In particular, the following velocity field is sought

Table 8.2 Fujita scales of tornadoes

F-scale	Wind speed (m/s)	Tornado	Waterspout	Dust devil	Path length (km)	Path width (km)
F5	≥ 117	rare	nil	nil	≥161	≥1.6
F4	93-116	2%	nil	nil	51-160	0.5-1.5
F3	71-92	8%	rare	nil	16-50	0.161-0.49
F2	51-70	24%	2%	nil	5.1-16	0.051-0.16
F1	32-50	42%	23%	rare	1.6-5	0.016-0.05
F0	<32	24%	75%	100%	<1.5	<0.015

$$u = -ar, \quad w = 2az, \quad v = \frac{\Gamma_\infty}{r}[1 - \exp(-\frac{ar^2}{2\nu})] \tag{8.310}$$

where a represents both the inward flow rate (indicating the suction strength) and the upward flow gradient, and as before, Γ_∞ is the circulation at the far field ($r \to \infty$). Physically, the radial flow given in (8.310) brings the far field vorticity to the vortex core. This is the first stretched vortex solution to model turbulent eddies. We are going to see that this solution satisfies (8.243). Since v given in (8.310) is not a function of z and t,

$$v \neq v(t, z) \tag{8.311}$$

Using (8.311), we can simplify (8.243), in conjunction with (8.278) and (8.279), as

$$u\frac{\partial v}{\partial r} + \frac{vu}{r} = \nu\frac{\partial}{\partial r}[\frac{1}{r}\frac{\partial (rv)}{\partial r}] \tag{8.312}$$

From (8.278), we get

$$\frac{\partial}{\partial r}[\frac{1}{r}\frac{\partial (rv)}{\partial r}] = \frac{\partial^2 v}{\partial r^2} + \frac{1}{r}\frac{\partial v}{\partial r} - \frac{v}{r^2} \tag{8.313}$$

Using (8.310), the left-hand side of (8.312) becomes

$$u\frac{\partial v}{\partial r} + \frac{vu}{r} = -\frac{\Gamma_\infty a^2 r}{\nu}\exp(-\frac{ar^2}{2\nu}) \tag{8.314}$$

On the other hand, we have

$$\frac{1}{r}\frac{\partial(rv)}{\partial r}=\frac{\Gamma_{\infty}a}{\nu}\exp(-\frac{ar^2}{2\nu}) \tag{8.315}$$

Differentiating (8.315), we obtain

$$\frac{\partial}{\partial r}[\frac{1}{r}\frac{\partial(rv)}{\partial r}]=-\frac{\Gamma_{\infty}a^2r}{\nu^2}\exp(-\frac{ar^2}{2\nu}) \tag{8.316}$$

Comparing (8.314) and (8.316), we see that tangential momentum equation (8.312) is identically satisfied by (8.310).

Next, we can use (8.242) to find the pressure distribution. First, we note that (8.244) can be rewritten as

$$u\frac{\partial u}{\partial r}-\frac{v^2}{r}=-\frac{1}{\rho}\frac{\partial p}{\partial r}+\nu\frac{\partial}{\partial r}[\frac{1}{r}\frac{\partial(ru)}{\partial r}] \tag{8.317}$$

The last term obtained in (8.317) can be validated by following the procedure in deriving (8.275) in the last section. First, we note from (8.310) that

$$\frac{1}{r}\frac{\partial(ru)}{\partial r}=-2a \tag{8.318}$$

Thus, we have that the last term in (8.317) is identically zero. Substitution of (8.310) into (8.317) gives

$$-\frac{1}{\rho}\frac{\partial p}{\partial r}=a^2r-\frac{\Gamma_{\infty}^2}{r^3}[1-\exp(-\frac{ar^2}{2\nu})]^2 \tag{8.319}$$

Integrating (8.319), we obtain

$$p=-\frac{\rho a^2}{2}r^2+\rho\Gamma_{\infty}^2\int_0^r\frac{1}{r^3}[1-\exp(-\frac{ar^2}{2\nu})]^2dr+f(z) \tag{8.320}$$

To simplify the integral in (8.320), we introduce the following change of variables

$$\frac{ar^2}{2\nu}=x \tag{8.321}$$

Substitution of (8.321) into the integral in (8.320) leads to

$$\int_0^r\frac{1}{r^3}[1-\exp(-\frac{ar^2}{2\nu})]^2dr=\frac{a}{4\nu}\int_0^{\frac{ar^2}{2\nu}}(\frac{1-e^{-x}}{x})^2dx \tag{8.322}$$

Substituting (8.322) into (8.320), we get

$$p=-\frac{\rho a^2}{2}r^2+\frac{\rho\Gamma_{\infty}^2a}{4\nu}\int_0^{\frac{ar^2}{2\nu}}(\frac{1-e^{-x}}{x})^2dx+f(z) \tag{8.323}$$

The unknown function $f(z)$ can be found by using the final momentum equation given in (8.244). For the solution form given in (8.310), we have

$$\frac{\partial w}{\partial t}=0,\quad \frac{\partial w}{\partial r}=0,\quad \frac{\partial w}{\partial z}=2a,\quad \frac{\partial^2 w}{\partial z^2}=0 \tag{8.324}$$

Using (8.324), (8.244) can be simplified to

$$4a^2z=-\frac{1}{\rho}\frac{\partial p}{\partial z} \tag{8.325}$$

Integrating (8.325), we obtain

$$p = -2\rho a^2 z^2 + g(r) \tag{8.326}$$

Comparing (8.323) and (8.326), we get

$$f(z) = -2\rho a^2 z^2 \tag{8.327}$$

Finally, the pressure is

$$p = -\frac{\rho a^2}{2}(r^2 + 4z^2) + \frac{\rho \Gamma_\infty^2 a}{4\nu}\int_0^{\frac{ar^2}{2\nu}} \left(\frac{1-e^{-x}}{x}\right)^2 dx + p(0,0) \tag{8.328}$$

where $p(0,0)$ is the pressure at the center of the vortex. In conclusion, we have shown that the Burgers-Rott vortex model does satisfy the Navier-Stokes equation.

To find the radius from the axis of the vortex at which the tangential velocity is maximum, we can differentiate (8.310) with respect to r and set the result to zero as

$$\frac{dv}{dr} = -\frac{\Gamma_\infty}{r}[1 - \exp(-\frac{ar^2}{2\nu})] + \frac{\Gamma_\infty a}{\nu}\exp(-\frac{ar^2}{2\nu}) = 0 \tag{8.329}$$

This can be rearranged as

$$(\frac{ar^2}{\nu} + 1)\exp(-\frac{ar^2}{2\nu}) = 1 \tag{8.330}$$

An analytical solution of r for (8.330) is not possible, but numerical calculation gives approximately

$$r_{\max} = 1.5813(\frac{\nu}{a})^{1/2} \tag{8.331}$$

The maximum velocity can be found accordingly (see Problem 8.6). Consider the values of the far-field circulation and the effective turbulent eddy viscosity adopted in the last section: $\Gamma_\infty \approx 10^4$ m^2/s and $\nu \approx 10$ m^2/s. Assuming that the maximum wind speed occurs at 100 m, (8.331) gives $a \approx 2.5\times10^{-3}$ s^{-1}. Using this set of parameters, (8.310) can be used to find the maximum velocity

$$v_{\max} = \frac{\Gamma_\infty}{r_{\max}}[1 - \exp(-\frac{ar_{\max}^2}{2\nu})] = 71.35 \text{ m/s} \tag{8.332}$$

This value is only 57% of the value predicted by (8.309). We can also revise this analysis by using $r_{max} = 80$ m, in order to be more comparable to that of (8.309). In doing so, we obtain $a \approx 1.56\times10^{-3}$ s^{-1}. Again using (8.310), we have

$$v_{\max} = \frac{\Gamma_\infty}{r_{\max}}[1 - \exp(-\frac{ar_{\max}^2}{2\nu})] = 49.2 \text{ m/s} \tag{8.333}$$

This value is only 39% of the prediction by the Rankine vortex model. This shows that the continuous solution of the Burgers-Rott vortex model smooths out the maximum velocity.

One main objection to the Burgers-Rott vortex model is that the second equation in (8.310) shows that the extent of the updraft region is infinite. This is clearly not possible. Because of this limitation, a more refined vortex model called the Sullivan vortex will be presented in Section 8.9.8. The Sullivan vortex model is the first two-celled model, that predicts a downdraft in the core region and an updraft region outside the core. In addition, the solution is time independent, and, thus, the process of tornado formation and decay can be considered. In the next section, we will consider a vortex model that allows for the time evolution, and this is the Oseen-Lamb vortex model.

8.9.7 Oseen-Lamb Vortex Model

In this section, a time-dependent vortex model is considered. This model was proposed by Oseen in 1912, and it is sometimes referred to as the viscous vortex. In particular, the following special solution form is sought:

$$u = 0, \quad v = v(r,t), \quad w = 0 \tag{8.334}$$

By noting that $\Gamma = rv$, we first rewrite (8.277) as

$$\frac{\partial \Gamma}{\partial t} + u\frac{\partial \Gamma}{\partial r} + \frac{v}{r}\frac{\partial \Gamma}{\partial \theta} + w\frac{\partial \Gamma}{\partial z} = \nu\{r\frac{\partial}{\partial r}(\frac{1}{r}\frac{\partial \Gamma}{\partial r}) + \frac{\partial^2 \Gamma}{\partial z^2}\} \tag{8.335}$$

This equation can also be rewritten as

$$\frac{\partial \Gamma}{\partial t} = \nu\{r\frac{\partial}{\partial r}(\frac{1}{r}\frac{\partial \Gamma}{\partial r}) + \frac{\partial^2 \Gamma}{\partial z^2}\} \tag{8.336}$$

This form can be found on p.193 of Wu et al. (2015). In view of (8.334), we can simplify (8.336) to

$$\frac{\partial \Gamma}{\partial t} = \nu\{r\frac{\partial}{\partial r}(\frac{1}{r}\frac{\partial \Gamma}{\partial r})\} \tag{8.337}$$

Expanding (8.337), we have

$$\frac{\partial \Gamma}{\partial t} = \nu\frac{\partial^2 \Gamma}{\partial r^2} - \frac{\nu}{r}\frac{\partial \Gamma}{\partial r} \tag{8.338}$$

To solve this PDE, we need both initial and boundary conditions. Consider the following initial condition

$$\Gamma(0,0) = \Gamma_\infty, \quad t, r = 0 \tag{8.339}$$

and boundary conditions:

$$\Gamma(0,t) = 0, \quad r = 0 \tag{8.340}$$

$$\Gamma(\infty,t) = \Gamma_\infty, \quad r \to \infty \tag{8.341}$$

We now introduce the following change of variables

$$\eta = \frac{r}{\sqrt{\nu t}}, \quad \Gamma = f(\eta) \tag{8.342}$$

The time and radial coordinate differentiations of (8.342) give

$$\frac{\partial \Gamma}{\partial t} = \frac{\partial \Gamma}{\partial \eta}\frac{\partial \eta}{\partial t} = -f'(\frac{1}{2}\frac{t^{-2/3}}{\sqrt{\nu}}r) \tag{8.343}$$

$$\frac{\partial \Gamma}{\partial r} = \frac{\partial \Gamma}{\partial \eta}\frac{\partial \eta}{\partial r} = \frac{1}{\sqrt{\nu t}}f' \tag{8.344}$$

$$\frac{\partial^2 \Gamma}{\partial r^2} = \frac{1}{\nu t}f'' \tag{8.345}$$

Substitution of (8.343) to (8.345) into (8.338) gives an ODE for f

$$\frac{d^2 f}{d\eta^2} + (\frac{\eta}{2} - \frac{1}{\eta})\frac{df}{d\eta} = 0 \tag{8.346}$$

Clearly, this ODE can be solved by reducing the order as discussed in Section 3.5.8 of Chau (2018) using:

$$\frac{df}{d\eta} = \zeta \tag{8.347}$$

Consequently, (8.346) is reduced to

$$\frac{d\zeta}{d\eta} + (\frac{\eta}{2} - \frac{1}{\eta})\zeta = 0 \tag{8.348}$$

This is a separable ODE as discussed in Section 3.2.1 of Chau (2018), and can be solved by direct integration as

$$\frac{d\zeta}{\zeta} = (\frac{1}{\eta} - \frac{\eta}{2})d\eta \tag{8.349}$$

Thus, the solution for ζ is

$$\zeta = C\eta e^{-\frac{\eta^2}{4}} = \frac{df}{d\eta} \tag{8.350}$$

The last equation of (8.350) results from (8.347) and the solution of f is

$$f = -4Ce^{-\eta^2/4} + B \tag{8.351}$$

The boundary condition given in (8.340) requires

$$B = 4C \tag{8.352}$$

The boundary condition $\Gamma = \Gamma_\infty$ corresponds to $r \to \infty$ given in (8.341) and leads to

$$B = \Gamma_\infty \tag{8.353}$$

Substitution of (8.353) into (8.352) and (8.351) gives

$$\Gamma(\eta) = -\Gamma_\infty e^{-\eta^2/4} + \Gamma_\infty \tag{8.354}$$

In terms of the original variables r and t, we have

$$\Gamma(r,t) = \Gamma_\infty[1 - \exp(-\frac{r^2}{4\nu t})] \tag{8.355}$$

This is the Oseen-Lamb vortex model. In order to agree with the Burgers-Rott vortex model in the limiting case, we can actually revise our analysis slightly without changing the mathematical structure of the solution form. In particular, we can use the following change of variables

$$\eta = \frac{\alpha r}{\sqrt{\nu t}}, \quad \Gamma = f(\eta) \tag{8.356}$$

where α is an arbitrary constant. By Problem 8.5 at the end of this chapter, we can show that the circulation for this change of variables becomes

$$\Gamma(r,t) = \Gamma_\infty[1 - \exp(-\frac{\alpha r^2}{4\nu t})] \tag{8.357}$$

The tangential velocity becomes

$$v(r,t) = \frac{\Gamma_\infty}{r}[1 - \exp(-\frac{\alpha r^2}{4\nu t})] \tag{8.358}$$

This particular form of Oseen-Lamb solution is given in Lewellen (1976) without proof. For the case of a thin vortex core, we have $r \to 0$ and Taylor's series expansion of (8.358) gives

$$v(r,t) = \frac{\Gamma_\infty}{r}[1-(1-\frac{\alpha r^2}{4\nu t}+...)] \tag{8.359}$$

The first term on the right-hand side of (8.361) closely resembles the inner core solution of the Rankine vortex. Equation (8.359) shows that the velocity decays with time. We can also recast (8.359) as

$$v(r,t) = \frac{\alpha\Gamma_\infty r}{4\nu t} = \frac{\alpha\Gamma_\infty r}{r_0^2} \tag{8.360}$$

where

$$r_0 = \sqrt{4\nu t} \tag{8.361}$$

This can be interpreted as an unsteady transitional zone of the vortex core. Once the core diffuses out (through eddy viscosity ν), the intensity of velocity drops correspondingly. The vorticity of the Oseen-Lamb vortex can be obtained by

$$\omega_z = \frac{1}{r}\frac{\partial(rv)}{\partial r} = \frac{\Gamma_\infty \alpha}{2\nu t}\exp(-\frac{\alpha r^2}{4\nu t}) \tag{8.362}$$

The vorticity distribution at various times is shown in Figure 8.13, which is for $\Gamma_\infty\alpha/\nu = 2.5$. As a function of time, the vorticity of the vortex core diffuses outside with increasing time.

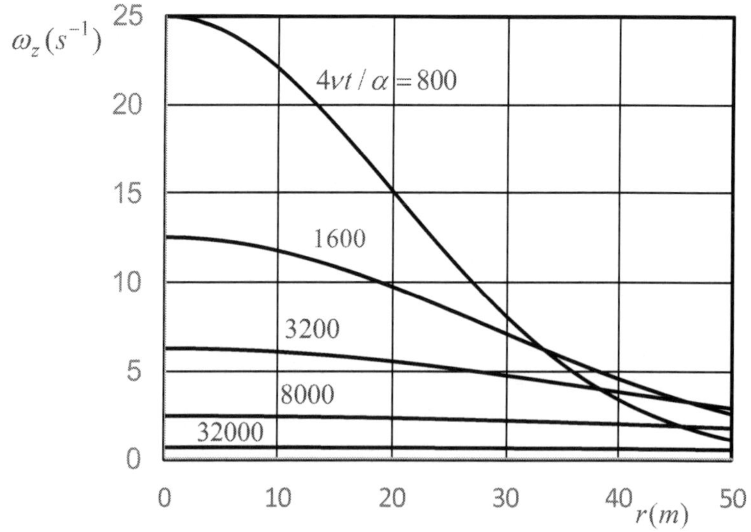

Figure 8.13 Diffusion process of the vorticity in the Oseen-Lamb vortex model

The total vorticity of the vortex is

$$\int_0^\infty \omega_z 2\pi r dr = \frac{\Gamma_\infty \alpha}{2\nu t} 2\pi \int_0^\infty r \exp(-\frac{\alpha r^2}{4\nu t}) dr$$
$$= -(\frac{2\nu t}{\alpha}) \frac{\Gamma_\infty \alpha \pi}{\nu t} [e^{-\infty} - e^0] \tag{8.363}$$
$$= 2\pi\Gamma_\infty$$

Thus, as expected, the total vorticity is an invariant and does not change with time during the evolution of the vortex decay.

Example 8.4 (i) Find the maximum velocity and the radius at which it occurs in the Oseen-Lamb vortex model. (ii) Assume that $\Gamma_\infty = 10^4\, m^2/s$ and $\nu = 10\, m^2/s$, the time that attains the maximum velocity is from 60 to 90 seconds, and the maximum velocity ranges from 50 m/s to about 125 m/s. Based on these data, find possible ranges for α in the Oseen-Lamb vortex model.

Solution: (i) Differentiating (8.360) and setting its value to zero, we obtain an equation for r where the maximum velocity occurs

$$\frac{dv}{dr} = -\frac{\Gamma_\infty}{r^2}[1 - \exp(-\frac{\alpha r^2}{4\nu t})] + \frac{\Gamma_\infty \alpha}{2\nu t}\exp(-\frac{\alpha r^2}{4\nu t}) = 0 \tag{8.364}$$

This can be rearranged as

$$(\frac{\alpha r^2}{2\nu t} + 1)\exp(-\frac{\alpha r^2}{4\nu t}) = 1 \tag{8.365}$$

This equation is the same as that for the Burgers-Rott vortex model found in (8.331) if the following identification is made:

$$\frac{2\nu t}{\alpha} \leftarrow \frac{\nu}{a} \tag{8.366}$$

Therefore, the result obtained numerically in (8.333) can be used readily as

$$r_{\max} = 1.5813(\frac{2\nu t}{\alpha})^{1/2} \tag{8.367}$$

This can be rewritten as

$$r_{\max} = 2.2363(\frac{\nu t}{\alpha})^{1/2} \tag{8.368}$$

According to Dergarabedian and Fendell (1976), tornadoes sometimes go through a sequence of transitions of cyclic decay and re-intensification of vortex strength for a period of about 5 minutes. The maximum increase rate of swirl speed may last for up to 90 seconds in each cycle. Thus, the given 60 to 90 seconds appear to be a reasonable estimate of the time at maximum velocity. From the previous calculations of speed given by using the Rankine and Burgers-Rott vortex models, the maximum velocities are in the order of 50 to 125 m/s, which again agree roughly with the given values.

Substitution of (8.368) into (8.358) gives

$$v_{\max} = \frac{\Gamma_\infty}{2.2363}\left(\frac{\alpha}{\nu t}\right)^{1/2}[1-\exp(-2.2363^2)]$$
$$= 0.3191\Gamma_\infty\left(\frac{\alpha}{\nu t}\right)^{1/2} \tag{8.369}$$

(ii) Adopting v_{max} = 125 m/s and t_{max} = 90 s, we have

$$\alpha = \left[\frac{v_{\max}}{0.3191\Gamma_\infty}\right]^2 \nu t = \left[\frac{125}{0.3191\times 10^4}\right]^2 10\times 90$$
$$= 1.3810 \tag{8.370}$$

Similarly, adopting v_{max} = 125 m/s and t_{max} = 60 s, we have

$$\alpha = 1.3810\times\frac{60}{90} = 0.9207 \tag{8.371}$$

For the lower bound velocity, we use v_{max} = 50 m/s and t_{max} = 60 s, and we have

$$\alpha = 0.9207\times\left(\frac{50}{125}\right)^2 = 0.1473 \tag{8.372}$$

Similarly, adopting v_{max} = 50 m/s and t_{max} = 90 s, we have

$$\alpha = 0.1473\times\frac{90}{60} = 0.2210 \tag{8.373}$$

Therefore, we conclude that the appropriate range of α for the Oseen-Lamb vortex model is from 0.14 to 1.38.

8.9.8 Sullivan Vortex Model

So far, we have only considered a single-cell vortex model. That is, a model with a one-celled vortex with central updraft. This is true when the swirl ratio S is small, which is defined as (Lewellen, 1976)

$$S = \frac{\pi\Gamma_\infty r_0}{q} \tag{8.374}$$

where q is the updraft flow rate, and r_0 is a characteristic outer radius of the vortex. The swirl ratio is the relative strength of circulation to the axial updraft. However, when the swirl ratio is large, an axial downdraft will develop in the core (also known as the "eye") such that a two-celled structure is developed, as illustrated in Figure 8.14. Field observations suggest that tornadoes were normally formed as a one-celled vortex with moderate strength, then under certain perturbations and subject to some not-fully-understood mechanisms transform into a more severe, but shorter-lived two-celled vortex. The appearance of an "eye" structure is considered as a signature or necessary condition of an intense tornado. The situation is very much similar to the case of hurricanes or tropical cyclones that strong hurricanes or typhoons are always accompanied by a well-defined "eye." Therefore, the studies of the two-celled vortex are of paramount importance.

In particular, Figure 8.14(a) shows the cross-section of the wind field of a one-celled vortex that consists of an updraft in the axis, whereas the two-celled vortex shown in Figure 8.14(b) consists of a funnel with a downdraft and updraft cell in the axial zone. Field observations and laboratory simulations both suggest that a tornado with a strong swirl ratio will lead to a two-celled structure as illustrated in Figure 8.14(b).

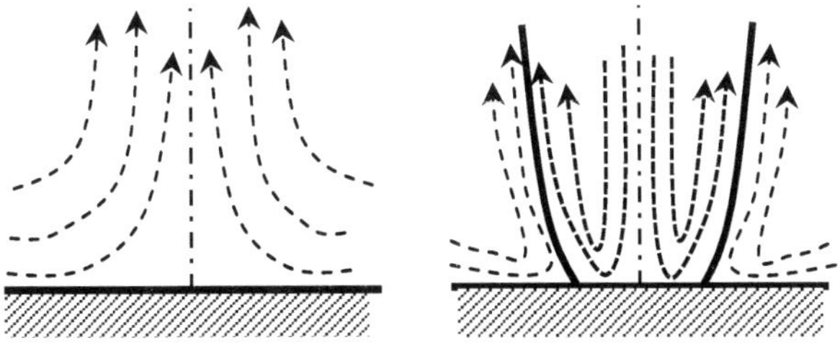
(a) One-celled tornado (b) Two-celled tornado

Figure 8.14 The cross-section of the wind fields for a one-celled vortex and a two-celled vortex

The Sullivan vortex model is the first two-celled vortex model, and will be discussed in this section. The price to pay is that the mathematical analysis involved is more complicated. In particular, the following velocity field is sought

$$u = u(r), \quad v = v(r), \quad w = f(r)g(z) \tag{8.375}$$

The axisymmetric components of vorticity given in (8.249) to (8.251) are

$$\omega_r = \frac{1}{r}\frac{\partial w}{\partial \theta} - \frac{\partial v}{\partial z} \tag{8.376}$$

$$\omega_\theta = \frac{\partial u}{\partial z} - \frac{\partial w}{\partial r} \tag{8.377}$$

$$\omega_z = \frac{1}{r}\frac{\partial (rv)}{\partial r} \tag{8.378}$$

In view of (8.375), these are further reduced to

$$\omega_r = 0\,, \ \omega_\theta = -f'(r)g(z)\,, \ \omega_z = \frac{1}{r}\frac{d(rv)}{dr} \tag{8.379}$$

The continuity given in (8.245) can be written as

$$\frac{1}{r}\frac{d(rv)}{dr} = -fg' \tag{8.380}$$

Since the left-hand side is only a function of r, we must require

$$g'(z) = 2a \tag{8.381}$$

where a is an unknown constant. Integrating from an initial value of z_0, we find

$$g(z) = 2a(z - z_0) \tag{8.382}$$

Substitution of (8.382) into (8.375) and (8.380) gives

$$w = 2a(z - z_0)f(r) \tag{8.383}$$

$$\frac{1}{r}\frac{d(ru)}{dr} = -2af(r) \tag{8.384}$$

Substitution of (8.382) into (8.379) gives the vorticity

$$\omega_r = 0, \ \omega_\theta = -2a(z-z_0)f'(r), \ \omega_z = \frac{1}{r}\frac{d(rv)}{dr} \tag{8.385}$$

In view of (8.375), the angular momentum equation given in (8.277) can be reduced to

$$u\frac{\partial(rv)}{\partial r} + w\frac{\partial(rv)}{\partial z} = \nu r\{\frac{\partial}{\partial r}(\frac{1}{r}\frac{\partial(rv)}{\partial r})\} \tag{8.386}$$

Using the definition of ω_z given in (8.385) into (8.386), we get

$$\frac{u}{\nu}\omega_z = \frac{d\omega_z}{dr} \tag{8.387}$$

This can be integrated as

$$\ln \omega_z = \frac{1}{\nu}\int^r u(\xi)d\xi + \bar{C} \tag{8.388}$$

Therefore, the vorticity is

$$\omega_z = C\exp\{\frac{1}{\nu}\int^r u(\xi)d\xi\} \tag{8.389}$$

where C is a constant. Equation (8.385) can be integrated to yield v

$$v = \frac{1}{r}\int^r \omega_z(\xi)\xi d\xi + \frac{C_1}{r} \tag{8.390}$$

Since we must have a finite solution for v as $r \to \infty$, this boundedness requirement yields $C_1 = 0$. Substitution of (8.389) into (8.390) leads to the following expression for v

$$v = \frac{C}{r}\int^r \xi \exp\{\frac{1}{\nu}\int^\xi u(\bar{\xi})d\bar{\xi}\}d\xi \tag{8.391}$$

The axisymmetric radial component of the Helmholtz equation given in (8.258) is

$$\frac{\partial\omega_r}{\partial t} + (u\frac{\partial}{\partial r} + w\frac{\partial}{\partial z})\omega_r = (\omega_r\frac{\partial}{\partial r} + \omega_z\frac{\partial}{\partial z})u + \nu[\frac{1}{r}\frac{\partial}{\partial r}(r\frac{\partial\omega_r}{\partial r}) + \frac{\partial^2\omega_r}{\partial z^2} - \frac{\omega_r}{r^2}] \tag{8.392}$$

It is straightforward to see that (8.392) is identically satisfied by (8.375) to (8.378). The θ-component of (8.258) reads

$$\begin{aligned}\frac{\partial\omega_\theta}{\partial t} + (u\frac{\partial}{\partial r} + w\frac{\partial}{\partial z})\omega_\theta + \frac{v\omega_r}{r} &= (\omega_r\frac{\partial}{\partial r} + \omega_z\frac{\partial}{\partial z})v \\ &\quad + \frac{u\omega_\theta}{r} + \nu[\frac{1}{r}\frac{\partial}{\partial r}(r\frac{\partial\omega_\theta}{\partial r}) + \frac{\partial^2\omega_\theta}{\partial z^2} - \frac{\omega_\theta}{r^2}]\end{aligned} \tag{8.393}$$

For the present case, (8.393) is reduced to

$$(u\frac{\partial}{\partial r} + w\frac{\partial}{\partial z})\omega_\theta = \frac{u\omega_\theta}{r} + \nu[\frac{1}{r}\frac{\partial}{\partial r}(r\frac{\partial\omega_\theta}{\partial r}) - \frac{\omega_\theta}{r^2}] \tag{8.394}$$

Differentiating the second equation of (8.385) gives

$$\frac{\partial\omega_\theta}{\partial r} = -2a(z-z_0)f'' \tag{8.395}$$

Plugging (8.395) into (8.394), we obtain an ODE for f and u as

$$uf'' + 2aff' = f'\frac{u}{r} + \frac{\nu}{r}(rf'')' - \frac{\nu}{r^2}f' \tag{8.396}$$

This equation can be recast as

$$(ru-v)\frac{d}{dr}(\frac{1}{r}\frac{df}{dr})-v\frac{d^3 f}{dr^3}+2af\frac{df}{dr}=0 \tag{8.397}$$

Recalling (8.380), we have another equation for u and f

$$\frac{1}{r}\frac{d(ru)}{dr}=-2af(r) \tag{8.398}$$

Equations (8.397) and (8.398) provide two coupled ODEs for u and f. This system is nonlinear and, in general, is not easy to solve. Let us proceed by assuming that f is given as

$$f=C+A\exp(-\alpha r^2) \tag{8.399}$$

where A and C are constants. Let us consider a special case that $A=0$ in (8.399), or

$$f=C \tag{8.400}$$

It is straightforward to see that (8.397) is identically satisfied and substitution of (8.400) into (8.398) gives

$$\frac{1}{r}\frac{d(ru)}{dr}=-2aC \tag{8.401}$$

Multiplying both sides by r and integrating the result, we have

$$u=-aCr+\frac{C_1}{r} \tag{8.402}$$

This radial velocity field indicates that for near field $r\to 0$ we have $u>0$ (outward), but for far field $r\to\infty$ we have $u<0$ (inward). This already resembles a two-celled structure as illustrated in Figure 8.14(b).

Let us consider a more general case of

$$f=1+A\exp(-\alpha r^2) \tag{8.403}$$

Substitution of (8.403) into (8.398) gives

$$\frac{1}{r}\frac{d(ru)}{dr}=-2a[1+A\exp(-\alpha r^2)] \tag{8.404}$$

Integrating both sides, we obtain

$$u=-ar+\frac{aA}{\alpha r}(e^{-\alpha r^2}-\beta) \tag{8.405}$$

where β is a constant. Similar to (8.402), this also indicates a two-celled structure. Differentiating (8.403), we get

$$\frac{df}{dr}=-2\alpha rAe^{-\alpha r^2} \tag{8.406}$$

$$\frac{d}{dr}(\frac{1}{r}\frac{df}{dr})=4\alpha^2 rAe^{-\alpha r^2} \tag{8.407}$$

$$\frac{d^3 f}{dr^3}=12\alpha^2 rAe^{-\alpha r^2}-8r^3\alpha^3 Ae^{-\alpha r^2} \tag{8.408}$$

Substitution of (8.406) to (8.408) into (8.397) gives

$$\alpha(2v\alpha-a)r^2-4\alpha v-a-aA\beta=0 \tag{8.409}$$

Thus, we must have

$$2\nu\alpha - a = 0 \tag{8.410}$$

$$4\alpha\nu + a + aA\beta = 0 \tag{8.411}$$

The solutions of (8.410) and (8.411) are

$$\alpha = \frac{a}{2\nu}, \quad A = -\frac{3}{\beta} \tag{8.412}$$

Substitution of (8.412) into (8.403) results in

$$f(r) = 1 - \frac{3}{\beta} e^{-\frac{ar^2}{2\nu}} \tag{8.413}$$

Substitution of (8.413) into (8.405) yields

$$u = -ar + \frac{6\nu}{r}[1 - \frac{1}{\beta} e^{-\frac{ar^2}{2\nu}}] \tag{8.414}$$

Finally, putting (8.413) into (8.383), we obtain the vertical velocity field as

$$w = 2a(z - z_0)(1 - \frac{3}{\beta} e^{-\frac{ar^2}{2\nu}}) \tag{8.415}$$

Substituting the function u found in (8.414) into (8.389), we have

$$\omega_z = C\exp\{-\frac{ar^2}{2\nu} + 6\int (1 - \frac{1}{\beta} e^{-\frac{ar^2}{2\nu}})\frac{1}{r} dr\} \tag{8.416}$$

To further evaluate the integral, we introduce the following change of variables

$$s = \frac{ar^2}{2\nu} \tag{8.417}$$

Using this new variable, (8.416) becomes

$$\omega_z = C\exp\{-s + 3\int (1 - \frac{e^{-s}}{\beta})\frac{1}{s} ds\} \tag{8.418}$$

Carrying out Taylor's expansion inside the integral, we have

$$\int (1 - \frac{e^{-s}}{\beta})\frac{1}{s} ds = \int (1 - \frac{1}{\beta} + \frac{s}{\beta} - \frac{s^2}{2\beta} + \ldots)\frac{1}{s} ds \tag{8.419}$$

We see that the integral diverges at the lower limit at $s = 0$. To remove the singularity, we extract the singular point as

$$\begin{aligned} \omega_z &= C\exp\{-s + 3\int (1 - \frac{1}{\beta})\frac{1}{s} ds + \frac{3}{\beta}\int^s (\frac{1 - e^{-\tau}}{\tau}) d\tau\} \\ &= C\exp\{-s + \frac{3}{\beta}\int^s (\frac{1 - e^{-\tau}}{\tau}) d\tau + 3(1 - \frac{1}{\beta})\ln s\} \\ &= Cs^{3(1-1/\beta)} \exp\{-s + \frac{3}{\beta}\int^s (\frac{1 - e^{-\tau}}{\tau}) d\tau\} \end{aligned} \tag{8.420}$$

To evaluate v, we can use (8.390) to get

$$v = \frac{1}{r}\int_0^r \xi\omega_z(\xi) d\xi + \frac{C_3}{r} \tag{8.421}$$

For a finite v at $r = 0$, we require $C_3 = 0$. With this result, we can substitute (8.416) into (8.421) to obtain

$$v = \frac{C}{r}\int_0^r \xi \exp\{-\frac{a\xi^2}{2\nu} + 6\int_0^\xi [1 - \frac{1}{\beta}\exp(-\frac{a\rho^2}{2\nu})]\frac{1}{\rho}d\rho\}d\xi \tag{8.422}$$

This can be simplified by introducing the following change of variables

$$\zeta = \frac{a\xi^2}{2\nu}, \quad s = \frac{a\rho^2}{2\nu} \tag{8.423}$$

In view of (8.420), the integral in (8.422) can be simplified to

$$v = \frac{C\nu}{ar} H_\beta(\frac{ar^2}{2\nu}) \tag{8.424}$$

$$H_\beta(x) = \int_0^x t^{3(1-1/\beta)} \exp[-t + \frac{3}{\beta}\int_0^t (\frac{1-e^{-s}}{s})ds]dt \tag{8.425}$$

To find the constant C, we can integrate v to get the far-field circulation as

$$\begin{aligned} 2\pi\Gamma_\infty &= \lim_{R\to\infty} \oint_{C_R} \boldsymbol{v}\cdot d\boldsymbol{r} = \lim_{R\to\infty} v(R)2\pi R \\ &= \lim_{R\to\infty} \frac{C\nu 2\pi}{a} H_\beta(\frac{aR^2}{2\nu}) = \frac{C\nu 2\pi}{a} H_\beta(\infty) \end{aligned} \tag{8.426}$$

Therefore, the constant C becomes

$$C = \frac{\Gamma_\infty a}{\nu} \frac{1}{H_\beta(\infty)} \tag{8.427}$$

Substitution of (8.427) into (8.424) gives the tangential velocity as

$$v = \frac{\Gamma_\infty}{r} \frac{H_\beta(\frac{ar^2}{2\nu})}{H_\beta(\infty)} \tag{8.428}$$

The Sullivan vortex model corresponds to β = 1. Thus, the Sullivan vortex model is given by the following velocity field

$$\frac{dr}{dt} = u = -ar + \frac{6\nu}{r}[1 - e^{-\frac{ar^2}{2\nu}}] \tag{8.429}$$

$$\frac{d\theta}{dt} = \frac{v}{r} = \frac{\Gamma_\infty}{r^2} \frac{H_1(\frac{ar^2}{2\nu})}{H_1(\infty)} \tag{8.430}$$

$$\frac{dz}{dt} = w = 2a(z - z_0)(1 - 3e^{-\frac{ar^2}{2\nu}}) \tag{8.431}$$

Numerical calculation shows that $H(\infty)$ is about 37.9044. As expected, for $r \to \infty$, (8.428) gives $\Gamma_\infty = rv$. The three-dimensional updraft from the far field in the Sullivan vortex model is illustrated in Figure 8.15.

To examine the two-celled structure, we can set

$$\frac{dz}{dt} = w = 0 \tag{8.432}$$

The critical radius at which the vertical velocity change from downward flow to upward flow is

$$1-\exp(-\frac{ar_*^2}{2\nu})=0 \quad (8.433)$$

Thus, we have

$$r_*=[\frac{2\nu}{a}\ln(3)]^{1/2}=1.4823(\frac{\nu}{a})^{1/2} \quad (8.434)$$

Similarly, to examine the radial flow structure, we set

$$\frac{dr}{dt}=u=0 \quad (8.435)$$

The critical radius for the change of sign of u can be found as

$$ar_0^2=6\nu[1-\exp(-\frac{ar_0^2}{2\nu})] \quad (8.436)$$

The actual value of r_0 depends on the ratio ν/a, and numerical calculation shows

$$r_0=2.375475(\frac{\nu}{a})^{1/2} \quad (8.437)$$

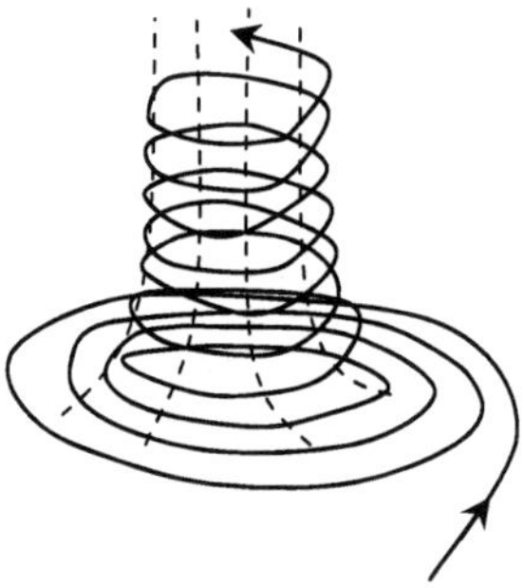

Figure 8.15 Illustration of three-dimensional updraft in the Sullivan vortex model

Mathematically, $r = r_0$ is a limit circle. That is, the velocity field will converge to the limit circle, or the core of the vortex. Comparison of (8.434) and (8.437) shows that r_0 is always larger than r_*. Let us consider the radial velocity at r_*

$$u(r=r_*)=-2\nu\ln(3)+4\nu=1.8028\nu>0 \quad (8.438)$$

In view of (8.429), (8.434), and (8.438), we see more generally that

$$u>0, \quad \text{when } r<r_0$$
$$u<0, \quad \text{when } r>r_0 \quad (8.439)$$

In view of (8.431) and (8.434), we have

$$w<0, \quad \text{when } r<r_*$$
$$w>0, \quad \text{when } r>r_* \quad (8.440)$$

Combining (8.439) and (8.440), the flow structure for the Sullivan vortex is illustrated in Figure 8.16.

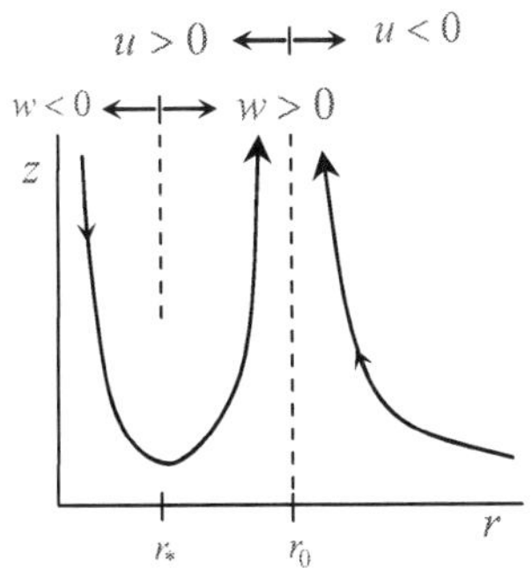

Figure 8.16 The direction of vertical and radial flows as a function of *r* in the Sullivan vortex

8.10 SUMMARY AND FURTHER READING

Geophysical flows dictate many different phenomena on Earth, leading to observed weather systems and ocean currents. More extreme events include hurricanes, tropical cyclones, tornadoes, storm surges, seiches, and tsunami. Our coverage is minimal here and only addresses those problems that allow simple analytic solutions to be derived. The main solutions considered here include those for the Ekman top layer in ocean, storm surge, and tornadoes. An excellent introduction to geophysical flows was given by Pedlosky (1987), by Cushman-Roisin (1994), and by Cushman-Roisin and Beckers (2009). Various kinds of topics have been left out in this chapter. Upper atmospheric currents in the form of Rossby waves are very important in controlling climate change and navigation of airplanes. Kelvin waves along coastal areas will influence the tides. Rogue waves in the ocean are of utmost importance for the safety of ocean liners and cruise ships, and this problem will be covered in a little more detail in Chapter 9 in the context of solitons through the discussion of the nonlinear Schrödinger's equation. Nonlinear terms in the Navier-Stokes equation, which controls all fluid on Earth, were originated from the advective force only. Yet this nonlinearity makes the fluid flows on Earth so unpredictable. For example, meteorological agencies around the globe issue only 14-day weather forecasts. Even with this short time window of forecast, we all have experienced that the forecast is often unreliable for any medium term of planning (such as whether we should cancel an open water swimming competition on the coming Sunday). Professor Edward Lorenz at MIT discovered in the 1960s that a tiny change in the initial parameters in weather simulations leads to a completely different weather scenario. His discovery of the chaotic behavior of atmospheric dynamics is one of the pioneering works on the development of deterministic chaos and strange attractors. The notion of the "butterfly effect" is associated with the unpredictability of the weather system, and says that the flipping of a butterfly's wings will subsequently induce a storm in another part of the globe.

For more discussions on storm surge, we refer to Pugh (1987), Murty (1984), and Gonnert et al. (2001). In the United States, the most popular programs used for storm surge simulations are SLOSH (the abbreviation of Sea, Lake, Overland Hurricane) by Chester Jelesnianski and ADCIRC (the abbreviation of Advanced Circulation) by Rick Luethich. SLOSH employs the finite difference method and it is adopted by the National Weather Service. ADCIRC employs the finite element method in space and the finite difference method in time. While SLOSH did not take into account tide, precipitation, river flow, and wind-driven waves, these effects are incorporated in ADCIRC. For this reason, ADCIRC has been adopted by many US government agencies, whereas SLOSH is mainly for the use of civilian activities.

For references on tornadoes, the 1976 Proceedings on the Symposium on Tornadoes: Assessment of Knowledge and Implications for Man compiled and surveyed most essential information up to 1976 (Peterson, 1976). For more updated books on vortices dynamics related to tornadoes, the reader can refer to Wu et al. (2015), Alekseenko et al. (2007), and Bluestein (2013). A more general introduction to dynamic meteorology was given by Holton (2004). Waves in oceans and atmosphere are covered by Pedlosky (2003). In general, waves in fluid are covered by Lighthill (1978) and by Mei et al. (2005).

Geophysical flows have become more important in recent years, as greenhouse effects, climate changes, rise of sea level, and more frequent occurrence of extreme weather is believed to be affecting the activities of human beings, including fisheries and tourists. This chapter serves as an introduction to this difficult but important topic.

8.11 PROBLEMS

Problem 8.1 For the case of the Ekman layer on the surface, show that the maximum flow velocity along the x-direction appears at the following depth

$$\frac{z}{d} = \frac{\pi}{4} + \tan^{-1}(\frac{\tau_x - \tau_y}{\tau_y + \tau_x}) \tag{8.441}$$

Problem 8.2 For the case of the Ekman layer on the surface, show that the maximum flow velocity along the y-direction appears at the following depth

$$\frac{z}{d} = \frac{\pi}{4} + \tan^{-1}(\frac{\tau_y - \tau_x}{\tau_y + \tau_x}) \tag{8.442}$$

Problem 8.3 Prove the validity of (8.214) for the divergence in the surface Ekman layer.

Problem 8.4 Prove the validity of (8.274) by the following steps:

(i) For the special case of $u = w = 0$ and $v \neq v(\theta, z)$, show that (8.277) can be simplified to

$$\frac{\partial(rv)}{\partial t} = \nu r\{\frac{\partial}{\partial r}(\frac{1}{r}\frac{\partial(rv)}{\partial r})\} \tag{8.443}$$

(ii) Let $\Gamma = rv$, and show that the governing equation for Γ is

$$\frac{\partial\Gamma}{\partial t} = \nu r\{\frac{\partial}{\partial r}(\frac{1}{r}\frac{\partial\Gamma}{\partial r})\} \tag{8.444}$$

(iii) Show that the vertical component of the vorticity given in (8.253) becomes

$$\omega_z = \frac{1}{r}\frac{\partial\Gamma}{\partial r} \tag{8.445}$$

(iv) Prove the following identities by differentiating (8.445)

$$\frac{\partial\omega_z}{\partial r} = -\frac{1}{r^2}\frac{\partial\Gamma}{\partial r} + \frac{1}{r}\frac{\partial^2\Gamma}{\partial r^2} \tag{8.446}$$

$$\frac{\partial^2\omega_z}{\partial r^2} = \frac{2}{r^3}\frac{\partial\Gamma}{\partial r} - \frac{2}{r^2}\frac{\partial^2\Gamma}{\partial r^2} + \frac{1}{r}\frac{\partial^3\Gamma}{\partial r^3} \tag{8.447}$$

(v) Show that the right-hand side of (8.444) is

$$\nu r\{\frac{\partial}{\partial r}(\frac{1}{r}\frac{\partial\Gamma}{\partial r})\} = \nu r\frac{\partial\omega_z}{\partial r} \tag{8.448}$$

(vi) Differentiate (8.448) with respect to r and the result divided by r to show the validity of the following formula:

$$\frac{1}{r}\frac{\partial}{\partial r}[\nu r\frac{\partial}{\partial r}(\frac{1}{r}\frac{\partial\Gamma}{\partial r})] = \nu\{\frac{1}{r^3}\frac{\partial\Gamma}{\partial r} - \frac{1}{r^2}\frac{\partial^2\Gamma}{\partial r^2} + \frac{1}{r}\frac{\partial^3\Gamma}{\partial r^3}\} \tag{8.449}$$

(vii) Finally, use (8.444) to (8.449) to prove

$$\frac{\partial\omega_z}{\partial t} = \nu(\frac{\partial^2\omega_z}{\partial r^2} + \frac{1}{r}\frac{\partial\omega_z}{\partial r}) \tag{8.450}$$

This completes the proof of (8.274).

Problem 8.5 Reconsider the derivation of the Burgers-Rott vortex model by considering the following change of variables:

$$\eta = \frac{\alpha r}{\sqrt{\nu t}} \tag{8.451}$$

$$\Gamma = f(\eta) \tag{8.452}$$

(i) Show that the ODE for f becomes

$$\frac{d^2 f}{d\eta^2} + (\frac{\eta}{2\alpha} - \frac{1}{\eta})\frac{df}{d\eta} = 0 \tag{8.453}$$

(ii) Show that Γ is

$$\Gamma(r,t) = \Gamma_\infty[1 - \exp(-\frac{\alpha r^2}{4\nu t})] \tag{8.454}$$

Problem 8.6 Show that the maximum velocity in the Burgers-Rott vortex model is approximately

$$v_{\max} = 0.45125\Gamma_\infty (\frac{a}{\nu})^{1/2} \tag{8.455}$$

Problem 8.7 Show that the axisymmetric components of vorticity of the Sullivan vortex model are

$$\omega_r = 0 \tag{8.456}$$

$$\omega_\theta = -\frac{6a^2 r}{\nu}(z - z_0)\exp(-\frac{ar^2}{2\nu}) \tag{8.457}$$

$$\omega_z = \frac{a\Gamma_\infty}{\nu H(\infty)} \exp[-\frac{ar^2}{2\nu} + 3\int_0^{\frac{ar^2}{2\nu}} (\frac{1-e^{-\varsigma}}{\varsigma}) d\varsigma] \tag{8.458}$$

Problem 8.8 Show that the axisymmetric components of vorticity of the Burgers-Rott vortex model are

$$\omega_r = 0 \tag{8.459}$$

$$\omega_\theta = 0 \tag{8.460}$$

$$\omega_z = \frac{a\Gamma_\infty}{\nu} \exp(-\frac{ar^2}{2\nu}) \tag{8.461}$$

Problem 8.9 The total dissipation per unit axial length is defined:

$$E_d = \rho\nu \int_0^\infty 2\pi r \omega_z^2 dr \tag{8.462}$$

Show for the Burgers-Rott vortex model that the total dissipation per unit axial length is independent of the effective turbulent eddy viscosity ν and is given by

$$E_d = \rho\pi a \Gamma_\infty^2 \tag{8.463}$$

Problem 8.10 For the following Helmholtz vorticity equation, there exists a similarity solution:

$$\frac{\partial \omega_z}{\partial t} = \nu(\frac{\partial^2 \omega_z}{\partial r^2} + \frac{1}{r}\frac{\partial \omega_z}{\partial r}) \tag{8.464}$$

In particular, we assume the following function form

$$\omega_z = T(t) f(\eta) \tag{8.465}$$

where

$$\eta = -\frac{r^2}{4\nu t} \tag{8.466}$$

(i) Show that

$$\frac{\partial \omega_z}{\partial t} = T'f - \frac{\eta}{t} Tf' \tag{8.467}$$

$$\frac{1}{r}\frac{\partial \omega_z}{\partial r} = -\frac{1}{2\nu t} Tf' \tag{8.468}$$

$$\frac{\partial^2 \omega_z}{\partial r^2} = -\frac{\eta}{\nu t} T f'' - \frac{1}{2\nu t} T f' \tag{8.469}$$

(ii) Substitute (8.467) to (8.469) into (8.464) to show

$$\frac{t}{T}\frac{dT}{dt} = -\frac{\eta}{f}\frac{d^2 f}{d\eta^2} + (\eta - 1)\frac{1}{f}\frac{df}{d\eta} = n \tag{8.470}$$

where n is a constant.

(iii) Show that the ODEs for T and f are

$$\frac{d^2 f}{d\eta^2} + (\frac{1}{\eta} - 1)\frac{df}{d\eta} + \frac{n}{\eta} f = 0 \tag{8.471}$$

$$\frac{dT}{dt} - \frac{n}{t} T = 0 \tag{8.472}$$

(iv) Show that the similarity solution for ω_z is

$$\omega_z = C t^n L_n(-\frac{r^2}{4\nu t}) \tag{8.473}$$

where the Laguerre polynomials $L_n(x)$ is the solution of the following equation

$$x\frac{d^2 L_n}{dx^2} + (1 - x)\frac{dL_n}{dx} + nL_n = 0 \tag{8.474}$$

(v) Finally, show that the tangential velocity v can be expressed as:

$$v = -\frac{2C\nu t^{n+1}}{r}\{L_n(-\frac{r^2}{4\nu t}) - \frac{L_{n+1}(-\frac{r^2}{4\nu t})}{n+1}\} + \frac{1}{r} f_1(t) \tag{8.475}$$

where f_1 is an arbitrary function of time.

Hints: Laguerre polynomials will be discussed in more detail in Chapter 11 on quantum mechanics. Note the following identity

$$\int_0^x L_n(t) dt = Ln(x) - \frac{L_{n+1}(x)}{n+1} \tag{8.476}$$

CHAPTER NINE

Nonlinear Wave and Solitons

9.1 INTRODUCTION

One of the first nonlinear ODEs considered analytically by mathematicians was the Riccati equation. The study of the nonlinearity of the Riccati equation reveals that it can be converted to a system of linear ODEs. It is the Riccati equation or other solvable nonlinear ODEs and PDEs that prompts mathematicians to ask a much deeper question. What is the condition that a nonlinear ODE needs to satisfy such that it is convertible to a linear ODE? Before we look into the problem more closely, the Riccati equation is first considered. In particular, we start with the following Riccati equation:

$$\frac{du}{dt} = a(t) + 2b(t)u + u^2 \tag{9.1}$$

To transform this nonlinear first order ODE to linear ODEs, we introduce the following change of variable

$$u = \frac{g}{f} \tag{9.2}$$

Differentiation of (9.2) with respect to t gives

$$u_t = \frac{g_t f - f_t g}{f^2} \tag{9.3}$$

Substitution of (9.2) and (9.3) into (9.1) gives

$$[g_t - a(t)f - b(t)g]f - [f_t + b(t)f + g]g = 0 \tag{9.4}$$

Equation (9.4) is identically satisfied if we set

$$g_t - a(t)f - b(t)g = \lambda(t)g \tag{9.5}$$

$$f_t + b(t)f + g = \lambda(t)f \tag{9.6}$$

Thus, the nonlinear Riccati equation given in (9.1) is converted to a linear system of two first order ODEs. In fact, the Riccati equation can also be converted to a linear second order ODE by the following change of variables (Section 3.2.8 of Chau, 2018):

$$y = \frac{1}{u}\frac{du}{dx} = \frac{u'}{u} \tag{9.7}$$

The resulting equation for u is

$$u'' - 2bu' - au = 0 \tag{9.8}$$

Thus, the Riccati equation is a special kind of nonlinear ODE, which is convertible to linear forms. The natural question to ask is what kinds of nonlinear ODEs and PDEs can be transformed to linear ODEs and PDEs? This intriguing question leads to a branch of mathematics focusing on the general development of exact solutions for nonlinear ODEs and PDEs. A major breakthrough was achieved by Painlevé in searching for the types of second order nonlinear ODEs that are convertible to linear ODEs. Painlevé discovered that six equations are not reducible to "known" differential equations, but are convertible to a linear system. These are called Painlevé transcendents. Among these equations are the KdV equation, Sine-Gordon equation, Boussinesq equation, and nonlinear Schrödinger equations, and their solutions can be expressed in the form of a solution called a soliton. The term soliton was proposed by Zabusky and Kruskal in 1965 to describe a localized solitary wave with a permanent form. This wave does not disperse with distance and does not amplify with distance. It violates the normal principle of superposition. This chapter discusses the soliton solution.

This kind of soliton was first observed in nature by Scottish engineer John Scott Russell in 1834. When Russell conducted experiments on the efficiency of canal boats at the Union Canal in Scotland, he discovered that if the boat was pulled at the "right" speed (corresponding to a Ursell number of about one) by a pair of horses, a wave of "translation" was generated in front of the boat. This wave maintained its form (i.e., did not disperse at all) and traveled with a constant speed upstream. Russell rode on horse to chase the wave for miles before it vanished. Russell gave a very detailed account of the observation of this wave. This wave is now known as a "soliton." It was known that dispersion causes a wave to attenuate along the travel distance, but nonlinearity effects on the other hand lead to shock wave formation or the eventual breaking wave. The formation of a soliton requires a delicate balance between dispersion and nonlinearity such that the shape of the soliton remains constant. Dutch professor Korteweg and his PhD student de Vries were the first to derive a nonlinear PDE that can successfully depict the wave observed by Russell in Scotland. Naturally, their PDE was eventually known as the KdV (short form from their last names) equation, which will be the main focus of this chapter. However, it was later discovered that the same PDE and soliton solution was published in a book by Boussinesq.

It turns out that the soliton was discovered and observed in many areas of engineering and science, not just in fluid mechanics. The following are examples of solitons: a wave traveling along a series of masses connecting by a nonlinear spring (Fermi-Pasta-Ulam problem), light traveling in fiber optics, electric nerve pulse in animals, propagation of deformation along the DNA double helix, propagation of waves in nonlinear medium, and magnetic flux propagation in Josephson junctions (gaps between superconductors).

A nonlinear wave in the form of a soliton remains an active area of research. Mathematical techniques in solving the N-soliton solution were discovered in the last few decades. Many more nonlinear PDEs were found to be solvable in terms of solitons using techniques developed originally for KdV, including the inverse scattering transform (an indirect method of solving scattering problem of Schrödinger equation), the Bäcklund transform, Lax pairs, the Darboux transform, and Hirota's bilinear form. A number of books have been published summarizing this exciting development. This chapter serves as a brief introduction to this rapidly developing subject. We will consider two other famous PDEs which are convertible to linear form, namely the modified KdV equation and the nonlinear Schrödinger equation, and their relation to rogue waves or freak waves observed in oceans. Before our discussion of KdV, we will consider two related equations in the next sections. One is a nonlinear transport equation and the other is a third order dispersive wave equation. Their physical meanings are discussed before they are combined to formulate the final KdV equation.

9.2 NONLINEAR TRANSPORT AND SHOCKS

The following nonlinear transport equation is known as the Poisson or Riemann equation:

$$u_t + uu_x = 0 \tag{9.9}$$

It was derived by Poisson and Riemann in 1820s. This is also a special form of the Burgers equation given in (2.57) of Chau (2018). This nonlinear PDE can be interpreted as a dispersionless KdV equation (a short form of the Korteweg-de Vries equation). The solution for this nonlinear transport equation can be solved by the method of characteristics (Chau, 2018). The solution of (9.9) is constant along the characteristics, which is given by

$$\frac{dx}{dt} = u(x,t) = C \tag{9.10}$$

where C is a constant. Taking the total differential of (9.10), we get

$$\frac{du}{dt} = \frac{\partial u}{\partial x}\frac{dx}{dt} + \frac{\partial u}{\partial t} = \frac{\partial u}{\partial x}u + \frac{\partial u}{\partial t} = 0 \tag{9.11}$$

Note that (9.11) is exactly (9.9). The solution of (9.10) gives the characteristics and, thus, the solution of u is

$$u(x,t) = f(x - ut) \tag{9.12}$$

This agrees with Chapter 6 of Chau (2018) that the general solution for the first order PDE can be an arbitrary function. The main feature of this solution is that the speed of propagation is the solution itself. Figure 9.1 illustrates that this solution will break down eventually. The higher the water surface u, the faster the wave will move. The figure on the left-hand side of Figure 9.1 shows an initial profile of the wave, and (9.12) predicts that the upper part of the initial profile moves faster than the lower part (indicated by the size of the arrows). Thus, we are expecting the wave shown in Figure 9.1 to break sooner or later. In other words, this nonlinear wave solution is not stable. We will, however, show that if dispersion is allowed for, this nonlinearity can be balanced by dispersion. The application of shock wave to traffic flow will be considered in Section 9.4.

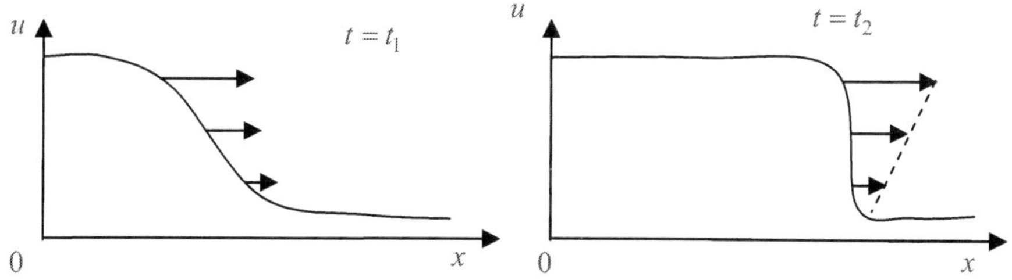

Figure 9.1 Evolution of the nonlinear transport equation

9.3 DISPERSIVE WAVES

A third order dispersive wave can be modeled by

$$u_t + u_{xxx} = 0 \tag{9.13}$$

A more intuitive approach to see the physical meaning of this dispersive wave is to be given in Section 9.5.3. Consider the following change of variables:

$$u = t^{-1/3}F(\zeta), \quad \zeta = \frac{x}{t^{1/3}} \tag{9.14}$$

Differentiating (9.14) with respect to t, we obtain

$$u_t = -\frac{1}{3}t^{-4/3}F(\zeta) - \frac{1}{3}t^{-4/3}\zeta F'(\zeta) \tag{9.15}$$

Differentiating (9.14) with respect to x, we obtain the following formulas

$$u_x = t^{-2/3}F'(\zeta) \tag{9.16}$$

$$u_{xx} = t^{-1}F''(\zeta) \tag{9.17}$$

$$u_{xxx} = t^{-4/3}F'''(\zeta) \tag{9.18}$$

Substitution of (9.18) and (9.15) into (9.13) gives

$$-\frac{1}{3}[F(\zeta) + \zeta F'(\zeta)] + F'''(\zeta) = 0 \tag{9.19}$$

It is straightforward to see that it can be rearranged as

$$-\frac{1}{3}[\zeta F(\zeta)]' + F'''(\zeta) = 0 \tag{9.20}$$

Integrating once with respect to ζ, we get

$$-\frac{1}{3}\zeta F(\zeta) + F''(\zeta) = 0 \tag{9.21}$$

We introduce another change of variables as

$$\zeta = (\frac{1}{3})^{-1/3}\xi \tag{9.22}$$

Using (9.22), we convert (9.21) to the following well-known Airy equation

$$F''(\xi) - \xi F(\xi) = 0 \tag{9.23}$$

Its solution is, of course, the Airy function (Abramowitz and Stegun, 1964)

$$F(\xi) = Ai(\xi) \tag{9.24}$$

Substitution of (9.24) into (9.14) gives the final solution

$$u(x,t) = t^{-1/3}Ai[\frac{x}{(3t)^{1/3}}] \tag{9.25}$$

The prediction of this Airy function is illustrated in Figure 9.2, indicating that a sharp initial undulation of the water profile will be dispersive with time. The time dependence in the solution given in (9.25) suggests that the solution actually decays to zero as $t \to \infty$.

In Section 9.5, we will see that when dispersion interacts with nonlinearity, unexpected things can happen. In the next section, we will first consider a one-dimensional shock wave and its application to a traffic wave.

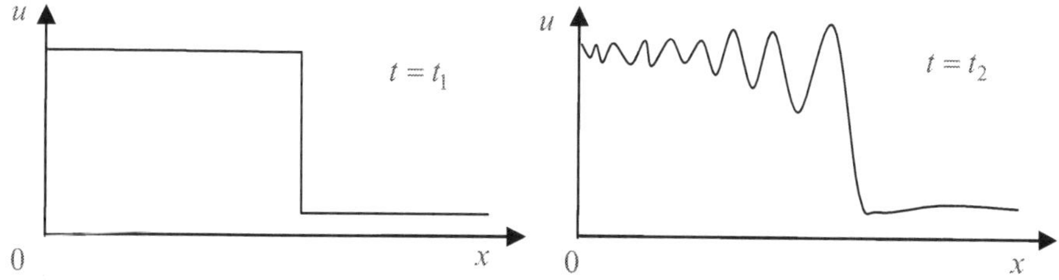

Figure 9.2 Evolution of the nonlinear transport equation

9.4 SHOCK WAVES IN TRAFFIC FLOW

In this section, we discuss the application of a shock wave to traffic flow problems. The presentation here follows that in the interesting book by Banks (1998). Highway traffic flow can be modeled as a fluid flow problem, as shown in Figure 9.3. The inflow of traffic Q is expressed in terms of car per time for highway section dx, whereas outflow is $Q+dQ$. The difference of the inflow and outflow of traffic must be balanced by the number of car changes per time per segment dx. Thus, we can formulate this as a one-dimensional continuity:

$$Q-(Q+\frac{\partial Q}{\partial x}dx)=\frac{\partial K}{\partial t}dx \tag{9.26}$$

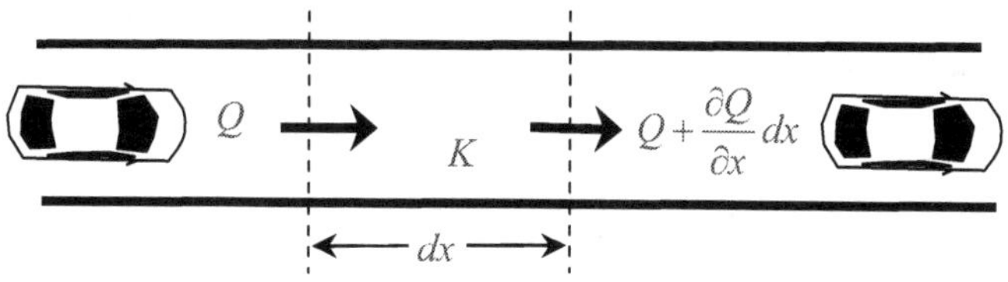

Figure 9.3 Traffic flow as a shock wave

Equation (9.26) can be simplified as

$$\frac{\partial Q}{\partial x}+\frac{\partial K}{\partial t}=0 \tag{9.27}$$

By continuity, the flow rate of the traffic can relate to K as:

$$Q=UK \tag{9.28}$$

Clearly, the flow speed of vehicles on the highway should decrease with the number of cars in the segment of highway K (it is restricted by the safe separation distance between cars). Figure 9.4 illustrates a simple admissible model for U and Q as a function of K, together with a typical plot of data observed on the highway for vehicles per time Q against vehicles per length K. Mathematically, the model shown in Figure 9.4 can be expressed as

$$U = U_*(1-\frac{K}{K_*}) \tag{9.29}$$

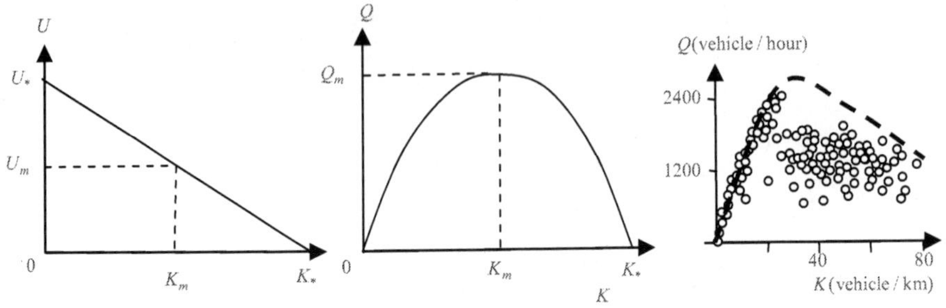

Figure 9.4 Optimum vehicle capacity and vehicle flow on a highway

Substitution of (9.29) into (9.28) gives

$$Q = U_* K(1-\frac{K}{K_*}) \tag{9.30}$$

This is the equation of state. Physically, if the number of cars per highway length attains a maximum capacity K^*, the cars are not moving. The maximum number of cars that can travel on the highway is attained when the car speed is U_m and the cars per length is K_m, as shown in Figure 9.4.

By setting the flow rate to zero, we can find the maximum traffic flow as:

$$\frac{dQ}{dK} = U_* - \frac{2K_m}{K_*}U_* = 0 \tag{9.31}$$

This gives K_m as

$$K_m = K_* / 2 \tag{9.32}$$

Substitution of (9.32) into (9.30) gives the maximum capacity of the highway as

$$Q_m = \frac{U_* K_*}{4} \tag{9.33}$$

We are going to show that when traffic speed changes at a certain segment of the highway, a kinematic wave of K will propagate along the highway. In particular, we consider the following kinematic wave condition:

$$\frac{\partial K}{\partial t} + \frac{dQ}{dK}\frac{\partial K}{\partial x} = 0 \tag{9.34}$$

This can be rewritten as (see Section 6.3.2 of Chau, 2018)

$$\frac{\partial K}{\partial t} + C(K)\frac{\partial K}{\partial x} = 0 \tag{9.35}$$

where $C(K)$ is the wave speed which is a function of the unknown K defined as

$$C(K) = \frac{dQ}{dK} = U_*(1-\frac{2K}{K_*}) \tag{9.36}$$

In obtaining (9.36), we have employed the simple model shown in Figure 9.4 and given in (9.30). More generally, we have the wave speed calculated as

$$C(K) = \frac{d(UK)}{dK} = U + K\frac{dU}{dK} < U \tag{9.37}$$

Note from Figure 9.4 that dU/dK is normally less than zero. Thus, $C(K)$ given in (9.37) is smaller than U. That is, a kinematic wave in traffic flow always moves backward because it actually reflects a braking wave. Compared to (9.12), we see that (9.35) has the characteristics of a shock wave. The solution of (9.35) can be solved using the method of characteristics as

$$K(x,t) = f(x - Ct) \tag{9.38}$$

where the characteristics are (e.g., Chau, 2018)

$$x - Ct = c_1 \tag{9.39}$$

Let us now consider the situation of a shock wave in traffic, shown in Figure 9.5. Suppose there is an accident that occurs at point 2 in Figure 9.5, leading to heavy traffic there; whereas at point 1 along the highway, the traffic is light and the car speed is fast. At point 1, we have

$$K < K_m = K_* / 2 \Rightarrow C = \frac{dQ}{dK} > 0 \tag{9.40}$$

At Point 2, we have

$$K > K_m = K_* / 2 \Rightarrow C = \frac{dQ}{dK} < 0 \tag{9.41}$$

If there is no kinematic wave, we have the stationary condition

$$C = \frac{dQ}{dK} = 0 \tag{9.42}$$

The slope at point 2 in Figure 9.5 represents the characteristics, which is shown as the hatched lines in the upper part of the right diagram. Similarly, the hatched lines in the lower part show the characteristics at Point 1. The intersection line joining Points 1 and 2 represents a shock wave. The speed of the shock wave can be calculated as

$$C_* = \frac{Q_2 - Q_1}{K_2 - K_1} \tag{9.43}$$

Since the traffic is slowing down backward, we can imagine that a shock wave of brake lights is traveling backward. This kind of shock wave in traffic was first studied by Lighthill and Whitham (1955).

Whitham was a PhD student of Lighthill at University of Manchester, and the author of the influential classic book *Linear and Nonlinear Waves* and had taught at MIT and Caltech. Whitham played a key role in developing the applied mathematics program at Caltech and was elected fellow of the Royal Society in 1965. Sir M.J. Lighthill was born in 1924 and passed away in 1998 at the age of 74 after nearly completing a nine-hour swim round the Channel Island of Sark near France for a record 7 times, battling against strong currents. The death resulted from the fact that Lighthill suffered from a mitral valve weakness which had never been diagnosed. Swimming around the Island Sark takes at least 12.5 km, without counting the current effect. This distance is longer than the 10-km Olympic Marathon Swimming Event, which has its inauguration at the 2008 Beijing

Olympics. Lighthill typically took the double-hand backstroke with reverse breaststroke kick. This stroke can be very comfortable and efficient, and world records in masters swimming have been broken by using this style. At the age of 49, he was the first person to swim round the island. Being a renowned expert in fluid mechanics, Sir Lighthill ended his life in fluid. He was a serious swimmer, completing many long adventure swims from age 40. Lighthill was a British applied mathematician and mechanician who made major contributions in fluid dynamics. He was known to others by his middle name James. Lighthill had a formidable memory (he played chess blindfolded at the age of 10) and was extremely good in music, languages and chess. He learned the three-volume Jordan's *Course d'Analyse* in French at the age of 14. He was a classmate of physicist and mathematician Freeman Dyson, and a student of Hardy and Littlewood at Cambridge. He succeeded Paul Dirac as the Lusian Professor of Applied Mathematics at Cambridge. He was the Vice President of the Royal Society, the Provost of University College, London, and President of IUTAM. He was awarded knighthood in 1971. Being a major in mathematics at Cambridge, he learned fluid dynamics during WWII while he was working at the National Physical Laboratory for the supersonic flow past thin wings, eventually leading to the making of the "Concorde." He developed the hodograph transformation method for transonic flow. The multiple-scale perturbation was named the PLK-method by Tsien (Poincaré-Lighthill-Kuo) (see Section 12.5 of Chau, 2018).

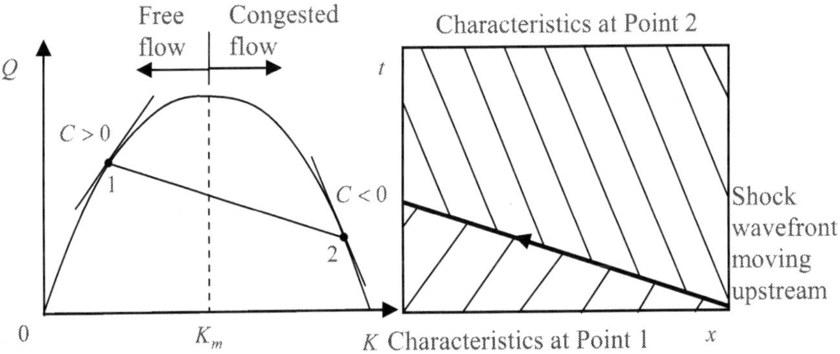

Figure 9.5 Kinematic wave generated by a traffic jam as characteristics

9.5 KDV EQUATION

In this section, we will first derive the KdV equation and consider its solution in the form of a soliton. The KdV equation is appropriate for shallow-water waves as shown in Figure 9.6.

9.5.1 Formulation of KdV

For inviscid flows, the velocity field can be expressed in terms of the flow potential as (see Section 12.10 of Chau, 2018):

$$\boldsymbol{u} = \nabla\phi \tag{9.44}$$

Substitution of (9.44) into continuity equations results in the Laplace equation

$$\nabla^2\phi = \frac{\partial^2\phi}{\partial x^2} + \frac{\partial^2\phi}{\partial z^2} = 0 \tag{9.45}$$

The kinematic condition at the sea bottom is (see (12.424) of Chau, 2018):

$$\frac{\partial\phi}{\partial z} = 0, \quad z = 0 \tag{9.46}$$

The dynamic conditions at the sea surface are (see (12.423) and (12.425) of Chau, 2018):

$$\frac{\partial\phi}{\partial t} + g\eta + \frac{1}{2}[(\frac{\partial\phi}{\partial x})^2 + (\frac{\partial\phi}{\partial z})^2] = 0, \quad z = h + \eta \tag{9.47}$$

$$\frac{\partial\eta}{\partial t} - \frac{\partial\phi}{\partial z} + \frac{\partial\phi}{\partial x}\frac{\partial\eta}{\partial x} = 0, \quad z = h + \eta \tag{9.48}$$

where h is the depth of the water and η is the free water surface shown in Figure 9.6. These equations could also be extracted from (12.452) to (12.455) of Chau (2018) by making the following identifications:

$$\eta_2 \leftarrow \eta, \quad \eta_1 \leftarrow 0, \quad \frac{\partial\eta_1}{\partial x} \leftarrow \frac{\partial\eta}{\partial x}, \quad \phi_2 \leftarrow -\phi, \quad \phi_1 \leftarrow -\phi \tag{9.49}$$

and by dropping the second derivative terms of ϕ. We then introduce the following normalizations:

$$\tilde{t} \leftarrow \frac{t}{\tau_0}, \quad \tilde{u} \leftarrow \frac{u}{k\alpha_1}, \quad \tilde{\eta} \leftarrow \frac{\eta}{a}, \quad \tilde{x} \leftarrow kx, \quad \tilde{z} \leftarrow \frac{z}{h} \tag{9.50}$$

where

$$\tau_0 = \frac{1}{kc} = \frac{\lambda}{2\pi c}, \quad \alpha_1 = \frac{ca}{kh} = \frac{c\alpha}{k}, \quad c = \sqrt{gh}, \quad \beta = (kh)^2 \tag{9.51}$$

Referring to Section 12.11 of Chau (2018) and Figure 9.6, physically we have k being the wave number, c the phase speed, λ the wavelength, a the amplitude of η from mean water level, and g the gravitational constant. Physically, the magnitude of α indicates nonlinearity and that of β reflects a measure of frequency dispersion. Making these substitutions in (9.45), and dropping the tilde sign, we have the Laplace equation becoming

$$\beta\frac{\partial^2\phi}{\partial x^2} + \frac{\partial^2\phi}{\partial z^2} = 0 \tag{9.52}$$

The conditions at the sea bottom and surface become

$$\frac{\partial\phi}{\partial z} = 0, \quad z = 0 \tag{9.53}$$

$$\frac{\partial\phi}{\partial t} + \eta + \frac{\alpha}{2}(\frac{\partial\phi}{\partial x})^2 + \frac{\alpha}{2\beta}(\frac{\partial\phi}{\partial z})^2 = 0, \quad z = 1 + \alpha\eta \tag{9.54}$$

$$\frac{\partial \eta}{\partial t}-\frac{1}{\beta}\frac{\partial \phi}{\partial z}+\alpha\frac{\partial \phi}{\partial x}\frac{\partial \eta}{\partial x}=0, \quad z=1+\alpha\eta \tag{9.55}$$

where α is defined in (9.51). We apply the following multiple-scale Galilean transform (i.e., a right-moving frame):

$$T=\varepsilon t, \quad X=x-t, \quad Z=z, \quad \frac{\partial}{\partial t}=-\frac{\partial}{\partial X}+\varepsilon\frac{\partial}{\partial T}, \quad \varepsilon=\frac{h}{\lambda} \tag{9.56}$$

Using these normalizations, we obtain the following set of four equations

$$\beta\frac{\partial^2 \phi}{\partial X^2}+\frac{\partial^2 \phi}{\partial Z^2}=0 \tag{9.57}$$

$$\frac{\partial \phi}{\partial Z}=0, \quad Z=0 \tag{9.58}$$

$$(-\frac{\partial}{\partial X}+\varepsilon\frac{\partial}{\partial T})\phi+\eta+\frac{\alpha}{2}(\frac{\partial \phi}{\partial X})^2+\frac{\alpha}{2\beta}(\frac{\partial \phi}{\partial Z})^2=0, \quad Z=1+\alpha\eta \tag{9.59}$$

$$(-\frac{\partial}{\partial X}+\varepsilon\frac{\partial}{\partial T})\eta-\frac{1}{\beta}\frac{\partial \phi}{\partial Z}+\alpha\frac{\partial \phi}{\partial X}\frac{\partial \eta}{\partial X}=0, \quad Z=1+\alpha\eta \tag{9.60}$$

We now apply a perturbation expansion for both ϕ and η in a series of small parameters β and α:

$$\phi=\phi_0+\beta\phi_1+\beta^2\phi_2+... \tag{9.61}$$

$$\eta=\eta_0+\alpha\eta_1+\alpha^2\eta_2+... \tag{9.62}$$

Substituting (9.61) into (9.57) and applying (9.58) for the constant order system and the first order system of β gives the following equations

$$\phi_0=\phi_0(X,T) \tag{9.63}$$

$$\phi_1=-\frac{1}{2}\frac{\partial^2 \phi_0}{\partial X^2}Z^2+\phi_{10}(X,T) \tag{9.64}$$

where ϕ_1 and ϕ_{10} are unknown functions to be determined. The proof of (9.63) and (9.64) are set as Problem 9.13. Substitution of (9.61) to (9.64) into (9.59) to (9.60) with $\alpha \approx \beta \approx \varepsilon$ leads to two equations in expansions of ε. Collecting the constant order terms in these equations, we obtain

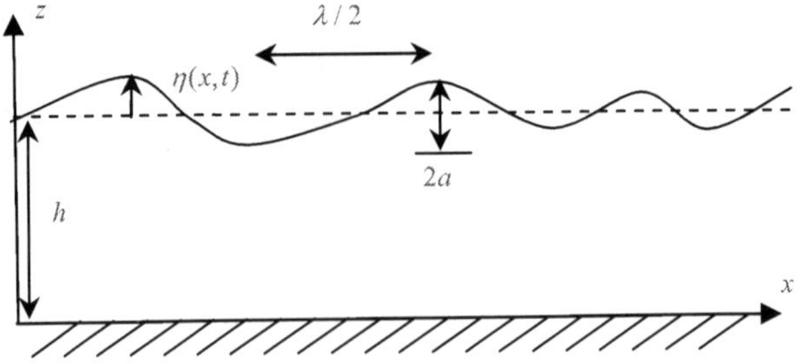

Figure 9.6 Shallow-water waves modeled by KdV

$$O(1): \quad \eta_0 = \frac{\partial \phi_0}{\partial X}$$
$$\frac{\partial^2 \phi_0}{\partial X^2} = \frac{\partial \eta_0}{\partial X} \tag{9.65}$$

Collecting the first order term of ε, we have

$$O(\varepsilon): \quad \eta_1 - \frac{\partial \phi_{10}}{\partial X} = -\frac{\partial \phi_0}{\partial T} - \frac{1}{2}\eta_0^2 - \frac{1}{2}\frac{\partial^2 \eta_0}{\partial X^2}$$
$$\frac{\partial \eta_1}{\partial X} - \frac{\partial^2 \phi_{10}}{\partial X^2} = \frac{\partial \eta_0}{\partial T} + 2\eta_0 \frac{\partial \eta_0}{\partial X} - \frac{1}{6}\frac{\partial^3 \eta_0}{\partial X^3} \tag{9.66}$$

Clearly, the solution of (9.65) is simply

$$\eta_0 = \frac{\partial \phi_0}{\partial X} \tag{9.67}$$

Differentiating the first part of (9.66) with respect to X and subtracting it from the second part of (9.66), we can eliminate η_1 to obtain

$$\frac{\partial \eta_0}{\partial T} + \frac{3}{2}\eta_0 \frac{\partial \eta_0}{\partial X} + \frac{1}{6}\frac{\partial^3 \eta_0}{\partial X^3} = 0 \tag{9.68}$$

We now can rewrite η_0 as u_0 to get the following KdV

$$\frac{\partial u_0}{\partial T} + \frac{3}{2}\eta_0 \frac{\partial u_0}{\partial X} + \frac{1}{6}\frac{\partial^3 u_0}{\partial X^3} = 0 \tag{9.69}$$

The constants in (9.69) can be scaled arbitrarily to the standard form, and this will be done in the next section.

9.5.2 Scale Invariance

To do the scaling, let us consider the following general form of (9.69):

$$u_t + \alpha u u_x + \beta u_{xxx} = 0 \tag{9.70}$$

where α and β are arbitrary constants. We now introduce the following change of variables:

$$x \to a\hat{x}, \quad t \to b\hat{t}, \quad u \to c\hat{u} \tag{9.71}$$

where a, b, and c are unknown constants to be fixed in scaling. Using these new variables, we have the first term in (9.70) becoming

$$u_t = \frac{\partial}{\partial t} c\hat{u} = \frac{c}{b}\frac{\partial \hat{u}}{\partial \hat{t}} \tag{9.72}$$

Similarly, the spatial derivative can be expressed as:

$$u_x = \frac{\partial}{\partial x} c\hat{u} = \frac{c}{a}\frac{\partial \hat{u}}{\partial \hat{x}} \tag{9.73}$$

Finally, the third order derivative can be obtained as

$$u_{xxx} = \frac{\partial^2}{\partial x^2}\left(\frac{c}{a}\frac{\partial \hat{u}}{\partial \hat{x}}\right) = \frac{c}{a^3}\frac{\partial^3 \hat{u}}{\partial \hat{x}^3} \tag{9.74}$$

Substitution of (9.72) to (9.74) into (9.70) leads to

$$\frac{1}{b}u_t + \frac{c}{a}\alpha u u_x + \frac{1}{a^3}\beta u_{xxx} = 0 \tag{9.75}$$

where the "^" has been dropped for simplicity. We now select a, b, and c such that the standard KdV can be obtained:

$$\frac{1}{b} = 1, \quad \frac{c}{a}\alpha = 6, \quad \frac{1}{a^3}\beta = 1 \tag{9.76}$$

Therefore, we can set the following constants

$$a = \beta^{1/3}, \quad b = 1, \quad c = \frac{6a}{\alpha} = \frac{6\beta^{1/3}}{\alpha} \tag{9.77}$$

such that (9.70) can be converted to the following standard form:

$$u_t + 6uu_x + u_{xxx} = 0 \tag{9.78}$$

Subsequently, (9.69) can also be transformed to the standard form given in (9.78). More general types of scaling invariance and Galilean invariance of KdV are considered in Problems 9.9 and 9.10.

9.5.3 Physical Interpretation of KdV

For a water wave of finite depth, the dispersion relation is obtained as (see (12.495) of Chau, 2018):

$$\omega^2 = gk \tanh kh \tag{9.79}$$

Using Taylor's series expansion, we find that

$$\tanh x = x - \frac{1}{3}x^3 + \frac{2}{13}x^5 - \frac{17}{315}x^7 + ... \tag{9.80}$$

Substitution of (9.80) into (9.79) yields

$$\omega = (gk \tanh kh)^{1/2} = [gk(kh - \frac{1}{3}k^3h^3 + ...)]^{1/2} \approx k\sqrt{gh}(1 - \frac{1}{6}k^2h^2) \tag{9.81}$$

The approximate dispersive relation (wave motions of small dispersion) is

$$\omega = c_0 k - \frac{c_0 h^2}{6}k^3 \tag{9.82}$$

where

$$c_0 = \sqrt{gh} \tag{9.83}$$

By definition, the phase velocity can be determined from the following formula (Achenbach, 1973)

$$c_p = \frac{\omega}{k} = c_0 - \frac{c_0 h^2}{6}k^2 \tag{9.84}$$

On the other hand, the group velocity is defined as (Achenbach, 1973)

$$c_g = \frac{d\omega}{dk} = c_0 - \frac{c_0 h^2}{2}k^2 \tag{9.85}$$

For the special case of long wavelength limit (i.e., $k \to 0$), we have the following special case

$$c_g = c_p = c_0 = \sqrt{gh} \tag{9.86}$$

That is, the wave velocity, group velocity, and particle velocity are the same, Actually, this is the non-dispersive wave limit. We can show that the following linearized KdV will lead to the dispersion relation given in (9.82):

$$u_t + c_0 u_x + \frac{c_0 h^2}{6} u_{xxx} = 0 \tag{9.87}$$

To see that, we seek the usual wave type solution as

$$u = Ae^{i(kx-\omega t)} \tag{9.88}$$

where k is the wave number and ω is the circular frequency. Substituting (9.88) into (9.87), we obtain (9.82). Thus, we see that the k^3 term is the simplest type of dispersion term as suggested by (9.82). In other words, the KdV equation contains the simplest type of dispersion. Adding the simplest type of nonlinearity, we have

$$u_t + c_0 (1 + \frac{3u}{2h}) u_x + \frac{c_0 h^2}{6} u_{xxx} = 0 \tag{9.89}$$

This is another popularly adopted mathematical form of KdV equation. Although it looks different from (9.78), (9.89) can actually be converted to the standard form given in (9.78). To do that, we use the following change of variables

$$x \to a\hat{x}, \quad t \to b\hat{t}, \quad u \to c\hat{u} \tag{9.90}$$

Similar to the discussions given in the last section we have

$$u_t = \frac{\partial}{\partial t} c\hat{u} = \frac{c}{b} \frac{\partial \hat{u}}{\partial \hat{t}} \tag{9.91}$$

$$u_x = \frac{\partial}{\partial x} c\hat{u} = \frac{c}{a} \frac{\partial \hat{u}}{\partial \hat{x}} \tag{9.92}$$

$$u_{xxx} = \frac{\partial^2}{\partial x^2} (\frac{c}{a} \frac{\partial \hat{u}}{\partial \hat{x}}) = \frac{c}{a^3} \frac{\partial^3 \hat{u}}{\partial \hat{x}^3} \tag{9.93}$$

With this change of variables, (9.89) becomes

$$\frac{1}{b} \hat{u}_t + \frac{1}{a} c_0 (1 + \frac{3c}{2h} \hat{u}) \hat{u}_x + \frac{1}{a^3} (\frac{c_0 h^2}{6}) \hat{u}_{xxx} = 0 \tag{9.94}$$

By choosing the following values for the constants involved in the change of variables given in (9.90),

$$a = (\frac{c_0}{6})^{1/2} h, \quad b = \frac{h}{\sqrt{6c_0}}, \quad c = \frac{2h}{3} \tag{9.95}$$

we obtain the following PDE after dropping the hat sign

$$u_t + (1+u) u_x + u_{xxx} = 0 \tag{9.96}$$

Next, we propose another round of change of variables

$$x \to \gamma \overline{x}, \quad t \to \beta \overline{t}, \quad 1+u \to \overline{u} \tag{9.97}$$

Substitution of (9.97) into (9.96) gives

$$\overline{u}_t + \frac{\beta}{\gamma} \overline{u}\,\overline{u}_x + \frac{\beta}{\gamma^3} \overline{u}_{xxx} = 0 \tag{9.98}$$

We now set the following values for γ and β

$$\gamma = \sqrt{6}, \quad \beta = 6^{3/2} \tag{9.99}$$

Finally, using (9.99) in (9.98), we obtain the standard KdV equation

$$\bar{u}_t + 6\bar{u}\bar{u}_x + \bar{u}_{xxx} = 0 \tag{9.100}$$

A slightly different proof is given in Problem 9.5. This gives a different approach in arriving at the KdV equation, and this current approach is more intuitive and instructive.

9.5.4 Dispersion versus Nonlinearity

We now recall the major results that we have obtained in Section 9.2 and 9.3, and discuss the interplay between the dispersion and the nonlinearity. In particular, the dispersive wave equation and its solution discussed in Section 9.3 are

$$u_t + u_{xxx} = 0 \tag{9.101}$$

$$u = t^{-1/3} A_i [\frac{x}{(3t)^{1/3}}] \tag{9.102}$$

The nonlinear wave equation and its solution discussed in Section 9.2 can be recast by adding a factor of 6 as:

$$u_t + 6uu_x = 0 \tag{9.103}$$

$$u = f(x - 6ut) \tag{9.104}$$

The validity of (9.104) can be demonstrated by differentiating (9.104) with respect to t and x as

$$u_t = f'(x-6ut)\frac{\partial(x-6ut)}{\partial t} = -6uf' \tag{9.105}$$

$$u_x = f'(x-6ut)\frac{\partial(x-6ut)}{\partial x} = f' \tag{9.106}$$

Substitution of (9.105) and (9.106) into the left-hand side of (9.103) leads to zero or the right-hand side of (9.103). The nonlinear term in (9.103) leads to wave breaking whereas the dispersion term in (9.101) leads to decay. Adding both terms, we have the standard KdV given in (9.100), as demonstrated in Figure 9.7. The nonlinear growth of wave cancels the dispersive decay of the wave resulting in a stable non-decaying shape-preserving soliton solution.

9.5.5 Soliton Solution

As we have seen from (9.100), the standard form of KdV is given as:

$$u_t + 6uu_x + u_{xxx} = 0 \tag{9.107}$$

There are also three other closely related forms of KdV that were commonly adopted in the literature. For example, if we make a substitution of $u \to -u$ in (9.107), we have

$$u_t - 6uu_x + u_{xxx} = 0 \tag{9.108}$$

On the other hand, if we make a substitution of $x \to -x$ in (9.107), we have

$$u_t - 6uu_x - u_{xxx} = 0 \tag{9.109}$$

Finally, if we make substitutions of $x \to -x$ and $u \to -u$ in (9.107), we have

$$u_t + 6uu_x - u_{xxx} = 0 \tag{9.110}$$

The mirror symmetry of these four solutions closely related to the KdV is illustrated in Figure 9.8. Solutions labeled 1, 2, 3, and 4 in Figure 9.8 correspond to equations (9.107), (9.108), (9.109) and (9.110).

We look for a wave-type solution as

$$u(x,t) = f(x-ct) = f(\xi) \tag{9.111}$$

where c is the speed of propagation of the soliton. Differentiation of (9.111) gives

$$u_x = f'(\xi) \tag{9.112}$$

$$u_{xxx} = f'''(\xi) \tag{9.113}$$

$$u_t = -cf'(\xi) \tag{9.114}$$

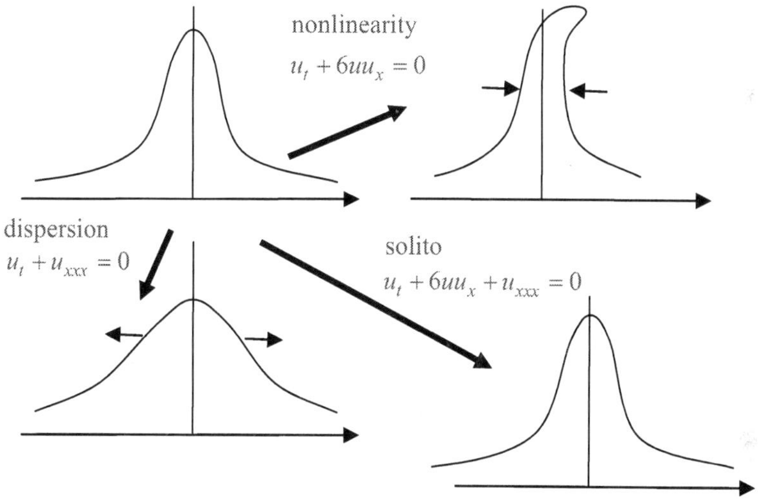

Figure 9.7 Delicate balance between dispersion and nonlinearity leads to the soliton (shape-preserving wave)

Substitution of these results into (9.107) gives

$$-cf' + 6ff' + f''' = 0 \tag{9.115}$$

Integrating once with respect to ξ, we get

$$-cf + 3f^2 + f'' = A \tag{9.116}$$

Next, we can multiply (9.116) by the first derivative of f to get

$$-cf\,f' + 3f'f^2 + f'f'' = Af' \tag{9.117}$$

We note that we can now integrate each term of (9.117) as

$$-\frac{1}{2}cf^2 + f^3 + \frac{1}{2}(f')^2 = Af + B \tag{9.118}$$

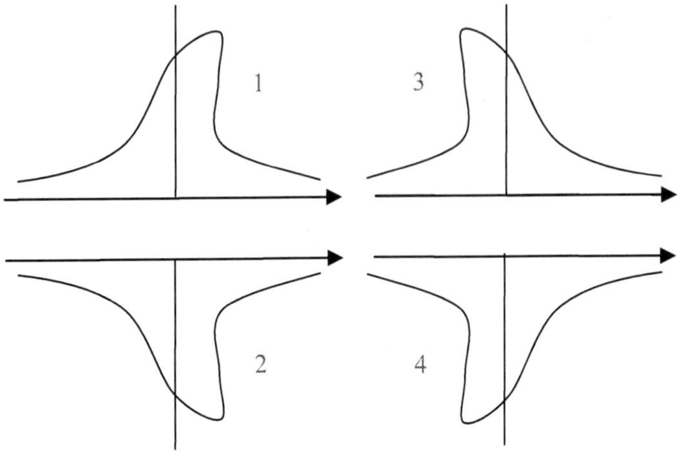

Figure 9.8 Mirror images of four types of KdV equations

Thus, after integration we obtain a first order ODE of f. We observe that the most crucial step in the solution technique is to multiply (9.116) by f'. Rearranging (9.118), we get a first order ODE as

$$(f')^2 = 2Af + 2B - 2f^3 + cf^2 \tag{9.119}$$

Since the soliton is localized spatially in the origin, we require the decay condition as

$$\xi \to \pm\infty,\ f' = 0,\ f'' = 0,\ f = 0 \tag{9.120}$$

This condition results in

$$A = 0,\ B = 0 \tag{9.121}$$

Using (9.121), we can rewrite (9.119) as

$$f' = \sqrt{cf^2 - 2f^3} \tag{9.122}$$

More specifically, we have

$$\frac{df}{d\xi} = f\sqrt{c-2f} \tag{9.123}$$

Integrating both sides of (9.123), we obtain

$$\xi - \xi_0 = \int \frac{df}{f\sqrt{c-2f}} \tag{9.124}$$

To integrate the right-hand side of (9.124), we introduce the following change of variables

$$v^2 = c - 2f \tag{9.125}$$

Differentiating (9.125), we get

$$2vdv = -2df \tag{9.126}$$

Rearranging (9.125), we have f being

$$f = \frac{1}{2}(c - v^2) \tag{9.127}$$

Substitution of (9.125) to (9.127) into (9.124) gives

$$\xi - \xi_0 = -\int \frac{vdv}{\frac{1}{2}(c - v^2)v} = -2\int \frac{dv}{c - v^2} \tag{9.128}$$

We now apply another round of change of variables

$$v = \sqrt{c} \tanh x \tag{9.129}$$

$$dv = \sqrt{c} \sec h^2 x dx \tag{9.130}$$

Substitution of (9.129) and (9.130) into (9.128) gives

$$\xi - \xi_0 = -2\int \frac{\sqrt{c} \sec h^2 x dx}{c(1 - \tanh^2 x)} \tag{9.131}$$

Noting the following identity for hyperbolic functions

$$\sec h^2 x + \tanh^2 x = 1, \tag{9.132}$$

we can integrate (9.131) as

$$\xi - \xi_0 = -\frac{2}{\sqrt{c}} x = -\frac{2}{\sqrt{c}} \tanh^{-1}(\frac{v}{\sqrt{c}}) \tag{9.133}$$

Inverting (9.133), we get

$$v = -\sqrt{c} \tanh[\frac{\sqrt{c}}{2}(\xi - \xi_0)] \tag{9.134}$$

Back substitution of (9.134) into (9.127) results in

$$f = \frac{1}{2}(c - v^2) = \frac{1}{2}\{c - c \tanh^2[\frac{\sqrt{c}}{2}(\xi - \xi_0)]\} \tag{9.135}$$

In view of the identity given in (9.132), we finally obtain the solution

$$u(x,t) = f(\xi) = \frac{c}{2} \sec h^2[\frac{\sqrt{c}}{2}(\xi - \xi_0)] \tag{9.136}$$

where

$$\xi = x - ct \tag{9.137}$$

In fact, according to Sattinger of the University of Minnesota, the KdV equation and the solution given in (9.136) was actually first derived by Boussinesq and reported in his 680-page book published in 1877, 18 years before the paper of D.J. Korteweg and G. de Vries.

One main feature of the soliton given in (9.136) is that the wave magnitude is proportional to the speed c; thus, a taller soliton moves faster than a shorter soliton. It will be of interest to ask what may happen when a taller soliton catches up with a shorter soliton. The problem will be considered after we discuss Hirota's direct method. Although there are other mathematical techniques in solving soliton problems, such as the inverse scattering transform, the Bäcklund transform, Lax pairs, and the Darboux transform, we will restrict our discussion to Hirota's direct method in the present chapter.

9.6 HIROTA'S DIRECT METHOD

There are various types of techniques for obtaining the soliton solution for KdV. In this chapter, we will only discuss Hirota's bilinear form or Hirota's direct method.

9.6.1 Bilinear Form of KdV Equation

We first recall the KdV given in (9.108) as

$$u_t - 6uu_x + u_{xxx} = 0 \tag{9.138}$$

The solution just obtained (9.136) can be modified as

$$u = -\frac{c}{2}\operatorname{sech}^2[\frac{\sqrt{c}}{2}(x-ct)+C] \tag{9.139}$$

Now, we can introduce a potential v as

$$u = \frac{\partial v}{\partial x} \tag{9.140}$$

To understand the physical meaning of v in (9.140), we quote the following gravitational potential U as

$$F = G\frac{mM}{r^2} = \frac{\partial U}{\partial r} \tag{9.141}$$

Integrating (9.141), we obtain U as

$$U = -G\frac{mM}{r} \tag{9.142}$$

Employing (9.140), we find that KdV becomes

$$v_{xt} - 6v_x v_{xx} + v_{xxxx} = 0 \tag{9.143}$$

Integrating once with respect to x, we get

$$v_t - 3v_x^2 + v_{xxx} = 0 \tag{9.144}$$

This equation is also called the potential KdV equation. Substitution of (9.139) into (9.140) and integration of the result gives

$$\begin{aligned} v &= \int u dx = -\frac{c}{2}\int \operatorname{sech}^2\{\frac{\sqrt{c}}{2}(x-ct)\}dx + C_0 \\ &= -\sqrt{c}\tanh[\frac{\sqrt{c}}{2}(x-ct)] + C_0 \end{aligned} \tag{9.145}$$

We look for a decay solution and this requires

$$x - ct \to -\infty \quad v \to 0 \tag{9.146}$$

Application of (9.146) to (9.145) gives the integration constant and back substitution of it into (9.145) gives the final solution for the potential v

$$v = -\sqrt{c}\{\tanh[\frac{\sqrt{c}}{2}(x-ct)]+1\} \tag{9.147}$$

In view of the definition of the hyperbolic tangent function, we can rewrite (9.147) as

$$v = -2\sqrt{c}\{\frac{e^{\sqrt{c}(x-ct)}}{e^{\sqrt{c}(x-ct)}+1}\} \tag{9.148}$$

We define the denominator as

$$\eta = 1 + e^{\sqrt{c}(x-ct)} \tag{9.149}$$

Differentiating (9.149), we get the numerator of (9.148)

$$\eta_x = \sqrt{c}e^{\sqrt{c}(x-ct)} \tag{9.150}$$

Therefore, it is obvious that we can rewrite (9.148) as

$$v = -2\frac{\eta_x}{\eta} \tag{9.151}$$

Differentiating (9.151) with respect to *t*, we find

$$v_t = -2(\frac{\eta\eta_{xt} - \eta_x\eta_t}{\eta^2}) \tag{9.152}$$

Differentiating (9.151) with respect to *x*, we obtain

$$v_x = -2(\frac{\eta\eta_{xx} - \eta_x^2}{\eta^2}) \tag{9.153}$$

$$v_{xx} = -\frac{2}{\eta^4}(\eta^3\eta_{xxx} - 3\eta^2\eta_x\eta_{xx} + 2\eta\eta_x^3) \tag{9.154}$$

$$v_{xxx} = -\frac{2}{\eta^4}(\eta^3\eta_{xxxx} - 4\eta^2\eta_x\eta_{xxx} + 12\eta\eta_x^2\eta_{xx} - 3\eta^2\eta_{xx}^2 - 6\eta_x^4) \tag{9.155}$$

Squaring (9.153), we obtain

$$v_x^2 = \frac{4}{\eta^4}\{\eta^2\eta_{xx}^2 - 2\eta_{xx}\eta_x^2\eta + \eta_x^4\} \tag{9.156}$$

Substitution of (9.152), (9.155) and (9.156) into (9.144) gives Hirota's bilinear form

$$\eta\eta_{xt} - \eta_x\eta_t = -\eta\eta_{xxxx} + 4\eta_x\eta_{xxx} - 3\eta_{xx}^2 \tag{9.157}$$

It is clear that Hirota's bilinear form is equivalent to (9.144) and subsequently (9.138). The trouble is that we find the transformation given in (9.151) because of the fact that we have its final solution. There is no systematic way to find the appropriate form of the transformation that is needed for other integrable nonlinear PDEs. This is the most difficult part of using Hirota's method.

9.6.2 One-Soliton Solution

Equation (9.149) suggests that the solution for η can be expressed as

$$\eta = 1 + e^{\theta} \tag{9.158}$$

where

$$\theta = Ax + Bt + C \tag{9.159}$$

Differentiation of (9.158) gives

$$\eta_x = Ae^{\theta} \tag{9.160}$$

$$\eta_{xt} = ABe^{\theta} \tag{9.161}$$

$$\eta_t = Be^{\theta} \tag{9.162}$$

$$\eta_{xxxx} = A^4 e^{\theta} \tag{9.163}$$

$$\eta_{xxx} = A^3 e^{\theta} \tag{9.164}$$

Substitution of these results into (9.157) gives

$$(1+e^{\theta})ABe^{\theta} - ABe^{2\theta} = -(1+e^{\theta})A^4 e^{\theta} + A^4 e^{2\theta} \tag{9.165}$$

Balancing terms on both sides of (9.165), we get

$$AB = -A^4 \tag{9.166}$$

$$B = -A^3 \tag{9.167}$$

To recover the KdV equation, we can set A, B, and C as

$$A = \sqrt{c}, \quad B = -c\sqrt{c}, \quad C = 0 \tag{9.168}$$

Substitution of (9.168) into (9.159) and (9.158) gives

$$\eta = 1 + e^{\sqrt{c}(x-ct)} \tag{9.169}$$

Of course, this is the same as (9.149). Differentiation of (9.169) gives

$$\eta_x = \sqrt{c}e^{\sqrt{c}(x-ct)} \tag{9.170}$$

Finally, using (9.169) and (9.170), we get

$$\begin{aligned} v &= -\frac{2\eta_x}{\eta} = -\frac{2\sqrt{c}e^{\sqrt{c}(x-ct)}}{e^{\sqrt{c}(x-ct)} + 1} \\ &= -\sqrt{c}\{\tanh[\frac{\sqrt{2}}{2}(x-ct)] + 1\} \end{aligned} \tag{9.171}$$

This is, of course, the same as (9.147). At this juncture, one may wonder that we are just going around a circle and getting nowhere. That is, we get a solution of the potential KdV before we identify the transformation given (9.151). Then, using (9.151), we finally get a more complicated PDE called Hirota's bilinear form given in (9.157). From the solution of (9.158), we finally find the mathematical form and eventually get back the solution in (9.171). On the surface, we get nothing out of this procedure. However, the story does not stop here. Since the KdV is a nonlinear PDE, it admits more than one soliton. In the next section, we will see that the procedure for obtaining the one-soliton solution here can be extended to obtain the more complicated two-soliton or even N-soliton solution.

9.6.3 Two-Soliton Solution

The solution form given in (9.158) suggests that we can look for a two-soliton solution as

$$\eta = 1 + e^{\theta_1} + e^{\theta_2} + ae^{\theta_1+\theta_2} \tag{9.172}$$

where

$$\theta_1 = A_1 x + B_1 + C_1 \tag{9.173}$$

$$\theta_2 = A_2 x + B_2 + C_2 \tag{9.174}$$

Differentiation of (9.172) gives

$$\eta_x = A_1 e^{\theta_1} + A_2 e^{\theta_2} + (A_1 + A_2) a e^{\theta_1+\theta_2} \tag{9.175}$$

$$\eta_{xt} = -A_1^4 e^{\theta_1} - A_2^4 e^{\theta_2} - (A_1 + A_2)(A_1^3 + A_2^3) a e^{\theta_1+\theta_2} \tag{9.176}$$

$$\eta_t = -A_1^3 e^{\theta_1} - A_2^3 e^{\theta_2} - (A_1^3 + A_2^3) a e^{\theta_1+\theta_2} \tag{9.177}$$

$$\eta_{xx} = A_1^2 e^{\theta_1} + A_2^2 e^{\theta_2} + (A_1 + A_2)^2 a e^{\theta_1+\theta_2} \tag{9.178}$$

Substitution of these equations into Hirota's bilinear form given in (9.157) gives

$$\begin{aligned}
&(1+e^{\theta_1}+e^{\theta_2}+ae^{\theta_1+\theta_2})(-A_1^4 e^{\theta_1} - A_2^4 e^{\theta_2} - a(A_1+A_2)(A_1^3+A_2^3)e^{\theta_1+\theta_2}) \\
&-(A_1 e^{\theta_1} + A_2 e^{\theta_2} + a(A_1+A_2)e^{\theta_1+\theta_2})(-A_1^3 e^{\theta_1} - A_2^3 e^{\theta_2} - a(A_1^3+A_2^3)e^{\theta_1+\theta_2}) \\
&= -(1+e^{\theta_1}+e^{\theta_2}+ae^{\theta_1+\theta_2})(A_1^4 e^{\theta_1} + A_2^4 e^{\theta_2} + a(A_1+A_2)^4 e^{\theta_1+\theta_2}) \\
&+4(A_1 e^{\theta_1} + A_2 e^{\theta_2} + a(A_1+A_2)e^{\theta_1+\theta_2})(A_1^3 e^{\theta_1} + A_2^3 e^{\theta_2} + a(A_1+A_2)^3 e^{\theta_1+\theta_2}) \\
&-3(A_1^2 e^{\theta_1} + A_2^2 e^{\theta_2} + a(A_1+A_2)^2 e^{\theta_1+\theta_2})^2
\end{aligned} \tag{9.179}$$

Balancing terms on both sides, we obtain

$$a = \left(\frac{A_1 - A_2}{A_1 + A_2}\right)^2 \tag{9.180}$$

$$A_1 = \sqrt{c_1}, \quad B_1 = -c_1\sqrt{c_1}, \quad C_1 = 0 \tag{9.181}$$

$$A_2 = \sqrt{c_2}, \quad B_2 = -c_2\sqrt{c_2}, \quad C_2 = 0 \tag{9.182}$$

This procedure is tedious, but symbolic manipulation software such as Mathematica and Maxima can be used to do the comparison.

Recall the fact that η relates to v given in (9.151) and v relates to u through (9.140). Thus, the bilinear transformation can also be expressed as

$$u = -2\left(\frac{\eta_x}{\eta}\right)_x = \frac{\partial^2}{\partial x^2}\ln\eta = -2\left(\frac{\eta\eta_{xx} - \eta_x^2}{\eta^2}\right) \tag{9.183}$$

Substitution of (9.180) to (9.182) into (9.172), and then differentiation of the result gives

$$\eta = 1 + e^{\sqrt{c_1}(x-c_1 t)} + e^{\sqrt{c_2}(x-c_2 t)} + \left(\frac{\sqrt{c_1}-\sqrt{c_2}}{\sqrt{c_1}+\sqrt{c_2}}\right)^2 e^{(\sqrt{c_1}+\sqrt{c_2})x-(c_1\sqrt{c_1}+c_2\sqrt{c_2})t} \tag{9.184}$$

$$\eta_{xx} = c_1 e^{\sqrt{c_1}(x-c_1 t)} + c_2 e^{\sqrt{c_2}(x-c_2 t)} + (\sqrt{c_1}-\sqrt{c_2})^2 e^{(\sqrt{c_1}+\sqrt{c_2})x-(c_1\sqrt{c_1}+c_2\sqrt{c_2})t} \tag{9.185}$$

$$\eta_x^2 = \left\{\sqrt{c_1} e^{\sqrt{c_1}(x-c_1 t)} + \sqrt{c_2} e^{\sqrt{c_2}(x-c_2 t)} + \frac{(\sqrt{c_1}-\sqrt{c_2})^2}{\sqrt{c_1}+\sqrt{c_2}} e^{(\sqrt{c_1}+\sqrt{c_2})x-(c_1\sqrt{c_1}+c_2\sqrt{c_2})t}\right\}^2 \tag{9.186}$$

Finally, substituting these results into (9.183), we obtain

$$u(x,t)=\frac{1}{2}\frac{(c_1-c_2)\left\{c_2\cosh^2[\frac{\sqrt{c_1}}{2}\zeta]+c_1\sinh^2[\frac{\sqrt{c_2}}{2}\zeta]\right\}}{\left\{\sqrt{c_2}\cosh[\frac{\sqrt{c_1}}{2}\zeta]\cosh[\frac{\sqrt{c_2}}{2}\zeta]-\sqrt{c_1}\sinh[\frac{\sqrt{c_1}}{2}\zeta]\sinh[\frac{\sqrt{c_2}}{2}\zeta]\right\}^2} \tag{9.187}$$

where

$$\zeta = x - ct \tag{9.188}$$

This is the two-soliton solution, and we see that we did come up with a new solution using Hirota's bilinear form. Figure 9.9 illustrates the prediction of (9.187).

Figure 9.9 shows that the taller soliton overtakes the shorter and slower one without changing its shape and speed. The soliton collision can be considered as linear collisions as compared to nonlinear collisions. Some researchers even considered this as a kind of "nonlinear superposition."

We should emphasize that the existence of Hirota's bilinear form does not imply that the corresponding PDE is automatically integrable. However, we will see from Section 9.8 that there are infinite conservation laws for KdV and this guarantees the integrability of KdV. Because of this, we can actually find higher order soliton solutions, i.e., the N-soliton solutions.

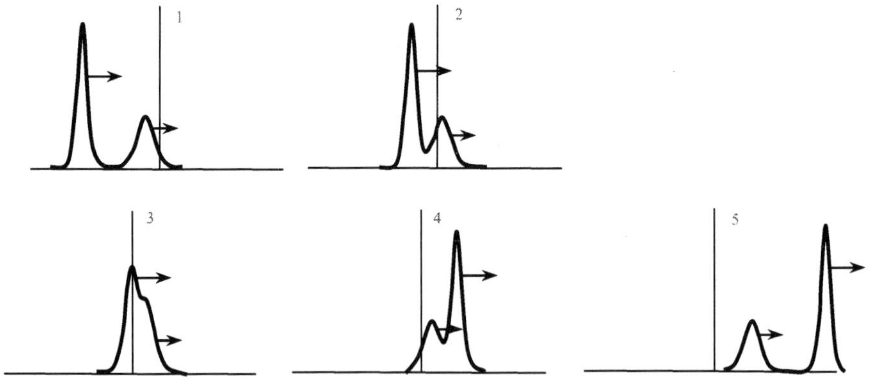

Figure 9.9 Two-soliton solutions with the taller and faster one overtaking the shorter and slower one

9.6.4 N-Soliton Solution

The procedure in obtaining the two-soliton solution in the last section can be extended to higher order solitons. For the three-soliton solution, we can set η as

$$\eta = 1 + e^{\theta_1} + e^{\theta_2} + e^{\theta_3} + a_{12}e^{(\theta_1+\theta_2)} + a_{13}e^{(\theta_1+\theta_3)} \\ + a_{23}e^{(\theta_2+\theta_3)} + a_{12}a_{13}a_{23}e^{(\theta_1+\theta_2+\theta_3)} \tag{9.189}$$

where factors a_{12}, a_{13}, and a_{23} are constants to be determined. It turns out that this is the exact mathematical form of (9.180), and θ_i is exactly the same form of (9.181):

$$\theta_i = A_i x - A_i^3 t \tag{9.190}$$

More generally, using this machinery, we can generate the N-soliton as

$$\eta = 1 + \sum_{n=1}^{N} \sum_{{}_N C_n} a(i_1, i_2, \dots, i_n) \exp(\theta_{i_1} + \theta_{i_2} + \dots + \theta_{i_n}) \tag{9.191}$$

$$a(i_1, i_2, \dots, i_n) = \prod_{k<l}^{(n)} a(i_k, i_l) \tag{9.192}$$

$$a(i_k, i_l) = (A_{i_k} - A_{i_l})^2 / (A_{i_k} + A_{i_l})^2 \tag{9.193}$$

Although getting the final solution can be tedious, it is systematic for generating the N-soliton. Symbolic manipulation software can be used in carrying out the calculation for the final solution for u.

9.6.5 Hirota's D-Operator

We have seen that the procedure of getting Hirota's bilinear form is tedious. Because of this, Hirota developed a new system of operator to simplify the calculation. In particular, Hirota found that the following operator is useful:

$$D_x^n(fg) = \left(\frac{\partial}{\partial x_1} - \frac{\partial}{\partial x_2}\right)^n [f(x_1)g(x_2)]\Big|_{x_1=x_2=x} \\ = f^{(n)}g^{(0)} + \frac{n}{1!}(-1)^1 f^{(n-1)}g^{(1)} + \frac{n(n-1)}{2!}(-1)^2 f^{(n-2)}g^{(2)} \\ + \frac{n(n-1)(n-2)}{3!}(-1)^3 f^{(n-3)}g^{(3)} + \dots + g^{(n)}f^{(0)} \tag{9.194}$$

This formula can be proved analytically, and it closely resembles Leibniz's rule for differentiation with the exception of an addition factor as a power of -1. Considering the lowest order, we get

$$D_x(fg) = D_x[f(x_1)g(x_2)]\Big|_{x_1=x_2=x} \\ = gf_x - fg_x \tag{9.195}$$

Similarly, we can also define the operator for variable t as

$$D_x D_t(fg) = D_t[gf_x - fg_x]\Big|_{x_1=x_2=t} \\ = gf_{xt} - g_t f_x - g_x f_t + fg_{xt} \tag{9.196}$$

Reapplying (9.195), we get

$$D_x^2(fg) = D_x[gf_x - fg_x]\big|_{x_1=x_2=x} \\ = gf_{xx} - 2g_x f_x + g_{xx} f \tag{9.197}$$

$$D_x^3(fg) = \left(\frac{\partial}{\partial x_1} - \frac{\partial}{\partial x_2}\right)^3 [f(x_1)g(x_2)]\big|_{x_1=x_2=x} \\ = gf_{xxx} - 3g_x f_{xx} + 3g_{xx} f_x - g_{xxx} f \tag{9.198}$$

$$D_x^4(fg) = \left(\frac{\partial}{\partial x_1} - \frac{\partial}{\partial x_2}\right)^4 [f(x_1)g(x_2)]\big|_{x_1=x_2=x} \\ = gf_{xxxx} - 4g_x f_{xxx} + 6g_{xx} f_{xx} - 4g_{xxx} f_x + g_{xxxx} f \tag{9.199}$$

$$D_t(fg) = \left(\frac{\partial}{\partial x_1} - \frac{\partial}{\partial x_2}\right)[f(x_1)g(x_2)]\big|_{x_1=x_2=t} = f_t g - f g_t \tag{9.200}$$

With these results, we can show that

$$(D_x^4 + D_x D_t)(\eta\eta) = 0 \tag{9.201}$$

In fact, in explicit form, (9.201) is actually the bilinear form of the KdV:

$$\eta\eta_{xt} - \eta_x\eta_t + \eta\eta_{xxxx} - 4\eta_x\eta_{xxx} + 3\eta_{xx}^2 = 0 \tag{9.202}$$

We should note that if Hirota's bilinear form exists, the PDE automatically has the one- and two-soliton solutions. However, there is no guarantee that a three-soliton solution exists.

The main problem with Hirota's operator technique is that there is no systematic way to convert a nonlinear PDE to Hirota's bilinear form. Several appropriate forms of change of variables in converting some well-known nonlinear PDEs to Hirota's bilinear form are considered in Problems 9.1 to 9.4. Hietarinta (2005) proposed a technique to find the appropriate change of variables for Hirota's method for any PDE by using Laurent series expansion.

9.7 KDV EQUATION AND OTHER NONLINEAR EQUATIONS

So far, we have focused our discussion on KdV. However, KdV turns out to be a very general type of nonlinear PDE. Many well-known nonlinear PDEs with soliton-type solutions relate to KdV in one way or the other. In this section, we will consider some of the more notable ones, such that the universality of KdV can be demonstrated.

9.7.1 KdV Equation and mKdV Equation

The modified KdV is defined as

$$u_t + 6u^2 u_x + u_{xxx} = 0 \tag{9.203}$$

This equation can be shown as equivalent to the KdV via the following complex Miura transformation

$$u = v^2 - iv_x \tag{9.204}$$

Differentiation of (9.204) gives

$$u_x = 2vv_x - iv_{xx} \tag{9.205}$$

$$u_{xx} = 2v_x^2 + 2vv_{xx} - iv_{xxx} \tag{9.206}$$

$$u_{xxx} = 6v_x v_{xx} + 2vv_{xxx} - iv_{xxx} \tag{9.207}$$

Substitution of these results into the KdV given in (9.203) gives

$$\begin{aligned} u_t + 6uu_x + u_{xxx} &= (2vv_t - iv_{tx}) + 6v_x v_{xx} + 2vv_{xxx} - iv_{xxx} \\ &\quad +6(v^2 - iv_x)(2vv_x - iv_{xx}) \\ &= (2v - i\frac{\partial}{\partial x})v_t + (2v - i\frac{\partial}{\partial x})v_{xxx} \\ &\quad +6(2v^3 v_x - iv^2 v_{xx} - 2ivv_x^2) \end{aligned} \tag{9.208}$$

We note that the last term of (9.208) can be grouped as

$$(2v - i\frac{\partial}{\partial x})v^2 v_x = 2v^3 v_x - iv^2 v_{xx} - 2ivv_x^2 \tag{9.209}$$

Finally, substituting (9.209) into (9.208), we obtain

$$u_t + 6uu_x + u_{xxx} = (2v - i\frac{\partial}{\partial x})(v_t + 6v^2 v_x + v_{xxx}) = 0 \tag{9.210}$$

Therefore, mKdV is equivalent to KdV.

9.7.2 KdV Equation and Boussinesq Equation

The Boussinesq equation is one of the most popularly used shallow water equations. However, many slightly different versions of the Boussinesq equation have been reported in the literature. The original Boussinesq equation was given in a set of two equations:

$$\eta_x + u_t + \alpha uu_x - \frac{1}{3}\beta u_{xxt} = 0 \tag{9.211}$$

$$\eta_t + u_x + \alpha(\eta u)_x = 0 \tag{9.212}$$

where

$$\alpha = \frac{a}{h}, \quad \beta = (kh)^2 \tag{9.213}$$

Among many different versions of Boussinesq equations, we consider the following particular form of Boussinesq equation

$$u_{tt} - c^2 u_{xx} = \varepsilon c^2 (u_x u_{xx} + \delta^2 u_{xxxx}) \tag{9.214}$$

This equation can be shown as equivalent to the KdV equation. To show this, the following multiple time scale is assumed

$$u(x,t) = f(\theta, T) + \varepsilon v(x,t) \tag{9.215}$$

where

$$\theta = x - ct, \quad T = \varepsilon t \tag{9.216}$$

Employing the chain rule, we find the differentiation of (9.215) with respect to t as

$$u_t = \frac{\partial f}{\partial \theta}\frac{\partial \theta}{\partial t} + \frac{\partial f}{\partial T}\frac{\partial T}{\partial t} + \varepsilon v_t = -cf_\theta + \varepsilon f_T + \varepsilon v_t \tag{9.217}$$

Repeating the differentiation one more time, we get

$$u_{tt} = c^2 f_{\theta\theta} - 2c\varepsilon f_{\theta T} + \varepsilon^2 f_{TT} + \varepsilon v_{tt} \tag{9.218}$$

The differentiation of (9.215) with respect to x is simpler as

$$u_x = f_\theta + \varepsilon v_x \tag{9.219}$$

$$u_{xx} = f_{\theta\theta} + \varepsilon v_{xx} \tag{9.220}$$

$$u_{xxxx} = f_{\theta\theta\theta\theta} + \varepsilon v_{xxxx} \tag{9.221}$$

Using these results in (9.214), we get

$$\begin{aligned} &c^2 f_{\theta\theta} - 2c\varepsilon f_{\theta T} + \varepsilon^2 f_{TT} + \varepsilon v_{tt} - c^2(f_{\theta\theta} + \varepsilon v_{xx}) \\ &= \varepsilon c^2 (f_\theta + \varepsilon v_x)(f_{\theta\theta} + \varepsilon v_{xx}) + \varepsilon c^2 \delta^2 (f_{\theta\theta\theta\theta} + \varepsilon v_{xxxx}) \end{aligned} \tag{9.222}$$

Collecting the first order term in ε, we get

$$\varepsilon(v_{tt} - c^2 v_{xx}) = \varepsilon[2cf_{\theta T} + c^2 f_\theta f_{\theta\theta} + c^2\delta^2 f_{\theta\theta\theta\theta}] + O(\varepsilon^2) \tag{9.223}$$

The nonhomogeneous terms on the right of (9.223) are, in general, functions of θ and T and the particular solution for v may not decay to zero as $\theta \to \infty$, and, thus, we require

$$2cf_{\theta T} + c^2 f_\theta f_{\theta\theta} + c^2 \delta^2 f_{\theta\theta\theta\theta} = 0 \tag{9.224}$$

We introduce the following change of variables

$$\tau = \frac{cT}{2}, \quad \frac{1}{6} f_\theta = q \tag{9.225}$$

Substitution of (9.225) into (9.224) gives

$$q_T + 6qq_\theta + \delta^2 q_{\theta\theta\theta} = 0 \tag{9.226}$$

As we have seen in Section 9.5.2, we can absorb δ to get the standard form of KdV. Thus, KdV is a special case of the Boussinesq equation given in (9.214) for small ε. In essence, the Boussinesq equation is more nonlinear than the KdV equation.

9.7.3 KdV Equation and Nonlinear Schrödinger Equation

In Section 9.9, we will consider another commonly encountered nonlinear PDE called the nonlinear Schrödinger equation (NLSE), for which a soliton can be found. In this section, we prove that the NLSE is equivalent to the KdV if scaling is applied and weak nonlinearity is imposed. Let us recall the following KdV

$$\varphi_t + \varphi\varphi_x + \varphi_{xxx} = 0 \tag{9.227}$$

Note that a factor of 6 can be added to the second term by scaling as discussed in Section 9.5.2. First, we can rescale the KdV to becoming weakly nonlinear by assuming the following change of variables

$$\varphi = \varepsilon u \tag{9.228}$$

Substitution of (9.228) into (9.227) gives

$$u_t + \varepsilon u u_x + u_{xxx} = 0 \tag{9.229}$$

Now we introduce multiple time and space scales as

$$u(x,t) = U(X_0, X_1, T_0, T_1, T_2) \tag{9.230}$$

where

$$X_j = \varepsilon^j x, \quad j = 0,1, \qquad T_k = \varepsilon^k t, \quad k = 0,1,2 \tag{9.231}$$

Substitution of (9.230) into (9.229) gives

$$\begin{aligned}&\frac{\partial U}{\partial T_0} + \varepsilon\frac{\partial U}{\partial T_1} + \varepsilon^2\frac{\partial U}{\partial T_2} + \varepsilon U\frac{\partial U}{\partial X_0} + \varepsilon^2 U\frac{\partial U}{\partial X_1} + \frac{\partial^3 U}{\partial X_0^3} + 3\varepsilon\frac{\partial^3 U}{\partial X_0^2 \partial X_1}\\ &+3\varepsilon^2\frac{\partial^3 U}{\partial X_0 \partial X_1^2} + \varepsilon^3\frac{\partial^3 U}{\partial X_1^3} = 0\end{aligned} \tag{9.232}$$

We introduce an asymptotic series of U in terms of the small parameter ε

$$U = U_0 + \varepsilon U_1 + \varepsilon^2 U_2 + \ldots \tag{9.233}$$

Substitution of (9.233) into (9.232) and collection of different order terms in ε gives

ε^0 order

$$\frac{\partial U_0}{\partial T_0} + \frac{\partial^3 U_0}{\partial X_0^3} = 0 \tag{9.234}$$

ε^1 order

$$\frac{\partial U_1}{\partial T_0} + \frac{\partial^3 U_1}{\partial X_0^3} = -\frac{\partial U_0}{\partial X_1} - U_0\frac{\partial U_0}{\partial X_0} - 3\frac{\partial^3 U_0}{\partial X_0^2 \partial X_1} \tag{9.235}$$

ε^2 order

$$\begin{aligned}\frac{\partial U_2}{\partial T_0} + \frac{\partial^3 U_2}{\partial X_0^3} &= -\frac{\partial U_1}{\partial T_1} - U_1\frac{\partial U_0}{\partial X_0} - U_0\frac{\partial U_1}{\partial X_0}\\ &-3\frac{\partial^3 U_1}{\partial X_0^2 \partial X_1} - \frac{\partial U_0}{\partial T_2} - U_0\frac{\partial U_0}{\partial X_1} - 3\frac{\partial^3 U_0}{\partial X_0 \partial X_1^2}\end{aligned} \tag{9.236}$$

We seek the following solution form for U_0

$$U_0 = A\exp\{i(kX_0 + k^3 T_0)\} \tag{9.237}$$

Note that (9.234) is automatically satisfied by (9.237). Substitution of (9.237) into (9.235) gives

$$\frac{\partial U_1}{\partial T_0} + \frac{\partial^3 U_1}{\partial X_0^3} = -[\frac{\partial A}{\partial T_1} - 3k^2\frac{\partial A}{\partial X_1}]e^{i(kX_0+k^3T_0)} - ikA^2 e^{2i(kX_0+k^3T_0)} + \ldots \tag{9.238}$$

For the decay solution, we set the first bracket term on the right-hand side of (9.238) to zero

$$\frac{\partial A}{\partial T_1} + \omega'(k)\frac{\partial A}{\partial X_1} = 0 \tag{9.239}$$

where $\omega = -k^3$ as recognized from (9.237). We first rename the last term in (9.238) as

$$W = ikA^2 e^{2i(kX_0+k^3T_0)} \tag{9.240}$$

Differentiation of (9.240) gives

$$\frac{\partial W}{\partial T_0} = 2k^3 iW \tag{9.241}$$

$$\frac{\partial^3 W}{\partial X_0^3} = -8k^3 iW \tag{9.242}$$

Equating (9.241) and (9.242) requires

$$2k^3 = -8k^3 \tag{9.243}$$

Equation (9.243) cannot be satisfied unless k is zero, but k being zero is ruled out. Thus, the nonhomogeneous term does not coincide with the homogeneous solution, and we can assume the particular solution as

$$U_1^{(p)} = Be^{2i(kX_0+k^3T_0)} \tag{9.244}$$

Substitution of (9.244) into (9.238) gives

$$B = \frac{A^2}{6k^2} \tag{9.245}$$

To solve U_2 in (9.236), the particular solution obtained in (9.244) is not enough and we assume the following solution form

$$U_1 = Be^{2i(kX_0+k^3T_0)} + M \tag{9.246}$$

Substitution of (9.246) into (9.236) gives

$$\begin{aligned} \frac{\partial U_2}{\partial T_0} + \frac{\partial^3 U_2}{\partial X_0^3} &= -[ikMA + \frac{i}{6k}|A|^2 A + \frac{\partial A}{\partial T_2} + 3ik\frac{\partial^2 A}{\partial X_1^2}]\mathrm{e}^{i(kX_0+k^3T_0)} \\ &\quad -\frac{\partial M}{\partial T_1} - A\frac{\partial \bar{A}}{\partial X_1} - \bar{A}\frac{\partial A}{\partial X_1} + \ldots \end{aligned} \tag{9.247}$$

Again, we can first set the bracket term in (9.247) to zero as

$$i\frac{\partial A}{\partial T_2} - \frac{1}{6k}|A|^2 A - 3k\frac{\partial^2 A}{\partial X_1^2} - kMA = 0 \tag{9.248}$$

Setting the last few terms in (9.247) to zero, we obtain

$$\frac{\partial M}{\partial T_1} = -\frac{\partial}{\partial X_1}|A|^2 \tag{9.249}$$

Equations (9.248) and (9.249) provide a system of two coupled PDEs for A and M. We can actually solve for M by using the following change of variables (using group velocity frame):

$$\xi = X_1 - \omega'(k)T_1, \quad \tau = T_1 \tag{9.250}$$

Using these new variables, we can rewrite the group velocity equation given in (9.239) as

$$\frac{\partial A}{\partial \tau} = 0 \tag{9.251}$$

To show this, we have

$$\frac{\partial A}{\partial X_1} = \frac{\partial A}{\partial \xi}\frac{\partial \xi}{\partial X_1} + \frac{\partial A}{\partial \tau}\frac{\partial \tau}{\partial X_1} = \frac{\partial A}{\partial \xi} \tag{9.252}$$

$$\frac{\partial A}{\partial T_1}=\frac{\partial A}{\partial \xi}\frac{\partial \xi}{\partial T_1}+\frac{\partial A}{\partial \tau}\frac{\partial \tau}{\partial T_1}=-\omega'(k)\frac{\partial A}{\partial \xi}+\frac{\partial A}{\partial \tau} \tag{9.253}$$

Substitution of (9.252) and (9.253) into (9.239) gives

$$\frac{\partial A}{\partial T_1}+\omega'(k)\frac{\partial A}{\partial X_1}=-\omega'(k)\frac{\partial A}{\partial \xi}+\frac{\partial A}{\partial \tau}+\omega'(k)\frac{\partial A}{\partial \xi}=\frac{\partial A}{\partial \tau}=0 \tag{9.254}$$

This completes the proof of (9.251). Expressing M and A in the group velocity frame, we have

$$\frac{\partial M}{\partial T_1}=\frac{\partial M}{\partial \xi}\frac{\partial \xi}{\partial T_1}+\frac{\partial M}{\partial \tau}\frac{\partial \tau}{\partial T_1}=-\omega'(k)\frac{\partial M}{\partial \xi}+\frac{\partial M}{\partial \tau} \tag{9.255}$$

$$\frac{\partial |A|^2}{\partial X_1}=\frac{\partial |A|^2}{\partial \xi} \tag{9.256}$$

Substitution of (9.255) and (9.256) into (9.249) gives

$$-\omega'(k)\frac{\partial M}{\partial \xi}+\frac{\partial M}{\partial \tau}=-\frac{\partial |A|^2}{\partial \xi} \tag{9.257}$$

Recall from (9.251) that we can see that the right-hand side of (9.257) is not a function of τ, and thus we can set

$$\frac{\partial M}{\partial \tau}=0 \tag{9.258}$$

Thus, we have

$$-\omega'(k)\frac{\partial M}{\partial \xi}=-\frac{\partial |A|^2}{\partial \xi} \tag{9.259}$$

Integrating both sides, we obtain

$$M=\frac{|A|^2}{\omega'(k)}=-\frac{|A|^2}{3k^2} \tag{9.260}$$

Substitution of (9.260) into (9.248) gives

$$i\frac{\partial A}{\partial T_2}+\frac{1}{6k}|A|^2 A-3k\frac{\partial^2 A}{\partial X_1^2}=0 \tag{9.261}$$

This can be rewritten as

$$i\frac{\partial A}{\partial T_2}+\frac{\omega''(k)}{2}\frac{\partial^2 A}{\partial X_1^2}+\frac{1}{6k}|A|^2 A=0 \tag{9.262}$$

This is the nonlinear Schrödinger equation (NLSE). Therefore, the NLSE is considered as a weakly nonlinear version of the KdV in the frame traveling with the group velocity of the waves. Solution of the NLSE is discussed in Section 9.9.

9.7.4 KdV Equation and First Painlevé Equation

We first use the scale invariance of Section 9.5.2 to rewrite the KdV as

$$u_t+uu_x+u_{xxx}=0 \tag{9.263}$$

Next, the following change of variable is introduced:

$$u(x,t) = w(z) + \frac{t}{a} \tag{9.264}$$

where

$$z = x - \frac{t^2}{2a} \tag{9.265}$$

Differentiating (9.264), we get

$$u_t = -\frac{t}{a} w_z + \frac{1}{a} \tag{9.266}$$

$$u_x = w_x = w_z \tag{9.267}$$

$$u_{xxx} = w_{xxx} = w_{zzz} \tag{9.268}$$

Substitution of (9.266) to (9.268) into (9.263) gives

$$-\frac{t}{a} w_z + \frac{1}{a} + w_z (w + \frac{t}{a}) + w_{zzz} = 0 \tag{9.269}$$

Simplification of (9.269) gives

$$w_{zzz} + w_z w + \frac{1}{a} = 0 \tag{9.270}$$

Integrating once with respect to z, we get

$$w_{zz} + \frac{1}{2} w^2 + \frac{z}{a} + c = 0 \tag{9.271}$$

This equation can be further reduced to the first Painlevé equation. To see this, we assume the following change of variables

$$v = \alpha w, \quad \zeta = \lambda z \tag{9.272}$$

Differentiation of v with respect to ζ gives

$$v_\zeta = \alpha \frac{dw}{d\zeta} = \alpha \frac{dw}{dz} \frac{dz}{d\zeta} = \frac{\alpha}{\lambda} w_z \tag{9.273}$$

$$v_{\zeta\zeta} = \frac{\alpha}{\lambda} w_{zz} \frac{1}{\lambda} = \frac{\alpha}{\lambda^2} w_{zz} \tag{9.274}$$

Substitution of (9.272) and (9.274) into (9.271) gives

$$v_{\zeta\zeta} + \frac{1}{2\lambda^2 \alpha} v^2 + \frac{\alpha}{\lambda^3 a} \zeta + \frac{c\alpha}{\lambda^2} = 0 \tag{9.275}$$

Next, we set the coefficients of (9.275) as follows:

$$\frac{1}{2\lambda^2 \alpha} = -6, \quad \frac{\alpha}{\lambda^3 a} = -1 \tag{9.276}$$

Solution of (9.276) gives the required change of variables

$$\alpha = \sqrt{\frac{a}{12}} (\frac{1}{12a})^{2/5}, \quad \lambda = (\frac{1}{12a})^{1/5} \tag{9.277}$$

Thus, (9.271) becomes

$$v_{\zeta\zeta} - 6v^2 - \zeta + \frac{c\alpha}{\lambda^2} = 0 \tag{9.278}$$

Finally, we introduce the following shifting of ζ

$$\zeta = \xi - \frac{c\alpha}{\lambda^2} \tag{9.279}$$

This change of variables converts (9.278) to

$$v_{\xi\xi} = 6v^2 + \xi \tag{9.280}$$

This is exactly the first Painlevé equation (see (4.647) of Chau, 2018).

9.7.5 mKdV Equation and Second Painlevé Equation

In this section, we will demonstrate that mKdV can be converted to the second Painleve equation. First, we recall the mKdV from (9.210) as

$$v_t + 6v^2 v_x + v_{xxx} = 0 \tag{9.281}$$

We introduce the following change of variables

$$v = \frac{1}{t^{1/3}} y(z), \quad z = \frac{x}{t^{1/3}} \tag{9.282}$$

The time derivative of v becomes

$$v_t = -\frac{1}{3t^{4/3}}[y + \frac{x}{t^{1/3}} y'] \tag{9.283}$$

Similarly, the first three spatial derivatives of v are

$$v_x = \frac{1}{t^{3/2}} y' \tag{9.284}$$

$$v_{xx} = \frac{1}{t} y'' \tag{9.285}$$

$$v_{xxx} = \frac{1}{t^{4/3}} y''' \tag{9.286}$$

The square of v can be rewritten in terms of y as:

$$v^2 = \frac{1}{t^{3/2}} y^2 \tag{9.287}$$

Substitution of (9.283), (9.284), (9.286) and (9.287) into (9.281) yields a governing equation for y

$$3y''' + 18y^2 y' - y - zy' = 0 \tag{9.288}$$

We recognize that this can be integrated as

$$\frac{d}{dz}(3y'' + 6y^3 - zy) = 0 \tag{9.289}$$

We now apply the following change of variables

$$\zeta = \lambda z, \quad w(\zeta) = i\alpha y \tag{9.290}$$

where λ and α are to be determined. The first two derivatives of w are

$$w' = i\frac{\alpha}{\lambda} y' \tag{9.291}$$

$$w'' = i\frac{\alpha}{\lambda^2} y'' \tag{9.292}$$

The cube of w can be found easily from (9.290) as

$$w^3 = -i\alpha^3 y^3 \tag{9.293}$$

Substitution of these results into (9.289) gives the governing equation for w

$$w'' - \frac{2\alpha^2}{\lambda^2} w^3 - \frac{1}{3\lambda^3}\zeta w = C_1 \tag{9.294}$$

In order to recover the second Painlevé equation, we can set

$$\frac{1}{3\lambda^3} = 1, \quad \frac{\alpha^2}{\lambda^2} = 1 \tag{9.295}$$

Solutions of (9.295) are

$$\alpha = \lambda = \frac{1}{3^{1/3}} \tag{9.296}$$

Using these values for λ and α, we can rewrite (9.294) as

$$w'' = 2w^3 + \zeta w + C \tag{9.297}$$

This is exactly the second Painlevé equation (see (4.648) of Chau, 2018).

9.8 CONSERVATION LAWS OF KDV

The integrability of a nonlinear PDE requires the existence of an infinite number of conservation laws. By definition, the conservation law is given as

$$\frac{\partial T}{\partial t} + \frac{\partial X}{\partial x} = 0 \tag{9.298}$$

where T is the density and X is the flux. If we integrate (9.298) with respect to x, we obtain

$$\frac{d}{dt}\int_{-\infty}^{\infty} T dx = -\int_{-\infty}^{\infty} \frac{\partial X}{\partial x} dx = -[X]_{-\infty}^{\infty} = 0 \tag{9.299}$$

We have assumed the flux at infinity is zero. Equation (9.299) requires

$$\int_{-\infty}^{\infty} T dx = C \tag{9.300}$$

Now, let us consider the following form of the KdV equation (see (9.108))

$$\frac{\partial u}{\partial t} - 6u\frac{\partial u}{\partial x} + \frac{\partial^3 u}{\partial x^3} = 0 \tag{9.301}$$

We recognize that it can be written as

$$\frac{\partial u}{\partial t} + \frac{\partial}{\partial x}(-3u^2 + \frac{\partial^2 u}{\partial x^2}) = 0 \tag{9.302}$$

Comparing (9.302) with the conservation law given in (9.298), we have

$$T = u, \quad X = -3u^2 + \frac{\partial^2 u}{\partial x^2} \tag{9.303}$$

Substituting the first equation of (9.303) into (9.300), we obtain the conservation of mass as

$$\int_{-\infty}^{\infty} u dx = \text{mass} \tag{9.304}$$

If we multiply the KdV equation given in (9.301) by u, we have

$$u\frac{\partial u}{\partial t} - 6u^2\frac{\partial u}{\partial x} + u\frac{\partial^3 u}{\partial x^3} = \frac{\partial}{\partial t}(\frac{1}{2}u^2) + \frac{\partial}{\partial x}\{-2u^3 + u\frac{\partial^2 u}{\partial x^2} - \frac{1}{2}(\frac{\partial u}{\partial x})^2\} = 0 \tag{9.305}$$

The last part of (9.305) is again in the structural form of (9.298), and by virtue of (9.300) we have

$$\int_{-\infty}^{\infty} T dx = \int_{-\infty}^{\infty} \frac{1}{2}u^2 dx = \text{conservation of momentum} \tag{9.306}$$

Physically, this is the conservation of momentum. We can continue the process by multiplying the KdV equation by $3u^2$ and add the result to the product of multiplying the spatial derivative of the KdV by $\partial u/\partial x$, and after grouping we get

$$\frac{\partial}{\partial t}(u^3 + \frac{1}{2}u_x^2) + \frac{\partial}{\partial x}\{-\frac{18}{4}u^4 + 3u^2 u_{xx} - 6uu_x^2 + u_x u_{xxx} - \frac{1}{2}(\frac{\partial^2 u}{\partial x^2})^2\} = 0 \tag{9.307}$$

This is again in the structural form of the conservation law given in (9.298), thus, we have

$$\int_{-\infty}^{\infty} T dx = \int_{-\infty}^{\infty} (u^3 + \frac{1}{2}u_x^2) dx = \text{conservation of energy} \tag{9.308}$$

Physically, this is the conservation of energy. We can continue this process but the procedure is getting less obvious. The integrability of the KdV requires the existence of an infinite number of functionally independent conservation laws. The numerical simulations on collisions of solitons by Zabusky and Kruskal did suggest that there exist infinite conservation laws. An ingenious way to solve the problem is proposed by Muira, Gardner and Kruskal. They assumed the solution of the KdV can be expressed as:

$$u = w + \varepsilon w_x + A\varepsilon^2 w^2 \tag{9.309}$$

where ε is a small parameter. This is actually a generalization of the Miura transformation that we used in (9.204). Differentiating (9.309) with respect to t, we find

$$u_t = w_t + \varepsilon w_{tx} + 2A\varepsilon^2 w w_t = (1 + \varepsilon\frac{\partial}{\partial x} + 2A\varepsilon^2 w) w_t \tag{9.310}$$

We then differentiate (9.309) with respect to x once and three times, and consider the following sum

$$\begin{aligned} uu_x + u_{xxx} &= ww_x + w_{xxx} + \varepsilon(ww_{xx} + w_x^2 + w_{xxxx}) \\ &+\varepsilon^2[3Aw^2 w_x + 2Aww_{xxx} + (1+6A)w_x w_{xx}] \\ &+\varepsilon^3(2Aww_x^2 + Aw^2 w_{xx}) + 2A^2\varepsilon^4 w^3 w_x \\ &= (1+\varepsilon\frac{\partial}{\partial x} + 2A\varepsilon^2 w)(ww_x + w_{xxx} + A\varepsilon^2 w^2 w_x) \\ &+\varepsilon^2(1+6A)w_x w_{xx} \end{aligned} \tag{9.311}$$

We now consider the special case that

$$A = -\frac{1}{6} \tag{9.312}$$

Substitution of (9.312) into the sum of (9.310) and (9.311) results in the following form:

$$u_t + uu_x + u_{xxx} = (1+\varepsilon\frac{\partial}{\partial x} - \frac{\varepsilon^2}{3}w)(w_t + ww_x + w_{xxx} - \frac{\varepsilon^2}{6}w^2 w_x) = 0 \tag{9.313}$$

Thus, the KdV equation leads to the following Gardner equation:

$$w_t + ww_x + w_{xxx} - \frac{\varepsilon^2}{6}w^2 w_x = 0 \tag{9.314}$$

Note that the constant 6 can always be scaled back to the second term on the left-hand side of (9.313) by the result of Section 9.5.2. If w satisfies the Gardner equation given in (9.314), (9.309) is also a solution of the KdV equation. It is straightforward to show that the Gardner equation can be rewritten as

$$\frac{\partial w}{\partial t} + \frac{\partial}{\partial x}\{\frac{1}{2}w^2 - \frac{\varepsilon^2}{18}w^3 + w_{xx}\} = 0 \tag{9.315}$$

Assuming a series expansion of w, we have

$$w = \sum_{n=0}^{\infty} \varepsilon^n w_n \tag{9.316}$$

Substitution of (9.316) into (9.309) gives

$$\begin{aligned} u &= w_0 + \varepsilon w_1 + \varepsilon^2 w_2 + \ldots + \varepsilon(w_{0,x} + \varepsilon w_{1,x} + \varepsilon^2 w_{2,x} + \ldots) \\ &-\frac{1}{6}[\varepsilon^2 w_0^2 + 2\varepsilon^3 w_0 w_1 + 2\varepsilon^4 w_2 w_1 + \ldots] \end{aligned} \tag{9.317}$$

Balancing the coefficients of ε on both sides of (9.317) for the first three order terms, we obtain

$$w_0 = u \tag{9.318}$$

$$w_1 = -w_{0,x} = -u_x \tag{9.319}$$

$$w_2 = -w_{1,x} + \frac{1}{6}w_0^2 = u_{xx} + \frac{1}{6}u^2 \tag{9.320}$$

For $n \geq 3$, we have a more general form:

$$w_n = -w_{n-1,x} + \frac{1}{3}uw_{n-2} + \frac{1}{6}\sum_{k=1}^{n-3} w_k w_{n-2-k} \tag{9.321}$$

Substitution of (9.316) into the Gardner equation in (9.314) and collection of terms for different order of ε leads to a generating function for the local conserved densities:

$$T(\varepsilon)=w=\sum_{n=0}^{\infty}\varepsilon^n T_n \tag{9.322}$$

where

$$T_0=u,\quad T_1=-u_x,\quad T_2=u_{xx}+\frac{1}{6}u^2,... \tag{9.323}$$

Note that the local densities in the conservation laws are all polynomials of u and its x derivatives only (i.e., does not involve variable x or t explicitly). Thus, w is the density itself. The generating functions for the flux X can be found as

$$X(\varepsilon)=\frac{1}{2}w^2-\frac{\varepsilon^2}{18}w^3+w_{xx}=\sum_{n=0}^{\infty}\varepsilon^n X_n \tag{9.324}$$

The first two terms are

$$X_0=\frac{1}{2}w_0^2+w_{0,xx}=\frac{1}{2}u^2+u_{xx} \tag{9.325}$$

$$X_1=w_0w_1+w_{1,xx}=-uu_x-u_{xxx} \tag{9.326}$$

Substitution of (9.322) into (9.300) yields

$$\int_{-\infty}^{\infty}Tdx=\sum_{n=0}^{\infty}\varepsilon^n\int_{-\infty}^{\infty}T_n dx=C \tag{9.327}$$

Clearly, there is an infinite number of conservation laws and this indicates the integrability of the KdV equation.

9.9 NONLINEAR SCHRÖDINGER EQUATION (NLSE)

The nonlinear Schrödinger equation (NLSE) has been formulated for a number of problems, including the weak pair repulsion between atoms in Bose gas, optical-potential scattering, a special case of the sine-Gordon equation (see Problem 9.1) for low amplitude, the vortex in superfluid helium, and rogue or freak waves in deep ocean. Note that the Schrödinger equation can be obtained from (11.45) in Chapter 11 (proof of the Schrödinger equation is derived there)

$$i\hbar\psi_t+\frac{\hbar^2}{2m}\nabla^2\psi-U\psi=0 \tag{9.328}$$

If the potential U is nonlinear and proportional to the square of the wave function, we have

$$i\hbar\psi_t+\frac{\hbar^2}{2m}\nabla^2\psi-g\left|\psi\right|^2\psi=0 \tag{9.329}$$

This can be normalized in the standard form to become the so-called nonlinear Schrödinger equation (NLSE):

$$i\psi_t + \nabla^2\psi + g|\psi|^2\psi = 0 \tag{9.330}$$

We are going to see that the NLSE can be related to either KdV or mKdV.

9.9.1 mKdV Equation and NLSE

In this section, we will show that the mKdV can be converted to the nonlinear Schrödinger equation. We recall the mKdV as

$$\varphi_t + \varphi^2\varphi_x + \varphi_{xxx} = 0 \tag{9.331}$$

Note that a factor of 6 can be added by scaling if we prefer. We first rescale the mKdV to become weakly nonlinear by assuming the following change of variables

$$\varphi = \varepsilon u \tag{9.332}$$

where, as usual, ε is a small parameter. Substitution of (9.332) into (9.331) results in

$$u_t + \varepsilon^2 u^2 u_x + u_{xxx} = 0 \tag{9.333}$$

We now apply a multiple time and space scale as:

$$u = U(X_0, X_1, T_0, T_1, T_2) \tag{9.334}$$

where the scaled space and time variables are defined as

$$X_j = \varepsilon^j x, \quad j = 0,1, \qquad T_k = \varepsilon^k t, \quad k = 0,1,2 \tag{9.335}$$

Substitution of (9.335) into (9.333) leads to

$$\begin{aligned}&\frac{\partial U}{\partial T_0} + \varepsilon\frac{\partial U}{\partial T_1} + \varepsilon^2\frac{\partial U}{\partial T_2} + \varepsilon^2 U^2\frac{\partial U}{\partial X_0} + \varepsilon^3 U^2\frac{\partial U}{\partial X_1} + \frac{\partial^3 U}{\partial X_0^3} + 3\varepsilon\frac{\partial^3 U}{\partial X_0^2\partial X_1}\\ &+3\varepsilon^2\frac{\partial^3 U}{\partial X_0\partial X_1^2} + \varepsilon^3\frac{\partial^3 U}{\partial X_1^3} = 0\end{aligned} \tag{9.336}$$

Next, we assume a series expansion of U as

$$U = U_0 + \varepsilon U_1 + \varepsilon^2 U_2 + \ldots \tag{9.337}$$

Substituting (9.337) into (9.336) and collecting the constant, linear and second order terms in ε, we find

ε^0 order:

$$\frac{\partial U_0}{\partial T_0} + \frac{\partial^3 U_0}{\partial X_0^3} = 0 \tag{9.338}$$

ε^1 order:

$$\frac{\partial U_1}{\partial T_0} + \frac{\partial^3 U_1}{\partial X_0^3} = -\frac{\partial U_0}{\partial X_1} - 3\frac{\partial^3 U_0}{\partial X_0^2\partial X_1} \tag{9.339}$$

ε^2 order:

$$\frac{\partial U_2}{\partial T_0} + \frac{\partial^3 U_2}{\partial X_0^3} = -\frac{\partial U_1}{\partial T_1} - 3\frac{\partial^3 U_1}{\partial X_0^2\partial X_1} - \frac{\partial U_0}{\partial T_2} - U_0^2\frac{\partial U_0}{\partial X_0} - 3\frac{\partial^3 U_0}{\partial X_0\partial X_1^2} \tag{9.340}$$

Similar to the standard perturbation method, we see that the mathematical structure of the left-hand side for (9.338) to (9.340) is the same.

It is straightforward to see the solution for (9.338) is

$$U_0 = A\exp\{i(kX_0 + k^3T_0)\} + ... \tag{9.341}$$

where

$$A = A(X_1, T_1, T_2) \tag{9.342}$$

To find the unknown function A, we have to go one order up. Substitution of (9.341) into (9.339) leads to

$$\frac{\partial U_1}{\partial T_0} + \frac{\partial^3 U_1}{\partial X_0^3} = -[\frac{\partial A}{\partial T_1} - 3k^2 \frac{\partial A}{\partial X_1}]e^{i(kX_0 + k^3T_0)} - ikA^2 e^{2i(kX_0 + k^3T_0)} + ... \tag{9.343}$$

We see that the first nonhomogeneous term on the right-hand side coincides with the homogeneous solution obtained in (9.341). Thus, the particular solution induced by this term is proportional to T_0 (similar to the discussion of Section 3.3.1 of Chau, 2018). Thus, the solution tends to infinity with time, and such solution is unphysical. To remove this unrealistic "resonant effect," we must set this bracket term to zero:

$$\frac{\partial A}{\partial T_1} - 3k^2 \frac{\partial A}{\partial X_1} = \frac{\partial A}{\partial T_1} + \omega'(k)\frac{\partial A}{\partial X_1} = 0 \tag{9.344}$$

Equation (9.344) is actually a conservation law that we have discussed. In the last equation of (9.344), we have replaced the dispersion relation of the linearized mKdV (i.e., (9.331) without the middle term) or $\omega = -k^3$. In particular, we can obtain this dispersion equation by substituting the following wave solution into the linearized mKdV (i.e., (9.331) without the middle term):

$$\varphi = A_0 \exp[i(\omega t - kx)] \tag{9.345}$$

Setting $U_1 = 0$ in (9.340) and neglecting the higher order terms, we have

$$\begin{aligned}\frac{\partial U_2}{\partial T_0} + \frac{\partial^3 U_2}{\partial X_0^3} &= -\frac{\partial U_0}{\partial T_2} - U_0^2 \frac{\partial U_0}{\partial X_1} - 3\frac{\partial^3 U_0}{\partial X_0 \partial X_1^2} \\ &= -[\frac{\partial A}{\partial T_2} + 3ik\frac{\partial^2 A}{\partial X_1^2} + ik|A|^2 A]e^{i(kX_0 + k^3T_0)} + ...\end{aligned} \tag{9.346}$$

To remove the unwanted resonance due to the nonhomogeneous term, again we must set the bracket term to zero:

$$i\frac{\partial A}{\partial T_2} - 3k\frac{\partial^2 A}{\partial X_1^2} - k|A|^2 A = 0 \tag{9.347}$$

This is the required NLSE. It can be rewritten in terms of the dispersion relation as before:

$$i\frac{\partial A}{\partial T_2} + \frac{\omega''(k)}{2}\frac{\partial^2 A}{\partial X_1^2} - k|A|^2 A = 0 \tag{9.348}$$

It was discovered that two different cases needed to be considered

$$\begin{aligned}-\frac{k\omega''(k)}{2} &> 0, \quad \text{focusing NLSE} \\ &< 0, \quad \text{defocusing NLSE}\end{aligned} \tag{9.349}$$

In the terminology of fibre optics, the soliton solution of the focusing NLSE is called a bright soliton and that of the defocusing NLSE is called a dark soliton. They are considered next.

9.9.2 Bright Soliton

The nonlinear Schrödinger equation is normally recast in the following form

$$i\psi_t + \psi_{xx} + 2\sigma|\psi|^2 \psi = 0 \quad (9.350)$$

For $\sigma >> 1$, the nonlinearity dominates, whereas for $\sigma << 1$, the dispersion phenomenon dominates. For a soliton solution to be possible, we need to have $|2\sigma| \approx 1$. We look for a separable type of solution

$$\psi(x,t) = u(x)e^{i\phi(t)} \quad (9.351)$$

Substitution of (9.351) into (9.350) results in

$$-u\phi_t + u_{xx} + 2\sigma u^3 = 0 \quad (9.352)$$

Since a function of time cannot be equal to a function of x, unless both of them equals the constant, this leads to

$$\frac{d\phi}{dt} = \frac{u_{xx}}{u} + 2\sigma u^2 = C \quad (9.353)$$

where C is a constant. The first and the last terms of (9.353) can be used to give

$$\phi = Ct \quad (9.354)$$

The last two terms of (9.353) can be rearranged as

$$u_{xx} = -2\sigma u^3 + Cu \quad (9.355)$$

Similar to the technique in integrating the KdV, we can multiply (9.355) by u_x to get

$$u_x u_{xx} = -2\sigma u_x u^3 + Cuu_x \quad (9.356)$$

By observation, this can be written as

$$\frac{d}{dt}[\frac{1}{2}(u_x)^2] = \frac{d}{dx}[-\frac{1}{2}\sigma u^4 + \frac{1}{2}Cu^2] \quad (9.357)$$

This is an exact differential and thus can be integrated immediately as

$$(u_x)^2 = -\sigma u^4 + Cu^2 + C_0 \quad (9.358)$$

The shape of the soliton, of course, depends on σ and, more precisely, the solution form of the soliton depends on the sign of σ.

For bright solitons, we have σ = 1 and we look for non-periodic solutions. Thus, the NLSE becomes focusing. The boundary conditions are that u and u' go to zero for $x \to \pm\infty$. This leads to $C_0 = 0$ and using this condition (9.358) becomes

$$u_x = u(C - u^2)^{1/2} \quad (9.359)$$

This is a separable first order ODE and can be integrated directly to give

$$\frac{du}{u(C-u^2)^{1/2}} = dx \quad (9.360)$$

For the simplest case of $C = 1$, we have

$$-\ln[\frac{1+\sqrt{1-u^2}}{u}] = x \quad (9.361)$$

Rearranging, we get

$$\sqrt{1-u^2} = (ue^{-x} - 1) \quad (9.362)$$

Squaring both sides, we have

$$u = \frac{2}{e^x + e^{-x}} = \frac{1}{\cosh x} = \operatorname{sech} x \tag{9.363}$$

Therefore, u is a sech function (recalling the soliton of the KdV is also expressible in such a function). Using $C = 1$, we can write (9.354) as

$$\phi = t \tag{9.364}$$

Finally, we can substitute (9.363) and (9.364) into (9.351) to give

$$\psi = (\operatorname{sech} x)e^{it} \tag{9.365}$$

To visualize the prediction of this solution, we can scale the unknown, the time, and the spatial variables by (see Problem 9.11)

$$t \to a^2 t, \quad x \to ax, \quad \psi \to \frac{\psi}{a} \tag{9.366}$$

By virtue of (9.366), we get the solution as

$$\psi = a \operatorname{sech}(ax)e^{ia^2 t} \tag{9.367}$$

Next, we introduce a moving coordinate or the Galilean transformation (or gauge transformation) (see Problem 9.12):

$$x \to x - ct, \quad \psi \to \psi e^{i(\frac{cx}{2} - \frac{c^2}{4}t)} \tag{9.368}$$

Finally, we obtain the standard form of the bright soliton solution found in the literature:

$$\psi = a \operatorname{sech}[a(x - ct)]e^{i\frac{cx}{2} + i(a^2 - \frac{c^2}{4})t} \tag{9.369}$$

The real part of the bright soliton is plotted against x for a fixed t in Figure 9.10 to illustrate the form of a deep water nonlinear wave that can be modeled by the bright soliton. Note that the envelope of the time-dependent function (dotted lines) is similar to the soliton for the KdV (sech function), whereas the bright soliton (solid line) is oscillating between the upper and lower envelopes. The number of oscillations in the soliton solution increases with the value of c and the magnitude increases with a. Thus, (9.369) can be considered as a two-parameter solution.

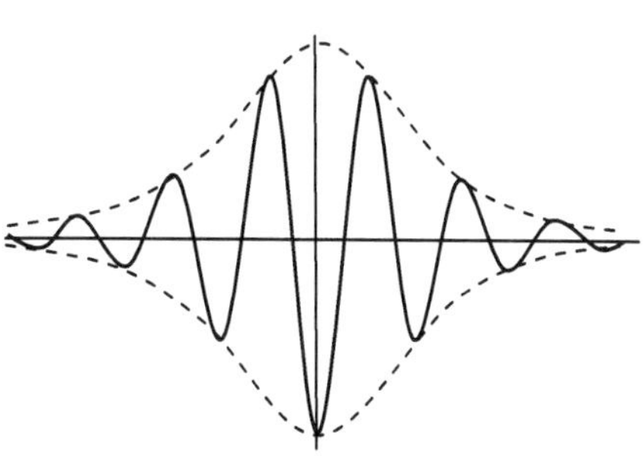

Figure 9.10 Bright soliton and its envelope

9.9.3 Dark Soliton

For dark solitons, we have $\sigma = -1$ and we look for non-periodic solutions. For this case, the NLSE becomes defocusing. The analysis presented in the previous section for the bright soliton applies equally to the dark soliton up to (9.358). If we take $\sigma = -1$, $C = -1$ and $C_0 = 1/4$, (9.358) can be written as:

$$(u_x)^2 = u^4 - u^2 + \frac{1}{4} = (u^2 - \frac{1}{2})^2 = (\frac{1}{2} - u^2)^2 \tag{9.370}$$

This corresponds to the simplest scenario. By using $C_0 = 1/4$, we have a complete square term in (9.370). Again, this is a separable first order ODE and can be rearranged as

$$\frac{du}{\frac{1}{2} - u^2} = dx \tag{9.371}$$

Direct integration gives

$$\sqrt{2}\tanh^{-1}(\sqrt{2}u) = x \tag{9.372}$$

This can be inverted to give u as

$$u = \frac{1}{\sqrt{2}}\tanh(\frac{x}{\sqrt{2}}) \tag{9.373}$$

With $\sigma = -1$, (9.354) becomes

$$\phi = -t \tag{9.374}$$

Substitution of (9.374) and (9.373) into (9.351) gives

$$\psi = \frac{1}{\sqrt{2}}\tanh(\frac{x}{\sqrt{2}})e^{-it} \tag{9.375}$$

Similarly, we can apply the following invariant scaling for the wave function ψ, the time t, and the spatial variable x to (9.375) (see Problem 9.11)

$$t \to a^2 t, \quad x \to ax, \quad \psi \to \frac{\psi}{a} \tag{9.376}$$

Using these changes of variables, we have

$$\psi = \frac{a}{\sqrt{2}}\tanh(\frac{ax}{\sqrt{2}})e^{-ia^2 t} \tag{9.377}$$

Next, we adopt an invariant Galilean transform as (see Problem 9.12)

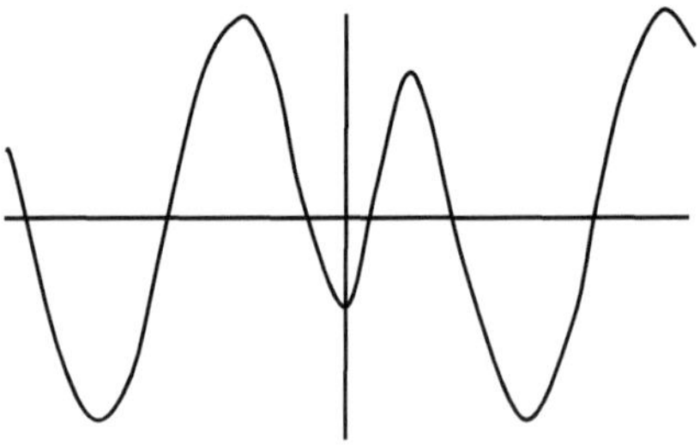

Figure 9.11 Dark soliton with dispersive behavior

$$x \to x - ct, \quad \psi \to \psi e^{i(\frac{cx}{2} - \frac{c^2}{4}t)} \tag{9.378}$$

Substitution of (9.378) into (9.377) gives the dark soliton as

$$\psi = \frac{a}{\sqrt{2}} \tanh[\frac{a}{\sqrt{2}}(x - ct)] e^{i[\frac{cx}{2} - (a^2 + \frac{c^2}{4})t]} \tag{9.379}$$

Figure 9.11 illustrates a typical dark soliton plotted against x at a fixed t, showing a dispersive behavior. The number of oscillations increases with c and amplitude increases with a. This is again a two-parameter solution, depending on a and c.

9.9.4 Rogue Waves in Oceans

We have shown that after proper change of variables, the KdV can be converted to the nonlinear Schrödinger equation. In 1968, Zakharov was the first to show that deep water waves are governed by the nonlinear Schrödinger equation (NLSE), and derived the breather-type solution through the Benjamin-Feir instability (FBI) (Benjamin and Feir, 1967). In particular, they showed that modulational instability of Stokes's waves occurs and waves traveling at different speeds can focus at a specific point. It was speculated that a periodic solution with infinite period appears as a form of rogue wave or freak wave. Such solution is referred to as the breather solution (probably because of its resemblance to the water surface during the breathing of whales). It has been reported that huge monster waves can come from nowhere in oceans and disappear without a trace. One notable event is the damage of the ocean-liner *Wilstar* in 1974 by such freak or rogue waves. It was recorded that it appears in the Agulhas current off the coast of South Africa when storm waves from Antarctica impact on the Agulhas current. Therefore, it is commonly believed that rogue waves were formed by wave-current interaction. Another possibility is that rogue waves result from geometrical focusing, or dispersion focusing of linear waves. It is likely that such interaction or focusing of linear waves leads to the formation of nonlinear waves (such as those modeled by the NLSE). On New Year's Day 1995, a single rogue wave of 26 m was recorded at the Draupner oil platform in the North Sea off the Norwegian coastline (compared to the 5-m waves in the background at that time). This provides the first solid scientific evidence of the existence of rogue waves. With the advance in satellite technology, it was found that rogue waves occur much more often than we normally anticipate. Radar records from the North Sea Gorm oil field show 466 rogue waves in a time span of 12 years from 1992. It was reported that over 200 supertankers longer than 200 m have sunk beneath such rogue waves from 1994 to 2004. This area remains an active area of research. In this section, we will only quote one of the so-called "breather" solutions of the NLSE that closely resemble the rogue wave.

We consider the normalized form of the nonlinear Schrödinger equation

$$i\psi_x + \psi_{tt} + |\psi|^2 \psi = 0 \tag{9.380}$$

Note that (9.380) is equivalent to (9.350) by setting $2\sigma = 1$ and reversing the roles of x and t. One of its solutions is called the Peregrine breather (a special form of the Akhmediev breather or Ma breather with infinite period) and is given as (Peregrine, 1983):

$$\psi(x,t) = [\frac{4(1+2ix)}{1+4x^2+2t^2} - 1]e^{ix} \tag{9.381}$$

Strictly speaking, the Peregrine breather is not a soliton. A three-dimensional plot of such a breather reveals that there are holes in the water in front and behind the monster-like rogue or freak wave. This seems to agree with eyewitness reports. This solution closely resembles the appearance of rogue waves in oceans. More discussion of the Peregrine breather as a potential rogue wave is given by Shrira and Geogjaev (2010). Rogue waves were also observed in fiber optics.

Figure 9.12 Rogue or freak waves observed in deep ocean (by NOAA)

To derive the Peregrine breather, we follow the displaced phase-amplitude variables method proposed by Karjanto and van Groesen (2007). In particular, we first rewrite (9.380) in a more general form as

$$A_\xi + i\beta A_{\tau\tau} + i\gamma |A|^2 A = 0 \tag{9.382}$$

We first note that a "continuous wave" or plane wave exists for (9.382) and is given by

$$A_0(\xi) = r_0 e^{-i\gamma r_0^2 \xi} \tag{9.383}$$

It can be shown by direct substitution that (9.383) indeed satisfies (9.382) (see Problem 9.8). We now seek a breather-type solution that arises from the background wave A_0 as

$$A(\xi,\tau) = A_0(\xi)[G(\xi,\tau)e^{i\phi(\xi)} - 1] \tag{9.384}$$

This "displaced" solution only depends on the phase $\phi(\xi)$ which is not a function of τ. Note that the solution in (9.384) has already subtracted the background plane wave. Differentiation of (9.384) gives

$$A_\xi = A_{0,\xi}[Ge^{i\phi} - 1] + A_0[G_\xi e^{i\phi} + i\phi_\xi Ge^{i\phi}] \tag{9.385}$$

$$A_{\tau\tau} = A_0 G_{\tau\tau} e^{i\phi} \tag{9.386}$$

The complex function A and its magnitude are

$$A = A_0[G\cos\phi - 1 + iG\sin\phi] \tag{9.387}$$

$$|A|^2 = A\bar{A} = A_0^2[G^2 - 2G\cos\phi + 1] \tag{9.388}$$

In view of the plane wave solution given in (9.383), these equations can be written as

$$\begin{aligned} A_\xi &= \{-i\gamma r_0^3(G\cos\phi - 1) + r_0(G_\xi\cos\phi - \phi_\xi G\sin\phi) \\ &+[\gamma r_0^3 G\sin\phi + ir_0(G_\xi\sin\phi + \phi_\xi G\cos\phi)]\}e^{-i\gamma r_0^2\xi} \end{aligned} \tag{9.389}$$

$$i\beta A_{\tau\tau} = \beta r_0 G_{\tau\tau}(i\cos\phi - \sin\phi)e^{-i\gamma r_0^2\xi} \tag{9.390}$$

$$i\gamma|A|^2 A = \gamma r_0^3[G(G - 2\cos\phi) + 1][i(G\cos\phi - 1) - G\sin\phi]e^{-i\gamma r_0^2\xi} \tag{9.391}$$

Substitution of (9.389) to (9.391) into the nonlinear Schrödinger equation given in (9.382) gives a complex equation. Collecting terms, we obtain the following real part of this nonlinear Schrödinger equation

$$\begin{aligned} &r_0(G_\xi\cos\phi - \phi_\xi G\sin\phi) + \gamma r_0^3 G\sin\phi - \beta r_0 G_{\tau\tau}\sin\phi \\ &-\gamma r_0^3[G(G - 2\cos\phi) + 1]G\sin\phi = 0 \end{aligned} \tag{9.392}$$

Similarly, if we collect the imaginary terms, we obtain the following imaginary part of the nonlinear Schrödinger equation

$$\begin{aligned} &-\gamma r_0^3(G\cos\phi - 1) + r_0(G_\xi\sin\phi + \phi_\xi G\cos\phi) + \beta r_0 G_{\tau\tau}\cos\phi \\ &+\gamma r_0^3[G(G - 2\cos\phi) + 1](G\cos\phi - 1) = 0 \end{aligned} \tag{9.393}$$

Multiplying (9.392) by $\cos\phi$, multiplying (9.393) by $\sin\phi$, and adding these results, we get

$$G_\xi + \gamma r_0^2 G\sin 2\phi - \gamma r_0^2\sin\phi G^2 = 0 \tag{9.394}$$

This is a nonlinear first order PDE. To solve for G, we assume the following change of variables

$$G = \frac{1}{H} \tag{9.395}$$

Substitution of (9.395) into (9.394) gives

$$-H_\xi + \gamma r_0^2 H\sin 2\phi - \gamma r_0^2\sin\phi = 0 \tag{9.396}$$

We have converted the nonlinear PDE to a linear first order PDE. We can rewrite (9.396) as

$$dH + [-\gamma r_0^2 H\sin 2\phi + \gamma r_0^2\sin\phi]d\xi = 0 \tag{9.397}$$

This equation can be recast as

$$N(H,\xi)dH + M(H,\xi)d\xi = 0 \tag{9.398}$$

Recognizing the function N and M from (9.397), we find that

$$\frac{\partial N}{\partial \xi} = 0, \quad \frac{\partial M}{\partial H} = -\gamma r_0^2\sin 2\phi \tag{9.399}$$

This first order PDE is not exact. Multiplying (9.398) by an integrating factor, we enforce the following condition

$$\frac{\partial \mu N}{\partial \xi} = \frac{\partial \mu M}{\partial H} \tag{9.400}$$

Assuming that the integrating factor is not a function of H, we get

$$\frac{d\mu}{\mu} = -\gamma r_0^2 \sin 2\phi d\xi \tag{9.401}$$

Integrating both sides, we find the integrating factor as

$$\mu = e^{-\gamma r_0^2 \int \sin 2\phi d\xi} \tag{9.402}$$

We can assume the following solution form

$$u(H, \xi) = C \tag{9.403}$$

Thus, we have

$$du = \frac{\partial u}{\partial H} dH + \frac{\partial u}{\partial \xi} d\xi = 0 \tag{9.404}$$

Comparing this to the result of equation (9.398) and multiplying by μ, we get

$$\frac{\partial u}{\partial H} = \mu \tag{9.405}$$

Integration gives

$$u = \mu(\xi)H + f(\xi) \tag{9.406}$$

Differentiating (9.406) with respect to ξ, we get

$$\begin{aligned} \frac{\partial u}{\partial \xi} &= \frac{d\mu}{d\xi}(\xi)H + \mu\frac{dH}{d\xi} + f'(\xi) = \mu M \\ &= \gamma r_0^2 \{-\mu H \sin 2\phi + \mu \sin \phi\} \end{aligned} \tag{9.407}$$

From the result from (9.401), we have

$$\frac{d\mu}{d\xi} = -\gamma r_0^2 \mu \sin 2\phi \tag{9.408}$$

Substituting (9.408) into (9.407), we get an ODE for f as

$$f'(\xi) = \gamma r_0^2 \mu \sin \phi \tag{9.409}$$

Integration gives

$$f(\xi) = \gamma r_0^2 \int \sin \phi \mu d\xi \tag{9.410}$$

Note that since G or H is a function of τ also, we need to set u equal to an arbitrary function of τ, instead of a constant as shown in (9.403). Therefore, the solution for H is (in view of (9.406) and (9.403))

$$u = \mu(\xi)H + \gamma r_0^2 \int \sin \phi \mu d\xi = -\zeta(\tau) \tag{9.411}$$

In particular, we have to replace the constant C by an arbitrary function $\zeta(\tau)$. Thus, we can determine H as

$$H(\xi, \tau) = \frac{-\gamma r_0^2 \int \sin \phi \mu d\xi - \zeta(\tau)}{\mu(\xi)} = \frac{\tilde{Q}(\xi) - \zeta(\tau)}{\mu(\xi)} \tag{9.412}$$

Substitution of (9.412) into (9.395) gives the solution for G

$$G(\xi,\tau)=\frac{\mu(\xi)}{\tilde{Q}(\xi)-\zeta(\tau)} \tag{9.413}$$

We can invert the phase as

$$\phi=\phi(\xi),\quad \xi=\xi(\phi) \tag{9.414}$$

Using (9.414), we can rewrite (9.413) as

$$G(\phi,\tau)=\frac{P(\phi)}{Q(\phi)-\zeta(\tau)} \tag{9.415}$$

This is the general solution for G. Before we consider the particular form of $\zeta(\tau)$, we derive another equation for G by subtracting the results of multiplying (9.392) by $\sin\phi$ and of multiplying (9.393) by $\cos\phi$:

$$\beta G_{\tau\tau}+[\phi_{\xi}+2\gamma r_0^2\cos^2\phi]G-3\gamma r_0^2\cos\phi G^2+\gamma r_0^2 G^3=0 \tag{9.416}$$

This equation will be used later for solving P and Q in (9.415).

As observed by Karjanto and van Groesen (2007), when we substitute different mathematical forms for ζ, we can obtain different breather solutions, including the Akhmediev breather, Ma breather, and Peregrine breather. In this present section, we will only consider the Peregrine breather, which is believed to resemble the rogue wave or freak wave more closely. In particular, we consider the following form for ζ:

$$\zeta(\tau)=1-\frac{1}{2}\nu^2\tau^2 \tag{9.417}$$

Substituting (9.417) into (9.415), we get

$$G(\phi,\tau)=\frac{P(\phi)}{Q(\phi)-1+\frac{1}{2}\nu^2\tau^2} \tag{9.418}$$

Differentiation of (9.418) gives

$$G_{\tau}=-\frac{P\nu^2\tau}{(Q-1+\frac{1}{2}\nu^2\tau^2)^2} \tag{9.419}$$

$$G_{\tau\tau}=\frac{2P(\nu^2\tau)^2}{(Q-1+\frac{1}{2}\nu^2\tau^2)^3}-\frac{P\nu^2}{(Q-1+\frac{1}{2}\nu^2\tau^2)^2} \tag{9.420}$$

From (9.418), we have the following identities

$$\frac{G}{P}=\frac{1}{Q-1+\frac{1}{2}\nu^2\tau^2} \tag{9.421}$$

$$\nu^2\tau^2=2(\frac{P}{G}+1-Q) \tag{9.422}$$

Substitution of (9.421) and (9.422) into (9.420) arrives at

$$G_{\tau\tau}=3\nu^2\frac{G^2}{P}-4\nu^2(Q-1)\frac{G^3}{P^2} \tag{9.423}$$

Comparison of (9.416) and (9.420) gives three equations

$$P = \frac{\beta v^2}{\gamma r_0^2 \cos\phi} = \frac{\tilde{v}^2}{\cos\phi} \tag{9.424}$$

$$Q = \frac{1}{4\cos\phi} + 1 \tag{9.425}$$

$$\phi_\xi = -2\gamma r_0^2 \cos^2\phi \tag{9.426}$$

where

$$\tilde{v}^2 = \frac{\beta v^2}{\gamma r_0^2} \tag{9.427}$$

Integrating (9.426) with respect to ξ, we find

$$\tan\phi = -2\gamma r_0^2 \xi \tag{9.428}$$

This result of ϕ can be used to find

$$\cos\phi = \frac{1}{\sqrt{1+(2\gamma r_0^2 \xi)^2}} \tag{9.429}$$

$$\sin\phi = \frac{-2\gamma r_0^2 \xi}{\sqrt{1+(2\gamma r_0^2 \xi)^2}} \tag{9.430}$$

With these results, (9.418) becomes

$$G = \frac{4\tilde{v}^2 \cos\phi}{\tilde{v}^2 + 2v^2\tau^2 \cos^2\phi} \tag{9.431}$$

Substituting (9.427) and (9.431) into (9.384), we have

$$\begin{aligned} A(\xi,\tau) &= A_0(\xi)[G(\xi,\tau)\cos\phi - 1 + iG(\xi,\tau)\sin\phi] \\ &= [\frac{4(1-2i\gamma r_0^2 \xi)}{1+4(\gamma r_0^2 \xi)^2 + 2\frac{\gamma}{\beta} r_0^2 \tau^2} - 1] r_0 e^{-i\gamma r_0^2 \xi} \end{aligned} \tag{9.432}$$

This is the Peregrine breather solution. To recover (9.380) from (9.382), we can set:

$$\gamma = -1, \quad \beta = -1, \quad r_0 = 1 \tag{9.433}$$

Thus, the solution becomes

$$A(\xi,\tau) = [\frac{4(1+2i\xi)}{1+4\xi^2 + 2\tau^2} - 1] e^{i\xi} \tag{9.434}$$

which agrees with (9.381). Note that (9.380) can also be written as

$$i\psi_t + \psi_{xx} + |\psi|^2 \psi = 0 \tag{9.435}$$

For this form, we have the Peregrine breather as:

$$\psi(x,t) = [\frac{4(1+2it)}{1+4t^2 + 2x^2} - 1] e^{it} \tag{9.436}$$

Equation (6.7) of Peregrine (1983) is recovered if the following identification is made:

$$t \leftarrow 2t \tag{9.437}$$

Figure 9.13 plots the Peregrine breather profile as a function of time. The soliton-like solution oscillates with time and decays to zero as $t \to \infty$. Figure 9.14 plots the real part, the imaginary part, and the absolute magnitude of the Peregrine breather versus time t and spatial distance x. A very distinct peak resembling a rogue wave with one hole in front and one hole behind was observed in Figure 9.14. A rogue wave is normally referred as a single wave with an unusually large magnitude, which is also known as a freak wave.

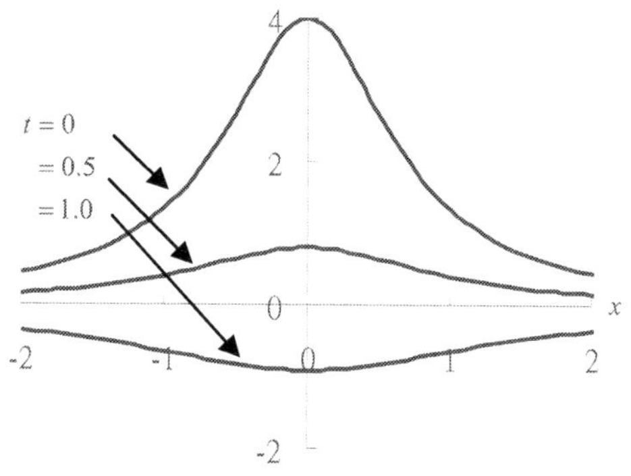

Figure 9.13 Peregrine breather as a function of time

Figure 9.15 shows the surface elevation of the Draupner oil platform in the North Sea recorded on January 1, 1995 (the first scientific record of the existence of the rogue wave) together with the Peregrine solution plus a harmonic wave solution. The wave magnitude of the Peregrine wave is given by

$$|\psi(x,t)| = [(\frac{4}{1+16t^2+2x^2}-1)^2+(\frac{16t}{1+16t^2+2x^2})^2]^{1/2} \tag{9.438}$$

In the plot of Figure 9.15, we have added 3.5 times this magnitude to a cosine wave, which is treated as a background wave. To the best of our knowlegde, this is the first numerical comparison of the Peregrine breather to the Draupner measurement.

9.10 OTHER NONLINEAR WAVE EQUATIONS

There are many different kinds of nonlinear wave equations which are similar to the KdV or nonlinear Schrödinger equation. The following are some of them:

Sawada-Koera equation

$$\varphi_t + 15(\varphi^3 + \varphi\varphi_{xx}) + \varphi_{xxxxx} = 0 \tag{9.439}$$

Boussineq equation

$$\varphi_{tt} - \varphi_{xx} - 3(\varphi^2)_{xx} - \varphi_{xxxx} = 0 \tag{9.440}$$

Lax's 5th order equation

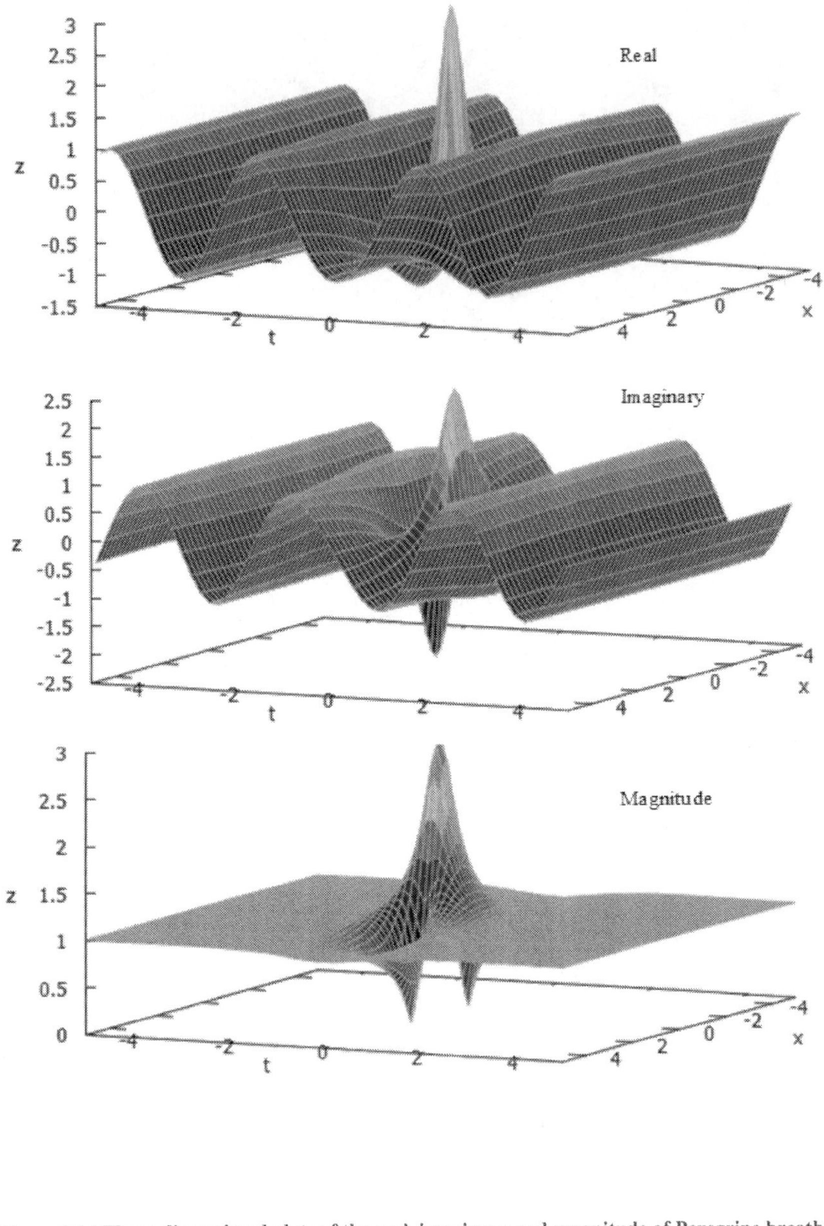

Figure 9.14 Three-dimensional plots of the real, imaginary and magnitude of Peregrine breather

$$\varphi_{xxxxx} + 10\varphi\varphi_{xxx} + 20\varphi_x\varphi_{xx} + 30\varphi^2\varphi_x + \varphi_t = 0 \tag{9.441}$$

Camassa-Holm (CH) equation

$$\varphi_t - \varphi_{xxt} + \kappa\varphi_x + 3\varphi\varphi_x - 2\varphi_x\varphi_{xx} - \varphi\varphi_{xxx} = 0 \tag{9.442}$$

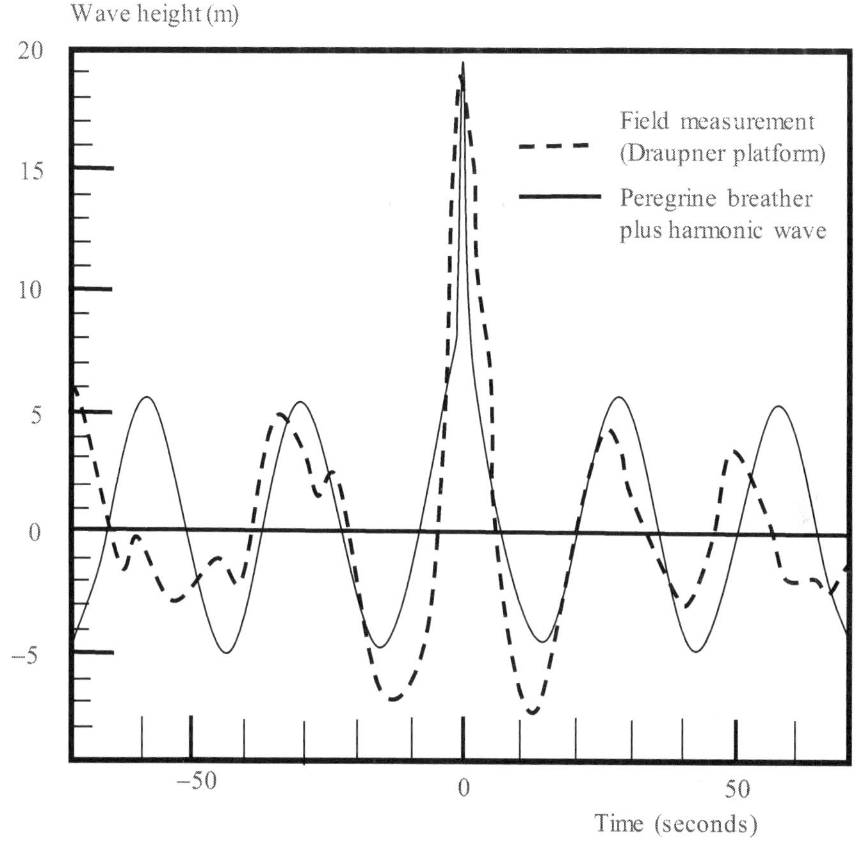

Figure 9.15 Wave height measurements at the Draupner platform on January 1, 1995 together with the superposition of a harmonic wave plus Peregrine breather solution

where κ is a constant relating to the critical shallow-water wave speed. A special feature of the CH equation is that it gives rise to a soliton solution with a non smooth peak. Therefore, the CH equation has been proposed to model wave breaking.

9.11 SUMMARY AND FURTHER READING

In this chapter, we have introduced the important concepts relating to nonlinear waves and solitons. Nonlinear transports, shocks and dispersive waves are discussed before the KdV is introduced. The physical interpretation of the KdV is considered as a balance of nonlinearity and dispersion. The direct method of Hirota is discussed through the bilinear form and D-operator, and applied to find the one-, two-, and N-soliton solutions. The relations between the KdV, mKdV,

Boussinesq equation, nonlinear Schrödinger equation (NLSE), first Painlevé equation, and second Painlevé equation are found. The integrability of KdV is demonstrated through conservation laws via the Miura-Gardner-Kruskal transformation. Finally the NLSE is considered, and solution for bright, dark, and Peregrine breather solitons are derived. The focus is on its relevance to rogue or freak waves.

There are many important topics related to the KdV and solitons are left out in the present short chapter. For example, nearly all textbooks cover the inverse scattering transform (IST) and its application in deriving soliton solutions. In the present chapter, we prefer to discuss the more direct approach by Hirota. The concept and solution procedure of IST is less straightforward compared to Hirota's approach. It was said that when Gardner, Greene, Kruskal and Miura thought that the Cole-Hopf transform used in the linearized Burgers equation can be generalized as for the KdV

$$u = -6\frac{\psi_{xx}}{\psi} \tag{9.443}$$

This is because the KdV involves a higher derivative term. Thus, (9.443) did not work. They recognized that they can add a constant to the solution in view of the Galilean invariance (see Problem 9.10). Thus, they proposed

$$u + E = -6\frac{\psi_{xx}}{\psi} \tag{9.444}$$

Rearranging, they got

$$-6\psi_{xx} - u\psi = E\psi \tag{9.445}$$

When they were looking at this equation written on the blackboard, someone walked by the corridor and said, "You guys are working on quantum mechanics, right?" Indeed, (9.444) can be recognized as a special case of the Schrödinger equation. This started the whole idea of relating solving the KdV to solving the eigenvalue spectrum of the Schrödinger equation for a given energy E. It turns out that there are two kinds of eigenvalues if a complex value of E is allowed: a discrete spectrum and a continuous spectrum. The wave function of the continuous spectrum case in the Schrödinger equation appears as an incoming wave and an outgoing wave with a transmission coefficient for $x \to \infty$ and with a reflection coefficient for $x \to -\infty$. However, the solution of the KdV u is a function of time and can be interpreted as $V = -u$. Thus, they allowed the potential V of the Schrödinger equation to be a function of time t and looked for the eigenvalues of E. For some initial data for u, they solved the eigenvalue problem of Schrödinger equation. The eigenfunctions are normalized by a constant, to which its time dependence was recognized from the KdV equation. Then, the inverse scattering of the time-evolving spectral data is used to determine $V = u(x,t)$. This is the reason why the technique is called the inverse scattering transform. It closely resembles the Fourier transform for linear problems.

For IST, we refer the reader to Ablowitz and Segur (1981) and Ablowitz and Clarkson (1991). The thesis by Druitt (2005) submitted to the University of Melbourne provided a clear coverage of the Hirota direct method. For the Peregrine breather, we refer to the coverage by Karjanto and van Groesen (2007).

9.12 PROBLEMS

Problem 9.1. The sine-Gordon equation is given by

$$\phi_{xx} - \phi_{tt} = \sin\phi \tag{9.446}$$

This equation relates to magnetic flux propagation in Josephson junctions between two superconductors, and to propagation of deformations along the double helix of DNA. Use the following change of variables to find Hirota's bilinear form

$$\phi = 4\tan^{-1}\left(\frac{g}{f}\right) \tag{9.447}$$

Ans:

$$\begin{aligned} &(f^2 - g^2)(f_{xx}g - 2f_x g_x + fg_{xx} - f_{tt}g + 2f_t g_t - fg_{tt} - fg) \\ &-2fg(ff_{xx} - f_x^2 - gg_{xx} + g_x^2 - ff_{tt} + f_t^2 + gg_{tt} - g_t^2) = 0 \end{aligned} \tag{9.448}$$

Problem 9.2. The Kadomtsev-Petviashvili (KP) equation is given by

$$(u_t - 6uu_x + u_{xxx})_x + 3u_{yy} = 0 \tag{9.449}$$

Use the following change of variables to find Hirota's bilinear form

$$u(x,t,y) = -2(\ln f)_{xx} \tag{9.450}$$

Ans:

$$ff_{xt} + f_x f_t + 3f_{xx}^2 + ff_{xxxx} - 4f_x f_{xxx} + 3f_{yy}f - 3f_y^2 = 0 \tag{9.451}$$

Problem 9.3. A particular form of the mKdV equation is given by

$$u_t + 24u^2 u_x + u_{xxx} = 0 \tag{9.452}$$

Use the following change of variables to find Hirota's bilinear form

$$u(x,t) = \frac{g_x f - gf_x}{g^2 + f^2} \tag{9.453}$$

Ans: There are two bilinear forms

$$g_t f - gf_t + g_{xxx}f - 3g_{xx}f_x + 3g_x f_{xx} - gf_{xxx} = 0 \tag{9.454}$$

$$ff_{xx} - f_x^2 + gg_{xx} - g_x^2 = 0 \tag{9.455}$$

Problem 9.4. The extended KP equation is given by

$$(u_t - 6uu_x + u_{xxx})_x + 3u_{yy} + \gamma u_{tt} + \beta u_{ty} = 0 \tag{9.456}$$

Use the following change of variables to find Hirota's bilinear form

$$u(x,t,y) = -2(\ln f)_{xx} \tag{9.457}$$

Ans:

$$\begin{aligned} &ff_{xt} - f_x f_t + f_{xxxx}f - 4f_x f_{xxx} + 3f_{xx}^2 + 3f_{yy}f - 3f_y^2 \\ &+\gamma ff_{tt} - \gamma f_t^2 + \beta f_{ty}f - \beta f_t f_y = 0 \end{aligned} \tag{9.458}$$

Problem 9.5. The KdV equation in fluid mechanics is normally given in dimensional form as

$$u_t + c_0 u_x + \frac{3c_0}{2h} u u_x + \frac{1}{6} h^2 c_0 u_{xxx} = 0 \tag{9.459}$$

where h is the depth of the water and c_0 is defined as

$$c_0 = \sqrt{gh} \tag{9.460}$$

(i) Applying the following change of variables

$$\bar{t} = \frac{t}{\tau_0}, \quad \bar{x} = kx, \quad \bar{u} = \frac{u}{k\alpha_1}, \quad \tau_0 = \frac{1}{kc_0} = \frac{\lambda}{2\pi c_0}, \quad \alpha_1 = \frac{c_0 a}{kh}, \tag{9.461}$$

where a is the magnitude of the wave, show that

$$\bar{u}_{\bar{t}} + \bar{u}_{\bar{x}} + \frac{3}{2}\alpha \bar{u}\bar{u}_{\bar{x}} + \frac{1}{6}\beta \bar{u}_{\bar{x}\bar{x}\bar{x}} = 0 \tag{9.462}$$

where

$$\alpha = \frac{a}{h}, \quad \beta = (kh)^2 \tag{9.463}$$

(ii) Consider another round of change of variables to show

$$X = \bar{x} - \bar{t}, \quad T = \varepsilon \bar{t}, \quad \bar{u}(x,t) = u(X,T), \quad \varepsilon = \frac{h}{\lambda} \tag{9.464}$$

$$\varepsilon u_T + \frac{3}{2}\alpha u u_X + \frac{1}{6}\beta u_{XXX} = 0 \tag{9.465}$$

(iii) Assume that $\alpha \approx \beta \approx \varepsilon$ and use the following change of variables

$$\tilde{x} \to 6^{1/3} x, \quad \tilde{u} \to 4(6)^{1/3} u, \quad t \to T, \tag{9.466}$$

to show that

$$\tilde{u}_t + 6\tilde{u}\tilde{u}_x + \tilde{u}_{xxx} = 0 \tag{9.467}$$

Problem 9.6. Show that if $u(x,t)$ is a solution of the KdV equation, the following function is also a solution of the KdV equation:

$$v(x,t) = u(x-a, t-b) \tag{9.468}$$

Problem 9.7. In Section 9.5.1, we considered the derivation of the KdV. In this problem, we consider another proof of the KdV equation. Recall from (9.45) to (9.48) but shifting the coordinate to the mean sea level as

$$\nabla^2 \phi = \frac{\partial^2 \phi}{\partial x^2} + \frac{\partial^2 \phi}{\partial z^2} = 0 \tag{9.469}$$

$$\frac{\partial \phi}{\partial z} = 0, \quad z = -h_0 \tag{9.470}$$

$$\frac{\partial \phi}{\partial t} + g\eta + \frac{1}{2}[(\frac{\partial \phi}{\partial x})^2 + (\frac{\partial \phi}{\partial z})^2] = 0, \quad z = \eta(x,t) \tag{9.471}$$

$$\frac{\partial \eta}{\partial t} + \frac{\partial \phi}{\partial x}\frac{\partial \eta}{\partial x} - \frac{\partial \phi}{\partial z} = 0, \quad z = \eta(x,t) \tag{9.472}$$

(i) First, assume the following expansion for ϕ

$$\phi(x,z,t) = \phi_0(x,t) + \sum_{n=2}^{\infty} \phi_n(x,t)(z+1)^n \tag{9.473}$$

Substitute (9.473) into the Laplace equation (9.469) to show

$$\phi_{2k}(x,t) = \frac{(-1)^k}{(2k)!} \frac{\partial^{2k}\phi_0(x,t)}{\partial x^{2k}} \tag{9.474}$$

(ii) Show that (9.470) is automatically satisfied by the choice of (9.473).

(iii) Rescale the variables as

$$\phi_0(x,t) = W(X,T), \quad X = \varepsilon^{1/2} x, \quad T = \varepsilon^{1/2} t \tag{9.475}$$

Substituting (9.475) into (9.473), show that

$$\phi(x,z,t) = W(X,T) + \frac{\varepsilon}{2}(1+z)^2 W_{XX} + \frac{\varepsilon^2}{24}(1+z)^4 W_{XXXX} + O(\varepsilon^2) \tag{9.476}$$

(iv) Make the solution of the KdV weakly nonlinear by assuming multiple scaling for the water level η and W in the flow potential ϕ:

$$\eta(x,t) = \varepsilon G(X,T), \quad W(X,T) = \varepsilon^{1/2} N(X,T) \tag{9.477}$$

Substituting (9.477) and (9.476) into (9.472), show that

$$G_T + N_{XX}(z+1) + G_X N_X \varepsilon - \frac{1}{6} N_{XXXX}(z+1)^3 \varepsilon + O(\varepsilon^2) = 0 \tag{9.478}$$

(v) Substituting (9.477) and (9.476) into (9.471), show that

$$N_T + G - \frac{1}{2} N_{XXT}(z+1)^2 \varepsilon + O(\varepsilon^2) = 0 \tag{9.479}$$

(vi) Substituting $z = \varepsilon G$ at the free surface boundary into (9.478) and (9.479), show that

$$G_T + N_{XX} + G N_{XX}\varepsilon + G_X N_X \varepsilon - \frac{1}{6} N_{XXXX}\varepsilon + O(\varepsilon^2) = 0 \tag{9.480}$$

$$N_T + G - \frac{1}{2} N_{XXT}\varepsilon + O(\varepsilon^2) = 0 \tag{9.481}$$

(vii) Eliminate G from (9.480) and (9.481) to show

$$N_{XX} - N_{TT} - N_T N_{XX}\varepsilon - N_{TX} N_X \varepsilon + \frac{1}{3} N_{XXXX}\varepsilon + O(\varepsilon^2) = 0 \tag{9.482}$$

(viii) Apply the following change of variables

$$\xi = X - T, \quad \tau = \varepsilon T \tag{9.483}$$

to show

$$2N_{\xi\tau} + \frac{1}{3} N_{\xi\xi\xi\xi} + 2N_\xi N_{\xi\xi} + O(\varepsilon) = 0 \tag{9.484}$$

(ix) Finally, apply the following definition

$$N_\xi = F \tag{9.485}$$

to show

$$F_\tau + FF_\xi + \frac{1}{6}F_{\xi\xi\xi} = 0 \qquad (9.486)$$

This is the KdV after a proper scaling. This completes the derivation of the KdV.

Problem 9.8 Show that (9.383) is a solution of (9.382).

Problem 9.9 This problem investigates the symmetries of the KdV equation. In particular, we consider the following form of the KdV:

$$u_t + 6uu_x + u_{xxx} = 0 \qquad (9.487)$$

We now introduce the following scaling substitutions:

$$u \to \mu u, \quad t \to \tau t, \quad x \to \chi x \qquad (9.488)$$

(i) Show that using (9.488), (9.487) can be converted to

$$\frac{1}{\tau}u_t + \frac{\mu}{\chi}6uu_x + \frac{1}{\chi^3}u_{xxx} = 0 \qquad (9.489)$$

(ii) Show that the scaled KdV is exactly the same as (9.487) if

$$\frac{\tau}{\chi^3} = 1, \quad \frac{\mu\tau}{\chi} = 1 \qquad (9.490)$$

(iii) Prove that the KdV is invariant by the following scaling

$$u \to \frac{1}{\lambda^2}u, \quad t \to \lambda^3 t, \quad x \to \lambda x \qquad (9.491)$$

Problem 9.10 This problem illustrates the Galilean invariance of the KdV equation. In particular, we consider the following form of the KdV:

$$u_t + 6uu_x + u_{xxx} = 0 \qquad (9.492)$$

We now consider the following Galilean transform in the solution

$$x \to x - \lambda t \qquad (9.493)$$

In particular, we assume

$$v(x,t) = u(x - \lambda t, t) + \mu = u(\xi, t) + \mu \qquad (9.494)$$

where λ and μ are constants.

(i) Show that

$$v_t = u_t - \lambda u_\xi \qquad (9.495)$$

$$v_{xxx} = u_{\xi\xi\xi} \qquad (9.496)$$

(ii) Further show that

$$v_t = -v_{xxx} - (6v + 6\mu + \lambda)v_x \qquad (9.497)$$

(iii) Finally, demonstrate that if u is a solution of KdV, the following v is also a solution of KdV

$$v(x,t) = u(x - \lambda t, t) - \frac{1}{6}\lambda \qquad (9.498)$$

Remarks: This is called the Galilean invariance of the KdV.

Problem 9.11 Following the procedure described in Problem 9.9, show that the NLSE given by

$$i\psi_t + \psi_{xx} + |\psi|^2 \psi = 0 \tag{9.499}$$

allows the following scaling invariance:

$$\psi \to a\psi, \quad t \to \frac{1}{a^2}t, \quad x \to \frac{1}{a}x \tag{9.500}$$

Problem 9.12 This problem derives the Galilean invariance of the NLSE. In particular, we consider the following NLSE

$$i\psi_t + \psi_{xx} + |\psi|^2 \psi = 0 \tag{9.501}$$

Consider the following function in terms of the Galilean transform

$$\tilde{\psi}(x,t) = \psi(x-ct,t)e^{i(\frac{cx}{2}-\frac{c^2}{4}t)} \tag{9.502}$$

(i) Show the following

$$\tilde{\psi}_x = [\psi_\xi + \frac{ic}{2}\psi]e^{i(\frac{cx}{2}-\frac{c^2}{4}t)} \tag{9.503}$$

$$\tilde{\psi}_{xx} = [\psi_{\xi\xi} + ic\psi_\xi - \frac{c^2}{4}\psi]e^{i(\frac{cx}{2}-\frac{c^2}{4}t)} \tag{9.504}$$

$$\tilde{\psi}_t = [\psi_t - c\psi_\xi - \frac{ic^2}{4}\psi]e^{i(\frac{cx}{2}-\frac{c^2}{4}t)} \tag{9.505}$$

$$|\tilde{\psi}| = |\psi| \tag{9.506}$$

where

$$\xi = x - ct \tag{9.507}$$

(ii) Demonstrate that

$$i\tilde{\psi}_t + \tilde{\psi}_{xx} + |\tilde{\psi}|\tilde{\psi} = 0 \tag{9.508}$$

Remarks: In conclusion, if ψ is a solution of the NLSE, (9.502) is also a solution of the NLSE.

CHAPTER TEN

Mathematical Theory for Maxwell Equations

10.1 INTRODUCTION

The development and derivation of the Maxwell equations are always considered as one of the major triumphs of human beings. It is because it links magnetism to electricity, and to electromagnetic waves. Seemingly unrelated in nature, visible light rays, X-rays, gamma rays, infrared, ultraviolet, radio wave, microwave, magnetic fields and electric fields are all linked together through a set of four vector differential equations called the Maxwell equations. The two main unknowns are the vector electric field and vector magnetic field. Denoted in terms of curl and divergence in vector calculus (see Chau, 2018), Maxwell equations can be expressed in a compact form. In fact, the whole set of Maxwell equations consists of four vector equations, namely, the Gauss law for electric fields, Gauss law for magnetism, the Maxwell-Faraday equation, and the Ampere circuital law with Maxwell correction. The vector calculus form of the Maxwell equations was proposed by Oliver Heaviside and Heinrich Hertz, and will be presented in the next section. The tensor or vector form of Maxwell equations has become the standard and they are much more compact than the original 20 equations by Maxwell. In 1864, James Clark Maxwell, a British physicist, suggested that accelerated charges generated linked electric and magnetic disturbances that can travel indefinitely through vacuum space. If the charges oscillate periodically, the disturbance propagates as waves with those electric and magnetic components perpendicular to each other and to the direction of propagation. Because of the dual components of the waves, they are called electromagnetic waves and the study of them is known as electrodynamics.

The idea behind Maxwell's theory came from a Faraday law that predicts a changing magnetic field can induce an electric current in a wire loop. When Maxwell examined the mathematical structure of the set of Gauss laws for electric fields, the Gauss law for magnetism, the Faraday equation, and the Ampere circuital law, he discovered that there is a dual symmetry in the mathematical structure in electric and magnetic fields in this set of equations (more mathematical detail of this will be discussed in later sections). That is, if a set of solutions for the electric and magnetic field is given, we can find another set of solutions by setting the new magnetic field as the given electric field divided by the speed of light (c) and by setting the new electric field as the given magnetic field multiplying by the negative of the speed of light ($-c$). This amazing discovery led Maxwell to the speculation that a changing electric field has a magnetic field associated with it (i.e., the reverse of the Faraday equation). If Maxwell's theory is correct, electromagnetic waves must be generated by varying electric and magnetic fields coupled through electromagnetic induction predicted by the Faraday law as well as the reverse mechanism Maxwell proposed. Maxwell then continued to show that the speed of these electromagnetic waves is exactly the speed of light. Maxwell knew that this cannot be coincidence, and thus, he concluded that light is a kind of electromagnetic wave. During Maxwell's lifetime his prediction of electromagnetic waves was never verified by experiments. In 1888, German physicist Heinrich Hertz conceived and conducted experiments to show that electromagnetic waves exist, and as Maxwell predicted these waves consisted of both electric and magnetic components. The whole spectra of ultraviolet to infrared are electromagnetic waves. Light is not the only electromagnetic wave; depending on its frequency, electromagnetic waves include high-frequency X-ray and gamma waves and low-frequency radio waves used in communication. We are going to see in coming sections that the set of Maxwell equations are linear and thus electromagnetic waves obey the principle of superposition. This linear nature also allows exact solutions to be found for electromagnetic waves.

Among physicists, Maxwell's equations are as important as Newton's second law $F = ma$, Einstein's equation $E = mc^2$, and Schrödinger's equation $H\psi = E\psi$. Without electromagnetic waves, our daily lives would no longer be the same. We would not have radio, television, cellular phone, satellite TV, GPS, etc. However, Maxwell never achieved the same level of fame in the general public as Newton, or Einstein. It is speculated that Maxwell's equations do not receive the deserved popularity because they involve the more obscure symbols of vector calculus (involving divergence and curl) and they are incomprehensible to the general public without training in advanced mathematics. I hope this chapter, at least, makes Maxwell's equations more accessible to engineers because of today's interdisciplinary nature of many new technologies, such as nanotechnology. At nano-scale, electromagnetic effects become more important.

In the present chapter, our main focus is to discuss the mathematical theory employed in solving Maxwell's equations, and practical applications will not be our focus.

10.2 MICROSCOPIC MAXWELL EQUATIONS

In general, Maxwell's equations can be written at a microscopic level and macroscopic level. At the microscopic level, we are concerned with the effect of individual charges in a vacuum (empty space except the charges themselves); whereas at the macroscopic scale we are concerning with the effect of charges in a material (like a conductor or just air). The original Maxwell equations are macroscopic, and the extension to the microscopic level was due to Clausius, Heaviside, Hertz and Lorentz. For the microscopic level, Maxwell's equations for the electric field $\boldsymbol{E}$ and magnetic field $\boldsymbol{B}$ can be expressed as:

$$\begin{gathered}\nabla \cdot \boldsymbol{E} = \frac{\rho}{\varepsilon_0}, \quad \nabla \cdot \boldsymbol{B} = 0, \\ \nabla \times \boldsymbol{E} = -\frac{\partial \boldsymbol{B}}{\partial t}, \quad \nabla \times \boldsymbol{B} = \mu_0 (\boldsymbol{J} + \varepsilon_0 \frac{\partial \boldsymbol{E}}{\partial t})\end{gathered} \tag{10.1}$$

where ε and μ are the electric permittivity and magnetic permeability in vacuum where the electric and magnetic field exits, ρ is the electric charges in vacuum, and $\boldsymbol{J}$ is the displacement current or total current density. The original equations by Maxwell are not easy to comprehend, and the vector calculus form given in (10.1) is due to Oliver Heaviside (recall his Heaviside step function in mathematics). The first and third equations describe how fields vary in space due to sources (the inhomogeneous terms in the equations). Note that the third gives the curl of $\boldsymbol{E}$ being proportional to $\boldsymbol{B}$, and this implies that $\boldsymbol{E}$ and $\boldsymbol{B}$ must be perpendicular to each other. Alternatively, the same conclusion would be drawn from the fourth term of (10.1). This is the reason why we mentioned in the introduction that the magnetic and electric fields are perpendicular to each other.

These four equations will be discussed in more detail one by one.

10.2.1 Gauss Law for Electric Field

The first of Maxwell's equations is actually the Gauss law of electric fields. It relates the distribution of electric charges to the resulting electric fields.

$$\nabla \cdot \boldsymbol{E} = \frac{\rho}{\varepsilon_0} \tag{10.2}$$

Physically, the net electric flux through any hypothetical closed surface equals $1/\varepsilon_0$ times the net electric charges within the closed surface. This physical meaning will be vivid when we present the integral form of it in a later section. This equation was actually derived independently by Lagrange in 1773 in the context of gravitation and by Gauss in 1813 in the context of electromagnetism. This has been historically known as Gauss's law. Its derivation relies on the divergence theorem. Gauss's law can be used to drive Coulomb's inverse square law of force between charges. The inverse square law was first deduced by John Priestley in 1767 from an experimental result of Benjamin Franklin, and examined experimentally again by Henry Cavendish in 1771 to 1781 (his work was not published and was discovered by Maxwell a century later). However, it was Coulomb's systematic and careful experiments using torsion balance in 1785 that led to the name of Coulomb's inverse square law. The mathematical structure of this Gauss law is similar to that of the Lagrange-Gauss law for gravity.

10.2.2 Gauss Law for Magnetism

Gauss's law for magnetism is given by

$$\nabla \cdot \boldsymbol{B} = 0 \tag{10.3}$$

Essentially, it states that the divergence of the magnetic field is zero. Mathematically, $\boldsymbol{B}$ is also known as a solenoidal vector field, and in fact this terminology came from magnetism. It is also closely related to Helmholtz decomposition of the vector field. For incompressible inviscid fluid flow (or the so-called potential flow), the velocity field will also be solenoidal or satisfying (10.3).

We can see from the last section that there can be a net positive charge or negative charges in a closed space such that the divergence of the electric field is proportional to it. The mathematical structure of (10.3) is identical to (10.2). The zero on the right-hand side actually implies a very important concept in magnetism. That is, magnetic monopoles do not exist. If they did exist, there would be a nonhomogeneous term on the right-hand side of (10.3). That is, we can only find a dipole in magnetism not a monopole that corresponds to charges in the electric field. In fact, this observation has been made by Matthew Hale, John Michell, and C.A. Coulomb. Poisson in 1824 developed a mathematical theory for magnetism and introduced the concept of the magnetic dipole moment, suggesting formally that a monopole does not exist. It was Gauss in 1832 who conducted careful experiments to verify the inverse square law for magnetic forces. Recognizing the formalism for deriving (10.2) also applicable to magnetism and the fact that monopoles do not exist, Gauss formulated the law of magnetism. We will show that for the case of no electric charge, there is a very nice dual symmetry in the Maxwell equations. Assuming the existence of mathematical beauty and simplicity in physical laws, many physicists believe that a monopole must exist in nature such that the Maxwell equations will possess true symmetry. So far, nobody has discovered the existence of a monopole. This idea of non-existence of the monopole can date back to 1269 by Petrus Peregrinus, and adopted by Gilbert and Faraday. We will discuss this more in a later section.

10.2.3 Maxwell-Faraday Law

The Maxwell-Faraday law is a generalization of Faraday's law of electromagnetic induction:

$$\nabla \times E = -\frac{\partial B}{\partial t} \tag{10.4}$$

Physically, it states that a time-varying magnetic field will always accompany a spatially varying non-conservative electric field. This is an important piece in the formulation of the Maxwell equations. This law was inspired by Faraday's experiment on magnetic coils in 1831. In fact, electromagnetic induction had been discovered by Joseph Henry in 1830, which was, however, unknown to Faraday. Faraday showed that the energy of a magnet is in the field around it and not in the magnet itself, and that magnetism can be converted to electricity. Without this discovery, he would not invent the dynamo and electric motor. Without such discoveries our daily lives would no longer be the same. Faraday was largely self-taught, and his life changed when he worked for Humphry Davy at the Royal Institution in 1813 as a laboratory assistant. Faraday also established the link between electricity and chemical affinity, or electrolysis. He also considered light as a form of the electromagnetic waves, which leads to the later development of Maxwell's theory.

10.2.4 Ampere Circuital Law (with Maxwell Correction)

The following Ampere circuital law with Maxwell Correction was derived by Maxwell in 1861:

$$\nabla \times B = \mu_0 (J + \varepsilon_0 \frac{\partial E}{\partial t}) \tag{10.5}$$

Physically, it relates the magnetic field associated with electric current. The original Ampere law was inspired by the experiments of Hans Oersted in 1819 (deflection of compass by current carrying wire). Oersted's experiment also led to the derivation of the Biot-Savart law, which relates the magnetic field at any point in space to the current in the wire that generates it. Maxwell first formulated the first term of total current density on the right of (10.5) in 1855, and included the second term on the time-varying electric field on the right of (10.5) in 1861. This second term on the right-hand side suggests that a changing electric field induces a magnetic field. This law actually was derived by Maxwell and was not by Ampere. Therefore, the commonly adopted name of Ampere circuital law with Maxwell correction seems not an appropriate choice. Without the second term on the right-hand side of (10.5), electromagnetic waves will not be generated. If the electromagnetic fields are inside a material, the total current density $\boldsymbol{J}$ includes the free current and bounded current. Electrons passing through a wire or battery is an example of free current $\boldsymbol{J}_f$. There are two sources of bounded currents. First, when a material is magnetized, microscopic current will circulate inside the magnetized object, leading to the so-called magnetization current $\boldsymbol{J}_M$. Second, when an external electric current is applied to a polarizable or dielectric material, bound charges (the charged elements are bounded and not free to move around) may displace at the atomic levels to create a polarized current $\boldsymbol{J}_P$. Therefore, we have

$$\boldsymbol{J} = \boldsymbol{J}_f + \boldsymbol{J}_M + \boldsymbol{J}_P \tag{10.6}$$

More discussion of these currents will be given in later sections. This is the most import piece of work within the framework of Maxwell's equations.

10.2.5 Dual Symmetry of Electromagnetic Waves in Vacuum Space

As remarked in the introduction, Maxwell recognized there is a dual symmetry in the Maxwell equations when he proposed the reverse mechanism of electromagnetic induction. In particular, for the case of vacuum space with no charge ($\rho = 0$) and no electric current ($\boldsymbol{J} = 0$), Maxwell's equations are reduced to

$$\nabla \cdot E = 0, \quad \nabla \cdot B = 0, \\ \nabla \times E = -\frac{\partial B}{\partial t}, \quad \nabla \times B = \mu_0 \varepsilon_0 \frac{\partial E}{\partial t} \tag{10.7}$$

where μ_0 and ε_0 are the electric permittivity and magnetic permeability in vacuum. The value of ε_0 is the capability of the vacuum to permit electric field lines, and it is given numerically as

$$\varepsilon_0 = 8.854187817... \times 10^{-12} F / \mathrm{s} \tag{10.8}$$

where F is Faraday and s is second. Similarly, the value of μ_0 is the capability of the vacuum to permit magnetic field lines, and it is given as

$$\mu_0 = 4\pi \times 10^{-7} s^2 / (Fm) \tag{10.9}$$

where m is meter. Maxwell found numerically that

$$c = \frac{1}{\sqrt{\mu_0 \varepsilon_0}} = 2.99792458 \times 10^8 \text{m/s} \tag{10.10}$$

This value coincides with the speed of light in vacuum. Maxwell realized that there is no coincidence and he, therefore, inferred that light consists of electromagnetic waves. If we can find a set of solutions for $\boldsymbol{B}$ and $\boldsymbol{E}$, we can make the following substitutions

$$\boldsymbol{E} \leftarrow c\boldsymbol{B}, \quad \boldsymbol{B} \leftarrow -\frac{\boldsymbol{E}}{c} \tag{10.11}$$

Then, substitution of (10.11) into the first part of (10.7) gives

$$\nabla \cdot (c\boldsymbol{B}) = 0, \quad \Rightarrow \quad \nabla \cdot \boldsymbol{B} = 0 \tag{10.12}$$

Substitution of (10.11) into the second part of (10.7) gives

$$\nabla \cdot (-\frac{\boldsymbol{E}}{c}) = 0 \quad \Rightarrow \quad \nabla \cdot \boldsymbol{E} = 0 \tag{10.13}$$

Substitution of (10.11) into the third part of (10.7) gives

$$\nabla \times (c\boldsymbol{B}) = -\frac{\partial}{\partial t}(-\frac{\boldsymbol{E}}{c}), \quad \Rightarrow \quad \nabla \times \boldsymbol{B} = \frac{1}{c^2}\frac{\partial \boldsymbol{E}}{\partial t} \tag{10.14}$$

Finally, substitution of (10.11) into the fourth equation of (10.7) gives

$$\nabla \times (-\frac{\boldsymbol{E}}{c}) = \frac{1}{c^2}\frac{\partial}{\partial t}(c\boldsymbol{B}) \quad \Rightarrow \quad \nabla \times \boldsymbol{E} = -\frac{\partial \boldsymbol{B}}{\partial t} \tag{10.15}$$

Amazingly, we get back the original set of Maxwell's equations. This is the so-called dual symmetry. This mathematical symmetry suggests that the electromagnetic induction effect of the magnetic field on the electric field also appears conversely. This is the main reason why Maxwell suspected the reverse mechanism of electromagnetic induction. More general discussion of duality or symmetry of electrodynamics in materials (i.e., nonzero ρ and $\boldsymbol{J}$) is considered in Section 10.14.

10.3 INTEGRAL VERSUS DIFFERENTIAL FORMS

The whole set of Maxwell's equations given in (10.1) can be converted to integral forms via the Kelvin-Stokes theorem and Gauss divergence theorem.

$$\oint_S \boldsymbol{E} \cdot d\boldsymbol{s} = \int_V \frac{\rho}{\varepsilon_0} dV \tag{10.16}$$

$$\oint_S \boldsymbol{B} \cdot d\boldsymbol{s} = 0 \tag{10.17}$$

$$\oint_{\partial \Sigma} \boldsymbol{E} \cdot d\boldsymbol{l} = -\frac{d}{dt}\int_\Sigma \boldsymbol{B} \cdot d\boldsymbol{s} \tag{10.18}$$

$$\oint_{\partial \Sigma} \boldsymbol{B} \cdot d\boldsymbol{l} = \int_\Sigma \mu_0(\boldsymbol{J} + \varepsilon_0 \frac{\partial \boldsymbol{E}}{\partial t}) \cdot d\boldsymbol{s} \tag{10.19}$$

The integral form of Maxwell's equations involves the ideas of flux and circulation. The area integrals in (10.16) to (10.19) represent flux, which gives an average outward component of a vector field multiplied by the surface area. The line integrals in (10.16) to (10.19) represent circulation that gives the average tangential component of a vector field multiplied by the path length. The integral form normally gives a better insight and is particular useful if there is symmetry in the electric and magnetic fields, whereas the differential form considers the effect of charges locally by neglected sources far away. Note that the total electric charge and electric current are given by

$$Q = \int_V \rho dV, \quad I = \int_\Sigma \boldsymbol{J} \cdot d\boldsymbol{s} \tag{10.20}$$

The integral form of Gauss's law given in (10.16) can be applied to find the total charge within the surface by calculating the electric flux through the closed surface. We will illustrate the use of this integral form in the following example.

Example 10.1 Refer to the case of a charge Q_1 at the center of a sphere and conduct the surface integral in (10.16) to derive Coulomb's law of the attractive force between two electric charges. As shown in Figure 10.1, the electric field $\boldsymbol{E}$ is symmetric with respect to any orientations.

Solution: The integral form given in (10.16) can be used to calculate the electric flux through the spherical surface. Assuming that there is a charge Q_1 locates at the center of a sphere, by symmetry we have a uniform electric field on the spherical surface. The electric flux through the spherical surface is then

$$\oint_S \boldsymbol{E} \cdot d\boldsymbol{s} = |\boldsymbol{E}| 4\pi r^2 \tag{10.21}$$

On the other hand, the integration of the density charge must be Q_1:

$$\int_V \frac{\rho}{\varepsilon_0} dV = \frac{Q_1}{\varepsilon_0} \tag{10.22}$$

Using Gauss's law for the electric field to equate (10.21) and (10.22), we find

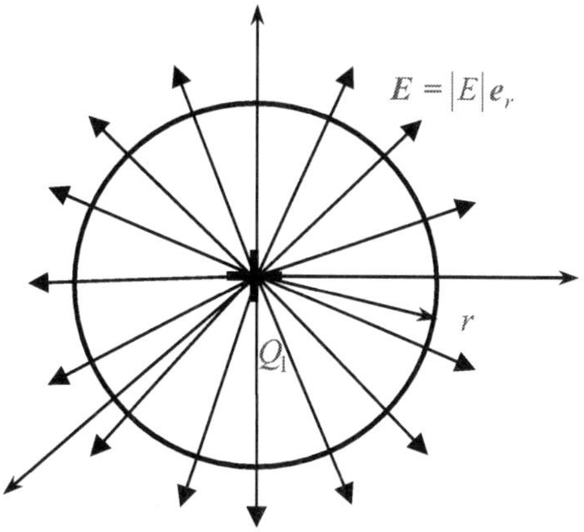

Figure 10.1 A charge with a uniform radial electric field on the spherical surface

$$|E| = \frac{Q}{4\pi\varepsilon_0 r^2} \tag{10.23}$$

This electric vector field could be used to find the Coulomb force associated with each point in space experienced by a test charge. For example, if there is another charge Q_2 at distance r from the center charge Q_1, it will experience a Coulomb force as

$$F = Q_2|E| = \frac{Q_2 Q_1}{4\pi\varepsilon_0 r^2} \tag{10.24}$$

In this case, the force is, of course, radial. More generally, in tensor or vector form Coulomb law can be expressed as

$$\boldsymbol{F} = Q_2\boldsymbol{E} = \frac{Q_2 Q_1}{4\pi\varepsilon_0 |\boldsymbol{r}|^3}\boldsymbol{r} \tag{10.25}$$

where $\boldsymbol{r}$ is the positive vector between the two charges. Thus, Coulomb's law has been incorporated in Maxwell's equations. Equation (10.25) also suggests another way of defining an electric field:

$$\boldsymbol{E} = \lim_{q\to 0}\frac{\boldsymbol{F}}{q} \tag{10.26}$$

where q is an arbitrary small charge experiencing a force $\boldsymbol{F}$ in the electric field. The existence of this charge actually will change the original field, and in turn $\boldsymbol{F}$ will be changed and so on. If the electric field $\boldsymbol{E}$ already incorporates the existence of charge q in the space, we can simply define it as:

$$\boldsymbol{F} = q\boldsymbol{E} \tag{10.27}$$

Physically, the integral form of the Maxwell-Faraday equation given in (10.18) says that if you are traveling on the closed path $\partial\Sigma$ of a surface Σ with the surface on your left, taking the dot product of the electric field $\boldsymbol{E}$ with the path vector $d\boldsymbol{l}$, and summing all these results of the dot product, the answer of this summation equals the negative of the time variation of magnetic flux through the surface Σ. We will illustrate the use of the Maxwell-Faraday equation in the following example.

Example 10.2 Consider the case of a closed integral loop moving through a uniform magnetic field, as shown in Figure 10.2. Employ the Maxwell-Faraday equation given in (10.18) to derive the Lorentz force law for the force experienced by a moving charge in the presence of both electric and magnetic fields.

Solution: In some classical textbooks on electrodynamics, it has been assumed that Maxwell's equations plus the Lorentz force equation form the complete set of governing equations in classic electrodynamics. In others, their relation is never discussed. More recently, Housner (2002) derived that the Lorentz force equation is a natural consequence of the Maxwell-Faraday equation. Housner's (2002) proof is based mainly on the shape independence of the magnetic flux (the result of the integral) on the right of (10.18).

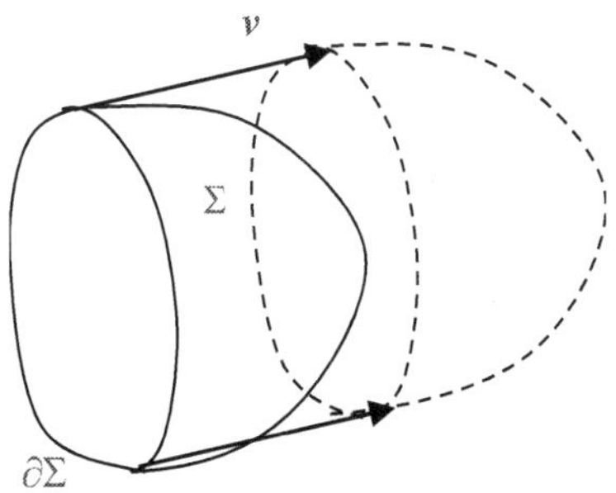

Figure 10.2 A closed integral loop moving in an electric and magnetic field with velocity *v*

Here, we will propose a new and different approach based upon the use of a moving field we can employ the idea by Euler for fluid flow that the control volume will experience a convective term in the material time derivative. Using this approach, the right-hand side of (10.18) can be expressed as:

$$\frac{d}{dt}\int_{\Sigma} \boldsymbol{B}\cdot d\boldsymbol{s} = \int_{\Sigma} \frac{d\boldsymbol{B}}{dt}\cdot d\boldsymbol{s} = \int_{\Sigma} [\frac{\partial \boldsymbol{B}}{\partial t} + (\boldsymbol{v}\cdot\nabla)\boldsymbol{B}]\cdot d\boldsymbol{s} \tag{10.28}$$

Next, we recall the following identity (see Problem 1.42 of Chau, 2018)

$$\begin{aligned}\nabla\times(\boldsymbol{B}\times\boldsymbol{v}) &= (\boldsymbol{v}\cdot\nabla)\boldsymbol{B} - (\boldsymbol{B}\cdot\nabla)\boldsymbol{v} + \boldsymbol{B}(\nabla\cdot\boldsymbol{v}) - \boldsymbol{v}(\nabla\cdot\boldsymbol{B}) \\ &= (\boldsymbol{v}\cdot\nabla)\boldsymbol{B}\end{aligned} \tag{10.29}$$

The last term on the right of (10.29) vanishes because of the second equation of Maxwell's equations given in (10.7), and the second and third terms on the right of (10.29) vanish because the velocity of the moving loop is not a function of position (i.e., there is no acceleration of the moving loop). Substitution of (10.29) into (10.28) gives

$$\frac{d}{dt}\int_{\Sigma} \boldsymbol{B}\cdot d\boldsymbol{s} = \int_{\Sigma} \frac{d\boldsymbol{B}}{dt}\cdot d\boldsymbol{s} = \int_{\Sigma} [\frac{\partial \boldsymbol{B}}{\partial t} + \nabla\times(\boldsymbol{B}\times\boldsymbol{v})]\cdot d\boldsymbol{s} \tag{10.30}$$

The Maxwell-Faraday equation given in (10.18) becomes

$$\begin{aligned}\oint_{\partial\Sigma} \boldsymbol{E}\cdot d\boldsymbol{l} &= -\int_{\Sigma} [\frac{\partial \boldsymbol{B}}{\partial t} + \nabla\times(\boldsymbol{B}\times\boldsymbol{v})]\cdot d\boldsymbol{s} \\ &= -\int_{\Sigma} \frac{\partial \boldsymbol{B}}{\partial t}\cdot d\boldsymbol{s} - \int_{\Sigma} \nabla\times(\boldsymbol{B}\times\boldsymbol{v})\cdot d\boldsymbol{s} \\ &= -\int_{\Sigma} \nabla\times(\boldsymbol{B}\times\boldsymbol{v})\cdot d\boldsymbol{s}\end{aligned} \tag{10.31}$$

The last of (10.31) is a consequence of the fact that the magnetic field is not a function of time. That is, only the closed loop is moving and the magnetic field is not changing. We now apply the Kelvin-Stokes theorem to the right-hand side of (10.31) to get

$$\oint_{\partial\Sigma} \boldsymbol{E}\cdot d\boldsymbol{l} = -\oint_{\partial\Sigma} (\boldsymbol{B}\times\boldsymbol{v})\cdot d\boldsymbol{l} \tag{10.32}$$

By the property of the cross product of vectors, we can rewrite (10.32) as

$$\oint_{\partial\Sigma} \boldsymbol{E}\cdot d\boldsymbol{l} = \oint_{\partial\Sigma} (\boldsymbol{v}\times\boldsymbol{B})\cdot d\boldsymbol{l} \tag{10.33}$$

Since we did not impose any condition on the path integral, it should apply to any closed loop integration. In other words, we require the equality be satisfied point-wise and we obtain

$$\boldsymbol{E}_m = \boldsymbol{v}\times\boldsymbol{B} \tag{10.34}$$

Note that we have added the subscript m to denote the electric field is due to the movement of the closed loop alone. We can now substitute this electric field into (10.27) derived in the last example to get

$$\boldsymbol{F}_m = q\boldsymbol{E}_m = q(\boldsymbol{v}\times\boldsymbol{B}) \tag{10.35}$$

This can now be combined with the force experienced by the charge due to the existence of the electric field to get

$$\boldsymbol{F} = \boldsymbol{F}_e + \boldsymbol{F}_m = q(\boldsymbol{E}+\boldsymbol{v}\times\boldsymbol{B}) \tag{10.36}$$

This is the Lorentz force equation. This completes the proof. This proof is evidently simpler than those given by Housner (2002). Nevertheless, the Lorentz force equation is within the framework of the Maxwell equation and does not need to be postulated as suggested by many textbooks.

10.4 MACROSCOPIC MAXWELL EQUATIONS

If we are not interested in the electric and magnetic fields of individual moving charge particles, we can rewrite Maxwell equation for macroscopic electric and magnetic fields. The macroscopic electric field is the average of the electric fields over a large volume. In particular, we have the macroscopic Maxwell equations as:

$$\begin{gathered}\nabla\cdot\boldsymbol{D} = \rho_f, \quad \nabla\cdot\boldsymbol{B} = 0, \\ \nabla\times\boldsymbol{E} = -\frac{\partial\boldsymbol{B}}{\partial t}, \quad \nabla\times\boldsymbol{H} = \boldsymbol{J}_f + \frac{\partial\boldsymbol{D}}{\partial t}\end{gathered} \tag{10.37}$$

where $\boldsymbol{D}$ is the displacement field or electric flux density, $\boldsymbol{H}$ is the magnetic flux density or the magnetizing field, and $\boldsymbol{J}_f$ is the free current density. The subscript f is used to indicate the free charge and free current. We are going to see in the next section that $\boldsymbol{D}$ and $\boldsymbol{H}$ can be related to $\boldsymbol{E}$ and $\boldsymbol{B}$ through the constitutive models. The corresponding integral forms of the macroscopic Maxwell equations are:

$$\begin{gathered}\oint_S \boldsymbol{D}\cdot d\boldsymbol{s} = \int_V \rho_f dV, \quad \oint_S \boldsymbol{B}\cdot d\boldsymbol{s} = 0, \\ \oint_{\partial\Sigma} \boldsymbol{E}\cdot d\boldsymbol{l} = -\frac{d}{dt}\int_\Sigma \boldsymbol{B}\cdot d\boldsymbol{s}, \quad \oint_{\partial\Sigma} \boldsymbol{H}\cdot d\boldsymbol{l} = \int_\Sigma (\boldsymbol{J}_f + \frac{\partial\boldsymbol{D}}{\partial t})\cdot d\boldsymbol{s}\end{gathered} \tag{10.38}$$

The whole set of Maxwell equations can be converted to integral forms via the Kelvin-Stokes theorem and the Gauss divergence theorem.

The set of Maxwell equations contains four equations as shown in (10.37), similarly for the case of microscopic Maxwell equations. In fact, not all four equations in the macroscopic Maxwell equations are independent. There are two choices in choosing the independent set of equations.

Independent Set 1:

Taking the divergence of the third equation of (10.37), we have

$$\nabla\cdot(\nabla\times\boldsymbol{E}) = -\frac{\partial(\nabla\cdot\boldsymbol{B})}{\partial t} = 0 \tag{10.39}$$

The last result obtained in (10.39) is the result of the vector identity (see (1.347) of Chau, 2018). Thus, we have

$$\frac{\partial(\nabla\cdot\boldsymbol{B})}{\partial t} = 0, \quad \text{or} \quad \nabla\cdot\boldsymbol{B} = 0 \tag{10.40}$$

This is precisely the second part of (10.37). Similarly, we can take the divergence of the fourth equation of (10.37) as

$$\nabla\cdot(\nabla\times\boldsymbol{H}) = \nabla\cdot\boldsymbol{J}_f + \frac{\partial(\nabla\cdot\boldsymbol{D})}{\partial t} = 0 \tag{10.41}$$

Substitution of the first equation of (10.37) into the right-hand side of (10.41) gives

$$\nabla\cdot\boldsymbol{J}_f = -\frac{\partial\rho_f}{\partial t} \tag{10.42}$$

This is a conservation equation for electric charges. Clearly, (10.42) and the first equation in (10.37) are not independent. In summary, we have a set of independent equations as

$$\nabla\cdot\boldsymbol{J}_f = -\frac{\partial\rho_f}{\partial t} \tag{10.43}$$

$$\nabla\times\boldsymbol{E} = -\frac{\partial\boldsymbol{B}}{\partial t} \tag{10.44}$$

$$\nabla\times\boldsymbol{H} = \boldsymbol{J}_f + \frac{\partial\boldsymbol{D}}{\partial t} \tag{10.45}$$

There is one scalar equation and two vector equations, and thus altogether (10.43) to (10.45) give seven equations.

Independent Set 2:

Alternatively, we can prefer the first part of (10.37) over (10.42) to get

$$\nabla\cdot\boldsymbol{D} = \rho_f \tag{10.46}$$

$$\nabla\times\boldsymbol{E} = -\frac{\partial\boldsymbol{B}}{\partial t} \tag{10.47}$$

$$\nabla \times \boldsymbol{H} = \boldsymbol{J}_f + \frac{\partial \boldsymbol{D}}{\partial t} \tag{10.48}$$

Similarly, this set of equations provides seven equations. However, there are altogether 16 unknowns (five vector unknowns $\boldsymbol{B}$, $\boldsymbol{D}$, $\boldsymbol{E}$, $\boldsymbol{J}_f$ and $\boldsymbol{H}$, and one scalar unknown ρ_f). Therefore, we still need nine more conditions. These additional conditions are constitutive equations:

$$\boldsymbol{D} = f_1(\boldsymbol{E}, \boldsymbol{H}) \tag{10.49}$$

$$\boldsymbol{B} = f_2(\boldsymbol{E}, \boldsymbol{H}) \tag{10.50}$$

$$\boldsymbol{J}_f = f_3(\boldsymbol{E}, \boldsymbol{H}) \tag{10.51}$$

Specific constitutive relations relating $\boldsymbol{B}$, $\boldsymbol{D}$, and $\boldsymbol{J}_f$ to $\boldsymbol{E}$ and $\boldsymbol{H}$ will be introduced next.

10.5 CONSTITUTIVE RELATION AND OHM'S LAW

Similar to the constitutive model between stress and strain for solids, the electric flux density $\boldsymbol{D}$ defined in the last section can be related to the electric field $\boldsymbol{E}$ and the magnetic flux density $\boldsymbol{H}$ can be related to the magnetic field $\boldsymbol{B}$:

$$\boldsymbol{D}(\boldsymbol{r},t) = \varepsilon_0 \boldsymbol{E}(\boldsymbol{r},t) + \boldsymbol{P}(\boldsymbol{r},t) \tag{10.52}$$

$$\boldsymbol{H}(\boldsymbol{r},t) = \frac{1}{\mu_0} \boldsymbol{B}(\boldsymbol{r},t) - \boldsymbol{M}(\boldsymbol{r},t) \tag{10.53}$$

where μ_0 and ε_0 are the electric permittivity and magnetic permeability in vacuum, and they are given in (10.8) and (10.9). Note that $\boldsymbol{E}$ and $\boldsymbol{B}$ are pure fields whereas $\boldsymbol{D}$ and $\boldsymbol{H}$ are fields containing median data. The symbols $\boldsymbol{P}$ and $\boldsymbol{M}$ denote the electric polarization in material and magnetization in material. To a certain extent, nearly all materials are polarizable and magnetizable, and the most recognizable polarizable material is dielectric (such as capacitors) and the most magnetizable material is ferrite (a kind of iron III oxide ceramic used as ferromagnets). In most crystalline solids (e.g., ferroelectric and ferromagnetic materials), $\boldsymbol{P}$ and $\boldsymbol{M}$ are nonzero. The study of $\boldsymbol{P}$ and $\boldsymbol{M}$ is very important for solid-state physics. In addition, there are also bound charges defined in terms of the electric polarization $\boldsymbol{P}$ as:

$$\rho_b = -\nabla \cdot \boldsymbol{P} \tag{10.54}$$

For the special case of vacuum, both $\boldsymbol{P}$ and $\boldsymbol{M}$ are zeros or

$$\boldsymbol{D} = \varepsilon_0 \boldsymbol{E} \tag{10.55}$$

$$\boldsymbol{H} = \frac{1}{\mu_0} \boldsymbol{B} \tag{10.56}$$

Equation (10.53) can also be recast as:

$$\boldsymbol{B}(\boldsymbol{r},t) = \mu_0[\boldsymbol{H}(\boldsymbol{r},t) + \boldsymbol{M}(\boldsymbol{r},t)] \tag{10.57}$$

For nonzero electric and magnetic fields inside the material, these constitutive relations can be recast as

$$\boldsymbol{D} = \varepsilon \boldsymbol{E} = \varepsilon_r \varepsilon_0 \boldsymbol{E} \tag{10.58}$$

$$H = \frac{1}{\mu}B = \frac{1}{\mu_r \mu_0}B \tag{10.59}$$

where ε_r is the relative dielectric permittivity and can be interpreted as the ratio of electrical energy stored in a material by an applied voltage to that stored in vacuum. It is also known as the dielectric constant, indicating the ratio of electrical capacitance of a material. The relative magnetic permeability μ_r can be interpreted in a similar fashion. Table 10.1 summarizes some values of ε_r and μ_r for various types of materials.

Table 10.1 Typical values of relative dielectric permittivity and relative magnetic permeability at 1 Hz

Material	ε_r	μ_r
vacuum	1	1
air	1.00058986	1.00000037
water	1.77 (visible light) 88, 80.1, 55.3 (0, 20, 100°C)	0.999992
concrete	4.5	1
ferrite	>640	640
copper	∞	0.999994

We can see that the relative dielectric permittivity and relative magnetic permeability for air is close to one. This is the reason why we can transmit electromagnetic waves through air without much energy loss, and we can also use a compass to navigate on the surface of the Earth, guided by the magnetic fields in air. In addition, we have tacitly assumed a scalar constant for both ε and μ to describe the dielectric permittivity and magnetic permeability. This is only true if the material is isotropic. For anisotropic material, instead of a scalar, we have to assume a tensor for both μ and ε. In addition, material can be dispersive, that is, both ε and μ can be functions of frequency. Note that the numerical values given in Table 10.1 are for 1 Hz. In the earlier discussion of the Ampere circuital law with Maxwell Correction, we have defined bounded current $\boldsymbol{J}_b$ and with the introduction of $\boldsymbol{P}$ and $\boldsymbol{M}$ it can now be determined as:

$$\boldsymbol{J}_b = \nabla \times \boldsymbol{M} + \frac{\partial \boldsymbol{P}}{\partial t} \tag{10.60}$$

For the free current, we have Ohm's law:

$$\boldsymbol{J}_f = \sigma \boldsymbol{E} \tag{10.61}$$

where σ is the electric conductivity. Alternatively, if there are external electric fields applied to the charges, Ohm's law can be revised as

$$\boldsymbol{J} = \sigma(\boldsymbol{E} + \boldsymbol{E}^e) \tag{10.62}$$

In short, the material constants now include ε μ, and σ. Equations (10.60) together with (10.52) and (10.53) provide a set of nine conditions required to solve for all unknowns as discussed in the last section.

Alternatively, the constitutive law can be incorporated into the independent set of Maxwell equations. For example, using (10.62), (10.58) and (10.59), the Set 1 Maxwell equations can rewritten as

$$\nabla \cdot [\sigma(\boldsymbol{E}+\boldsymbol{E}^e)] = -\frac{\partial \rho}{\partial t} \tag{10.63}$$

$$\nabla \times \boldsymbol{E} = -\frac{\partial}{\partial t}(\mu \boldsymbol{H}) \tag{10.64}$$

$$\nabla \times \boldsymbol{H} = \sigma(\boldsymbol{E}+\boldsymbol{E}^e) + \frac{\partial}{\partial t}(\varepsilon \boldsymbol{E}) \tag{10.65}$$

For crystalline ferroelectric or ferromagnetic solids with given $\boldsymbol{P}$, $\boldsymbol{J}$ and $\boldsymbol{M}$, using (10.52) and (10.53) we can write the Set 1 Maxwell equations as:

$$\nabla \times \boldsymbol{B} = \mu_0 (\varepsilon_0 \frac{\partial \boldsymbol{E}}{\partial t} + \boldsymbol{J} + \frac{\partial \boldsymbol{P}}{\partial t} + \nabla \times \boldsymbol{M}) \tag{10.66}$$

$$\nabla \times \boldsymbol{E} = -\frac{\partial \boldsymbol{B}}{\partial t} \tag{10.67}$$

$$\nabla \cdot \boldsymbol{E} = \frac{1}{\varepsilon_0}(\rho - \nabla \cdot \boldsymbol{P}) \tag{10.68}$$

Note that we have added the bound charge to the free charge in (10.68). If $\boldsymbol{P}$, $\boldsymbol{J}$ and $\boldsymbol{M}$ are prescribed, there will be changes to both the electric and magnetic fields.

10.6 ELECTROMAGNETIC WAVES IN VACUUM

We will now look at the wave phenomenon of the Maxwell equation for vacuum space. We first recall the following vector identity from (1.362) of Chau (2018):

$$\nabla \times (\nabla \times \boldsymbol{F}) = \nabla(\nabla \cdot \boldsymbol{F}) - \nabla^2 \boldsymbol{F} \tag{10.69}$$

Taking the curl of the third equation of (10.7) gives

$$\nabla \times (\nabla \times \boldsymbol{E}) = -\frac{\partial (\nabla \times \boldsymbol{B})}{\partial t} \tag{10.70}$$

Substitution of (10.69) into the left-hand side of (10.70) and (10.7) into the right-hand side of (10.13) gives

$$\nabla(\nabla \cdot \boldsymbol{E}) - \nabla^2 \boldsymbol{E} = -\frac{\partial (\nabla \times \boldsymbol{B})}{\partial t} = -\frac{\partial}{\partial t}[\frac{1}{c^2}\frac{\partial \boldsymbol{E}}{\partial t}] = -\frac{1}{c^2}\frac{\partial^2 \boldsymbol{E}}{\partial t^2} \tag{10.71}$$

However, the first term on the left of (10.71) is zero by the first equation of (10.7); thus, we get a vector wave equation for $\boldsymbol{E}$:

$$\nabla^2 \boldsymbol{E} = \frac{1}{c^2}\frac{\partial^2 \boldsymbol{E}}{\partial t^2} \tag{10.72}$$

Following the same procedure, we can take the curl of the fourth equation of (10.7) to yield

$$\nabla \times (\nabla \times \boldsymbol{B}) = \mu_0 \varepsilon_0 \frac{\partial (\nabla \times \boldsymbol{E})}{\partial t} \tag{10.73}$$

$$\nabla(\nabla \cdot \boldsymbol{B}) - \nabla^2 \boldsymbol{B} = \frac{1}{c^2}\frac{\partial (\nabla \times \boldsymbol{E})}{\partial t} = -\frac{1}{c^2}\frac{\partial^2 \boldsymbol{B}}{\partial t^2} \tag{10.74}$$

In view of the second equation of (10.7), we get another vector wave equation for $\boldsymbol{B}$

$$\nabla^2 \boldsymbol{B} = \frac{1}{c^2}\frac{\partial^2 \boldsymbol{B}}{\partial t^2} \tag{10.75}$$

This is the reason why time-oscillating electric and magnetic fields propagate as electromagnetic waves with the speed of light. If the changing electric and magnetic fields are not in vacuum (say in our atmosphere), we need to modify the wave speed as

$$c = \frac{1}{\sqrt{\mu_0 \mu_r \varepsilon_0 \varepsilon_r}} \tag{10.76}$$

where μ_r and ε_r are the relative permittivity and relative permeability respectively. They have been discussed in the last section.

10.7 MAXWELL EQUATIONS IN GAUSS UNIT

Another popular choice in writing out the Maxwell equations is using Gaussian units, which is also known as the cgs unit (i.e., centimeter-gram-second). The definition for the electric field is not the same. So we have to be careful in using any results of Maxwell equations. In particular, the Gaussian electric field relates to the electric field in SI units as:

$$\boldsymbol{E}_G = \frac{1}{c}\boldsymbol{E}_{SI} \tag{10.77}$$

The set of Maxwell equations can be obtained by making the following substitutions:

$$\varepsilon_0 \leftarrow \frac{1}{4\pi c}, \quad \mu_0 \leftarrow \frac{4\pi}{c} \tag{10.78}$$

More specifically, the microscopic Maxwell equation in Gaussian units can be expressed as (by dropping the G for Gaussian):

$$\begin{gathered} \nabla \cdot \boldsymbol{E} = 4\pi\rho, \quad \nabla \cdot \boldsymbol{B} = 0, \\ \nabla \times \boldsymbol{E} = -\frac{1}{c}\frac{\partial \boldsymbol{B}}{\partial t}, \quad \nabla \times \boldsymbol{B} = \frac{1}{c}(4\pi \boldsymbol{J} + \frac{\partial \boldsymbol{E}}{\partial t}) \end{gathered} \tag{10.79}$$

The macroscopic Maxwell equation in Gaussian units can be expressed as (by dropping the G for Gaussian):

$$\begin{gathered} \nabla \cdot \boldsymbol{D} = 4\pi\rho_f, \quad \nabla \cdot \boldsymbol{B} = 0, \\ \nabla \times \boldsymbol{E} = -\frac{1}{c}\frac{\partial \boldsymbol{B}}{\partial t}, \quad \nabla \times \boldsymbol{H} = \frac{1}{c}(4\pi \boldsymbol{J}_f + \frac{\partial \boldsymbol{D}}{\partial t}) \end{gathered} \tag{10.80}$$

10.8 BOUNDARY CONDITIONS

Now, we recall the original set of macroscopic Maxwell equations in S.I. units as:

$$\nabla \cdot \boldsymbol{D} = \rho, \quad \nabla \cdot \boldsymbol{B} = 0,$$
$$\nabla \times \boldsymbol{E} = -\frac{\partial \boldsymbol{B}}{\partial t}, \quad \nabla \times \boldsymbol{H} = \boldsymbol{J} + \frac{\partial \boldsymbol{D}}{\partial t} \tag{10.81}$$

Figure 10.3 shows the given boundary conditions at the interface of two materials.

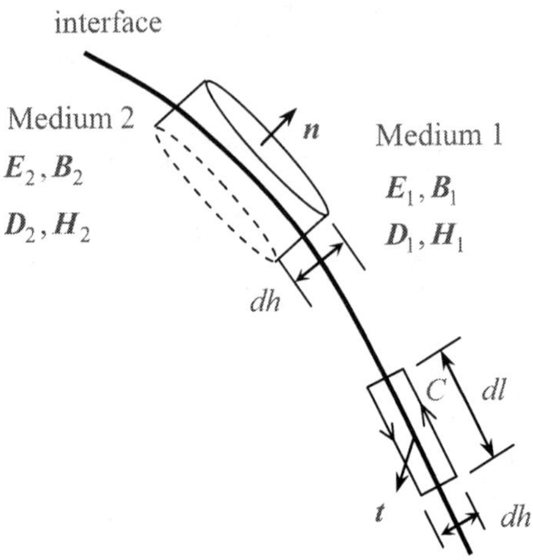

Figure 10.3 Boundary between two materials

The corresponding continuity conditions at the interface between the two materials are

$$\boldsymbol{n} \cdot (\boldsymbol{D}_1 - \boldsymbol{D}_2) = \rho_s, \quad \boldsymbol{n} \cdot (\boldsymbol{B}_1 - \boldsymbol{B}_2) = 0,$$
$$\boldsymbol{n} \times (\boldsymbol{E}_1 - \boldsymbol{E}_2) = 0, \quad \boldsymbol{n} \times (\boldsymbol{H}_1 - \boldsymbol{H}_2) = \boldsymbol{J}_s \tag{10.82}$$

where ρ_s is the surface charge density at the interface and $\boldsymbol{J}_s$ is the current density at the interface. For the case of no surface current, the fourth of (10.82) shows that the tangential component of $\boldsymbol{H}$ is continuous. These boundary conditions can be obtained by using the following identifications in (10.81):

$$\boldsymbol{E} \leftarrow \boldsymbol{E}_1 - \boldsymbol{E}_2 \quad \boldsymbol{H} \leftarrow \boldsymbol{H}_1 - \boldsymbol{H}_2, \quad \boldsymbol{D} \leftarrow \boldsymbol{D}_1 - \boldsymbol{D}_2,$$
$$\boldsymbol{B} \leftarrow \boldsymbol{B}_1 - \boldsymbol{B}_2, \quad \boldsymbol{J} \leftarrow \boldsymbol{J}_s, \quad \nabla \leftarrow \boldsymbol{n}, \quad \rho \leftarrow \rho_s \tag{10.83}$$

If both materials are not conductors, there will be no current at the boundary or $\boldsymbol{J}_s = 0$ in (10.82). If the electric and magnetic fields at the boundary are discontinuous as shown in (10.82), the differential form of Maxwell's equations is not valid but instead we should adopt the integral form of Maxwell's equations. This idea is similar to the finite volume method (expressed in integral form) in fluid mechanics dealing with shock waves discussed in Section 15.10 of Chau (2018). Let us illustrate the idea by considering the left-hand side of the first equation in (10.38):

$$\oint_S \boldsymbol{D} \cdot d\boldsymbol{s} = \oint_S \boldsymbol{D} \cdot \boldsymbol{n} dA = (\boldsymbol{D}_1 - \boldsymbol{D}_2) \cdot \boldsymbol{n} \Delta A \tag{10.84}$$

The right-hand side of it is

$$\int_V \rho dV = \lim_{dh\to 0}(\rho dh dA) = \rho_s \Delta A \tag{10.85}$$

where ρ_s is the surface charge density at the interface. Combining (10.84) and (10.85), we get

$$(\boldsymbol{D}_1 - \boldsymbol{D}_2)\cdot \boldsymbol{n} = \rho_s \tag{10.86}$$

This is precisely the boundary condition given in the first part of (10.82).

If both materials are perfect dielectric, the boundary conditions become

$$\begin{aligned} &\boldsymbol{n}\cdot(\boldsymbol{D}_1 - \boldsymbol{D}_2) = 0, \quad \boldsymbol{n}\cdot(\boldsymbol{B}_1 - \boldsymbol{B}_2) = 0, \\ &\boldsymbol{n}\times(\boldsymbol{E}_1 - \boldsymbol{E}_2) = 0, \quad \boldsymbol{n}\times(\boldsymbol{H}_1 - \boldsymbol{H}_2) = 0 \end{aligned} \tag{10.87}$$

If material 1 is perfect dielectric and the material is a perfect conductor, we have

$$\begin{aligned} &\boldsymbol{n}\cdot\boldsymbol{D}_1 = \rho_s, \quad \boldsymbol{n}\cdot\boldsymbol{B}_1 = 0, \\ &\boldsymbol{n}\times\boldsymbol{E}_1 = 0, \quad \boldsymbol{n}\times\boldsymbol{H}_1 = \boldsymbol{J}_s \end{aligned} \tag{10.88}$$

These boundary conditions for Maxwell's equations were considered by Lorentz.

10.9 MAXWELL'S VECTOR AND SCALAR POTENTIALS

We now look at the solution technique for Maxwell equations. In particular, Maxwell proposed the potential fields

$$\boldsymbol{B} = \nabla\times\boldsymbol{A} \tag{10.89}$$

$$\boldsymbol{E} = -\nabla\varphi - \frac{\partial \boldsymbol{A}}{\partial t} \tag{10.90}$$

The choice of the vector potential field $\boldsymbol{A}$ in (10.88) is obvious. By vector identity, we must have (Chau, 2018)

$$\nabla\cdot\boldsymbol{B} = \nabla\cdot(\nabla\times\boldsymbol{A}) = 0 \tag{10.91}$$

Thus, the Gauss law for magnetism in Maxwell equations is automatically satisfied, if (10.89) is assumed. We now substitute (10.89) into the Maxwell-Faraday equation, and we have

$$\nabla\times\boldsymbol{E} + \frac{\partial\boldsymbol{B}}{\partial t} = \nabla\times\boldsymbol{E} + \frac{\partial(\nabla\times\boldsymbol{A})}{\partial t} = \nabla\times(\boldsymbol{E} + \frac{\partial\boldsymbol{A}}{\partial t}) = 0 \tag{10.92}$$

We now recall another vector identity that for any scalar function (see (1.346) of Chau, 2018)

$$\nabla\times(\nabla\varphi) = 0 \tag{10.93}$$

Therefore, this suggests that we can set

$$\boldsymbol{E} + \frac{\partial\boldsymbol{A}}{\partial t} = -\nabla\varphi \tag{10.94}$$

Rearranging (10.94), we have the electric field expressed in terms of a vector potential $\boldsymbol{A}$ and a scalar potential φ:

$$\boldsymbol{E} = -\nabla\varphi - \frac{\partial\boldsymbol{A}}{\partial t} \tag{10.95}$$

This is precisely (10.90). The remaining two Maxwell equations will provide governing equations for both $\boldsymbol{A}$ and φ. Substitution of (10.89) into the first part of (10.1) gives

$$\nabla \cdot \boldsymbol{E} = \nabla \cdot (-\nabla \varphi - \frac{\partial \boldsymbol{A}}{\partial t}) = \frac{\rho}{\varepsilon_0} \tag{10.96}$$

Thus, Gauss's law for electricity gives

$$\nabla^2 \varphi + \frac{\partial}{\partial t}(\nabla \cdot \boldsymbol{A}) = -\frac{\rho}{\varepsilon_0} \tag{10.97}$$

Finally, the left-hand side of Ampere's circuital law with Maxwell correction can be written in terms of $\boldsymbol{A}$ as

$$\nabla \times \boldsymbol{B} = \nabla \times (\nabla \times \boldsymbol{A}) = \nabla(\nabla \cdot \boldsymbol{A}) - \nabla^2 \boldsymbol{A} \tag{10.98}$$

Using (10.90), the right-hand side of Ampere's circuital law with Maxwell correction can also be expressed in terms of these potentials:

$$\mu_0 (\boldsymbol{J} + \varepsilon_0 \frac{\partial \boldsymbol{E}}{\partial t}) = \mu_0 \boldsymbol{J} + \mu_0 \varepsilon_0 (-\nabla \frac{\partial \varphi}{\partial t} - \frac{\partial^2 \boldsymbol{A}}{\partial t^2}) \tag{10.99}$$

Equating (10.98) and (10.99), Ampere's circuital law with Maxwell correction becomes

$$(\nabla^2 \boldsymbol{A} - \frac{1}{c^2}\frac{\partial^2 \boldsymbol{A}}{\partial t^2}) - \nabla(\nabla \cdot \boldsymbol{A} + \frac{1}{c^2}\frac{\partial \varphi}{\partial t}) = -\mu_0 \boldsymbol{J} \tag{10.100}$$

Now, we have two equations (10.97) and (10.100) for $\boldsymbol{A}$ and φ. Actually (10.100) is a vector equation and (10.97) is a scalar equation, and we have exactly 4 equations for 4 unknowns (3 components of vector $\boldsymbol{A}$ and 1 scalar function φ).

10.10 GAUGE FREEDOM

Before we proceed to solve these equations, we first report that there is a gauge freedom in the potentials. To see that, we can rewrite (10.90) as

$$\begin{aligned} \boldsymbol{E} &= -(\nabla \varphi + \frac{\partial \boldsymbol{A}}{\partial t}) = -(\nabla \varphi - \nabla \frac{\partial \lambda}{\partial t} + \nabla \frac{\partial \lambda}{\partial t} + \frac{\partial \boldsymbol{A}}{\partial t}) \\ &= -[\nabla(\varphi - \frac{\partial \lambda}{\partial t}) + \frac{\partial}{\partial t}(\boldsymbol{A} + \nabla \lambda)] \\ &= -[\nabla \bar{\varphi} + \frac{\partial \bar{\boldsymbol{A}}}{\partial t}] \end{aligned} \tag{10.101}$$

where

$$\bar{\varphi} = (\varphi - \frac{\partial \lambda}{\partial t}), \quad \bar{\boldsymbol{A}} = \boldsymbol{A} + \nabla \lambda \tag{10.102}$$

We see that we can arbitrarily add a function λ to the original vector potential and scalar potential according to the rules given in (10.102) such that we end up with the original mathematical form of (10.90). In other words, the choice of the potentials is not unique. In experiments, we are measuring $\boldsymbol{E}$ and $\boldsymbol{B}$, but not $\boldsymbol{A}$ and φ. Thus, it is argued that adding function λ to the potentials as shown in (10.102) won't affect the physical quantities of electric and magnetic fields. Well, this may be true, but may not be the whole truth. In a later section on Aharonov-Bohm effect, we will see that the potentials $\boldsymbol{A}$ and φ might have physical significance. We will postpone our discussion until then.

Since we can add an arbitrary function λ (as long as it enters potentials $\boldsymbol{A}$ and φ through (10.102)), there is a gauge freedom. Two commonly adopted gauges are the Coulomb gauge and Lorenz gauge. They are discussed next.

10.10.1 Coulomb Gauge

Coulomb proposed to gauge the potentials in the following way. Considering the divergence of the second equation of (10.102), we have

$$\nabla \cdot \overline{\boldsymbol{A}} = \nabla \cdot \boldsymbol{A} + \nabla \cdot (\nabla \lambda) = \nabla \cdot \boldsymbol{A} + \nabla^2 \lambda \tag{10.103}$$

Since the scalar function λ is arbitrary, Coulomb set

$$\nabla^2 \lambda = -\nabla \cdot \boldsymbol{A} \tag{10.104}$$

That is, we can always gauge the potential using (10.103) such that

$$\nabla \cdot \overline{\boldsymbol{A}} = 0 \tag{10.105}$$

With this freedom, we can actually put (10.102) into (10.97) and (10.100) to get

$$\nabla^2 \overline{\varphi} = -\frac{\rho}{\varepsilon_0} \tag{10.106}$$

$$\nabla^2 \overline{\boldsymbol{A}} - \frac{1}{c^2}\frac{\partial^2 \overline{\boldsymbol{A}}}{\partial t^2} = -\mu_0 \boldsymbol{J} + \frac{1}{c^2}\nabla\left(\frac{\partial \overline{\varphi}}{\partial t}\right) \tag{10.107}$$

Effectively, we can actually drop the bar in the notation. Equation (10.106) is the classical Poisson equation (or nonhomogeneous Laplace equation), and equation (10.107) is a nonhomogeneous vector wave equation in which part of the nonhomogeneous term came from the solution of the scalar potential φ from (10.106). There are time derivative terms for $\boldsymbol{A}$ on the left and also time derivative term for φ on the right, suggesting that electric potentials are changing instantly everywhere. Another more popular choice of gauge freedom is due to Lorenz and is considered next.

10.10.2 Lorenz Gauge

Danish mathematician, L.V. Lorenz (1829-1891), proposed another popular choice of gauge to be discussed in this section. Lorenz should not be confused with H.A. Lorentz, who derived the Lorentz force law (see Example 10.2) and was a Nobel Prize winner in Physics in 1902. Some textbooks in classical electromagnetic dynamics even mistakenly called it the Lorentz gauge. In fact, Lorenz's work in 1867 on the gauge theory is 25 years ahead of that of Lorentz in 1892. According to Jackson and Okun (2001), Lorenz also independently derived the "Maxwell equations" but was never recognized properly. In particular, the Lorenz gauge is defined by

$$\nabla \cdot \overline{\boldsymbol{A}} = -\frac{1}{c^2}\frac{\partial \overline{\varphi}}{\partial t} \tag{10.108}$$

Using the definition of gauge freedom given in (10.108), we have

$$\nabla^2 \lambda - \frac{1}{c^2}\frac{\partial^2 \lambda}{\partial t^2} = -\nabla \cdot \boldsymbol{A} - \frac{1}{c^2}\frac{\partial \varphi}{\partial t} \tag{10.109}$$

That is, Coulomb proposed to choose λ according to (10.109). Using the first equation of (10.102), the left-hand side of (10.96) can be rewritten as

$$\nabla^2 \bar{\varphi} + \frac{\partial}{\partial t}\nabla^2 \lambda + \frac{\partial}{\partial t}(\nabla \cdot \boldsymbol{A}) = -\frac{\rho}{\varepsilon_0} \tag{10.110}$$

Substitution of (10.109) into (10.110) gives

$$\nabla^2 \bar{\varphi} + \frac{\partial}{\partial t}\nabla^2 \lambda + [-\frac{\partial}{\partial t}\nabla^2 \lambda - \frac{1}{c^2}\frac{\partial^2 \varphi}{\partial t^2} + \frac{1}{c^2}\frac{\partial^3 \lambda}{\partial t^3}] = -\frac{\rho}{\varepsilon_0} \tag{10.111}$$

Differentiating the first equation of (10.102), we find

$$-\frac{1}{c^2}\frac{\partial^2 \bar{\varphi}}{\partial t^2} = -\frac{1}{c^2}\frac{\partial^2 \varphi}{\partial t^2} + \frac{1}{c^2}\frac{\partial^3 \lambda}{\partial t^3} \tag{10.112}$$

Substituting (10.112) into (10.111), we finally get

$$\nabla^2 \bar{\varphi} - \frac{1}{c^2}\frac{\partial^2 \bar{\varphi}}{\partial t^2} = -\frac{\rho}{\varepsilon_0} \tag{10.113}$$

This is a nonhomogeneous scalar wave equation. We now substitute (10.109) into (10.100) to get

$$(\nabla^2 \boldsymbol{A} - \frac{1}{c^2}\frac{\partial^2 \boldsymbol{A}}{\partial t^2}) - \nabla(\nabla^2 \lambda - \frac{1}{c^2}\frac{\partial \lambda}{\partial t}) = -\mu_0 \boldsymbol{J} \tag{10.114}$$

Taking the Laplacian and the second equation in (10.102), we find

$$\nabla^2 \bar{\boldsymbol{A}} = \nabla^2 \boldsymbol{A} + \nabla(\nabla^2 \lambda) \tag{10.115}$$

$$\frac{\partial^2 \bar{\boldsymbol{A}}}{\partial t^2} = \frac{\partial^2 \boldsymbol{A}}{\partial t^2} + \nabla(\frac{\partial^2 \lambda}{\partial t^2}) \tag{10.116}$$

Substitution of (10.115) and (10.116) into (10.114) gives

$$\nabla^2 \bar{\boldsymbol{A}} - \frac{1}{c^2}\frac{\partial^2 \bar{\boldsymbol{A}}}{\partial t^2} = -\mu_0 \boldsymbol{J} \tag{10.117}$$

In summary, we have uncoupled the vector potential and scalar potential, and two nonhomogeneous wave equations are obtained (i.e., (10.113) and (10.117)). Thus, using the Lorenz gauge in solving Maxwell equations is equivalent to solving nonhomogeneous wave equations.

Section 9.3 of Chau (2018) derived the solution of the nonhomogeneous scalar wave equation in terms of retarded potential. Using (9.205) of Chau (2018), we have

$$\bar{\varphi}(\boldsymbol{r}, t) = \frac{1}{4\pi\varepsilon_0}\int_V \frac{\rho(\boldsymbol{r}', t - |\boldsymbol{r} - \boldsymbol{r}'|/c)}{|\boldsymbol{r} - \boldsymbol{r}'|} dV(\boldsymbol{r}') \tag{10.118}$$

Following the same procedure for the vector, we can write the solution for the nonhomogeneous vector wave equation in terms of retarded potential as:

$$\bar{\boldsymbol{A}}(\boldsymbol{r}, t) = \frac{\mu_0}{4\pi}\int_V \frac{\boldsymbol{J}(\boldsymbol{r}', t - |\boldsymbol{r} - \boldsymbol{r}'|/c)}{|\boldsymbol{r} - \boldsymbol{r}'|} dV(\boldsymbol{r}') \tag{10.119}$$

Therefore, the effect of the wave potentials propagates with the speed of light. The concept of retarded potentials can date back to Riemann in 1858 and Lorenz in 1861. In principle, once we have the solutions given in (10.118) and (10.119), we can use (10.89) and (10.90) to give the electric and magnetic fields. Because both φ and $\boldsymbol{A}$ depend on the retarded time, the Lorenz gauge is also known as retarded potentials. The solutions for $\boldsymbol{B}$ and $\boldsymbol{E}$ are called Jefimenko's (1992) equations and will be discussed in Section 10.11.

10.10.3 Aharonov-Bohm Effect (Physical Meaning of Wave Potentials)

We have shown that a gauge function can be added to the potentials (both scalar and vector potentials) such that the electric and magnetic fields remain the same. This non-uniqueness of the potentials leads to the belief that potentials are nonphysical. The Aharonov-Bohm effect is a quantum mechanical phenomenon in which an electrically charged particle is affected by the electromagnetic potential φ and $\boldsymbol{A}$, despite being confined to a region of zero electric and magnetic fields. This discovery is amazing in the sense that potential cannot be measured directly in experiments and potentials are not unique. Yet, they can be used to explain the phase shift in the wave function of a charged particle in the absence of electric and magnetic fields. In this special case, experimental observation seems to suggest the physical meaning of mathematical tools.

10.11 SOLUTIONS OF MAXWELL EQUATIONS: JEFIMENKO'S EQUATIONS

Although Jefimenko's solution was derived in 1966 in his textbook *Electricity and Magnetism*, the solution is surprisingly not well known among physicists. It becomes more well known after Jefimenko's (1992) paper appeared in the *American Journal of Physics* and another "Letters and Comments" in the *European Journal of Physics* in 1998 (Jefimenko, 1998). Subsequently, a number of papers discussed its importance (McDonald, 1997; Heras, 1994, 1995).

Returning to the technical side and substituting (10.118) and (10.119) into (10.89) and (10.90), we have solutions for the electric and magnetic fields as

$$\boldsymbol{B} = \frac{\mu_0}{4\pi}\int_V \frac{\nabla\times\boldsymbol{J}(\boldsymbol{r}',t-|\boldsymbol{r}-\boldsymbol{r}'|/c)}{|\boldsymbol{r}-\boldsymbol{r}'|}dV(\boldsymbol{r}') \tag{10.120}$$

$$\begin{aligned}\boldsymbol{E} = &-\frac{1}{4\pi\varepsilon_0}\int_V \frac{\nabla\rho(\boldsymbol{r}',t-|\boldsymbol{r}-\boldsymbol{r}'|/c)}{|\boldsymbol{r}-\boldsymbol{r}'|}dV(\boldsymbol{r}')\\ &-\frac{\mu_0}{4\pi}\int_V \frac{1}{|\boldsymbol{r}-\boldsymbol{r}'|}\frac{\partial}{\partial t}\boldsymbol{J}(\boldsymbol{r}',t-|\boldsymbol{r}-\boldsymbol{r}'|/c)dV(\boldsymbol{r}')\end{aligned} \tag{10.121}$$

In the following, we are going to employ a clever mathematical trick of Jefimenko (1992) to remove the spatial derivatives involved in the gradient of ρ and the curl of $\boldsymbol{J}$, expressing the final results in their time derivative only. The present approach is, however, different from that of Jefimenko (1992) in the sense that we use vector and scalar potential, whereas the formulation of Jefimenko (1992) starts from the Poisson integral and Helmholtz decomposition theorem. Nevertheless, the final result is the same as that given by Jefimenko (1992).

In particular, the following identities of Jefimenko (1966) can be used:

$$\nabla\left\{\frac{[\rho]}{r}\right\}=-\frac{\boldsymbol{r}_u}{r^2}[\rho]-\frac{\boldsymbol{r}_u}{rc}\frac{\partial[\rho]}{\partial t} \tag{10.122}$$

$$\nabla\times\left\{\frac{[\boldsymbol{J}]}{r}\right\}=\{\frac{[\boldsymbol{J}]}{r^2}+\frac{1}{rc}\frac{\partial[\boldsymbol{J}]}{\partial t}\}\times\boldsymbol{r}_u \tag{10.123}$$

where the bracket [...] means that the function inside it is evaluated at the retarded time

$$t_r(\boldsymbol{r},t)=t-|\boldsymbol{r}-\boldsymbol{r}'|/c=t-r/c \tag{10.124}$$

And the vector $\boldsymbol{r}_u$ is defined as:

$$\boldsymbol{r}_u(\boldsymbol{r},t)=\frac{\boldsymbol{r}-\boldsymbol{r}'}{r} \tag{10.125}$$

The derivation of (10.122) and (10.123) will be deferred to later sections. Substituting (10.122) and (10.123) into (10.120) and (10.121), we obtain the Jefimenko equations:

$$\begin{aligned}\boldsymbol{E}(\boldsymbol{r},t)=\frac{1}{4\pi\varepsilon_0}\int_V\{[\frac{\rho(\boldsymbol{r}',t_r)}{|\boldsymbol{r}-\boldsymbol{r}'|^3}+\frac{1}{|\boldsymbol{r}-\boldsymbol{r}'|^2c}\frac{\partial\rho(\boldsymbol{r}',t_r)}{\partial t}](\boldsymbol{r}-\boldsymbol{r}')\\ -\frac{1}{|\boldsymbol{r}-\boldsymbol{r}'|c^2}\frac{\partial\boldsymbol{J}(\boldsymbol{r}',t_r)}{\partial t}\}dV(\boldsymbol{r}')\end{aligned} \tag{10.126}$$

$$\boldsymbol{B}(\boldsymbol{r},t)=\frac{\mu_0}{4\pi}\int_V[\frac{\boldsymbol{J}(\boldsymbol{r}',t_r)}{|\boldsymbol{r}-\boldsymbol{r}'|^3}+\frac{1}{|\boldsymbol{r}-\boldsymbol{r}'|^2c}\frac{\partial\boldsymbol{J}(\boldsymbol{r}',t_r)}{\partial t}]\times(\boldsymbol{r}-\boldsymbol{r}')dV(\boldsymbol{r}') \tag{10.127}$$

where the retarded time is defined in (10.124). For the case that $\boldsymbol{J}$ is independent of time, (10.126) reduced to the Biot-Savart law.

The remaining task is to prove (10.122) and (10.123), and this is the objective for the next two sections.

10.11.1 Gradient Identity of Jefimenko

To prove (10.122), we first report the following identities using tensor notation (e.g., Chau, 2018):

$$r=\sqrt{(x-x')^2+(y-y')^2+(z-z')^2}=\{(x_i-x_i')(x_i-x_i')\}^{1/2} \tag{10.128}$$

where i = 1,2,3 and we have used Einstein's convention in dropping the summation sign for repeated indices (in this case is i). The observation point is denoted by $\boldsymbol{x}$ whereas the source point is denoted by $\boldsymbol{x}'$. Differentiation of r with respect to the observation point gives

$$\begin{aligned}\frac{\partial r}{\partial x_j}&=\frac{1}{2}\{(x_k-x_k')(x_k-x_k')\}^{-1/2}2(x_i-x_i')\delta_{ij}\\&=\frac{1}{r}(x_j-x_j')\end{aligned} \tag{10.129}$$

Note that once an index is repeated, it becomes a dummy for the summation and cannot be used anymore. Similarly, we have

$$\frac{\partial r}{\partial x_j{}'} = -\frac{1}{2}\{(x_k - x_k{}')(x_k - x_k{}')\}^{-1/2} 2(x_i - x_i{}')\delta_{ij} \\ = -\frac{1}{r}(x_j - x_j{}') \tag{10.130}$$

In addition, we have

$$\nabla(\frac{1}{r}) = \boldsymbol{e}_k \frac{\partial}{\partial x_k}(\frac{1}{r}) = -\frac{1}{r^2}\frac{\partial r}{\partial x_k}\boldsymbol{e}_k = -\frac{(x_k - x_k{}')}{r^3}\boldsymbol{e}_k = -\frac{\boldsymbol{r}_u}{r^2} \tag{10.131}$$

We have used (10.125) in obtaining the last part of (10.131). Similarly, we also have

$$\nabla'(\frac{1}{r}) = \boldsymbol{e}_k \frac{\partial}{\partial x_k'}(\frac{1}{r}) = -\frac{1}{r^2}\frac{\partial r}{\partial x_k'}\boldsymbol{e}_k = \frac{(x_k - x_k')}{r^3}\boldsymbol{e}_k = \frac{\boldsymbol{r}_u}{r^2} \tag{10.132}$$

where the gradient operator with the prime means differentiation with respect to the source point $\boldsymbol{x}'$. We now consider the charge density function ρ which depends on both source point $\boldsymbol{x}'$ and retarded time t_r:

$$\left.\frac{\partial[\rho(\boldsymbol{x}', t - r/c)]}{\partial x_j'}\right|_t = \left.\frac{\partial[\rho]}{\partial x_j'}\right|_{t-r/c} + \left.\frac{\partial[\rho]}{\partial(t - r/c)}\right|_{x_j'} \frac{\partial(t - r/c)}{\partial x_j'} \tag{10.133}$$

Note that since $\boldsymbol{x}'$ also enters the retarded time implicitly through r, there are two derivative terms as shown in (10.133). The subscript after the vertical bar indicates that this variable is treated as constant in the partial differentiation. Recall one more time that the square bracket indicates the function being evaluated at the retarded time. The last partial differentiation in (10.133) can be evaluated as

$$\frac{\partial(t - r/c)}{\partial x_k'} = -\frac{1}{c}\frac{\partial r}{\partial x_k'} = \frac{(x_k - x_k')}{rc} \tag{10.134}$$

We have obviously employed (10.130) in obtaining the last equation of (10.134). Multiplying (10.133) by the base vector, we have

$$\nabla'[\rho] = \boldsymbol{e}_k \frac{\partial[\rho]}{\partial x_j'} = [\nabla'\rho] + \frac{\boldsymbol{r}_u}{c}\frac{\partial[\rho]}{\partial t} \tag{10.135}$$

Similarly, we have the following differentiation with respect to $\boldsymbol{x}$

$$\nabla[\rho] = \boldsymbol{e}_k \frac{\partial[\rho]}{\partial x_j} = [\nabla\rho] - \frac{\boldsymbol{r}_u}{c}\frac{\partial[\rho]}{\partial t} \tag{10.136}$$

However, the charge density function in (10.113) is only a function of $\boldsymbol{x}'$, thus, the first term on the right of (10.136) is identically zero or

$$\nabla[\rho] = \boldsymbol{e}_k \frac{\partial[\rho]}{\partial x_j} = -\frac{\boldsymbol{r}_u}{c}\frac{\partial[\rho]}{\partial t} \tag{10.137}$$

Adding (10.135) and (10.137), we find

$$\nabla'[\rho] = [\nabla'\rho] - \nabla[\rho] \tag{10.138}$$

Equation (10.138) is a very important identity that apparently has been missed by many others, except Jefimenko (1966). We are now ready to consider the left-hand side of (10.122)

$$\nabla\left\{\frac{[\rho]}{r}\right\} = \nabla(\frac{1}{r})[\rho] + \frac{1}{r}\nabla[\rho] \tag{10.139}$$

Substituting (10.136) into (10.139), we find

$$\nabla\left\{\frac{[\rho]}{r}\right\} = -\frac{\boldsymbol{r}_u}{r^2}[\rho] + \frac{1}{r}\nabla[\rho] \tag{10.140}$$

Similarly, we have

$$\nabla'\left\{\frac{[\rho]}{r}\right\} = \nabla'(\frac{1}{r})[\rho] + \frac{1}{r}\nabla'[\rho] \tag{10.141}$$

Putting (10.132) into (10.141), we find

$$\nabla'\left\{\frac{[\rho]}{r}\right\} = \frac{\boldsymbol{r}_u}{r^2}[\rho] + \frac{1}{r}\nabla'[\rho] \tag{10.142}$$

Comparing (10.140) and (10.142), we have

$$\nabla\left\{\frac{[\rho]}{r}\right\} + \nabla'\left\{\frac{[\rho]}{r}\right\} = \frac{1}{r}\nabla[\rho] + \frac{1}{r}\nabla'[\rho] \tag{10.143}$$

Substituting (10.138) into (10.143), we finally get

$$\frac{1}{r}[\nabla'\rho] = \nabla\left\{\frac{[\rho]}{r}\right\} + \nabla'\left\{\frac{[\rho]}{r}\right\} \tag{10.144}$$

Rearranging (10.144) and substituting (10.142) and (10.135) into the resulting equation, we obtain

$$\begin{aligned}
\nabla\left\{\frac{[\rho]}{r}\right\} &= \frac{1}{r}[\nabla'\rho] - \nabla'\left\{\frac{[\rho]}{r}\right\} \\
&= \frac{1}{r}[\nabla'\rho] - \frac{\boldsymbol{r}_u}{r^2}[\rho] - \frac{1}{r}\nabla'[\rho] \\
&= \frac{1}{r}[\nabla'\rho] - \frac{\boldsymbol{r}_u}{r^2}[\rho] - \frac{1}{r}[\nabla'\rho] - \frac{\boldsymbol{r}_u}{rc}\frac{\partial[\rho]}{\partial t} \\
&= -\frac{\boldsymbol{r}_u}{r^2}[\rho] - \frac{\boldsymbol{r}_u}{rc}\frac{\partial[\rho]}{\partial t}
\end{aligned} \tag{10.145}$$

This completes the proof for (10.122).

10.11.2 Curl Identity of Jefimenko

The analysis given in the previous section can be extended to consider the curl of $\boldsymbol{J}$. First of all, we recall the definition of curl using tensor notation (Chau, 2018):

$$\nabla\times\left\{\frac{[\boldsymbol{J}]}{r}\right\} = e_{ijk}\frac{\partial}{\partial x_j}\{\frac{[J_k]}{r}\}\boldsymbol{e}_i \tag{10.146}$$

In view of this, we consider the following differentiation

$$\begin{aligned}\frac{\partial}{\partial x_j}\{\frac{[J_k]}{r}\} &= \frac{\partial}{\partial x_j}(\frac{1}{r})[J_k] + \frac{1}{r}\frac{\partial [J_k]}{\partial x_j} \\ &= -\frac{(x_j - x'_j)}{r^3}[J_k] + \frac{1}{r}\frac{\partial [J_k]}{\partial x_j}\end{aligned} \tag{10.147}$$

For differentiation with respect to the source point $\boldsymbol{x}'$, we have

$$\frac{\partial}{\partial x'_j}\{\frac{[J_k]}{r}\} = \frac{(x_j - x'_j)}{r^3}[J_k] + \frac{1}{r}\frac{\partial [J_k]}{\partial x'_j} \tag{10.148}$$

The last term in (10.148) can be considered as

$$\begin{aligned}\left.\frac{\partial [J_k]}{\partial x'_j}\right|_t &= \left.\frac{\partial [J_k]}{\partial x'_j}\right|_{t-r/c} + \frac{\partial [J_k]}{\partial (t - r/c)}\frac{\partial (t - r/c)}{\partial x'_j} \\ &= \left.\frac{\partial [J_k]}{\partial x'_j}\right|_{t-r/c} + \frac{1}{c}\frac{(x_j - x'_j)}{r}\frac{\partial [J_k]}{\partial t}\end{aligned} \tag{10.149}$$

Multiplying the permutation tensor and the base vector to (10.149), we obtain

$$e_{ijk}\frac{\partial [J_k]}{\partial x'_j}\boldsymbol{e}_i = e_{ijk}\boldsymbol{e}_i\left.\frac{\partial [J_k]}{\partial x'_j}\right|_{t-r/c} + e_{ijk}\boldsymbol{e}_i\frac{(x_j - x'_j)}{rc}\frac{\partial [J_k]}{\partial t} \tag{10.150}$$

In symbolic form, (10.150) can be rewritten as

$$\nabla' \times [\boldsymbol{J}] = [\nabla' \times \boldsymbol{J}] + \frac{\boldsymbol{r}_u \times}{c}\frac{\partial [\boldsymbol{J}]}{\partial t} \tag{10.151}$$

Similarly, we also have

$$\left.\frac{\partial [J_k]}{\partial x_j}\right|_t = \left.\frac{\partial [J_k]}{\partial x_j}\right|_{t-r/c} - \frac{1}{c}\frac{(x_j - x'_j)}{r}\frac{\partial [J_k]}{\partial t} \tag{10.152}$$

Multiplying the permutation tensor and the base vector to (10.152), we obtain

$$e_{ijk}\boldsymbol{e}_i\left.\frac{\partial [J_k]}{\partial x_j}\right|_t = e_{ijk}\boldsymbol{e}_i\left.\frac{\partial [J_k]}{\partial x_j}\right|_{t-r/c} - \frac{1}{c}e_{ijk}\boldsymbol{e}_i\frac{(x_j - x'_j)}{r}\frac{\partial [J_k]}{\partial t} \tag{10.153}$$

In symbolic form, (10.153) can be rewritten as

$$\nabla \times [\boldsymbol{J}] = [\nabla \times \boldsymbol{J}] - \frac{\boldsymbol{r}_u \times}{c}\frac{\partial [\boldsymbol{J}]}{\partial t} \tag{10.154}$$

Since $\boldsymbol{J}$ is only a function of $\boldsymbol{x}'$, (10.154) becomes

$$\nabla \times [\boldsymbol{J}] = -\frac{\boldsymbol{r}_u \times}{c}\frac{\partial [\boldsymbol{J}]}{\partial t} \tag{10.155}$$

Adding (10.151) and (10.155), we find

$$\nabla' \times [\boldsymbol{J}] + \nabla \times [\boldsymbol{J}] = [\nabla' \times \boldsymbol{J}] \tag{10.156}$$

Multiplying (10.147) by the Levi-Civita symbol and base vector, we have

$$e_{ijk}\boldsymbol{e}_i\frac{\partial}{\partial x_j}\{\frac{[J_k]}{r}\} = -e_{ijk}\boldsymbol{e}_i\frac{(x_j - x'_j)}{r^3}[J_k] + \frac{1}{r}e_{ijk}\boldsymbol{e}_i\frac{\partial [J_k]}{\partial x_j} \tag{10.157}$$

Thus, we get

$$\nabla \times \left\{ \frac{[\boldsymbol{J}]}{r} \right\} = -\frac{\boldsymbol{r}_u \times [\boldsymbol{J}]}{r^2} + \frac{1}{r} \nabla \times [\boldsymbol{J}] \tag{10.158}$$

Similarly, multiplying (10.148) by the Levi-Civita symbol and base vector, we have

$$e_{ijk} \boldsymbol{e}_i \frac{\partial}{\partial x'_j} \{ \frac{[J_k]}{r} \} = e_{ijk} \boldsymbol{e}_i \frac{(x_j - x'_j)}{r^3} [J_k] + \frac{1}{r} e_{ijk} \boldsymbol{e}_i \frac{\partial [J_k]}{\partial x'_j} \tag{10.159}$$

Finally, we find another identity

$$\nabla' \times \left\{ \frac{[\boldsymbol{J}]}{r} \right\} = \frac{\boldsymbol{r}_u \times [\boldsymbol{J}]}{r^2} + \frac{1}{r} \nabla' \times [\boldsymbol{J}] \tag{10.160}$$

Substituting (10.156) into (10.160), we arrive at

$$\nabla' \times \left\{ \frac{[\boldsymbol{J}]}{r} \right\} = \frac{\boldsymbol{r}_u \times [\boldsymbol{J}]}{r^2} + \frac{1}{r} [\nabla' \times \boldsymbol{J}] - \frac{1}{r} \nabla \times [\boldsymbol{J}] \tag{10.161}$$

Combining (10.158) and (10.160), we obtain a very important identity

$$\nabla \times \left\{ \frac{[\boldsymbol{J}]}{r} \right\} + \nabla' \times \left\{ \frac{[\boldsymbol{J}]}{r} \right\} = \frac{1}{r} [\nabla' \times \boldsymbol{J}] \tag{10.162}$$

Substitution of (10.160) into (10.162) yields

$$\nabla \times \left\{ \frac{[\boldsymbol{J}]}{r} \right\} = \frac{1}{r} [\nabla' \times \boldsymbol{J}] - \frac{\boldsymbol{r}_u \times [\boldsymbol{J}]}{r^2} - \frac{1}{r} \nabla' \times [\boldsymbol{J}] \tag{10.163}$$

This can be further simplified using (10.160) and (10.155) as

$$\begin{aligned} \nabla \times \left\{ \frac{[\boldsymbol{J}]}{r} \right\} &= \frac{1}{r} [\nabla' \times \boldsymbol{J}] - \frac{\boldsymbol{r}_u \times [\boldsymbol{J}]}{r^2} - \frac{\boldsymbol{r}_u}{rc} \times \frac{\partial [\boldsymbol{J}]}{\partial t} - \frac{1}{r} [\nabla' \times \boldsymbol{J}] \\ &= -\frac{\boldsymbol{r}_u \times [\boldsymbol{J}]}{r^2} - \frac{\boldsymbol{r}_u}{rc} \times \frac{\partial [\boldsymbol{J}]}{\partial t} \\ &= \{ \frac{[\boldsymbol{J}]}{r^2} + \frac{1}{rc} \frac{\partial [\boldsymbol{J}]}{\partial t} \} \times \boldsymbol{r}_u \end{aligned} \tag{10.164}$$

The last part of (10.164) results from the fact that reversing the order of the cross-product will lead to negation of the original cross-product. This proved the curl identity of Jefimenko (1966) given in (10.123).

This concludes the complete solution of Maxwell equations for vacuum.

10.12 ELECTROMAGNETIC WAVES IN MATERIALS

The Jefimenko (1992) equations presented in the last section are restricted to the case of vacuum and, in addition, we consider the solution through the use of nonhomogeneous wave equation for a vector potential and a scalar potential resulting from the Lorenz gauge. In fact, both electric and magnetic fields can be shown as satisfying the nonhomogeneous wave equation.

Using the constitutive model given in Section 10.5, for crystalline ferroelectric or ferromagnetic solids with the given $\boldsymbol{P}$, $\boldsymbol{J}$ and $\boldsymbol{M}$, Maxwell equations in SI units are expressed as (see (10.66)-(10.68)):

$$\nabla\times\boldsymbol{B}=\mu_0(\varepsilon_0\frac{\partial\boldsymbol{E}}{\partial t}+\boldsymbol{J}+\frac{\partial\boldsymbol{P}}{\partial t}+\nabla\times\boldsymbol{M}) \tag{10.165}$$

$$\nabla\times\boldsymbol{E}=-\frac{\partial\boldsymbol{B}}{\partial t} \tag{10.166}$$

$$\nabla\cdot\boldsymbol{B}=0 \tag{10.167}$$

$$\nabla\cdot\boldsymbol{E}=\frac{1}{\varepsilon_0}(\rho-\nabla\cdot\boldsymbol{P}) \tag{10.168}$$

Taking the curl of (10.165), we find

$$\nabla\times\nabla\times\boldsymbol{B}=\mu_0(\varepsilon_0\frac{\partial\nabla\times\boldsymbol{E}}{\partial t}+\nabla\times\boldsymbol{J}+\frac{\partial\nabla\times\boldsymbol{P}}{\partial t}+\nabla\times\nabla\times\boldsymbol{M}) \tag{10.169}$$

We recall the following identity (Chau, 2018):

$$\nabla\times\nabla\times\boldsymbol{B}=\nabla(\nabla\cdot\boldsymbol{B})-\nabla^2\boldsymbol{B}=-\nabla^2\boldsymbol{B} \tag{10.170}$$

The last part is obtained by using (10.167). Substitution of (10.170) into (10.169) gives

$$-\nabla^2\boldsymbol{B}=-\frac{1}{c^2}\frac{\partial^2\boldsymbol{B}}{\partial t^2}+\mu_0\nabla\times\boldsymbol{J}+\mu_0\frac{\partial\nabla\times\boldsymbol{P}}{\partial t}+\mu_0\nabla\times\nabla\times\boldsymbol{M} \tag{10.171}$$

The first term on the right-hand side results from (10.166). Thus, we obtain the following nonhomogeneous wave equation:

$$\nabla^2\boldsymbol{B}-\frac{1}{c^2}\frac{\partial^2\boldsymbol{B}}{\partial t^2}=-\mu_0(\nabla\times\boldsymbol{J}+\frac{\partial\nabla\times\boldsymbol{P}}{\partial t}+\nabla\times\nabla\times\boldsymbol{M}) \tag{10.172}$$

This is a nonhomogeneous wave equation for $\boldsymbol{B}$ with given $\boldsymbol{P}$, $\boldsymbol{J}$ and $\boldsymbol{M}$. Similarly, we can take the curl of (10.166)

$$\nabla\times\nabla\times\boldsymbol{E}=-\frac{\partial\nabla\times\boldsymbol{B}}{\partial t}=-\frac{\partial}{\partial t}[\frac{1}{c^2}\frac{\partial\boldsymbol{E}}{\partial t}+\mu_0(\boldsymbol{J}+\frac{\partial\boldsymbol{P}}{\partial t}+\nabla\times\boldsymbol{M})] \tag{10.173}$$

The last part of (10.173) is obtained by using (10.165). Using vector identity, we have

$$\begin{aligned}\nabla\times\nabla\times\boldsymbol{E}&=\nabla(\nabla\cdot\boldsymbol{E})-\nabla^2\boldsymbol{E}\\&=\frac{1}{\varepsilon_0}[\nabla\rho-\nabla(\nabla\cdot\boldsymbol{P})]-\nabla^2\boldsymbol{E}\end{aligned} \tag{10.174}$$

The last equation of (10.174) results from (10.168). Thus, substituting (10.174) into (10.173), we get

$$\frac{1}{\varepsilon_0}[\nabla\rho-\nabla(\nabla\cdot\boldsymbol{P})]-\nabla^2\boldsymbol{E}=-\frac{\partial}{\partial t}[\frac{1}{c^2}\frac{\partial\boldsymbol{E}}{\partial t}+\mu_0(\boldsymbol{J}+\frac{\partial\boldsymbol{P}}{\partial t}+\nabla\times\boldsymbol{M})] \tag{10.175}$$

Rearrangement of (10.175) gives

$$\nabla^2\boldsymbol{E}-\frac{1}{c^2}\frac{\partial^2\boldsymbol{E}}{\partial t^2}=\frac{1}{\varepsilon_0}[\nabla\rho-\nabla(\nabla\cdot\boldsymbol{P})]+\mu_0[\frac{\partial\boldsymbol{J}}{\partial t}+\frac{\partial^2\boldsymbol{P}}{\partial t^2}+\frac{\partial}{\partial t}(\nabla\times\boldsymbol{M})] \tag{10.176}$$

Again, we find that the electric field satisfies a nonhomogeneous wave equation with given charge density, electric current, polarization, and magnetism. Therefore, from (10.172) and (10.176) the wave mechanism is obvious, with or without the introduction of potentials.

Without going through the details, we simply report the solution in this case as (Jefimenko, 1992):

$$\boldsymbol{E}(\boldsymbol{r},t)=\frac{1}{4\pi\varepsilon_0}\int_V \{[\frac{\rho(\boldsymbol{r}',t_r)-\nabla\cdot\boldsymbol{P}}{|\boldsymbol{r}-\boldsymbol{r}'|^3}+\frac{1}{|\boldsymbol{r}-\boldsymbol{r}'|^2 c}\frac{\partial\{\rho(\boldsymbol{r}',t_r)-\nabla\cdot\boldsymbol{P}\}}{\partial t}](\boldsymbol{r}-\boldsymbol{r}')$$
$$-\frac{1}{|\boldsymbol{r}-\boldsymbol{r}'|c^2}\frac{\partial[\boldsymbol{J}(\boldsymbol{r}',t_r)+\nabla\times\boldsymbol{M}]}{\partial t}+\frac{\partial^2\boldsymbol{P}}{\partial t^2}\}dV(\boldsymbol{r}') \tag{10.177}$$

$$\boldsymbol{B}(\boldsymbol{r},t)=\frac{\mu_0}{4\pi}\int_V \{[\frac{\boldsymbol{J}(\boldsymbol{r}',t_r)+\nabla\times\boldsymbol{M}}{|\boldsymbol{r}-\boldsymbol{r}'|^3}+\frac{1}{|\boldsymbol{r}-\boldsymbol{r}'|^2 c}\frac{\partial[\boldsymbol{J}(\boldsymbol{r}',t_r)+\nabla\times\boldsymbol{M}]}{\partial t}]$$
$$\times(\boldsymbol{r}-\boldsymbol{r}')+\frac{1}{|\boldsymbol{r}-\boldsymbol{r}'|}\frac{\partial\nabla\times\boldsymbol{P}}{\partial t}\}dV(\boldsymbol{r}') \tag{10.178}$$

where the retarded time is defined in (10.124).

10.13 MATHEMATICAL THEORY FOR LORENZ GAUGE

Mathematically, Chau (2018) showed that there is close resemblance between three-dimensional dynamic elasticity and the Maxwell equations. Both of them can be solved in terms of a nonhomogeneous wave equation. In three-dimensional elasticity, Helmholtz decomposition reduces the problem mathematically to solving biharmonic equations of a vector potential and a scalar potential (see Section 4.2.1 of Chau, 2018). This is very similar to the Lorenz gauge situation that we encountered in an earlier section. Alternatively, in three-dimensional elasticity, the Galerkin vector reduces the problem to solving a biharmonic equation of a single vector function (see Section 4.2.3 of Chau, 2013). In fact, instead of using $\boldsymbol{A}$ and ρ, in electrodynamics we can also employ a single vector function, and it is called the Hertz vector (having a role similar to the Galerkin vector in three-dimensional elasticity). We will consider the Hertz vector in this section. The beginning of Hertz's career coincided with the death of Maxwell, so they never had correspondence. As we mentioned in the introduction, it was Hertz who first conceived and conducted experiments that showed the existence of electromagnetic waves as predicted by Maxwell, and this led to the general acceptance of Maxwell equations. Hertz was awarded the Rumford Medal of the Royal Society because of this. Unfortunately, Maxwell did not live to see this. Theoretically, the electric and magnetic Hertz vector approach was proposed by Hertz, and is considered next.

10.13.1 Hertz Vector for Electric Field

For non-conducting materials, let us recall the conservation of charge given in (10.42)

$$\nabla \cdot \boldsymbol{J} + \frac{\partial \rho}{\partial t} = 0 \tag{10.179}$$

We see that if the charge density is time-independent, the current field must be solenoidal or irrotational:

$$\nabla \cdot \boldsymbol{J} = 0 \tag{10.180}$$

Mathematically, we see from (10.60) that the irrotational condition implies

$$\boldsymbol{J} = \frac{\partial \boldsymbol{P}}{\partial t} \tag{10.181}$$

Taking the divergence of (10.181), we have

$$\nabla \cdot \boldsymbol{J} = \frac{\partial \nabla \cdot \boldsymbol{P}}{\partial t} \tag{10.182}$$

Substitution of (10.182) into (10.179) gives

$$\frac{\partial \nabla \cdot \boldsymbol{P}}{\partial t} + \frac{\partial \rho}{\partial t} = 0 \tag{10.183}$$

Equation (10.183) yields

$$\rho = -\nabla \cdot \boldsymbol{P} \tag{10.184}$$

Using (10.181), we now recall the result of the Lorenz gauge given in (10.117) as

$$\nabla^2 \boldsymbol{A} - \frac{1}{c^2}\frac{\partial^2 \boldsymbol{A}}{\partial t^2} = -\mu \boldsymbol{J} = -\mu \frac{\partial \boldsymbol{P}}{\partial t} \tag{10.185}$$

Now, Hertz wanted to get rid of the time derivative on the right side by letting the following electric Hertz vector

$$\boldsymbol{A} = \varepsilon \mu \frac{\partial \boldsymbol{\Pi}_e}{\partial t} \tag{10.186}$$

Using this definition, (10.185) becomes

$$(\nabla^2 - \frac{1}{c^2}\frac{\partial^2}{\partial t^2})\boldsymbol{\Pi}_e = -\frac{\boldsymbol{P}}{\varepsilon} \tag{10.187}$$

We now consider the Lorenz condition. In particular, substituting (10.186) into (10.108) we get

$$\nabla \cdot \boldsymbol{A} = \mu \varepsilon \frac{\partial \nabla \cdot \boldsymbol{\Pi}_e}{\partial t} = -\frac{1}{c^2}\frac{\partial \varphi}{\partial t} \tag{10.188}$$

Therefore, we have the following relation between Maxwell's scalar potential and the electric Hertz vector

$$\varphi = -\nabla \cdot \boldsymbol{\Pi}_e \tag{10.189}$$

Substitution of (10.186) and (10.189) into (10.90) give

$$\boldsymbol{E} = -\nabla \varphi - \frac{\partial \boldsymbol{A}}{\partial t} = \nabla(\nabla \cdot \boldsymbol{\Pi}_e) - \varepsilon \mu \frac{\partial^2 \boldsymbol{\Pi}_e}{\partial t^2} \tag{10.190}$$

Substitution of (10.186) into (10.89) give

$$\boldsymbol{B} = \varepsilon \mu \frac{\partial \nabla \times \boldsymbol{\Pi}_e}{\partial t} = \varepsilon \mu \nabla \times \frac{\partial \boldsymbol{\Pi}_e}{\partial t} \tag{10.191}$$

In summary, the governing equation for the electric Hertz vector is given in (10.187) and again it is another nonhomogeneous vector wave equation. The $\boldsymbol{E}$ and $\boldsymbol{B}$ fields are given in terms of $\boldsymbol{\Pi}_e$ in (10.190) and (10.191).

10.13.2 Gauge Invariance of Hertz Vector

There is also a gauge invariance for the electric Hertz vector. Recalling the gauge invariance of $\boldsymbol{A}$ given in (10.102) and the fact that $\boldsymbol{\Pi}_e$ is proportional to $\boldsymbol{A}$, we let

$$\Pi_e = \Pi_e' + \nabla\chi \tag{10.192}$$

That is, the Hertz vector is expressed as another Hertz vector plus an arbitrary function. We will determine under what condition the $\boldsymbol{E}$ and $\boldsymbol{B}$ fields remain the same. Taking the divergence of (10.192) gives

$$\nabla\cdot\Pi_e = \nabla\cdot\Pi_e' + \nabla^2\chi \tag{10.193}$$

Then, the vector $\boldsymbol{A}$ becomes

$$\boldsymbol{A} = \varepsilon\mu\left(\frac{\partial}{\partial t}\Pi_e' + \frac{\partial}{\partial t}\nabla\chi\right) \tag{10.194}$$

Substitution of (10.193) into (10.187) gives

$$\left(\nabla^2 - \frac{1}{c^2}\frac{\partial^2}{\partial t^2}\right)\left(\frac{\partial}{\partial t}\Pi_e' + \frac{\partial}{\partial t}\nabla\chi\right) = -\frac{1}{\varepsilon}\frac{\partial \boldsymbol{P}}{\partial t} \tag{10.195}$$

Expanding (10.195) gives

$$\left(\nabla^2 - \frac{1}{c^2}\frac{\partial^2}{\partial t^2}\right)\frac{\partial \Pi_e'}{\partial t} + \frac{1}{\varepsilon}\frac{\partial \boldsymbol{P}}{\partial t} + \left(\nabla^2 - \frac{1}{c^2}\frac{\partial^2}{\partial t^2}\right)\frac{\partial}{\partial t}\nabla\chi = 0 \tag{10.196}$$

In order to satisfy this, we must have

$$\left(\nabla^2 - \frac{1}{c^2}\frac{\partial^2}{\partial t^2}\right)\chi = 0 \tag{10.197}$$

That is, the function χ satisfies the scalar wave equation. Substituting (10.194) into (10.108) gives

$$\varepsilon\mu\frac{\partial}{\partial t}(\nabla\cdot\Pi_e') + \varepsilon\mu\frac{\partial}{\partial t}\nabla^2\chi + \varepsilon\mu\frac{\partial\varphi}{\partial t} = 0 \tag{10.198}$$

Therefore, this gives the scalar potential as

$$\varphi = -\nabla\cdot\Pi_e' - \nabla^2\chi \tag{10.199}$$

Substituting (10.194) and (10.199) into (10.90), we get the electric field $\boldsymbol{E}$ as

$$\boldsymbol{E} = -\nabla\varphi - \frac{\partial \boldsymbol{A}}{\partial t} = \nabla(\nabla\cdot\Pi_e') + \nabla(\nabla^2\chi) - \varepsilon\mu\frac{\partial^2\Pi_e'}{\partial t^2} - \varepsilon\mu\frac{\partial^2\nabla\chi}{\partial t^2} \tag{10.200}$$

In view of (10.197), we obtained

$$\boldsymbol{E} = \nabla(\nabla\cdot\Pi_e') - \varepsilon\mu\frac{\partial^2\Pi_e'}{\partial t^2} \tag{10.201}$$

This is exactly the same as (10.190). Thus, arbitrarily adding a scalar wave function χ according to (10.192) will not affect the electric field. Finally, substitution of (10.194) into (10.89)

$$\boldsymbol{B} = \varepsilon\mu\frac{\partial\nabla\times\Pi_e'}{\partial t} + \varepsilon\mu\frac{\partial}{\partial t}(\nabla\times\nabla\chi) \tag{10.202}$$

It is well known that the curl of the gradient of any scalar function must be zero. Thus, we have

$$B = \varepsilon\mu \frac{\partial \nabla \times \Pi_e'}{\partial t} \tag{10.203}$$

Again, the magnetic field is not affected by the introduction of χ. Thus, the gauge invariance is established as (10.192) provided that χ satisfies (10.197).

10.13.3 Hertz Vector for Magnetic Polarization

The electric Hertz vector is established using (10.181) (i.e., electric polarization). We now consider that the case of material with magnetic polarization, and in this case, we have from (10.60):

$$J = \nabla \times M \tag{10.204}$$

The nonhomogeneous wave equation for A becomes

$$\nabla^2 A - \frac{1}{c^2}\frac{\partial^2 A}{\partial t^2} = -\mu J = -\mu \nabla \times M \tag{10.205}$$

Similar to our previous discussion, we want to remove the curl on the right-hand side by introducing:

$$A = \mu \nabla \times \Pi_m \tag{10.206}$$

where Π_m is called the magnetic Hertz vector. Substitution of (10.206) into (10.205) gives

$$(\nabla^2 - \frac{1}{c^2}\frac{\partial^2}{\partial t^2})\nabla \times \Pi_m = -\nabla \times M \tag{10.207}$$

Thus, we have

$$(\nabla^2 - \frac{1}{c^2}\frac{\partial^2}{\partial t^2})\Pi_m = -M \tag{10.208}$$

The divergence of A becomes

$$\nabla \cdot A = \mu \nabla \cdot (\nabla \times \Pi_m) = 0 \tag{10.209}$$

In view of (10.206), the last equality of (10.209) results from the fact that the divergence of the curl of any vector must be zero (Chau, 2013, 2018). Using this information, the Lorenz condition given in (10.108) implies

$$\nabla \cdot A = -\frac{1}{c^2}\frac{\partial \varphi}{\partial t} = 0 \tag{10.210}$$

Thus, φ is independent of time. Finally, substitution of (10.206) into (10.89) and (10.90) yields

$$B = \nabla \times A = \mu \nabla \times \nabla \times \Pi_m \tag{10.211}$$

$$E = -\nabla\varphi - \frac{\partial A}{\partial t} = -\mu \nabla \times \frac{\partial \Pi_m}{\partial t} \tag{10.212}$$

We have set $\varphi = 0$ in (10.212). The governing equation for the magnetic Hertz vector Π_m is given in (10.208) and the electric and magnetic fields can be found in terms of the Hertz vector in (10.211) and (10.212). Therefore, we can use a single Hertz vector to solve the Maxwell equations. To consider the gauge invariance of the magnetic Hertz vector, we assume

$$\Pi_m = \Pi_m' + \nabla \chi_m \tag{10.213}$$

Substitution of (10.213) into (10.207) gives

$$(\nabla^2 - \frac{1}{c^2}\frac{\partial^2}{\partial t^2})\nabla \times (\Pi_m' + \nabla \chi_m) = -\nabla \times \boldsymbol{M} \tag{10.214}$$

Recalling again the following vector identity for arbitrary scalar function, we get

$$\nabla \times \nabla \chi_m = 0 \tag{10.215}$$

Finally, we arrive at

$$(\nabla^2 - \frac{1}{c^2}\frac{\partial^2}{\partial t^2})\Pi_m' = -\boldsymbol{M} \tag{10.216}$$

This is of course the same as (10.208). Finally, using (10.213) we have the electric and magnetic fields as:

$$\boldsymbol{B} = \mu\nabla \times \nabla \times \Pi_m = \mu\nabla \times \nabla \times (\Pi_m' + \nabla \chi_m) = \mu\nabla \times \nabla \times \Pi_m' \tag{10.217}$$

$$\boldsymbol{E} = -\mu\nabla \times \frac{\partial}{\partial t}(\Pi_m' + \nabla \chi_m) = -\mu\nabla \times \frac{\partial \Pi_m'}{\partial t} \tag{10.218}$$

Thus, both $\boldsymbol{E}$ and $\boldsymbol{B}$ fields remain the same, and thus χ_m is arbitrary.

More generally, if both polarization and magnetization exist in the material, we can use superposition to get

$$\boldsymbol{B} = \nabla \times \boldsymbol{A} = \mu\nabla \times \nabla \times \Pi_m + \varepsilon\mu\nabla \times \frac{\partial \Pi_e}{\partial t} \tag{10.219}$$

$$\boldsymbol{E} = \nabla(\nabla \cdot \Pi_e) - \mu\nabla \times \frac{\partial \Pi_m}{\partial t} - \varepsilon\mu\frac{\partial^2 \Pi_e}{\partial t^2} \tag{10.220}$$

For linear material, we have the magnetic flux density $\boldsymbol{H}$ as

$$\boldsymbol{H} = \nabla \times \nabla \times \Pi_m + \varepsilon\frac{\partial}{\partial t}\nabla \times \frac{\partial \Pi_e}{\partial t} \tag{10.221}$$

10.13.4 Debye Potential Function for Transverse Magnetic Waves

One problem with the use of the Hertz vector is that it is not useful in solving problems in a spherical coordinate. However, one of the most fundamental problems in electromagnetic dynamics is the problem of radiation by an oscillating electric charge. This problem was solved by Peter Debye, who received the Nobel Prize in Chemistry in 1936, by formulating spherical problems in the so-called Debye potentials. In Chau (2018), we have seen that Debye discovered Riemann's method of steepest descent or the Debye saddle point method. He also derived the integral representation of Bessel functions for large argument and large order.

Figure 10.4 shows the oscillating charge problem and the associated spherical coordinate. Note that the main complication in working in the spherical coordinate arise from the fact that the base vectors are changing directions from point to point (see Section 1.23 of Chau, 2018).

In view of the spherical symmetry, we have the electric Hertz vector being:

$$\Pi_e = \Pi_e \boldsymbol{e}_r \tag{10.222}$$

In the absence of external current, the electric Hertz vector satisfies the wave equation (see (10.187))

$$(\nabla^2 - \frac{1}{c^2}\frac{\partial^2}{\partial t^2})\mathbf{\Pi}_e = 0 \tag{10.223}$$

Recall the vector differential operator in polar form as

$$\nabla = \left(\boldsymbol{e}_r \frac{\partial}{\partial r} + \boldsymbol{e}_\theta \frac{1}{r}\frac{\partial}{\partial \theta} + \boldsymbol{e}_\phi \frac{1}{r\sin\theta}\frac{\partial}{\partial \phi} \right) \tag{10.224}$$

By virtue of the vector identity (1.362) of Chau (2018), we can express the Laplacian of the Hertz vector as

$$\nabla^2\mathbf{\Pi}_e = \nabla(\nabla\cdot\mathbf{\Pi}_e) - \nabla\times(\nabla\times\mathbf{\Pi}_e) \tag{10.225}$$

Substitution of (10.225) into (10.223) gives

$$\nabla\times(\nabla\times\mathbf{\Pi}_e) - \nabla(\nabla\cdot\mathbf{\Pi}_e) + \frac{1}{c^2}\frac{\partial^2\mathbf{\Pi}_e}{\partial t^2} = 0 \tag{10.226}$$

Spherical components of (10.226) will be considered explicitly. In particular, using the formulas obtained in Problems 1.28 and 1.30 of Chau (2018) and (10.222), we find

$$\nabla\times\mathbf{\Pi}_e = \mathbf{e}_\theta \frac{1}{r\sin\theta}\frac{\partial\mathbf{\Pi}_e}{\partial\phi} - \mathbf{e}_\phi \frac{1}{r}\frac{\partial\mathbf{\Pi}_e}{\partial\theta} \tag{10.227}$$

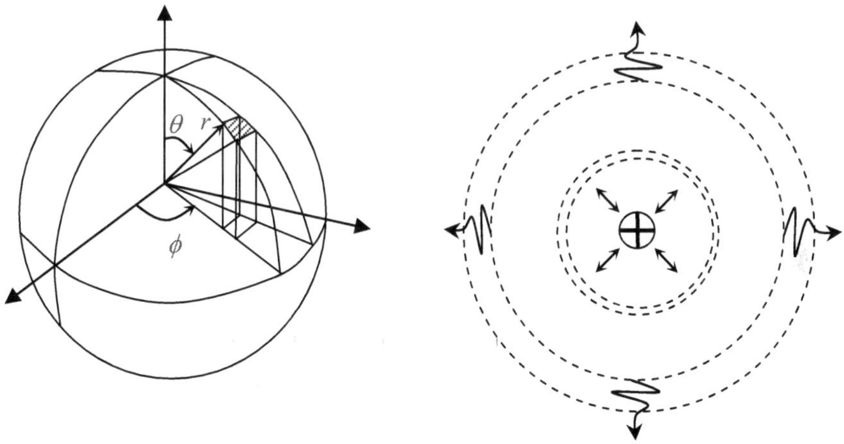

Figure 10.4 Spherical coordinate and the oscillating charge

$$\nabla(\nabla\cdot\mathbf{\Pi}_e) = \boldsymbol{e}_r(\frac{\partial^2 \Pi_e}{\partial r^2} + \frac{2}{r}\frac{\partial \Pi_e}{\partial r} - \frac{2\Pi_e}{r^2}) + \boldsymbol{e}_\theta(\frac{1}{r}\frac{\partial^2 \Pi_e}{\partial\theta\partial r} + \frac{2}{r^2}\frac{\partial \Pi_e}{\partial\theta})$$
$$+\boldsymbol{e}_\phi(\frac{1}{r\sin\theta}\frac{\partial^2 \Pi_e}{\partial\phi\partial r} + \frac{2}{r^2\sin\theta}\frac{\partial \Pi_e}{\partial\phi}) \tag{10.228}$$

$$\nabla\times(\nabla\times\Pi_e) = -e_r(\frac{1}{r^2}\frac{\partial^2 \Pi_e}{\partial\theta^2}+\frac{\cot\theta}{r^2}\frac{\partial \Pi_e}{\partial\theta}+\frac{1}{r^2\sin^2\theta}\frac{\partial^2 \Pi_e}{\partial\phi^2})$$
$$+e_\theta\frac{1}{r}\frac{\partial^2 \Pi_e}{\partial r\partial\theta}+e_\phi\frac{1}{r\sin\theta}\frac{\partial^2 \Pi_e}{\partial r\partial\phi} \quad (10.229)$$

We are now ready to consider Debye potential. Recall from (10.191) that

$$B = \varepsilon\mu\nabla\times\frac{\partial\Pi_e}{\partial t} \quad (10.230)$$

Taking the curl to (10.230) and substituting the result into (10.165) for the case of zero $\boldsymbol{P}$, $\boldsymbol{M}$, and $\boldsymbol{J}$, we find

$$\nabla\times B = \varepsilon\mu\nabla\times\nabla\times\frac{\partial\Pi_e}{\partial t} = \varepsilon\mu\frac{\partial E}{\partial t} \quad (10.231)$$

We find an alternative form for (10.220).

$$E = \nabla\times\nabla\times\Pi_e \quad (10.232)$$

We will use this result later. On the other hand, substituting (10.230) into (10.166), we get

$$\nabla\times E = -\varepsilon\mu\nabla\times\frac{\partial^2\Pi_e}{\partial t^2} \quad (10.233)$$

Rearranging (10.233), we get

$$\nabla\times(E+\varepsilon\mu\frac{\partial^2\Pi_e}{\partial t^2}) = 0 \quad (10.234)$$

By the well-known vector identity given in ((1.346) of Chau, 2018; (1.46) of Chau, 2013), the bracket term in (10.234) can be expressed as:

$$E+\varepsilon\mu\frac{\partial^2\Pi_e}{\partial t^2} = -\nabla\psi \quad (10.235)$$

where ψ is a scalar function to be determined. Solving for $\boldsymbol{E}$, we have

$$E = -\varepsilon\mu\frac{\partial^2\Pi_e}{\partial t^2} - \nabla\psi \quad (10.236)$$

Expressed explicitly in polar components, we have

$$E_r = -\varepsilon\mu\frac{\partial^2 \Pi_e}{\partial t^2} - \frac{\partial\psi}{\partial r} \quad (10.237)$$

$$E_\theta = -\frac{1}{r}\frac{\partial\psi}{\partial\theta} \quad (10.238)$$

$$E_\phi = -\frac{1}{r\sin\theta}\frac{\partial\psi}{\partial\phi} \quad (10.239)$$

Alternatively, using (10.232) and vector identity (10.229), we can find the component form of $\boldsymbol{E}$

$$E_r = -(\frac{1}{r^2}\frac{\partial^2 \Pi_e}{\partial\theta^2}+\frac{\cot\theta}{r^2}\frac{\partial \Pi_e}{\partial\theta}+\frac{1}{r^2\sin^2\theta}\frac{\partial^2 \Pi_e}{\partial\phi^2}) \quad (10.240)$$

$$E_\theta = \frac{1}{r}\frac{\partial^2 \Pi_e}{\partial r \partial \theta} \tag{10.241}$$

$$E_\phi = \frac{1}{r\sin\theta}\frac{\partial^2 \Pi_e}{\partial r \partial \phi} \tag{10.242}$$

Equating these components of the electric field, we have three equations for finding ψ

$$E_r = -\left(\frac{1}{r^2}\frac{\partial^2 \Pi_e}{\partial \theta^2} + \frac{\cot\theta}{r^2}\frac{\partial \Pi_e}{\partial \theta} + \frac{1}{r^2\sin^2\theta}\frac{\partial^2 \Pi_e}{\partial \phi^2}\right) = -\varepsilon\mu\frac{\partial^2 \Pi_e}{\partial t^2} - \frac{\partial \psi}{\partial r} \tag{10.243}$$

$$E_\theta = \frac{1}{r}\frac{\partial^2 \Pi_e}{\partial r \partial \theta} = -\frac{1}{r}\frac{\partial \psi}{\partial \theta} \tag{10.244}$$

$$E_\phi = \frac{1}{r\sin\theta}\frac{\partial^2 \Pi_e}{\partial r \partial \phi} = -\frac{1}{r\sin\theta}\frac{\partial \psi}{\partial \phi} \tag{10.245}$$

By inspection, we can set

$$\psi = -\frac{\partial \Pi_e}{\partial r} \tag{10.246}$$

Both (10.244) and (10.245) are satisfied identically. Substitution of (10.243) into (10.243) gives

$$\frac{\partial^2 \Pi_e}{\partial r^2} + \left(\frac{1}{r^2}\frac{\partial^2 \Pi_e}{\partial \theta^2} + \frac{\cot\theta}{r^2}\frac{\partial \Pi_e}{\partial \theta} + \frac{1}{r^2\sin^2\theta}\frac{\partial^2 \Pi_e}{\partial \phi^2}\right) - \varepsilon\mu\frac{\partial^2 \Pi_e}{\partial t^2} = 0 \tag{10.247}$$

Note that the second term on the left-hand side can be rewritten as

$$\frac{1}{r^2}\frac{\partial^2 \Pi_e}{\partial \theta^2} + \frac{\cot\theta}{r^2}\frac{\partial \Pi_e}{\partial \theta} + \frac{1}{r^2\sin^2\theta}\frac{\partial^2 \Pi_e}{\partial \phi^2} = \frac{1}{r^2\sin\theta}\frac{\partial}{\partial \theta}\left(\sin\theta\frac{\partial \Pi_e}{\partial \theta}\right) + \frac{1}{r^2\sin^2\theta}\frac{\partial^2 \Pi_e}{\partial \phi^2} \tag{10.248}$$

Thus, substitution of (10.248) into (10.247) yields

$$\frac{\partial^2 \Pi_e}{\partial r^2} + \frac{1}{r^2\sin\theta}\frac{\partial}{\partial \theta}\left(\sin\theta\frac{\partial \Pi_e}{\partial \theta}\right) + \frac{1}{r^2\sin^2\theta}\frac{\partial^2 \Pi_e}{\partial \phi^2} - \varepsilon\mu\frac{\partial^2 \Pi_e}{\partial t^2} = 0 \tag{10.249}$$

We further note that the Laplacian of Π_e is (see (1.469) of Chau, 2018)

$$\nabla^2 \Pi_e = \frac{1}{r^2}\frac{\partial}{\partial r}\left(r^2\frac{\partial \Pi_e}{\partial r}\right) + \frac{1}{r^2\sin\theta}\frac{\partial}{\partial \theta}\left(\sin\theta\frac{\partial \Pi_e}{\partial \theta}\right) + \frac{1}{r^2\sin^2\theta}\frac{\partial^2 \Pi_e}{\partial \phi^2} \tag{10.250}$$

Debye found that it is possible to reduce the first three terms on the left of (10.249) to Laplacian. In particular, the most important step is that Debye assumed the following new function P

$$\Pi_e = rP \tag{10.251}$$

Note the following identity

$$\frac{\partial^2 (rP)}{\partial r^2} = 2\frac{\partial P}{\partial r} + r\frac{\partial^2 P}{\partial r^2} = \frac{1}{r}\frac{\partial}{\partial r}\left(r^2\frac{\partial P}{\partial r}\right) \tag{10.252}$$

In view of (10.252), we finally get

$$\frac{1}{r^2}\frac{\partial}{\partial r}(r^2\frac{\partial P}{\partial r})+\frac{1}{r^2\sin\theta}\frac{\partial}{\partial\theta}(\sin\theta\frac{\partial P}{\partial\theta})+\frac{1}{r^2\sin^2\theta}\frac{\partial^2 P}{\partial\phi^2}-\varepsilon\mu\frac{\partial^2 P}{\partial t^2}=0 \quad (10.253)$$

Comparing (10.250) with (10.253), we get

$$(\nabla^2-\varepsilon\mu\frac{\partial^2}{\partial t^2})P=0 \quad (10.254)$$

The function P is called the Debye scalar potential for transverse magnetic (TM) waves. This is also called TM waves. To see the transverse nature of the induced magnetic waves, we first generalize (10.251) in an arbitrary direction as:

$$\boldsymbol{\Pi}_e=\boldsymbol{r}P \quad (10.255)$$

where $\boldsymbol{r}$ is the position vector. Substitution (10.255) into (10.232) gives

$$\boldsymbol{E}=\nabla\times\nabla\times(\boldsymbol{r}P) \quad (10.256)$$

The magnetic field $\boldsymbol{B}$ can be found by substitution of (10.255) into (10.230)

$$\boldsymbol{B}=\varepsilon\mu\nabla\times\frac{\partial(\boldsymbol{r}P)}{\partial t}=\varepsilon\mu\frac{\partial}{\partial t}\nabla\times(\boldsymbol{r}P) \quad (10.257)$$

First of all, we note the following vector identity

$$\nabla\times(\boldsymbol{r}P)=-\boldsymbol{r}\times(\nabla P) \quad (10.258)$$

To prove (10.258), we use the Cartesian tensor notation (e.g., see (1.428) of Chau, 2018):

$$\begin{aligned}\nabla\times(\boldsymbol{r}P)&=e_{ijk}\frac{\partial}{\partial x_j}(x_k P)\boldsymbol{e}_i=e_{ijk}(\delta_{kj}+x_k P_{,j})\boldsymbol{e}_i=e_{ijk}x_k P_{,j}\boldsymbol{e}_i\\&=(\nabla P)\times\boldsymbol{r}=-\boldsymbol{r}\times(\nabla P)\end{aligned} \quad (10.259)$$

Substitution of (10.258) into (10.257) leads to

$$\boldsymbol{B}=-\varepsilon\mu\frac{\partial}{\partial t}[\boldsymbol{r}\times(\nabla P)] \quad (10.260)$$

We are going to show that the radial component of $\boldsymbol{B}$ given in (10.257) is zero. In the Cartesian coordinate, we find

$$\boldsymbol{r}\times(\nabla P)=(x_2P_{,3}-x_3P_{,2})\boldsymbol{e}_1+(-x_1P_{,3}+x_3P_{,1})\boldsymbol{e}_2+(x_1P_{,2}-x_2P_{,1})\boldsymbol{e}_3 \quad (10.261)$$

To rewrite (10.261) in spherical coordinates, we note that (see (1.457) of Chau, 2018)

$$x_1=r\sin\theta\cos\phi,\quad x_2=r\sin\theta\sin\varphi,\quad x_3=r\cos\theta \quad (10.262)$$

The base vectors in polar form can be expressed in terms of the base vectors in Cartesian form as (e.g., (1.461) of Chau, 2018)

$$\begin{aligned}\boldsymbol{e}_1&=\sin\theta\cos\phi\boldsymbol{e}_r+\cos\theta\cos\phi\boldsymbol{e}_\theta-\sin\theta\boldsymbol{e}_\phi\\\boldsymbol{e}_2&=\sin\theta\sin\phi\boldsymbol{e}_r+\cos\theta\sin\phi\boldsymbol{e}_\theta+\cos\theta\boldsymbol{e}_\phi\\\boldsymbol{e}_3&=\cos\theta\boldsymbol{e}_r-\sin\theta\boldsymbol{e}_\theta\end{aligned} \quad (10.263)$$

Substitution of (10.262) and (10.263) into (10.261) gives the radial component as

$$\begin{aligned}[\boldsymbol{r}\times(\nabla P)]_r &= (r\sin^2\theta\cos\phi\sin\phi - r\sin^2\theta\cos\phi\sin\phi)P_{,3} \\ &+(-r\sin\theta\cos\theta\cos\phi + r\sin\theta\cos\theta\cos\phi)P_{,2} \\ &+(r\sin\theta\cos\theta\sin\phi - r\sin\theta\cos\theta\sin\phi)P_{,1} \\ &= 0\end{aligned} \tag{10.264}$$

This is identically zero regardless of the function P. Therefore, the radial component of $\boldsymbol{E}$ is nonzero, whereas the radial component of the magnetic field is zero. Thus, they represent TM waves.

10.13.5 Debye Potential Function for Transverse Electric Waves

To express spherical waves in terms of the magnetic Hertz vector, we define

$$\boldsymbol{\Pi}_m = \Pi_m \boldsymbol{e}_r \tag{10.265}$$

Recall from (10.211) and (10.212) that the electric and magnetic fields can be expressed in terms of the magnetic Hertz vector as

$$\boldsymbol{B} = \nabla\times\boldsymbol{A} = \mu\nabla\times\nabla\times\boldsymbol{\Pi}_m \tag{10.266}$$

$$\boldsymbol{E} = -\mu\nabla\times\frac{\partial\boldsymbol{\Pi}_m}{\partial t} \tag{10.267}$$

Substituting (10.267) into (10.165) for the case of zero $\boldsymbol{P}$, $\boldsymbol{M}$, and $\boldsymbol{J}$, we find

$$\nabla\times\boldsymbol{B} = -\mu^2\varepsilon\nabla\times\frac{\partial^2\boldsymbol{\Pi}_m}{\partial t^2} \tag{10.268}$$

Rearranging (10.268), we have

$$\nabla\times(\boldsymbol{B} + \mu^2\varepsilon\frac{\partial^2\boldsymbol{\Pi}_m}{\partial t^2}) = 0 \tag{10.269}$$

Following the same argument in the previous section, by the well-known vector identity given in ((1.346) of Chau, 2018; (1.46) of Chau, 2013), the bracket term in (10.269) can be expressed as:

$$\boldsymbol{B} + \varepsilon\mu^2\frac{\partial^2\boldsymbol{\Pi}_m}{\partial t^2} = -\nabla\psi^* \tag{10.270}$$

where ψ^* is a scalar function to be determined. Solving for $\boldsymbol{B}$, we have

$$\boldsymbol{B} = -\varepsilon\mu^2\frac{\partial^2\boldsymbol{\Pi}_m}{\partial t^2} - \nabla\psi^* \tag{10.271}$$

Equating (10.266) and (10.271), we have

$$\frac{1}{\mu}B_r = -\frac{1}{r^2\sin\theta}[\frac{\partial}{\partial\theta}(\sin\theta\frac{\partial\Pi_m}{\partial\theta}) + \frac{1}{\sin\theta}\frac{\partial^2\Pi_m}{\partial\phi^2}] = -\varepsilon\mu\frac{\partial^2\Pi_m}{\partial t^2} - \frac{\partial\psi^*}{\partial r} \tag{10.272}$$

$$\frac{1}{\mu}B_\theta = \frac{1}{r}\frac{\partial^2\Pi_m}{\partial r\partial\theta} = -\frac{1}{r}\frac{\partial\psi^*}{\partial\theta} \tag{10.273}$$

$$\frac{1}{\mu}B_\phi = \frac{1}{r\sin\theta}\frac{\partial^2\Pi_m}{\partial r\partial\phi} = -\frac{1}{r\sin\theta}\frac{\partial\psi^*}{\partial\phi} \tag{10.274}$$

By inspection, we can set

$$\psi^* = -\frac{\partial \Pi_m}{\partial r} \tag{10.275}$$

Both (10.273) and (10.274) are satisfied identically. Substitution of (10.275) into (10.272) gives

$$\frac{\partial^2 \Pi_m}{\partial r^2} + \frac{1}{r^2 \sin\theta}\left[\frac{\partial}{\partial \theta}(\sin\theta \frac{\partial \Pi_m}{\partial \theta}) + \frac{1}{\sin\theta}\frac{\partial^2 \Pi_m}{\partial \phi^2}\right] - \varepsilon\mu \frac{\partial^2 \Pi_m}{\partial t^2} = 0 \tag{10.276}$$

Finally, the following Debye potential is assumed:

$$\Pi_m = rQ \tag{10.277}$$

Consequently, we finally get

$$\frac{1}{r^2}\frac{\partial}{\partial r}(r^2 \frac{\partial Q}{\partial r}) + \frac{1}{r^2 \sin\theta}\frac{\partial}{\partial \theta}(\sin\theta \frac{\partial Q}{\partial \theta}) + \frac{1}{r^2 \sin^2\theta}\frac{\partial^2 Q}{\partial \phi^2} - \varepsilon\mu \frac{\partial^2 Q}{\partial t^2} = 0 \tag{10.278}$$

In view of the definition of Laplacian given in (10.250), we have

$$(\nabla^2 - \varepsilon\mu \frac{\partial^2}{\partial t^2})Q = 0 \tag{10.279}$$

The function Q is called the Debye scalar potential for transverse electric (TE) waves. To see the transverse nature of the induced magnetic waves, we first generalize (10.277) in an arbitrary direction as:

$$\mathbf{\Pi}_m = \mathbf{r}Q \tag{10.280}$$

where $\mathbf{r}$ is the position vector. Substitution of (10.280) into (10.267) gives

$$\mathbf{E} = -\mu \frac{\partial}{\partial t}[\nabla \times (\mathbf{r}Q)] = \mu \frac{\partial}{\partial t}[\mathbf{r} \times (\nabla Q)] \tag{10.281}$$

Substituting (10.280) into (10.266) and in view of (10.271), (10.275) and (10.277), we finally get

$$\begin{aligned} \mathbf{B} &= \mu \nabla \times \nabla \times \mathbf{\Pi}_m = \mu \nabla \times \nabla \times (\mathbf{r}Q) = -\varepsilon\mu^2 \frac{\partial^2 (\mathbf{r}Q)}{\partial t^2} - \mu \nabla \psi^* \\ &= -\varepsilon\mu^2 \frac{\partial^2 (\mathbf{r}Q)}{\partial t^2} + \mu \nabla \frac{\partial (rQ)}{\partial r} \end{aligned} \tag{10.282}$$

Thus, with a nonzero radial component of $\mathbf{B}$, we arrive at a zero radial component of the electric field $\mathbf{E}$, as suggested by (10.281). The proof of this statement follows directly from (10.260) and (10.264), except for reversing the roles of $\mathbf{E}$ and $\mathbf{B}$. Thus, Π_m will lead to the only transverse component of the electric waves.

For more general situations for nonzero Π_m and Π_e, we use superposition to get

$$\begin{aligned} \mathbf{E} &= \nabla \times \nabla \times (\mathbf{r}P) + \mu \frac{\partial}{\partial t}[\mathbf{r} \times (\nabla Q)] \\ &= -\varepsilon\mu \frac{\partial^2 (\mathbf{r}P)}{\partial t^2} + \nabla \frac{\partial (rP)}{\partial r} + \mu \frac{\partial}{\partial t}[\mathbf{r} \times (\nabla Q)] \end{aligned} \tag{10.283}$$

$$\begin{aligned} \boldsymbol{B} &= -\varepsilon\mu\frac{\partial}{\partial t}[\boldsymbol{r}\times(\nabla P)] + \mu\nabla\times\nabla\times(\boldsymbol{r}Q) \\ &= -\varepsilon\mu\frac{\partial}{\partial t}[\boldsymbol{r}\times(\nabla P)] - \varepsilon\mu^2\frac{\partial^2(\boldsymbol{r}Q)}{\partial t^2} + \mu\nabla\frac{\partial(\boldsymbol{r}Q)}{\partial r} \end{aligned} \tag{10.284}$$

where

$$(\nabla^2 - \varepsilon\mu\frac{\partial^2}{\partial t^2})P = 0 \tag{10.285}$$

$$(\nabla^2 - \varepsilon\mu\frac{\partial^2}{\partial t^2})Q = 0 \tag{10.286}$$

Both of the Debye potentials satisfy homogeneous scalar wave equations, and the technique discussed in Chau (2018) can be used to solve for P and Q. Therefore, for the case of zero $\boldsymbol{P}$, $\boldsymbol{M}$, and $\boldsymbol{J}$, only two independent scalar potentials are needed to represent the electromagnetic wave fields.

10.14 DUALITY AND SYMMETRY

As shown in Section 10.2.5 for the case of vacuum space with zero charge ($\rho = 0$) and electric current ($\boldsymbol{J} = 0$), there is dual symmetry in the Maxwell equations. We will extend the discussion here to the case of electromagnetic waves in materials with the presence of both ρ and $\boldsymbol{J}$. We will see here the interplay between mathematics and physics.

First, let us recast (10.7) slightly by using $\boldsymbol{H}$ instead of $\boldsymbol{B}$ as (recall the constitutive model (10.59)):

$$\begin{aligned} &\nabla\cdot\boldsymbol{E} = 0, \quad \nabla\cdot\boldsymbol{H} = 0, \\ &\nabla\times\boldsymbol{E} = -\mu_0\frac{\partial\boldsymbol{H}}{\partial t}, \quad \nabla\times\boldsymbol{H} = \varepsilon_0\frac{\partial\boldsymbol{E}}{\partial t} \end{aligned} \tag{10.287}$$

where μ_0 and ε_0 are the electric permittivity and magnetic permeability in vacuum. If we can find a set of solutions for $\boldsymbol{H}$ and $\boldsymbol{E}$, we propose to make the following identifications:

$$\boldsymbol{E} \leftarrow \boldsymbol{H}, \quad \boldsymbol{H} \leftarrow \boldsymbol{E}, \quad \mu_0 \leftarrow -\varepsilon_0, \quad \varepsilon_0 \leftarrow -\mu_0 \tag{10.288}$$

We again see that the role of $\boldsymbol{H}$ and $\boldsymbol{E}$ reverses, but the same set of Maxwell equations results. The symmetry between electric and magnetic is intriguing.

Many theoretical physicists believed that natural laws must be beautiful mathematically. This leads to the hypothesis of symmetrizing the Maxwell equation by introducing the existence of magnetic charges ρ_m and this magnetic charge will produce magnetic current density $\boldsymbol{J}_m$ (this should not be confused with $\boldsymbol{J}_M$ discussed in Section 10.2.4). In addition, they are proposed to satisfy the conservation law as the electric charge does:

$$\nabla\cdot\boldsymbol{J}_m + \frac{\partial\rho_m}{\partial t} = 0 \tag{10.289}$$

With this hypothesis, the symmetrized Maxwell equations will look like:

$$\nabla\cdot\boldsymbol{E}=\frac{\rho_e}{\varepsilon_0},\quad \nabla\cdot\boldsymbol{B}=\mu_0\rho_m,$$
$$\nabla\times\boldsymbol{E}=-\mu_0\boldsymbol{J}_m-\frac{\partial\boldsymbol{B}}{\partial t},\quad \nabla\times\boldsymbol{B}=\mu_0\boldsymbol{J}_e+\frac{1}{c^2}\frac{\partial\boldsymbol{E}}{\partial t} \tag{10.290}$$

where subscript m denotes the magnetic charge and associated current, whereas subscript e denotes the electric charge and associated current. The most interesting feature of this hypothesized theory is that it allows the following duality transformation rules parameterized by an angle θ:

$$\boldsymbol{E}'=\boldsymbol{E}\cos\theta+c\boldsymbol{B}\sin\theta,\quad c\boldsymbol{B}'=-\boldsymbol{E}\sin\theta+c\boldsymbol{B}\cos\theta \tag{10.291}$$

$$c\rho_e'=c\rho_e\cos\theta+\rho_m\sin\theta,\quad \rho_m'=-c\rho_e\sin\theta+\rho_m\cos\theta \tag{10.292}$$

$$c\boldsymbol{J}_e'=c\boldsymbol{J}_e\cos\theta+\boldsymbol{J}_m\sin\theta,\quad \boldsymbol{J}_m'=-c\boldsymbol{J}_e\sin\theta+\boldsymbol{J}_m\cos\theta \tag{10.293}$$

It is straightforward to show that

$$c^2(\rho_e')^2+(\rho_m')^2=c^2\rho_e^2+\rho_m^2 \tag{10.294}$$

To check the duality, we take the divergence of the first definition in (10.291) as

$$\begin{aligned}\nabla\cdot\boldsymbol{E}'&=\nabla\cdot\boldsymbol{E}\cos\theta+c\nabla\cdot\boldsymbol{B}\sin\theta\\&=\frac{\rho_e}{\varepsilon_0}\cos\theta+c\mu_0\rho_m\sin\theta\\&=\frac{\rho_e'}{\varepsilon_0}\end{aligned} \tag{10.295}$$

Similarly, taking the divergence of the second definition in (10.291), we have

$$\begin{aligned}\nabla\cdot\boldsymbol{B}'&=-\frac{1}{c}\nabla\cdot\boldsymbol{E}\sin\theta+\nabla\cdot\boldsymbol{B}\cos\theta\\&=-c\mu_0\rho_e\sin\theta+\mu_0\rho_m\cos\theta\\&=\mu_0\rho_m'\end{aligned} \tag{10.296}$$

Taking the curl of the first equation in (10.291), we find

$$\begin{aligned}\nabla\times\boldsymbol{E}'&=\nabla\times\boldsymbol{E}\cos\theta+c\nabla\times\boldsymbol{B}\sin\theta\\&=-(\mu_0\boldsymbol{J}_m-\frac{\partial\boldsymbol{B}}{\partial t})\cos\theta+(c\mu_0\boldsymbol{J}_e+\frac{1}{c}\frac{\partial\boldsymbol{E}}{\partial t})\sin\theta\\&=-\mu_0\boldsymbol{J}_m'-\frac{\partial\boldsymbol{B}'}{\partial t}\end{aligned} \tag{10.297}$$

Taking the curl of the second equation in (10.291), we find

$$\begin{aligned}\nabla\times\boldsymbol{B}'&=-\frac{1}{c}\nabla\times\boldsymbol{E}\sin\theta+\nabla\times\boldsymbol{B}\cos\theta\\&=\frac{1}{c}(\mu_0\boldsymbol{J}_m+\frac{\partial\boldsymbol{B}}{\partial t})\sin\theta+(\mu_0\boldsymbol{J}_e+\frac{1}{c^2}\frac{\partial\boldsymbol{E}}{\partial t})\cos\theta\\&=\mu_0\boldsymbol{J}_e'+\frac{1}{c^2}\frac{\partial\boldsymbol{E}'}{\partial t}\end{aligned} \tag{10.298}$$

Thus, the transformed set of variables satisfies exactly the same symmetrized Maxwell equations given (10.290). Although ρ_m and $\boldsymbol{J}_m$ are fictitious, due to this duality and symmetry, Dirac argued the existence of a monopole or magnetic monopole (i.e., the magnetic charges that we hypothesized above). Paul Dirac, is a British engineer and theoretical physicist, and shared Nobel Prize in Physics in 1933 with Erwin Schrödinger. However, magnetism in a bar magnet does not come from a magnetic monopole. Although various experiments have been set up to search for a magnetic monopole, there is no experimental evidence for its existence yet. This remains a hypothesis that a natural phenomenon must obey law with mathematical beauty (in this case, duality or symmetry in the mathematical structure for $\boldsymbol{B}$ and $\boldsymbol{E}$).

10.15 MATHEMATICAL THEORY FOR COULOMB GAUGE

We now look at the Coulomb gauge more closely in this section. Physically, it is also known as the radiation gauge or transverse gauge.

10.15.1 Scalar and Vector Potentials

Recalling from our earlier discussion, we have the following condition for Coulomb gauge:

$$\nabla \cdot \boldsymbol{A} = 0 \tag{10.299}$$

The differential equations for Maxwell's potentials become (see (10.106) and (10.107))

$$\nabla^2 \varphi = -\frac{\rho}{\varepsilon_0} \tag{10.300}$$

$$\nabla^2 \boldsymbol{A} - \frac{1}{c^2}\frac{\partial^2 \boldsymbol{A}}{\partial t^2} = -\mu_0 \boldsymbol{J} + \frac{1}{c^2}\nabla(\frac{\partial \varphi}{\partial t}) \tag{10.301}$$

Before our discussion, we first recall the solution for the Poisson equation ((9.207) of Chau, 2018):

$$\varphi(\boldsymbol{r},t) = \frac{1}{4\pi\varepsilon_0}\int_V \frac{\rho(\boldsymbol{r}')}{|\boldsymbol{r}-\boldsymbol{r}'|} dV(\boldsymbol{r}') \tag{10.302}$$

This formula will be proved in Section 10.15.3 and it would be used in later discussions.

10.15.2 Transverse Waves or Radiation Gauge

To solve for (10.301), we first note that the current $\boldsymbol{J}$ can be decomposed into a transverse or solenoidal part and a longitudinal or irrotational part. More specifically, by definition we have

$$\boldsymbol{J} = \boldsymbol{J}_t + \boldsymbol{J}_l \tag{10.303}$$

where $\boldsymbol{J}_l$ is the irrotational part and $\boldsymbol{J}_t$ is the solenoidal part. These components satisfy

$$\nabla \cdot \boldsymbol{J}_t = 0 \tag{10.304}$$

$$\nabla \times \boldsymbol{J}_l = 0 \tag{10.305}$$

In fact, this is a mathematical truth that any vector can always be decomposed into a solenoidal part and an irrotational part (see Section 1.16 of Chau, 2018). These solenoidal and irrotational parts can be expressed in terms of integrals involving $\boldsymbol{J}$. To see this, let us consider the following property of the Dirac delta function:

$$\boldsymbol{J}(\boldsymbol{r},t) = \int_V \boldsymbol{J}(\boldsymbol{r}',t)\delta(\boldsymbol{r}-\boldsymbol{r}')dV(\boldsymbol{r}') \tag{10.306}$$

Note from Section 8.10 of Chau (2018) that (10.306) is the basic definition of the Dirac delta function. In the present context, it is given in three-dimensional space. As shown in Chau (2018), the three-dimensional fundamental function of the Laplace equation satisfies (see Section 8.3 of Chau, 2018)

$$\nabla^2 \left(\frac{1}{|\boldsymbol{r}-\boldsymbol{r}'|}\right) = -4\pi\delta(\boldsymbol{r}-\boldsymbol{r}') \tag{10.307}$$

Substitution of (10.307) into (10.306) gives

$$\boldsymbol{J}(\boldsymbol{r},t) = -\int_V \frac{\boldsymbol{J}(\boldsymbol{r}',t)}{4\pi}\nabla^2 \left(\frac{1}{|\boldsymbol{r}-\boldsymbol{r}'|}\right)dV(\boldsymbol{r}') \tag{10.308}$$

Since the current $\boldsymbol{J}$ inside the integral is independent of the position vector $\boldsymbol{r}$, we can reverse the order of Laplacian and integration. Thus, we have

$$\boldsymbol{J}(\boldsymbol{r},t) = -\nabla^2 \int_V \frac{\boldsymbol{J}(\boldsymbol{r}',t)}{4\pi R}dV(\boldsymbol{r}') \tag{10.309}$$

where

$$R = |\boldsymbol{r}-\boldsymbol{r}'| \tag{10.310}$$

We now apply the following vector identity given in (1.362) of Chau (2018)

$$\nabla\times\nabla\times\boldsymbol{J} = \nabla(\nabla\cdot\boldsymbol{J}) - \nabla^2\boldsymbol{J} \tag{10.311}$$

Applying (10.311) to (10.309), we obtain

$$\begin{aligned}\boldsymbol{J}(\boldsymbol{r},t) &= \nabla\times\nabla\times\int_V \frac{\boldsymbol{J}(\boldsymbol{r}',t)}{4\pi R}dV(\boldsymbol{r}') - \nabla[\nabla\cdot\int_V \frac{\boldsymbol{J}(\boldsymbol{r}',t)}{4\pi R}dV(\boldsymbol{r}')] \\ &= \boldsymbol{J}_l + \boldsymbol{J}_t\end{aligned} \tag{10.312}$$

where

$$\boldsymbol{J}_t = \nabla\times\nabla\times\int_V \frac{\boldsymbol{J}(\boldsymbol{r}',t)}{4\pi R}dV(\boldsymbol{r}') \tag{10.313}$$

$$\boldsymbol{J}_l = -\nabla[\nabla\cdot\int_V \frac{\boldsymbol{J}(\boldsymbol{r}',t)}{4\pi R}dV(\boldsymbol{r}')] \tag{10.314}$$

The divergence of $\boldsymbol{J}/R$ can be evaluated as

$$\nabla\cdot[\frac{\boldsymbol{J}(\boldsymbol{r}',t)}{R}] = \boldsymbol{J}(\boldsymbol{r}',t)\cdot\nabla(\frac{1}{R}) + \frac{1}{R}\nabla\cdot\boldsymbol{J}(\boldsymbol{r}',t) = \boldsymbol{J}(\boldsymbol{r}',t)\cdot\nabla(\frac{1}{R}) \tag{10.315}$$

$$\nabla'\cdot[\frac{\boldsymbol{J}(\boldsymbol{r}',t)}{R}] = \boldsymbol{J}(\boldsymbol{r}',t)\cdot\nabla'(\frac{1}{R}) + \frac{1}{R}\nabla'\cdot\boldsymbol{J}(\boldsymbol{r}',t) \tag{10.316}$$

In addition, we note that

$$\boldsymbol{J}(\boldsymbol{r}',t)\nabla'(\frac{1}{R}) = -\boldsymbol{J}(\boldsymbol{r}',t)\nabla(\frac{1}{R}) \tag{10.317}$$

Combining (10.315) to (10.317), we get

$$\nabla\cdot[\frac{J(r',t)}{R}] = -\nabla'\cdot[\frac{J(r',t)}{R}] + \frac{1}{R}\nabla'\cdot J(r',t) \tag{10.318}$$

Substitution of (10.318) into (10.314) yields

$$J_l = \nabla\int_V \nabla'\cdot[\frac{J(r',t)}{4\pi R}]dV(r') - \nabla\int_V \frac{\nabla'\cdot J(r',t)}{4\pi R}dV(r') \tag{10.319}$$

Applying the Gauss divergence theorem, we obtain

$$J_l = \nabla\oint_S \frac{J(r',t)\cdot n}{4\pi R}dS' - \nabla\int_V \frac{\nabla'\cdot J(r',t)}{4\pi R}dV(r') \tag{10.320}$$

All the normal components of current through a closed surface must add to zero. Thus, (10.320) is reduced to

$$J_l = -\nabla\int_V \frac{\nabla'\cdot J(r',t)}{4\pi R}dV(r') \tag{10.321}$$

The conservation of charges gives

$$\nabla'\cdot J = -\frac{\partial\rho(r',t)}{\partial t} \tag{10.322}$$

Using (10.322), we rewrite (10.321) as

$$J_l = \nabla\frac{\partial}{\partial t}\int_V \frac{\rho(r',t)}{4\pi R}dV(r') = \varepsilon_0\nabla\frac{\partial}{\partial t}\int_V \frac{\rho(r',t)}{4\pi\varepsilon_0 R}dV(r') = \varepsilon_0\nabla\frac{\partial\varphi}{\partial t} \tag{10.323}$$

The last equation of (10.323) is obtained by employing (10.302). This result can be substituted into (10.301) to get

$$\begin{aligned}\nabla^2 A - \frac{1}{c^2}\frac{\partial^2 A}{\partial t^2} &= -\mu_0 J + \mu_0 J_l \\ &= -\mu_0 J_t\end{aligned} \tag{10.324}$$

This result shows that A is determined by the transverse component of J. Because of this, the Coulomb gauge is also known as the "transverse gauge." In addition, using (10.321), we can rewrite (10.301) as

$$\nabla^2 A - \frac{1}{c^2}\frac{\partial^2 A}{\partial t^2} = -\mu_0 J - \mu_0\nabla\int_V \frac{\nabla'\cdot J(r',t)}{4\pi R}dV(r') \tag{10.325}$$

Clearly, this is a nonhomogeneous wave equation for A. Suppose the charges only locate close to the origin. Thus, for $R\to\infty$ and $V\to\infty$, the charge density ρ will drop to zero and by (10.302) φ also drops to zero. Thus, for $R\to\infty$, we have

$$\nabla^2 A - \frac{1}{c^2}\frac{\partial^2 A}{\partial t^2} = -\mu_0 J \tag{10.326}$$

Thus, the wave functions of A radiate with distance. Therefore, the scalar potential φ in the Coulomb gauge only contributes to near fields, and the transverse radiating fields are given by the vector potential A alone. For this reason, the Coulomb gauge is also known as the "radiation gauge."

10.15.3 General Solution for Poisson Equation

We have seen from (10.300) that the scalar potential satisfies the Poisson equation if we adopt the Coulomb gauge. In this section, we will establish the general solution for the Poisson equation using Green's function method for a finite region with boundary conditions (see Chau, 2018). We start with the following Gauss divergence theorem for a vector function $\boldsymbol{\Phi}$.

$$\int_V \nabla\cdot\boldsymbol{\Phi}dV = \oint_S \boldsymbol{\Phi}\cdot\boldsymbol{n}dA \tag{10.327}$$

Let us consider the following function

$$\boldsymbol{\Phi} = \phi\nabla\psi \tag{10.328}$$

The divergence of $\boldsymbol{\Phi}$ becomes

$$\nabla\cdot\boldsymbol{\Phi} = \nabla\cdot(\phi\nabla\psi) = \phi\nabla^2\psi + \nabla\cdot\phi\nabla\cdot\psi \tag{10.329}$$

In addition, we note that

$$\phi\nabla\psi\cdot\boldsymbol{n} = \phi\frac{\partial\psi}{\partial n} \tag{10.330}$$

Substitution of (10.330) and (10.329) into (10.327) gives

$$\int_V (\phi\nabla^2\psi + \nabla\cdot\phi\nabla\cdot\psi)dV = \oint_S \phi\frac{\partial\psi}{\partial n}dA \tag{10.331}$$

This is Green's first identity (Chau, 2018). Interchanging the role of ϕ and ψ, we get

$$\int_V (\psi\nabla^2\phi + \nabla\cdot\phi\nabla\cdot\psi)dV = \oint_S \psi\frac{\partial\phi}{\partial n}dA \tag{10.332}$$

Subtracting (10.332) from (10.331) gives

$$\int_V (\phi\nabla^2\psi - \psi\nabla^2\phi)dV = \oint_S (\phi\frac{\partial\psi}{\partial n} - \psi\frac{\partial\phi}{\partial n})dA \tag{10.333}$$

This is Green's second identity (Chau, 2018). Now consider the following set of equations:

$$\nabla^2(\frac{1}{R}) = -4\pi\delta(\boldsymbol{x}-\boldsymbol{x}') \tag{10.334}$$

$$\nabla^2\varphi = -\frac{\rho(\boldsymbol{x})}{\varepsilon_0} \tag{10.335}$$

Physically, $1/R$ is the Green's function for unit excitation for the Poisson equation, whereas (10.335) is the governing equation for the scalar potential φ using the Coulomb gauge. Now we make the following substitutions:

$$\phi = \varphi, \quad \psi = \frac{1}{R} = \frac{1}{|\boldsymbol{x}-\boldsymbol{x}'|} \tag{10.336}$$

Substitution of (10.336) into (10.333) gives

$$\int_V (\varphi\nabla^2(\frac{1}{R}) - \frac{1}{R}\nabla^2\varphi)dV = \oint_S [\varphi\frac{\partial}{\partial n}(\frac{1}{R}) - \frac{1}{R}\frac{\partial\varphi}{\partial n}]dA \tag{10.337}$$

Note that

$$\int_V \varphi\nabla^2(\frac{1}{R})dV = -\int_V \varphi 4\pi\delta(\boldsymbol{x}-\boldsymbol{x}')dV = -4\pi\varphi(\boldsymbol{x}) \tag{10.338}$$

In view of (10.338) and (10.335), we finally obtain

$$\varphi(\boldsymbol{x}) = \frac{1}{4\pi\varepsilon_0}\int_V \frac{\rho(\boldsymbol{x}')}{R}dV' + \frac{1}{4\pi}\oint_S [\frac{1}{R}\frac{\partial\varphi}{\partial n} - \varphi\frac{\partial}{\partial n}(\frac{1}{R})]dA \tag{10.339}$$

This is the solution to the Poisson equation. If ρ is inside V, and φ and $\partial\varphi/\partial n$ on the boundary S are known, according to (10.339), we can find φ everywhere in the domain. As we mentioned earlier, for the case that $R \to \infty$, we have φ and $\partial\varphi/\partial n \to 0$, thus we recover the case for infinite domain:

$$\varphi(\boldsymbol{x}) = \frac{1}{4\pi\varepsilon_0}\int_V \frac{\rho(\boldsymbol{x}')}{R} dV' \tag{10.340}$$

If we set $\rho = 0$ in (10.339), we recover the formula reported in (8.52) of Chau (2018). Equation (10.339) can also be obtained by substituting (10.335) into (8.55) of Chau (2018).

10.15.4 Single- and Doble-Layer Potentials

As discussed in Example 8.1 of Chau (2018), physically the single-layer potential (or surface density charges) and double-layer potential are given respectively by

$$\sigma = \frac{1}{4\pi}\frac{\partial\varphi}{\partial n}, \quad \mu = -\frac{\varphi}{4\pi} \tag{10.341}$$

Note that φ and $\partial\varphi/\partial n$ cannot be given on the same boundary. If only φ is given on the whole boundary, it is the Dirichlet problem; if only $\partial\varphi/\partial n$ is given on the whole boundary, it is the Dirichlet problem; otherwise, it is a mixed boundary problem.

10.16 KIRCHHOFF INTEGRAL FORMULA FOR WAVES

We have seen that both the Lorenz gauge and Coulomb gauge lead to a nonhomogeneous wave equation. In this section, we will consider the Kirchhoff integral formula for harmonic waves. In particular, we consider the following nonhomogeneous wave equation:

$$(\nabla^2 - \varepsilon\mu\frac{\partial^2}{\partial t^2})\psi(\boldsymbol{r},t) = f(\boldsymbol{r},t) \tag{10.342}$$

Any time-dependent wave function can be represented by superposition of harmonic waves. Thus, we assume

$$\psi(\boldsymbol{r},t) = \psi(\boldsymbol{r})e^{i\omega t}, \quad f(\boldsymbol{r},t) = f(\boldsymbol{r})e^{i\omega t} \tag{10.343}$$

Substitution of (10.343) into (10.342) results in

$$(\nabla^2 + k^2)\psi(\boldsymbol{r}) = f(\boldsymbol{r}) \tag{10.344}$$

where

$$k^2 = \omega^2\mu\varepsilon = \omega^2 / c^2 \tag{10.345}$$

Equation (10.344) is the Helmholtz equation. Now consider the Green's function problem with spherical symmetry:

$$(\nabla^2 + k^2)G(\boldsymbol{r},\boldsymbol{r}') = -\delta(\boldsymbol{r}-\boldsymbol{r}') \tag{10.346}$$

To simplify the problem, we set the source point at the origin and rewrite (10.346) as:

$$\frac{1}{R}\frac{d^2}{dR^2}(RG)+k^2G=-\delta(R) \tag{10.347}$$

For $R \neq 0$, we can rearrange (10.347) as

$$\frac{d^2}{dR^2}(RG)+k^2(RG)=0 \tag{10.348}$$

The solutions of (10.348) are

$$G=\frac{A}{R}e^{\pm ikR} \tag{10.349}$$

This is the three-dimensional Green's function for the Helmholtz equation. To consider the effect at $R \to 0$, we integrate (10.347) over a spherical surface about the origin:

$$\int_V (\nabla^2 G+k^2G)dV=-\int_V \delta(R)dV \tag{10.350}$$

For $kR << 1$ and $R \to 0$, we have

$$G \approx \frac{A}{R} \tag{10.351}$$

With this approximation, (10.350) becomes

$$A\int_V [\nabla^2(\frac{1}{R})+\frac{k^2}{R}]dV=-1 \tag{10.352}$$

We note that the second term on the left is

$$\int_V \frac{k^2}{R}dV=\int_V \frac{k^2}{R}R^2\sin\theta dRd\theta d\phi=\int_V k^2R\sin\theta dRd\theta d\phi \tag{10.353}$$

Thus, this integral goes to zero as $R \to 0$. Thus, the left-hand side of (10.352) becomes

$$A\int_V \nabla^2(\frac{1}{R})dV=A\int_V -4\pi\delta(R)dV=-4\pi A \tag{10.354}$$

Substitution of (10.354) into (10.352) gives

$$A=\frac{1}{4\pi} \tag{10.355}$$

Finally, Green's function becomes

$$G(\boldsymbol{r},\boldsymbol{r}')=\frac{e^{\pm ikR}}{4\pi R} \tag{10.356}$$

where R is the distance between the observation point and source point. The positive sign represents an outgoing wave whereas the minus sign represents an incoming wave. We now apply the Fourier transform for the function ψ:

$$\psi(\boldsymbol{r},t)=\frac{1}{2\pi}\int_{-\infty}^{\infty}\psi(\boldsymbol{r},\omega)e^{-i\omega t}d\omega \tag{10.357}$$

$$\psi(\boldsymbol{r},\omega)=\int_{-\infty}^{\infty}\psi(\boldsymbol{r},t)e^{i\omega t}dt \tag{10.358}$$

In a sense, this Fourier transform pair is a mathematical expression for saying that the arbitrary time function can be represented as a sum of harmonic time series of different ω. Using this approach, the wave equation can be written as a function of frequency ω:

$$(\nabla^2 + k^2)\psi(\boldsymbol{r},\omega) = f(\boldsymbol{r},\omega) \tag{10.359}$$

Green's function becomes

$$G(\boldsymbol{r},\boldsymbol{r}',t,t') = \frac{1}{2\pi}\int_{-\infty}^{\infty} G(\boldsymbol{r},\boldsymbol{r}')e^{-i\omega(t-t')}d\omega \tag{10.360}$$

Substitution of (10.356) into (10.360) results in

$$G(\boldsymbol{r},\boldsymbol{r}',t,t') = (\frac{1}{4\pi R})\frac{1}{2\pi}\int_{-\infty}^{\infty} e^{-i\omega(t-t'\mp kR/\omega)}d\omega \tag{10.361}$$

As proved in (8.216) of Chau (2018), the Fourier integral representation of the delta function is

$$\delta(t) = \frac{1}{2\pi}\int_{-\infty}^{\infty} e^{i\omega t}d\omega \tag{10.362}$$

Thus, Green's function becomes

$$G(\boldsymbol{r},\boldsymbol{r}',t,t') = \frac{\delta(t'-t \pm kR/\omega)}{4\pi R} \tag{10.363}$$

The retarded Green's function is given by

$$G^{(+)}(\boldsymbol{r},\boldsymbol{r}',t,t') = \frac{\delta(t'-\{t-kR/\omega\})}{4\pi R} \tag{10.364}$$

This represents an outgoing wave, whereas the advanced Green's function is given by

$$G^{(-)}(\boldsymbol{r},\boldsymbol{r}',t,t') = \frac{\delta(t'-\{t+kR/\omega\})}{4\pi R} \tag{10.365}$$

Clearly, for the present problem we should look at the outgoing waves or

$$G(\boldsymbol{r},\boldsymbol{r}',t,t') = \frac{\delta(t'-[t-kR/\omega])}{4\pi R} \tag{10.366}$$

Now we can apply Green's function method given in (10.333) by using the following identification:

$$\phi \leftarrow G(\boldsymbol{r},\boldsymbol{r}'), \quad \psi \leftarrow \psi(\boldsymbol{r}',\omega) \tag{10.367}$$

Using (10.367), we can rewrite (10.333) as

$$\begin{aligned}&\int_V (G(\boldsymbol{r},\boldsymbol{r}')\nabla^2\psi(\boldsymbol{r}',\omega) - \psi(\boldsymbol{r}',\omega)\nabla^2 G(\boldsymbol{r},\boldsymbol{r}'))dV \\ &= \oint_S (G(\boldsymbol{r},\boldsymbol{r}')\frac{\partial\psi(\boldsymbol{r}',\omega)}{\partial n} - \psi(\boldsymbol{r}',\omega)\frac{\partial G(\boldsymbol{r},\boldsymbol{r}')}{\partial n})dA\end{aligned} \tag{10.368}$$

Substituting (10.346) and (10.359) into (10.368), we get

$$\begin{aligned}&\int_V (G(\boldsymbol{r},\boldsymbol{r}')f(\boldsymbol{r}',\omega)dV' + \psi(\boldsymbol{r},\omega) \\ &= \oint_S (G(\boldsymbol{r},\boldsymbol{r}')\frac{\partial\psi(\boldsymbol{r}',\omega)}{\partial n} - \psi(\boldsymbol{r}',\omega)\frac{\partial G(\boldsymbol{r},\boldsymbol{r}')}{\partial n})dA\end{aligned} \tag{10.369}$$

Rearranging (10.369), we find

$$\psi(\boldsymbol{r},\omega) = -\int_V (G(\boldsymbol{r},\boldsymbol{r}')f(\boldsymbol{r}',\omega)dV' + \oint_S (G(\boldsymbol{r},\boldsymbol{r}')\frac{\partial\psi(\boldsymbol{r}',\omega)}{\partial n} - \psi(\boldsymbol{r}',\omega)\frac{\partial G(\boldsymbol{r},\boldsymbol{r}')}{\partial n})dA \tag{10.370}$$

Substituting (10.356) into (10.370), we obtain

$$\psi(\boldsymbol{r},\omega) = -\int_V \frac{e^{ikR}}{4\pi R} f(\boldsymbol{r}',\omega)dV' + \oint_S [\frac{e^{ikR}}{4\pi R}\frac{\partial\psi(\boldsymbol{r}',\omega)}{\partial n} - \psi(\boldsymbol{r}',\omega)\frac{\partial}{\partial n}(\frac{e^{ikR}}{4\pi R})]dA \tag{10.371}$$

Thus, using (10.343), we find

$$\psi(\boldsymbol{r}) = -\int_V \frac{e^{ikR}}{4\pi R} f(\boldsymbol{r}')dV' + \oint_S [\frac{e^{ikR}}{4\pi R}\frac{\partial\psi(\boldsymbol{r}')}{\partial n} - \psi(\boldsymbol{r}')\frac{\partial}{\partial n}(\frac{e^{ikR}}{4\pi R})]dA \tag{10.372}$$

If $f = 0$, we have

$$\psi(\boldsymbol{r}) = -\frac{1}{4\pi}\oint_S [\psi(\boldsymbol{r}')\nabla'(\frac{e^{ikR}}{R}) - \frac{e^{ikR}}{R}\nabla'\psi(\boldsymbol{r}')]\cdot \boldsymbol{n}dA \tag{10.373}$$

Therefore, for given ψ and $\partial\psi/\partial n$ on the boundary, (10.373) can be used to evaluate $\psi(\boldsymbol{r})$ everywhere. We will now consider the radiation condition that needs to be satisfied by the wave field. For an infinite domain with two spherical boundaries S_1 and S_2, the source point is on S_2 shown by the dotted line and the observation point is fixed within the domain S_1 (see Figure 10.5). For such a domain, the integral given in (10.373) can be split into two parts:

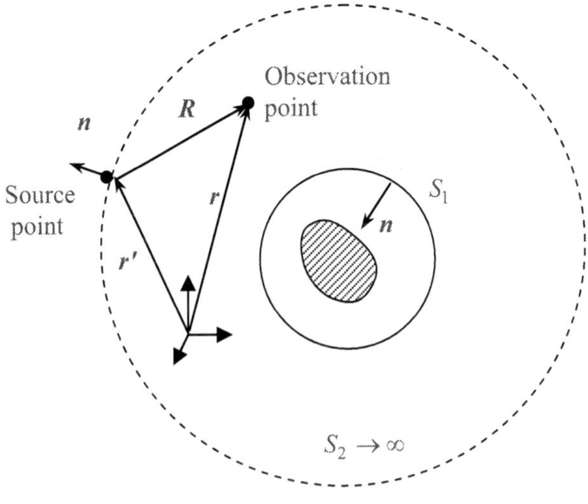

Figure 10.5 Domain for Kirchhoff integral formula for waves

$$\psi(\mathbf{r}) = -\frac{1}{4\pi}\oint_{S_1}[\psi(\mathbf{r}')\nabla'(\frac{e^{ikR}}{R}) - \frac{e^{ikR}}{R}\nabla'\psi(\mathbf{r}')]\cdot\mathbf{n}dA$$
$$-\frac{1}{4\pi}\oint_{S_2}[\psi(\mathbf{r}')\nabla'(\frac{e^{ikR}}{R}) - \frac{e^{ikR}}{R}\nabla'\psi(\mathbf{r}')]\cdot\mathbf{n}dA \quad (10.374)$$

For boundary S_2, we have the second integral in (10.374) becoming

$$I_2 = \oint_{S_2}[\psi(\mathbf{r}')\frac{\partial}{\partial R}(\frac{e^{ikR}}{R}) - \frac{e^{ikR}}{R}\frac{\partial}{\partial R}\psi(\mathbf{r}')]dA \quad (10.375)$$

Note that we have replaced the normal derivative by the R derivative. Note further that

$$\psi\frac{\partial}{\partial R}(\frac{e^{ikR}}{R}) = \psi ik\frac{e^{ikR}}{R} - \frac{1}{R^2}\psi e^{ikR} = \frac{\psi}{R}e^{ikR}(ik - \frac{1}{R}) \quad (10.376)$$

As $R \to \infty$, we can use (10.376) to get

$$I_2 = \oint_{S_2\to\infty}[\frac{\psi}{R}e^{ikR}(ik - \frac{1}{R}) - \frac{e^{ikR}}{R}\frac{\partial\psi}{\partial R}]R^2\sin\theta d\theta d\phi = 0 \quad (10.377)$$

This gives the radiation condition (i.e., at infinity this integral must approach zero). It can be further simplified as:

$$I_2 = \oint_{S_2\to\infty}[\psi Rik - \psi - R\frac{\partial\psi}{\partial R}]e^{ikR}\sin\theta d\theta d\phi = 0 \quad (10.378)$$

For radiating fields, we must have

$$\psi \sim \frac{e^{ikR}}{R}, \quad R \to \infty \quad (10.379)$$

Using this condition, we require

$$\lim_{R\to\infty} R(\psi ik - \frac{\partial\psi}{\partial R}) = 0 \quad (10.380)$$

This condition was derived by Sommerfeld (1949) and is called Sommerfeld's radiation condition. More generally, for n-dimension waves, we have

$$\lim_{R\to\infty} R^{\frac{n-1}{2}}(ik\psi - \frac{\partial\psi}{\partial R}) = 0 \quad (10.381)$$

This condition ensures that there are only outgoing waves. Consequently, (10.374) is reduced to

$$\psi(\mathbf{r}) = -\frac{1}{4\pi}\oint_{S_1}[\psi(\mathbf{r}')\nabla'(\frac{e^{ikR}}{R}) - \frac{e^{ikR}}{R}\nabla'\psi(\mathbf{r}')]\cdot\mathbf{n}dA \quad (10.382)$$

Note that

$$\nabla'(\frac{e^{ikR}}{R}) = -\nabla(\frac{e^{ikR}}{R}) = -\frac{\partial}{\partial R}(\frac{e^{ikR}}{R})\frac{\mathbf{R}}{R} = -(\frac{ike^{ikR}}{R} - \frac{e^{ikR}}{R^2})\frac{\mathbf{R}}{R}$$
$$= -\frac{e^{ikR}}{R}ik(1 + \frac{i}{kR})\frac{\mathbf{R}}{R} \quad (10.383)$$

Substitution of (10.383) into (10.382) leads to the so-called Kirchhoff integral formula for harmonic waves

$$\psi(\boldsymbol{r}) = \frac{1}{4\pi}\oint_{S_1} \frac{e^{ikR}}{R}[\nabla'\psi(\boldsymbol{r}') + ik(1+\frac{i}{kR})\frac{\boldsymbol{R}}{R}\psi]\cdot \boldsymbol{n}dA \tag{10.384}$$

10.17 SUMMARY AND FURTHER READING

In this chapter, we have summarized various mathematical theories leading to the solution of the well-known Maxwell's equations. Both the microscopic and macroscopic Maxwell's equations are presented, in both SI and Gaussian units. They are also expressed in differential forms as well as integral forms. Maxwell equations demonstrate that there is coupling between the magnetic and electric fields if both of them are functions of time, whereas such coupling disappears in case of static problems. Following the idea of Housner (2002), we propose a new proof that the Lorentz force law can be deduced from Maxwell equations, instead of treating it as an additional law to Maxwell equations to form the complete set of electromagnetic dynamics. However, this viewpoint is subject to further discussion.

The introduction of Maxwell's scalar and vector potentials leads to two coupled PDEs. This set of coupled PDEs can further be simplified by observing the gauge freedom in choosing these potentials. The gauge freedom is the freedom of adding various functions to the potentials such that the remaining electric and magnetic fields remain unchanged. We found that the coupled system can be reduced to two nonhomogeneous wave equations for the case of the Lorenz gauge and to one Poisson equation and one nonhomogeneous vector wave equation for the case of the Coulomb gauge. We also demonstrated why the Coulomb gauge is also known as the radiation gauge or the transverse gauge. For the case of the Lorenz gauge, we discuss the solution of two nonhomogeneous wave equations and the associated Jefimenko equations for the electric and magnetic fields (which are actually simplified forms for $\boldsymbol{B}$ and $\boldsymbol{E}$ in terms of the solution of nonhomogeneous wave equations). Using constitutive models, we showed the wave nature of the electric and magnetic fields of moving charges. For the Lorenz gauge, we discussed the introduction of the electric Hertz vector and the magnetic Hertz vector, and for each case only one nonhomogeneous wave equation is needed. For more general cases, we can superimpose these two types of waves. For problems with spherical symmetry, Debye potentials for transverse electric (TE) waves and for transverse magnetic (TM) waves are discussed. In view of the importance of the Poisson equation and nonhomogeneous wave equations in Maxwell's equations, we introduce the Green's function method for the Poisson equation for a finite domain with prescribed boundary conditions, and the Kirchhoff integral formula for harmonic waves.

The idea of gauge invariance apparently originated from the study of the Maxwell equations but its application to physics has been extended to the study of electroweak and strong interactions of elementary particles in quantum mechanics. More discussions of gauge theory, we refer to Jackson and Okun (2001) and Jackson (2002). In particular, in addition to the most popular Coulomb and Lorenz gauges, other gauges have been proposed. They include the Poincare gauge (see Problems 10.1 to 10.3), velocity gauge, Kirchhoff gauge, Gibbs gauge, and static-voltage gauge.

The mathematical theory for solving Maxwell equations is a not an easy subject. However, there are many excellent textbooks available in the literature. Although Maxwell's original textbook was made available in a Dover edition (Maxwell, 1954), it was not recommended for beginners because it did not use the modern tensor form and became obsolete in a sense. Griffith (1981) is a popular undergraduate- to intermediate-level textbook used by many, and Feynman et al. (1964a) is another. Fleisch (2008) and Purcell (1985) give an elementary introduction to electrodynamics and Maxwell equations. From the German school, Sommerfeld (1952) again provided a solid coverage in electrodynamics. Jackson (1999) is one of the standard graduate-level textbooks although some readers complained that it is not easy to read. Zangwill (2013) provided a modern view of electrodynamics.

10.18 PROBLEMS

Problem 10.1 Find the electric and magnetic fields for the following vector and scalar potentials:

$$A = -\frac{1}{2}\boldsymbol{r}\times\boldsymbol{B}_0, \quad \varphi = -\boldsymbol{r}\cdot\boldsymbol{E}_0 \tag{10.385}$$

where $\boldsymbol{B}_0$ and $\boldsymbol{E}_0$ are constant vectors.

Hint: Recall the following identity first

$$\nabla\times(\boldsymbol{r}\times\boldsymbol{B}) = (\boldsymbol{B}\cdot\nabla)\boldsymbol{r} - (\boldsymbol{r}\cdot\nabla)\boldsymbol{B} + \boldsymbol{r}(\nabla\cdot\boldsymbol{B}) - \boldsymbol{B}(\nabla\cdot\boldsymbol{r}) \tag{10.386}$$

Ans:

$$\boldsymbol{B} = \boldsymbol{B}_0, \quad \boldsymbol{E} = \boldsymbol{E}_0 \tag{10.387}$$

Problem 10.2

(i) Show that

$$\frac{d\boldsymbol{A}(\lambda\boldsymbol{r})}{d\lambda} = \frac{1}{\lambda}(\boldsymbol{r}\cdot\nabla)\boldsymbol{A}(\lambda\boldsymbol{r}) \tag{10.388}$$

(ii) Show that the vector and scalar potentials can be found in terms of the electric and magnetic fields by the following so-called Poincare gauge (Zangwill, 2013):

$$\boldsymbol{A}(\boldsymbol{r},t) = -\int_0^1 \lambda\boldsymbol{r}\times\boldsymbol{B}(\lambda\boldsymbol{r},t)d\lambda \tag{10.389}$$

$$\varphi(\boldsymbol{r},t) = -\boldsymbol{r}\cdot\int_0^1 \boldsymbol{E}(\lambda\boldsymbol{r},t)d\lambda \tag{10.390}$$

Problem 10.3 Show that for the vector potential given in (10.389), we have

$$\boldsymbol{r}\cdot\boldsymbol{A}(\boldsymbol{r},t) = 0 \tag{10.391}$$

This is the condition of the so-called Poincare gauge.

CHAPTER ELEVEN

Quantum Mechanics and Schrödinger Equation

11.1 INTRODUCTION

Quantum mechanics is one of the most difficult topics for first-time learners because it differs from our daily experience. Quantum means "how much" in Latin. It was originally proposed for modeling the wave behavior of moving particles. Quantum mechanics can only provide the probability of finding moving particles at the atomic level. The Schrödinger equation provides the information of a wavefunction, which gives only the quantitative prediction of probability of finding a moving particle at a particular location. This differs from what we learn from classical mechanics. Heisenberg's uncertainty principle, one of the backbones of quantum mechanics, states that we cannot find the position and momentum of any moving particle accurately at the same time. In quantum mechanics, an individual event cannot be determined, and only the average value (or the expectation value) of a quantity (like position and momentum) can be determined. If you find this subject difficult, you are not alone. Einstein never believed quantum mechanics. Even Richard Feynman, one of the main contributors to quantum mechanics, said: "I think I can safely say that nobody understands quantum mechanics." The now-accepted probabilistic view of quantum mechanics proposed by Max Born was opposed by Schrödinger himself. Although quantum mechanics appears as a strange theory, it can provide a powerful tool to explain many observed phenomena at the atomic level. Therefore, it has been well accepted as a valid theory in physics.

Historically, quantum mechanics originated from the physical observation that energy emitted in a black body is in quantum form (i.e., a multiple of some threshold quantum energy). This quantum energy is found proportional to Planck's constant. de Broglie's theory of wave-particle duality of matter forms the backbone of modern quantum mechanics. In fact, Schrödinger derived his renowned Schrödinger equation based on both the quantum form of energy as well as the wavelength of de Broglie's wave (or called the matter wavelength).

Modeling of quantum mechanics using the Schrödinger equation is only one of the choices. Born, Jordan, and Heisenberg provided an alternative methodology by proposing matrix mechanics. Subsequently, Schrödinger showed that the matrix mechanics is equivalent to the Schrödinger equation. A more recent approach was the path integral approach developed by Richard Feynman in 1950s. Unlike the Schrödinger equation, statistical mechanics was built into the path integral method in the formulation at the very beginning.

The wave and particle behaviors of light have been proposed by Huygens and Newton respectively. It is the wave behavior of matter (or more precisely moving particles) that provided the basis for the wave-like Schrödinger equation for quantum mechanics. For "matter waves," the momentum of the de Broglie wave of matter relates to the wavelength. A major difference between the classical mechanics (Newtonian mechanics) and quantum mechanics is the uncertain nature of the quantum state. Only the probability of finding moving particles (such as photons) at a particular position and at a particular time can be found using quantum mechanics through the wavefunction. In a sense, quantum mechanics can only provide information about moving particles in a statistical sense. This kind of indeterministic nature has been ill-received by many, including Albert Einstein. Einstein did not believe quantum mechanics, although he did not come up with a better theory than quantum mechanics.

One distinct feature of the present chapter from most textbooks on quantum mechanics is that full mathematical details in deriving the exact solution of the wavefunction are covered. These steps are normally neglected in textbooks in quantum mechanics, especially for the radial dependent function that involves the associated Laguerre polynomials. The Laguerre polynomials and associated Laguerre polynomials are relatively less known. For example, they are not covered in the classic mathematical handbook of Abramowitz and Stegun (1964), and also not covered in the classic book of analysis by Whittaker and Watson (1927). Two subsections will be devoted exclusively to discussing the essential properties of the Laguerre polynomials and the associated Laguerre polynomials, including the orthogonality of the associated Laguerre polynomials. It is these properties that eventually lead to the quantized state of the admissible energy levels of electrons. We will also demonstrate that this result agrees with the simple hydrogen atom model proposed by Bohr.

11.2 BLACK BODY RADIATION AND QUANTIZED ENERGY

The black body radiation problem was posed by Kirchhoff in 1859, aiming to find how the radiation energy of a black body depends on the frequency of radiation and the temperature of the black body. Experiments were conducted to solve the problem. In 1896, Wien proposed the following formula

$$u(\nu) = \frac{8\pi h\nu^3}{c^3} e^{-h\nu/kT} \tag{11.1}$$

where u is the radiation energy density or radiation energy per mass, ν is the frequency of the radiation, T is the temperature of the black body in Kelvin, k is Boltzmann's constant (or 1.382×10^{-23}J/K), h is Planck's constant (6.626×10^{-34}J·s) and c is the speed of light. This formula accurately predicts the radiation at high frequency but fails to predict radiation at low frequency.

In 1900, Rayleigh proposed another formula for black body radiation based upon both physical argument and empirical facts. In terms of radiation energy density or radiation energy per mass u, the Rayleigh-Jeans radiation formula can be expressed as:

$$u(\nu) = \frac{8\pi kT}{c^3}\nu^2 \tag{11.2}$$

where ν is the frequency of the radiation, T is the temperature of the black body, k is Boltzmann's constant (or 1.382×10^{-23}J/K), and c is the speed of light. This Rayleigh-Jeans law agreed with experiments well for low frequency but does not perform well for high-frequency radiation. In the same year, Planck proposed the well-known Planck's law, which can be expressed as:

$$u(\nu) = \frac{8\pi h}{c^3}\frac{\nu^3}{e^{h\nu/kT}-1} \tag{11.3}$$

As we can see, for high frequency $\nu \to \infty$, we have

$$e^{h\nu/kT} - 1 \approx e^{h\nu/kT} \tag{11.4}$$

Substitution of (11.4) into (11.3) gives Wien's law as the high-frequency limit. For the low-frequency limit $\nu \to 0$, we have

$$e^{h\nu/kT} - 1 = 1 + \frac{h\nu}{kT} + \ldots - 1 \approx \frac{h\nu}{kT} \tag{11.5}$$

Substitution of (11.5) into (11.3) recovers the Rayleigh-Jeans law as the low-frequency limit.

The central assumption behind Planck's derivation was the supposition, now known as the Planck postulate, that electromagnetic energy could be emitted only in quantized form. That is, energy radiation could only be a multiple of an elementary unit:

$$E = h\nu \tag{11.6}$$

where h is the Planck's constant and ν is the frequency of the radiation. This quantization of energy is a major breakthrough in our understanding in the atomic world. Planck was awarded the Nobel Prize in Physics in 1918 for his revolutionary idea. Thus, this value of E is the quantum magnitude of the energy emission. It is also this idea that the modern quantum mechanics is based upon.

Let us demonstrate how Planck arrived at the quantum form of energy given in (11.6). By 1900, Planck already knew, by fitting experimental data, that the empirical formula for the change of radiation energy per unit volume with respect to the change in frequency is:

$$\frac{dU(\nu,T)}{d\nu} = \frac{8\pi h}{c^3}\frac{\nu^3}{(e^{h\nu/kT}-1)} \tag{11.7}$$

Theoretically, for the frequency interval ν to $\nu+d\nu$, Planck formulated the radiation energy density increment per unit volume U as

$$dU(\nu,T) = \frac{4\pi\nu^2}{c^3}d\nu \times 2 \times \bar{\bar{E}} \tag{11.8}$$

The first factor on the right-hand side of (11.8) is the number of degrees of freedom per unit volume in the electromagnetic radiation in the frequency range ν to $\nu+d\nu$, the second factor 2 counts the number of polarizations of the emitted photons, and the last factor is the average energy per degree of freedom. Based on the Maxwell-Boltzmann distribution of the energy, Planck found

$$\bar{E} = \frac{\sum_{E} E e^{-E/kT}}{\sum_{E} e^{-E/kT}} \tag{11.9}$$

If the energy for the black body radiation is continuous, we can replace the summation in (11.9) by integration. The numerator of (11.9) becomes:

$$\begin{aligned}
\sum_{E} E e^{-E/kT} &= \int_0^{\infty} E e^{-E/kT} dE = -kT \int_0^{\infty} E de^{-E/kT} \\
&= -kT\{E\, e^{-E/kT}\Big|_0^{\infty} - \int_0^{\infty} e^{-E/kT} dE\} \\
&= -(kT)^2 \int_0^{\infty} e^{-E/kT} d(-\frac{E}{kT}) \qquad (11.10) \\
&= -(kT)^2 [0-1] \\
&= (kT)^2
\end{aligned}$$

Similarly, the denominator of (11.9) is obviously:

$$\begin{aligned}
\sum_{E} e^{-E/kT} &= \int_0^{\infty} e^{-E/kT} dE \\
&= -kT \int_0^{\infty} de^{-E/kT} \qquad (11.11) \\
&= -kT[0-1] \\
&= kT
\end{aligned}$$

Substitution of (11.10) and (11.11) into (11.9) gives

$$\bar{E} = kT \tag{11.12}$$

Employing this result, we find that (11.8) becomes

$$dU(v,T) = \frac{8\pi v^2 kT}{c^3} dv \tag{11.13}$$

This is, in fact, the Rayleigh-Jeans formula given in (11.2) by recognizing:

$$\frac{dU(v,T)}{dv} = u(v,T) \tag{11.14}$$

This clearly does not agree with experimental fitting obtained in (11.7). Planck hypothesized that the energy is in multiple of quantum energy, or

$$E = nhv \tag{11.15}$$

where $n = 1,2,....$ Since the energy is assumed to be discrete, we cannot replace it by integration. More specifically, substitution of (11.15) into (11.9) gives

$$\bar{E} = \frac{\sum_{n=0}^{\infty} nhv e^{-nhv/kT}}{\sum_{n=0}^{\infty} e^{-nhv/kT}} = \frac{hv(e^{-hv/kT} + 2e^{-2hv/kT} + 3e^{-3hv/kT} + ...)}{(1 + e^{-hv/kT} + e^{-2hv/kT} + e^{-3hv/kT} + ...)} \tag{11.16}$$

We see the series in the numerator is an arithmetic-geometric series and can be summed analytically as (Spiegel, 1964):

$$a+(a+d)r+(a+2d)r^2+...+[a+(n-1)d]r^{n-1}=\frac{a(1-r^n)}{1-r}+\frac{rd\{1-nr^{n-1}+(n-1)r^n\}}{(1-r)^2} \tag{11.17}$$

with

$$a=0, \quad d=1, \quad r=e^{-h\nu/kT} \tag{11.18}$$

Thus, the series in the numerator equals

$$e^{-h\nu/kT}+2e^{-2h\nu/kT}+3e^{-3h\nu/kT}+... = \lim_{n\to\infty}\frac{e^{-h\nu/kT}\{1-ne^{-(n-1)h\nu/kT}+(n-1)e^{-nh\nu/kT}\}}{(1-e^{-h\nu/kT})^2} \tag{11.19}$$

From L'Hôpital's rule, we have

$$\lim_{n\to\infty} ne^{-(n-1)h\nu/kT}=\lim_{n\to\infty}\frac{n}{e^{(n-1)h\nu/kT}}=\lim_{n\to\infty}\frac{1}{\frac{h\nu}{kT}e^{(n-1)h\nu/kT}}=0 \tag{11.20}$$

Similarly, we also have

$$\lim_{n\to\infty}(n-1)e^{-nh\nu/kT}=\lim_{n\to\infty}\frac{(n-1)}{e^{nh\nu/kT}}=\lim_{n\to\infty}\frac{1}{\frac{h\nu}{kT}e^{nh\nu/kT}}=0 \tag{11.21}$$

Substitution of (11.20) and (11.21) into (11.19) gives

$$e^{-h\nu/kT}+2e^{-2h\nu/kT}+3e^{-3h\nu/kT}+...=\frac{e^{-h\nu/kT}}{(1-e^{-h\nu/kT})^2} \tag{11.22}$$

The series in the denominator of (11.16) is a geometric series

$$a+ar+ar^2+ar^3+...=\frac{a}{1-r} \tag{11.23}$$

with

$$a=1, \quad r=e^{-h\nu/kT} \tag{11.24}$$

Thus, we have

$$1+e^{-h\nu/kT}+e^{-2h\nu/kT}+e^{-3h\nu/kT}+...=\frac{1}{1-e^{-h\nu/kT}} \tag{11.25}$$

Substitution of (11.22) and (11.25) into (11.16) gives

$$\bar{E}=\frac{h\nu}{e^{h\nu/kT}-1} \tag{11.26}$$

Combining (11.26) and (11.8), we finally get

$$\frac{dU(\nu,T)}{d\nu}=\frac{8\pi h}{c^3}\left(\frac{\nu^3}{e^{h\nu/kT}-1}\right) \tag{11.27}$$

This is exactly (11.7). Thus, the experimental data suggests that the black body radiation is emitting quanta of energy. This provides the basis of quantum mechanics. The predictions of Wien's law, the Rayleigh-Jeans law, and Planck's law are compared in Figure 11.1. We can see that Planck's law approaches the exact lower limit for $\nu \to 0$ predicted by Wien's law and the exact upper limit for $\nu \to \infty$ predicted by the Rayleigh-Jeans law for black body radiation.

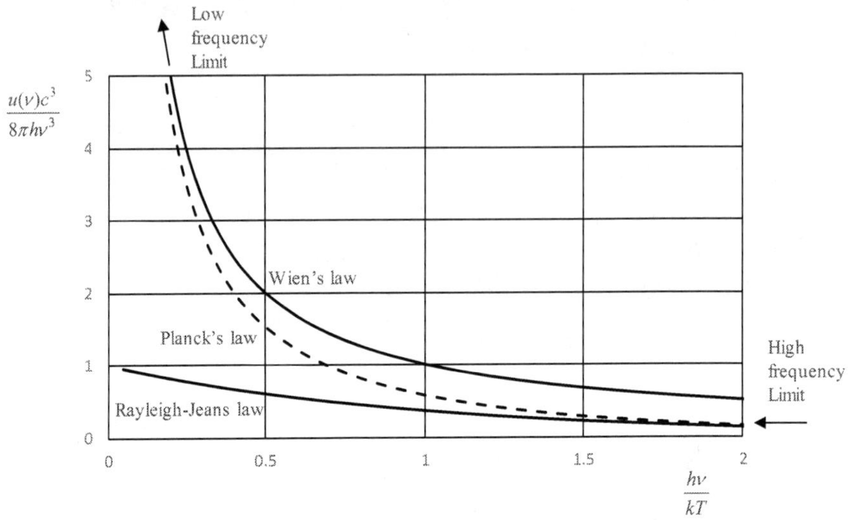

Figure 11.1 Comparison of Wien's law, Rayleigh-Jeans law, and Planck's law of black body radiation

11.3 SCHRÖDINGER EQUATION

11.3.1 One-Dimensional Schrödinger Equation

We now consider the wave property of a moving atomic particle. At the atomic level, the de Broglie wave theory asserts that there is a dual property of a moving particle, i.e., both wave and particle behaviors. De Broglie is a Nobel Prize winner in Physics in 1929. In particular, the wave function of a moving particle can be expressed in terms of velocity v, circular frequency ω, position x, frequency ν and time t as:

$$\Psi = Ae^{-i\omega(t-x/V)} \tag{11.28}$$

The frequency ν and circular frequency ω is related by:

$$\omega = 2\pi\nu \tag{11.29}$$

On the other hand, the velocity V can be related to the frequency ν as:

$$V = \lambda\nu \tag{11.30}$$

where λ is the wavelength of the wave function. Substitution of (11.30) into (11.28) gives

$$\Psi = Ae^{-i2\pi\nu(t-x/\lambda\nu)} = Ae^{-i2\pi(\nu t-x/\lambda)} \tag{11.31}$$

The quantum energy defined in (11.6) can be rewritten in terms of a basic unit of angular momentum:

$$E = h\nu = 2\pi\hbar\nu \tag{11.32}$$

where

$$\hbar = \frac{h}{2\pi} \tag{11.33}$$

We now consider the de Broglie wave theory of particles such that the de Broglie wavelength is defined as

$$\lambda = \frac{h}{p} = \frac{2\pi\hbar}{p} \tag{11.34}$$

where the momentum of the particle p is defined as

$$p = mV \tag{11.35}$$

where m is the mass and V is the velocity of the particle. For the case of "light waves," (11.34) defines the wavelength of a photon of light. This is also known as the quantum theory of light. For the case of atomic particles, the de Broglie wavelength is defined as

$$\lambda = \frac{h}{mV}\sqrt{1-\frac{V^2}{c^2}} \tag{11.36}$$

We now use (11.32) and (11.34) to rewrite the frequency ν and wavelength λ in terms of the E and p as

$$\Psi = Ae^{-i2\pi(\frac{E}{2\pi\hbar}t-\frac{px}{2\pi\hbar})} = Ae^{-\frac{i}{\hbar}(Et-px)} \tag{11.37}$$

The two main features of this wave function Ψ is that we have used de Brogie wave theory to express its wavelength in terms of the momentum of the particle, and to express the frequency in terms of the quantized energy from black body radiation. This is the wave function of a free particle, and it provides the basis for deriving the Schrödinger equation. Differentiating this wave function with respect to x, we have

$$\frac{\partial^2\Psi}{\partial x^2} = A(\frac{ip}{\hbar})^2 e^{-i(Et-px)/\hbar} = -(\frac{p}{\hbar})^2\Psi \tag{11.38}$$

This can be rearranged as

$$p^2\Psi = -\hbar^2\frac{\partial^2\Psi}{\partial x^2} \tag{11.39}$$

The time differentiation of the wave function is

$$\frac{\partial\Psi}{\partial t} = -\frac{iE}{\hbar}Ae^{-i(Et-px)/\hbar} = -\frac{iE}{\hbar}\Psi \tag{11.40}$$

This can be rearranged as

$$E\Psi = -\frac{\hbar}{i}\frac{\partial\Psi}{\partial t} \tag{11.41}$$

For moving particles, the total energy E can also be evaluated as the kinetic energy plus the potential energy as:

$$E = \frac{p^2}{2m} + U(x,t) \tag{11.42}$$

where U is the potential energy from the influence of the rest of the universe on the moving particle. For the case of a hydrogen atom, U is the potential due to the electric field of the nucleus of the atom. Multiplying (11.42) by the wavefunction Ψ, we get

$$E\Psi = \frac{p^2\Psi}{2m} + U\Psi \tag{11.43}$$

Substitution of (11.41) and (11.39) into (11.43) gives

$$-\frac{\hbar}{i}\frac{\partial\Psi}{\partial t} = -\frac{\hbar^2}{2m}\frac{\partial^2\Psi}{\partial x^2} + U\Psi \tag{11.44}$$

This step of substitution, however, cannot be justified theoretically. The question is why it is appropriate to express frequency in terms of quantized energy and to express wavelength by using the de Broglie wave theory of matter. Equation (11.44) can be rewritten as

$$i\hbar\frac{\partial\Psi}{\partial t} = -\frac{\hbar^2}{2m}\frac{\partial^2\Psi}{\partial x^2} + U\Psi \tag{11.45}$$

If Ψ is real, the right-hand side of (11.45) is real, but then the left-hand side is imaginary. This is impossible; thus, Ψ must be complex. Therefore, Ψ is not a measurable quantity. The physical meaning of Ψ will be discussed in Section 11.3.3. Equation (11.45) is the 1-D Schrödinger equation, which can be rewritten into a slightly different form using the Hamiltonian operator:

$$i\hbar\frac{\partial\Psi}{\partial t} = \hat{H}\Psi \tag{11.46}$$

where the Hamiltonian operator is

$$\hat{H} = -\frac{\hbar^2}{2m}\frac{\partial^2}{\partial x^2} + U \tag{11.47}$$

The Schrödinger equation was derived by Erwin Schrödinger in 1926 in an attempt to describe the wave behavior of a moving particle. Schrödinger received the 1933 Nobel Prize in Physics on this work. This equation cannot be obtained from known principles in physics. This is a linear second order PDE with constant coefficients (although one of them is a complex constant). The derivation of it is also not totally justifiable. It turns out that the results from solving the Schrödinger equation agree remarkable well with experiments; thus, it has been generally accepted as a physical law. This is a situation that mathematical speculation prevails in the physical world. Note also that although we call Ψ a wave function, the Schrödinger equation differs from the traditional wave equation (see for example Chapter 9 of Chau, 2018). To most students, the most intimidating feature of the Schrödinger equation is the appearance of the imaginary constant i in its coefficient, and this is resulting from the wave type solution that we assumed in (11.28). In the next section, we will generalize the 1-D Schrödinger equation to the 3-D situation. After all, electrons are moving in the three-dimensional space around atoms.

11.3.2 Three-Dimensional Schrödinger Equation

The 1-D Schrödinger equation derived in (11.45) can be generalized to 3-D Schrödinger equation as:

$$i\hbar\frac{\partial \Psi}{\partial t} = -\frac{\hbar^2}{2m}\nabla^2\Psi + U\Psi \tag{11.48}$$

To prove (11.48), we can first extend (11.37) to the following form:

$$\Psi = A\exp[-\frac{i}{\hbar}(Et - \boldsymbol{p}\cdot\boldsymbol{x})] = A\exp\{-\frac{i}{\hbar}(Et - p_1x_1 - p_2x_2 - p_3x_3)\} \tag{11.49}$$

where $\boldsymbol{p}$ is the momentum vector. The differentiation of the wavefunction given in (11.49) gives

$$\frac{\partial^2\Psi}{\partial x_1^2} = (\frac{i}{\hbar}p_1)^2\Psi = -\frac{p_1^2}{\hbar^2}\Psi \tag{11.50}$$

Similarly, we have

$$\frac{\partial^2\Psi}{\partial x_2^2} = -\frac{p_2^2}{\hbar^2}\Psi \tag{11.51}$$

$$\frac{\partial^2\Psi}{\partial x_3^2} = -\frac{p_3^2}{\hbar^2}\Psi \tag{11.52}$$

Combining these results, we have the Laplacian of Ψ as:

$$\nabla^2\Psi = \frac{\partial^2\Psi}{\partial x_1^2} + \frac{\partial^2\Psi}{\partial x_2^2} + \frac{\partial^2\Psi}{\partial x_3^2} = -\frac{p_1^2 + p_2^2 + p_3^2}{\hbar^2}\Psi = -\frac{p^2}{\hbar^2}\Psi \tag{11.53}$$

By multiplying an appropriate factor, we get

$$-\frac{\hbar^2}{2m}\nabla^2\Psi = \frac{p^2}{2m}\Psi \tag{11.54}$$

The total energy is the sum of kinetic energy plus the potential energy

$$E = \frac{p^2}{2m} + U \tag{11.55}$$

Substitution of (11.55) into (11.54) gives

$$-\frac{\hbar^2}{2m}\nabla^2\Psi = (E - U)\Psi \tag{11.56}$$

This is the three-dimensional steady-state Schrödinger equation (note that this PDE is real). The time differentiation of (11.49) results in

$$\frac{\partial \Psi}{\partial t} = -\frac{iE}{\hbar}\Psi \tag{11.57}$$

This is equivalent to

$$i\hbar\frac{\partial \Psi}{\partial t} = E\Psi \tag{11.58}$$

Substitution of (11.55) and (11.54) into (11.58) gives

$$i\hbar\frac{\partial\Psi}{\partial t}=-\frac{\hbar^2}{2m}\nabla^2\Psi+U\Psi \tag{11.59}$$

This is the three-dimensional Schrödinger equation given in (11.48).

11.3.3 Wave Functions of Particles

In water waves, the elevation of the water surface appears as a wave function. In sound waves, pressure distribution in air appears as a wave function. In light waves, electric and magnetic fields propagate as a wave function. In quantum mechanics, the wave function in the Schrödinger equation is associated with the likelihood of finding the particle at a particular time and at a particular place. The wave function Ψ does not have a definite physical meaning in quantum mechanics, but the square of the absolute magnitude of Ψ is proportional to the probability of finding the particle at a particular place at a particular time. This physical interpretation of the wavefunction Ψ is given by Max Born in 1926, who was awarded the Nobel Prize in Physics in 1954. From (11.48), it is clear that Ψ is a complex function. Thus, the square of its magnitude is

$$|\Psi|^2=\Psi\bar{\Psi}=[\mathrm{Re}(\Psi)]^2+[\mathrm{Im}(\Psi)]^2 \tag{11.60}$$

If we set the value of (11.60) equals the probability P of finding the particle described by the wave function Ψ, we must have the following normalization:

$$\int_{-\infty}^{\infty}|\Psi|^2 dV=\int_{-\infty}^{\infty}P dV=1 \tag{11.61}$$

We also expect that the wave function Ψ is continuous and single-valued. Its partial derivatives with respect to spatial variables must also be continuous and single-valued. The wave function must decay to zero as x, y, and $z \to \pm\infty$. For particles that can only move along the x-axis, the probability of finding the particle between x_1 and x_2 is

$$P_{x_1x_2}=\int_{x_1}^{x_2}|\Psi|^2 dx \tag{11.62}$$

Since (11.59) is a second order PDE, we have two independent solutions:

$$\Psi=A\Psi_1+B\Psi_2 \tag{11.63}$$

Let us consider the case that an electron beam is passing two slits and is being diffracted as shown in Figure 11.2. The probability density for opening slit 1 only, slit 2 only, superposition of results of slit 1 and 2, and finally the actual probability density of a pair of slits opening are also shown in Figure 11.2.

Note that for slit 1 open only, we have the probability of electron beam intensity as:

$$P_1=|\Psi_1|^2=\Psi_1\bar{\Psi}_1 \tag{11.64}$$

Note that for slit 2 open only, we have the probability of electron beam intensity as:

$$P_2=|\Psi_2|^2=\Psi_2\bar{\Psi}_2 \tag{11.65}$$

If we assume the probability can be added, the probability of electron beam intensity for both slits open would have been:

$$P_1 + P_2 = |\Psi_1|^2 + |\Psi_2|^2 \tag{11.66}$$

However, in reality, superposition applies to the wave function not to its magnitude squared. Thus, we have

$$\Psi = \Psi_1 + \Psi_2 \tag{11.67}$$

The probability density is

$$\begin{aligned} P &= |\Psi|^2 = |\Psi_1 + \Psi_2|^2 = (\Psi_1 + \Psi_2)(\bar{\Psi}_1 + \bar{\Psi}_2) \\ &= \Psi_1\bar{\Psi}_1 + \Psi_2\bar{\Psi}_2 + \bar{\Psi}_1\Psi_2 + \Psi_1\bar{\Psi}_2 = P_1 + P_2 + \bar{\Psi}_1\Psi_2 + \Psi_1\bar{\Psi}_2 \end{aligned} \tag{11.68}$$

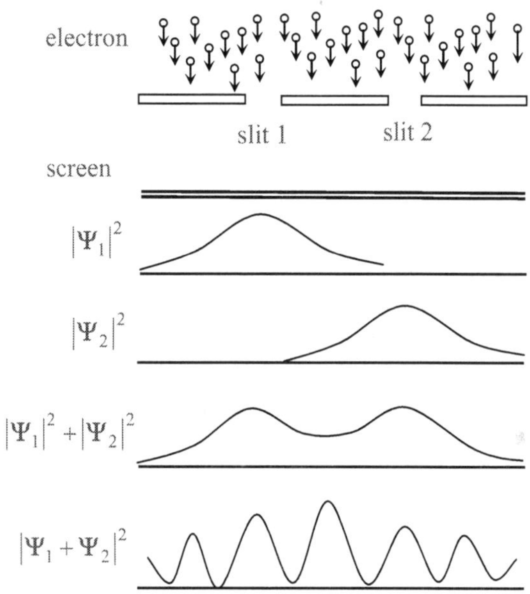

Figure 11.2 The probability of finding electrons in the two slit experiments

This result given in (11.68) is plotted as the last probability intensity shown in Figure 12.2. This also illustrates the wave nature of the beam of particles. The pattern resembles wave interference.

11.3.4 Expectation Values

In quantum mechanics, the wave property of a moving particle leads to very peculiar concepts. One of them is Heisenberg's uncertainty principle discovered in

1927 which was formulated under the suggestion of Wolfgang Pauli. Heisenberg received the Nobel Prize in Physics in 1932. Heisenberg remained in Germany during the Nazi period and was reportedly responsible for developing the atomic bomb for the Nazis. This news led to the US government sending the former Boston Red Sox catcher Moe Berg to shoot him during a lecture in Switzerland in 1944, but Berg did not shoot Heisenberg after the lecture because he was uncertain about their progress. In fact, not much progress was made by the end of World War II. The uncertainty principle can be stated as:

> "It is impossible to know both the exact position and exact momentum of an object at the same time."

Heisenberg initially argued that at the atomic level, any measurement process on a particle's position and momentum will inevitably change its position and momentum. Thus, they cannot be known at the same time. However, Bohr thought that the uncertainty is a result of the wave-particle duality, and has nothing to do with the measurement method. Eventually, Bohr's view was accepted by Heisenberg. Since a moving particle is regarded as a wave group, it is impossible to know its exact position and momentum at the same time. This view agrees with the physical meaning of the wave function that it is associated with the probability of finding the particle at a particular place and at a particular time. Once Ψ is solved from the Schrödinger equation, the wave function contains all information about the particle that is permitted by the uncertainty principle.

Consider the case of a particle that can only exist along the x-axis. If we would measure the positions of many particles described by the same wave function at a particular time t, we take the average of this position. This mean value can be expressed as an expectation value. In simple terms, after many observations, if we found a particle at position x_1 N_1 times, at position x_2 N_2 times, and so on, we have the mean value as:

$$\overline{x} = \frac{x_1 N_1 + x_2 N_2 + \ldots}{N_1 + N_2 + \ldots} = \frac{\sum x_i N_i}{\sum N_i} \tag{11.69}$$

Using the wave function, the probability of finding the particle at x_i can be evaluated as

$$P_i = |\Psi_i|^2 \, dx \tag{11.70}$$

where the point x_i is within the interval dx. Replacing the summation by the integration given in (11.69), we have the expectation value defined as

$$\langle x \rangle = \frac{\int_{-\infty}^{\infty} x|\Psi|^2 dx}{\int_{-\infty}^{\infty} |\Psi|^2 dx} \tag{11.71}$$

If the wave function has been normalized as given in (11.61), we have

$$\langle x \rangle = \int_{-\infty}^{\infty} x|\Psi|^2 dx \tag{11.72}$$

This procedure can be extended to find the expectation value of any function $G(x)$:

$$\langle G(x)\rangle = \int_{-\infty}^{\infty} G(x)|\Psi|^2 dx \tag{11.73}$$

However, this formula cannot be extended to find the expectation value of momentum for a specified position x and the expectation value of energy for a specified time t. These expectation values will be considered next.

11.3.5 Stationary State of Energy E

Recalling (11.37), we can express the wavefunction as

$$\Psi(x,t) = e^{-iEt/\hbar}\psi(x) \tag{11.74}$$

This time-dependent wavefunction is called the stationary state of energy E. Recall from our earlier discussion that we cannot measure the wavefunction but we can measure the probability of observing moving particles at a particular place. According to (11.61), the probability P can be expressed as:

$$\begin{aligned} P = |\Psi|^2 &= \bar{\Psi}(x,t)\Psi(x,t) = e^{iEt/\hbar}\bar{\psi}(x)e^{iEt/\hbar}\psi(x) \\ &= \bar{\psi}(x)\psi(x) = |\psi(x)|^2 \end{aligned} \tag{11.75}$$

Note that we start with a time-dependent function $\Psi(x,t)$ but the final probability obtained in (11.75) is only a function of x and, thus, is time independent. This is the reason why it is called a stationary state. Substitution of (11.74) into (11.46) gives

$$i\hbar(-i\frac{E}{\hbar})e^{-iEt/\hbar}\psi(x) = e^{-iEt/\hbar}\hat{H}\psi(x) \tag{11.76}$$

Using the definition of the Hamiltonian operator given in (11.47), we find

$$E\psi = -\frac{\hbar^2}{2m}\frac{d^2\psi}{dx^2} + U\psi \tag{11.77}$$

This can be rearranged as

$$\frac{d^2\psi}{dx^2} + \frac{2m}{\hbar^2}(E-U)\psi = 0 \tag{11.78}$$

This is a second order ODE. In general, the solution depends on the mathematical form of potential $U(x)$. If $U(x)$ is continuous, all ψ, ψ', and ψ'' are continuous. If $U(x)$ has finite jumps, ψ and ψ' are continuous but ψ'' is not smooth. If $U(x)$ contains Dirac delta functions, ψ'' also contains the Dirac delta function and ψ' has finite jumps. To solve (11.78), we need to impose both ψ and ψ' at some point x_0. This is an eigenvalue problem that we may not satisfy (11.78) with an arbitrary energy E. Similar to the discussions by Chau (2018), the discrete spectrum of eigenvalues is normally found for finite volume and the continuous spectrum of eigenvalues is found for an infinite domain. Figure 11.3 illustrates the regions of eigenvalue of energies. Note that discrete eigenstates only appear in a potential well between U_- and U_+.

Thus, the problem becomes finding all admissible energies E for which the problem can be solved for the given boundary conditions. The solution $\psi_n(x)$

corresponding to a certain energy level E_n is called the eigenstate. These energy levels E_n form the spectrum of the Hamiltonian operator. Let the eigenstates of energy be

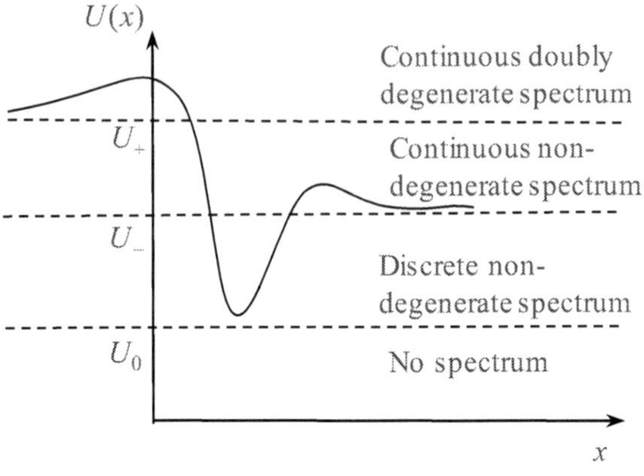

Figure 11.3 Illustration of potential $U(x)$ as a function of x and its eigenstates

$$E_1 \leq E_2 \leq E_3 \leq \ldots \tag{11.79}$$

These discrete energies agree with the basic requirement of quantum states of the energy. The eigenstate clearly satisfies

$$\hat{H}\psi_n(x) = E_n\psi_n(x) \tag{11.80}$$

This system closely resembles the eigenvalue problem that we encounter in mathematics with the eigenvalues being E_n and eigenfunctions ψ_n. The eigenstate wavefunctions are orthonormal such that

$$\int_{-\infty}^{\infty} \bar{\psi}_n(x)\psi_m(x)dx = \delta_{nm} \tag{11.81}$$

This is the orthogonality of the eigenstate wavefunctions. Similar to the eigenfunction expansion discussed in Chapter 10 of Chau (2018), we can, in general, express any wavefunction as

$$\psi(x) = \sum_{m=1}^{\infty} b_m\psi_m(x) \tag{11.82}$$

Suppose that the initial condition is given as

$$\Psi(x, t=0) = \psi(x) = \sum_{m=1}^{\infty} b_m\psi_m(x) \tag{11.83}$$

From the normalization requirement, we have

$$\int_{-\infty}^{\infty} \bar{\psi}(x)\psi(x)dx = 1 \tag{11.84}$$

Substitution of (11.83) into (11.84) gives

$$\sum_{m=1}^{\infty} |b_m|^2 = 1 \tag{11.85}$$

This expression is similar to Parseval's equation for the Fourier coefficients in Fourier series expansion. Multiplying both sides of (11.83) by the complex conjugate of $\psi_n(x)$ and integrating the result from $-\infty$ to ∞, we have

$$\begin{aligned} \int_{-\infty}^{\infty} \psi(x)\overline{\psi}_n(x) &= \sum_{m=1}^{\infty} b_m \int_{-\infty}^{\infty} \psi_m(x)\overline{\psi}_n(x)dx \\ &= \sum_{m=1}^{\infty} b_m \delta_{mn} = b_n \end{aligned} \tag{11.86}$$

Thus, the wavefunction subject to initial condition (11.83) can be found as:

$$\Psi(x,t) = \sum_{n=1}^{\infty} b_n e^{-iE_n t/\hbar} \psi_n(x) \tag{11.87}$$

where

$$b_n = \int_{-\infty}^{\infty} \psi(x)\overline{\psi}_n(x) \tag{11.88}$$

Reconsidering (11.83), we can rewrite it as:

$$\psi(x) = \sum_{n=1}^{\infty} [\int_{-\infty}^{\infty} \psi(\xi)\overline{\psi}_n(\xi)d\xi]\psi_n(x) \tag{11.89}$$

Reversing the order of the summation and the integration of (11.89), we obtain

$$\psi(x) = \int_{-\infty}^{\infty} [\sum_{n=1}^{\infty} \overline{\psi}_n(\xi)\psi_n(x)]\psi(\xi)d\xi \tag{11.90}$$

This can be recast in a general form as:

$$f(x) = \int_{-\infty}^{\infty} K(\xi,x)f(\xi)d\xi \tag{11.91}$$

This is a Fredholm integral equation of the second kind. Consider the special case that

$$f(x) = \delta(x - x_0) \tag{11.92}$$

Substitution of (11.92) into (11.91) gives

$$\delta(x - x_0) = \int_{-\infty}^{\infty} K(\xi,x)\delta(\xi - x_0)d\xi = K(x_0, x) \tag{11.93}$$

Thus, comparing (11.93) and (11.90) we have

$$\sum_{n=1}^{\infty} \overline{\psi}_n(\xi)\psi_n(x) = \delta(x - \xi) \tag{11.94}$$

This provides an alternate form of the Dirac delta function. This is also known as the completeness condition. Equations (11.84) and (11.94) provide a dual pair of properties of the eigenfunctions or the eigenstates of energy.

For any differential operator, we can define the expectation value as:

$$\langle \hat{A} \rangle = \int_{-\infty}^{\infty} \bar{\Psi}_n(x,t)[\hat{A}\Psi_n(x,t)]dx \tag{11.95}$$

We may also normalize the expectation of any operator as

$$\langle \hat{A} \rangle = \frac{\int_{-\infty}^{\infty} \bar{\Psi}_n(x,t)[\hat{A}\Psi_n(x,t)]dx}{\int_{-\infty}^{\infty} \bar{\Psi}_n(x,t)\Psi_n(x,t)dx} \tag{11.96}$$

Recalling the Hamiltonian operator given in (11.80) and multiplying it by the conjugate of the wavefunction, we find

$$\langle E \rangle = \langle \hat{H} \rangle = \int_{-\infty}^{\infty} \bar{\Psi}_n(x,t)[\hat{H}\Psi_n(x,t)]dx \tag{11.97}$$

Substitution of the wavefunction given in (11.87) into (11.97) leads to

$$\begin{aligned}\langle \hat{H} \rangle &= \sum_{m=1}^{\infty}\sum_{n=1}^{\infty} \int_{-\infty}^{\infty} \bar{b}_n b_m e^{iE_n t/\hbar} \bar{\psi}_n(x) e^{-iE_m t/\hbar} [\hat{H}\psi_m(x)]dx \\ &= \sum_{m=1}^{\infty}\sum_{n=1}^{\infty} \bar{b}_n b_m E_m e^{i(E_n - E_m)t/\hbar} \int_{-\infty}^{\infty} \bar{\psi}_n(x)\psi_m(x)]dx\end{aligned} \tag{11.98}$$

We have used (11.80) in obtaining the result in (11.98). Using orthogonality given in (11.81), (11.98) is reduced to

$$\langle E \rangle = \langle \hat{H} \rangle = \sum_{m=1}^{\infty}\sum_{n=1}^{\infty} \bar{b}_n b_m E_m e^{i(E_n - E_m)t/\hbar} \delta_{mn} = \sum_{n=1}^{\infty} |b_n|^2 E_n \tag{11.99}$$

Therefore the expectation of a Hamiltonian or energy is time independent, but the expectation of other observables is generally time dependent. According to the Heisenberg uncertainty principle, the position of the electron cannot be found exactly, and the coefficients b_n actually give participations of each energy level to the overall energy level of the electron. Equation (11.99) can be considered as the quantum version of energy conservation. The energy state is stationary. For other observables, their expectation values are generally time dependent.

11.4 OPERATORS AND EXPECTATION VALUES

For the case of a free particle, the wave function is given in (11.38). The time and spatial differentiations of Ψ give

$$\frac{\partial \Psi}{\partial t} = -\frac{i}{\hbar} E\Psi \tag{11.100}$$

$$\frac{\partial \Psi}{\partial x} = \frac{i}{\hbar} p\Psi \tag{11.101}$$

We can rearrange these equations as

$$E\Psi = i\hbar \frac{\partial \Psi}{\partial t} \tag{11.102}$$

$$p\Psi = \frac{\hbar}{i}\frac{\partial \Psi}{\partial x} \tag{11.103}$$

Equation (11.102) suggests that the dynamical quantity E on the left of (11.102) in some sense plays the role of temporal differential operator on the right of (11.102). Similarly, the dynamical quantity p plays the role of the spatial differential operator on the left of (11.103). These differential operators, defined in (11.102) and (11.103), are customarily written as:

$$\hat{E} = i\hbar\frac{\partial}{\partial t} \tag{11.104}$$

$$\hat{p} = \frac{\hbar}{i}\frac{\partial}{\partial x} \tag{11.105}$$

Because of their correspondence with the energy and momentum, they are called the total-energy operator and momentum operator. Although (11.104) and (11.105) are obtained by assuming the wavefunction of free particles, they are entirely general results. For more general cases, we can replace the total energy E as:

$$\hat{E} = \hat{KE} + \hat{U} \tag{11.106}$$

The kinetic energy is defined in terms of momentum as

$$KE = \frac{p^2}{2m} \tag{11.107}$$

Thus, we have the kinetic-energy operator being

$$\hat{KE} = \frac{\hat{p}^2}{2m} = \frac{1}{2m}\left(\frac{\hbar}{i}\frac{\partial}{\partial x}\right)^2 = -\frac{\hbar^2}{2m}\frac{\partial^2}{\partial x^2} \tag{11.108}$$

Combining (11.104), (11.106) and (11.108), we get

$$i\hbar\frac{\partial}{\partial t} = -\frac{\hbar^2}{2m}\frac{\partial^2}{\partial x^2} + \hat{U} \tag{11.109}$$

If we operate this equation on wavefunction Ψ, we get

$$i\hbar\frac{\partial \Psi}{\partial t} = -\frac{\hbar^2}{2m}\frac{\partial^2 \Psi}{\partial x^2} + U\Psi \tag{11.110}$$

This is precisely the Schrödinger equation. Therefore, postulating (11.104) and (11.105) is equal to postulating the Schrödinger equation. Because E and p can be replaced by their corresponding operators, we can evaluate the expectation value of p as

$$\langle p \rangle = \int_{-\infty}^{\infty} \bar{\Psi}\,\hat{p}\,\Psi dx = \int_{-\infty}^{\infty} \bar{\Psi}\frac{\hbar}{i}\frac{\partial}{\partial x}\Psi dx = \frac{\hbar}{i}\int_{-\infty}^{\infty} \bar{\Psi}\frac{\partial \Psi}{\partial x}dx \tag{11.111}$$

Similarly, we can find the expectation value of E as

$$\langle E \rangle = \int_{-\infty}^{\infty} \bar{\Psi}\,\hat{E}\,\Psi dx = \int_{-\infty}^{\infty} \bar{\Psi} i\hbar\frac{\partial}{\partial t}\Psi dx = i\hbar\int_{-\infty}^{\infty} \bar{\Psi}\frac{\partial \Psi}{\partial t}dx \tag{11.112}$$

For the case of position, we have the expectation value as:

$$\langle x(t) \rangle = \int_{-\infty}^{\infty} \bar{\Psi}(x,t)\,x\,\Psi(x,t)dx \tag{11.113}$$

We can extend this result to any observable quantity as

$$\langle G(x,p)\rangle = \int_{-\infty}^{\infty} \bar{\Psi}\hat{G}(x,-i\hbar\frac{\partial}{\partial x})\Psi dx \tag{11.114}$$

Formula (11.114) can be generalized to 3-D easily:

$$\langle G(x,y,z,p_x,p_y,p_z)\rangle = \int_{-\infty}^{\infty} \bar{\Psi}\hat{G}(x,y,z,-i\hbar\frac{\partial}{\partial x},-i\hbar\frac{\partial}{\partial y},-i\hbar\frac{\partial}{\partial z})\Psi dx \tag{11.115}$$

To find the operator for G, we can first express G in terms of x and p, then substitute p with the operator defined in (11.115). In quantum mechanics, for every physical observable parameter, there corresponds a linear operator (like G in (11.115)) in Hilbert space, which is a vector space particularly suitable for the analysis of quantum mechanics. The states of a quantum mechanical system are vectors in a certain Hilbert vector space, which is symmetric self-adjoint (or Hermitian) and of infinite dimension. The term Hilbert space was coined by John von Neumann. Therefore, an individual event cannot be determined exactly, we only determine its average (or expected) value for certain observable parameters.

11.5 CLASSICAL MECHANICS VERSUS QUANTUM MECHANICS

By observing the role of (11.104) and (11.105), we can actually use the corresponding substitution of them in the classical mechanics formulation to get the quantum mechanics expressions. This is called the Bohr's correspondence principle, which can be stated as:

> Between classical physics and quantum mechanics exists a formal analogy. A classical dynamic variable corresponds to a quantum mechanical Hermitian operator.

Table 11.1 summarizes the correspondence between classical mechanics and quantum mechanics. In particular, quantum mechanics can be obtained by classical mechanics if the identification made by the third column of Table 11.1 is used. The physical meaning behind this mathematical coincidence is not fully understood yet.

11.6 HYDROGEN-LIKE ATOM MODEL

In this section, we will model the wavefunctions for hydrogen-like atoms by solving Schrödinger's equation. The hydrogen-like atoms mean there is only one electron orbiting around the nucleus as shown in Figure 11.4. The charge and mass of the electron are denoted as e^- and m, respectively. The mass and charge of the nucleus are Ze^+ and m_p. For hydrogen, we have only one proton and, thus, $Z = 1$. For ionized Helium (or He^+), we have $Z = 2$. There are two electrons in the normal state of Helium. If one of the electrons is removed (or so-called ionized), the net charge of the atom is positive one e (electric charge of an electron), and it is traditionally represented by the superscripted "+." Similarly, for ionized Lithium (Li^{++}), we have $Z = 3$. Of course, we can continue to consider atoms with higher atomic number, but it would be more difficult experimentally to remove all electrons but one. For the case of multiple electrons orbiting around a nucleus, a brief formulation will be given at the end of this section.

Table 11.1 A summary for correspondence between classical mechanics and quantum mechanics.

Quantities	Classical Mechanics	Quantum Mechanics
Momentum	p_x, p_y, p_z	$\hat{p}_x = -i\hbar\dfrac{\partial}{\partial x}, \hat{p}_y = -i\hbar\dfrac{\partial}{\partial y}, \hat{p}_z = -i\hbar\dfrac{\partial}{\partial z}$
Position	x, y, z	$\hat{x} = x, \hat{y} = y, \hat{z} = z$
Energy	E	$\hat{E} = -i\hbar\dfrac{\partial}{\partial t}$
Hamiltonian	$H = \dfrac{1}{2m}(p_x^2 + p_y^2 + p_y^2) + U(\boldsymbol{r})$	$\hat{H} = -\dfrac{\hbar^2}{2m}\nabla^2 + U(\boldsymbol{r})$
Angular momentum	$\boldsymbol{L} = \boldsymbol{r} \times \boldsymbol{p}$	$\hat{\boldsymbol{L}} = \hat{\boldsymbol{r}} \times \hat{\boldsymbol{p}} = -i\hbar\hat{\boldsymbol{r}} \times \nabla$
General	$Q(\boldsymbol{r}, \boldsymbol{p})$	$\hat{Q}(\hat{\boldsymbol{r}}, \hat{\boldsymbol{p}}) = \hat{Q}(\hat{\boldsymbol{r}}, -i\hbar\nabla)$

One main feature of our presentation compared to the standard textbooks in quantum mechanics is that full mathematical detail in arriving at the solution of the wavefunction is covered. These steps are normally neglected in textbooks in quantum mechanics, especially for the radial dependent function resulting from the separation of variables. The exact solution is expressed in terms of Laguerre polynomials, which is relatively less known (e.g., it is not covered in the handbook of Abramowitz and Stegun, 1964, or in the classic book by Whittaker and Watson, 1927). Two subsections will be devoted to discussing the essential properties of the Laguerre polynomials and the associated Laguerre polynomials, including the orthogonality of the associated Laguerre polynomials. This knowledge is essential for eigenfunction expansion of the wavefunctions in terms of the associated Laguerre polynomials.

11.6.1 Schrödinger Equation in Polar Form

Recalling from (11.56), we have the three-dimensional Schrödinger equation in Cartesian coordinates

$$\frac{\partial^2\psi}{\partial x^2} + \frac{\partial^2\psi}{\partial y^2} + \frac{\partial^2\psi}{\partial z^2} + \frac{2m}{\hbar^2}(E - U)\psi = 0 \tag{11.116}$$

However, for modeling hydrogen-like atoms, it is more convenient to use spherical coordinates. In particular, the conversion between Cartesian coordinates and spherical coordinates is

$$x = r\sin\theta\cos\phi, \quad y = r\sin\theta\sin\phi, \quad z = r\cos\theta \tag{11.117}$$

where the angles ϕ and θ are defined in Figure 11.4. As shown in Chau (2018), the spherical form of the Laplacian can be expressed as:

$$\frac{1}{r^2}\frac{\partial}{\partial r}(r^2\frac{\partial \psi}{\partial r})+\frac{1}{r^2\sin\theta}\frac{\partial}{\partial \theta}(\sin\theta\frac{\partial \psi}{\partial \theta})+\frac{1}{r^2\sin^2\theta}\frac{\partial^2\psi}{\partial\phi^2}+\frac{2m}{\hbar^2}(E-U)\psi=0 \quad (11.118)$$

Normally, the electric force is much larger than the gravitational force in atoms. If we neglect the gravitational force between the nucleus and the electron, the potential depends only on the electric Coulomb force. Thus, the potential U can be approximated as

$$U=-\frac{Ze^2}{4\pi\varepsilon_0 r} \quad (11.119)$$

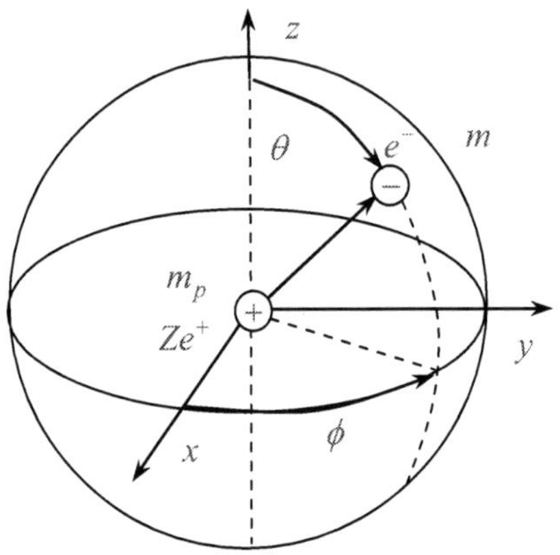

Figure 11.4 Illustration of a hydrogen-like atom

Substituting (11.119) into (11.118), we obtain

$$\sin^2\theta\frac{\partial}{\partial r}(r^2\frac{\partial \psi}{\partial r})+\sin\theta\frac{\partial}{\partial \theta}(\sin\theta\frac{\partial \psi}{\partial \theta})+\frac{\partial^2\psi}{\partial\phi^2}+\frac{2mr^2\sin^2\theta}{\hbar^2}(\frac{Ze}{4\pi\varepsilon_0 r}+E)\psi=0$$

(11.120)

11.6.2 Separation of Variables

It is well known in mathematics that the Laplacian operator is separable in spherical coordinates (see Chau, 2018). Thus, we can assume the following separation of variables

$$\psi(r,\theta,\phi)=R(r)\Theta(\theta)\Phi(\phi) \quad (11.121)$$

Partial differentiations of the wavefunctions become total differentiations of the corresponding coordinate dependent functions as

$$\frac{\partial \psi}{\partial r}=\Theta\Phi\frac{dR}{dr} \tag{11.122}$$

$$\frac{\partial \psi}{\partial \theta}=R\Phi\frac{d\Theta}{d\theta} \tag{11.123}$$

$$\frac{\partial^2 \psi}{\partial \phi^2}=R\Theta\frac{d^2\Phi}{d\phi^2} \tag{11.124}$$

Using (11.121) to (11.124), we can rewrite (11.120) as

$$\begin{aligned}&\frac{\sin^2\theta}{R}\frac{d}{dr}(r^2\frac{dR}{dr})+\frac{\sin\theta}{\Theta}\frac{d}{d\theta}(\sin\theta\frac{d\Theta}{d\theta})\\&+\frac{2mr^2\sin^2\theta}{\hbar^2}(\frac{Ze^2}{4\pi\varepsilon_0 r}+E)=-\frac{1}{\Phi}\frac{d^2\Phi}{d\phi^2}=m_l^2\end{aligned} \tag{11.125}$$

Clearly, each term in (11.125) must be constant, as a function of ϕ cannot equal to a function of θ, unless both them are constants. The constant of separation is denoted by m_l, and it is related to the magnetic property of the atom and this will be discussed later. Another constant of separation of variables l is introduced as

$$\frac{1}{R}\frac{d}{dr}(r^2\frac{dR}{dr})+\frac{2mr^2}{\hbar^2}(\frac{Ze^2}{4\pi\varepsilon_0 r}+E)=\frac{m_l^2}{\sin^2\theta}-\frac{1}{\Theta\sin\theta}\frac{d}{d\theta}(\sin\theta\frac{d\Theta}{d\theta})=l(l+1) \tag{11.126}$$

The reason why the constant of separation of variables expressed as $l(l+1)$ will be discussed shortly. It can be shown later that l relates to the angular momentum of the electron (see later subsection). Finally, we arrive at three ODEs instead of dealing with the original PDE. More specifically, three uncoupled ODEs for R, Θ, and Φ are

$$\frac{d^2\Phi}{d\phi^2}+m_l^2\Phi=0 \tag{11.127}$$

$$\frac{1}{\sin\theta}\frac{d}{d\theta}(\sin\theta\frac{d\Theta}{d\theta})+[l(l+1)-\frac{m_l^2}{\sin^2\theta}]\Theta=0 \tag{11.128}$$

$$\frac{1}{r^2}\frac{d}{dr}(r^2\frac{dR}{dr})+[\frac{2m}{\hbar^2}(\frac{Ze^2}{4\pi\varepsilon_0 r}+E)-\frac{l(l+1)}{r^2}]R=0 \tag{11.129}$$

Equation (11.127) is the familiar equation of the harmonic oscillator. The solutions of (11.127) are, of course, sine and cosine, but since the wavefunction ψ is complex anyway, we can simply express it as

$$\Phi_{m_l}=A\exp(im_l\phi) \tag{11.130}$$

Note that we have added a subscript to the azimuth function to indicate its dependence on the magnetic quantum number m_l. Physically, we can see from Figure 11.4 that if we go around the spherical coordinate one complete revolution, we will end up at the same point. Mathematically, we require

$$\exp(im_l\phi)=\exp[im_l(\phi+2\pi)] \tag{11.131}$$

where

$$m_l = 0, \pm 1, \pm 2, \pm 3, \ldots \tag{11.132}$$

This is called the magnetic quantum number and the reason for its name will be discussed later. Equation (11.128) is a Legendre equation if l is an integer, and this is the reason why we assume this particular form of the constant in (11.126). Traditionally, we assume the following change of variables

$$x = \cos\theta, \quad \sin\theta = \sqrt{1-x^2}, \quad d\theta = \frac{dx}{\sqrt{1-x^2}} \tag{11.133}$$

Substitution of (11.133) into (11.128) yields

$$(1-x^2)\frac{d^2\Theta}{dx^2} - 2x\frac{d\Theta}{dx} + [l(l+1) - \frac{m_l^2}{1-x^2}]\Theta = 0 \tag{11.134}$$

The solution is the associated Legendre polynomials

$$\Theta_{lm_l} = BP_l^{m_l}(\cos\theta) \tag{11.135}$$

As shown in Chau (2018), the associated Legendre polynomials is a finite series. The first few terms are given here:

$$\begin{aligned} &P_0^0 = 1, \quad P_1^0 = 2\cos\theta, \quad P_1^1 = \sin\theta, \\ &P_2^0 = 4(3\cos^2\theta - 1), \quad P_2^1 = 4\sin\theta\cos\theta, \quad P_2^2 = \sin^2\theta, \\ &P_3^0 = 24(5\cos^3\theta - 3\cos\theta), \quad P_3^1 = 6\sin\theta(5\cos^2\theta - 1), \\ &P_3^2 = 6\sin^2\theta\cos\theta, \quad P_3^3 = \sin^3\theta \end{aligned} \tag{11.136}$$

In addition, the associated Legendre polynomials have the important property

$$P_l^{m_l}(\cos\theta) = 0, \quad m_l > l \tag{11.137}$$

This provides a constraint on the magnetic quantum number. The solution of (11.129) is less straightforward. We can write it in a slightly different form

$$\frac{d^2R}{dr^2} + \frac{2}{r}\frac{dR}{dr} + \frac{2m}{\hbar^2}[(\frac{Ze^2}{4\pi\varepsilon_0 r} + E) - \frac{\hbar^2 l(l+1)}{2mr^2}]R = 0 \tag{11.138}$$

First, we will normalize the coordinate r by the factor κ involving the admissible energy E, and another normalized parameter λ. We will see later that the solution form of R will impose the condition that λ must be an integer. It is precisely this condition that leads to the quantized form of E. In particular, we have

$$\frac{d^2R}{d\rho^2} + \frac{2}{\rho}\frac{dR}{dr} + [\frac{\lambda}{\rho} - \frac{1}{4} - \frac{l(l+1)}{\rho^2}]R = 0 \tag{11.139}$$

where

$$\rho = \kappa r, \quad \lambda = \frac{2mZe^2}{4\pi\varepsilon_0\kappa\hbar^2}, \quad \kappa^2 = -\frac{8mE}{\hbar^2} \tag{11.140}$$

To solve (11.138), we assume the following form

$$R = e^{-\rho/2}\rho^l w(\rho) \tag{11.141}$$

Differentiation of (11.141) gives

$$\frac{dR}{d\rho} = e^{-\rho/2}\rho^l\{w' + (\frac{l}{\rho} - \frac{1}{2})w\} \tag{11.142}$$

$$\frac{d^2R}{d\rho^2} = e^{-\rho/2}\rho^l \left\{ w'' + [(\frac{2l}{\rho} - 1)w' + [\frac{1}{4} - \frac{l}{\rho} + \frac{l(l-1)}{\rho^2}]w \right\} \tag{11.143}$$

Substitution of (11.141) to (11.143) into (11.139) leads to

$$\rho w'' + (2l + 2 - \rho)w' + (\lambda - 1 - l)w = 0 \tag{11.144}$$

This is recognized as the associated Laguerre equation if the following identification is made:

$$\rho w'' + (m + 1 - \rho)w' + \alpha w = 0 \tag{11.145}$$

with

$$m = 2l + 1, \quad \alpha = \lambda - 1 - l \tag{11.146}$$

The solution of (11.145) is called the associated Laguerre polynomials and is given as

$$w = L_\alpha^m(\rho) \tag{11.147}$$

Full discussion of this solution in the form of an finite series can be found in Section 11.6.4. Finally, we get the following radial function of the wave function

$$R_{nl}(r) = e^{-\kappa r/2}(\kappa r)^l L_{\lambda-l-1}^{2l+1}(\kappa r) \tag{11.148}$$

Further simplification of λ and κ will be made after we discuss the property of the associated Laguerre polynomials. In short, the wavefunction becomes

$$\begin{aligned} \psi_{nlm_l}(r,\theta,\phi) &= e^{-\kappa r/2}(\kappa r)^l L_{\lambda-l-1}^{2l+1}(\kappa r) P_l^{m_l}(\cos\theta)\exp(im_l\phi) \\ &= e^{-\kappa r/2}(\kappa r)^l L_{\lambda-l-1}^{2l+1}(\kappa r) Y_{lm_l}(\theta,\phi) \end{aligned} \tag{11.149}$$

where the spherical harmonics are defined as (e.g., Hobson, 1955)

$$Y_{lm_l}(\theta,\phi) = P_l^{m_l}(\cos\theta)\exp(im_l\phi) \tag{11.150}$$

It seems that we have obtained the final solution in (11.149), but this is not the whole story. We are going to see that there are restrictions on the possible choices of λ and, thus, in turn, the energy levels E.

Before we continue, we mention a bit more about Prof. Hobson (see our citation of his book on spherical harmonics above). In Chau (2018), we told the interesting story of Ramanujan. In fact, Ramanujan first wrote to E.W. Hobson before he wrote to G.H. Hardy. Fortunately or unfortunately, Hobson ignored Ramanujan's his letter with unproved formulas.

11.6.3 Constraints Imposed by Wavefunctions

Before we discuss the mathematical properties of the associated Laguerre polynomials, it is instructive to explore the structural form of the wavefunction first. In order to illustrate the restrictions on the possible solution for the wavefunction, we take one step back to see that the energy level E must be negative in order to fulfill the boundedness condition. Let us consider the far-field solution for the radial wavefunction for $r \to \infty$ in the following example.

Example 11.1 By solving the radial wavefunctions governed by (11.138) for $r \to \infty$, show that the energy E must be negative.

Solution: For $r \to \infty$, the governing equation for R given in (11.138) is reduced to:

$$\frac{d^2R}{dr^2} + \frac{2mE}{\hbar^2}R = 0 \tag{11.151}$$

If the energy state $E > 0$, we have the solution as

$$R = C_1 \sin(\frac{\sqrt{2mE}}{\hbar}r) + C_2 \cos(\frac{\sqrt{2mE}}{\hbar}r) \tag{11.152}$$

This is a finite non-zero solution. However, physically the wavefunction must decay to zero as $r \to \infty$. This solution is physically inadmissible. If the energy state $E < 0$, the governing equation becomes

$$\frac{d^2R}{dr^2} - \frac{2m|E|}{\hbar^2}R = 0 \tag{11.153}$$

The solution then becomes an exponential function as

$$R = C_1 \exp(-\frac{\sqrt{2m|E|}}{\hbar}r) + C_2 \exp(\frac{\sqrt{2m|E|}}{\hbar}r) \tag{11.154}$$

The boundedness condition of $R \to 0$ for $r \to \infty$ requires $C_2 = 0$. Finally, we have

$$R = C_1 \exp(-\frac{\sqrt{2m|E|}}{\hbar}r) \tag{11.155}$$

Therefore, it is obvious that we must have $E < 0$. We will consider the lowest level of energy E in the following example. Physically, $E < 0$ indicates that energy is needed to bond protons and electrons together in the atom.

Example 11.2 Solve the wavefunction governed by the Schrödinger equation for the case of $n = 1$, $l = 0$ and $m_l = 0$. Find the corresponding lowest energy state.

Solution: For this case, (11.138) is reduced to

$$\frac{d^2R}{dr^2} + \frac{2}{r}\frac{dR}{dr} + \frac{2m}{\hbar^2}(\frac{Ze^2}{4\pi\varepsilon_0 r} + E)R = 0 \tag{11.156}$$

Although this ODE is not of constant coefficients, we still want to seek an exponential function as the solution

$$R = e^{\alpha r} \tag{11.157}$$

Substitution of (1.157) into (11.156) gives

$$\alpha^2 + \frac{2}{r}\alpha + \frac{2m}{\hbar^2}(\frac{Ze^2}{4\pi\varepsilon_0 r} + E) = 0 \tag{11.158}$$

By setting the coefficients of the constant order term and the $1/r$ term to zero, we have

$$\alpha^2 = -\frac{2mE}{\hbar^2}, \quad \alpha = -\frac{mZe^2}{4\pi\varepsilon_0\hbar^2} \tag{11.159}$$

Squaring the second equation of (11.159) and equating it to the first equation in (11.159), we get

$$\frac{m^2 Z^2 e^4}{16\pi^2 \varepsilon_0^2 \hbar^4} = -\frac{2mE}{\hbar^2} \tag{11.160}$$

Solving for E, we find

$$E = -\frac{mZ^2 e^4}{32\pi^2 \varepsilon_0^2 \hbar^2} \tag{11.161}$$

In view of (11.157), we see that the unit of α must be proportional to one over length. In particular, we set α as:

$$\alpha = -\frac{mZe^2}{4\pi\varepsilon_0 \hbar^2} = -\frac{1}{a_0} \tag{11.162}$$

where a_0 is radius of the innermost orbit of Bohr's hydrogen atom model. For the case of hydrogen or $Z = 1$, we have the radius being 5.292×10^{-11}m. We are going to show that this radius agrees with the radius of the innermost orbit of Bohr's hydrogen atom model, which does not depend on solving the Schrödinger equation. For $l = 0$ and $m_l = 0$, it is obvious from (11.127) and (11.128) that both Φ and Θ are constants (see Problem 11.7). Traditionally, they are normalized as:

$$\Phi = \frac{1}{\sqrt{2\pi}}, \quad \Theta = \frac{1}{\sqrt{2}} \tag{11.163}$$

Thus, the wavefunction is traditionally normalized as:

$$\psi(r,\theta,\phi) = \frac{1}{\sqrt{\pi} a_0^{3/2}} \exp(-r/a_0) \tag{11.164}$$

This is the final wavefunction for the lowest mode.

Example 11.3 Show that the probability of finding the electron in the full space for the wavefunction found in Example 11.2 is indeed unity (or one).

Solution: The probability of finding the electron in the full space can be evaluated as

$$\begin{aligned} P = \int |\psi(r,\theta,\phi)|^2 dV &= \frac{1}{\pi a_0^3} \int_0^\infty \int_0^{2\pi} \int_0^\pi \exp(-2r/a_0) r^2 \sin dr d\phi d\theta \\ &= \frac{1}{\pi a_0^3} \int_0^{2\pi} d\phi \int_0^{2\pi} \sin\theta d\theta \int_0^\infty r^2 \exp(-2r/a_0) dr \\ &= \frac{1}{\pi a_0^3} (2\pi)(2)(-\frac{a_0}{2}) \int_0^\infty r^2 d(e^{-2r/a_0}) \end{aligned} \tag{11.165}$$

Applying integration by parts twice, we get

$$P = \frac{2}{a_0}(-\frac{a_0}{2}) \int_0^\infty d(e^{-2r/a_0}) = -[e^{-2r/a_0}]_0^\infty = 1 \tag{11.166}$$

This illustrates that the wavefunction given in (11.164) is normalized properly.

Example 11.4 Referring to Figure 11.5, derive the radius orbit in Bohr's model of the hydrogen atom, which assumed that the electron is orbiting the proton in a circular motion. Show that the energy for the lowest state in this model is the same as that obtained in (11.161).

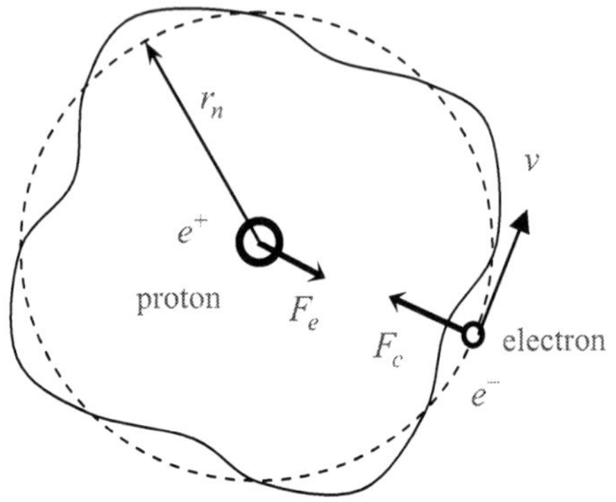

Figure 11.5 Sketch of Bohr's model of the hydrogen atom with $n = 4$

Hints: (i) Orbit stability is ensured by allowing an integral number of waves in the orbit; (ii) the de Broglie wavelength of matter waves should be used; (iii) centrifugal force should be balanced by Coulomb's electric force; (iv) the energy of the electron can be considered by summing the kinetic energy and potential energy.

Solution: The circumference of the circular orbit must accommodate a complete wave number, and we have the orbit stability as

$$n\lambda = 2\pi r_n, \quad n = 1, 2, 3, ... \tag{11.167}$$

In view of the particle-wave duality of the matter waves, we have the wavelength expressed in terms of the momentum as (see (11.33) to (11.35)):

$$\lambda = \frac{h}{mV} \tag{11.168}$$

The centrifugal force of the electron must balance the Coulomb force from the electric charge between the electron and proton:

$$F_c = \frac{mV^2}{r_n}, \quad F_e = \frac{1}{4\pi\varepsilon_0}\frac{e^2}{r_n^2} \tag{11.169}$$

Equating these forces, we obtain the velocity of the electron as:

$$V = \frac{e}{\sqrt{4\pi\varepsilon_0 m r_n}} \tag{11.170}$$

Substitution of (11.170) into (11.168) gives

$$\lambda = \frac{h}{e}\sqrt{\frac{4\pi\varepsilon_0 r_n}{m}} \tag{11.171}$$

Substitution of (11.171) into (11.167) yields

$$\frac{nh}{e}\sqrt{\frac{4\pi\varepsilon_0 r_n}{m}} = 2\pi r_n \tag{11.172}$$

Squaring both sides, we find

$$r_n = \frac{n^2 h^2 \varepsilon_0}{\pi m e^2}, \quad n = 1, 2, 3, \ldots \tag{11.173}$$

Putting $n = 1$, we recover the radius a_0 as

$$a_0 = r_1 = \frac{h^2 \varepsilon_0}{\pi m e^2} = \frac{4\pi \hbar^2 \varepsilon_0}{m e^2} \tag{11.174}$$

For the case of hydrogen $Z = 1$, (11.174) is the same as that given in (11.62). Physically, a_0 represents the typical spatial extension of an electron in the ground state, which is about 0.52Å (1 Å $=10^{-10}$m). In terms of energy, the energy of an electron can be found as

$$E_n = KE + PE = \frac{mV^2}{2} - \frac{e^2}{4\pi\varepsilon_0 r_n} = \frac{e^2}{8\pi\varepsilon_0 r_n} - \frac{e^2}{4\pi\varepsilon_0 r_n} = -\frac{e^2}{8\pi\varepsilon_0 r_n} \tag{11.175}$$

We have applied (11.169) to obtain the result in (11.175). Finally, substituting (11.173) into (11.175), we find

$$E_n = -\frac{me^4}{8\varepsilon_0^2 h^2}\left(\frac{1}{n^2}\right) = -\frac{me^4}{32\pi^2 \varepsilon_0^2 \hbar^2}\left(\frac{1}{n^2}\right) \tag{11.176}$$

Setting $n = 1$, we recover the energy given in (11.161). We see that the energy states given in (11.176) are discrete. We are going to show in Section 11.6.6 that (11.76) can be obtained by solving the Schrödinger equation alone. Before we do this, we will discuss Laguerre and associated Laguerre polynomials in the next two subsections.

11.6.4 Laguerre and Associated Laguerre Polynomials

We have seen from (11.149) that the radial portion of the wavefunction is found to be proportional to the associated Laguerre function. We will consider some essential results related to the associated Laguerre polynomials. Surprisingly, associated Laguerre polynomials are normally not covered in standard books on special functions. For example, it was not discussed in Whittaker and Watson (1927) and not covered in Abramowitz and Stegun (1964). Associated Laguerre polynomials find applications in solving Helmholtz's equation in parabolic coordinates, in the hydrogen atom model (as we have seen in the previous section), and in propagation of electromagnetic waves along transmission lines.

Historically, Laguerre polynomials had been considered by Lagrange in 1762, Abel in 1826, Murphy in 1835, and Chebyshev in 1859 before its properties were studied more extensively by Laguerre in 1879. In Russian literature, they were often referred as Chebyshev-Laguerre polynomials to reflect the priority of Chebyshev over Laguerre. Laguerre mainly considered the case of Laguerre polynomials, whereas the associated Laguerre polynomials were considered extensively by Sonine in 1880. However, associated Laguerre polynomials were not covered in classical textbooks on special functions (with the exception of Courant and Hilbert, 1962). Thus, it is relatively less known. It was the development of quantum mechanics that made the Laguerre polynomials and associated Laguerre polynomials more popular in the areas of physical science. In analyzing the hydrogen atom model, Schrödinger and Dirac derived the approximate solution of the Schrödinger equation. In 1928, Walter Gordon solved the hydrogen atom model analytically, in terms of the confluent hypergeometric functions. We should note that the associated Laguerre polynomials represent a special case of the confluent hypergeometric function. It was Pidduck who solved the Schrödinger equation in 1928 using the associated Laguerre polynomials for the hydrogen atom. It was subsequently adopted as the standard solution since then. I will summarize the essential results needed to understand the Laguerre polynomials in this section. For more detailed discussions, we refer readers to Bell (1968) and Lebedev (1972).

Laguerre polynomials

Let us consider the following Laguerre equation

$$x\frac{d^2L_\alpha}{dx^2}+(1-x)\frac{dL_\alpha}{dx}+\alpha L_\alpha=0 \tag{11.177}$$

We look for a series solution as

$$L_\alpha=\sum_{s=0}^{\infty}a_s x^s \tag{11.178}$$

Differentiation of (11.178) gives

$$\frac{dL_\alpha}{dx}=\sum_{s=1}^{\infty}s a_s x^{s-1} \tag{11.179}$$

$$\frac{d^2L_\alpha}{dx^2}=\sum_{s=2}^{\infty}s(s-1)a_s x^{s-2} \tag{11.180}$$

Substitution of (11.179) and (11.180) into (11.177) gives

$$\sum_{s=2}^{\infty}s(s-1)a_s x^{s-1}+\sum_{s=1}^{\infty}s a_s x^{s-1}-\sum_{s=1}^{\infty}s a_s x^{s}+\alpha\sum_{s=0}^{\infty}a_s x^{s}=0 \tag{11.181}$$

Because of the infinite summation, we can shift the indices such that the powers of x are the same for all summations

$$\sum_{m=1}^{\infty}m(m+1)a_{m+1}x^m+\sum_{m=0}^{\infty}(m+1)a_{m+1}x^m-\sum_{m=1}^{\infty}m a_m x^m+\alpha\sum_{m=0}^{\infty}a_m x^m=0 \tag{11.182}$$

For $m=0$, we have

$$a_1 + \alpha a_0 = 0 \tag{11.183}$$

For $m > 0$, we have

$$m(m+1)a_{m+1} + (m+1)a_{m+1} - ma_m + \alpha a_m = 0 \tag{11.184}$$

This can be simplified as

$$a_{m+1} = -\frac{\alpha - m}{(m+1)^2} a_m \tag{11.185}$$

Rewriting the left-hand side for index m, we have

$$\begin{aligned} a_m &= -\frac{\alpha - m + 1}{m^2} a_{m-1} \\ &= (-1)^2 \frac{(\alpha - m + 1)(\alpha - m + 2)}{m^2 (m-1)^2} a_{m-2} \\ &= (-1)^m \frac{(\alpha - m + 1)(\alpha - m + 2)\cdots\alpha}{m^2 (m-1)^2 \cdots 1^2} a_0 \\ &= (-1)^m \frac{\Gamma(\alpha + 1)}{m!m!\Gamma(\alpha - m + 1)} a_0 \end{aligned} \tag{11.186}$$

Substituting (11.186) into (11.178), we have

$$L_\alpha(x) = a_0 \sum_{m=0}^{\infty} (-1)^m \frac{\Gamma(\alpha + 1)}{m!m!\Gamma(\alpha - m + 1)} x^m \tag{11.187}$$

It looks like we get the final solution, but this is not true. It is because this solution behaves like an exponential function for large m and the solution diverges for $x \to \infty$. In particular, we must truncate the infinite series in (11.187). For $m \to \infty$, we have

$$\begin{aligned} a_m &= (-1)^m \frac{(\alpha - m + 1)(\alpha - m + 2)\cdots\alpha}{m^2 (m-1)^2 \cdots 1^2} a_0 \\ &= \frac{(m - \alpha - 1)(m - \alpha - 2)\cdots[m + (\alpha - 1) + 1 - m]}{m^2 (m-1)^2 \cdots 1^2} a_0 \\ &\approx \frac{m(m-1)\cdots 1}{m!m!} a_0 = \frac{1}{m!} a_0 \end{aligned} \tag{11.188}$$

Thus, for $m \to \infty$, we have

$$L_\alpha(x) \approx a_0 \sum_{m=0}^{\infty} \frac{x^m}{m!} = a_0 e^x \tag{11.189}$$

The wavefunction becomes

$$\begin{aligned} \psi(r,\theta,\phi) &= R_{nl}(r) Y_{lm_l}(\theta,\phi) \approx e^{-\kappa r/2} (\kappa r)^l a_0 e^{\kappa r} Y_{lm_l}(\theta,\phi) \\ &= a_0 e^{\kappa r/2} (\kappa r)^l Y_{lm_l}(\theta,\phi) \end{aligned} \tag{11.190}$$

This solution clearly diverges for large r, thus we have $\psi \to \infty$ for $r \to \infty$ (or $x \to \infty$). This is not allowed by physical consideration. For the special case when α is an integer, say n, we can terminate the series at n such that

$$a_m = (-1)^m \frac{(n-m+1)(n-m+2)\cdots n}{m^2(m-1)^2\cdots 1^2} a_0 \tag{11.191}$$

$$a_n = (-1)^n \frac{1}{n!} a_0 \neq 0, \quad a_{n+1} = 0 \tag{11.192}$$

Clearly, this is a natural choice and this is called the Laguerre polynomials of order n and is defined by setting $a_0 = 1$ as

$$L_n(x) = \sum_{m=0}^{n} \frac{n!}{(m!)^2(n-m)!}(-1)^m x^m \tag{11.193}$$

It is well known that there is a Rodrigues' formula for the Legendre function (see Section 4.10.2 of Chau, 2018). In general, this type of function also exists for other special functions, such as Bessel functions, Hermit polynomials, Chebyshev polynomials, Gegenbauer polynomials, and Jacobi polynomials. They are called generating functions. The generating function for Laguerre polynomials is

$$\frac{\exp\{-xt/(1-t)\}}{1-t} = \sum_{n=0}^{\infty} L_n(x)t^n \tag{11.194}$$

The left-hand side of (11.194) can be written in infinite series by using the Taylor series expansion as:

$$\begin{aligned} \frac{\exp\{-xt/(1-t)\}}{1-t} &= \frac{1}{1-t}\sum_{r=0}^{\infty}\frac{1}{r!}\left(-\frac{xt}{1-t}\right)^r \\ &= \sum_{r=0}^{\infty}\frac{(-1)^r}{r!}\frac{x^r t^r}{(1-t)^{r+1}} \end{aligned} \tag{11.195}$$

Using the binominal theorem, we have

$$\begin{aligned} \frac{1}{(1-t)^{r+1}} &= 1+(r+1)t+\frac{(r+1)(r+2)}{2!}t^2+\frac{(r+1)(r+2)(r+3)}{3!}t^3+\ldots \\ &= \sum_{s=0}^{\infty}\frac{(r+s)!}{r!s!}t^s \end{aligned} \tag{11.196}$$

Substitution of (11.196) into (11.195) gives

$$\frac{\exp\{-xt/(1-t)\}}{1-t} = \sum_{s=0}^{\infty}\sum_{r=0}^{\infty}\frac{(-1)^r(r+s)!}{(r!)^2 s!}x^r t^{r+s} \tag{11.197}$$

Let $r+s = n$ or $s = n-r$, we have

$$\frac{\exp\{-xt/(1-t)\}}{1-t} = \sum_{n=0}^{\infty}\sum_{r=0}^{n}\frac{(-1)^r (n)!}{(r!)^2 (n-r)!}x^r t^n$$
$$= \sum_{n=0}^{\infty}\{\sum_{r=0}^{n}\frac{(-1)^r (n)!}{(r!)^2 (n-r)!}x^r\}t^n \tag{11.198}$$
$$= \sum_{n=0}^{\infty} L_n(x)t^n$$

This completes the proof of the generating function given in (11.194). The first few Laguerre polynomials are

$$L_0 = 1,$$
$$L_1 = -x+1,$$
$$L_2 = x^2 - 4x + 2, \tag{11.199}$$
$$L_3 = -x^3 + 9x^2 - 18x + 6,$$
$$L_4 = x^4 - 16x^3 + 72x^2 - 96x + 24$$

Note that the technique used here in deriving (11.194) is similar to that in obtaining Ramanujan's master theorem in Appendix C.1 of Chau (2018).

Associated Laguerre polynomials

We first recall the associated Laguerre equation

$$x\frac{d^2 y}{dx^2} + (k+1-x)\frac{dy}{dx} + ny = 0 \tag{11.200}$$

To find the solution of this associated Laguerre equation, we write the Laguerre equation given (11.177) for order $n+k$ as

$$x\frac{d^2 z}{dx^2} + (1-x)\frac{dz}{dx} + (n+k)z = 0 \tag{11.201}$$

We know that the solution of this Laguerre equation is

$$z = L_{n+k}(x) \tag{11.202}$$

Differentiating (11.201) k times, we have

$$\frac{d^k}{dx^k}(x\frac{d^2 z}{dx^2}) + \frac{d^k}{dx^k}[(1-x)\frac{dz}{dx}] + (n+k)\frac{d^k}{dx^k}z = 0 \tag{11.203}$$

The first two terms in (11.203) can be determined easily by using the Leibniz's rule for differentiation as

$$\frac{d^k}{dx^k}(fg) = \sum_{r=0}^{k} C_k^r \frac{d^r f}{dx^r}\frac{d^{k-r} g}{dx^{k-r}} \tag{11.204}$$

Thus, we have

$$x\frac{d^{k+2} z}{dx^{k+2}} + k\frac{d^{k+1} z}{dx^{k+1}} + (1-x)\frac{d^{k+1} z}{dx^{k+1}} - k\frac{d^k z}{dx^k} + (n+k)\frac{d^k z}{dx^k} = 0 \tag{11.205}$$

This can be simplified as

$$x\frac{d^2}{dx^2}(\frac{d^k z}{dx^k})+(k+1-x)\frac{d}{dx}(\frac{d^k z}{dx^k})+n\frac{d^k z}{dx^k}=0 \tag{11.206}$$

Comparing (11.200) with (11.206) gives the solution of the associated Laguerre equation as (in view of and (11.200) and (11.201))

$$y=\frac{d^k z}{dx^k}=\frac{d^k}{dx^k}L_{n+k}(x) \tag{11.207}$$

However, multiplication of any constant by (11.207) also leads to another solution of the original associated Laguerre equation. Traditionally, the associated Laguerre polynomials have been defined as

$$L_n^k(x)=(-1)^k\frac{d^k}{dx^k}L_{n+k}(x) \tag{11.208}$$

Clearly, we can also express it as a finite series in view of (11.193). First, from (11.193) we get

$$L_{n+k}(x)=\sum_{r=0}^{n+k}\frac{(n+k)!}{(r!)^2(n+k-r)!}(-1)^r x^r \tag{11.209}$$

Substitution of (11.209) into (11.208) yields

$$\begin{aligned} L_n^k(x)&=(-1)^k\frac{d^k}{dx^k}\sum_{r=0}^{n+k}\frac{(n+k)!}{(r!)^2(n+k-r)!}(-1)^r x^r \\ &=(-1)^k\frac{d^k}{dx^k}\sum_{r=k}^{n+k}\frac{(n+k)!}{(r!)^2(n+k-r)!}(-1)^r x^r \end{aligned} \tag{11.210}$$

The change of the lower limit of the summation results from the fact that the differential operator applied on any power series of order less than k gives zero. The differentiation of the series of x gives

$$\frac{d^k x^r}{dx^k}=r(r-1)\cdots(r-k+1)x^{r-k}=\frac{r!}{(r-k)!}x^{r-k} \tag{11.211}$$

Substitution of (11.211) into (11.210) gives

$$\begin{aligned} L_n^k(x)&=(-1)^k\sum_{r=k}^{n+k}(-1)^r\frac{(n+k)!}{r!(n+k-r)!(r-k)!}x^{r-k} \\ &=\sum_{s=0}^{n}(-1)^s\frac{(n+k)!}{(n-s)!(s+k)!s!}x^s \end{aligned} \tag{11.212}$$

The last line in (11.212) results from shifting the summation index from r to $s = r-k$. This is the series representation of the associated Laguerre polynomials. For the case when $k=\alpha$ is not an integer, it is straightforward to see that

$$L_n^\alpha(x)=\sum_{s=0}^{n}\frac{\Gamma(n+\alpha+1)}{\Gamma(s+\alpha+1)(n-s)!s!}(-x)^s \tag{11.213}$$

Note also that

$$L_n^{\alpha}(x) = 0, \quad \alpha > n \tag{11.214}$$

If α equals a half integer, Hermite polynomials are recovered as a special case. In other words, we must have $\alpha \leq n$. Next, we will prove the following generating function for the associated Laguerre polynomials

$$\frac{\exp\{-xt/(1-t)\}}{(1-t)^{k+1}} = \sum_{n=0}^{\infty} L_n^k(x)t^n \tag{11.215}$$

To prove this we start with differentiating the generating function of the Laguerre polynomials shown in (11.194) as

$$\frac{d^k}{dx^k}\frac{\exp\{-xt/(1-t)\}}{1-t} = \frac{d^k}{dx^k}\sum_{n=k}^{\infty} L_n(x)t^n \tag{11.216}$$

Note that the lower index of the summation on the right of (11.216) has been shifted to k because of the differential operator that we apply. Carrying out the differentiation on the left-hand side and shifting the lower index of the summation back to zero, we get

$$(-\frac{t}{1-t})^k \frac{\exp\{-xt/(1-t)\}}{1-t} = \frac{d^k}{dx^k}\sum_{m=0}^{\infty} L_{m+k}(x)t^{m+k} \tag{11.217}$$

This can be simplified as

$$\begin{aligned} \frac{\exp\{-xt/(1-t)\}}{(1-t)^{k+1}} &= (-1)^k \frac{d^k}{dx^k}\sum_{m=0}^{\infty} L_{m+k}(x)t^m \\ &= \sum_{m=0}^{\infty}(-1)^k \frac{d^k}{dx^k} L_{m+k}(x)t^m \\ &= \sum_{m=0}^{\infty} L_m^k(x)t^m \end{aligned} \tag{11.218}$$

This is exactly (11.215) and the proof is completed. Some of the lower order associated Laguerre polynomials are:

$$\begin{aligned} &L_1^1 = -1, \quad L_2^1 = 2x-4, \quad L_2^2 = 2, \\ &L_3^1 = -3x^2+18x-18, \quad L_3^2 = -6x+18, \quad L_3^3 = -6, \\ &L_4^1 = 4x^3-48x^2+144x-96, \quad L_4^2 = 12x^2-96x+144, \\ &L_4^3 = 24x-96, \quad L_4^4 = 24 \end{aligned} \tag{11.219}$$

11.6.5 Orthogonality of Associated Laguerre Polynomials

We have learned from eigenfunction expansion of arbitrary functions that we need to find the orthogonality of the eigenmodes (see Chapter 10 of Chau, 2018). For the case of hydrogen-like atoms, the eigenmodes are expressed in terms of the associated Laguerre polynomials. To prove the orthogonality of the associated

Laguerre polynomials, we first convert the associated Laguerre equation to a slightly different ODE. In particular, we consider the following ODE:

$$xu''+(\alpha+1-2\nu)u'+[n+\frac{\alpha+1}{2}-\frac{x}{4}+\frac{\nu(\nu-\alpha)}{x}]u=0 \tag{11.220}$$

Let us now assume the following solution form:

$$u=e^{-x/2}x^{\nu}\varphi(x) \tag{11.221}$$

Differentiation of (11.271) gives

$$u'=e^{-x/2}x^{\nu}\{(\frac{\nu}{x}-\frac{1}{2})\varphi+\varphi'\} \tag{11.222}$$

$$u''=e^{-x/2}x^{\nu}\{\varphi''+(\frac{2\nu}{x}-1)\varphi'+[\frac{1}{4}-\frac{\nu}{x}+\frac{\nu(\nu-1)}{x^2}]\varphi\} \tag{11.223}$$

Substitution of (11.222) and (11.223) into (11.220) gives the ODE for φ

$$x\varphi''+(\alpha+1-x)\varphi'+n\varphi=0 \tag{11.224}$$

This is exactly the associated Laguerre equation, and thus, the solution of (11.220)

$$u_n=e^{-x/2}x^{\nu}L_n^{\alpha}(x) \tag{11.225}$$

Now let us consider two solutions (or two eigenfunctions of different mode numbers in the case of our atom problem) with the order of the associated Laguerre polynomials of order m and n. For $\nu=\alpha/2$ and mode n, (11.220) can be simplified to

$$xu_n''+u_n'+[n+\frac{\alpha+1}{2}-\frac{x}{4}-\frac{\alpha^2}{4x}]u_n=0 \tag{11.226}$$

This can be further reduced to

$$\frac{d}{dx}(x\frac{du_n}{dx})+[n+\frac{\alpha+1}{2}-\frac{x}{4}-\frac{\alpha^2}{4x}]u_n=0 \tag{11.227}$$

Similarly, the governing equation for the m-th mode is

$$\frac{d}{dx}(x\frac{du_m}{dx})+[m+\frac{\alpha+1}{2}-\frac{x}{4}-\frac{\alpha^2}{4x}]u_m=0 \tag{11.228}$$

Multiplying (11.227) by u_m, subtracting the result of (11.228), and multiplying by u_n, we obtain

$$\int_0^{\infty}[u_m\frac{d}{dx}(x\frac{du_n}{dx})-u_n\frac{d}{dx}(x\frac{du_m}{dx})]dx+(n-m)\int_0^{\infty}u_mu_ndx=0 \tag{11.229}$$

Using integration by parts, we have

$$\int_0^{\infty}[u_m\frac{d}{dx}(x\frac{du_n}{dx})dx=\int_0^{\infty}[u_m d(x\frac{du_n}{dx})=xu_m\frac{du_n}{dx}\Big|_0^{\infty}-\int_0^{\infty}x\frac{du_n}{dx}\frac{du_m}{dx}dx \tag{11.230}$$

Using this and the expression of reversing n and m in (11.230), we have the first term of (11.229) equal to

$$\int_0^{\infty}[u_m\frac{d}{dx}(x\frac{du_n}{dx})-u_n\frac{d}{dx}(x\frac{du_m}{dx})]dx=x(u_m\frac{du_n}{dx}-u_m\frac{du_n}{dx})\Big|_0^{\infty} \tag{11.231}$$

To consider (11.231), we note that

$$
\begin{aligned}
u_n &= e^{-x/2}x^{\alpha/2}\sum_{s=0}^{n}\frac{\Gamma(n+\alpha+1)}{\Gamma(s+\alpha+1)(n-s)!s!}(-x)^s \\
&= e^{-x/2}x^{\alpha/2}\{\frac{\Gamma(n+\alpha+1)}{\Gamma(\alpha+1)n!}-\frac{\Gamma(n+\alpha+1)}{\Gamma(\alpha+2)(n-1)!}x+...\}
\end{aligned} \tag{11.232}
$$

$$
\begin{aligned}
\frac{du_n}{dx} &= e^{-x/2}x^{\alpha/2}\{-\frac{\Gamma(n+\alpha+1)}{\Gamma(\alpha+2)(n-1)!}+...\} \\
&-\frac{1}{2}e^{-x/2}x^{\alpha/2}\{\frac{\Gamma(n+\alpha+1)}{\Gamma(\alpha+1)n!}-\frac{\Gamma(n+\alpha+1)}{\Gamma(\alpha+2)(n-1)!}x+...\} \\
&+\frac{\alpha}{2}e^{-x/2}x^{\alpha/2-1}\{\frac{\Gamma(n+\alpha+1)}{\Gamma(\alpha+1)n!}-\frac{\Gamma(n+\alpha+1)}{\Gamma(\alpha+2)(n-1)!}x+...\}
\end{aligned} \tag{11.233}
$$

Similarly, we can get corresponding expressions for u_m by replacing n by m:

$$
u_m = e^{-x/2}x^{\alpha/2}\{\frac{\Gamma(m+\alpha+1)}{\Gamma(\alpha+1)m!}-\frac{\Gamma(m+\alpha+1)}{\Gamma(\alpha+2)(m-1)!}x+...\} \tag{11.234}
$$

$$
\begin{aligned}
\frac{du_m}{dx} &= e^{-x/2}x^{\alpha/2}\{-\frac{\Gamma(m+\alpha+1)}{\Gamma(\alpha+2)(m-1)!}+...\} \\
&-\frac{1}{2}e^{-x/2}x^{\alpha/2}\{\frac{\Gamma(m+\alpha+1)}{\Gamma(\alpha+1)m!}-\frac{\Gamma(m+\alpha+1)}{\Gamma(\alpha+2)(m-1)!}x+...\} \\
&+\frac{\alpha}{2}e^{-x/2}x^{\alpha/2-1}\{\frac{\Gamma(m+\alpha+1)}{\Gamma(\alpha+1)m!}-\frac{\Gamma(m+\alpha+1)}{\Gamma(\alpha+2)(m-1)!}x+...\}
\end{aligned} \tag{11.235}
$$

Finally, we have

$$
\begin{aligned}
x\frac{du_m}{dx}u_n &= e^{-x}x^{\alpha+1}\{-\frac{\Gamma(n+\alpha+1)}{\Gamma(\alpha+1)(n)!}[\frac{\Gamma(m+\alpha+1)}{\Gamma(\alpha+2)(m-1)!} \\
&+\frac{1}{2}\frac{\Gamma(m+\alpha+1)}{\Gamma(\alpha+1)(m)!}]+...\}+e^{-x}x^{\alpha}\{\frac{\Gamma(n+\alpha+1)}{\Gamma(\alpha+1)n!}\frac{\Gamma(m+\alpha+1)}{\Gamma(\alpha+1)m!}+...\}
\end{aligned} \tag{11.236}
$$

$$
\begin{aligned}
x\frac{du_n}{dx}u_m &= e^{-x}x^{\alpha+1}\{-\frac{\Gamma(m+\alpha+1)}{\Gamma(\alpha+1)(m)!}[\frac{\Gamma(n+\alpha+1)}{\Gamma(\alpha+2)(n-1)!}+\frac{1}{2}\frac{\Gamma(n+\alpha+1)}{\Gamma(\alpha+1)(n)!}]+...\} \\
&+e^{-x}x^{\alpha}\{\frac{\Gamma(m+\alpha+1)}{\Gamma(\alpha+1)m!}\frac{\Gamma(n+\alpha+1)}{\Gamma(\alpha+1)n!}+...\}
\end{aligned} \tag{11.237}
$$

Substitution of (11.236) and (11.237) into (11.231) gives

$$
\begin{aligned}
x(\frac{du_n}{dx}u_m-\frac{du_m}{dx}u_n) &= e^{-x}x^{\alpha+1}\{\frac{\Gamma(n+\alpha+1)!}{\Gamma(\alpha+1)(n)!}\frac{\Gamma(m+\alpha+1)!}{\Gamma(\alpha+2)(m-1)!} \\
&-\frac{\Gamma(m+\alpha+1)!}{\Gamma(\alpha+1)m!}\frac{\Gamma(n+\alpha+1)!}{\Gamma(\alpha+2)(n-1)!}+...\}
\end{aligned} \tag{11.238}
$$

Note the coefficient for the term $e^{-x}x^{\alpha}$ is zero. Therefore, we have

$$
\lim_{x\to 0}[x(u_m\frac{du_n}{dx}-u_n\frac{du_m}{dx})]\sim\lim_{x\to 0}x^{\alpha+1}=0,\quad \alpha>-1 \tag{11.239}
$$

On the other hand, the upper limit on the right-hand side of (11.231) is

$$\lim_{x\to\infty}[x(u_m \frac{du_n}{dx} - u_n \frac{du_m}{dx})] \sim \lim_{x\to\infty} e^{-x} x^{\alpha+1} = 0 \tag{11.240}$$

The last result of (11.240) can be proved by applying L'Hôpital's rule k times with $k > \alpha+1$. Substitution of (11.239), (11.240) into (11.231), then the result into (11.229) yields

$$(m-n)\int_0^\infty u_m u_n dx = 0 \tag{11.241}$$

If $m \neq n$, we must require

$$\int_0^\infty u_m u_n dx = 0 \tag{11.242}$$

Substituting (11.225) into (11.242) gives the

$$\int_0^\infty e^{-x} x^\alpha L_n^\alpha(x) L_m^\alpha(x) dx = 0 \tag{11.243}$$

If $m = n$, it can be shown that the integral is:

$$\int_0^\infty e^{-x} x^\alpha L_n^\alpha(x) L_n^\alpha(x) dx = \frac{\Gamma(n+\alpha+1)}{n!} \tag{11.244}$$

Combining these two results, the orthogonality of the associated Laguerre polynomials is

$$\int_0^\infty e^{-x} x^\alpha L_n^\alpha(x) L_m^\alpha(x) dx = \frac{\Gamma(n+\alpha+1)}{n!} \delta_{mn} \tag{11.245}$$

The remaining task is to prove the validity of (11.244). To prove (11.244), we need to derive two identities first.

The first one is

$$(1-t)^2 \frac{\partial w}{\partial t} + [x-(1-t)(1+\alpha)]w = 0 \tag{11.246}$$

where w is the generating function defined in (11.215) or

$$w(x,t) = (1-t)^{-\alpha-1} \exp\{-xt/(1-t)\} = \sum_{m=0}^\infty L_m^\alpha(x) t^m \tag{11.247}$$

The second identity is the 3-term recursive formula of the associated Laguerre polynomials:

$$(n+1)L_{n+1}^\alpha(x) + (x-1-\alpha-2n)L_n^\alpha(x) + (n+\alpha)L_{n-1}^\alpha(x) = 0 \tag{11.248}$$

In deriving (11.246), we first differentiate (11.247) with respect to t as

$$\frac{\partial w}{\partial t} = (1-t)^{-\alpha-2} \exp\{-xt/(1-t)\}[\alpha+1-\frac{x}{1-t}] \tag{11.249}$$

Thus, multiplying (11.249) by $(1-t)^2$, we get

$$\begin{aligned}(1-t)^2 \frac{\partial w}{\partial t} &= (1-t)^{-\alpha-1} \exp\{-xt/(1-t)\}[(1-t)(\alpha+1)-x] \\ &= [(1-t)(\alpha+1)-x]w\end{aligned} \tag{11.250}$$

This is clearly the same as (11.246).

We now proceed to prove (11.248). We can substitute the right-hand side of the generating function (11.247) into (11.246):

$$(1-t)^2\sum_{n=1}^{\infty} nL_n^{\alpha}(x)t^{n-1} + [x-(1-t)(1+\alpha)]\sum_{n=0}^{\infty} L_n^{\alpha}(x)t^n = 0 \qquad (11.251)$$

Expanding (11.251), we find

$$\begin{aligned}&\sum_{n=1}^{\infty} nL_n^{\alpha}(x)t^{n-1} - 2\sum_{n=1}^{\infty} nL_n^{\alpha}(x)t^n + \sum_{n=1}^{\infty} nL_n^{\alpha}(x)t^{n+1} + (x-1-\alpha)\sum_{n=0}^{\infty} L_n^{\alpha}(x)t^n \\ &+(1+\alpha)\sum_{n=0}^{\infty} L_n^{\alpha}(x)t^{n+1} = 0\end{aligned} \qquad (11.252)$$

Shifting the summation index for each sum in (11.252) such that they have the same power series for t, we obtain

$$\begin{aligned}&\sum_{n=0}^{\infty} (n+1)L_{n+1}^{\alpha}(x)t^n - 2\sum_{n=1}^{\infty} nL_n^{\alpha}(x)t^n + \sum_{n=2}^{\infty} (n-1)L_{n-1}^{\alpha}(x)t^n \\ &+(x-1-\alpha)\sum_{n=0}^{\infty} L_n^{\alpha}(x)t^n + (1+\alpha)\sum_{n=1}^{\infty} L_{n-1}^{\alpha}(x)t^n = 0\end{aligned} \qquad (11.253)$$

Writing out the two zero terms explicitly and grouping all the summations, we obtain

$$\begin{aligned}&L_1^{\alpha}(x) + (x-1-\alpha)L_0^{\alpha}(x) \\ &+\sum_{n=1}^{\infty} \{(n+1)L_{n+1}^{\alpha} + (x-1-\alpha-2n)L_n^{\alpha} + (n+\alpha)L_{n-1}^{\alpha}\}t^n = 0\end{aligned} \qquad (11.254)$$

Using the definition of the associated Laguerre polynomials, we have the following identities

$$L_0^{\alpha} = 1, \quad L_1^{\alpha} = 1+\alpha-x, \quad L_2^{\alpha} = \frac{x^2}{2} - (\alpha+2)x + \frac{(\alpha+2)(\alpha+1)}{2}, \ldots \qquad (11.255)$$

The first terms of (11.254) clearly cancel one another. Thus, we obtain the following 3-term recursive formula for the associated Laguerre polynomials:

$$(n+1)L_{n+1}^{\alpha} + (x-1-\alpha-2n)L_n^{\alpha} + (n+\alpha)L_{n-1}^{\alpha} = 0 \qquad (11.256)$$

for $n = 1,2,3,\ldots$ This completes the proof of (11.242). There is a so-called Shohat-Favard theorem that when functions satisfy a 3-term recursive formula similar to (11.256), then this function must be orthogonal.

We are now ready to prove (11.244). First, we can rewrite (11.256) for order $m = n-1$ and multiply the result by the associated Laguerre polynomials of order m, and we obtain

$$mL_m^{\alpha}L_m^{\alpha} + (x-1-\alpha-2m+2)L_{m-1}^{\alpha}L_m^{\alpha} + (m-1+\alpha)L_{m-2}^{\alpha}L_m^{\alpha} = 0 \qquad (11.257)$$

In a similar fashion, we can rewrite (11.256) replacing n by m and multiply the result by the associated Laguerre polynomials of order $m-1$ to get

$$(m+1)L_{m+1}^{\alpha}L_{m-1}^{\alpha} + (x-1-\alpha-2m)L_m^{\alpha}L_{m-1}^{\alpha} + (m+\alpha)L_{m-1}^{\alpha}L_{m-1}^{\alpha} = 0 \quad (11.258)$$

Subtracting (11.258) from (11.257), we get

$$m[L_m^\alpha]^2-(m+\alpha)[L_{m-1}^\alpha]^2+2L_{m-1}^\alpha L_m^\alpha+(m-1+\alpha)L_{m-2}^\alpha L_m^\alpha+(m+1)L_{m+1}^\alpha L_{m-1}^\alpha=0 \tag{11.259}$$

We now can multiply (11.259) by $e^{-x}x^\alpha$ and integrate the product from 0 to ∞ to yield

$$m\int_0^\infty e^{-x}x^\alpha[L_m^\alpha]^2\,dx=(m+\alpha)\int_0^\infty e^{-x}x^\alpha[L_{m-1}^\alpha]^2\,dx \tag{11.260}$$

Note that the last three integrals from (11.259) are zero because of the orthogonal property of (11.243). Alternatively, it can be written as

$$\int_0^\infty e^{-x}x^\alpha[L_m^\alpha]^2\,dx=\frac{(m+\alpha)}{m}\int_0^\infty e^{-x}x^\alpha[L_{m-1}^\alpha]^2\,dx \tag{11.261}$$

This is actually a recursive formula for the integral itself. Reapplying this formula to the right-hand side of (11.261) m times, we get

$$\int_0^\infty e^{-x}x^\alpha[L_m^\alpha]^2\,dx=\frac{(m+\alpha)(m+\alpha-1)\cdots(\alpha+1)}{m(m-1)\cdots 2\cdot 1}\int_0^\infty e^{-x}x^\alpha[L_0^\alpha]^2\,dx \tag{11.262}$$

In view of (11.255), the last integral can be evaluated as

$$\int_0^\infty e^{-x}x^\alpha[L_0^\alpha]^2\,dx=\int_0^\infty e^{-x}x^\alpha\,dx=\Gamma(\alpha+1) \tag{11.263}$$

The last equation in (11.263) results from the definition of the gamma function (see Chau, 2018). Substitution of (11.263) into (11.262) gives

$$\int_0^\infty e^{-x}x^\alpha[L_m^\alpha]^2\,dx=\frac{\Gamma(m+\alpha+1)}{m!},\quad m=2,3,\ldots \tag{11.264}$$

This result can be extended to cover the case of m = 1, 0. For m = 0, the left-hand side of (11.264) is actually the left-hand side of (11.263), and, thus, (11.264) is valid. For m = 1, the left-hand side of (11.264) is

$$\begin{aligned}\int_0^\infty e^{-x}x^\alpha[L_1^\alpha]^2\,dx&=\int_0^\infty e^{-x}x^\alpha(1+\alpha-x)^2\,dx\\&=(1+\alpha)^2\int_0^\infty e^{-x}x^\alpha dx+\int_0^\infty e^{-x}x^{\alpha+2}dx-2(1+\alpha)\int_0^\infty e^{-x}x^{\alpha+1}dx\\&=(1+\alpha)^2\Gamma(\alpha+2)+\Gamma(\alpha+3)-2(1+\alpha)\Gamma(\alpha+2)\\&=\Gamma(\alpha+2)\end{aligned} \tag{11.265}$$

Thus, (11.264) is extended to m = 0,1,2,3,... In conclusion, the orthogonality of the associated Laguerre polynomials is

$$\int_0^\infty e^{-x}x^\alpha L_n^\alpha(x)L_m^\alpha(x)dx=\frac{\Gamma(n+\alpha+1)}{n!}\delta_{mn} \tag{11.266}$$

Our discussion of the associated Laguerre polynomials ends here. We will see how these properties of the associated Laguerre polynomials lead to the quantization of energy.

11.6.6 Admissible Form of the Wavefunctions

We now return to the atom problem. Recall from (11.145) and (11.146) that

$$\rho w'' + [(2l+1)+1-\rho]w' + (\lambda - 1 - l)w = 0 \tag{11.267}$$

We see from the last section that for the associated Laguerre polynomials to be finite, $\lambda - l - 1$ must be an integer. In turn, we can also interpret that λ must be an integer as well, say n:

$$\lambda = n \tag{11.268}$$

This mathematical restriction is of utmost importance in the quantum states of energy E. In particular, recall from (11.140) that

$$\lambda = \frac{2mZe^2}{4\pi\varepsilon_0 \kappa \hbar^2} = n \tag{11.269}$$

Rearranging (11.269), we get

$$\kappa = \frac{2mZe^2}{4n\pi\varepsilon_0 \hbar^2} \tag{11.270}$$

Squaring of this result and equating it to the last equation of (11.140), we obtain

$$\kappa^2 = \frac{4m^2 Z^2 e^4}{16n^2 \pi^2 \varepsilon_0^2 \hbar^4} = -\frac{8mE}{\hbar^2} \tag{11.271}$$

Then, we have the energy states as

$$E = -\frac{mZ^2 e^4}{32\pi^2 \varepsilon_0^2 \hbar^2}\left(\frac{1}{n^2}\right) = -E_1\left(\frac{1}{n^2}\right) \tag{11.272}$$

This is exactly (11.176), predicted by Bohr's model for the case of hydrogen (i.e., $Z = 1$). It also agrees with our earlier result that $E < 0$ (see Example 11.1). The coincidence is not fully understood yet, because the Schrödinger equation is completely different from the classical mechanics-based Bohr's hydrogen model (see Example 11.4). Recall from (11.162) that

$$\kappa = \frac{mZe^2}{2n\pi\varepsilon_0 \hbar^2} = \frac{1}{a_0}\left(\frac{2Z}{n}\right) \tag{11.273}$$

where

$$a_0 = -\frac{4\pi\varepsilon_0 \hbar^2}{Zme^2} \tag{11.274}$$

Using this result, we can simplify our solution given in (11.149) as

$$\begin{aligned} \psi_{nlm_l}(r,\theta,\phi) &= e^{-Zr/(na_0)}\left(\frac{Zr}{na_0}\right)^l L_{n-l-1}^{2l+1}\left(\frac{Zr}{na_0}\right) P_l^{m_l}(\cos\theta)\exp(im_l\phi) \\ &= e^{-Zr/(na_0)}\left(\frac{Zr}{na_0}\right)^l L_{n-l-1}^{2l+1}\left(\frac{Zr}{na_0}\right) Y_{lm_l}(\theta,\phi) \end{aligned} \tag{11.275}$$

For the case of the hydrogen atom (i.e., $Z = 1$), this wavefunction can be simplified to

$$\psi_{nlm_l}(r,\theta,\phi)=e^{-r/(na_0)}(\frac{r}{na_0})^l L_{n-l-1}^{2l+1}(\frac{r}{na_0})Y_{lm_l}(\theta,\phi) \quad (11.276)$$

There are three quantum numbers, namely n, l, and m_l. Mathematically, these three constants result from the separation of variables. Recalling the wavefunction, we have

$$\psi_{nlm_l}(r,\theta,\phi)=R_{nl}(r)\Theta_{lm_l}(\theta)\Phi_{m_l}(\phi) \quad (11.277)$$

We can see that the radial function R depends on both n and l. The angular function Θ depends on both l and m_l. The azimuth function Φ depends only on m_l. We have seen that the angular periodicity of Φ around the spherical coordinate requires that the magnetic quantum number m_l is an integer, the boundedness condition of the associated Laguerre polynomials at infinity requires that the principal quantum number n is an integer and (11.272) shows that n reflects the energy levels, and finally the associated Laguerre polynomials given in (11.214) require $n > l$. In summary, we have the following quantum numbers:

$$\begin{aligned}&\text{Principal}: \quad n=1,2,3,...\\&\text{Orbital}: \quad l=1,2,3,...,(n-1)\\&\text{Magnetic}: \quad m_l=0,\pm1,\pm2,\pm3,...,\pm l\end{aligned} \quad (11.278)$$

Physically, we have seen in (11.272) that n is related to the quantization of energy levels. To see the physical meaning of l, the total energy E is

$$E=KE_{radial}+KE_{orbital}+U \quad (11.279)$$

Recalling the governing equation for R given in (11.129) but without assuming (11.119), we have

$$\frac{1}{r^2}\frac{d}{dr}(r^2\frac{dR}{dr})+\frac{2m}{\hbar^2}[KE_{radial}+KE_{orbital}-\frac{\hbar^2 l(l+1)}{2mr^2}]R=0 \quad (11.280)$$

The last two terms in (11.278) are related to angular momentum, but this equation for R is for radial dependence only. Thus, we must have

$$KE_{orbital}=\frac{\hbar^2 l(l+1)}{2mr^2} \quad (11.281)$$

The orbital kinetic energy is

$$KE_{orbital}=\frac{1}{2}mv_0^2 \quad (11.282)$$

The angular momentum is defined as

$$L=mv_0 r \quad (11.283)$$

Using this information, we get

$$\frac{L^2}{2mr^2}=\frac{\hbar^2 l(l+1)}{2mr^2} \quad (11.284)$$

This yields the angular momentum in terms of the orbital quantum number l:

$$L=\hbar\sqrt{l(l+1)} \quad (11.285)$$

Therefore, the angular momentum of the electron is also quantized. The z-component of the angular momentum is defined as

$$L_z=\hbar m_l < L=\hbar\sqrt{l(l+1)} \quad (11.286)$$

where

$$m_l = 0, \pm 1, \pm 2, \pm 3, ..., \pm l \tag{11.287}$$

In the presence of a magnetic field, the z-component of angular momentum is found to be quantized as given in (11.286). This is the reason why it was labelled as the magnetic quantum number in (11.278). Figure 11.6 demonstrates the case of $l = 2$, and thus 5 possible values of m_l.

In the presence of a magnetic field, the z-component of the angular momentum is quantized as shown in Figure 11.6. We see that L_z is fixed, but according to Heisenberg's uncertainty principle, we must have the other components of the angular momentum indeterministic such that

$$\langle L_x \rangle = \langle L_y \rangle = 0 \tag{11.288}$$

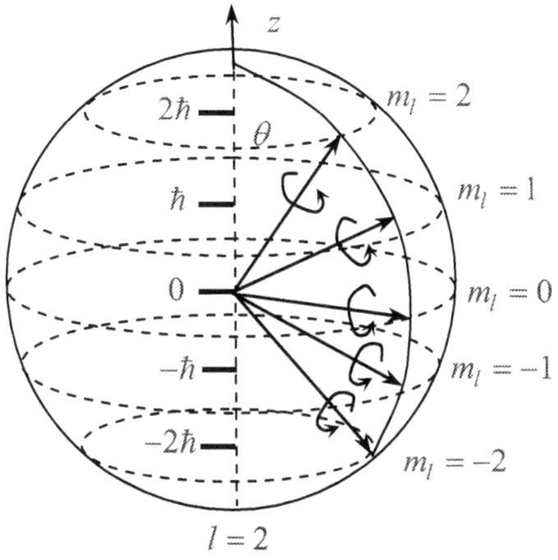

Figure 11.6 Illustration of the magnetic quantum number

We put forward an analogy here that it resembles the unsteady whopping of a coin with normal being as the z-axis. The magnitude of the tilting of the coin depends on the angular momentum. In the context of quantum mechanics, such whopping motions can be stable only at certain amplitudes, depending on l.

Table 11.2 tabulates the number system that has adopted in naming the admissible orbits for different energy levels n. The capital letters are referred to as shells, such as the K-shell.

Table 11.2 Energy levels and their associated shell letters in atoms

n	1	2	3	4	5	6	7
Shell	K	L	M	N	O	P	Q

Table 11.3 tabulates the letter system that has been adopted in naming the admissible orbits for different angular momenta l. The number of degeneracy (i.e., number of orbits of the same energy level) of each level l can be calculated by the Aufbau principle:

$$D(l) = 2(2l+1) \tag{11.289}$$

This is given in the third row in Table 11.3. The letter system given in Table 11.3 was adopted before quantum mechanics was developed. In fact, they are from spectroscopic notation in atomic spectra. When a gas tube of gaseous hydrogen is excited by a 500-volt electrical discharge, four lines are observed in the visible part of red, blue-green, blue-violet and violet, and they are the so-called Balmer series (electrons jumping from $n \geq 3$ to $n = 2$). It turns out that these observed frequencies of these observed lines satisfy the following formula established by Rydberg in 1890:

$$\frac{1}{\lambda} = R(\frac{1}{n_1^2} - \frac{1}{n_2^2}) \tag{11.290}$$

Table 11.3 Orbital angular quantum number and its associated spectroscopic notation, degeneracy, and angular momentum

$l =$	0	1	2	3	4	5	6
Subshell	s	p	d	f	g	h	i
$D(l)$	2	6	10	14	18	22	26
L	0	$\sqrt{2}\hbar$	$\sqrt{6}\hbar$	$2\sqrt{3}\hbar$	$2\sqrt{5}\hbar$	$\sqrt{30}\hbar$	$\sqrt{42}\hbar$

where R equals 109,677 cm^{-1}. Another series is called the Lyman series arising from electrons jumping from energy level $n \geq 2$ to $n = 1$. The letter s means "sharp," p means "principal," d means "diffuse," and f means "fundamental." These lower case letters are referred to as subshells, such as the p-subshell.

Combining the two systems given in Tables 11.2 and 11.3, we have the system in naming the atomic electron states used in the literature. These atomic electron states are compiled in Table 11.4. The periodic table was formulated by Mendeleev, and whether the shells are closed (completed filled) has profound importance in the chemical behavior of elements. For example, the atomic numbers of the noble gases Helium, Neon, Argon, Xenon, and Radon are 2, 10, 18, 36, 54, and 86 respectively. It turns out that they are highly stable elements and the shells are closed (or fully filled up by electrons). The difference between these atomic numbers are 8, 8, 18, and 32. This leads to the observations, including the first 2, that

$$2 = 2(1)^2, \quad 8 = 2(2)^2, \quad 18 = 2(3)^2, \quad 32 = 2(4)^2 \tag{11.291}$$

This suggests that the number of electrons that can be accommodated in shells is of the form:

$$\text{electron number in shell} = 2(n^2) \tag{11.292}$$

Actually, this can also be explained by the wavefunctions that we have obtained. The maximum number of elements in a shell appears to relate to the probability of magnetic angular momentum states for each energy level n (see (11.287))

$$\begin{aligned} N_{max} &= \sum_{l=}^{n-1} 2(2l+1) = 2[1+3+5+...+2(n-1)+1] \\ &= 2[1+3+5+...+2n-1] \\ &= 2\frac{n}{2}[1+(2n-1)] = 2n^2 \end{aligned} \tag{11.293}$$

The factor 2 in the summation accounts for two spins of the electrons, which will be discussed in Section 11.7. This agrees exactly with our observation for noble gas or inert gas. The reason is that atoms with closed shells have no dipole moment and they do not tend to react with other elements. Therefore, the closed K shell holds 2 electrons, a closed L shell holds 8 electrons, a closed M holds 18 electrons, and so on.

Table 11.4 Atomic electron states

$n=$	Atomic electron states					
	$l=0$	$l=1$	$l=2$	$l=3$	$l=4$	$l=5$
1	1*s*					
2	2*s*	2*p*				
3	3*s*	3*p*	3*d*			
4	4*s*	4*p*	4*d*	4*f*		
5	5*s*	5*p*	5*d*	5*f*	5*g*	
6	6*s*	6*p*	6*d*	6*f*	6*g*	6*h*

The first few wavefunctions predicted by (11.274) are compiled in Table 11.5. Higher order wavefunctions are given in Problems 11.6.

Table 11.5 Atomic electron states

n	l	m_l	wavefunctions
1	0	0	$\psi = \frac{1}{\sqrt{\pi}}(\frac{Z}{a_0})^{3/2}\exp(-\frac{Zr}{a_0})$
2	0	0	$\psi = \frac{1}{4\sqrt{2\pi}}(\frac{Z}{a_0})^{3/2}(1-\frac{Zr}{a_0})\exp(-\frac{Zr}{2a_0})$
2	1	0	$\psi = \frac{1}{4\sqrt{2\pi}}(\frac{Z}{a_0})^{3/2}\frac{Zr}{a_0}\exp(-\frac{Zr}{2a_0})\cos\theta$
2	1	±1	$\psi = \frac{1}{8\sqrt{\pi}}(\frac{Z}{a_0})^{3/2}\frac{Zr}{a_0}\exp(-\frac{Zr}{2a_0})\sin\theta\exp(\pm i\phi)$

Example 11.5 Find the probability of finding an electron at r for the $1s$ orbital.

Solution: For this case we have

$$n=1, \quad l=0, \quad m_l=0 \tag{11.294}$$

From Table 11.5, we find

$$\psi_{1,0,0}=\frac{1}{\sqrt{\pi}}(\frac{Z}{a_0})^{3/2}\exp(-\frac{Zr}{a_0}) \tag{11.295}$$

The probability of finding the electron is given by

$$P_{nlm_l}(r,\theta,\phi)=\left|\psi_{nlm_l}(r,\theta,\phi)\right|^2=\psi_{nlm_l}(r,\theta,\phi)\overline{\psi}_{nlm_l}(r,\theta,\phi) \tag{11.296}$$

Substitution of (11.295) into (11.296) gives the probability as

$$P_{1,0,0}(r,\theta,\phi)=[\frac{1}{\sqrt{\pi}}(\frac{Z}{a_0})^{3/2}\exp(-\frac{Zr}{a_0})]^2=\frac{1}{\pi}(\frac{Z}{a_0})^3\exp(-\frac{2Zr}{a_0}) \tag{11.297}$$

For the wavefunctions adopted in chemistry, we normally adopt a slightly different notation and only real wavefunctions are used. For example, the p-orbitals for $n = 2$ from Table 11.5 are

$$\psi_{2,1,0}=\frac{1}{4\sqrt{2\pi}}(\frac{Z}{a_0})^{3/2}\frac{Zr}{a_0}\exp(-\frac{Zr}{2a_0})\cos\theta \tag{11.298}$$

$$\psi_{2,1,\pm1}=\frac{1}{8\sqrt{\pi}}(\frac{Z}{a_0})^{3/2}\frac{Zr}{a_0}\exp(-\frac{Zr}{2a_0})\sin\theta\exp(\pm i\phi) \tag{11.299}$$

By observing from the spherical coordinate that

$$z=r\cos\theta, \quad x=r\sin\theta\cos\phi, \quad y=r\sin\theta\sin\phi, \tag{11.300}$$

we can define a new wavefunction as

$$\psi_{2pz}=\psi_{2,1,0}=\frac{1}{4\sqrt{2\pi}}(\frac{Z}{a_0})^{5/2}z\exp(-\frac{Zr}{2a_0}) \tag{11.301}$$

The two other wavefunctions given in (11.299) are complex. To remove the imaginary parts, we can define two independent wavefunctions as:

$$\psi_{2px}=\frac{1}{\sqrt{2}}(\psi_{2,1,-1}+\psi_{2,1,+1}) \tag{11.302}$$

$$\psi_{2py}=\frac{-i}{\sqrt{2}}(\psi_{2,1,+1}-\psi_{2,1,-1}) \tag{11.303}$$

Substituting (11.299) into (11.302) and (11.303) and in view of (11.300) and Euler's formula (see Chau, 2018), we obtain

$$\psi_{2py}=\frac{1}{4\sqrt{2\pi}}(\frac{Z}{a_0})^{5/2}y\exp(-\frac{Zr}{2a_0}) \tag{11.304}$$

$$\psi_{2px}=\frac{1}{4\sqrt{2\pi}}(\frac{Z}{a_0})^{5/2}x\exp(-\frac{Zr}{2a_0}) \tag{11.305}$$

In fact, it turns out that (11.301), (11.304) and (11.305) are general, and all wavefunctions for p-orbitals for $n \geq 3$ can also be cast into the form

$$\psi_{npz} = zf_n(r), \quad \psi_{npx} = xf_n(r), \quad \psi_{npy} = yf_n(r) \tag{11.306}$$

These functions are real and can be visualized in 3-D plotting easily. Graphically, the wavefunctions (11.301), (11.304) and (11.305) are illustrated in Figure 11.7, and they appear symmetrically along the z, y, and x-axes, respectively.

11.7 ELECTRON SPINS

The vertical lines for hydrogen observed in spectroscopic plots are split into more closely spaced lines when a magnetic field is applied to the hydrogen gas. This is called the Zeeman effect, which was first observed in 1896 by Dutch physicist Pieter Zeeman. The splitting of atomic lines due to an electric field is called the Stark effect. By considering the moving electron around the nucleus as current, the radiation was shifted left and right in the frequency spectrum (depending on the magnetic quantum number m_l) or so-called splitting. However, the calculation from the orbital magnetic quantum number does not match the number of splitting as well as the magnitude of splitting. This problem is called the Anomalous Zeeman effect. In 1925, two Dutch graduate students (S. Goudsmit and G. Uhlenbeck) proposed the idea that all electrons have intrinsic angular momentum (called spin). However, the effects of this spin are too weak to be observed in experiments. In 1929, Paul Dirac proved analytically the existence of such spin using relativistic quantum mechanics. The spin angular momentum of an electron was found by Dirac as

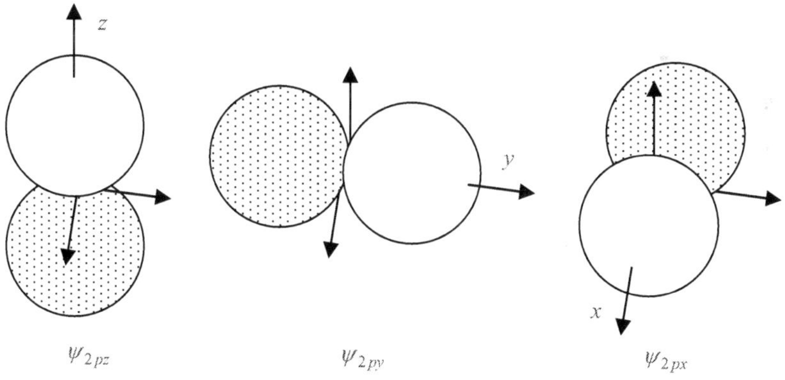

Figure 11.7 Illustration of the magnetic quantum number

$$S = \hbar\sqrt{s(s+1)} = \frac{\sqrt{3}}{2}\hbar \tag{11.307}$$

where

$$s = \frac{1}{2} \tag{11.308}$$

This value of spin agrees with the observation of spectral lines. Note that this angular momentum is the same as that given in (11.285) except for s not being an integer. Therefore, we need altogether a set of four quantum numbers by adding a spin magnetic quantum number m_s equal to

$$m_s = -\frac{1}{2}, +\frac{1}{2} \tag{11.309}$$

The first one is referred to as spin up whereas the second one is referred to as spin down. It turns out that all atomic particles have spin. If the spin number s of certain particles is of half integral (1/2, 3/2,...), they are called fermions and their statistical distributions follow the Fermi-Dirac distribution law. Fermions include protons, neutrons, and electrons. If the spin number s of certain particles is 0 or an integer, they are called bosons and their statistical distributions follow the Bose-Einstein distribution law. Bosons include photons, alpha particles, and helium atoms.

11.8 SCHRÖDINGER EQUATION FOR GENERAL ATOMS

For general atoms with Z electrons and Z protons, the Hamiltonian is

$$H\psi = \sum_{j=1}^{Z} \left\{ -\frac{\hbar^2}{2m} \nabla_j^2 \psi - \frac{1}{4\pi\varepsilon_0} \frac{Ze^2}{r_j} \psi \right\} + \frac{1}{2} \frac{1}{4\pi\varepsilon_0} \sum_{j \neq k}^{Z} \frac{e^2}{|\vec{r}_j - \vec{r}_k|} \psi \tag{11.310}$$

where the first summation is the kinetic energy and potential energy of the j-th electron produced by the nucleus, and the second summation is potential energy associated with mutual repulsion of the electrons. The factor 1/2 in front of the second sum is to cancel the double count for each electron pair.

Thus, the Schrödinger equation can be written as

$$\sum_{j=1}^{Z} \left\{ -\frac{\hbar^2}{2m} \nabla_j^2 \psi - \frac{1}{4\pi\varepsilon_0} \frac{Ze^2}{r_j} \psi \right\} + \frac{1}{2} \frac{1}{4\pi\varepsilon_0} \sum_{j \neq k}^{Z} \frac{e^2}{|\vec{r}_j - \vec{r}_k|} \psi = E\psi \tag{11.311}$$

This is a very difficult mathematical problem to solve. If we can find the wavefunction, the probability of finding electrons can be estimated. For example, for the case of Helium, the probability of finding the two electrons at two specific positions is

$$P(\vec{r}_1, \vec{r}_2) = |\psi(\vec{r}_1, \vec{r}_2)|^2 \, dV(\vec{r}_1) dV(\vec{r}_2) \tag{11.312}$$

A similar expression can be written for atoms with a larger atomic number.

To simplify the problem given in (11.311), if repulsion between electrons is ignored, we can write

$$\sum_{j=1}^{Z} \left\{ -\frac{\hbar^2}{2m} \nabla_j^2 - \frac{1}{4\pi\varepsilon_0} \frac{Ze^2}{r_j} \right\} \psi(\vec{r}_1, \vec{r}_2, \ldots, \vec{r}_Z) = E\psi(\vec{r}_1, \vec{r}_2, \ldots, \vec{r}_Z) \tag{11.313}$$

With this approximation, the solution can be expressed as separable:

$$\psi(\vec{r}_1, \vec{r}_2, \ldots, \vec{r}_Z) = \psi(\vec{r}_1)\psi(\vec{r}_2) \cdots \psi(\vec{r}_Z) \tag{11.314}$$

The solution of each of the wavefunctions is

$$\psi(\vec{r}_j) = \psi_{nlm_l}(r_j, \theta, \phi) \tag{11.315}$$

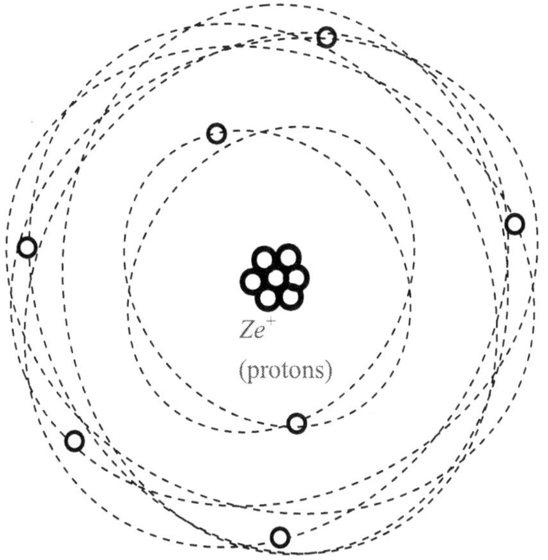

Figure 11.8 Atom models

for the (n,l,m_l) orbitals. Once the orbital is determined for each electron, we can use (11.312) or its extension for higher atomic numbers to find the probability of finding Z electrons at current locations. But, we have to fill in the shells one by one, 2 electrons in the K-shell, 8 electrons in the L-shell, and 18 electrons in M-shell and so on. For electrons in higher shells, the electron-electron repulsion becomes important and cannot be neglected as we did in (11.313). However, the same state cannot be occupied by two or more different electrons, and we need to fill the states sequentially. This rule is imposed by Pauli's exclusion principle, which is considered next.

If two particles a and b are identical and there is no interaction between the two particles, we can assume separation of variables as

$$\psi(\vec{r}_1, \vec{r}_2) = \psi_a(\vec{r}_1)\psi_b(\vec{r}_2) \tag{11.316}$$

However, there is no way that we can distinguish two electrons from one another. According to the uncertainty principle, we cannot identify the electron without disturbing it. Thus, (11.316) is not valid. Instead, we can take the linear combination of the two different states and each of these states is of the same probability as:

$$\psi(\vec{r}_1, \vec{r}_2) = A_1\psi_a(\vec{r}_1)\psi_b(\vec{r}_2) + A_2\psi_b(\vec{r}_1)\psi_a(\vec{r}_2) \tag{11.317}$$

with

$$|A_1| = |A_2| = \frac{1}{\sqrt{2}} \tag{11.318}$$

Therefore, (11.317) can be expressed as two possibilities

$$\psi_f(\vec{r}_1,\vec{r}_2)=\frac{1}{\sqrt{2}}[\psi_a(\vec{r}_1)\psi_b(\vec{r}_2)-\psi_b(\vec{r}_1)\psi_a(\vec{r}_2)] \tag{11.319}$$

$$\psi_b(\vec{r}_1,\vec{r}_2)=\frac{1}{\sqrt{2}}[\psi_a(\vec{r}_1)\psi_b(\vec{r}_2)+\psi_b(\vec{r}_1)\psi_a(\vec{r}_2)] \tag{11.320}$$

The wavefunction given in (11.319) applies to fermions whereas the second one given in (11.320) applies to bosons. As discussed in the last section, electrons are fermions. If the electrons are at the same state, we have

$$\psi_a(\vec{r})=\psi_b(\vec{r}) \tag{11.321}$$

Equation (11.319) implies the wavefunctions for fermions (including electrons) are zero, and this is physically impossible. In other words, two fermions can occupy the same state. That is, no two electrons can occupy the same state. This is Pauli's exclusion principle.

11.9 RADIATIVE TRANSITIONS FROM ATOMS

Experiments showed that the frequency of radiation emitted from atom when an electron jumps from one energy state to another satisfies the following equation

$$\nu=\frac{E_m-E_n}{h} \tag{11.322}$$

Actually, this formula can be proved by wavefunctions. Figure 11.8 illustrates the transition of an electron from one energy state to another after being excited by an external input (e.g., in the form of collision by other particles). A one-dimensional state is assumed in the present section. Recall the stationary state of the wavefunction given in (11.74) that

$$\Psi_n=\psi_n\exp(-iE_nt/\hbar) \tag{11.323}$$

Employing (11.323), we obtain the following expectation value of an electron at a particular position x as

$$\begin{aligned}\langle x\rangle&=\int_{-\infty}^{\infty}x\Psi_n\bar{\Psi}_n dx\\&=\int_{-\infty}^{\infty}x\psi_n\exp(-iE_nt/\hbar)\bar{\psi}_n\exp(iE_nt/\hbar)dx\\&=\int_{-\infty}^{\infty}x\psi_n\bar{\psi}_n dx\end{aligned} \tag{11.324}$$

For this case, we can see that the expectation is independent of time. This implies the electron does not oscillate with time. Therefore, a specific quantum state does not radiate any energy and this, of course, agrees with experimental observation. If the electron is in the transition from state n to state m, the probability of finding the electron at a particular position is

$$\Psi=a\Psi_n+b\Psi_m \tag{11.325}$$

Thus, the probability that the electron is in state n is

$$\langle\Psi_n\rangle=a\bar{a} \tag{11.326}$$

whereas the probability that the electron is in state m is

$$\langle \Psi_m \rangle = b\bar{b} \tag{11.327}$$

Of course, we must have

$$b\bar{b} + a\bar{a} = 1 \tag{11.328}$$

At the start of transition, since the electron is at state n, we have $a = 1$ and $b = 0$ at time zero. At the excited state m, the electron is at state m, and we have $a = 0$ and $b = 1$. In the midst of transition from state n to state m, we have

$$b \neq 0, \quad a \neq 0 \tag{11.329}$$

Thus, the expectation value of finding an electron at a particular position x is

$$\begin{aligned} \langle x \rangle &= \int_{-\infty}^{\infty} x(a\Psi_n + b\Psi_m)(\bar{a}\bar{\Psi}_n + \bar{b}\bar{\Psi}_m)dx \\ &= \int_{-\infty}^{\infty} x(a^2\Psi_n\bar{\Psi}_n + \bar{b}a\bar{\Psi}_m\Psi_n + \bar{a}b\bar{\Psi}_n\Psi_m + b^2\Psi_m\bar{\Psi}_m)dx \end{aligned} \tag{11.330}$$

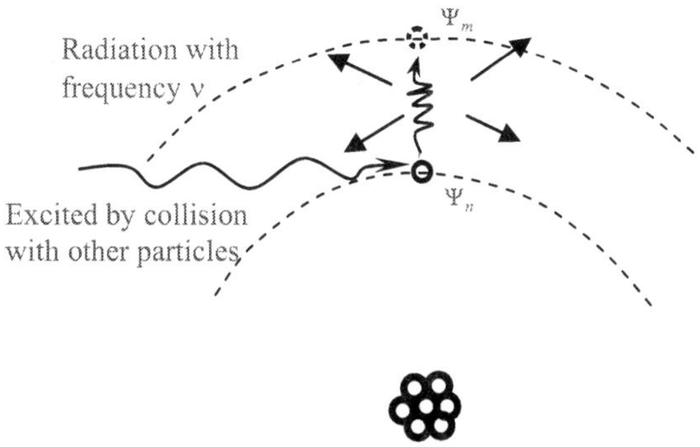

Figure 11.9 Radiation emitted from an electron jumping from one energy level to another

From (11.324), it is clear that the first term and the last term in (11.330) are time-independent. Substituting the stationary state given in (11.323) into (11.330), we obtain

$$\begin{aligned} \langle x \rangle &= \int_{-\infty}^{\infty} x(a^2\Psi_n\bar{\Psi}_n + b^2\Psi_m\bar{\Psi}_m)dx + \bar{b}a\int_{-\infty}^{\infty} x\bar{\psi}_m\psi_n e^{i(E_m-E_n)t/\hbar}dx \\ &+\bar{a}b\int_{-\infty}^{\infty} x\bar{\psi}_n\psi_m e^{i(E_n-E_m)t/\hbar}dx \end{aligned} \tag{11.331}$$

Recall Euler's formulas for both positive and negative arguments

$$e^{i\theta} = \cos\theta + i\sin\theta, \quad e^{-i\theta} = \cos\theta - i\sin\theta \tag{11.332}$$

Application of (11.332) to the second integral on the right-hand side of (11.331) gives

$$\bar{b}a\int_{-\infty}^{\infty} x\bar{\psi}_m\psi_n e^{i(E_m-E_n)t/\hbar}dx = \bar{b}a\int_{-\infty}^{\infty} x\bar{\psi}_m\psi_n[\cos(\frac{E_m-E_n}{\hbar}t) + i\sin(\frac{E_m-E_n}{\hbar}t)]dx \tag{11.333}$$

Similarly, the third integral in (11.332) is

$$\bar{a}b\int_{-\infty}^{\infty} x\bar{\psi}_n\psi_m e^{-i(E_m-E_n)t/\hbar}dx = \bar{a}b\int_{-\infty}^{\infty} x\bar{\psi}_n\psi_m[\cos(\frac{E_m-E_n}{\hbar}t)-i\sin(\frac{E_m-E_n}{\hbar}t)]dx \tag{11.334}$$

The sum of these two terms is

$$\begin{aligned}\langle x\rangle &= \int_{-\infty}^{\infty} x(a^2\Psi_n\bar{\Psi}_n + b^2\Psi_m\bar{\Psi}_m)dx \\ &+\cos(\frac{E_m-E_n}{\hbar}t)\int_{-\infty}^{\infty} x[\bar{\psi}_m\psi_n a\bar{b} + \bar{\psi}_n\psi_m\bar{a}b]dx \\ &+\sin(\frac{E_m-E_n}{\hbar}t)\int_{-\infty}^{\infty} x[\bar{\psi}_m\psi_n a\bar{b} - \bar{\psi}_n\psi_m\bar{a}b]dx\end{aligned} \tag{11.335}$$

Therefore, (11.335) shows that the position of the electron oscillates at the frequency ν

$$\nu = \frac{E_m - E_n}{\hbar} \tag{11.336}$$

The oscillating electron will act like an electric dipole and, thus, according to Maxwell's equations, radiate electromagnetic waves at the same frequency. This result agrees with (11.322) obtained by Bohr and agrees with experimental observation. This is another major success of quantum mechanics. The transition of an electron between two different states is only possible if

$$\int_{-\infty}^{\infty} x\bar{\psi}_m\psi_n dx > 0 \tag{11.337}$$

The transition of an electron between two different states is forbidden if

$$\int_{-\infty}^{\infty} x\bar{\psi}_m\psi_n dx = 0 \tag{11.338}$$

The general rule is that transition is possible only for

$$\Delta l = \pm 1, \quad \Delta m_l = 0, \pm 1 \tag{11.339}$$

Equivalently, the change of angular momentum between states is exactly $\pm\hbar$.

11.10 SUMMARY AND FURTHER READING

In this chapter, we have derived and discussed the Schrödinger equation and its application to quantum mechanics. The Schrödinger equation is a linear PDE and is considered as a wave equation, although mathematically it is quite different from the classical wave equation. This surprisingly simple equation, when applied appropriately, can model the essential observations in quantum mechanics. Both one-dimensional and three-dimensional cases are discussed. The physical significance of the Schrödinger equation is discussed through the meaning of the wave function of particles and the interpretation of the expectation of encountering an electron in an atom. In a statistical sense, operators and expectation values are discussed. The mathematical similarity between classical mechanics and quantum mechanics is summarized. A hydrogen-like atom is considered through the three-dimensional spherical polar form of the Schrödinger equation. The solution of the Schrödinger equation in polar form is obtained in terms of the associated Laguerre polynomials. The concept of electron spin is considered and the Schrödinger equation is then applied to model general atoms. Radiative transition from atoms is discussed. In view of the recent development of nano-mechanics and nano-technology, we should be equipped with the basic mathematical tools for analyzing problems in quantum mechanics. This chapter serves as an introduction to this important development of mechanics at small scales, in contrast to the classical mechanics.

There are many good textbooks on quantum mechanics. Some of them are more on conceptual and intuitive and suitable for beginners, and some of them are more mathematical and more ideal for the prepared reader. Older textbooks are normally presented within the historical context, and all arguments were used to justify the necessity and correctness of quantum mechanics, whereas the modern quantum mechanics textbooks simply compile the postulates of the theory and quantum mechanics is tacitly accepted as fact. Mathematical formulations in newer textbooks are more abstract but more powerful. Some good textbooks are Dirac (1947), Feynman et al. (1964b), Landau and Lifshitz (1965), Messiah (1967), Liboff (1980), Sakurai (1994), Shankar (1994), Griffiths (1995), Townsend (2000), Fitts (2002), Rae (2002), and McIntyre (2012). Beiser (2003) covers a wide variety of topics in modern physics, but its coverage on quantum mechanics is highly recommended for beginners.

11.11 PROBLEMS

Problem 11.1 Prove the following identity for the Laguerre polynomials:

$$\left.\frac{d^2 L_n(x)}{dx^2}\right|_{x=0} = \frac{1}{2}n(n-1) \tag{11.340}$$

Problem 11.2 Show the validity of the following equation for the radial dependent functions of the wavefunction

$$R_{30}(r) = (\frac{Z}{3a_0})^{3/2} 2[1 - \frac{2Zr}{3a_0} + \frac{2}{3}(\frac{Zr}{a_0})^2]\exp(-\frac{Zr}{3a_0}) \tag{11.341}$$

$$R_{31}(r) = (\frac{Z}{3a_0})^{3/2}(\frac{4\sqrt{2}}{3})(\frac{Zr}{3a_0})[1 - \frac{1}{2}(\frac{Zr}{3a_0})]\exp(-\frac{Zr}{3a_0}) \tag{11.342}$$

$$R_{32}(r) = (\frac{Z}{3a_0})^{3/2}(\frac{2\sqrt{2}}{3\sqrt{5}})(\frac{Zr}{3a_0})^2 \exp(-\frac{Zr}{3a_0}) \tag{11.343}$$

Problem 11.3 Show the validity of the following spherical harmonics of the wavefunction

$$Y_{30} = \frac{1}{4}\sqrt{\frac{7}{\pi}}(5\cos^3\theta - 3\cos\theta) \tag{11.344}$$

$$Y_{3,\pm 1} = \mp\frac{1}{8}\sqrt{\frac{21}{\pi}}\sin\theta(5\cos^3\theta - 1)\exp(\pm i\phi) \tag{11.345}$$

$$Y_{3,\pm 2} = \frac{1}{4}\sqrt{\frac{105}{2\pi}} \sin^2\theta \cos\theta \exp(\pm 2i\phi) \tag{11.346}$$

$$Y_{3,\pm 3} = \mp\frac{1}{8}\sqrt{\frac{35}{\pi}} \sin^3\theta \exp(\pm 3i\phi) \tag{11.347}$$

Problem 11.4 Find the probability of finding an electron in the three 2*s* orbitals.

Ans:

$$P_{2,0,0}(r,\theta,\phi) = \frac{1}{32\pi}(\frac{Z}{a_0})^3(1-\frac{Zr}{a_0})^2 \exp(-\frac{Zr}{a_0}) \tag{11.348}$$

Problem 11.5 Find the probability of finding an electron in the three 2*p* orbitals.

Ans:

$$P_{2,1,0}(r,\theta,\phi) = \frac{1}{32\pi}(\frac{Z}{a_0})^3(\frac{Zr}{a_0})^2 \exp(-\frac{Zr}{a_0})\cos^2\theta \tag{11.349}$$

$$P_{2,1,+1}(r,\theta,\phi) = P_{2,1,-1}(r,\theta,\phi) = \frac{1}{64\pi}(\frac{Z}{a_0})^3(\frac{Zr}{a_0})^2 \exp(-\frac{Zr}{a_0})\sin^2\theta \tag{11.350}$$

Problem 11.6 Show the validity of the following wavefunctions:

$$\psi_{3,0,0}(r,\theta,\phi) = \frac{1}{81\sqrt{3\pi}}(\frac{Z}{a_0})^{3/2}(27-18\frac{Zr}{a_0}+2\frac{Z^2r^2}{a_0^2})\exp(-\frac{Zr}{3a_0}) \tag{11.351}$$

$$\psi_{3,1,0}(r,\theta,\phi) = \frac{\sqrt{2}}{81\sqrt{\pi}}(\frac{Z}{a_0})^{3/2}(6-\frac{Zr}{a_0})(\frac{Zr}{a_0})\exp(-\frac{Zr}{3a_0})\cos\theta \tag{11.352}$$

$$\psi_{3,2,0}(r,\theta,\phi) = \frac{1}{81\sqrt{\pi}}(\frac{Z}{a_0})^{3/2}(\frac{Zr}{a_0})^2 \exp(-\frac{Zr}{3a_0})(3\cos^2\theta-1) \tag{11.353}$$

$$\psi_{3,1,\pm 1}(r,\theta,\phi) = \frac{1}{81\sqrt{\pi}}(\frac{Z}{a_0})^{3/2}(6-\frac{Zr}{a_0})(\frac{Zr}{a_0})\exp(-\frac{Zr}{3a_0})\sin\theta\exp(\pm i\phi) \tag{11.354}$$

$$\psi_{3,2,\pm 1}(r,\theta,\phi) = \frac{1}{81\sqrt{\pi}}(\frac{Z}{a_0})^{3/2}(\frac{Zr}{a_0})^2 \exp(-\frac{Zr}{3a_0})\sin\theta\cos\theta\exp(\pm i\phi) \tag{11.355}$$

Problem 11.7 For $l = 0$ and $m_l = 0$, show that the solutions for Φ and Θ governed by (11.127) and (11.128) are both constants.

Problem 11.8 It was mentioned in Section 10.7 of Chau (2018) that the Laguerre equation can be recast as the Sturm-Liouville problem. Show that the Laguerre equation can be converted to

$$\frac{d}{dx}(e^{-x}x\frac{dy}{dx}) + ne^{-x}y = 0 \tag{11.356}$$

CHAPTER TWELVE

Celestial Mechanics and Astrodynamics

12.1 INTRODUCTION

One of the "oldest" problems in mechanics and dynamics is the investigation of the motion of celestial bodies in the sky. One of the measures in judging ancient civilizations is how detailed their observations were of the heavenly bodies and how well they applied such knowledge in setting calendars and in making farming systematic. The regularity of the orbits of the moon, planets, and stars makes their appearance predictable, but on the other hand "less unpredictable" infrequent events such as a solar eclipse, meteors, comets, or asteroids appear to be omens to ancient civilizations. In the East, the ancient Chinese use cosmic phenomena to forecast natural disasters on Earth and even to forecast the fall or rise of a dynasty. For example, the disappearance of a bright star suggests the fall of an empire, and it gives rebels the timing for overthrowing a dynasty. In the West, a systematic fortune-telling theory was established based on the position of constellations when a person was born. In ancient Greece, cosmology is blended with myths and gods. It was the belief of Aristotle that the Earth is fixed and at the center of the solar system, versus the heliocentric theory of Nicolaus Copernicus. The dispute could only be resolved by observations.

The major breakthrough came from the invention of the telescope, with which Galelio was able to observe the circular orbits of the four large moons of Jupiter, and theorized that smaller bodies are orbiting the larger one. Tycho Brahe, a Danish aristocrat, who pursued his hobby in observational astronomy, built one of the finest observatories in the world on the Island of Hven near Copenhagen with his wealth. He made accurate measurement of positions of the Sun, Moon, planets, and stars for more than 20 years. Three years before his death, Brahe was exiled from Denmark in 1598, and Emperor Rudoff II invited Brahe to become the astronomer to the Imperial Court in Prague. At the same time, Johannes Kepler was a professor of mathematics at the University of Graz, Austria, was not successful in formulating his theory of celestial bodies, and decided to apply for the post as Brahe's assistant. Brahe passed away shortly after Kepler's arrival, and Kepler fought a lengthy legal battle to get access to Brahe's data. Eventually, Kepler was successful and began his great work on data analysis. Kepler discovered that the orbit of Mars is actually elliptical and adopted Copernicus's idea that the Sun is at the center of the solar system. Finally, Kepler formulated his three laws of planetary motions. Then came Isaac Newton, who discovered that all Kepler's laws could be explained if a single law of universal gravitation is formulated (i.e., the attractive force between masses). In our solar system, 99.8% of the mass is from the Sun, and, therefore, Newton's formulation for two bodies (i.e., the Sun and the corresponding planet) provides a good approximation for the orbits of planets. It was Edmund Halley (the Halley comet was named after him) who insisted and helped Newton to publish the theory of gravitation in his *Principia*. When the problem of gravitational interaction between three bodies was considered, the mathematics became more complicated and Newton was not able to solve the problem with satisfaction. In fact, Newton complained that it was this problem that gave him headaches. Based on the discovery of calculus by Newton and Leibniz and the discovery of gravitation law, Euler established the lunar theory in 1767, which was used by both the Russian and British Navy in navigation. For two fixed massive bodies, the orbit of a third body was solved by Euler in terms of an elliptic integral. Lagrange found the existence of Lagrangian Points in the restricted three-body problem, which will be covered briefly in the last section in this chapter. Lagrange also formulated calculus of variations and Lagrange multiplier technique in solving problems of celestial mechanics, and summarized his findings in his celebrated book *Analytical Mechanics*. For many body problems, Laplace proposed the existence of potential and the use of perturbation theory. The so-called Laplace equation was formulated. Laplace published five volumes of *Celestial Mechanics* from 1799 to 1825. Laplace's perturbation theory was instrumental in the discovery of planet Neptune. Based upon this theory, both J.C. Adams and U.J.J. Leverier predicted in 1845 the existence of another planet beyond Uranus. It was observed in 1846 by J.F. Encke and H.I. d'Arrest at the expected location. Henri Poincaré examined the stability of the three-body problem. He discovered that solutions of the deterministic equations turn out to be quite chaotic, and highly dependent on the initial conditions of the system. There remain a lot of unsolved problems in celestial mechanics (especially related to the stability of our solar system), and the existence of dark matter in spiral galaxies. The human ambition of exploring space led to many problems in studying artificial satellites, planetary travels, transfer orbits, and space navigation. A new branch of celestial mechanics called astrodynamics resulted. It deals mainly with the orbit and motion of artificial bodies, like spacecraft and satellites, in space. In this chapter, we will not distinguish between celestial mechanics and astrodynamics.

Space travel of mankind poses a lot of issues. Although the gravitation forces from the Sun and the planets appear to be predictable, space travel has to be conducted at high speed (a few times of the speed of a bullet) with limited fuel. Therefore, significant changes in direction and speed are limited. The flight of spacecraft consists of two phases: coasting flight affected by gravity only in a time scale of days or months and accelerated flight supported by firing an engine in a time scale of only minutes. Spacecraft must rendezvous with the landing body at the right time and at the right place. For a lunar landing, the spacecraft must pass the Moon with the correct speed, altitude, and direction of flight. In view of the chaotic behavior of orbits of the n-body problem, the initial conditions of liftoff must be accurate. In addition, the prediction by Newton's second law may not be accurate enough as the "exact" size, shape, density, and mass of all planets were not known. Continuous monitoring of the flight path must be done by optical or radar measurements from the spacecraft or from the Earth with respect to the directions and locations of the stars, and corrections to flight path need to be made continuously. The spacecraft needs to slow down when it approaches the Moon, speed up again to go into the Moon's orbit, decelerate during landing, and return to the Earth after the mission or in case of mission abortion. The re-entry problem is probably the most critical part of space travel. If the re-entry angle is too steep, excessive deceleration will burn the spacecraft. If the re-entry angle is too shallow, the spacecraft may skip out of the atmosphere.

The use of gunpowder-fueled rockets in battles was reported in China more than a thousand years ago. British inventor Sir William Congreve designed the first modern rocket in 1806, K. Tsiolkovsky derived the rocket equation in 1903, American Robert H. Goddard fired the world's first liquid-fueled rocket in Auburn, Massachusetts in 1926, and Hermann Oberth of Romania published his book *The Way to Space Travel* in 1923 (second edition in 1926); these men provided all the technical details required for space travel. The first artificial satellite was sent to Earth's orbit in 1957 by the Russians. Four classes of spacecraft were used by the US: Mariners, Pioneers, Voyagers, and Vikings. In 1973, Mariner 10 used the first gravity-assisted trajectory, flying past both Venus and Mercury. Mariner 9 was the first spacecraft ever put in the orbit of another planet (Mars). In 1979, Viking was the first to make in situ analysis of the soils of another planet (Mars). In 1989, Voyager 2 flew past Neptune. The nuclear-powered Pioneer 10 became the first man-made object to leave our solar system.

Nearly all major applied mathematicians have made contributions to celestial mechanics, including Newton, Leibniz, Euler, Lagrange, Lambert, DÁlembert, Laplace, Gauss, Poincare, Hilbert, and Maxwell. In terms of space travel, modern rocketry was founded by Tsiolkovsky, Goddard, and Oberth. In this chapter, we will demonstrate the use of differential equations and mathematics in solving problems in celestial mechanics and in astrodynamics.

12.2 EQUATION OF MOTION FOR A RIGID MASS

For the dynamics of a rigid mass, we describe the motion of a body by its position vector $\boldsymbol{r}$ defined as:

$$\boldsymbol{r} = (x, y, z) \tag{12.1}$$

The velocity vector is defined as the time derivative of the position vector defined in (12.1)

$$\boldsymbol{v} = \frac{d\boldsymbol{r}}{dt} = (u, v, w) = (\frac{dx}{dt}, \frac{dy}{dt}, \frac{dz}{dt}) \tag{12.2}$$

where t is the time variable. The acceleration $\boldsymbol{a}$ is the time derivative of the velocity vector defined in (12.2):

$$\boldsymbol{a} = \frac{d\boldsymbol{v}}{dt} = (\frac{d^2x}{dt^2}, \frac{d^2y}{dt^2}, \frac{d^2z}{dt^2}) \tag{12.3}$$

Newton's second law relates the acceleration $\boldsymbol{a}$ to the applied force $\boldsymbol{F}$ on the rigid body as:

$$\boldsymbol{F} = m\boldsymbol{a} \tag{12.4}$$

In component forms of Cartesian coordinates, Newton's second law under external applied force is expressed as

$$m\frac{d^2x}{dt^2} = X, \quad m\frac{d^2y}{dt^2} = Y, \quad m\frac{d^2z}{dt^2} = Z \tag{12.5}$$

where X, Y, and Z are external applied forces. The special case of a mass under gravitational pull is considered next.

12.3 MASS UNDER GRAVITATIONAL PULL

In this section, we consider the motion of a rigid body under gravitational pull such as a projectile of a mass near the surface of the Earth. If the gravitational pull is along the y axis (or the so-called vertical direction), we can simplify the equation of motion given in (12.5) to

$$m\frac{d^2x}{dt^2}=0, \quad m\frac{d^2y}{dt^2}=-mg, \quad m\frac{d^2z}{dt^2}=0 \tag{12.6}$$

Suppose that at zero time, the rigid body is released from an altitude of h above the center of mass of another body which attracts the mass m:

$$\boldsymbol{r}\ (t=0)=\boldsymbol{r}_0=(0,h,0) \tag{12.7}$$

The initial velocity of the rigid body is assumed to be zero:

$$\boldsymbol{v}(t=0)=\boldsymbol{v}_0=(0,0,0) \tag{12.8}$$

Integration of (12.6) with respect to time gives

$$\frac{dx}{dt}=C_1, \quad \frac{dy}{dt}=-gt+C_5, \quad \frac{dz}{dt}=C_3 \tag{12.9}$$

Integrating for the second time, we obtain the displacement field as

$$x=C_1t+C_2, \quad y=-\frac{1}{2}gt^2+C_5t+C_6, \quad z=C_3t+C_4 \tag{12.10}$$

Using the initial velocity field imposed in (12.8), we find

$$C_1=C_5=C_3=0 \tag{12.11}$$

Using the initial position imposed in (12.7), we obtain the remaining constants as:

$$C_4=C_2=0, \quad C_6=h \tag{12.12}$$

This gives the familiar results of the path of a falling mass under gravitational pull from a height of h:

$$x=0, \quad y=-\frac{1}{2}gt^2+h, \quad z=0 \tag{12.13}$$

The following example considers the more familiar problem of the projectile of a cannon fired from an initial position of altitude h above the ground and under an initial inclined velocity.

Example 12.1 Consider the projectile problem shown in Figure 12.1. A mass m is shot up in the air with an initial velocity of v_0 and with an inclination angle α with the horizontal axis.

Solution: We first recall the velocity and displacement fields derived in (12.9) and (12.10) as

$$\frac{dx}{dt}=C_1, \quad \frac{dy}{dt}=-gt+C_5, \quad \frac{dz}{dt}=C_3 \tag{12.14}$$

$$x=C_1t+C_2, \quad y=-\frac{1}{2}gt^2+C_5t+C_6, \quad z=C_3t+C_4 \tag{12.15}$$

The initial position and initial velocity at $t=0$ depicted in Figure 12.1 can be expressed as

$$\boldsymbol{r}_0 = (0, h, 0), \quad \boldsymbol{v}_0 = (v_0 \cos\alpha, v_0 \sin\alpha, 0) \tag{12.16}$$

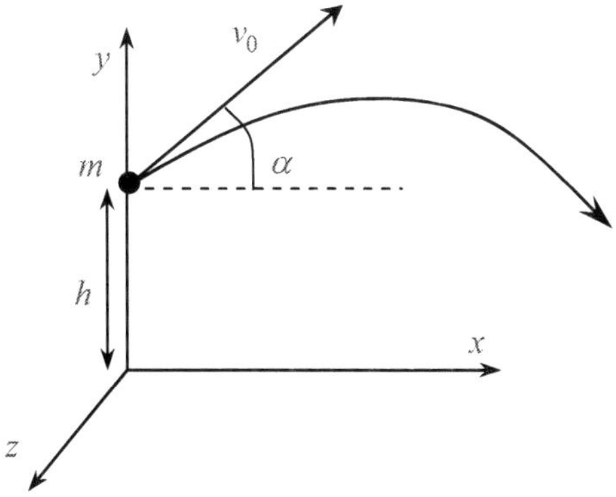

Figure 12.1 A mass at an elevation *h* was shot up initially at velocity v_0 with an inclination of α

Imposing the initial velocity given in the second equation of (12.16) onto (12.14), we get

$$C_1 = v_0 \cos\alpha, \quad C_5 = v_0 \sin\alpha, \quad C_3 = 0 \tag{12.17}$$

Similarly, applying the initial position given in the first equation of (12.16) onto (12.15), we obtain

$$x(0) = C_2 = 0, \quad y(0) = C_6 = h, \quad z(0) = C_4 = 0 \tag{12.18}$$

The final solution of the projectile is

$$x = v_0 (\cos\alpha) t \tag{12.19}$$

$$y = -\frac{1}{2} g t^2 + (\sin\alpha) v_0 t + h \tag{12.20}$$

$$z = 0 \tag{12.21}$$

A major assumption that we have made so far is that the gravitational pull is always parallel to the y-axis. This is a good approximation if the motion is very close the surface of the Earth and the surface curvature of the Earth can be ignored. More generally, the gravitational force between two masses must be modeled as a central force as described in Figure 12.2. The center of mass of the gravitational field locates at the origin of the coordinate system. In particular, the gravitational pull must always be along the radial direction of the motion from the rigid body depicted as a spherical body in Figure 12.2. We now reformulate the problem of gravitational pull using the central force concept.

Newton's second law in vector form can be expressed as

$$\boldsymbol{F} = m \frac{d\boldsymbol{v}}{dt} \tag{12.22}$$

Taking the dot product of (12.22) with velocity vector $\boldsymbol{v}$, we get

$$F \cdot v = m\frac{dv}{dt} \cdot v \tag{12.23}$$

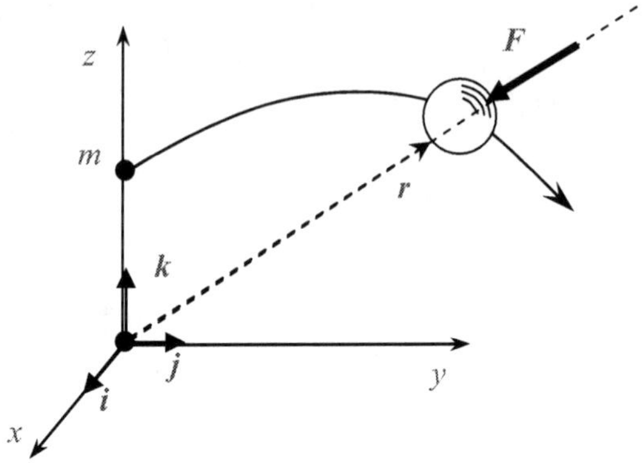

Figure 12.2 A mass under central force

Using the definition of velocity, we can rewrite this scalar equation as

$$\frac{d}{dt}(\frac{m}{2} v \cdot v) = F \cdot \frac{dr}{dt} \tag{12.24}$$

Dividing through the dt, the left-hand side of (12.24) can be recognized as the total change of kinetic energy:

$$d(\frac{m}{2} v^2) = F \cdot dr \tag{12.25}$$

where v is defined as

$$v^2 = v \cdot v = (\frac{dx}{dt})^2 + (\frac{dy}{dt})^2 + (\frac{dz}{dt})^2 \tag{12.26}$$

Assuming the existence of a potential U for force F, we get

$$F \propto \frac{\partial U}{\partial r} \tag{12.27}$$

Whenever (12.27) is satisfied, the applied force is called conservative (i.e., energy is conserved during the motion of the body). A typical example is, of course, the body force F induced by the gravitational field. In Cartesian coordinates, we have the components of the conservative body force as

$$X = \frac{\partial U}{\partial x}, \quad Y = \frac{\partial U}{\partial y}, \quad Z = \frac{\partial U}{\partial z} \tag{12.28}$$

Using the principle of the total differential for an arbitrary function, we have a relation between the change of potential and the body force components

$$dU = F \cdot dr = \frac{\partial U}{\partial x} dx + \frac{\partial U}{\partial y} dy + \frac{\partial U}{\partial z} dz = Xdx + Ydy + Zdz \tag{12.29}$$

In view of (12.25), we obtain

$$d(\frac{m}{2}v^2) = dU \tag{12.30}$$

Integration gives

$$\frac{m}{2}v^2 - U = C \tag{12.31}$$

Physically, this implies that the sum of kinetic energy and the potential energy is a constant. Thus, there is a conservation or interchange of kinetic energy and potential energy during the motion of the mass m, and this is the reason why we call the applied force conservative.

We now consider another important property of the gravitational force between two masses. Taking the cross product of (12.22) with position vector $\boldsymbol{r}$, we have

$$\boldsymbol{F} \times \boldsymbol{r} = m\frac{d\boldsymbol{v}}{dt} \times \boldsymbol{r} \tag{12.32}$$

Since the mass is a constant, we can rewrite (12.32) as

$$\boldsymbol{F} \times \boldsymbol{r} = \frac{d}{dt}(m\boldsymbol{v} \times \boldsymbol{r}) \tag{12.33}$$

Under gravitational pull, the force $\boldsymbol{F}$ always points to the origin, thus, we have the force $\boldsymbol{F}$ being parallel to the position vector $\boldsymbol{r}$ such that the cross product between them must be zero

$$\begin{aligned} \boldsymbol{F} \times \boldsymbol{r} &= (X\boldsymbol{i} + Y\boldsymbol{j} + Z\boldsymbol{k})(x\boldsymbol{i} + y\boldsymbol{j} + z\boldsymbol{k}) \\ &= (Yz - Zy)\boldsymbol{i} + (Zx - Xz)\boldsymbol{j} + (Xy - Yx)\boldsymbol{k} = \boldsymbol{0} \end{aligned} \tag{12.34}$$

Thus, all components of the vector must be zero.

$$(Yz - Zy) = (Zx - Xz) = (Xy - Yx) = 0 \tag{12.35}$$

These results can be rearranged as

$$\frac{X}{x} = \frac{Y}{y} = \frac{Z}{z} = c \tag{12.36}$$

Substitution of (12.34) into (12.33) gives

$$\frac{d}{dt}(m\boldsymbol{v} \times \boldsymbol{r}) = 0 \tag{12.37}$$

Integration gives

$$m\boldsymbol{v} \times \boldsymbol{r} = \boldsymbol{C} \tag{12.38}$$

where $\boldsymbol{C}$ is a constant vector. In component form, (12.38) can be expressed as

$$m(y\frac{dz}{dt} - z\frac{dy}{dt}) = C_1 \tag{12.39}$$

$$m(z\frac{dx}{dt} - x\frac{dz}{dt}) = C_2 \tag{12.40}$$

$$m(x\frac{dy}{dt} - y\frac{dx}{dt}) = C_3 \tag{12.41}$$

We now consider a very important condition for mass under gravitational pull:

$$C_1x + C_2y + C_3z = m(xy\frac{dz}{dt} - zx\frac{dy}{dt} + yz\frac{dx}{dt} - xy\frac{dz}{dt} + xz\frac{dy}{dt} - yz\frac{dx}{dt}) = 0 \tag{12.42}$$

Therefore, the motion of a mass under gravitational pull must be within a three-dimensional plane. This will greatly simplify the subsequent discussion. The problem of mass under central force (or gravitational pull) was first considered and solved by Jacob Hermann in 1710, a student of Jacob Bernoulli.

12.4 ORBITAL EQUATIONS FOR AN ARTIFICIAL SATELLITE

Consider that a satellite of mass m is fired from the surface of the Earth of mass M, as shown in Figure 12.3, and the inclination at firing is α. By neglecting the gravitational effects from other planets, the motion of the satellite is solely controlled by the gravitational pull of the Earth. According to Newton's law of universal gravitational force, we have:

$$F = \frac{GMm}{r^2} = \frac{GMm}{x^2 + y^2} \tag{12.43}$$

Projecting the force F onto the coordinates x and y at the location of firing, we have

$$X = -F\cos\alpha = -\frac{GMm}{x^2 + y^2}\frac{x}{\sqrt{x^2 + y^2}} \tag{12.44}$$

$$Y = -F\sin\alpha = -\frac{GMm}{x^2 + y^2}\frac{y}{\sqrt{x^2 + y^2}} \tag{12.45}$$

The mass of the Earth was estimated as

$$M = 5.98\times 10^{27}\, g \tag{12.46}$$

The universal gravitational constant was estimated experimentally as:

$$G = 6.685\times 10^{-23}(km)^3 / (gs^2) \tag{12.47}$$

Substituting (12.44) and (12.45) into Newton's second law on force equilibrium given in (12.5), we obtain two equations for the two unknowns x and y which describe the orbit of the mass as:

$$m\frac{d^2 x}{dt^2} = -\frac{GmMx}{(x^2 + y^2)^{3/2}}, \quad m\frac{d^2 y}{dt^2} = -\frac{GmMy}{(x^2 + y^2)^{3/2}} \tag{12.48}$$

Next, we differentiate the following term

$$\begin{aligned} \frac{d}{dt}[(\frac{dy}{dt})^2 + (\frac{dx}{dt})^2] &= 2(\frac{dx}{dt})\frac{d^2 x}{dt^2} + 2(\frac{dy}{dt})\frac{d^2 y}{dt^2} \\ &= -\frac{2GM}{(x^2 + y^2)^{3/2}}(x\frac{dx}{dt} + y\frac{dy}{dt}) \end{aligned} \tag{12.49}$$

The last of (12.49) results from (12.48). This can further be simplified in view of the following identity:

$$\frac{d}{dt}(x^2 + y^2) = 2(x\frac{dx}{dt} + y\frac{dy}{dt}) \tag{12.50}$$

Substitution of (12.50) into (12.49) gives

$$\frac{d}{dt}[(\frac{dy}{dt})^2+(\frac{dx}{dt})^2]=-\frac{GM}{(x^2+y^2)^{3/2}}\frac{d}{dt}(x^2+y^2) \quad (12.51)$$

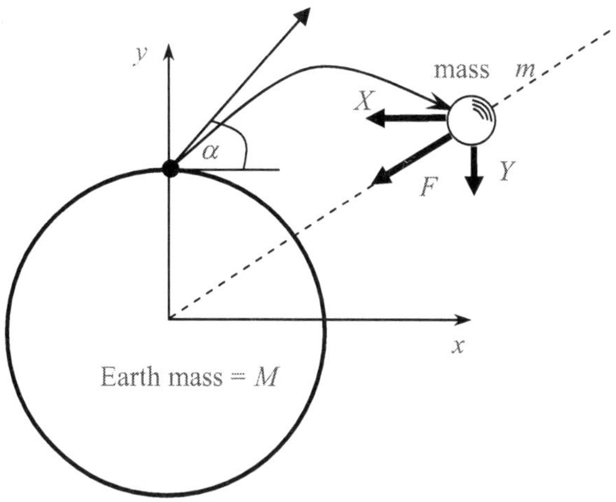

Figure 12.3 A mass under central force was shot up initially at velocity v_0 with an inclination of α from Earth's surface

Integration on both sides gives

$$(\frac{dy}{dt})^2+(\frac{dx}{dt})^2=\frac{2GM}{(x^2+y^2)^{1/2}}+C_2 \quad (12.52)$$

This equation can also be rewritten as a slightly different form:

$$v^2=\frac{2GM}{r}+C_2 \quad (12.53)$$

Let us recall (12.41) that gives another governing equation for x and y as

$$\frac{d}{dt}(x\frac{dy}{dt}-y\frac{dx}{dt})=x\frac{d^2y}{dt^2}-y\frac{d^2x}{dt^2}=\frac{d}{dt}(\frac{C_3}{m})=0 \quad (12.54)$$

For constant mass m, it can be written as

$$x\frac{dy}{dt}-y\frac{dx}{dt}=C_1 \quad (12.55)$$

Equations (12.52) and (12.55) provide a system of nonlinear ODEs for unknown x and y. This coupled nonlinear system appears to be difficult to solve. However, it can be solved exactly if we consider the problem in polar form.

12.5 ORBITAL EQUATIONS IN POLAR FORM

It is more natural to use polar coordinates in solving the system of nonlinear ODEs given in (12.52) and (12.55). In particular, we introduce

$$x = r\cos\theta, \quad y = r\sin\theta \tag{12.56}$$

Substitution of (12.56) into (12.55) gives

$$\begin{aligned} x\frac{dy}{dt} - y\frac{dx}{dt} &= r\cos\theta\left(\frac{dr}{dt}\sin\theta + r\cos\theta\frac{d\theta}{dt}\right) \\ &\quad - r\sin\theta\left(\frac{dr}{dt}\cos\theta - r\sin\theta\frac{d\theta}{dt}\right) \\ &= r^2\frac{d\theta}{dt} = C_1 \end{aligned} \tag{12.57}$$

Substitution of (12.56) into (12.52) gives another equation in polar form

$$\begin{aligned} \left(\frac{dy}{dt}\right)^2 + \left(\frac{dx}{dt}\right)^2 &= \left(\frac{dr}{dt}\cos\theta - r\sin\theta\frac{d\theta}{dt}\right)^2 + \left(\frac{dr}{dt}\sin\theta + r\cos\theta\frac{d\theta}{dt}\right)^2 \\ &= \left(\frac{dr}{dt}\right)^2\cos^2\theta - 2r\sin\theta\cos\theta\frac{d\theta}{dt}\frac{dr}{dt} + r^2\sin^2\theta\left(\frac{d\theta}{dt}\right)^2 \\ &\quad + \left(\frac{dr}{dt}\right)^2\sin^2\theta + 2r\sin\theta\cos\theta\frac{d\theta}{dt}\frac{dr}{dt} + r^2\cos^2\theta\left(\frac{d\theta}{dt}\right)^2 \\ &= \left(\frac{dr}{dt}\right)^2 + r^2\left(\frac{d\theta}{dt}\right)^2 = \frac{2GM}{r} + C_2 \end{aligned} \tag{12.58}$$

In summary, we obtain a system of governing equations for the new unknowns r and θ as

$$\left(\frac{dr}{dt}\right)^2 + r^2\left(\frac{d\theta}{dt}\right)^2 = \frac{2GM}{r} + C_2 \tag{12.59}$$

$$r^2\frac{d\theta}{dt} = C_1 \tag{12.60}$$

Although (12.59) remains nonlinear, the structural form of (12.60) allows (12.59) to be solved exactly. This will be considered next, and the exact orbital solution is known as Kepler's first law.

12.6 KEPLER'S 1ST LAW

Substitution of (12.60) into (12.59) gives

$$\left(\frac{dr}{dt}\right)^2 + \left(\frac{C_1}{r}\right)^2 = \frac{2GM}{r} + C_2 \tag{12.61}$$

Equation (12.61) can be solved as

$$\frac{dr}{dt} = \sqrt{C_2 + \frac{2GM}{r} - \left(\frac{C_1}{r}\right)^2} \tag{12.62}$$

We see that the variable time t does not appear explicitly in (12.62), and thus we can use (12.60) to rewrite r as a function of θ

$$\frac{dr}{d\theta}=\frac{r^2}{C_1}\sqrt{C_2+\frac{2GM}{r}-(\frac{C_1}{r})^2} \tag{12.63}$$

We now propose a change of variables as:

$$r=\frac{1}{u},\quad \frac{dr}{d\theta}=-\frac{1}{u^2}\frac{du}{d\theta} \tag{12.64}$$

Substitution of (12.64) into (12.63) gives

$$-\frac{1}{u^2}\frac{du}{d\theta}=\frac{1}{C_1u^2}\sqrt{C_2+2GMu-C_1^2u^2} \tag{12.65}$$

Rearranging (12.65), we get

$$d\theta=\frac{-du}{\sqrt{\frac{C_2}{C_1^2}+\frac{2GM}{C_1^2}u-u^2}} \tag{12.66}$$

The denominator can be grouped as a complete square involving u

$$d\theta=\frac{-du}{\sqrt{\frac{C_2}{C_1^2}+(\frac{GM}{C_1^2})^2-(\frac{GM}{C_1^2}-u)^2}} \tag{12.67}$$

Obviously, we can change variables as:

$$\cos\alpha=\frac{(GM/C_1^2-u)}{\sqrt{\frac{C_2}{C_1^2}+(\frac{GM}{C_1^2})^2}},\quad \sin\alpha d\alpha=\frac{-du}{\sqrt{\frac{C_2}{C_1^2}+(\frac{GM}{C_1^2})^2}} \tag{12.68}$$

Substitution of (12.68) into (12.67) gives

$$d\theta=\frac{\sin\alpha d\alpha}{\sqrt{1-\cos^2\alpha}}=d\alpha \tag{12.69}$$

Integrating both sides and substituting the definition of α from (12.68), we find the solution for θ

$$\theta=\alpha+C=\cos^{-1}\{\frac{u-GM/C_1^2}{\sqrt{\frac{C_2}{C_1^2}+(\frac{GM}{C_1^2})^2}}\}+C \tag{12.70}$$

Since r and u are related through (12.64), we can rearrange (12.70) to give u as a function of θ:

$$\begin{aligned}\frac{1}{r}=u&=\frac{GM}{C_1^2}+\sqrt{\frac{C_2}{C_1^2}+(\frac{GM}{C_1^2})^2}\cos(\theta-C)\\&=\frac{GM}{C_1^2}+\frac{GM}{C_1^2}\sqrt{1+\frac{C_2C_1^2}{(GM)^2}}\cos(\theta-C)\end{aligned} \tag{12.71}$$

This solution for r in terms of θ can be recast as:

$$r=\frac{p}{1+e\cos(\theta-C)} \tag{12.72}$$

where

$$p = \frac{C_1^2}{GM}, \quad e = \sqrt{1 + \frac{C_2 C_1^2}{(GM)^2}} \tag{12.73}$$

The constants C, C_1, and C_2 can be determined from the conditions at the moment of firing of the satellite. Equation (12.72) is also known as Kepler's first law, which states that all planets move in ellipse with the Sun as one of its focuses. In fact, Kepler's laws (two more to be discussed) were established in 1618 whereas the universal gravitational law given in (12.46) was not published until 1687 in *Principia* by Newton. It was Kepler's laws on celestial bodies that led to the gravitational theory of Newton. Physically, (12.72) gives the orbit of the motion of a mass under gravitational pull. A number of special cases can be observed from (12.72), depending on the values of e. In short, we have

$$\begin{aligned} &e = 0, && \text{circular orbit,} \\ &0 < e^2 < 1, && \text{elliptic orbit,} \\ &e^2 = 1, && \text{parabolic orbit,} \\ &e^2 > 1, && \text{hyperbolic orbit} \end{aligned} \tag{12.74}$$

For the case of planets moving around the Sun, we have elliptic orbits. The eccentricity of the ellipse is given by e.

12.7 FIRST ESCAPE VELOCITY (ORBITAL SPEED)

We now consider the first escape velocity, which is the minimum velocity of a satellite that can stay on a circular orbit around the Earth. Let us consider that at $t = 0$, the satellite is located at $x = 0$ and $y = R$ or $\theta = \pi/2$ and $r = R$ (which is the radius of the Earth). We have tacitly assumed that the Earth is a perfect sphere. The initial velocities are given by

$$\frac{dx}{dt} = v_0 \cos\alpha, \quad \frac{dy}{dt} = v_0 \sin\alpha \tag{12.75}$$

Substitution of these initial conditions into (12.55) gives

$$x\frac{dy}{dt} - y\frac{dx}{dt} = -Rv_0 \cos\alpha = C_1 \tag{12.76}$$

Substitution of (12.75) into (12.52) gives

$$v_0^2 = \frac{2GM}{R} + C_2 \tag{12.77}$$

Substituting $\theta = \pi/2$ and $r = R$ into (12.72) gives

$$R = \frac{p}{1 + e\cos(\pi/2 - C)} = \frac{p}{1 + e\sin C} \tag{12.78}$$

Solving for C, we get

$$C = \sin^{-1}\left(\frac{p/R - 1}{e}\right) \tag{12.79}$$

All constants have been found in terms of the initial conditions. From (12.73), the eccentricity e can be found as

$$e=\sqrt{1+\frac{(v_0^2-\frac{2GM}{R})(Rv_0\cos\alpha)^2}{(GM)^2}} \tag{12.80}$$

Squaring both sides of (12.80) we get

$$e^2=1+\frac{v_0^4R^2\cos^2\alpha}{(GM)^2}-\frac{2Rv_0^2\cos^2\alpha}{GM} \tag{12.81}$$

This can be regrouped as

$$\begin{aligned} e^2&=(1-\frac{Rv_0^2\cos^2\alpha}{GM})^2+\frac{v_0^4R^2\cos^2\alpha(1-\cos^2\alpha)}{(GM)^2} \\ &=(1-\frac{Rv_0^2\cos^2\alpha}{GM})^2+\frac{v_0^4R^4\cos^4\alpha}{(GM)^2}\frac{\sin^2\alpha}{R^2\cos^2\alpha} \end{aligned} \tag{12.82}$$

Substituting (12.76) into the first equation of (12.73), we obtain

$$p=\frac{R^2v_0^2\cos^2\alpha}{GM} \tag{12.83}$$

Substituting (12.83) into (12.82), we find

$$e^2=(1-\frac{p}{R})^2+\frac{p^2}{R^2}\tan^2\alpha\geq(\frac{p}{R}-1)^2 \tag{12.84}$$

This gives

$$\frac{p/R-1}{e}\leq 1 \tag{12.85}$$

From (12.79), we see that C exists. Let us consider the special case of a circular orbit (i.e., $e=0$). Equation (12.84) and (12.85) gives

$$p/R=1,\quad \alpha=0 \tag{12.86}$$

Substitution of (12.86) into (12.83) gives

$$v_0^2=\frac{GM}{R} \tag{12.87}$$

Plugging in the values of G and M from equations (12.47) and (12.46) and R = 6370 km, we obtain

$$v_0^2=\frac{GM}{R}=\frac{(6.685\times10^{-23})(5.98\times10^{27})}{6370}=62.76\ (km/s)^2 \tag{12.88}$$

Thus, the first escape velocity is given by

$$v_0=7.92\ km/s \tag{12.89}$$

This is the minimum speed on a circular orbit. For any velocity less than this value, the mass will return to the surface of the Earth. It is interesting to note that for this case we have to fire the satellite horizontally. This is, of course, not possible. However, the rocket will tend to turn parallel to the ground once it is fired off the ground.

12.8 SECOND ESCAPE VELOCITY (FROM EARTH)

The second escape velocity is defined as the initial velocity of the mass leading to its escape from Earth's gravitational pull. This occurs when $e = 1$ and the satellite has a parabolic orbit and thus the rocket will approach infinity. Using this value of e in (12.73), we get

$$\frac{C_2 C_1^2}{(GM)^2} = 0 \tag{12.90}$$

Equivalently, we have

$$C_2 = 0 \tag{12.91}$$

Substitution of (12.91) into (10.53) leads to

$$v_0 = \sqrt{\frac{2GM}{R}} = \sqrt{2} \times 7.92 = 11.2 km/s \tag{12.92}$$

This is the velocity for a rocket to escape from Earth's gravitational field. For a firing velocity larger than (12.92), we have $e > 1$; the orbit becomes a hyperbola and in principle the satellite will fly to infinity. However, in reality, the satellite will only escape from the gravitational pull from the Earth but it may or may not be able to escape from the gravitational pull of the Sun. The escape velocity from the solar system is called the third escape velocity and will be considered in the next section.

When $e < 1$, we expect an elliptic orbit for the satellite. We see from the orbital solution given in (12.72) that the value of r is given by

$$r = \frac{p}{1 + e\cos(\theta - C)} \tag{12.93}$$

If for some values of θ we find $r \leq R$, the fired rocket will have a crash landing at the Earth's surface. If we set $0 < e < 1$ and $p \geq R$, (12.83) gives the minimum value of r

$$R \leq p \leq \frac{R^2 v_0^2}{GM} \tag{12.94}$$

This implies

$$v_0 \geq (\frac{GM}{R})^{1/2} \tag{12.95}$$

Therefore, for firing a satellite in an elliptic orbit around the Earth, we expect the initial velocity to be larger than the first escape velocity but smaller than the second escape velocity.

Example 12.2 Recently, it was proposed by a privately owned space travel company (such as *Virgin Galactic*) that spacecraft should be carried to high altitude by an airplane before it is fired outward to the orbit, as shown in Figure 12.4. Suppose that the satellite is fired at an altitude of 200 km. Find the first and second escape velocities for firing at this altitude. For the case of circular orbits, find the firing angle α at which crash landing can be prevented.

Solution: The initial point at firing is

$$x = 0, \quad y = R + h \tag{12.96}$$

Initial condition at $t = 0$

$$\frac{dx}{dt} = v_0 \cos\alpha, \quad \frac{dy}{dt} = v_0 \sin\alpha \tag{12.97}$$

Substitution of these initial conditions into (12.55) and (12.53) gives

$$x\frac{dy}{dt} - y\frac{dx}{dt} = -(R+h)v_0 \cos\alpha = C_1 \tag{12.98}$$

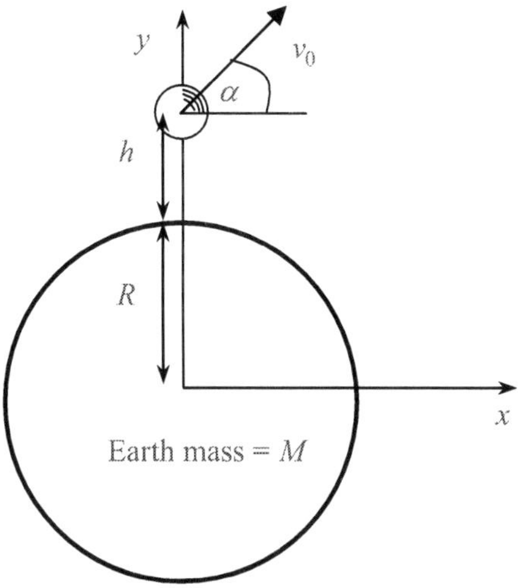

Figure 12.4 A mass at an elevation of *h* was shot up initially at velocity v_0 with an inclination of α

$$v_0^2 = \frac{2GM}{R+h} + C_2 \tag{12.99}$$

Thus, we have

$$C_1 = -(R+h)v_0 \cos\alpha \tag{12.100}$$

$$C_2 = v_0^2 - \frac{2GM}{R+h} \tag{12.101}$$

Substituting (12.100) into the first equation of (12.73) and (12.72), we obtain

$$p = \frac{C_1^2}{GM} = \frac{(R+h)^2 v_0^2 \cos^2\alpha}{GM} \tag{12.102}$$

$$R + h = \frac{p}{1 + e\cos(\pi/2 - C)} = \frac{p}{1 + e\sin C} \tag{12.103}$$

Solving for C from (12.103), we get

$$C=\sin^{-1}[\frac{p/(R+h)-1}{e}] \tag{12.104}$$

By now, all constants have been found. From (12.73), the eccentricity e can be found as

$$e=\sqrt{1+\frac{(v_0^2-\frac{2GM}{R+h})(R+h)^2 v_0^2\cos^2\alpha}{(GM)^2}} \tag{12.105}$$

Following the procedure in obtaining (12.84), we finally get

$$e^2=(1-\frac{p}{R+h})^2+\frac{p^2}{(R+h)^2}\tan^2\alpha \ge (\frac{p}{R+h}-1)^2 \tag{12.106}$$

This gives

$$\frac{p/(R+h)-1}{e}\le 1 \tag{12.107}$$

Equation (12.104) shows that C exists. Let us consider the special case of a circular orbit (i.e., $e = 0$). Equation (12.107) and (12.106) gives

$$p/(R+h)=1, \quad \alpha=0 \tag{12.108}$$

Substituting (12.100) into the first part of (12.73) and plugging (12.108) into its result, we find

$$v_0=\sqrt{\frac{GM}{R+h}}=\sqrt{\frac{(6.685\times10^{-23})(5.98\times10^{27})}{6370+200}}=7.80\ km/s \tag{12.109}$$

This value is smaller than the first escape velocity at the ground surface at 7.92 km/s (see (12.89)), as expected. In view of (12.73) and $e = 1$, the second escape velocity is obtained by setting C_2 into (12.101) as

$$v_0=\sqrt{\frac{2GM}{R+h}}=\sqrt{2}\times7.80\ km/s=11.03 km/s \tag{12.110}$$

which is smaller than the value found in (12.92), as expected. If we set $0 < e < 1$ and $r = R$ for crash landing, we find

$$R=\frac{(R+h)^2 v_0^2\cos^2\alpha}{(GM)[1+e\cos(\theta-C)]} \tag{12.111}$$

Using $\theta = \pi/2$ and $e = 0$, we have

$$\alpha=\pm\cos^{-1}\{\frac{\sqrt{RGM}}{(R+h)v_0}\} \tag{12.112}$$

Substitution of (12.109) into (12.112) gives

$$\alpha=\pm\cos^{-1}\{\sqrt{\frac{R}{R+h}}\}=\pm14.17^\circ \tag{12.113}$$

Therefore, we can fire the satellite at an angle of 14.17° from horizontal to avoid a crash landing. The current value is for an elevation of 200 km above the ground only. This critical angle avoiding crashing on the Earth clearly depends on the elevation from the ground.

12.9 THIRD ESCAPE VELOCITY (FROM OUR SOLAR SYSTEM)

If we want to send a spacecraft to explore beyond our solar system, we need to escape from the gravitational pull of the Sun. To do that, we have to first calculate the speed of the Earth revolving around the Sun. We need this assistance to accelerate beyond the solar system. We need to follow three steps: first we have to calculate the second escape speed to escape from the Sun's gravitational pull from the Earth; secondly, we need to find the revolving speed of the Earth in its orbit; thirdly, the additional velocity exceeding the second escape velocity from the Earth needs to be larger the difference between the results in step one and step two.

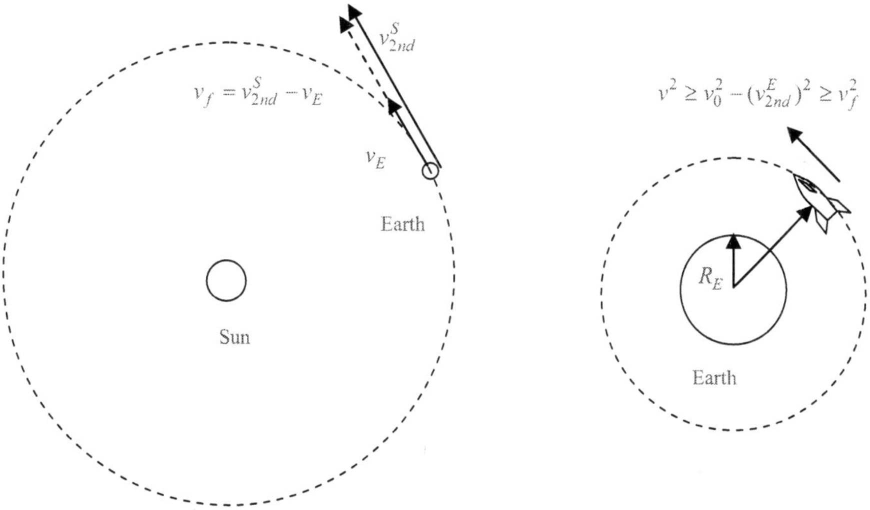

Figure 12.5 The third escape velocity from the solar system

Step 1: Note the mass of the Sun and the distance between the Sun and the Earth as:

$$M_s = 1.983\times10^{30}\,kg, \quad D_{SE} = 1.495\times10^{11}\,m \tag{12.114}$$

Using the formula given in (12.95), we have

$$v_{2nd}^{S} = \sqrt{\frac{2GM_s}{D_{SE}}} = \sqrt{\frac{(2)(6.685\times10^{-23})(1.983\times10^{33})}{1.495\times10^{8}}} \tag{12.115}$$
$$= 42.11\,km/s$$

This is the second escape velocity from the solar system needed from a distance of 1.495×10^{11} m from the Sun.

Step 2: The period of Earth's orbit is

$$T = \frac{2\pi}{\omega} = 365.3 \times 24 \times 60 \times 60 s \tag{12.116}$$

The tangential velocity of circular motion can be estimated as

$$v_E = \omega D_{SE} \tag{12.117}$$

Combining (12.116) and (12.117), we have the Earth's velocity as

$$v_E = \frac{2\pi D_{SE}}{T} = \frac{(2)(3.142592654)(1.495 \times 10^{11})}{365.3 \times 24 \times 60 \times 60} = 29.76 km/s \tag{12.118}$$

If we fire the rocket in the direction of the rotation of the Earth around the Sun, the additional velocity of firing from the orbit of the Earth is:

$$\begin{aligned} v_f &= v_{2nd}^S - v_E = 42.11 - 29.76 \\ &= 12.35 km/s \end{aligned} \tag{12.119}$$

This is the required additional velocity to be gained in the orbit of the Earth. We still need to fire the rocket from the Earth's surface and to escape from the Earth's gravity. This leads to the final step in the calculation.

Step 3: This is the most subtle step in the whole analysis. Recall from (12.53) that the velocity of a moving mass on an orbit satisfies the following equation

$$v^2 - \frac{2GM_E}{r} = C_2 = v_0^2 - \frac{2GM_E}{R_E} \tag{12.120}$$

with $r > R_E$, where r is the radial distance on the orbit of the satellite from the Earth. Thus, the velocity on the orbit is not a constant. Rearranging this gives

$$v^2 = \frac{2GM_E}{r} + v_0^2 - \frac{2GM_E}{R_E} > v_0^2 - \frac{2GM_E}{R_E} = v_0^2 - (11.2)^2 \tag{12.121}$$

Note that the last term is the second escape velocity from the surface of the Earth obtained in (12.92). The left-hand side on (12.121) is the velocity of the satellite on the orbit. It is this velocity that we need to achieve an additional velocity needed from Step 2. Recall from Step 2 that we need to achieve an additional velocity of 12.35 km/s. Thus, we require

$$v_0^2 - (v_{2nd}^E)^2 \geq v_f^2 \tag{12.122}$$

Substitution of the second escape velocity from the Earth and the firing velocity required from Step 2 gives

$$v_0^2 - (11.2)^2 \geq (12.35)^2 \tag{12.123}$$

Thus, we have

$$v_0^2 \geq (12.35)^2 + (11.2)^2 = 277.96\ (\text{km/s})^2 \tag{12.124}$$

Finally, the third escape velocity is

$$v_0 \geq 16.67\,\text{km/s} \tag{12.125}$$

That is, instead of achieving a velocity of 29.76 km/s from zero speed, we only need to accelerate to an additional 12.35 km/s from the orbit of the Earth. Equation (12.124) shows that the firing velocity on the Earth's surface needs to be higher than 16.67 km/s in order to escape our solar system. To benefit the most from this gravitational slingshot, it has been proposed that the takeoff should align with the last quarter-moon. In the next section, we will examine the energy hump between the Moon and the Earth that we need to overcome to be able to travel to the Moon. However, a more energy-efficient way to travel to the Moon will be covered in Section 12.14 using the Hohmann transfer orbit.

12.10 TRAVEL TO THE MOON

In this section, we consider the required initial velocity for a spacecraft to travel from the Earth to the Moon. As shown in Figure 12.6, the distance between the Earth and the Moon is D, the gravitational potential due to the Earth is U_1 and that due to the Moon is U_2, and the total potential on a spacecraft is also shown (see (12.27) for the definition of U). There is a potential hump that the spacecraft needs to overcome.

The mass and radius of the moon are

$$M_m = 7.35\times10^{22}\, kg, \quad R_m = 1.74\times10^{6}\, m \tag{12.126}$$

The distance between the Earth and Moon D is 3.84×10^8 m. The ratios between various parameters between the Earth and the Moon are:

$$\frac{M_E}{M_m} = 81.22, \quad \frac{D}{R_E} = 60.2, \quad \frac{R_E}{R_m} = 3.667 \tag{12.127}$$

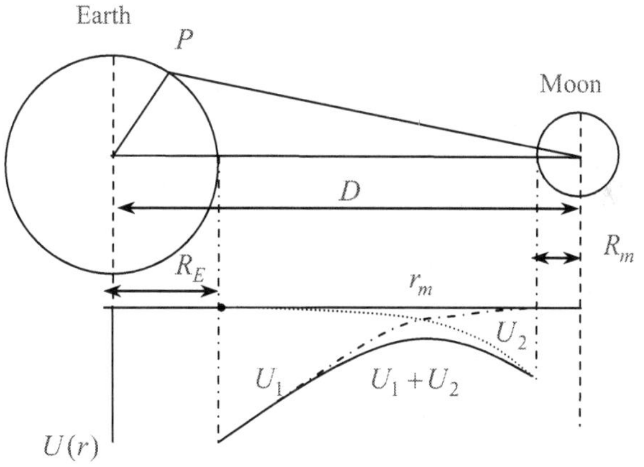

Figure 12.6 Energy hump between the Earth and the Moon

The force and potential between two masses m and M can be expressed as (see (12.27))

$$F = \frac{dU}{dr} = \frac{GMm}{r^2} \tag{12.128}$$

Integration gives

$$U = -\frac{GMm}{r} \tag{12.129}$$

For a spacecraft of mass m traveling between the Earth and the Moon, the total gravitational potential is

$$U(r) = -\frac{GM_E m}{r} - \frac{GM_m m}{D-r} \tag{12.130}$$

To find the location of the maximum potential, we differentiate (12.130) with respect to r

$$\left(\frac{dU}{dr}\right)_{r_m} = \frac{GM_E m}{r^2} - \frac{GM_m m}{(D-r)^2} = 0 \tag{12.131}$$

Thus, we have

$$\left(\frac{M_E}{M_m}\right)^{1/2} = \frac{r_m}{D-r_m} \tag{12.132}$$

Rearranging (12.132), we get

$$r_m = \frac{D}{1+\left(\frac{M_E}{M_m}\right)^{1/2}} \tag{12.133}$$

Using the mass ratio between the Earth and the Moon given in (12.127), we get

$$r_m = \frac{D}{1+\left(\frac{1}{81.22}\right)^{1/2}} = 0.9D \tag{12.134}$$

The point with minimum potential is very close to the Moon. On the Earth's surface, we have $r = R_E$ and the potential is

$$U(R_E) = -\frac{GM_E m}{R_E} - \frac{GM_m m}{D-R_E} \tag{12.135}$$

The ratio of the potential due to the Earth to that due to the Moon is

$$\frac{U_1}{U_2} = \frac{M_E / M_m}{R_E / (D-R_E)} = 4808 \tag{12.136}$$

On the surface of the Moon, we have $r = D - R_m$ and the potential is

$$U(D-R_m) = -\frac{GM_E m}{D-R_m} - \frac{GM_m m}{R_m} \tag{12.137}$$

The ratio of the potential due to the Earth to that due to the Moon is

$$\frac{U_1}{U_2} = \frac{M_E / M_m}{(D-R_m)/R_m} = 0.3697 \tag{12.138}$$

Inversely, we have

$$\frac{U_2}{U_1} = 2.7 \tag{12.139}$$

At the point with maximum potential, we have $r = r_m$ and

$$U(r_m) = -\frac{GM_E m}{0.9D} - \frac{GM_m m}{0.1D} = -\frac{GM_E m}{0.9D}\left(1+\frac{9M_m}{M_E}\right) = -\frac{1.23GM_E m}{D} \tag{12.140}$$

Considering the total energy of the spacecraft (sum of the kinetic energy and potential energy), we have

$$\frac{1}{2}mv_0^2 - \frac{GM_E m}{R_E} = U(r_m) = -\frac{1.23GM_E m}{D} \tag{12.141}$$

More discussion of the total energy of the spacecraft in the gravitational field will be given in a later section. Rearranging (12.141), we get

$$v_0 = \sqrt{\frac{2GM_E}{R_E}}(1 - \frac{1.23R_E}{D})^{1/2} = 11.08km/s \tag{12.142}$$

We see that this initial velocity is slower but very close to the second escape velocity found in (12.92). Once the maximum energy hump between the Earth and the Moon is surpassed, the spacecraft will fall within the gravitational pull of the Moon. The landing velocity of the spacecraft is about 1.76 km/s (see Problem 12.2). Therefore, additional fuel is needed to stop the spacecraft for safe landing on the Moon.

However, it is not energy efficient to travel to the Moon using an initial speed as discussed here. In a later section, we will see that the Hohmann transfer orbit is normally used (see Section 12.14). It is also known as the translunar orbit. It may be even more energy efficient to get into various stages of various elliptic orbits before getting into the actual translunar orbit. It is also more practical to send the spacecraft to a low Earth orbit before getting into a translunar orbit. This will be demonstrated in a later section. Another option is to station an spacecraft at the so-called Lagrangian Points between the Moon and the Earth, before landing on the Moon. There are five Lagrangian Points, at which the combined gravitational pull of two massive bodies (or the Moon and the Earth) precisely equals the centrifugal force of the small body to orbit with them. Therefore, the spacecraft at the Lagrangian Points will appear stationary relative to the motions of the Moon-Earth system. Three of the Lagrangian Points are collinear with the two large masses and were found by Euler and two other Lagrangian Points each form an equilateral triangle with the two masses that were found by Lagrange in 1772 in his prize memoir submitted to the Paris Academy. The collinear Lagrangian Points were shown by Joseph Liouville in 1845 to be unstable. The two triangular Lagrangian Points are, however, stable if the mass ratio between the two large masses is greater than 24.96, which is the case for the Earth-Moon system (see (12.127)).

12.11 KEPLER'S SECOND LAW

Kepler observed that the line joining the Sun to a given planet sweeps out an equal area in equal times. This is called Kepler's second law. To prove this, the hatched area that sweeps out by the position vector from Point P to Point Q shown in Figure 12.7 can be estimated as

$$dA = \frac{1}{2}(r + dr)rd\theta \tag{12.143}$$

Figure 12.7 also demonstrates that the area that sweeps out by time dt is the same regardless of the current position of the orbit. Dividing through by dt, we get

$$\frac{dA}{dt} = \frac{1}{2} r^2 \frac{d\theta}{dt} \tag{12.144}$$

Recalling from (12.60), one of the governing equations for motion of mass under gravitation pull is

$$r^2 \frac{d\theta}{dt} = C_1 \tag{12.145}$$

Substitution of (12.145) into (12.144) gives

$$\frac{dA}{dt} = \frac{1}{2} C_1 \tag{12.146}$$

Thus, the constant C_1 derived earlier has a physical meaning that it equals twice the rate of area swept out by the position vector $\boldsymbol{r}$. We see that Kepler's second law is a natural consequence of gravitational pull. In fact, chronologically, this second law is the first law discovered by Kepler.

12.12 KEPLER'S THIRD LAW (NEWTON'S LAW)

Kepler's third law states that the square of a planet's year, divided by the cube of its mean distance from the Sun, is the same for all planets. Mathematically, it states that

$$T \propto r^{3/2} \tag{12.147}$$

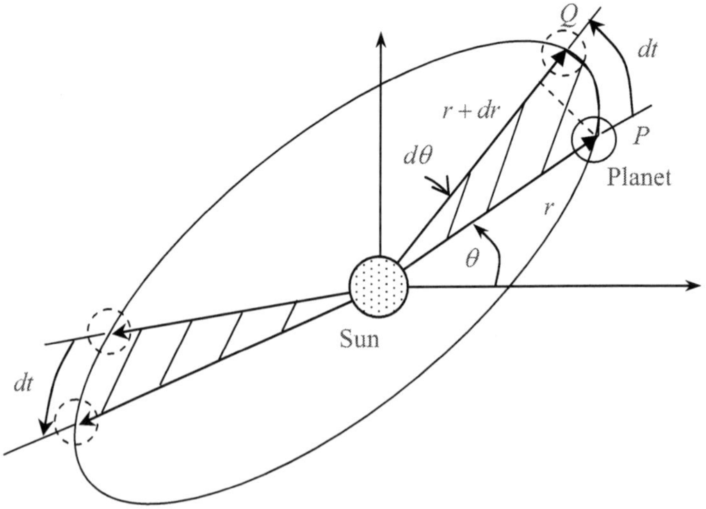

Figure 12.7 Kepler's second law: equal area sweeping in equal time

Table 12.1 compiles the periods and radii of the orbits of 6 planets and the prediction by (12.147). In the table, AU is called an astronomical unit and is a length in terms of the distance of the Sun from the Earth or 1.496×10^{11} m. We see from the last column that (12.147) gives a value very close to a constant.

For circular motions, we have the radial acceleration being

$$a_r = -\frac{v^2}{r} \tag{12.148}$$

On the other hand, the velocity relates to period as:

$$v = \frac{2\pi r}{T} \tag{12.149}$$

Table 12.1 Radii of orbits and periods of various planets around the Sun

Planet	Radius r of orbit (AU)	T (days)	r^3/T^2 (AU)^3/day^2 $\times 10^{-6}$
Mercury	0.389	87.77	7.64
Venus	0.724	224.70	7.52
Earth	1.000	365.25	7.50
Mars	1.524	686.98	7.50
Jupiter	5.200	4332.62	7.49
Saturn	9.510	10,759.20	7.43

Substitution of (12.149) into (12.148) gives

$$a_r = -\frac{4\pi^2 r}{T^2} \tag{12.150}$$

Newton's second law requires

$$F_r = ma_r = -\frac{4\pi^2 mr}{T^2} \tag{12.151}$$

Now if we assume the validity of the Kepler's third law, we have

$$\frac{r^3}{T^2} = K \tag{12.152}$$

Substitution of (12.152) into (12.151) gives

$$F_r = -\frac{4K\pi^2 m}{r^2} \tag{12.153}$$

This suggests that the force between the Sun and the planet satisfies the following law:

$$F_r \propto \frac{1}{r^2} \tag{12.154}$$

More generally, the attractive force between any two masses are given by:

$$F = -\frac{Gm_1 m_2}{r^2} \tag{12.155}$$

where G is the universal gravitational constant. Therefore, Kepler's third law leads to the gravitational law or it is a consequence of the gravitational law.

Example 12.3 The first human-made satellite was fired into space in October 1957 and was called Sputnik I by the former USSR. It followed an elliptic orbit with the minimum and maximum heights of about 228 km and 947 km. Find the period of the satellite in revolving the Earth and the velocity of the satellite to stay on the orbit.

Solution: Recall the first escape velocity that

$$v = (\frac{GM_E}{r})^{1/2} \tag{12.156}$$

At the surface of the Earth, we have

$$mg = \frac{GM_E m}{R_E^2} \tag{12.157}$$

From (12.157), we get

$$GM_E = gR_E^2 \tag{12.158}$$

Substitution of (12.158) into (12.156) gives

$$v = (\frac{gR_E^2}{r})^{1/2} \tag{12.159}$$

The period of a satellite can be estimated as

$$T = \frac{2\pi r}{v} = \frac{2\pi r^{3/2}}{g^{1/2} R_E} \tag{12.160}$$

Using the value of the radius of the Earth and the gravitational constant, we have the period in seconds as

$$T = 3.14529 \times 10^{-7} r^{3/2} \quad (r \text{ in m}) \tag{12.161}$$

For the case of Sputnik I, we have

$$r = R_E + h_{ave} = 6378 + (\frac{228 + 947}{2}) = 6965.5 \text{ km} \tag{12.162}$$

Substitution of (12.162) into (12.161) gives

$$T = 96.37 \text{ min} \tag{12.163}$$

The revolving velocity is

$$v = (\frac{gR_E^2}{r})^{1/2} = \{\frac{(9.81)(6378 \times 10^3)^2}{6965.5 \times 10^3}\}^{1/2} = 7.569 \text{ km/s} \tag{12.164}$$

Example 12.4 For satellites used for telecommunications, GPS positioning, and other applications, we may want to have the satellites be geosynchronous with our daily life, as shown in Figure 12.8. That is, these satellites appear to be fixed in the sky relative to a particular location on Earth. The first such satellite is the Syncom II launched in July 1963. Find the altitude and the speed of these geosynchronous satellites.

Solution: First, we recall (12.161) as

$$T = 3.14529 \times 10^{-7} r^{3/2} \quad (r \text{ in m}) \tag{12.165}$$

For geosynchronous satellites, the period is 1 day and (12.165) can be rearranged as:

$$r = \frac{1}{10^3}\{\frac{24\times 60\times 60}{3.14529\times 10^{-7}}\}^{2/3} = 42,257 \text{ km} \tag{12.166}$$

Therefore, the elevation of the satellite is 35,879 km. The ratio of r to the radius of the Earth is about 6.625. That is, these satellites have to be located more than 1/10th of the distance to the Moon. The speed can be estimated by using (12.164) as

$$v = (\frac{gR_E^2}{r})^{1/2} = \{\frac{(9.81)(6378\times 10^3)^2}{42,257\times 10^3}\}^{1/2} = 3.073 \text{ km/s} \tag{12.167}$$

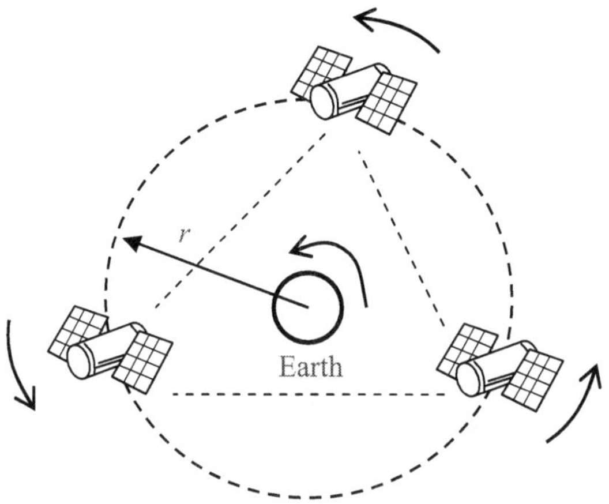

Figure 12.8 Geosynchronous satellites in Earth's orbit

12.13 ENERGY IN AN ELLIPTIC ORBIT

All planets revolve around the Sun in an elliptic orbit. In this section, we will examine, in more detail, the energy for a mass in an elliptic orbit. Figure 12.9 shows that the Sun of M is located at one of the focuses of the ellipse of size a and b (with $a > b$) and a planet of mass m is revolving around the Sun. Figure 12.8 shows that the maximum value of r is attained when the planet locates on the same level of the focus points F and F' but closer to focus F', whereas the minimum value of r is attained when the planet is on the other end closer to focus F. The maximum and minimum values of r are

$$r_{\max} = a(1+e), \quad r_{\min} = a(1-e) \tag{12.168}$$

From right-angle triangle FNO, we have

$$a^2 - b^2 = a^2e^2 \tag{12.169}$$

Thus, b is

$$b = a(1-e^2)^{1/2} \tag{12.170}$$

The maximum potential energy is attained at $r = r_{\max}$ and equals

$$U(r_{\max}) = -\frac{GMm}{a(1+e)} \tag{12.171}$$

The kinetic energy is

$$K = \frac{1}{2}mv_\theta^2 = \frac{1}{2}mr^2(\frac{d\theta}{dt})^2 \tag{12.172}$$

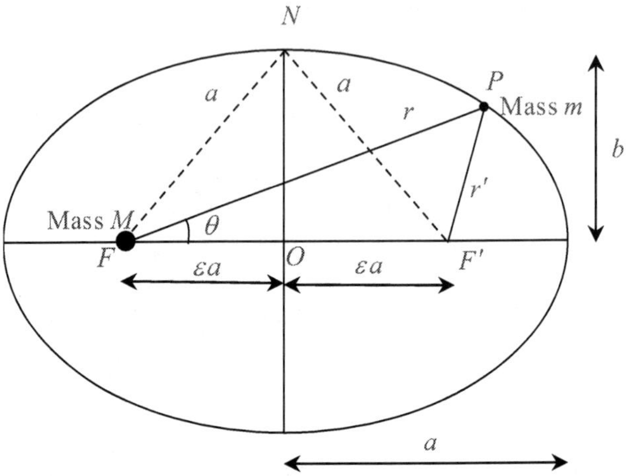

Figure 12.9 Elliptic orbit of the solar system

In view of (12.60), we can rearrange the result as

$$K = \frac{1}{2}mv_\theta^2 = \frac{1}{2}\frac{m}{r^2}r^4(\frac{d\theta}{dt})^2 = \frac{1}{2}\frac{m}{r^2}C_1^2 \tag{12.173}$$

However, we have shown in (12.146) in Section 12.11 that physically the constant C_1 equals twice the rate of change of the area that sweeps out by the line of joining the Sun and the planet. Thus, for the case of an ellipse, we have

$$C_1 = \frac{2\pi ab}{T} \tag{12.174}$$

Thus,

$$C_1^2 = \frac{4\pi^2a^2b^2}{T^2} = \frac{4\pi^2a^4(1-e^2)}{T^2} \tag{12.175}$$

Now, referring to Figure 12.8, we have

$$r + r' = 2a \tag{12.176}$$

Applying the cosine rule to triangle FPF', we find

$$r'^2 = r^2 + (2ea)^2 - 4ear\cos\theta \tag{12.177}$$

From (12.176), we have

$$r'^2 = (2a-r)^2 = 4a^2 - 4ar + r^2 \tag{12.178}$$

Substitution of (12.178) into (12.177) gives

$$4a^2 - 4ar + r^2 = r^2 + (2ea)^2 - 4ear\cos\theta \tag{12.179}$$

Simplification leads to

$$r = \frac{a(1-e^2)}{1-e\cos\theta} \tag{12.180}$$

This expression was apparently first derived by Euler. Substitution of (12.170) into (12.180) results in

$$\frac{1}{r} = \frac{a}{b^2}(1-\varepsilon\cos\theta) \tag{12.181}$$

Differentiation of (12.181) with respect to t gives

$$-\frac{1}{r^2}\frac{dr}{dt} = \frac{\varepsilon a}{b^2}\sin\theta\frac{d\theta}{dt} \tag{12.182}$$

Using (12.60) in (12.182), we get

$$\frac{dr}{dt} = -\frac{C_1\varepsilon a}{b^2}\sin\theta \tag{12.183}$$

Differentiating (12.183) with respect to t, we find

$$\frac{d^2r}{dt^2} = -\frac{C_1\varepsilon a}{b^2}\cos\theta\frac{d\theta}{dt} = -\frac{C_1^2\varepsilon a}{b^2}\frac{\cos\theta}{r^2} \tag{12.184}$$

It can be proved that the radial acceleration is (see Problem 12.3):

$$a_r = \frac{d^2r}{dt^2} - r(\frac{d\theta}{dt})^2 \tag{12.185}$$

Substitution of (12.184) into (12.185) gives

$$\begin{aligned} a_r &= -\frac{C_1^2\varepsilon a}{b^2}\frac{\cos\theta}{r^2} - r(\frac{d\theta}{dt})^2 = -\frac{C_1^2\varepsilon a}{b^2}\frac{\cos\theta}{r^2} - r(\frac{C_1}{r^2})^2 \\ &= -\frac{C_1^2}{r^2}(\frac{\varepsilon a\cos\theta}{b^2} + \frac{1}{r}) \end{aligned} \tag{12.186}$$

In view of (12.181), (12.186) can be simplified as

$$a_r = -\frac{aC_1^2}{b^2}(\frac{1}{r^2}) \tag{12.187}$$

We again obtain a $1/r^2$ law for any central force directed toward one of the focuses of the elliptic orbit. This result is, of course, consistent with Kepler's third law. More specifically, we can substitute (12.175) into (12.187) to get

$$a_r = -\frac{4\pi^2a^3}{T^2}(\frac{1}{r^2}) \tag{12.188}$$

The attractive force between the Sun and the planet is

$$F_r = ma_r = -\frac{GMm}{r^2} = -\frac{4m\pi^2a^3}{T^2}(\frac{1}{r^2}) \tag{12.189}$$

Thus, we obtain the following formula for the period T

$$T^2 = \frac{4\pi^2a^3}{GM} \tag{12.190}$$

This is precisely Kepler's third law considered in Section 12.12. However, the discussion given in Section 12.12 is restricted to circular orbit. We see from

(12.190) that the period is a function of the major axis a but not the minor axis b of the ellipse. It is also independent of the eccentricity e but only the major axis a. In fact, the major axis a can also be interpreted as the mean distance of the planet from the Sun as:

$$\frac{r_{max}+r_{min}}{2}=\frac{a(1+e)+a(1-e)}{2}=a \tag{12.191}$$

All elliptic orbits illustrated in Figure 12.10 have the same period.

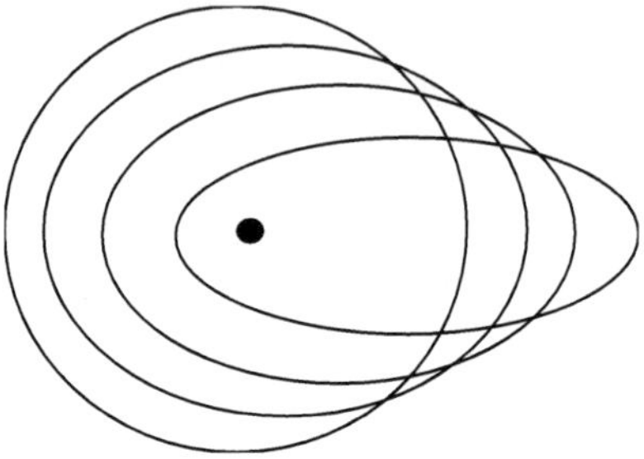

Figure 12.10 Various elliptic orbits having the same period

Substitution of (12.190) into (12.175) gives

$$C_1^2 = GMa(1-e^2) \tag{12.192}$$

Finally, with this result of C_1 and (12.168), the kinetic energy given in (12.173) can be simplified as

$$K(r_{max})=\frac{GMm(1-e^2)}{2a(1+e)^2}=\frac{GMm(1-e)}{2a(1+e)} \tag{12.193}$$

The total energy is

$$E=K(r_{max})+U(r_{max})=\frac{GMm(1-e)}{2a(1+e)}-\frac{GMm}{a(1+e)}=-\frac{GMm}{2a} \tag{12.194}$$

From the conservation of energy, we have the velocity v versus distance r as

$$E=\frac{1}{2}mv^2-\frac{GMm}{r}=-\frac{GMm}{2a} \tag{12.195}$$

Rearranging (12.195), we find the kinetic energy as a function of r

$$\frac{1}{2}mv^2=\frac{GMm}{r}-\frac{GMm}{2a} \tag{12.196}$$

This can be rewritten to give the velocity as

$$v=\sqrt{GM}\,(\frac{2}{r}-\frac{1}{a})^{1/2} \tag{12.197}$$

As illustrated in Figure 12.11, the kinetic energy of the planet on the left-hand side of (12.195) can be interpreted as the loss of potential energy from point R on a hypothetical circular orbit for the planet with a radius $2a$ to point S on the actual elliptic orbit with a major axis of a. We have demonstrated that the physical meaning of C_1, which equals twice the rate of change of area swept by the line joining the Sun and the planet, or it was given in (12.173). In fact, the physical meaning of C_2 can also be found. Multiplying (12.59) by $m/2$, we get

$$\frac{1}{2}m[(\frac{dr}{dt})^2+r^2(\frac{d\theta}{dt})^2]=\frac{GMm}{r}+\frac{1}{2}mC_2 \tag{12.198}$$

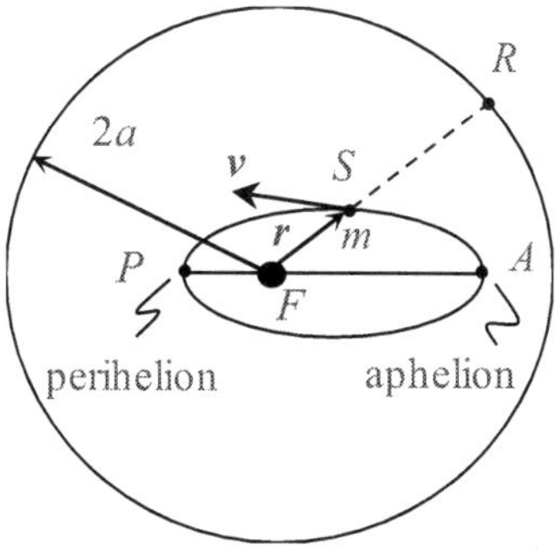

Figure 12.11 Hypothetical circular orbit of radius $2a$ of an elliptic orbit around the Sun

We recognize that the total energy E is

$$E=\frac{1}{2}m[(\frac{dr}{dt})^2+r^2(\frac{d\theta}{dt})^2]-\frac{GMm}{r}=\frac{1}{2}mC_2 \tag{12.199}$$

The last part of (12.199) is a result of (12.198). Thus, we can find C_2 in terms of the total energy as:

$$C_2=\frac{2E}{m} \tag{12.200}$$

Using (12.200) and (12.174), the eccentricity given in (12.73) can be written as:

$$e=\sqrt{1+\frac{C_2C_1^2}{(GM)^2}}=\sqrt{1+\frac{2E}{m}(\frac{2\pi ab}{GMT})^2} \tag{12.201}$$

Thus, (12.201) gives another physical meaning of the total energy. That is, the types of orbits are totally controlled by the sign of E:

$$\begin{aligned} &E>0, \quad e>1, \quad \text{hyperbolic orbit} \\ &E=0, \quad e=1, \quad \text{parabolic orbit} \\ &E<0, \quad e<1, \quad \text{elliptic or circular orbit} \end{aligned} \tag{12.202}$$

We will apply this energy equation to study the interplanetary transfer orbits.

12.14 INTERPLANETARY TRAVEL

The most efficient way in terms of least amount of energy needed to travel between planets is to use the transfer orbit, which was proposed by German engineer and scientist, Walter Hohmann, in 1925. Therefore, it is also known as the Hohmann transfer orbit. This is not the most direct or the fastest way of conducting interplanetary travel, but it requires the least energy. Thus, it has been used in most space travel so far.

12.14.1 Hohmann Transfer Orbit

Figure 12.12 shows a proposed transfer from circular orbit 1 to circular orbit 2 through an elliptic Hohmann transfer orbit 3 for translunar travel (or interplanetary travel). We assume here that the orbits of the planets are roughly circular. The path of the transfer orbit is shown by a solid line with an arrow. Let us recall the speeds on the circular orbits (i.e., (12.87)):

$$v_{c1} = \sqrt{\frac{GM}{r_1}}, \quad v_{c2} = \sqrt{\frac{GM}{r_2}} \tag{12.203}$$

In Section 12.7, we also call these orbital speeds of the first escape. A Hohmann transfer orbit 3 with its perigee (if the mass at the center is the Sun, it becomes the perihelion) overlaps with circular orbit 1 and with its apogee (if the mass at the focus is the Sun, it becomes the aphelion) overlaps with circular orbit 2. We have to accelerate the spacecraft at both the perigee (or the perihelion) and the apogee (or the aphelion) of the Hohmann transfer orbit. The required velocity increment at the perigee (or perihelion) is

$$\Delta v_p = v_p - v_{c1} \tag{12.204}$$

where v_p is the velocity of the spacecraft at the perigee (or perihelion) of the transfer orbit. By using (12.197), it can be calculated as:

$$v_p = \sqrt{2GM}\left(\frac{1}{r_1} - \frac{1}{r_1 + r_2}\right)^{1/2} \tag{12.205}$$

In obtaining (12.205), we have used the following formula (see Figure 12.12):

$$a = \frac{r_1 + r_2}{2} \tag{12.206}$$

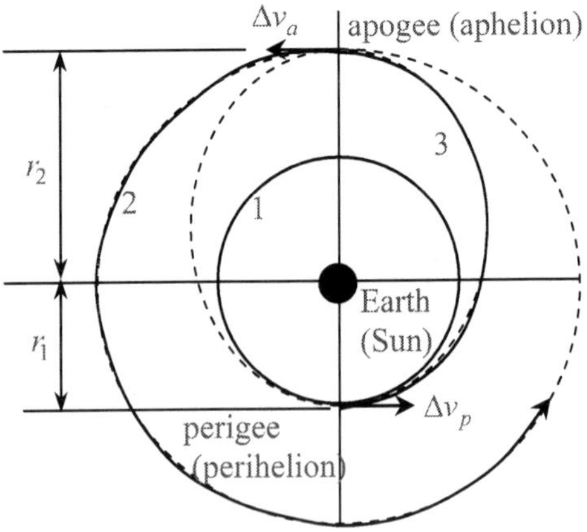

Figure 12.12 Hohmann transfer orbit for translunar travel (or interplanetary travel)

Substitution of the first equation of (12.203) and (12.205) into (12.204) gives

$$\Delta v_p = \sqrt{2GM}\left(\frac{1}{r_1} - \frac{1}{r_1 + r_2}\right)^{1/2} - \sqrt{\frac{GM}{r_1}} = \sqrt{\frac{GM}{r_1}}\left(\frac{2r_2}{r_1 + r_2}\right)^{1/2} - \sqrt{\frac{GM}{r_1}} \tag{12.207}$$

Therefore, the velocity increment needed at the perigee (or perihelion) of the transfer orbit is

$$\Delta v_p = \sqrt{\frac{GM}{r_1}}\left\{\sqrt{\frac{2r_2}{r_1 + r_2}} - 1\right\} \tag{12.208}$$

We assumed that this velocity is applied instantaneously. In reality, it normally takes a few minutes. After the spacecraft travels from the perigee (or perihelion) to the apogee (or aphelion) of the elliptic transfer orbit, another acceleration is needed to push the spacecraft to the circular orbit 2. In particular, we have

$$\Delta v_a = v_{c2} - v_a \tag{12.209}$$

where

$$v_a = \sqrt{2GM}\left(\frac{1}{r_2} - \frac{1}{r_1 + r_2}\right)^{1/2} \tag{12.210}$$

Therefore, substitution of the second equation of (12.203) and (12.20) into (12.209) gives

$$\Delta v_a = \sqrt{\frac{GM}{r_2}} - \sqrt{2GM}\left(\frac{1}{r_2} - \frac{1}{r_1 + r_2}\right)^{1/2} = \sqrt{\frac{GM}{r_2}}\left\{1 - \sqrt{\frac{2r_1}{r_1 + r_2}}\right\} \tag{12.211}$$

This is the second velocity increment needed for the spacecraft to transfer to circular orbit 2. For interplanetary travel, we have to use the mass of the Sun for M in the above equations:

$$M = M_s \tag{12.212}$$

For travel to the Moon, we have to use the mass of the Earth for M in the above equations:

$$M = M_e \tag{12.213}$$

If we travel from the Earth to the outer planets (i.e., $r_2 > r_1$), we need to increase velocities at both the perihelion and aphelion of the transfer orbit. Conversely, if we travel from the Earth to the planets closer to the Sun (i.e., $r_1 > r_2$), we need to decrease velocities at both the perihelion and aphelion of the transfer orbit. Both deceleration and acceleration require energy.

As remarked earlier, one problem with the Hohmann transfer orbit is the long travel time. Therefore, it would be informative to estimate the travel time on the transfer orbit. For planetary travel from the Earth to other planets, we can use Kepler's third law to write:

$$\left(\frac{T_c}{T_E}\right)^2 = \left(\frac{a_c}{r_E}\right)^3 \tag{12.214}$$

where T_c and T_E are the periods of the transfer orbit and of the Earth around the Sun. As shown in Figure 12.12, the travel time on the Hohmann transfer orbit is clearly half of the period and thus equals

$$T = \frac{1}{2}T_c = \frac{1}{2}T_E(\frac{a_c}{r_E})^{3/2} = \frac{1}{2}T_E(\frac{r_1 + r_2}{2r_1})^{3/2} \tag{12.215}$$

Since the period of the orbit of the Earth can be expressed as

$$T_E^2 = \frac{4\pi^2 r_E^3}{GM}, \tag{12.216}$$

substitution of this into (12.216) gives the travel time on the Hohmann transfer orbit

$$T = \pi\sqrt{\frac{(r_1 + r_2)^3}{8GM}} \tag{12.217}$$

Let us consider some examples.

Example 12.5 Consider interplanetary travel from the Earth to Venus using an elliptic Hohmann transfer orbit. The relative distance of Venus to the Sun is 0.72AU, as shown in Figure 12.13.

Solution: For Earth's orbit, we have the velocity as

$$v_0 = \sqrt{\frac{GM_s}{r_E}} \tag{12.218}$$

In view of the period given in (12.217), we have the period being

$$T_E = \frac{2\pi r_E^{3/2}}{\sqrt{GM_s}} \tag{12.219}$$

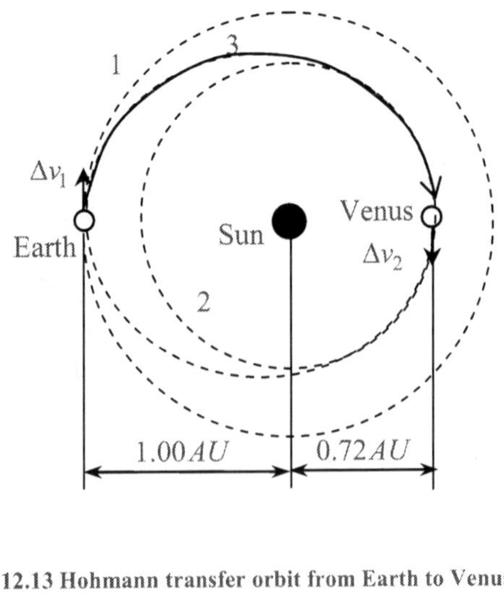

Figure 12.13 Hohmann transfer orbit from Earth to Venus

Employing (12.219), we get the following equation, as expected,

$$v_0 = \frac{2\pi r_E}{T_E} = \frac{(2)(3.141592654)(1.49\times 10^8)}{(356.25)(24)(60)(60)} = 29.6264 \text{ km/s} \tag{12.220}$$

The speed of the spacecraft must be suddenly reduced to

$$\begin{aligned}\Delta v_p &= \sqrt{\frac{GM_s}{r_E}}\{\sqrt{\frac{2r_V}{r_E + r_V}} - 1\} \\ &= (29.6264)\{\sqrt{\frac{(2)(0.72)}{1+0.72}} - 1\} = -2.5185 \text{ km/s}\end{aligned} \tag{12.221}$$

Therefore, we need to reduce the speed from 29.6264 km/s to 27.1079 km/s at the perihelion of the elliptic orbit. Using (12.215), we get the travel time from Earth's orbit to Venus's orbit as:

$$T = \frac{1}{2}T_E(\frac{r_1 + r_2}{2r_1})^{3/2} = \frac{1}{2}(1)(\frac{1.72}{2})^{3/2} = 0.3988\text{year} = 145.6\text{days} \tag{12.222}$$

Once the spacecraft enters the transfer orbit, under the gravitational field of the Sun its speed will increase. More specifically, using the conservation of total energy we have

$$E = \frac{1}{2}mv_a^2 - \frac{GM_s m}{0.72 r_E} = -\frac{GM_s m}{1.72 r_E} \tag{12.223}$$

Rearranging (12.223), we obtain

$$v_a^2 = \frac{GM_s}{r_E}2(\frac{1}{0.72} - \frac{1}{1.72}) = 1.61498 v_0^2 \tag{12.224}$$

Thus, we have the speed of the spacecraft at the aphelion

$$v_a = 1.2708 v_0 = 37.6498 \text{km/s} \tag{12.225}$$

On the orbit of Venus, we have the speed as

$$v_V = \sqrt{\frac{GM_s}{r_V}} = \frac{2\pi r_V}{T_V} \tag{12.226}$$

From Table 12.1, we find that the period of Venus is about $0.6152T_E$ or 0.6152 year. If we use the approximate circular orbit, using Kepler's third law we find 0.6109 year. In the following calculation, we use the period calculated from Kepler's third law for the sake of consistency. Thus, from the last equation in (12.238) we have

$$v_V == \frac{(2)(3.141592654)(0.72)(1.49\times 10^8)}{(0.6109)(3.16\times 10^7)} = 34.9174 \text{km/s} \tag{12.227}$$

The speed reduction at this point is

$$\Delta v_a = v_V - v_a = 34.9174 - 37.6498 = -2.7324 \text{km/s} \tag{12.228}$$

Alternatively, we have the speed reduction as

$$\Delta v_a = \sqrt{\frac{GM}{r_E}(\frac{r_E}{r_V})}\{1-\sqrt{\frac{2r_E}{r_E+r_V}}\}$$
$$= (29.6264)\sqrt{\frac{1}{0.72}}\{1-\sqrt{\frac{2}{1+0.72}}\} = -2.7348 \text{ km/s} \tag{12.229}$$

These values should be the same if we use exact and consistent values of parameters. Impulse is needed to reduce the speed of the spacecraft, but these values given in (12.228) and (12.229) are a relatively small expenditure of energy for orbit transfer. In reality, human maneuvering of the spacecraft may be needed for final adjustment.

12.14.2 Launching Time Window

So far, we have only considered the speed increments needed to be applied to the spacecraft at the perihelion and aphelion of the Hohmann transfer orbit planned for the interplanetary travels. However, such interplanetary travel cannot be launched arbitrarily at any time. The target planet needs to be at the same location of the orbit when the spacecraft arrives from the transfer orbit. Since the revolving speed of different planets depends on its distance from the Sun, only a certain launching time window will allow for interplanetary travel by the Hohmann transfer orbit. Figure 12.14 illustrates the relative position of Venus (V_0) and the Earth (E_0) at launch. The head angle between the Earth and Venus at launch is labeled as α_0.

The travel time of the spacecraft on the Hohmann transfer orbit is (see (12.217))

$$T_{travel} = \frac{\pi}{\sqrt{GM_s}}a^{3/2} = \frac{\pi}{\sqrt{GM_s}}(\frac{R_{SE}+R_{SV}}{2})^{3/2} \tag{12.230}$$

The travel time of Venus can be evaluated by using Kepler's third law as:

$$T_V = (\alpha_0 + \pi)\frac{R_{SV}^{3/2}}{\sqrt{GM_s}} \tag{12.231}$$

Equating (12.230) and (12.231), we have

$$\alpha_0 = \pi\{(\frac{R_{SE}+R_{SV}}{2R_{SV}})^{3/2} - 1\} \tag{12.232}$$

Plugging in the values of the planet's distances to the Sun, we obtain

$$\alpha_0 = \pi\{(\frac{1+0.7223}{2\times 0.7223})^{3/2} - 1\} = 54^\circ \tag{12.233}$$

If we want to travel to Mars, we can show that the head angle can be evaluated as

$$\alpha_0 = \pi\{1-(\frac{R_{SE}+R_{SM}}{2R_{SM}})^{3/2}\} = \pi\{1-(\frac{1+1.523679}{2\times 1.523679})^{3/2}\} = 44.3^\circ \tag{12.234}$$

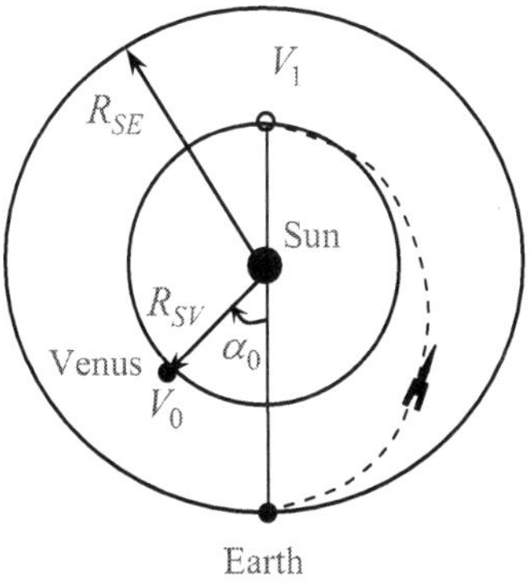

Figure 12.14 The Earth's head angle α_0 at launch

We can see that this desirable setting of the head angle α_0 will occur periodically. This period of suitable launching is also known as the synodic period, and the corresponding values for travel to Venus and Mars can be calculated as:

$$T_{Vsyn} = \frac{1}{\dfrac{1}{T_V} - \dfrac{1}{T_E}} = \frac{1}{\dfrac{1}{224.7} - \dfrac{1}{365.25}} = 584 \text{ days} \tag{12.235}$$

$$T_{Msyn} = \frac{1}{\dfrac{1}{T_E} - \dfrac{1}{T_M}} = \frac{1}{\dfrac{1}{365.25} - \dfrac{1}{686.98}} = 780 \text{ days} \tag{12.236}$$

To derive (12.235) and (12.236), we first note that the angle θ of a general elliptic orbit shown in Figure 12.15 can be expressed as:

$$\cos\theta = \frac{1}{e}\left(\frac{p}{r} - 1\right) \tag{12.237}$$

At the launching position, the angle θ_1 is given by

$$\cos\theta_1 = \frac{1}{e}\left(\frac{p}{r_e} - 1\right) \tag{12.238}$$

At the transfer position, the angle θ_2 can be expressed as

$$\cos\theta_2 = \frac{1}{e}\left(\frac{p}{r_t} - 1\right) \tag{12.239}$$

Note that for the special case of the Hohmann transfer orbit, we have

$$\theta_1 = 0, \quad \theta_2 = \pi \tag{12.240}$$

More generally, the head angles of the target planet at launching and at transfer are denoted by ψ_1 and ψ_2. They are functions of θ_1 and θ_2 as well as the orbital angular motion of the target planet ω_t and the flight time T. From Figure 12.15, we have

$$\psi_1 = \theta_2 - \theta_1 - \omega_t T \tag{12.241}$$

Similarly, at the transfer position, the angle ψ_2 can be expressed as

$$\psi_2 = \theta_2 - \theta_1 + (2\pi - \omega_e T) \tag{12.242}$$

where ω_e is the angular frequency of the Earth. For the case of the Hohmann transfer orbit, we have

$$\psi_{1H} = \pi - \omega_t T_H \tag{12.243}$$

$$\psi_{2H} = \pi - \omega_e T_H \tag{12.244}$$

The relative position between the Earth and the target planet at the moment of transfer is important. If ψ_{2H} is about π, the Sun will locate between the Earth and the target planet and will block the communication during landing on the target planet. This situation should be avoided. Another important parameter is the distance between the Earth and the target planet at the time of transfer. Using the cosine law for a triangle, we have:

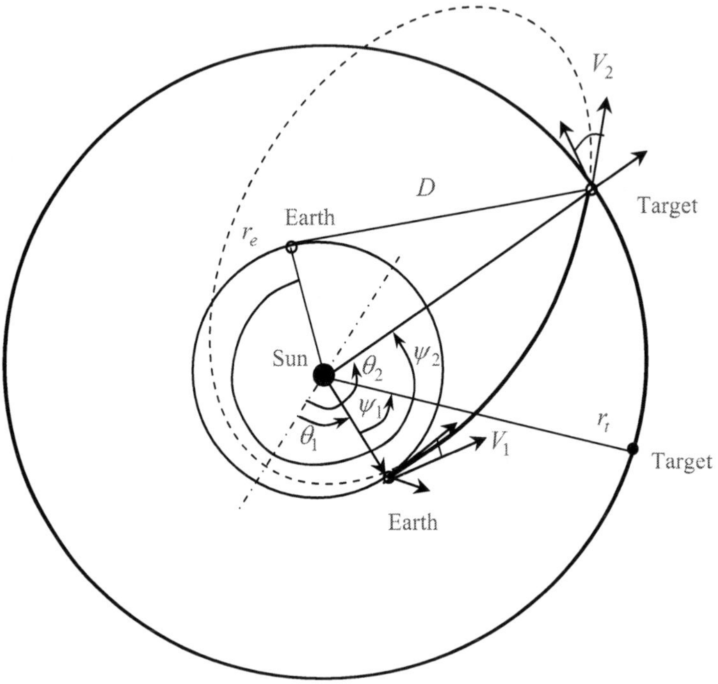

Figure 12.15 Non-Hohmann transfer orbit

$$D^2 = r_t^2 + r_e^2 - 2r_t r_e \cos\psi_2 \tag{12.245}$$

This distance D will control the energy required to send the electromagnetic waves as telecommunication from the spacecraft to the Earth.

To derive the synodic period, we observe that the future time window t relates to the current launching time t_0 by

$$t = t_0 \pm kT_{syn} \tag{12.246}$$

where k is an integer. The relative positions between the Earth and the target planet at time t and time t_0 are the same. Thus, we have

$$\psi_1(t) = \psi_1(t_0) \pm 2\pi \tag{12.247}$$

On the other hand, their relative angle at launching can be found as

$$\psi_1(t) = \psi_1(t_0) - \omega_e(t - t_0) + \omega_t(t - t_0) \tag{12.248}$$

Equating (12.247) and (12.248), we have

$$\begin{aligned}\psi_1(t) - \psi_1(t_0) = 2\pi &= -\omega_e(t - t_0) + \omega_t(t - t_0) \\ &= -\omega_e T_{syn} + \omega_t T_{syn}\end{aligned} \tag{12.249}$$

The periods of revolving around the Sun relate to the rate of angular motion as

$$\omega_e = \frac{2\pi}{T_e}, \quad \omega_t = \frac{2\pi}{T_t} \tag{12.250}$$

Substitution of (12.250) into (12.249) leads to

$$T_{syn} = \frac{1}{1/T_t - 1/T_e} \tag{12.251}$$

This completes the proof for (12.235). We can also derive (12.236) following a similar procedure. Table 12.2 compiles the parameters of the Hohmann transfer orbit from the Earth to other planets. The launching time window is the smallest for Mars. We can make the journey once per two years.

Table 12.2 Parameters of the Hohmann orbit from Earth to other planets

Planet	a (AU)	e	T (yr)	T_{syn} (yr)	ψ_1(°)	ψ_2(°)	$D(AU)$
Mercury	0.6935	0.4419	0.289	0.317	108.3	76.0	7.64
Venus	0.8617	0.1605	0.400	1.599	−54.0	36.0	7.52
Mars	1.2618	0.2075	0.709	2.135	44.3	−75.1	7.50
Jupiter	3.1012	0.6775	0.731	1.092	97.2	−83.0	7.49
Saturn	5.2741	0.8104	6.056	1.035	106.1	159.8	7.50
Uranus	10.0674	0.9007	15.972	1.012	111.3	−169.7	7.49
Neptune	15.5055	0.9356	30.529	1.006	113.2	−10.1	7.50
Pluto	20.1446	0.9504	45.208	1.004	113.9	105.3	7.43

12.15 STRIKING SPEED OF METEORS ON EARTH

By observing the Moon's surface, it is quite obvious that it has been subject to numerous meteor bombardments in its history. There are a number of well-preserved meteor impact craters on the surface of the Earth, including Tenoumer Crater, Mauritania, and Arizona's meteor crater in the United States. More recently, it has been inferred that much larger meteors or asteroids have impacted the Earth. Examples are the giant craters at Vredefort in South Africa (diameter of 300 km), at the northern coastline of Chicxulub in Mexico (diameter of 150 km), and at Sudbury in Canada (diameter of 140 km). It is necessary to estimate the striking speed of meteors on Earth. If the meteor is coming from infinity, it must have a parabolic orbit around the Sun. That is, we have

$$e = \sqrt{1 + \frac{C_2 C_1^2}{(GM)^2}} = 1 \tag{12.252}$$

Thus, from (12.202) we must have

$$E = 0 \tag{12.253}$$

For $e = 1$, the second escape velocity can be used to evaluate the impact speed. That is, the meteor impact is treated as a reverse event of a meteor flying to infinity from the orbit of the Earth. In particular, we have

$$v = \sqrt{\frac{2GM_s}{r_E}} = \sqrt{\frac{(2)(6.685\times10^{-23})(1.99\times10^{33})}{1.495\times10^{8}}} = 42.1863 \text{ km/s} \tag{12.254}$$

As illustrated in Figure 12.16, the impact can be coming from the opposite direction or the same direction as the revolving orbit of the Earth. Thus, the maximum and minimum impact velocity can be estimated as:

$$v_{impact}^{\max} = 42.1863 + 29.6264 = 71.8127 \text{ km/s} \tag{12.255}$$

$$v_{impact}^{\min} = 42.1863 - 29.6264 = 12.5599 \text{ km/s} \tag{12.256}$$

Based on thousands of data, Bjork (1961) and many others have interpreted that the impact velocity of meteor on Earth is indeed between 11 km/s and 72 km/s. These values agree very well with the results given in (12.245) and (12.246).

12.16 PRECESSION OF THE PERIHELION OF MERCURY

Apparently, Newton's universal law of gravitation works fine most of the time in predicting the motions of celestial bodies. However, there are a number of exceptions. In particular, in the study of dynamics of spiral galaxies, Newton's law of gravitation fails to explain why the speed of the spiral arms of galaxies does not decay as $r^{-1/2}$ as expected from Newton's law of gravitation because most of the bright stars in galaxies are concentrated near the center (see discussion in Lin and Segel, 1988). Gravitation density wave theory was proposed in the 1920s by B. Lindblad to explain this peculiar observation. These gravity density waves are supposed to be caused by periodic temporary concentrations of stars caused by their self-gravitation. This was revived by C.C. Lin and F.H. Shu in 1960s (e.g., Lin and Segel, 1988). On the other hand, this also leads to the speculation that there is lots of unobservable dark matter in the universe. The most recent Shaw Prize in Astronomy was awarded to Simon White for his study on dark matter distribution. Recently, this observation has also been regarded as partial evidence of the existence of "black holes." It will be discussed in the next section that the black hole is linked to general relativity. Nevertheless, this remains an unsolved mystery in astronomy.

An expedition headed by Eddington and Dyson measured the light deflection during the total solar eclipse of May 29, 1919 (right after the World War I) simultaneously in Brazil and on the west coast of Africa. The results seem to favor the prediction by general relativity rather than by the Newtonian gravitation law (their predictions differ by a factor of two). This result was considered spectacular news and made the front page in many newspapers. Similar measurements have been conducted in 1922, 1953, and 1973 solar eclipse.

Another triumph of general relativity is the prediction of the perihelion precession of Mercury. This is the main topic of the present section. The eccentricity of the elliptic orbit of Mercury is 0.2056, which is the largest in all planets in our solar system except Pluto (with a value of 0.2491). The point of closest approach to the Sun is called the perihelion, which is fixed. Under the influence of other planets in the solar system, the perihelia of planets precess (or rotate) around the Sun, as illustrated in Figure 12.17.

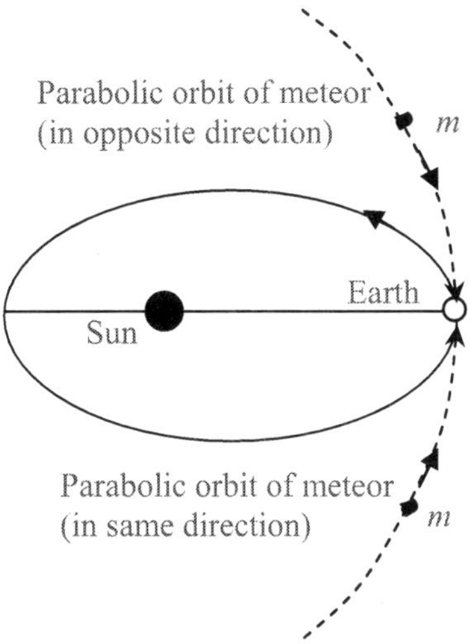

Figure 12.16 Parabolic orbit of meteors

The observed amount of precession of the perihelion of Mercury is about 574 arc seconds per century. Using Newton's law of gravitation, the effects of all other planets is 531 arc seconds. Therefore, 43 arc seconds cannot be explained by gravitation alone. In the following we will discuss the amount of precession due to the effect of general relativity. We will first introduce the Schwarzschild metric, consider the amount of energy change due to relativity, and the amount precession due to this additional energy.

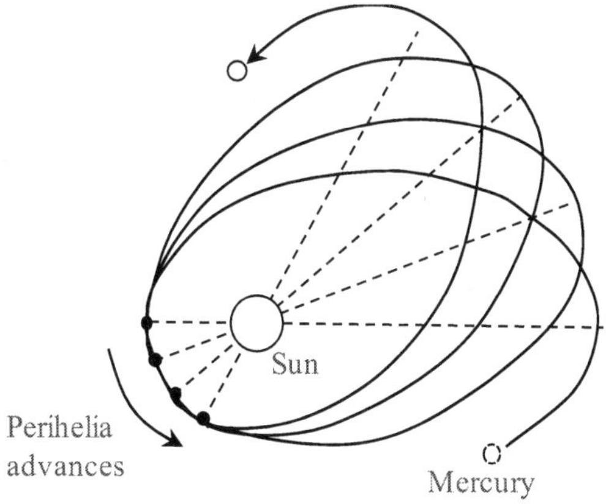

Figure 12.17 Precession of the perihelion of Mercury

12.16.1 Schwarzschild Metric for Curved Space-Time

Einstein's general theory of relativity speculated that the space-time around a massive body is distorted and curved, and such curved space-time can be described mathematically by Riemann geometry, as illustrated in Figure 12.18. In 1915, Einstein published his celebrated paper on the advance of perihelion precession of Mercury using general relativity based on the approximate gravitational field around a spherically symmetric, non-rotating, non-charged mass. On December 22, 1915, while still serving a war stationed on the Russia front of World War I, Karl Schwarzschild sent a letter to Einstein and presented his exact solution of Einstein's field equation in polar form. In 1916, Einstein replied:

> I have read your paper with the utmost interest. I had not expected that one could formulate the exact solution of the problem in such a simple way. I liked very much your mathematical treatment of the subject. Next Thursday I shall present the work to the Academy with a few words of explanation. Albert Einstein

Schwarzschild managed to publish two papers on general relativity before he passed away in 1916. This exact solution is also referred to as Schwarzschild geometry or the Schwarzschild metric. The singularity in the metric at a particular radius, called the Schwarzschild radius, was the basis of the subsequent theory of the non-rotating black hole by David Finkelstein in 1958. Schwarzschild's black hole is also called the "dead" black hole versus the "live" rotating Kerr black hole and the rotating and charged Kerr-Newman black hole. Recently, LIGO (the Laser Interferometer Gravitation-Wave Observatory in the United States) and VIRO (Virgo Interferometer by France, Italy, the Netherlands, Poland and Hungary) announced on February 11, 2016 that gravitational waves were observed and matched the theoretical predictions of general relativity for a gravitational wave from the merger of a pair of stellar mass black holes. Drever, Thorne and Weiss received the 1916 Shaw Prize in Astronomy for such discovery. It is of interest to note that the paper "Observation of Gravitational Waves from a Binary Black Hole Merger" was published in *Physical Review Letters* in 2016 and was co-authored by 1026 authors from 133 institutions! The number is so huge that the authorship order of the paper is actually listed alphabetically! Nevertheless, this observation on gravity waves seems to provide more evidence for the validity of the general relativity theory of Einstein. This is, however, out of the scope of the present discussion.

The Schwarzschild metric for curved space-time can be expressed

$$ds^2 = g_{\alpha\beta} dx^\alpha dx^\beta \tag{12.257}$$

where Einstein's notation has been used (i.e., summation over α and β for 1,2,3 and 4) and $g_{\alpha\beta}$ is the covariant metric tensor component for the curved space and x^α is the contravariant component of the coordinate. For curvilinear coordinates, it is important to distinguish covariant and contravariant components because the base vectors of the coordinates are not perpendicular.

Physically, the metric tensor can be used to define the length of and the angle between tangent vectors on a manifold. Integration of the metric tensor allows us to calculate the length of curves on the manifold. In Euclidean space, the components of the metric tensor can be found by dot product between the tangent vectors. Riemann geometry is positive definite in the sense that

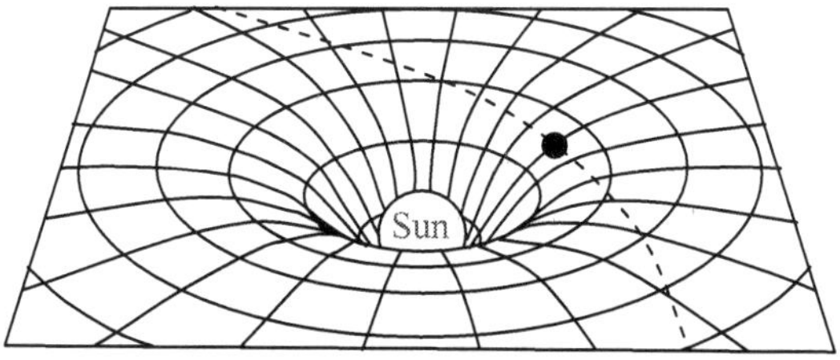

The mass of the Sun distort the space-time
Sun
Object moves along the shortest path allowed by the curved space-time

Figure 12.18 Distorted space-time around the Sun (a heavy mass)

$$g_{\alpha\alpha} > 0 \tag{12.258}$$

for all values of α. For spherical coordinates, we can write (12.257) explicitly as

$$ds^2 = g_{rr}dr^2 + g_{\theta\theta}d\theta^2 + g_{\varphi\varphi}d\varphi^2 + g_{tt}dt^2 \tag{12.259}$$

Note that the base vectors for spherical coordinates are perpendicular, and thus we do not need to use subscript and superscript to distinguish the components. In addition, (12.259) illustrates that in general relativity we treat the time variable as another coordinate in a four-dimensional space-time, instead of the three spatial variables and one time variable. We will introduce another proper time scale later. In view of the classical length of the line element in an ordinary spherical metric without the effect of relativity, we are prone to assume

$$ds^2 = A(r)dr^2 + r^2d\theta^2 + r^2\sin^2\theta d\varphi^2 + B(r)dt^2 \tag{12.260}$$

By spherical symmetry, we expect that g_{rr} and g_{tt} can only be a function of radial direction. Instead of starting from Einstein's field equation for vacuum, which involves curvilinear tensors, we will follow a more intuitive approach in obtaining the functions $A(r)$ and $B(r)$. It is instructive to recall that for the case of special relativity, the gravitational effect is unimportant. In this case, the Minkowski space-time applies:

$$ds^2 = dr^2 + r^2d\theta^2 + r^2\sin^2\theta d\varphi^2 - c^2dt^2 \tag{12.261}$$

Let us consider the case of spherical symmetry and temporarily stationary space-time, and define the following functional

$$\begin{aligned} F &= (\frac{ds}{d\tau})^2 = -c^2 = A(r)(\frac{dr}{d\tau})^2 + r^2(\frac{d\varphi}{d\tau})^2 + B(r)(\frac{dt}{d\tau})^2 \\ &= A(r)\dot{r}^2 + r^2\dot{\varphi}^2 + B(r)\dot{t}^2 \end{aligned} \tag{12.262}$$

Similar to the discussion of geodesics and suggested by Figure 12.18, we consider the first variation of the arc-length in the four-dimensional space:

$$J = \int_{r_1}^{r_2} \sqrt{-(\frac{ds}{d\tau})^2}\, d\tau \tag{12.263}$$

Since $ds/d\tau$ is stationary, the Euler-Lagrange of this function is exactly the same for the following function:

$$I = \int_{r_1}^{r_2} (\frac{ds}{d\tau})^2 d\tau \tag{12.264}$$

Applying calculus of variations (see Section 13.6 of Chau, 2018), for the case of three dependent functions and one independent variable, we have

$$\frac{\partial F}{\partial r} - \frac{d}{d\tau}(\frac{\partial F}{\partial \dot{r}}) = 0 \tag{12.265}$$

$$\frac{\partial F}{\partial \varphi} - \frac{d}{d\tau}(\frac{\partial F}{\partial \dot{\varphi}}) = 0 \tag{12.266}$$

$$\frac{\partial F}{\partial t} - \frac{d}{d\tau}(\frac{\partial F}{\partial \dot{t}}) = 0 \tag{12.267}$$

Substitution of (12.262) into these equations gives

$$2A\ddot{r} + A'\dot{r}^2 = 2r\dot{\varphi}^2 + B'\dot{t}^2 \tag{12.268}$$

$$4r\dot{r}\dot{\varphi} + 2r^2\ddot{\varphi} = 0 \tag{12.269}$$

$$2B'\dot{r}\dot{t} + 2B\ddot{t} = 0 \tag{12.270}$$

In obtaining these results, we have noted the following

$$\frac{\partial B}{\partial \tau} = \frac{\partial B}{\partial r}\frac{\partial r}{\partial \tau} = B'\dot{r}, \quad \frac{\partial A}{\partial \tau} = \frac{\partial A}{\partial r}\frac{\partial r}{\partial \tau} = A'\dot{r} \tag{12.271}$$

Some steps of the derivation are given in Problem 12.13. For circular motions, the radius is a constant and we get

$$\dot{r} = \ddot{r} = 0 \tag{12.272}$$

Thus, (12.268) gives

$$B' = -\frac{2r\dot{\varphi}^2}{\dot{t}^2} = -2r(\frac{d\varphi}{dt})^2 \tag{12.273}$$

From Kepler's third law, we have

$$\frac{T^2}{r^3} = \frac{1}{K} \tag{12.274}$$

where K is a constant. From (12.189), we have K being

$$F_r = -\frac{4\pi^2 Km}{r^2} = -\frac{GMm}{r^2} \tag{12.275}$$

Comparing the coefficients in (12.275), we get

$$K = \frac{GM}{4\pi^2} \tag{12.276}$$

Substitution of (12.276) into (12.274) gives

$$T^2 = \frac{4\pi^2}{GM} r^3 \tag{12.277}$$

By definition, the period relates to the rotation speed as

$$T = \frac{2\pi}{(d\varphi / dt)} \tag{12.278}$$

Comparing (12.277) and (12.278), we obtain

$$(\frac{d\varphi}{dt})^2 = \frac{GM}{r^3} \tag{12.279}$$

Substitution of (12.279) into (12.273) yields

$$B' = -\frac{2GM}{r^2} \tag{12.280}$$

Integration of (12.269) gives

$$B = \frac{2GM}{r} + C_3 \tag{12.281}$$

For the case of flat space-time, we have $M = 0$ or $r \to \infty$ and (12.261) for Minkowski space-time gives

$$B(\infty) = -c^2 = C_3 \tag{12.282}$$

Substitution of this result into (12.281) gives the final solution for B

$$B(r) = c^2(\frac{2GM}{c^2 r} - 1) = c^2(\frac{r_s}{r} - 1) \tag{12.283}$$

where the Schwarzschild radius is defined as

$$r_s = \frac{2GM}{c^2} \tag{12.284}$$

For the case of a stationary point mass, we have

$$\dot{r} = 0, \quad \dot{\varphi} = 0 \tag{12.285}$$

Using (12.285), (12.262) becomes

$$\dot{t}^2 = -\frac{c^2}{B(r)} \tag{12.286}$$

Substituting (12.285) into (12.268), we get

$$A = \frac{B'\dot{t}^2}{2\ddot{r}} \tag{12.287}$$

However, for the point mass being temporarily stationary, we have the radial force equilibrium as

$$\ddot{r} = -\frac{MG}{r^2} \tag{12.288}$$

Therefore, in view of (12.286) and (12.288), the unknown function A in (12.287) becomes

$$A = (-\frac{2GM}{r^2})[-\frac{c^2}{c^2(\frac{2GM}{c^2 r} - 1)}](-\frac{r^2}{2GM}) = \frac{1}{1 - \frac{2GM}{c^2 r}} \tag{12.289}$$

Finally, the well-known Schwarzschild metric is obtained as

$$ds^2 = \frac{1}{1 - \frac{2GM}{c^2 r}} dr^2 + r^2 d\theta^2 + r^2 \sin^2\theta d\varphi^2 - c^2(1 - \frac{2GM}{c^2 r})dt^2 \tag{12.290}$$

This is the first analytic solution for Einstein's vacuum field equation. For the special case $M = 0$, Minkowski space-time is recovered as a special case. We see that if r equals the Schwarzschild radius, the metric g_{rr} becomes infinite and g_{tt} becomes zero. For our Sun, it can be shown that the Schwarzschild radius is about 2.959 km (using (12.284)), which is a very small value. Therefore, there is a singularity in the Schwarzschild metric at the Schwarzschild radius. This eventually leads to the study of collapsing stars and black holes. In other words, if our Sun were squeezed in a sphere of radius 2.959 km, a black hole would be formed. Even light cannot escape from it because of its immense gravitation force, so it was called the black hole. Various kinds of transformations have been proposed to remove the singularity and to study the dynamics of Schwarzschild geometry. Novikov coordinates have been proposed to study the dynamics of a falling particle, whereas Eddington-Finkelstein coordinates and Kruskal-Szekeres coordinates have been proposed for studying the dynamics of a photon falling into a black hole. All kinds of strange consequences have been obtained based on the dynamics of Schwarzschild geometry, including the event horizon, a wormhole connecting two asymptotically flat universes, Einstein-Rosen bridges (two singularities join to form a non-singular bridge), and time travel. However, there is no reason whatsoever to believe that these strange things exist in the real universe. We refer readers to more advanced textbooks on gravitation by Weinberg (1972) and Misner et al. (1973).

12.16.2 Energy Term Due to Relativity

To investigate the effect of relativity on the energy, we employ variational principle. In particular, we consider world line along the Schwarzschild geometry. Following the ideas of geodesics discussed in Chapter 13 of Chau (2018) and from (12.290), the shortest path on the equator (or $\theta = \pi/2$ in (12.290)) is

$$I = \int ds = c\int d\tau = \int \sqrt{c^2(1-\frac{2GM}{c^2 r})dt^2 - (1-\frac{2GM}{c^2 r})^{-1} dr^2 - r^2 d\varphi^2} \quad (12.291)$$

By introducing a time parameter τ which is the proper time read by a clock carried along by the moving object, we rewrite (12.291) as

$$I = \int \sqrt{c^2(1-\frac{2GM}{c^2 r})\dot{t}^2 - (1-\frac{2GM}{c^2 r})^{-1} \dot{r}^2 - r^2 \dot{\varphi}^2}\, d\tau \quad (12.292)$$

Note again that t is regarded as a coordinate in relativity. From (12.291), we have

$$ds^2 = -c^2 d\tau^2 \quad (12.293)$$

Thus, along the world line of the moving object, we have

$$c^2(1-\frac{2GM}{c^2 r})\dot{t}^2 - (1-\frac{2GM}{c^2 r})^{-1} \dot{r}^2 - r^2 \dot{\varphi}^2 = c^2 \quad (12.294)$$

Again, in view of (12.264) we have I defined in (12.292) being stationary, also implying J below being stationary:

$$\delta J = \delta \int L d\tau = 0 \quad (12.295)$$

where

$$L = c^2(1-\frac{2GM}{c^2 r})\dot{t}^2 - (1-\frac{2GM}{c^2 r})^{-1} \dot{r}^2 - r^2 \dot{\varphi}^2 \quad (12.296)$$

In the dynamics of a rigid body, L is also known to be Lagrangian. For circular orbits, we have t and φ being the variables and associated conjugate forces can be defined as

$$p^t = \frac{\partial L}{\partial \dot{t}} = -2c^2(1-\frac{2GM}{c^2 r})\dot{t} = -2c^2\mathscr{E} \tag{12.297}$$

$$p^\varphi = \frac{\partial L}{\partial \dot{\varphi}} = -2r^2\dot{\varphi} = 2\mathscr{L} \tag{12.298}$$

where $\mathscr{E}$ and $\mathscr{L}$ are constants. Physically, we can see from (12.298) that

$$\mathscr{L} = r^2\dot{\varphi}, \quad or \quad \mathscr{L} = \frac{l}{m} \tag{12.299}$$

where l is the angular momentum which is constant in view of conservation of angular momentum. Comparing (12.284) and (12.286), we find that L is conserved or

$$L = c^2 \tag{12.300}$$

Thus, we can substitute (12.297) and (12.298) into (12.294) to get

$$c^2(1-\frac{2GM}{c^2 r})\frac{\mathscr{E}^2}{(1-\frac{2GM}{c^2 r})^2} - (1-\frac{2GM}{c^2 r})^{-1}\dot{r}^2 - \frac{\mathscr{L}}{r^2} = c^2 \tag{12.301}$$

Rearranging (12.301), we find

$$\dot{r}^2 - \frac{2GM}{r} + \frac{\mathscr{L}^2}{r^2}(1-\frac{2GM}{c^2 r}) = c^2(\mathscr{E}^2 - 1) \tag{12.302}$$

Multiplying $m/2$ to (12.302), we have the familiar form:

$$\frac{1}{2}m\dot{r}^2 - \frac{GMm}{r} + \frac{l^2}{2mr^2}(1-\frac{2GM}{c^2 r}) = E \tag{12.303}$$

This is precisely the energy of the object in the gravitation field. For the case of non-relativistic formation, we have the energy being

$$E = \frac{1}{2}m(\dot{r}^2 + r^2\dot{\varphi}^2) + U(r) = \frac{1}{2}m(\dot{r}^2 + r^2\dot{\varphi}^2) - \frac{GMm}{r} \tag{12.304}$$

Comparison of (12.303) and (12.304) gives the additional energy term due to relativity in (12.303) as

$$\Delta E = -\frac{GMl^2}{mc^2 r^3} \tag{12.305}$$

12.16.3 Contribution to Perihelion Precession

It is important to recognize that the angular momentum l conserves and thus we need to express the time derivative of r in terms of angular momentum first. Recall the change of variables $r = 1/u$ and (12.298) and we have

$$\dot{r} = \frac{dr}{d\tau} = \frac{dr}{du}\frac{du}{d\varphi}\frac{d\varphi}{d\tau} = -r^2\frac{du}{d\varphi}\frac{l}{mr^2} = -\frac{l}{m}\frac{du}{d\varphi} \tag{12.306}$$

In terms of the new variable and using (12.303), we have

$$E=\frac{l^2}{2m}[(\frac{du}{d\varphi})^2+u^2]-GMmu-\frac{GMl^2}{mc^2}u^3 \quad (12.307)$$

where l is a constant. Since we have the energy at the perihelion being the minimum, differentiation of (12.307) with respect to φ gives

$$\frac{dE}{d\varphi}=\frac{l^2}{2m}[2(\frac{du}{d\varphi})\frac{d^2u}{d\varphi^2}+2u\frac{du}{d\varphi}]-GMm\frac{du}{d\varphi}-\frac{3GMl^2}{mc^2}u^2\frac{du}{d\varphi}=0 \quad (12.308)$$

Simplification of (12.308) gives

$$\frac{l^2}{m}[\frac{d^2u}{d\varphi^2}+u]-GMm-\frac{3GMl^2}{mc^2}u^2=0 \quad (12.309)$$

This can be rearranged as

$$\frac{d^2u}{d\varphi^2}+u=\frac{GMm^2}{l^2}+\frac{3GM}{c^2}u^2 \quad (12.310)$$

This is a nonlinear second order nonhomogeneous ODE. By using the perturbation method, we seek a solution form composed of two parts:

$$u=u_0+u_1 \quad (12.311)$$

with

$$\frac{d^2u_0}{d\varphi^2}+u_0=\frac{GMm^2}{l^2} \quad (12.312)$$

With the definition given in (12.311), it is clear that the effect of relativity only enters the solution through u_1. Substitution of (12.311) into (12.310) and in view of (12.312) gives

$$\frac{d^2u_1}{d\varphi^2}+u_1=\frac{3GM}{c^2}(u_0+u_1)^2\approx\frac{3GM}{c^2}u_0^2 \quad (12.313)$$

We drop the last term because we are expecting that the contribution from u_1 is much smaller than that of u_0. We seek the solution of (12.312) in the following form

$$u_0=A+B\cos\varphi=A(1+e\cos\varphi) \quad (12.314)$$

Substitution of (12.314) into (12.312) gives

$$A=\frac{GMm^2}{l^2} \quad (12.315)$$

On the other hand, at $\varphi=\pi/2$, we have $r=b$ or

$$u_0(\pi/2)=\frac{1}{b}=A=\frac{1}{a(1-e^2)} \quad (12.316)$$

Substitution of (12.314) and (12.316) into (12.313) gives

$$\begin{aligned}\frac{d^2u_1}{d\varphi^2}+u_1&=\frac{3GM}{c^2}A^2(1+e\cos\varphi)^2\\&=D^2[(1+\frac{e^2}{2})+2e\cos\varphi+\frac{e^2}{2}\cos 2\varphi]\end{aligned} \quad (12.317)$$

where

$$D^2 = (\frac{3GM}{c^2})A^2 \tag{12.318}$$

Because of the linearity of (12.317), we can find the nonhomogeneous solution term by term. It is not difficult to show that the nonhomogeneous solution of (12.317) is

$$u_1 = D^2[(1+\frac{e^2}{2}) - (\frac{e^2}{6})\cos 2\varphi + e\varphi \sin\varphi] \tag{12.319}$$

For each complete revolution, we have

$$\varphi \leftarrow \varphi + 2\pi \tag{12.320}$$

We see that only the last term in (12.310) can cause perihelion precession, whereas the other two terms can only give a slight change in shape. Thus, the zeroth-order solution and the secular term give

$$u = u_0 + u_1 \approx A(1+e\cos\varphi) + D^2 e\varphi \sin\varphi \tag{12.321}$$

In addition, the homogeneous solution of u_1 is the same as that of u_0, thus, it adds nothing new to the solution in (12.321). At the perihelion, we have r being minimum or

$$\frac{\partial u}{\partial \varphi} = 0 = -Ae\sin\varphi + D^2 e(\sin\varphi + \varphi\cos\varphi) \tag{12.322}$$

Thus, this gives $\varphi = 0$ but $\varphi = 2\pi$ is not a solution. We look for a solution of the form $\varphi = 2\pi + \delta$ for small δ. Physically, δ is the precession. For small δ, we have

$$\sin(2\pi+\delta) = \sin\delta \approx \delta, \quad \cos(2\pi+\delta) = \cos\delta \approx 1 \tag{12.323}$$

At the end of one revolution, (12.322) gives

$$-Ae\delta + D^2 e[\delta + (2\pi+\delta)] = 0 \tag{12.324}$$

However, $\delta << 2\pi$ and we can approximate δ as:

$$\delta = \frac{2\pi D^2}{A} = 2\pi(\frac{3GM}{c^2})\frac{1}{a(1-e^2)} \tag{12.325}$$

We can substitute the following parameters for Mercury into (12.325):

$$\begin{aligned} &a = 57.91\times 10^6 \text{km}, \quad \varepsilon = 0.2056, \\ &M_S = 1.9891\times 10^{30} \text{kg}, \quad c = 2.99792458\times 10^8 \text{m/s} \end{aligned} \tag{12.326}$$

At the perihelion, we have r being minimum or

$$\begin{aligned} \delta &= \frac{6\pi(1.3297\times 10^{11})}{(299792.458)^2(57.91\times 10^6)(1-0.2056^2)} \\ &= 5.028262\times 10^{-7} \text{rad/revolution} \end{aligned} \tag{12.327}$$

The period of Mercury is 87.969 days per revolution. We now convert this value to arc seconds per century as:

$$\begin{aligned} \delta &= 5.028262\times 10^{-7}(\frac{360}{2\pi})(3600)(\frac{1}{87.969})(365)(100) \\ &= 43.03 \text{sec/century} \end{aligned} \tag{12.328}$$

This value agrees very well with the 43 seconds per century that was unaccounted for by Newton's law of gravitation. Finally, a long-standing problem in the study of the orbit of Mercury was resolved by the theory of general relativity (Weinberg,

1972). This is one of the very first validations of Einstein's theory of general relativity. However, the theory of general relativity is out of the scope of the present chapter.

12.17 MOTION NEAR THE EARTH'S SURFACE

In this section, we will extend our analysis of the elliptic orbit to body travel very close to the surface of the Earth. A commonly encountered example of this kind is the study of trajectory of missiles, which will be considered in the next section. As shown in Figure 12.19, an object is fired at Point A and lands on Point B. We will examine its orbit in view of our results presented in the previous sections. The apogee and perigee of the elliptic orbits are denoted as a and p on Figure 12.19. Note that the center of the Earth must locate at one of the focuses of the elliptic orbit. The elevation climb of the object is shown as h in the figure.

The object is launched by a velocity v at Point A. From (12.195), the total energy of a body of mass m on the orbit is

$$\frac{1}{2}mv^2 - \frac{GM_E m}{R_E} = E = -\frac{GM_E m}{2a} \tag{12.329}$$

From Figure 12.19, we see that twice the semi-major axis is:

$$2a = R_E + H \tag{12.330}$$

Substitution of (12.330) into (12.329) gives

$$\begin{aligned} E &= -\frac{GM_E m}{(H+R_E)} = -\frac{GM_E m}{R_E(1+\frac{H}{R_E})} \\ &= -\frac{GM_E m}{R_E}(1-\frac{H}{R_E}+\ldots) = -\frac{GM_E m}{R_E} + \frac{GM_E mH}{R_E^2} + \ldots \end{aligned} \tag{12.331}$$

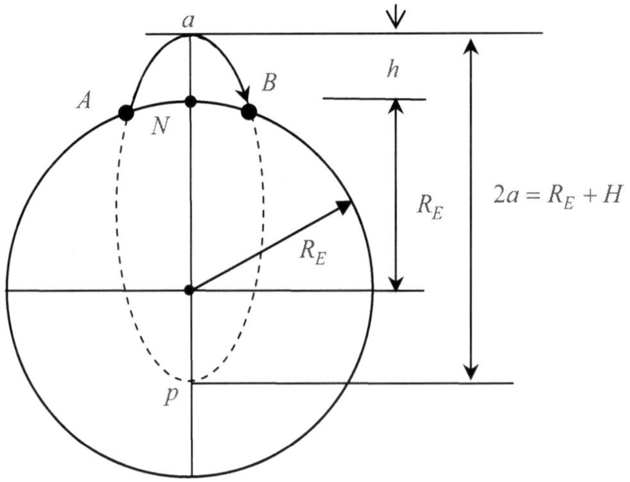

Figure 12.19 Motion near Earth's surface

Comparing (12.329) and (12.331), we find

$$\frac{1}{2}mv^2 = \frac{GM_E mH}{R_E^2} + ... \tag{12.332}$$

Rearranging (12.332), we find the following approximation for H

$$H \approx \frac{v^2}{2GM_E / R_E^2} \tag{12.333}$$

On the surface of the Earth, we must have the weight of the object equal to the gravitational pull of the Earth as:

$$mg = \frac{GM_E m}{R_E^2} \tag{12.334}$$

Cancellation of mass m on both sides gives

$$g = \frac{GM_E}{R_E^2} \tag{12.335}$$

In view of (12.335), we have the following approximation for H

$$H \approx \frac{v^2}{2GM_E / R_E^2} = \frac{v^2}{2g} \tag{12.336}$$

It is clear from Figure 12.19 that $H > h$. For the special case that the object is launched vertically, we have the perigee located at the center of the Earth. Thus, we have

$$h = \frac{v^2}{2g} \tag{12.337}$$

In high school, we learn that the conservation of energy leads to

$$\frac{1}{2}mv^2 = mgh \tag{12.338}$$

This gives the classic solution that

$$v = \sqrt{2gh} \tag{12.339}$$

This, of course, agrees with (12.337).

12.18 ROCKET AND MISSILE PROBLEM

Let us now consider the general solution for the orbit of an object with a specified initial launching condition. As shown in Figure 12.20, if a missile is fired from the surface of the Earth with an initial velocity of v_0 and inclination with horizontal of θ, the total energy of the orbit is

$$E = \frac{1}{2}mv_0^2 - \frac{GM_E m}{R_E} = -\frac{GM_E m}{2a} \tag{12.340}$$

The first step is to check with whether E is smaller than zero. If yes, we indeed have a bounded elliptic orbit. For missile problems, we must have $E < 0$. The initial orbital angular momentum is

$$l = |\boldsymbol{r} \times m\boldsymbol{v}| = mR_E v_0 \sin\varphi \tag{12.341}$$

The perigee and apogee are located at

$$r = r_1 \text{ (perigee)}, \quad r = r_2 \text{ (apogee)} \tag{12.342}$$

These points are labeled as p and a in Figure 12.20, and their corresponding velocities are denoted by v_1 and v_2. For a central force field under gravitational pull, angular momentum conserves. In addition, at perigee and apogee the velocities are perpendicular to the displacement vector $\boldsymbol{r}$. That is, we have $\varphi = \pi/2$ and thus

$$l = mv_1 r_1 = mv_2 r_2 \tag{12.343}$$

Plunging these into the energy equation given in (12.340), we get

$$\frac{1}{2}mv_1^2 - \frac{GM_E m^2 v_1}{l} = \frac{1}{2}mv_2^2 - \frac{GM_E m^2 v_2}{l} = -\frac{GM_E m}{2a} \tag{12.344}$$

Both v_1 and v_2 are found satisfying the following quadratic equations:

$$lamv^2 - GM_E 2am^2 v + GM_E ml = 0 \tag{12.345}$$

The solutions of v_1 and v_2 at the perigee and apogee are

$$v_1 = \frac{GM_E m}{l}\{1 + \sqrt{1 - \frac{l^2}{GM_E m^2 a}}\} \tag{12.346}$$

$$v_2 = \frac{GM_E m}{l}\{1 - \sqrt{1 - \frac{l^2}{GM_E m^2 a}}\} \tag{12.347}$$

However, for the missile problem, perigee is likely inside the Earth, so v_1 is not a physically meaningful result. For more general situations of higher orbital travel, these equations do provide the speeds at perigee and apogee in terms of the initial angular momentum.

Equation (12.343) gives

$$r_1 = \frac{l}{mv_1}, \quad r_2 = \frac{l}{mv_2} \tag{12.348}$$

For the missile problem, the second part of (12.348) gives the maximum altitude of flight of the missile as:

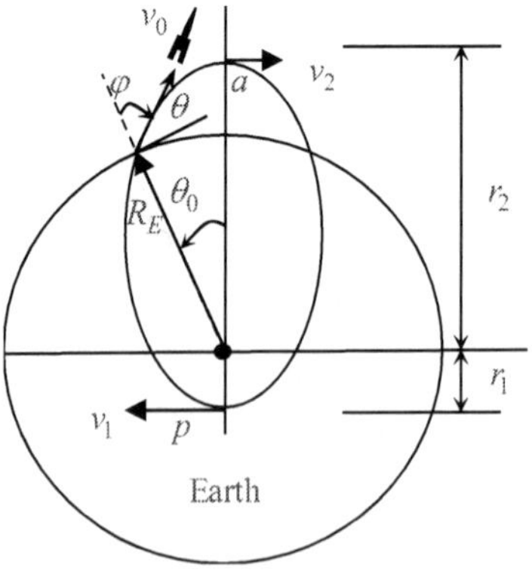

Figure 12.20 Orbit of missile

$$h = \frac{l^2}{GM_E m^2} \frac{1}{\{1 - \sqrt{1 - \frac{l^2}{GM_E m^2 a}}\}} - R_E \tag{12.349}$$

To further simplify (12.349), we note that the angular momentum is

$$l = mr_E v_0 \sin\varphi = mr_E v_0 \sin(\pi/2 - \theta) = mr_E v_0 \cos\theta \tag{12.350}$$

where θ is defined in Figure 12.20. On the other hand, a can be solved by using (12.340) as:

$$a = \frac{1}{2(\frac{1}{R_E} - \frac{v_0^2}{2GM_E})} = \frac{R_E}{2(1-\alpha)} \tag{12.351}$$

where the dimensionless parameter α is defined as

$$\alpha = \frac{R_E v_0^2}{2GM_E} \tag{12.352}$$

Substituting (12.351) and (12.352) into (12.349) gives

$$h = R_E\{\frac{2\alpha\cos^2\theta}{1 - \sqrt{1 - 4\alpha(1-\alpha)\cos^2\theta}} - 1\} \tag{12.353}$$

Note from the equation of an ellipse that we have

$$r_1 = a(1-e), \quad r_2 = a(1+e) \tag{12.354}$$

Subtracting these equations, we get the eccentricity of the elliptic orbit as

$$e = \frac{r_2 - r_1}{2a} \tag{12.355}$$

Recall from (12.180) that

$$r_E = \frac{a(1-e^2)}{1 - e\cos\theta_0} \tag{12.356}$$

Solving for the angle θ_0, we have

$$\theta_0 = \cos^{-1}\{\frac{1}{e}[1 - \frac{a}{R_E}(1-e^2)]\} \tag{12.357}$$

In summary, if the initial speed v and inclination angle θ of the missile of mass m are given, we can use (12.351) to find a, (12.346) and (12.347) to find v_1 and v_2, (12.348) to find r_1 and r_2, (12.353) to find maximum altitude h, (12.355) to find the eccentricity e, and finally (12.357) to find θ_0. Therefore, the orbit of flight is completely determined.

Example 12.6 On July 28, 2017, North Korea fired an intercontinental ballistic missile into the Sea of Japan at 11:41 p.m. It was reported that the missile was fired to an altitude of 3700 km. It lasted for 47 minutes before landing in the Sea of Japan with a travel distance of about 1000 km (see Figure 12.21). The missile is assumed to be 700 kg. A weather webcam at Muroran City on Hokkaido Island appeared to capture the missile before it landed in the Sea of Japan. The Korean Central Television released a video the next day showing the process of launching of Hwasong-13 missile. The launching angle is very close to vertical, and a crude estimation from the video is about $\theta = 85°$. Using this information, (i) estimate the angle θ_0, (ii) estimate the initial speed of the missile v_0, and (iii) find the maximum striking distance of this missile at $\theta = 45°$.

Solution:

(i) The angle θ_0 shown in Figure 12.13 covers only half of the travel distance. Thus, we have

$$\theta_0 = \frac{d}{R_E} = \frac{500}{6378} = 0.0783945 = 4.49^\circ \tag{12.358}$$

(ii) We note that (12.357) can be rearranged as

$$e\cos\theta_0 = 1 - \frac{r_2}{(1+e)R_E}(1-e^2) = 1 - \frac{r_2}{R_E}(1-e) \tag{12.359}$$

Eccentricity of the orbit can be solved from (12.359) as

$$e = \frac{\dfrac{r_2}{R_E} - 1}{\dfrac{r_2}{R_E} - \cos\theta_0} = \frac{(6378+3700)/6378-1}{(6378+3700)/6378-\cos(4.49^\circ)} = 0.9947 \tag{12.360}$$

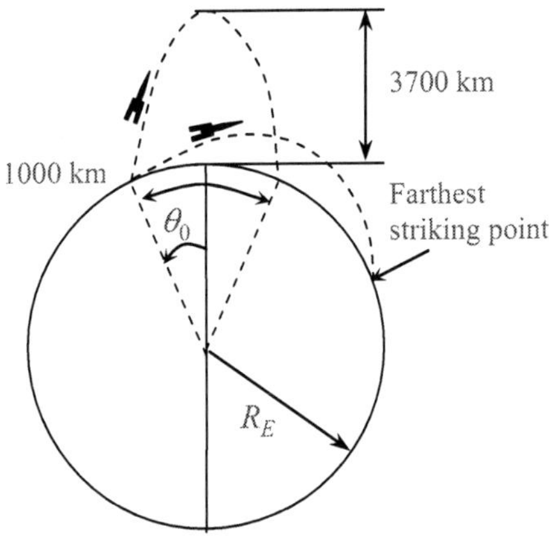

Figure 12.21 Striking distance of ballistic missile

The semi-axis a is

$$a = \frac{r_2}{1+e} = \frac{6378+3700}{1+0.9947} = 5052 \text{ km} \tag{12.361}$$

The apogee distance r_1 is

$$r_1 = a(1-e) = 5052(1-0.9947) = 26.77 \text{ km} \tag{12.362}$$

Substituting (12.348) into the total energy given in (12.329), we find

$$l^2 \frac{a}{mr_2^2} - \frac{2aGM_E m}{r_2} + GM_E m = 0 \tag{12.363}$$

Solving for the angular momentum, we get

$$\begin{aligned} l &= mr_2 \sqrt{\frac{GM_E}{a}(\frac{2a}{r_2} - 1)} \\ &= (700)(10078 \times 10^3)\sqrt{\frac{3.99763 \times 10^{14}}{5052 \times 10^3}(\frac{2 \times 5052}{10078} - 1)} \\ &= 3.187436 \times 10^{12} \text{kgm}^2 / \text{s} \end{aligned} \tag{12.364}$$

Finally, the initial velocity can be found as

$$v_0 = \frac{l}{mR_E \cos\theta} = \frac{3.187436 \times 10^{12}}{(700)(6378 \times 10^3)\cos(85^\circ)} = 8191.7 \text{m/s} \tag{12.365}$$

This value is, of course, larger than the first escape velocity but smaller than the second escape velocity of the Earth.

(iii) For a launching angle of $\theta = \pi/4$, the angular momentum is

$$l = mR_E v_0 \cos\theta = \frac{mR_E v_0}{\sqrt{2}} \tag{12.366}$$

Next, we consider the velocities at perigee and apogee. Let us simplify the result by noting:

$$\frac{GM_E m}{l} = \frac{GM_E}{R_E v_0}\sqrt{2} = \frac{v_0}{\sqrt{2\alpha}} \tag{12.367}$$

$$\frac{l^2}{GM_E m^2 a} = \frac{m^2 R_E^2 v_0^2}{2GM_E m^2 a} = \frac{m^2 R_E^2 v_0^2}{2GM_E m^2}[\frac{2(1-\alpha)}{R_E}] = 2\alpha(1-\alpha) \tag{12.368}$$

where α has been defined in (12.352). In obtaining the last equation of (12.368), we have used (12.351). Substituting (12.367) and (12.368) into (12.346), we obtain

$$v_1 = \frac{v_0}{\sqrt{2\alpha}}\{1 + \sqrt{1 - 2\alpha(1-\alpha)}\} \tag{12.369}$$

$$v_2 = \frac{v_0}{\sqrt{2\alpha}}\{1 - \sqrt{1 - 2\alpha(1-\alpha)}\} \tag{12.370}$$

In view of (12.348) and (12.366), we find

$$r_1 = \frac{\alpha R_E}{1 + \sqrt{1 - 2\alpha(1-\alpha)}} \tag{12.371}$$

$$r_2 = \frac{\alpha R_E}{1 - \sqrt{1 - 2\alpha(1-\alpha)}} \tag{12.372}$$

With these results, we can determine the eccentricity by using (12.355)

$$e = \frac{r_2 - r_1}{2a} = \frac{1}{2}[\frac{\sqrt{1 - 2\alpha(1-\alpha)}}{1 - \alpha(1-\alpha)}] \tag{12.373}$$

The main parameter is

$$\alpha = \frac{R_E v_0^2}{2GM_E} = \frac{6378\times10^3(8191.7)^2}{2(3.99763\times10^{14})} = 0.5353 \tag{12.374}$$

Using this value of α, eccentricity becomes

$$e = \frac{r_2 - r_1}{2a} = \frac{1}{2}[\frac{\sqrt{1-2(0.5353)(0.4647)}}{1-(0.5353)(0.4647)}] = 0.4718 \tag{12.375}$$

Finally, using (12.351), we can rewrite (12.357) as

$$\begin{aligned} \theta_0 &= \cos^{-1}\{\frac{1}{e}[1-\frac{1-e^2}{2(1-\alpha)}]\} \\ &= \cos^{-1}\{\frac{1}{0.4718}[1-\frac{1-0.4718^2}{2(0.4647)}]\} = 69.72^\circ \end{aligned} \tag{12.376}$$

Finally, the maximum striking distance becomes

$$d = 2R_E\theta_0 = 2(6378)(\frac{69.72}{360})2\pi = 15521.7 \text{ km} \tag{12.377}$$

This striking distance covers most of the places in the United States, as speculated by many.

12.19 DYNAMICS OF ATMOSPHERIC RE-ENTRY

Space travel mainly consists of three main phases, the launch, placing the spacecraft into orbit, and the re-entry of spacecraft. Among these, re-entry is probably the most dangerous and difficult phase. The total energy of the spacecraft at re-entry is high because of the high potential energy (high orbital altitude) and high kinetic energy (high orbital speed). The friction between the Earth's atmosphere and the spacecraft at high speed leads to deceleration of the spacecraft and to heat generation. The surface of the spacecraft may rise to 1400°C during re-entry. The whole deceleration may take up to 30 minutes. The speed typically decreases from 11 km/s (hypersonic speed) down to 0.044 km/s (landing speed). The re-entry design depends on whether the mission is manned or unmanned. The re-entry design may also depend on safety issues (manned spacecraft), reusability of the spacecraft (such as the space shuttle), and the special mission requirement. Four types of forces are applied on the spacecraft during re-entry, including the draft force, lift force, gravitational force, and other forces. These forces depend on the velocity of the spacecraft, the cross-section area, the air density, shape of the spacecraft (including the airfoil and body shape), and the angle of attack (or the flight path angle). The first application of re-entry problems was in the area of inter-continental ballistic missile development after the Second World War. The missile head is normally a blunted head covered by thick metal. The metal is designed to allow melting and vaporization, and through this process, a large amount of heat is dissipated. This mechanism of heat dissipation is called ablation. For missile re-entry, it is normally a steep flight path angle and the re-entry time is as short as possible such that the landing location is as accurate as possible. For spacecraft re-entry, the angle of the flight path is normally small (say less than 12°) so the control range distance and entry time are long. Heat dissipation per time is minimized.

Figure 12.22 shows the re-entry of a spacecraft to the Earth. Six main variables are r, θ, ϕ, V, γ, and ψ. The first three variables change according to kinetic conditions, where the latter three according to the equation of motion.

12.19.1 Formulation

By kinetic conditions, the position of the spacecraft as seen from the rotating planet is determined by

$$\frac{dr}{dt} = V \sin\gamma \tag{12.378}$$

$$\frac{d\theta}{dt} = \frac{V\cos\gamma\cos\psi}{r\cos\phi} \tag{12.379}$$

$$\frac{d\phi}{dt} = \frac{V\cos\gamma\sin\psi}{r} \tag{12.380}$$

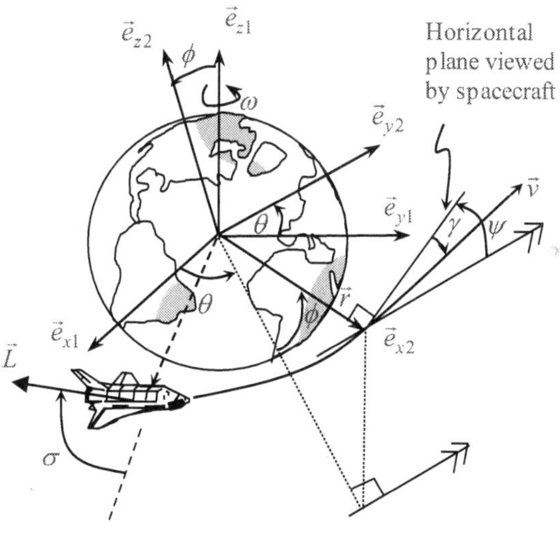

Figure 12.22 The re-entry of a spacecraft to the Earth

where the involved parameters are defined in Figure 12.22. The r is the distance of the spacecraft from the center of the Earth, θ the longitude of the spacecraft, and ϕ the latitude of the spacecraft. Force equilibrium conditions lead to the following three equations

$$\frac{dV}{dt} = -\frac{D}{m} - g\sin\gamma + A_{R\omega} \tag{12.381}$$

$$V\frac{d\gamma}{dt}=\frac{L}{m}\cos\sigma-g\cos\gamma+\frac{V^2}{r}\cos\gamma+A_{\gamma\omega} \tag{12.382}$$

$$V\frac{d\psi}{dt}=\frac{L\sin\sigma}{m\cos\gamma}-\frac{V^2}{r}\cos\gamma\cos\psi\tan\phi+A_{\psi\omega} \tag{12.383}$$

where V is the velocity of the spacecraft, γ is the flight-path angle between the local horizontal viewed from the spacecraft and the velocity vector, and ψ is the heading angle between the local parallel of latitude and the projection of the velocity vector on the local horizontal plane (see Figure 12.22). The bank angle σ is the angle between the lift force L and the r-v plane. The bank angle is normal imposed by the spacecraft to steer the course of re-entry and the final destination on landing.

The three additional terms at the end of (12.381) and (12.383) are due to the rotation of the planet Earth

$$A_{R\omega}=r\omega^2\cos\phi(\cos\phi\sin\gamma-\sin\phi\sin\psi\cos\gamma) \tag{12.384}$$

$$A_{\gamma\omega}=2V\omega\cos\phi\cos\psi+r\omega^2\cos\phi(\cos\phi\cos\gamma+\sin\phi\sin\psi\sin\gamma) \tag{12.385}$$

$$A_{\psi\omega}=2V\omega(\sin\psi\cos\phi\tan\gamma-\sin\phi)-\frac{r\omega^2}{\cos\gamma}\sin\phi\cos\phi\cos\psi \tag{12.386}$$

The full derivation of this set of six nonlinear coupled ODEs is tedious and is not so straightforward. We refer the reader to Hicks (2009) for complete derivation. The gravity and density of air as a function of the elevation are given by

$$g=g_s\left(\frac{R_E}{r}\right)^2 \tag{12.387}$$

$$\rho=\rho_s e^{-\beta(r-R_E)} \tag{12.388}$$

The lift and drag forces can be expressed in terms of the lift and drag coefficients as

$$L=\frac{1}{2}\rho C_L S V^2 \tag{12.389}$$

$$D=\frac{1}{2}\rho C_D S V^2 \tag{12.390}$$

where S is the reference area in calculating the drag and lift coefficients. In general, C_D and C_L are functions of the flight velocity and the attached angle (or flight path angle γ in Figure 12.22). For hypersonic flight, these coefficients are essentially functions of attack angle only. The system of (12.381) to (12.383) needs to be solved by numerical integration. In particular, the fourth order Runge-Kutta method is discussed in Section 15.5 of Chau (2018). In doing so, the bank angle should be given as a function of time.

However, for highly idealized situations a simple analytical solution has been obtained for the re-entry problem. These solutions are useful in the stage of conceptual design of the spacecraft and in the effects of various parameters of the problem. These solutions have been very helpful in making preliminary planning decisions. Let us consider the case that the re-entry path is along a great circle drawn from the center of the Earth as shown in Figure 12.23. Only three variables remain for this case: they are r, γ, and V.

In addition to planar re-entry, we ignore the heading angle and bank angle

$$\sigma = 0, \quad \psi = 0, \quad \theta = const., \quad \phi = const. \tag{12.391}$$

The remaining three equations, one kinematic and two equations of motion, are

$$\dot{r} = V \sin \gamma \tag{12.392}$$

$$\dot{V} = -\frac{D}{m} - g \sin \gamma \tag{12.393}$$

$$V\dot{\gamma} = \frac{L}{m}\cos \sigma - g \cos \gamma + \frac{V^2}{r} \tag{12.394}$$

For $\gamma = \pi/2$, the re-entry is steep and fast, and it is applicable to the case of ballistic missile re-entry, where the trajectory is designed to pass through the atmosphere as quickly and as straight as possible. In this relatively straight down manner, the vehicle minimizes the time spent in the atmosphere and, as a result, minimizes the uncertainty in the trajectory. In this section, we consider the shallow ballistic re-entry and the analytic solution by Yaroshevsky.

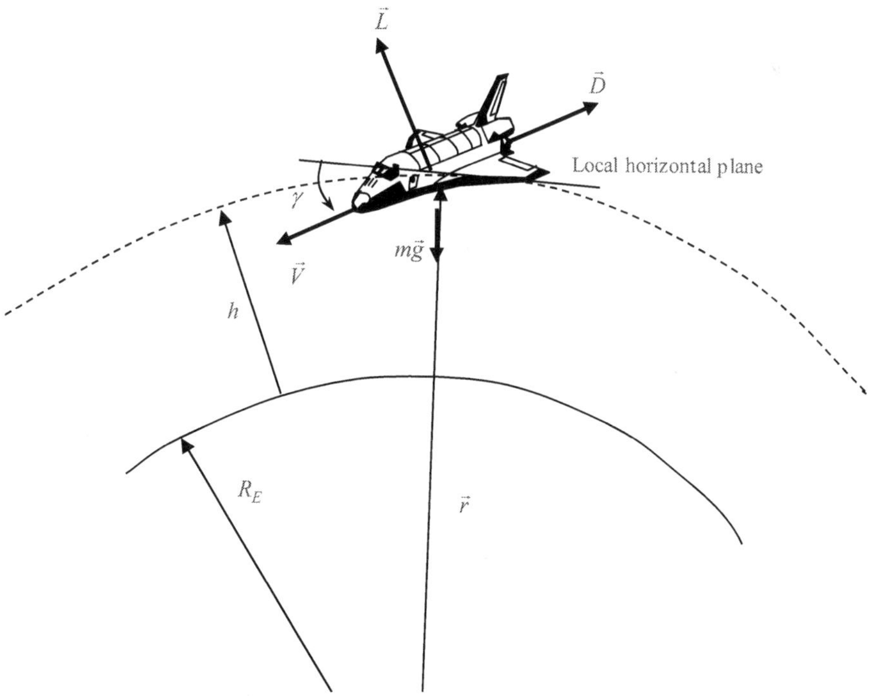

Figure 12.23 Flight-path angle at re-entry

12.19.2 Yaroshevsky Solution

The first man who traveled in the Earth's orbit of 300 km altitude was a Russian astronaut, Yuri Gagarin, on April 12, 1961. The Yaroshevsky solution was first published in 1964. Apparently, the Yaroshevsky solution may have helped to put the first man in space and returned him safely. More specifically, it is for shallow ballistic entry (i.e., γ is small) and for a vehicle with no wing (i.e., no lift force applied) such as the capsule in the Vostok 1, which took Yuri Gagarin on the first spaceflight by man. In particular, for shallow ballistic entry at near circular entry speed, we have

$$\sin\gamma \approx \gamma, \quad \cos\gamma \approx 1, \quad L \approx 0 \tag{12.395}$$

Substitution of (12.387), (12.404), and (12.390) into (12.392) to (12.394) results in

$$\dot{r} = V\gamma \tag{12.396}$$

$$\dot{V} = -\frac{1}{2m}\rho C_D S V^2 - g_0\left(\frac{r_0}{r}\right)^2 \gamma \tag{12.397}$$

$$V\dot{\gamma} = \frac{V^2}{r} - g_0\left(\frac{r_0}{r}\right)^2 \tag{12.398}$$

The first term on the right-hand side of (12.398) is the centrifugal force term. This can be further simplified for the case of thin atmosphere; the gravity can be assumed as the ground surface value and for shallow entry we further neglect the tangential gravity force term on the right-hand side of (12.397). Thus, we have

$$\dot{r} = V\gamma \tag{12.399}$$

$$\dot{V} = -\frac{1}{2m}\rho C_D S V^2 \tag{12.400}$$

$$V\dot{\gamma} = \frac{V^2}{R_0} - g_0 \tag{12.401}$$

Since the variable time does not appear explicitly in this system of ODEs, we can change the variable from time to velocity V such that (12.399) becomes:

$$\frac{dr}{dt} = \frac{dr}{dV}\frac{dV}{dt} = V\gamma \tag{12.402}$$

Substitution of (12.402) into (12.400) gives

$$\frac{dr}{dV}\frac{dV}{dt} = V\gamma = -\frac{1}{2m}\rho C_D S V^2 \frac{dr}{dV} \tag{12.403}$$

Rearranging (12.403), we find

$$\frac{dr}{dV} = -\frac{2m\gamma}{\rho C_D S V} \tag{12.404}$$

Similarly, (12.401) can be rewritten as

$$V\frac{d\gamma}{dV}\frac{dV}{dt} = \left(\frac{V^2}{R_0} - g_0\right) \tag{12.405}$$

Substitution of (12.400) into (12.405) gives

$$\frac{d\gamma}{dV} = \frac{2m}{\rho C_D S V^3}\left(g_0 - \frac{V^2}{R_0}\right) \tag{12.406}$$

Using the air density model given in (12.388), we can rewrite (12.406) as

$$\frac{d\gamma}{dV} = \frac{2m}{\rho_s C_D S V^3}(g_0 - \frac{V^2}{R_0})e^{\beta(r-R_E)} \tag{12.407}$$

The system is reduced to (12.404) and (12.407) with unknown r and γ. However, the system is still nonlinear and difficult to solve. Instead of working on r, Yaroshevsky ingeniously proposed the following change of variables x and y replacing γ and r (note that ρ is a function of r) as

$$x = -\ln \bar{V} \tag{12.408}$$

$$y = \frac{\rho C_D S}{2m}\sqrt{\frac{R_0}{\beta}} \tag{12.409}$$

where

$$\bar{V} = \frac{V}{\sqrt{R_0 g_0}} \tag{12.410}$$

Differentiating (12.409), we can relate r to y through ρ as

$$\frac{dy}{y} = \frac{d\rho}{\rho} = -\beta dr \tag{12.411}$$

Differentiating (12.408), we find that x can replace variable V as

$$dx = -\frac{d\bar{V}}{\bar{V}} = -\frac{dV}{V} \tag{12.412}$$

Dividing (12.411) by (12.412), we get

$$V\beta\frac{dr}{dV} = \frac{dy}{ydx} = -\frac{2m\gamma\beta}{\rho C_D S} \tag{12.413}$$

Rearranging (12.413), we find

$$\frac{dy}{dx} = -y\frac{2m\gamma\beta}{\rho C_D S} \tag{12.414}$$

From the definition of (12.409), we see that y is proportional to ρ and thus (12.414) is reduced to

$$\frac{dy}{dx} = -\gamma\sqrt{R_0\beta} \tag{12.415}$$

It is important to recognize that one of the unknown γ is found proportional to dy/dx. From the definition given in (12.412), we can rewrite (12.406) as

$$\frac{d\gamma}{dV} = -\frac{1}{V}\frac{d\gamma}{dx} = \frac{2mg_0}{\rho C_D S V^3}(1 - \frac{V^2}{R_0 g_0}) \tag{12.416}$$

Simplification of (12.416) gives

$$\frac{d\gamma}{dx} = -\frac{1}{y}\frac{1}{\sqrt{R_0\beta}}(\frac{1}{\bar{V}^2} - 1) \tag{12.417}$$

Finally, substitution of (12.417) into the differentiation of (12.415) with respect to x gives a single second order ODE for y with variable x

$$\frac{d^2 y}{dx^2} = \frac{1}{y}(\frac{1}{\bar{V}^2} - 1) = \frac{1}{y}(e^{2x} - 1) \tag{12.418}$$

To solve this ODE, we have to figure out the initial condition for y. At the initial time $t = 0$, the spacecraft is at high attitude (typically 24 km above ground)

$$V = \sqrt{R_0 g_0}, \quad \bar{V} = 1, \quad \rho = 0 \tag{12.419}$$

Note that at high attitude the air density is roughly zero. Thus, we have

$$x(0) = 0, \quad y(0) = 0, \quad \gamma(0) = 0, \quad \left.\frac{dy}{dx}\right|_{t=0} = 0 \tag{12.420}$$

The last condition given in (12.420) results from the assumption of an initial circular orbit. We will first find an approximation of y for small x. In particular, we first note that

$$e^{2x} = 1 + 2x + \frac{(2x)^2}{2!} + \frac{(2x)^3}{3!} + ... \tag{12.421}$$

Thus, the right-hand side of (12.418) becomes

$$e^{2x} - 1 = 2x + \frac{(2x)^2}{2!} + \frac{(2x)^3}{3!} + ... \tag{12.422}$$

Retaining the first term in this expansion for small x, we get

$$\frac{d^2 y}{dx^2} = \frac{2x}{y} \tag{12.423}$$

This ODE is singular at $y = 0$, unless y is proportional to x. Thus, we assume

$$y = Ax^p \tag{12.424}$$

where p is a constant. Substitution of (12.424) into (12.423) gives

$$Ap(p-1)x^{p-2} = \frac{2x}{Ax^p} \tag{12.425}$$

Equating the power of x on both sides, we require

$$p = 3/2 \tag{12.426}$$

Using this value of p in (12.424), we get

$$A = \sqrt{8/3} \tag{12.427}$$

Thus, the first approximation of (12.418) is

$$y = \sqrt{8/3} x^{3/2} \tag{12.428}$$

We can generalize this series solution to higher order:

$$y = \sqrt{8/3} x^{3/2} (C_0 + C_1 x + C_2 x^2 + ...) \tag{12.429}$$

Differentiating (12.429) twice, we find

$$\begin{aligned} y'' &= \frac{3}{4}\sqrt{\frac{8}{3}} x^{-1/2} (C_0 + C_1 x + C_2 x^2 + ...) + 3\sqrt{\frac{8}{3}} x^{1/2} (C_1 + 2C_2 x + ...) \\ &+ \sqrt{\frac{8}{3}} x^{3/2} (2C_2 x + 3C_3 x^2 ...) \end{aligned} \tag{12.430}$$

Note that (12.418) can be rearranged as

$$yy'' = e^{2x} - 1 \tag{12.431}$$

Substitution of (12.430) into the left-hand side of (12.431) and collection to third order of power of x gives

$$yy'' = \frac{8}{3}(\frac{3}{4})x[C_0^2 + 2C_0C_1x + (2C_0C_2 + C_1^2)x^2 + ...]$$
$$+3(\frac{8}{3})x^2[(C_0C_1 + (2C_0C_2 + C_1^2)x + ...] + \frac{8}{3}x^3(2C_0C_2 + ...) \quad (12.432)$$
$$= \frac{8}{3}[(\frac{3}{4}C_0^2)x + (\frac{9}{2}C_0C_1)x^2 + (\frac{19}{2}C_0C_2 + \frac{15}{4}C_1^2)x^3] + ...$$

On the other hand, the series expansion on the right-hand side of (12.431) gives

$$e^{2x} - 1 = 2x + 2x^2 + \frac{4}{3}x^3 + ... \quad (12.433)$$

Substitution of (12.432) and (12.433) into (12.431) gives

$$C_0 = 1, \quad C_1 = \frac{1}{6}, \quad C_2 = \frac{1}{24} \quad (12.434)$$

This gives a series solution as

$$y = \sqrt{8/3}x^{3/2}(1 + \frac{1}{6}x + \frac{1}{24}x^2 + ...) \quad (12.435)$$

We define the normalized deceleration as

$$\frac{a_d}{g_0} = -\frac{1}{g_0}\frac{dV}{dt} = \frac{1}{2mg_0}\rho C_D S V^2 = \frac{1}{g_0}(\frac{\rho C_D S}{2m}\sqrt{\frac{R_0}{\beta}})\sqrt{\frac{\beta}{R_0}}V^2$$
$$= \frac{y}{g_0}\sqrt{\frac{\beta}{R_0}}(\bar{V}^2 g_0 R_0) = y\bar{V}^2\sqrt{\beta R_0} \quad (12.436)$$

In obtaining the result in (12.436), we have used (12.400) and (12.409). Substitution of (12.435) into (12.436) results in

$$\frac{a_d}{g_0} = \sqrt{\frac{8\beta R_0}{3}}x^{3/2}(1 + \frac{1}{6}x + \frac{1}{24}x^2)e^{-2x} \quad (12.437)$$

The maximum deceleration can be evaluated by differentiating (12.437):

$$\frac{d}{dx}(\frac{a_d}{g_0}) = 0 \quad (12.438)$$

In view of (12.437), (12.438) leads to

$$(\frac{3}{2}x^{1/2} - \frac{19}{12}x^{3/2} - \frac{3}{16}x^{5/2} - \frac{1}{12}x^{7/2})e^{-2x} = 0 \quad (12.439)$$

If x is neither zero nor infinity, (12.448) gives

$$4x^3 + 9x^2 + 76x - 72 = 0 \quad (12.440)$$

By Decartes' rule of sign, there is only one change of sign for its coefficient in (12.440), and, consequently, there is one real root and it equals approximately 0.835. Substitution of this value of x into (12.408) and (12.435) gives

$$\bar{V} = e^{-x} = 0.434,$$
$$y = \sqrt{\frac{8}{3}}(0.835)^{3/2}[1 + \frac{1}{6}(0.835) + \frac{1}{24}(0.835)^2 + ...] = 1.46 \quad (12.441)$$

For planet Earth, we have approximately

$$\beta = 0.14 \text{ km}^{-1}, \quad R_0 = 6378 \text{ km} \tag{12.442}$$

The maximum deceleration can be evaluated by substituting these values into (12.436):

$$\frac{a_d}{g_0} = y\bar{V}^2\sqrt{\beta R_0} = (1.46)(0.434)^2\sqrt{(0.14)(6378)} = 8.2 \tag{12.443}$$

Therefore, regardless of the design of the spacecraft or capsule we have the maximum deceleration of about 8.2 times the ground gravitational acceleration:

$$a_d = 8.2 g_0 \tag{12.444}$$

This value is close to the maximum deceleration that can be tolerated by a human being (normally is about $12g_0$). Astronauts need to train in centrifuge to withstand such deceleration. The maximum velocity of the re-entry capsule is

$$V_{\max} = \bar{V}\sqrt{R_0 g_0} = 0.434\sqrt{(6378000)(9.81)} = 3.433 \text{ km/s} \tag{12.445}$$

This solution by Yaroshevsky might have played an important role in the design of the first human spaceflight in 1961.

12.20 RESTRICTED PROBLEM OF THREE BODIES

So far, we have only considered the gravitational interaction between two masses. Clearly, there are many planets and heavenly bodies in our solar system. The existence of a third body clearly will change the orbit of any mass. The so-called three-body problem in celestial mechanics was formulated by Newton, but he confessed that the problem was too difficult for him to solve it. In this section, we will present one of the few problems that can be solved analytically, called the restricted three-body problem obtained by Euler and Lagrange. The solutions are called Lagrangian Points. The restriction imposed in the formulation is that the motion of the three bodies is all within the same plane and the orbits are approximately circular. This celestial mechanics problem has been considered by the greatest mathematicians in history, including Newton, Euler, Lagrange, Laplace, Jacobi, Hill, Poincare, Gauss, Delaunay, Birkhoff, Kolmogorov, Painleve, Bendixson, Darwin, Moulton, Levi-Civita, Lyapunov, Sundman, Moser, and Arnold. Henri Poincaré's stability theory of dynamic systems is, in fact, motivated by the three-body problem. One of the very first three-body problems is the lunar theory of the moon under the gravitation pull of the Earth and the Sun. In this section, we will examine the particular solution to the three-body problem called Lagrangian Points.

12.20.1 Formulation of the Three-Body Problem

If there are n bodies rotating about their center of mass with a circular frequency of ω, the centrifugal force for each body can be formulated as:

$$\boldsymbol{f}_i = -m_i\omega^2\boldsymbol{r}_i, \quad i = 1, 2, ..., n \tag{12.446}$$

where $\boldsymbol{r}_i$ is the position vector of the i-th body measured from the center of mass, and m_i is the mass of the i-th body. We have the total gravitational force applied on the i-th mass from the other $n-1$ masses being:

$$f_i = G\sum_{j=1}^{n} \frac{m_i m_j}{r_{ij}^3}(\boldsymbol{r}_j - \boldsymbol{r}_i) \quad i = 1,2,...,n \;\&\; j \neq i \tag{12.447}$$

where G is the universal gravitational constant, and r_{ij} is the distance between the i-th and the j-th bodies. However, the resultant force of the n-body system must be zero:

$$\sum_{i=1}^{n} f_i = 0 \tag{12.448}$$

Substitution of (12.446) into (12.448) gives

$$\omega^2 (m_1 r_1 + m_2 r_2 + ... + m_n r_n) = 0 \tag{12.449}$$

The center of mass of the n-body system must coincide with the axis of rotation. We now formulate the three-body case, or $n = 3$. Equating (12.446) and (12.447) for the case of $i = 1$, we have

$$f_1 = G[\frac{m_1 m_2}{r_{12}^3}(r_2 - r_1) + \frac{m_1 m_3}{r_{13}^3}(r_3 - r_1)] = -m_1 \omega^2 r_1 \tag{12.450}$$

Rearranging (12.450), we obtain

$$(\frac{\omega^2}{G} - \frac{m_2}{r_{12}^3} - \frac{m_3}{r_{13}^3})r_1 + \frac{m_2}{r_{12}^3} r_2 + \frac{m_3}{r_{13}^3} r_3 = 0 \tag{12.451}$$

Similarly, we obtain for the second mass the following equation

$$\frac{m_1}{r_{12}^3} r_1 + (\frac{\omega^2}{G} - \frac{m_1}{r_{12}^3} - \frac{m_3}{r_{23}^3})r_2 + \frac{m_3}{r_{23}^3} r_3 = 0 \tag{12.452}$$

Instead of using force equilibrium for the third mass, we use (12.449) for the center of mass to form a system of three equations

$$m_1 r_1 + m_2 r_2 + m_3 r_3 = 0 \tag{12.453}$$

In summary, we have a system of equations (i.e., (12.451) to (12.453)) for three position vectors.

In the next two sections, we are going to show that there are five points (called Lagrangian Points) in the restricted three-body system at which the third body appears to be stationary with respect to the other two bodies. Assuming that two of the three bodies are more massive, once the third body is brought to the Lagrangian Points, it is held fixed at these points relative to the three-body system. Three of the Lagrangian Points are collinear with the two heavier masses, whereas two Lagrangian Points form two equilateral triangles with the other two masses. Figure 12.24 shows the locations of the five Lagrangian Points, labeled as L_1, L_2, L_3, L_4 and L_5. Triangular Lagrangian Points will be considered next, before we discuss the collinear Lagrangian Points and their potential applications in space exploration and in deploying telescopes. It turns out that the triangular Lagrangian Points coincide with observed locations of two groups of Trojan asteroids of the Sun-Jupiter system. Other bodies were also found at the triangular Lagrangian Points of the Sun-Mars system, and the Sun-Neptune system.

12.20.2 Triangular Lagrangian Points

We will demonstrate in this section that the three bodies located at the vertices of a equilateral triangle is a solution of the three-body system. To see this, we first express the distance between the bodies as ρ:

$$r_{12} = r_{13} = r_{23} = \rho \tag{12.454}$$

In view of (12.454), (12.451) to (12.453) becomes

$$\left(\frac{\omega^2}{G} - \frac{m_2}{\rho^3} - \frac{m_3}{\rho^3}\right) r_1 + \frac{m_2}{\rho^3} r_2 + \frac{m_3}{\rho^3} r_3 = 0 \tag{12.455}$$

$$\frac{m_1}{\rho^3} r_1 + \left(\frac{\omega^2}{G} - \frac{m_1}{\rho^3} - \frac{m_3}{\rho^3}\right) r_2 + \frac{m_3}{\rho^3} r_3 = 0 \tag{12.456}$$

$$m_1 r_1 + m_2 r_2 + m_3 r_3 = 0 \tag{12.457}$$

Equations (12.45) to (12.457) provide a homogeneous system of equations. For nonzero position vectors of the bodies to be possible, we must have the determinant of the coefficient be zero. This results in

$$\frac{\omega^4}{G^2} m_3 - \frac{\omega^2}{G} \frac{2m_3}{\rho^3} (m_1 + m_2 + m_3) + \frac{m_3}{\rho^6} (m_1 + m_2 + m_3)^2 = 0 \tag{12.458}$$

This is a quadratic equation in ω^2. It is straightforward to see that for the solution to exist, we must have the rotation frequency of the three-body system be

$$\frac{\omega^2}{G} = \frac{1}{\rho^3} (m_1 + m_2 + m_3) \tag{12.459}$$

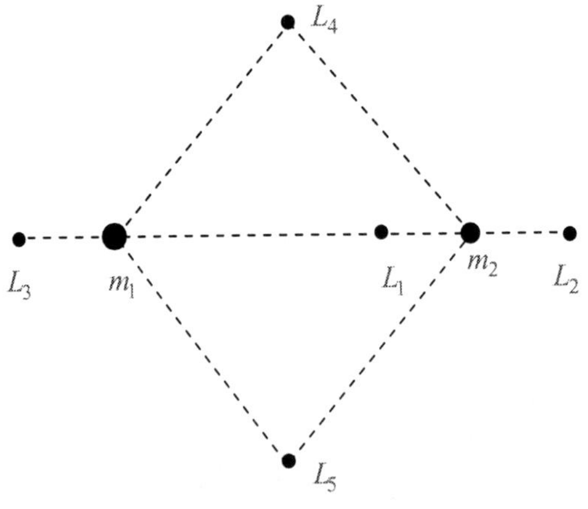

Figure 12.24 The locations of the five Lagrangian Points for the third smaller body in the restricted three-body problem

Since the determinant of the system has been set to zero, the three equations are now dependent, or actually there is only one independent equation for (12.455) to (12.457). Once an initial position vector of a body is fixed and the system starts to rotate with ω given in (12.459), the other two position vectors are arbitrary. In other words, the equilateral arrangement of the three bodies with a rotating frequency given in (12.459) is a particular solution of the three-body problem. This problem was first obtained by Lagrange in 1772 in his memoir submitted to the Paris Academy for a prize. It has been shown that these two Lagrangian Points are stable subject to small perturbation in motion (e.g., Battin, 1999). Therefore, it has been suggested that the triangular Lagrangian Points of the Sun-Earth system can be used for artificial space colonies for future space exploration.

For the Sun-Jupiter system, among the several thousand asteroids, two groups of Trojan asteroids have their centroids at the Lagrangian Points L_4 and L_5. By 2004, there are known to have 1051 Trojan asteroids at L_4 and 628 Trojan asteroids at L_5. The first Trojan asteroid was discovered by German astronomer Max Wolf in 1907 at the Sun-Jupiter L_4, and he named it 588 Achilles. In the Sun-Mars system, there is a 5261 Eureka at L_5. There are four bodies around Sun-Neptune L_4. Saturnian moon Tethys also has two smaller moons located at their L_4 and L_5.

12.20.3 Three Collinear Lagrangian Points

To prove the existence of three collinear Lagrangian Points, we first assume that the locations of the three bodies are described by

$$\boldsymbol{r}_1 = \xi_1 \boldsymbol{i}_\xi \tag{12.460}$$

$$\boldsymbol{r}_2 = \xi_2 \boldsymbol{i}_\xi = (\xi_1 + r_{12}) \boldsymbol{i}_\xi \tag{12.461}$$

$$\boldsymbol{r}_3 = \xi_3 \boldsymbol{i}_\xi = (\xi_1 + r_{12} + r_{23}) \boldsymbol{i}_\xi \tag{12.462}$$

with $\xi_3 > \xi_2 > \xi_1$. Substitution of these straight-line solutions into (12.451) gives

$$\frac{\omega^2}{G} r_{12}^3 \frac{\xi_1}{r_{12}} + m_2 + (\frac{1}{1+\chi})^2 m_3 = 0 \tag{12.463}$$

where

$$\chi = \frac{r_{23}}{r_{12}} \tag{12.464}$$

Similarly, (12.452) becomes

$$\frac{\omega^2}{G} r_{12}^3 (1 + \frac{\xi_1}{r_{12}}) - m_1 + \frac{1}{\chi^2} m_3 = 0 \tag{12.465}$$

Finally, (12.453) gives

$$(m_1 + m_2 + m_3) \frac{\xi_1}{r_{12}} + m_2 + (1+\chi) m_3 = 0 \tag{12.466}$$

Solving for ξ_1 from (12.466), we obtain

$$\xi_1 = -(\frac{m_2 + (1+\chi) m_3}{m_1 + m_2 + m_3}) r_{12} \tag{12.467}$$

Substituting (12.467) into (12.463) gives

$$\omega^2 = \frac{G}{r_{12}^3} [\frac{m_1 + m_2 + m_3}{m_2 + (1+\chi) m_3}][\frac{(1+\chi)^2 m_2 + m_3}{(1+\chi)^2}] \tag{12.468}$$

Finally, we can substitute (12.467) and (12.468) into (12.465) to obtain

$$\begin{aligned}&[(1+2\chi+\chi^2)\chi^2 m_2+m_3\chi^2](m_1+m_2+m_3)\\&-[(1+\chi)^2(m_2+m_1)+m_3][m_2+(1+\chi)m_3]\chi^2\\&+(1+2\chi+\chi^2)[m_2+(1+\chi)m_3]m_3=0\end{aligned} \tag{12.469}$$

Collecting terms in the power of χ, we obtain

$$\begin{aligned}&(m_1+m_2)\chi^5+(3m_1+2m_2)\chi^4+(3m_1+m_2)\chi^3-(m_2+3m_3)\chi^2\\&-(2m_2+3m_3)\chi-(m_2+m_3)=0\end{aligned} \tag{12.470}$$

This is called Lagrange's quintic equation. By using Decartes' rule of sign for roots of algebraic equations, we see that there is only one change of sign for the successive coefficients of (12.470) implying one real root. However, by changing the masses in (12.470) in cyclic order, we have two more roots for a straight-line solution. Approximations to these solutions are considered next.

12.20.4 Approximate Solution to Lagrange's Quintic Equation

If one of the three bodies is much lighter than the other two bodies, we can obtain a very good approximation of Lagrange's quintic equation given in (12.470). Three different combinations of the three body masses are considered:

Case (a): $m_3 = 0$, $m_2 < m_1$ (Lagrangian Point L_2)

Substituting $m_3 = 0$ into (12.470), we find

$$(\frac{1}{3}+\sigma^3)\chi^5+(1+2\sigma^3)\chi^4+(1+\sigma^3)\chi^3-\sigma^3\chi^2-2\sigma^3\chi-\sigma^3=0 \tag{12.471}$$

where

$$\sigma^3=\frac{m_2}{3m_1} \tag{12.472}$$

Rearranging (12.471), we have

$$\sigma^3=\frac{\chi^3(\chi^2+3\chi+3)}{3(1-\chi^3)(1+\chi)^2} \tag{12.473}$$

Applying Taylor's series expansion for the terms in the denominator of (12.481), we obtain

$$\sigma^3=\chi^3\{1-\chi+\frac{4}{3}\chi^2-\frac{2}{3}\chi^3+\chi^4-\chi^5+2\chi^6-2\chi^7+...\} \tag{12.474}$$

Taking the cube root of (12.474), we have

$$\sigma=\chi-\frac{1}{3}\chi^2+\frac{1}{3}\chi^3+\frac{1}{81}\chi^4+\frac{47}{243}\chi^5-\frac{43}{243}\chi^6+\frac{2549}{6561}\chi^7+...\} \tag{12.475}$$

We can apply a technique called series reversion to solve for χ. In particular, we consider the following series:

$$\sigma=\chi+b\chi^2+c\chi^3+d\chi^4+e\chi^5+f\chi^6+... \tag{12.476}$$

Next, we assume that χ can be expressed in series of σ as

$$\chi = \sigma + B\sigma^2 + C\sigma^3 + D\sigma^4 + E\sigma^5 + F\sigma^6 + ... \tag{12.477}$$

Substitution of (12.477) into (12.476) and comparison of coefficients on both sides gives

$$B = -b \tag{12.478}$$

$$C = -c + 2b^2 \tag{12.479}$$

$$D = -d + 5bc - 5b^3 \tag{12.480}$$

$$E = -e + 6bd - 21b^2c + 3c^2 + 14b^4 \tag{12.481}$$

$$F = -f + 7be - 28b^2d - 28bc^2 + 84b^3c - 42b^5 + 7cd \tag{12.482}$$

Therefore, the reversion of series given in (12.475) results in

$$\chi = \sigma + \frac{1}{3}\sigma^2 - \frac{1}{9}\sigma^3 - \frac{31}{81}\sigma^4 - \frac{119}{243}\sigma^5 - \frac{1}{9}\sigma^6 + \frac{3089}{6561}\sigma^7 + ... \tag{12.483}$$

<u>Case (b): $m_2 = 0$, $m_1 < m_3$</u> (Lagrangian Point L_1)

Substituting $m_2 = 0$ into (12.470), we find

$$\varphi^3\chi^5 + 3\varphi^3\chi^4 + 3\varphi^3\chi^3 - \chi^2 - \chi - \frac{1}{3} = 0 \tag{12.484}$$

where

$$\varphi^3 = \frac{m_1}{3m_3} \tag{12.485}$$

Solving for φ_3 from (12.484), we have

$$\varphi^3 = \frac{3\chi^2 + 3\chi + 1}{3\chi^2(3 + 3\chi + \chi^2)} \tag{12.486}$$

Next, we define the following parameter

$$\alpha = \frac{1}{1+\chi}, \quad \text{or} \quad \chi = \frac{1-\alpha}{\alpha} \tag{12.487}$$

Substitution of (12.487) into (12.486) gives

$$\begin{aligned} \varphi^3 &= \frac{\alpha^3(3 - 3\alpha + \alpha^2)}{3(1-\alpha^3)(1-\alpha)^2} \\ &= \alpha^3(1 + \alpha + \frac{4}{3}\alpha^2 + \frac{8}{3}\alpha^3 + 3\alpha^4 + \frac{11}{3}\alpha^5 + \frac{16}{3}\alpha^6 + 6\alpha^7 ...) \end{aligned} \tag{12.488}$$

Applying reversion of series to (12.488), we obtain

$$\alpha = \frac{1}{1+\chi} = \varphi - \frac{1}{3}\varphi^2 - \frac{1}{9}\varphi^3 - \frac{23}{81}\varphi^4 + \frac{151}{243}\varphi^5 - \frac{1}{9}\varphi^6 - \frac{691}{6561}\varphi^7 + ... \tag{12.489}$$

<u>Case (c): $m_1 = 0$, $m_3 < m_2$</u> (Lagrangian Point L_3)

Substituting $m_1 = 0$ into (12.470), we find

$$m_2(\chi^5 + 2\chi^4 + \chi^3 - \chi^2 - 2\chi - 1) = m_3(3\chi^2 + 3\chi + 1) \tag{12.490}$$

Rearranging (12.490), we obtain

$$\frac{m_3}{m_2} = \frac{\chi^5 + 2\chi^4 + \chi^3 - \chi^2 - 2\chi - 1}{3\chi^2 + 3\chi + 1} = \frac{(\chi^3 - 1)(1+\chi)^2}{3\chi^2 + 3\chi + 1} \tag{12.491}$$

Similar to the Case (b), we define the following parameter

$$\chi = \frac{1}{1+\beta}, \quad \text{or} \quad \beta = \frac{1-\chi}{\chi} \tag{12.492}$$

Using the following mass ratio parameter δ, we find

$$\delta = \frac{m_3}{m_2 + m_3} = -\frac{\beta(12 + 24\beta + 19\beta^2 + 7\beta^3 + \beta^4)}{7 + 14\beta + 13\beta^2 + 6\beta^3 + \beta^4} \tag{12.493}$$

Expanding (12.493) in power series of β, we obtain

$$\delta = -\frac{12}{7}\beta + \frac{23}{49}\beta^3 - \frac{23}{49}\beta^4 + \frac{58}{343}\beta^5 + \frac{45}{343}\beta^6 - \frac{579}{2401}\beta^7 + \frac{386}{2401}\beta^8 + \ldots \tag{12.494}$$

Applying series reversion to (12.494), we obtain an approximation of χ as

$$\frac{1-\chi}{\chi} = -\frac{7}{12}\delta - \frac{1127}{20736}\delta^3 - \frac{7889}{248832}\delta^4 - \frac{261023}{11943936}\delta^5 + \ldots \tag{12.495}$$

Example 12.7 Find the locations of the collinear Lagrangian Points of the Earth-Moon system. The masses of the Moon and the Earth can be taken as 0.07346×10^{24} kg and 5.9724×10^{24} kg with a mass ratio of the Earth to the Moon of about 81.3008 (this value is suggested by NASA). The distance between the Moon and the Earth can be taken as 384,400 km and denoted as r. The gravitational constant G can be taken as 6.67428×10^{11} Nm²/kg².

(i) Find the mass centroid of the rotating Earth-Moon system.
(ii) Find the circular frequency and period of the rotating Earth-Moon system.
(iii) Find the distance ratio of χ for the collinear Lagrangian Points L_1, L_2 and L_3.

Solution:

(i) Let the distance of the Earth from the mass centroid of the Earth-Moon system be x_M and that of the Moon to the centroid be x_m. The masses of the Earth and the Moon are denoted by M and m. The center of mass is defined as

$$Mx_M = mx_m = m(r - x_M) \tag{12.496}$$

Solving for x_M, we have

$$x_M = \frac{m}{m+M}r = \frac{1}{M/m+1}r = \frac{1}{81.30081+1}384400 = 4670.67km \tag{12.497}$$

Noting the fact that the radius of the Earth is about 6380 km, the mass centroid is actually within the Earth!

(ii) To find the circular frequency of the rotating system, we can balance the centrifugal force about the mass centroid by the gravitation force between the two bodies:

$$M\omega^2 x_M = \frac{GmM}{r^2} \tag{12.498}$$

Solving for ω, we have

$$\omega = \sqrt{\frac{Gm}{r^2 x_M}} = \sqrt{\frac{G(m+M)}{r^3}} = 2.665362\times 10^{-6} \text{ rad / s} \tag{12.499}$$

The period of rotation can be determined as

$$T = \frac{2\pi}{\omega} = \frac{2\pi}{\sqrt{\frac{G(m+M)}{r^3}}} = 27.284 \text{ days} \tag{12.500}$$

This is the sidereal month or the lunar month.

(iii) For L_2 or Case (a), we have from (12.472) that

$$\sigma^3 = \frac{m_2}{3m_1} = \frac{1}{3\times 81.3007} = 4.1\times 10^{-3} \tag{12.501}$$

This gives σ being 0.160052. Substitution of this value into (12.483) gives χ = 0.168135, 0.167884, 0.167833, 0.167831, and 0.167832, respectively, for taking 3, 4, 5, 6, and 7 terms in (12.483), respectively. On the other hand, we can set $m_3 = 0$ in Lagrange's Quintic equation given in (12.470) and use the numerical method to search for the real root. The solution for χ that we obtained is 0.167832, which suggests that the 7-term approximation derived in (12.483) is accurate up to six digits.

For L_1 or Case (b), we have from (12.485) that

$$\varphi^3 = \frac{m_1}{3m_3} = 4.1\times 10^{-3} \tag{12.502}$$

Therefore, substitution of φ = 0.160052 into (12.489) gives χ = 5.619993, 5.628169, 5.625303, and 5.625385 for taking 3, 4, 5, and 6, respectively.

For L_3 or Case (c), we have from (12.493) that

$$\delta = \frac{m_3}{m_2 + m_3} = 0.012151267 \tag{12.503}$$

Substitution of δ = 0.012151267 into (12.495) gives χ = 1.007138398, 1.0071389405, and 1.0071389405 for taking 2, 3, and 4 terms, respectively. The locations of these collinear Lagrangian Points of the Earth Moon system are illustrated in Figure 12.25. It appears from the figure that both Lagrangian Points L_1 and L_2 can provide a service hub for Moon landing. Although Joseph Liouville in 1845 showed that these collinear Lagrangian Points are unstable, the gravitational pull to move the spacecraft from the Lagrangian Point is, however, very small so it is not too energy consuming to stay at these unstable Lagrangian Points. In addition, it is possible to find stable periodic orbits around these unstable collinear Lagrangian Points for the case of a three-body system. These orbits are called Halo orbits. Even for the case of the n-body problem (i.e., for the case where gravitation effects from other planets are not negligible), we can find quasi-periodic (bounded but not precisely periodic) orbits called Lissajous orbits around the Lagrangian Points.

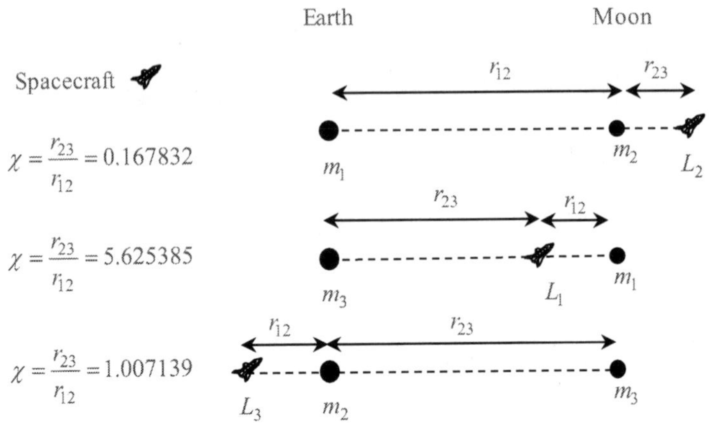

Figure 12.25 Locations of the collinear Lagrangian Points for the Earth-Moon system

For the case of the restricted three-body problem of the Sun-Earth system, Lagrangian Point L_1 is ideal for observing the Sun, since there is no obstruction by the Earth and the Moon. The Solar and Heliospheric Observatory (SOHO) is stationed in a Halo orbit at L_1. The Wilkinson Microwave Anisotropy Probe (WMAP) is already in orbit around the Sun-Earth L_2 and observes the full sky every six months, as it orbits around the Sun behind the Earth. The future Hershel Space infra-red Observatory as well as the James Webb Space Telescope will be placed at the Sun-Earth L_2, since they will be behind the Earth and orbit the Sun. As shown in Problem 12.10, the Sun-Earth L_2 is about 1,505,564 km behind the Earth. More detailed calculations of the collinear Lagrangian Points of the Sun-Earth system are found in Problem 12.10. When the mass difference between the two massive bodies is large, the distances between the L_1 and L_2 Lagrangian Points and the less massive body are about the same. For the Sun-Earth system, the distance between L_1 and L_2 and the Earth is about the same with a value of 1,500,578 km (see Problem 12.12). This distance can be estimated in terms of the so-called Hill sphere or Roche sphere defined as:

$$R = r\sqrt[3]{\frac{m_S}{3m_L}} \tag{12.504}$$

where r is the distance between the two masses denoted by m_S (smaller mass) and m_L (larger mass). The accuracy of this formula for the Earth-Moon and the Sun-Earth system are considered in Problems 12.11 and 12.12. Physically, the Hill sphere or Roche sphere is the gravitational sphere of influence. It is named after American astronomer G.W. Hill and French astronomer E. Roche as they derived the formula independently. Clearly, the Hill sphere extends somewhere between the L_1 and L_2 Lagrangian Points.

12.21 SUMMARY AND FURTHER READING

In this chapter, we demonstrated that all three of Kepler's laws are a natural consequence of the universal law of gravitation. Application of gravitational law to the studies of the orbits of artificial satellites is considered. The first, second, and third escape velocities from Earth's surface are considered. The Hohmann transfer orbit is studied in the context of interplanetary travels, as an energy-efficient way of space travel. The striking speed of meteors on Earth is studied as a special case of the orbital equations. Einstein's general relativity is employed in considering the precession of the perihelion of Mercury, through the use of Schwarzschild geometry of four-dimensional space-time. The problem of the flight of a rocket or missile near Earth's surface is considered as a special case of elliptic orbits. The dynamics of re-entry of spacecraft back to the Earth through the atmosphere is introduced and Yarochevsky analytical solution is derived for the special case of shallow entry with constant gravitational acceleration. The Lagrangian Points of the restricted three-body problem are considered.

For applications of Newtonian mechanics to celestial mechanics, we recommend the highly readable books by French (1971), by Fowles and Cassiday (2005), and by Hua (2009, 2012). There are many good textbooks on celestial mechanics, including Moulton (1914), Brouwer and Clemence (1961), and Szebehely and Mark (2004). A very comprehensive historical development of celestial mechanics and astrodynamics was given by Szebehely and Mark (2004). More specialized textbooks on astrodynamics are Bates et al. (1971) and Battin (1999). I highly recommend the excellent book by Battin (1999) which provided mathematical details on problems of astrodynamics. Gurfil and Seidelmann (2016) covered both celestial mechanics and astrodynamics. Olivier (1925) gave a good introduction to meteors. Rosser et al. (1947) provided a mathematical theory of rocket flight. For re-entry engineering, Hicks (2009) provided the most comprehensive mathematical coverage. Gallais (2007) provided a more technical consideration of re-entry problems.

12.21 PROBLEMS

Problem 12.1 Find the first and second escape velocities from the surface of the Moon. It is given that the mass and radius of the Moon are 7.35×10^{25} g and 1740 km.

Ans: 1.67854 km/s and 2.3738 km/s

Problem 12.2 Find the minimum landing velocity of a Moon-landing spacecraft without applying deceleration.

Ans: 1.76 km/s

Hint: For spacecraft under the gravitational pull of the Moon, we have

$$\frac{1}{2}mv^2 - \frac{GM_m m}{R_m} = U(r_m) = -\frac{1.23GM_E m}{D} \tag{12.505}$$

Problem 12.3 Rederive (12.60) by referring to Figure 12.26 in polar form.

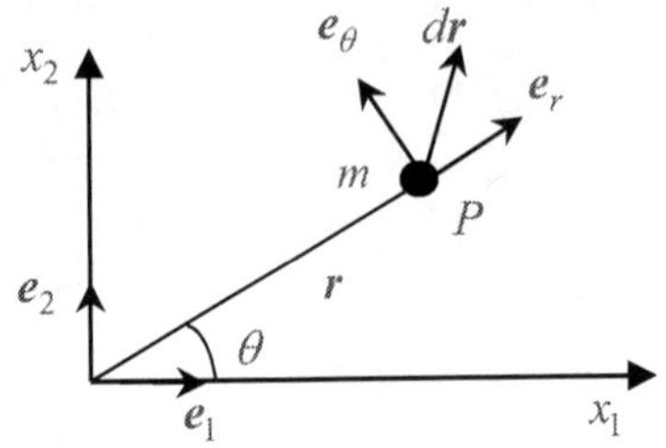

Figure 12.26 Polar coordinate for a rigid mass

(i) It is given that the position vector is

$$\boldsymbol{r} = r\boldsymbol{e}_r \tag{12.506}$$

Use (1.462) of Chau (2018) to show that the velocity and acceleration vectors are

$$\boldsymbol{v} = \frac{d\boldsymbol{r}}{dt} = \frac{dr}{dt}\boldsymbol{e}_r + r\frac{d\theta}{dt}\boldsymbol{e}_\theta \tag{12.507}$$

$$\boldsymbol{a} = \frac{d\boldsymbol{v}}{dt} = [\frac{d^2 r}{dt^2} - r(\frac{d\theta}{dt})^2]\boldsymbol{e}_r + [r\frac{d^2\theta}{dt^2} + 2\frac{dr}{dt}\frac{d\theta}{dt}]\boldsymbol{e}_\theta \tag{12.508}$$

(ii) Show that for the central force case (i.e., only nonzero radial component of force) we have

$$r^2(\frac{d\theta}{dt}) = C_1 \tag{12.509}$$

Hints: The base vectors in polar form are, in general, a function of time.

Problem 12.4 Halley's comet is a comet that approaches the Earth on a current return period of about 75.3 years (which changes over the centuries since its orbit is easily influenced by the gravitational pull of other planets, such as Jupiter and Saturn). It was named after Sir Edmund Halley, who was the first to calculate its orbit. It was observed that the perihelion distance r_p of it is about 0.586 AU (which is somewhere between Mercury and Venus). It follows an elliptic orbit with a very large value of eccentricity e, and its orbit inclines at 18° to the orbit of the Earth (see Figure 12.27). Calculate the following parameters of Halley's comet: (i) the semi-major axis a; (ii) the aphelion distance r_a; and (iii) the eccentricity of its orbit.

Hint: Clearly, it is under the gravitational pull of the Sun (this is the reason why Halley's comet keeps coming back). Use Kepler's third law to compare its period with that of the Earth.

Ans: (i) 17.8 AU; (ii) 35.0 AU; (iii) 0.967

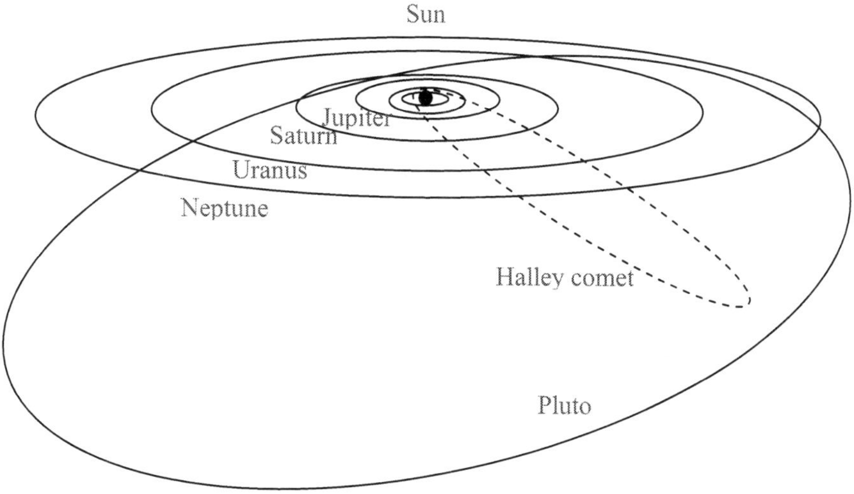

Figure 12.27 Orbit of Halley's comet

Problem 12.5 For an elliptic Hohmann transfer orbit, show that the velocities at the perigee and apogee can also be expressed as:

$$v_p = \sqrt{\frac{GM(1+\varepsilon)}{a(1-\varepsilon)}} \tag{12.510}$$

$$v_a = \sqrt{\frac{GM(1-\varepsilon)}{a(1+\varepsilon)}} \tag{12.511}$$

Problem 12.6 Find the orbital speeds of comet Halley at the perihelion and aphelion.

Hints: The results of Problem 12.5 could be useful.

Ans:

$$v_a = 0.913\text{km}/\text{s}, \quad v_p = 54.4\text{km}/\text{s} \tag{12.512}$$

Problem 12.7 Using the results of Problem 12.4, verify that the return period of the comet Halley is about 75.5 years. The last observance of the comet Halley was in 1986. Find the next return of Halley's comet.

Hint: Kepler's third law may be useful.

Ans: 2061

Problem 12.8 Repeat the analysis in Section 12.18 for the case of arbitrary launching angle of θ, and show the validity of the following formulas:

$$v_1 = \frac{v_0}{2\alpha\cos\theta}\{1+\sqrt{1-4\alpha(1-\alpha)\cos^2\theta}\} \tag{12.513}$$

$$v_2 = \frac{v_0}{2\alpha\cos\theta}\{1-\sqrt{1-4\alpha(1-\alpha)\cos^2\theta}\} \tag{12.514}$$

$$r_1 = \frac{2\alpha R_E \cos^2\theta}{1+\sqrt{1-4\alpha(1-\alpha)\cos^2\theta}} \tag{12.515}$$

$$r_2 = \frac{2\alpha R_E \cos^2\theta}{1-\sqrt{1-4\alpha(1-\alpha)\cos^2\theta}} \tag{12.516}$$

where α is defined in (12.352).

Problem 12.9 We have seen that Hohmann transfer orbit for space travel is the most energy efficient but it also takes very long travel time. A simple alternative is to use a Hohmann transfer orbit bigger than necessary (i.e., a higher orbit energy level) such that the travel time will be faster in the expense of more energy increment at transfer. As shown in Figure 12.28, the alternative transfer orbit to travel to the Moon from a circular orbit of the Earth, and the major axis of the Hohmann transfer orbit is $2a = r_1+r_3$.

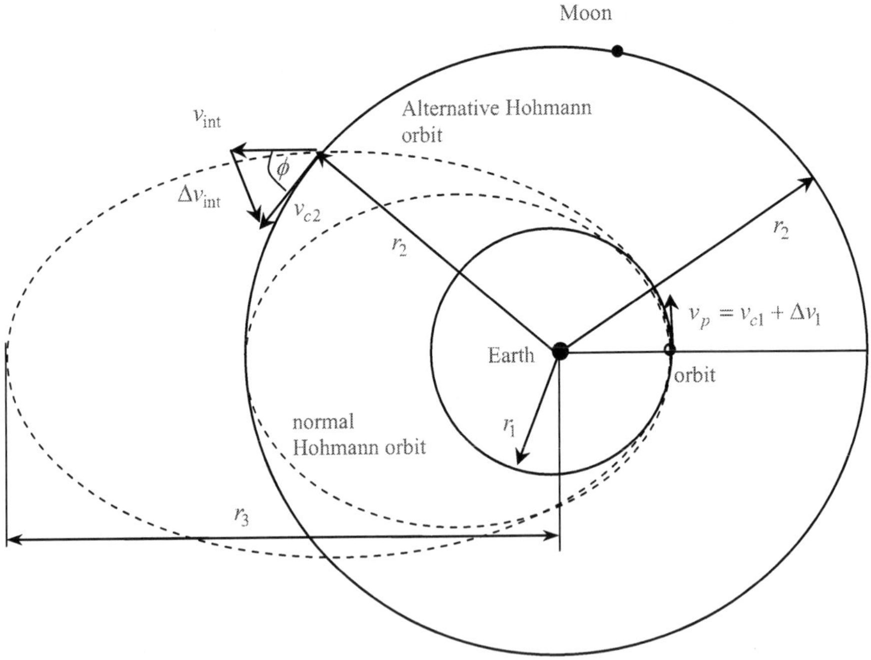

Figure 12.28 Alternative transfer orbit to the Moon

(i) Show that the velocity at the perigee of the Hohmann transfer orbit is

$$v_p = \sqrt{GM_E}\,(\frac{2}{r_1} - \frac{2}{r_1 + r_3})^{1/2} \tag{12.517}$$

(ii) Show that the velocity increment at the perigee should be

$$\Delta v_1 = \sqrt{GM_E}\,(\frac{2}{r_1} - \frac{2}{r_1 + r_3})^{1/2} - \sqrt{\frac{GM_E}{r_1}} \tag{12.518}$$

(iii) Show that the velocity of the spacecraft at the interception of the Hohmann transfer orbit and the Moon's orbit is

$$v_{\text{int}} = \sqrt{GM_E}\,(\frac{2}{r_2} - \frac{2}{r_1 + r_3})^{1/2} \tag{12.519}$$

(iv) Use the principle of conservation of angular momentum to show that

$$\phi = \cos^{-1}[\frac{(v_{\text{int}})_\theta}{v_{\text{int}}}] \tag{12.520}$$

where

$$(v_{\text{int}})_\theta = \frac{r_1}{r_2}\sqrt{GM_E}\,(\frac{2}{r_1} - \frac{2}{r_1 + r_3})^{1/2} \tag{12.521}$$

(v) Use the cosine law to show that the velocity increment at transfer is given by

$$\Delta v_{\text{int}}^2 = v_{c2}^2 + v_{\text{int}}^2 - 2v_{c2}v_{\text{int}}\cos\phi \tag{12.522}$$

Problem 12.10 Calculate the distance ratio χ for the collinear Lagrangian Points for the Sun-Earth system of the restricted three-body problem. The masses of the Sun and Earth can be taken as 1988500×10^{24} kg and 5.9724×10^{24} kg, respectively. The distance between the Sun and the Earth can be taken as 1.5×10^{11} m, and the gravitational constant G can be taken as 6.67428×10^{11} Nm^2/kg^2.

Ans:
L_1 Lagrangian Point: $\chi = 99.29707042$
L_2 Lagrangian Point: $\chi = 0.0100370979097$
L_3 Lagrangian Point: $\chi = 1.00000175202195$

Problem 12.11 For the Earth-Moon system of the restricted three-body problem: (i) find the distance between the Moon and L_1; (ii) find the distance between the Moon and L_2; (iii) find the size of the Hill sphere given by (12.504).

Ans: (i) 58,019 km; (ii) 64,536 km; (iii) 61,524 km

Problem 12.12 For the Sun-Earth system of the restricted three-body problem: (i) find the distance between the Earth and L_1; (ii) find the distance between the Earth and L_2; (iii) find the size of the Hill sphere given by (12.504).

Hints: Data given in Problem 12.10 could be useful.

Ans: (i) 1,510,618 km; (ii) 1,505,564 km; (iii) 1,500,578 km

Problem 12.13 For F defined in (12.262), prove the following

$$\frac{\partial F}{\partial r} = A'\dot{r}^2 + 2r\dot{\varphi}^2 + B'\dot{t}^2, \quad \frac{\partial F}{\partial \varphi} = 0, \quad \frac{\partial F}{\partial t} = 0 \tag{12.523}$$

$$\frac{\partial F}{\partial \dot{r}} = 2A\dot{r}, \quad \frac{\partial F}{\partial \dot{\varphi}} = 2r^2\dot{\varphi}, \quad \frac{\partial F}{\partial \dot{t}} = 2B\dot{t} \tag{12.524}$$

$$\frac{d}{d\tau}(\frac{\partial F}{\partial \dot{r}}) = 2A\ddot{r} + 2A'\dot{r}^2, \quad \frac{d}{d\tau}(\frac{\partial F}{\partial \dot{\varphi}}) = 4r\dot{r}\dot{\varphi} + 2r^2\ddot{\varphi}, \quad \frac{d}{d\tau}(\frac{\partial F}{\partial \dot{t}}) = 2B'\dot{r}\dot{t} + 2B\ddot{t} \tag{12.525}$$

CHAPTER THIRTEEN

Fracture Mechanics and Dynamics

13.1 INTRODUCTION

This chapter considers two main analyses, namely fracture mechanics analysis using dual integral equations, and dynamic fracture analysis using wave equations.

Many crack problems can be formulated as mixed boundary value problems and this in turn can be expressed in dual integral equations. In particular, for the case of a crack under internal pressure, the shear stress on the plane of the crack turns out to be identically zero. For such cases, the internal crack problem can be considered as a half-space or half-plane problem subject to different boundary conditions within the crack surface and outside the crack surface. More specifically, a special form of Papkovitch-Neuber potentials can be used. For both axisymmetric problems as well as 2-D plane problems, a single potential is found sufficient and in addition this potential satisfies Laplace equation. Thus, elastic problems become potential problems. The mixed boundary value problem can be formulated as dual integral equations. If the appropriate form of the solution is assumed, the dual integrals can be converted to a single integral equation of Abel type. We will consider the axisymmetric penny-shaped crack problem as well as the plane strain Griffith problem as examples. For the special case that the internal pressure is uniform, classical results for both the penny-shaped crack and the Griffith crack are recovered as special cases.

For dynamic fracture mechanics, we will derive the asymptotic stress field for modes I, II, and III for a crack tip moving at a speed of v. The problems for modes I and II are governed by two wave equations, one dilatation wave and one shear wave. The mode III problem is governed by a shear wave. The Galilean transform and coordinate scaling are used to convert both of them to Laplace equations. The traction-free boundary condition leads to an eigenvalue problem. Only one of the infinite eigenvalues leads to classical stress singularity at the crack tip. The polar variation of the stress field is also a function of the crack speed v. These asymptotic stress fields remain valid for a non-uniformly propagating crack tip. For a crack propagating faster than shear wave speed, or so-called subsonic speed, it will be shown that only mode II crack is possible whereas intersonic mode I crack propagation is prohibited. The shear wave equation remains as hyperbolic type even when the Galilean transform is applied. The classical stress singularity is recovered only for the case that the crack speed equals $2^{1/2}c_s$, where c_s is the shear wave speed. The asymptotic crack tip field is derived analytically for the intersonic case as well.

13.2 PAPKOVITCH-NEUBER POTENTIALS FOR AXISYMMETRIC ELASTICITY

In this section, we review the use of the "Papkovitch–Neuber displacement potential" in 3-D elasticity and its specialization for 2-D problems. The following formulation follows that of Chau (2013). It can be shown that the equilibrium equation in terms of stress is (e.g., Chau, 2013)

$$\nabla \bullet \boldsymbol{\sigma} + \boldsymbol{F} = 0 \tag{13.1}$$

For small strain and displacement, the displacement-strain equation is written as:

$$\boldsymbol{\varepsilon} = \tfrac{1}{2}(\nabla \boldsymbol{u} + \boldsymbol{u}\nabla) \tag{13.2}$$

For elastic isotropic solids, the stress and strain are related by Hooke's law:

$$\boldsymbol{\sigma} = 2\mu\boldsymbol{\varepsilon} + \lambda\varepsilon_{kk}\boldsymbol{I} \tag{13.3}$$

where μ and λ are called Lamé constants. Substitution of (13.2) and (13.3) into (13.1) gives the equilibrium in terms of displacement

$$\mu\nabla^2\boldsymbol{u} + (\lambda+\mu)\nabla\nabla \bullet \boldsymbol{u} = 0 \tag{13.4}$$

To derive the Papkovitch–Neuber displacement potential, we start with Helmholtz decomposition for the displacement vector, which is defined as:

$$\boldsymbol{u} = \nabla\phi + \nabla\times\boldsymbol{\psi} \tag{13.5}$$

Taking the divergence of (13.5) gives

$$\nabla \bullet u = \nabla \bullet \nabla\phi + \nabla \bullet \nabla\times\psi = \nabla^2\phi \tag{13.6}$$

Substitution of (13.6) into (13.4) yields

$$\nabla^2[(\lambda+\mu)\nabla\phi + \mu\boldsymbol{u}] = 0 \tag{13.7}$$

Integrating (13.7) gives

$$(\lambda+\mu)\nabla\phi + \mu\boldsymbol{u} = \mu\boldsymbol{\Phi} \tag{13.8}$$

where $\boldsymbol{\Phi}$ is a harmonic vector function. Rearranging (13.8) gives

$$\boldsymbol{u} = \boldsymbol{\Phi} - (1+\lambda/\mu)\nabla\phi \tag{13.9}$$

Taking the divergence of (13.9) and in view of (13.6), we obtain

$$\nabla^2\phi = \nabla \bullet \boldsymbol{\Phi} - (1+\lambda/\mu)\nabla^2\phi \tag{13.10}$$

Solving for ϕ from (13.10) gives

$$(2+\frac{\lambda}{\mu})\nabla^2\phi = \nabla \bullet \boldsymbol{\Phi} = \frac{1}{2}\nabla^2(\boldsymbol{r} \bullet \boldsymbol{\Phi}) \tag{13.11}$$

The last equation of (13.11) can be shown by using the fact that $\boldsymbol{r}$ is a position vector and the following identity:

$$\begin{aligned}\nabla^2(\boldsymbol{r} \bullet \boldsymbol{\Phi}) &= (\Phi_i x_i)_{,kk} = (\Phi_{i,k}x_i + \Phi_i x_{i,k})_{,k} = (\Phi_{i,k}x_i + \Phi_i\delta_{ik})_{,k} \\ &= (\Phi_{i,kk}x_i + \Phi_{i,k}\delta_{ik} + \Phi_{i,k}\delta_{ik}) = (2\Phi_{k,k} + \Phi_{i,kk}x_i) \\ &= 2\nabla \bullet \boldsymbol{\Phi} + \boldsymbol{r} \bullet \nabla^2\boldsymbol{\Phi} = 2\nabla \bullet \boldsymbol{\Phi}\end{aligned} \tag{13.12}$$

The last term of (13.12) is a result of the fact that $\boldsymbol{\Phi}$ is a harmonic vector function. The scalar potential given in (13.11) can be rearranged as

$$\nabla^2\phi = \frac{\mu}{2(\lambda+2\mu)}\nabla^2(\boldsymbol{r} \bullet \boldsymbol{\Phi}) \tag{13.13}$$

Integrating (13.12) finally gives

$$\phi = \frac{\mu}{2(\lambda+2\mu)}(\Phi_0 + \boldsymbol{r} \bullet \boldsymbol{\Phi}) = \frac{1-2\nu}{4(1-\nu)}(\Phi_0 + \boldsymbol{r} \bullet \boldsymbol{\Phi}) \tag{13.14}$$

where Φ_0 is another harmonic function and ν is the Poisson's ratio. The last equation of (13.14) can be verified by using (2.50) and (2.51) given in Chapter 2 of Chau (2013). Back substitution of (13.14) into (13.9) gives the Papkovitch-Neuber or the so-called P-N displacement potentials as

$$\boldsymbol{u} = \boldsymbol{\Phi} - \frac{1}{4(1-\nu)}\nabla(\Phi_0 + \boldsymbol{r} \bullet \boldsymbol{\Phi}), \quad \nabla^2\Phi_0 = 0, \quad \nabla^2\boldsymbol{\Phi} = 0 \tag{13.15}$$

More discussion on P-N displacement potentials can be found in Chau (2013). The advantage of using the P-N displacement potentials is that we do not raise the order of the governing differential equations.

Substitution of (13.15) into (13.3) and (13.2) gives the stress in terms of P-N potentials (see (4.212) of Chau, 2013):

$$\sigma_{ij} = \frac{\mu}{2(1-\nu)}\left\{2\nu\Phi_{k,k}\delta_{ij} + (1-2\nu)(\Phi_{i,j} + \Phi_{j,i}) - \Phi_{0,ij} - x_k\Phi_{k,ij}\right\} \tag{13.16}$$

For the axisymmetric case, it can be shown that only Φ_0 and Φ_3 are nonzero. The corresponding shear stresses are

$$\sigma_{31} = \frac{\mu}{2(1-\nu)}\left\{(1-2\nu)\Phi_{3,1} - \Phi_{0,13} - x_3\Phi_{3,31}\right\} \tag{13.17}$$

$$\sigma_{32} = \frac{\mu}{2(1-\nu)}\left\{(1-2\nu)\Phi_{3,2} - \Phi_{0,23} - x_3\Phi_{3,32}\right\} \tag{13.18}$$

We now impose the condition that the shear stresses given in (13.17) and (13.18) vanish on $x_3 = 0$ (we will see that this corresponds to both indentation problems by circular punch and the penny-shaped crack problem). In particular, from (13.17) and (13.18) we have on $x_3 = 0$:

$$\frac{\partial}{\partial x_1}[(1-2\nu)\Phi_3 - \Phi_{0,3}] = 0 \tag{13.19}$$

$$\frac{\partial}{\partial x_2}[(1-2\nu)\Phi_3 - \Phi_{0,3}] = 0 \tag{13.20}$$

By integrating (13.19) and (13.20) independently, it is straightforward to see that the admissible solution is

$$(1-2\nu)\Phi_3 - \Phi_{0,3} = C \tag{13.21}$$

Let us consider the elastic half-space problems. We note that if (13.21) is satisfied identically everywhere in the half-space, all shear stress on the surface of a half-space will always be identically zero. Thus, (13.21) corresponds to a special case of half-space problems that all shear tractions on the surface of the half-space is zero. For example, frictionless rigid indentation on the half-space is a typical case of such kind of problem. Another problem is a penny-shaped crack subject to internal normal stress. Due to symmetry of deformation of the penny-shaped crack, shear stress will be automatically zero on the plane of the penny-shaped crack. Consequently, we can actually formulate the penny-shaped problem by considering elastic half-space. The main objective of this section is to consider such a crack problem formulation.

Now, recall from (13.15) that both these potentials are harmonic functions

$$\nabla^2\Phi_0 = 0, \quad \nabla^2\Phi_3 = 0 \tag{13.22}$$

We now will further explore the structure of (13.21). First we rewrite (13.21) as

$$(1-2\nu)\Phi_3 = \frac{\partial \Phi_0}{\partial x_3} + C = \frac{\partial \bar{\Phi}_0}{\partial x_3} \tag{13.23}$$

Clearly, we have

$$\bar{\Phi}_0 = \Phi_0 + Cx_3 \tag{13.24}$$

Let us define a new potential such that

$$\bar{\Phi}_0 = (1-2\nu)\Psi \tag{13.25}$$

Substitution of (13.25) into (13.23) gives

$$\Phi_3 = \frac{\partial \Psi}{\partial x_3} \tag{13.26}$$

It is clear from (13.22) and (13.26) that Ψ is also a harmonic function, or it satisfies

$$\nabla^2\Psi = 0 \tag{13.27}$$

Both Φ_0 and Φ_3 can be expressed in terms of it. Therefore, for a shear traction-free problem we need only one harmonic function. In other words, a penny-shaped crack problem is equivalent to solving harmonic functions in half-space. For zero surface traction half-spaces, elastic problems become potential problems. In terms of this new displacement potential, the stresses are

$$\sigma_{33} = \frac{\mu}{2(1-\nu)}\left\{\frac{\partial^2\Psi}{\partial x_3^2} - x_3\frac{\partial^3\Psi}{\partial x_3^3}\right\} \tag{13.28}$$

$$\sigma_{31} = -\frac{\mu}{2(1-\nu)}x_3\frac{\partial^3\Psi}{\partial x_3^3} \tag{13.29}$$

$$\sigma_{32} = -\frac{\mu}{2(1-\nu)}x_3\frac{\partial^3\Psi}{\partial x_3^2\partial x_2} \tag{13.30}$$

$$\sigma_{11} = \frac{\mu}{2(1-\nu)}\left\{2\nu\frac{\partial^2\Psi}{\partial x_3^2} - (1-2\nu)\frac{\partial^2\Psi}{\partial x_1^2} - x_3\frac{\partial^3\Psi}{\partial x_1^2\partial x_3}\right\} \tag{13.31}$$

$$\sigma_{22} = \frac{\mu}{2(1-\nu)}\left\{2\nu\frac{\partial^2\Psi}{\partial x_2^2} - (1-2\nu)\frac{\partial^2\Psi}{\partial x_1^2} - x_3\frac{\partial^3\Psi}{\partial x_2^2\partial x_3}\right\} \tag{13.32}$$

$$\sigma_{12} = -\frac{\mu}{2(1-\nu)}\{(1-2\nu)\frac{\partial^2\Psi}{\partial x_1\partial x_2} + x_3\frac{\partial^3\Psi}{\partial x_1\partial x_2\partial x_3}\} \tag{13.33}$$

The corresponding displacements are

$$u_1 = \frac{1}{4(1-\nu)}[(1-2\nu)\frac{\partial\Psi}{\partial x_1} - x_3\frac{\partial^2\Psi}{\partial x_3\partial x_1}] \tag{13.34}$$

$$u_2 = \frac{1}{4(1-\nu)}[(1-2\nu)\frac{\partial\Psi}{\partial x_2} - x_3\frac{\partial^2\Psi}{\partial x_3\partial x_2}] \tag{13.35}$$

$$u_3 = \frac{1}{4(1-\nu)}[2(1-\nu)\frac{\partial \Psi}{\partial x_3} - x_3 \frac{\partial^2 \Psi}{\partial x_3^2}] \tag{13.36}$$

Note from (13.29) and (13.30) that the condition of zero shear tractions on $x_3 = 0$ is satisfied by:

$$\frac{\partial \Psi}{\partial x_3} = 0 \tag{13.37}$$

The two most typical elastic half-space problems are summarized in Figures 13.1 and 13.2, namely the penny-shaped crack problem and the contact problem respectively. The main feature of this half-space problem is discussed next.

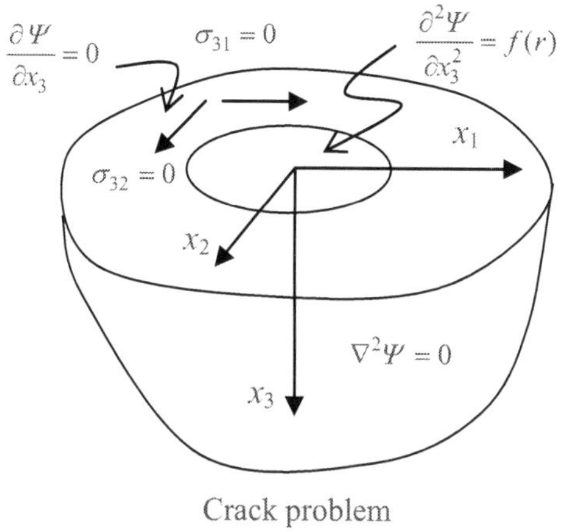

Figure 13.1 Penny-shaped crack modeled as an elastic half-space or potential problem

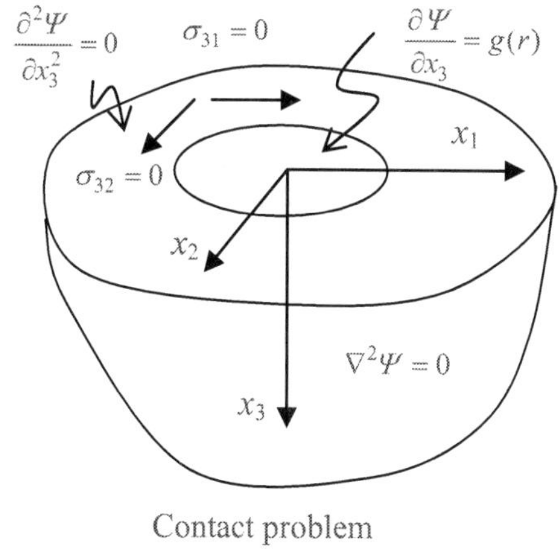

Figure 13.2 Circular contact modeled as an elastic half-space or potential problem

13.3 MIXED BOUNDARY VALUE PROBLEMS AS POTENTIAL PROBLEMS

The main feature of the half-space problems given in Figures 13.1 and 13.2 is that both of them are called mixed boundary value problems. In particular, the only boundary of the half-space is on the surface $x_3 = 0$, but the value of our displacement potential Ψ or its derivative is not uniformly imposed everywhere on this surface. As shown in Figure 13.1, the second derivative of Ψ (corresponding to vertical normal stress in the elastic problem) is imposed on the circular area with center at the origin of the coordinate system (see (13.28)), whereas the first derivative of Ψ (corresponding to vertical displacement in the elastic problem) is imposed elsewhere (see (13.36)). Figure 13.2 shows that the reverse is true for the circular rigid indentation problem formulated in terms of the mixed boundary value problems.

13.4 FORMULATION OF DUAL INTEGRAL EQUATIONS

The mixed boundary value problem is not an easy problem and a standard technique is the use of dual integral equations, which will be discussed in this section. More specifically, the boundary problem for Figure 13.2 for circular contact can be formulated as the potential problem or Laplace equation:

$$\nabla^2 \Psi = 0 \tag{13.38}$$

with boundary conditions on $x_3 = 0$

$$u_3 = \frac{1}{2}\frac{\partial \Psi}{\partial x_3} = g(r) \quad 0 \le r \le a \tag{13.39}$$

$$\sigma_{33} = 0, \quad r > a \tag{13.40}$$

$$\sigma_{31} = \sigma_{32} = 0, \quad 0 \le r \le \infty \tag{13.41}$$

Note that (13.41) has been satisfied identically as (13.29) to (13.30). The mixed boundary condition of (13.39) and (13.40) can be formulated by assuming:

$$\frac{\partial^2 \Psi}{\partial x_3^2} = -\sum_{n=0}^{\infty} \int_0^{\infty} A_n(\xi)\xi e^{-\xi x_3} J_n(\xi r) d\xi \cos n\theta \tag{13.42}$$

where r and θ are the cylindrical coordinates, which is adopted by virtue of the axisymmetry of the problem. This can be found formally by assuming separation of variables to the Laplace equation in cylindrical coordinates or by applying the Hankel transform to the polar form of the Laplace equation. Here, we just want to demonstrate that (13.39) indeed satisfies the Laplace equation. We can easily see that (13.42) leads to

$$\frac{\partial \Psi}{\partial x_3} = \sum_{n=0}^{\infty} \int_0^{\infty} A_n(\xi) e^{-\xi x_3} J_n(\xi r) d\xi \cos n\theta \tag{13.43}$$

Taking differentiation with respect to r twice, we have

$$\frac{\partial^2}{\partial r^2}\left(\frac{\partial \Psi}{\partial x_3}\right) = \sum_{n=0}^{\infty} \int_0^{\infty} A_n(\xi) e^{-\xi x_3} \frac{d^2 J_n(\xi r)}{dr^2} d\xi \cos n\theta \tag{13.44}$$

$$\frac{1}{r}\frac{\partial}{\partial r}(\frac{\partial \Psi}{\partial x_3}) = \sum_{n=0}^{\infty}\int_0^{\infty} A_n(\xi)e^{-\xi x_3}\frac{1}{r}\frac{dJ_n(\xi r)}{dr}d\xi\cos n\theta \tag{13.45}$$

$$\frac{1}{r^2}\frac{\partial^2}{\partial \theta^2}(\frac{\partial \Psi}{\partial x_3}) = -\sum_{n=0}^{\infty}\frac{n^2}{r^2}\int_0^{\infty} A_n(\xi)e^{-\xi x_3}J_n(\xi r)d\xi\cos n\theta \tag{13.46}$$

Substitution of these results into the polar form of the Laplace equation gives

$$\frac{\partial^2}{\partial r^2}(\frac{\partial \Psi}{\partial x_3}) + \frac{1}{r}\frac{\partial}{\partial r}(\frac{\partial \Psi}{\partial x_3}) + \frac{1}{r^2}\frac{\partial^2}{\partial \theta^2}(\frac{\partial \Psi}{\partial x_3})$$

$$= \sum_{n=0}^{\infty}\int_0^{\infty} A_n(\xi)e^{-\xi x_3}\{\frac{d^2J_n(\xi r)}{dr^2} + \frac{1}{r}\frac{dJ_n(\xi r)}{dr} + (\frac{n^2}{r^2} - \xi^2)J_n(\xi r)\}d\xi\cos n\theta \tag{13.47}$$

$$= 0$$

Note that the bracket term in the integrand of (13.47) is zero as it is the Bessel equation for $J_n(\xi r)$. Thus, we have

$$\nabla^2(\frac{\partial \Psi}{\partial x_3}) = 0 \tag{13.48}$$

This also implies that (13.43) is harmonic.

For the axisymmetric problem, we can take $n = 0$. The vertical normal stress given in (13.28) leads to

$$\sigma_{33} = \frac{\mu}{2(1-\nu)}\left\{\frac{\partial^2 \Psi}{\partial x_3^2} - x_3\frac{\partial^3 \Psi}{\partial x_3^3}\right\} = -\frac{\mu}{2(1-\nu)}\int_0^{\infty}\xi(1+\xi x_3)e^{-\xi x_3}A_0(\xi)J_0(\xi r)d\xi \tag{13.49}$$

On the surface ($x_3 = 0$), zero normal traction outside the circular area leads to

$$\int_0^{\infty}\xi A_0(\xi)J_0(\xi r)d\xi = 0 \quad r > a \tag{13.50}$$

Within the circular contact area, we have

$$u_3 = \frac{1}{2}\int_0^{\infty} A_0(\xi)J_0(\xi r)d\xi = g(r) \quad 0 \le r \le a \tag{13.51}$$

Therefore, the mixed boundary value problem is governed by a pair of dual integral equations

$$\int_0^{\infty} A_0(\xi)J_0(\xi r)d\xi = g(r) \quad 0 \le r \le a \tag{13.52}$$

$$\int_0^{\infty}\xi A_0(\xi)J_0(\xi r)d\xi = 0 \quad r > a \tag{13.53}$$

This system can be normalized as

$$\int_0^{\infty} A_0(\xi)J_0(\xi\rho)d\xi = g(\rho) \quad 0 \le \rho \le 1 \tag{13.54}$$

$$\int_0^{\infty}\xi A_0(\xi)J_0(\xi\rho)d\xi = 0 \quad \rho > 1 \tag{13.55}$$

where the normalized radial coordinate is given by

$$\rho = r / a \tag{13.56}$$

The solution of this pair of dual integral equations has been obtained by Titchmarsh (1948) using the Mellin transform and summarized by Sneddon (1951). Once the solution for $A_0(\xi)$ is obtained, the stress component given in (13.49) can be expressed in integral form. The Bessel function is an oscillating function (see Figure 4.5 of Chau, 2018), therefore, as long as the prescribed displacement $g(r)$ is a regular finite function, the integration can be conducted numerically. For the case of a flat punch contact, the integral can be evaluated analytically. This problem is, however, out of the scope of the present chapter.

We now return to our main focus of formulating the penny-shaped crack problem using a similar procedure.

13.5 PENNY-SHAPED CRACK PROBLEM

We will formulate the penny-shaped crack problem in this section for given normal stress within the circular region and zero normal displacement outside the circular region, as shown in Figure 13.1. In view of (13.49) and (13.51), the pair of dual integral equations can be formulated as

$$\begin{aligned}\sigma_{33} &= \frac{\mu}{2(1-\nu)}\frac{\partial^2 \Psi}{\partial x_3^2} = -\frac{\mu}{2(1-\nu)}\int_0^\infty \xi A_0(\xi) J_0(\xi\rho)d\xi \\ &= -\frac{\mu}{2(1-\nu)}\bar{f}(\rho) = f(\rho), \quad 0 \le \rho \le 1\end{aligned} \tag{13.57}$$

$$\int_0^\infty A_0(\xi) J_0(\xi\rho)d\xi = 0 \quad 1 < \rho < \infty \tag{13.58}$$

In essence, the role of displacement and stress in the contact problem have been reversed in (13.54) and (13.55) in this penny-shaped problem.

We now follow a solution procedure given by Sneddon (1951). In particular, a special form of the solution is assumed such that the dual integral equations can be reduced to a single Abel integral equation.

13.5.1 Reduction of Dual Integral Equations to Abel Integral

To reduce the dual integral equation to a single integral equation, we propose the following solution form:

$$A_0(\xi) = \int_0^1 \varphi(t)\sin(\xi t)dt \tag{13.59}$$

with

$$\varphi(0) = 0 \tag{13.60}$$

Substitution of (13.59) into the left-hand side of (13.58) gives

$$LHS = \int_0^1 \varphi(t)\int_0^\infty \sin(\xi t) J_0(\xi\rho)d\xi dt \tag{13.61}$$

From page 730, Equation 6.671.2 of Section 6.67-6.68 in Gradshteyn and Ryzhik (1980), we find that

$$\int_0^{\infty} \sin(\xi t) J_0(\xi \rho) d\xi = \frac{H(t-\rho)}{\sqrt{t^2-\rho^2}} \tag{13.62}$$

where H is the Heaviside step function. Substitution of (13.62) into (13.61) gives

$$\begin{aligned}\int_0^1 \varphi(t) \int_0^{\infty} \sin(\xi t) J_0(\xi \rho) d\xi dt &= \int_0^1 \frac{\varphi(t) H(t-\rho)}{\sqrt{t^2-\rho^2}} dt \\ &= \int_{\rho}^1 \frac{\varphi(t)}{\sqrt{t^2-\rho^2}} dt\end{aligned} \tag{13.63}$$

This integral is defined only for $\rho < 1$ and is zero for $\rho > 1$, and thus (13.58) is satisfied.

Secondly, we apply integration by parts to

$$\begin{aligned}A_0(\xi) &= \int_0^1 \varphi(t) \sin(\xi t) dt \\ &= -\frac{1}{\xi} \int_0^1 \varphi(t) d\cos(\xi t) \\ &= -\frac{1}{\xi} [\varphi(t)\cos(\xi t)|_0^1 - \int_0^1 \varphi'(t) \cos(\xi t) d\xi]\end{aligned} \tag{13.64}$$

Substitution of (13.64) into the right-hand side of (13.57) gives

$$\begin{aligned}\int_0^{\infty} \xi A_0(\xi) J_0(\xi\rho) d\xi &= -\varphi(1) \int_0^{\infty} \cos(\xi) J_0(\xi\rho) d\xi + \varphi(0) \int_0^{\infty} J_0(\xi\rho) d\xi \\ &+ \int_0^1 \varphi'(t) \int_0^{\infty} \cos(\xi) J_0(\xi\rho) d\xi dt\end{aligned} \tag{13.65}$$

From page 730, Equation 6.671.2 of Section 6.67-6.68 in Gradshteyn and Ryzhik (1980), we have

$$\int_0^{\infty} \cos(\xi t) J_0(\xi\rho) d\xi = \frac{H(\rho - t)}{\sqrt{\rho^2 - t^2}} \tag{13.66}$$

Substitution of (13.60) and (13.66) into (13.65) gives

$$\begin{aligned}\int_0^{\infty} \xi A_0(\xi) J_0(\xi\rho) d\xi &= -\varphi(1) \frac{H(\rho-1)}{\sqrt{\rho^2-t^2}} + \int_0^1 \varphi'(t) \frac{H(\rho-t)}{\sqrt{\rho^2-t^2}} dt \\ &= \int_0^{\rho} \frac{\varphi'(t)}{\sqrt{\rho^2-t^2}} dt = f(\rho)\end{aligned} \tag{13.67}$$

The step function $H(\rho-1)$ in the first term on the right of (13.67) vanishes by virtue of the fact that (13.57) is for $0 \le \rho \le 1$.

Equation (13.58) has been satisfied, and thus, the dual integral equations have been reduced to a single integral equation given in the last part of (13.67) or

$$\int_0^{\rho} \frac{\varphi'(t)}{\sqrt{\rho^2-t^2}} dt = f(\rho) \tag{13.68}$$

This integral is known as the Abel integral. The solution of it is well known and is given by

$$\varphi'(s) = \frac{2}{\pi}\frac{d}{ds}\int_0^s \frac{\rho f(\rho) d\rho}{\sqrt{s^2-\rho^2}} \tag{13.69}$$

This equation appears naturally in crack problems (e.g., Mura, 1987). This integral was first considered by Abel in 1823 when he studied a mass sliding along a frictionless curve vertical plane under the influence of gravity. It turns out that the arrival time to the lowest point is independent of its starting point on this curve. The solution is called a tautochrone curve. The tautochrone curve is also the brachistochrone (see Section 13.8 and 13.9 of Chau, 2018) or the curve takes the shortest time to travel to the lowest point.

Niels Henrik Abel was a Norwegian mathematician who was born in 1802 and passed away in 1828 at the age of 26. Abel had shown that it is impossible to solve the quintic equation in radical forms. He also discovered the elliptic function, which was subsequently improved by Jacobi and is now called the Jacobi elliptic function (see Chapter 1 and the appendix of Chau, 2018). He lived in poverty for his whole life, and passed away just two days short of receiving the good news from his friend Crelle that he was appointed as professor at University of Berlin.

To prove (13.69), we investigate the following generalized integral:

$$\int_a^x \frac{g(t)dt}{[h(x)-h(t)]^{\alpha}} = f(x) \tag{13.70}$$

where

$$0<\alpha<1, \quad a<x<b \tag{13.71}$$

Assume that $h(t)$ is strictly monotonically increasing and its first derivative is nonzero for all t between a and b. Note that $h(t) = t^2$ in the Abel integral given in (13.69).

We first consider the following related integral:

$$I(x) = \int_a^x \frac{h'(u) f(u) du}{[h(x)-h(u)]^{1-\alpha}} \tag{13.72}$$

Substitution of (13.70) into (13.72) gives

$$\begin{aligned} I(x) &= \int_a^x \frac{h'(u)}{[h(x)-h(u)]^{1-\alpha}} \int_a^u \frac{g(t)dt}{[h(u)-h(t)]^{\alpha}} du \\ &= \int_a^x g(t) \int_a^x \frac{h'(u)du}{[h(x)-h(u)]^{1-\alpha}[h(u)-h(t)]^{\alpha}} dt \end{aligned} \tag{13.73}$$

The last part of (13.73) results from reversing the order of integration. Note that the upper limit for variable t has been set to x (instead of using u).

Applying the following change of variables, we get

$$h(u) = \xi, \quad \rho_2 = h(x), \quad \rho_1 = h(t) \tag{13.74}$$

$$\begin{aligned} \int_a^x \frac{h'(u)}{[h(x)-h(u)]^{1-\alpha}[h(u)-h(t)]^{\alpha}} du &= \int_{h(\xi)}^{h(x)} \frac{d\xi}{(\rho_2-\xi)^{1-\alpha}(\xi-\rho_1)^{\alpha}} \\ &= \int_{\rho_1}^{\rho_2} \frac{d\xi}{(\rho_2-\xi)^{1-\alpha}(\xi-\rho_1)^{\alpha}} \end{aligned} \tag{13.75}$$

This integral cannot be found in most handbooks on tables of integrations, including the comprehensive book by Gradshteyn and Ryzhik (1980). However, the integration given in (13.75) can be evaluated through an integral covered in Section 1.7.7 of Chau (2018) and it is also reported on p. 118 of Whittaker and Watson (1927), In particular, we recall the following result from (1.200) of Chau (2018):

$$I_1 = \int_0^{\infty} \frac{x^{\alpha-1}dx}{1+x} = \frac{\pi}{\sin(\alpha\pi)} \tag{13.76}$$

We introduce a new variable defined as:

$$x = \frac{z-y}{y-\xi} \tag{13.77}$$

Differentiating (13.79), we obtain

$$dx = \frac{\xi - z}{(y-\xi)^2} dy \tag{13.78}$$

Noting that $x = 0$ gives $y = z$ and $x \to \infty$ gives $y = \xi$, we can convert (13.76) to

$$I_1 = \int_{\xi}^{z} \frac{dy}{(z-y)^{1-\alpha}(y-\xi)^{\alpha}} = \frac{\pi}{\sin(\alpha\pi)} \tag{13.79}$$

We see that (13.79) is the same as (13.75). Substitution of (13.79) into (13.73) and (13.72) gives

$$I(x) = \int_a^x g(t)dt \frac{\pi}{\sin(\pi\alpha)} = \int_a^x \frac{h'(u)f(u)du}{[h(x)-h(u)]^{1-\alpha}} \tag{13.80}$$

Differentiating (13.80) with respect to x and using the Leibniz rule of differentiation on the integral, we obtain

$$g(x)\frac{\pi}{\sin(\pi\alpha)} = \frac{d}{dx}\int_a^x \frac{h'(u)f(u)du}{[h(x)-h(u)]^{1-\alpha}} \tag{13.81}$$

This is the solution for (13.70). We now make the following identifications:

$$h(u) = u^2, \quad g(\rho) = \varphi'(\rho), \quad \alpha = 1/2, \quad x = \rho, \quad a = 0 \tag{13.82}$$

We finally obtained

$$\varphi'(\rho) = \frac{2}{\pi}\frac{d}{d\rho}\int_0^{\rho} \frac{\xi f(\xi)d\xi}{\sqrt{\rho^2 - \xi^2}} \tag{13.83}$$

This is the solution given in (13.69). Integrating on both sides, we have

$$\varphi(\rho) = \frac{2}{\pi}\int_0^{\rho} \frac{\xi f(\xi)d\xi}{\sqrt{\rho^2 - \xi^2}} \tag{13.84}$$

With this solution, we can substitute (13.84) into (13.59) to give

$$A_0(\xi) = \frac{2}{\pi}\int_0^1 \int_0^t \frac{\eta f(\eta)d\eta}{\sqrt{t^2 - \eta^2}} \sin(\xi t)dt \tag{13.85}$$

In other words, we have found A_0 in terms of $f(\rho)$. Next, we can consider another round of change of variables as

$$\eta = t\zeta \tag{13.86}$$

Thus, (13.85) becomes

$$\begin{aligned} A_0(\xi) &= \frac{2}{\pi}\int_0^1\int_0^1 \frac{t^2\zeta f(t\zeta)d\zeta}{\sqrt{t^2-t^2\zeta^2}}\sin(\xi t)dt \\ &= \frac{2}{\pi}\int_0^1 t\sin(\xi t)[\int_0^1 \frac{\zeta f(t\zeta)d\zeta}{\sqrt{1-\zeta^2}}]dt \end{aligned} \tag{13.87}$$

Note that this result, of course, agrees with p. 489 of Sneddon (1951).

13.5.2 Displacement Field Due to Uniform Pressure

We now consider the vertical displacement on the plane of the penny-shaped crack

$$\begin{aligned} \left[u_3\right]_{x_3=0} &= \frac{1}{2}\frac{\partial \Psi}{\partial x_3} = \frac{1}{2}\int_0^\infty A_0(\xi)J_0(\xi r)d\xi \\ &= \frac{1}{\pi}\int_0^\infty \int_0^1 t\sin(\xi t)[\int_0^1 \frac{\zeta f(t\zeta)d\zeta}{\sqrt{1-\zeta^2}}]dtJ_0(\xi r)d\xi \end{aligned} \tag{13.88}$$

Recalling the result from (13.62), we have

$$\left[u_3\right]_{x_3=0} = \frac{1}{\pi}\int_\rho^1 \frac{t}{\sqrt{t^2-\rho^2}}\int_0^1 \frac{\zeta f(t\zeta)d\zeta}{\sqrt{1-\zeta^2}}dt \tag{13.89}$$

We now consider a particular form of the internal pressure inside the penny-shaped cracked that can be expressed in power series of ρ:

$$f(\rho) = \frac{2(1-\nu)}{\mu} p_0 \sum_{n=0}^\infty \alpha_n \rho^n \tag{13.90}$$

Substitution of (13.90) into (13.89) gives

$$\left[u_3\right]_{x_3=0} = \frac{2(1-\nu)}{\pi\mu} p_0 \sum_{n=0}^\infty \alpha_n \int_\rho^1 \frac{t^{n+1}}{\sqrt{t^2-\rho^2}}\int_0^1 \frac{\zeta^{n+1}d\zeta}{\sqrt{1-\zeta^2}}dt \tag{13.91}$$

We note the following result of integral (Sneddon, 1951):

$$\int_0^1 \frac{\zeta^{n+1}d\zeta}{\sqrt{1-\zeta^2}} = \frac{\Gamma(1/2)\Gamma(1+n/2)}{\Gamma(3/2+n/2)} \tag{13.92}$$

Using this result and noting $\Gamma(1/2) = \pi^{1/2}$, (13.91) can further be simplified to

$$\begin{aligned} \left[u_3\right]_{x_3=0} &= \frac{(1-\nu)}{\sqrt{\pi}\,\mu} p_0 \sum_{n=0}^\infty \alpha_n \frac{\Gamma(1+n/2)}{\Gamma(3/2+n/2)}\int_\rho^1 \frac{t^{n+1}}{\sqrt{t^2-\rho^2}}dt \\ &= \frac{(1-\nu)}{\sqrt{\pi}\,\mu} p_0 \sum_{n=0}^\infty \alpha_n \frac{\Gamma(1+n/2)}{\Gamma(3/2+n/2)}\left[\sqrt{1-\rho^2} - n\int_0^1 t^{n-1}\sqrt{t^2-\rho^2}\,dt\right] \end{aligned} \tag{13.93}$$

For uniform stress we have $n = 0$ and thus (13.93) is simplified to

$$\left[u_3\right]_{x_3=0}=\frac{2(1-\nu)}{\pi\mu}\alpha_0 p_0\sqrt{1-\rho^2} \tag{13.94}$$

In obtaining (13.94), we have used the following identities

$$\Gamma(1)=1,\quad \Gamma(m+1/2)=\frac{1\cdot 3\cdots(2m-1)}{2^m}\sqrt{\pi} \tag{13.95}$$

Rewriting (13.94) in physical coordinates, we obtain

$$\left[u_3\right]_{x_3=0}=\frac{2(1-\nu)}{\pi\mu}\frac{\alpha_0 p_0}{a}\sqrt{a^2-r^2} \tag{13.96}$$

Here we have retained the constant α_0 for dimension scaling purposes. This result agrees with the classical result of crack opening displacement for a penny-shaped crack subject to uniform pressure. For more general loading, we can sum (13.93) accordingly. Using dimensional analysis on both sides of (13.96), we see that $\alpha_0 = a$ and this final result agrees with (6.283) of Chau (2013) if the typo in Chau (2013) is corrected (i.e., $4(1-\nu)/(2-\nu)$ being replaced by $2(1-\nu)$ in (6.283)).

13.5.3 Energy Change Due to Crack Presence

Physically the maximum crack opening occurs at $r = 0$, and we have

$$u_{3\max}=\frac{2(1-\nu)}{\pi\mu}\alpha_0 p_0 \tag{13.97}$$

Inversely, the pressure required to open the crack to a certain amount of maximum displacement is

$$p_0=\frac{u_{3\max}\pi\mu}{2(1-\nu)\alpha_0} \tag{13.98}$$

Substitution of (13.98) into (13.96) gives

$$u_3=\frac{u_{3\max}}{a}\sqrt{a^2-r^2} \tag{13.99}$$

Rearranging this gives

$$\left(\frac{u_3}{u_{3\max}}\right)^2+\left(\frac{r}{a}\right)^2=1 \tag{13.100}$$

This is an equation for ellipsoidal shape. Thus, uniform pressure will open the crack in an ellipsoidal shape, but we should be cautious that this conclusion is only true for a penny-shaped crack under uniform pressure.

We now consider the energy required to open up a penny-shaped crack with a maximum crack opening of $u_{\max}$ (at $r = 0$). Consider a pressure change corresponding to a crack opening increment as

$$p_0\to p_0+dp_0,\quad u_{3\max}\to u_{3\max}+du_{3\max} \tag{13.101}$$

Substituting (13.101) into (13.98), we have

$$p_0+dp_0=\frac{(u_{3\max}+du_{3\max})\pi\mu}{2(1-\nu)\alpha_0} \tag{13.102}$$

Using the average pressure

$$p_{0ave} = \frac{(u_{3\max} + \frac{1}{2} du_{3\max})\pi\mu}{2(1-\nu)\alpha_0} \tag{13.103}$$

Considering the work done in deforming the surface for a circular ring on the half-space, we have

$$\Delta W_1 = 2\pi r p_{0ave} dr du_3 = \frac{(u_{3\max} + \frac{1}{2} du_{3\max})\pi\mu}{2(1-\nu)\alpha_0} 2\pi r dr du_3$$

$$= \frac{(u_{3\max} + \frac{1}{2} du_{3\max})\pi\mu}{2(1-\nu)\alpha_0} 2\pi r (1-\frac{r^2}{a^2})^{1/2} du_{3\max} dr \tag{13.104}$$

$$\approx \frac{\pi^2\mu}{(1-\nu)\alpha_0} u_{3\max} du_{3\max} (1-\frac{r^2}{a^2})^{1/2} r dr$$

We have employed (13.99) in obtaining (13.104). The energy required to form a depression on a circular surface of a half-space from zero to maximum displacement is

$$W_1 = \frac{\pi^2\mu}{(1-\nu)\alpha_0} \int_0^{u_{\max}} \zeta d\zeta \int_0^a (1-\frac{r^2}{a^2})^{1/2} r dr = \frac{\pi^2\mu}{6(1-\nu)\alpha_0} u_{\max}^2 a^2 \tag{13.105}$$

By symmetry, the work done that is required to open the penny-shaped crack in an infinite domain is twice this value or

$$W = 2W_1 = \frac{\pi^2\mu}{3(1-\nu)\alpha_0} a^2 u_{\max}^2 \tag{13.106}$$

Substitution of (13.97) into (13.106) gives

$$W = \frac{4(1-\nu)\alpha_0}{3\mu} a^2 p_0^2 \tag{13.107}$$

We remarked earlier that by dimension analysis we find that $\alpha_0 = a$, and thus

$$W = \frac{4(1-\nu)}{3\mu} p_0^2 a^3 \tag{13.108}$$

This result agrees with the classical result given by Sneddon (1951). This also agrees with (6.284) of Chau (2013) where a different technique is used.

13.6 PAPKOVITCH-NEUBER POTENTIALS FOR PLANE ELASTICITY

Let us recall from (13.16) that the stress in terms of the Papkovitch-Neuber (P-N) potentials is given by (i.e., Chau, 2013)

$$\sigma_{ij} = \frac{\mu}{2(1-\nu)} \left\{ 2\nu \Phi_{k,k}\delta_{ij} + (1-2\nu)(\Phi_{i,j} + \Phi_{j,i}) - \Phi_{0,ij} - x_k \Phi_{k,ij} \right\} \tag{13.109}$$

In addition, the component form of (13.15) can be written as (see (4.211) of Chau, 2013):

$$u_i = \frac{1}{4(1-\nu)}[(3-4\nu)\Phi_i - \Phi_{0,i} - x_k \Phi_{k,i}] \tag{13.110}$$

For plane problems, we can specify the P-N potentials as

$$\Phi_3 = \Phi_1 = 0, \quad \Phi_0, \Phi_2 = f(x_1, x_2) \tag{13.111}$$

That is, two of the functions are zero whereas the nonzero two are only a function of coordinates x_1 and x_2. For these specializations, we have

$$u_1 = -\frac{1}{4(1-\nu)}[\frac{\partial \Phi_0}{\partial x_1} + x_2 \frac{\partial \Phi_2}{\partial x_1}] \tag{13.112}$$

$$u_2 = \frac{1}{4(1-\nu)}[(3-4\nu)\Phi_2 - \frac{\partial \Phi_0}{\partial x_2} - x_2 \frac{\partial \Phi_2}{\partial x_2}] \tag{13.113}$$

$$u_3 = 0 \tag{13.114}$$

$$\sigma_{11} = \frac{\mu}{2(1-\nu)}\left\{2\nu \frac{\partial \Phi_2}{\partial x_2} - \frac{\partial^2 \Phi_0}{\partial x_1^2} - x_2 \frac{\partial^2 \Phi_2}{\partial x_1^2}\right\} \tag{13.115}$$

$$\sigma_{22} = \frac{\mu}{2(1-\nu)}\left\{2(1-\nu) \frac{\partial \Phi_2}{\partial x_2} - \frac{\partial^2 \Phi_0}{\partial x_2^2} - x_2 \frac{\partial^2 \Phi_2}{\partial x_2^2}\right\} \tag{13.116}$$

$$\sigma_{12} = \frac{\mu}{2(1-\nu)}\left\{(1-2\nu) \frac{\partial \Phi_2}{\partial x_1} - \frac{\partial^2 \Phi_0}{\partial x_1 \partial x_2} - x_2 \frac{\partial^2 \Phi_2}{\partial x_1 \partial x_2}\right\} \tag{13.117}$$

These expressions are plane strain conditions. For plane stress problems, we can make the following substitution:

$$\nu = \frac{\bar{\nu}}{1+\bar{\nu}^2} \tag{13.118}$$

Similar to the formulation for axisymmetric problems, we want to simplify the potentials further such that the shear stress on the x_1 axis is zero identically. Mathematically, we have

$$\sigma_{12} = 0, \quad \text{on } x_2 = 0 \tag{13.119}$$

We find that this is automatically satisfied if

$$(1-2\nu)\Phi_2 = \frac{\partial \Phi_0}{\partial x_2} \tag{13.120}$$

Clearly, (13.120) is identically satisfied if we set

$$\Phi_2 = \frac{\partial \Psi}{\partial x_2}, \quad \Phi_0 = (1-2\nu)\Psi, \quad \nabla_1^2 \Psi = \frac{\partial^2 \Psi}{\partial x_1^2} + \frac{\partial^2 \Psi}{\partial x_2^2} = 0 \tag{13.121}$$

The last part of (13.122) results from the fact that Φ_0 is a harmonic function (see (13.22)). Again, we need only one potential Ψ, which is a harmonic function.

In terms of this new potential, the displacements and stresses are

$$u_1 = -\frac{1}{4(1-\nu)}[(1-2\nu)\frac{\partial \Psi}{\partial x_1} + x_2 \frac{\partial^2 \Psi}{\partial x_1 \partial x_2}] \tag{13.122}$$

$$u_2 = \frac{1}{4(1-\nu)}[2(1-\nu)\frac{\partial \Psi}{\partial x_2} - x_2 \frac{\partial^2 \Psi}{\partial x_2^2}] \tag{13.123}$$

$$u_3 = 0 \tag{13.124}$$

$$\sigma_{11} = \frac{\mu}{2(1-\nu)}\left\{2\nu \frac{\partial^2 \Psi}{\partial x_2^2} - (1-2\nu)\frac{\partial^2 \Psi}{\partial x_1^2} - x_2 \frac{\partial^3 \Psi}{\partial x_1^2 \partial x_2}\right\} \tag{13.125}$$

$$\sigma_{22} = \frac{\mu}{2(1-\nu)}\left\{\frac{\partial^2 \Psi}{\partial x_2^2} - x_2 \frac{\partial^3 \Psi}{\partial x_2^3}\right\} \tag{13.126}$$

$$\sigma_{12} = -\frac{\mu}{2(1-\nu)} x_2 \frac{\partial^3 \Psi}{\partial x_1 \partial x_2^2} \tag{13.127}$$

13.7 FORMULATION OF DUAL INTEGRAL EQUATIONS

Due to symmetry, the crack problem can be considered a half-plane problem with surface central traction on the region of $2c$ around the origin and zero displacement outside the central region on the surface, as shown in Figure 13.3. The solution should converge when we set $x_2 \to \infty$. On the surface of half-plane ($x_2 = 0$), we have

$$u_2 = \frac{1}{2}\frac{\partial \Psi}{\partial x_2} = 0, \quad |x_1| > c \tag{13.128}$$

$$\sigma_{22} = \frac{\mu}{2(1-\nu)}\frac{\partial^2 \Psi}{\partial x_2^2} = -f(x_1), \quad 0 < |x_1| < c \tag{13.129}$$

We can now recast the problem as a potential problem for a half-plane

$$\nabla_1^2 \Psi = \frac{\partial^2 \Psi}{\partial x_1^2} + \frac{\partial^2 \Psi}{\partial x_2^2} = 0 \tag{13.130}$$

$$\frac{\partial^2 \Psi}{\partial x_2^2} = -\frac{2(1-\nu)}{\mu} f(x_1) = -\kappa f(x_1), \quad 0 < |x_1| < c \tag{13.131}$$

$$\frac{\partial \Psi}{\partial x_2} = 0, \quad |x_1| > c \tag{13.132}$$

The 2-D elastic problem becomes a 2-D potential problem. To formulate the dual integral equations, we assume

$$\frac{\partial \Psi}{\partial x_2} = \sqrt{\frac{2}{\pi}}\int_0^\infty A(\xi)e^{-\xi x_2}\cos(\xi x_1)d\xi \tag{13.133}$$

The boundary conditions (13.131) and (13.132) become

$$\sqrt{\frac{2}{\pi}}\int_0^\infty \xi A(\xi)\cos(\xi x_1)d\xi = \kappa f(x_1), \quad c > x_1 > 0 \tag{13.134}$$

$$\sqrt{\frac{2}{\pi}}\int_0^{\infty} A(\xi)\cos(\xi x_1)d\xi = 0, \quad x_1 > c \tag{13.135}$$

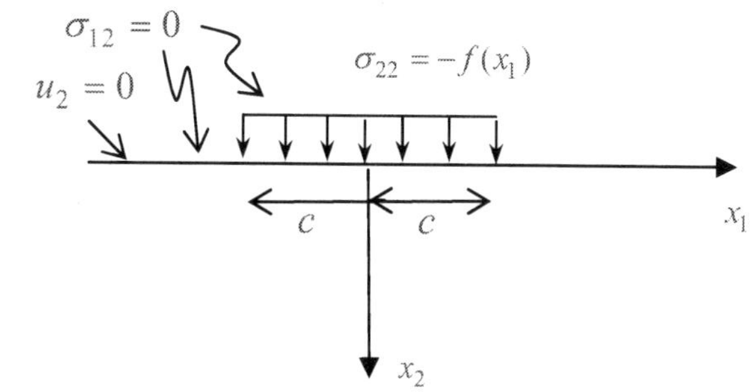

Figure 13.3 Mixed boundary value problem on a half-plane

where κ is defined in (13.132). Thus, again the mixed boundary value problem becomes the solution of dual integral equations.

13.8 GRIFFITH CRACK PROBLEM

By combining an upper and a lower half-plane, we get a full plane problem of having a central crack of length $2c$ subject to internal pressure. The negative sign of the applied stress implies compression. This problem actually corresponds to the classical Griffith crack problem, as shown in Figure 13.4, and the elastic fracture mechanics problem becomes a potential problem governed by the Laplace equation as summarized in (13.130) to (13.132).

13.8.1 Reduction of Dual Integral Equations to Abel Integral

To solve the dual integral equations given in (13.134) and (13.135), we introduce

$$A(\xi) = \sqrt{\frac{\pi}{2}}\int_0^{c} \varphi(t)J_0(\xi t)dt \tag{13.136}$$

where $\varphi(t)$ is an unknown function to be determined. Substitution of (13.136) into the left-hand side of (13.135) gives

$$LHS = \int_0^{\infty}\int_0^{c} \varphi(t)J_0(\xi t)dt\cos(\xi x_1)d\xi \tag{13.137}$$

Reversing the order of integration and recalling the integration formula given in (13.66) gives

$$LHS = \int_0^{c} \frac{\varphi(t)H(t-x_1)}{\sqrt{t^2-x_1^2}}dt = \int_{x_1}^{c} \frac{\varphi(t)}{\sqrt{t^2-x_1^2}}dt \tag{13.138}$$

However, (13.138) is valid for $x_1 < t < c$ or otherwise it is zero. However, (13.135) is defined as $x_1 > c$. Thus, (13.138) is identically zero and (13.135) is satisfied.

We see that (13.134) can be rewritten as

$$\sqrt{\frac{2}{\pi}}\frac{d}{dx_1}\int_0^\infty A(\xi)\sin(\xi x_1)d\xi = \sqrt{\frac{2}{\pi}}\int_0^\infty \xi A(\xi)\cos(\xi x_1)d\xi = \kappa f(x_1), \quad c > x_1 > 0 \tag{13.139}$$

Using this mathematical trick, we now can substitute (13.136) into (13.134) and reverse the order of integration to yield

$$\frac{d}{dx_1}\left\{\int_0^c \varphi(t)[\int_0^\infty [J_0(\xi t)\sin(\xi x_1)d\xi]dt\right\} = \kappa f(x_1) \tag{13.140}$$

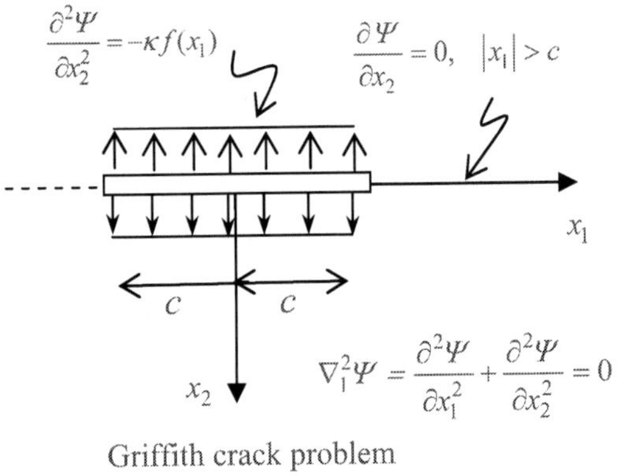

Figure 13.4 Mixed boundary value problem for the Griffith crack

The integration for variable ξ can be integrated using formula (13.62) and leads to

$$\frac{d}{dx_1}\left\{\int_0^c \frac{\varphi(t)H(x_1 - t)dt}{\sqrt{x_1^2 - t^2}}\right\} = \kappa f(x_1) \tag{13.141}$$

Integrating both sides, we obtain

$$\int_0^{x_1} \frac{\varphi(t)dt}{\sqrt{x_1^2 - t^2}} = \kappa\int_0^{x_1} f(t)dt = \kappa F(x_1) \tag{13.142}$$

This is again the Abel integral equation and the solution is considered next.

13.8.2 Solutions

We again obtain the Abel integral equation in (13.142). Applying the result in formula (13.69), we get

$$\varphi(t)=\frac{2\kappa}{\pi}\frac{d}{dt}\int_0^t\frac{xF(x)dx}{\sqrt{t^2-x^2}} \tag{13.143}$$

Integration by parts gives

$$\varphi(t)=\frac{2\kappa}{\pi}\frac{d}{dt}\left[-\sqrt{t^2-x^2}\,F(x)\Big|_0^t+\int_0^t\sqrt{t^2-x^2}\,f(x)dx\right] \tag{13.144}$$

Note that

$$F(0)=0 \tag{13.145}$$

We thus have

$$\varphi(t)=\frac{2\kappa}{\pi}\frac{d}{dt}\int_0^t\sqrt{t^2-x^2}\,f(x)dx \tag{13.146}$$

Applying Leibniz's rule of differentiation on an integral gives

$$\varphi(t)=\frac{2\kappa}{\pi}t\int_0^t\frac{f(x)}{\sqrt{t^2-x^2}}dx \tag{13.147}$$

Now consider the case of uniform applied pressure, or $f(x_1)=p_0$. The integral in (13.147) can be integrated exactly:

$$\varphi(t)=\frac{2\kappa}{\pi}p_0t\int_0^t\frac{dx}{\sqrt{t^2-x^2}}=\kappa p_0t \tag{13.148}$$

The normal stress on $x_2=0$ is

$$\begin{aligned}\sigma_{22}(x_1,0)&=\frac{\mu}{2(1-\nu)}\frac{\partial^2\Psi}{\partial x_2^2}(x_1,0)=\frac{d}{dx_1}\int_0^{x_1}\frac{\varphi(t)dt}{\sqrt{x_1^2-t^2}} \qquad 0<x_1<a\\&=\frac{d}{dx_1}\int_0^{a}\frac{\varphi(t)dt}{\sqrt{x_1^2-t^2}} \qquad a<x_1<\infty\end{aligned} \tag{13.149}$$

For uniform applied pressure, using (13.148) and considering the normal stress ahead of the crack tip we have

$$\begin{aligned}\sigma_{22}(x_1,0)&=p_0\frac{d}{dx_1}\int_0^a\frac{tdt}{\sqrt{x_1^2-t^2}}=p_0\frac{d}{dx_1}[-\sqrt{x_1^2-a^2}+x_1]\\&=p_0\left\{1-\frac{x_1}{\sqrt{x_1^2-a^2}}\right\}\end{aligned} \tag{13.150}$$

We note that the normal stress approaches infinity as $x_1\to a$, and approaches zero as $x_1\to\infty$. The opening displacement can be evaluated as

$$u_2=\frac{1}{2}\frac{\partial\Psi}{\partial x_2}=\frac{1}{2}\sqrt{\frac{2}{\pi}}\int_0^\infty A(\xi)\cos(\xi x_1)d\xi=\frac{1}{2}\int_{x_1}^a\frac{\varphi(t)dt}{\sqrt{t^2-x_1^2}} \tag{13.151}$$

Substituting (13.149) into (13.152), we have

$$u_2(x_1,0)=\frac{\kappa}{2}p_0\int_{x_1}^a\frac{tdt}{\sqrt{t^2-x_1^2}}=\frac{\kappa}{2}p_0\sqrt{a^2-x_1^2} \tag{13.152}$$

By symmetry, the crack opening can be evaluated as

$$\Delta u_2 = u_2^+ - u_2^- = \frac{2(1-\nu)}{\mu} p_0 \sqrt{a^2 - x_1^2}, \quad 0 < x_1 < a \tag{13.153}$$

This gives the crack opening for the case of uniform pressure. More generally, for other internal pressure we may have to evaluate (13.151) numerically.

13.9 FRACTURE DYNAMICS IN WAVE EQUATIONS

Section 4.2.1 of Chau (2013) discussed the use of the Helmholtz decomposition for 3-D elastic solids under quasi-static conditions, and Section 9.4 of Chau (2013) considered its usage under dynamic conditions. In this section, we will summarize the use of the Helmholtz decomposition and will specialize its usage in 2-D elastic solids. In particular, the equation of motion for 3-D dynamic problems is (Section 9.3 of Chau, 2013):

$$(\lambda + \mu)\nabla(\nabla \bullet \boldsymbol{u}) + \mu \nabla^2 \boldsymbol{u} + \rho \boldsymbol{f} = \rho \ddot{\boldsymbol{u}} \tag{13.154}$$

For the case of zero body forces, we use the following Helmholtz decomposition for displacement as

$$\boldsymbol{u} = \nabla \varphi + \nabla \times \boldsymbol{\psi} \tag{13.155}$$

Substitution of (13.154) into (13.155) results in the following scalar and vector wave equations (the details are referred to Chau, 2013):

$$c_d^2 \nabla^2 \varphi = \frac{\partial^2 \varphi}{\partial t^2} \tag{13.156}$$

$$c_s^2 \nabla^2 \boldsymbol{\psi} = \frac{\partial^2 \boldsymbol{\psi}}{\partial t^2} \tag{13.157}$$

where the dilatation and shear wave speeds are defined as

$$c_d = \sqrt{\frac{\lambda + 2\mu}{\rho}}, \quad c_s = \sqrt{\frac{\mu}{\rho}} \tag{13.158}$$

We now consider the 2-D plane strain condition. In this case, we can set

$$\phi = \phi(x_1, x_2, t), \quad \boldsymbol{\psi} = \psi(x_1, x_2, t)\boldsymbol{e}_3 \tag{13.159}$$

That is, the displacement is only a function of x_1 and x_2, not a function of x_3. In addition, only one component of the vector potential exists as shown in the second part of (13.159). The wave equations given in (13.156) and (13.157) can be rewritten as:

$$\nabla^2 \varphi = \frac{1}{c_d^2} \frac{\partial^2 \varphi}{\partial t^2}, \quad \nabla^2 \psi = \frac{1}{c_s^2} \frac{\partial^2 \psi}{\partial t^2} \tag{13.160}$$

By referring to Section 4.2.1 of Chau (2013), we can express the displacements and stresses as:

$$u_1 = \frac{\partial \varphi}{\partial x_1} + \frac{\partial \psi}{\partial x_2} \tag{13.161}$$

$$u_2 = \frac{\partial \varphi}{\partial x_2} - \frac{\partial \psi}{\partial x_1} \tag{13.162}$$

$$\sigma_{11} = \lambda\nabla^2\varphi + 2\mu(\frac{\partial^2\varphi}{\partial x_1^2} + \frac{\partial^2\psi}{\partial x_1\partial x_2}) \quad (13.163)$$

$$\sigma_{22} = \lambda\nabla^2\varphi + 2\mu(\frac{\partial^2\varphi}{\partial x_2^2} - \frac{\partial^2\psi}{\partial x_1\partial x_2}) \quad (13.164)$$

$$\sigma_{33} = \lambda(\frac{\partial^2\varphi}{\partial x_2^2} + \frac{\partial^2\varphi}{\partial x_1^2}) \quad (13.165)$$

$$\sigma_{12} = \mu[2\frac{\partial^2\varphi}{\partial x_1\partial x_2} + \frac{\partial^2\psi}{\partial x_2^2} - \frac{\partial^2\psi}{\partial x_1^2}] \quad (13.166)$$

Note that Equation (13.167) differs from that of Ravi-Chandar (2004), which has a typo in the shear stress expression.

13.10 REDUCTION OF WAVE TO HARMONIC PROBLEM BY GALILEAN TRANSFORM

We now consider a moving coordinate aligning with the moving crack tip as shown in Figure 13.5. This transformation is also called Galilean transformation, which is defined as:

$$\xi_1 = x_1 - vt, \quad \xi_2 = x_2 \quad (13.167)$$

where v is the velocity of the moving crack which is, in general, a function of time.

Substitution of this change of variables into the wave equations given in (13.161) gives

$$(1-\frac{v^2}{c_d^2})\frac{\partial^2\varphi}{\partial\xi_1^2} + \frac{\partial^2\varphi}{\partial\xi_2^2} = 0, \quad (1-\frac{v^2}{c_s^2})\frac{\partial^2\psi}{\partial\xi_1^2} + \frac{\partial^2\psi}{\partial\xi_2^2} = 0 \quad (13.168)$$

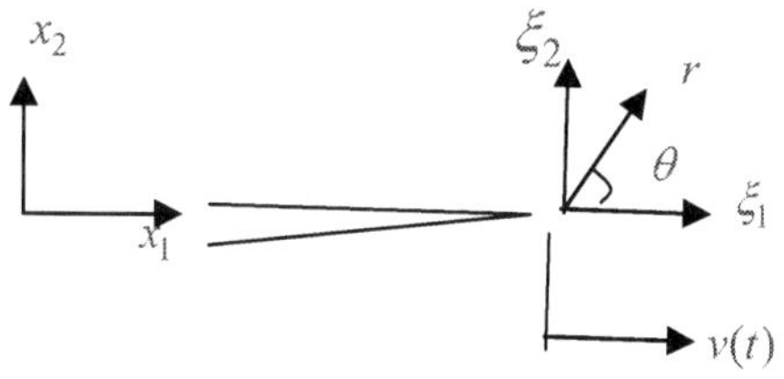

Figure 13.5 Asymptotic field near a moving crack tip

Note that for a moving velocity v smaller than the dilatational and shear wave speed, we have to convert the two wave equations into two elliptic equations. The case of an intersonic moving crack (i.e., $c_s < v < c_d$) will be considered it in a later section. For $v < c_s < c_d$, we can use a scaled coordinate such that (13.169) and (13.170) become Laplace equations:

$$\frac{\partial^2 \varphi}{\partial \xi_1^2} + \frac{1}{\alpha_d^2}\frac{\partial^2 \varphi}{\partial \xi_2^2} = \frac{\partial^2 \varphi}{\partial \xi_1^2} + \frac{\partial^2 \varphi}{\partial \zeta_d^2} = \frac{\partial^2 \varphi}{\partial r_d^2} + \frac{1}{r_d}\frac{\partial \varphi}{\partial r_d} + \frac{1}{r_d^2}\frac{\partial^2 \varphi}{\partial \theta_d^2} = 0 \tag{13.169}$$

$$\frac{\partial^2 \psi}{\partial \xi_1^2} + \frac{1}{\alpha_s^2}\frac{\partial^2 \psi}{\partial \xi_2^2} = \frac{\partial^2 \psi}{\partial \xi_1^2} + \frac{\partial^2 \psi}{\partial \zeta_s^2} = \frac{\partial^2 \psi}{\partial r_s^2} + \frac{1}{r_s}\frac{\partial \psi}{\partial r_s} + \frac{1}{r_s^2}\frac{\partial^2 \psi}{\partial \theta_s^2} = 0 \tag{13.170}$$

where

$$\xi_1 = r_d \cos\theta_d, \quad \zeta_d = \alpha_d \xi_2 = r_d \sin\theta_d, \quad \alpha_d = \sqrt{1 - \frac{v^2}{c_d^2}} \tag{13.171}$$

$$\xi_1 = r_s \cos\theta_s, \quad \zeta_s = \alpha_s \xi_2 = r_s \sin\theta_s, \quad \alpha_s = \sqrt{1 - \frac{v^2}{c_s^2}} \tag{13.172}$$

$$\begin{gathered} r_s = \sqrt{\xi_1^2 + \alpha_s^2 \xi_2^2}, \quad r_d = \sqrt{\xi_1^2 + \alpha_d^2 \xi_2^2}, \quad r = \sqrt{\xi_1^2 + \xi_2^2} \\ \theta_d = \tan^{-1}\left(\frac{\alpha_d \xi_2}{\xi_1}\right), \quad \theta_s = \tan^{-1}\left(\frac{\alpha_s \xi_2}{\xi_1}\right), \quad \theta = \tan^{-1}\left(\frac{\xi_2}{\xi_1}\right) \end{gathered} \tag{13.173}$$

Since the method of analysis is the same for both equations, we will discuss only (13.169) here. Similar to the static crack tip analysis by William in 1957, we can assume (see Section 6.5 of Chau, 2013):

$$\varphi = r_d^{\lambda} f(\theta_d, \lambda) \tag{13.174}$$

where λ is a constant to be determined. Substitution of (13.174) into (13.169), we find the following ODE for f

$$\lambda^2 f + f'' = 0 \tag{13.175}$$

The general solution for this equation becomes

$$\varphi = r_d^{\lambda}(A\cos\lambda\theta_d + B\sin\lambda\theta_d) \tag{13.176}$$

Without repeating the details, we can also obtain the following result for ψ

$$\psi = r_s^{\lambda}(C\cos\lambda\theta_s + D\sin\lambda\theta_s) \tag{13.177}$$

Whether we retain the cosine or sine functions, depends on the symmetry of displacement and stress fields that we are expecting. It turns out that for the case of mode I cracking, there is symmetry in the stress σ_{22} and anti-symmetry in displacement u_2; whereas for the case of mode II there is anti-symmetry in both displacement and stresses. These situations will be discussed separately next.

Before we consider the boundary conditions for modes I and II separately, let us first rewrite the stress expressions using α_d and α_s instead of in terms of Lamé constants. In particular, using (13.167), we can rewrite (13.164) as

$$\begin{aligned} \sigma_{22} &= \lambda\frac{\partial^2 \varphi}{\partial \xi_1^2} + (\lambda + 2\mu)\frac{\partial^2 \varphi}{\partial \xi_2^2} - 2\mu\frac{\partial^2 \psi}{\partial \xi_1 \partial \xi_2} \\ &= \mu\left\{\frac{\lambda}{\mu}\frac{\partial^2 \varphi}{\partial \xi_1^2} - \alpha_d^2 \frac{c_d^2}{c_s^2}\frac{\partial^2 \varphi}{\partial \xi_1^2} - 2\frac{\partial^2 \psi}{\partial \xi_1 \partial \xi_2}\right\} \end{aligned} \tag{13.178}$$

We note that

$$\frac{\lambda}{\mu}-\alpha_d^2\frac{c_d^2}{c_s^2}=\frac{v^2}{c_s^2}\{\frac{2\alpha_d^2-\alpha_s^2-1}{(1-\alpha_d^2)(1-\alpha_s^2)}\}-\alpha_d^2\frac{v^2}{c_s^2(1-\alpha_d^2)}$$
$$=-\frac{v^2}{c_s^2}\{\frac{1+\alpha_s^2}{1-\alpha_s^2}\} \tag{13.179}$$

Substituting (13.179) into (13.178), we find

$$\sigma_{22}=-\mu\left\{(1+\alpha_s^2)\frac{\partial^2\varphi}{\partial\xi_1^2}+2\frac{\partial^2\psi}{\partial\xi_1\partial\xi_2}\right\} \tag{13.180}$$

Similarly, (13.166) can be written as

$$\sigma_{12}=\mu\left\{2\frac{\partial^2\varphi}{\partial\xi_1\partial\xi_2}+\frac{\partial^2\psi}{\partial\xi_2^2}-\frac{\partial^2\psi}{\partial\xi_1^2}\right\}=\mu\left\{2\frac{\partial^2\varphi}{\partial\xi_1\partial\xi_2}+\frac{\partial^2\psi}{\partial\xi_2^2}+\frac{1}{\alpha_s^2}\frac{\partial^2\psi}{\partial\xi_2^2}\right\}$$
$$=\mu\left\{2\frac{\partial^2\varphi}{\partial\xi_1\partial\xi_2}+(\frac{1+\alpha_s^2}{\alpha_s^2})\frac{\partial^2\psi}{\partial\xi_2^2}\right\} \tag{13.181}$$

Substituting (13.167) into (13.163), we have

$$\sigma_{11}=\mu\left\{\frac{\lambda}{\mu}\frac{\partial^2\varphi}{\partial\xi_2^2}+(\frac{\lambda+2\mu}{\mu})\frac{\partial^2\varphi}{\partial\xi_1^2}+2\frac{\partial^2\psi}{\partial\xi_1\partial\xi_2})\right\} \tag{13.182}$$

The first of (13.168) can be rearranged as

$$\frac{\partial^2\varphi}{\partial\xi_2^2}=-\alpha_d^2\frac{\partial^2\varphi}{\partial\xi_1^2} \tag{13.183}$$

Substituting (13.183) into (13.182) and noting the following identity:

$$\frac{\lambda+2\mu}{\mu}-\alpha_d^2\frac{\lambda}{\mu}=\frac{c_d^2}{c_s^2}-\alpha_d^2\frac{v^2}{c_s^2}\left\{\frac{2\alpha_d^2-\alpha_s^2-1}{(1-\alpha_d^2)(1-\alpha_s^2)}\right\}$$
$$=\frac{v^2}{c_s^2}\left\{\frac{1+2\alpha_d^2-\alpha_s^2}{(1-\alpha_s^2)}\right\}=1+2\alpha_d^2-\alpha_s^2 \tag{13.184}$$

we finally get

$$\sigma_{11}=\mu\left\{(1+2\alpha_d^2-\alpha_s^2)\frac{\partial^2\phi}{\partial\xi_1^2}+2\frac{\partial^2\psi}{\partial\xi_1\partial\xi_2})\right\} \tag{13.185}$$

In the next section, we will first consider the asymptotic field at a mode I moving crack tip.

13.11 MODE I ASYMPTOTIC FIELD AT MOVING CRACK TIP

For the case of mode I cracking, the boundary conditions are given as:

$$\sigma_{22}(r,\pm\pi)=0,\quad \sigma_{12}(r,\pm\pi)=0 \tag{13.186}$$

In view of symmetry, we only retain the following nonzero terms in (13.177) and (13.178):

$$\varphi = Ar_d^{\lambda} \cos \lambda\theta_d\,, \quad \psi = Br_s^{\lambda} \sin \lambda\theta_s \tag{13.187}$$

Substitution of (13.187) into these stress expressions gives stresses in terms of the unknowns A and B. To find the derivative of potential φ with respect to ξ_1 and ξ_2, the following chain rules are used

$$\frac{\partial\varphi}{\partial\xi_1} = \frac{\partial\varphi}{\partial r_d}\frac{\partial r_d}{\partial\xi_1} + \frac{\partial\varphi}{\partial\theta_d}\frac{\partial\theta_d}{\partial\xi_1} \tag{13.188}$$

$$\frac{\partial r_d}{\partial\xi_1} = \cos\theta_d\,, \quad \frac{\partial\theta_d}{\partial\xi_1} = -\frac{1}{r_d}\sin\theta_d \tag{13.189}$$

$$\frac{\partial\varphi}{\partial\xi_2} = \frac{\partial\varphi}{\partial r_d}\frac{\partial r_d}{\partial\xi_2} + \frac{\partial\varphi}{\partial\theta_d}\frac{\partial\theta_d}{\partial\xi_2} \tag{13.190}$$

$$\frac{\partial r_d}{\partial\xi_2} = \alpha_d \sin\theta_d\,, \quad \frac{\partial\theta_d}{\partial\xi_2} = \frac{\alpha_d}{r_d}\cos\theta_d \tag{13.191}$$

Applying the chain rules given in (13.188) to (13.191), we obtain the following results

$$\frac{\partial\varphi}{\partial\xi_1} = \lambda A r_d^{\lambda-1}\cos(\lambda-1)\theta_d \tag{13.192}$$

$$\frac{\partial\varphi}{\partial\xi_2} = -\lambda A r_d^{\lambda-1}\alpha_d \sin(\lambda-1)\theta_d \tag{13.193}$$

$$\frac{\partial^2\varphi}{\partial\xi_1^2} = \lambda(\lambda-1)A r_d^{\lambda-2}\cos(\lambda-2)\theta_d \tag{13.194}$$

$$\frac{\partial^2\varphi}{\partial\xi_2^2} = -\lambda(\lambda-1)A\alpha_d^2 r_d^{\lambda-2}\cos(\lambda-2)\theta_d \tag{13.195}$$

$$\frac{\partial^2\varphi}{\partial\xi_2\partial\xi_1} = -\lambda(\lambda-1)A\alpha_d r_d^{\lambda-2}\sin(\lambda-2)\theta_d \tag{13.196}$$

Similarly, we find the following expressions for ψ:

$$\frac{\partial\psi}{\partial\xi_1} = \lambda B r_s^{\lambda-1}\sin(\lambda-1)\theta_s \tag{13.197}$$

$$\frac{\partial\psi}{\partial\xi_2} = \lambda B r_s^{\lambda-1}\alpha_s \cos(\lambda-1)\theta_s \tag{13.198}$$

$$\frac{\partial^2\psi}{\partial\xi_1^2} = \lambda(\lambda-1)B r_s^{\lambda-2}\sin(\lambda-2)\theta_s \tag{13.199}$$

$$\frac{\partial^2\psi}{\partial\xi_2^2} = -\lambda(\lambda-1)B\alpha_s^2 r_s^{\lambda-2}\sin(\lambda-2)\theta_s \tag{13.200}$$

$$\frac{\partial^2 \psi}{\partial \xi_2 \partial \xi_1} = \lambda(\lambda-1)B\alpha_s r_s^{\lambda-2}\cos(\lambda-2)\theta_s \tag{13.201}$$

Applying these results to the stress expressions given in (13.180) and (13.181), we obtain

$$\sigma_{22} = -\mu\lambda(\lambda-1)\left\{A(1+\alpha_s^2)r_d^{\lambda-2}\cos(\lambda-2)\theta_d + 2B\alpha_s r_s^{\lambda-2}\cos(\lambda-2)\theta_s\right\} \tag{13.202}$$

$$\sigma_{12} = -\mu\lambda(\lambda-1)\left\{B(1+\alpha_s^2)r_s^{\lambda-2}\sin(\lambda-2)\theta_s + 2A\alpha_d r_d^{\lambda-2}\sin(\lambda-2)\theta_d\right\} \tag{13.203}$$

We are now ready to consider the boundary conditions given in (13.187).

13.11.1 Eigenvalue Problem

Now we can specify the stress to satisfy the traction-free conditions on the crack face given in (13.186). Note that on the crack face, we have $\xi_2 = 0$ and $\theta_d = \theta_s = \pi$ and $r_d = r_s = r$. In view of these, (13.186) leads to two equations for constants A and B as:

$$\begin{bmatrix} (1+\alpha_s^2)\cos(\lambda-2)\pi & 2\alpha_s\cos(\lambda-2)\pi \\ 2\alpha_d\sin(\lambda-2)\pi & (1+\alpha_s^2)\sin(\lambda-2)\pi \end{bmatrix}\begin{Bmatrix} A \\ B \end{Bmatrix} = \begin{Bmatrix} 0 \\ 0 \end{Bmatrix} \tag{13.204}$$

Mathematically, finding the condition under which this homogeneous system is satisfied is an eigenvalue problem. For nonzero solutions for the stresses, we require the determinant of (13.204) to vanish and this leads to

$$[(1+\alpha_s^2)^2 - 4\alpha_s\alpha_d]\cos(\lambda-2)\pi\sin(\lambda-2)\pi = 0 \tag{13.205}$$

Note that the Rayleigh wave speed is given by the condition (Chau, 2013):

$$R(v) = [(1+\alpha_s^2(v))^2 - 4\alpha_s(v)\alpha_d(v)] = 0 \tag{13.206}$$

Therefore, it is clear that the stress analysis breaks down when the movement is at the Rayleigh wave speed. In other words, we can also argue that the mode I crack would only move at a speed slower than the Rayleigh wave speed.

For $v < v_R$ (where v_R is the Rayleigh wave speed), (13.205) is satisfied if

$$\lambda = \frac{n}{2} + 1, \quad n = 1, 2, 3, \ldots \tag{13.207}$$

As we expect from eigenvalue problems, the solution for n is discrete and there are infinite eigenvalues.

Substituting (13.207) into (13.187), we find the following displacement potentials for the crack tip problems

$$\varphi = A_n r_d^{(n/2+1)}\cos(n/2+1)\theta_d, \quad \psi = B_n r_s^{n/2+1}\sin(n/2+1)\theta_s \tag{13.208}$$

With this eigenvalue, the components of the displacement are:

$$u_1 = (\frac{n}{2}+1)A_n\left\{r_d^{n/2}\cos(\frac{n\theta_d}{2}) + \kappa_1(n,v)\alpha_s r_s^{n/2}\cos(\frac{n\theta_s}{2})\right\} \tag{13.209}$$

$$u_2 = -(\frac{n}{2}+1)A_n\left\{r_d^{n/2}\alpha_d\sin(\frac{n\theta_d}{2}) + \kappa_1(n,v)r_s^{n/2}\sin(\frac{n\theta_s}{2})\right\} \tag{13.210}$$

where

$$\kappa_1 = -(\frac{2\alpha_d}{1+\alpha_s^2}) \quad n = \text{odd}$$
$$= -(\frac{1+\alpha_s^2}{2\alpha_s}) \quad n = \text{even} \quad , \quad n = 1,2,3,... \tag{13.211}$$

The corresponding stress components can be obtained by substituting (13.208) into (13.163), (13.164), and (13.166) as:

$$\sigma_{12} = \frac{\mu A_n}{2\alpha_s}\frac{n}{2}(\frac{n}{2}+1)\left\{-4\alpha_s\alpha_d r_d^{n/2-1}\sin(\frac{n-2}{2})\theta_d + \kappa_2(n) r_s^{n/2-1}\sin(\frac{n-2}{2})\theta_s\right\} \tag{13.212}$$

$$\sigma_{22} = \frac{\mu A_n}{(1+\alpha_s^2)}\frac{n}{2}(\frac{n}{2}+1)\{-(1+\alpha_s^2)^2 r_d^{n/2-1}\cos(\frac{n-2}{2})\theta_d + \kappa_2(n) r_s^{n/2-1}\cos(\frac{n-2}{2})\theta_s\} \tag{13.213}$$

$$\sigma_{11} = \frac{\mu A_n}{1+\alpha_s^2}\frac{n}{2}(\frac{n}{2}+1)\{(1+\alpha_s^2)(1+2\alpha_d^2-\alpha_s^2) r_d^{n/2-1}\cos(\frac{n-2}{2})\theta_d - \kappa_2(n) r_s^{n/2-1}\cos(\frac{n-2}{2})\theta_s\} \tag{13.214}$$

where

$$\kappa_2 = 4\alpha_d\alpha_s \quad n = odd$$
$$= (1+\alpha_s^2)^2 \quad n = even \quad , \quad n = 1,2,3,... \tag{13.215}$$

If we specialize the case of $n = 1$, which is the most singular case for the stresses, we recover the singular stress field near the crack tip. Thus, for $n = 1$, the stress expression given by Rice (1968) is recovered. These expressions are also the same as (3.29) of Ravi-Chandar (2004).

13.11.2 Asymptotic Fields

As we remarked earlier that the most singular case corresponds to $n = 1$, we can define the mode I stress intensity factor as

$$K_I^d = \lim_{\xi_1 \to 0} \sqrt{2\pi\xi_1}\,\sigma_{22}(r, 0^{\pm}) \tag{13.216}$$

Note that the stress intensity depends on the loading, crack size and body geometries, and the boundary condition of the body, and cannot be obtained by the current asymptotic analysis and it must be found independently for the particular problem. In other words, the asymptotic crack field to be obtained here is universal and it is independent of the crack problem itself. For $n = 1$ and for $\theta \to 0$, we also have

$$\theta_d \to 0, \quad \theta_s \to 0, \quad \xi_2 \to 0, \quad \xi_1 \to r, \quad r_s = r_d = r \tag{13.217}$$

Considering these limits with $n = 1$, we can reduce (13.213) to

$$\sigma_{22}(r,0^{\pm})=\frac{\mu A}{1+\alpha_s^2}(\frac{3}{4})R(v)r^{-1/2} \tag{13.218}$$

where we have write A_1 as A for simplicity. Substituting (13.218) into (13.216), we find

$$A=\frac{4(1+\alpha_s^2)}{3\mu\sqrt{2\pi}}\frac{K_I^d}{R(v)} \tag{13.219}$$

Note also that

$$\begin{aligned} r_s^2 &= \xi_1^2+\alpha_s^2\xi_2^2=\xi_1^2+\xi_2^2+(\alpha_s^2-1)\xi_2^2=r^2+(\alpha_s^2-1)\xi_2^2 \\ &= r^2+r^2\sin^2\theta(\alpha_s^2-1)=r^2-r^2\sin^2\theta\frac{v^2}{c_s^2} \end{aligned} \tag{13.220}$$

To simplify the presentation, we define γ_s as

$$r_s^2=r^2(1-\sin^2\theta\frac{v^2}{c_s^2})=r^2\gamma_s^2 \tag{13.221}$$

Similarly, we can define γ_d as

$$r_d^2=r^2(1-\sin^2\theta\frac{v^2}{c_d^2})=r^2\gamma_d^2 \tag{13.222}$$

Now, we can express the crack tip asymptotic stress field as

$$\sigma_{ij}(r,\theta)=\frac{K_I^d}{\sqrt{\pi r}}f_{ij}^I(\theta,v) \tag{13.223}$$

where

$$f_{22}^I(\theta,v)=\frac{1}{R(v)}\left\{-(1+\alpha_s^2)^2\frac{1}{\sqrt{\gamma_d}}\cos(\frac{\theta_d}{2})+4\alpha_s\alpha_d\frac{1}{\sqrt{\gamma_s}}\cos(\frac{\theta_s}{2})\right\} \tag{13.224}$$

$$f_{11}^I(\theta,v)=\frac{1}{R(v)}\left\{(1+\alpha_s^2)(1+2\alpha_d^2-\alpha_s^2)\frac{1}{\sqrt{\gamma_d}}\cos(\frac{\theta_d}{2})-4\alpha_s\alpha_d\frac{1}{\sqrt{\gamma_s}}\cos(\frac{\theta_s}{2})\right\} \tag{13.225}$$

$$f_{12}^I(\theta,v)=\frac{2\alpha_d(1+\alpha_s^2)}{R(v)}\left\{\frac{1}{\sqrt{\gamma_d}}\sin(\frac{\theta_d}{2})-\frac{1}{\sqrt{\gamma_s}}\sin(\frac{\theta_s}{2})\right\} \tag{13.226}$$

$$\gamma_d=\sqrt{1-\frac{v^2\sin^2\theta}{c_d^2}},\quad \gamma_s=\sqrt{1-\frac{v^2\sin^2\theta}{c_s^2}} \tag{13.227}$$

These equations agree with those given in Section 9.10.2 of Chau (2013), by Ravi-Chandar (2004), and by Freund (1998). This mode I dynamic stress intensity factor has been employed by Yoffe (1951) to argue that when the crack speed is about $0.6c_s$ (see the conclusion of Yoffe, 1951) dynamic crack branching occurs because the maximum tensile hoop is no longer at $\theta = 0°$ but roughly $\theta = 60°$ (see Figure 1 of Yoffe, 1951). These values of crack speed and angle of maximum hoop stress have been taken "too" seriously. However, the numerical results of Yoffe (1951) are for a Poisson ratio of 0.25 only, and they are not accurate even for $\nu = 0.25$. To see this, we plot in Figure 13.6, the hoop stress as a function of θ for various values of crack speed v for a Poisson ratio of 0.25. The formula for the hoop stress in the neighborhood of the crack tip is given in Problem 13.1. Figure 13.6 shows that the hoop stress attains a maximum at an angle of about 70° (instead of 60° obtained by Yoffe, 1951) when $v/c_s \geq 0.7417$ or with $v/c_R \geq 0.8083$ (note that c_R/c_s =0.9176 for $\nu = 0.25$, see for example Eq. (9.35) of Chau, 2013). Our results for the critical velocity and critical angle are generally larger than those recorded in the literature. For example, it has been quoted in p. 836 of Rice at al. (1994) that this critical value given by Yoffe (1951) is $v = 0.60c_R$ but, in fact, Yoffe quoted a value of $0.6c_s$ without giving the details (see the Conclusion of Yoffe, 1951). For ν = 0.3, Gao (1993) plotted in his Figure 5(a) the hoop stress versus angle θ. But Gao (1993) then quoted that for $v > 0.6c_s$, dynamic crack branching would occur at 60°. This value is, actually, not consistent with his Figure 5(a), which did not show any peak of hoop stress at $v = 0.6\ c_s$ other than $\theta = 0°$. We have calculated the critical speed for ν = 0.3 and we get the critical value of $v = 0.7635\ c_s$ or $v =$ 0.82438 c_R (from (9.35) of Chau, 2013, we have $c_R/c_s = 0.92615$ for $\nu = 0.3$). Nevertheless, these values are too large compared to experimental observations of crack surface roughness (which is considered as micro-branching instability in the literature) for glass and a PMMA of about $v = 0.42\text{-}0.49\ c_R$ (see Rice et al., 1994).

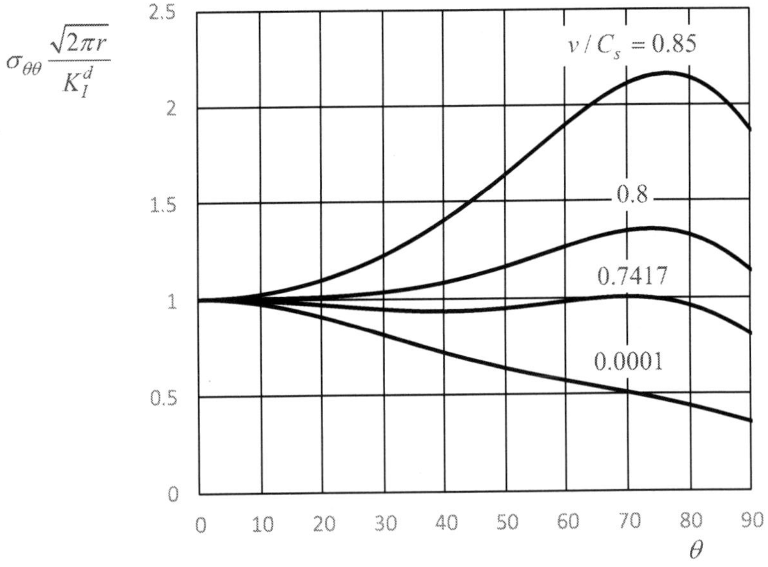

Figure 13.6 Hoop distribution of a mode I dynamic crack

Similarly, the displacements at the crack tip are

$$u_i(r,\theta) = \frac{K_I^d \sqrt{r}}{\sqrt{2\pi}} g_i^I(\theta, v) \tag{13.228}$$

$$g_1^I(\theta, v) = \frac{2}{\mu R(v)}\left\{(1+\alpha_s^2)\sqrt{\gamma_d}\cos(\frac{\theta_d}{2}) - 2\alpha_s\alpha_d\sqrt{\gamma_s}\cos(\frac{\theta_s}{2})\right\} \quad (13.229)$$

$$g_2^I(\theta, v) = \frac{2\alpha_d}{\mu R(v)}\left\{(1+\alpha_s^2)\sqrt{\gamma_d}\sin(\frac{\theta_d}{2}) - 2\sqrt{\gamma_s}\sin(\frac{\theta_s}{2})\right\} \quad (13.230)$$

where γ_d and γ_s are given in (13.227).

13.12 MODE II ASYMPTOTIC FIELD AT MOVING CRACK TIP

For the case of mode II cracking, the boundary conditions are again given as:

$$\sigma_{22}(r, \pm\pi) = 0, \quad \sigma_{12}(r, \pm\pi) = 0 \quad (13.231)$$

which are the same as those for mode I. However, we are looking for anti-symmetry in the shear stress, and the corresponding displacement potentials are:

$$\varphi = Ar_d^\lambda \sin\lambda\theta_d, \quad \psi = Br_s^\lambda \cos\lambda\theta_s \quad (13.232)$$

The following identities can be established similar to the case of mode I discussed in the last section:

$$\frac{\partial\varphi}{\partial\xi_1} = \lambda Ar_d^{\lambda-1}\sin(\lambda-1)\theta_d \quad (13.233)$$

$$\frac{\partial\varphi}{\partial\xi_2} = \lambda Ar_d^{\lambda-1}\alpha_d\cos(\lambda-1)\theta_d \quad (13.234)$$

$$\frac{\partial^2\varphi}{\partial\xi_1^2} = \lambda(\lambda-1)Ar_d^{\lambda-2}\sin(\lambda-2)\theta_d \quad (13.235)$$

$$\frac{\partial^2\varphi}{\partial\xi_2^2} = -\lambda(\lambda-1)A\alpha_d^2 r_d^{\lambda-2}\sin(\lambda-2)\theta_d \quad (13.236)$$

$$\frac{\partial^2\varphi}{\partial\xi_2\partial\xi_1} = \lambda(\lambda-1)A\alpha_d r_d^{\lambda-2}\cos(\lambda-2)\theta_d \quad (13.237)$$

The proofs of these identities will be left for readers. Similarly, the derivatives of ψ given in (13.232) can be found as

$$\frac{\partial\psi}{\partial\xi_1} = \lambda Br_s^{\lambda-1}\cos(\lambda-1)\theta_s \quad (13.238)$$

$$\frac{\partial\psi}{\partial\xi_2} = -\lambda Br_s^{\lambda-1}\alpha_s\sin(\lambda-1)\theta_s \quad (13.239)$$

$$\frac{\partial^2\psi}{\partial\xi_1^2} = \lambda(\lambda-1)Br_s^{\lambda-2}\cos(\lambda-2)\theta_s \quad (13.240)$$

$$\frac{\partial^2\psi}{\partial\xi_2^2} = -\lambda(\lambda-1)B\alpha_s^2 r_s^{\lambda-2}\cos(\lambda-2)\theta_s \quad (13.241)$$

$$\frac{\partial^2 \psi}{\partial \xi_2 \partial \xi_1} = -\lambda(\lambda-1)B\alpha_s r_s^{\lambda-2}\sin(\lambda-2)\theta_s \tag{13.242}$$

Employing these identities, the following stresses can be derived:

$$\sigma_{22} = -\mu\lambda(\lambda-1)\left\{A(1+\alpha_s^2)r_d^{\lambda-2}\sin(\lambda-2)\theta_d - 2B\alpha_s r_s^{\lambda-2}\sin(\lambda-2)\theta_s\right\} \tag{13.243}$$

$$\sigma_{12} = \mu\lambda(\lambda-1)\left\{2A\alpha_d r_d^{\lambda-2}\cos(\lambda-2)\theta_d - B(1+\alpha_s^2)r_s^{\lambda-2}\cos(\lambda-2)\theta_s\right\} \tag{13.244}$$

$$\sigma_{11} = \mu\lambda(\lambda-1)\left\{(1+2\alpha_d^2-\alpha_s^2)Ar_d^{\lambda-2}\sin(\lambda-2)\theta_d - 2B\alpha_s r_s^{\lambda-2}\sin(\lambda-2)\theta_s\right\} \tag{13.245}$$

The proofs of these expressions are set as Problem 13.4.

13.12.1 Eigenvalue Problem

Substitution of (13.243) and (13.244) into the boundary conditions given in (13.231) leads to:

$$\begin{bmatrix} (1+\alpha_s^2)\sin(\lambda-2)\pi & -2\alpha_s\sin(\lambda-2)\pi \\ 2\alpha_d\cos(\lambda-2)\pi & -(1+\alpha_s^2)\cos(\lambda-2)\pi \end{bmatrix} \begin{Bmatrix} A \\ B \end{Bmatrix} = \begin{Bmatrix} 0 \\ 0 \end{Bmatrix} \tag{13.246}$$

Setting the determinant of the matrix in (13.246) to zero, we get

$$[(1+\alpha_s^2)^2 - 4\alpha_s\alpha_d]\cos(\lambda-2)\pi\sin(\lambda-2)\pi = 0 \tag{13.247}$$

which is exactly the same as (13.205) obtained in the mode I case. Thus, we have the same eigenvalue for λ:

$$\lambda = \frac{n}{2}+1, \quad n = 1,2,3,... \tag{13.248}$$

The corresponding displacement potentials given in (13.232) become

$$\varphi = Ar_d^{n/2+1}\sin(\frac{n+2}{2})\theta_d, \quad \psi = Br_s^{n/2+1}\cos(\frac{n+2}{2})\theta_s \tag{13.249}$$

Employing (13.249), we find the displacement components in (13.161) and (13.162) being:

$$u_1 = (\frac{n}{2}+1)A_n\left\{r_d^{n/2}\sin(\frac{n\theta_d}{2}) - \chi(n)\alpha_s r_s^{n/2}\sin(\frac{n\theta_s}{2})\right\} \tag{13.250}$$

$$u_2 = (\frac{n}{2}+1)A_n\left\{r_d^{n/2}\alpha_d\cos(\frac{n\theta_d}{2}) - \chi(n)r_s^{n/2}\cos(\frac{n\theta_s}{2})\right\} \tag{13.251}$$

where

$$\chi(n) = \begin{cases} \dfrac{2\alpha_d}{1+\alpha_s^2} & n = even \\[2ex] \dfrac{1+\alpha_s^2}{2\alpha_s} & n = odd \end{cases}, \quad n = 1,2,3,... \tag{13.252}$$

Substitution of (13.249) into (13.180), (13.181) and (13.185) gives the corresponding stresses (see Problem 13.5)

$$\sigma_{12} = \mu A_n \frac{n}{2}(\frac{n}{2}+1)\left\{2\alpha_d r_d^{n/2-1}\cos(\frac{n-2}{2})\theta_d - (1+\alpha_s^2)\chi(n) r_s^{n/2-1}\cos(\frac{n-2}{2})\theta_s\right\} \tag{13.253}$$

$$\sigma_{22} = -\mu A_n \frac{n}{2}(\frac{n}{2}+1)\{(1+\alpha_s^2) r_d^{n/2-1}\sin(\frac{n-2}{2})\theta_d - 2\alpha_s \chi(n) r_s^{n/2-1}\sin(\frac{n-2}{2})\theta_s\} \tag{13.254}$$

$$\sigma_{11} = \mu A_n \frac{n}{2}(\frac{n}{2}+1)\{(1+2\alpha_d^2-\alpha_s^2) r_d^{n/2-1}\sin(\frac{n-2}{2})\theta_d - 2\chi(n) r_s^{n/2-1}\sin(\frac{n-2}{2})\theta_s\} \tag{13.255}$$

For the most singular stress field, we have to consider the case of $n=1$.

13.12.2 Asymptotic Fields

For a mode II crack, we can define the dynamic stress intensity factor as

$$K_{II}^d = \lim_{\xi_1 \to 0} \sqrt{2\pi\xi_1}\,\sigma_{12}(r, 0^{\pm}) \tag{13.256}$$

Considering the limits given in (13.217) into (13.253) and setting $n=1$, we have

$$\sigma_{12}(r, 0^{\pm}) = \frac{\mu A}{2\alpha_s}(\frac{3}{4}) R(v) r^{-1/2} \tag{13.257}$$

where we have dropped the subscript of A_1. Substitution of (13.257) into (13.256) results in

$$A = \frac{8\alpha_s}{3\mu\sqrt{2\pi}}\frac{K_{II}^d}{R(v)} \tag{13.258}$$

Substitution of (13.258) into (13.253) and (13.255) gives the asymptotic stress field at the crack tip as

$$\sigma_{ij}(r,\theta) = \frac{K_{II}^d}{\sqrt{2\pi r}} f_{ij}^{II}(\theta, v) \tag{13.259}$$

where

$$f_{22}^{II}(\theta, v) = \frac{2\alpha_s(1+\alpha_s^2)}{R(v)}\left\{\frac{1}{\sqrt{\gamma_d}}\sin(\frac{\theta_d}{2}) - \frac{1}{\sqrt{\gamma_s}}\sin(\frac{\theta_s}{2})\right\} \tag{13.260}$$

$$f_{11}^{II}(\theta, v) = -\frac{2\alpha_s}{R(v)}\left\{(1+2\alpha_d^2-\alpha_s^2)\frac{1}{\sqrt{\gamma_d}}\sin(\frac{\theta_d}{2}) - (1+\alpha_s^2)\frac{1}{\sqrt{\gamma_s}}\sin(\frac{\theta_s}{2})\right\} \tag{13.261}$$

$$f_{12}^{II}(\theta, v) = \frac{1}{R(v)}\left\{4\alpha_d\alpha_s \frac{1}{\sqrt{\gamma_d}}\cos(\frac{\theta_d}{2}) - (1+\alpha_s^2)\frac{1}{\sqrt{\gamma_s}}\cos(\frac{\theta_s}{2})\right\} \tag{13.262}$$

These equations agree with those given in Section 9.10.2 of Chau (2013), by Ravi-Chandar (2004), and by Freund (1998).

Substitution of (13.258) into (13.250) and (13.251) gives the following displacements at the crack tip

$$u_i(r,\theta) = \frac{K_{II}^d \sqrt{r}}{\sqrt{2\pi}} g_i^{II}(\theta, v) \tag{13.263}$$

$$g_1^{II}(\theta,v) = \frac{2\alpha_s}{\mu R(v)} \left\{ 2\sqrt{\gamma_d} \sin(\frac{\theta_d}{2}) - (1+\alpha_s^2)\sqrt{\gamma_s} \sin(\frac{\theta_s}{2}) \right\} \tag{13.264}$$

$$g_2^{II}(\theta,v) = \frac{2}{\mu R(v)} \left\{ 2\alpha_s \alpha_d \sqrt{\gamma_d} \cos(\frac{\theta_d}{2}) + (1+\alpha_s^2)\sqrt{\gamma_s} \cos(\frac{\theta_s}{2}) \right\} \tag{13.265}$$

where γ_d and γ_s are given in (13.227).

13.13 MODE III ASYMPTOTIC FIELD AT MOVING CRACK TIP

For the case of a mode III crack, the elastic problem is called the anti-plane problem, a term coined by Filon (see Chau, 2013). For this case, the only nonzero displacement is u_3, which is not a function of coordinate x_3. That is,

$$u_3 = u_3(x_1, x_2, t), \quad u_1 = u_2 = 0 \tag{13.266}$$

The only equation of motion is expressed as (Freund, 1998):

$$\nabla^2 u_3 = \frac{1}{c_s^2} \frac{\partial^2 u_3}{\partial t^2} \tag{13.267}$$

Similar to the case of modes I and II, a Galilean transform can be used to convert this wave equation to an elliptic type. In particular, we assume

$$\xi_1 = x_1 - vt, \quad \xi_2 = x_2 \tag{13.268}$$

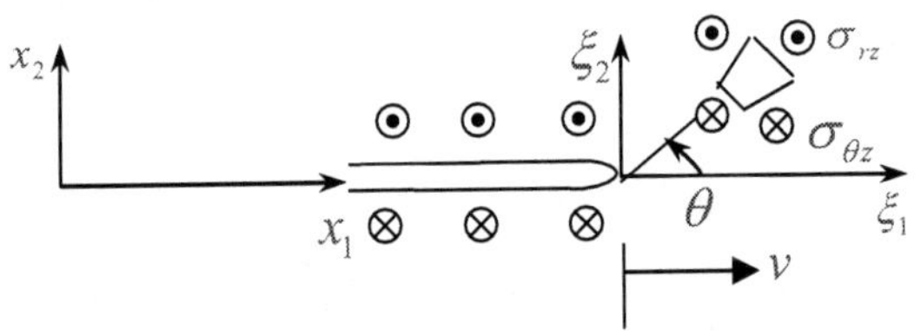

Figure 13.7 Asymptotic field near a moving mode III crack

Substitution of (13.268) into (13.267) yields

$$(1 - \frac{v^2}{c_s^2}) \frac{\partial^2 u_3}{\partial \xi_1^2} + \frac{\partial^2 u_3}{\partial \xi_2^2} = 0 \tag{13.269}$$

Rewriting it in a more compact form, we have

$$\frac{\partial^2 u_3}{\partial \xi_1^2} + \frac{1}{\alpha_s^2} \frac{\partial^2 u_3}{\partial \xi_2^2} = 0 \tag{13.270}$$

where

$$\alpha_s = \sqrt{1 - \frac{v^2}{c_s^2}} \tag{13.271}$$

It can expressed as a Laplace equation by introducing a new variable

$$\zeta_2 = \alpha_s \xi_2 \tag{13.272}$$

such that

$$\frac{\partial^2 u_3}{\partial \xi_1^2} + \frac{\partial^2 u_3}{\partial \zeta_2^2} = \nabla_d^2 u_3 = 0 \tag{13.273}$$

Rewriting this Laplace equation in polar form, we get

$$(\frac{\partial^2}{\partial r_s^2} + \frac{1}{r_s}\frac{\partial}{\partial r_s} + \frac{1}{r_s^2}\frac{\partial^2}{\partial \theta_s^2})u_3 = 0 \tag{13.274}$$

Similar to the earlier analysis for modes I and II, we seek a solution of the form:

$$u_3 = r_s^\lambda f(\theta_s; \lambda) \tag{13.275}$$

Substitution of (13.275) into (13.274) results in

$$f'' + \lambda^2 f = 0 \tag{13.276}$$

Its solution is obviously sine and cosine (see Section 1.11 of Chau, 2018)

$$f = A \sin \lambda\theta_s + B \cos \lambda\theta_s \tag{13.277}$$

Substituting (13.277) into (13.275), we have the out-of-plane displacement as

$$u_3 = r_s^\lambda (A \sin \lambda\theta_s + B \cos \lambda\theta_s) \tag{13.278}$$

The boundary conditions will be considered next.

13.13.1 Eigenvalue Problem

Due to symmetry, the boundary conditions of the anti-plane problem can be set as

$$u_3(r_s, 0) = 0, \quad \sigma_{z\theta}(r_s, \pi) = 0 \tag{13.279}$$

By Hooke's law in polar form, we have

$$\sigma_{z\theta} = \frac{1}{2}(\frac{1}{r_s}\frac{\partial u_3}{\partial \theta_s} + \frac{\partial u_\theta}{\partial z}) = \frac{1}{2r_s}\frac{\partial u_3}{\partial \theta_s} \tag{13.280}$$

Substitution of (13.280) and (13.278) into (13.279) leads to

$$B = 0, \quad \cos \lambda\pi = 0 \tag{13.281}$$

Thus, the eigenvalues are

$$\lambda = \pm(\frac{1}{2} + n) \quad n = 0, 1, 2, 3, ... \tag{13.282}$$

There are an infinite number of discrete eigenvalues. Similar to the quasi-static case (e.g., Section 6.7 of Chau, 2013), the boundedness of the strain energy at the crack tip leads to

$$\int_0^R \int_{-\pi}^{\pi} \mathcal{E}(r, \theta) r dr d\theta = \frac{1}{2\mu}\int_0^R \int_{-\pi}^{\pi} (\sigma_{rz}^2 + \sigma_{z\theta}^2) r dr d\theta < \infty \tag{13.283}$$

As all shear stresses are proportional to $r^{\lambda-1}$, we have

$$\int_0^R r^{2\lambda-1} dr < \infty \tag{13.284}$$

The only possibility is 2λ–1 > –1, or λ > 0. Therefore, only λ = 1/2 gives stress singularity and at the same time results in finite energy at the crack tip. Thus, the mode III displacement near the crack tip is

$$u_3(r_s,\theta_s) = 2Ar_s^{1/2}\sin\frac{\theta_s}{2}+... \tag{13.285}$$

Note that we have scaled the constant A here compared to (13.278).

13.13.2 Asymptotic Fields

We can define the dynamic stress intensity factor as:

$$K_{III}^d = \lim_{\xi_1\to 0}\sqrt{2\pi\xi_1}\,\sigma_{32}(r,0^{\pm}) \tag{13.286}$$

The shear stress can be determined from the displacement field obtained in (13.285) as

$$\begin{aligned}\sigma_{32}(r_s,\theta_s) &= \mu\frac{\partial u_3}{\partial \xi_2} = \mu(\frac{\partial u_3}{\partial r_s}\frac{\partial r_s}{\partial \xi_2}+\frac{\partial u_3}{\partial \theta_s}\frac{\partial \theta_s}{\partial \xi_2}) \\ &= \frac{\mu A}{r_s^{1/2}}\alpha_s\{\text{os}\,\theta_s\cos\frac{\theta_s}{2}+\sin\theta_s\sin\frac{\theta_s}{2}\} = \frac{\mu A}{r_s^{1/2}}\alpha_s\cos\frac{\theta_s}{2}\end{aligned} \tag{13.287}$$

Thus, the shear stress at $\theta_s = 0$ and $r_s \to \xi_1$ becomes

$$\sigma_{32}(r_s,0) = \frac{\mu A}{r_s^{1/2}}\alpha_s \tag{13.288}$$

Substitution of (13.288) into (13.286) yields

$$A = \frac{K_{III}^d}{\mu\alpha_s\sqrt{2\pi}} \tag{13.289}$$

Employing (13.289), we finally have the shear stress in (13.288) being

$$\sigma_{32}(r_s,\theta_s) = \frac{K_{III}^d}{\sqrt{2\pi r}}\frac{1}{\sqrt{\gamma_s}}\cos\frac{\theta_s}{2} \tag{13.290}$$

where K_{III}^d needs to be determined independently from the actual crack problem. Recall from (6.139) of Chau (2013) that

$$\begin{aligned}\boldsymbol{\sigma} &= \sigma_{32}\mathbf{e}_3\mathbf{e}_2+\sigma_{31}\mathbf{e}_3\mathbf{e}_1 = \sigma_{zr}\mathbf{e}_z\mathbf{e}_r+\sigma_{z\theta}\mathbf{e}_z\mathbf{e}_\theta \\ &= \sigma_{zr}\mathbf{e}_3(\mathbf{e}_1\cos\theta_s+\mathbf{e}_2\sin\theta_s)+\sigma_{z\theta}\mathbf{e}_3(-\mathbf{e}_1\sin\theta_s+\mathbf{e}_2\cos\theta_s)\end{aligned} \tag{13.291}$$

Comparing terms with the same base vectors, we find

$$\sigma_{32} = \sigma_{zr}\sin\theta_s+\sigma_{z\theta}\cos\theta_s, \quad \sigma_{31} = \sigma_{zr}\cos\theta_s-\sigma_{z\theta}\sin\theta_s \tag{13.292}$$

The shear stresses in polar coordinates can be determined easily from Hooke's law as

$$\sigma_{zr} = \mu \frac{\partial u_3}{\partial r_s} = \frac{\mu A}{r_s^{1/2}} \sin \frac{\theta_s}{2}, \quad \sigma_{z\theta} = \frac{\mu}{r_s} \frac{\partial u_3}{\partial \theta_s} = \frac{\mu A}{r_s^{1/2}} \cos \frac{\theta_s}{2} \tag{13.293}$$

Combining the result in (13.292) and (13.293), we finally obtain

$$\sigma_{31}(r_s, \theta_s) = -\frac{\mu A}{r_s^{1/2}} \sin \frac{\theta_s}{2} = -\frac{K_{III}^d}{\sqrt{2\pi r}} \frac{1}{\sqrt{\gamma_s}} \sin \frac{\theta_s}{2} \tag{13.294}$$

where

$$\gamma_s = \sqrt{1 - (\frac{v \sin \theta}{c_s})^2}, \quad \theta = \tan^{-1}(\frac{\xi_2}{\xi_1}), \quad \theta_s = \tan^{-1}(\frac{\alpha_s \xi_2}{\xi_1}) \tag{13.295}$$

Note that these definitions are the same as those defined in (13.171), (13.171) and (13.227). This can be combined to give a compact form of the crack tip asymptotic field as

$$\sigma_{ij}(r_s, \theta_s) = \frac{K_{III}^d}{\sqrt{2\pi r}} f_{ij}^{III}(\theta, v) \tag{13.296}$$

$$f_{32}^{III}(\theta, v) = \frac{1}{\sqrt{\gamma_s}} \cos \frac{\theta_s}{2}, \quad f_{31}^{III}(\theta, v) = -\frac{1}{\sqrt{\gamma_s}} \sin \frac{\theta_s}{2} \tag{13.297}$$

This result agrees with that given in Chau (2013), Freund (1998), and Ravi-Chandar (2004).

13.14 ASYMPTOTIC FIELD OF TRANSIENT CRACK GROWTH

If the moving velocity of the crack tip is not uniform or $v = v(t)$, the mathematics involved is more complicated. This problem is solved by Freund and Rosakis (1992) and Rosakis et al. (1991). We will summarize the main mathematical steps in this section.

Let us consider the plane problem, and recall the equations of motions from (13.160) as

$$\nabla^2 \varphi = \frac{1}{c_d^2} \frac{\partial^2 \varphi}{\partial t^2}, \quad \nabla^2 \psi = \frac{1}{c_s^2} \frac{\partial^2 \psi}{\partial t^2} \tag{13.298}$$

The main difference compared to (13.167) lies in the following Galilean transformation, which is defined as:

$$\xi_1 = x_1 - v(t)t, \quad \xi_2 = x_2 \tag{13.299}$$

Note that the velocity v is now a function of t. We first rewrite the scalar potential in these new variables as

$$\varphi(x_1, x_2, t) = \Phi(\xi_1, \xi_2, t), \quad \psi(x_1, x_2, t) = \Psi(\xi_1, \xi_2, t) \tag{13.300}$$

The differential of φ is

$$d\varphi = \frac{\partial \Phi}{\partial \xi_1} \frac{\partial \xi_1}{\partial t} dt + \frac{\partial \Phi}{\partial t} dt = -v \frac{\partial \Phi}{\partial \xi_1} dt + \frac{\partial \Phi}{\partial t} dt \tag{13.301}$$

Equation (13.302) can be rewritten as

$$\frac{\partial \varphi}{\partial t} = \frac{\partial \Phi}{\partial \xi_1} \frac{\partial \xi_1}{\partial t} + \frac{\partial \Phi}{\partial t} = -v \frac{\partial \Phi}{\partial \xi_1} + \frac{\partial \Phi}{\partial t} \tag{13.302}$$

Note that the left-hand side is written as a partial derivative as we recognize that it is also a function of position (ξ_1, ξ_2). Note also that we have to treat v and t as two independent variables when we take the differential.

Application of differentiation one more time gives

$$\begin{aligned}\frac{\partial^2\varphi}{\partial t^2} &= -\dot{v}\frac{\partial\Phi}{\partial\xi_1} - v\frac{\partial}{\partial t}[\frac{\partial\Phi}{\partial\xi_1}] + \frac{\partial}{\partial t}[\frac{\partial\Phi}{\partial t}] \\ &= -\dot{v}\frac{\partial\Phi}{\partial\xi_1} - v\left\{-v\frac{\partial^2\Phi}{\partial\xi_1^2} + \frac{\partial^2\Phi}{\partial t\partial\xi_1}\right\} - v\frac{\partial^2\Phi}{\partial t\partial\xi_1} + \frac{\partial^2\Phi}{\partial t^2} \\ &= -\dot{v}\frac{\partial\Phi}{\partial\xi_1} + v^2\frac{\partial^2\Phi}{\partial\xi_1^2} - 2v\frac{\partial^2\Phi}{\partial t\partial\xi_1} + \frac{\partial^2\Phi}{\partial t^2}\end{aligned} \tag{13.303}$$

Note that the partial differentiation of square bracket terms in (13.303) follows from (13.302). Using the result in (13.303), the first (13.298) becomes

$$(1-\frac{v^2}{c_d^2})\frac{\partial^2\Phi}{\partial\xi_1^2} + \frac{\partial^2\Phi}{\partial\xi_2^2} + \frac{\dot{v}}{c_d^2}\frac{\partial\Phi}{\partial\xi_1} + 2\frac{v}{c_d^2}\frac{\partial^2\Phi}{\partial t\partial\xi_1} - \frac{1}{c_d^2}\frac{\partial^2\Phi}{\partial t^2} = 0 \tag{13.304}$$

Similarly, we also have the second of (13.298) as

$$(1-\frac{v^2}{c_s^2})\frac{\partial^2\Psi}{\partial\xi_1^2} + \frac{\partial^2\Psi}{\partial\xi_2^2} + \frac{\dot{v}}{c_s^2}\frac{\partial\Psi}{\partial\xi_1} + 2\frac{v}{c_s^2}\frac{\partial^2\Psi}{\partial t\partial\xi_1} - \frac{1}{c_s^2}\frac{\partial^2\Psi}{\partial t^2} = 0 \tag{13.305}$$

Note that these PDEs have a non-constant coefficient. Its method of solution is not straightforward. The following technique is proposed by Freund and Rosakis (1992) and Rosakis et al. (1991), and it is similar to the perturbation technique for nonlinear differential equations. In particular, a parameter is introduced to scale the spatial coordinates and the solution is assumed as an infinite series of this new scaling parameter ε

$$\Phi(\xi_1,\xi_2,t) = \sum_{m=0}^{\infty}\varepsilon^{\frac{m+3}{2}}\Phi_m(\eta_1,\eta_2,t), \tag{13.306}$$

$$\Psi(\xi_1,\xi_2,t) = \sum_{m=0}^{\infty}\varepsilon^{\frac{m+3}{2}}\Psi_m(\eta_1,\eta_2,t) \tag{13.307}$$

where the scaled variables are defined as

$$\eta_1 = \xi_1/\varepsilon, \quad \eta_2 = \xi_2/\varepsilon \tag{13.308}$$

Since we are interested in the crack tip region, our focus is on

$$r = \sqrt{\xi_1^2 + \xi_2^2} \to 0 \tag{13.309}$$

Physically, ε is supposed to be a small parameter that enlarges the crack tip region, and consequently the leading terms in the series are the dominant terms of the stress singularity.

Substitution of (13.306) into (13.304) results in

$$\sum_{m=0}^{\infty}\varepsilon^{\frac{m+3}{2}}\left\{\frac{1}{\varepsilon^2}(1-\frac{v^2}{c_d^2})\frac{\partial^2\Phi_m}{\partial\eta_1^2} + \frac{1}{\varepsilon^2}\frac{\partial^2\Phi_m}{\partial\eta_2^2} + \frac{1}{\varepsilon}\frac{\dot{v}}{c_d^2}\frac{\partial\Phi_m}{\partial\eta_1} + \frac{2}{\varepsilon}\frac{v}{c_d^2}\frac{\partial^2\Phi_m}{\partial t\partial\eta_1} - \frac{1}{c_d^2}\frac{\partial^2\Phi_m}{\partial t^2}\right\} = 0 \tag{13.310}$$

If this differential equation is going to be identically zero, we require every order terms of the power series in ε to be zero as well. Expanding the first six terms explicitly, we get

$$
\begin{aligned}
&\left\{\frac{1}{\varepsilon^{1/2}}(1-\frac{v^2}{c_d^2})\frac{\partial^2\Phi_0}{\partial\eta_1^2}+\frac{1}{\varepsilon^{1/2}}\frac{\partial^2\Phi_0}{\partial\eta_2^2}+\varepsilon^{1/2}\frac{\dot{v}}{c_d^2}\frac{\partial\Phi_0}{\partial\eta_1}+\varepsilon^{1/2}\frac{2v}{c_d^2}\frac{\partial^2\Phi_0}{\partial t\partial\eta_1}-\frac{\varepsilon^{3/2}}{c_d^2}\frac{\partial^2\Phi_0}{\partial t^2}\right\}\\
&+\left\{(1-\frac{v^2}{c_d^2})\frac{\partial^2\Phi_1}{\partial\eta_1^2}+\frac{\partial^2\Phi_1}{\partial\eta_2^2}+\varepsilon\frac{\dot{v}}{c_d^2}\frac{\partial\Phi_1}{\partial\eta_1}+\varepsilon\frac{2v}{c_d^2}\frac{\partial^2\Phi_1}{\partial t\partial\eta_1}-\frac{\varepsilon^2}{c_d^2}\frac{\partial^2\Phi_1}{\partial t^2}\right\}\\
&+\left\{\varepsilon^{1/2}(1-\frac{v^2}{c_d^2})\frac{\partial^2\Phi_2}{\partial\eta_1^2}+\varepsilon^{1/2}\frac{\partial^2\Phi_2}{\partial\eta_2^2}+\varepsilon^{3/2}\frac{\dot{v}}{c_d^2}\frac{\partial\Phi_2}{\partial\eta_1}+\varepsilon^{3/2}\frac{2v}{c_d^2}\frac{\partial^2\Phi_2}{\partial t\partial\eta_1}-\frac{\varepsilon^{5/2}}{c_d^2}\frac{\partial^2\Phi_2}{\partial t^2}\right\}\\
&+\left\{\varepsilon(1-\frac{v^2}{c_d^2})\frac{\partial^2\Phi_3}{\partial\eta_1^2}+\varepsilon\frac{\partial^2\Phi_3}{\partial\eta_2^2}+\varepsilon^2\frac{\dot{v}}{c_d^2}\frac{\partial\Phi_3}{\partial\eta_1}+\varepsilon^2\frac{2v}{c_d^2}\frac{\partial^2\Phi_3}{\partial t\partial\eta_1}-\frac{\varepsilon^3}{c_d^2}\frac{\partial^2\Phi_3}{\partial t^2}\right\}\\
&+\left\{\varepsilon^{3/2}(1-\frac{v^2}{c_d^2})\frac{\partial^2\Phi_4}{\partial\eta_1^2}+\varepsilon^{3/2}\frac{\partial^2\Phi_4}{\partial\eta_2^2}+\varepsilon^{5/2}\frac{\dot{v}}{c_d^2}\frac{\partial\Phi_4}{\partial\eta_1}+\varepsilon^{5/2}\frac{2v}{c_d^2}\frac{\partial^2\Phi_4}{\partial t\partial\eta_1}-\frac{\varepsilon^{7/2}}{c_d^2}\frac{\partial^2\Phi_4}{\partial t^2}\right\}\\
&+\left\{\varepsilon^2(1-\frac{v^2}{c_d^2})\frac{\partial^2\Phi_5}{\partial\eta_1^2}+\varepsilon^2\frac{\partial^2\Phi_5}{\partial\eta_2^2}+\varepsilon^3\frac{\dot{v}}{c_d^2}\frac{\partial\Phi_5}{\partial\eta_1}+\varepsilon^3\frac{2v}{c_d^2}\frac{\partial^2\Phi_5}{\partial t\partial\eta_1}-\frac{\varepsilon^4}{c_d^2}\frac{\partial^2\Phi_5}{\partial t^2}\right\}\\
&+\ldots=0
\end{aligned}
\tag{13.311}
$$

Collecting the coefficients of the order of $\varepsilon^{-1/2}$, constant, $\varepsilon^{1/2}$, ε, $\varepsilon^{3/2}$, and ε^2, we get six differential equations for the first six power series in ε:

$$(1-\frac{v^2}{c_d^2})\frac{\partial^2\Phi_0}{\partial\eta_1^2}+\frac{\partial^2\Phi_0}{\partial\eta_2^2}=0 \tag{13.312}$$

$$(1-\frac{v^2}{c_d^2})\frac{\partial^2\Phi_1}{\partial\eta_1^2}+\frac{\partial^2\Phi_1}{\partial\eta_2^2}=0 \tag{13.313}$$

$$(1-\frac{v^2}{c_d^2})\frac{\partial^2\Phi_2}{\partial\eta_1^2}+\frac{\partial^2\Phi_2}{\partial\eta_2^2}=-(\frac{\dot{v}}{c_d^2}\frac{\partial\Phi_0}{\partial\eta_1}+\frac{2v}{c_d^2}\frac{\partial^2\Phi_0}{\partial t\partial\eta_1}) \tag{13.314}$$

$$(1-\frac{v^2}{c_d^2})\frac{\partial^2\Phi_3}{\partial\eta_1^2}+\frac{\partial^2\Phi_3}{\partial\eta_2^2}=-(\frac{\dot{v}}{c_d^2}\frac{\partial\Phi_1}{\partial\eta_1}+\frac{2v}{c_d^2}\frac{\partial^2\Phi_1}{\partial t\partial\eta_1}) \tag{13.315}$$

$$(1-\frac{v^2}{c_d^2})\frac{\partial^2\Phi_4}{\partial\eta_1^2}+\frac{\partial^2\Phi_4}{\partial\eta_2^2}=-(\frac{\dot{v}}{c_d^2}\frac{\partial\Phi_2}{\partial\eta_1}+\frac{2v}{c_d^2}\frac{\partial^2\Phi_2}{\partial t\partial\eta_1})+\frac{1}{c_d^2}\frac{\partial^2\Phi_0}{\partial t^2} \tag{13.316}$$

$$(1-\frac{v^2}{c_d^2})\frac{\partial^2\Phi_5}{\partial\eta_1^2}+\frac{\partial^2\Phi_5}{\partial\eta_2^2}=-(\frac{\dot{v}}{c_d^2}\frac{\partial\Phi_3}{\partial\eta_1}+\frac{2v}{c_d^2}\frac{\partial^2\Phi_3}{\partial t\partial\eta_1})+\frac{1}{c_d^2}\frac{\partial^2\Phi_1}{\partial t^2} \tag{13.317}$$

It is observed that the leading order pair of PDEs (13.312) and (13.313) are exactly the same as (13.168) for the case of a uniform moving crack. We have already obtained the asymptotic stress field corresponding to them in previous sections. The nonhomogeneous terms on the right-hand sides of the second pair of PDEs given in (13.314) and (13.315) depend on the solution of the lower order solutions Φ_0 and Φ_1 respectively. Similar conclusions can also be drawn for (13.316) and (13.317), except that one more nonhomogeneous term occurs on the right-hand side. It can be easily demonstrated that all PDEs for higher order terms of ε possess the mathematical structures of (13.316) and (13.317).

The general PDE for any order can be put into a universal form as (Rosakis et al., 1991):

$$\alpha_d^2(t)\frac{\partial^2\Phi_m}{\partial\eta_1^2}+\frac{\partial^2\Phi_m}{\partial\eta_2^2}=-\frac{2v^{1/2}}{c_d^2}\frac{\partial}{\partial t}(v^{1/2}\frac{\partial\Phi_{m-2}}{\partial\eta_1})+\frac{1}{c_d^2}\frac{\partial^2\Phi_{m-4}}{\partial t^2} \tag{13.318}$$

provided that

$$\begin{aligned}\Phi_m &= \Phi_m \qquad m\ge 0\\ &=0 \qquad m<0\end{aligned} \tag{13.319}$$

Note that α_d has been defined in (13.170) except that $v = v(t)$. Without repeating the details, it is obvious that we can also find the general PDE for the series expansion of ε as

$$\alpha_s^2(t)\frac{\partial^2\Psi_m}{\partial\eta_1^2}+\frac{\partial^2\Psi_m}{\partial\eta_2^2}=-\frac{2v^{1/2}}{c_s^2}\frac{\partial}{\partial t}(v^{1/2}\frac{\partial\Psi_{m-2}}{\partial\eta_1})+\frac{1}{c_s^2}\frac{\partial^2\Psi_{m-4}}{\partial t^2} \tag{13.320}$$

provided that

$$\begin{aligned}\Psi_m &= \Psi_m \qquad m\ge 0\\ &=0 \qquad m<0\end{aligned} \tag{13.321}$$

Again, α_s has been defined in (13.171) except that $v = v(t)$. As for the case of a constant moving crack, we express the potentials in terms of polar coordinates. For $m = 0$ and 1, the solution can be assumed as:

$$\Phi_m(\eta_1,\eta_2,t)=\tilde{\Phi}_m(r_d,\theta_d,t)=A_m(t)r_d^{(m+3)/2}\cos(\frac{m+3}{2})\theta_d \tag{13.322}$$

where

$$r_d=\sqrt{\eta_1^2+\alpha_d^2\eta_2^2},\quad \theta_d=\tan^{-1}(\frac{\alpha_d\eta_2}{\eta_1}) \tag{13.323}$$

These translating polar coordinates are distorted from the conventional polar coordinates by an amount determined by the crack speed v. Similarly, the solution for Ψ with $m = 0$ and 1 can be expressed as

$$\Psi_m(\eta_1,\eta_2,t)=\tilde{\Psi}_m(r_s,\theta_s,t)=B_m(t)r_s^{(m+3)/2}\sin(\frac{m+3}{2})\theta_s \tag{13.324}$$

where

$$r_s=\sqrt{\eta_1^2+\alpha_s^2\eta_2^2},\quad \theta_s=\tan^{-1}(\frac{\alpha_s\eta_2}{\eta_1}) \tag{13.325}$$

Equation (13.324) is the same as (13.208) if we identify $n = m+1$. Therefore, for the case of $m = 0$, we can show by referring to the previous section that

$$A_0(t)=-\frac{4K_I^d(t)}{3\mu\sqrt{2\pi}}\frac{1+\alpha_s^2}{R(v)}=-\frac{1+\alpha_s^2}{2\alpha_d}B_0(t) \tag{13.326}$$

where again $R(v)$ is

$$R(v) = 4\alpha_d \alpha_s - (1+\alpha_s^2)^2 \tag{13.327}$$

For the case of $m = 1$ (or $n = 2$ in the Section 13.11), we have from (13.208)

$$\hat{\Phi}_1(r_d, \theta_d, t) = A_1(t) r_d^2 \cos(2\theta_d) \tag{13.328}$$

$$\hat{\Psi}_1(r_s, \theta_s, t) = B_1(t) r_s^2 \sin(2\theta_s) \tag{13.329}$$

The displacements derived in (13.209) and (13.210) can be used to obtain

$$u_1 = \frac{3}{2} A_1 \left\{ r_d \cos\theta_d - \frac{(1+\alpha_s^2)}{2\alpha_s} \alpha_s r_s \cos\theta_s \right\} \tag{13.330}$$

$$u_2 = -\frac{3}{2} A_1 \left\{ r_d \alpha_d \sin\theta_d - \frac{(1+\alpha_s^2)}{2\alpha_s} r_s \sin\theta_s \right\} \tag{13.331}$$

The stresses derived in (13.212) to (13.215) can be specified as

$$\sigma_{12} = \frac{\mu A_1}{\alpha_s} \left\{ -4\alpha_s \alpha_d \sin 0 + (1+\alpha_s^2) \sin 0 \right\} = 0 \tag{13.332}$$

$$\sigma_{22} = \frac{2\mu A_n}{(1+\alpha_s^2)} \{ -(1+\alpha_s^2)^2 + (1+\alpha_s^2)^2 \} = 0 \tag{13.333}$$

$$\sigma_{11} = \frac{2\mu A_1}{1+\alpha_s^2} \{ (1+\alpha_s^2)(1+2\alpha_d^2 - \alpha_s^2) - (1+\alpha_s^2)^2 \} = 4\mu A_1 (\alpha_d^2 - \alpha_s^2) \tag{13.334}$$

Note that we have already imposed the boundary condition on the crack face to arrive at the above results. The only nonzero stress is given in (13.334) and as discussed in Broberg (1999) and Ravi-Chandar (2004), this is the so-called T-stress, which is the non-singular stress at the crack tip. This T-stress is thought to be an important factor in controlling the size and mechanism of the process zone ahead of the crack tip.

For $m = 2$ and 3, both (13.314) and (13.315) are nonhomogeneous. The right-hand side of the PDE can be rewritten by recognizing

$$\begin{aligned} \frac{\partial \Phi_m}{\partial \eta_1} &= \frac{\partial \hat{\Phi}_m}{\partial r_d} \frac{\partial r_d}{\partial \eta_1} + \frac{\partial \hat{\Phi}_m}{\partial \theta_d} \frac{\partial \theta_d}{\partial \eta_1} \\ &= \frac{m+3}{2} r_d^{(m+1)/2} \cos(\frac{m+1}{2}) \theta_d A_m(t) \end{aligned} \tag{13.335}$$

Thus, in view of (13.335) we have the following differentiation

$$\frac{\partial}{\partial t} \left\{ v^{1/2} \frac{\partial \Phi_m}{\partial \eta_1} \right\} = \frac{m+3}{2} r_d^{(m+1)/2} \cos(\frac{m+1}{2}) \theta_d \frac{d}{dt} \{ v^{1/2}(t) A_m(t) \} \tag{13.336}$$

Consequently, the nonhomogeneous term on the right-hand side of (13.318) or (13.320) becomes

$$-\frac{2v^{1/2}}{\alpha_d^2 c_d^2} \frac{\partial}{\partial t} (v^{1/2} \frac{\partial \Phi_{m-2}}{\partial \eta_1}) = -\frac{2v^{1/2}(t)}{\alpha_d^2 c_d^2} (\frac{m+3}{2}) r_d^{(m+1)/2} \cos(\frac{m+1}{2}) \theta_d \frac{d}{dt} \{ v^{1/2}(t) A_m(t) \} \tag{13.337}$$

Following Freund and Rosakis (1992), we introduce the following differential operator as

$$D_d^1[A_m(t)] = -\frac{(m+3)v^{1/2}(t)}{\alpha_d^2 c_d^2}\frac{d}{dt}\{v^{1/2}(t)A_m(t)\} \tag{13.338}$$

Finally, the PDE for Φ_m is simplified to

$$\frac{\partial^2\bar{\Phi}_m}{\partial r_d^2} + \frac{1}{r_d}\frac{\partial\bar{\Phi}_m}{\partial r_d} + \frac{1}{r_d^2}\frac{\partial^2\bar{\Phi}_m}{\partial\theta_d^2} = D_d^1[A_{m-2}(t)]r_d^{(m-1)/2}\cos(\frac{m-1}{2})\theta_d \tag{13.339}$$

The general solution of (13.339) is the homogeneous solution plus the particular solution. The homogeneous solution is obviously the same as that given in (13.322). For the case of constant v, the particular solution for (13.339) can be shown equal to (Freund and Rosakis, 1992):

$$\bar{\Phi}_{mp} = \frac{1}{2m+2}r_d^{(m+3)/2}D_d^1[A_{m-2}(t)]\cos(\frac{m-1}{2})\theta_d \tag{13.340}$$

To illustrate the validity of this solution, one can first show the following identities:

$$\frac{1}{r_d}\frac{\partial\bar{\Phi}_{mp}}{\partial r_d} = (\frac{m+3}{2})\frac{1}{2m+2}r_d^{(m-1)/2}D_d^1[A_{m-2}(t)]\cos(\frac{m-1}{2})\theta_d \tag{13.341}$$

$$\frac{\partial^2\bar{\Phi}_{mp}}{\partial r_d^2} = (\frac{m+3}{2})(\frac{m+1}{2})\frac{1}{2m+2}r_d^{(m-1)/2}D_d^1[A_{m-2}(t)]\cos(\frac{m-1}{2})\theta_d \tag{13.342}$$

$$\frac{1}{r_d^2}\frac{\partial^2\bar{\Phi}_{mp}}{\partial\theta_d^2} = -\frac{(m-1)^2}{4}(\frac{1}{2m+2})r_d^{(m-1)/2}D_d^1[A_{m-2}(t)]\cos(\frac{m-1}{2})\theta_d \tag{13.343}$$

Substitution of (13.341) and (13.343) into the left-hand side of (13.339) will result in the right-hand side of (13.190). Therefore, the general solution is

$$\bar{\Phi}_m = r_d^{(m+3)/2}\left\{A_m(t)\cos(\frac{m+3}{2})\theta_d + \frac{D_d^1[A_{m-2}(t)]}{2m+2}\cos(\frac{m-1}{2})\theta_d\right\} \tag{13.344}$$

For non-constant v, the solution is more complicated, and we refer the reader to Rosaki et al. (1991) for details.

Freund and Rosakis (1992) also gave the particular solution for (13.167) and (13.168) as:

$$\bar{\Phi}_{mp} = r_d^{(m+3)/2}\{\frac{1}{2m+2}\left[D_d^1[A_{m-2}(t)] + \frac{D_d^2[A_{m-2}(t)]}{(m-1)^2} + \ddot{A}_{m-4}(t)\right]\cos(\frac{m-1}{2})\theta_d$$
$$+\frac{D_d^2[A_{m-4}(t)]}{8(m-1)^2}\cos(\frac{m-2}{2})\theta_d\} \tag{13.345}$$

where

$$D_d^2() = D_d^1 D_d^1(), \quad \ddot{A}_m = \frac{1}{\alpha_d^2 c_d^2}\frac{d^2 A_m}{dt^2} \tag{13.346}$$

Although Freund and Rosakis (1992) demonstrated that a six-term solution of (13.306) and (13.307) leads to a better experimental comparison for fringe patterns away from the crack tip (see their Fig. 4), only the first order term is singular and the second order term corresponds to the T-stress, and higher order terms are only important when we are away from the tips.

The stress field at the crack tip can be summarized as:

$$\sigma_{ij}(r,\theta)=\frac{K_I^d(t)}{\sqrt{\pi r}}f_{ij}^I(\theta,v)+T_{ij}^I+O(r^{1/2}) \tag{13.347}$$

where the spatial variation function has been given in (13.224) to (13.226) and the constant T-stress is

$$T_{ij}^I(\theta,v)=4\mu A_1(\alpha_d^2-\alpha_s^2)\delta_{i1}\delta_{j1} \tag{13.348}$$

Therefore, we can conclude that the crack tip asymptotic stress field for an unsteady moving crack is the same as that for a crack with constant speed. The main difference is that the stress intensity factor can be a function of time.

13.15 CRACK GROWTH WITH INTERSONIC SPEED

So far, we have seen that the crack tip cannot move at a speed higher than the Rayleigh wave speed, otherwise our mathematical theory breaks down. However, there are both field observations after earthquakes and experimental studies by the experimental team at Caltech that a crack can indeed move at a wave speed faster than Rayleigh wave speed and shear wave speed c_s. Its implication on ground motions induced by earthquakes can be tremendous. If the fault rupturing speed is faster than that of shear waves, it means that there will be no clear signal of an S-wave after a P-wave before the arrival of the most devastating waves of Rayleigh type. Then, if you do not feel the shaking due to P-waves (which is normally small in magnitude unless for near-field strong earthquakes), you pretty much do not have any time to respond (say, go under the tables in your houses) before the arrival of Rayleigh waves.

In this section, we will look at the possibility of intersonic speed for dynamic fracture defined as:

$$c_s<v<c_d \tag{13.349}$$

13.15.1 Formulation

Again, the two wave equations for dilatation and shear waves given in (13.160) remain valid, and we still can use the Galilean transform to simplify the two wave equations. In particular, using

$$\xi_1=x_1-v(t)t,\quad \xi_2=x_2 \tag{13.350}$$

we get

$$(1-\frac{v^2}{c_d^2})\frac{\partial^2\varphi}{\partial\xi_1^2}+\frac{\partial^2\varphi}{\partial\xi_2^2}=0,\quad -(\frac{v^2}{c_s^2}-1)\frac{\partial^2\psi}{\partial\xi_1^2}+\frac{\partial^2\psi}{\partial\xi_2^2}=0 \tag{13.351}$$

In this way, the governing equations become

$$\frac{\partial^2\varphi}{\partial\xi_1^2}+\frac{1}{\alpha_d^2}\frac{\partial^2\varphi}{\partial\xi_2^2}=0,\quad \frac{\partial^2\psi}{\partial\xi_1^2}-\frac{1}{\bar{\alpha}_s^2}\frac{\partial^2\psi}{\partial\xi_2^2}=0 \tag{13.352}$$

where

$$\alpha_d = \sqrt{1-\frac{v^2}{c_d^2}}, \quad \hat{\alpha}_s = \sqrt{\frac{v^2}{c_s^2}-1} \tag{13.353}$$

Although the PDE for φ remains elliptic, the PDE for ψ becomes hyperbolic. Following the analysis used in the previous sections, we can obtain the corresponding stress expressions. The main difference lies on:

$$c_s^2 = \frac{v^2}{1+\hat{\alpha}_s^2} \tag{13.354}$$

Consequently, we find the following identities

$$\frac{\lambda}{\mu} - \alpha_d^2 \frac{c_d^2}{c_s^2} = -(1-\hat{\alpha}_s^2) \tag{13.355}$$

$$\frac{\lambda+2\mu}{\mu} - \alpha_d^2 \frac{\lambda}{\mu} = 1 + 2\alpha_d^2 + \hat{\alpha}_s^2 \tag{13.356}$$

Using these identities, we find

$$\sigma_{11} = \mu\left\{(1+2\alpha_d^2+\hat{\alpha}_s^2)\frac{\partial^2\varphi}{\partial\xi_1^2} + 2\frac{\partial^2\psi}{\partial\xi_1\partial\xi_2})\right\} \tag{13.357}$$

$$\sigma_{22} = -\mu\left\{(1-\hat{\alpha}_s^2)\frac{\partial^2\varphi}{\partial\xi_1^2} + 2\frac{\partial^2\psi}{\partial\xi_1\partial\xi_2}\right\} \tag{13.358}$$

$$\sigma_{12} = \mu\left\{2\frac{\partial^2\varphi}{\partial\xi_1\partial\xi_2} - (1-\hat{\alpha}_s^2)\frac{\partial^2\psi}{\partial\xi_1^2}\right\} \tag{13.359}$$

Note that these stress expressions are different from (3.55) of Ravi-Chandar (2004), and there are clearly typos in Ravi-Chandar (2004).

We seek the following solution forms

$$\varphi(r_d,\theta_d) = r_d^{\lambda}(A\cos\lambda\theta_d + B\sin\lambda\theta_d), \tag{13.360}$$

$$\psi(r_s,\theta_s) = Cf_L(\xi_1+\hat{\alpha}_s\xi_2) + Df_R(\xi_1-\hat{\alpha}_s\xi_2) \tag{13.361}$$

where θ_d, θ_s, r_d, and r_s have been defined as (13.172). The solutions of hyperbolic-type equations are expressed in terms of arbitrary functions of characteristics, one characteristic is a left-going wave and the other is a right-going wave. Note that (13.360) differs from (3.56) of Ravi-Chandra (2004). In particular, λ is missing in the cosine and sine functions in (3.56) of Ravi-Chandar (2004). The solution of this wave in terms of characteristics is demonstrated in Figure 13.8. For the upper half-plane, we only have a left-going wave non-zero whereas in the lower half plane, we only have the right-going waves non-zero. Basically, the shear wave front is chasing behind the crack tip, which moves faster than the characteristics.

The boundary conditions for the problem can be written as

$$\sigma_{22}(\xi_1<0,\xi_2=0)=0, \quad \sigma_{12}(\xi_1<0,\xi_2=0)=0 \tag{13.362}$$

This is also equivalent to

$$\sigma_{22}(r_d,\theta_d=\pi)=0, \quad \sigma_{12}(r_d,\theta_d=\pi)=0 \tag{13.363}$$

We can choose either one of (13.362) or (13.363), depending on whether the stress is given in terms of the distorted coordinates or the polar coordinates.

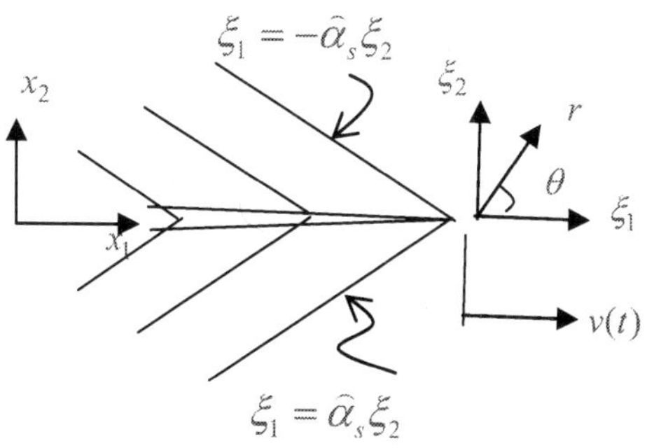

Figure 13.8 Characteristics lines for shear waves behind the crack tip moving with intersonic speed

13.15.2 Mode I

For a mode I crack, we have the symmetric case and if the upper half-plane is considered, we have $B = D = 0$ in (13.360) and (13.361). In particular, the displacement potentials for mode I are

$$\varphi(r_d, \theta_d) = r_d^{\lambda} A \cos \lambda \theta_d , \tag{13.364}$$

$$\psi(r_s, \theta_s) = C f_L(\xi_1 + \hat{\alpha}_s \xi_2) \tag{13.365}$$

Note that for the shear stress, we have

$$\frac{\partial \psi}{\partial \xi_1} = C f_L{}' , \quad \frac{\partial^2 \psi}{\partial \xi_1^2} = C f_L{}'' \tag{13.366}$$

Substitution of (13.366) and (13.364) into (13.359) and (13.358) gives

$$\sigma_{12} = -\mu\{2\alpha_d \lambda(\lambda - 1) r_d^{\lambda-2} A \sin(\lambda - 2)\theta_d + H(|\theta| - \theta_f)(1 - \hat{\alpha}_s^2) C f_L{}''\} \tag{13.367}$$

$$\sigma_{22} = -\mu\left\{(1 - \hat{\alpha}_s^2)\lambda(\lambda - 1) A r_d^{\lambda-2} \cos(\lambda - 2)\theta_d + H(|\theta| - \theta_f) 2C\hat{\alpha}_s f_L{}''\right\} \tag{13.368}$$

where

$$\sin \theta_f = \frac{c_s}{v} \tag{13.369}$$

Note however that the characteristics solution corresponding to the wave equation only exists for the space-time behind the characteristics. It is well known that only the region behind the wave front is affected by waves and it is known as the region of influence by the waves (see Section 9.2.2 of Chau, 2018). Thus, a Heaviside step function is added to the wave solution because of this causality of characteristics or waves.

To see the validity of (13.369), we refer to Figure 13.9. The step function given in (13.367) and (13.368) can also be written as

$$H(|\theta|-\theta_f)=H(-\xi_1-\hat{\alpha}_s\xi_2) \tag{13.370}$$

For nonzero wave influence, we must have the argument of the step function larger than zero

$$\begin{aligned}-\xi_1-\hat{\alpha}_s\xi_2>0\\ -r\cos\theta-\hat{\alpha}_s r\sin\theta>0\end{aligned} \tag{13.371}$$

Since $r > 0$, we must have

$$1+\hat{\alpha}_s\tan\theta<0 \tag{13.372}$$

Therefore, we have

$$\tan\theta<-\frac{1}{\hat{\alpha}_s} \tag{13.373}$$

For the upper plane, we must have $\pi/2 < \theta < \pi$. This critical angle θ_f can be expressed in a more appealing form in terms of its physical meaning. In particular, it is straightforward to see from (13.354) that

$$\sin\theta_f=\frac{1}{\sqrt{1+\hat{\alpha}_s^2}}=\frac{c_s}{v} \tag{13.374}$$

where $\pi/2 < \theta_f < \pi$. This completes the proof of (13.369).

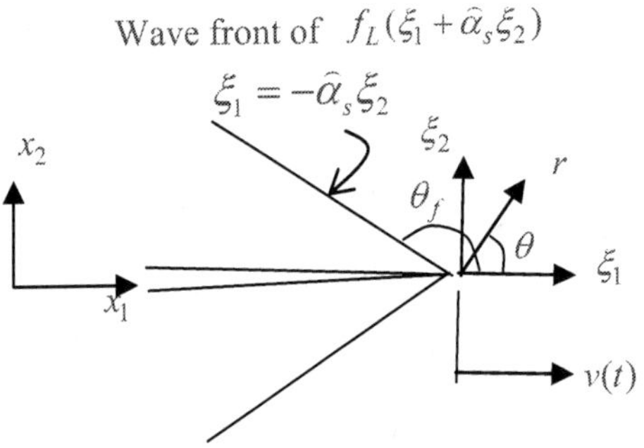

Figure 13.9 The definition of θ_f on the wave front

Substitution of (13.367) and (13.368) into the boundary conditions (13.363) gives

$$\begin{bmatrix} 2\alpha_d\lambda(\lambda-1)r^{\lambda-2}\sin(\lambda-2)\pi & (1-\hat{\alpha}_s^2)f_L'' \\ (1-\hat{\alpha}_s^2)\lambda(\lambda-1)r^{\lambda-2}\cos(\lambda-2)\pi & 2\hat{\alpha}_s f_L'' \end{bmatrix}\begin{Bmatrix} A \\ C \end{Bmatrix}=\begin{Bmatrix} 0 \\ 0 \end{Bmatrix} \tag{13.375}$$

Note that we have dropped the Heaviside step function since the crack line is clearly behind the crack front (see Figure 13.9). Setting the determinant of the coefficient matrix to zero, we have

$$2\lambda(\lambda-1)f_L''\left\{4\alpha_d\hat{\alpha}_s\sin(\lambda-2)\pi-(1-\hat{\alpha}_s^2)^2\cos(\lambda-2)\pi\right\}=0 \tag{13.376}$$

The eigenvalue equation becomes

$$\tan(\lambda-2)\pi = \frac{(1-\bar{\alpha}_s^2)^2}{4\alpha_d\bar{\alpha}_s} \tag{13.377}$$

Consequently, we have

$$\lambda-2 = \frac{1}{\pi}\tan^{-1}\{\frac{(1-\bar{\alpha}_s^2)^2}{4\alpha_d\bar{\alpha}_s}\}+n = \varepsilon(v)+n \tag{13.378}$$

Because the periodicity of the tangent function is π, from (13.378) we have

$$0<\varepsilon(v)<1/2 \tag{13.379}$$

We can see that the stress and displacement are of the order:

$$\sigma \sim r_d^{\lambda-2} = r_d^{\varepsilon(v)+n} \tag{13.380}$$

$$u \sim r_d^{\lambda-1} = r_d^{\varepsilon(v)+n+1} \tag{13.381}$$

As the crack tip is approached, we must have the displacement remain finite. This requires:

$$n \geq -2 \tag{13.382}$$

Otherwise, the displacement at the crack tip will approach infinity. On the other hand, we must have

$$n<0 \tag{13.383}$$

Otherwise, the stress field will go to zero as $r \to 0$, and this implies there is no crack propagation. Only two possibilities remain, that is, $n = 0$ and -1. Consider the case of $n = -1$; the strain energy, which is a product of stress and strain, is of the order of

$$\mathcal{E}(r,\theta) \sim r_d^{2\varepsilon(v)-2} \tag{13.384}$$

Because $\varepsilon(v)$ must be less than 1/2, we have that the strain energy goes to infinity as $r \to 0$. This is again physically unacceptable (i.e., we need infinite amount of energy to drive the crack).

Finally, for the remaining possibility that $n = 0$, the stress field is of the order of

$$\sigma \sim r_d^{2\varepsilon(v)} \tag{13.385}$$

Thus, the stress field at the crack tip is bounded, and therefore this is not a crack problem. Our final conclusion is that a mode I crack cannot propagate at an intersonic speed of propagation. This observation also agrees with experimental observations.

13.15.3 Mode II

For a mode II crack, we have the anti-symmetric case and if the upper half-plane is considered, we have $A = D = 0$ in (13.360) and (13.361). Thus, the displacement potentials for mode II are

$$\varphi(r_d,\theta_d) = r_d^{\lambda} B\sin\lambda\theta_d, \tag{13.386}$$

$$\psi(r_s,\theta_s) = Cf_L(\xi_1+\bar{\alpha}_s\xi_2) \tag{13.387}$$

Substitution of (13.386) and (13.387) into (13.359) and (13.358) gives

$$\sigma_{12} = \mu\{2\lambda(\lambda-1)B\alpha_d r_d^{\lambda-2}\cos(\lambda-2)\theta_d-(1-\bar{\alpha}_s^2)H(|\theta|-\theta_f)Cf_L''\} \tag{13.388}$$

$$\sigma_{22} = -\mu\left\{(1-\hat{\alpha}_s^2)\lambda(\lambda-1)Br_d^{\lambda-2}\sin(\lambda-2)\theta_d + 2C\hat{\alpha}_s H(|\theta|-\theta_f)f_L''\right\} \tag{13.389}$$

where θ_f is defined in (13.369). The two equations from the boundary conditions given in (13.363) can be put in a matrix form as

$$\begin{bmatrix} 2\alpha_d\lambda(\lambda-1)r^{\lambda-2}\cos(\lambda-2)\pi & -(1-\hat{\alpha}_s^2)f_L'' \\ (1-\hat{\alpha}_s^2)\lambda(\lambda-1)r^{\lambda-2}\sin(\lambda-2)\pi & 2\hat{\alpha}_s f_L'' \end{bmatrix}\begin{Bmatrix} B \\ C \end{Bmatrix} = \begin{Bmatrix} 0 \\ 0 \end{Bmatrix} \tag{13.390}$$

Setting the determinant of the coefficient matrix to zero gives

$$4\hat{\alpha}_s\alpha_d\cos(\lambda-2)\pi + (1-\hat{\alpha}_s^2)^2\sin(\lambda-2)\pi = 0 \tag{13.391}$$

This leads to

$$\tan(\lambda-2)\pi = -\frac{4\hat{\alpha}_s\alpha_d}{(1-\hat{\alpha}_s^2)^2} \tag{13.392}$$

The eigenvalue is

$$\lambda - 2 = -\frac{1}{\pi}\tan^{-1}\left\{\frac{4\hat{\alpha}_s\alpha_d}{(1-\hat{\alpha}_s^2)^2}\right\} + n = \delta(v) + n \tag{13.393}$$

Note from Figure 13.10 that we have the following property of the tangent function

$$\theta = \tan^{-1}\left\{\frac{a}{b}\right\}, \quad \frac{\pi}{2} - \theta = \tan^{-1}\left\{\frac{b}{a}\right\} \tag{13.394}$$

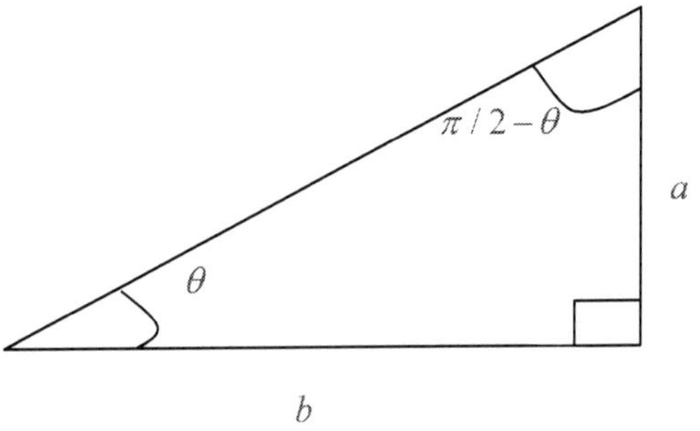

Figure 13.10 Tangent function for adjacent angles in a right-angle triangle

With this identity, we can relate $\varepsilon(v)$ defined in (13.378) and $\delta(v)$ defined in (13.393) as

$$\lambda - 2 = \delta(v) + n = \varepsilon(v) + n - \frac{1}{2} \tag{13.395}$$

For the range of $\varepsilon(v)$ given in (13.379), we have the following range for $\delta(v)$

$$-1/2 \le \delta(v) \le 0 \tag{13.396}$$

The displacement is of the order

$$u \sim r_d^{\delta(v)+n+1} \tag{13.397}$$

We require that

$$\delta(v)+n+1>0 \tag{13.398}$$

otherwise the displacement tends to infinity as $r \to 0$. This leads to

$$n>-1 \tag{13.399}$$

On the other hand, the stress field is of the order

$$\sigma \sim r_d^{\delta(v)+n} \tag{13.400}$$

Thus, in order to have singular stress at the crack tip, we require

$$\delta(v)+n \le 0 \tag{13.401}$$

Consequently, we need to have

$$n \le 0 \tag{13.402}$$

Therefore, the only possibility is $n = 0$. However, for $n = 0$, the stress singularity differs from the classical singularity of $r^{-1/2}$, unless the denominator of the function in the arctangent is zero:

$$\delta(v) = -\frac{1}{\pi}\tan^{-1}\left\{\frac{4\hat{\alpha}_s\alpha_d}{(1-\hat{\alpha}_s^2)^2}\right\} \tag{13.403}$$

Therefore, we find that

$$\hat{\alpha}_s^2 = 1 = \frac{v^2}{c_s^2} - 1 \tag{13.404}$$

Thus, the crack speed becomes

$$v = \sqrt{2}c_s = 1.414c_s \tag{13.405}$$

which is clearly an intersonic speed. This result agrees with that of Broberg (1999). This solution seems to agree with experimental observation (see Figure 3.10 of Ravi-Chandar, 2004). That is, a sharp wave front at 45° was observed for a shear crack propagating along a weak plane and for intact homogeneous materials.

Rosakis (2002) provided a summary of both experimental observations and field evidence observed during earthquakes, supporting the existence of intersonic shear crack (e.g., 1999 Izmit and Duzce earthquakes in Turkey, 1979 Imperial valley earthquake, and 1989 Landers earthquake).

13.15.4 Asymptotic Field for Mode II Crack

We will now consider the asymptotic stress field for the case of $n = 0$ obtained in the last section. The eigenvalue thus becomes

$$\lambda = \delta(v)+2 \tag{13.406}$$

Substitution of (13.406) into (13.388), (13.389) and (13.357) yields

$$\sigma_{12} = \mu\{2\alpha_d(\delta+2)(\delta+1)Br_d^{\delta}\cos\delta\theta_d - H(|\theta|-\theta_f)(1-\hat{\alpha}_s^2)Cf_L''\} \tag{13.407}$$

$$\sigma_{22} = -\mu\left\{(1-\hat{\alpha}_s^2)(\delta+2)(\delta+1)Br_d^\delta \sin\delta\theta_d + H(|\theta|-\theta_f)2C\hat{\alpha}_s f_L''\right\} \tag{13.408}$$

$$\sigma_{11} = \mu\left\{(1+2\alpha_d^2+\hat{\alpha}_s^2)(\delta+2)(\delta+1)Br_d^\delta \sin\delta\theta_d + H(|\theta|-\theta_f)2C\hat{\alpha}_s f_L''\right\} \tag{13.409}$$

However, the boundary condition on the crack line (i.e., $\theta = \pi$) given in (13.363) can be used to relate the constants B and C as:

$$Cf_L''(r) = \frac{2\alpha_d}{(1-\hat{\alpha}_s^2)}(\delta+2)(\delta+1)r^\delta \cos\delta\pi B \tag{13.410}$$

In addition, we can define the intersonic mode II stress intensity factor as

$$K_{II}^*(v) = \lim_{\xi_1\to 0}\sqrt{2\pi\xi_1}\,\sigma_{12}(r,0) = \sqrt{2\pi}\{2\mu\alpha_d(\delta+2)(\delta+1)B\} \tag{13.411}$$

In obtaining (13.411), we have noted that for $\theta_d \to 0$, we have $\xi_1 \to r$ and the Heaviside function is zero ahead of the crack tip at $\theta = 0$. Using these two results, the asymptotic stress field can be grouped in the following form:

$$\sigma_{ij}(r,\theta) = K_{II}^*(v)\left\{\frac{l_{ij}(\theta_d,\alpha_d,\hat{\alpha}_s)}{r_d^{-\delta}} - \frac{m_{ij}(\theta_d,\alpha_d,\hat{\alpha}_s)}{r^{-\delta}}H(|\theta|-\theta_f)\right\} \tag{13.412}$$

where

$$m_{11} = \frac{(1-\hat{\alpha}_s^2)}{2\alpha_d\sqrt{2\pi}}\frac{f_L''(\xi_1+\hat{\alpha}_s\xi_2)}{f_L''(r)}\sin\delta\pi \tag{13.413}$$

$$m_{12} = \frac{1}{\sqrt{2\pi}}\frac{f_L''(\xi_1+\hat{\alpha}_s\xi_2)}{f_L''(r)}\cos\delta\pi \tag{13.414}$$

$$m_{22} = -\frac{(1-\hat{\alpha}_s^2)}{2\alpha_d\sqrt{2\pi}}\frac{f_L''(\xi_1+\hat{\alpha}_s\xi_2)}{f_L''(r)}\sin\delta\pi \tag{13.415}$$

$$l_{11} = \frac{(1+2\alpha_d^2+\hat{\alpha}_s^2)}{2\alpha_d\sqrt{2\pi}}\sin\delta\theta_d \tag{13.416}$$

$$l_{12} = \frac{1}{\sqrt{2\pi}}\cos\delta\theta_d \tag{13.417}$$

$$l_{22} = -\frac{(1-\hat{\alpha}_s^2)}{2\alpha_d\sqrt{2\pi}}\sin\delta\theta_d \tag{13.418}$$

$$\delta(v) = -\frac{1}{\pi}\tan^{-1}\left\{\frac{4\hat{\alpha}_s\alpha_d}{(1-\hat{\alpha}_s^2)^2}\right\} \tag{13.419}$$

$$\alpha_d = \sqrt{1-\frac{v^2}{c_d^2}}, \quad \hat{\alpha}_s = \sqrt{\frac{v^2}{c_s^2}-1} \tag{13.420}$$

The corresponding displacements can be grouped as

$$u_i(r,\theta) = K_{II}^*(v)\left\{r_d^{\delta+1}f_i(\theta_d,\alpha_d,\hat{\alpha}_s) - r^\delta g_i(\theta_d,\alpha_d,\hat{\alpha}_s)H(|\theta|-\theta_f)\right\} \tag{13.421}$$

where

$$f_1 = \frac{1}{\sqrt{2\pi}} \frac{\sin(\delta+1)\theta_d}{2\mu\alpha_d(1+\delta)} \tag{13.422}$$

$$f_2 = \frac{1}{\sqrt{2\pi}} \frac{\cos(\delta+1)\theta_d}{2\mu(1+\delta)} \tag{13.423}$$

$$g_1 = \frac{(1-\hat{\alpha}_s^2)}{\sqrt{2\pi}} \frac{\sin\delta\pi}{4\mu\alpha_d} \frac{f_L'(\zeta)}{f_L''(r)} \tag{13.424}$$

$$g_2 = \frac{1}{\sqrt{2\pi}} \frac{\cos\delta\pi}{\mu\alpha_d(1-\hat{\alpha}_s^2)} \frac{f_L'(\zeta)}{f_L''(r)} \tag{13.425}$$

The steps for deriving (13.421) are given in Problem 13.2. Figure 13.11 plots the index of singularity versus the crack speed for various values of Poisson's ratios. It is clear that the maximum value is 1/2 at $v = 2^{1/2}c_s$ regardless of the value of Poisson's ratio.

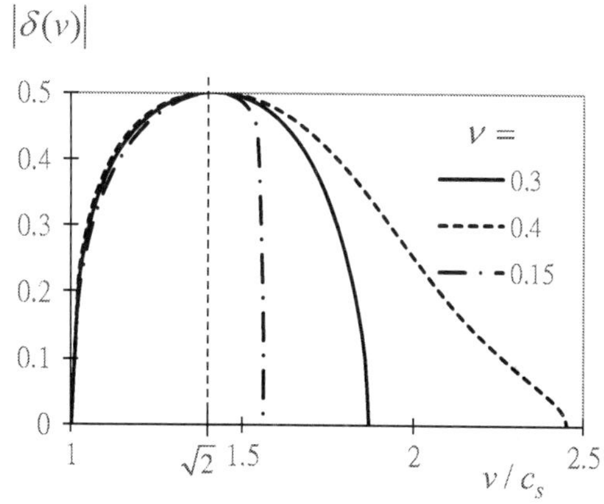

Figure 13.11 Singularity index $\delta(v)$ versus v/c_s for various ν

There are various theoretical issues related to this intersonic crack propagation. First, the energy release is not well defined if $\delta \neq -1/2$. It is also not clear why the crack propagation can jump from its theoretical maximum value at Rayleigh wave speed, which is smaller than the shear wave speed c_s, to shear wave speed or even to its maximum value of $2^{1/2}c_s$. Various models based on a crack tip cohesive zone have been proposed to explain the mechanism of velocity jumping. Numerical results suggest that there is a possibility of the creation of a secondary rupture just ahead of the crack tip due to a stress peak ahead of the tip as illustrated in Figure 13.12. This joining mechanism of the main rupture with the secondary crack ahead of the crack makes it possible for the rupture to accelerate to intersonic speed. This mechanism is normally referred as the Burridge-Andrews mechanism.

13.16 SUMMARY AND FURTHER READING

The most comprehensive coverage of the fracture mechanics problem using dual integral equations is given by Sneddon (1951). However, the book is not a target for beginners and thus students will find it difficult to read. The solution of dual integral equations discussed by Sneddon (1951) follows from Titchmarsh (1948) and Busbridge (1938). The derivation is given in Section 12 of Sneddon (1951). However, many students and researchers alike found that it is not easy to follow, as it involves the use of the Mellin transform of complex functions. In contrast, the coverage here is given in full detail and follows a different path of analysis via the Abel integral equation. In particular, the solution of the dual integral equations is given through the reduction of the pair of dual integral equations to the Abel integral equation and its solution can be solved readily. This analysis is somewhat similar to that given in Section 32 of Mura (1987) for the Dugdale-Barenblatt crack problem, although the name of dual integral equations is never mentioned in Mura (1987). They are actually mathematically equivalent. In this sense, the dual integral analysis can be extended to a cohesive crack model, such as the Dugdale-Barenblatt model discussed by Mura (1987). In this kind of model, a yielding zone is allowed around the crack tip. Mura (1987) also extended the Dugdale model to mode III cracking using a similar approach.

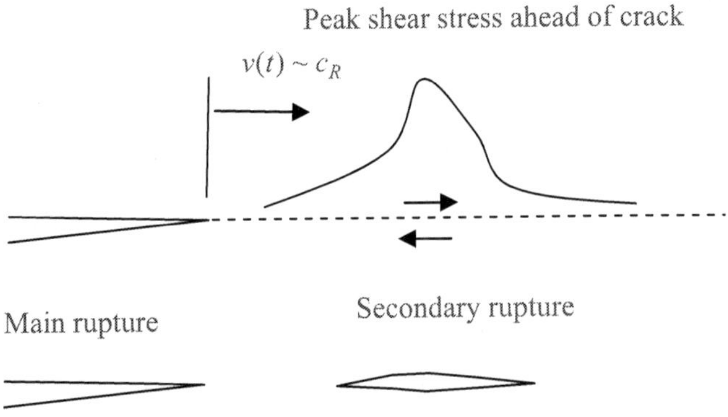

Figure 13.12 Burridge-Andrews mechanism for intersonic crack propagation

Sneddon (1951) also demonstrated that dual integral equations can be used to solve fluid mechanics problems. That is not surprising as we have already seen that the dual integral equations are for harmonic functions that satisfy the Laplace equation. Naturally, ideal incompressible flow of fluids (or so-called potential flow) is governed by the Laplace equation through the use of velocity potential. Section 31.3 of Sneddon (1951) discussed the potential flow through a circular aperture in a plane rigid screen using dual integral formulation. The recent book by

Duffy (2008) provided a list of mixed boundary value problems that can be solved using dual integral formulation. Sneddon (1951) also solved the indentation problems (both plane and 3-D punches) using dual integral formulation. More discussion of dual integral equations can be found in Keer (1968).

The second part of this chapter considered the asymptotic stress field at a moving crack tip through the solution of wave equations. It can be shown as in Chau (2013) that all dynamic problems in elastic isotropic solids can be solved through two wave equations through Helmholtz decomposition of vectors, one for the dilatation wave and one for the shear wave. One is a scalar wave equation and the other is a vector wave equation (its mathematical equivalence to the Maxwell equation formulation for electromagnetic waves is considered in Chapter 2 of Chau (2018). For our dynamic wave problems, one component of the shear wave is found sufficient to model the problems. For a steadily moving crack tip, the use of Galilean transformation of the coordinate system reduces the wave problem to a potential problem. The requirement of a traction-free condition posed an eigenvalue problem. Although there are infinite eigenvalues, there is only one that corresponds to stress singularity at the crack tip, while all others lead to a finite stress field. It turns out that the eigenvalues for mode I and mode II are identical. Thus, we can obtain the crack tip fields for moving cracks under all modes I, II and III. The classical $r^{-1/2}$ stress singularity is obtained, but the polar function of the stress field also depends on the crack speed v, and the Rayleigh wave speed quotient appears naturally in the derivation of the stress field. Thus, this imposes the condition that Rayleigh wave speed should not be exceeded. Thus, the crack speed must be subsonic (or less than the shear wave speed) since Rayleigh wave speed is always less than the shear wave speed. We also presented the analyses by Freund and Rosakis (1992) and Rosakis et al. (1991) for the case of a non-uniform propagating crack tip. A technique similar to perturbation analysis for nonlinear differential equations is proposed and it turns out that the crack tip field remains the same for a non-uniformly moving crack tip. The possibility of intersonic speed of crack growth is investigated (i.e., the crack speed is between shear and dilatation wave speeds). In contrast to the subsonic case, only the dilatation wave is converted to the Laplace equation through Galilean transformation, whilst the shear wave equation remains of hyperbolic type. The method of characteristics is used to represent the wave type solution. The eigenvalue problem leads to an eigenvalue equation in terms of the arctangent function of crack speed and Poisson's ratio of the solid (see Figure 13.11). This eigenvalue equation predicts that only when the crack speed equals $2^{1/2}c_s$, where c_s is the shear wave speed, the classical stress singularity is recovered. This solution predicts that the characteristics lines of the hyperbolic equation for the shear wave is trailing the crack tip at 45°, which has been verified by experiments conducted by Ravi-Chandar and Rosakis at California Institute of Technology. The intersonic crack propagation is only possible for a shear crack and is theoretically impossible for a tensile crack. There is also strong evidence that faulting speed during an earthquake can be intersonic and is summarized in Rosakis et al. (1991) and Rosakis (2002).

For more coverage of dynamic fracture mechanics, we refer to the books by Freund (1998), Ravi-Chandar (2004), Slepyan (2002) and Broberg (1999).

13.17 PROBLEMS

Problem 13.1 Show that the hoop stress in the vicinity of a mode I crack propagating at a constant speed of v is given by:

$$\sigma_{\theta\theta}(r,\theta)\frac{\sqrt{2\pi r}}{K_I^d} = \{f_{11}^I(\theta,v)\sin^2\theta + f_{22}^I(\theta,v)\cos^2\theta - 2\sin\theta\cos\theta f_{12}^I(\theta,v)\} \tag{13.426}$$

Problem 13.2 In deriving the displacement of mode II intersonic cracks given in (13.421), we note that following identities are useful.

(i) Show that

$$\frac{\partial\varphi}{\partial\xi_1} = \lambda B_d^{\lambda-1}\sin(\lambda-1)\theta_d \tag{13.427}$$

$$\frac{\partial\varphi}{\partial\xi_2} = \lambda B r_d^{\lambda-1}\alpha_d\cos(\lambda-1)\theta_d \tag{13.428}$$

$$\frac{\partial\psi}{\partial\xi_1} = Cf_L' \tag{13.429}$$

$$\frac{\partial\psi}{\partial\xi_2} = C\widehat{\alpha}_s f_L' \tag{13.430}$$

(ii) Note the following formulas for displacement

$$u_1 = \frac{\partial\varphi}{\partial x_1} + \frac{\partial\psi}{\partial x_2} = \frac{\partial\varphi}{\partial\xi_1} + \frac{\partial\psi}{\partial\xi_2} \tag{13.431}$$

$$u_2 = \frac{\partial\varphi}{\partial x_2} - \frac{\partial\psi}{\partial x_1} = \frac{\partial\varphi}{\partial\xi_2} - \frac{\partial\psi}{\partial\xi_1} \tag{13.432}$$

Show that

$$u_1 = \lambda B_d^{\lambda-1}\sin(\lambda-1)\theta_d + C\widehat{\alpha}_s f_L' \tag{13.433}$$

$$u_2 = \lambda B r_d^{\lambda-1}\alpha_d\cos(\lambda-1)\theta_d - Cf_L' \tag{13.434}$$

(iii) Show that

$$C\widehat{\alpha}_s = -\frac{(1-\widehat{\alpha}_s^2)(\delta+2)(\delta+1)r^\delta\sin\delta\pi}{2f_L''(r)}B \tag{13.435}$$

$$C = \frac{2\alpha_d(\delta+2)(\delta+1)r^\delta\cos\delta\pi}{(1-\widehat{\alpha}_s^2)f_L''(r)}B \tag{13.436}$$

(iv) Show that

$$u_1 = B(\delta+2)\{r^{\delta+1}\gamma_d^{(\delta+1)/2}\sin(\delta+1)\theta_d - \frac{(1-\widehat{\alpha}_s^2)(\delta+1)r^\delta\sin\delta\pi}{2}\frac{f_L'(\zeta)}{f_L''(r)}H(|\theta|-\theta_f)\} \tag{13.437}$$

$$u_2 = B(\delta+2)\alpha_d\{r^{\delta+1}\gamma_d^{(\delta+1)/2}\cos(\delta+1)\theta_d - \frac{2(\delta+1)r^{\delta}\cos\delta\pi}{(1-\widehat{\alpha}_s^2)}\frac{f_L'(\zeta)}{f_L''(r)}\} \tag{13.438}$$

(v) Note the following condition on the boundary of the crack

$$\sigma_{12}(r,0) = \frac{K_{II}^*}{\sqrt{2\pi\xi_1}} \tag{13.439}$$

By using this condition, show that

$$B = \frac{K_{II}^*}{\sqrt{2\pi}}\left(\frac{1}{2\mu\alpha_d(1+\delta)(2+\delta)}\right) \tag{13.440}$$

(vi) Finally, show the validity of (13.421).

Problem 13.3 Show the validity of (13.212) to (13.215).

Problem 13.4 Show the validity of (13.243) to (13.245).

Problem 13.5 Show the validity of (13.253) to (13.255).

Problem 13.6 Prove (13.335).

REFERENCES

Ablowitz, M.J. and Clarkson, P.A., 1991, *Solitons, Nonlinear Evolution Equations and Inverse Scattering,* London Mathematical Society Lecture Note Series 149 (Cambridge: Cambridge University Press).

Ablowitz, M.J. and Segur, H., 1981, *Solitons and the Inverse Scattering Transform,* SIAM Studies in Applied Mathematics (Philadelphia: SIAM).

Abramowitz, M. and Stegun, I.A., 1964, *Handbook of Mathematical Functions* (New York: Dover).

Achenbach, J.D., 1973, *Wave Propagation in Elastic Solids* (Amsterdam: North-Holland).

Adhikari, S., 2006, Damping modeling using generalized proportional damping. *Journal of Sound and Vibration*, **293**, 156-170.

Aki, K. and Richards, P.G., 1980, *Quantitative Seismology—Theory and Methods* (San Francisco: W.H. Freeman).

Alekseenko, S.V., Kuibin, P.A., and Okulov, V.L., 2007, *Theory of Concentrated Vortices: An Introduction* (Berlin: Springer).

Arboleda-Monsalve, L.G., Zapata-Medina, D.G., and Aristizabal-Ochoa, J.D., 2007, Stability and natural frequencies of a weakened Timoshenko beam-column with generalized end conditions under constant axial load. *Journal of Sound and Vibration*, **307**, 89-112.

Banks, R.B., 1994, *Growth and Diffusion Phenomena: Mathematical Frameworks and Applications* (Berlin: Springer-Verlag).

Banks, R.B., 1998, *Towing Icebergs, Falling Dominoes, and Other Adventures in Applied Mathematics* (Princeton: Princeton University Press).

Banks, R.B., 1999, *Slicing Pizzas, Racing Turtles and Further Adventures in Applied Mathematics* (Princeton: Princeton University Press).

Bates, R.R., Mueller, D.D., and White, J.E., 1971, *Fundamentals of Astrodynamics* (New York: Dover).

Bathe, K.J., 1982, *Finite Element Procedures in Engineering Analysis* (Englewood Cliffs: Prentice-Hall).

Battin, R.H., 1999, *An Introduction to the Mathematics and Methods of Astrodynamics,* revised edition (Reston: American Institute of Aeronautics and Astronautics).

Bauer, F., Keller, H.B., and Reiss, E.L., 1967, Imperfections in the buckling of spherical caps. *SIAM Journal on Applied Mathematics*, **15**(2), 273-283.

Bazant, Z.P., and Cedolin, L., 1991, *Stability of Structures* (New York: Oxford University Press).

Beiser, A., 2003, *Concepts of Modern Physics,* 6th edition (New York: McGraw-Hill).

Bell, W.W., 1968, *Special Functions for Scientists and Engineers* (Mineola: Dover).

Benjamin, T.B., and Feir, J.E., 1967, The disintegration of wavetrains in deep water. Part I. *Journal of Fluid Mechanics*, **27**, 417-430.

Bjork, R.L., 1961, Analysis of the formation of meteor crater, Arizona: A preliminary report. *Journal of Geophysical Research*, **66**(10), 3379-3387.

Blaauwendraad, J., and Hoeffaker, J.H., 2014, *Structural Shells Analysis: Understanding and Application* (Dordrecht: Springer).

Bleich, F., McCullough, C.B., Rosecrans, R., and Vincent, G.S., 1950, *The Mathematical Theory of Vibration in Suspension Bridges* (Washington DC: United States Government Printing Office).

Bloom, F., and Coffin, D., 2001, *Handbook of Thin Plate Buckling and Postbuckling* (Boca Raton: Chapman & Hall).

Bluestein, H.B., 2013, *Severe Convective Storms and Tornadoes: Observations and Dynamics* (Heidelberg: Springer).

Bolotin, V.V., 1963, *Nonconservative Problems of the Theory of Elastic Stability* translated from Russian by T.K. Lusher (Oxford: Pergamon).

Bolotin, V.V., 1964, *The Dynamic Stability of Elastic Systems* translated from Russian by V.I. Weingarten, L.B. Greszczuk, K.N. Trirogoff and K.D. Gallegos (San Francisco: Holden-Day).

Britton, J.R., Kriegh, R.B., and Rutland, L.W., 1966, *Calculus and Analytic Geometry* (San Francisco: Freeman and Company).

Broberg, K.B., 1999, *Cracks and Fracture* (San Diego: Academic Press).

Brouwer, D., and Clemence, G.M., 1961, *Methods of Celestial Mechanics* (New York: Academic Press).

Brush, D.O., and Almroth, B.O., 1975, *Buckling of Bars, Plates, and Shells* (New York: McGraw Hill).

Busbridge, I.W., 1938, Dual integral equations. *Proceedings of the London Mathematical Society*, **44**(1), 115-129.

Calladine, C.R., 1983, *Theory of Shell Structure* (Cambridge: Cambridge University Press).

Carslaw, H.S. and Jaeger, J.C., 1959. *Conduction of Heat in Solids*, 2nd edition (Oxford: Oxford University Press).

Caughey, T. K. 1960, Classical normal modes in damped linear dynamic systems. *Journal of Applied Mechanics ASME*, **27**(2), 269-271.

Caughey, T. K., and O'Kelly, M. E. J. 1965, Classical normal modes in damped linear dynamic systems. *Journal of Applied Mechanics ASME* **32**(3), 583-588.

Charney, F.A., 2008, Unintended consequences of modeling damping in structures. *Journal Structural Engineering ASCE*, **134**, 581-592.

Chau, K.T., 1993, Anti-symmetric bifurcation in a compressible pressure-sensitive circular cylinder under axisymmetric tension and compression. *Journal of Applied Mechanics ASME*, **60**, 282-289.

Chau, K.T., 1994, Half-space instabilities and short wavelength bifurcations in cylinders and rectangular blocks. *Journal of Applied Mechanics ASME,* **61,** 742-744.

Chau, K.T., 1995, Buckling, barrelling, and surface instabilities of a finite, transversely isotropic circular cylinder. *Quarterly of Applied Mathematics*, **53**, 225-244.

Chau, K.T., 2013, *Analytic Methods in Geomechanics* (Boca Raton: CRC Press).

Chau, K.T., 2018, *Theory of Differential Equations in Engineering and Mechanics* (Boca Raton: CRC Press).

Cheo, L.S and Reiss, E.L., 1973, Unsymmetric wrinkling of circular plates. *Quarterly of Applied Mathematics*, **31**, 75-91.

Cheo, L.S and Reiss, E.L., 1974, Secondary buckling of circular plates. *SIAM Journal on Applied Mathematics*, **26**(3), 490-495.

Clough, R.W., and Penzien, J., 2003, *Dynamics of Structures*, 3rd edition (Berkeley: Computer & Structures Inc.).

Coman, C.D., and Bassom, A.P., 2016, Asymptotic limits and wrinkling patterns in a pressurized shallow spherical cap. *International Journal of Nonlinear Mechanics*, **81**, 8-18.

Courant, R., and Hilbert, D., 1962, *Methods of Mathematical Physics: Vol. II, Partial Differentiation Equations* (New York: John Wiley & Sons).

Crank, J., 1975, *The Mathematics of Diffusion* (Oxford: Clarendon Press).

Cushman-Roisin, B., 1994, *Introduction to Geophysical Fluid Dynamics* (Englewood Cliffs: Prentice Hall).

Cushman-Roisin, B., and Beckers, J.-M., 2009, *Introduction to Geophysical Fluid Dynamics: Physical and Numerical Aspects* (New York: Academic Press).

Dergarabedian, P., and Fendell, F., 1976, One-and two-cell structure in tornadoes. In *Proceedings of the Symposium on Tornadoes: Assessment of Knowledge and Implications for Man,* ed. by R.E. Peterson, June 22-24, 1976, Texas Tech University, pp. 501-521.

Dirac, P.A.M., 1947, *The Principle of Quantum Mechanics* (Oxford: Clarendon Press).

Druitt, F., 2005, *Hirota's Direct Method and Sato's Formalism in Soliton Theory*, Honors thesis, Melbourne University.

Duffy, D.G., 2008, *Mixed Boundary Value Problems* (Boca Raton: CRC Press).

Esmailzadeh, E., and Ohadi, A.R., 2000, Vibration and stability analysis of non-uniform Timoshenko beams under axial and distributed tangential loads. *Journal of Sound and Vibration*, **236**(3), 443-456.

Farquharson, F.B., 1950, *Aerodynamic Stability of Suspension Bridges with Special Reference to the Tacoma Narrows Bridge*. Part I: Investigations Prior to October, 1941 (Washington DC: United States Government Printing Office).

Feynman, R.P., Leighton, R.B., and Sands, M.L., 1964a, *Feynman Lectures on Physics*, Vol. 2, Electricity and Magnetism (Reading: Addison-Wesley Publishing).

Feynman, R.P., Leighton, R.B., and Sands, M.L., 1964b, *Feynman Lectures on Physics*, Vol. 3, Quantum Mechanics (Reading: Addison-Wesley Publishing).

Fischer, H.B., List, E.J., Koh, R.C.Y., Imberger, J., and Brooks, N.H., 1979, *Mixing in Inland and Coastal Waters* (San Diego: Academic Press).

Fitts, D.D., 2002, *Principles of Quantum Mechanics as Applied to Chemistry and Chemical Physics* (Cambridge: Cambridge University Press).

Fleisch, D., 2008, *A Student's Guide to Maxwell's Equations* (Cambridge: Cambridge University Press).

Flugge, W., 1960, *Stresses in Shells* (Berlin: Springer).

Flugge, W., 1962, Shells. Chapter 40 in *Handbook of Engineering Mechanics*, ed. by W. Flugge (New York: McGraw-Hill).

Forbes, G.S. 1976, E. Photogrammetric characteristics of the Parker Tornado of April 3, 1974. In *Proceedings of the Symposium on Tornadoes: Assessment of Knowledge and Implications for Man,* ed. by R.E. Peterson, June 22-24, 1976, Texas Tech University, pp. 58-77.

Fowles, G.R., and Cassiday, G.L., 2005, *Analytical Mechanics*, 7th edition (Belmont: Brooks/Cole).

French, A.P., 1971, *Newtonian Mechanics*, MIT Introductory Physics Series (New York: W.W. Norton & Company).

Freund, L.B., 1998, *Dynamic Fracture Mechanics* (Cambridge: Cambridge University Press).

Freund, L.B., and Rosakis, A.J., 1992, The structure of the near-tip field during transient elastodynamics crack growth. *Journal of the Mechanics and Physics of Solids*, **40**(3), 699-719.

Friedrichs, K.O., and Stoker, J.J., 1941, The nonlinear boundary value problem of the buckled plate. *American Journal of Mathematics*, **63**, 839-888.

Fujino, Y., 2002, Vibration, control and monitoring of long-span bridges-recent research, developments and practice in Japan. *Journal of Constructional Steel Research*, **58**, 71-97.

Fujita, T.T., and Pearson, A.D., 1976, B. Glossary of whirlwinds. In *Proceedings of the Symposium on Tornadoes: Assessment of Knowledge and Implications for Man,* ed. by R.E. Peterson, June 22-24, 1976, Texas Tech University, pp. 45-47.

Gallais, P., 2007, *Atmospheric Re-entry Vehicle Mechanics* (Berlin: Springer-Verlag).

Gao, H., 1993, Surface roughening and branching instabilities in dynamic fracture. *Journal of the Mechanics and Physics of Solids*, **41**(3), 457-486.

Gere, J.M., and Timoshenko, S.P., 1990, *Mechanics of Materials,* 3rd edition (Boston: PWS-KENT).

Gimsing, N.J., and Georgakis, C.T., 2011, *Cable Supported Bridges: Concept and Design,* 3rd edition (Chichester: John Wiley & Sons).

Gonnert, G., Dube, S.K., Murty, T., and Siefert, W., 2001, *Global Storm Surges,* ed. by German Coastal Engineering Research Council (Westholsteinische: Verlagsanstalt Boyens & Co).

Gradshteyn, I.S. and Ryzhik, I.M., 1980, *Table of Integrals, Series, and Products* (New York: Academic Press).

Griffiths, D.J., 1981, *Introduction to Electrodynamics* (Upper Saddle River: Prentice- Hall).

Griffiths, D.J., 1995, *Introduction to Quantum Mechanics* (Upper Saddle River: Prentice- Hall).

Gurfil, P., and Seidelmann, P.K., 2016, *Celestial Mechanics and Astrodynamics: Theory and Practice* (Berlin: Springer-Verlag).

Hasselsman, T. K., 1976, Modal coupling in lightly damped structures. *AIAA Journal*, **14**, 1627-1628.

Heras, J.A., 1994, Jefimenko's formulas with magnetic monopoles and the Lienard-Wiechert fields of a dual-charged particle. *American Journal of Physics*, **62**, 525-531.

Heras, J.A., 1995, Time-dependent generalizations of the Biot-Savart and Coulomb laws: A formal derivation. *American Journal of Physics*, **63**, 928-932.

Hetenyi, M., 1946, *Beams on Elastic Foundation: Theory with Applications in the Field in Civil Engineering and Mechanics* (Ann Arbor: Michigan University Press).

Heyman, J., 1999, *The Science of Structural Engineering* (London: Imperial College Press).

Hicks, K.D., 2009, *Introduction to Astrodynamic Reentry*, AFIT/EN/TR-09-03 (Wright-Patterson Air Force Base: Air Force Institute of Technology) www.dtic.mil/dtic/tr/fulltext/u2/a505342.pdf.

Hietarinta, J., 2005, Hirota's bilinear method and soliton solutions. *Physics AUC*, **15**(part I), 31-37.

Hobson, E.W., 1955, *The Theory of Spherical Harmonics* (New York: Chelsea).

Hoff, N.J., 1951, The dynamics of the buckling of elastic columns. *Journal of Applied Mechanics,* **18**, 68-74.

Holton, J.R., 2004, *An Introduction to Dynamic Meteorology*, 4th edition (Amsterdam: Elsevier).

Housner, W.P., 2002, Deriving the Lorentz force equation from Maxwell's equations. *Proceedings IEEE Southeast Conference* 2002, pp. 422-425.

Hua, L.K., 2009, *An Introduction to Higher Mathematics,* Vols. 1–4 (Beijing: Higher Education Press) (in Chinese).

Hua, L.K., 2012, *An Introduction to Higher Mathematics,* translated by P. Shiu (Cambridge: Cambridge University Press).

Hutchinson, J.R., 2001, Shear coefficients for Timoshenko beam theory. *Journal of Applied Mechanics*, **68**, 87-92.

Hutchinson, J.W. and Budiansky, B., 1966, Dynamic buckling estimates. *AIAA Journal*, **4**(3), 525-530.

Irvine, H.M., 1981, *Cable Structures* (Cambridge: MIT Press).

Jackson, J.D., 1999, *Classical Electrodynamics,* 3rd edition (New York: John Wiley & Sons).

Jackson, J.D., 2002, From Lorenz to Coulomb and other explicit gauge transformations. *American Journal of Physics*, **70**, 917-928.

Jackson, J.D., and Okun, L.B., 2001, Historical roots of gauge invariance. *Review of Modern Physics*, **73**, 663-680.

Jefimenko C.D., 1966, *Electricity and Magnetism: An Introduction to the Theory of Electric and Magnetic Fields* (New York: Appleton-Century-Crofts).

Jefimenko C.D., 1992, Solutions of Maxwell's equations for electric and magnetic fields in arbitrary media. *American Journal of Physics*, **60**, 899-902.

Jefimenko C.D., 1998, On Maxwell's displacement current. *European Journal of Physics,* **19**, 469-470.

Johnson, N.L., Kotz, S. and Balakrishnan, N., 1994, *Continuous Univariate Distributions* (New York: John Wiley & Sons).

Kanok-Nukulchai, W., Yiu, P.K.A., and Brotton, D.M., 1992, Mathematical modeling of cable-stayed bridges. *Science and Technology, Structural Engineering International*, **2/92**, 108-113.

Kaneko, T., 1975, On Timoshenko's correction for shear in vibrating beams. *Journal of Physics D: Applied Physics*, **8**, 1927-1936.

Karjanto, N., and van Groesen, E., 2007, Derivation of the NLS breather solutions using displaced phase-amplitude variables. *Proceedings of SEAMS-GMU Conference* 2007, pp. 1-12.

Keer, L.M., 1968, Coupled pairs of dual integral equations. *Quarterly of Applied Mathematics*, **25**(4), 453-457.

Krylov, V., and Ferguson, C., 1994, Calculation of low-frequency ground vibrations from railway trains. *Applied Acoustics*, **42**, 199-213.

Landau, L.D. and Lifshitz, E.M., 1965, *Quantum Mechanics* (*Non-Relativistic Theory*), 2nd edition (Oxford: Pergamon).

Lebedev, N.N., Sakalskaya, I.P. and Uflyand, Y.S., 1965, *Worked Problems in Applied Mathematics,* translated by R.A. Silverman and supplemented by E.L. Reiss (New York: Dover).

Lebedev, N.N., 1972, *Special Functions & Their Applications,* translated and edited by R.A. Silverman (New York: Dover).

Leissa, A.W., 1969, *Vibration of Plates*, NASA SP-160 (Washington: NASA).

Lewellen, W.S., 1976, Theoretical models of the tornado vortex. In *Proceedings of the Symposium on Tornadoes: Assessment of Knowledge and Implications for Man,* ed. by R.E. Peterson, June 22-24, 1976, Texas Tech University, pp. 103-143.

Liboff, R.L., 1980, *Introductory Quantum Mechanics* (Reading: Addison-Wesley).

Lighthill, J., 1978, *Waves in Fluids* (Cambridge: Cambridge University Press).

Lighthill, J., 1998, Fluid Mechanics of Tropical Cyclone. *Theoretical and Computational Fluid Dynamics*, **10**, 3-21.

Lighthill, M.J., and Whitham, G.B., 1955, On kinematic waves: II A theory of traffic flow on long crowded highways. *Proceedings of the Royal Society of London A*, **229**, 317-345.

Lin, C.C. and Segel, L.A., 1988, *Mathematics Applied to Deterministic Problems in the Natural Sciences* (Philadelphia: Society for Industrial and Applied Mathematics).

Lin, Y.K., 1966, Discussion on "Classical normal modes in damped linear dynamic systems." *Journal of Applied Mechanics ASME*, **33**, 471-472.

Liu, M, and Gorman, D.G., 1995, Formulation of Rayleigh damping and its extension. *Computers and Structures*, **57**(2), 277-285.

Maxwell, J.C., 1954, *A Treatise on Electricity and Magnetism,* 3rd edition (New York: Dover).

McDonald, K.T., 1997, The relation between expressions for time-dependent electromagnetic fields given by Jefimenko and by Panofsky and Philips. *American Journal of Physics*, **65**, 1074-1076.

McIntyre, D.H., 2012, *Quantum Mechanics: A Paradigms Approach* (Boston: Pearson).

Mei, C.C., Stiassnie, M., and Yue, D.K.-P., 2005, *Theory and Applications of Ocean Surface Waves* (Singapore: World Scientific).

Messiah, A., 1967, *Quantum Mechanics,* Vol. I, translated from French by G.M. Temmer (Amsterdam: North-Holland).

Mindlin, R.D., 2001, *An Introduction to the Mathematical Theory of Vibrations of Elastic Plates* (New Jersey: World Scientific).

Minor, J.E., 1976, Applications of tornado technology in professional practice. In *Proceedings of the Symposium on Tornadoes: Assessment of Knowledge and Implications for Man,* ed. by R.E. Peterson, June 22-24, 1976, Texas Tech University, pp. 375-392.

Minorsky, N., 1962, *Nonlinear Oscillations* (Princeton: Van Nostrand).

Misner, C.W., Thorne, K.S., and Wheeler, J.A., 1973, *Gravitation* (San Francisco: Freeman and Company).

Moulton, F.R., 1914, *An Introduction to Celestial Mechanics,* 2nd revised edition (New York: Macmillan).

Mura, T., 1987, *Micromechanics of Defeats in Solids* (Dordrecht: Martinus Nijhoff).

Murty, T.S., 1984, *Storm Surges: Meteorological Ocean Tides* (Ottawa: Department of Fisheries and Oceans).

Naguleswara, S., 2004, Transverse vibration of an uniform Euler-Bernoulli beam under linearly varying axial force. *Journal of Sound and Vibrations*, **275**, 47-57.

Nevel, D.E., 1959, *Tables of Kelvin Functions and Their Derivatives,* Technical Report 67 (Wilmette, IL: U.S. Army Snow Ice and Permafrost Research Establishment).

Novozhilov, V.V., 1953, *Foundations of the Nonlinear Theory of Elasticity* (Rochester: Graylock Press).

Ogata, A., and Banks, R.B., 1961, *A Solution of the Differential Equation of Longitudinal Dispersion in Porous Media,* Geological Survey Professional Paper 411-A (Washington DC: US Department of the Interiors).

Okukawa, A., Suzuki, S., and Harazaki, I., 2000, Suspension bridges. Chapter 18 in *Bridge Engineering Handbook*, ed. W.F. Chen and L. Duan (Boca Raton: CRC Press).

Olivier, C.P., 1925, *Meteors* (Baltimore: Williams & Wilkins).

Ozisik, M.N., 1968, *Boundary Value Problems of Heat Conduction* (New York, Dover).

Parkus, H., 1962, Thermal Stresses. Chapter 43 in *Handbook of Engineering Mechanics*, ed. by W. Flugge (New York: McGraw-Hill).

Pedlosky J., 1987, *Geophysical Fluid Dynamics* (New York: Springer-Verlag).

Pedlosky J., 2003, *Waves in Oceans and Atmosphere* (Berlin: Springer-Verlag).

Peregrine, D.H., 1983, Water waves, nonlinear Schrodinger equations and their solutions. *Journal of Australian Mathematical Society Series B*, **25**, 16-43.

Peterson, R.E. (ed.), 1976, *Proceedings of the Symposium on Tornadoes: Assessment of Knowledge and Implications for Man*, June 22-24, 1976 (Lubbock: Texas Tech University).

Press, W.H., Flannery, B.P., Teukolsky, S.A. and Vetterling, W.T., 1992, *Numerical Recipes: The Art of Scientific Computing,* 2nd edition (New York: Cambridge University Press).

Pugh, D.T., 1987, *Tides, Surges and Mean Sea-Level: A Handbook for Engineers and Scientists* (New York: John Wiley and Sons).

Pugsley, A., 1968, *The Theory of Suspension Bridges* (London: Edward Arnold).

Purcell, E.M., 1985, *Electricity and Magnetism,* Berkeley Physics Course-Volume 2 (New York: McGraw-Hill).

Rae, A.I.M., 2002, *Quantum Mechanics*, 4th edition (Bristol: Institute of Physics Publishing).

Ravi-Chandar, K., 2004, *Dynamic Fracture* (Amsterdam: Elsevier).

Rayleigh, Lord, 1877, *The Theory of Sound* (London: Macmillan).

Reid, W.P., 1962, Free vibrations of a circular plate. *Journal of the Society for Industrial and Applied Mathematics*, **10**(4), 668-674.

Reiss, E.L., 1969, Column buckling: An elementary example of bifurcation. *Bifurcation Theory and Nonlinear Eigenvalue Problems,* ed. J.B. Keller and S. Antman (New York: Benjamin), pp. 1-16.

Reiss, E.L., 1980a, A modified two-time method for the dynamic transitions of bifurcation. *SIAM Journal of Applied Mathematics,* **38**(2), 249-260.

Reiss, E.L., 1980b, A new asymptotic method for jump phenomena. *SIAM Journal of Applied Mathematics,* **39**(3), 440-455.

Reiss, E.L., 1984, A nonlinear structural concept for drag-reducing compliant walls. *AIAA Journal,* **22**(3), 399-402.

Reiss, E.L. and Matkowsky, B.J., 1971, Nonlinear dynamic buckling of a compressed elastic column. *Quarterly of Applied Mathematics,* **29**, 245-260.

Rice, J.R., 1968, Mathematical analysis in mechanics of fracture. In *Fracture: An Advanced Treatise,* Vol. 2, edited by H. Liebowitz, pp. 191-311 (New York: Academic Press).

Rice, J.R., Ben-Zion, Y., and Kim, Y.-S., 1994, Three-dimensional perturbation solution for a dynamic planar crack moving unsteadily in a model elastic solid. *Journal of the Mechanics and Physics of Solids*, **42**(5), 813-843.

Rocard, Y., 1957, *Dynamic Instability: Automobiles, Aircraft, Suspension Bridges* (London: Crosby Lockwood & Son).

Rosakis, A.J., 2002, Intersonic shear cracks and fault ruptures. *Advances in Physics,* **51**(4), 1189-1257.

Rosakis, A.J., Liu, C., and Freund, L.B., 1991, A note on the asymptotic stress field of a non-uniformly propagating dynamic crack. *International Journal of Fracture*, **50**, R39-R45.

Rosser, J.B., Newton, R.R., Gross, G.L., 1947, *Mathematical Theory of Rocket Flight* (New York: McGraw-Hill).

Sakurai, J.J., 1994, *Modern Quantum Mechanics,* Revised edition (Reading: Addison-Wesley).

Segel, L.A., 1987, *Mathematics Applied to Continuum Mechanics* (New York: Dover).

Shankar, R., 1994, *Principles of Quantum Mechanics* (New York: Plenum Press).

Shilkrut, D.I., 2002, *Stability of Nonlinear Shells* (Amsterdam: Elsevier Science).

Shrira, V.I., and Geogjaev, V.V., 2010, What makes the Peregrine soliton so special as a prototype of freak waves? *Journal of Engineering Mathematics*, **67**, 11-22.

Slepyan, L.I., 2002, *Models and Phenomena in Fracture Mechanics* (Berlin: Springer).

Sneddon, I.N., 1951, *Fourier Transforms* (New York: McGraw-Hill).

Sommerfeld, A., 1949, *Partial Differential Equations in Physics.* Lectures on Theoretical Physics, Vol. VI, Translated by E. G. Straus (New York: Academic Press).

Sommerfeld, A., 1952, *Electrodynamics.* Lectures on Theoretical Physics, Vol. III, Translated by E. G. Ramberg (New York: Academic Press).

Song, Z., and Su, C., 2017, Computation of Rayleigh damping coefficients for the seismic analysis of a hydro-powerhouse. *Shock and Vibration*, Vol. 2017, Article ID 2046345, 11 pages, https://doi.org/10.1155/2017/2046345.

Spiegel, M.R., 1968, *Schaum's Outline Series: Mathematical Handbook* (New York: McGraw-Hill).

Stoker, J.J., 1950, *Nonlinear Vibrations of Mechanical and Electrical Systems* (New York: Wiley, Interscience).

Szebehely, V.G., and Mark, H., 2004, *Adventures in Celestial Mechanics*, 2nd edition (Weinheim: Wiley-VCH).

Takahashi, K., 1980, Eigenvalue problem of a beam with a mass and spring at the end subjected to an axial force. *Journal of Sound and Vibration*, **71**(3), 453-457.

Tang, M.-C., 2000, Cable-Stayed Bridges. Chapter 19 in *Bridge Engineering Handbook*, ed. by W.F. Chen and L. Duan (Boca Raton: CRC Press).

Taylor, G.I., 1954, The dispersion of matter in turbulent flow through a pipe. *Proceedings of Royal Society of London A* **223**, 446-468.

Timoshenko, S.P., 1921, On the correction for shear of the differential equation for transverse vibrations of prismatic bars. *Philosophical Magazine*, **41**(245), 744-746.

Timoshenko, S.P., 1956, *Strength of Materials*, 3rd edition (Princeton: Van Nostrand).

Timoshenko, S.P. and Gere, J.M., 1961, *Theory of Elastic Stability*, 2nd edition (New York: McGraw-Hill).

Timoshenko, S.P. and Woinowsky-Krieger, S., 1959, *Theory of Plates and Shells*, 2nd edition (New York: McGraw-Hill).

Timoshenko, S.P. and Young, D.H., 1965, *Theory of Structures*, 2nd edition (New York: McGraw-Hill).

Titchmarsh, E.C., 1948, *Introduction to the Theory of Fourier Integrals*, 2nd edition (Oxford: Clarendon Press).

Townsend, J.S., 2000, *A Modern Approach to Quantum Mechanics* (Sausalito: University Science Books).

TRACOR, Inc., 1971, *Estuarine Modeling: An Assessment.* Water Pollution Control Research Series, 16070 DZV 02/71 (Washington DC: Water Quality Office, Environmental Protection Agency).

Troitsky, M.S., 1988, *Cable-Stayed Bridges* (Boston: BSP Professional Books).

Ventsel, E., and Krauthammer, T., 2001, *Thin Plates and Shells: Theory, Analysis and Applications* (New York: Marcel Dekker).

Vlasov, V.S., 1951, *Basic Differential Equations in General Theory of Elastic Shells* NACA TM 1241 (Washington: NACA) (English translation from Russian).

Vlasov, V.S., and Leontev, N.N., 1966, *Beams, Plates and Shells on Elastic Foundations* (Washington: NASA) (English translation from Russian by Israel Program for Scientific Translation).

Walther, R., Houriet, B., Isler, W., Moia, P., Klein, J.-F., 1999, *Cable Stayed Bridges*, 2nd edition (London: Thomas Telford).

Wang, C.M., Wang, C.Y., and Reddy, J.N., 2005, *Exact Solutions for Buckling of Structural Members* (Boca Raton: CRC Press).

Weinberg, S., 1972, *Gravitation and Cosmology: Principles and Applications of the General Theory of Relativity* (New York: John Wiley & Sons).

Weitsman, Y., 1970, On foundations that react in compression only. *Journal of Applied Mechanics*, Vol. 92, 1019-1030.

Wexler, E.J., 1992, *Analytical Solution for One-, Two, and Three-Dimensional Solute Transport in Ground Water System with Uniform Flows,* Ch B7, Book 3 Applications of Hydraulics (Denver: USGS).

Whittaker, E. T. and Watson, G. N., 1927, *A Course of Modern Analysis,* 4-th edition (London: Cambridge University Press).

Wu, J.-Z., Ma, H.-Y., and Zhou, M.-D., 2015, *Vortical Flows* (Heidelberg: Springer).

Wyman, M., 1950, Deflections of an infinite plate. *Canadian Journal of Research,* A**28**, 293–302.

Xu, Y.L., 2013, *Wind Effects on Cable-Supported Bridges* (Singapore: John Wiley & Sons).

Yoffe, E.H., 1951, The moving Griffith crack. *Philosophical Magazine*, **42**, 739-750.

Yokoyama, T., 1990, Vibrations of a hanging Timoshenko beam under gravity. *Journal of Sound and Vibration* **141**(2), 245-258.

Zangwill, A., 2013, *Modern Electrodynamics* (Cambridge: Cambridge University Press).

Author Index

A

Abel, N.H., 629, 731, 738-740, 747-748, 780
Abramowitz, M., 129, 132, 398, 498, 604, 621, 629
Adams, J.C., 656
Airy, G.B., 104, 334, 498

B

Banks, R.B., 398, 405-406, 408, 410-412, 420, 428-429, 499
Bathe, K.J., 337
Bazant, Z.P., 334, 378, 386, 391
Berg, M., 614
Bernoulli, D., 1-2
Bernoulli, Jacob, 1
Bernoulli, Johann (or John), 291
Bohr, N.H.D., 604, 614, 620, 627-628, 641, 652
Boussinesq, J.V., 80, 239, 496, 511, 519-520, 544,
Brahe, T., 655
Broberg, K.B., 769, 777, 781
Busbridge, I.W., 780

C

Calladine, C.R., 230
Carslaw, H.S.,398, 428
Caughey, K.T., 265, 271, 273-274, 283, 286, 289
Chau, K.T., 10, 13-16, 34, 38, 43, 46, 50, 54, 59-60, 80-81, 89, 95, 104, 112-113, 117, 125, 131-132, 177, 187, 195-196, 213-214, 218, 220-221, 245-247, 256-257, 265-267, 293, 295-297, 299, 303-305, 307, 316, 318, 322, 331-334, 342, 344-345, 347, 350, 353, 358-360, 362-365, 381, 398-399, 401, 406, 414, 417, 423, 428, 431, 433, 442, 464, 468-470, 472, 479, 495, 497, 500-503, 506, 525-526, 531, 551, 559, 561, 564, 566-567, 570, 572, 574, 577-578, 581-587, 589, 591-592, 595, 597, 610, 615-616, 624, 632-633, 635, 640, 646, 654, 696, 698, 710, 726, 732-733, 738, 740-741, 743-744, 749-750, 752, 757-758, 761-765, 773, 781
Clairaut, A.C., 464
Clough, R., 284
Copernicus, N., 655
Courant, R., 630
Cushman-Roisin, B., 449, 489

D

D'Alembert, J.L.R., 239-240,
da Vinci, L., 1,2
de Broglie, L., 603-604, 608-610, 628
Debye, P.J.W., 582, 584-589, 600
De Vries, G., 496-497, 511
Dirac, P.A.M., 344, 414, 502, 590-592, 615, 618, 630, 647-648, 653
Dyson, F., 502, 692

E

Eddington, A., 692, 698
Einstein, A., 552, 572, 603, 648, 694-695, 697-698, 702, 725
Euler, L., 1-5, 8, 31, 34, 37-38, 40, 42-44, 47, 53, 55, 58, 65, 67, 71, 73-77, 79, 83, 164, 177, 239-240, 331-332, 346, 351-352, 377, 441, 460, 559, 646, 651, 656-657, 675, 681, 695, 716

F

Feynman, R., 601, 603, 653
Filon, L.N.G., 762,
Flugge, W., 138, 230
Fourier, J.B.J., 79, 88, 106, 108, 228, 358, 405, 445, 544, 596-597, 617,
Freund, B., 757, 761-762, 765-766, 769, 770, 781

Friedrichs, K.O., 334
Fujino, Y., 327

G

Gagarin, Y., 711-712
Galerkin, B.G., 80, 95, 97-98, 134, 571
Galilei, G., 1, 239, 655
Gao, H., 749
Gauss, C.F., 84, 139, 230, 266, 271, 398-399, 401, 415, 417, 425, 551, 553-554, 556-557, 560, 565, 567-568, 593-594, 600, 657, 716
Gere, J.M., 43, 75, 167, 230, 242, 265, 334, 336-337, 391
Goddard, R.H., 657
Goudsmit, S., 647
Gradshteyn, I.S., 408, 738-740

H

Halley, E., 655, 726-727
Hamilton, W.R., 53-54, 74, 272, 350, 376, 610, 615-616, 618, 621, 648
Heaviside, O., 255, 432, 551-552, 739, 773-774, 778
Heisenberg, W.K., 603, 614, 618, 643
Hermite, C., 634
Hertz, H., 31, 80, 109, 122, 130, 134, 284, 551-552, 578-583, 587, 600
Heyman, J., 2, 302
Hicks, K.D., 710, 725
Hilbert, D., 620, 630, 657
Hua, L.K., 725
Hutchinson, J.R., 53
Hutchinson, J.W., 229, 341

I

Irvine, H.M., 292, 327

J

Jackson, J.D., 569, 600
Jacobi, C.G.J., 632, 716, 740
Jefimenko, C.D., 571-574, 576-577, 600

K

Kaneko, T., 53
Keer, L.M., 781
Kelvin, Lord, 79-80, 86, 125, 128-132, 134, 213, 215-216, 230, 271, 445, 469, 489, 556, 559, 604
Kepler, J., 655, 664, 666, 675-677, 681, 685, 687-688, 696, 725-727
Kirchhoff, G.R., 79, 86, 126, 138-139, 174, 239, 334, 346, 595, 598-600, 604
Korteweg, D.J., 496-497, 511
Krylov/Kriloff, A.N., 80, 239-240,
Krylov, V., 37

L

Lagrange, J.L., 55, 74, 79-80, 85, 239 351, 553, 629, 656-657, 675, 695, 716, 719-720, 723
Lamb, H., 239, 434, 478-482
Lambert, J.H., 657
Laplace, P.S., 214, 239, 399, 401, 406-407, 415, 417, 469, 503, 547, 569, 592, 656-657, 716, 731, 736-737, 747, 751, 762, 780-781
Lebedev, N.N., 25, 630
Legendre, A.M., 624, 632
Leibniz, G.W., 256, 298, 407, 517, 633, 656-657, 741, 749
Leverier, U.J.J., 656
Levi-Civita, T., 575-576, 716
L'Hôpital, G.F.A., 111, 132, 607, 637
Lighthill, M.J., 447, 490, 501-502
Lin, C.C., 692,
Liouville, J., 654, 675, 723
Love, A.E.H., 138, 139, 174, 271, 346

M

Matkowsky, B.J., 331, 361, 363, 365-366, 369, 372, 391, 394
Maxwell, J.C., 205, 551-555, 651-652, 657, 781
Mellin, H., 738, 780
Mendeleev, D.I., 644
Misner, C.W., 698

Moulton, F.R., 716, 725
Mura, T., 740, 780

N

Nevel, D.E., 216
Newton, I., 240, 552, 604, 655-657, 662, 666, 676-677, 692-693, 701, 716, 725

O

Oberth, H., 657
O'Kelly, E.M.J., 271

P

Painleve, P., 495, 523-526, 544, 716
Papkovitch, P.F., 731-733, 744
Pauli, W., 614, 649-650
Poincaré, H., 239, 342, 363-364, 502, 600-601, 656-657, 716
Poisson, S.D., 53, 79-80, 86, 107, 180, 182, 239, 398, 497, 554, 569, 572, 591, 593-595, 600, 733, 757, 779, 781
Press, W.H., 22, 27

R

Ramanujan, S., 625, 633
Ravi-Chandar, K. 743, 748-749, 753, 757, 761, 764, 769, 773
Rayleigh, Lord, 1, 53, 56, 60, 64, 67, 80, 112-113, 116, 126-127, 134-136, 205, 239, 265, 268, 270-274, 276-277, 282-283, 288-289, 331, 398, 403, 430, 604-606, 608, 755, 771, 779, 781
Reiss, E.L., 331, 333-334, 361, 363, 365-366, 369, 372-373, 378, 391, 394
Rice, J.R., 756-758
Riemann, G.F.B., 497, 571, 582, 694
Ritz, W., 80, 113, 126-127, 134
Rosakis, A.J., 765-766, 768-770, 777, 781
Ryzhik, I.M., 408, 738-740

S

Schrodinger, J., 194, 489, 496, 520, 524, 529, 535, 541, 544, 552, 591, 603-604, 608, 610-614, 619, 621, 626-627, 629-630, 641, 648, 652
Schwarzschild, K., 693-695, 697, 698, 725
Segel, L.A., 398, 430, 692
Sneddon, I.N., 772, 729-730, 734, 737
Sommerfeld, A., 239, 599, 601
Slepyan, L.I., 781
Spiegel, M.R., 91, 300, 341, 462, 607
Stegun, I.A., 129, 132, 398, 498, 604, 621, 629
Stoker, J.J., 334, 339, 363
Stokes, G.G., 239, 433, 468, 471, 489, 556

T

Taylor, G.I., 397, 414-416, 418-419, 430, 464,
Timoshenko, S.P., 1-2, 5, 29, 43, 52-53, 55-56, 58, 60-61, 63, 65, 67, 72, 74-75, 77, 97, 107, 134, 138, 149, 167, 199, 207, 209, 217, 226, 228, 230, 239, 257, 265, 285, 292, 331, 334, 336-337, 346, 377, 391
Titchmarsh, E.T., 738, 780
Tsiolkovsky, K., 657

U

Uhlenbeck, G., 647

V

Vlasov, V.S., 109, 134, 138, 219-221, 224, 228, 230, 233
von Karman, T., 104, 176, 183, 230, 257, 292, 331, 334, 457

W

Watson, G.N., 604, 621, 629, 741
Weinberg, S., 698, 702
Weitsman, Y., 36

Whitham, G.B., 501
Whittaker. E. T., 604, 621, 629, 741
Wyman, M., 130, 133,

Y
Yaroshevsky, V.A., 711-713, 716
Yoffe, E.H., 757

Z

Zeeman, P., 647

Subject Index

A

Abel integral, 738-740, 747-748, 780
Aharonov-Bohm effect, 568, 571
Akhmediev breather, 535, 539
Ampere circuital law, 551, 554-555, 563
Angular momentum, 699, 703-705, 707, 729
Apogee, 684-658, 702, 704, 706-707, 727,
Applied end displacement, 354, 391
Arithmetic-geometric series, 607
Artificial Satellite, 662
Associated Laguerre polynomials,
  Generating function, 632, 634, 638
  Orthogonality, 604, 616, 618, 621, 635, 640,
Astrodynamics, 655-656, 725
Asymptotic field, 751, 753, 756, 758, 761-762, 764-765, 777
Atmospheric re-entry, 656-657, 708-711, 716, 725
Axisymmetric Indentation Problem, 774
Axisymmetric shell, 141, 145, 173, 175, 229
Aufbau principle, 643

B

Ballistic missile, 705-708, 711
Beam bending,
  Cable load, 10
  Cantilever, 2, 8-10, 12, 18, 21, 23-25, 28, 32, 44-45, 48, 51, 63, 71
  Green's function, 13
  Simply-supported, 7, 14, 38, 40, 42, 61
Beam on elastic foundation, 6, 33-35, 67, 173, 175-176, 230
  Under concentrated load, 35
Beam vibrations, 1, 13, 17, 21, 28, 70
  Orthogonality, 28, 50-51
  Rigid beam as seismograph, 29
  Rocket launch pad, 31
  Suddenly removed load, 21
  Subject to impulse, 27
  Tip lump mass, 25
  Under axial loads, 45
Benjamin-Feir instability, 535
Biochemical oxygen demand (BOD), 410
Black body radiation, 604, 606-609
Black holes, 692, 694, 698
Bohr's model of hydrogen atom, 620
Buckling,
  Stability of buckled state, 360
  Stability of straight state, 358
  Static buckling of beams, 332, 341-342, 355
Burridge-Andrews mechanism,779

C

Cable stay bridges, 308
  Non-overlapping cable, 309-310
  Overlapping cable, 309-310
Cable suspension bridges, 305
  Anti-symmetric modes, 319
  Extensible cable, 314-315
  Inextensible cable, 314, 317
  Symmetric modes, 316
  With flexible deck, 311
  With stiffened truss, 321
Catenary, 292, 298
Caughey proportional damping, 271
Celestial Mechanics, 655-656, 716, 725
Chebyshev-Laguerre polynomials, 629
Circular contact, 735-737
Classical mechanics, 603-604, 620-621, 652
Collinear Lagrangian point, 675, 717, 719, 722-724, 729
Complementary error function, 398, 401, 407, 409, 430

Coriolis force, 434
Coulomb gauge,
Radiation gauge, 591, 593, 600
Crack pressure, 747
Cylindrical roofs, 170, 228, 230
Cylindrical shells
Vlasov's stress function, 220

D

Dark matter, 656, 692
Decartes' rule of sign, 720
Diffusion equation,
Decaying pollutants, 410
Dimensional analysis, 416
Dilatation wave, 731, 781
Dispersive waves, 498, 543
Displaced phase-amplitude method, 536
Draupner oil platform, 535, 541
Dual Integral Equations, 728, 738
Dual symmetry in electrodynamics, 551, 554-556, 589
Debye potentials
For transverse electric waves, 587
For transverse magnetic waves, 582
Decaying pollutants, 410
Duffing equation, 362-363, 391
Duhamel integral, 253, 255, 406

E

Effiel Tower, 2
Edge disturbance, 161-162, 180
Ekman number, 450
Ekman pumping, 463
Ekman transport, 433, 458, 462-463
Eigenvalue problem,
Mode I, 753, 755-758
Mode II, 758
Mode III, 762-763, 780
Einstein vacuum field equation, 697
Elastic arches, 331, 378
Electric polarization, 562, 581
Electrodynamics, 551, 556, 558, 578, 601
Electromagnetic waves,
In material, 563, 576
In vacuum, 555, 564
Electron spins, 647
Energy in an Elliptic Orbit, 672
Energy change, 693
Error function, 397-402, 407, 409, 427, 430
Escape velocity,
First, 666-667, 670, 678, 707
Second, 668, 670-672, 675, 692, 707, 725
Third, 668, 671-672, 725
Euler-Bernoulli beam,
Calculus of variations, 350
Hamiltonian principle, 350
Strain energy function, 347
Euler buckling, 40
Eccentric loads, 42
Subject to end moment, 41
Expectation value, 603, 614-615, 618-619, 650, 652

F

Femi-Dirac distribution, 648
Fick's first law, 405
Fick's second law, 405, 414, 422, 424-425, 430
First Escape Velocity, 666-667, 670, 678, 707
Fourier transform, 596
Fracture Dynamics, 749
Fracture Mechanics, 731
Fredholm integral equation, 617

G

Galerkin method, 80, 95, 97-98, 134
Galilean Transform, 504, 533-534, 548-549, 731, 751, 762, 765, 771, 781
Gauge freedom, 568-569, 600
Coulomb gauge, 569, 591, 593-595, 600
Lorenz gauge, 569-571, 576, 578-579, 595, 600
Poincare gauge, 600-601
Gauss law for electricity, 553, 568
Gauss law for magnetism, 551, 553, 567

Gaussian normal distribution, 398
Generating function, 529, 632, 634, 638
Geophysical flows,
Constitutive law, 443
Continuity equation, 440
Energy equation, 444
Mass conservation, 442
Momentum equation, 441
Scale consideration, 446
State equation, 445
Geostrophic flows, 464
Taylor-Proudman theorem,464
Geosynchronous satellite, 677-678
Gravitational pull, 658-659, 661-662, 666, 668, 671, 675, 703-704, 723, 725
Gravity wave, 694
Green's strain, 344
Griffith Crack Problem, 747

H

Hagia Sophia, 138
Half-power bandwidth method, 250, 286
Halley's comet, 726-727
Halo orbit, 724
Hamiltonian, 350, 376, 610, 615-616, 618, 621, 648
Hammer test, 249, 286
Hanging chains, 293-295
Harmonic function, 213, 243, 266, 733-734, 745, 780
Harmonic problem, 751
Heliospheric Observatory (SOHO), 724
Helium, 529, 621, 644, 648
Helmholtz decomposition, 553, 572, 578, 732, 749-750, 781
Hertz problem, 80, 109, 122
Rayleigh-Ritz method, 127
Series solution, 123
Solution in Kelvin functions, 128
Variational principle, 126
Wyman's solution, 133
Hertz vector, 579-580
For electric field, 578
For magnetic polarization, 581
Gauge invariance, 580
Hilbert space, 620
Hirota's direct method, 511-512
Bilinear form, 512
D-operator, 517
Hohmann Transfer Orbit, 675, 683-686, 688-691, 725, 727-729
Hong Kong Exhibition and Convention Center, 137
Hydrogen atom, 604, 610, 627-630, 641
Hydrogen-like atom, 620, 622, 635, 652

I

Inglis solution, 303
Interplanetary Travel, 683-686, 688, 725,
Intersonic moving crack, 751, 771
Mode I, 773
Mode II, 775
Inverted catenary, 301
Gateway Arch, 291
Stone arches, 302

J

Jefimenko's equation, 571
Jefimenko's identities
Curl identity, 574
Gradient identity, 572

K

KdV equation, 496, 502
Conservation laws, 516, 526, 544
Dispersion versus nonlinearity, 508
Physical interpretation, 506
Scale invariance, 505
Versus Boussinesq equation, 519
Versus First Painleve equation, 523
Versus mKdV, 518
Versus NLSE, 520
Kelvin function, 80, 128-131, 134, 213, 215-216, 230
Kepler's 1st Law, 656

Kepler's 2nd Law, 667
Kepler's 3rd Law, 667-669, 678, 680, 689, 720
Kirchhoff integral formula, 595, 598, 600
Kresge Auditorium, 138
Kruskal-Szekeres coordinate, 698

L

Lagrangian points, 648, 667, 710-711, 715-717, 722
  Collinear, 675, 717, 719, 722-724, 729
  Triangular, 675, 717, 719
Lagrange's Quintic equation, 720, 723
Lagrangian strain, 375
Laguerre polynomials, 470, 493, 604, 621, 629-633
Laplace equation, 729
Launching Time Window, 688, 691,
Legendre equation, 624
L'Hôpital's rule, 111, 132, 607, 637
Liu-Gorman proportional damping, 271-274, 283, 286, 289
Lorenz gauge,
  Hertz vector, 578-583, 587, 600
  Debye potentials, 582, 584, 587-589, 600
Lorentz force, 569, 558-560, 600
Lyman series, 644

M

Ma breather, 535, 539
Magnetization,555, 562, 582
Matter waves, 604, 628
Maxwell-Boltzmann distribution, 605
Maxwell equations,
  Boundary conditions, 567
  Differential form, 552
  Integral form, 556
  In Gaussian unit, 565
  In S.I. unit, 566
  Macroscopic, 560
  Microscopic, 552
Maxwell-Faraday law, 554
Maxwell potentials, 567
Mellin transform, 729
Meteor
  Striking speed, 691, 725
Meteor crater
  Chicxulub (Mexico), 691
  Sudbury (Canada), 691
  Vredefort (South Africa), 691
Method of characteristics, 497, 501, 781
mKdV versus Second Painleve equation, 525
Minkowski space-time, 695, 697
Missile, 702-706, 708, 711, 725
Mixed Boundary Value Problems, 728
  Half plane, 739
Modal analysis, 265, 274-276, 283
Moving Crack,
  Mode I, 746
  Mode II, 751
  Mode III, 754
Multi-story building, 240, 263
Multi-time perturbation, 332, 368

N

Navier-Stokes equation, 433, 489
Non-Hohmann orbit, 690,
Nonlinear buckling, 1, 7, 328
  Column buckling, 332
  Plate buckling, 334
  Shell buckling, 336
  Snap through, 336-337, 339, 377-378, 383-387, 391
Nonlinear Schrodinger equation
  Bright soliton, 532
  Dark soliton, 533
  Versus mKdV, 529
Nonlinear transport, 496

O

Ogata-Banks solution, 406, 408, 410-412, 429
Orbit
  Elliptic, 655, 666, 668, 675, 677, 679-687, 689, 692, 702-703, 705, 725-727,
  Hyperbolic, 666, 683
  Parabolic, 666, 683, 691-692
Ohm's law, 562-563

P

Painleve transcendents, 496
Pantheon, 137, 197, 236
Papkovitch-Neuber potential, 724, 737
Pauli's exclusion principle, 649-650
Penny-shaped crack, 731, 733-735, 738, 742-744
Penny-shaped crack, 731, 733-735, 738, 742-744
  Subject to uniform Pressure, 734
  Energy Change, 735
Peregrine breather, 536
Perigee, 684-685, 702-704, 706, 727, 729
Perihelion, 676, 679-680, 684-686, 692-695, 697, 717, 719-720
Perturbation method, 700
Pipe,137, 166-167, 178, 181, 183-184, 230
  Effective length, 183, 231
  Infinite, 181
  Semi-infinite, 235
PMMA, 758
Planck's constant, 604
Plate, 79-136
  Anisotropic, 80, 107-108, 134
  Buckling, 103, 104-106, 134
  Circular, 80, 98-101, 108, 116, 122, 125-127, 133-136
  Clamped, 86, 95-96, 98-102, 113-116, 126
  Galerkin method, 98, 134
  Levy's solution, 88-89, 95, 134
  Navier's solution, 88, 108, 134
  On elastic foundation, 108, 122, 134
  Rectangular, 79-80, 89-90, 92, 95-96, 103, 105, 108, 110-111, 114
  Simply-supported, 88, 90, 91, 95, 102, 105, 108
  von Karman equation, 104
Plate theories,
  Kirchhoff theory, 79-80, 86, 138
Point source of pollutants
  1-D solution, 415
  2-D solution, 422, 426
  3-D solution, 425
Poisson equation, 569, 591, 593-595, 600
Pollutants in river, 403
Potential problem, 731, 734-736, 746-747, 781
Precession of perihelion of Mercury, 693-693, 699, 701, 725
Pseudo-response spectrum, 259-260

Q

Quantized energy, 604, 609-610
Quantum mechanics, 603

R

Radiative transitions from atom, 650
Ramanujan's master theorem, 633
Rayleigh damping, 265, 268, 270-273, 276-277, 282-283
Rayleigh-Jeans law, 605, 608
Rayleigh quotient technique, 80, 112, 134
Rayleigh wave speed, 747, 763, 771-773
Relativity, 692-695, 698-702, 725
Response spectrum, 240, 257
Reynolds number, 403, 433, 449-450
Riccati equation, 495
Riemann equation, 497
Rigid Mass, 657, 726
Rocket, 657, 667-668, 672, 703, 725
Rodrigues' formula, 632
Rogue waves, 496, 535-536, 489
Rolex library, 137
Rossby number, 449-450

S

Sagitta, 236
Satellite, 657, 662, 666-668, 670-671, 677-678, 725
Schwarzschild geometry, 693-694, 697-698,
Schwarzschild metric, 693-695, 697-698
Schwarzschild radius, 694, 697-698,
SDOF,
  Damped, 246

Forced responses, 250,
Free responses, 243
Undamped, 243
Secant formula, 43
Second escape velocity, 668, 670-672, 675, 692, 707, 725
Separation of variables, 14, 21, 46, 58, 89, 116, 295, 423, 425, 621-622, 641, 649, 736
Series reversion, 720, 722
Shallow water equation, 433, 465, 519
Shallow spherical shells,
Hetenyi approximation, 206, 232
Geckeler-Staerman approximation, 138, 201, 206-207, 209, 230
Shear wave, 732, 750-751, 771-773, 779, 781
Shell, 137-238
Axisymmetric, 172-173, 187, 217
Buckling, 176, 184-187
Conical, 156-159
Cylindrical,160-164, 170-187, 217-229
Dome, 145-147, 149-151, 197
Liquid storage tank, 157
Of revolution, 150, 187
Pipe, 166-167, 178-184
Ring foundation, 148
Roof, 221
Spherical, 210, 210
Truncated dome, 145-150
Shell theories,
Bending theory, 187, 217
Membrane theory, 141, 160
Reissner formulation, 191
Shocks, 496-502
Snap through buckling
Elastic arches, 377-390
Two-bar system, 337-339
Solid state physics, 562
Soliton, 495-549
N-soliton solution, 516
One-soliton solution, 513
Two-soliton solutions, 514
Sommerfeld radiation condition, 599
Spacecraft, 656-657, 668, 671, 673-675, 684-685, 687-688, 690, 708-710, 713, 716, 723-725, 729,
Sputnik I, 677-678
Stationary state of energy, 615
Storm surge, 450-457
Typhoon Hato, 456
Structural dynamics, 239
Sturm-Liouville problem, 654
Syncom II, 678

T

Tacoma Narrows Bridge, 292, 311, 313, 318-319, 321, 327
Tautochrone curve, 740
Taylor's point source solution, 414
Thin shallow spherical shells, 210
Kelvin function, 213
Third escape velocity, 668, 671-672, 725
Three-body problem, 716, 718-719, 724, 729
Timoshenko beam,
Free vibration, 58
Static solution, 56
Variational formulation, 52
Tornado dynamics, 467
Traffic flow, 499
Transient crack growth, 758
Transitions of snap through buckling, 387
Transitions to the buckled states, 361
Travel to the Moon, 673, 675, 685,
Triangular Lagrangian point, 675, 717, 719
Turbulent diffusion, 397

U

Uncertainty principle, 603, 614, 618, 643, 649
Universal law of gravitation, 692, 725

V

Vibrations of cable suspension bridge, 311
Vibrations of carriage, 240-242
Vibrations of plates,
Clamped circular, 116
Free, 109
Forced, 110

Rayleigh quotient, 112
Rayleigh-Ritz method, 113
von Karman constant, 457
Vortex model,
Burgers-Rott vortex model, 475
Oseen-Lamb vortex model, 478
Potential vortex model, 472
Rankine vortex model, 473
Sullivan vortex model, 482
Vorticity 467

W

Wavefunctions, 603-604, 610-612, 615-621, 623, 625-627, 629, 631, 640-641, 644-646, 648-650, 653
Wave-particle duality, 603, 614
Wien's law, 605, 608

Y

Yaroshevsky solution, 711-713, 716
Yoffe problem, 757

Z

Zeeman effect, 647
Zeiss planetarium, 138